Body systems maintain homeostasis

ENDOCRINE SYSTEM
Acts by means of hormones secreted into the blood to regulate processes that require duration rather than speed—e.g., metabolic activities and water and electrolyte balance
See Chapters 4, 18, and 19.

HOMEOSTASIS
A dynamic steady state of the constituents in the internal fluid environment that surrounds and exchanges materials with the cells
See Chapter 1.
Factors homeostatically maintained:
- Concentration of nutrient molecules
  See Chapters 16, 17, 18, and 19.
- Concentration of O_2 and CO_2
  See Chapter 13.
- Concentration of waste products
  See Chapter 14.
- pH *See Chapter 15.*
- Concentration of water, salts, and other electrolytes
  See Chapters 14, 15, 18, and 19.
- Temperature *See Chapter 17.*
- Volume and pressure
  See Chapters 10, 14, and 15.

INTEGUMENTARY SYSTEM
Serves as a protective barrier between the external environment and the remainder of the body; the sweat glands and adjustments in skin blood flow are important in temperature regulation
See Chapters 12 and 17.

Keeps internal fluids in

Keeps foreign material out

IMMUNE SYSTEM
Defends against foreign invaders and cancer cells; paves way for tissue repair
See Chapter 12.

Protects against foreign invaders

Homeostasis is essential for survival of cells

MUSCULAR AND SKELETAL SYSTEMS
Support and protect body parts and allow body movement; heat-generating muscle contractions are important in temperature regulation; calcium is stored in the bone
See Chapters 8, 17, and 19.

Enables the body to interact with the external environment

CELLS
Need homeostasis for their own survival and for performing specialized functions essential for survival of the whole body
See Chapters 1, 2, and 3.
Need a continual supply of nutrients and O_2 and ongoing elimination of acid-forming CO_2 to generate the energy needed to power life-sustaining cellular activities as follows:
Food + $O_2 \rightarrow CO_2 + H_2O$ + energy
See Chapter 17.

Exchanges with all other systems

Cells make up body systems

SEVENTH EDITION

Human Physiology

From Cells to Systems

Lauralee Sherwood

Department of Physiology and Pharmacology

School of Medicine

West Virginia University

BROOKS/COLE

CENGAGE Learning

Australia • Brazil • Japan • Korea • Mexico • Singapore • Spain • United Kingdom • United States

BROOKS/COLE
CENGAGE Learning™

Human Physiology: From Cells to Systems,
  Seventh Edition
Lauralee Sherwood

Publisher: Yolanda Cossio

Development Editor: Mary Arbogast

Assistant Editor: Lauren Oliveira

Editorial Assistant: Samantha Arvin

Technology Project Manager: Lauren Tarson

Marketing Manager: Stacy Best

Marketing Coordinator: Elizabeth Wong

Marketing Communications Manager: Linda Yip

Project Manager, Editorial Production: Trudy Brown

Creative Director: Rob Hugel

Art Director: John Walker

Print Buyer: Judy Inouye

Permissions Editor, Text: Bob Kauser

Production Service: Graphic World Inc.

Text Designer: Carolyn Deacy

Art Editor: Suzannah Alexander

Photo Researcher: Linda Sykes

Illustrator: Graphic World Inc. and
  Dragonfly Media Group

Cover Designer: John Walker

Cover Image: © Getty Images, Inc.

Compositor: Graphic World Inc.

For product information and technology assistance, contact us at
Cengage Learning Customer & Sales Support, 1-800-354-9706.

For permission to use material from this text or product,
submit all requests online at **www.cengage.com/permissions.**
Further permissions questions can be e-mailed to
permissionrequest@cengage.com.

Library of Congress Control Number: 2008940733

ISBN-13: 978-0-495-39184-5

ISBN-10: 0-495-39184-0

Brooks/Cole
10 Davis Drive
Belmont, CA 94002-3098
USA

Cengage Learning is a leading provider of customized learning solutions with office locations around the globe, including Singapore, the United Kingdom, Australia, Mexico, Brazil, and Japan. Locate your local office at **www.cengage.com/international**

Cengage Learning products are represented in Canada by Nelson Education, Ltd.

To learn more about Brooks/Cole, visit **www.cengage.com/brookscole**

Purchase any of our products at your local college store or at our preferred online store **www.ichapters.com**

Printed in Canada
1 2 3 4 5 6 7 12 11 10 09 08

With love to my family,
for all that they mean to me
and all that they have done for me:

My parents,
Larry (in memoriam) and Lee Sherwood

My husband,
Peter Marshall

My daughters,
Melinda Marple
Allison Tadros

My grandchildren,
Lindsay Marple
Emily Marple
Alexander Tadros

Brief Contents

Contents

Chapter 12

Body Defenses 417

Chapter 15

Fluid and Acid–Base Balance   557

Chapter 20

The Reproductive System   741

Uniqueness of the Reproductive System   741

Male Reproductive Physiology   749

Sexual Intercourse Between Males and Females   759

Female Reproductive Physiology   764

Preface

Goals, Philosophy, and Theme

I have been teaching physiology since the mid-1960s and remain awestruck at the miraculous intricacies and efficiency of body function. No machine can perform even a portion of natural body function as effectively. My goal in writing physiology textbooks is not only to help students learn about how the body works but also to share my enthusiasm for the subject matter. Most of us, even infants, have a natural curiosity about how our bodies work. When a baby first discovers it can control its own hands, it is fascinated and spends many hours manipulating them in front of its face. By capitalizing on students' natural curiosity about themselves, I try to make physiology a subject they can enjoy learning.

Even the most tantalizing subject can be difficult to comprehend if not effectively presented, however. Therefore, this book has a logical, understandable format with an emphasis on how each concept is an integral part of the entire subject. Too often, students view the components of a physiology course as isolated entities; by understanding how each component depends on the others, a student can appreciate the integrated functioning of the human body. The text focuses on the mechanisms of body function from cells to systems and is organized around the central theme of *homeostasis*—how the body meets changing demands while maintaining the internal constancy necessary for all cells and organs to function.

This text is written with undergraduate students preparing for health-related careers in mind. Its approach and depth are appropriate for other undergraduates as well. Because it is intended as an introduction and, for most students, may be their only exposure to a formal physiology text, all aspects of physiology receive broad coverage, yet depth, where needed, is not sacrificed. The scope of this text has been limited by judicious selection of pertinent content that a student can reasonably be expected to assimilate in a one-semester physiology course. Materials were selected for inclusion on a "need-to-know" basis, so the book is not cluttered with unnecessary detail. Instead, content is restricted to relevant information needed to understand basic physiologic concepts and to serve as a foundation for future careers in the health professions.

Because most students have a clinical orientation, research methodologies and data are not emphasized, although the material is based on up-to-date evidence. New information based on recent discoveries has been included in all chapters. Students can be assured of the timeliness and accuracy of the material presented. This edition is the most extensive revision yet, taking into account new discoveries in the field as well as clarifying and modifying as needed based on reviewers' comments. Some controversial ideas and hypotheses are presented to illustrate that physiology is a dynamic, changing discipline.

To keep pace with today's rapid advances in the health sciences, students in the health professions must be able to draw on their conceptual understanding of physiology instead of merely recalling isolated facts that soon may be outdated. This text is designed to promote understanding of the basic principles and concepts of physiology rather than memorization of details. The text is written in simple, straightforward language, and every effort has been made to assure smooth reading through good transitions, logical reasoning, and integration of ideas throughout the text.

Because the function of an organ depends on the organ's construction, enough relevant anatomy is provided to make the inseparable relation between form and function meaningful.

Text Features and Learning Aids

Implementing the homeostasis theme

A unique, easy-to-follow, pictorial homeostatic model showing the relationship among cells, systems, and homeostasis is developed in the introductory chapter and presented on the inside front cover as a quick reference. Each chapter begins with a specially tailored version of this model, accompanied by a brief description emphasizing how the body system considered in the chapter functionally fits in with the body as a whole. This opening feature orients students to the homeostatic aspects of the material that follows. At the close of each chapter, **Chapter in Perspective: Focus on Homeostasis** helps students put into perspective how the part of the body just discussed contributes to homeostasis. This closing feature, the opening homeostatic model, and the introductory comments work together to facilitate students' comprehension of the interactions and interdependency of body systems, even though each system is discussed separately.

Pedagogical illustrations

Anatomic illustrations, schematic representations, photographs, tables, and graphs complement and reinforce the written material. Over 90% of the art has been upgraded in this edition, with more three-dimensional art; many conceptually redesigned or new figures to enhance student understanding; brighter, more contemporary, and more visually appealing colors; and more consistent style throughout. More electron micrographs are included than ever before.

Widespread use of integrated descriptions within figures, including numerous process-oriented figures with step-by-step descriptions, allows visually oriented students to review processes through figures.

Flow diagrams are used extensively to help students integrate the written information. In the flow diagrams, lighter and darker shades of the same color denote a decrease or increase in a controlled variable, such as blood pressure or the concentration of blood glucose. Physical entities, such as body structures and chemicals, are distinguished visually from actions. New to this edition, icons of physical entities are incorporated into the flow diagrams to aid students in learning what structures are involved in specific actions.

Also, integrated color-coded figure/table combinations help students better visualize what part of the body is responsible for what activities. For example, anatomic depiction of the brain is integrated with a table of the functions of the major brain components, with each component shown in the same color in the figure and the table.

A unique feature of this book is that people depicted in the various illustrations are realistic representatives of a cross-section of humanity (they were drawn from photographs of real people). Sensitivity to various races, sexes, and ages should enable all students to identify with the material being presented.

Analogies

Many analogies and frequent references to everyday experiences are included to help students relate to the physiology concepts presented. These useful tools have been drawn in large part from my over four decades of teaching experience. Knowing which areas are likely to give students the most difficulty, I have tried to develop links that help them relate the new material to something with which they are already familiar.

Pathophysiology and clinical coverage

Another effective way to keep students' interest is to help them realize they are learning worthwhile and applicable material. Because most students using this text will have health-related careers, frequent references to pathophysiology and clinical physiology demonstrate the content's relevance to their professional goals. Clinical Note icons flag clinically relevant material, which is integrated throughout the text.

Boxed features

Two types of boxed features are integrated within the chapters. **Concepts, Challenges, and Controversies** boxes expose students to high-interest information on such diverse topics as stem cell research, acupuncture, new discoveries regarding common diseases such as strokes, historical perspectives, and body responses to new environments such as those encountered in space flight and deep-sea diving.

A Closer Look at Exercise Physiology boxes are included for three reasons: increasing national awareness of the importance of physical fitness, increasing recognition of the value of prescribed therapeutic exercise programs for a variety of conditions, and growing career opportunities related to fitness and exercise.

Feedforward statements as subsection titles

Instead of traditional short topic titles for each major subsection (for example, "Heart valves"), feedforward statements alert students to the main point of the subsection to come (for example, "Pressure-operated heart valves ensure that blood flows in the right direction through the heart"). These headings also break up large concepts into smaller, more manageable pieces for the student, and as an added bonus, the listing of these headings in the **Contents** at the beginning of the book serves as a set of objectives for each chapter.

Key terms and word derivations

Key terms are defined as they appear in the text. Because physiology is laden with new vocabulary words, many of which are rather intimidating at first glance, word derivations are provided to enhance understanding of new words.

Review and Self-Evaluation Tools in the Text

New to this edition, tear-out **Study Cards** are available for each chapter. Students can carry these handy chapter summaries instead of the book to conveniently review key concepts for exams. The study cards present the major points of each chapter in concise, section-by-section bulleted lists, including cross-references for page numbers, figures, and tables. The tear-out design lets students more efficiently review material even with the book in hand because they can see the written summary and visual information side-by-side without having to flip pages back and forth. This feature enables students to easily review main concepts before moving on.

The **Review Exercises** at the end of each chapter include a variety of question formats for students to self-test their knowledge and application of the facts and concepts presented. Also available are **Quantitative Exercises,** which provide the students with an opportunity to practice calculations that will enhance their understanding of complex relationships. A **Points to Ponder** section features thought-provoking problems that encourage students to analyze what they have learned, and the **Clinical Consideration,** a mini case history, challenges them to apply their knowledge to a patient's specific symptoms. Answers and explanations for all of these questions are found in Appendix F.

Appendixes and Glossary

The appendixes are designed for the most part to help students who need to brush up on some foundation materials that they are assumed to already have studied in prerequisite courses.

- *Appendix A,* **The Metric System,** is a conversion table between metric measures and their English equivalents.

- Most undergraduate physiology texts have a chapter on chemistry, yet physiology instructors rarely teach basic chemistry concepts. Knowledge of chemistry beyond that introduced in secondary schools is not required for understanding this text. Therefore, I provide instead *Appendix B,* **A Review of Chemical Principles,** as a handy reference for students who need a brief review of basic chemistry concepts that apply to physiology.

- Likewise, *Appendix C,* **Storage, Replication, and Expression of Genetic Information,** serves as a reference for students or as assigned material if the instructor deems appropriate. It includes a discussion of DNA and chromosomes, protein synthesis, cell division, and mutations.

- *Appendix D,* **Principles of Quantitative Reasoning,** will help students become more comfortable working with equations and translating back and forth between words, concepts, and equations. This appendix supports the Quantitative Exercises at each chapter's end.

- *Appendix E,* **Text References to Exercise Physiology,** provides an index of all relevant content on this topic.

- *Appendix F,* **Answers to End-of-Chapter Objective Questions, Quantitative Exercises, Points to Ponder, and Clinical Considerations,** provides answers to all objective learning activities, solutions to the Quantitative Exercises, and explanations for the Points to Ponder and Clinical Considerations.

- The **Glossary,** which offers a way to review the meaning of key terminology, includes phonetic pronunciations of the entries.

Organization

There is no ideal organization of physiologic processes into a logical sequence. In the sequence I chose, most chapters build on material presented in immediately preceding chapters, yet each chapter is designed to stand on its own, allowing the instructor flexibility in curriculum design. This flexibility is facilitated by cross-references to related material in other chapters. The cross-references let students quickly refresh their memory of material already learned or proceed, if desired, to a more in-depth coverage of a particular topic.

The general flow is from introductory background information to cells to excitable tissue (nerve and muscle) to organ systems, with logical transitions from one chapter to the next. For example, Chapter 8, "Muscle Physiology," ends with a discussion of cardiac muscle, which is carried forward in Chapter 9, "Cardiac Physiology." Even topics that seem unrelated in sequence, such as Chapter 12, "Body Defenses," and Chapter 13, "The Respiratory System," are linked together, in this case by ending Chapter 12 with a discussion of respiratory defense mechanisms.

Several organizational features warrant specific mention. The most difficult decision in organizing this text was placement of the endocrine material. There is merit in placing the chapters on the nervous and endocrine (hormone-secreting) systems in close proximity because they are the body's two major regulatory

systems. However, discussing details of the endocrine system immediately after the nervous system would disrupt the logical flow of material related to excitable tissue. In addition, the endocrine system cannot be covered in the depth its importance warrants if it is discussed before students have the background to understand this system's roles in maintaining homeostasis.

My solution to this dilemma is Chapter 4, "Principles of Neural and Hormonal Communication." This chapter introduces the underlying mechanisms of neural and hormonal action before the nervous system and specific hormones are mentioned in later chapters. It contrasts how nerve cells and endocrine cells communicate with other cells in carrying out their regulatory actions. Building on the different modes of action of nerve and endocrine cells, the last section of this new chapter compares, in a general way, how the nervous and endocrine systems differ as regulatory systems. Chapter 5 then begins with the nervous system, providing a good link between Chapters 4 and 5. Chapters 5, 6, and 7 are devoted to the nervous system. Specific hormones are introduced in appropriate chapters, such as hormonal control of the heart and blood vessels in maintaining blood pressure in Chapters 9 and 10 and hormonal control of the kidneys in maintaining fluid balance in Chapters 14 and 15. The body's processing of absorbed energy-rich nutrient molecules is largely under endocrine control, providing a link from digestion (Chapter 16) and energy balance (Chapter 17) to the endocrine chapters, Chapters 18 and 19. These endocrine chapters pull together the source, functions, and control of specific endocrine secretions and serve as a summarizing/unifying capstone for homeostatic body function. Finally, building on the hormones that control the gonads (testes and ovaries) introduced in the endocrine chapters, the last chapter, Chapter 20, diverges from the theme of homeostasis to focus on reproductive physiology.

Besides the novel handling of hormones and the endocrine system, other organizational features are unique to this book. For example, unlike other physiology texts, the skin is covered in the chapter on defense mechanisms of the body, in consideration of the skin's recently recognized immune functions. Bone is also covered more extensively in the endocrine chapter than in most undergraduate physiology texts, especially with regard to hormonal control of bone growth and bone's dynamic role in calcium metabolism.

Departure from traditional groupings of material in several important instances has permitted more independent and more extensive coverage of topics that are frequently omitted or buried within chapters concerned with other subject matter. For example, a separate chapter (Chapter 15) is devoted to fluid balance and acid–base regulation, topics often tucked within the kidney chapter. Another example is the grouping of the autonomic nervous system, motor neurons, and the neuromuscular junction in an independent chapter on the efferent division of the peripheral nervous system, which serves as a link between the nervous system chapters and the muscle chapter. Energy balance and temperature regulation are also grouped into an independent chapter.

Although there is a rationale for covering the various aspects of physiology in the order given here, it is by no means the only logical way of presenting the topics. Because each chapter is able to stand on its own, especially with the cross-references provided, instructors can vary the sequence of presentation at their discretion. Some chapters may even be omitted, depending on the students' needs and interests and the time constraints of the course. For example, the defense role of the leukocytes is explained briefly in Chapter 11 on blood, so an instructor can choose to omit the more detailed explanations of immune defense in Chapter 12.

Ancillaries for Students

CengageNOW with Personal Tutor and InfoTrac® College Edition

CengageNOW is an easy-to-use online resource that helps students study in less time to get the grade they want—NOW. Class-tested and student-praised, CengageNOW offers a variety of features that support course objectives and interactive learning. Helping students to be better prepared for class and learn difficult physiology concepts, this web-based resource offers self-paced learning modules with tutorials, interactive quizzes, and animations. With a Personalized Study plan including pre- and post-tests, animations, and interactive exercises, students can focus on difficult topics in which they need help most.

The animations help bring to life some of the physiologic processes most difficult to visualize, enhancing the understanding of complex sequences of events. Better than ever, the collection of animations has been expanded to include some new, more integrative, realistic, and technologically sophisticated animations.

Cengage Learning has partnered with a leading service provider to give students the best in online tutoring. **Personal Tutor** provides one-on-one tutoring and on-demand assignment help from experienced tutors in an easy-access online environment.

Ideal for selecting topics and gathering information for papers, **InfoTrac College Edition** is a fully searchable database offering 20 years' worth of full-length articles (more than 20 million!) from nearly 6000 diverse sources, such as academic journals, newsletters, and up-to-the-minute periodicals, including *The New York Times, Time, Newsweek, Science, Forbes,* and *USA Today.* It includes access to **InfoMarks,** stable URLs that can be linked to articles, journals, and searches, and to **InfoWrite,** a complete set of online critical thinking and paper writing tools.

ISBN: 0495826502
ISBN: 0495826375 (with **CLeBook**)

Study Guide

Each chapter of this student-oriented supplementary manual, which is correlated with the corresponding chapter in *Human Physiology: From Cells to Systems,* Seventh Edition, contains a chapter overview, a detailed chapter outline, a list of Key Terms, and Review Exercises (multiple-choice, true/false, fill-in-the-blanks, and matching). This learning resource also offers Points to Ponder, questions that stimulate use of material in the chapter as a starting point to critical thinking and further learning.

Clinical Perspectives (common applications of the physiology under consideration), Experiments of the Day, and simple hands-on activities enhance the learning process. Answers to the Review Questions are provided at the back of the Study Guide.

ISBN: 0495826251

Coloring Book

This helpful study tool contains key pieces of art from the book and provides opportunities for students to interact with the material and explain the processes associated with the figures in their own words.

ISBN: 0495826391

Photo Atlas for Anatomy and Physiology

This full-color atlas (with more than 600 photographs) depicts structures in the same colors as they would appear in real life or in a slide. Labels as well as color differentiations within each structure are employed to facilitate identification of the structure's various components. The atlas includes photographs of tissue and organ slides, the human skeleton, commonly used models, cat dissections, cadavers, some fetal pig dissections, and some physiology materials.

ISBN: 0534517161

Acknowledgments

I gratefully acknowledge the many people who helped with the first six editions or this edition of the textbook. A special thank-you goes to four people who contributed substantially to the content of the book: Rachel Yeater (Professor Emeritus and past Chairwoman, Exercise Physiology, School of Medicine, West Virginia University), who contributed the original boxed features titled "A Closer Look at Exercise Physiology"; Spencer Seager (Chairman, Chemistry Department, Weber State University), who prepared Appendix B, "A Review of Chemical Principles"; and Kim Cooper (Associate Professor, Midwestern University) and John Nagy (Professor, Scottsdale Community College), who provided the Quantitative Exercises at the ends of chapters and prepared Appendix D, "Principles of Quantitative Reasoning."

During the book's creation and revision, many colleagues at West Virginia University shared resource materials, answered my questions, and offered suggestions for improvement. I thank each of them for helping ensure the accuracy and currency of the book.

In addition to the over 150 reviewers who carefully evaluated the forerunner books for accuracy, clarity, and relevance, I express sincere appreciation to the following individuals who served as reviewers for this edition:

Erwin Bautista, University of California, Davis

Eric Blough, Marshall University

Eric Hall, Rhode Island College

Jon Hunter, Texas A&M University

Valerie Kalter, Wilkes University

David Mallory, Marshall University

John McReynolds, University of Michigan, Ann Arbor

Susan Mounce, Eastern Illinois University

Steven Price, Virginia Commonwealth University

Nida Sehweil-Elmuti, Eastern Illinois University

Donal Skinner, University of Wyoming

Mark Taylor, Baylor University

I also thank those who participated as a virtual focus group via e-mail to provide immediate feedback regarding their preference on content, style, and pedagogical issues about which we were debating during development. Art survey participants included:

Daniel Castellanos, Florida International University

Norman Eugene Garrison, James Madison University

Jon Hunter, Texas A&M University

Erika Kancler Michaels, James Madison University

Nida Sehweil-Elmuti, Eastern Illinois University

Roy Silcox, Brigham Young University

Rachel Smetanka, Southern Utah University

Alan Sved, Pittsburgh University

Study card survey participants were:

Daniel Castellanos, Florida International University

Jon Hunter, Texas A&M University

Najma Javed, Ball State University

Erika Kancler Michaels, James Madison University

Paul Pillitteri, Southern Utah University

Nida Sehweil-Elmuti, Eastern Illinois University

Roy Silcox, Brigham Young University

Rachel Smetanka, Southern Utah University

Dixon Woodbury, Brigham Young University

Also, I am grateful to the users of the textbook who have taken time to send helpful comments.

I have been fortunate to work with a highly competent, dedicated team from Brooks/Cole. It has been a source of comfort and inspiration to know that so many people have been working so diligently in so many ways to bring this book to fruition.

Yolanda Cossio, Publisher, deserves warm thanks for her vision, creative ideas, leadership, and ongoing helpfulness. Yolanda was a strong advocate for making this the best edition yet. She secured the financial and human resources to make this a reality. I especially appreciated the conference calls she held with the team to exchange ideas. These frequent conversations kept us all excited about what could be and motivated us to get it done. Above all, Yolanda's decisions were guided by what is best for the instructors and students who will use the textbook and ancillary package. Thanks also to Editorial Assistant Samantha Arvin, who trafficked paperwork and coordinated many tasks during the development process. Furthermore, I appreciate the efforts of Senior Developmental Editor Mary Arbogast for facilitating and offering valuable insights during development and production. She had great ideas for improving the flow of the writing and for making many of the art pieces more effective. I can always count on Mary, the one consistent member of my publishing team at Brooks/Cole. She's a valuable resource because she "knows the ropes" and is always thinking of ways to make the book better and the process smoother.

Major efforts for this edition were to upgrade the art and enhance the media package. I am grateful for the creative insight of Brooks/Cole Senior Art Director John Walker, who oversaw the overall artistic design of the text and found the powerful yet graceful cover image. Freelancer Suzannah Alexander, Art Development Editor, developed the new art style and carefully evaluated and proofed every figure to ensure that the visual aspects of the text are aesthetically pleasing, consistent, contemporary, and meaningful.

The technology-enhanced learning tools in the media package were updated under the guidance of Senior Media Editor Lauren Tarson. These include the online interactive tutorials, media exercises, and other learning opportunities in CengageNOW. Melinda Newfarmer went above and beyond the call of duty to negotiate with the animation developers and get this time-consuming and important part of the media package launched when, early in the project, Lauren had to go on complete bed rest before the arrival of her twins. Shelley Ryan, Managing Media Editor, also contributed to the media package. Senior Assistant Editor Lauren Oliveira oversaw the development of the multiple hard-copy components of the ancillary package, making sure it was a cohesive whole. A hearty note of gratitude is extended to all of them for the high-quality multimedia package that accompanies this edition.

On the production side, Trudy Brown, Senior Content Project Manager, closely monitored every step of the production process while simultaneously overseeing the complex production process of multiple books. I felt confident knowing that she was making sure that everything was going according to plan. Trudy was especially helpful in coming up with efficient ways of compressing the production schedule without compromising quality. I also thank Permissions Editor Bob Kauser for tracking down permissions for the art and other copyrighted materials incorporated in the text, an absolutely essential task. With everything finally coming together, Print/Media Buyer Judy Inouye oversaw the manufacturing process, coordinating the actual printing of the book.

No matter how well a book is conceived, produced, and printed, it would not reach its full potential as an educational tool without being efficiently and effectively marketed. Marketing Managers Mandy Jellerichs and Stacy Best played the lead roles in marketing this text, for which I am most appreciative.

Brooks/Cole also did an outstanding job in selecting highly skilled vendors to carry out particular production tasks. First and foremost, it has been my personal and professional pleasure to work with Carol O'Connell, Production Editor at Graphic World, who coordinated the day-to-day management of production. In her competent hands lay the responsibility of seeing that all art, typesetting, page layout, and other related details got done right and in a timely fashion. Thanks to her, the production process went smoothly despite a compressed time schedule. Carol always established reasonable deadlines and helped set priorities when multiple tasks needed my attention, which was most of the time! I also want to extend a hearty note of gratitude to compositor Graphic World for their accurate typesetting, execution of many of the art revisions, and attractive, logical layout. Thanks as well to Dragonfly Media Group for taking my most complex, new, and extensively revised sketches and instructions and turning them into attractive, pedagogically relevant works of art. Designer Carolyn Deacy deserves thanks for the fresh and attractive, yet space-conscious appearance of the book's interior, and for envisioning the book's visually appealing exterior.

Finally, my love and gratitude go to my family for the sacrifices in family life as this seventh edition was being developed and produced. The schedule for this book was especially hectic because it came at a time when a lot else was going on in our lives. I want to thank my husband, children, grandchildren, and mother for their patience and understanding during the times I was working on the book instead of being there with them or for them. My husband, Peter Marshall, deserves special appreciation and recognition for retiring in order to free up time for me to work on the text while at the same time enabling our daughters to pursue their careers. He has become a very capable "nanny" for our preschool grandson, a "soccer mom" for our tween and teen granddaughters, and a great "househusband," taking over my share of household responsibilities. I could not have done this, or any of the preceding books, without his help, support, and encouragement.

Thanks to all!

Lauralee Sherwood

Human Physiology

FROM CELLS TO SYSTEMS

During the minute that it will take you to read this page:

Your eyes will convert the image from this page into electrical signals (nerve impulses) that will transmit the information to your brain for processing.

Besides receiving and processing information such as visual input, your brain will provide output to your muscles to help maintain your posture, move your eyes across the page as you read, and turn the page as needed. Chemical messengers will carry signals between your nerves and muscles to trigger appropriate muscle contraction.

Your heart will beat 70 times, pumping 5 liters (about 5 quarts) of blood to your lungs and another 5 liters to the rest of your body.

Approximately 150 million old red blood cells will die and be replaced by newly produced ones.

You will breathe in and out about 12 times, exchanging 6 liters of air between the atmosphere and your lungs.

More than 1 liter of blood will flow through your kidneys, which will act on the blood to conserve the "wanted" materials and eliminate the "unwanted" materials in the urine. Your kidneys will produce 1 ml (about a thimbleful) of urine.

Your cells will consume 250 ml (about a cup) of oxygen and produce 200 ml of carbon dioxide.

Your digestive system will be processing your last meal for transfer into your bloodstream for delivery to your cells.

You will use about 2 calories of energy derived from food to support your body's "cost of living," and your contracting muscles will burn additional calories.

Introduction to Physiology and Homeostasis

CENGAGENOW™ Log on to CengageNOW at **http://www.cengage.com/sso/** for an opportunity to explore animations and interactive quizzes to help you learn, review, and master physiology concepts.

Introduction to Physiology

The activities described on the preceding page are a sampling of the processes that occur in our bodies all the time just to keep us alive. We usually take these life-sustaining activities for granted and do not really think about "what makes us tick," but that's what physiology is about. **Physiology** is the study of the functions of living things. Specifically, we will focus on how the human body works.

Physiology focuses on mechanisms of action.

There are two approaches to explaining events that occur in the body: one emphasizing the *purpose* of a body process and the other the underlying *mechanism* by which this process occurs. In response to the question "Why do I shiver when I am cold?" one answer would be "to help warm up, because shivering generates heat." This approach, which explains body functions in terms of meeting a bodily need, emphasizes "why" body processes occur. Physiologists, however, explain "how" processes occur in the body. They view the body as a machine whose mechanisms of action can be explained in terms of cause-and-effect sequences of physical and chemical processes—the same types of processes that occur throughout the universe. A physiologist's explanation of shivering is that when temperature-sensitive nerve cells detect a fall in body temperature, they signal the area in the brain responsible for temperature regulation. In response, this brain area activates nerve pathways that ultimately bring about involuntary, oscillating muscle contractions (that is, shivering).

Structure and function are inseparable.

Physiology is closely related to **anatomy,** the study of the structure of the body. Physiological mechanisms are made possible by the structural design and relationships of the various body parts that carry out each of these functions. Just as the functioning of an automobile depends on the shapes, organization, and interactions of its various parts, the structure and function of the human body are inseparable. Therefore, as we tell the story of how the body works, we will provide sufficient anatomic background for you to understand the function of the body part being discussed.

Some structure–function relationships are obvious. For example, the heart is well designed to receive and pump blood, the teeth to tear and grind food, and the hingelike elbow joint to permit bending of the arm. In other situations, the interdependence of form and function is more subtle but equally important. Consider the interface between air and blood in the lungs as an example: The respiratory airways, which carry air from the outside into the lungs, branch extensively when they reach the lungs. Tiny air sacs cluster at the ends of the huge number of airway branches. The branching is so extensive that the lungs contain about 300 million air sacs. Similarly, the vessels carrying blood into the lungs branch extensively and form dense networks of small vessels that encircle each air sac (see ● Figure 13-2, p. 463). Because of this structural relationship, the total surface area forming an interface between the air in the air sacs and the blood in the small vessels is about the size of a tennis court. This tremendous interface is crucial for the lungs' ability to efficiently carry out their function: the transfer of needed oxygen from the air into the blood and the unloading of the waste product carbon dioxide from the blood into the air. The greater the surface area available for these exchanges, the faster oxygen and carbon dioxide can move between the air and the blood. This large functional interface packaged within the confines of your lungs is possible only because both the air-containing and blood-containing components of the lungs branch extensively.

Levels of Organization in the Body

We now turn our attention to how the body is structurally organized into a total functional unit, from the chemical level to the whole body (● Figure 1-1). These levels of organization make possible life as we know it.

The chemical level: Various atoms and molecules make up the body.

Like all matter, both living and nonliving, the human body is a combination of specific *atoms,* which are the smallest building blocks of matter. The most common atoms in the body—oxygen, carbon, hydrogen, and nitrogen—make up approximately 96% of the total body chemistry. These common atoms and a few others combine to form the *molecules* of life, such as proteins, carbohydrates, fats, and nucleic acids (genetic material, such as deoxyribonucleic acid, or DNA). These important atoms and molecules are the inanimate raw ingredients from which all living things arise. (See Appendix B for a review of this chemical level.)

The cellular level: Cells are the basic units of life.

The mere presence of a particular collection of atoms and molecules does not confer the unique characteristics of life. Instead, these nonliving chemical components must be arranged and packaged in very precise ways to form a living entity. The **cell,** the fundamental unit of both structure and function in a living being, is the smallest unit capable of carrying out the processes associated with life. Cell physiology is the focus of Chapter 2.

An extremely thin, oily barrier, the *plasma membrane,* encloses the contents of each cell and controls the movement of materials into and out of the cell. Thus, the cell's interior contains a combination of atoms and molecules that differs from the mixture of chemicals in the environment surrounding the cell. Given the importance of the plasma membrane and its associated functions for carrying out life processes, Chapter 3 is devoted entirely to this structure.

Organisms are independent living entities. The simplest forms of independent life are single-celled organisms such as bacteria and amoebas. Complex multicellular organisms, such as trees and humans, are structural and functional aggregates of trillions of cells (*multi* means "many"). In the simpler multicellular forms of life—for example, a sponge—the cells of the organism are all similar. However, more complex organisms, such as humans, have many different kinds of cells, such as muscle cells, nerve cells, and gland cells.

Each human organism begins when an egg and sperm unite to form a single new cell, which multiplies and forms a growing mass through myriad cell divisions. If cell multiplication were the only process involved in development, all the body cells would be essentially identical, as in the simplest multicellular life-forms. However, during development of complex multicellular organisms such as humans, each cell also **differentiates,** or becomes specialized to carry out a particular function. As a result of cell differentiation, your body is made up of about 200 different specialized types of cells.

BASIC CELL FUNCTIONS All cells, whether they exist as solitary cells or as part of a multicellular organism, perform certain basic functions essential for their own survival. These basic cell functions include the following:

1. Obtaining food (nutrients) and oxygen (O_2) from the environment surrounding the cell.

2. Performing chemical reactions that use nutrients and O_2 to provide energy for the cells, as follows:

$$\text{Food} + O_2 \rightarrow CO_2 + H_2O + \text{energy}$$

3. Eliminating to the cell's surrounding environment carbon dioxide (CO_2) and other by-products, or wastes, produced during these chemical reactions.

4. Synthesizing proteins and other components needed for cell structure, for growth, and for carrying out particular cell functions.

5. Controlling to a large extent the exchange of materials between the cell and its surrounding environment.

6. Moving materials internally from one part of the cell to another, with some cells also being able to move themselves through their surrounding environment.

7. Being sensitive and responsive to changes in the surrounding environment.

8. In the case of most cells, reproducing. Some body cells, most notably nerve cells and muscle cells, lose the ability to reproduce soon after they are formed. This is the reason why strokes, which result in lost nerve cells in the brain, and heart attacks, which bring about death of heart muscle cells, can be so devastating.

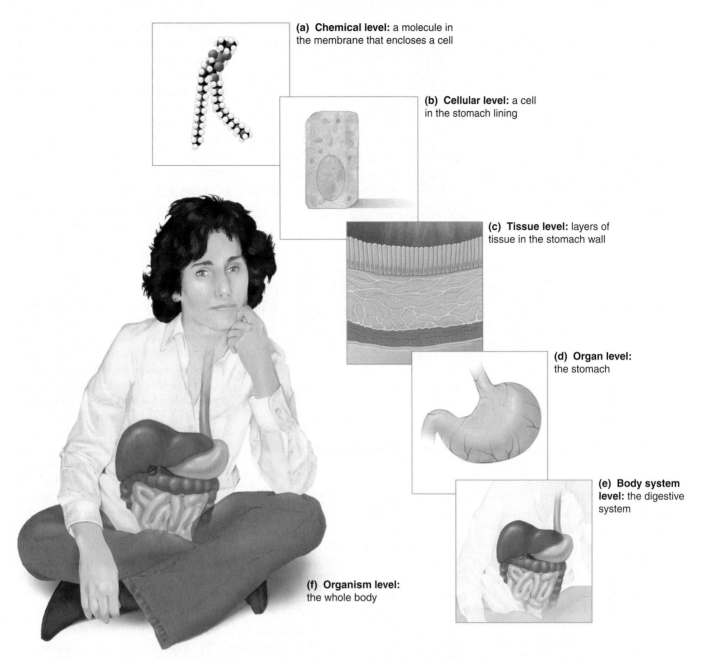

(a) Chemical level: a molecule in the membrane that encloses a cell

(b) Cellular level: a cell in the stomach lining

(c) Tissue level: layers of tissue in the stomach wall

(d) Organ level: the stomach

(e) Body system level: the digestive system

(f) Organism level: the whole body

● **FIGURE 1-1 Levels of organization in the body, showing an example for each level.**

Cells are remarkably similar in the ways they carry out these basic functions. Thus, all cells share many common characteristics.

SPECIALIZED CELL FUNCTIONS In multicellular organisms, each cell also performs a specialized function, which is usually a modification or elaboration of a basic cell function. Here are a few examples:

■ By taking special advantage of their protein-synthesizing ability, the gland cells of the digestive system secrete digestive enzymes that break down ingested food; **enzymes** are specialized proteins that speed up particular chemical reactions in the body.

■ Certain kidney cells are able to selectively retain the substances needed by the body while eliminating unwanted sub-stances in the urine because of their highly specialized ability to control exchange of materials between the cell and its environment.

■ Muscle contraction, which involves selective movement of internal structures to generate tension in the muscle cells, is an elaboration of the inherent ability of these cells to produce intracellular movement (*intra* means "within").

■ Capitalizing on the basic ability of cells to respond to changes in their surrounding environment, nerve cells generate and transmit to other regions of the body electrical impulses that relay information about changes to which the nerve cells are responsive. For example, nerve cells in the ear can relay information to the brain about sounds in the body's surroundings.

Organ:
Body structure that integrates different tissues and carries out a specific function

Stomach

Epithelial tissue:
Protection, secretion, and absorption

Connective tissue:
Structural support

Muscle tissue:
Movement

Nervous tissue:
Communication, coordination, and control

● **FIGURE 1-2 The stomach as an organ made up of all four primary tissue types.**

Each cell performs these specialized activities in addition to carrying on the unceasing, fundamental activities required of all cells. The basic cell functions are essential for survival of each individual cell, whereas the specialized contributions and interactions among the cells of a multicellular organism are essential for survival of the whole body.

Just as a machine does not function unless all its parts are properly assembled, the cells of the body must be specifically organized to carry out the life-sustaining processes of the body as a whole, such as digestion, respiration, and circulation. Cells are progressively organized into tissues, organs, body systems, and finally the whole body.

The tissue level: Tissues are groups of cells of similar specialization.

Cells of similar structure and specialized function combine to form **tissues,** of which there are four *primary types:* muscle, nervous, epithelial, and connective (● Figure 1-2). Each tissue consists of cells of a single specialized type, along with varying amounts of extracellular material (*extra* means "outside of").

■ **Muscle tissue** consists of cells specialized for contracting, which generates tension and produces movement. There are three types of muscle tissue: *skeletal muscle,* which moves the skeleton; *cardiac muscle,* which pumps blood out of the heart; and *smooth muscle,* which controls movement of contents through hollow tubes and organs, such as movement of food through the digestive tract.

■ **Nervous tissue** consists of cells specialized for initiating and transmitting electrical impulses, sometimes over long distances. These electrical impulses act as signals that relay information from one part of the body to another. Such signals are important in communication, coordination, and control in the body. Nervous tissue is found in the brain, spinal cord, nerves, and special sense organs.

■ **Epithelial tissue** consists of cells specialized for exchanging materials between the cell and its environment. Any substance that enters or leaves the body proper must cross an epithelial barrier. Epithelial tissue is organized into two general types of structures: *epithelial sheets* and *secretory glands.* Epithelial sheets are layers of very tightly joined cells that cover and line various parts of the body. For example, the outer layer of the skin is epithelial tissue, as is the lining of the digestive tract. In general, epithelial sheets serve as boundaries that separate the body from its surroundings and from the contents of cavities that open to the outside, such as the digestive tract lumen. (A **lumen** is the cavity within a hollow organ or tube.) Only selective transfer of materials is possible between regions separated by an epithelial barrier. The type and extent of controlled exchange vary, depending on the location and function of the epithelial tissue. For example, the skin can exchange very little between the body and surrounding environment, making it a protective barrier. By contrast the epithelial cells lining the small intestine of the digestive tract are specialized for absorbing nutrients that have come from outside the body.

Glands are epithelial tissue derivatives specialized for secreting. **Secretion** is the release from a cell, in response to appropriate stimulation, of specific products that have been produced by the cell. Glands are formed during embryonic development by pockets of epithelial tissue that invaginate (dip inward from the surface) and develop secretory capabilities. There are two categories of glands: *exocrine* and *endocrine* (● Figure 1-3). During development, if the connecting cells between the epithelial surface cells and the secretory gland cells within the invaginated pocket remain intact as a duct between the gland and the surface, an exocrine gland is formed. **Exocrine glands** secrete through ducts to the outside of the body (or into a cavity that communicates with the outside) (*exo* means "external"; *crine* means "secretion"). Examples are sweat glands and glands that secrete digestive juices. If, in contrast, the connecting cells disappear during development and the secretory gland cells are isolated from the surface, an endocrine gland is formed. **Endocrine glands** lack ducts and release their secretory products, known as *hormones,* internally into the blood (*endo* means "internal"). For example, the pancreas secretes insulin into the blood, which transports this hormone to its sites of action throughout the body. Most cell types depend on insulin for taking up glucose (sugar).

■ **Connective tissue** is distinguished by having relatively few cells dispersed within an abundance of extracellular material. As its name implies, connective tissue connects, supports, and anchors various body parts. It includes such diverse structures as the loose connective tissue that attaches epithelial tissue to underlying structures; tendons, which attach skeletal muscles to bones; bone, which gives the body shape, support, and protection; and blood, which transports materials from one part of the body to another. Except for blood, the cells within connective tissue produce specific structural molecules that they release into the extracellular spaces between the cells. One such molecule is the rubber band–like protein fiber *elastin;* its presence facilitates the stretching and recoiling of structures such as the lungs, which alternately inflate and deflate during breathing.

Muscle, nervous, epithelial, and connective tissue are the primary tissues in a classical sense; that is, each is an integrated collection of cells with the same specialized structure and function. The term *tissue* is also often used, as in clinical medicine, to mean the aggregate of various cellular and extracellular components that make up a particular organ (for example, lung tissue or liver tissue).

The organ level: An organ is a unit made up of several tissue types.

Organs consist of two or more types of primary tissue organized together to perform a particular function or functions. The stomach is an example of an organ made up of all four primary tissue types (see ● Figure 1-2). The tissues of the stomach function collectively to store ingested food, move it forward into the rest of the digestive tract, and begin the digestion of protein. The stomach is lined with epithelial tissue that restricts the transfer of harsh digestive chemicals and undigested food from the stomach lumen into the blood. Epithelial gland cells in the stomach include exocrine cells, which secrete protein-digesting juices into the lumen, and endocrine cells, which secrete a hormone that helps regulate the stomach's exocrine secretion and muscle contraction. The wall of the stomach contains smooth muscle tissue, whose contractions mix ingested food with the digestive juices and push the mixture out of the stomach and into the intestine. The stomach wall also contains nervous tissue, which, along with hormones, controls muscle contraction and gland secretion. Connective tissue binds together all these various tissues.

The body system level: A body system is a collection of related organs.

Groups of organs are further organized into **body systems.** Each system is a collection of organs that perform related functions and interact to accomplish a common activity that is essential for survival of the whole body. For example, the

(a) Invagination of surface epithelium during gland formation

(b) Exocrine gland

(c) Endocrine gland

● **FIGURE 1-3 Exocrine and endocrine gland formation during development.** (a) Glands arise from the formation of pocket-like invaginations of surface epithelial cells. (b) If the cells at the deepest part of the invagination become secretory and release their product through the connecting duct to the surface, an exocrine gland is formed. (c) If the connecting cells are lost and the deepest secretory cells release their product into the blood, an endocrine gland is formed.

digestive system consists of the mouth, salivary glands, pharynx (throat), esophagus, stomach, pancreas, liver, gallbladder, small intestine, and large intestine. These digestive organs cooperate to break food down into small nutrient molecules that can be absorbed into the blood for distribution to all cells.

The human body has 11 systems: circulatory, digestive, respiratory, urinary, skeletal, muscular, integumentary, immune, nervous, endocrine, and reproductive (● Figure 1-4). Chapters 4 through 20 cover the details of these systems.

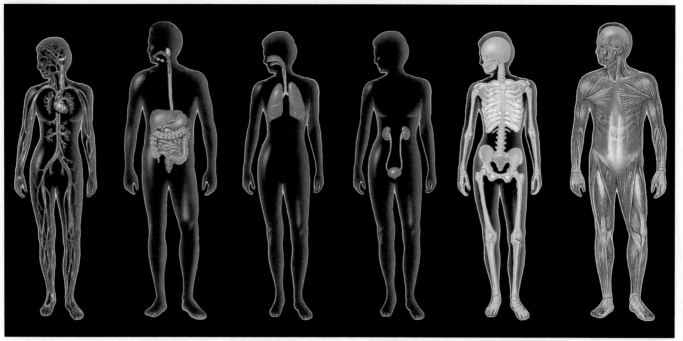

Circulatory system
heart, blood vessels, blood

Digestive system
mouth, pharynx, esophagus, stomach, small intestine, large intestine, salivary glands, exocrine pancreas, liver, gallbladder

Respiratory system
nose, pharynx, larynx, trachea, bronchi, lungs

Urinary system
kidneys, ureters, urinary bladder, urethra

Skeletal system
bones, cartilage, joints

Muscular system
skeletal muscles

● **FIGURE 1-4 Components of the body systems.**

The organism level: The body systems are packaged together into a functional whole body.

Each body system depends on the proper functioning of other systems to carry out its specific responsibilities. The whole body of a multicellular organism—a single, independently living individual—consists of the various body systems structurally and functionally linked as an entity that is separate from its surrounding environment. Thus, the body is made up of living cells organized into life-sustaining systems.

The different body systems do not act in isolation from one another. Many complex body processes depend on the interplay among multiple systems. For example, regulation of blood pressure depends on coordinated responses among the circulatory, urinary, nervous, and endocrine systems. Even though physiologists may examine body functions at any level from cells to systems (as indicated in the title of this book), their ultimate goal is to integrate these mechanisms into the big picture of how the entire organism works as a cohesive whole.

Currently, researchers are hotly pursuing several approaches to repairing or replacing tissues or organs that can no longer adequately perform vital functions because of disease, trauma, or age-related changes. (See the boxed feature on pp. 8 and 9, ■ Concepts, Challenges, and Controversies. Each chapter has similar boxed features that explore in greater depth high-interest, tangential information on such diverse topics as environmental impact on the body, aging, ethical issues, new discoveries regarding common diseases, historical perspectives, and so on.)

We next focus on how the different body systems normally work together to maintain the internal conditions necessary for life.

Concept of Homeostasis

If each cell has basic survival skills, why can't the body's cells live without performing specialized tasks and being organized according to specialization into systems that accomplish functions essential for the whole organism's survival? The cells in a multicellular organism cannot live and function without contributions from the other body cells because the vast majority of cells are not in direct contact with the external environment. The **external environment** is the surrounding environment in which an organism lives. A single-celled organism such as an amoeba obtains nutrients and O_2 directly from its immediate external surroundings and eliminates wastes back into those surroundings. A muscle cell or any other cell in a multicellular organism has the same need for life-supporting nutrient and O_2 uptake and waste elimination, yet the muscle cell is isolated from the external environment surrounding the body. How can

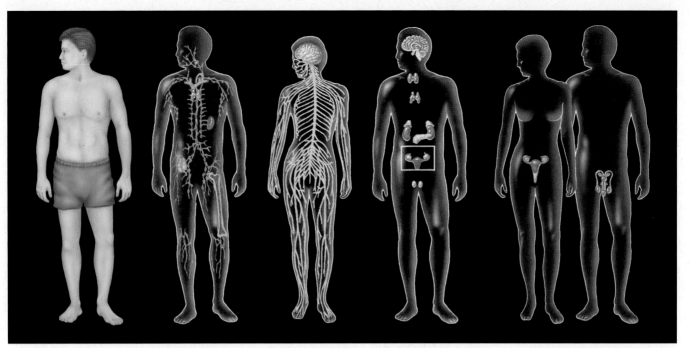

Integumentary system
skin, hair, nails

Immune system
lymph nodes, thymus, bone marrow, tonsils, adenoids, spleen, appendix, and, not shown, white blood cells, gut-associated lymphoid tissue, and skin-associated lymphoid tissue

Nervous system
brain, spinal cord, peripheral nerves, and, not shown, special sense organs

Endocrine system
all hormone-secreting tissues, including hypothalamus, pituitary, thyroid, adrenals, endocrine pancreas, gonads, kidneys, pineal, thymus, and, not shown, parathyroids, intestine, heart, skin, and adipose tissue

Reproductive system
Male: testes, penis, prostate gland, seminal vesicles, bulbourethral glands, and associated ducts

Female: ovaries, oviducts, uterus, vagina, breasts

it make vital exchanges with the external environment with which it has no contact? The key is the presence of a watery internal environment. The **internal environment** is the fluid that surrounds the cells and through which they make life-sustaining exchanges.

Body cells are in contact with a privately maintained internal environment.

The fluid collectively contained within all body cells is called **intracellular fluid (ICF).** The fluid outside the cells is called **extracellular fluid (ECF).** The ECF is the internal environment of the body. Note that the ECF is outside the cells but inside the body. You live in the external environment; your cells live within the body's internal environment.

Extracellular fluid is made up of two components: the **plasma,** the fluid portion of the blood; and the **interstitial fluid,** which surrounds and bathes the cells (*inter* means "between"; *stitial* means "that which stands") (● Figure 1-5).

No matter how remote a cell is from the external environment, it can make life-sustaining exchanges with its own surrounding fluids. In turn, particular body systems accomplish the transfer of materials between the external environment and the internal environment so that the composition of the internal environment is appropriately maintained to support

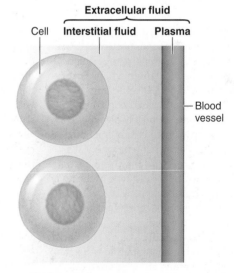

● **FIGURE 1-5 Components of the extracellular fluid (internal environment).**

the life and functioning of the cells. For example, the digestive system transfers the nutrients required by all body cells from the external environment into the plasma. Likewise, the respiratory system transfers O_2 from the external environment into the plasma. The circulatory system distributes these nutrients and O_2 throughout the body. Materials are

Stem-Cell Science and Tissue Engineering: The Quest to Make Defective Body Parts Like New Again

Liver failure, stroke, paralyzing spinal-cord injury, diabetes mellitus, damaged heart muscle, arthritis, extensive burns, surgical removal of a cancerous breast, an arm mangled in an accident. Although our bodies are remarkable and normally serve us well, sometimes a body part is defective, injured beyond repair, or lost in such situations. In addition to the loss of quality of life for affected individuals, the cost of treating patients with lost, permanently damaged, or failing organs accounts for about half of the total health-care expenditures in the United States. Ideally, when the body suffers an irreparable loss, new, permanent replacement parts could be substituted to restore normal function and appearance. Fortunately, this possibility is moving rapidly from the realm of science fiction to the reality of scientific progress.

The Medical Promise of Stem Cells

Stem cells offer exciting medical promise for repairing or replacing organs that are diseased, damaged, or worn out. **Stem cells** are versatile cells that are not specialized for a specific function but can divide to give rise to highly specialized cells while at the same time maintaining a supply of new stem cells. Two categories of stem cells are being explored: *embryonic stem cells* and *tissue-specific stem cells* from adults. **Embryonic stem cells (ESCs)** result from the early divisions of a fertilized egg. These undifferentiated cells ultimately give rise to all the mature, specialized cells of the body, while simultaneously self-renewing. ESCs are *pluripotent,* meaning they have the potential to generate any cell type in the body if given the appropriate cues.

During development, the undifferentiated ESCs give rise to many partially differentiated **tissue-specific stem cells,** each of which becomes committed to generating the highly differentiated, specialized cell types that compose a particular tissue. For example, tissue-specific muscle stem cells give rise to specialized muscle cells. Some tissue-specific stem cells remain in adult tissues where they serve as a continual source of new specialized cells to maintain or repair structure and function in that particular tissue. Tissue-specific stem cells have even been found in adult brain and muscle tis-

sue. Even though mature nerve and muscle cells cannot reproduce themselves, to a limited extent, adult brains and muscles can grow new cells throughout life by means of these persisting stem cells. However, this process is too slow to keep pace with major losses, as in a stroke or heart attack. Some investigators are searching for drugs that might spur a person's own tissue-specific stem cells into increased action to make up for damage or loss of that tissue—a feat that is not currently feasible.

More hope is pinned on nurturing stem cells outside the body for possible transplant into the body. In 1998, for the first time, researchers succeeded in isolating ESCs and maintaining them indefinitely in an undifferentiated state in culture. With **cell culture**, cells isolated from a living organism continue to thrive and reproduce in laboratory dishes when supplied with appropriate nutrients and supportive materials.

The medical promise of ESCs lies in their potential to serve as an all-purpose material that can be coaxed into whatever cell types are needed to patch up the body. Studies in the last decade demonstrate that these cells have the ability to differentiate into particular cells when exposed to the appropriate chemical signals. As scientists gradually learn to prepare the right cocktail of chemical signals to direct the undifferentiated cells into the desired cell types, they will have the potential to fill in deficits in damaged or dead tissue with healthy cells. Scientists even foresee the ability to grow customized tissues and eventually whole, made-to-order replacement organs, a process known as **tissue engineering.**

The Medical Promise of Tissue Engineering

The era of tissue engineering is being ushered in by advances in cell biology, plastic manufacturing, and computer graphics. Using computer-aided designs, very pure, biodegradable plastics are shaped into three-dimensional molds or scaffoldings that mimic the structure of a particular tissue or organ. The plastic mold is then "seeded" with the desired cell types, which are coaxed, by applying appropriate nourishing and stimulatory chemicals, into multiplying and assembling into the desired body part.

After the plastic scaffolding dissolves, only the newly generated tissue remains, ready to be implanted into a patient as a permanent, living replacement part.

More recently, some investigators are experimenting with *organ printing*. Based on the principle used in desktop printers, organ printing involves computer-aided layer-by-layer deposition of "biological ink." Biological ink consists of cells, scaffold materials, and supportive growth factors that are simultaneously printed in thin layers in a highly organized pattern based on the anatomy of the organ under construction. Fusion of these living layers forms a three-dimensional structure that mimics the body part the printed organ is designed to replace.

What about the source of cells to seed the plastic mold or print the organ? The immune system is programmed to attack foreign cells, such as invading bacteria or cells transplanted into the body from another individual. Such an attack brings about rejection of transplanted cells unless the transplant recipient is treated with *immunosuppressive drugs* (drugs that suppress the immune system's attack on the transplanted material). An unfortunate side effect of these drugs is the reduced ability of the patient's immune system to defend against potential disease-causing bacteria and viruses. To prevent rejection by the immune system and avoid the necessity of lifelong immunosuppressive drugs, tissue engineers could use appropriate specialized cells harvested from the recipient if these cells were available. However, because of the very need for a replacement part, the patient often does not have any appropriate cells for seeding or printing the replacement. This is what makes ESCs potentially so exciting. Through genetic engineering, these stem cells could be converted into "universal" seed cells that would be immunologically acceptable to any recipient; that is, they could be genetically programmed to not be rejected by any body.

Here are some of the tissue engineers' early accomplishments and future predictions:

- Engineered skin patches are being used to treat victims of severe burns.
- Laboratory-grown cartilage and bone graft substitutes are already in use.

- Lab-grown bladders are the first organs successfully implanted in humans.
- Replacement blood vessels are almost ready for clinical trials.
- Heart-muscle patches are being developed for repairing damaged hearts.
- Progress has been made on building artificial heart valves and teeth.
- Tissue-engineered scaffolding to promote nerve regeneration is being tested in animals.
- Progress has been made on growing two complicated organs, the liver and the pancreas.
- Engineered joints will be used as a living, more satisfactory alternative to the plastic and metal devices used as replacements today.
- Ultimately, complex body parts such as arms and hands will be produced in the laboratory for attachment as needed.

Tissue engineering thus holds the promise that damaged body parts can be replaced with the best alternative, a laboratory-grown version of "the real thing."

Ethical Concerns and Political Issues

Despite this potential, ESC research is fraught with controversy because of the source of these cells: They are isolated from discarded embryos from abortion clinics and in-vitro fertility ("test-tube baby") clinics. Opponents of using ESCs are morally and ethically concerned because embryos are destroyed in the process of harvesting these cells. Proponents argue that these embryos were destined to be destroyed anyway—a decision already made by the parents of the embryos—and that these stem cells have great potential for alleviating human suffering. Thus, ESC science has become inextricably linked with stem-cell politics.

Federal policy currently prohibits use of public funding to support research involving human embryos, so the scientists who isolated existing cultured ESCs relied on private money. Public policy makers, scientists, and bioethicists now face balancing a host of ethical issues against the tremendous potential clinical application of ESC research. Such research will proceed at a much faster pace if federal money is available to support it. In a controversial decision, President George W. Bush issued guidelines in August 2001 permitting the use of government funds for studies using a dwindling pool of established lines of human ESCs but not for deriving new cell lines. Increasingly, where a candidate stands on ESC science is becoming a hot campaign issue. Federal support for ESC research will be forthcoming under the leadership of President Barack Obama.

In the meantime, some states, with California leading the way, have passed legislation to provide taxpayer funds to support ESC research, with the hope of bringing sizable profits into these states as scientists discover new ways of treating a variety of diseases.

The Search for Noncontroversial Stem Cells

In a different approach, some researchers have been searching for alternative ways to obtain stem cells. Some are exploring the possibility of using tissue-specific stem cells from adult tissues as a substitute for pluripotent ESCs. Until recently, most investigators believed these adult stem cells could give rise only to the specialized cells of a particular tissue. However, although these partially differentiated adult stem cells do not have the complete developmental potential of ESCs, they have been coaxed into producing a wider variety of cells than originally thought possible. To name a few examples, provided the right supportive environment, stem cells from the brain have given rise to blood cells, bone-marrow stem cells to liver and nerve cells, and fat-tissue stem cells to bone, cartilage, and muscle cells. Thus, researchers may be able to tap into the more limited but still versatile developmental potential of adult tissue-specific stem cells. Although ESCs hold greater potential for developing treatments for a broader range of diseases, adult stem cells are more accessible than ESCs, and their use is not controversial. For example, researchers dream of being able to take fat stem cells from a person and transform them into a needed replacement knee joint. However, interest in these cells is dwindling because they have failed to live up to expectations in recent scientific studies.

Political setbacks have inspired still other scientists to search for new ways to obtain the more versatile ESCs for culturing new cell lines without destroying embryos. Following are among the recent breakthroughs:

- Turning back the clock on adult mouse skin cells, converting them to their embryonic state, by inserting four key regulatory genes active only in early embryos. These reprogrammed cells, called *induced pluripotent stem cells (iPSCs)*, have the potential to differentiate into any cell type in the body. By using iPSCs from a person's own body, scientists could manipulate these patient-specific, genetically matched cells for treating the individual, thus avoiding the issue of transplant rejection. Currently one big problem precluding the use of iPSCs is that the viruses used to insert the genes can cause cancer. Researchers are now looking for safer ways to shunt adult cells back to their embryonic state.
- Harvesting ESCs from "virgin" embryos. These are embryo-like structures formed when an egg is prodded to divide on its own without having been fertilized by sperm. These virgin embryos are not considered potential human lives because they never develop beyond a few days, but they can yield ESCs.
- Obtaining ESCs from "dead" embryos, embryos that have stopped dividing on their own in fertility labs. These embryos are normally discarded because their further development is arrested.
- Taking one cell from an eight-cell human embryo, a procedure that does not harm the embryo and is routinely done in fertility labs to check for genetic abnormalities. Now scientists are using the plucked cells to generate more ESCs.
- Nuclear transfer, in which DNA is removed from the nucleus of a collected egg and replaced with the DNA from a patient's body cell. By coaxing the egg into dividing, researchers can produce ESCs that genetically match the patient.

Whatever the source of the cells, stem-cell research promises to revolutionize medicine. According to the Center for Disease Control's National Center for Health Statistics, an estimated 3000 Americans die every day from conditions that may in the future be treatable with stem cell derivatives.

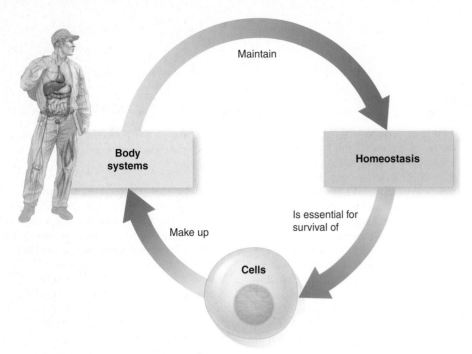

Maintain

Body systems

Homeostasis

Is essential for survival of

Make up

Cells

● **FIGURE 1-6 Interdependent relationship of cells, body systems, and homeostasis.** Homeostasis is essential for the survival of cells, body systems maintain homeostasis, and cells make up body systems. This relationship serves as the foundation for modern-day physiology.

thoroughly mixed and exchanged between the plasma and interstitial fluid across the capillaries, the smallest and thinnest of the blood vessels. As a result, the nutrients and O_2 originally obtained from the external environment are delivered to the interstitial fluid, from which the body cells pick up these needed supplies. Similarly, wastes produced by the cells are released into the interstitial fluid, picked up by the plasma, and transported to the organs that specialize in eliminating these wastes from the internal environment to the external environment. The lungs remove CO_2 from the plasma, and the kidneys remove other wastes for elimination in the urine.

Thus, a body cell takes in essential nutrients from its watery surroundings and eliminates wastes into these same surroundings, just as an amoeba does. The main difference is that each body cell must help maintain the composition of the internal environment so that this fluid continuously remains suitable to support the existence of all the body cells. In contrast, an amoeba does nothing to regulate its surroundings.

Body systems maintain homeostasis, a dynamic steady state in the internal environment.

Body cells can live and function only when the ECF is compatible with their survival; thus, the chemical composition and physical state of this internal environment must be maintained within narrow limits. As cells take up nutrients and O_2 from the internal environment, these essential materials must constantly be replenished. Likewise, wastes must constantly be removed from the internal environment so they do not reach toxic levels. Other aspects of the internal environment important for main-

taining life, such as temperature, also must be kept relatively constant. Maintenance of a relatively stable internal environment is termed **homeostasis** (*homeo* means "the same"; *stasis* means "to stand or stay").

The functions performed by each body system contribute to homeostasis, thereby maintaining within the body the environment required for the survival and function of all the cells. Cells, in turn, make up body systems. This is the central theme of physiology and of this book: *Homeostasis is essential for the survival of each cell, and each cell, through its specialized activities, contributes as part of a body system to the maintenance of the internal environment shared by all cells* (● Figure 1-6).

The fact that the internal environment must be kept relatively stable does not mean that its composition, temperature, and other characteristics are absolutely unchanging. Both external and internal factors continuously threaten to disrupt homeostasis. When any factor starts to move the internal environment away from optimal conditions, the body systems initiate appropriate counterreactions to minimize the change. For example, exposure to a cold environmental temperature (an external factor) tends to reduce the body's internal temperature. In response, the temperature control center in the brain initiates compensatory measures, such as shivering, to raise body temperature to normal. By contrast, production of extra heat by working muscles during exercise (an internal factor) tends to raise the body's internal temperature. In response, the temperature control center brings about sweating and other compensatory measures to reduce body temperature to normal.

Thus, homeostasis is not a rigid, fixed state but a dynamic steady state in which the changes that do occur are minimized by compensatory physiological responses. The term *dynamic* refers to the fact that each homeostatically regulated factor is marked by continuous change, whereas *steady state* implies that these changes do not deviate far from a constant, or steady, level. This situation is comparable to the minor steering adjustments you make as you drive a car along a straight stretch of highway. Small fluctuations around the optimal level for each factor in the internal environment are normally kept, by carefully regulated mechanisms, within the narrow limits compatible with life.

Some compensatory mechanisms are immediate, transient responses to a situation that moves a regulated factor in the internal environment away from the desired level, whereas others are more long-term adaptations that take place in response to prolonged or repeated exposure to a situation that disrupts homeostasis. Long-term adaptations make the body more efficient in responding to an ongoing or repetitive challenge. The body's reaction to exercise includes examples of both short-

What Is Exercise Physiology?

Exercise physiology is the study of both the functional changes that occur in response to a single session of exercise and the adaptations that occur as a result of regular, repeated exercise sessions. Exercise initially disrupts homeostasis. The changes that occur in response to exercise are the body's attempt to meet the challenge of maintaining homeostasis when increased demands are placed on the body. Exercise often requires prolonged coordination among most body systems, including the muscular, skeletal, nervous, circulatory, respiratory, urinary, integu-mentary (skin), and endocrine (hormone-producing) systems.

Heart rate is one of the easiest factors to monitor that shows both an immediate response to exercise and long-term adaptation to a regular exercise program. When a person begins to exercise, the active muscle cells use more O_2 to support their increased energy demands. Heart rate increases to deliver more oxygenated blood to the exercising muscles. The heart adapts to regular exercise of sufficient intensity and duration by increasing its strength and efficiency so that it pumps more blood per beat. Because of increased pumping ability, the heart does not have to beat as rapidly to pump a given quantity of blood as it did before physical training.

Exercise physiologists study the mechanisms responsible for the changes that occur as a result of exercise. Much of the knowledge gained from studying exercise is used to develop appropriate exercise programs that increase the functional capacities of people ranging from athletes to the infirm. The importance of proper and sufficient exercise in disease prevention and rehabilitation is becoming increasingly evident.

---

term compensatory responses and long-term adaptations among the different body systems. (See the accompanying boxed feature ■ A Closer Look at Exercise Physiology. Most chapters have a boxed feature focusing on exercise physiology. Also, we will be mentioning issues related to exercise physiology throughout the text. Appendix E will help you locate all the references to this important topic.)

FACTORS HOMEOSTATICALLY REGULATED Many factors of the internal environment must be homeostatically maintained. They include the following:

1. *Concentration of nutrients.* Cells need a constant supply of nutrient molecules for energy production. Energy, in turn, is needed to support life-sustaining and specialized cell activities.

2. *Concentration of O_2 and CO_2.* Cells need O_2 to carry out energy-yielding chemical reactions. The CO_2 produced during these reactions must be removed so that acid-forming CO_2 does not increase the acidity of the internal environment.

3. *Concentration of waste products.* Some chemical reactions produce end products that have a toxic effect on the body's cells if these wastes are allowed to accumulate.

4. *pH.* Changes in the pH (relative amount of acid; see p. A-10 and p. 570) of the ECF adversely affect nerve cell function and wreak havoc with the enzyme activity of all cells.

5. *Concentration of water, salt, and other electrolytes.* Because the relative concentrations of salt (NaCl) and water in the ECF influence how much water enters or leaves the cells, these concentrations are carefully regulated to maintain the proper volume of the cells. Cells do not function normally when they are swollen or shrunken. Other electrolytes (chemicals that form ions in solution and conduct electricity; see p. A-5 and A-9) perform a variety of vital functions. For example, the rhythmic beating of the heart depends on a relatively constant concentration of potassium (K^+) in the ECF.

6. *Volume and pressure.* The circulating component of the internal environment, the plasma, must be maintained at adequate volume and blood pressure to ensure bodywide distribution of this important link between the external environment and the cells.

7. *Temperature.* Body cells function best within a narrow temperature range. If cells are too cold, their functions slow down too much; if they get too hot, their structural and enzymatic proteins are impaired or destroyed.

CONTRIBUTIONS OF THE BODY SYSTEMS TO HOMEOSTASIS The 11 body systems contribute to homeostasis in the following important ways (● Figure 1-7):

1. The *circulatory system* (heart, blood vessels, and blood) transports materials such as nutrients, O_2, CO_2, wastes, electrolytes, and hormones from one part of the body to another.

2. The *digestive system* (mouth, esophagus, stomach, intestines, and related organs) breaks down dietary food into small nutrient molecules that can be absorbed into the plasma for distribution to the body cells. It also transfers water and electrolytes from the external environment into the internal environment. It eliminates undigested food residues to the external environment in the feces.

3. The *respiratory system* (lungs and major airways) gets O_2 from and eliminates CO_2 to the external environment. By adjusting the rate of removal of acid-forming CO_2, the respiratory system is also important in maintaining the proper pH of the internal environment.

4. The *urinary system* (kidneys and associated "plumbing") removes excess water, salt, acid, and other electrolytes from the plasma and eliminates them in the urine, along with waste products other than CO_2.

5. The *skeletal system* (bones, joints) provides support and protection for the soft tissues and organs. It also serves as a stor-

---

BODY SYSTEMS
Made up of cells organized according to specialization to maintain homeostasis
See Chapter 1.

NERVOUS SYSTEM
Acts through electrical signals to control rapid responses of the body; also responsible for higher functions—e.g., consciousness, memory, and creativity
See Chapters 4, 5, 6, and 7.

Information from the external environment relayed through the nervous system

RESPIRATORY SYSTEM
Obtains O_2 from and eliminates CO_2 to the external environment; helps regulate pH by adjusting the rate of removal of acid-forming CO_2
See Chapters 13 and 15.

O_2

CO_2

URINARY SYSTEM
Important in regulating the volume, electrolyte composition, and pH of the internal environment; removes wastes and excess water, salt, acid, and other electrolytes from the plasma and eliminates them in the urine
See Chapters 14 and 15.

Urine containing wastes and excess water and electrolytes

DIGESTIVE SYSTEM
Obtains nutrients, water, and electrolytes from the external environment and transfers them into the plasma; eliminates undigested food residues to the external environment
See Chapter 16.

Nutrients, water, electrolytes

Feces containing undigested food residue

REPRODUCTIVE SYSTEM
Not essential for homeostasis, but essential for perpetuation of the species
See Chapter 20.

Sperm leave male

Sperm enter female

EXTERNAL ENVIRONMENT

Regulate

Exchanges with all other systems

CIRCULATORY SYSTEM
Transports nutrients, O_2, CO_2, wastes, electrolytes, and hormones throughout the body
See Chapters 9, 10, and 11.

● **FIGURE 1-7 Role of the body systems in maintaining homeostasis.**

ENDOCRINE SYSTEM
Acts by means of hormones secreted into the blood to regulate processes that require duration rather than speed—e.g., metabolic activities and water and electrolyte balance
See Chapters 4, 18, and 19.

INTEGUMENTARY SYSTEM
Serves as a protective barrier between the external environment and the remainder of the body; the sweat glands and adjustments in skin blood flow are important in temperature regulation
See Chapters 12 and 17.

Keeps internal fluids in

Keeps foreign material out

IMMUNE SYSTEM
Defends against foreign invaders and cancer cells; paves way for tissue repair
See Chapter 12.

Protects against foreign invaders

MUSCULAR AND SKELETAL SYSTEMS
Support and protect body parts and allow body movement; heat-generating muscle contractions are important in temperature regulation; calcium is stored in the bone
See Chapters 8, 17, and 19.

Enables the body to interact with the external environment

Exchanges with all other systems

Body systems maintain homeostasis

HOMEOSTASIS
A dynamic steady state of the constituents in the internal fluid environment that surrounds and exchanges materials with the cells
See Chapter 1.
Factors homeostatically maintained:
• Concentration of nutrient molecules
  See Chapters 16, 17, 18, and 19.
• Concentration of O_2 and CO_2
  See Chapter 13.
• Concentration of waste products
  See Chapter 14.
• pH   *See Chapter 15.*
• Concentration of water, salts, and other electrolytes
  See Chapters 14, 15, 18, and 19.
• Temperature   *See Chapter 17.*
• Volume and pressure
  See Chapters 10, 14, and 15.

Homeostasis is essential for survival of cells

CELLS
Need homeostasis for their own survival and for performing specialized functions essential for survival of the whole body
See Chapters 1, 2, and 3.
Need a continual supply of nutrients and O_2 and ongoing elimination of acid-forming CO_2 to generate the energy needed to power life-sustaining cellular activities as follows:
Food + $O_2 \rightarrow CO_2 + H_2O$ + energy
See Chapter 17.

Cells make up body systems

age reservoir for calcium (Ca^{2+}), an electrolyte whose plasma concentration must be maintained within very narrow limits. Together with the muscular system, the skeletal system also enables movement of the body and its parts. Furthermore, the bone marrow—the soft interior portion of some types of bone—is the ultimate source of all blood cells.

6. The *muscular system* (skeletal muscles) moves the bones to which the skeletal muscles are attached. From a purely homeostatic view, this system enables an individual to move toward food or away from harm. Furthermore, the heat generated by muscle contraction is important in temperature regulation. In addition, because skeletal muscles are under voluntary control, a person can use them to accomplish myriad other movements of his or her own choice. These movements, which range from the fine motor skills required for delicate needlework to the powerful movements involved in weight lifting, are not necessarily directed toward maintaining homeostasis.

7. The *integumentary system* (skin and related structures) serves as an outer protective barrier that prevents internal fluid from being lost from the body and foreign microorganisms from entering. This system is also important in regulating body temperature. The amount of heat lost from the body surface to the external environment can be adjusted by controlling sweat production and by regulating the flow of warm blood through the skin.

8. The *immune system* (white blood cells, lymphoid organs) defends against foreign invaders such as bacteria and viruses and against body cells that have become cancerous. It also paves the way for repairing or replacing injured or worn-out cells.

9. The *nervous system* (brain, spinal cord, nerves, and sense organs) is one of the body's two major regulatory systems. In general, it controls and coordinates bodily activities that require swift responses. It is especially important in detecting changes in the external environment and initiating reactions to them. Furthermore, it is responsible for higher functions that are not entirely directed toward maintaining homeostasis, such as consciousness, memory, and creativity.

10. The *endocrine system* (all hormone-secreting glands) is the other major regulatory system. In contrast to the nervous system, the endocrine system in general regulates activities that require duration rather than speed, such as growth. It is especially important in controlling the concentration of nutrients and, by adjusting kidney function, controlling the volume and electrolyte composition of the ECF.

11. The *reproductive system* (male and female gonads and related organs) is not essential for homeostasis and therefore is not essential for survival of the individual. It is essential, however, for perpetuating the species.

As we examine each of these systems in greater detail, always keep in mind that the body is a coordinated whole even though each system provides its own special contributions. It is easy to forget that all the body parts actually fit together into a functioning, interdependent whole body. Accordingly, each chapter begins with a figure and discussion that focus on how the body system to be described fits into the body as a whole. In addition,

each chapter ends with a brief review of the homeostatic contributions of the body system. As a further tool to help you keep track of how all the pieces fit together, ● Figure 1-7 is duplicated on the inside front cover as a handy reference.

Also be aware that the functioning whole is greater than the sum of its separate parts. Through specialization, cooperation, and interdependence, cells combine to form an integrated, unique, single living organism with more diverse and complex capabilities than are possessed by any of the cells that make it up. For humans, these capabilities go far beyond the processes needed to maintain life. A cell, or even a random combination of cells, obviously cannot create an artistic masterpiece or design a spacecraft, but body cells working together permit those capabilities in an individual.

Now that you have learned what homeostasis is and how the functions of different body systems maintain it, let's look at the regulatory mechanisms by which the body reacts to changes and controls the internal environment.

Homeostatic Control Systems

A **homeostatic control system** is a functionally interconnected network of body components that operate to maintain a given factor in the internal environment relatively constant around an optimal level. To maintain homeostasis, the control system must be able to (1) detect deviations from normal in the internal environmental factor that needs to be held within narrow limits; (2) integrate this information with any other relevant information; and (3) make appropriate adjustments in the activity of the body parts responsible for restoring this factor to its desired value.

Homeostatic control systems may operate locally or bodywide.

Homeostatic control systems can be grouped into two classes—intrinsic and extrinsic controls. **Intrinsic,** or **local, controls** are built into or are inherent in an organ (*intrinsic* means "within"). For example, as an exercising skeletal muscle rapidly uses up O_2 to generate energy to support its contractile activity, the O_2 concentration within the muscle falls. This local chemical change acts directly on the smooth muscle in the walls of the blood vessels that supply the exercising muscle, causing the smooth muscle to relax so that the vessels dilate, or open widely. As a result, increased blood flows through the dilated vessels into the exercising muscle, bringing in more O_2. This local mechanism helps maintain an optimal level of O_2 in the fluid immediately around the exercising muscle's cells.

Most factors in the internal environment are maintained, however, by **extrinsic,** or **systemic, controls,** which are regulatory mechanisms initiated outside an organ to alter the organ's activity (*extrinsic* means "outside of"). Extrinsic control of the organs and body systems is accomplished by the nervous and endocrine systems, the two major regulatory systems. Extrinsic control permits coordinated regulation of several organs toward a common goal; in contrast, intrinsic controls serve only the organ in which they occur. Coordinated, overall regulatory mechanisms are crucial for maintaining the dynamic steady state in the internal environment as a whole. To restore blood pressure to the proper level when it falls too low, for example, the nervous system acts simultaneously on the heart and blood vessels throughout the body to increase the blood pressure to normal.

To stabilize the physiological factor being regulated, homeostatic control systems must be able to detect and resist change. The term **feedback** refers to responses made after a change has been detected; the term **feedforward** is used for responses made in anticipation of a change. Let us look at these mechanisms in more detail.

Negative feedback opposes an initial change and is widely used to maintain homeostasis.

Homeostatic control mechanisms operate primarily on the principle of negative feedback. In **negative feedback,** a change in a homeostatically controlled factor triggers a response that seeks to restore the factor to normal by moving the factor in the opposite direction of its initial change. That is, a corrective adjustment opposes the original deviation from the normal desired level.

A common example of negative feedback is control of room temperature. Room temperature is a **controlled variable,** a factor that can vary but is held within a narrow range by a control system. In our example, the control system includes a thermostat, a furnace, and all their electrical connections. The room temperature is determined by the activity of the furnace, a heat source that can be turned on or off. To switch on or off appropriately, the control system as a whole must "know" what the *actual* room temperature is, "compare" it with the *desired* room temperature, and "adjust" the output of the furnace to bring the actual temperature to the desired level. A thermometer in the thermostat provides information about the actual room temperature. The thermometer is the **sensor,** which monitors the magnitude of the controlled variable. The sensor typically converts the original information regarding a change into a "language" the control system can "understand." For example, the thermometer converts the magnitude of the air temperature into electrical impulses. This message serves as the input into the control system. The thermostat setting provides the desired temperature level, or **set point.** The thermostat acts as an **integrator,** or **control center:** It compares the sensor's input with the set point and adjusts the heat output of the furnace to bring about the appropriate response to oppose a deviation from the set point. The furnace is the **effector,** the component of the control system commanded to bring about the desired effect. These general components of a negative-feedback control system are summarized in ● Figure 1-8a. Carefully examine this figure and its key; the symbols and conventions introduced here are used in comparable flow diagrams throughout the text.

Let us look at a typical negative-feedback loop. For example, if the room temperature falls below the set point because it is cold outside, the thermostat, through connecting circuitry, activates the furnace, which produces heat to raise the room temperature (● Figure 1-8b). Once the room temperature reaches the set point, the thermometer no longer detects a deviation from that point. As a result, the activating mechanism in the thermostat and the furnace are switched off. Thus, the heat from the furnace counteracts, or is "negative" to, the original fall in temperature. If the heat-generating pathway were not shut off once the target temperature was reached, heat production would continue and the room would get hotter and hotter. Overshooting the set point does not occur, because the heat "feeds back" to shut off the thermostat that triggered its output. Thus, a negative-feedback control system detects a change away from the ideal value in a controlled variable, initiates mechanisms to correct the situation, then shuts itself off. In this way, the controlled variable does not drift too far above or below the set point.

What if the original deviation is a rise in room temperature above the set point because it is hot outside? A heat-producing furnace is of no use in returning the room temperature to the desired level. An opposing control system involving a cooling air conditioner is needed to reduce the room temperature. In this case, the thermostat, through connecting circuitry, activates the air conditioner, which cools the room air, the opposite effect from that of the furnace. In negative-feedback fashion, once the set point is reached, the air conditioner is turned off to prevent the room from becoming too cold. Note that if the controlled variable can be deliberately adjusted to oppose a change in one direction only, the variable can move in uncontrolled fashion in the opposite direction. For example, if the house is equipped only with a furnace that produces heat to oppose a fall in room temperature, no mechanism is available to prevent the house from getting too hot in warm weather. However, the room temperature can be kept relatively constant through two opposing mechanisms, one that heats and one that cools the room, despite wide variations in the temperature of the external environment.

Homeostatic negative-feedback systems in the human body operate in the same way. For example, when temperature-monitoring nerve cells detect a decrease in body temperature below the desired level, these sensors signal the temperature control center in the brain, which begins a sequence of events that ends in responses, such as shivering, that generate heat and raise the temperature to the proper level (● Figure 1-8c). When the body temperature reaches the set point, the temperature-monitoring nerve cells turn off the stimulatory signal to the skeletal muscles. As a result, the body temperature does not continue to increase above the set point. Conversely, when the temperature-monitoring nerve cells detect a rise in body temperature above normal, cooling mechanisms such as sweating are called into play to lower the temperature to normal. When the temperature reaches the set point, the cooling mechanisms are shut off. As with body temperature, opposing mechanisms can move most homeostatically controlled variables in either direction as needed.

Positive feedback amplifies an initial change.

In negative feedback, a control system's output is regulated to resist change so that the controlled variable is kept at a relatively steady set point. With **positive feedback,** by contrast, the

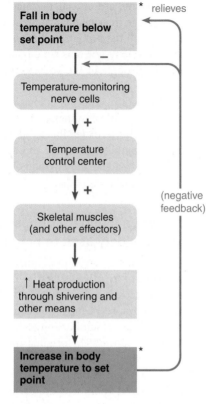

(a) Components of a negative-feedback control system

(b) Negative-feedback control of room temperature

(c) Negative-feedback control of body temperature

KEY

Flow diagrams throughout the text
+ = Stimulates or activates
− = Inhibits or shuts off
⬭ = Physical entity, such as body structure or a chemical
☐ = Actions
❙ = Compensatory pathway
❙ = Turning off of compensatory pathway (negative feedback)
* Note that lighter and darker shades of the same color are used to denote, respectively, a decrease or an increase in a controlled variable.

● **FIGURE 1-8 Negative feedback.**

output enhances or amplifies a change so that the controlled variable continues to move in the direction of the initial change. Such action is comparable to the heat generated by a furnace triggering the thermostat to call for even *more* heat output from the furnace so that the room temperature would continuously rise.

Because the major goal in the body is to maintain stable, homeostatic conditions, positive feedback does not occur nearly as often as negative feedback. Positive feedback does play an important role in certain instances, however, as in birth of a baby. The hormone oxytocin causes powerful contractions of the uterus. As the contractions push the baby against the cervix (the exit from the uterus), the resultant stretching of the cervix triggers a sequence of events that brings about the release of

even more oxytocin, which causes even stronger uterine contractions, triggering the release of more oxytocin, and so on. This positive-feedback cycle does not stop until the baby is finally born. Likewise, all other normal instances of positive-feedback cycles in the body include some mechanism for stopping the cycle.

Feedforward mechanisms initiate responses in anticipation of a change.

In addition to feedback mechanisms, which bring about a reaction to a change in a regulated variable, the body less frequently employs feedforward mechanisms, which respond in anticipation of a change in a regulated variable. For example, when a meal is still in the digestive tract, a feedforward mechanism increases secretion of a hormone (insulin) that will promote the cellular uptake and storage of ingested nutrients after they have been absorbed from the digestive tract. This anticipatory response helps limit the rise in blood nutrient concentration after nutrients have been absorbed.

Disruptions in homeostasis can lead to illness and death.

Despite control mechanisms, when one or more of the body's systems malfunction, homeostasis is disrupted, and all the cells suffer because they no longer have an optimal environment in which to live and function. Various pathophysiological states develop, depending on the type and extent of the disruption. The term **pathophysiology** refers to the abnormal functioning of the body (altered physiology) associated with disease. When a homeostatic disruption becomes so severe that it is no longer compatible with survival, death results.

Chapter in Perspective: Focus on Homeostasis

In this chapter, you learned what homeostasis is: a dynamic steady state of the constituents in the internal fluid environment (the extracellular fluid) that surrounds and exchanges materials with the cells. Maintenance of homeostasis is essential for the survival and normal functioning of cells. Each cell, through its specialized activities, contributes as part of a body system to the maintenance of homeostasis.

This relationship is the foundation of physiology and the central theme of this book. We have described how cells are organized according to specialization into body systems. How homeostasis is essential for cell survival and how body systems maintain this internal constancy are the topics covered in the rest of this book. Each chapter concludes with this capstone feature to facilitate your understanding of how the system under discussion contributes to homeostasis, as well as of the interactions and interdependency of the body systems.

REVIEW EXERCISES

Objective Questions (Answers on p. A-39)

1. Which of the following activities is *not* carried out by every cell in the body?
   a. obtaining O_2 and nutrients
   b. performing chemical reactions to acquire energy for the cell's use
   c. eliminating wastes
   d. controlling to a large extent exchange of materials between the cell and its external environment
   e. reproducing

2. Which of the following is the proper progression of the levels of organization in the body?
   a. chemicals, cells, organs, tissues, body systems, whole body
   b. chemicals, cells, tissues, organs, body systems, whole body
   c. cells, chemicals, tissues, organs, whole body, body systems
   d. cells, chemicals, organs, tissues, whole body, body systems
   e. chemicals, cells, tissues, body systems, organs, whole body

3. Which of the following is *not* a type of connective tissue?
   a. bone
   b. blood

   c. the spinal cord
   d. tendons
   e. the tissue that attaches epithelial tissue to underlying structures

4. The term *tissue* can apply either to one of the four primary tissue types or to a particular organ's aggregate of cellular and extracellular components. *(True or false?)*

5. Cells in a multicellular organism have specialized to such an extent that they have little in common with single-celled organisms. *(True or false?)*

6. Cell specializations are usually a modification or elaboration of one of the basic cell functions. *(True or false?)*

7. The four primary types of tissue are ____, ____, ____, and ____.

8. The term ____ refers to the release from a cell, in response to appropriate stimulation, of specific products that have in large part been synthesized by the cell.

9. ____ glands secrete through ducts to the outside of the body, whereas ____ glands release their secretory products, known as ____, internally into the blood.

10. ____ controls are inherent to an organ, whereas ____ controls are regulatory mechanisms initiated outside an organ that alter the activity of the organ.

11. Match the following:

___ 1. circulatory system
___ 2. digestive system
___ 3. respiratory system
___ 4. urinary system
___ 5. muscular and skeletal systems
___ 6. integumentary system
___ 7. immune system
___ 8. nervous system
___ 9. endocrine system
___ 10. reproductive system

(a) obtains O_2 and eliminates CO_2
(b) support, protect, and move body parts
(c) controls, via hormones it secretes, processes that require duration
(d) acts as transport system
(e) removes wastes and excess water, salt, and other electrolytes
(f) perpetuates the species
(g) obtains nutrients, water, and electrolytes
(h) defends against foreign invaders and cancer
(i) acts through electrical signals to control body's rapid responses
(j) serves as outer protective barrier

Essay Questions

1. Define *physiology.*

2. What are the basic cell functions?

3. Distinguish between the external environment and the internal environment. What constitutes the internal environment?

4. What fluid compartments make up the internal environment?

5. Define *homeostasis.*

6. Describe the interrelationships among cells, body systems, and homeostasis.

7. What factors must be homeostatically maintained?

8. Define and describe the components of a homeostatic control system.

9. Compare negative and positive feedback.

POINTS TO PONDER

(Explanations on p. A-39)

1. Considering the nature of negative-feedback control and the function of the respiratory system, what effect do you predict that a decrease in CO_2 in the internal environment would have on how rapidly and deeply a person breathes?

2. Would the O_2 levels in the blood be (a) normal, (b) below normal, or (c) elevated in a patient with severe pneumonia resulting in impaired exchange of O_2 and CO_2 between the air and blood in the lungs? Would the CO_2 levels in the same patient's blood be (a) normal, (b) below normal, or (c) elevated? Because CO_2 reacts with H_2O to form carbonic acid (H_2CO_3), would the patient's blood (a) have a normal pH, (b) be too acidic, or (c) not be acidic enough (that is, be too alkaline), assuming that other compensatory measures have not yet had time to act?

3. The hormone insulin enhances the transport of glucose (sugar) from the blood into most body cells. Its secretion is controlled by a negative-feedback system between the concentration of glucose in the blood and the insulin-secreting cells. Therefore, which of the following statements is correct?
   a. A decrease in blood glucose concentration stimulates insulin secretion, which in turn further lowers blood glucose concentration.
   b. An increase in blood glucose concentration stimulates insulin secretion, which in turn lowers blood glucose concentration.
   c. A decrease in blood glucose concentration stimulates insulin secretion, which in turn increases blood glucose concentration.
   d. An increase in blood glucose concentration stimulates insulin secretion, which in turn further increases blood glucose concentration.
   e. None of the preceding is correct.

4. Given that most AIDS victims die from overwhelming infections or rare types of cancer, what body system do you think HIV (the AIDS virus) impairs?

5. Body temperature is homeostatically regulated around a set point. Given your knowledge of negative feedback and homeostatic control systems, predict whether narrowing or widening of the blood vessels of the skin will occur when a person exercises strenuously. (*Hints:* Muscle contraction generates heat. Narrowing of the vessels supplying an organ decreases blood flow through the organ, whereas vessel widening increases blood flow through the organ. The more warm blood flowing through the skin, the greater is the loss of heat from the skin to the surrounding environment.)

CLINICAL CONSIDERATION

(Explanation on p. A-39)

Jennifer R. has the "stomach flu" that is going around campus and has been vomiting profusely for the past 24 hours. Not only has she been unable to keep down fluids or food but she has also lost the acidic digestive juices secreted by the stomach that are normally reabsorbed back into the blood farther down the digestive tract. In what ways might this condition threaten to disrupt homeostasis in Jennifer's internal environment? That is, what homeostatically maintained factors are moved away from normal by her profuse vomiting? What body systems respond to resist these changes?

Body Systems

Body systems maintain homeostasis

Homeostasis
The specialized activities of the cells that make up the body systems are aimed at maintaining homeostasis, a dynamic steady state of the constituents in the internal fluid environment.

Homeostasis is essential for survival of cells

Cells

Nucleus

Organelles

Plasma membrane

Cytosol

Cells make up body systems

Cells are the highly organized, living building blocks of the body. A cell has three major parts: the **plasma membrane,** which encloses the cell; the **nucleus,** which houses the cell's genetic material; and the **cytoplasm.** The cytoplasm consists of the cytosol, organelles, and cytoskeleton. The *cytosol* is a gel-like liquid within which the organelles and cytoskeleton are suspended. *Organelles* are discrete, well-organized structures that carry out specialized functions. The *cytoskeleton* is a protein scaffolding that extends throughout the cell and serves as the cell's "bone and muscle."

Through the coordinated action of these components, every cell performs certain basic functions essential to its own survival and a specialized task that helps maintain homeostasis. Cells are organized according to their specialization into body systems that maintain the stable internal environment essential for the whole body's survival. All body functions ultimately depend on the activities of the individual cells that make up the body.

Cell Physiology

CENGAGENOW™ Log on to CengageNOW at **http://www .cengage.com/sso/** for an opportunity to explore a learning module that illustrates difficult concepts with self-study tutorials, animations, and interactive quizzes to help you learn, review, and master physiology concepts.

Cell Theory

Although the same chemicals that make up living cells are also found in nonliving matter, researchers have not been able to organize these chemicals into living cells in a laboratory. Life stems from the unique and complex organization and interactions of these inanimate chemicals within the cell. Cells, the smallest living entities, serve as the living building blocks for the immensely complicated whole body. Thus, cells are the bridge between chemicals and humans (and all other living organisms). All body functions of a multicellular organism ultimately depend on the collective structural and functional capabilities of its individual cells. Furthermore, all new cells and all new life arise from the division of preexisting cells, not from nonliving sources. Because of this continuity of life, the cells of all organisms are fundamentally similar in structure and function. ▲ Table 2-1 summarizes these principles, which are known collectively as the **cell theory.** By probing deeper into the molecular structure and organization of the cells that make up the body, modern physiologists are unraveling many of the broader mysteries of how the body works.

Observations of Cells

The cells that make up the human body are so small they cannot be seen by the unaided eye. The smallest visible particle is 5 to 10 times larger than a typical human cell, which averages about 10 to 20 micrometers (μm) in diameter (1 μm = 1 millionth of a meter). About 100 average-sized cells lined up side by side would stretch a distance of only 1 mm (1 mm = 1 thousandth of a meter; 1 m = 39.37 in.). (See Appendix A for a comparison of metric units and their English equivalents. This appendix also provides a visual comparison of the size of cells in relation to other selected structures.)

Until the microscope was invented in the middle of the 17th century, scientists did not know that cells existed. With the development of better light microscopes in the early 19th century, they learned that all plant and animal tissues consist of individual cells. The cells of a hummingbird, a human, and a whale are all about the same size. Larger species have more cells, not larger cells. These early investigators also discovered that cells are filled with a fluid, which, given the microscopic capabilities of the time, appeared to be a rather uniform, soupy mixture

▲ TABLE 2-1	**Principles of the Cell Theory**

- The cell is the smallest structural and functional unit capable of carrying out life processes.
- The functional activities of each cell depend on the specific structural properties of the cell.
- Cells are the living building blocks of all multicellular organisms.
- An organism's structure and function ultimately depend on the collective structural characteristics and functional capabilities of its cells.
- All new cells and new life arise only from preexisting cells.
- Because of this continuity of life, the cells of all organisms are fundamentally similar in structure and function.

about 100 times more powerful than light microscopes.) Now that scientists have even more sophisticated microscopes, biochemical techniques, cell culture technology, and genetic engineering, the concept of the cell as a microscopic bag of formless fluid has given way to our current understanding of the cell as a complex, highly organized, compartmentalized structure. (See the accompanying boxed feature, ■ Concepts, Challenges, and Controversies, for a glimpse into an interesting chapter in the history of cell culture.)

An Overview of Cell Structure

The trillions of cells in a human body are classified into about 200 different types based on specific variations in structure and function. Yet, despite their diverse structural and functional specializations, different cells share many common features. Most cells have three major subdivisions: the *plasma membrane*, which encloses the cells; the *nucleus*, which contains the cell's genetic material; and the *cytoplasm*, the portion of the cell's interior not occupied by the nucleus (● Figure 2-1). Here, we provide an overview of each subdivision, then we will focus primarily on the cytoplasm in this chapter. The plasma membrane and nucleus are described in further detail later.

believed to be the elusive "stuff of life." In the 1940s, when researchers first employed electron microscopes to observe living matter, they began to realize the great diversity and complexity of the internal structure of cells. (Electron microscopes are

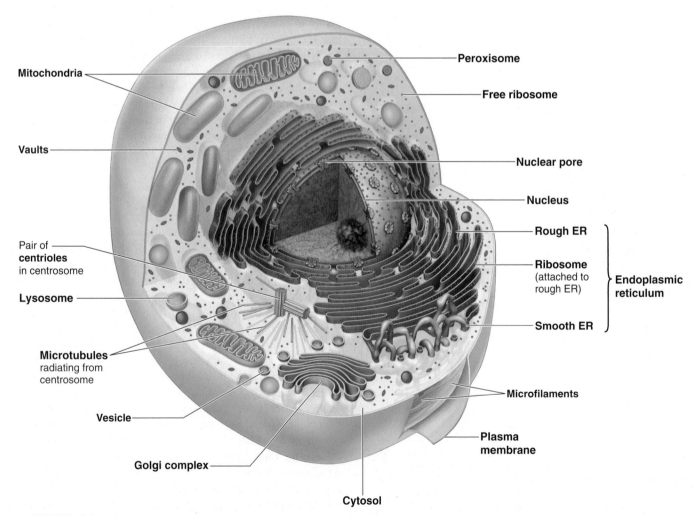

● **FIGURE 2-1 Diagram of cell structures visible under an electron microscope.**

HeLa Cells: Problems in a "Growing" Industry

Many basic advances in cell physiology, genetics, and cancer research have come about through the use of cells grown, or cultured, outside the body. In the middle of the last century, many attempts were made to culture human cells using tissues obtained from biopsies or surgical procedures. These early attempts usually met with failure; the cells died after a few days or weeks in culture, most without having divided. These difficulties continued until February 1951, when a researcher at Johns Hopkins University received a sample of a cervical cancer from a patient named Henrietta Lacks. Following convention, the culture was named HeLa by combining the first two letters of the donor's first and last names. This cell line not only grew but prospered under culture conditions and represented one of the earliest cell lines successfully grown outside the body.

Researchers were eager to have human cells available on demand to study the effects of drugs, toxic chemicals, radiation, and viruses on human tissue. For example, poliomyelitis virus reproduced well in HeLa cells, providing a breakthrough in the development of a polio vaccine. As cell-culture techniques improved, human cell lines were started from other cancers and from normal tissues, including heart, kidney, and liver tissues. By the early 1960s, a central collection of cell lines had been established in Washington, D.C., and cultured human cells were an important tool in many areas of biological research.

But in 1966, geneticist Stanley Gartler made a devastating discovery. He analyzed 18 different human cell lines and found that all had been contaminated and taken over by HeLa cells. Over the next two years, scientists confirmed that 24 of the 32 cell lines in the central repository were actually HeLa cells. Researchers who had spent years studying what they thought were heart or kidney cells had in reality been working with a cervical cancer cell instead. Gartler's discovery meant that hundreds of thousands of experiments performed in laboratories around the world were invalid.

As painful as this lesson was, scientists started over, preparing new cell lines and using new, stricter rules of technique to prevent contamination with HeLa cells. Unfortunately, the problem did not end. In 1974, Walter Nelson-Rees published a paper demonstrating that 5 cell lines extensively used in cancer research were in fact all HeLa cells. In 1976, 11 additional cell lines, each widely used in research, were also found to be HeLa cells; and in 1981, Nelson-Rees listed 22 more cell lines that were contaminated with HeLa. In all, one third of all cell lines used in cancer research were apparently really HeLa cells. The result was an enormous waste of dollars and resources. In 2005 Roland Nardone, frustrated by the still ongoing problem of cross-contamination, took up the battle. He wrote a "call for action" white paper arguing for two broad changes: increasing education efforts on how to avoid contamination and having scientific grant-awarding agencies and journals mandate that the cell lines used in the research they fund or publish be checked for authenticity. However, to date, agencies and journals have been reluctant to impose regulations in the belief that the solution instead should reside with professional societies and scientists themselves.

The invasion of other cultures by HeLa cells is a testament to the ferocious and aggressive nature of some cancer cells. In cell-culture laboratories, rules of sterile technique are supposed to ensure that cross-contamination of one culture with another does not occur. But researchers are only human, and they sometimes make mistakes. For example, perhaps a bottle of culture medium was contaminated through improper handling or a cell line was mislabeled. Whatever the case, clearly at some point one or more HeLa cells were introduced into cultures where they did not belong.

HeLa cells divide more rapidly than most other human cells, whether normal or cancerous, and thus they use up nutrients more quickly than other cell types. What's more, cells grown in culture are not readily distinguished from one another just by looking; usually, biochemical tests are needed for identification. As a result, within a few cycles of transfer, a culture that started out as kidney cells or some other human cell type can be taken over completely by the rapidly growing HeLa cells, which crowd out the original cell line just as cancer cells do in the body.

Henrietta Lacks died long ago from the cervical cancer that started the HeLa line, yet these potent cells live on. Their spread throughout human cell cultures underscores the unrelenting nature of cancer, a disease of uncontrolled growth and resource consumption.

The plasma membrane bounds the cell.

The **plasma membrane** is a very thin membranous structure that encloses each cell and is composed mostly of lipid (fat) molecules and studded with proteins. This oily barrier separates the cell's contents from its surroundings; it keeps the intracellular fluid (ICF) within the cells from mingling with the extracellular fluid (ECF) outside the cells. The plasma membrane is not simply a mechanical barrier to hold in the contents of a cell; its proteins selectively control movement of molecules between the ICF and ECF. The plasma membrane can be likened to the gated walls that enclosed ancient cities. Through this structure, the cell controls the entry of nutrients and other needed supplies and the export of products manufactured within, while at the same time guarding against unwanted traffic into or out of the cell. The plasma membrane is discussed thoroughly in Chapter 3.

The nucleus contains the DNA.

The two major parts of the cell's interior are the nucleus and the cytoplasm. The **nucleus,** which is typically the largest single organized cell component, can be seen as a distinct spherical or oval structure, usually located near the center of the cell. It is surrounded by a double-layered membrane, the **nuclear envelope,** which separates the nucleus from the rest of the cell. The nuclear envelope is pierced by many **nuclear pores** that allow necessary traffic to move between the nucleus and the cytoplasm.

The nucleus houses the cell's genetic material, **deoxyribonucleic acid (DNA),** which has two important functions: (1) directing protein synthesis and (2) serving as a genetic blueprint during cell replication. DNA provides codes, or "instructions," for directing synthesis of specific structural and enzymatic proteins within the cell. By specifying the kinds and amounts of proteins that are produced, the nucleus indirectly governs most cell activities and serves as the cell's control center.

Three types of **ribonucleic acid (RNA)** play roles in protein synthesis. First, DNA's genetic code for a particular protein is transcribed into a **messenger RNA (mRNA)** molecule, which exits the nucleus through the nuclear pores. Within the cytoplasm, mRNA delivers the coded message to *ribosomes,* which "read" the code and translate it into the appropriate amino acid sequence for the designated protein being synthesized. **Ribosomal RNA (rRNA)** is an essential component of ribosomes. **Transfer RNA (tRNA)** delivers the appropriate amino acids within the cytoplasm to their designated site in the protein under construction.

Besides providing codes for protein synthesis, DNA also serves as a genetic blueprint during cell replication to ensure that the cell produces additional cells just like itself, thus continuing the identical type of cell line within the body. Furthermore, in the reproductive cells (eggs and sperm), the DNA blueprint serves to pass on genetic characteristics to future generations. (See Appendix C for further details of DNA and RNA function and protein synthesis.)

The cytoplasm consists of various organelles, the cytoskeleton, and cytosol.

The **cytoplasm** is that portion of the cell interior not occupied by the nucleus. It contains a number of discrete, specialized *organelles* (the cell's "little organs") and the *cytoskeleton* (a scaffolding of proteins that serve as the cell's "bone and muscle") dispersed within the *cytosol* (a complex, gel-like liquid).

Organelles are distinct, highly organized structures that perform specialized functions within the cell. On average, nearly half of the total cell volume is occupied by two categories of organelles—*membranous organelles* and *nonmembranous organelles.* Each **membranous organelle** is a separate compartment within the cell that is enclosed by a membrane similar to the plasma membrane. Thus, the contents of a membranous organelle are separated from the surrounding cytosol and from the contents of other organelles. Nearly all cells contain five main types of membranous organelles—the *endoplasmic reticulum, Golgi complex, lysosomes, peroxisomes,* and *mitochondria.* Membranous organelles are like intracellular "specialty shops." Each is a separate internal compartment that contains a specific set of chemicals for carrying out a particular cellular function. This compartmentalization permits chemical activities that would not be compatible with one another to occur simultaneously within the cell. For example, the enzymes that destroy unwanted proteins in the cell operate within the protective confines of the lysosomes without the risk of destroying essential cell proteins. The **nonmembranous organelles** are not surrounded by membrane and thus are in direct contact with the cytosol. They include *ribosomes, vaults,* and *centrioles.* Like

membranous organelles, nonmembranous organelles are organized structures that carry out specific functions within the cell. Organelles are similar in all cells, although there are some variations depending on the specialized capabilities of each cell type. Just as each organ plays a role essential for survival of the whole body, each organelle performs a specialized activity necessary for survival of the whole cell.

The **cytoskeleton** is an interconnected system of protein fibers and tubes that extends throughout the cytosol. This elaborate protein network gives the cell its shape, provides for its internal organization, and regulates its various movements.

The remainder of the cytoplasm not occupied by organelles and cytoskeleton consists of the **cytosol** ("cell liquid"). The cytosol is a semiliquid, gel-like mass. Many of the chemical reactions that are compatible with one another are conducted in the cytosol. (For clarification, the intracellular fluid encompasses all the fluid inside the cell, including that within the cytosol, the organelles, and the nucleus.)

In this chapter, we examine each of the cytoplasmic components in more detail, concentrating first on the membranous organelles.

Endoplasmic Reticulum and Segregated Synthesis

The **endoplasmic reticulum (ER)** is an elaborate fluid-filled membranous system distributed extensively throughout the cytosol. It is primarily a protein- and lipid-producing factory. Two distinct types of endoplasmic reticulum—the rough ER and the smooth ER—can be distinguished. The **rough ER** consists of stacks of relatively flattened interconnected sacs, whereas the **smooth ER** is a meshwork of tiny interconnected tubules (● Figure 2-2). Even though these two regions differ considerably in appearance and function, they are connected to each other. In other words, the ER is one continuous organelle with many interconnected channels. The relative amount of rough and smooth ER varies between cells, depending on the activity of the cell.

The rough endoplasmic reticulum synthesizes proteins for secretion and membrane construction.

The outer surface of the rough ER membrane is studded with small particles that give it a "rough" or granular appearance under a light microscope. These particles are *ribosomes,* which are the "workbenches" where protein synthesis takes place. We will discuss the structure and function of ribosomes in more detail later. Not all ribosomes in the cell are attached to the rough ER. Unattached or "free" ribosomes are dispersed throughout the cytosol.

The rough ER, in association with its ribosomes, synthesizes and releases a variety of new proteins into the ER lumen, the fluid-filled space enclosed by the ER membrane. These proteins serve one of two purposes: (1) Some proteins are destined for export to the cell's exterior as secretory products, such as protein hormones or enzymes. (2) Other proteins are transported

Rough ER

Smooth ER

Rough ER lumen

Smooth ER lumen

Ribosomes

Sacs

Tubules

© Don W. Fawcett/Visuals Unlimited

© Don W. Fawcett/Visuals Unlimited

Rough ER lumen   Ribosomes

Smooth ER lumen

(a) Rough ER

(b) Smooth ER

● **FIGURE 2-2 Endoplasmic reticulum (ER).** (a) Diagram and electron micrograph of the rough ER, which consists of stacks of relatively flattened interconnected sacs studded with ribosomes. (b) Diagram and electron micrograph of the smooth ER, which is a meshwork of tiny interconnected tubules. The rough ER and smooth ER are connected, making one continuous organelle.

to sites within the cell for use in constructing new cellular membrane (either plasma membrane or organelle membrane) or other protein components of organelles. The membranous wall of the ER also contains enzymes essential for the synthesis of nearly all the lipids needed to produce new membranes. These newly synthesized lipids enter the ER lumen along with the proteins. Predictably, the rough ER is most abundant in cells specialized for protein secretion (for example, cells that secrete digestive enzymes) or in cells that require extensive membrane synthesis (for example, rapidly growing cells such as immature egg cells).

Within the ER lumen, a newly synthesized protein is folded into its final conformation; it may also be modified in other ways, such as being pruned or having carbohydrates attached to

it. After this processing, a new protein cannot pass out of the lumen through the ER membrane and therefore becomes permanently separated from the cytosol as soon as it has been synthesized. In contrast to the rough ER ribosomes, free ribosomes synthesize proteins that are used within the cytosol. In this way, newly produced molecules that are destined for export out of the cell or for synthesis of new cell components (those synthesized by the ER) are physically separated from those that belong in the cytosol (those produced by the free ribosomes).

How do the molecules newly synthesized within the ER lumen get to their destinations at other sites inside the cell or to the outside of the cell if they cannot pass out through the ER membrane? They do so through the action of the smooth endoplasmic reticulum.

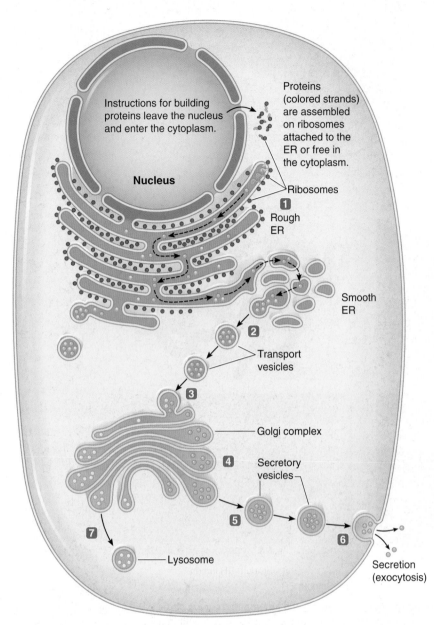

Instructions for building proteins leave the nucleus and enter the cytoplasm.

Proteins (colored strands) are assembled on ribosomes attached to the ER or free in the cytoplasm.

Nucleus

Ribosomes

1 Rough ER

Smooth ER

2

Transport vesicles

3

Golgi complex

4

Secretory vesicles

5

6

7

Lysosome

Secretion (exocytosis)

1 The rough ER synthesizes proteins to be secreted to the exterior or to be incorporated into cellular membrane.

2 The smooth ER packages the secretory product into transport vesicles, which bud off and move to the Golgi complex.

3 The transport vesicles fuse with the Golgi complex, open up, and empty their contents into the closest Golgi sac.

4 The newly synthesized proteins from the ER travel by vesicular transport through the layers of the Golgi complex, which modifies the raw proteins into final form and sorts and directs the finished products to their final destination by varying their wrappers.

5 Secretory vesicles containing the finished protein products bud off the Golgi complex and remain in the cytosol, storing the products until signaled to empty.

6 On appropriate stimulation, the secretory vesicles fuse with the plasma membrane, open, and empty their contents to the cell's exterior. Secretion has occurred by exocytosis, with the secretory products never having come into contact with the cytosol.

7 Lysosomes also bud from the Golgi complex.

● **FIGURE 2-3 Overview of the secretion process for proteins synthesized by the endoplasmic reticulum.** Note that the secretory product never comes into contact with the cytosol.

The smooth endoplasmic reticulum packages new proteins in transport vesicles.

The smooth ER does not contain ribosomes, so it is "smooth." Lacking ribosomes, it is not involved in protein synthesis. Instead, it serves other purposes that vary in different cell types.

In the majority of cells, the smooth ER is rather sparse and serves primarily as a central packaging and discharge site for molecules to be transported from the ER. Newly synthesized proteins and lipids move within the continuous lumen from the rough ER to gather in the smooth ER. Portions of the smooth ER then "bud off" (that is, balloon outward on the surface, then are pinched off), forming **transport vesicles** that enclose the new molecules in a spherical capsule derived from the smooth ER membrane (● Figure 2-3). (A **vesicle** is a fluid-filled, membrane-enclosed intracellular cargo container.) Newly syn-

thesized membrane components are rapidly incorporated into the ER membrane itself to replace the membrane used to form the transport vesicle. Transport vesicles move to the Golgi complex, described in the next section, for further processing of their cargo.

In contrast to the sparseness of the smooth ER in most cells, some specialized types of cells have an extensive smooth ER, which has additional functions as follows:

■ The smooth ER is abundant in cells that specialize in lipid metabolism—for example, cells that secrete lipid-derived steroid hormones. The membranous wall of the smooth ER, like that of the rough ER, contains enzymes for synthesis of lipids. By themselves, the lipid-producing enzymes in the rough ER membrane cannot carry out enough lipid synthesis to maintain adequate steroid-hormone secretion levels, so these cells have an expanded smooth-ER compartment that houses the additional enzymes necessary to keep pace with demands for hormone secretion.

■ In liver cells, the smooth ER contains enzymes specialized for detoxifying harmful substances produced within the body by metabolism or substances that enter the body from the outside in the form of drugs or other foreign compounds. These detoxification enzymes alter the toxic substances so that the latter can be eliminated more readily in the urine. The amount of smooth ER available in liver cells for the task of detoxification can vary dramatically, depending on the need. For example, if phenobarbital, a sedative drug, is taken in large quantities, the amount of smooth ER in liver cells doubles within a few days, only to return to normal within 5 days after drug administration ceases.

■ Muscle cells have an elaborate, modified smooth ER known as the *sarcoplasmic reticulum*, which stores calcium used in the process of muscle contraction (see p. 264).

Golgi Complex and Exocytosis

The **Golgi complex** is closely associated with the endoplasmic reticulum. Each Golgi complex consists of a stack of flattened, slightly curved, membrane-enclosed sacs (● Figure 2-4). The sacs within each Golgi stack do not come into physical contact with one another. Note that the flattened sacs are thin in the middle but have dilated, or bulging, edges. The number of Golgi complexes varies, depending on the cell type. Some cells have only one Golgi stack, whereas cells highly specialized for protein secretion may have hundreds of stacks.

Transport vesicles carry their cargo to the Golgi complex for further processing.

The majority of newly synthesized molecules that have just budded off from the smooth ER enter a Golgi stack. When a transport vesicle reaches a Golgi stack, the vesicle membrane fuses with the membrane of the sac closest to the center of the cell. The vesicle membrane opens up and becomes integrated into the Golgi membrane, and the contents of the vesicle are released to the interior of the sac (see ● Figure 2-3).

These newly synthesized raw materials from the ER travel by means of vesicle formation through the layers of the Golgi stack, from the innermost sac closest to the ER to the outermost sac near the plasma membrane. During this transit, two important, interrelated functions take place:

1. *Processing the raw materials into finished products.* Within the Golgi complex, the "raw" proteins from the ER are modified into their final form, for example, by having a carbohydrate attached. The biochemical pathways that the proteins follow during their passage through the Golgi complex are elaborate, precisely programmed, and specific for each final product.

2. *Sorting and directing the finished products to their final destinations.* The Golgi complex is responsible for sorting and segregating different types of products according to their function and destination, namely, products (1) to be secreted to the cell's exterior, (2) to be used for construction of new plasma

Golgi complex

Transport vesicle from ER, about to fuse with the Golgi membrane

Golgi lumen

Golgi sacs

Vesicles containing finished product

Golgi complex

Dr. Don Fawcett & R. Bollender/Visuals Unlimited

● **FIGURE 2-4 Golgi complex.** Diagram and electron micrograph of a Golgi complex, which consists of a stack of slightly curved, membrane-enclosed sacs. The vesicles at the dilated edges of the sacs contain finished protein products packaged for distribution to their final destination.

membrane, or (3) to be incorporated into other organelles, especially lysosomes.

The Golgi complex packages secretory vesicles for release by exocytosis.

How does the Golgi complex sort and direct finished proteins to the proper destinations? Finished products collect within the dilated edges of the Golgi complex's sacs. The edge of the outermost sac then pinches off to form a membrane-enclosed vesicle containing a selected product. For each type of product to reach its appropriate site of function, each distinct type of vesicle takes up a specific product before budding off. Vesicles with their selected cargo destined for different sites are wrapped in membranes containing distinctly different surface proteins. Each type of surface protein serves as a specific **docking marker**

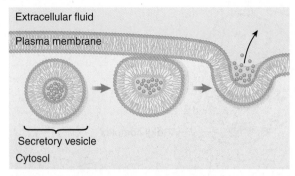

Extracellular fluid

Plasma membrane

Secretory vesicle

Cytosol

(a) Exocytosis: A secretory vesicle fuses with the plasma membrane, releasing the vesicle contents to the cell exterior. The vesicle membrane becomes part of the plasma membrane.

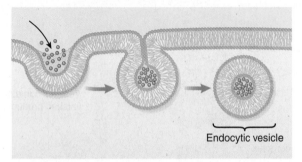

Endocytic vesicle

(b) Endocytosis: Materials from the cell exterior are enclosed in a segment of the plasma membrane that pockets inward and pinches off as an endocytic vesicle.

● **FIGURE 2-5** Exocytosis and endocytosis.

(like an address on an envelope). A vesicle can "dock" and "unload" its selected cargo only at the appropriate **docking-marker acceptor,** a protein located only at the proper destination within the cell (like a house address). Thus, Golgi products reach their appropriate site of function because they are sorted and delivered like addressed envelopes containing particular pieces of mail being delivered only to the appropriate house addresses.

Specialized secretory cells include endocrine cells, which secrete protein hormones, and digestive gland cells, which secrete digestive enzymes. In secretory cells, numerous large **secretory vesicles,** which contain proteins to be secreted, bud off from the Golgi stacks. Secretory vesicles, which are about 200 times larger than transport vesicles, store the secretory proteins until the cell is stimulated by a specific signal that indicates a need for release of that particular secretory product. On the appropriate signal, a vesicle moves to the cell's periphery, fuses with the plasma membrane, opens, and empties its contents to the outside (● Figures 2-3 and 2-5a). This mechanism—release to the exterior of substances originating within the cell—is referred to as **exocytosis** (*exo* means "out of"; *cyto* means "cell"). Exocytosis is the primary mechanism for accomplishing secretion. Secretory vesicles fuse only with the plasma membrane and not with any of the internal membranes that enclose organelles, thereby preventing fruitless or even dangerous discharge of secretory products into the organelles.

Let's now look in more detail at how secretory vesicles take up specific products in the Golgi stacks for release into the ECF

and why they are able to dock only at the plasma membrane (● Figure 2-6):

■ The newly finished proteins destined for secretion contain a unique sequence of amino acids known as a *sorting signal*, and the interior surface of the Golgi membrane contains *recognition markers,* proteins that recognize and attract specific sorting signals. Recognition of the right protein's sorting signal by the complementary membrane marker ensures that the proper cargo is captured and packaged into the secretory vesicle.

■ *Coat proteins* called *coatomer* from the cytosol bind with another specific protein facing the outer surface of the membrane. The linking of these coat proteins causes the surface membrane of the Golgi sac to curve and form a dome-shaped bud around the captured cargo. Eventually, the surface membrane closes and pinches off the vesicle.

■ After budding off, the vesicle sheds its coat proteins and exposes the docking markers, known as *v-SNAREs,* which face the outer surface of the vesicle membrane.

■ A v-SNARE can link lock-and-key fashion only with its docking marker acceptor, called a *t-SNARE,* on the targeted membrane. In the case of secretory vesicles, the targeted membrane is the plasma membrane, the designated site for secretion to take place. Thus, the v-SNAREs of secretory vesicles fuse only with the t-SNAREs of the plasma membrane. Once a vesicle has docked at the appropriate membrane by means of matching SNAREs, the two membranes completely fuse; then the vesicle opens up and empties its contents at the targeted site.

Note that the contents of secretory vesicles never come into contact with the cytosol. From the time these products are first synthesized in the ER until they are released from the cell by exocytosis, they are always wrapped in membrane and thus isolated from the remainder of the cell. By manufacturing its particular secretory protein ahead of time and storing this product in secretory vesicles, a secretory cell has a readily available reserve from which to secrete large amounts of this product on demand. If a secretory cell had to synthesize all its product on the spot as needed for export, the cell would be more limited in its ability to meet varying levels of demand.

Secretory vesicles are formed only by secretory cells. However, the Golgi complexes of these and other cell types also sort and package newly synthesized products for different destinations within the cell. In each case, a particular vesicle captures a specific kind of cargo from among the many proteins in the Golgi lumen, then addresses each shipping container for a distinct destination.

Lysosomes and Endocytosis

Lysosomes are small, membrane-enclosed organelles that break down organic molecules (*lys* means "breakdown"; *some* means "body"). Instead of having a uniform structure, as is characteristic of all other organelles, lysosomes vary in size and shape, depending on the contents they are digesting. Most commonly, lysosomes are small (0.2 to 0.5 μm in diameter) oval or spherical bodies (● Figure 2-7). On average, a cell contains about 300 lysosomes.

1 Secretory vesicle formation　　**2** Budding from Golgi　　**3** Uncoating　　**4** Docking at plasma membrane　　**5** Exocytosis

Golgi lumen　Membrane of outer-most Golgi sac　Cytosol　Plasma membrane　ECF

KEY

> t-SNARE (docking-marker acceptor)

Recognition marker
Coat-protein acceptor
v-SNARE (docking marker)

Sorting signal

Cargo proteins

Coatomer (coat protein that causes membrane to curve)

1 Recognition markers in the membrane of the outermost Golgi sac capture the appropriate cargo from the Golgi lumen by binding only with the sorting signals of the protein molecules to be secreted. The membrane that will wrap the vesicle is coated with coatomer, which causes the membrane to curve, forming a bud.

2 The membrane closes beneath the bud, pinching off the secretory vesicle.

3 The vesicle loses its coating, exposing v-SNARE docking markers on the vesicle surface.

4 The v-SNAREs bind only with the t-SNARE docking-marker acceptors of the targeted plasma membrane, ensuring that secretory vesicles empty their contents to the cell's exterior.

● **FIGURE 2-6 Packaging, docking, and release of secretory vesicles.** The diagram series illustrates secretory vesicle formation and budding with the aid of a coat protein and docking with the plasma membrane by means of v-SNAREs and t-SNARES. The transmission electron micrograph shows secretion by exocytosis.

Lysosomes digest extracellular material brought into the cell by phagocytosis.

Lysosomes are formed by budding from the Golgi complex. A lysosome contains more than 30 different powerful **hydrolytic enzymes** that are synthesized in the ER, then transported to the Golgi complex for packaging in the budding lysosome (see ● Figure 2-3). These enzymes catalyze **hydrolysis,** reactions that break down organic molecules by the addition of water at a bond site (see p. A-15) (*hydrolysis* means "splitting with water"). In lysosomes the organic molecules are cell debris and foreign material, such as bacteria, that have been brought into the cell. Lysosomal enzymes are similar to the hydrolytic enzymes that the digestive system secretes to digest food. Thus, lysosomes serve as the intracellular "digestive system."

Extracellular material to be attacked by lysosomal enzymes is brought into the cell through the process of phagocytosis, a type of endocytosis. **Endocytosis,** the reverse of exocytosis, refers to the internalization of extracellular material within a cell (*endo* means "within") (see ● Figure 2-5b). Endocytosis can be accomplished in three ways—*pinocytosis, receptor-mediated endocytosis,* and *phagocytosis*—depending on the contents of the internalized material (● Figure 2-8).

Peroxisome

Lysosome

Oxidative enzymes

Hydrolytic enzymes

Peroxisome

Peroxisome

Lysosome

● **FIGURE 2-7 Lysosomes and peroxisomes.** Diagram and electron micrograph of lysosomes, which contain hydrolytic enzymes, and peroxisomes, which contain oxidative enzymes.

1 Solute molecules and water molecules are outside the plasma membrane.

2 Membrane pockets inward, enclosing solute molecules and water molecules.

3 Pocket pinches off as endocytic vesicle containing sample of ECF.

(a) Pinocytosis

1 Substances attach to membrane receptors.

2 Membrane pockets inward.

3 Pocket pinches off as endocytic vesicle containing target molecule.

(b) Receptor-mediated endocytosis

1 Pseudopods begin to surround prey.

2 Pseudopods close around prey.

3 Prey is enclosed in endocytic vesicle that sinks into cytoplasm.

4 Lysosome fuses with vesicle, releasing enzymes that attack material inside vesicle.

(c) Phagocytosis

● **FIGURE 2-8 Forms of endocytosis.** (a) Diagram and electron micrograph of pinocytosis. The surface membrane dips inward to form a pouch, then seals the surface, forming an intracellular endocytic vesicle that nonselectively internalizes a bit of ECF. (b) Diagram and electron micrograph of receptor-mediated endocytosis. When a large molecule such as a protein attaches to a specific surface receptor, the membrane pockets inward with the aid of a coat protein, forming a coated pit, then pinches off to selectively internalize the molecule in an endocytic vesicle. (c) Diagram and scanning electron micrograph series of phagocytosis. White blood cells internalize multimolecular particles such as bacteria or old red blood cells by extending pseudopods that wrap around and seal in the targeted material. A lysosome fuses with and degrades the vesicle contents.

PINOCYTOSIS In **pinocytosis** ("cell drinking"), a droplet of extracellular fluid is taken up nonselectively. First, the plasma membrane dips inward, forming a pouch that contains a small bit of ECF (● Figure 2-8a). The plasma membrane then seals at the surface of the pouch, trapping the contents in a small, intracellular **endocytic vesicle.** *Dynamin,* the protein responsible for pinching off an endocytic vesicle, forms rings that wrap around and "wring the neck" of the pouch, severing the vesicle from the surface membrane. Besides bringing ECF into a cell, pinocytosis provides a means to retrieve extra plasma membrane that has been added to the cell surface during exocytosis.

RECEPTOR-MEDIATED ENDOCYTOSIS Unlike pinocytosis, which involves the nonselective uptake of the surrounding fluid, **receptor-mediated endocytosis** is a highly selective process that enables cells to import specific large molecules that it needs from its environment. Receptor-mediated endocytosis is triggered by the binding of a specific target molecule such as a protein to a surface membrane receptor specific for that molecule (● Figure 2-8b). This binding causes the plasma membrane at that site to pocket inward, then seal at the surface, trapping the bound molecule inside the cell. The pouch is formed by the linkage of *clathrin* molecules, which are membrane-deforming coat proteins on the inner surface of the plasma membrane. Clathrin is a different coat protein than the one used for exocytosis. The resulting pouch is known as a *coated pit,* because it is coated with clathrin. Cholesterol complexes, vitamin B_{12}, the hormone insulin, and iron are examples of substances selectively taken into cells by receptor-mediated endocytosis.

Unfortunately, some viruses can sneak into cells by exploiting this mechanism. For instance, flu viruses and HIV, the virus that causes AIDS (see p. 440), gain entry to cells via receptor-mediated endocytosis. They do so by binding with membrane receptors normally designed to trigger the internalization of a needed molecule.

PHAGOCYTOSIS During **phagocytosis** ("cell eating"), large multimolecular particles are internalized. Most body cells perform pinocytosis, many carry out receptor-mediated endocytosis, but only a few specialized cells are capable of phagocytosis, the most notable being certain types of white blood cells that play an important role in the body's defense mechanisms. When a white blood cell encounters a large particle, such as a bacterium or tissue debris, it extends surface projections known as **pseudopods** ("false feet") that completely surround or engulf the particle and trap it within an internalized vesicle (● Figure 2-8c). A lysosome fuses with the membrane of the phagocytic vesicle and releases its hydrolytic enzymes into the vesicle, where they safely attack the bacterium or other trapped material without damaging the remainder of the cell. The enzymes largely break down the engulfed material into raw ingredients, such as amino acids, glucose, and fatty acids, that the cell can use.

Lysosomes remove worn out organelles.

Lysosomes can also fuse with aged or damaged organelles to remove these useless parts of the cell. This selective self-digestion, known as **autophagy** (*auto* means "self"; *phag* means

"eating") makes way for new replacement parts. In most cells all the organelles are renewable.

Some individuals lack the ability to synthesize one or more of the lysosomal enzymes. The result is massive accumulation within the lysosomes of the compound normally digested by the missing enzyme. Clinical manifestations often accompany such disorders, because the engorged lysosomes interfere with normal cell activity. The nature and severity of the symptoms depend on the type of substance accumulating, which in turn depends on what lysosomal enzyme is missing. Among these so-called *storage diseases* is **Tay-Sachs disease,** which is characterized by abnormal accumulation of complex molecules found in nerve cells. As the accumulation continues, profound symptoms of progressive nervous-system degeneration result.

Peroxisomes and Detoxification

Peroxisomes are membranous organelles that produce and decompose hydrogen peroxide in the process of degrading potentially toxic molecules (*peroxi* refers to "hydrogen peroxide"). Typically, several hundred small peroxisomes that are about one third to one half the average size of lysosomes are present in a cell (see ● Figure 2-7). They too arise from the ER/Golgi complex.

Peroxisomes house oxidative enzymes that detoxify various wastes.

Peroxisomes are similar to lysosomes in that they are membrane-enclosed sacs containing enzymes, but unlike the lysosomes, which contain hydrolytic enzymes, peroxisomes house several powerful *oxidative enzymes* and contain most of the cell's *catalase.*

Oxidative enzymes, as the name implies, use oxygen (O_2), in this case to strip hydrogen from certain organic molecules. This reaction helps detoxify various wastes produced within the cell or foreign toxic compounds that have entered the cell, such as alcohol consumed in beverages.

The major product generated in the peroxisome, *hydrogen peroxide* (H_2O_2), is formed by molecular oxygen and the hydrogen atoms stripped from the toxic molecule. Hydrogen peroxide is potentially destructive if allowed to accumulate or escape from the confines of the peroxisome. However, peroxisomes also contain an abundance of **catalase,** an enzyme that decomposes potent H_2O_2 into harmless H_2O and O_2. This latter reaction is an important safety mechanism that destroys the potentially deadly peroxide at its site of production, thereby preventing its possible devastating escape into the cytosol.

Mitochondria and ATP Production

Mitochondria are the energy organelles, or "power plants," of the cell; they extract energy from the nutrients in food and transform it into a usable form for cell activities. Mitochondria generate about 90% of the energy that cells—and, accordingly, the whole body—need to survive and function. A single cell may contain as few as a hundred or as many as several thousand mitochondria, depending on the energy needs of each particular cell type. In some cell types, the mitochondria are densely compacted into the regions of the cell that use the most energy. For example, mitochondria are packed between the contractile units in the muscle cells of the heart.

Mitochondria are enclosed by two membranes.

Mitochondria are rod-shaped or oval structures about the size of bacteria. In fact, mitochondria are descendants of bacteria that invaded or were engulfed by primitive cells early in evolutionary history and that subsequently became permanent organelles. As part of their separate heritage, mitochondria possess their own DNA, distinct from the DNA housed in the cell's nucleus. Mitochondrial DNA contains the genetic codes for producing many of the molecules the mitochondria need to generate energy.

Clinical Note Flaws gradually accumulate in mitochondrial DNA over a person's lifetime; these flaws have been implicated in aging as well as in an array of disorders. Mitochondrial diseases affect an estimated 1 in 5000 people. The clinical manifestations of these disorders are highly diverse, depending on the location and extent of mitochondrial DNA mutations. Prominent among mitochondrial diseases are those that become debilitating in later life, such as some forms of chronic degenerative muscle and nervous-system diseases, an example being **Charcot-Marie-Tooth type 2a.** The more common symptoms of mitochondrial diseases include muscle weakness, incoordination, seizures, blindness, and hearing loss.

Each mitochondrion is enclosed by a double membrane—a smooth outer membrane that surrounds the mitochondrion itself, and an inner membrane that forms a series of infoldings or shelves called **cristae,** which project into an inner cavity filled with a gel-like solution known as the **matrix** (● Figure 2-9). The two membranes are separated by a narrow intermembrane space. The cristae contain proteins that ultimately are responsible for converting much of the energy in food into a usable form. The generous folds of the inner membrane greatly increase the surface area available for housing these important proteins. The matrix consists of a concentrated mixture of hundreds of different dissolved enzymes that prepare nutrient molecules for the final extraction of usable energy by the cristae proteins.

Mitochondria play a major role in generating ATP.

The source of energy for the body is the chemical energy stored in the carbon bonds of ingested food. Body cells are not equipped to use this energy directly. Instead, the cells must extract energy from food nutrients and convert it into a form they can use—namely, the high-energy phosphate bonds of **adenosine triphosphate (ATP),** which consists of adenosine with three phosphate groups attached (*tri* means "three") (see p. A-17). When a high-energy bond such as that binding the terminal phosphate to adenosine is split, a substantial amount of energy is released.

Adenosine triphosphate is the universal energy carrier—the common energy "currency" of the body. Cells can "cash in" ATP to pay the energy "price" for running the cell machinery. To

● **FIGURE 2-9 Mitochondrion.** Diagram and electron micrograph of a mitochondrion. Note that the outer membrane is smooth while the inner membrane forms folds known as cristae that extend into the matrix. An intermembrane space separates the outer and inner membranes. The electron transport proteins embedded in the cristae are ultimately responsible for converting much of the energy of food into a usable form.

obtain immediate usable energy, cells split the terminal phosphate bond of ATP, which yields **adenosine diphosphate (ADP)**—adenosine with two phosphate groups attached (*di* means "two")—plus inorganic phosphate (P_i) plus energy:

$$ATP \xrightarrow{\text{splitting}} ADP + P_i + \text{energy for use by the cell}$$

In this energy scheme, food can be thought of as the "crude fuel" and ATP as the "refined fuel" for operating the body's machinery. The food is digested, or broken down by the digestive system into smaller absorbable units that can be transferred from the digestive tract lumen into the blood (see Chapter 16). For example, dietary carbohydrates are broken down primarily into glucose, which is absorbed into the blood. No usable energy is released during the digestion of food. When delivered to the cells by the blood, the nutrient molecules are transported across the plasma membrane into the cytosol. (Details of how materials cross the membrane are covered in Chapter 3.)

We now turn our attention to the steps involved in ATP production within the cell and the role of the mitochondria in these steps. **Cellular respiration** refers collectively to the intracellular reactions in which energy-rich molecules are broken down to form ATP, using O_2 and producing CO_2 in the process. In most cells, ATP is generated from the sequential dismantling of absorbed nutrient molecules in three stages: *glycolysis* in the cytosol, the *citric acid cycle* in the mitochondrial matrix, and *oxidative phosphorylation* at the mitochondrial inner membrane (● Figure 2-10). (Muscle cells use an additional cytosolic pathway for immediately generating energy at the onset of exercise; see p. 276.) We will use glucose as an example to describe these stages.

GLYCOLYSIS Among the thousands of enzymes in the cytosol are the 10 responsible for **glycolysis**, a chemical process involving 10 sequential reactions that break down a six-carbon sugar molecule, glucose, into two pyruvate molecules, each with three carbons (*glyc-* means "sweet"; *lysis* means "breakdown") (● Figure 2-11). During this process, two hydrogens are released and transferred to 2 NADH molecules for later use. (You will learn more about NADH shortly.) Some of the energy from the broken chemical bonds of glucose is used directly to convert ADP into ATP. However, glycolysis is not very efficient in terms of energy extraction: the net yield is only two molecules of ATP per glucose molecule processed. Much of the energy originally contained in the glucose molecule is still locked in the chemical bonds of the pyruvate molecules. The low-energy yield of glycolysis is not enough to support the body's demand for ATP. This is where the mitochondria come into play.

CITRIC ACID CYCLE The pyruvate produced by glycolysis in the cytosol is selectively transported into the mitochondrial matrix. Here, one of its carbons is enzymatically removed in the form of carbon dioxide (CO_2) and later eliminated from the body as an end product, or waste (● Figure 2-12). Also, another hydrogen is released and transferred to another NADH. The two-carbon molecule remaining after the breakdown process, acetate, combines with coenzyme A (CoA), a derivative of pantothenic acid (a B vitamin), producing the compound acetyl coenzyme A (acetyl-CoA).

Acetyl-CoA then enters the **citric acid cycle**, a cyclical series of eight biochemical reactions that are catalyzed by the enzymes of the mitochondrial matrix. This cycle of reactions can be compared to one revolution around a Ferris wheel, except that the molecules themselves are not physically moved around in a cycle. On the top of the Ferris wheel, acetyl-

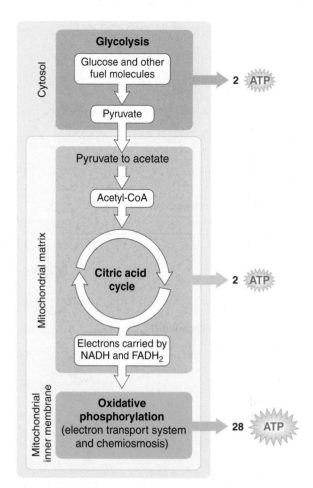

● **FIGURE 2-10 Stages of cellular respiration.** The three stages of cellular respiration are (1) glycolysis, (2) citric acid cycle, and (3) oxidative phosphorylation.

● **FIGURE 2-11 Glycolysis in the cytosol.** Glycolysis splits glucose (six carbons) into two pyruvate molecules (three carbons each), with a net yield of 2 ATP plus 2 NADH (available for further energy extraction by the electron transport system).

CoA, a two-carbon molecule, enters a seat already occupied by oxaloacetate, which has four carbons. These two molecules link to form a six-carbon citrate molecule (at intracellular pH, citric acid exists in an ionized form, citrate), and the trip around the

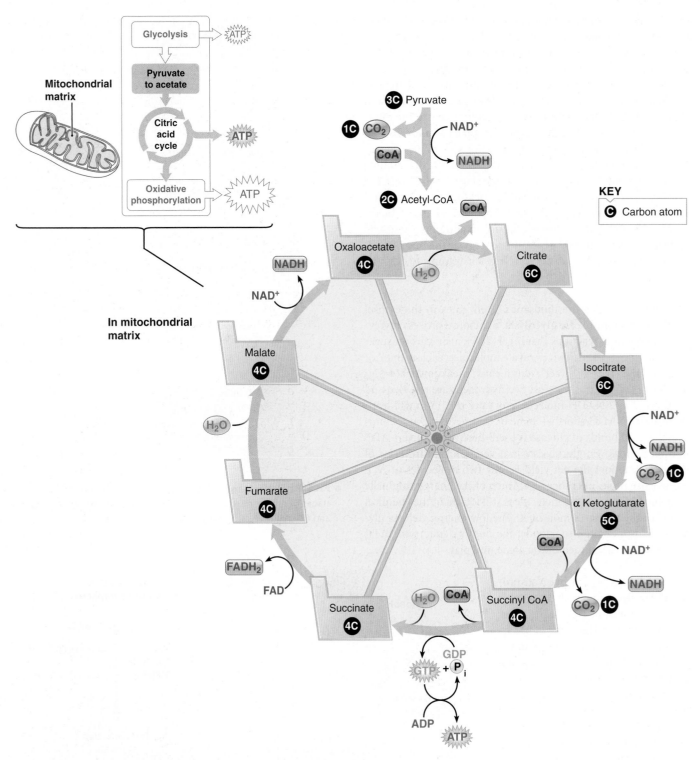

● **FIGURE 2-12 Citric acid cycle in the mitochondrial matrix.** The two carbons entering the cycle by means of acetyl-CoA are eventually converted to CO_2, with oxaloacetate, which accepts acetyl-CoA, being regenerated at the end of the cyclical pathway. The hydrogens released at specific points along the pathway bind to the hydrogen carrier molecules NAD^+ and FAD for further processing by the electron transport system. One molecule of ATP is generated for each molecule of acetyl-CoA that enters the citric acid cycle, for a total of two molecules of ATP for each molecule of processed glucose.

citric acid cycle begins. (This cycle is alternatively known as the **Krebs cycle,** in honor of its principal discoverer, or the **tricarboxylic acid cycle,** because citrate contains three carboxylic acid groups.) At each step in the cycle, matrix enzymes modify the passenger molecule to form a slightly different molecule (shown in ● Figure 2-12). These molecular alterations have the following important consequences:

1. Two carbons are "kicked off the ride"—released one at a time from six-carbon citrate, converting it back into four-carbon oxaloacetate, which is now available at the top of the cycle to pick up another acetyl-CoA for another revolution through the cycle. CoA is recycled too; it is released at the end of the cycle, making it available to bind with a new acetate to form another acetyl-CoA.

2. The released carbons are converted into two molecules of CO_2. Note that two carbon atoms enter the cycle in the form of acetyl-CoA and two carbon atoms leave the cycle in the form of two CO_2 molecules. This CO_2, as well as the CO_2 produced during the formation of acetate from pyruvate, passes out of the mitochondrial matrix and subsequently out of the cell to enter the blood. The blood carries the CO_2 to the lungs, where it is eliminated to the atmosphere through breathing. The oxygen used to make CO_2 from these released carbons is derived from the molecules involved in the reactions, not from free molecular oxygen supplied by breathing.

3. Hydrogens are also "bumped off" during the cycle at four of the chemical conversion steps. The key purpose of the citric acid cycle is to produce these hydrogens for entry into the electron transport system in the inner mitochondrial membrane. The hydrogens are transferred to two different hydrogen carrier molecules—**nicotinamide adenine dinucleotide (NAD^+),** a derivative of the B vitamin niacin, and **flavine adenine dinucleotide (FAD),** a derivative of the B vitamin riboflavin. The transfer of hydrogen converts these compounds to NADH and $FADH_2$, respectively. Three NADH and one $FADH_2$ are produced for each turn of the citric acid cycle.

4. One more molecule of ATP is produced for each molecule of acetyl-CoA processed. Actually, ATP is not directly produced by the citric acid cycle. The released energy directly links inorganic phosphate to **guanosine diphosphate (GDP),** forming **guanosine triphosphate (GTP),** a high-energy molecule similar to ATP. The energy from GTP is then transferred to ATP as follows:

$$ADP + GTP \rightleftharpoons ATP + GDP$$

Because each glucose molecule is converted into two acetate molecules, fueling two turns of the citric acid cycle, two more ATP molecules are produced from each glucose molecule.

So far, the cell still does not have much of an energy profit. However, the citric acid cycle is important in preparing the hydrogen carrier molecules for their entry into the final stage, oxidative phosphorylation, which produces far more energy than the sparse amount of ATP produced by the cycle itself.

OXIDATIVE PHOSPHORYLATION Considerable untapped energy is still stored in the released hydrogens, which contain electrons at high energy levels. **Oxidative phosphorylation** refers to the process by which ATP is synthesized using the energy released by electrons as they are transferred to O_2. This process involves two groups of proteins, both located at the inner mitochondrial membrane: the *electron transport system* and *ATP synthase*.

The "big payoff" in energy capture begins when NADH and $FADH_2$ enter the electron transport system. The **electron transport system** consists of electron carriers found in four large stationary protein complexes, numbered I, II, III, and IV, along with two small highly mobile electron carriers, cytochrome *c* and ubiquinone (also known as coenzyme Q or CoQ), which shuttle electrons between the major complexes (● Figure 2-13).

As the electron transport system begins, high-energy electrons are extracted from the hydrogens held in NADH and $FADH_2$ and are transferred through a series of steps from one electron-carrier molecule to another in an assembly line (step **1**). As a result of giving up hydrogen ions (H^+) and electrons within the electron transport system, NADH and $FADH_2$ are converted back to NAD^+ and FAD (step **2**), freeing them to pick up more hydrogen atoms released during glycolysis and the citric acid cycle. Thus, NAD^+ and FAD serve as the link between the citric acid cycle and the electron transport system. The electron carriers are arranged in a specific order in the inner membrane so that the high-energy electrons fall to successively lower energy levels as they are transferred from carrier to carrier through a chain of reactions (step **3**). Ultimately, when they are in their lowest energy state, the electrons are bound to molecular oxygen (O_2) derived from the air we breathe. Oxygen enters the mitochondria to serve as the final electron acceptor of the electron transport system. This negatively charged oxygen (negative because it has acquired additional electrons) then combines with the positively charged hydrogen ions (positive because they have donated electrons at the beginning of the electron transport system) to form water, H_2O (step **4**).

As electrons move through this chain of reactions, they release free energy. Part of the released energy is lost as heat, but some is harnessed by the mitochondrion to synthesize ATP. At three sites in the electron transport system (Complexes I, III, and IV), the energy released during the transfer of electrons is used to transport hydrogen ions (H^+) across the inner mitochondrial membrane from the matrix to the space between the inner and outer mitochondrial membranes, the *intermembrane space* (step **5**). As a result, hydrogen ions are more heavily concentrated in the intermembrane space than in the matrix. This H^+ gradient generated by the electron transport system (step **6**) supplies the energy that drives ATP synthesis by the membrane-bound mitochondrial enzyme ATP synthase.

ATP synthase consists of a *basal unit* embedded in the inner membrane, connected by a stalk to a *headpiece* located in the matrix, with the *stator* bridging the basal unit and headpiece. Because H^+ ions are more heavily concentrated in the intermembrane space than in the matrix, they have a strong tendency to flow back into the matrix through the inner membrane via channels formed between the basal units and stators of the ATP synthase complexes (step **7**). This flow of H^+ ions activates ATP synthase and powers ATP synthesis by the headpiece, a process called **chemiosmosis.** Passage of H^+ ions through the channel makes the headpiece and stalk spin like a

● **FIGURE 2-13 Oxidative phosphorylation at the mitochondrial inner membrane.** Oxidative phosphorylation involves the electron transport system (steps 1–6) and chemiosmosis by ATP synthase (steps 7–9). The pink circles in the electron transport system represent specific electron carriers.

1 The high-energy electrons extracted from the hydrogens in NADH and FADH$_2$ are transferred from one electron-carrier molecule to another.

2 The NADH and FADH$_2$ are converted to NAD$^+$ and FAD, which frees them to pick up more hydrogen atoms released during glycolysis and the citric acid cycle.

3 The high-energy electrons fall to successively lower energy levels as they are transferred from carrier to carrier through the electron transport system.

4 The electrons are passed to O$_2$, the final electron acceptor of the electron transport system. This oxygen, now negatively charged because it has acquired additional electrons, combines with H$^+$ ions, which are positively charged because they donated electrons at the beginning of the electron transport system, to form H$_2$O.

5 As electrons move through the electron transport system, they release free energy. Part of the released energy is lost as heat, but some is harnessed by the mitochondrion to transport H$^+$ across the inner mitochondrial membrane from the matrix to the intermembrane space at Complexes I, III, and IV .

6 As a result, H$^+$ ions are more heavily concentrated in the intermembrane space than in the matrix. This H$^+$ gradient supplies the energy that drives ATP synthesis by ATP synthase.

7 Because of this gradient, H$^+$ ions have a strong tendency to flow into the matrix across the inner membrane via channels between the basal units and stators of the ATP synthase complexes.

8 This flow of H$^+$ ions activates ATP synthase and powers ATP synthesis by the headpiece, a process called **chemiosmosis**. Passage of H$^+$ ions through the channel makes the headpiece and stalk spin like a top.

9 As a result of changes in its shape and position as it turns, the headpiece picks up ADP and P$_i$, combines them, and releases the ATP product.

Electron tranport system
Electrons flow through a series of electron carriers from high-energy to low-energy levels; the energy released builds an H$^+$ gradient across the inner mitochondrial membrane.

Chemiosmosis
ATP synthase catalyzes ATP synthesis using energy from the H$^+$ gradient across the membrane.

Oxidative phosphorylation

top (step **8**), similar to the flow of water making a waterwheel turn. As a result of the changes in its shape and position as it turns, the headpiece is able to sequentially pick up ADP and P_i, combine them, and release the ATP product (step **9**).

Oxidative phosphorylation encompasses the entire process by which ATP synthase synthesizes ATP by phosphorylating (adding a phosphate to) ADP using the energy released by electrons as they are transferred to O_2 by the electron transport system. The harnessing of energy into a useful form as the electrons tumble from a high-energy state to a low-energy state can be likened to a power plant converting the energy of water tumbling down a waterfall into electricity.

When activated, ATP synthase provides a rich yield of 28 more ATP molecules for each glucose molecule processed (● Figure 2-14). The most recent research indicates that approximately 2.5 ATP are synthesized as a pair of electrons released by NADH travels through the entire electron transport system to oxygen. The shorter pathway followed by an electron pair released from $FADH_2$ (see ● Figure 2-13) synthesizes about 1.5 ATP. This means a total of 32 molecules of ATP are produced when a glucose molecule is completely dismantled in cellular respiration: 2 during glycolysis, 2 during the citric acid cycle, and 28 during oxidative phosphorylation. The ATP is transported out of the mitochondrion into the cytosol for use as the cell's energy source.

The steps leading to oxidative phosphorylation might at first seem an unnecessary complication. Why not just directly oxidize, or "burn," food molecules to release their energy? When this process occurs outside the body, all the energy stored in the food molecule is released explosively in the form of heat (● Figure 2-15). Think about what happens when a marshmallow you're roasting accidentally catches on fire. The burning marshmallow gets very hot quickly as a result of the rapid oxidation of sugar. In the body, food molecules are oxidized within the mitochondria in many small, controlled steps so that their chemical energy is gradually released in small quantities that can be more efficiently captured in ATP bonds and stored in a form that is useful to the cell. In this way, much less of the energy is converted to heat. The heat that is produced is not completely wasted energy; it is used to help maintain body temperature, with any excess heat being eliminated to the environment.

The cell generates more energy in aerobic than in anaerobic conditions.

The cell is a much more efficient energy converter when oxygen is available (● Figure 2-16). In an **anaerobic** ("lack of air," specifically "lack of O_2") condition, the degradation of glucose cannot proceed beyond glycolysis, which takes place in the cytosol and yields only two molecules of ATP per molecule of glucose. The untapped energy of the glucose molecule remains

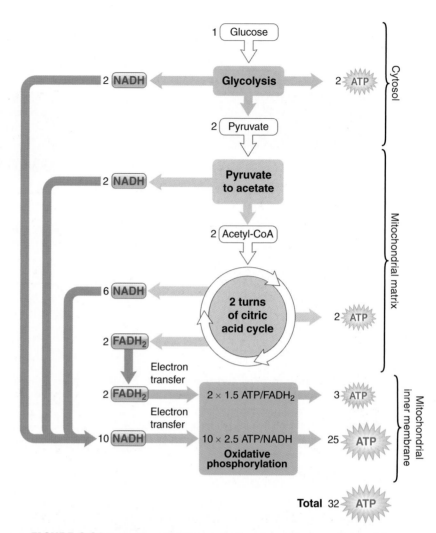

● **FIGURE 2-14 Summary of ATP production from the complete oxidation of one molecule of glucose.** The total of 32 ATP assumes that electrons carried by each NADH yield 2.5 ATP and those carried by each $FADH_2$ yield 1.5 ATP during oxidative phosphorylation.

locked in the bonds of the pyruvate molecules, which are eventually converted to **lactate** if they do not enter the pathway that ultimately leads to oxidative phosphorylation.

However, when sufficient O_2 is present—an **aerobic** ("with air" or "with O_2") condition—mitochondrial processing (that is, the citric acid cycle in the matrix and the electron transport system and ATP synthase at the inner membrane) harnesses enough energy to generate 28 more molecules of ATP, for a total net yield of 32 ATPs per molecule of glucose processed. (For a description of aerobic exercise, see the boxed feature on p. 39, ■ A Closer Look at Exercise Physiology.) The overall reaction for the oxidation of food molecules to yield energy during cellular respiration is as follows:

$$\text{Food} + \underset{\substack{\text{(necessary} \\ \text{for oxidative} \\ \text{phosphorylation)}}}{O_2} \rightleftharpoons \underset{\substack{\text{(produced} \\ \text{primarily} \\ \text{by the citric} \\ \text{acid cycle)}}}{CO_2} + \underset{\substack{\text{(produced by} \\ \text{the electron} \\ \text{transport} \\ \text{system)}}}{H_2O} + \underset{\substack{\text{(produced} \\ \text{primarily by} \\ \text{ATP synthase)}}}{\text{ATP}}$$

Glucose, the principal nutrient derived from dietary carbohydrates, is the fuel preference of most cells. However, nutrient

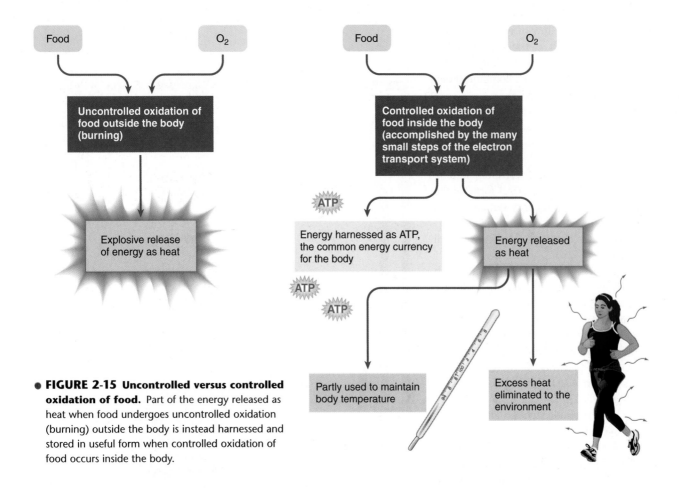

● **FIGURE 2-15 Uncontrolled versus controlled oxidation of food.** Part of the energy released as heat when food undergoes uncontrolled oxidation (burning) outside the body is instead harnessed and stored in useful form when controlled oxidation of food occurs inside the body.

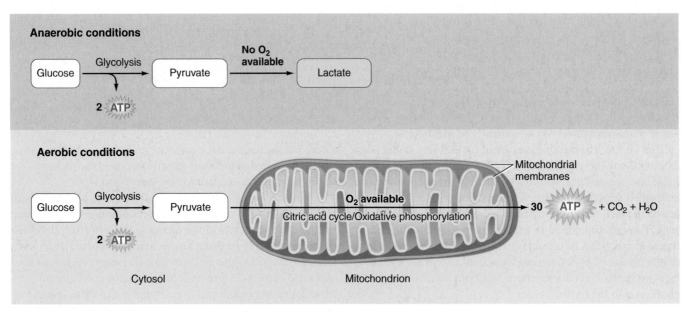

● **FIGURE 2-16 Comparison of energy yield and products under anaerobic and aerobic conditions.** In anaerobic conditions, only 2 ATP are produced for every glucose molecule processed, but in aerobic conditions a total of 32 ATP are produced per glucose molecule.

molecules derived from fats (fatty acids) and, if necessary, from proteins (amino acids) can also participate at specific points in this overall chemical reaction to eventually produce energy. Amino acids are usually used for protein synthesis rather than energy production, but they can be used as fuel if insufficient glucose and fat are available (see Chapter 17).

Note that the oxidative reactions within the mitochondria generate energy, unlike the oxidative reactions controlled by

Aerobic Exercise: What For and How Much?

Aerobic ("with O_2") **exercise** involves large muscle groups and is performed at a low-enough intensity and for a long-enough period of time that fuel sources can be converted to ATP by using the citric acid cycle and oxidative phosphorylation as the predominant metabolic pathway. Aerobic exercise can be sustained from 15 to 20 minutes to several hours at a time. Short-duration, high-intensity activities such as weight training and the 100-meter dash, which last for a matter of seconds and rely solely on energy stored in the muscles and on glycolysis, are forms of **anaerobic** ("without O_2") **exercise.**

Inactivity is associated with increased risk of developing both hypertension (high blood pressure) and coronary artery disease (blockage of the arteries that supply the heart). The American College of Sports Medicine recommends that an individual participate in aerobic exercise a minimum of three times per week for 20 to 60 minutes to reduce the risk of hypertension and coronary artery disease and to improve physical work capacity. Recent studies have shown the same health benefits are derived whether the exercise is accomplished in one long stretch or is broken down into multiple shorter stints. This is good news, because many individuals find it easier to stick with brief bouts of exercise sprinkled throughout the day.

The intensity of the exercise should be based on a percentage of the individual's maximal capacity to work. The easiest way to establish the proper intensity of exercise and to monitor intensity levels is by checking the heart rate. The estimated maximal heart rate is determined by subtracting the person's age from 220. Significant benefits can be derived from aerobic exercise performed between 70% and 80% of maximal heart rate. For example, the estimated maximal heart rate for a 20-year-old is 200 beats per minute. If this person exercised three times per week for 20 to 60 minutes at an intensity that increased the heart rate to 140 to 160 beats per minute, the participant should significantly improve his or her aerobic work capacity and reduce the risk of cardiovascular disease.

---

the peroxisome enzymes. Both organelles use O_2, but for different purposes.

The energy stored within ATP is used for synthesis, transport, and mechanical work.

Once formed, ATP is transported out of the mitochondria and is then available as an energy source in the cell. Cell activities that require energy expenditure fall into three main categories:

1. *Synthesis of new chemical compounds,* such as protein synthesis by the endoplasmic reticulum. Some cells, especially cells with a high rate of secretion and cells in the growth phase, use up to 75% of the ATP they generate just to synthesize new chemical compounds.

2. *Membrane transport,* such as the selective transport of molecules across the kidney tubules during the process of urine formation. Kidney cells can expend as much as 80% of their ATP currency to operate their selective membrane-transport mechanisms.

3. *Mechanical work,* such as contraction of the heart muscle to pump blood or contraction of skeletal muscles to lift an object. These activities require tremendous quantities of ATP.

As a result of cell energy expenditure to support these various activities, large quantities of ADP are produced. These energy-depleted molecules enter the mitochondria for "recharging" and then cycle back into the cytosol as energy-rich ATP molecules after participating in oxidative phosphorylation. In this recharging–expenditure cycle, a single ADP/ATP molecule may shuttle between the mitochondria and cytosol thousands of times per day. On average a person recycles the equivalent of his or her body weight of ATP every day!

The high demands for ATP make glycolysis alone an insufficient as well as inefficient supplier of power for most cells. Were it not for the mitochondria, which house the metabolic machinery for oxidative phosphorylation, the body's energy capability would be very limited. However, glycolysis does provide cells with a sustenance mechanism that can produce at least some ATP under anaerobic conditions. Skeletal muscle cells in particular take advantage of this ability during short bursts of strenuous exercise, when energy demands for contractile activity outstrip the body's ability to bring adequate O_2 to the exercising muscles to support oxidative phosphorylation. Also, red blood cells, which are the only cells that do not contain mitochondria, rely solely on glycolysis for their limited energy production. The energy needs of red blood cells are low, however, because they also lack a nucleus and therefore are not capable of synthesizing new substances, the biggest energy expenditure for most noncontractile cells.

Having completed our discussion of the membranous organelles, we now turn attention to the nonmembranous organelles, which include ribosomes, vaults, and centrioles.

Ribosomes and Protein Synthesis

Ribosomes carry out protein synthesis by translating mRNA into chains of amino acids in the ordered sequence dictated by the original DNA code. Ribosomes bring together all components that participate in protein synthesis—mRNA, tRNA, and

(a) Ribosome

Ribosomes

Where a messenger
RNA fits through
a ribosome

Large
ribosomal
subunit

Small
ribosomal
subunit

Growing
amino acid
chain

Released
tRNA with no
amino acid

Incoming tRNA,
with amino acid
attached, reads
mRNA instructions
and introduces
amino acid to be
added next.

mRNA

E   P   A

Ribosome
movement

(b) Protein synthesis in ribosome

● **FIGURE 2-17 Ribosomes.** (a) Diagram of an assembled ribosome consisting of a large and a small ribosomal subunit, with mRNA fitting through a groove formed between these subunits. (b) Schematic representation of the ribosome interior showing the A, P, and E sites where tRNA molecules interact with mRNA during protein synthesis.

amino acids—and provide the enzymes and energy required for linking the amino acids together. The nature of the protein synthesized by a given ribosome is determined by the mRNA being translated. Each mRNA serves as a code for only one protein.

Ribosomes exist free in the cytosol or attached to the rough ER. A finished ribosome is about 20 nm in diameter and is made up of two parts of unequal size, a *large* and a *small ribosomal subunit* (● Figure 2-17a). Each subunit is composed of rRNA and ribosomal proteins. These subunits are brought together when a protein is being synthesized. When the two subunits unite, a groove is formed. In translation, mRNA moves through this groove. The ribosome also has three binding sites where tRNAs interact with mRNA (● Figure 2-17b). The *A site* is where incoming tRNA, bearing a specified amino acid, binds to mRNA. The *P site* is where the tRNA carrying the growing chain of amino acids is bound to mRNA. At each step in translation, the amino acid chain from the tRNA in the P site is cleaved and linked to the amino acid on the incoming tRNA in the A site. The *E site* is where an empty tRNA is bound prior to release from the ribosome. After the amino acid on the incoming tRNA has been added to the amino acid chain, the ribosome moves forward along mRNA, repositioning all tRNAs at the next site. The tRNA formerly in the E site exits the ribosome. The now empty tRNA that had been in the P site moves into the E site. The tRNA that had been in the A site and now bears the growing amino acid chain moves into the P site. A new incoming tRNA, carrying the next amino acid in the protein sequence, enters the vacant A site. This process is repeated until all amino acids specified by the mRNA instructions for the protein under construction have been linked together. See pp. A-24–A-26 for further details.

Vaults as Cellular Trucks

Vaults, which are three times as large as ribosomes, are shaped like octagonal barrels (● Figure 2-18). Their name comes from their multiple arches, which reminded their discoverers of vaulted or cathedral ceilings. Just like barrels, vaults have a hollow interior. When open, they appear like pairs of unfolded flowers with each half of the vault bearing eight "petals" attached to a central ring. A cell may contain thousands of vaults, yet they were not discovered until the 1990s. Why would the presence of these numerous, relatively large organelles have been elusive until recently? The reason is that they do not show up with ordinary staining techniques.

Vaults may serve as cellular transport vehicles.

Currently, the function of vaults is uncertain, but their octagonal shape and their hollow interior may provide clues. Nuclear pores are also octagonal and the same size as vaults, leading to speculation that vaults may be cellular "trucks." According to this proposal, vaults would dock at or enter nuclear pores, pick up molecules synthesized in the nucleus, and deliver their cargo elsewhere in the cell. Ongoing research supports the role of vaults in nucleus-to-cytoplasm transport, but their cargo has not been determined. One possibility is that vaults carry mRNA from the nucleus to the ribosomal sites of protein synthesis within the cytoplasm. Another possibility is that vaults transport the two subunits that make up ribosomes from the nucleus, where they are produced, to their sites of action—either attached to the rough ER or in the cytosol. The interior of a vault is the right size to accommodate these ribosomal subunits.

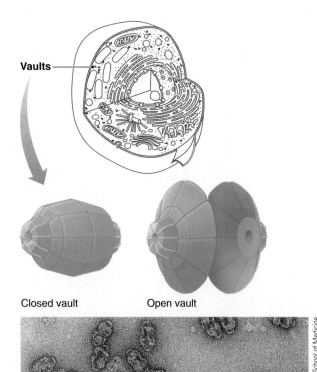

Vaults

Closed vault   Open vault

FIGURE 2-18 Vaults. Diagram of closed and open vaults and electron micrograph of vaults, which are octagonal barrel-shaped nonmembranous organelles believed to transport either messenger RNA or the ribosomal subunits from the nucleus to cytoplasmic ribosomes.

Clinical Note Vaults may play an undesirable role in bringing about the multidrug resistance sometimes displayed by cancer cells. Chemotherapy drugs designed to kill cancer cells tend to accumulate in the nuclei of these cells, but some cancer cells develop resistance to a wide variety of these drugs. This broad resistance is a major cause of cancer treatment failure. Researchers have shown that some cancer cells resistant to chemotherapy produce up to 16 times more than normal quantities of the major vault protein. If further investigation confirms that vaults play a role in drug resistance—perhaps by transporting the drugs from the nucleus to sites for exocytosis from the cancer cells—the exciting possibility exists that interference with this vault activity could improve the sensitivity of cancer cells to chemotherapeutic drugs.

Centrosome, Centrioles, and Microtubule Organization

The **centrosome,** or **cell center,** located near the nucleus, consists of the centrioles surrounded by an amorphous mass of proteins. The **centrioles** are a pair of short cylindrical structures that lie at right angles to each other at the centrosome's center

Centrioles

Microtubule triplet

FIGURE 2-19 Centrioles. The two cylindrical centrioles of the pair lie at right angles to each other as shown in the diagram. The electron micrograph shows a centriole in cross section. Note that a centriole is made up of nine microtubule triplets that form a ring.

(● Figure 2-19). The centrosome is the cell's main microtubule organizing center. Microtubules are one of the components of the cytoskeleton. When a cell is not dividing, microtubules are formed from the amorphous mass and radiate outward in all directions from the centrosome (see ● Figure 2-1). These microtubules anchor many of the membranous organelles and also serve as "highways" along which vesicles are transported within the cell by "molecular motors." In some cells, the centrioles form cilia and flagella, which are elongated, slender motile structures made up of bundles of microtubules. During cell division, the centrioles form a mitotic spindle out of microtubules to direct movement of chromosomes. You will learn about the microtubules and their functions in further detail when we examine the cytoskeleton following a brief description of the cytosol.

Cytosol: Cell Gel

Occupying about 55% of the total cell volume, the cytosol is the semiliquid portion of the cytoplasm that surrounds the organelles. Its nondescript appearance under an electron microscope gives the false impression that the cytosol is a liquid mixture of uniform consistency, but it is actually a highly organized, gel-like mass with differences in composition and consistency from one part of the cell to another.

The cytosol is important in intermediary metabolism, ribosomal protein synthesis, and nutrient storage.

Three general categories of activities are associated with the cytosol: (1) enzymatic regulation of intermediary metabolism, (2) ribosomal protein synthesis, and (3) storage of fat, carbohydrate, and secretory vesicles.

ENZYMATIC REGULATION OF INTERMEDIARY METABOLISM The term **intermediary metabolism** refers collectively to the large set of chemical reactions inside the cell that involve the degradation, synthesis, and transformation of small organic molecules such as simple sugars, amino acids, and fatty acids. These reactions are critical for ultimately capturing the energy used for cell activities and for providing the raw materials needed to maintain the cell's structure, function, and growth. All intermediary metabolism occurs in the cytoplasm, with most of it being accomplished in the cytosol. The cytosol contains thousands of enzymes involved in intermediary biochemical reactions.

RIBOSOMAL PROTEIN SYNTHESIS Also dispersed throughout the cytosol are the free ribosomes, which synthesize proteins for use in the cytosol itself. In contrast, recall that the rough-ER ribosomes synthesize proteins for secretion and for construction of new cell components.

STORAGE OF FAT, GLYCOGEN, AND SECRETORY VESICLES Excess nutrients not immediately used for ATP production are converted in the cytosol into storage forms that are readily visible under a light microscope. Such nonpermanent masses of stored material are known as **inclusions.** Inclusions are not surrounded by membrane, and they may or may not be present, depending on the type of cell and the circumstances. The largest and most important storage product is fat. Small fat droplets are present within the cytosol in various cells. In **adipose tissue,** the tissue specialized for fat storage, stored fat molecules can occupy almost the entire cytosol, where they merge to form one large fat droplet (● Figure 2-20a). The other visible storage product is **glycogen,** the storage form of glucose, which appears as clusters or granules dispersed throughout the cell (● Figure 2-20b). Cells vary in their ability to store glycogen, with liver and muscle cells having the greatest stores. When food is not available to provide fuel for the citric acid cycle and electron transport system, stored glycogen and fat are broken down to release glucose and fatty acids, respectively, which can feed the mitochondrial energy-producing machinery. An average adult human stores enough glycogen to provide energy for about a

Fat droplet   Nucleus of adipose cell

(a) Fat storage in adipose cells

Glycogen granules   Liver cell

(b) Glycogen storage in liver cells

● **FIGURE 2-20 Inclusions.** (a) Light micrograph showing fat storage in adipose cells. A fat droplet occupies almost the entire cytosol of each cell. (b) Light micrograph showing glycogen storage in liver cells. The red-stained granules throughout the cytosol of each liver cell are glycogen deposits.

day of normal activities, and typically has enough fat stored to provide energy for two months.

Secretory vesicles that have been processed and packaged by the endoplasmic reticulum and Golgi complex also remain in the cytosol, where they are stored until signaled to empty their contents to the outside. In addition, transport and endocytic vesicles move through the cytosol.

Cytoskeleton: Cell "Bone and Muscle"

Different cell types in the body have distinct shapes, structural complexities, and functional specializations. These unique characteristics are maintained by the **cytoskeleton,** an elaborate protein scaffolding dispersed throughout the cytosol that acts as the "bone and muscle" of the cell by supporting and organizing the cell components and controlling their movements.

The cytoskeleton has three distinct elements: (1) *microtubules,* (2) *microfilaments*, and (3) *intermediate filaments* (● Figure 2-21). These elements are structurally linked and functionally coordinated to provide certain integrated functions for the cell. These functions, along with the functions of all other components of the cytoplasm, are summarized in ▲ Table 2-2.

Tubulin
subunit

Polypeptide
strand

Actin
subunit

(a) Microtubule

(b) Microfilament

(c) Keratin, an intermediate filament

● **FIGURE 2-21 Components of the cytoskeleton.** (a) Microtubules, the largest of the cytoskeletal elements, are long, hollow tubes formed by two slightly different variants of globular-shaped tubulin molecules. (b) Most microfilaments, the smallest of the cytoskeletal elements, consist of two chains of actin molecules wrapped around each other. (c) The intermediate filament keratin, found in skin, is made of three polypeptide strands wound around one another. The composition of intermediate filaments, which are intermediate in size between the microtubules and microfilaments, varies among different cell types.

Microtubules help maintain asymmetric cell shapes and play a role in complex cell movements.

Microtubules are the largest of the cytoskeletal elements. They are slender (22 nm in diameter), long, hollow, unbranched tubes composed primarily of **tubulin,** a small, globular, protein molecule (● Figure 2-21a) (1 nanometer (nm) = 1 billionth of a meter).

Microtubules position many of the cytoplasmic organelles, such as the ER, Golgi complex, lysosomes, and mitochondria. They are also essential for maintaining the shape of asymmetric cells, such as nerve cells, whose elongated axons may extend up to a meter in length from where the cell body originates in the spinal cord to where the axon ends at a muscle (● Figure 2-22). Along with specialized intermediate filaments, microtubules stabilize this asymmetric axonal extension.

Microtubules also play an important role in some complex cell movements, including (1) transport of secretory vesicles or other materials from one region of the cell to another, (2) movement of specialized cell projections such as cilia and flagella, and (3) distribution of chromosomes during cell division through formation of a mitotic spindle. Let's examine each of these roles.

VESICLE TRANSPORT Axonal transport provides a good example of the importance of an organized system for moving secretory vesicles. In a nerve cell, specific chemicals are released from the terminal end of the elongated axon to influence a muscle or another structure that the nerve cell controls. These chemicals are largely produced within the cell body where the nuclear DNA blueprint, endoplasmic reticular factory, and Golgi packaging and distribution outlet are located. If they had to diffuse on their own from the cell body to a distant axon terminal, it would take the chemicals about 50 years to get there—obviously an impractical solution. Instead, the microtubules that extend from the beginning to the end of the axon provide a "highway" for vesicular traffic along the axon.

Molecular motors are the transporters. A molecular motor is a protein that attaches to the particle to be transported, then uses energy harnessed from ATP to "walk" along the microtubule with the particle riding in "piggyback" fashion (*motor* means "movement"). **Kinesin,** the molecular motor that carries secretory vesicles in axons, consists of two "feet," a stalk, and a fanlike tail (● Figure 2-22). The tail binds to the secretory vesicle to be moved, and the feet move one at a time, as if walking (● Figure 2-23, p. 46). They alternately attach to one tubulin molecule on the microtubule, bend and push forward, then let go. During this process, the back foot swings ahead of what was the front foot and then attaches to the next tubulin molecule farther down the microtubule. The process is repeated over and over as kinesin moves its cargo to the end of the axon by using the tubulin molecules as stepping-stones.

Reverse vesicular traffic also occurs along these microtubular highways. Vesicles that contain debris are transported by a different ATP-driven molecular motor, **dynein,** from the axon terminal to the cell body for degradation by lysosomes, which are confined within the cell body. The two ends of a microtubule are different, and each molecular motor can travel in only one direction along the microtubule toward a specific end. Dynein always moves toward the centrosome (or "minus") end of the microtubule and kinesin always walks toward the outermost (or "plus") end, ensuring that their cargo is moved in the right direction.

Reverse axonal transport may also serve as a pathway for the movement of some infectious agents, such as herpes viruses (the ones that cause cold sores, genital herpes, and shingles), poliomyelitis virus, and rabies virus. These viruses travel backward along nerves from their surface site of contamination, such as a break in the skin or an animal bite, to the central nervous system (brain and spinal cord).

MOVEMENT OF CILIA AND FLAGELLA Microtubules are also the dominant structural and functional components of cilia and flagella. These specialized motile protrusions from the cell sur-

Summary of Cytoplasm Components

Cytoplasm Component	Structure	Function
Membranous organelles		
Endoplasmic reticulum	Extensive, continuous membranous network of fluid-filled tubules and flattened sacs, partially studded with ribosomes	Forms new cell membrane and other cell components and manufactures products for secretion
Golgi complex	Sets of stacked, flattened membranous sacs	Modifies, packages, and distributes newly synthesized proteins
Lysosomes	Membranous sacs containing hydrolytic enzymes	Serve as cell's digestive system, destroying foreign substances and cellular debris
Peroxisomes	Membranous sacs containing oxidative enzymes	Perform detoxification activities
Mitochondria	Rod- or oval-shaped bodies enclosed by two membranes, with the inner membrane folded into cristae that project into an interior matrix	Act as energy organelles; major site of ATP production; contain enzymes for citric acid cycle, proteins of electron transport system, and ATP synthase
Nonmembranous organelles		
Ribosomes	Granules of RNA and proteins—some attached to rough ER, some free in cytosol	Serve as workbenches for protein synthesis
Vaults	Shaped like hollow octagonal barrels	Serve as cellular trucks for transport from nucleus to cytoplasm
Centrosome/centrioles	A pair of cylindrical structures at right angles to each other (centrioles) surrounded by an amorphous mass	Form and organize the microtubule cytoskeleton
Cytosol		
Intermediary metabolism enzymes	Dispersed within the cytosol	Facilitate intracellular reactions involving degradation, synthesis, and transformation of small organic molecules
Transport, secretory, and endocytic vesicles	Transiently formed, membrane-enclosed products synthesized within or engulfed by the cell	Transport and/or store products being moved within, out of, or into the cell, respectively
Inclusions	Glycogen granules, fat droplets	Store excess nutrients
Cytoskeleton		As an integrated whole, serves as the cell's "bone and muscle"
Microtubules	Long, slender, hollow tubes composed of tubulin molecules	Maintain asymmetric cell shapes and coordinate complex cell movements, specifically serving as highways for transport of secretory vesicles within cell, serving as main structural and functional component of cilia and flagella, and forming mitotic spindle during cell division
Microfilaments	Intertwined helical chains of actin molecules; microfilaments composed of myosin molecules also present in muscle cells	Play a vital role in various cellular contractile systems, including muscle contraction and amoeboid movement; serve as a mechanical stiffener for microvilli
Intermediate filaments	Irregular, threadlike proteins	Help resist mechanical stress

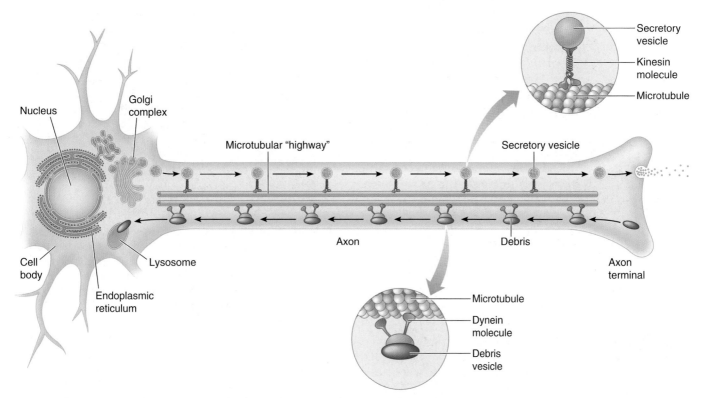

● **FIGURE 2-22 Two-way vesicular axonal transport facilitated by the microtubular "highway" in a neuron.** Secretory vesicles are transported from the site of production in the cell body along a microtubule "highway" to the terminal end for secretion. Vesicles containing debris are transported in the opposite direction for degradation in the cell body. The top enlargement depicts kinesin, a molecular motor, carrying a secretory vesicle down the microtubule by using its "feet" to "step" on one tubulin molecule after another. The bottom enlargement depicts another molecular motor, dynein, transporting the debris up the microtubule.

face allow a cell to move materials across its surface (in the case of a stationary cell) or to propel itself through its environment (in the case of a motile cell). **Cilia** (meaning "eyelashes"; singular, *cilium*) are short, tiny, hairlike protrusions usually found in large numbers on the surface of a ciliated cell. **Flagella** (meaning "whips"; singular, *flagellum*) are long, whiplike appendages; typically, a cell has one or a few flagella at most. Even though they project from the surface of the cell, cilia and flagella are both intracellular structures—they are covered by the plasma membrane.

Cilia beat or stroke in unison in a given direction, much like the coordinated efforts of a rowing team. In humans, ciliated cells line the respiratory tract, the oviduct of the female reproductive tract, and the fluid-filled ventricles (chambers) of the brain. The coordinated stroking of the thousands of respiratory cilia help keep foreign particles out of the lungs by sweeping outward dust and other inspired (breathed-in) particles (● Figure 2-24). In the female reproductive tract, the sweeping action of the cilia lining the oviduct draws the egg (ovum) released

from the ovary during ovulation into the oviduct and then guides it toward the uterus (womb). In the brain, the ciliated cells lining the ventricles produce cerebrospinal fluid, which flows through the ventricles and around the brain and spinal cord, cushioning and bathing these fragile neural structures. Beating of the cilia helps promote circulation of this supportive fluid.

In addition to the multiple motile cilia found in cells in these specific locations, almost all cells in the human body possess a single nonmotile *primary cilium*. Until recently, primary cilia were considered useless vestiges, but growing evidence suggests that they act as microscopic sensory organs that sample the extracellular environment. They may be critical for receiving regulatory signals involved in controlling growth, cell differentiation, and cell proliferation (expansion of a given cell type). Defects in primary and motile cilia have been implicated in a range of human disorders, including a form of abnormal kidney development (polycystic kidney disease) and chronic respiratory disease, respectively.

● **FIGURE 2-23 How a kinesin molecule "walks."** A kinesin molecule walks along the surface of a microtubule by alternately attaching and releasing its "feet" as it cyclically swings the rear foot ahead of the front foot.

In humans, the only cells that have a flagellum are sperm (see ● Figure 20-9, p. 754). The whiplike motion of its single flagellum, or "tail," enables a sperm to move through its environment, which is crucial for maneuvering into position to fertilize the female ovum.

Cilia and flagella have the same basic internal structure, their only difference being their length: cilia are short, flagella are long. Both consist of nine fused pairs of microtubules (doublets) arranged in an outer ring around two single unfused microtubules in the center (● Figure 2-25). This characteristic "9 + 2" grouping of microtubules extends throughout the

Cilia   Goblet cell

● **FIGURE 2-24 Cilia in the respiratory tract.** Scanning electron micrograph of cilia on cells lining the human respiratory tract. The respiratory airways are lined by goblet cells, which secrete a sticky mucus that traps inspired particles, and epithelial cells that bear numerous hairlike cilia. The cilia all beat in the same direction to sweep inspired particles up and out of the airways.

length of the motile appendage. Spokelike accessory proteins hold the structure together.

Cilia and flagella arise from centrioles. Each cylinder of the centriole pair contains a bundle of microtubules similar to the 9 + 2 complex, except that the central pair of single microtubules is missing and the outer ring has nine fused triplets rather than doublets of microtubules (see ● Figure 2-19). During formation of a cilium or flagellum, a duplicated centriole moves to a position just under the plasma membrane, where microtubules grow outward from the centriole in an orderly pattern to form the motile appendage. The centriole remains at the base of the developed cilium or flagellum as the **basal body** of the structure.

In addition to the accessory proteins that maintain the microtubule's organization, another accessory protein, dynein, plays an essential part in the microtubular movement that causes the entire structure to bend. Dynein, a molecular motor, forms armlike projections from each doublet of microtubules (● Figure 2-25b and c). These dynein arms walk along the adjacent microtubule doublets, causing the doublets to slide past each other, bringing about the bending and stroking (● Figure 2-25d). Groups of cilia working together are oriented to beat in the same direction and contract in a synchronized manner through poorly understood control mechanisms involving the single microtubules at the cilium's center.

FORMATION OF THE MITOTIC SPINDLE Cell division involves two discrete but related activities: *mitosis* (nuclear division), which depends on microtubules, and *cytokinesis* (cytoplasmic division), which depends on microfilaments and will be described in the next section. During **mitosis,** the DNA-containing chromosomes of the nucleus are replicated, resulting in two identical sets. These duplicate sets of chromosomes are separated and drawn apart to opposite sides of the cell so that the genetic material is evenly distributed in the two halves of the cell (see p. A-28).

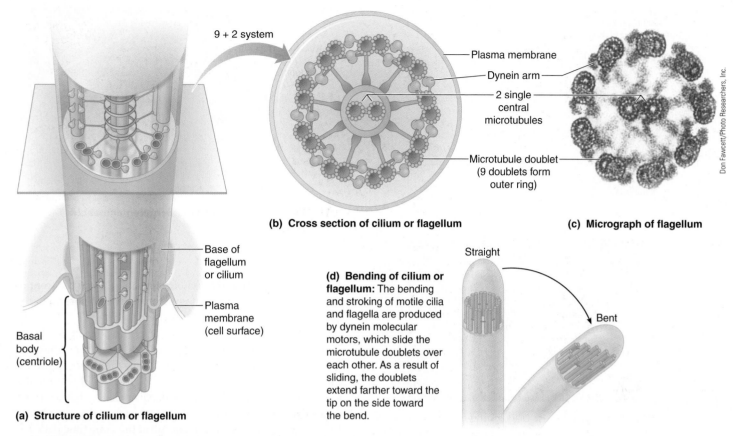

9 + 2 system

Plasma membrane

Dynein arm

2 single central microtubules

Microtubule doublet (9 doublets form outer ring)

(b) Cross section of cilium or flagellum

(c) Micrograph of flagellum

Base of flagellum or cilium

Plasma membrane (cell surface)

Basal body (centriole)

(a) Structure of cilium or flagellum

(d) Bending of cilium or flagellum: The bending and stroking of motile cilia and flagella are produced by dynein molecular motors, which slide the microtubule doublets over each other. As a result of sliding, the doublets extend farther toward the tip on the side toward the bend.

Straight

Bent

● **FIGURE 2-25 Internal structure of a cilium or flagellum.** (a) The relationship between the microtubules and the centriole turned basal body of a cilium or flagellum. (b) Diagram of a cilium or flagellum in cross section showing the characteristic "9 + 2" arrangement of microtubules along with the dynein arms and other accessory proteins that hold the system together. (c) Electron micrograph of a flagellum in cross section; individual tubulin molecules are visible in the microtubule walls. (d) Depiction of bending of a cilium or flagellum caused by microtubule sliding brought about by dynein "walking."

The replicated chromosomes are pulled apart by a cellular apparatus called the **mitotic spindle,** which is transiently assembled from microtubules only during cell division (see ● Figure C-10, p. A-29). The microtubules of the mitotic spindle are formed by the centrioles. As part of cell division, the centrioles first duplicate themselves; then, the new centriole pairs move to opposite ends of the cell and form the spindle apparatus between them through a precisely organized assemblage of microtubules. Importantly, some anticancer drugs prevent cancer cells from reproducing by interfering with the microtubules that ordinarily pull the chromosomes to opposite poles during cell division.

Microfilaments are important to cellular contractile systems and as mechanical stiffeners.

Microfilaments are the smallest (6 nm in diameter) elements of the cytoskeleton. The most obvious microfilaments in most cells are those composed of **actin,** a protein molecule that has a globular shape similar to that of tubulin. Unlike tubulin, which forms a hollow tube, actin assembles into two strands, which twist around each other to form a microfilament (see ● Figure 2-21b). In muscle cells, the protein **myosin** forms a different kind of microfilament (see ● Figure 8-5). In most cells, myosin is not as abundant and does not form such distinct filaments.

Microfilaments serve two functions: (1) They play a vital role in various cell contractile systems, and (2) they act as mechanical stiffeners for several specific cell projections.

MICROFILAMENTS IN CELL CONTRACTILE SYSTEMS Actin-based assemblies are involved in muscle contraction, cell division, and cell locomotion. The most obvious, best-organized, and most clearly understood cell contractile system is that found in muscle. Muscle contains an abundance of actin and myosin microfilaments, which bring about muscle contraction through the ATP-powered sliding of actin microfilaments in relation to stationary myosin microfilaments. Myosin is a molecular motor that has heads that walk along the actin microfilaments, pulling them inward between the myosin microfilaments. Microfilament sliding and force development are triggered by a complex sequence of electrical, biochemical, and mechanical events initiated when the muscle cell is stimulated to contract (see Chapter 8 for details).

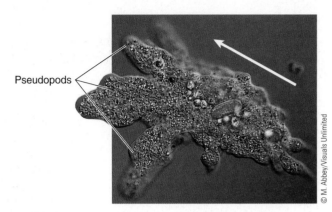

● **FIGURE 2-27 An amoeba undergoing amoeboid movement.**

● **FIGURE 2-26 Cytokinesis.** Diagram and micrograph of a cell undergoing cytokinesis, in which a contractile ring composed of actin filaments tightens, squeezing apart the two duplicate cell halves formed by mitosis.

Nonmuscle cells may also contain "musclelike" assemblies. Some of these microfilament contractile systems are transiently assembled to perform a specific function when needed. A good example is the contractile ring that forms during **cytokinesis,** the process by which the two halves of a dividing cell separate into two new daughter cells, each with a full complement of chromosomes. The ring consists of a beltlike bundle of actin filaments located just beneath the plasma membrane in the middle of the dividing cell. When this ring of fibers contracts and tightens, it pinches the cell in two (● Figure 2-26).

Complex actin-based assemblies are also responsible for most cell locomotion. Four types of human cells are capable of moving on their own—sperm, white blood cells, fibroblasts, and skin cells. Sperm move by the flagellar mechanism already described. Motility for the other cells is accomplished by **amoeboid movement,** a cell-crawling process that depends on the activity of their actin filaments, in a mechanism similar to that used by amoebas to maneuver through their environment. When crawling, the motile cell forms fingerlike *pseudopods* at the "front" or leading edge of the cell in the direction of the target. For example, the target that triggers amoeboid movement might be the proximity of food in the case of an amoeba or a bacterium in the case of a white blood cell (see ● Figure 2-8c). Pseudopods are formed as a result of the organized assembly and disassembly of branching actin networks. During amoeboid movement, actin filaments continuously grow at the cell's lead-

ing edge through the addition of actin molecules at the front of the actin chain. This filament growth pushes that portion of the cell forward as a pseudopod protrusion (● Figure 2-27). Simultaneously, actin molecules at the rear of the filament are being disassembled and transferred to the front of the line. Thus, the filament does not get any longer; it stays the same length but moves forward through the continuous transfer of actin molecules from the rear to the front of the filament in what is termed *treadmilling* fashion. The cell attaches the advancing pseudopod to surrounding connective tissue and at the same time detaches from its older adhesion site at the rear. The cell uses the new adhesion site at the leading edge as a point of traction to pull the bulk of its body forward through cytoskeletal contraction.

White blood cells are the most active crawlers in the body. These cells leave the circulatory system and travel by amoeboid movement to areas of infection or inflammation, where they engulf and destroy microorganisms and cellular debris. Amazingly, it is estimated that the total distance traveled collectively per day by all your white blood cells while they roam the tissues in their search-and-destroy tactic would circle Earth twice!

Fibroblasts ("fiber formers"), another type of motile cell, move amoeboid fashion into a wound from adjacent connective tissue to help repair the damage; they are responsible for scar formation. Skin cells, which are ordinarily stationary, can become modestly mobile and move by amoeboid motion toward a cut to restore the skin surface.

MICROFILAMENTS AS MECHANICAL STIFFENERS Besides their role in cellular contractile systems, actin filaments serve as mechanical supports or stiffeners for several cellular extensions, of which the most common are microvilli. **Microvilli** are microscopic, nonmotile, hairlike projections from the surface of epithelial cells lining the small intestine and kidney tubules (● Figure 2-28). Their presence greatly increases the surface area available for transferring material across the plasma membrane. In the small intestine, the microvilli increase the area available for absorbing digested nutrients. In the kidney tubules, microvilli enlarge the absorptive surface that salvages useful substances passing through the kidney so that these materials are saved for the body instead of being eliminated in the urine. Within each microvillus, a core consisting of parallel linked

Microvilli

● **FIGURE 2-28 Microvilli in the small intestine.** Scanning electron micrograph showing microvilli on the surface of a small-intestine epithelial cell.

<small>Dr. R. G. Kessel & Dr. R. H. Kardon/Visuals Unlimited</small>

actin filaments forms a rigid mechanical stiffener that keeps these valuable surface projections intact.

Intermediate filaments are important in cell regions subject to mechanical stress.

Intermediate filaments are intermediate in size between microtubules and microfilaments (7 to 11 nm in diameter)—hence their name. The proteins that compose the intermediate filaments vary between cell types, but in general they appear as irregular, threadlike molecules. These proteins form tough, durable fibers that play a central role in maintaining the structural integrity of a cell and in resisting mechanical stresses externally applied to a cell.

Different types of intermediate filaments are tailored to suit their structural or tension-bearing role in specific cell types. In general, only one class of intermediate filament is found in a particular cell type. Two important examples follow:

■ *Neurofilaments* are intermediate filaments found in nerve cell axons. Together with microtubules, neurofilaments strengthen and stabilize these elongated cellular extensions.

■ Skin cells contain irregular networks of intermediate filaments made of the protein **keratin** (see ● Figure 2-21c). These intracellular filaments connect with the extracellular filaments that tie adjacent cells together, creating a continuous filamentous network that extends throughout the skin and gives it strength. When surface skin cells die, their tough keratin skeletons persist, forming a protective, waterproof outer layer. Hair and nails are also keratin structures.

Intermediate filaments account for up to 85% of the total protein in nerve cells and keratin-producing skin cells, whereas these filaments constitute only about 1% of other cells' total protein on average.

Clinical Note Neurofilament abnormalities contribute to some neurological disorders. An important example is **amyotrophic lateral sclerosis (ALS),** more familiarly known as **Lou Gehrig's disease.** ALS is characterized by progressive degeneration and death of motor neurons, the type of nerve cells that control skeletal muscles. This adult-onset condition leads to gradual loss of control of skeletal muscles, including the muscles of breathing, and ultimately to death, as it did for baseball legend Lou Gehrig. One underlying problem may be an abnormal accumulation and disorganization of neurofilaments. Motor neurons, which have the most neurofilaments, are the most affected. The disorganized neurofilaments are believed to block the axonal transport of crucial materials along the microtubular highways, thus choking off vital supplies from the cell body to the axon terminal.

The cytoskeleton functions as an integrated whole and links other parts of the cell.

Collectively, the cytoskeletal elements and their interconnections support the plasma membrane and are responsible for the particular shape, rigidity, and spatial geometry of each different cell type. Furthermore, growing evidence suggests that the cytoskeleton serves as a lattice to organize groups of enzymes for many cellular activities. This internal framework thus acts as the cell's "skeleton."

New studies hint that the cytoskeleton as a whole is not merely a supporting structure that maintains the tensional integrity of the cell but may serve as a mechanical communications system as well. Various components of the cytoskeleton behave as if they are structurally connected or "hardwired" to each other as well as to the plasma membrane and the nucleus. This force-carrying network may serve as the mechanism by which mechanical forces acting on the cell surface reach all the way from the plasma membrane through the cytoskeleton to ultimately influence gene regulation in the nucleus.

Furthermore, as you have learned, the coordinated action of the cytoskeletal elements is responsible for directing intracellular transport and for regulating numerous cellular movements and thereby also serves as the cell's "muscle."

Chapter in Perspective: Focus on Homeostasis

The ability of cells to perform functions essential for their own survival as well as specialized tasks that help maintain homeostasis within the body ultimately depends on the successful, cooperative operation of the intracellular components. For example, to support life-sustaining activities, all cells must generate energy, in a usable form, from nutrient molecules. Energy is generated intracellularly by chemical reactions in the cytosol and mitochondria.

In addition to being essential for basic cell survival, the organelles and cytoskeleton participate in many cells' specialized tasks that contribute to homeostasis. Here are several examples:

■ Both nerve and endocrine cells release protein chemical messengers (neurotransmitters in nerve cells and hormones in endocrine cells) that are important in regulatory activities aimed at maintaining homeostasis. For example, neurotransmitters stimulate the respiratory muscles, which accomplish life-sustaining exchanges of O_2 and CO_2 between the body and atmosphere through breathing. These protein chemical messengers are all produced by the endoplasmic reticulum and Golgi complex and released by exocytosis from the cell when needed.

- The ability of muscle cells to contract depends on their highly developed cytoskeletal microfilaments sliding past one another. Muscle contraction is responsible for many homeostatic activities, including (1) contracting the heart muscle, which pumps life-supporting blood throughout the body; (2) contracting the muscles attached to bones, which enables the body to procure food; and (3) contracting the muscle in the walls of the stomach and intestine, which moves the food along the digestive tract so that ingested nutrients can be progressively broken down into a form that can be absorbed into the blood for delivery to the cells.

- White blood cells help the body resist infection by making extensive use of lysosomal destruction of engulfed particles as they police the body for microbial invaders. These white blood cells are able to roam the body by means of amoeboid movement, a cell-crawling process accomplished by coordinated assembly and disassembly of actin, one of their cytoskeletal components.

As we begin to examine the various organs and systems, keep in mind that proper cell functioning is the foundation of all organ activities.

REVIEW EXERCISES

Objective Questions (Answers on p. A-39)

1. The barrier that separates and controls movement between the cell contents and the extracellular fluid is the _____.

2. The chemical that directs protein synthesis and serves as a genetic blueprint is _____, which is found in the _____ of the cell.

3. The cytoplasm consists of _____, which are discrete specialized, intracellular compartments, a gel-like mass known as _____, and an elaborate protein scaffolding called the _____.

4. Transport vesicles from the _____ fuse with and enter the _____ for modification and sorting.

5. The (what kind of) _____ enzymes within the peroxisomes primarily detoxify various wastes produced within the cell or foreign compounds that have entered the cell.

6. The universal energy carrier of the body is _____.

7. The largest cells in the human body can be seen by the unaided eye. (True or false?)

8. Amoeboid movement is accomplished by the coordinated assembly and disassembly of microtubules. (True or false?)

9. Using the following answer code, indicate which type of ribosome is being described:
   ___ 1. synthesizes proteins used to construct new cell membrane
   ___ 2. synthesizes proteins used intracellularly within the cytosol
   ___ 3. synthesizes secretory proteins such as enzymes or hormones
   ___ 4. synthesizes the hydrolytic enzymes incorporated in lysosomes

   (a) free ribosome
   (b) rough ER-bound ribosome

10. Using the following answer code, indicate which form of energy production is being described:
    ___ 1. takes place in the mitochondrial matrix
    ___ 2. produces H_2O as a by-product
    ___ 3. results in a rich yield of ATP
    ___ 4. takes place in the cytosol
    ___ 5. processes acetyl-CoA
    ___ 6. takes place in the mitochondrial inner-membrane cristae
    ___ 7. converts glucose into two pyruvate molecules
    ___ 8. uses molecular oxygen
    ___ 9. accomplished by the electron transport system and ATP synthase

    (a) glycolysis
    (b) citric acid cycle
    (c) oxidative phosphorylation

Essay Questions

1. What are a cell's three major subdivisions?

2. State an advantage of organelle compartmentalization.

3. List the five types of membranous organelles and the three types of nonmembranous organelles.

4. Describe the structure of the endoplasmic reticulum, distinguishing between the rough and the smooth ER. What is the function of each?

5. Compare exocytosis and endocytosis. Define secretion, pinocytosis, receptor-mediated endocytosis, and phagocytosis.

6. Which organelles serve as the intracellular digestive system? What type of enzymes do they contain? What functions do these organelles serve?

7. Compare lysosomes with peroxisomes.

8. Distinguish among cellular respiration, oxidative phosphorylation, and chemiosmosis.

9. Describe the structure of mitochondria, and explain their role in cellular respiration.

10. Distinguish between the oxidative enzymes found in peroxisomes and those found in mitochondria.

11. Cells expend energy on what three categories of activities?

12. List and describe the functions of each component of the cytoskeleton.

Quantitative Exercises (Solutions on p. A-40)

(See Appendix D, "Principles of Quantitative Reasoning.")

1. Each "turn" of the Krebs cycle
   a. generates 3 NAD^+, 1 $FADH_2$, and 2 CO_2
   b. generates 1 GTP, 2 CO_2, and 1 $FADH_2$
   c. consumes 1 pyruvate and 1 oxaloacetate
   d. consumes an amino acid

2. Let's consider how much ATP you synthesize in a day. Assume that you consume 1 mole of O_2 per hour or 24 moles/day (a mole is the number of grams of a chemi-

cal equal to its molecular weight). About 6 moles of ATP are produced per mole of O_2 consumed. The molecular weight of ATP is 507. How many grams of ATP do you produce per day at this rate? Given that 1000 g equal 2.2 lb, how many pounds of ATP do you produce per day at this rate? (This is under relatively inactive conditions!)

3. Under resting circumstances a person produces about 144 moles of ATP per day (73,000 g ATP/day). The amount of *free energy* represented by this amount of ATP can be calculated as follows. Cleavage of the terminal phosphate bond from ATP results in a decrease in free energy of approximately 7300 cal/mole. This is a crude measure of the energy available to do work that is contained in the terminal phosphate bond of the ATP molecule. How many calories, in the form of ATP, are produced per day by a resting individual, crudely speaking?

4. Calculate the number of cells in the body of an average 68 kg (150 lb) adult. (This will only be accurate to about 1 part in 10 but should give you an idea how scientists estimate this commonly quoted number.) Assume all cells are spheres 20 μm in diameter. The volume of a sphere can be determined by the equation $v = 4/3 \ \pi r^3$. (*Hint:* We know that about two thirds of the water in the body is intracellular, and the density of cells is nearly 1 g/ml. The proportion of the mass made up of water is about 60%.)

5. If sucrose is injected into the bloodstream, it tends to stay out of the cells (cells do not use sucrose directly). If it does not go into cells, where does it go? In other words, how much "space" is in the body that is not inside some cell? Sucrose can be used to determine this space. Suppose 150 mg of sucrose is injected into a 55 kg woman. If the concentration of sucrose in her blood is 0.015 mg/ml, what is the volume of her extracellular space, assuming that no metabolism is occurring and that the blood sucrose concentration is equal to the sucrose concentration throughout the extracellular space?

POINTS TO PONDER

(Explanations on p. A-40)

1. The stomach has two types of exocrine secretory cells: *chief cells,* which secrete an inactive form of the protein-digesting enzyme *pepsinogen;* and *parietal cells,* which secrete *hydrochloric acid (HCl)* that activates pepsinogen. Both cell types have an abundance of mitochondria for ATP production—the chief cells need energy to synthesize pepsinogen, and the parietal cells need energy to transport H^+ and Cl^- from the blood into the stomach lumen. Only one of these cell types also has an extensive rough endoplasmic reticulum and abundant Golgi stacks. Would this type be the chief cells or the parietal cells? Why?

2. The poison *cyanide* acts by binding irreversibly to one component of the electron transport system, blocking its action. As a result, the entire electron-transport process comes to a screeching halt, and the cells lose more than 94% of their ATP-producing capacity. Considering the types of cell activities that depend on energy expenditure, what would be the consequences of cyanide poisoning?

3. Hydrogen peroxide, which belongs to a class of very unstable compounds known as *free radicals,* can bring about drastic, detrimental changes in a cell's structure and function by reacting with almost any molecule with which it comes in contact, including DNA. The resultant cellular changes can lead to genetic mutations, cancer, or other serious consequences. Furthermore, some researchers speculate that cumulative effects of more subtle cellular damage resulting from free radical reactions over a period of time might contribute to the gradual deterioration associated with aging. Related to this speculation, studies have shown that longevity decreases in fruit flies in direct proportion to a decrease in a specific chemical found in one of the cellular organelles. Based on your knowledge of how the body rids itself of dangerous hydrogen peroxide, what do you think this chemical in the organelle is?

4. Why do you think a person is able to perform anaerobic exercise (such as lifting and holding a heavy weight) only briefly but can sustain aerobic exercise (such as walking or swimming) for long periods? (*Hint:* Muscles have limited energy stores.)

5. One type of the affliction *epidermolysis bullosa* is caused by a genetic defect that results in production of abnormally weak keratin. Based on your knowledge of the role of keratin, what part of the body do you think would be affected by this condition?

CLINICAL CONSIDERATION

(Explanation on p. A-40)

Kevin S. and his wife have been trying to have a baby for the past three years. On seeking the help of a fertility specialist, Kevin learned that he has a hereditary form of male sterility involving nonmotile sperm. His condition can be traced to defects in the cytoskeletal components of the sperm's flagella. As a result of this finding, the physician suspected that Kevin also has a long history of recurrent respiratory tract disease. Kevin confirmed that indeed he has had colds, bronchitis, and influenza more frequently than his friends. Why would the physician suspect that Kevin probably had a history of frequent respiratory disease based on his diagnosis of sterility due to nonmotile sperm?

Body Systems

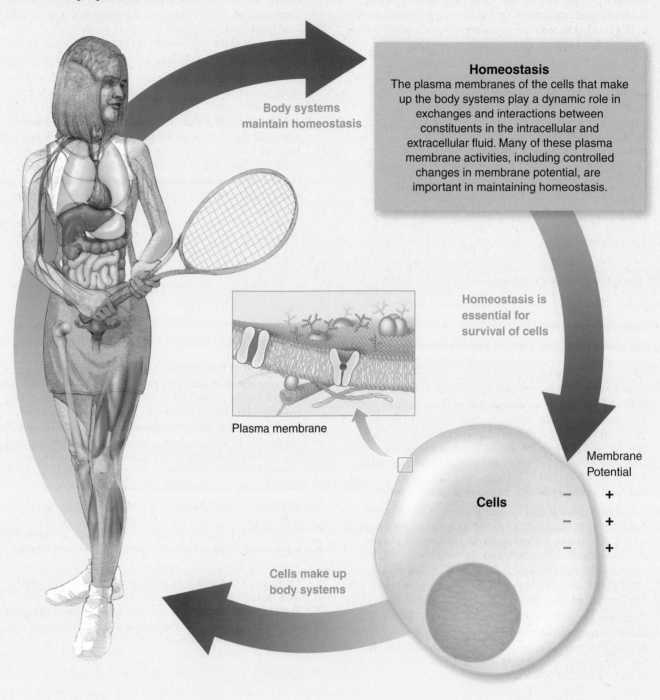

Body systems maintain homeostasis

Homeostasis
The plasma membranes of the cells that make up the body systems play a dynamic role in exchanges and interactions between constituents in the intracellular and extracellular fluid. Many of these plasma membrane activities, including controlled changes in membrane potential, are important in maintaining homeostasis.

Homeostasis is essential for survival of cells

Plasma membrane

Cells

Membrane Potential

− +
− +
− +

Cells make up body systems

All cells are enveloped by a **plasma membrane,** a thin, flexible, lipid barrier that separates the contents of the cell from its surroundings. To carry on life-sustaining and specialized activities, each cell of the body must exchange materials across this membrane with the homeostatically maintained internal fluid environment that surrounds it. This discriminating barrier contains specific proteins, some of which enable selective passage of materials. Other membrane proteins serve as receptor sites

for interaction with specific chemical messengers in the cell's environment. These messengers control many cell activities crucial to homeostasis.

Cells have a membrane potential, a slight excess of negative charges lined up along the inside of the membrane and a slight excess of positive charges on the outside. The specialization of nerve and muscle cells depends on the ability of these cells to alter their potential when appropriately stimulated.

3

The Plasma Membrane and Membrane Potential

CENGAGENOW™ Log on to CengageNOW at **http://www .cengage.com/sso/** for an opportunity to explore a learning module that illustrates difficult concepts with self-study tutorials, animations, and interactive quizzes to help you learn, review, and master physiology concepts.

Membrane Structure and Functions

To survive, every cell must maintain a specific composition of its contents unique for that cell type despite the remarkably different composition of the extracellular fluid surrounding it. This difference in fluid composition inside and outside a cell is maintained by the **plasma membrane,** the extremely thin layer of lipids and proteins that forms the outer boundary of every cell and encloses the intracellular contents. Besides acting as a mechanical barrier that traps needed molecules within the cell, the plasma membrane helps determine the cell's composition by selectively permitting specific substances to pass between the cell and its environment. The plasma membrane controls the entry of nutrient molecules and the exit of secretory and waste products. In addition, it maintains differences in ion concentrations inside and outside the cell, which are important in the membrane's electrical activity. The plasma membrane also participates in the joining of cells to form tissues and organs. Finally, it plays a key role in enabling a cell to respond to changes, or signals, in the cell's environment; this ability is important in communication between cells. No matter what the cell type, these common membrane functions are crucial to the cell's survival, to its ability to perform specialized homeostatic activities, and to its ability to coordinate its functions with those of other cells. Many of the functional differences between cell types are due to subtle variations in the composition of their plasma membranes, which in turn enable different cells to interact in different ways with essentially the same extracellular fluid environment.

The plasma membrane is a fluid lipid bilayer embedded with proteins.

The plasma membrane of every cell consists mostly of lipids and proteins plus small amounts of carbohydrate. It is too thin to be seen under an ordinary light microscope, but with an electron microscope it appears as a trilaminar (meaning "three-layered") structure consisting of two dark layers separated by a light middle layer (● Figure 3-1) (*tri* means "three"; *lamina* means "layer"). The specific arrangement of the molecules that make up the plasma membrane is responsible for this "sandwich" appearance.

The most abundant membrane lipids are phospholipids, with lesser amounts of cholesterol. An estimated billion phospholipid

Cell 1 — Plasma membranes

Intercellular space — Cell 2

© Don W. Fawcett/Visuals Unlimited

● **FIGURE 3-1 Trilaminar appearance of a plasma membrane in an electron micrograph.** Depicted are the plasma membranes of two adjacent cells. Note that each membrane appears as two dark layers separated by a light middle layer.

Head (negatively charged, polar, hydrophilic)

Tails (uncharged, nonpolar, hydrophobic)

(a) Phospholipid molecule

ECF (water)

Polar heads (hydrophilic)

Nonpolar tails (hydrophobic)

Polar heads (hydrophilic)

Lipid bilayer

ICF (water)

(b) Organization of phospholipids into a bilayer in water

Lipid bilayer

Intracellular fluid

Extracellular fluid

(c) Separation of ECF and ICF by the lipid bilayer

● **FIGURE 3-2 Structure and organization of phospholipid molecules in a lipid bilayer.** (a) Phospholipid molecule. (b) In water, phospholipid molecules organize themselves into a lipid bilayer with the polar heads interacting with the polar water molecules at each surface and the nonpolar tails all facing the interior of the bilayer. (c) An exaggerated view of the plasma membrane enclosing a cell, separating the ICF from the ECF.

molecules are present in the plasma membrane of a typical human cell. **Phospholipids** have a polar (electrically charged; see p. A-7) head containing a negatively charged phosphate group and two nonpolar (electrically neutral) fatty-acid chain tails (● Figure 3-2a). The polar end is hydrophilic ("water loving") because it can interact with water molecules, which are also polar; the nonpolar end is hydrophobic ("water fearing") and will not mix with water. In water, phospholipids self-assemble into a **lipid bilayer,** a double layer of lipid molecules (● Figure 3-2b) (*bi* means "two"). The hydrophobic tails bury themselves in the center of the bilayer away from the water, and the hydrophilic heads line up on both sides in contact with the water. The outer surface of the layer is exposed to extracellular fluid (ECF), whereas the inner surface is in contact with the intracellular fluid (ICF) (● Figure 3-2c).

The lipid bilayer is fluid, not rigid, with a consistency more like liquid cooking oil than solid shortening. The phospholipids, which are not held together by strong chemical bonds, are constantly moving. They can twirl, vibrate, and move around within their own half of the bilayer, exchanging places millions of times a second. This phospholipid movement accounts in large part for membrane fluidity.

Cholesterol contributes to both the fluidity and the stability of the membrane. Cholesterol molecules are tucked in between the phospholipid molecules, where they prevent the fatty acid chains from packing together and crystallizing, a process that would drastically reduce membrane fluidity. Through their spatial relationship with phospholipid molecules, cholesterol molecules also help stabilize the phospholipids' position.

Because of its fluidity, the plasma membrane has structural integrity but at the same time is flexible, enabling the cell to change shape. For example, muscle cells change shape as they contract, and red blood cells must change shape considerably as they squeeze their way single file through the capillaries, the narrowest blood vessels.

Membrane proteins are inserted within or attached to the lipid bilayer (● Figure 3-3). **Integral proteins** are embedded in the lipid bilayer, with most extending through the entire thickness of the membrane, in which case they are alternatively called **transmembrane proteins** (*trans* means "across"). Like phospholipids, integral proteins also have hydrophilic and hydrophobic regions. **Pe-**

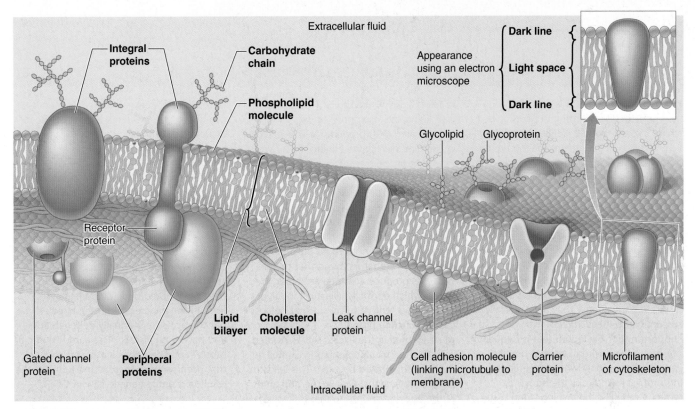

Extracellular fluid

Integral proteins

Carbohydrate chain

Phospholipid molecule

Receptor protein

Lipid bilayer

Cholesterol molecule

Leak channel protein

Gated channel protein

Peripheral proteins

Intracellular fluid

Appearance using an electron microscope

Dark line

Light space

Dark line

Glycolipid   Glycoprotein

Cell adhesion molecule (linking microtubule to membrane)

Carrier protein

Microfilament of cytoskeleton

● **FIGURE 3-3 Fluid mosaic model of plasma membrane structure.** The plasma membrane is composed of a lipid bilayer embedded with proteins. Integral proteins extend through the thickness of the membrane or are partially submerged in the membrane, and peripheral proteins are loosely attached to the surface of the membrane. Short carbohydrate chains attach to proteins or lipids on the outer surface only.

ripheral proteins** are usually polar molecules that do not penetrate the membrane. They stud only the outer or more commonly the inner surface, anchored by relatively weak chemical bonds with the polar parts of integral membrane proteins or membrane lipids. The plasma membrane has about 50 times more lipid molecules than protein molecules. However, proteins account for nearly half of the membrane's mass because they are much larger than the lipids. The fluidity of the lipid bilayer enables many membrane proteins to float freely like "icebergs" in a moving "sea" of lipid, although the cytoskeleton restricts the mobility of proteins that perform a specialized function in a specific area of the cell. This view of membrane structure is known as the **fluid mosaic model,** in reference to the membrane fluidity and the ever-changing mosaic pattern of the proteins embedded in the lipid bilayer. (A mosaic is a surface decoration made by inlaying small pieces of variously colored tiles to form patterns or pictures.)

A small amount of **membrane carbohydrate** is located on the outer surface of cells, "sugar coating" them. Short carbohydrate chains protrude like tiny antennas from the outer surface, bound primarily to membrane proteins and, to a lesser extent, to lipids. These sugary combinations are known as *glycoproteins* and *glycolipids,* respectively (● Figure 3-3), and the coating they form is called the *glycocalyx* (*glyco* means "sweet"; *calyx* means "husk").

This proposed structure accounts for the trilaminar appearance of the plasma membrane. When stains are used to help visualize the plasma membrane under an electron microscope (as in ● Figure 3-1), the two dark lines represent the hydrophilic polar regions of the lipid and protein molecules that have taken up the stain. The light space between corresponds to the poorly stained hydrophobic core formed by the nonpolar regions of these molecules.

The different components of the plasma membrane carry out a variety of functions. The lipid bilayer forms the primary barrier to diffusion, the proteins perform most of the specific membrane functions, and the carbohydrates play an important role in "self-recognition" processes and cell-to-cell interactions. We now examine the functions of these membrane components in more detail.

The lipid bilayer forms the basic structural barrier that encloses the cell.

The lipid bilayer serves three important functions:

1. It forms the basic structure of the membrane. The phospholipids can be visualized as the "pickets" that form the "fence" around the cell.

2. Its hydrophobic interior serves as a barrier to passage of water-soluble substances between the ICF and ECF. Water-soluble substances cannot dissolve in and pass through the lipid bilayer. By means of this barrier, the cell can maintain different mixtures and concentrations of solutes (dissolved substances) inside and outside the cell.

3. It is responsible for the fluidity of the membrane.

Cystic Fibrosis: A Fatal Defect in Membrane Transport

Cystic fibrosis (CF), the most common fatal genetic disease in the United States, strikes 1 in every 2000 Caucasian children. It is characterized by the production of abnormally thick, sticky mucus. Most dramatically affected are the respiratory airways and the pancreas.

Respiratory Problems

The presence of thick, sticky mucus in the respiratory airways makes it difficult to get adequate air in and out of the lungs. Also, because bacteria thrive in the accumulated mucus, CF patients suffer from repeated respiratory infections. They are especially susceptible to *Pseudomonas aeruginosa,* an "opportunistic" bacterium that is frequently present in the environment but usually causes infection only when some underlying problem handicaps the body's defenses. Gradually, the involved lung tissue becomes scarred (fibrotic), making the lungs harder to inflate. This complication increases the work of breathing beyond the extra effort required to move air through the clogged airways.

Underlying Cause

Cystic fibrosis is caused by any one of several different genetic defects that lead to production of a flawed version of a protein known as *cystic fibrosis transmembrane conductance regulator (CFTR).* CFTR normally helps form and regulate the chloride (Cl^-) channels in the plasma membrane. With CF, the defective CFTR "gets stuck" in the endoplasmic reticulum/Golgi system, which normally manufactures and processes this product and ships it to the plasma membrane (see p. 27). That is, in CF patients the mutated version of CFTR is only partially processed and never makes it to the cell surface. The resultant absence of CFTR protein in the plasma membrane's Cl^- channels makes the membrane impermeable to Cl^-. Because Cl^- transport across the mem-brane is closely linked to Na^+ transport, cells lining the respiratory airways cannot absorb salt (NaCl) properly. As a result, salt accumulates in the fluid lining the airways.

What puzzles researchers is how this Cl^- channel defect and resultant salt accumulation lead to the excess mucus problem. Two recent discoveries may provide an answer, although these proposals remain to be proven and research into other possible mechanisms continues to be pursued. One group of investigators found that the airway cells produce a natural antibiotic, *defensin,* which normally kills most inhaled airborne bacteria. It turns out that defensin cannot function properly in a salty environment. Bathed in the excess salt associated with CF, the disabled antibiotic cannot rid the lungs of inhaled bacteria. This leads to repeated infections. One of the outcomes of the body's response to these infections is production of excess mucus, which serves as a breeding ground for more bacterial growth. The cycle continues as the lung-clogging mucus accumulates and lung infections become more frequent. To make matters worse, the excess mucus is especially thick and sticky, making it difficult for the normal ciliary defense mechanisms of the lungs to sweep up the bacteria-laden mucus (see pp. 24 and 457). The mucus is thick and sticky because it is underhydrated (has too little water), a problem believed to be linked to the defective salt transport.

The second new study found an additional complicating factor in the CF story. These researchers demonstrated that CFTR appears to serve a dual role as a Cl^- channel and as a membrane receptor that binds to and destroys *P. aeruginosa* (and perhaps other bacteria). Because CFTR is absent from the airway cell membranes of CF patients, *P. aeruginosa* is not cleared from the airways as usual. Besides causing infection, these bacteria also trigger the airway cells to produce unusually large amounts of ab-normal, thick, sticky mucus. This mucus promotes more bacterial growth as the vicious cycle continues.

Pancreatic Problems

Furthermore, in CF patients the pancreatic duct, which carries secretions from the pancreas to the small intestine, becomes plugged with thick mucus. Because the pancreas produces enzymes important in the digestion of food, malnourishment eventually results. In addition, as the pancreatic digestive secretions accumulate behind the blocked duct, fluid-filled cysts form in the pancreas, with the affected pancreatic tissue gradually degenerating and becoming fibrotic. The name "cystic fibrosis" aptly describes long-term changes that occur in the pancreas and lungs as the result of a single genetic flaw in CFTR.

Treatment and New Research Directions

Treatment consists of physical therapy to help clear the airways of the excess mucus and antibiotic therapy to combat respiratory infections, plus special diets and administration of supplemental pancreatic enzymes to maintain adequate nutrition. Despite this supportive treatment, most CF victims do not survive beyond their early 30s, with most dying from lung complications.

Having identified the genetic defect responsible for the majority of CF cases, investigators are hopeful of developing a means to correct or compensate for the defective gene. Another potential cure being studied is development of drugs that induce the mutated CFTR to be "finished off" and inserted in the plasma membrane. Furthermore, several promising new drug therapies, such as a mucus-thinning aerosol drug that can be inhaled, offer hope of reducing the number of lung infections and extending the life span of CF victims until a cure can be found.

The membrane proteins perform various specific membrane functions.

Different types of membrane proteins serve the following specialized functions:

1. Some transmembrane proteins form water-filled pathways, or **channels,** through the lipid bilayer (● Figure 3-3). Water-soluble substances small enough to enter a channel can pass through the membrane by this means without coming into direct contact with the hydrophobic lipid interior. Channels are highly selective. The small diameter of channels prevents particles larger than 0.8 nm (40 billionths of an inch) in diameter from entering. Only small ions can fit through channels. Furthermore, a given channel selectively admits particular ions.

For example, only sodium ions (Na^+) can pass through Na^+ channels, and only potassium ions (K^+) can pass through K^+ channels. This channel selectivity is due to specific arrangements of chemical groups on the interior surfaces of the channels. Some channels are *leak channels* that always permit passage of their selected ion. Others are *gated channels* that may be open or closed to their specific ion as a result of changes in channel shape in response to controlling mechanisms, described later. This is a good example of function depending on structural details. Cells vary in the number, kind, and activity of channels they possess. Some drugs target channels, for example Ca^{2+} *channel blockers* that are widely used in the management of high blood pressure and abnormal heart rhythms. More than 60 genetic mutations in channels have been linked to human diseases. To learn how a specific channel defect can lead to a devastating disease, see the accompanying boxed feature, ■ Concepts, Challenges, and Controversies.

2. Other proteins that span the membrane serve as **carrier,** or **transport, molecules;** they transfer across the membrane specific substances that are unable to cross on their own. The means by which carriers accomplish this transport will be described later. (Thus, channels and carrier molecules are both important in selective movement of substances between the ECF and ICF.) Each carrier can transport only a particular molecule (or ion) or group of closely related molecules. Cells of different types have different kinds of carriers. As a result, they vary as to which substances they can selectively transport across their membranes. For example, thyroid gland cells are the only cells to use iodine. Appropriately, only the plasma membranes of thyroid gland cells have carriers for iodine, so only these cells can transport iodine from the blood into the cell interior.

3. Other proteins, located on the inner membrane surface, serve as **docking-marker acceptors;** they bind lock-and-key fashion with the docking markers of secretory vesicles (see p. 28). Secretion is initiated as stimulatory signals trigger fusion of the secretory vesicle membrane with the inner surface of the plasma membrane through interactions between these matching labels. The secretory vesicle subsequently opens up and empties its contents to the outside by exocytosis.

4. Some proteins located on either the inner or the outer cell surface function as **membrane-bound enzymes** that control specific chemical reactions. Cells are specialized in the types of membrane-bound enzymes they have. For example, the outer plasma membrane surface of skeletal muscle cells contains an enzyme that destroys the chemical messenger responsible for triggering muscle contraction, thus allowing the muscle to relax.

5. Many proteins on the outer surface serve as **receptors,** sites that "recognize" and bind with specific molecules in the cell's environment. This binding initiates a series of membrane and intracellular events (to be described later) that alter the activity of the particular cell. In this way, chemical messengers in the blood, such as water-soluble hormones, influence only the specific cells that have receptors for a given messenger. Even though every cell is exposed to the same messenger via the blood circulation, a given messenger has no effect on cells lacking receptors for this specific messenger. To illustrate, the anterior pituitary gland secretes into the blood thyroid-stimulating hormone (TSH), which attaches only to the surface of thyroid gland cells to stimulate secretion of thyroid hormone. No other cells have receptors for TSH, so only thyroid cells are influenced by TSH despite its widespread distribution.

6. Still other proteins serve as **cell adhesion molecules (CAMs).** Many CAMs protrude from the outer membrane surface and form loops or hooks by which cells grip each other or grasp the connective tissue fibers between cells. For example, *cadherins,* a type of CAM found on the surface of adjacent cells, interlock in zipper fashion to help hold the cells within tissues and organs together. Other CAMs, such as the *integrins,* span the plasma membrane, where they serve as a structural link between the outer membrane surface and its extracellular surroundings and also connect the inner membrane surface to the intracellular cytoskeletal scaffolding. Besides mechanically linking the cell's external environment and intracellular components, integrins also relay regulatory signals through the plasma membrane in either direction. Some CAMs participate in signaling cells to grow or in signaling immune system cells to interact with the right kind of other cells in inflammatory responses and wound healing, among other things.

7. Finally, still other proteins on the outer membrane surface, especially in conjunction with carbohydrates (as glycoproteins), are important in the cells' ability to recognize "self" (that is, cells of the same type).

The membrane carbohydrates serve as self-identity markers.

The short carbohydrate chains on the outer membrane surface serve as self-identity markers that enable cells to identify and interact with one another in the following ways:

1. Different cell types have different markers. The unique combination of sugar chains projecting from the surface membrane proteins serves as the "trademark" of a particular cell type, enabling a cell to recognize others of its own kind. These carbohydrate chains play an important role in recognition of "self" and in cell-to-cell interactions. Cells can recognize other cells of the same type and join to form tissues. This is especially important during embryonic development. If cultures of embryonic cells of two different types, such as nerve cells and muscle cells, are mixed, the cells sort themselves into separate aggregates of nerve cells and muscle cells.

2. Carbohydrate-containing surface markers are also involved in tissue growth, which is normally held within certain limits of cell density. Cells do not "trespass" across the boundaries of neighboring tissues; that is, they do not overgrow their own territory. The exception is the uncontrolled spread of cancer cells, which have been shown to bear abnormal surface carbohydrate markers.

Cell-to-Cell Adhesions

In multicellular organisms such as humans, the plasma membrane not only serves as the outer boundary of all cells but also participates in cell-to-cell adhesions. These adhesions bind groups of cells together into tissues and package them further into organs. The life-sustaining activities of the body systems depend not only on the functions of the individual cells of which they are made but also on how these cells live and work together in tissue and organ communities.

At least in part, similar cells organize into appropriate groupings by means of the carbohydrate markers on the membrane surface. Once arranged, cells are held together by three different means: (1) CAMs, (2) the extracellular matrix, and (3) specialized cell junctions. You're already familiar with CAMs. We now examine the extracellular matrix, then specialized junctions.

The extracellular matrix serves as the biological "glue."

Tissues are not made up solely of cells, and many cells within a tissue are not in direct physical contact with neighboring cells. Instead, they are held together by a biological "glue" called the **extracellular matrix (ECM)**. The ECM is an intricate meshwork of fibrous proteins embedded in a watery, gel-like substance composed of complex carbohydrates. The watery gel, usually called the interstitial fluid (see p. 7), provides a pathway for diffusion of nutrients, wastes, and other water-soluble traffic between the blood and tissue cells. The three major types of protein fibers woven through the gel are *collagen, elastin,* and *fibronectin.*

1. **Collagen** forms flexible but nonelastic fibers or sheets that provide tensile strength (resistance to longitudinal stress). Collagen is the most abundant protein in the body, making up nearly half of the total body protein by weight.

Clinical Note In **scurvy**, a condition caused by vitamin C deficiency, collagen fibers are not properly formed. As a result, the tissues, especially those of the skin and blood vessels, become very fragile. This leads to bleeding in the skin and mucous membranes, which is especially noticeable in the gums.

2. **Elastin** is a rubbery protein fiber most plentiful in tissues that must easily stretch and then recoil after the stretching force is removed. It is found, for example, in the lungs, which stretch and recoil as air moves in and out.

3. **Fibronectin** promotes cell adhesion and holds cells in position. Reduced amounts of this protein have been found within certain types of cancerous tissue, possibly accounting for the fact that cancer cells do not adhere well to each other but tend to break loose and metastasize (spread elsewhere in the body).

The ECM is secreted by local cells present in the matrix. The relative amount of ECM compared to cells varies greatly among tissues. For example, the ECM is scant in epithelial tissue but is the predominant component of connective tissue. Most of this abundant matrix in connective tissue is secreted by **fibroblasts** ("fiber formers"). The exact composition of the ECM also varies for different tissues, thus providing distinct local environments for the various cell types in the body. In some tissues, the matrix becomes highly specialized to form such structures as cartilage or tendons or, on appropriate calcification, the hardened structures of bones and teeth.

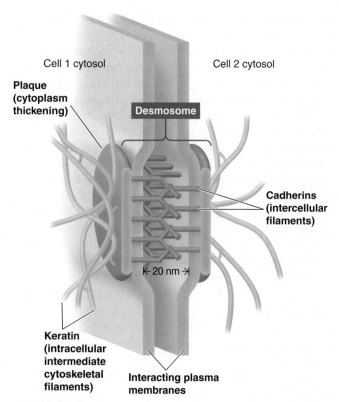

● **FIGURE 3-4 Desmosome.** Desmosomes are adhering junctions that spot-rivet cells, anchoring them together in tissues subject to considerable stretching.

Contrary to long-held belief, the ECM is not just a passive scaffolding for cellular attachment but also helps regulate the behavior and functions of the cells with which it interacts. Cells are able to function normally and indeed even to survive only when associated with their normal matrix components. The matrix is especially influential in cell growth and differentiation. In the body, only circulating blood cells are designed to survive and function without attaching to the ECM.

Some cells are directly linked by specialized cell junctions.

In tissues where the cells lie in close proximity to each other, CAMs provide some tissue cohesion as they "Velcro" adjacent cells to each other. In addition, some cells within given types of tissues are directly linked by one of three types of specialized cell junctions: (1) *desmosomes* (adhering junctions), (2) *tight junctions* (impermeable junctions), or (3) *gap junctions* (communicating junctions).

DESMOSOMES Desmosomes act like "spot rivets" that anchor together two adjacent but nontouching cells. A desmosome consists of two components: (1) a pair of dense, buttonlike cytoplasmic thickenings known as *plaques* located on the inner surface of each of the two adjacent cells; and (2) strong glycoprotein filaments containing cadherins (a type of CAM) that extend across the space between the two cells and attach to the plaque on both sides (● Figure 3-4). These intercellular filaments bind adjacent plasma membranes together so that they resist being pulled apart. Thus, desmosomes are adhering junctions. They are the strongest cell-to-cell connections.

Desmosomes are most abundant in tissues that are subject to considerable stretching, such as those found in the skin, the heart, and the uterus. In these tissues, functional groups of cells are riveted together by desmosomes. Furthermore, intermediate cytoskeletal filaments, such as tough keratin filaments in the skin (see p. 49), stretch across the interior of these cells and attach to the desmosome plaques located on opposite sides of the cell's inner surface. This arrangement forms a continuous network of strong fibers throughout the tissue, both through and between the cells, much like a continuous line of people firmly holding hands. This interlinking fibrous network provides tensile strength, reducing the chances of the tissue being torn when stretched.

TIGHT JUNCTIONS At **tight junctions,** adjacent cells bind firmly with each other at points of direct contact to seal off the passageway between the two cells. Tight junctions are found primarily in sheets of epithelial tissue, which cover the surface of the body and line its internal cavities. All epithelial sheets serve as highly selective barriers between two compartments having considerably different chemical compositions. For example, the epithelial sheet lining the digestive tract separates the food and potent digestive juices within the inner cavity (lumen) from the blood vessels on the other side. It is important that only completely digested food particles and not undigested food particles or digestive juices move across the epithelial sheet from the lumen to the blood. Accordingly, the lateral (side) edges of adjacent cells in the epithelial sheet are joined in a tight seal near their luminal border by "kiss" sites, at which *junctional proteins* on the outer surfaces of the two interacting plasma membranes fuse directly (● Figure 3-5). These tight junctions are impermeable and thus prevent materials from passing between the cells. Passage across the epithelial barrier, therefore, must take place *through* the cells, *not between* them. This traffic across the cell is regulated by channel and carrier proteins. If the cells were not joined by tight junctions, uncontrolled exchange of molecules could take place between the compartments by unpoliced traffic through the spaces between adjacent cells. Tight junctions thus prevent undesirable leaks within epithelial sheets.

GAP JUNCTIONS At a **gap junction,** as the name implies, a gap exists between adjacent cells, which are linked by small, connecting tunnels formed by connexons. A **connexon** is made up of six protein subunits arranged in a hollow tubelike structure. Two connexons, one from each of the plasma membranes of two adjacent cells, extend outward and join end-to-end to form a connecting tunnel between the two cells (● Figure 3-6). Gap junctions are communicating junctions. The small diameter of the tunnels permits small, water-soluble particles to pass between the connected cells but precludes passage of large molecules, such as vital intracellular proteins. Ions and small molecules can be directly exchanged between interacting cells through gap junctions without ever entering the ECF.

Gap junctions are especially abundant in cardiac muscle and smooth muscle. In these tissues, movement of ions through gap junctions transmits electrical activity throughout an entire

● **FIGURE 3-5 Tight junction.** Tight junctions are impermeable junctions that join the lateral edges of epithelial cells near their luminal borders, thus preventing movement of materials between the cells. Only regulated passage of materials can occur through these cells, which form highly selective barriers that separate two compartments of highly different chemical composition.

muscle mass. Because this electrical activity brings about contraction, the presence of gap junctions enables synchronized contraction of a whole muscle mass, such as the pumping chamber of the heart.

Cell 1 cytosol

Cell 2 cytosol

Connexon

Gap junction

Longitudinal section of connexon

Diameter of channel = 1.5 nm

PASSAGE OF IONS AND SMALL MOLECULES

NO PASSAGE OF LARGE MOLECULES

2–4 nm

Interacting plasma membranes

● **FIGURE 3-6 Gap junction.** Gap junctions are communicating junctions made up of connexons, which form tunnels that permit movement of charge-carrying ions and small molecules between two adjacent cells.

Gap junctions are also found in some nonmuscle tissues, where they permit unrestricted passage of small nutrient molecules between cells. For example, glucose, amino acids, and other nutrients pass through gap junctions to a developing egg cell from surrounding cells within the ovary, thus helping the egg stockpile these essential nutrients.

Gap junctions also serve as avenues for the direct transfer of small signaling molecules from one cell to the next. Such transfer permits cells connected by gap junctions to communicate with each other directly. This communication provides one possible mechanism by which cooperative cell activity may be coordinated. In the next chapter we will examine other means by which cells "talk to each other."

We now turn our attention to the topic of membrane transport, focusing on how the plasma membrane selectively controls what enters and exits the cell.

Overview of Membrane Transport

Anything that passes between a cell and the surrounding extracellular fluid must be able to penetrate the plasma membrane. If a substance can cross the membrane, the membrane is said to be **permeable** to that substance; if a substance cannot pass, the membrane is **impermeable** to it. The plasma membrane is **selectively permeable** in that it permits some particles to pass through while excluding others.

Two properties of particles influence whether they can permeate the plasma membrane without assistance: (1) the relative solubility of the particle in lipid and (2) the size of the particle. Highly lipid-soluble particles can dissolve in the lipid bilayer and pass through the membrane. Uncharged or nonpolar molecules (such as O_2, CO_2, and fatty acids) are highly lipid soluble and readily permeate the membrane. Charged particles (ions such as Na^+ and K^+) and polar molecules (such as glucose and proteins) have low lipid solubility but are very soluble in water. The lipid bilayer serves as an impermeable barrier to particles poorly soluble in lipid. For water-soluble (and thus lipid-insoluble) ions less than 0.8 nm in diameter, the protein channels serve as an alternative route for passage across the membrane. Only ions for which specific channels are available and open can permeate the membrane.

Particles that have low lipid solubility and are too large for channels cannot permeate the membrane on their own. Yet some of these particles, for example glucose, must cross the membrane for the cell to survive and function. (Most cells use glucose as their fuel of choice to produce ATP.) Cells have several means of assisted transport to move particles across the membrane that must enter or leave the cell but cannot do so unaided, as you will learn shortly.

Even if a particle can permeate the membrane by virtue of its lipid solubility or its ability to fit through a channel, some force is needed to move it across the membrane. Two general types of forces accomplish transport of substances across the membrane: (1) **passive forces,** which do not require the cell to expend energy to produce movement, and (2) **active forces,** which do require the cell to expend energy (ATP) in transporting a substance across the membrane.

We now examine the various methods of membrane transport, indicating whether each is an unassisted or assisted means of transport and whether each is a passive- or active-transport mechanism.

Unassisted Membrane Transport

Molecules and ions that can penetrate the plasma membrane on their own are passively driven across the membrane by two forces: diffusion down a concentration gradient and/or movement along an electrical gradient. We will first examine diffusion down a concentration gradient.

Particles that can permeate the membrane diffuse passively down their concentration gradient.

All molecules and ions are in continuous random motion at temperatures above absolute zero as a result of thermal (heat) energy. This motion is most evident in liquids and gases, where the individual molecules (or ions) have more room to move

⬤ **FIGURE 3-7 Diffusion.** (a) Diffusion down a concentration gradient. (b) Dynamic equilibrium,
with no net diffusion occurring.

before colliding with another molecule. Each molecule moves separately and randomly in any direction. As a consequence of this haphazard movement, the molecules frequently collide, bouncing off each other in different directions like billiard balls striking each other.

Solutions are homogeneous mixtures containing a relatively large amount of one substance called the **solvent** (the dissolving medium, which in the body is water) and smaller amounts of one or more dissolved substances called **solutes.** The **concentration** of a solution refers to the amount of solute dissolved in a specific amount of solution. The greater the concentration of solute molecules (or ions), the greater the likelihood of collisions. Consequently, molecules within a particular space tend to become evenly distributed over time. Such uniform spreading out of molecules due to their random intermingling is known as **simple diffusion** (*diffusere* means "to spread out"). The term *diffusion* without a qualifier implies simple diffusion, as distinct from *facilitated diffusion*, which is diffusion facilitated by a carrier protein in the plasma membrane. You will learn more about facilitated diffusion later.

To illustrate simple diffusion, in ⬤ Figure 3-7a, the concentration of the solute in a solution differs between area A and area B. Such a difference in concentration between two adjacent areas is called a **concentration gradient** (or **chemical gradient**). Random molecular collisions will occur more frequently in area A because of its greater concentration of solute molecules. For this reason, more molecules will bounce from area A into area B than in the opposite direction. In both areas, individual molecules will move randomly and in all directions, but the net movement of molecules by diffusion will be from the area of higher concentration to the area of lower concentration.

The term **net diffusion** refers to the difference between two opposing movements. If 10 molecules move from area A to area B while 2 molecules simultaneously move from B to A, the net diffusion is 8 molecules moving from A to B. Molecules will spread in this way until the substance is uniformly distributed between the two areas and a concentration gradient no longer exists (⬤ Figure 3-7b). At this point, even though movement is still taking place, no net diffusion is occurring because the opposing movements exactly counterbalance each other, that is, are in **dynamic equilibrium** (*dynamic* in reference to the continuous movement, *equilibrium* in reference to the exact balance between opposing forces). Movement of molecules from area A to area B will be exactly matched by movement of molecules from B to A.

What happens if a plasma membrane separates different concentrations of a substance? If the substance can permeate the membrane, net diffusion of the substance will occur through the membrane down its concentration gradient from the area of high concentration to the area of low concentration until the concentration gradient is abolished, unless there's some opposing force (⬤ Figure 3-8a). No energy is required for this movement, so it is a *passive* mechanism of membrane transport. The process of diffusion is crucial to the survival of every cell and plays an important role in many specialized homeostatic activities. As an example, O_2 is transferred across the lung membrane by diffusion. The blood carried to the lungs is low in O_2, having given up O_2 to the body tissues for cell metabolism. The air in the lungs, in contrast, is high in O_2 because it is continuously exchanged with fresh air during breathing. Because of this concentration gradient, net diffusion of O_2 occurs from the lungs into the blood as blood flows through the lungs. Thus, as blood leaves the lungs for delivery to the tissues, it is high in O_2.

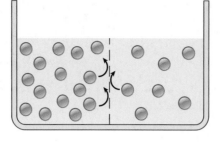

If a substance can permeate the membrane:

Membrane

If the membrane is impermeable to a substance:

(a) Diffusion occurs

(b) No diffusion occurs

KEY

= Penetrating solute

= Nonpenetrating solute

● **FIGURE 3-8 Diffusion through a membrane.** (a) Net diffusion of a penetrating solute across the membrane down a concentration gradient. (b) No diffusion of a non-penetrating solute through the membrane despite the presence of a concentration gradient.

Of course, if the membrane is impermeable to the substance, no diffusion can take place across the membrane, even though a concentration gradient may exist (● Figure 3-8b). For example, because the plasma membrane is impermeable to the vital intracellular proteins, they are unable to escape from the cell, even though they are in much greater concentration in the ICF than in the ECF.

FICK'S LAW OF DIFFUSION Several factors in addition to the concentration gradient influence the rate of net diffusion across a membrane. The effects of these factors collectively make up **Fick's law of diffusion** (▲ Table 3-1):

1. *The magnitude (or steepness) of the concentration gradient.* If a substance can permeate the membrane, its rate of simple diffusion is always directly proportional to its concentration gradient; that is, the greater the difference in concentration, the faster is the rate of net diffusion (see ● Figure 3-15, p. 69). For example, during exercise the working muscles produce CO_2 more rapidly than usual because they are burning additional fuel to produce the extra ATP they need to power the stepped-up, energy-demanding contractile activity. The increase in CO_2 level in the muscles creates a greater-than-normal difference in CO_2 between the muscles and the blood supplying the muscles. Because of this steeper gradient, more CO_2 than usual enters the blood. When this blood with its increased CO_2 load reaches the lungs, a greater-than-normal CO_2 gradient exists between the blood and the air sacs in the lungs. Accordingly, more CO_2 than normal diffuses from the blood into the air sacs. This extra CO_2 is subsequently breathed out to the environment. Thus, any additional CO_2 produced by exercising muscles is eliminated from the body through the lungs as a result of the increase in CO_2 concentration gradient.

2. *The surface area of the membrane across which diffusion is taking place.* The larger the surface area available, the greater is the rate of diffusion it can accommodate. Various strategies are used throughout the body for increasing the membrane surface area across which diffusion and other types of transport take place. For example, absorption of nutrients in the small intestine is enhanced by the presence of microvilli, which greatly increase the available absorptive surface in contact with the nutrient-rich contents of the small intestine lumen (see p. 49). Conversely, abnormal loss of membrane surface area decreases the rate of net diffusion. For example, in *emphysema*, O_2 and CO_2 exchange between the air and blood in the lungs is reduced because the walls of the air sacs break down, resulting in less surface area available for diffusion of these gases.

3. *The lipid solubility of the substance.* The greater the lipid solubility of a substance, the more rapidly the substance can diffuse through the membrane's lipid bilayer down its concentration gradient.

4. *The molecular weight of the substance.* Heavier molecules do not bounce as far on collision as lighter molecules such as O_2 and CO_2 do. Consequently, O_2 and CO_2 diffuse rapidly, permit-

▲ **TABLE 3-1**   **Factors Influencing the Rate of Net Diffusion of a Substance across a Membrane (Fick's Law of Diffusion)**

Factor	Effect on Rate of Net Diffusion
↑ Concentration gradient of substance (ΔC)	↑
↑ Surface area of membrane (A)	↑
↑ Lipid solubility (β)	↑
↑ Molecular weight of substance (MW)	↓
↑ Distance (thickness) (ΔX)	↓

Modified Fick's equation:

$$\text{Net rate of diffusion } (Q) = \frac{\Delta C \cdot A \cdot \beta}{\sqrt{MW} \cdot \Delta X}$$

$$\left[\text{diffusion coefficient } (D) \propto \frac{1}{\sqrt{MW}} \right]$$

$$\left[\text{permeability } (P) = \frac{D\beta}{\Delta X} \right]$$

Restated $Q \propto \Delta C \cdot A \cdot P$

ting rapid exchanges of these gases across the lung membranes. As molecular weight increases, the rate of diffusion decreases.

5. *The distance through which diffusion must take place.* The greater the distance, the slower is the rate of diffusion. Accordingly, membranes across which diffusing particles must travel are normally relatively thin, such as the membranes separating the air and blood in the lungs. Thickening of this air–blood interface (as in *pneumonia,* for example) slows down exchange of O_2 and CO_2. Furthermore, diffusion is efficient only for short distances between cells and their surroundings. It becomes an inappropriately slow process for distances of more than a few millimeters. To illustrate, it would take months or even years for O_2 to diffuse from the surface of the body to the cells in the interior. Instead, the circulatory system provides a network of tiny vessels that deliver and pick up materials at every "block" of a few cells, with diffusion accomplishing short local exchanges between the blood and surrounding cells.

Ions that can permeate the membrane also move passively along their electrical gradient.

In addition to their concentration gradient, movement of ions is also affected by their electrical charge. Ions with like charges (those with the same kind of charge) repel each other, and ions with opposite charges attract each other. If a relative difference in charge exists between two adjacent areas, positively charged ions (*cations*) tend to move toward the more negatively charged area, and negatively charged ions (*anions*) tend to move toward the more positively charged area. A difference in charge between two adjacent areas thus produces an **electrical gradient** that promotes movement of ions toward the area of opposite charge. Because a cell does not have to expend energy for ions to move into or out of it along an electrical gradient, this method of membrane transport is *passive.* When an electrical gradient exists between the ICF and ECF, only ions that can permeate the plasma membrane can move along this gradient.

Both an electrical and a concentration (chemical) gradient may be acting on a particular ion at the same time. The net effect of simultaneous electrical and concentration gradients on this ion is called an **electrochemical gradient.** Later in this chapter you will learn how electrochemical gradients contribute to the electrical properties of the plasma membrane.

Osmosis is the net diffusion of water down its own concentration gradient.

Water molecules can readily permeate the plasma membrane. Even though water molecules are strongly polar, they are small enough to slip through momentary spaces created between the phospholipid molecules' tails as they sway and move within the lipid bilayer. However, this type of water movement across the membrane is relatively slow. In many cell types, membrane proteins form **aquaporins,** which are channels specific for the passage of water (*aqua* means "water"). This avenue greatly increases membrane permeability to water. Different cell types vary in their density of aquaporins and thus in their water permeability. About a billion water molecules can pass in single file through an aquaporin channel in one second. The driving force

100% water concentration
0% solute concentration

(a) Pure water

90% water concentration
10% solute concentration

(b) Solution

KEY

= Water molecule

= Solute molecule

● **FIGURE 3-9** Relationship between solute and water concentration in a solution.

for net movement of water across the membrane is the same as for any other diffusing molecule, namely, its concentration gradient. Usually the term *concentration* refers to the density of the solute in a given volume of water. It is important to recognize, however, that adding a solute to pure water in essence decreases the water concentration. In general, one molecule of a solute displaces one molecule of water.

Compare the water and solute concentrations in the two containers in ● Figure 3-9. The container in part (a) is full of pure water, so the water concentration is 100% and the solute concentration is 0%. In part (b), solute has replaced 10% of the water molecules. The water concentration is now 90%, and the solute concentration is 10%—a lower water concentration and a higher solute concentration than in part (a). Note that as the solute concentration increases, the water concentration decreases correspondingly. We will now examine what happens if pure water is separated from a solution by a membrane permeable to water but not to the solute.

MOVEMENT OF WATER WHEN A SELECTIVELY PERMEABLE MEMBRANE SEPARATES PURE WATER FROM A SOLUTION OF A NONPENETRATING SOLUTE If as in ● Figure 3-10 pure water (side 1) and a solution containing a nonpenetrating solute (side 2) are separated by a selectively permeable membrane that permits passage of water but not the solute, water will move passively down its own concentration gradient from the area of higher water concentration (lower solute concentration) to the area of lower water concentration (higher solute concentration). This net diffusion of water down its concentration gradient through a selectively permeable membrane is known as **osmosis.** Because solutions are always referred to in terms of concentration of solute, *water moves by osmosis to the area of higher solute concentration.* Despite the impression that the solutes are "pulling," or attracting, water, osmosis is nothing more than diffusion of water down its own concentration gradient across the membrane.

Osmosis occurs from side 1 to side 2, but the concentrations between the two compartments can never become equal. No

Membrane (permeable to H_2O but impermeable to solute)

Side 1   Side 2

H_2O
Osmosis
Solute

Pure water   Lower H_2O concentration, higher solute concentration

H_2O moves from side 1 to side 2 down its concentration gradient = osmosis

Solute unable to move from side 2 to side 1 down its concentration gradient

Side 1   Side 2

Hydrostatic (fluid) pressure difference
Osmosis
Hydrostatic pressure

Original level of solutions

- **Water concentrations not equal**
- **Solute concentrations not equal**
- **Tendency for water to diffuse by osmosis into side 2 is exactly balanced by opposing tendency for hydrostatic pressure difference to push water into side 1**
- **Osmosis ceases; dynamic equilibrium exists**

KEY

= Water molecule

= Nonpenetrating solute

● **FIGURE 3-10 Osmosis when pure water is separated from a solution containing a nonpenetrating solute.**

matter how dilute side 2 becomes because of water diffusing into it, it can never become pure water, nor can side 1 ever acquire any solute. Because equilibrium is impossible to achieve, does net diffusion of water (osmosis) continue until all the water has left side 1? No. As the volume expands in compartment 2, a difference in hydrostatic pressure between the two compartments is created, and it opposes osmosis. **Hydrostatic (fluid) pressure** is the pressure exerted by a standing, or stationary, fluid on an object—in this case the membrane (*hydro* means "fluid"; *static* means "standing"). The hydrostatic pressure exerted by the larger volume of fluid on side 2 is greater than the hydrostatic pressure exerted on side 1. This difference in hydrostatic pressure tends to push fluid from side 2 to side 1.

The **osmotic pressure** of a solution is a measure of the tendency for osmotic flow of water into that solution because of its relative concentration of nonpenetrating solutes and water. Net movement of water by osmosis continues until the opposing hydrostatic pressure exactly counterbalances the osmotic pressure. You can think of the osmotic pressure of a solution as a "pulling pressure" that tends to pull water into the solution whereas the hydrostatic pressure of a solution is a "pushing pressure" that tends to push water out of the solution. The magnitude of the osmotic pressure is equal to the magnitude of the opposing hydrostatic pressure necessary to completely stop osmosis. The greater the concentration of nonpenetrating solute → the lower the concentration of water → the greater the drive for water to move by osmosis from pure water into the solution → the greater the opposing pressure required to stop the osmotic flow → the greater the osmotic pressure of the solution. Therefore, a solution with a high concentration of nonpenetrating solute exerts greater osmotic pressure than a solution with a lower concentration of nonpenetrating solute does.

Osmotic pressure is an indirect measure of solute concentration, expressed in units of pressure. A more direct means of expressing solute concentration is the **osmolarity** of a solution, which is a measure of its total solute concentration given in terms of the *number* of particles (molecules or ions). Osmolarity is expressed in *osmoles/liter* (or *Osm*), the number of moles of solute particles in 1 liter of solution (see p. A-9). Because glucose remains as an intact molecule when in solution, 1 mole of glucose equals 1 osmole, that is, 1 mole of solute particles. By contrast, because a molecule of NaCl dissociates (separates) into 2 ions—Na^+ and Cl^-—when in solution, 1 mole of NaCl equals 2 osmoles—1 mole of Na^+ and 1 mole of Cl^-, or 2 moles of solute particles. The osmolarity of body fluids is typically expressed in *milliosmoles/liter* (*mOsm*) (1/1000 of an osmole) because the solutes in body fluids are too dilute to conveniently use the osmole unit. Because osmolarity depends on the number, not the nature, of particles, any mixture of particles can contribute to the osmolarity of a solution. The normal osmolarity of body fluids is 300 mOsm.

Thus far in our discussion of osmosis, we have considered movement of water when pure water is separated from a solution by a membrane permeable to water but not to nonpenetrating solutes. However in the body, the plasma membrane separates the ICF and ECF, both of which contain solutes, some that can and others that cannot penetrate the membrane. Let us compare the results of water movement when solutions of differing osmolarities are separated by a selectively permeable membrane that permits movement of water and some but not all solutes.

MOVEMENT OF WATER AND SOLUTE WHEN A MEMBRANE SEPARATES UNEQUAL SOLUTIONS OF A PENETRATING SOLUTE Assume that solutions of *unequal* concentration of *penetrating* solute (differing osmolarities) are separated by a membrane that is permeable to both water and the solute (● Figure 3-11). In this situation, the solute moves down its own concentration gradient in the opposite direction of the net water movement. The movement continues until both solute and water are evenly distributed across the membrane. With all concentration gradients gone, net movement ceases. The final volume of the com-

Membrane (permeable to both H₂O and solute)

Side 1   Side 2

H_2O

Solute

Higher H₂O concentration,   Lower H₂O concentration,
lower solute concentration   higher solute concentration

**H₂O moves from side 1 to side 2
down its concentration gradient**

**Solute moves from side 2 to side 1
down its concentration gradient**

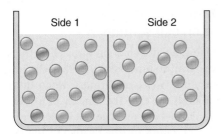

Side 1   Side 2

- **Water concentrations equal**
- **Solute concentrations equal**
- **No further net diffusion**
- **Dynamic equilibrium exists**
- **No change in volume of
  the two sides**

KEY

 = Water molecule

= Penetrating solute

● **FIGURE 3-11 Movement of water and a penetrating solute unequally distributed across a membrane.**

Membrane (permeable to H₂O
but impermeable to solute)

Side 1   Side 2

H_2O
Osmosis

Solute

Higher H₂O concentration,   Lower H₂O concentration,
lower solute concentration   higher solute concentration

**H₂O moves from side 1 to side 2
down its concentration gradient**

**Solute unable to move from side 2 to
side 1 down its concentration gradient**

New location   Original location
of membrane   of membrane

Side 1   Side 2

- **Water concentrations equal**
- **Solute concentrations equal**
- **Osmosis ceases; dynamic
  equilibrium exists**
- **Volume of side 1 has decreased
  and volume of side 2 has increased**

KEY

= Water molecule

= Nonpenetrating solute

● **FIGURE 3-12 Osmosis in the presence of an unequally distributed nonpenetrating solute.**

partments when dynamic equilibrium is achieved is the same as at the onset. Water and solute molecules merely exchange places between the two compartments until their distributions are equalized; that is, an equal number of water molecules move from side 1 to side 2 as solute molecules move from side 2 to side 1. Therefore solutes that can penetrate the plasma membrane do not contribute to osmotic differences between the ICF and ECF and do not affect cell volume (although before equilibrium is achieved, transient changes in volume may occur as a result of differing rates of diffusion of water and the solute across the membrane.)

MOVEMENT OF WATER WHEN A MEMBRANE SEPARATES EQUAL AND UNEQUAL SOLUTIONS OF A NONPENETRATING SOLUTE If solutions of *equal* concentration of *nonpenetrating* solute (the same osmolarities) are separated by a membrane that is perme-

able to water but impermeable to the solute, no concentration differences will exist so no net movement of water will occur across the membrane. Of course, the solute will not move because the membrane is impermeable to it and no concentration gradient exists for it. This is the usual situation in body fluids. Body cells normally do not experience any net gain (swelling) or loss (shrinking) of volume, because the concentration of nonpenetrating solutes in the ECF is normally carefully regulated (primarily by the kidneys) so that the ECF osmolarity is the same as the osmolarity within the cells. Intracellular osmolarity is normally 300 mOsm, and all intracellular solutes are assumed to be nonpenetrating.

Now assume that solutions of *unequal* concentration of *nonpenetrating* solute (differing osmolarities) are separated by a membrane that is permeable to water but impermeable to the solute (● Figure 3-12). Osmotic movement of water across the

Normal cell volume
Intracellular fluid: 300 mOsm
nonpenetrating solutes

300 mOsm
nonpenetrating solutes

200 mOsm
nonpenetrating solutes

H_2O

400 mOsm
nonpenetrating solutes

H_2O

M. Sheetz, R. Painter, & S. Singer. *Journal of Cell Biology.* 70:493, 1976. By permission of Rockefeller University Press.

No net movement of water; no change in cell volume.

Water diffuses into cells; cells swell.

Water diffuses out of cells; cells shrink.

(a) Isotonic conditions

(b) Hypotonic conditions

(c) Hypertonic conditions

● **FIGURE 3-13 Tonicity and osmotic water movement.**

membrane is driven by the difference in osmotic pressure of the two solutions. At first the concentration gradients are identical to those in ● Figure 3-11. Net diffusion of water takes place from side 1 to side 2, but the solute cannot cross the membrane down its concentration gradient. As a result of water movement alone, the volume of side 2 increases while the volume of side 1 correspondingly decreases. Loss of water from side 1 increases the solute concentration on side 1, whereas addition of water to side 2 reduces the solute concentration on that side. If the membrane is free to move so that side 2 can expand without an opposing hydrostatic pressure developing, eventually the concentrations of water and solute on the two sides of the membrane become equal, and net diffusion of water ceases. This situation is similar to what happens across plasma membranes in the body. Within the slight range of changes in ECF osmolarity that occur physiologically, if water moves by osmosis into the cells, their plasma membranes normally accommodate the increase in cell volume with no significant change in hydrostatic pressure inside the cells. Likewise, in the reverse situation if water

moves by osmosis out of the cells, the ECF compartment expands without a change in its hydrostatic pressure. Therefore, osmosis is the major force responsible for the net movement of water into or out of cells, without having to take hydrostatic pressure into consideration. At the end point when osmosis ceases, the volume has increased on the side that originally had the higher solute concentration, and the volume has decreased on the side with the lower solute concentration. Therefore, osmotic movement of water across the plasma membrane always results in a change in cell volume, and cells, especially brain cells, do not function properly when they swell or shrink.

Tonicity refers to the effect the concentration of nonpenetrating solutes in a solution has on cell volume.

The **tonicity** of a solution is the effect the solution has on cell volume—whether the cell remains the same size, swells, or shrinks—when the solution surrounds the cell. The tonicity of a solution has no units and is a reflection of its concentration of nonpenetrating solutes relative to the cell's concentration of nonpenetrating solutes. (By contrast, the osmolarity of a solution is a measure of its total concentration of both penetrating and nonpenetrating solutes expressed in units of osmoles/liter.) The easiest way to demonstrate this phenomenon is to place red blood cells in solutions with varying concentrations of a nonpenetrating solute (● Figure 3-13).

Normally, the plasma in which red blood cells are suspended has the same osmotic activity as the fluid inside these cells, so the cells maintain a constant volume. An **isotonic solution** (*iso* means "equal") has the same concentration of nonpenetrating solutes as normal body cells do. When a cell is bathed in an isotonic solution, no water enters or leaves the cell by osmosis, so cell volume remains constant. For this reason, the ECF is normally maintained isotonic so that no net diffusion of water occurs across the plasma membranes of body cells.

If red blood cells are placed in a dilute or **hypotonic solution** (*hypo* means "below"), a solution with a below-normal concentration of nonpenetrating solutes (and therefore a higher concentration of water), water enters the cells by osmosis. Net gain of water by the cells causes them to swell, perhaps to the point of rupturing or *lysing*. If, in contrast, red blood cells are placed in a concentrated, or **hypertonic, solution** (*hyper* means "above"), a solution with an above-normal concentration of nonpenetrating solutes (and therefore a lower concentration of water), the cells shrink as they lose water by osmosis. When a red blood cell decreases in volume, its surface area does not decrease correspondingly, so the cell assumes a *crenated*, or spiky, shape (● Figure 3-13c). Because cells change volume when surrounded by fluid that is not isotonic, it is crucial that the concentration of nonpenetrating solutes in the ECF quickly be restored to normal should the ECF become hypotonic (as with ingesting too much water) or hypertonic (as with losing too much water through severe diarrhea). (See pp. 565–569 for further details about the important homeostatic mechanisms that maintain the normal concentration of nonpenetrating solutes in the ECF.) For the same reason, fluids injected intravenously should be isotonic, to prevent unwanted movement of water into or out of the cells. For example, isotonic saline (0.9% NaCl solution) is used as a vehicle for delivering drugs intravenously or for expanding plasma volume without impacting the cells. (Sometimes hypotonic or hypertonic fluids are injected therapeutically to correct osmotic imbalances.)

Assisted Membrane Transport

All the kinds of transport we have discussed thus far—diffusion down concentration gradients, movement along electrical gradients, and osmosis—produce net movement of solutes capable of permeating the plasma membrane by virtue of their lipid solubility (nonpolar molecules of any size) or their ability to fit through channels (selected ions and water). Poorly lipid-soluble polar molecules that are too big for channels, such as proteins, glucose, and amino acids, cannot cross the plasma membrane on their own no matter what forces are acting on them. This impermeability ensures that large, polar intracellular proteins stay in the cell where they belong and can carry out their life-sustaining functions, for example, serving as metabolic enzymes.

However, because poorly lipid-soluble molecules cannot cross the plasma membrane on their own, the cell must provide mechanisms for deliberately transporting these types of molecules into or out of the cell as needed. For example, the cell must usher in essential nutrients, such as glucose for energy and amino acids for the synthesis of proteins, and transport out metabolic wastes and secretory products, such as water-soluble protein hormones. Furthermore, passive diffusion alone cannot always account for the movement of ions. Some ions move through the membrane passively in one direction and actively in the other direction. Cells use two different mechanisms to accomplish these selective transport processes: *carrier-mediated transport* for transfer of small to moderate-sized water-soluble substances across the membrane and *vesicular transport* for movement of large molecules and multimolecular particles between the ECF and ICF. We will examine each of these methods of assisted membrane transport in turn.

Carrier-mediated transport is accomplished by a membrane carrier changing its shape.

A carrier protein spans the thickness of the plasma membrane and can change its conformation (shape) so that specific binding sites within the carrier are alternately exposed to the ECF and the ICF. ● Figure 3-14 shows how this **carrier-mediated**

transport works. Step **1** shows the carrier open to the ECF. The molecule to be transported attaches to a carrier's binding site on one side of the membrane, in this case on the ECF side (step **2**). Then as a result of the carrier changing shape, the same site is now exposed to the other side of the membrane (step **3**). Having been moved in this way from one side of the membrane to the other, the bound molecule detaches from the carrier (step **4**). Next the carrier reverts to its original shape (back to step **1**).

Both channels and carriers are transmembrane proteins that serve as selective avenues for the movement of water-soluble substances across the membrane, but there are notable differences between them. (1) Only ions fit through the narrow channels, whereas small polar molecules such as glucose and amino acids are transported across the membrane by carriers. (2) Channels can be open or closed, but carriers are always "open for business" (although the number and kinds of carriers in the plasma membrane can be regulated). (3) Movement through channels is considerably faster than carrier-mediated transport is. When open for traffic, channels are open at both sides of the membrane at the same time, permitting continuous, rapid movement of ions between the ECF and ICF through these nonstop passageways. By contrast, carriers are never open to both the ECF and ICF simultaneously. They must change shape to alternately pick up passenger molecules on one side of the membrane and then drop them off on the other side, a time-consuming process. Whereas a carrier may move up to 5000 particles per second across the membrane, 5 million ions may pass through a channel in a second.

Carrier-mediated transport systems display three important characteristics that determine the kind and amount of material that can be transferred across the membrane: specificity, saturation, and competition.

1. **Specificity.** Each carrier protein is specialized to transport a specific substance or, at most, a few closely related chemical compounds. For example, amino acids cannot bind to glucose carriers, although several similar amino acids may be able to use the same carrier. Cells vary in the types of carriers they have, thus permitting transport selectivity among cells.

 A number of inherited diseases involve defects in transport systems for a particular substance. **Cysteinuria** (cysteine in the urine) is such a disease involving defective cysteine carriers in the kidney membranes. This transport system normally removes cysteine from the fluid destined to become urine and returns this essential amino acid to the blood. When this carrier malfunctions, large quantities of cysteine remain in the urine, where it is relatively insoluble and tends to precipitate. This is one cause of urinary stones.

2. **Saturation.** A limited number of carrier binding sites are available within a particular plasma membrane for a specific substance. Therefore, there is a limit to the amount of a substance a carrier can transport across the membrane in a given time. This limit is known as the **transport maximum (T_m)**. Until the T_m is reached, the number of carrier binding sites occupied by a substance and, accordingly, the substance's rate of transport across the membrane are directly related to its concentration. The more of a substance available for transport, the

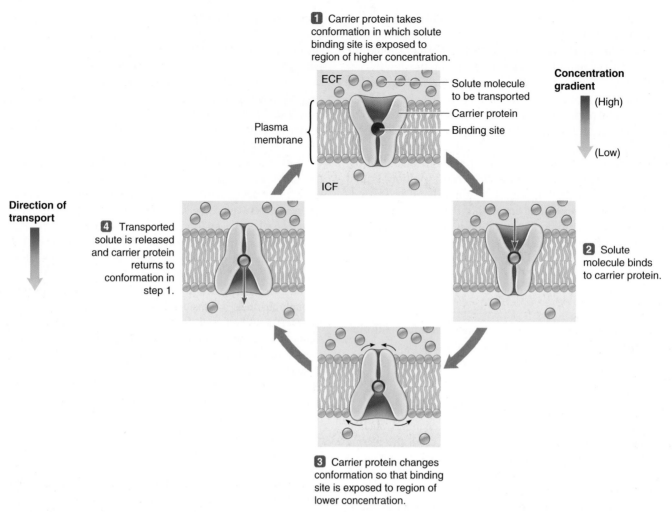

1 Carrier protein takes conformation in which solute binding site is exposed to region of higher concentration.

ECF

Solute molecule to be transported

Carrier protein

Binding site

Plasma membrane

ICF

Concentration gradient

(High)

(Low)

Direction of transport

4 Transported solute is released and carrier protein returns to conformation in step 1.

2 Solute molecule binds to carrier protein.

3 Carrier protein changes conformation so that binding site is exposed to region of lower concentration.

● **FIGURE 3-14 Model for facilitated diffusion, a passive form of carrier-mediated transport.**

more will be transported. When the T_m is reached, the carrier is saturated (all binding sites are occupied) and the rate of the substance's transport across the membrane is maximal. Further increases in the substance's concentration are no longer accompanied by corresponding increases in the rate of transport (● Figure 3-15).

As an analogy, consider a ferry boat that can maximally carry 100 people across a river during one trip in an hour. If 25 people are on hand to board the ferry, 25 will be transported that hour. Doubling the number of people on hand to 50 will double the rate of transport to 50 people that hour. Such a direct relationship will exist between the number of people waiting to board (the concentration) and the rate of transport until the ferry is fully occupied (its T_m is reached). Even if 150 people are waiting to board, only 100 can be transported per hour.

Saturation of carriers is a critical rate-limiting factor in the transport of selected substances across the kidney membranes during urine formation and across the intestinal membranes during absorption of digested foods. Furthermore, it is sometimes possible to regulate the rate of carrier-mediated transport

by varying the affinity (attraction) of the binding site for its passenger or by varying the number of binding sites. For example, the hormone insulin greatly increases the carrier-mediated transport of glucose into most cells of the body by promoting an increase in the number of glucose carriers in the cell's plasma membrane. Deficient insulin action (*diabetes mellitus*) drastically impairs the body's ability to take up and use glucose as the primary energy source.

3. **Competition.** Several closely related compounds may compete for a ride across the membrane on the same carrier. If a given binding site can be occupied by more than one type of molecule, the rate of transport of each substance is less when both molecules are present than when either is present by itself. To illustrate, assume the ferry has 100 seats (binding sites) that can be occupied by either men or women. If only men are waiting to board, up to 100 men can be transported during each trip; the same holds true if only women are waiting to board. If both men and women are waiting to board, however, they will compete for the available seats. Fifty of each might make the

trip, although the total number of people transported will still be the same, 100 people. In other words, when a carrier can transport two closely related substances, such as the amino acids glycine and alanine, the presence of both diminishes the rate of transfer for either.

Facilitated diffusion is passive carrier-mediated transport.

Carrier-mediated transport takes two forms, depending on whether energy must be supplied to complete the process: *facilitated diffusion* (not requiring energy) and *active transport* (requiring energy). **Facilitated diffusion** uses a carrier to facilitate (assist) the transfer of a particular substance across the membrane "downhill" from high to low concentration. This process is *passive* and does not require energy because movement occurs naturally down a concentration gradient. **Active transport**, however, requires the carrier to expend energy to transfer its passenger "uphill" against a concentration gradient, from an area of lower concentration to an area of higher concentration. An analogous situation is a car on a hill. To move the car downhill requires no energy; it will coast from the top down. Driving the car uphill, however, requires the use of energy (generated by the burning of gasoline).

The most notable example of facilitated diffusion is the transport of glucose into cells. Glucose is in higher concentration in the blood than in the tissues. Fresh supplies of this nutrient are regularly added to the blood by eating and by using reserve energy stores in the body. Simultaneously, the cells metabolize glucose almost as rapidly as it enters from the blood. As a result, a continuous gradient exists for net diffusion of glucose into the cells. However, glucose cannot cross cell membranes on its own because it is not lipid soluble and is too large to fit through a channel. Without glucose carrier molecules (called *glucose transporters,* or *GLUT*; see p. 716) to facilitate membrane transport of glucose, the cells would be deprived of their preferred source of fuel. (The boxed feature on p. 70, ■ A Closer Look at Exercise Physiology, describes the effect of exercise on glucose carriers in skeletal muscle cells.)

The binding sites on facilitated diffusion carriers can bind with their passenger molecules when exposed on either side of the membrane. As a result of thermal energy, these carriers undergo spontaneous changes in shape, alternately exposing their binding sites to the ECF or ICF. After picking up the passenger on one side, when the carrier changes its conformation, it drops off the passenger on the opposite side of the membrane. Because passengers are more likely to bind with the carrier on the high-concentration side than on the low-concentration side, the net movement always proceeds down the concentration gradient from higher to lower concentration (see ● Figure 3-14). As is characteristic of all types of mediated transport, the rate of facilitated diffusion is limited by saturation of the carrier binding sites, unlike the rate of simple diffusion, which

● **FIGURE 3-15 Comparison of carrier-mediated transport and simple diffusion down a concentration gradient.** With simple diffusion of a molecule down its concentration gradient, the rate of transport of the molecule into the cell is directly proportional to the extracellular concentration of the molecule. With carrier-mediated transport of a molecule down its concentration gradient, the rate of transport of the molecule into the cell is directly proportional to the extracellular concentration of the molecule until the carrier is saturated, at which time the rate of transport reaches the transport maximum (T_m). After T_m is reached, the rate of transport levels off despite further increases in the ECF concentration of the molecule.

is always directly proportional to the concentration gradient (● Figure 3-15).

Active transport is carrier-mediated transport that uses energy and moves a substance against its concentration gradient.

Active transport also involves the use of a carrier protein to transfer a specific substance across the membrane, but in this case the carrier transports the substance uphill *against* its concentration gradient. Active transport comes in two forms. In **primary active transport,** energy is *directly* required to move a substance against its concentration gradient; the carrier splits ATP to power the transport process. In **secondary active transport,** energy is required in the entire process, but it is *not directly* used to produce uphill movement. That is, the carrier does not split ATP; instead, it moves a molecule uphill by using "secondhand" energy stored in the form of an **ion concentration gradient** (most commonly a Na^+ gradient). This ion gradient is built up by primary active transport of the ion by a different carrier. Let's examine each process in more detail.

PRIMARY ACTIVE TRANSPORT In primary active transport, energy in the form of ATP is required to vary the affinity of the binding site when it is exposed on opposite sides of the plasma membrane. In contrast, in facilitated diffusion the affinity of the binding site is the same when exposed to either the outside or the inside of the cell.

Exercising Muscles Have a "Sweet Tooth"

Cells take up glucose from the blood by facilitated diffusion via glucose carriers in the plasma membrane. The cells maintain an intracellular pool of these carriers that can be inserted into the plasma membrane as the need for glucose uptake increases. In many cells, including resting muscle cells, glucose uptake depends on the hormone insulin, which promotes the insertion of glucose carriers in the plasma membranes of insulin-dependent cells.

During exercise, muscle cells use more glucose and other nutrient fuels than usual to power their increased contractile activity. The rate of glucose transport into exercising muscle may increase more than 10-fold during moderate or intense physical activity. Insulin is not responsible for the increased transport of glucose into exercising muscles, however, because blood insulin levels fall during exercise. Researchers have shown that muscle cells instead insert more glucose carriers in their plasma membranes in direct response to exercise. This has been demonstrated in rats that have undergone physical training.

Exercise influences glucose transport into cells in yet another way. Regular aerobic exercise (see p. 39) has been shown to increase both the affinity (degree of attraction) and number of plasma membrane receptors that bind specifically with insulin. This adaptation results in an increase in insulin sensitivity; that is, the cells are more responsive than normal to a given level of circulating insulin.

Because insulin enhances the facilitated diffusion of glucose into most cells, an exercise-induced increase in insulin sensitivity is one of the factors that makes exercise a beneficial therapy for controlling diabetes mellitus. In this disorder, glucose entry into most cells is impaired as a result of inadequate insulin action (see Chapter 19). Plasma levels of glucose become elevated because glucose remains in the plasma instead of being transported into the cells. In Type 1 diabetes, too little insulin is produced to meet the body's need for glucose uptake. Regular aerobic exercise reduces the amount of insulin that must be injected to promote glucose uptake and lower the blood glucose level toward normal. In Type 2 diabetes, insulin is produced, but insulin's target cells have decreased sensitivity to its presence. By increasing the cells' responsiveness to the insulin available, regular aerobic exercise helps drive glucose into the cells, where it can be used for energy production, instead of remaining in the plasma, where it leads to detrimental consequences for the body.

With primary active transport, the binding site has a greater affinity for its passenger (always an ion) on the low-concentration side as a result of **phosphorylation** of the carrier on this side (● Figure 3-16, step **1**). The carrier acts as an enzyme that has ATPase activity, which means it splits the terminal phosphate from an ATP molecule to yield ADP and inorganic phosphate plus free energy (see p. 32). (Do not confuse *ATPase*, which splits ATP, with *ATP synthase*, which synthesizes ATP.) The phosphate group then attaches to the carrier, increasing the affinity of its binding site for the ion. As a result the ion to be transported binds to the carrier on the low-concentration side (step **2**). In response to this binding, the carrier changes its conformation so that the ion is now exposed to the high-concentration side of the membrane (step **3**). The change in carrier shape reduces the affinity of the binding site for the passenger, so the ion is released on the high-concentration side. Simultaneously the change in shape is also accompanied by **dephosphorylation;** that is, the phosphate group detaches from the carrier (step **4**). The carrier then returns to its original conformation (step **5**). Thus, ATP energy is used in the phosphorylation–dephosphorylation cycle of the carrier. It alters the affinity of the carrier's binding sites on opposite sides of the membrane so that transported ions are moved uphill from an area of low concentration to an area of higher concentration. These active-transport mechanisms are frequently called "pumps," analogous to water pumps that require energy to lift water against the downward pull of gravity.

Primary active-transport pumps all move positively charged ions, namely Na^+, K^+, H^+, or Ca^{2+}, across the membrane. The simplest primary active-transport systems pump a single type of passenger. For example, the Ca^{2+} pump in the plasma membrane transports Ca^{2+} out of the cell, keeping the Ca^{2+} concentration in the cytosol very low. These Ca^{2+} transporters are particularly abundant in the plasma membrane of neuron terminals that store chemical messengers (neurotransmitters) in secretory vesicles (see p. 105). An electrical signal in a neuron terminal triggers the opening of Ca^{2+} channels in the terminal's plasma membrane. Entry of Ca^{2+} down its concentration gradient through these open channels promotes the secretion of neurotransmitter by exocytosis of the secretory vesicles. By keeping the intracellular Ca^{2+} concentration very low, the active Ca^{2+} pump helps maintain a large concentration gradient for the entry of secretion-inducing Ca^{2+} from the ECF into the neuron terminal.

Other more complicated primary active-transport mechanisms involve the transfer of two different passengers in opposite directions. The most important example is a sequentially active **Na^+–K^+ ATPase pump (Na^+–K^+ pump** for short) found in the plasma membrane of all cells. This carrier transports Na^+ out of the cell, concentrating it in the ECF, and picks up K^+ from the outside, concentrating it in the ICF (● Figure 3-17). The pump has high affinity for Na^+ on the ICF side. Binding of Na^+ to the carrier triggers the splitting of ATP through ATPase activity and subsequent phosphorylation of the carrier on the intracellular side, inducing a change in carrier shape, which moves Na^+ to the exterior. Simultaneously the change in shape increases the carrier's affinity for K^+ on the ECF side. Binding of K^+ leads to dephosphorylation of the carrier, inducing a second change in carrier shape, which transfers K^+ into the cytoplasm. There is not a direct exchange of Na^+ for K^+, however. The Na^+–K^+

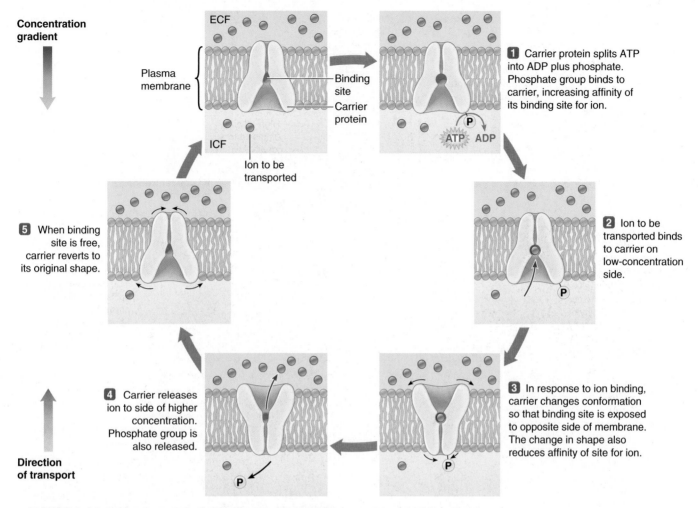

Concentration gradient

ECF

Plasma membrane

Binding site

Carrier protein

ICF

Ion to be transported

1 Carrier protein splits ATP into ADP plus phosphate. Phosphate group binds to carrier, increasing affinity of its binding site for ion.

ATP → ADP, P

2 Ion to be transported binds to carrier on low-concentration side.

P

5 When binding site is free, carrier reverts to its original shape.

Direction of transport

4 Carrier releases ion to side of higher concentration. Phosphate group is also released.

P

3 In response to ion binding, carrier changes conformation so that binding site is exposed to opposite side of membrane. The change in shape also reduces affinity of site for ion.

P

● **FIGURE 3-16 Model for active transport.** The energy of ATP is required in the phosphorylation–dephosphorylation cycle of the carrier to transport the molecule uphill from a region of low concentration to a region of high concentration.

pump moves three Na^+ out of the cell for every two K^+ it pumps in. (To appreciate the magnitude of active Na^+–K^+ pumping that takes place, consider that a single nerve cell membrane contains perhaps 1 million Na^+–K^+ pumps capable of transporting about 200 million ions per second.)

The Na^+–K^+ pump plays three important roles:

1. It establishes Na^+ and K^+ concentration gradients across the plasma membrane of all cells; these gradients are critically important in the ability of nerve and muscle cells to generate electrical signals essential to their functioning (a topic discussed more thoroughly later).

2. It helps regulate cell volume by controlling the concentrations of solutes inside the cell and thus minimizing osmotic effects that would induce swelling or shrinking of the cell.

3. The energy used to run the Na^+–K^+ pump also indirectly serves as the energy source for secondary active transport, a topic to which we now turn attention.

SECONDARY ACTIVE TRANSPORT With secondary active transport, the carrier does not directly split ATP to move a substance against its concentration gradient. Instead, the movement of

Na^+ into the cell down its concentration gradient (established by the ATP-splitting Na^+–K^+ pump) drives the uphill transport of another solute by a secondary active-transport carrier. This is very efficient, because Na^+ must be pumped out anyway to maintain the electrical and osmotic integrity of the cell.

In secondary active transport, the transfer of the solute across the membrane is always coupled (occurs together) with the transfer of the ion that supplies the driving force. We use Na^+ as the main example. Secondary active-transport carriers have two binding sites: one for the solute being moved against its concentration gradient and one for Na^+. Secondary active transport occurs by two mechanisms—*symport* and *antiport*—depending on the direction the transported solute moves in relation to Na^+ movement. In **symport** (also called **cotransport**), the solute and Na^+ move through the membrane in the same direction, that is, into the cell (*sym* means "together;" *co* means "with"). Glucose and amino acids are examples of molecules transported by symport in intestinal and kidney cells. We'll discuss the importance of these carriers in more detail shortly. In **antiport** (also known as **countertransport** or **exchange**), the solute and Na^+ move through the membrane in opposite directions, that is Na^+ into and the solute out of the

1 Pump has 3 high-affinity sites for Na^+ and 2 low-affinity sites for K^+ when exposed to ICF.

Na⁺ concentration gradient

ECF

High Na^+ Low K^+

Na^+–K^+ pump

Plasma membrane

High-affinity binding site for Na^+

Low-affinity binding site for K^+

Low Na^+ High K^+

ICF 3 Na^+

K⁺ concentration gradient

Direction of K⁺ transport

6 Two K^+ are released to ICF (where K^+ concentration is high) as affinity of K^+ binding sites markedly decreases during change in shape. At same time, affinity of Na^+ binding sites greatly increases, returning process to step 1.

2 K^+

2 When 3 Na^+ from ICF (where Na^+ concentration is low) bind to pump, it splits ATP into ADP plus phosphate; phosphate group binds to pump.

P

ATP ADP

5 When 2 K^+ from ECF (where K^+ concentration is low) bind to pump, it releases phosphate group. Dephosphorylation causes pump to revert to its original conformation.

P

3 Na^+

Low-affinity binding site for Na^+

High-affinity binding site for K^+

3 Phosphorylation causes pump to change conformation so that Na^+ binding sites are exposed to opposite side of membrane and 3 Na^+ are released to ECF (where Na^+ concentration is high) as affinity of Na^+ binding sites greatly decreases.

Direction of Na⁺ transport

2 K^+

P

4 Change in shape also exposes pump's binding sites for K^+ to ECF and greatly increases affinity of K^+ sites.

● **FIGURE 3-17 Na⁺–K⁺ pump.** The plasma membrane of all cells contains an active-transport carrier, the Na^+–K^+ pump, which uses energy in the carrier's phosphorylation–dephosphorylation cycle to sequentially transport Na^+ out of the cell and K^+ into the cell against these ions' concentration gradient. This pump moves three Na^+ out and two K^+ in for each ATP split.

cell (*anti* means "opposite;" *counter* means "against") (● Figure 3-18). For example, cells exchange Na^+ and H^+ by means of antiport. This carrier plays an important role in maintaining the appropriate pH inside the cells (a fluid becomes more acidic as its H^+ concentration rises).

Let's examine the Na^+ and glucose (or amino acid) symport in more detail as an example of secondary active transport. Unlike most cells of the body, the intestinal and kidney cells actively transport glucose and amino acids by moving them uphill from low to high concentration. The intestinal cells transport these nutrients from the intestinal lumen into the blood, concentrating them in the blood, until none are left in the lumen to be lost in the feces. The kidney cells save these nutrient molecules for the body by transporting them out of the fluid that is to become urine, moving them against a concentration gradient into the blood. The symport carriers that transport glucose against its concentration gradient from the lumen in the intestine and kidneys are distinct from the glucose facilitated-diffusion carriers that transport glucose down its concentration gradient.

Here we focus specifically on the symport carrier that co-transports Na^+ and glucose in intestinal epithelial cells. This carrier, known as the **sodium and glucose cotransporter** or **SGLT,** is located in the luminal membrane (the membrane facing the intestinal lumen) (● Figure 3-19). The Na^+–K^+ pump in these cells is located in the basolateral membrane (the membrane on the side of the cell opposite the lumen and along the lateral edge of the cell below the tight junction; see ● Figure 3-5, p. 59). More Na^+ is present in the lumen than inside the cells because the energy-requiring Na^+–K^+ pump transports Na^+ out of the cell at the basolateral membrane, keeping the intracellular Na^+ concentration low (● Figure 3-19, step **1**). Because of this Na^+ concentration difference, more Na^+ binds to the SGLT when it is exposed to the lumen than when it is exposed to the ICF. Binding of Na^+ to this carrier increases its affinity for glucose, so glucose binds to the SGLT when it is open to the lumen side where glucose concentration is low (step **2a**). When both Na^+ and glucose are bound to it, the SGLT changes shape and opens to the inside of the cell (step **2b**). Both Na^+ and glucose are released to the interior—Na^+ because of the lower intracellular Na^+ concentration and glucose because of the reduced affinity of the binding site on release of Na^+ (step **2c**). The movement of Na^+ into the cell by this cotransport carrier is downhill because the intracellular Na^+ concentration is low, but the movement of glucose is uphill because glucose becomes concentrated in the cell.

The released Na^+ is quickly pumped out by the active Na^+–K^+ transport mechanism, keeping the level of intracellular Na^+ low. The energy expended in this process is not used directly to run the SGLT because phosphorylation is not required to alter the affinity of the binding site to glucose. Instead, the establishment of a Na^+ concentration gradient by the Na^+–K^+ pump—a primary active-transport mechanism—

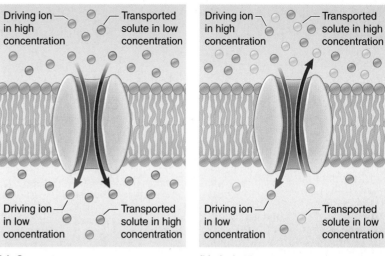

(a) Symport **(b)** Antiport

● **FIGURE 3-18 Secondary active transport, in which an ion concentration gradient is used as the energy source for active transport of a solute.** (a) In symport, the transported solute moves in the same direction as the gradient of the driving ion. (b) In antiport, the transported solute moves in the direction opposite from the gradient of the driving ion.

drives the SGLT—a secondary active-transport mechanism—to move glucose against its concentration gradient.

The glucose carried across the luminal membrane into the cell by secondary active transport then moves passively out of the cell across the basolateral membrane by facilitated diffusion down its concentration gradient and enters the blood (step **3**). This facilitated diffusion is mediated by a passive glucose transporter or GLUT identical to the one that transports glucose into other cells, but in intestinal and kidney cells it transports glucose out of the cell. The difference depends on the direction of the glucose concentration gradient. In the case of intestinal and kidney cells, the glucose concentration is higher inside the cells. Note that in this sequence of events, secondary active transport refers only to the cotransport of glucose uphill across the luminal membrane driven by the Na^+ concentration gradient; that is, the transport accomplished by SGLT.

Before leaving the topic of carrier-mediated transport, think about all the activities that rely on carrier assistance. All cells depend on carriers for the uptake of glucose and amino acids, which serve as the major energy source and structural building blocks, respectively. Na^+–K^+ pumps are essential for generating cellular electrical activity and for ensuring that cells have an appropriate intracellular concentration of osmotically active solutes. Primary and secondary active transport are used extensively to accomplish the specialized functions of the nervous and digestive systems, as well as the kidneys and all types of muscle.

With vesicular transport, material is moved into or out of the cell wrapped in membrane.

The special carrier-mediated transport systems embedded in the plasma membrane selectively transport ions and small polar molecules. But how do large polar molecules, such as the pro-

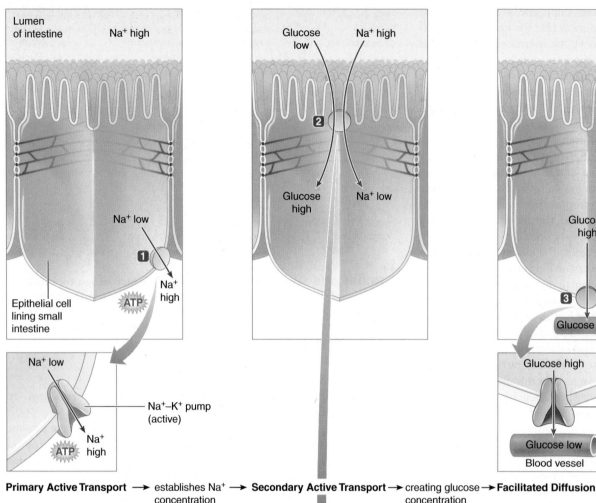

Lumen
of intestine   Na⁺ high

Epithelial cell
lining small
intestine

Na⁺ low

1

Na⁺
high

ATP

Na⁺ low

Na⁺–K⁺ pump
(active)

Na⁺
high

ATP

Glucose
low   Na⁺ high

2

Glucose
high   Na⁺ low

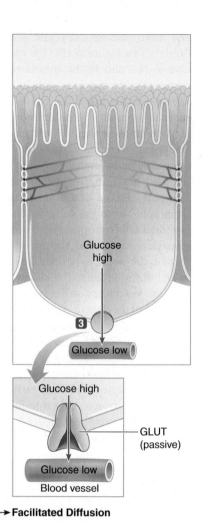

Glucose
high

3

Glucose low

Glucose high

GLUT
(passive)

Glucose low

Blood vessel

Primary Active Transport → establishes Na⁺ → **Secondary Active Transport** → creating glucose → **Facilitated Diffusion**
concentration   concentration
gradient from   gradient from cell
lumen to cell,   to blood used for
which drives

1 Na⁺–K⁺ pump uses energy   **2** SGLT uses Na⁺ concentration   **3** GLUT passively moves
to drive Na⁺ *uphill* out of cell.   gradient to simultaneously move   glucose *downhill* out of cell
                                          Na⁺ *downhill* and glucose *uphill*   into blood.
                                          from lumen into cell.

Glucose low   Na⁺ high

SGLT

Glucose high   Na⁺ low

Glucose low   Na⁺ high

Glucose high   Na⁺ low

2a Binding of Na⁺ on luminal side,   **2b** When both Na⁺ and glucose   **2c** SGLT releases Na⁺ to cell interior,
where Na⁺ concentration is higher,   are bound, SGLT changes   where Na⁺ concentration is lower.
increases affinity of SGLT for glucose.   shape, opening to cell interior.   Because affinity of SGLT for glucose
Therefore, glucose also binds to SGLT   decreases on release of Na⁺, SGLT
on luminal side, where glucose   also releases glucose to cell interior,
concentration is lower.   where glucose concentration is higher.

● **FIGURE 3-19 Symport of glucose.** Glucose is transported across intestinal and kidney cells
against its concentration gradient by means of secondary active transport mediated by the sodium and glucose cotransporter (SGLT) at the cells' luminal membrane.

tein hormones secreted by endocrine cells, or even multimolecular materials, such as the bacteria ingested by white blood cells, leave or enter the cell? These materials are unable to cross the plasma membrane, even with assistance: They are much too big for channels, and no carriers exist for them (they would not even fit into a carrier molecule). These large particles are transferred between the ICF and ECF not by crossing the membrane but by being wrapped in a membrane-enclosed vesicle, a process known as **vesicular transport.** Vesicular transport requires energy expenditure by the cell, so this is an *active* method of membrane transport. Energy is needed to accomplish vesicle formation and vesicle movement within the cell. Transport into the cell in this manner is termed *endocytosis,* whereas transport out of the cell is called *exocytosis* (see ● Figure 2-5, p. 28).

ENDOCYTOSIS To review, in **endocytosis** the plasma membrane surrounds the substance to be ingested, then fuses over the surface, pinching off a membrane-enclosed vesicle so that the engulfed material is trapped within the cell. Recall that there are three forms of endocytosis, depending on the nature of the material internalized: pinocytosis (nonselective uptake of ECF), receptor-mediated endocytosis (selective uptake of a large molecule), and phagocytosis (selective uptake of a multimolecular particle) (see ● Figure 2-8, p. 30).

Once inside the cell, an engulfed vesicle has two possible destinies:

1. In most instances, lysosomes fuse with the vesicle, degrading and releasing its contents into the intracellular fluid.

2. In some cells, the endocytic vesicle bypasses the lysosomes and travels to the opposite side of the cell, where it releases its contents by exocytosis. This provides a pathway to shuttle intact particles through the cell. Such vesicular traffic is one means by which materials are transferred through the thin cells lining the capillaries, across which exchanges are made between the blood and surrounding tissues.

EXOCYTOSIS In **exocytosis,** almost the reverse of endocytosis occurs. A membrane-enclosed vesicle formed within the cell fuses with the plasma membrane, then opens up and releases its contents to the exterior (see ● Figure 2-6, p. 29). Materials packaged for export by the endoplasmic reticulum and Golgi complex are externalized by exocytosis.

Exocytosis serves two different purposes:

1. It provides a mechanism for secreting large polar molecules, such as protein hormones and enzymes that are unable to cross the plasma membrane. In this case, the vesicular contents are highly specific and are released only on receipt of appropriate signals.

2. It enables the cell to add specific components to the membrane, such as selected carriers, channels, or receptors, depending on the cell's needs. In such cases, the composition of the membrane surrounding the vesicle is important, and the contents may be merely a sampling of ICF.

BALANCE BETWEEN ENDOCYTOSIS AND EXOCYTOSIS The rates of endocytosis and exocytosis must be kept in balance to maintain a constant membrane surface area and cell volume. In a cell actively involved in endocytosis, more than 100% of the plasma membrane may be used in an hour to wrap internalized vesicles, necessitating rapid replacement of surface membrane by exocytosis. In contrast, when a secretory cell is stimulated to secrete, it may temporarily insert up to 30 times its surface membrane through exocytosis. This added membrane must be specifically retrieved by an equivalent level of endocytic activity. Thus, through exocytosis and endocytosis, portions of the membrane are constantly being restored, retrieved, and generally recycled.

Our discussion of membrane transport is now complete; ▲ Table 3-2 summarizes the pathways by which materials can pass between the ECF and ICF. Cells are differentially selective in what enters or leaves because they have varying numbers and kinds of channels, carriers, and mechanisms for vesicular transport. Large polar molecules (too large for channels and not lipid soluble) for which there are no special transport mechanisms are unable to permeate.

The selective transport of K^+ and Na^+ is responsible for the electrical properties of cells. We turn our attention to this topic next.

Membrane Potential

The plasma membranes of all living cells have a membrane potential, or are polarized electrically.

Membrane potential is a separation of opposite charges across the plasma membrane.

The term **membrane potential** refers to a separation of opposite charges across the membrane or to a difference in the relative number of cations and anions in the ICF and ECF. Recall that opposite charges tend to attract each other and like charges tend to repel each other. Work must be performed (energy expended) to separate opposite charges after they have come together. Conversely, when oppositely charged particles have been separated, the electrical force of attraction between them can be harnessed to perform work when the charges are permitted to come together again. This is the basic principle underlying electrically powered devices. A separation of charges across the membrane is called a *membrane potential* because separated charges have the potential to do work. Potential is measured in volts (the same unit used for the voltage in electrical devices), but because the membrane potential is relatively low, the unit used is the **millivolt (mV)** (1 mV = 1/1000 volt).

Because the concept of potential is fundamental to understanding much of physiology, especially nerve and muscle physiology, it is important to understand clearly what this term means. The membrane in ● Figure 3-20a is electrically neutral; with an equal number of positive (+) and negative (−) charges on each side of the membrane, no membrane potential exists. In ● Figure 3-20b, some of the positive charges from the right side have been moved to the left. Now the left side has an excess of positive charges, leaving an excess of negative charges on the right. In other words, there is a separation of opposite charges

| ▲ TABLE 3-2 | **Methods of Membrane Transport and Their Characteristics** |

Method of Transport	Substances Involved	Energy Requirements and Force-Producing Movement	Limit to Transport
Simple Diffusion			
Diffusion through lipid bilayer	Nonpolar molecules of any size (e.g., O_2, CO_2, fatty acids)	Passive; molecules move down concentration gradient (from high to low concentration)	Continues until gradient is abolished (dynamic equilibrium with no net diffusion)
Diffusion through protein channel	Specific small ions (e.g., Na^+, K^+, Ca^{2+}, Cl^-)	Passive; ions move down electrochemical gradient through open channels (from high to low concentration and by attraction of ion to area of opposite charge)	Continues until there is no net movement and dynamic equilibrium is established
Osmosis	Water only	Passive; water moves down its own concentration gradient (to area of lower water concentration, i.e., higher solute concentration)	Continues until concentration difference is abolished or until stopped by opposing hydrostatic pressure or until cell is destroyed
Carrier-Mediated Transport			
Facilitated diffusion	Specific polar molecules for which carrier is available (e.g., glucose)	Passive; molecules move down concentration gradient (from high to low concentration)	Displays a transport maximum (T_m); carrier can become saturated
Primary active transport	Specific cations for which carriers are available (e.g., Na^+, K^+, H^+, Ca^{2+})	Active; ions move against concentration gradient (from low to high concentration); requires ATP	Displays a transport maximum; carrier can become saturated
Secondary active transport (symport or antiport)	Specific polar molecules and ions for which coupled transport carriers are available (e.g., glucose, amino acids for symport; some ions for antiport)	Active; substance moves against concentration gradient (from low to high concentration); driven directly by ion gradient (usually Na^+) established by ATP-requiring primary pump. In symport, cotransported molecule and driving ion move in same direction; in antiport, transported solute and driving ion move in opposite directions	Displays a transport maximum; coupled transport carrier can become saturated
Vesicular Transport			
Endocytosis			
Pinocytosis	Small volume of ECF fluid; also important in membrane recycling	Active; plasma membrane dips inward and pinches off at surface, forming internalized vesicle	Control poorly understood
Receptor-mediated endocytosis	Specific large polar molecule (e.g., protein)	Active; plasma membrane dips inward and pinches off at surface, forming internalized vesicle	Necessitates binding to specific receptor on membrane surface
Phagocytosis	Multimolecular particles (e.g., bacteria and cellular debris)	Active; cell extends pseudopods that surround particle, forming internalized vesicle	Necessitates binding to specific receptor on membrane surface
Exocytosis	Secretory products (e.g., hormones and enzymes) as well as large molecules that pass through cell intact; also important in membrane recycling	Active; increase in cytosolic Ca^{2+} induces fusion of secretory vesicle with plasma membrane; vesicle opens up and releases contents to outside	Secretion triggered by specific neural or hormonal stimuli; other controls involved in transcellular traffic and membrane recycling not known

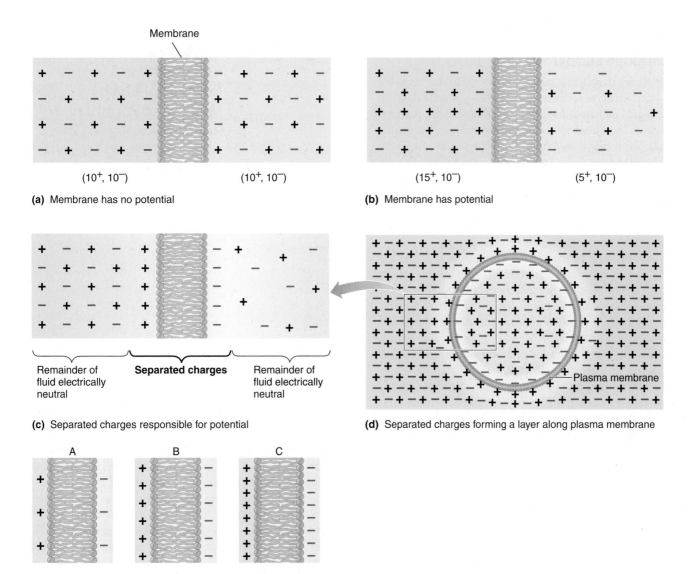

(a) Membrane has no potential

(b) Membrane has potential

(c) Separated charges responsible for potential

(d) Separated charges forming a layer along plasma membrane

(e) Magnitude of potential: membrane B has more potential than membrane A and less potential than membrane C

● **FIGURE 3-20 Determination of membrane potential by unequal distribution of positive and negative charges across the membrane.** (a) When the positive and negative charges are equally balanced on each side of the membrane, no membrane potential exists. (b) When opposite charges are separated across the membrane, membrane potential exists. (c) The unbalanced charges responsible for the potential accumulate in a thin layer along opposite surfaces of the membrane. (d) The vast majority of the fluid in the ECF and ICF is electrically neutral. The unbalanced charges accumulate along the plasma membrane. (e) The greater the separation of charges across the membrane, the larger the potential.

across the membrane, or a difference in the relative number of positive and negative charges between the two sides. That is, now a membrane potential exists. The attractive force between the separated charges causes them to accumulate in a thin layer along the outer and inner surfaces of the plasma membrane (● Figure 3-20c). These separated charges represent only a small fraction of the total number of charged particles (ions) present in the ICF and ECF, however, and the vast majority of the fluid inside and outside the cells is electrically neutral (● Figure 3-20d). The electrically balanced ions can be ignored, because they do not contribute to membrane potential. Thus, an almost

insignificant fraction of the total number of charged particles present in the body fluids is responsible for the membrane potential.

Note that the membrane itself is not charged. The term *membrane potential* refers to the difference in charge between the wafer-thin regions of ICF and ECF lying next to the inside and outside of the membrane, respectively. The magnitude of the potential depends on the number of opposite charges separated: The greater the number of charges separated, the larger the potential. Therefore, in ● Figure 3-20e membrane B has more potential than A and less potential than C.

Membrane potential is due to differences in the concentration and permeability of key ions.

All cells have membrane potential. The cells of **excitable tissues**—namely, nerve cells and muscle cells—have the ability to produce rapid, transient changes in their membrane potential when excited. These brief fluctuations in potential serve as electrical signals. The constant membrane potential present in the cells of nonexcitable tissues and those of excitable tissues when they are at rest—that is, when they are not producing electrical signals—is known as the **resting membrane potential.** Here we will concentrate on the generation and maintenance of the resting membrane potential and in later chapters will examine the changes that take place in excitable tissues during electrical signaling.

The unequal distribution of a few key ions between the ICF and ECF and their selective movement through the plasma membrane are responsible for the electrical properties of the membrane. In the body, electrical charges are carried by ions. The ions primarily responsible for the generation of the resting membrane potential are Na^+, K^+, and A^-. The last refers to the large, negatively charged (anionic) intracellular proteins. Other ions (calcium, magnesium, chloride, bicarbonate, and phosphate, to name a few) do not contribute directly to the resting electrical properties of the plasma membrane in most cells, even though they play other important roles in the body.

The concentrations and relative permeabilities of the ions critical to membrane electrical activity are compared in ▲ Table 3-3. Note that *Na^+ is more concentrated in the extracellular fluid and K^+ is more concentrated in the intracellular fluid.* These concentration differences are maintained by the Na^+–K^+ pump at the expense of energy. Because the plasma membrane is virtually impermeable to A^-, these large, negatively charged proteins are found only inside the cell. After they have been synthesized from amino acids transported into the cell, they remain trapped within the cell.

In addition to the active carrier mechanism, Na^+ and K^+ can passively cross the membrane through protein channels specific for them. It is usually much easier for K^+ than for Na^+ to get through the membrane because the membrane typically has many more channels open for passive K^+ traffic than for passive Na^+ traffic. At resting potential in a nerve cell, the membrane is typically about 25 to 30 times more permeable to K^+ than to Na^+.

Armed with knowledge of the relative concentrations and permeabilities of these ions, we can analyze the forces acting across the plasma membrane. We will consider first the direct contributions of the Na^+–K^+ pump to membrane potential; second, the effect that the movement of K^+ alone would have on membrane potential; third, the effect of Na^+ alone; and finally, the situation that exists in the cells when both K^+ and Na^+ effects are taking place concurrently. Remember throughout this discussion that *the concentration gradient for K^+ will always be outward and the concentration gradient for Na^+ will always be inward,* because the Na^+–K^+ pump maintains a higher concentration of K^+ inside the cell and a higher concentration of Na^+ outside the cell. Also, note that because K^+ and Na^+ are both cations (positively charged), *the electrical gradient for both of these ions will always be toward the negatively charged side of the membrane.*

▲ **TABLE 3-3** **Concentration and Permeability of Ions Responsible for Membrane Potential in a Resting Nerve Cell**

ION	CONCENTRATION (Millimoles/liter; mM)		Relative Permeability
	Extracellular	**Intracellular**	
Na^+	150	15	1
K^+	5	150	25–30
A^-	0	65	0

EFFECT OF THE SODIUM–POTASSIUM PUMP ON MEMBRANE POTENTIAL The Na^+–K^+ pump transports three Na^+ out for every two K^+ it transports in. Because Na^+ and K^+ are both positive ions, this unequal transport separates charges across the membrane, with the outside becoming relatively more positive and the inside becoming relatively more negative as more positive ions are transported out than in. However, this active transport mechanism only separates enough charges to generate an almost negligible membrane potential of 1 mV to 3 mV, with the interior negative to the exterior of the cell. The vast majority of the membrane potential results from the passive diffusion of K^+ and Na^+ down concentration gradients. Thus, the main role of the Na^+–K^+ pump in producing membrane potential is indirect, through its critical contribution to maintaining the concentration gradients directly responsible for the ion movements that generate most of the potential.

EFFECT OF THE MOVEMENT OF POTASSIUM ALONE ON MEMBRANE POTENTIAL: EQUILIBRIUM POTENTIAL FOR K^+ Let's consider a hypothetical situation characterized by (1) the concentrations that exist for K^+ and A^- across the plasma membrane, (2) free permeability of the membrane to K^+ but not to A^-, and (3) no potential as yet present. The concentration gradient for K^+ would tend to move these ions out of the cell (● Figure 3-21). Because the membrane is permeable to K^+, these ions would readily pass through, carrying their positive charge with them, so more positive charges would be on the outside. At the same time, negative charges in the form of A^- would be left behind on the inside, similar to the situation shown in ● Figure 3-20b. (Remember that the large protein anions cannot diffuse out, despite a tremendous concentration gradient.) A membrane potential would now exist. Because an electrical gradient would also be present, K^+ would be attracted toward the negatively charged interior and repelled by the positively charged exterior. Thus, two opposing forces would now be acting on K^+: the concentration gradient tending to move K^+ out of the cell and the electrical gradient tending to move these same ions into the cell.

1 The concentration gradient for K$^+$ tends to move this ion out of the cell.

2 The outside of the cell becomes more positive as K$^+$ ions move to the outside down their concentration gradient.

3 The membrane is impermeable to the large intracellular protein anion (A$^-$). The inside of the cell becomes more negative as K$^+$ ions move out, leaving behind A$^-$.

4 The resulting electrical gradient tends to move K$^+$ into the cell.

5 No further net movement of K$^+$ occurs when the inward electrical gradient exactly counterbalances the outward concentration gradient. The membrane potential at this equilibrium point is the equilibrium potential for K$^+$ (E_{K^+}) at −90 mV.

$E_{K^+} = -90$ mV

● **FIGURE 3-21 Equilibrium potential for K$^+$.**

Initially, the concentration gradient would be stronger than the electrical gradient, so net movement of K$^+$ out of the cell would continue, and the membrane potential would increase. As more and more K$^+$ moved out of the cell, however, the opposing electrical gradient would also become stronger as the outside became increasingly more positive and the inside more negative. One might think that the outward concentration gradient for K$^+$ would gradually decrease as K$^+$ leaves the cell down this gradient. Surprisingly, however, the K$^+$ concentration gradient would remain essentially constant despite the outward movement of K$^+$. The reason is that even infinitesimal movement of K$^+$ out of the cell would bring about rather large changes in membrane potential. Accordingly, such an extremely small number of K$^+$ ions would have to leave the cell to establish an opposing electrical gradient that the K$^+$ concentration inside and outside the cell would remain essentially unaltered. As K$^+$ continued to move out down its unchanging concentration gradient, the inward electrical gradient would continue to increase in strength. Net outward movement would gradually be reduced as the strength of the electrical gradient approached that of the concentration gradient. Finally, when these two forces exactly balanced each other (that is, when they were in equilibrium), no further net movement of K$^+$ would occur. The potential that would exist at this equilibrium is known as the **equilibrium potential for K$^+$ (E_{K^+})**. At this point, a large concentration gradient for K$^+$ would still exist, but no more net movement of K$^+$ out of the cell would occur down this concentration gradient because of the exactly equal opposing electrical gradient (● Figure 3-21).

The membrane potential at E_{K^+} is −90 mV. By convention, *the sign always designates the polarity of the excess charge on the inside of the membrane*. A membrane potential of −90 mV means that the potential is of a magnitude of 90 mV, with the inside being negative relative to the outside. A potential of +90 mV would have the same strength, but the inside would be more positive than the outside.

The equilibrium potential for a given ion with differing concentrations across a membrane can be calculated by means of the **Nernst equation** as follows:

$$E_{ion} = \frac{61}{z} \log \frac{C_o}{C_i}$$

where

E_{ion} = equilibrium potential for ion in mV

61 = a constant that incorporates the universal gas constant (R), absolute temperature (T), and an electrical constant known as Faraday (F), along with the conversion of the natural logarithm (*ln*) to the logarithm to base 10 (*log*); 61 = RT/F.

z = the ion's valence; z = 1 for K$^+$ and Na$^+$, the ions that contribute to membrane potential

C_o = concentration of the ion outside the cell in millimoles/liter (millimolars; mM)

C_i = concentration of the ion inside the cell in mM

Given that the ECF concentration of K$^+$ is 5 mM and the ICF concentration is 150 mM (see ▲ Table 3-3),

$$E_{K^+} = 61 \log \frac{5 \text{ mM}}{150 \text{ mM}}$$

$$= 61 \log \frac{1}{30}$$

Because the 61 log of 1/30 = −1.477,

$$E_{K^+} = 61(-1.477) = -90 \text{ mV}$$

Because 61 is a constant, the equilibrium potential is essentially a measure of the membrane potential (that is, the magnitude of the electrical gradient) that exactly counterbalances the concentration gradient for the ion (that is, the ratio between the ion's concentration outside and inside the cell). The larger the con-

Plasma membrane

ECF   ICF

Na+

Concentration gradient for Na+

Electrical gradient for Na+

ECF anions, mostly

Cl−

Na+

$E_{Na^+} = +60$ mV

1 The concentration gradient for Na+ tends to move this ion into the cell.

2 The inside of the cell becomes more positive as Na+ ions move to the inside down their concentration gradient.

3 The outside becomes more negative as Na+ ions move in, leaving behind in the ECF unbalanced negatively charged ions, mostly Cl−.

4 The resulting electrical gradient tends to move Na+ out of the cell.

5 No further net movement of Na+ occurs when the outward electrical gradient exactly counterbalances the inward concentration gradient. The membrane potential at this equilibrium point is the equilibrium potential for Na+ (E_{Na^+}) at +60 mV.

● **FIGURE 3-22 Equilibrium potential for Na+.**

centration gradient is for an ion, the greater the ion's equilibrium potential. A comparably greater opposing electrical gradient would be required to counterbalance the larger concentration gradient.

EFFECT OF MOVEMENT OF SODIUM ALONE ON MEMBRANE POTENTIAL: EQUILIBRIUM POTENTIAL FOR Na+ A similar hypothetical situation could be developed for Na+ alone (● Figure 3-22). The concentration gradient for Na+ would move this ion into the cell, producing a buildup of positive charges on the interior of the membrane and leaving negative charges unbalanced outside (primarily in the form of chloride, Cl−; Na+ and Cl−—that is, salt—are the predominant ECF ions). Net inward movement would continue until equilibrium was established by the development of an opposing electrical gradient that exactly counterbalanced the concentration gradient. At this point, given the concentrations for Na+, the **equilibrium potential for Na+** (E_{Na^+}) as calculated by the Nernst equation would be +61 mV. Given that the ECF concentration of Na+ is 150 mM and the ICF concentration is 15 mM,

$$E_{Na^+} = 61 \log \frac{150 \text{ mM}}{15 \text{ mM}}$$

$$= 61 \log 10$$

Because the log of 10 = 1,

$$E_{Na^+} = 61(1) = 61 \text{ mV}$$

In this case the inside of the cell would be positive, in contrast to the equilibrium potential for K+. The magnitude of E_{Na^+} is somewhat less than for E_{K^+} (61 mV compared to 90 mV) because the concentration gradient for Na+ is not as large (see ▲ Table 3-3); thus, the opposing electrical gradient (membrane potential) is not as great at equilibrium. (For convenience in representing the magnitude of charge in figures, we will round E_{Na^+} to +60 mV.)

CONCURRENT POTASSIUM AND SODIUM EFFECTS ON MEMBRANE POTENTIAL Neither K+ nor Na+ exists alone in the body fluids, so equilibrium potentials are not present in body cells. They exist only in hypothetical or experimental conditions. In a living cell, the effects of both K+ and Na+ must be taken into account. *The greater the permeability of the plasma membrane for a given ion, the greater is the tendency for that ion to drive the membrane potential toward the ion's own equilibrium potential.* Because the membrane at rest is 25 to 30 times more permeable to K+ than to Na+, K+ passes through more readily than Na+; thus, K+ influences the resting membrane potential to a much greater extent than Na+ does. Recall that K+ acting alone would establish an equilibrium potential of −90 mV. The membrane is somewhat permeable to Na+, however, so some Na+ enters the cell in a limited attempt to reach its equilibrium potential. This Na+ entry neutralizes, or cancels, some of the potential that would have been produced by K+ alone if Na+ were not present.

To better understand this concept, assume that each separated pair of charges in ● Figure 3-23 represents 10 mV of potential. (This is not technically correct, because in reality many separated charges must be present to account for a potential of 10 mV.) In this simplified example, nine separated pluses and minuses, with the minuses on the inside, represent the E_{K^+} of −90 mV. Superimposing the slight influence of Na+ on this K+-dominated membrane, assume that two sodium ions enter the cell down the Na+ concentration and electrical gradients. (Note that the electrical gradient for Na+ is now inward in contrast to the outward electrical gradient for Na+ at E_{Na^+}. At E_{Na^+}, the inside of the cell is positive as a result of the inward movement of Na+ down its concentration gradient. In a resting nerve cell, however, the inside is negative because of the dominant influence of K+ on membrane potential. Thus, both the concentration and electrical gradients now favor the inward movement of Na+.) The inward movement of these two positively charged sodium ions neutralizes some of the potential established by K+, so now only seven pairs of charges are separated, and the

1 The Na⁺–K⁺ pump actively transports Na⁺ out of and K⁺ into the cell, keeping the concentration of Na⁺ high in the ECF and the concentration of K⁺ high in the ICF.

2 Given the concentration gradients that exist across the plasma membrane, K⁺ tends to drive membrane potential to the equilibrium potential for K⁺ (−90 mV), whereas Na⁺ tends to drive membrane potential to the equilibrium potential for Na⁺ (+60 mV).

3 However, K⁺ exerts the dominant effect on resting membrane potential because the membrane is more permeable to K⁺. As a result, resting potential (−70 mV) is much closer to E_{K^+} than to E_{Na^+}.

4 During the establishment of resting potential, the relatively large net diffusion of K⁺ outward does not produce a potential of −90 mV because the resting membrane is slightly permeable to Na⁺ and the relatively small net diffusion of Na⁺ inward neutralizes (in gray shading) some of the potential that would be created by K⁺ alone, bringing resting potential to −70 mV, slightly less than E_{K^+}.

5 The negatively charged intracellular proteins (A⁻) that cannot cross the membrane remain unbalanced inside the cell during the net outward movement of the positively charged ions, so the inside of the cell is more negative than the outside.

• **FIGURE 3-23 Effect of concurrent K⁺ and Na⁺ movement on establishing the resting membrane potential.**

potential is −70 mV. This is the resting membrane potential of a typical nerve cell. The resting potential is much closer to E_{K^+} than to E_{Na^+} because of the greater permeability of the membrane to K⁺, but it is slightly less than E_{K^+} (−70 mV is a lower potential than −90 mV) because of the weak influence of Na⁺.

Membrane potential can be measured directly in experimental conditions by recording the voltage difference between the inside and outside of the cell, or it can be calculated using the **Goldman-Hodgkin-Katz equation (GHK equation)**, which takes into account the relative permeabilities and concentration gradients of all permeable ions. The stable, resting membrane is permeable to K⁺, Na⁺, and Cl⁻, but for reasons to be described later, Cl⁻ does not directly contribute to potential in most cells. Therefore, we can ignore it when calculating membrane potential, making the simplified GHK equation:

$$V_m = 61 \log \frac{P_{K^+}[K^+]_o + P_{Na^+}[Na^+]_o}{P_{K^+}[K^+]_i + P_{Na^+}[Na^+]_i}$$

where

V_m = membrane potential in mV

61 = a constant representing RT/zF, when z = 1, as it does for K⁺ and Na⁺

P_{K^+}, P_{Na^+} = permeabilities for K⁺ and Na⁺, respectively

$[K^+]_o, [Na^+]_o$ = concentration of K⁺ and Na⁺ outside the cell in mM, respectively

$[K^+]_i, [Na^+]_i$ = concentration of K⁺ and Na⁺ inside the cell in mM, respectively.

The GHK equation is basically an expanded version of the Nernst equation. The Nernst equation can only be used to calculate the potential generated by a specific ion, but the GHK equation takes into account the combined contributions to potential of all ions moving across the membrane. Assuming the resting membrane is 25 times more permeable to K⁺ than to Na⁺, then the relative permeabilities are $P_{K^+} = 1.0$ and $P_{Na^+} = 0.04$ (1/25 of 1.0). Given these permeabilities and the concentrations for K⁺ and Na⁺ in the ECF and ICF listed in ▲ Table 3-3,

$$V_m = 61 \log \frac{(1)(5) + (0.04)(150)}{(1)(150) + (0.04)(15)}$$

$$= 61 \log \frac{5 + 6}{150 + 0.6}$$

$$= 61 \log 0.073$$

Because the log of 0.073 is −1.137,

$$V_m = 61 (-1.137) = -69 \text{ mV}$$

Adding −1 mV of potential generated directly by the Na⁺–K⁺ pump to this value totals −70 mV for the resting membrane potential.

BALANCE OF PASSIVE LEAKS AND ACTIVE PUMPING AT RESTING MEMBRANE POTENTIAL At resting potential, neither K⁺ nor Na⁺ is at equilibrium. A potential of −70 mV does not exactly counterbalance the concentration gradient for K⁺; it takes a potential of −90 mV to do that. Thus, K⁺ slowly continues to passively exit through its leak channels down this small concentration gradient. **Leak channels** are channels that are open all the time, thus permitting unregulated leakage of their chosen ion down electrochemical gradients. In the case of Na$^+$, the concentration and electrical gradients do not even oppose each other; they both favor the inward movement of Na$^+$. Therefore, Na$^+$ continually leaks inward down its electrochemical gradient, but only slowly, because of its low permeability, that is, because of the scarcity of Na$^+$ leak channels.

Because such leaking goes on all the time, why doesn't the intracellular concentration of K$^+$ continue to fall and the concentration of Na$^+$ inside the cell progressively increase? The reason is that the Na$^+$–K$^+$ pump counterbalances the rate of passive leakage. At resting potential, this pump transports back into the cell essentially the same number of potassium ions that have leaked out and simultaneously transports to the outside the sodium ions that have leaked in. At this point, a **steady state** exists: No net movement of any ions takes place, because all passive leaks are exactly balanced by active pumping. No net change takes place in either a steady state or a dynamic equilibrium (see p. 61), but in a *steady state* energy must be used to maintain the constancy whereas in a *dynamic equilibrium* no energy is needed to maintain the constancy. That is, opposing passive and active forces counterbalance each other in a steady state and opposing passive forces counterbalance each other in a dynamic equilibrium. Because in the steady state across the membrane the active pump offsets the passive leaks, the concentration gradients for K$^+$ and Na$^+$ remain constant. Thus, not only is the Na$^+$–K$^+$ pump initially responsible for the Na$^+$ and K$^+$ concentration differences across the membrane but it also maintains these differences.

As just discussed, the magnitude of these concentration gradients, together with the difference in permeability of the membrane to these ions, accounts for the magnitude of the membrane potential. Because the concentration gradients and permeabilities for Na$^+$ and K$^+$ remain constant in the resting state, the resting membrane potential established by these forces remains constant.

CHLORIDE MOVEMENT AT RESTING MEMBRANE POTENTIAL Thus far, we have largely ignored one other ion present in high concentration in the ECF, namely, Cl$^-$. Chloride is the principal ECF anion. Its equilibrium potential is −70 mV, exactly the same as the resting membrane potential. Movement alone of negatively charged Cl$^-$ into the cell down its concentration gradient would produce an opposing electrical gradient, with the inside negative compared to the outside. When physiologists were first examining the ionic effects that could account for the membrane potential, they were tempted to think that Cl$^-$ movements and establishment of the Cl$^-$ equilibrium potential could be solely responsible for producing the identical resting membrane potential. Actually, the reverse is the case. The membrane potential is responsible for driving the distribution of Cl$^-$ across the membrane.

Most cells are highly permeable to Cl$^-$ but have no active-transport mechanisms for this ion. With no active forces acting on it, Cl$^-$ passively distributes itself to achieve an individual state of equilibrium. In this case, Cl$^-$ is driven out of the cell, establishing an inward concentration gradient that exactly counterbalances the outward electrical gradient (that is, the resting membrane potential) produced by K$^+$ and Na$^+$ movement. Thus, the concentration difference for Cl$^-$ between the

ECF and ICF is brought about passively by the presence of the membrane potential rather than maintained by an active pump, as is the case for K$^+$ and Na$^+$. Therefore, in most cells Cl$^-$ does not influence resting membrane potential; instead, membrane potential passively influences the Cl$^-$ distribution.

SPECIALIZED USE OF MEMBRANE POTENTIAL IN NERVE AND MUSCLE CELLS Nerve and muscle cells have developed a specialized use for membrane potential. They can rapidly and transiently alter their membrane permeabilities to the involved ions in response to appropriate stimulation, thereby bringing about fluctuations in membrane potential. The rapid fluctuations in potential are responsible for producing nerve impulses in nerve cells and for triggering contraction in muscle cells. These activities are the focus of the next five chapters. Even though all cells display a membrane potential, its significance in other cells is uncertain, although changes in membrane potential of some secretory cells, for example insulin-secreting cells, have been linked to their level of secretory activity.

Chapter in Perspective: Focus on Homeostasis

All cells of the body must obtain vital materials, such as nutrients and O$_2$, from the surrounding ECF; they must also eliminate wastes to the ECF and release secretory products, such as chemical messengers and digestive enzymes. Thus, transport of materials across the plasma membrane between the ECF and ICF is essential for cell survival, and the constituents of the ECF must be homeostatically maintained to support these life-sustaining exchanges.

Many cell types use membrane transport to carry out their specialized activities geared toward maintaining homeostasis. Here are several examples:

1. Absorption of nutrients from the digestive tract lumen involves the transport of these energy-giving molecules across the membranes of the cells lining the tract.
2. Exchange of O$_2$ and CO$_2$ between the air and blood in the lungs involves the transport of these gases across the membranes of the cells lining the air sacs and blood vessels of the lungs.
3. Urine is formed by the selective transfer of materials between the blood and the fluid within the kidney tubules across the membranes of the cells lining the tubules.
4. The beating of the heart is triggered by cyclic changes in the transport of Na$^+$, K$^+$, and Ca^{2+} across the heart cells' membranes.
5. Secretion of chemical messengers such as neurotransmitters from nerve cells and hormones from endocrine cells involves the transport of these regulatory products to the ECF on appropriate stimulation.

In addition to providing selective transport of materials between the ECF and ICF, the plasma membrane contains receptors for binding with specific chemical messengers that regulate various cell activities, many of which are specialized activities aimed toward maintaining homeostasis. For example, the hormone vasopressin, which is secreted in response to a water deficit in the body, binds with receptors in the plasma membrane of a specific type of kidney cell. This binding triggers the cells to conserve water during urine formation by promoting the insertion of additional aquaporins (water channels) in the plasma membrane of these cells, thus helping alleviate the water deficit that initiated the response.

All living cells have a membrane potential, with the cell's interior being slightly more negative than the fluid surrounding the cell when the cell is electrically at rest. The specialized activities of nerve and muscle cells depend on these cells' ability to change their membrane potential rapidly on appropriate stimulation. The transient, rapid changes in potential in nerve cells serve as electrical signals or nerve impulses, which provide a means to transmit information along nerve pathways. This information is used to accomplish homeostatic adjustments, such as restoring blood pressure to normal when signaled that it has fallen too low.

Rapid changes in membrane potential in muscle cells trigger muscle contraction, the specialized activity of muscle. Muscle contraction contributes to homeostasis in many ways, including the pumping of blood by the heart and the movement of food through the digestive tract.

REVIEW EXERCISES

Objective Questions (answers on p. A-40)

1. The nonpolar tails of the phospholipid molecules bury themselves in the interior of the plasma membrane. (*True or false?*)

2. Cells shrink when in contact with a hypertonic solution. (*True or false?*)

3. Channels are open to both sides of the membrane at the same time but carriers are open to only one side of the membrane at a time. (*True or false?*)

4. At resting membrane potential, there is a slight excess of negative charges on the inside of the membrane, with a corresponding slight excess of positive charges on the outside. (*True or false?*)

5. Using the answer code on the right, indicate which membrane component is responsible for the function in question:
   ___ 1. channel formation   (a) lipid bilayer
   ___ 2. barrier to passage of   (b) proteins
          water-soluble substances   (c) carbohydrates
   ___ 3. receptors
   ___ 4. membrane fluidity
   ___ 5. recognition of "self"
   ___ 6. membrane-bound enzymes
   ___ 7. structural boundary
   ___ 8. carriers

6. Using the answer code on the right, indicate the direction of net movement in each case:
   ___ 1. simple diffusion   (a) movement from
   ___ 2. facilitated diffusion   high to low
   ___ 3. primary active transport   concentration
   ___ 4. Na$^+$ during symport   (b) movement from
          or antiport   low to high
   ___ 5. transported solute during   concentration
          symport or antiport
   ___ 6. water with regard to the water concentration
          gradient during osmosis
   ___ 7. water with regard to the solute concentration
          gradient during osmosis

7. Using the answer code on the right, indicate the type of cell junction described:
   ___ 1. adhering junction   (a) gap junction
   ___ 2. impermeable junction   (b) tight junction
   ___ 3. communicating junction   (c) desmosome
   ___ 4. made up of connexons, which permit passage of ions and small molecules between cells
   ___ 5. consists of interconnecting fibers, which spot-rivet adjacent cells
   ___ 6. formed by fusion of proteins on the outer surfaces of two interacting cells
   ___ 7. important in tissues subject to mechanical stretching
   ___ 8. important in synchronizing contractions within heart and smooth muscle by allowing spread of electrical activity between the cells composing the muscle mass
   ___ 9. important in preventing passage of materials between cells in epithelial sheets that separate compartments containing different chemical compositions

Essay Questions

1. Describe the fluid mosaic model of membrane structure.

2. What are the functions of the three major types of protein fibers in the extracellular matrix?

3. What two properties of a particle influence whether it can permeate the plasma membrane?

4. List and describe the methods of membrane transport. Indicate what types of substances are transported by each method, and state whether each means of transport is passive or active and unassisted or assisted.

5. According to Fick's law of diffusion, what factors influence the rate of net diffusion across a membrane?

6. State three important roles of the Na$^+$–K$^+$ pump.

7. Describe the contribution of each of the following to establishing and maintaining membrane potential: (a) the Na$^+$–K$^+$ pump; (b) passive movement of K$^+$ across the membrane; (c) passive movement of Na$^+$ across the membrane; and (d) the large intracellular anions.

Quantitative Exercises (Solutions on p. A-40)

(See Appendix D, "Principles of Quantitative Reasoning.")

1. When using the Nernst equation for an ion that has a valence other than 1, you must divide the potential by the valence. Using the Nernst equation, calculate the equilibrium potential for Ca^{2+} from the following sets of data:
   a. Given $[Ca^{2+}]_o = 1$ mM, $[Ca^{2+}]_i = 100$ nM, find $E_{Ca^{2+}}$
   b. Given $[Cl^-]_o = 110$ mM, $[Cl^-]_i = 10$ mM, find E_{Cl^-}

2. One of the important uses of the Nernst equation is in describing the flow of ions across plasma membranes. Ions move under the influence of two forces, the concentration gradient (given in electrical units by the Nernst equation) and the electrical gradient (given by the membrane voltage). This is summarized by *Ohm's law:*

$$I_x = G_x (V_m - E_x)$$

   which describes the movement of ion x across the membrane. I is the current in amperes (A); G is the conductance, a measure of the permeability of x, in Siemens (S), which is $\Delta I / \Delta V$; V_m is the membrane voltage; and E_x is the equilibrium potential of ion x. Not only does this equation tell how large the current is but it also tells what direction the current is flowing. By convention, a negative value of the current represents either a positive ion entering the cell or a negative ion leaving the cell. The opposite is true of a positive value of the current.
   a. Using the following information, calculate the magnitude of I_{Na^+}.

$[Na^+]_o = 145$ mM, $[Na^+]_i = 15$ mM, $G_{Na^+} = 1$ nS, $V_m = -70$ mV

   b. Is Na^+ entering or leaving the cell?
   c. Is Na^+ moving with or against the concentration gradient? Is it moving with or against the electrical gradient?

3. Using the Goldman-Hodgkin-Katz equation, determine what happens to the resting membrane potential if the ECF K^+ concentration doubles to 10 mM.

POINTS TO PONDER

(Explanations on p. A-41)

1. Assume that a membrane permeable to Na^+ but not to Cl^- separates two solutions. The concentration of sodium chloride on side 1 is much higher than on side 2. Which of the following ionic movements would occur?
   a. Na^+ would move until its concentration gradient is dissipated (until the concentration of Na^+ on side 2 is the same as the concentration of Na^+ on side 1).
   b. Cl^- would move down its concentration gradient from side 1 to side 2.
   c. A membrane potential, negative on side 1, would develop.
   d. A membrane potential, positive on side 1, would develop.
   e. None of the above are correct.

2. A solution may have the same osmolarity as normal body fluids, yet not be isotonic. Explain why.

3. Compared to resting potential, would the membrane potential become more negative or more positive if the membrane were more permeable to Na^+ than to K^+?

4. Which of the following methods of transport is being used to transfer the substance into the cell in the accompanying graph?

a. diffusion down a concentration gradient
b. osmosis
c. facilitated diffusion
d. active transport
e. vesicular transport
f. It is impossible to tell with the information provided.

5. Colostrum, the first milk that a mother produces, contains an abundance of antibodies, large protein molecules. These maternal antibodies help protect breast-fed infants from infections until the babies are capable of producing their own antibodies. By what means would you suspect these maternal antibodies are transported across the cells lining a newborn's digestive tract into the bloodstream?

6. The rate at which the Na^+–K^+ pump operates is not constant but is controlled by a combined effect of changes in ICF Na^+ concentration and ECF K^+ concentration. Do you think an increase in both ICF Na^+ and ECF K^+ concentrations would accelerate or slow down the Na^+–K^+ pump? What would be the benefit of this response? Before you reply, consider the following additional information about Na^+ and K^+ movement across the membrane. Not only do Na^+ and K^+ slowly and passively leak through their channels in a resting cell, but during an electrical impulse, known as an action potential, Na^+ rapidly and passively enters the cell; this movement is followed by a rapid, passive outflow of K^+. (These ion movements, which result from rapid changes in membrane permeability, bring about rapid, pronounced changes in membrane potential. This sequence of rapid potential changes—an *action potential*—serves as an electrical signal for conveying information along a nerve pathway.)

CLINICAL CONSIDERATION

(Explanation on p. A-41)

When William H. was helping victims following a devastating earthquake in a region that was not prepared to swiftly set up adequate temporary shelter, he developed severe diarrhea. He was diagnosed as having *cholera,* a disease transmitted through unsanitary water supplies that have been contaminated by fecal material from infected individuals. The toxin produced by cholera bacteria causes Cl^- channels in the luminal membranes of the intestinal cells to open, thereby increasing the secretion of Cl^- from the cells into the intestinal tract lumen. By what mechanisms would Na^+ and water be secreted into the lumen in conjunction with Cl^- secretion? How does this secretory response account for the severe diarrhea that is characteristic of cholera?

Nervous and Endocrine Systems

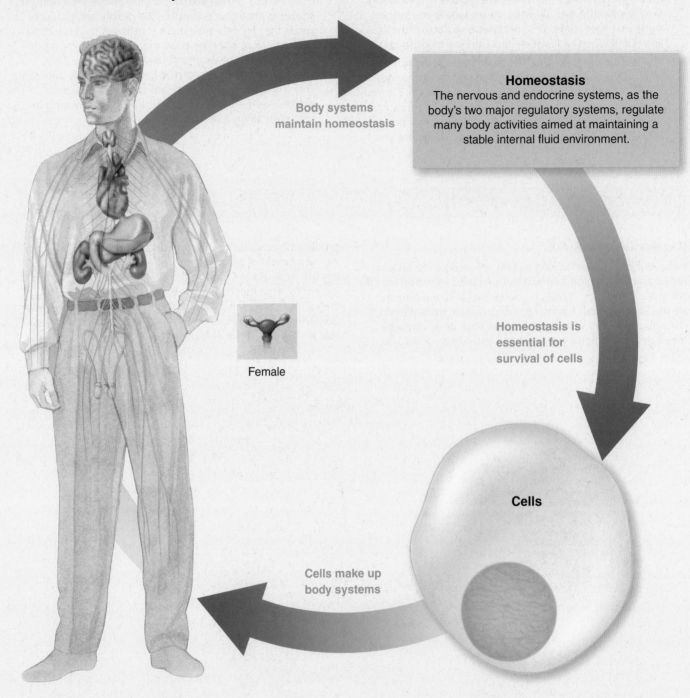

Body systems maintain homeostasis

Homeostasis
The nervous and endocrine systems, as the body's two major regulatory systems, regulate many body activities aimed at maintaining a stable internal fluid environment.

Homeostasis is essential for survival of cells

Female

Cells

Cells make up body systems

To maintain homeostasis, cells must work in a coordinated fashion toward common goals. The two major regulatory systems of the body that help ensure life-sustaining coordinated responses are the nervous and endocrine systems.

Neural communication is accomplished by nerve cells, or neurons, which are specialized for rapid electrical signaling and for secreting neurotransmitters, short-distance chemical messengers that act on nearby target organs. The nervous system exerts rapid control over most of the body's muscular and glandular activities, most of which are directed toward main-

taining homeostasis. Furthermore, many higher-level (thought-related) activities carried out by the nervous system also contribute to homeostasis.

Hormonal communication is accomplished by hormones, which are long-distance chemical messengers secreted by the endocrine glands into the blood. The blood carries the hormones to distant target sites, where they regulate processes that require duration rather than speed, such as metabolic activities, water and electrolyte balance, and growth.

CHAPTER 4

Principles of Neural and Hormonal Communication

CENGAGENOW˙ Log on to CengageNOW at **http://www .cengage.com/sso/** for an opportunity to explore a learning module that illustrates difficult concepts with self-study tutorials, animations, and interactive quizzes to help you learn, review, and master physiology concepts.

Introduction to Neural Communication

All body cells display a membrane potential, which is a separation of positive and negative charges across the membrane, as discussed in the preceding chapter (see pp. 25–32). This potential is related to the uneven distribution of Na$^+$, K$^+$, and large intracellular protein anions between the intracellular fluid (ICF) and extracellular fluid (ECF), and to the differential permeability of the plasma membrane to these ions.

Nerve and muscle are excitable tissues.

The constant membrane potential present when a cell is electrically at rest, that is, not producing electrical signals, is referred to as the *resting membrane potential.* Two types of cells, neurons (nerve cells) and muscle cells, have developed a specialized use for membrane potential. They can undergo transient, rapid fluctuations in their membrane potentials, which serve as electrical signals.

Nerve and muscle are considered *excitable tissues* because they produce electrical signals when excited. Neurons use these electrical signals to receive, process, initiate, and transmit messages. In muscle cells, these electrical signals initiate contraction. Thus, electrical signals are critical to the function of the nervous system and all muscles. In this chapter, we examine how neurons undergo changes in potential to accomplish their function. Muscle cells are discussed in later chapters.

Membrane potential becomes less negative during depolarization and more negative during hyperpolarization.

Before you can understand what electrical signals are and how they are created, you must become familiar with several terms used to describe changes in potential, which are graphically represented in ● Figure 4-1:

1. **Polarization:** Charges are separated across the plasma membrane, so the membrane has potential. Any time the value of the membrane potential is other than 0 mV, in either the positive or negative direction, the membrane is in a state of polarization. Recall that the magnitude of the potential is directly proportional to the number of positive and negative

● **FIGURE 4-1 Types of changes in membrane potential.**

charges separated by the membrane and that the sign of the potential (+ or −) always designates whether excess positive or excess negative charges are present, respectively, on the inside of the membrane. At resting potential, the membrane is polarized at −70 mV in a typical neuron (see p. 31).

2. Depolarization: The membrane becomes less polarized; the inside becomes less negative than at resting potential, with the potential moving closer to 0 mV (for example, a change from −70 to −60 mV); fewer charges are separated than at resting potential. This term also refers to the inside even becoming positive as it does during an action potential (a major type of electrical signal) when the membrane potential reverses itself (for example, becoming +30 mV).

3. Repolarization: The membrane returns to resting potential after having been depolarized.

4. Hyperpolarization: The membrane becomes more polarized; the inside becomes more negative than at resting potential, with the potential moving even farther from 0 mV (for instance, a change from −70 to −80 mV).

One possibly confusing point should be clarified. On the device used for recording rapid changes in potential, during a depolarization when the inside becomes *less* negative than at resting, this *decrease* in the magnitude of the potential is represented as an *upward* deflection. By contrast, during a hyperpolarization when the inside becomes *more* negative than at resting, this *increase* in the magnitude of the potential is represented by a *downward* deflection.

Electrical signals are produced by changes in ion movement across the plasma membrane.

Changes in membrane potential are brought about by changes in ion movement across the membrane. For example, if the net inward flow of positively charged ions increases compared to the resting state, the membrane depolarizes (becomes less negative inside). By contrast, if the net outward flow of positively charged ions increases compared to the resting state, the membrane hyperpolarizes (becomes more negative inside).

Changes in ion movement are brought about by changes in membrane permeability in response to *triggering events*. Depending on the type of electrical signal, a triggering event might be (1) a change in the electrical field in the vicinity of an excit-

able membrane; (2) an interaction of a chemical messenger with a surface receptor on a nerve or muscle cell membrane; (3) a stimulus, such as sound waves stimulating specialized neurons in the ear; or (4) a change of potential caused by inherent cyclical changes in channel permeability. (You will learn more about the nature of these various triggering events as our discussion of electrical signals continues.)

Because the water-soluble ions responsible for carrying charge cannot penetrate the plasma membrane's lipid bilayer, these charges can cross the membrane only through channels specific for them or by carrier-mediated transport. Membrane channels may be either *leak channels* or *gated channels*. As described in Chapter 3, **leak channels,** which are open all the time, permit unregulated leakage of their specific ion across the membrane through the channels. **Gated channels,** in contrast, have gates that can be open or closed, permitting ion passage through the channels when open and preventing ion passage through the channels when closed. Gate opening and closing results from a change in the conformation (shape) of the protein that forms the gated channel. There are four kinds of gated channels, depending on the factor that causes the change in channel conformation: (1) **voltage-gated channels** open or close in response to changes in membrane potential; (2) **chemically gated channels** change conformation in response to binding of a specific extracellular chemical messenger to a surface membrane receptor; (3) **mechanically gated channels** respond to stretching or other mechanical deformation; and (4) **thermally gated channels** respond to local changes in temperature (heat or cold).

Triggering events alter membrane permeability and consequently alter ion flow across the membrane by opening or closing the gates guarding particular ion channels. These ion movements redistribute charge across the membrane, causing membrane potential to fluctuate.

There are two basic forms of electrical signals: (1) *graded potentials,* which serve as short-distance signals; and (2) *action potentials,* which signal over long distances. We next examine these types of signals in more detail, beginning with graded potentials, and then explore how neurons use these signals to convey messages.

Graded Potentials

Graded potentials are local changes in membrane potential that occur in varying grades or degrees of magnitude or strength. For example, membrane potential could change from −70 to −60 mV (a 10 mV graded potential) or from −70 to −50 mV (a 20 mV graded potential).

The stronger a triggering event is, the larger the resultant graded potential.

Graded potentials are usually produced by a specific triggering event that causes gated ion channels to open in a specialized region of the excitable cell membrane. The resultant ion movement produces the graded potential, which most commonly is a depolarization resulting from net Na^+ entry. The graded potential is confined to this small, specialized region of the total plasma membrane.

(a) Entire membrane at resting potential

(b) Inward movement of Na⁺ depolarizes membrane, producing a graded potential

Spread of depolarization

(c) Depolarization spreads by local current flow to adjacent inactive areas, away from point of origin

● **FIGURE 4-2 Current flow during a graded potential.** (a) The membrane of an excitable cell at resting potential. (b) A triggering event opens ion channels, usually leading to net Na⁺ entry that depolarizes the membrane at this site. The adjacent inactive areas are still at resting potential. (c) Local current flows between the active and adjacent inactive areas, resulting in depolarization of the previously inactive areas. In this way, the depolarization spreads away from its point of origin.

The magnitude of the initial graded potential (that is, the difference between the new potential and resting potential) is related to the magnitude of the triggering event: *The stronger the triggering event, the more gated channels that open, the greater is the positive charge entering the cell, and the larger is the depolarizing graded potential at the point of origin. Also, the longer the duration of the triggering event, the longer is the duration of the graded potential.*

Graded potentials spread by passive current flow.

When a graded potential occurs locally in a nerve or muscle cell membrane, the rest of the membrane remains at resting potential. The temporarily depolarized region is called an *active area.*

Note from ● Figure 4-2 that inside the cell, the active area is relatively more positive than the neighboring *inactive areas* that are still at resting potential. Outside the cell, the active area is relatively less positive than adjacent inactive areas. Because of this difference in potential, electrical charges, which are carried by ions, passively flow between the active and adjacent resting regions on both the inside and outside of the membrane.

Any flow of electrical charges is called a **current.** By convention, the direction of current flow is always expressed as the direction in which the positive charges are moving (● Figure 4-2c). Inside the cell, positive charges flow through the ICF away from the relatively more positive depolarized active region toward the more negative adjacent resting regions. Outside the cell, positive charges flow through the ECF from the more positive adjacent inactive regions toward the relatively more

(a) Current loss across the membrane

* Numbers refer to the local potential in mV at various points along the membrane.

(b) Decremental spread of graded potentials

● **FIGURE 4-3 Current loss across the plasma membrane leading to decremental spread of a graded potential.** (a) Leakage of charge-carrying ions across the plasma membrane results in progressive loss of current with increasing distance from the initial site of the change in potential. (b) Because of leaks in current, the magnitude of a graded potential continues to decrease as it passively spreads from the initial active area. The potential dies out altogether within a few millimeters of its site of initiation.

negative active region. Ion movement (that is, current) is occurring *along* the membrane between regions next to each other on the same side of the membrane. This flow is in contrast to ion movement *across* the membrane through ion channels or by means of carriers.

As a result of local current flow between an active depolarized area and an adjacent inactive area, the potential changes in the previously inactive area. Because positive charges have flowed simultaneously into the adjacent inactive area on the inside and out of this area on the outside, the adjacent area is now more positive (or less negative) on the inside than before and less positive (or more negative) on the outside (● Figure 4-2c). Stated differently, the previously inactive adjacent area has been depolarized, so the graded potential has spread. This area's potential now differs from that of the inactive region immediately next to it on the other side, inducing further current flow at this new site, and so on. In this manner, current spreads in both directions away from the initial site of the change in potential.

The amount of current that flows between two areas depends on the difference in potential between the areas and on the resistance of the material through which the charges are moving. **Resistance** is the hindrance to electrical charge movement. The greater the difference in potential, the greater is the current flow; and the lower the resistance, the greater is the current flow. *Conductors* have low resistance, providing little hindrance to current flow. Electrical wires and the ICF and ECF are all good conductors, so current readily flows through them. *Insulators* have high resistance and greatly hinder movement of charge. The plastic surrounding electrical wires has high resistance, as do body lipids. Thus, current does not flow across the plasma membrane's lipid bilayer. Current, carried by ions, can move across the membrane only through ion channels.

Graded potentials die out over short distances.

The passive current flow between active and adjacent inactive areas is similar to the means by which current is carried through electrical wires. We know from experience that current leaks out of an electrical wire with dangerous results unless the wire is covered with an insulating material such as plastic. (People can get an electric shock if they touch a bare wire.) Likewise, current is lost across the plasma membrane as charge-carrying ions in the form of K^+ leak out through the "uninsulated" parts of the membrane, that is, by diffusing outward down their electrochemical gradient through open channels. Because of this current loss, the magnitude of the local current—and thus the magnitude of the graded potential—progressively diminishes the farther it moves away from the initial active area (● Figure 4-3a). Another way of saying this is that the spread of a graded potential is *decremental* (gradually decreases) (● Figure 4-3b). Note that in this example, the magnitude of the initial change in potential is 15 mV (a change from the resting state of −70 to −55 mV); the change in potential decreases as it moves along the membrane to a potential of 10 mV (from −70 to −60 mV) and continues to diminish the farther it moves away from the initial active area until there is no longer a change in potential. In this way, local currents die out within a few millimeters from the initial site of change in potential and consequently can function as signals for only very short distances.

Although graded potentials have limited signaling distance, they are critically important to the body's function, as explained in later chapters. The following are all graded potentials: *postsynaptic potentials, receptor potentials, end-plate potentials, pacemaker potentials,* and *slow-wave potentials.* These terms are unfamiliar to you now, but you will become well acquainted with them as we continue discussing nerve and muscle physiology. We are including this list here because it is the only place all these types of graded potentials are listed together. For now it's enough to say that most excitable cells produce one of these types of graded potentials in response to a triggering event. In turn, graded potentials can initiate *action potentials,* the long-distance signals, in an excitable cell.

Action Potentials

Action potentials are brief, rapid, large (100 mV) changes in membrane potential during which the potential actually reverses, so that the inside of the excitable cell transiently becomes more positive than the outside. As with a graded potential, a single action potential involves only a small portion of the total excitable cell membrane. Unlike graded potentials, however, action potentials are conducted, or propagated, throughout the entire membrane in *nondecremental* fashion; that is, they do not diminish in strength as they travel from their site of initiation throughout the remainder of the cell membrane. Thus, action potentials can serve as faithful long-distance signals.

Think about the neuron that causes the muscle cells in your big toe to contract (see ● Figure 4-8). If you want to wiggle your big toe, commands are sent from your brain down your spinal cord to initiate an action potential at the beginning of this neuron, which is located in the spinal cord. The action potential travels all the way down the neuron's long axon, which runs through your leg to terminate on your big-toe muscle cells. The signal does not weaken or die off, being instead preserved at full strength from beginning to end.

Let us now consider the changes in potential during an action potential, and the permeability and ion movements responsible for generating this change in potential, before we turn our attention to the means by which action potentials spread throughout the cell membrane without diminishing.

During an action potential, the membrane potential rapidly, transiently reverses.

If a graded potential is large enough, it can initiate an action potential before the graded change dies off. (Later you will discover how this initiation is accomplished for the various types of graded potentials.) Typically, the region of the excitable membrane where graded potentials are produced in response to a triggering event does not undergo action potentials. Instead, passive current flow from the region where a graded potential is taking place depolarizes adjacent portions of the membrane where action potentials can occur.

Depolarization from the resting potential of −70 mV proceeds slowly at first, until it reaches a critical level known as **threshold potential,** typically between −50 and −55 mV (● Figure 4-4). At threshold potential, an explosive depolarization takes place. A recording of the potential at this time shows a sharp upward deflection as the potential rapidly reverses itself so that the inside of the cell becomes positive compared to the outside. Peak potential is usually + 30 to + 40 mV, depending on the excitable cell. Just as rapidly, the membrane repolarizes, dropping back to resting potential. Often the forces that repolarize the membrane push the potential too far, causing a brief **after hyperpolarization,** during which the inside of the mem-

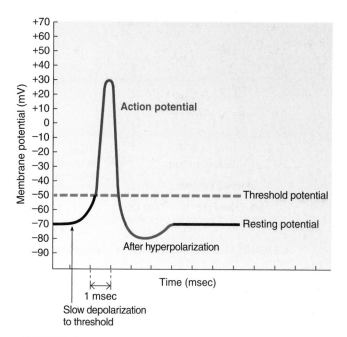

● **FIGURE 4-4 Changes in membrane potential during an action potential.**

brane briefly becomes even more negative than normal (for example, −80 mV) before the resting potential is restored.

The action potential is the entire rapid change in potential from threshold to peak and then back to resting. Unlike the variable duration of a graded potential, the duration of an action potential is always the same in a given excitable cell. In a neuron, an action potential lasts for only 1 msec (0.001 sec). It lasts longer in muscle, with the duration depending on the muscle type. Often an action potential is referred to as a **spike,** because of its spikelike recorded appearance. Alternatively, when an excitable membrane is triggered to undergo an action potential, it is said to **fire.** Thus, the terms *action potential, spike,* and *firing* all refer to the same phenomenon of rapid reversal of membrane potential. If the initial triggered depolarization does not reach threshold potential, no action potential takes place. Thus, threshold is a critical all-or-none point. Either the membrane is depolarized to threshold and an action potential takes place, or threshold is not reached in response to the depolarizing event and no action potential occurs.

Marked changes in membrane permeability and ion movement lead to an action potential.

How is the membrane potential, which is usually maintained at a constant resting level, altered to such an extent as to produce an action potential? Recall that K^+ makes the greatest contribution to the establishment of the resting potential, because the membrane at rest is considerably more permeable to K^+ than to Na^+ (see p. 28). During an action potential, marked changes in membrane permeability to Na^+ and K^+ take place, permitting rapid fluxes of these ions down their electrochemical gradients. These ion movements carry the current responsible for the potential changes that occur during an action potential. Action potentials take place as a result of the triggered opening and

VOLTAGE-GATED SODIUM CHANNEL

Na⁺

ECF

Plasma membrane

ICF

Activation gate

Inactivation gate

Na⁺

Na⁺

Rapid opening triggered at threshold

Na⁺

Slow closing triggered at threshold

(a) Closed but capable of opening

(b) Open (activated)

(c) Closed and not capable of opening (inactivated)

VOLTAGE-GATED POTASSIUM CHANNEL

Activation gate

K⁺

K⁺

Delayed opening triggered at threshold

K⁺

(d) Closed

(e) Open

● **FIGURE 4-5 Conformations of voltage-gated sodium and potassium channels.**

subsequent closing of two specific types of channels: voltage-gated Na⁺ channels and voltage-gated K⁺ channels.

VOLTAGE-GATED Na⁺ AND K⁺ CHANNELS Voltage-gated membrane channels consist of proteins that have a number of charged groups. The electrical field (potential) surrounding the channels can distort the channel structure as charged portions of the channel proteins are electrically attracted or repelled by charges in the fluids around the membrane. Unlike the majority of membrane proteins, which remain stable despite fluctuations in membrane potential, voltage-gated channel proteins are especially sensitive to voltage changes. Small distortions in shape induced by changes in potential can cause the channels to change their conformation. Here again is an example of how subtle changes in structure can profoundly influence function.

The voltage-gated Na⁺ channel has two gates: an *activation gate* and an *inactivation gate* (● Figure 4-5). The activation gate guards the channel interior by opening and closing like a sliding door. The inactivation gate consists of a ball-and-chain–like sequence of amino acids at the channel opening facing the ICF. This gate is open when the ball is dangling free on its chain and closed when the ball binds to the channel opening, thus blocking the opening. Both gates must be open to permit passage of Na⁺ through the channel, and closure of either gate prevents passage. This voltage-gated Na⁺ channel can exist in three different conformations: (1) *closed but capable of opening* (activation gate closed, inactivation gate open, ● Figure 4-5a); (2) *open,* or *activated* (both gates open, ● Figure 4-5b); and (3) *closed and not capable of opening,* or *inactivated* (activation gate open, inactivation gate closed, ● Figure 4-5c).

The voltage-gated K⁺ channel is simpler. It has only an activation gate, which can be either closed (● Figure 4-5d) or open (● Figure 4-5e). These voltage-gated Na⁺ and K⁺ channels exist in addition to the Na⁺–K⁺ pump and the leak channels for these ions (described in Chapter 3).

CHANGES IN PERMEABILITY AND ION MOVEMENT DURING AN ACTION POTENTIAL At resting potential (-70 mV), all the voltage-gated Na⁺ and K⁺ channels are closed, with the activation gates of the Na⁺ channels being closed and their inactivation gates being open; that is, the voltage-gated Na⁺ channels are in their "closed but capable of opening" conformation. Therefore, Na⁺ and K⁺ cannot pass through these voltage-gated channels at resting potential. However, because many K⁺ leak channels and very few Na⁺ leak channels are present, the resting membrane is 25 to 30 times more permeable to K⁺ than to Na⁺.

When current spreads passively from an adjacent site already depolarized into a new region at resting potential, the new region of membrane starts to depolarize toward threshold, causing the activation gates of some of its voltage-gated Na⁺ channels to open, so that both gates of these activated channels are now open. Because both the concentration and electrical gradients for Na⁺ favor its movement into the cell, Na⁺ starts to move in. The inward movement of positively charged Na⁺ depolarizes the membrane further, opening even more voltage-gated Na⁺ channels and allowing more Na⁺ to enter, and so on, in a positive-feedback cycle (● Figure 4-6).

At threshold potential, there is an explosive increase in Na⁺ permeability, which is symbolized as P_{Na^+}, as the membrane swiftly becomes about 600 times more permeable to Na⁺ than to K⁺. Each individual channel is either closed or open and cannot be partially open. However, the delicately poised gating mechanisms of the various voltage-gated Na⁺ channels are jolted open by slightly different voltage changes. During the early depolarizing phase, more and more Na⁺ channels open as the potential progressively decreases. At threshold, enough Na⁺ gates have opened to set off the positive-feedback cycle that rapidly causes the remaining Na⁺ gates to open. Now Na⁺ permeability dominates the membrane, in contrast to the K⁺ domination at resting potential. Thus, at threshold Na⁺ rushes into the cell, rapidly eliminating the internal negativity and even making the inside of the cell more positive than the outside in an attempt to drive the membrane potential to the Na⁺ equilibrium potential (which is $+60$ mV; see p. 30). The potential reaches approximately $+30$ mV, close to the Na⁺ equilibrium potential. The potential does not become any more positive, because, at the peak of the action potential, the Na⁺ channels start to close to the inactivated state, and P_{Na^+} starts to fall to its low resting value (● Figure 4-7).

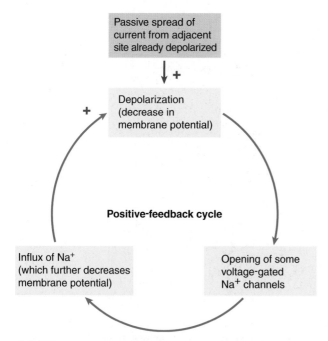

Passive spread of current from adjacent site already depolarized

↓ +

Depolarization (decrease in membrane potential)

+

Positive-feedback cycle

Influx of Na⁺ (which further decreases membrane potential)

Opening of some voltage-gated Na⁺ channels

● **FIGURE 4-6 Positive-feedback cycle responsible for opening Na⁺ channels at threshold.**

What causes the Na^+ channels to close? When the membrane potential reaches threshold, two closely related events take place in the gates of each Na^+ channel. First, the activation gates are triggered to *open rapidly* in response to the depolarization, converting the channel to its open (activated) conformation. Surprisingly, the conformational change that opens the channel also allows the inactivation gate's ball to bind to the channel opening, thereby physically blocking the mouth of the channel. However, this closure process takes time, so the inactivation gate *closes slowly* compared to the rapidity of channel opening (see ● Figure 4-5c). Meanwhile, during the 0.5 msec delay after the activation gate opens and before the inactivation gate closes, both gates are open and Na^+ rushes into the cell through these open channels, bringing the action potential to its peak. Then the inactivation gate closes, membrane permeability to Na^+ plummets to its low resting value, and further Na^+ entry is prevented. The channel remains in this inactivated conformation until the membrane potential has been restored to its resting value.

Simultaneous with inactivation of Na^+ channels, the voltage-gated K^+ channels start to slowly open at the peak of the action potential. Opening of the K^+ channel gate is a delayed voltage-gated response triggered by the initial depolarization to threshold (see ● Figures 4-5e and 4-7). Thus, three action potential-related events occur at threshold: (1) the rapid opening of the Na^+ activation gates, which permits Na^+ to enter, moving the potential from threshold to its positive peak; (2) the slow closing of the Na^+ inactivation gates, which halts further Na^+ entry after a brief time delay, thus keeping the potential from rising any further; and (3) the slow opening of the K^+ gates, which is in large part responsible for the potential plummeting from its peak back to resting.

The membrane potential would gradually return to resting after closure of the Na^+ channels as K^+ continued to leak out but no further Na^+ entered. However, the return to resting is hastened by the opening of K^+ gates at the peak of the action potential. Opening of the voltage-gated K^+ channels greatly increases K^+ permeability (designated P_{K^+}) to about 300 times the resting P_{Na^+}. This marked increase in P_{K^+} causes K^+ to rush out of the cell down its electrochemical gradient, carrying positive charges back to the outside. Note that at the peak of the action potential, the positive potential inside the cell tends to repel the positive K^+ ions, so the electrical gradient for K^+ is outward, unlike at resting potential. Of course the concentration gradient for K^+ is always outward. The outward movement of K^+ rapidly restores the negative resting potential.

To review (● Figure 4-7), *the rising phase of the action potential* (from threshold to +30 mV) *is due to Na^+ influx* (Na^+ entering the cell) induced by an explosive increase in P_{Na^+} at threshold. The *falling phase* (from +30 mV to resting potential) *is brought about largely by K^+ efflux* (K^+ leaving the cell) caused by the marked increase in P_{K^+} occurring simultaneously with the inactivation of the Na^+ channels at the peak of the action potential.

As the potential returns to resting, the changing voltage shifts the Na^+ channels to their "closed but capable of opening" conformation, with the activation gate closed and the inactivation gate open. Now the channel is reset, ready to respond to another triggering event. The newly opened voltage-gated K^+ channels also close, so the membrane returns to the resting number of open K^+ leak channels. Typically, the voltage-gated K^+ channels are slow to close. As a result of this persistent increased permeability to K^+, more K^+ may leave than is necessary to bring the potential to resting. This slight excessive K^+ efflux makes the interior of the cell transiently even more negative than resting potential, causing the after hyperpolarization.

The Na^+–K^+ pump gradually restores the concentration gradients disrupted by action potentials.

At the completion of an action potential, the membrane potential has been restored to its resting condition, but the ion distribution has been altered slightly. Sodium entered the cell during the rising phase, and a comparable amount of K^+ left during the falling phase. The Na^+–K^+ pump restores these ions to their original locations in the long run, but not after each action potential.

The active pumping process takes much longer to restore Na^+ and K^+ to their original locations than it takes for the passive fluxes of these ions during an action potential. However, the membrane does not need to wait until the concentration gradients are slowly restored before it can undergo another action potential. Actually, the movement of relatively few Na^+ and K^+ ions causes the large swings in membrane potential that occur during an action potential. Only about 1 out of 100,000 K^+ ions present in the cell leaves during an action potential, while a comparable number of Na^+ ions enter from the ECF. The movement of this extremely small proportion of the total Na^+ and K^+ during a single action potential produces dramatic 100 mV changes in potential (between −70 and +30 mV) but only infinitesimal changes in the ICF and ECF concentrations of these ions. Much more K^+ is still inside the cell than outside, and Na^+ is still predominantly an extracellular cation. Consequently, the Na^+ and K^+ concentration gradients still exist, so

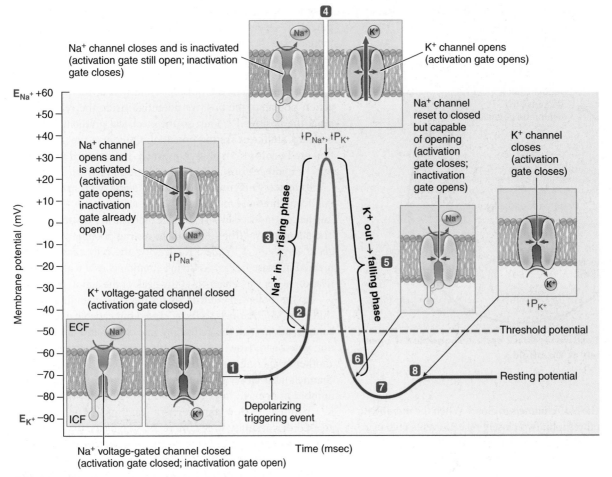

1 Resting potential: all voltage-gated channels closed.

2 At threshold, Na⁺ activation gate opens and P_{Na^+} rises.

3 Na⁺ enters cell, causing explosive depolarization to +30 mV, which generates rising phase of action potential.

4 At peak of action potential, Na⁺ inactivation gate closes and P_{Na^+} falls, ending net movement of Na⁺ into cell. At the same time, K⁺ activation gate opens and P_{K^+} rises.

5 K⁺ leaves cell, causing its repolarization to resting potential, which generates falling phase of action potential.

6 On return to resting potential, Na⁺ activation gate closes and inactivation gate opens, resetting channel to respond to another depolarizing triggering event.

7 Further outward movement of K⁺ through still-open K⁺ channel briefly hyperpolarizes membrane, which generates after hyperpolarization.

8 K⁺ activation gate closes, and membrane returns to resting potential.

● **FIGURE 4-7 Permeability changes and ion fluxes during an action potential.**

repeated action potentials can occur without the pump having to keep pace to restore the gradients.

Were it not for the pump, of course, even tiny fluxes accompanying repeated action potentials would eventually "run down" the concentration gradients so that further action potentials would be impossible. If the concentrations of Na⁺ and K⁺ were equal between the ECF and ICF, changes in permeability to these ions would not bring about ion fluxes, so no change in potential would occur. Thus, the Na⁺–K⁺ pump is critical to maintaining the concentration gradients in the long run. How-ever, it does not have to perform its role between action potentials, nor is it directly involved in the ion fluxes or potential changes that occur during an action potential.

Action potentials are propagated from the axon hillock to the axon terminals.

A single action potential involves only a small patch of the total surface membrane of an excitable cell. But if action potentials are to serve as long-distance signals, they cannot be merely

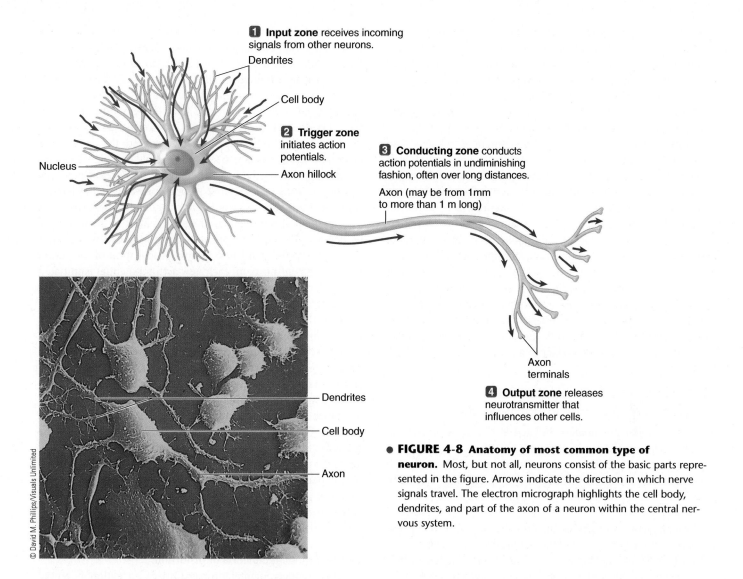

1 Input zone receives incoming signals from other neurons.

Dendrites

Cell body

2 Trigger zone initiates action potentials.

3 Conducting zone conducts action potentials in undiminishing fashion, often over long distances.

Nucleus

Axon hillock

Axon (may be from 1mm to more than 1 m long)

Axon terminals

4 Output zone releases neurotransmitter that influences other cells.

Dendrites

Cell body

Axon

© David M. Phillips/Visuals Unlimited

● **FIGURE 4-8 Anatomy of most common type of neuron.** Most, but not all, neurons consist of the basic parts represented in the figure. Arrows indicate the direction in which nerve signals travel. The electron micrograph highlights the cell body, dendrites, and part of the axon of a neuron within the central nervous system.

isolated events occurring in a limited area of a nerve or muscle cell membrane. Mechanisms must exist to conduct or spread the action potential throughout the entire cell membrane. Furthermore, the signal must be transmitted from one cell to the next (for example, along specific nerve pathways). To explain these mechanisms, we will first begin with a brief look at neuronal structure. Then we will examine how an action potential (nerve impulse) is conducted throughout a neuron before we turn our attention to how the signal is passed to another cell.

A single **neuron** typically consists of three basic parts—the *cell body,* the *dendrites,* and the *axon*—although there are variations in structure, depending on the location and function of the neuron. The nucleus and organelles are housed in the **cell body,** from which numerous extensions known as **dendrites** typically project like antennae to increase the surface area available for receiving signals from other neurons (● Figure 4-8). Some neurons have up to 400,000 dendrites. In most neurons the plasma membrane of the dendrites and cell body contains protein receptors that bind chemical messengers from other neurons. Therefore, the dendrites and cell body are the neuron's *input zone,* because these components receive and integrate *incoming* signals. This is the region where graded potentials are

produced in response to triggering events, in this case, incoming chemical messengers.

The **axon,** or **nerve fiber,** is a single, elongated, tubular extension that conducts action potentials *away from* the cell body and eventually terminates at other cells. The first portion of the axon plus the region of the cell body from which the axon leaves is known as the **axon hillock.** The axon hillock is the neuron's *trigger zone,* because it is the site where action potentials are triggered, or initiated, by the graded potential if it is of sufficient magnitude. The action potentials are then conducted along the axon from the axon hillock to the typically highly branched ending at the **axon terminals.** These terminals release chemical messengers that simultaneously influence numerous other cells with which they come into close association. Functionally, therefore, the axon is the *conducting zone* of the neuron, and the axon terminals constitute its *output zone.* (The major exceptions to this typical neuronal structure and functional organization are neurons specialized to carry sensory information, a topic described in a later chapter.)

Axons vary in length from less than a millimeter in neurons that communicate only with neighboring cells to longer than a meter in neurons that communicate with distant parts of the nervous system or with peripheral organs. For example, the axon of the neuron innervating your big toe must traverse the distance from the origin of its cell body within the spinal cord in the lower region of your back all the way down your leg to your toe.

Action potentials can be initiated only in portions of the membrane with abundant voltage-gated Na^+ channels that can be triggered to open by a depolarizing event. Typically, regions of excitable cells where graded potentials take place do not undergo action potentials because voltage-gated Na^+ channels are sparse there. Therefore, sites specialized for graded potentials do not undergo action potentials, even though they might be considerably depolarized. However, before dying out, graded potentials can trigger action potentials in adjacent portions of the membrane by bringing these more sensitive regions to threshold through local current flow spreading from the site of the graded potential. In a typical neuron, for example, graded potentials are generated in the dendrites and cell body in response to incoming chemical signals. If these graded potentials have sufficient magnitude by the time they have spread to the axon hillock, they initiate an action potential at this triggering zone.

Once initiated, action potentials are conducted throughout a nerve fiber.

Once an action potential is initiated at the axon hillock, no further triggering event is necessary to activate the remainder of the nerve fiber. The impulse is automatically conducted throughout the neuron without further stimulation by one of two methods of propagation: *contiguous conduction* or *saltatory conduction*.

Contiguous conduction involves the spread of the action potential along every patch of membrane down the length of the axon (*contiguous* means "touching" or "next to in sequence"). This process is illustrated in ● Figure 4-9, which represents a longitudinal section of the axon hillock and the portion of the axon immediately beyond it. The membrane at the axon hillock is at the peak of an action potential. The inside of the cell is positive in this active area, because Na^+ has already rushed in here. The remainder of the axon, still at resting potential and negative inside, is considered inactive. For the action potential to spread from the active to the inactive areas, the inactive areas must somehow be depolarized to threshold. This depolarization is accomplished by local current flow between the area already undergoing an action potential and the adjacent inactive area, similar to the current flow responsible for the spread of graded potentials. Because opposite charges attract, current can flow locally between the active area and the neighboring inactive area on both the inside and the outside of the membrane. This local current flow in effect neutralizes or eliminates some of the unbalanced charges in the inactive area; that is, it reduces the number of opposite charges separated across the membrane, reducing the potential in this area. This depolarizing effect quickly brings the involved inactive area to threshold, at which time the voltage-gated Na^+ channels in this region of the membrane are all thrown open, leading to an action potential in this previously inactive area. Meanwhile, the original active area returns to resting potential as a result of K^+ efflux.

Beyond the new active area is another inactive area, so the same thing happens again. This cycle repeats itself in a chain reaction until the action potential has spread to the end of the axon. *Once an action potential is initiated in one part of a neuron's cell membrane, a self-perpetuating cycle is initiated so that the action potential is propagated along the rest of the fiber automatically.* In this way, the axon is like a firecracker fuse that needs to be lit at only one end. Once ignited, the fire spreads down the fuse; it is not necessary to hold a match to every separate section of the fuse. Therefore a new action potential can be initiated in two ways, in both cases involving passive spread of current from an adjacent site already depolarized. An action potential is initiated in the axon hillock in the first place by depolarizing current spreading from a graded potential in the cell body and dendrites. During propagation of the action potential down the axon, each new action potential is initiated by depolarizing local current flow spreading from the preceding site undergoing an action potential.

Note that the original action potential does not travel along the membrane. Instead, it triggers an identical new action potential in the bordering area of the membrane, with this process being repeated along the axon's length. An analogy is the "wave" at a stadium. Each section of spectators stands up (the rising phase of an action potential), then sits down (the falling phase) in sequence one after another as the wave moves around the stadium. The wave, not individual spectators, travels around the stadium. Similarly, new action potentials arise sequentially down the axon. Each new action potential is a fresh local event that depends on induced permeability changes and electrochemical gradients that are virtually identical down the length of the axon. Therefore, the last action potential at the end of the axon is identical to the original one, no matter how long the axon is. In this way, action potentials can serve as long-distance signals without attenuation or distortion.

This nondecremental propagation of an action potential contrasts with the decremental spread of a graded potential, which dies out over a very short distance because it cannot regenerate itself. ▲ Table 4-1 summarizes the differences between graded potentials and action potentials, some of which are yet to be discussed.

The refractory period ensures one-way propagation of action potentials and limits their frequency.

What ensures the one-way propagation of an action potential away from the initial site of activation? Note from ● Figure 4-10 that once the action potential has been regenerated at a new neighboring site (now positive inside) and the original active area has returned to resting (once again negative inside), the close proximity of opposite charges between these two areas is conducive to local current flow in the backward direction, as well as in the forward direction into as yet unexcited portions of the membrane. If such backward current flow were able to bring the just inactivated area to threshold, another action potential would be initiated here, which would spread both forward and backward, initiating still other action potentials, and so on. But if action potentials were to move in both directions,

● FIGURE 4-9 Contiguous conduction. Local current flow between the active area at the peak of an action potential and the adjacent inactive area still at resting potential reduces the potential in this contiguous inactive area to threshold, which triggers an action potential in the previously inactive area. The original active area returns to resting potential, and the new active area induces an action potential in the next adjacent inactive area by local current flow as the cycle repeats itself down the length of the axon.

the situation would be chaotic, with numerous action potentials bouncing back and forth along the axon until the neuron eventually fatigued. Fortunately, neurons are saved from this fate of oscillating action potentials by the **refractory period,** during which a new action potential cannot be initiated by normal events in a region that has just undergone an action potential.

Because of the changing status of the voltage-gated Na^+ and K^+ channels during and after an action potential, the refractory period has two components: the *absolute refractory period* and the *relative refractory period* (● Figure 4-11). During the time that a particular patch of axonal membrane is undergoing an action potential, it cannot initiate another action potential, no

Comparison of Graded Potentials and Action Potentials

Property	Graded Potentials	Action Potentials
Triggering Events	Triggered by stimulus, by combination of neurotransmitter with receptor, or by inherent shifts in channel permeability	Triggered by depolarization to threshold, usually through passive spread of depolarization from adjacent area undergoing graded potential or action potential
Ion Movement Producing Change in Potential	Produced by net movement of Na^+, K^+, Cl^-, or Ca^{2+} across plasma membrane by various means	Produced by sequential movement of Na^+ into and K^+ out of cell through voltage-gated channels
Coding of Magnitude of Triggering Event	Graded potential change; magnitude varies with magnitude of triggering event	All-or-none membrane response; magnitude of triggering event coded in frequency rather than amplitude of action potentials
Duration	Varies with duration of triggering event	Constant
Magnitude of Potential Change with Distance from Initial Site	Decremental conduction; magnitude diminishes with distance from initial site	Propagated throughout membrane in undiminishing fashion; self-regenerated in neighboring inactive areas of membrane
Refractory Period	None	Relative, absolute
Summation	Temporal, spatial	None
Direction of Potential Change	Can be depolarization or hyperpolarization	Always depolarization and reversal of charges
Location	Occurs in specialized regions of membrane designed to respond to triggering event	Occurs in regions of membrane with abundance of voltage-gated Na^+ channels

Previous active area returned to resting potential

New active area at peak of action potential

New adjacent inactive area into which depolarization is spreading; will soon reach threshold

+ + + + + − − − − − − − + + + + + + + + + + + +

− − − − − + + + + + − − − − − − − − −

"Backward" current flow does not reexcite previously active area because this area is in its refractory period

"Forward" current flow excites new inactive area

Direction of propagation of action potential

● **FIGURE 4-10 Value of the refractory period.** The refractory period prevents "backward" current flow. During an action potential and slightly afterward, an area cannot be restimulated by normal events to undergo another action potential. Thus, the refractory period ensures that an action potential can be propagated only in the forward direction along the axon.

matter how strong the depolarizing triggering event is. This time period when a recently activated patch of membrane is completely refractory (meaning "stubborn," or unresponsive) to further stimulation is known as the **absolute refractory period.** Once the voltage-gated Na^+ channels have opened, they cannot open again in response to another depolarizing triggering event, no matter how strong, until resting potential is restored and the channels are reset to their original conformations. Accordingly, the absolute refractory period lasts the entire time from opening of the Na^+ channels' activation gates at threshold, through closure of their inactivation gates at the peak of the action potential, until the return to resting potential when the activation gates close and inactivation gates open once again; that is, until the channels are reset to their "closed but capable of opening" conformation. Only then can they respond to another depolarization with an explosive increase in P_{Na^+} to initiate another action potential. Because of the absolute refractory period, one action potential must be over before another can be initiated at the same site. Action potentials cannot overlap or be added one on top of another "piggyback fashion."

Following the absolute refractory period is a **relative refractory period,** during which a second action potential can be produced only by a triggering event considerably stronger than usual. The relative refractory period occurs after the action potential is completed because of a twofold effect: lingering inacti-

● FIGURE 4-11 Absolute and relative refractory periods.
During the absolute refractory period, the portion of the membrane that has just undergone an action potential cannot be restimulated. This period corresponds to the time during which the Na⁺ gates are not in their resting conformation. During the relative refractory period, the membrane can be restimulated only by a stronger stimulus than is usually necessary. This period corresponds to the time during which the K⁺ gates opened during the action potential have not yet closed, coupled with lingering inactivation of the voltage-gated Na⁺ channels.

vation of the voltage-gated Na^+ channels and slowness to close of the voltage-gated K^+ channels that opened at the peak of the action potential. During this time, fewer voltage-gated Na^+ channels than normal are in a position to be jolted open by a depolarizing triggering event. Simultaneously, K^+ is still leaving through its slow-to-close channels during the after hyperpolarization. The less-than-normal Na^+ entry in response to another triggering event is opposed by a persistent hyperpolarizing outward leak of K^+ through its not-yet-closed channels, and thus a greater depolarizing triggering event than normal is needed to bring the membrane to threshold during the relative refractory period.

By the time the original site has recovered from its refractory period and is capable of being restimulated by normal current flow, the action potential has been propagated in the forward direction only and is so far away that it can no longer influence the original site. Thus, *the refractory period ensures the one-way propagation of the action potential down the axon away from the initial site of activation.*

The refractory period is also responsible for setting an upper limit on the frequency of action potentials; that is, it determines the maximum number of new action potentials that can be initiated and propagated along a fiber in a given period of time. The original site must recover from its refractory period before a new action potential can be triggered to follow the preceding action potential. The length of the refractory period varies for different types of neurons. The longer the refractory period, the greater the delay before a new action potential can be initiated and the lower the frequency with which a neuron can respond to repeated or ongoing stimulation.

Action potentials occur in all-or-none fashion.

If any portion of the neuronal membrane is depolarized to threshold, an action potential is initiated and relayed along the membrane in undiminished fashion. Furthermore, once threshold has been reached, the resultant action potential always goes to maximal height. The reason for this effect is that the changes in voltage during an action potential result from ion movements down concentration and electrical gradients, and these gradients are not affected by the strength of the depolarizing triggering event. A triggering event stronger than is necessary to bring the membrane to threshold does not produce a larger action potential. However, a triggering event that fails to depolarize the membrane to threshold does not trigger an action potential at all. Thus, *an excitable membrane either responds to a triggering event with a maximal action potential that spreads nondecrementally throughout the membrane, or it does not respond with an action potential at all.* This property is called the **all-or-none law.**

The all-or-none concept is analogous to firing a gun. Either the trigger is not pulled sufficiently to fire the bullet (threshold is not reached), or it is pulled hard enough to elicit the full firing response of the gun (threshold is reached). Squeezing the trigger harder does not produce a greater explosion. Just as it is not possible to fire a gun halfway, it is not possible to cause a halfway action potential.

The threshold phenomenon allows some discrimination between important and unimportant stimuli or other triggering events. Stimuli too weak to bring the membrane to threshold do not initiate action potentials and therefore do not clutter up the nervous system by transmitting insignificant signals.

The strength of a stimulus is coded by the frequency of action potentials.

How is it possible to differentiate between two stimuli of varying strengths when both stimuli bring the membrane to threshold and generate action potentials of the same magnitude? For example, how can one distinguish between touching a warm object or touching a very hot object if both trigger identical action potentials in a nerve fiber relaying information about skin temperature to the central nervous system (CNS)? The answer lies in part on the *frequency* with which the action potentials are generated. A stronger stimulus does not produce a larger action potential, but it does trigger a greater *number* of action potentials per second. For an illustration, see ● Figure 10-36, p. 378, in which changes in blood pressure are coded by corresponding changes in the frequency of action potentials generated in the neurons monitoring blood pressure.

In addition, a stronger stimulus in a region causes more neurons to reach threshold, increasing the total information sent to the CNS. For example, lightly touch this page with your finger and note the area of skin in contact with the page. Now, press down more firmly and note that a larger surface area of skin is in contact with the page. Therefore, more neurons are brought to threshold with this stronger touch stimulus.

Once initiated, the velocity, or speed, with which an action potential travels down the axon depends on two factors: (1) whether the fiber is myelinated and (2) the diameter of the fiber. Contiguous conduction occurs in unmyelinated fibers. In

Nodes of Ranvier

Myelin sheath

←1 mm→

Myelin sheath

Axon

Plasma membrane

© C. Raines/Visuals Unlimited

Axon of neuron

(a) Myelinated fiber

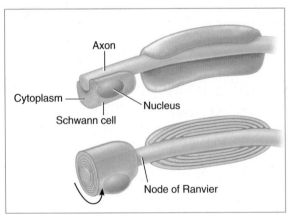

Axon

Cytoplasm

Nucleus

Schwann cell

Node of Ranvier

(b) Schwann cells in peripheral nervous system

Myelin sheath

Axon

Nodes of Ranvier

Oligodendrocyte

(c) Oligodendrocytes in central nervous system

● **FIGURE 4-12 Myelinated fibers.** (a) A myelinated fiber is surrounded by myelin at regular intervals. The intervening bare, unmyelinated regions are known as nodes of Ranvier. The electron micrograph shows a myelinated fiber in cross section at a myelinated region. (b) In the PNS each patch of myelin is formed by a separate Schwann cell that wraps itself jelly-roll fashion around the nerve fiber. (c) In the CNS each of the several processes ("arms") of a myelin-forming oligodendrocyte forms a patch of myelin around a separate nerve fiber.

this case, as you just learned, each individual action potential initiates an identical new action potential in the contiguous (bordering) segment of the axon membrane so that every portion of the membrane undergoes an action potential as this electrical signal is conducted from the beginning to the end of the axon. A faster method of propagation, *saltatory conduction,* takes place in myelinated fibers. Next, we compare myelinated fibers with unmyelinated fibers, and then see how saltatory conduction compares with contiguous conduction.

Myelination increases the speed of conduction of action potentials.

Myelinated fibers are axons covered with **myelin,** a thick layer composed primarily of lipids, at regular intervals along their length (● Figure 4-12a). Because the water-soluble ions responsible for carrying current across the membrane cannot permeate this myelin coating, it acts as an insulator, just like plastic around an electrical wire, to prevent leakage of current across the myelin-

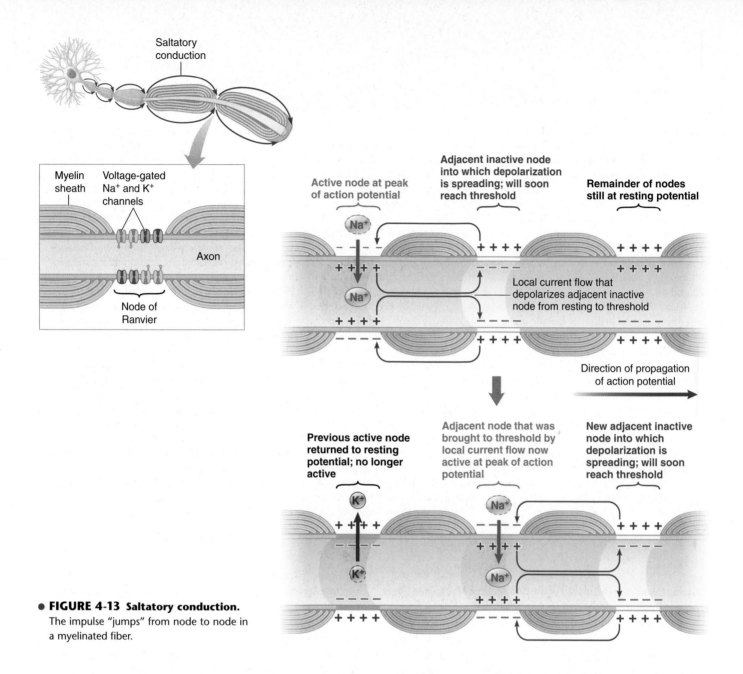

Saltatory
conduction

Myelin
sheath

Voltage-gated
Na⁺ and K⁺
channels

Axon

Node of
Ranvier

Active node at peak
of action potential

Adjacent inactive node
into which depolarization
is spreading; will soon
reach threshold

Remainder of nodes
still at resting potential

Na⁺

Na⁺

Local current flow that
depolarizes adjacent inactive
node from resting to threshold

Direction of propagation
of action potential

Previous active node
returned to resting
potential; no longer
active

Adjacent node that was
brought to threshold by
local current flow now
active at peak of action
potential

New adjacent inactive
node into which
depolarization is
spreading; will soon
reach threshold

K⁺

Na⁺

K⁺

Na⁺

● **FIGURE 4-13 Saltatory conduction.**
The impulse "jumps" from node to node in
a myelinated fiber.

ated portion of the membrane. Myelin is not actually a part of the neuron but consists of separate myelin-forming cells that wrap themselves around the axon in jelly-roll fashion. These myelin-forming cells are **Schwann cells** in the peripheral nervous system (PNS) (● Figure 4-12b), the nerves running between the CNS and the various regions of the body, and **oligodendrocytes** in the CNS (the brain and spinal cord) (● Figure 4-12c). Each patch of lipid-rich myelin consists of multiple layers of the myelin-forming cell's plasma membrane (predominantly the lipid bilayer) as the cell wraps itself around and around the axon. A patch of myelin might be made up of as many as 300 layers of wrapped lipid bilayers.

Between the myelinated regions, at the **nodes of Ranvier,** the axonal membrane is bare and exposed to the ECF. Current can flow across the membrane only at these bare spaces to produce action potentials. Voltage-gated Na⁺ channels are concentrated at the nodes, whereas the myelin-covered regions are almost devoid of these special passageways. By contrast, an unmyelinated fiber has a high density of voltage-gated Na⁺

channels along its entire length. As you now know, action potentials can be generated only at portions of the membrane furnished with an abundance of these channels.

The distance between the nodes is short enough that local current can flow between an active node and an adjacent inactive node before dying off. When an action potential occurs at one node, local current flow between this node and the oppositely charged adjacent node reduces the adjacent node's potential to threshold so that it undergoes an action potential, and so on. Consequently, in a myelinated fiber, the impulse "jumps" from node to node, skipping over the myelinated sections of the axon (● Figure 4-13); this process is called **saltatory conduction** (*saltare* means "to jump or leap"). Saltatory conduction propagates action potentials more rapidly than contiguous conduction does, because the action potential does not have to be regenerated at myelinated sections but must be regenerated within every section of an unmyelinated axonal membrane from beginning to end. In myelinated fibers, local current generated at an active node travels a longer distance, depo-

Multiple Sclerosis: Myelin—Going, Going, Gone

Multiple sclerosis (MS) is a pathophysiologic condition in which nerve fibers in various locations throughout the nervous system lose their myelin. MS is an autoimmune disease (*auto* means "self"; *immune* means "defense against"), in which the body's defense system erroneously attacks the myelin sheath surrounding myelinated nerve fibers. The condition afflicts about 1 in 1000 people in the United States. MS typically begins between the ages of 20 and 40.

Many investigators believe that MS arises from a combination of genetic and environmental factors. Relatives of those with MS have a 6 to 10 times greater chance of developing the disease themselves than the general population does. Because of their genetic predisposition, these relatives have increased susceptibility to environmental factors that may trigger the disease. Various environmental triggers have been proposed, including viral infections, environmental toxins, and vitamin D deficiency, but no evidence has been conclusive. Among the leading theories is that MS may be a result of an earlier infection with

a particular form of herpes virus, *HHV-6.* This virus causes *roseola,* a relatively mild babyhood condition characterized by a fever and rash. More than 90% of infants get roseola, after which HHV-6 may remain dormant in nerve fibers. In a recent study, researchers found that more than 70% of the MS patients studied showed evidence of active infection with HHV-6, which apparently became reactivated years later in these individuals. Interestingly, HHV-6 may act as a "molecular mimic" of myelin in those genetically predisposed; that is, the virus shares some common structural features with the person's myelin. This means that antibodies produced against this virus may mistakenly attack and inflame myelin instead.

Loss of myelin as a result of a misguided immune attack slows transmission of impulses in the affected neurons. A hardened scar known as a *sclerosis* (meaning "hard") forms at the multiple sites of myelin damage. These scars interfere with and can eventually block the propagation of action potentials in the underlying axons. Furthermore, the inflammatory phase char-

acterized by myelin destruction sets off a subsequent degenerative phase characterized by deterioration of the affected axons.

The symptoms of MS vary considerably, depending on the extent and location of myelin damage and axon degeneration. The most common symptoms include fatigue, visual problems, tingling and numbness, muscle weakness, impaired balance and coordination, and gradual paralysis. The early stage of the disease is often characterized by cycles of relapse and recovery, while the later chronic stage is marked by slow, progressive worsening of symptoms. MS is debilitating but not fatal.

Currently there is no cure and no really effective treatment for MS, although researchers are scrambling to find means to prevent, halt, or even reverse its debilitating symptoms. Among recent efforts are development of an experimental vaccine that calms the myelin-attacking immune cells, treatment with a drug originally designed to prevent immune rejection of transplanted organs, and strategies to promote remyelination.

larizing the next node instead of the next section. Myelinated fibers conduct impulses about 50 times faster than unmyelinated fibers of comparable size. You can think of myelinated fibers as the "superhighways" and unmyelinated fibers as the "back roads" of the nervous system when it comes to the speed with which information can be transmitted. Thus, the most urgent types of information are transmitted via myelinated fibers, whereas nerve pathways carrying less urgent information are unmyelinated.

Besides permitting action potentials to travel faster, myelination also conserves energy. Because the ion fluxes associated with action potentials are confined to the nodal regions, the energy-consuming Na^+–K^+ pump must restore fewer ions to their respective sides of the membrane following propagation of an action potential.

The accompanying boxed feature, ■ Concepts, Challenges, and Controversies, examines the myelin-destroying disease multiple sclerosis.

Fiber diameter also influences the velocity of action potential propagation.

Besides the effect of myelination, fiber diameter influences the speed with which an axon can conduct action potentials. The magnitude of current flow (that is, the amount of charge that

moves) depends not only on the difference in potential between two adjacent electrically charged regions but also on the resistance or hindrance to electrical charge movement between the two regions. When fiber diameter increases, the resistance to local current decreases. Thus, the larger the fiber diameter, the faster action potentials can be propagated.

Large myelinated fibers, such as those supplying skeletal muscles, can conduct action potentials at a velocity of up to 120 m/sec (268 mi/hr), compared with a conduction velocity of 0.7 m/sec (2 mi/hr) in small unmyelinated fibers such as those supplying the digestive tract. This difference in speed of propagation is related to the urgency of the information being conveyed. A signal to skeletal muscles to execute a particular movement (for example, to prevent you from falling as you trip on something) must be transmitted more rapidly than a signal to modify a slow-acting digestive process. Without myelination, axon diameters within urgent nerve pathways would have to be very large and cumbersome to achieve the necessary conduction velocities. Indeed, many invertebrates have large axons. In the course of vertebrate evolution, an efficient alternative to very large nerve fibers was development of the myelin sheath, which allows economic, rapid, long-distance signaling. For example, in humans the optic nerve leading from the eye to the brain is only 3 mm in diameter but is packed with more than a

Best Bets for Regenerating Nerve Fibers

Nerve fibers may be damaged by being severed or crushed (as during a traumatic event, such as a vehicle wreck, a gunshot wound, or a diving accident) or by being deprived of their blood supply (as during a stroke). When damaged, the affected axons can no longer conduct action potentials to convey messages. Using a severed axon as an example, the portion of the axon farthest from the cell body degenerates. Whether or not the lost portion of the axon regenerates depends on its location. Cut axons in the peripheral nervous system (PNS) can regenerate, whereas those in the central nervous system (CNS) cannot.

Regeneration of Peripheral Axons

In the case of a cut axon in a peripheral nerve, when the detached part of the axon degenerates, the surrounding Schwann cells phagocytize the debris. The Schwann cells themselves remain, then form a **regeneration tube** that guides the regenerating nerve fiber to its proper destination. The remaining portion of the axon connected to the cell body starts to grow and move forward within the Schwann cell column by amoeboid movement (see p. 48). The growing axon tip "sniffs" its way forward in the proper direction, guided by a chemical secreted into the regeneration tube by the Schwann cells. Successful fiber regeneration is responsible for the eventual return of sensation and movement after traumatic peripheral nerve injuries, although regeneration is not always successful.

Inhibited Regeneration of Central Axons

Fibers in the CNS, which are myelinated by oligodendrocytes, do not have this regenerative ability. Actually, the axons themselves have the ability to regenerate, but the oligodendrocytes surrounding them synthesize proteins that inhibit axonal growth, in sharp contrast to the nerve growth–promoting action of the Schwann cells that myelinate pe-

ripheral axons. Nerve growth in the brain and spinal cord is controlled by a delicate balance between *nerve growth–enhancing* and *nerve growth–inhibiting proteins*. During fetal development, nerve growth in the CNS is possible as the brain and spinal cord are being formed. Researchers speculate that nerve growth inhibitors, which are produced late in fetal development in the myelin sheaths surrounding central nerve fibers, may normally serve as "guardrails" that keep new nerve endings from straying outside their proper paths. The growth-inhibiting action of oligodendrocytes may thus serve to stabilize the enormously complex structure of the CNS.

Growth inhibition is a disadvantage, however, when CNS axons need to be mended, as when the spinal cord has been severed accidentally. Damaged central fibers show immediate signs of repairing themselves after an injury, but within several weeks they start to degenerate, and scar tissue forms at the site of injury, halting any recovery. Therefore, damaged neuronal fibers in the brain and spinal cord never regenerate.

Research on Regeneration of Central Axons

In the future, however, it may be possible to promote significant regeneration of damaged fibers in the CNS. Investigators are exploring several promising ways of spurring repair of central axonal pathways, with the goal of enabling victims of spinal-cord injuries to walk again. Here are some current lines of research:

- Scientists have been able to induce significant nerve regeneration in rats with severed spinal cords by *chemically blocking the nerve growth inhibitors*, thereby allowing the nerve growth enhancers to promote abundant sprouting of new nerve fibers at the site of injury. One of the nerve growth inhibitors, dubbed

Nogo, was recently identified. Now investigators are trying to encourage axon regrowth in experimental animals with spinal cord injuries by using an antibody to Nogo.

- Other researchers are experimentally *using peripheral nerve grafts* to bridge the defect at an injured spinal-cord site. These grafts contain the nurturing Schwann cells, which release nerve growth–enhancing proteins. Alternatively, some groups are using *biological scaffolding* (specifically engineered fibers made up of amino-acid chains) to bridge the gap and promote neuron growth.
- Another promising avenue under study involves *transplanting olfactory ensheathing glia* into the damaged site. Olfactory neurons, the cells that carry information about odors to the brain, are replaced on a regular basis, unlike most neurons. The growing axons of these newly generated neurons enter the brain and form functional connections with appropriate neurons in the brain. This capability is promoted by the special olfactory ensheathing glia, which wrap around and myelinate the olfactory axons. Early experimental evidence suggests that transplants of these special myelin-forming cells might help induce axonal regeneration in the CNS.
- Another avenue of hope is the discovery of *neural stem cells* (see pp. 8 and 138). These cells might someday be implanted into a damaged spinal cord and coaxed into multiplying and differentiating into mature, functional neurons that will replace those lost.
- Yet another new strategy under investigation is *enzymatically breaking down inhibitory components in the scar tissue* that naturally forms at the injured site and prevents sprouting nerve fibers from crossing this barrier.

---

million myelinated axons. If those axons were unmyelinated, each would have to be about 100 times thicker to conduct impulses at the same velocity, resulting in an optic nerve about 300 mm (12 inches) in diameter.

The presence of myelinating cells can be either a tremendous benefit or a tremendous detriment when an axon is cut, depending on whether the damage occurs in a peripheral

nerve or in the CNS. See the accompanying boxed feature, ■ Concepts, Challenges, and Controversies, to learn more about the regeneration of damaged nerve fibers, a matter of crucial importance in spinal-cord injuries or other trauma affecting nerves.

You have now seen how an action potential is propagated along the axon and learned about the factors influencing the

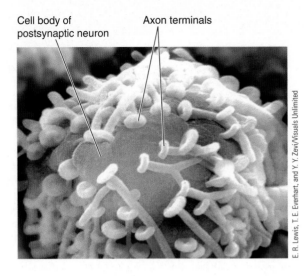

Dendrites   Synaptic inputs
(presynaptic axon terminals)   See Figure 4-15

Cell body of   Axon terminals
postsynaptic neuron

Cell body of   Axon   Myelinated
postsynaptic neuron   hillock   axon

E. R. Lewis, T. E. Everhart, and Y. Y. Zevi/Visuals Unlimited

● **FIGURE 4-14 Synaptic inputs (presynaptic axon terminals) to the cell body and dendrites of a single postsynaptic neuron.** The drying process used to prepare the neuron for the electron micrograph has toppled the presynaptic axon terminals and pulled them away from the postsynaptic cell body.

speed of this propagation. But what happens when an action potential reaches the end of the axon?

Synapses and Neuronal Integration

A neuron may terminate on one of three structures: a *muscle*, a *gland*, or *another neuron*. Therefore, depending on where a neuron terminates, it can cause a muscle cell to contract, a gland cell to secrete, another neuron to convey an electrical message along a nerve pathway, or some other function. When a neuron terminates on a muscle or a gland, the neuron is said to **innervate,** or supply, the structure. The junctions between nerves and the muscles and glands that they innervate will be described later. For now we will concentrate on the junction between two neurons—a **synapse** (*synapsis* means "juncture"). (Sometimes the term *synapse* is used to describe a junction between any two excitable cells, but we will reserve this term for the junction between two neurons.)

Synapses are typically junctions between presynaptic and postsynaptic neurons.

There are two types of synapses: *electrical synapses* and *chemical synapses*, depending on how information is transferred between the two neurons.

ELECTRICAL SYNAPSES In an **electrical synapse,** two neurons are connected by gap junctions (see p. 9), which allow charge-carrying ions to flow directly between the two cells in either direction. Although electrical synapses lead to unbroken transmission of electrical signals and are extremely rapid, this type of connection is essentially "on" or "off" and is unregulated. An action potential in one neuron always leads to an action potential in the connected neuron. Electrical synapses are relatively

rare in the human nervous system. They have been identified in the CNS, where they synchronize electrical activity in groups of neurons interconnected by gap junctions, and in specialized locations, such as the pulp of a tooth and the retina of the eye. Gap junctions are more numerous in smooth muscle and cardiac muscle, where their function is better understood.

CHEMICAL SYNAPSES The vast majority of synapses in the human nervous system are **chemical synapses** at which a chemical messenger transmits information one way across a space separating the two neurons. A chemical synapse involves a junction between an axon terminal of one neuron, known as the *presynaptic neuron,* and the dendrites or cell body of a second neuron, known as the *postsynaptic neuron.* (*Pre* means "before," and *post* means "after"; the presynaptic neuron lies before the synapse, and the postsynaptic neuron lies after the synapse.) The dendrites and, to a lesser extent, the cell body of most neurons receive thousands of synaptic inputs, which are axon terminals from many other neurons. Some neurons in the CNS receive as many as 100,000 synaptic inputs (● Figure 4-14).

The anatomy of one of these thousands of chemical synapses is shown in ● Figure 4-15. The axon terminal of the **presynaptic neuron,** which conducts its action potentials *toward* the synapse, ends in a slight swelling, the **synaptic knob.** The synaptic knob contains **synaptic vesicles,** which store a specific chemical messenger, a **neurotransmitter** that has been synthesized and packaged by the presynaptic neuron. The synaptic knob comes into close proximity to, but does not actually touch, the **postsynaptic neuron,** the neuron whose action potentials are propagated *away* from the synapse. The space between the presynaptic and postsynaptic neurons is called the **synaptic cleft.** Recently discovered fingerlike CAMs extend partway across the synaptic cleft from the surfaces of both the presynaptic and postsynaptic neurons. These projections are "Velcroed"

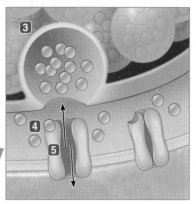

1 Action potential reaches axon terminal of presynaptic neuron.

2 Ca²⁺ enters synaptic knob (presynaptic axon terminal).

3 Neurotransmitter is released by exocytosis into synaptic cleft.

4 Neurotransmitter binds to receptors that are an integral part of chemically gated channels on subsynaptic membrane of postsynaptic neuron.

5 Binding of neurotransmitter to receptor opens that specific channel.

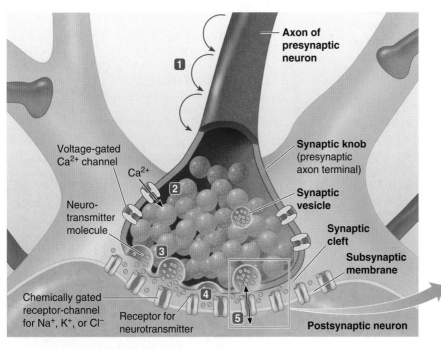

Axon of presynaptic neuron

Voltage-gated Ca²⁺ channel

Ca²⁺

Neurotransmitter molecule

Chemically gated receptor-channel for Na⁺, K⁺, or Cl⁻

Receptor for neurotransmitter

Synaptic knob (presynaptic axon terminal)

Synaptic vesicle

Synaptic cleft

Subsynaptic membrane

Postsynaptic neuron

© Dennis Kunkel/Visuals Unlimited

Axon terminal of presynaptic neuron

Dendrite of postsynaptic neuron

Synaptic vesicles   Synaptic cleft

● **FIGURE 4-15 Structure and function of a single synapse.** The numbered steps designate the sequence of events that take place at a synapse. The blowup depicts the release by exocytosis of neurotransmitter from the presynaptic axon terminal and its subsequent binding with receptors specific for it on the subsynaptic membrane of the postsynaptic neuron. The electron micrograph shows a synapse between a presynaptic axon terminal and a dendrite of a postsynaptic cell.

together where they meet and overlap in the middle of the cleft, much as if you interlock the extended fingers from both hands together. This physical tethering stabilizes the close proximity of the pre- and postsynaptic neurons at the synapse.

Current does not spread directly from the presynaptic to the postsynaptic neuron at a chemical synapse. Instead, an action potential in the presynaptic neuron alters the postsynaptic neuron's potential by chemical means. Synapses operate in one direction only; that is, the presynaptic neuron brings about changes in the membrane potential of the postsynaptic neuron, but the postsynaptic neuron does not directly influence the potential of the presynaptic neuron. The reason for this becomes readily apparent when you examine the events that occur at a synapse.

A neurotransmitter carries the signal across a synapse.

When an action potential in a presynaptic neuron has been propagated to the axon terminal (● Figure 4-15, step 1), this local change in potential triggers the opening of voltage-gated

Ca²⁺ channels in the synaptic knob. Because Ca²⁺ is much more highly concentrated in the ECF (see p. 70), this ion flows into the synaptic knob through the open channels (step 2). Ca²⁺ promotes the release of a neurotransmitter from some of the synaptic vesicles into the synaptic cleft (step 3). The release is accomplished by exocytosis (see p. 28). The released neurotransmitter diffuses across the cleft and binds with specific protein receptors on the **subsynaptic membrane,** the portion of the postsynaptic membrane immediately underlying the synaptic knob (*sub* means "under") (step 4). These receptors are an integral part of specific ion channels. Binding of neurotransmitter to the receptor-channels causes the channels to open, changing the ion permeability of the postsynaptic neuron (step 5). These are chemically gated channels, in contrast to the voltage-gated channels responsible for the action potential and for Ca²⁺ influx into the synaptic knob. Because the presynaptic terminal releases the neurotransmitter and the subsynaptic membrane of the postsynaptic neuron has receptor-channels for the neurotransmitter, the synapse can operate only in the direction from presynaptic to postsynaptic neuron.

Conversion of the electrical signal in the presynaptic neuron (an action potential) to an electrical signal in the postsynaptic neuron by chemical means (via the neurotransmitter–receptor combination) takes time. This **synaptic delay** is usually about 0.5 to 1 msec. In a neural pathway, chains of neurons often must be traversed. The more complex the pathway is, the more synaptic delays and the longer the *total reaction time* (the time required to respond to a particular event).

Each presynaptic neuron typically releases only one neurotransmitter; however, different neurons vary in the neurotransmitter they release. On binding with their subsynaptic receptor-channels, different neurotransmitters cause different ion permeability changes. There are two types of synapses, depending on the permeability changes induced in the postsynaptic neuron by the combination of a specific neurotransmitter with its receptor-channels: *excitatory synapses* and *inhibitory synapses*.

EXCITATORY SYNAPSES At an **excitatory synapse,** the response to the binding of a neurotransmitter to the receptor-channel is the opening of nonspecific cation channels in the subsynaptic membrane that permit simultaneous passage of Na^+ and K^+ through them. (These are a different type of channel from those you have encountered before.) Thus, permeability to both these ions is increased at the same time. How much of each ion diffuses through an open cation channel depends on their electrochemical gradients. At resting potential, both the concentration and electrical gradients for Na^+ favor its movement into the postsynaptic neuron, whereas only the concentration gradient for K^+ favors its movement outward. Therefore, the permeability change induced at an excitatory synapse results in the movement of a few K^+ ions out of the postsynaptic neuron, while a larger number of Na^+ ions simultaneously enter this neuron. The result is net movement of positive ions into the cell. This makes the inside of the membrane slightly less negative than at resting potential, thus producing a *small depolarization* of the postsynaptic neuron.

Activation of one excitatory synapse can rarely depolarize the postsynaptic neuron enough to bring it to threshold. Too few channels are involved at a single subsynaptic membrane to permit adequate ion flow to reduce the potential to threshold. This small depolarization, however, does bring the membrane of the postsynaptic neuron closer to threshold, increasing the likelihood that threshold will be reached (in response to further excitatory input) and that an action potential will occur. That is, the membrane is now more excitable (easier to bring to threshold) than when at rest. Accordingly, the change in postsynaptic potential occurring at an excitatory synapse is called an **excitatory postsynaptic potential,** or EPSP (● Figure 4-16a).

INHIBITORY SYNAPSES At an **inhibitory synapse,** the binding of a different released neurotransmitter with its receptor-channels increases the permeability of the subsynaptic membrane to either K^+ or Cl^-. The resulting ion movements typically bring about a *small hyperpolarization* of the postsynaptic neuron—that is, the inside of the neuron becomes slightly more negative. In the case of increased P_{K^+}, more positive charges leave the cell via K^+ efflux, leaving more negative charges behind on the inside. In

● **FIGURE 4-16 Postsynaptic potentials.** (a) An excitatory postsynaptic potential (EPSP) brought about by activation of an excitatory presynaptic input brings the postsynaptic neuron closer to threshold potential. (b) An inhibitory postsynaptic potential (IPSP) brought about by activation of an inhibitory presynaptic input moves the postsynaptic neuron farther from threshold potential.

the case of increased P_{Cl^-}, because the concentration of Cl^- is higher outside the cell, more negative charges enter the cell in the form of Cl^- ions than are driven out by the opposing electrical gradient established by the resting membrane potential (see p. 31). In either case, this small hyperpolarization moves the membrane potential even farther away from threshold (● Figure 4-16b), lessening the likelihood that the postsynaptic neuron will reach threshold and undergo an action potential. That is, the membrane is now less excitable (harder to bring to threshold by excitatory input) than when it is at resting potential. The membrane is said to be inhibited under these circumstances, and the small hyperpolarization of the postsynaptic cell is called an **inhibitory postsynaptic potential,** or IPSP.

In cells where the equilibrium potential for Cl^- exactly equals the resting potential, an increased P_{Cl^-} does not result in a hyperpolarization because there is no driving force to produce Cl^- movement. Opening of Cl^- channels in these cells tends to hold the membrane at resting potential, reducing the likelihood that threshold will be reached.

Note that EPSPs and IPSPs are produced by opening of chemically gated channels, unlike action potentials, which are produced by opening of voltage-gated channels.

Each neurotransmitter–receptor combination always produces the same response.

Many different chemicals serve as neurotransmitters (▲ Table 4-2). Even though neurotransmitters vary from synapse to synapse, the same neurotransmitter is always released at a particular synapse. Furthermore, at a given synapse, binding of a neurotransmitter with its appropriate subsynaptic receptor-channels always leads to the same change in permeability and resultant change in potential of the postsynaptic membrane. That is, the response to a given neurotransmitter–receptor combination is always the same; the combination does not generate an EPSP under one circumstance and an IPSP under another. Some neurotransmitters (for example, *glutamate,* the most common excitatory neurotransmitter in the brain) typically bring about EPSPs, whereas others (for example, *gamma-aminobutyric acid,* or *GABA,* the brain's main inhibitory neurotransmitter) always cause IPSPs. Still other neurotransmitters (for example, *norepinephrine*) are quite variable, producing EPSPs at one synapse and IPSPs at a different synapse; that is, different permeability changes in the postsynaptic neuron can occur in response to the binding of the same neurotransmitter to the subsynaptic receptor-channels of different postsynaptic neurons.

Most of the time, each axon terminal releases only one neurotransmitter. Recent evidence suggests, however, that in some cases two different neurotransmitters can be released simultaneously from a single axon terminal. For example, *glycine* and GABA, both of which produce inhibitory responses, can be packaged and released from the same synaptic vesicles. Scientists speculate that the fast-acting glycine and more slowly acting GABA may complement each other in the control of activities that depend on precise timing—for example, coordination of complex movements.

Neurotransmitters are quickly removed from the synaptic cleft.

As long as the neurotransmitter remains bound to the receptor-channels, the alteration in membrane permeability responsible for the EPSP or IPSP continues. For the postsynaptic neuron to be ready to receive additional messages from the same or other presynaptic inputs, the neurotransmitter must be inactivated or removed from the postsynaptic cleft after it has produced the appropriate response in the postsynaptic neuron; that is, the postsynaptic "slate" must be "wiped clean." Thus, after combining with the postsynaptic receptor-channel, chemical transmitters are removed and the response is terminated.

Several mechanisms can remove the neurotransmitter: It may diffuse away from the synaptic cleft, be inactivated by specific enzymes within the subsynaptic membrane, or be actively taken back up into the axon terminal by transport mechanisms in the presynaptic membrane. Once the neurotransmitter is taken back up, it can be stored and released another time (recycled) in response to a subsequent action potential or de-

▲ TABLE 4-2	Some Common Neurotransmitters
Acetylcholine	Histamine
Dopamine	Glycine
Norepinephrine	Glutamate
Epinephrine	Aspartate
Serotonin	Gamma-aminobutyric acid (GABA)

stroyed by enzymes within the synaptic knob. The method employed depends on the particular synapse.

Some drugs work by interfering with removal of specific neurotransmitters from synapses. For example, **selective serotonin reuptake inhibitors (SSRIs),** as their name implies, selectively block the reuptake of *serotonin* into presynaptic axon terminals, thereby prolonging the action of this neurotransmitter at synapses that use this messenger. SSRIs, such as *Prozac* and *Paxil,* are prescribed to treat depression, which is characterized by a deficiency of serotonin, among other things. Serotonin is involved in neural pathways that regulate mood and behavior.

The grand postsynaptic potential depends on the sum of the activities of all presynaptic inputs.

EPSPs and IPSPs are graded potentials. Unlike action potentials, which behave according to the all-or-none law, graded potentials can be of varying magnitude, have no refractory period, and can be summed (added on top of one another). What are the mechanisms and significance of summation?

The events that occur at a single synapse result in either an EPSP or an IPSP at the postsynaptic neuron. But if a single EPSP is inadequate to bring the postsynaptic neuron to threshold and an IPSP moves it even farther from threshold, how can an action potential be initiated in the postsynaptic neuron? The answer lies in the thousands of presynaptic inputs that a typical neuronal cell body receives from many other neurons. Some of these presynaptic inputs may be carrying sensory information from the environment; some may be signaling internal changes in homeostatic balance; others may be transmitting signals from control centers in the brain; and still others may arrive carrying other bits of information. At any given time, any number of these presynaptic neurons (probably hundreds) may be firing and thus influencing the postsynaptic neuron's level of activity. The total potential in the postsynaptic neuron, the **grand postsynaptic potential (GPSP),** is a composite of all EPSPs and IPSPs occurring at approximately the same time.

The postsynaptic neuron can be brought to threshold by either *temporal summation* or *spatial summation.* To illustrate these methods of summation, we will examine the possible interactions of three presynaptic inputs—two excitatory inputs (Ex1 and Ex2) and one inhibitory input (In1)—on a hypothetical postsynaptic neuron (● Figure 4-17). The recording shown in the figure represents the potential in the postsynaptic cell. Bear in mind during our discussion of this simplified version that many thousands of synapses are actually interacting in the same way on a single cell body and its dendrites.

TEMPORAL SUMMATION Suppose that Ex1 has an action potential that causes an EPSP in the postsynaptic neuron. After this EPSP has died off, if another action potential occurs in Ex1, an EPSP of the same magnitude takes place before dying off (● Figure 4-17a). Next, suppose that Ex1 has two action potentials in close succession (● Figure 4-17b). The first action potential in Ex1 produces an EPSP in the postsynaptic membrane. While the postsynaptic membrane is still partially depolarized from this first EPSP, the second action potential in Ex1 produces a second EPSP. Because graded potentials do not have a refractory period, the second EPSP can add to the first, bringing the membrane to threshold and initiating an action potential in the postsynaptic neuron. EPSPs can add together or sum because an EPSP lasts longer than the action potential that caused it. The presynaptic neuron (Ex1) can recover from its refractory period following the first action potential and have a second action potential, causing a second EPSP in the postsynaptic neuron, before the first EPSP is finished.

The summing of several EPSPs occurring very close together in time because of successive firing of a single presynaptic neuron is known as **temporal summation** (*tempus* means "time"). In reality, up to 50 EPSPs might be needed to bring the postsynaptic membrane to threshold. Each action potential in a single presynaptic neuron triggers the emptying of a certain number of synaptic vesicles. The amount of neurotransmitter released and the resultant magnitude of the change in postsynaptic potential are thus directly related to the frequency of presynaptic action potentials. One way in which the postsynaptic membrane can be brought to threshold, then, is through rapid, repetitive excitation from a single persistent input.

SPATIAL SUMMATION Let us now see what happens in the postsynaptic neuron if both excitatory inputs are stimulated simultaneously (● Figure 4-17c). An action potential in either Ex1 or Ex2 will produce an EPSP in the postsynaptic neuron; however, neither of these alone brings the membrane to threshold to elicit a postsynaptic action potential. But simultaneous action potentials in Ex1 and Ex2 produce EPSPs that add to each other, bringing the postsynaptic membrane to threshold, so an action potential does occur. The summation of EPSPs originating simultaneously from several different presynaptic inputs (that is, from different points in "space") is known as **spatial summation.** A second way to elicit an action potential in a postsynaptic cell, therefore, is through concurrent activation of several excitatory inputs. Again, in reality up to 50 simultaneous EPSPs are required to bring the postsynaptic membrane to threshold.

Similarly, IPSPs can undergo both temporal and spatial summation. As IPSPs add together, however, they progressively move the potential farther from threshold.

CANCELLATION OF CONCURRENT EPSPS AND IPSPS If an excitatory and an inhibitory input are simultaneously activated, the concurrent EPSP and IPSP more or less cancel each other out. The extent of cancellation depends on their respective magnitudes. In most cases, the postsynaptic membrane potential remains close to resting (● Figure 4-17d).

IMPORTANCE OF POSTSYNAPTIC NEURONAL INTEGRATION The magnitude of the GPSP depends on the sum of activity in all the presynaptic inputs and, in turn, determines whether or not the postsynaptic neuron will undergo an action potential to pass information on to the cells on which the neuron terminates. The following oversimplified real-life example demonstrates the benefits of this neuronal integration. The explanation is not completely accurate technically, but the principles of summation are accurate.

Assume for simplicity's sake that urination is controlled by a postsynaptic neuron that innervates the urinary bladder. When this neuron fires, the bladder contracts. (Actually, voluntary control of urination is accomplished by postsynaptic integration at the neuron controlling the external urethral sphincter rather than the bladder itself.) As the bladder fills with urine and becomes stretched, a reflex is initiated that ultimately produces EPSPs in the postsynaptic neuron responsible for causing bladder contraction. Partial filling of the bladder does not cause enough excitation to bring the neuron to threshold, so urination does not take place. That is, action potentials do not occur frequently enough in presynaptic neuron Ex1, which fires reflexly in response to the degree of bladder stretching, to generate EPSPs close enough together in the postsynaptic neuron to bring the latter to threshold (● Figure 4-17a). As the bladder progressively fills, the frequency of action potentials progressively increases in presynaptic neuron Ex1, leading to more rapid formation of EPSPs in the postsynaptic neuron. Thus the frequency of EPSP formation arising from Ex1 activity signals the postsynaptic neuron of the extent of bladder filling. When the bladder becomes sufficiently stretched that the Ex1-generated EPSPs are temporally summed to threshold, the postsynaptic neuron undergoes an action potential that stimulates bladder contraction (● Figure 4-17b).

What if the time is inopportune for urination to take place? Presynaptic inputs originating in higher levels of the brain responsible for voluntary control can produce IPSPs at the bladder postsynaptic neuron (In1 in ● Figure 4-17d). These "voluntary" IPSPs in effect cancel out the "reflex" EPSPs triggered by stretching of the bladder. Thus, the postsynaptic neuron remains at resting potential and does not have an action potential, so the bladder is prevented from contracting and emptying even though it is full.

What if a person's bladder is only partially filled so that the presynaptic input from this source (Ex1) is insufficient to bring the postsynaptic neuron to threshold to cause bladder contraction, yet he or she needs to supply a urine specimen for laboratory analysis? The person can voluntarily activate another excitatory presynaptic neuron originating in higher brain levels (Ex2 in ● Figure 4-17c). The "voluntary" EPSPs arising from Ex2 activity and the "reflex" EPSPs arising from Ex1 activity are spatially summed to bring the postsynaptic neuron to thresh-

(a) If an excitatory presynaptic input (Ex1) is stimulated a second time after the first EPSP in the postsynaptic cell has died off, a second EPSP of the same magnitude will occur.

(b) If, however, Ex1 is stimulated a second time before the first EPSP has died off, the second EPSP will add onto, or sum with, the first EPSP, resulting in *temporal summation*, which may bring the postsynaptic cell to threshold.

(c) The postsynaptic cell may also be brought to threshold by *spatial summation* of EPSPs that are initiated by simultaneous activation of two (Ex1 and Ex2) or more excitatory presynaptic inputs.

(d) Simultaneous activation of an excitatory (Ex1) and inhibitory (In1) presynaptic input does not change the postsynaptic potential, because the resultant EPSP and IPSP cancel each other out.

● **FIGURE 4-17 Determination of the grand postsynaptic potential by the sum of activity in the presynaptic inputs.** Two excitatory (Ex1 and Ex2) and one inhibitory (In1) presynaptic inputs terminate on this hypothetical postsynaptic neuron. The potential of the postsynaptic neuron is being recorded. For simplicity in the figure, summation of two EPSPs brings the postsynaptic neuron to threshold, but in reality many EPSPs must sum to reach threshold.

old. This achieves the action potential necessary to stimulate bladder contraction, even though the bladder is not full.

This example illustrates the importance of postsynaptic neuronal integration. Each postsynaptic neuron in a sense "computes" all the input it receives and makes a "decision" about whether to pass the information on (that is, whether threshold is reached and an action potential is transmitted down the axon). In this way, neurons serve as complex computational devices, or integrators. The dendrites function as the primary processors of incoming information. They receive and tally the signals from all the presynaptic neurons. Each neuron's output in the form of frequency of action potentials to other cells (muscle cells, gland cells, or other neurons) reflects the balance of activity in the inputs it receives via EPSPs or IPSPs from the thousands of other neurons that terminate on it. Each postsynaptic neuron filters out information that is not significant enough to bring it to threshold and does not pass it on. If every action potential in every presynaptic neuron that impinges on a particular postsynaptic neuron were to cause an action potential in the postsynaptic neuron, the neuronal pathways would be overwhelmed with trivia. Only if an excitatory presynaptic signal is reinforced by other supporting signals through summation will the informa-

tion be passed on. Furthermore, interaction of EPSPs and IPSPs provides a way for one set of signals to offset another, allowing a fine degree of discrimination and control in determining what information will be passed on. Thus, unlike an electrical synapse, a chemical synapse is more than a simple on-off switch because many factors can influence the generation of a new action potential in the postsynaptic cell. Whether or not the postsynaptic neuron has an action potential or not depends on the relative balance of information coming in via presynaptic neurons at all of its excitatory and inhibitory synapses.

Let us now see why action potentials are initiated at the axon hillock.

Action potentials are initiated at the axon hillock because it has the lowest threshold.

Threshold potential is not uniform throughout the postsynaptic neuron. The axon hillock has the lowest threshold, because this region has a much higher density of voltage-gated Na^+ channels than anywhere else in the neuron. For this reason, the axon hillock is considerably more responsive than the dendrites or remainder of the cell body to changes in potential. Because of

Comparison of Classical Neurotransmitters and Neuropeptides

Characteristic	Classical Neurotransmitters	Neuropeptides
Size	Small (one amino acid or similar chemical)	Large (2 to 40 amino acids)
Site of Synthesis	Cytosol of synaptic knob	Endoplasmic reticulum and Golgi complex in cell body; moved to synaptic knob by axonal transport
Site of Storage	Small synaptic vesicles in axon terminal	Large dense-core vesicles in axon terminal
Site of Release	Axon terminal	Axon terminal; may be co-secreted with neurotransmitter
Amount of Release	Variable, depending on synapse	Much lower concentration than classical neurotransmitter
Speed and Duration of Action	Rapid, brief response	Slow, prolonged response
Site of Action	Subsynaptic membrane of postsynaptic cell	Nonsynaptic sites on either presynaptic or postsynaptic cell
Effect	Usually alter potential of postsynaptic cell by opening specific ion channels	Modulate synaptic effectiveness by long-term changes in neurotransmitter synthesis or postsynaptic receptors

local current flow, EPSPs or IPSPs occurring anywhere on the dendrites or cell body spread throughout the dendrites, cell body, and axon hillock. When summation of EPSPs takes place, the lower threshold of the axon hillock is reached first, whereas the dendrites and cell body at the same potential are still considerably below their own, much higher thresholds. Therefore, an action potential originates in the axon hillock and is propagated from there to the end of the axon.

Neuropeptides act primarily as neuromodulators.

In addition to the classical neurotransmitters described earlier, some neurons also release *neuropeptides*. Neuropeptides differ from classical neurotransmitters in several important ways (▲ Table 4-3). Classical neurotransmitters are small, rapid-acting molecules that typically trigger the opening of specific ion channels to bring about a change in potential in the postsynaptic neuron (an EPSP or IPSP) within a few milliseconds or less. Most classical neurotransmitters are synthesized and packaged locally in synaptic vesicles in the cytosol of the axon terminal. These chemical messengers are primarily amino acids or closely related compounds.

Neuropeptides are larger molecules made up of anywhere from 2 to about 40 amino acids. They are synthesized in the endoplasmic reticulum and Golgi complex (see p. 26) of the neuronal cell body and are moved by axonal transport along the microtubular highways to the axon terminal (see p. 45). Neuropeptides are not stored in small synaptic vesicles with the classical neurotransmitters but instead are packaged in large **dense-core vesicles,** which are also present in the axon termi-

nal. The dense-core vesicles undergo Ca^{2+}-induced exocytosis and release neuropeptides at the same time that the neurotransmitter is released from the synaptic vesicles. An axon terminal typically releases only a single classical neurotransmitter, but the same terminal may also contain one or more neuropeptides that are co-secreted with the neurotransmitter.

Even though neuropeptides are currently the subject of intense investigation, our knowledge about their functions and control is still sketchy. They are known to diffuse locally and act on adjacent neurons at much lower concentrations than classical neurotransmitters do, and they bring about slower, more prolonged responses. Most neuropeptides are believed to function as neuromodulators. **Neuromodulators** are chemical messengers that do not cause the formation of EPSPs or IPSPs but rather bring about long-term changes that subtly *modulate*—depress or enhance—the action of the synapse. The neuronal receptors to which they bind are not located on the subsynaptic membrane, and they do not directly alter membrane permeability and potential. Neuromodulators may act at either presynaptic or postsynaptic sites. For example, a neuromodulator may influence the level of an enzyme involved in the synthesis of a neurotransmitter by a presynaptic neuron, or it may alter the sensitivity of the postsynaptic neuron to a particular neurotransmitter by causing long-term changes in the number of subsynaptic receptors for the neurotransmitter. Thus, neuromodulators delicately fine-tune the synaptic response. The effect may last for days, months, or even years. Whereas neurotransmitters are involved in rapid communication between neurons, neuromodulators are involved with more long-lasting events, such as learning and motivation.

Interestingly, the synaptically released neuromodulators include many substances that also have distinctly different roles as hormones released into the blood from endocrine tissues. *Cholecystokinin (CCK)* is an example. As a hormone, CCK is released from the small intestine following a meal and causes the gallbladder to contract and release bile into the intestine, among other digestive actions (see Chapter 16). CCK has also been found in axon terminal vesicles in the brain, where it is believed to function as a neuromodulator, causing the sensation of no longer being hungry. In many instances, neuropeptides are named for their first-discovered role as hormones, as is the case with cholecystokinin (*chole* means "bile"; *cysto* means "bladder"; *kinin* means "contraction"). A number of chemical messengers are quite versatile and can assume different roles, depending on their source, distribution, and interaction with different cell types.

● **FIGURE 4-18 Presynaptic inhibition.** A, an excitatory terminal ending on postsynaptic cell C, is itself innervated by inhibitory terminal B. Stimulation of terminal A alone produces an EPSP in cell C, but simultaneous stimulation of terminal B prevents the release of excitatory neurotransmitter from terminal A. Consequently, no EPSP is produced in cell C despite the fact that terminal A has been stimulated. Such presynaptic inhibition selectively depresses activity from terminal A without suppressing any other excitatory input to cell C. Stimulation of excitatory terminal D produces an EPSP in cell C even though inhibitory terminal B is simultaneously stimulated because terminal B only inhibits terminal A.

Presynaptic inhibition or facilitation can selectively alter the effectiveness of a presynaptic input.

Besides neuromodulation, another means of depressing or enhancing synaptic effectiveness is presynaptic inhibition or facilitation. Sometimes, a third neuron influences activity between a presynaptic ending and a postsynaptic neuron. The presynaptic axon terminal (labeled A in ● Figure 4-18) may itself be innervated by another axon terminal (labeled B). The neurotransmitter released from modulatory terminal B binds with receptors on terminal A. This binding alters the amount of neurotransmitter released from terminal A in response to action potentials. If the amount of neurotransmitter released from A is reduced, the phenomenon is known as **presynaptic inhibition.** If the release of neurotransmitter is enhanced, the effect is called **presynaptic facilitation.**

Let us look more closely at how this process works. You know that Ca^{2+} entry into an axon terminal causes the release of neurotransmitter by exocytosis of synaptic vesicles. The amount of neurotransmitter released from terminal A depends on how much Ca^{2+} enters this terminal in response to an action potential. Ca^{2+} entry into terminal A, in turn, can be influenced by activity in modulatory terminal B. We will use presynaptic inhibition to illustrate (● Figure 4-18). The amount of neurotransmitter released from presynaptic terminal A, an excitatory input in our example, influences the potential in the postsynaptic neuron at which it terminates (labeled C in the

figure). Firing of A by itself generates an EPSP in postsynaptic neuron C. Now consider that B is stimulated simultaneously with A. Neurotransmitter from terminal B binds on terminal A, reducing Ca^{2+} entry into terminal A. Less Ca^{2+} entry means less neurotransmitter release from A. Note that modulatory neuron B can suppress neurotransmitter release from A only when A is firing. If this presynaptic inhibition by B prevents A from releasing its neurotransmitter, the formation of EPSPs on postsynaptic membrane C from input A is specifically prevented. As a result, no change in the potential of the postsynaptic neuron occurs despite action potentials in A.

Would the simultaneous production of an IPSP through activation of an inhibitory input to negate an EPSP produced by excitatory input A achieve the same result? Not quite. Activation of an inhibitory input to cell C would produce an IPSP in cell C, but this IPSP could cancel out not only an EPSP from excitatory input A but also any EPSPs produced by other excitatory terminals, such as terminal D in the figure. The entire postsynaptic membrane is hyperpolarized by IPSPs, thereby negating (canceling) excitatory information fed into any part of the cell from any presynaptic input. By contrast, presynaptic inhibition (or presynaptic facilitation) works in a much more specific way. Presynaptic inhibition provides a means by which certain inputs to the postsynaptic neuron can be *selectively* inhibited without affecting the contributions of any other inputs. For example, firing of B specifically prevents the formation of an EPSP in the postsynaptic neuron from excitatory presynaptic neuron A but does not influence other excitatory presynaptic inputs. Excitatory input D can still produce an EPSP in the postsynaptic neuron even when B is firing. This type of neuronal integration is another means by which electrical signaling between neurons can be carefully fine-tuned.

Drugs and diseases can modify synaptic transmission.

Clinical Note The vast majority of drugs that influence the nervous system function by altering synaptic mechanisms. Synaptic drugs may block an undesirable effect or enhance a desirable effect. Possible drug actions include (1) altering the synthesis, storage, or release of a neurotransmitter; (2) modifying neurotransmitter interaction with the postsynaptic receptor; (3) influencing neurotransmitter reuptake or destruction; and (4) replacing a deficient neurotransmitter with a substitute transmitter.

You already learned about SSRIs. As another example, the socially abused drug **cocaine** blocks the reuptake of the neurotransmitter *dopamine* at presynaptic terminals. It does so by binding competitively with the dopamine reuptake transporter, which is a protein molecule that picks up released dopamine from the synaptic cleft and shuttles it back to the axon terminal. With cocaine occupying the dopamine transporter, dopamine remains in the synaptic cleft longer than usual and continues to interact with its postsynaptic receptors. The result is prolonged activation of neural pathways that use this chemical as a neurotransmitter, especially pathways that play a role in feelings of pleasure. In essence, when cocaine is present, the neural switches in the pleasure pathway are locked in the "on" position.

Cocaine is addictive because the involved neurons become *desensitized* to the drug. After the postsynaptic cells are incessantly stimulated for an extended time, they can no longer transmit normally across synapses without increasingly larger doses of the drug. Specifically, with prolonged use of cocaine, the number of dopamine receptors in the brain is reduced in response to the glut of the abused substance. As a result of this desensitization, the user must steadily increase the dosage of the drug to get the same "high," or sensation of pleasure, a phenomenon known as *drug tolerance*. When the cocaine molecules diffuse away, the sense of pleasure evaporates, because the normal level of dopamine activity no longer "satisfies" the overly needy demands of the postsynaptic cells for stimulation. Cocaine users reaching this low become frantic and profoundly depressed. Only more cocaine makes them feel good again. But repeated use of cocaine often modifies responsiveness to the drug; the user no longer derives pleasure from the drug but suffers unpleasant *withdrawal symptoms* once its effect has worn off. The user typically becomes *addicted* to the drug, compulsively seeking it out, first to experience the pleasurable sensations and later to avoid the negative withdrawal symptoms. Cocaine is abused by millions who have become addicted to its mind-altering properties, with devastating social and economic effects.

Whereas cocaine abuse leads to excessive dopamine activity, **Parkinson's disease (PD)** is attributable to a deficiency of dopamine in the *basal nuclei,* a region of the brain involved in controlling complex movements. This movement disorder is characterized by muscular rigidity and involuntary tremors at rest, such as involuntary rhythmic shaking of the hands or head. The standard treatment for PD is the administration of *levodopa (L-dopa),* a precursor of dopamine. Dopamine itself cannot be administered because it is unable to cross the blood–brain barrier (discussed in the following chapter), but L-dopa can enter the brain from the blood. Once inside the brain, L-dopa is converted into dopamine, thus substituting for the deficient neurotransmitter. In contrast to cocaine, which leads to excessive dopamine activity, L-dopa brings reduced dopamine levels up toward normal. This therapy greatly alleviates the symptoms associated with the deficit in most patients. You will learn more about this condition when we discuss the basal nuclei in the next chapter.

Synaptic transmission is also vulnerable to neural toxins, which may cause nervous system disorders by acting at either presynaptic or postsynaptic sites. For example, two different neural poisons, tetanus toxin and strychnine, act at different synaptic sites to block inhibitory impulses while leaving excitatory inputs unchecked. Tetanus toxin prevents the presynaptic release of a specific inhibitory neurotransmitter, whereas strychnine blocks specific postsynaptic inhibitory receptors.

Tetanus toxin prevents the release of the inhibitory neurotransmitter GABA from inhibitory presynaptic inputs terminating at neurons that supply skeletal muscles. Unchecked excitatory inputs to these neurons result in uncontrolled muscle spasms. These spasms occur especially in the jaw muscles early in the disease, giving rise to the common name of *lockjaw* for this condition. Later they progress to the muscles responsible for breathing, at which point death occurs.

Strychnine competes with another inhibitory neurotransmitter, glycine, at the postsynaptic receptor. This poison combines with the receptor but does not directly alter the potential of the postsynaptic cell in any way. Instead, it blocks the receptor so that it is not available for interaction with glycine when the latter is released from the inhibitory presynaptic ending. Thus, strychnine abolishes postsynaptic inhibition (formation of IPSPs) in nerve pathways that use glycine as an inhibitory neurotransmitter. Unchecked excitatory pathways lead to convulsions, muscle spasticity, and death.

Many other drugs and diseases influence synaptic transmission, but as these examples illustrate, any site along the synaptic pathway is vulnerable to either pharmacologic (drug induced) or pathological (disease induced) interference.

Neurons are linked through complex converging and diverging pathways.

Two important relationships exist between neurons: convergence and divergence. A given neuron may have many other neurons synapsing on it. Such a relationship is known as **convergence** (● Figure 4-19). Through converging input, a single cell is influenced by thousands of other cells. This single cell, in turn, influences the level of activity in many other cells by divergence of output. The term **divergence** refers to the branching of axon terminals so that a single cell synapses with and influences many other cells.

Note that a particular neuron is postsynaptic to the neurons converging on it but presynaptic to the other cells at which it terminates. Thus, the terms *presynaptic* and *postsynaptic* refer only to a single synapse. Most neurons are presynaptic to one group of neurons and postsynaptic to another group.

There are an estimated 100 billion neurons and 10^{14} (100 quadrillion) synapses in the brain alone! A single neuron may be connected to between 5000 and 10,000 other

Convergence of input
(one cell is influenced
by many others)

Divergence of output
(one cell influences
many others)

● **FIGURE 4-19 Convergence and divergence.** Arrows indicate the direction in which information is being conveyed.

neurons. When you consider the vast and intricate interconnections possible between these neurons through converging and diverging pathways, you can begin to imagine how complex the wiring mechanism of our nervous system really is. Even the most sophisticated computers are far less complex than the human brain. The "language" of the nervous system—that is, all communication between neurons—is in the form of graded potentials, action potentials, neurotransmitter signaling across synapses, and other nonsynaptic forms of chemical chatter. All activities for which the nervous system is responsible—every sensation, every command to move a muscle, every thought, every emotion, every memory, every spark of creativity—depend on the patterns of electrical and chemical signaling between neurons along these complexly wired neural pathways.

A neuron communicates with the cells it influences by releasing a neurotransmitter, but this is only one means of intercellular ("between cell") communication. We will now consider all the ways by which cells can "talk with one another."

Intercellular Communication and Signal Transduction

Coordination of the diverse activities of cells throughout the body to accomplish life-sustaining and other desired responses depends on the ability of cells to communicate with one another.

Communication between cells is largely orchestrated by extracellular chemical messengers.

Intercellular communication can take place either directly or indirectly (● Figure 4-20). *Direct* intercellular communication involves physical contact between the interacting cells:

■ *Through gap junctions.* The most intimate means of intercellular communication is through gap junctions, the minute tunnels that bridge the cytoplasm of neighboring cells in some types of tissues. Through gap junctions, small ions and molecules are directly exchanged between interacting cells without ever entering the extracellular fluid (see p. 59).

■ *Through transient direct linkup of surface markers.* Some cells, such as those of the immune system, have specialized markers on the surface membrane that allow them to directly link with certain other cells that have compatible markers for transient interactions. This is how the phagocytes of the body's defense system specifically recognize and selectively destroy only undesirable cells, such as cancer cells, while leaving the body's own healthy cells alone (see p. 448).

The most common means by which cells communicate with one another is *indirectly* through **extracellular chemical messengers** or **signal molecules,** of which there are four types: *paracrines, neurotransmitters, hormones,* and *neurohormones.* In each case, a specific chemical messenger, the signal molecule, is synthesized by specialized controlling cells to serve a designated purpose. On being released into the ECF by appropriate stimulation, these extracellular chemical messengers act on other particular cells, the messenger's **target cells,** in a prescribed manner. To exert its effect, an extracellular chemical messenger must bind with target cell receptors specific for it. Receptors are all transmembrane glycoproteins, that is, integral proteins that extend through the thickness of the plasma membrane and have carbohydrate chains attached on their outer surface (see p. 55). A given cell may have thousands up to as many as a few million receptors, of which hundreds to as many as 100,000 may be for the same chemical messenger. Different cell types have distinct combinations of receptors, allowing them to react individually to various regulatory extracellular chemical messengers. Nearly 5% of all genes in humans code for synthesis of these membrane recep-

DIRECT INTERCELLULAR COMMUNICATION

Small molecules and ions

(a) Gap junctions

(b) Transient direct linkup of cells' surface markers

INDIRECT INTERCELLULAR COMMUNICATION VIA EXTRACELLULAR CHEMICAL MESSENGERS

Secreting cell

Local target cell

Paracrine

(c) Paracrine secretion

Local target cell

Electrical signal

Secreting cell
(neuron)

Neurotransmitter

(d) Neurotransmitter secretion

Secreting cell
(endocrine cell)

Blood

Hormone

Distant target cell

Nontarget cell
(no receptors)

(e) Hormonal secretion

Neurohormone   Blood

Electrical signal

Secreting cell
(neuron)

Distant target cell

Nontarget cell
(no receptors)

(f) Neurohormone secretion

● **FIGURE 4-20 Types of intercellular communication.** Gap junctions and transient direct linkup of cells by means of complementary surface markers are both means of direct communication between cells. Paracrines, neurotransmitters, hormones, and neurohormones are all extracellular chemical messengers that accomplish indirect communication between cells. These chemical messengers differ in their source and the distance they travel to reach their target cells.

tors, indicative of the importance of this means of intercellular communication.

The four types of chemical messengers differ in their source and the distance and means by which they get to their site of action.

■ **Paracrines** are local chemical messengers whose effect is exerted only on neighboring cells in the immediate environment of their site of secretion. Because paracrines are distributed by simple diffusion within the interstitial fluid, their action is restricted to short distances. They do not gain entry to the blood in any significant quantity because they are rapidly inactivated by locally existing enzymes. One example of a paracrine is *histamine,* which is released from a specific type of connective tissue cell during an inflammatory response within an invaded or injured tissue (see p. 420). Among other things, histamine dilates (opens more widely) the blood vessels in the vicinity to increase blood flow to the tissue. This action brings additional blood-borne combat supplies into the affected area.

Paracrines must be distinguished from chemicals that influence neighboring cells after being nonspecifically released during the course of cellular activity. For example, an increased local concentration of CO_2 in an exercising muscle is among the factors that promote local dilation of the blood vessels supplying the muscle. The resultant increased blood flow helps to meet the more active tissue's increased metabolic demands. However, CO_2 is produced by all cells and is not specifically released to accomplish this response, so it and similar nonspecifically released chemicals are not considered paracrines.

■ As you just learned, neurons communicate directly with the cells they innervate (their target cells) by releasing **neurotransmitters,** which are very short-range chemical messengers, in response to electrical signals (action potentials). Like paracrines, neurotransmitters diffuse from their site of release across a narrow extracellular space to act locally on an adjoining target cell, which may be another neuron, a muscle, or a gland. Neurons themselves may carry electrical signals long distances (the length of the axon), but the chemical messenger released at the axon terminal acts at short range—just across the synaptic cleft.

■ **Hormones** are long-range chemical messengers specifically secreted into the blood by endocrine glands in response to an appropriate signal. The blood carries the messengers to other sites in the body, where they exert their effects on their target cells some distance away from their site of release. Only the target cells of a particular hormone have membrane receptors for binding with this hormone. Nontarget cells are not influenced by any blood-borne hormones that reach them.

■ **Neurohormones** are hormones released into the blood by *neurosecretory neurons.* Like ordinary neurons, neurosecretory neurons can respond to and conduct electrical signals. Instead of directly innervating target cells, however, a neurosecretory neuron releases its chemical messenger, a neurohormone, into the blood on appropriate stimulation. The neurohormone is then distributed through the blood to distant target cells. An example is *vasopressin,* a hormone produced by nerve cells in the brain that promotes water conservation by the kidneys during urine formation. Thus, like endocrine cells, neurosecretory neurons release blood-borne chemical messengers, whereas ordinary neurons secrete short-range neurotransmitters into a confined space. In the future, the general term *hormone* will tacitly include both blood-borne hormonal and neurohormonal messengers.

In every case, extracellular chemical messengers are released from one cell type and interact with other target cells to bring about a desired effect in the target cells. We now turn attention to how these chemical messengers cause the right cellular response.

Extracellular chemical messengers bring about cell responses primarily by signal transduction.

The term **signal transduction** refers to the process by which incoming signals (instructions from extracellular chemical messengers) are conveyed into the target cell where they are transformed into the dictated cellular response. Binding of the extracellular chemical messenger amounts to a signal for the cell to get a certain job done. During signal transduction, the extracellular signal is transduced, or changed into a form necessary to modify intracellular activities to accomplish the desired outcome. (A *transducer* is a device that receives energy from one system and transmits it in a different form to another system. For example, your radio receives radio waves sent out from the broadcast station and transmits these signals in the form of sound waves that can be detected by your ears.)

Signal transduction occurs by different mechanisms, depending on the messenger and receptor type. Lipid-soluble extracellular chemical messengers, such as cholesterol-derived steroid hormones, gain entry into the cell by dissolving in and passing through the lipid bilayer of the target cell's plasma membrane. Thus, these extracellular chemical messengers bind to receptors inside the target cell to initiate the desired intracellular response themselves, usually by changing gene activity. By contrast, water-soluble extracellular chemical messengers cannot gain entry to the target cell because they are poorly soluble in lipid and cannot dissolve in the plasma membrane. The major water-soluble extracellular messengers are protein hormones delivered by the blood and neurotransmitters released from nerve endings. These messengers signal the cell to perform a given response by first binding with surface membrane receptors specific for that given messenger. This binding triggers a sequence of intracellular events that ultimately controls a particular cellular activity, such as membrane transport, secretion, metabolism, contraction, or division and differentiation.

Despite the wide range of possible responses, binding of an extracellular messenger (also known as the **first messenger**) to its matching surface membrane receptor brings about the desired intracellular response primarily by three general means: (1) by opening or closing chemically gated *receptor-channels,* (2) by activating *receptor-enzymes,* or (3) by activating second-messenger pathways via *G-protein-coupled receptors.* Because of the universal nature of these events, let us examine each more closely (● Figure 4-21).

Some extracellular chemical messengers open chemically gated receptor-channels.

Some extracellular messengers carry out the assigned task by opening or closing specific chemically gated receptor-channels to regulate movement of particular ions across the membrane. In this case, the *receptor itself serves as an ion channel* (● Figure 4-21a). When the appropriate extracellular messenger binds to the **receptor-channel,** the channel opens or closes, depending on the signal. (In the future, for convenience when discussing receptor-channels in general we refer only to the more common opening of channels.) An example is the opening of chemically gated receptor-channels in the subsynaptic membrane in response to neurotransmitter binding. The resultant small, short-lived movement of given charge-carrying ions across the membrane through these open channels generates electrical signals—in this example, EPSPs and IPSPs.

On completion of the response, the extracellular messenger is removed from the receptor site, and the chemically gated channels close once again. The ions that moved across the membrane through opened channels to trigger the response are returned to their original location by special membrane carriers.

Some extracellular chemical messengers activate receptor-enzymes via the tyrosine kinase pathway.

Most extracellular chemical messengers that cannot enter their target cells do not act on chemically gated receptor-channels to bring about the desired intracellular response. Instead these first messengers issue their orders by triggering a "Psst, pass it on" process.

In most instances, the signal is relayed inside the cell by **protein kinases,** the name for any enzyme that transfers a phos-

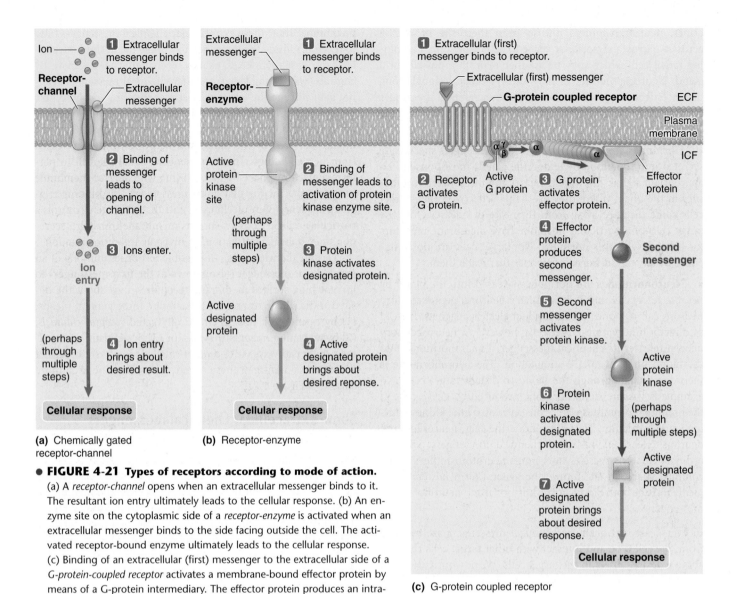

● FIGURE 4-21 Types of receptors according to mode of action.
(a) A *receptor-channel* opens when an extracellular messenger binds to it. The resultant ion entry ultimately leads to the cellular response. (b) An enzyme site on the cytoplasmic side of a *receptor-enzyme* is activated when an extracellular messenger binds to the side facing outside the cell. The activated receptor-bound enzyme ultimately leads to the cellular response. (c) Binding of an extracellular (first) messenger to the extracellular side of a *G-protein-coupled receptor* activates a membrane-bound effector protein by means of a G-protein intermediary. The effector protein produces an intracellular second messenger, which ultimately leads to the cellular response.

phate group from ATP to a particular intracellular protein. In response to phosphorylation, these proteins alter their shape and function, that is, are *activated,* to ultimately accomplish the cellular response dictated by the first messenger. Transduction may occur in a single step, although typically phosphorylation of a single kind of protein does not get the job done. Usually, protein kinases act in a chain of reactions, called a **cascade,** to pass along the signal to the final designated proteins capable of carrying out the desired effect. Protein kinase is activated on binding of the signal molecule to the surface receptor in one of two ways: the *tyrosine kinase pathway* or the *second-messenger pathway.*

In the tyrosine kinase pathway, the simplest of the two pathways, the *receptor itself functions as an enzyme,* a so-called **receptor-enzyme,** which has a protein kinase site on its portion that faces the cytoplasm (see ● Figure 4-21b). To activate the protein kinase, appropriate extracellular messengers must

bind with two of these receptors, which assemble into a pair. On activation, the receptor's protein kinase adds phosphate groups to the receptor itself (● Figure 4-22). Designated proteins inside the cell recognize and bind to the phosphorylated receptor. Then the receptor's protein kinase adds phosphate groups to the bound proteins. As a result of phosphorylation, the proteins change shape and function (are activated), enabling them to bring about the desired cellular response. In this pathway, the portions of the proteins that are specifically phosphorylated in both the receptor and designated proteins contain the amino acid tyrosine. Because of this specificity, such receptor-enzymes are called **tyrosine kinases.** The hormone insulin, which plays a major role in maintaining glucose homeostasis, exerts its effects via tyrosine kinases. Also, many growth factors that help regulate cell growth and division, such as *nerve growth factor* and *epidermal growth factor,* act via this pathway.

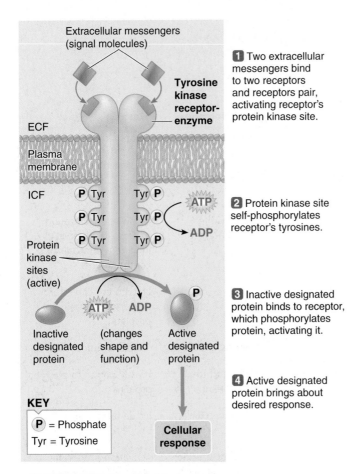

1 Two extracellular messengers bind to two receptors and receptors pair, activating receptor's protein kinase site.

2 Protein kinase site self-phosphorylates receptor's tyrosines.

3 Inactive designated protein binds to receptor, which phosphorylates protein, activating it.

4 Active designated protein brings about desired response.

KEY

$\boxed{P}$ = Phosphate

Tyr = Tyrosine

● **FIGURE 4-22 Tyrosine kinase pathway.**

Most extracellular chemical messengers activate second-messenger pathways via G-protein-coupled receptors.

The second-messenger pathway is initiated by binding of the first messenger (alias extracellular chemical messenger, alias signal molecule) to a surface membrane receptor specific for it. In this pathway, the *receptor is coupled with a G protein*, appropriately called a **G-protein-coupled receptor,** which snakes through the membrane seven times (see ● Figure 4-21c). Binding of the first messenger to the receptor activates the **G protein,** which serves as a membrane-bound intermediary that shuttles along the membrane to alter the activity of a nearby membrane protein called the **effector protein.** Once altered, the effector protein leads to an increased concentration of an intracellular messenger, known as the **second messenger.** The second messenger relays the orders through a cascade of biochemical reactions inside the cell that ultimately cause a change in the shape and function of designated proteins. These activated designated proteins accomplish the cellular response dictated by the first messenger. Most commonly, the second messenger activates an intracellular protein kinase, which leads to phosphorylation and thereby altered function of designated proteins. The intracellular pathways activated by a second messenger are remarkably similar among different cells despite the diversity of ultimate re-

sponses. The variability in response depends on the specialization of the cell, not on the mechanism used.

About half of all drugs prescribed today act on G-protein-coupled receptors. These receptors participate in some way in most body functions, so they are important targets for a variety of drugs used to treat diverse disorders. For example, they include drugs used to reduce high blood pressure, to treat congestive heart failure, to suppress stomach acid, to open airways in asthmatics, to ease symptoms of enlarged prostate, to block histamine-induced allergic responses, to relieve pain, and to treat hormone-dependent cancers.

The effects of protein kinases in the tyrosine kinase and second-messenger signal transduction pathways are reversed by another group of enzymes called **protein phosphatases,** which remove phosphate groups from the designated proteins. Unlike the protein kinases, which are active only when an extracellular messenger binds to a surface membrane receptor, most of the protein phosphatases are continuously active in cells. By continually removing phosphate groups from designated proteins, the protein phosphatases quickly shut off a signal transduction pathway if its signal molecule is no longer bound at the cell surface.

Some neurotransmitters function through intracellular second-messenger systems. Most, but not all, neurotransmitters function by changing the conformation of chemically gated receptor-channels, thereby altering membrane permeability and ion fluxes across the postsynaptic membrane, a process with which you are already familiar. Synapses involving these rapid responses are considered **"fast" synapses.** However, another mode of synaptic transmission used by some neurotransmitters, such as *serotonin,* involves the activation of intracellular second messengers. Synapses that lead to responses mediated by second messengers are known as **"slow" synapses,** because these responses take longer and often last longer than those accomplished by fast synapses. For example, neurotransmitter-activated second messengers may trigger long-term postsynaptic cellular changes believed to be linked to neuronal growth and development, as well as possibly play a role in learning and memory.

Depending on the cell type, the first messenger can either be released from the target cell and ultimately degraded by the liver and/or excreted in the urine, or alternatively the first messenger and receptor complex can be removed from action by receptor-mediated endocytosis, in which both the receptor and extracellular chemical messenger are internalized by the target cell (see p. 31).

Second-messenger systems are widely used throughout the body, including being one of the key means by which most water-soluble hormones ultimately bring about their effects. Let's now turn attention to hormonal communication, where we will examine specific second-messenger systems in more detail.

Introduction to Hormonal Communication

Endocrinology is the study of homeostatic chemical adjustments and other activities accomplished by hormones, which are secreted into the blood by endocrine glands. Earlier we de-

Chemical Classification of Hormones

Properties	Peptides	AMINES		Steroids
		Catecholamines and Indoleamines	**Thyroid Hormone**	
Solubility	Hydrophilic	Hydrophilic	Lipophilic	Lipophilic
Structure	Chains of specific amino acids	Tyrosine derivative (catecholamines) or tryptophan derivative (indoleamines)	Iodinated tyrosine derivative	Cholesterol derivative
Synthesis	In rough endoplasmic reticulum; packaged in Golgi complex	In cytosol	In colloid within thyroid gland (see p. 691)	Stepwise modification of cholesterol molecule in various intracellular compartments
Storage	Large amounts in secretory granules	In secretory granules	In colloid	Not stored; cholesterol precursor stored in lipid droplets
Secretion	Exocytosis of granules	Exocytosis of granules	Endocytosis of colloid	Simple diffusion
Transport in Blood	As free hormone	Half bound to plasma proteins	Mostly bound to plasma proteins	Mostly bound to plasma proteins
Receptor Site	Surface of target cell	Surface of target cell	Inside target cell	Inside target cell
Mechanism of Action	Activation of second-messenger pathway to alter activity of preexisting proteins that produce the effect	Activation of second-messenger pathway to alter activity of preexisting proteins that produce the effect	Activation of specific genes to make new proteins that produce the effect	Activation of specific genes to make new proteins that produce the effect
Hormones of This Type	Majority of hormones	Catecholamines: hormones from the adrenal medulla, dopamine from hypothalamus. Indoleamines: melatonin from pineal	Only hormones from the follicular cells of the thyroid	Hormones from the adrenal cortex and gonads and some placental hormones; vitamin D (a hormone) is steroidlike

scribed the underlying molecular and cellular mechanisms of the nervous system—electrical signaling within neurons and chemical transmission of signals between neurons. We now focus on the molecular and cellular features of hormonal action and compare the similarities and differences in how neurons and endocrine cells communicate with other cells in carrying out their regulatory actions.

Hormones are classified chemically as hydrophilic or lipophilic.

Hormones fall into two distinct chemical groups based on their solubility properties: *hydrophilic* and *lipophilic hormones.* Hormones can also be classified according to their biochemical structure, namely *peptides, amines,* and *steroids,* and/or source as follows (▲ Table 4-4):

1. Hydrophilic ("water-loving") **hormones** are highly water soluble and have low lipid solubility. Most hydrophilic hormones are peptide or protein hormones consisting of specific amino acids arranged in a chain of varying length. The shorter chains are peptides, and the longer ones are proteins. For convenience, we refer to this entire category as **peptides.** Insulin from the pancreas is a peptide hormone. The **amines** are so called because they are amino-acid derivatives. The amine hormones include two types of hydrophilic hormones (catecholamines and indoleamines) and one type of lipophilic hormone (thyroid hormone). *Catecholamines* are derived from the amino acid tyrosine and are largely secreted by the adrenal medulla. The adrenal gland consists of an inner adrenal medulla surrounded by an outer adrenal cortex. (You will learn more about the location and structure of the endocrine glands and the functions of specific hormones in later chapters.) Epinephrine is the major catecholamine hormone. *Indoleamines* are derived from the amino acid tryptophan and are secreted by the pineal gland. Melatonin is the only indoleamine hormone. Some neurotransmitters are also amines, such as dopamine (a catecholamine) and serotonin (an indoleamine). Dopamine also acts as a neurohormone, and serotonin is the precursor for melatonin, serving as examples of the overlapping activities of the nervous and endocrine systems.

2. **Lipophilic** ("lipid-loving") **hormones** have high lipid solubility and are poorly soluble in water. Lipophilic hormones include thyroid hormone and the steroid hormones. *Thyroid hormone,* as its name implies, is secreted exclusively by the thyroid gland; it is an iodinated tyrosine derivative. Even though catecholamines and thyroid hormone are both derived from tyrosine, they behave very differently because of their solubility properties. **Steroids** are neutral lipids derived from cholesterol. They include hormones secreted by the adrenal cortex, such as cortisol, and the sex hormones (testosterone in males and estrogen in females) secreted by the reproductive organs.

Minor differences in chemical structure between hormones within each category often result in profound differences in biological response. For example, in ● Figure 4-23, note the subtle difference between the steroid hormones *testosterone,* the male sex hormone responsible for inducing the development of masculine characteristics, and *estradiol,* a form of estrogen, which is the feminizing female sex hormone.

The solubility properties of a hormone determine (1) how the hormone is processed by the endocrine cell, (2) how the hormone is transported in the blood, and (3) how the hormone exerts its effects at the target cell. We first consider the different ways in which these hormone types are processed at their site of origin, the endocrine cell, before comparing their means of transport and their mechanisms of action.

The mechanisms of synthesis, storage, and secretion of hormones vary according to their chemical differences.

Because of their chemical differences, the means by which the various types of hormones are synthesized, stored, and secreted differ.

PROCESSING OF HYDROPHILIC PEPTIDE HORMONES Peptide hormones are synthesized and secreted by the same steps used for manufacturing any protein that is exported from the cell (see ● Figure 2-3, p. 26). From the time peptide hormones are synthesized until they are secreted, they are always segregated from intracellular proteins within membrane-enclosed compartments. Here is a brief outline of these steps:

1. Large precursor proteins, or **preprohormones,** are synthesized by ribosomes on the rough endoplasmic reticulum (ER).

Testosterone, a masculinizing hormone

Estradiol, a feminizing hormone

● **FIGURE 4-23 Comparison of two steroid hormones, testosterone and estradiol.**

They then migrate to the Golgi complex in membrane-enclosed transport vesicles that pinch off from the smooth ER.

2. During their journey through the ER and Golgi complex, the preprohormones are pruned to active hormones.

3. The Golgi complex then packages the finished hormones into secretory vesicles that are pinched off and stored in the cytoplasm until an appropriate signal triggers their secretion.

4. On appropriate stimulation, the secretory vesicles fuse with the plasma membrane and release their contents to the outside by exocytosis (see p. 28). Such secretion usually does not go on continuously; it is triggered only by specific stimuli. The blood then picks up the secreted hormones for distribution.

PROCESSING OF LIPOPHILIC STEROID HORMONES All steroidogenic (steroid-producing) cells perform the following steps to produce and release their hormonal product:

1. Cholesterol is the common precursor for all steroid hormones.

2. Synthesis of the various steroid hormones from cholesterol requires a series of enzymatic reactions that modify the basic cholesterol molecule—for example, by varying the type and position of side groups attached to the cholesterol framework. Each conversion from cholesterol to a specific steroid hormone requires the help of a number of enzymes that are limited to certain steroidogenic organs. Thus, each steroidogenic organ can produce only the steroid hormone or hormones for which it has a complete set of appropriate enzymes. For example, a key enzyme necessary for producing cortisol is found only in the adrenal cortex, so no other steroidogenic organ can produce this hormone.

3. Unlike peptide hormones, steroid hormones are not stored. Once formed, the lipid-soluble steroid hormones immediately diffuse through the steroidogenic cell's lipid plasma membrane to enter the blood. Only the hormone precursor cholesterol is stored in significant quantities within steroidogenic cells. Accordingly, the rate of steroid hormone secretion is controlled entirely by the rate of hormone synthesis. In contrast, peptide hormone secretion is controlled primarily by regulating the release of presynthesized stored hormone.

4. Following their secretion into the blood, some steroid hormones as well as thyroid hormone undergo further interconver-

sions within the blood or other organs, where they are changed into more potent or different hormones.

The adrenomedullary catecholamines and thyroid hormone have unique synthetic and secretory pathways that will be described when addressing each of these hormones specifically in the endocrine chapters (Chapters 18 and 19).

Hydrophilic hormones dissolve in the plasma; lipophilic hormones are transported by plasma proteins.

All hormones are carried by the blood, but they are not all transported in the same manner:

■ The hydrophilic peptide hormones are transported simply dissolved in the blood.

■ Lipophilic steroids and thyroid hormone, which are poorly soluble in water, cannot dissolve to any extent in the watery blood. Instead, the majority of the lipophilic hormones circulate to their target cells reversibly bound to plasma proteins in the blood. Some plasma proteins carry only one type of hormone, whereas others, such as albumin, indiscriminately pick up any "hitchhiking" hormone.

Only the small, unbound, freely dissolved fraction of a lipophilic hormone is biologically active (that is, free to cross capillary walls and bind with target cell receptors to exert an effect). Once a hormone has interacted with a target cell, it is rapidly inactivated or removed so it is no longer available to interact with another target cell. Because the carrier-bound hormone is in dynamic equilibrium with the free hormone pool, the bound form of steroid and thyroid hormones provides a large reserve of these lipophilic hormones that can be used to replenish the active free pool. To maintain normal endocrine function, the magnitude of the small free, effective pool, rather than the total blood concentration of a particular lipophilic hormone, is monitored and adjusted.

■ Catecholamines are unusual in that only about 50% of these hydrophilic hormones circulate as free hormone; the other 50% are loosely bound to the plasma protein albumin. Because catecholamines are water soluble, the importance of this protein binding is unclear.

Clinical Note The chemical properties of a hormone dictate not only the means by which blood transports it but also how it can be artificially introduced into the blood for therapeutic purposes. Because the digestive system does not secrete enzymes that can digest steroid and thyroid hormones, these hormones, such as the sex steroids contained in birth control pills, can when taken orally be absorbed intact from the digestive tract into the blood. No other type of hormones can be taken orally, because protein-digesting enzymes would attack and convert them into inactive fragments. Therefore, these hormones must be administered by nonoral routes; for example, insulin deficiency is typically treated with daily injections of insulin.

Next, we examine how the hydrophilic and lipophilic hormones vary in their mechanisms of action at their target cells.

Hormones generally produce their effect by altering intracellular proteins.

To induce their effect, hormones must bind with target cell receptors specific for them. Each interaction between a particular hormone and a target cell receptor produces a highly characteristic response that differs among hormones and among different target cells influenced by the same hormone. Both the location of the receptors within the target cell and the mechanism by which binding of the hormone with the receptors induces a response vary, depending on the hormone's solubility characteristics.

LOCATION OF RECEPTORS FOR HYDROPHILIC AND LIPOPHILIC HORMONES Hormones can be grouped into two categories based on the primary location of their receptors:

1. The hydrophilic peptides and catecholamines, which are poorly soluble in lipid, cannot pass through the lipid membrane barriers of their target cells. Instead, they bind with specific receptors on the *outer plasma membrane surface* of the target cell.

2. The lipophilic steroids and thyroid hormone easily pass through the surface membrane to bind with specific receptors located *inside* the target cell.

GENERAL MEANS OF HYDROPHILIC AND LIPOPHILIC HORMONE ACTION Even though hormones elicit a wide variety of biological responses, they ultimately influence their target cells by altering the cell's proteins in one of two major ways:

1. Surface-binding hydrophilic hormones function largely by *activating second-messenger pathways* within the target cell. This activation directly *alters the activity of preexisting intracellular proteins, usually enzymes,* to produce the desired effect.

2. Lipophilic hormones function mainly by *activating specific genes* in the target cell *to cause formation of new intracellular proteins,* which in turn produce the desired effect. The new proteins may be enzymatic or structural.

Let us examine the two major mechanisms of hormonal action (activation of second-messenger pathways and activation of genes) in more detail.

Hydrophilic hormones alter preexisting proteins via second-messenger systems.

Most hydrophilic hormones (peptides and catecholamines) bind to G-protein-coupled surface membrane receptors and produce their effects in their target cells by acting through a second-messenger system to alter the activity of preexisting proteins. There are two major second-messenger pathways: One uses **cyclic adenosine monophosphate (cyclic AMP, or cAMP)** as a second messenger, and the other employs Ca^{2+} in this role.

Both pathways use a G protein, which is found on the inner surface of the plasma membrane, as an intermediary between the receptor and effector protein (● Figure 4-24). G proteins are so named because they are bound to guanine nucleotides— *guanosine triphosphate (GTP)* when active or *guanosine diphosphate (GDP)* when inactive. An inactive G protein consists of a complex of alpha (α), beta (β), and gamma (γ) subunits, with a

1 Binding of extra-cellular messenger to receptor activates a G protein, the α subunit of which shuttles to and activates adenylyl cyclase.

2 Adenylyl cyclase converts ATP to cAMP

3 cAMP activates protein kinase A.

4 Protein kinase A phosphorylates inactive designated protein, activating it.

5 Active designated protein brings about desired response.

KEY

P = Phosphate

● **FIGURE 4-24 Mechanism of action of hydrophilic hormones via activation of the cyclic AMP second-messenger pathway.**

sengers through the membrane to effector proteins by means of G proteins.

CYCLIC AMP SECOND-MESSENGER PATHWAY Cyclic AMP is the most widely used second messenger. In the following description of the cAMP pathway, the numbered steps correlate to the numbered steps in ● Figure 4-24.

Binding of the first messenger to its surface membrane receptor activates the associated G protein, which in turn activates the effector protein, in this case the enzyme **adenylyl cyclase** (step **1**), which is located on the cytoplasmic side of the plasma membrane. Adenylyl cyclase induces the conversion of intracellular ATP to cAMP by cleaving off two of the phosphates (step **2**). (This is the same ATP used as the common energy currency in the body.) Acting as the intracellular second messenger, cAMP triggers a pre-programmed series of biochemical steps within the cell to bring about the response dictated by the first messenger. To begin, cyclic AMP activates a specific intracellular enzyme, **protein kinase A** (step **3**). Protein kinase A, in turn, phosphorylates a designated preexisting intracellular protein, such as an enzyme important in a particular metabolic pathway. Phosphorylation causes the protein to change its shape and function, thereby activating it (step **4**). This activated protein brings about the target cell's ultimate response to the first messenger (step **5**). For example, the activity of a particular enzymatic protein that regulates a specific metabolic event may be increased or decreased.

Note that in this signal transduction pathway, the steps involving the extracellular first messenger, the receptor, the G-protein complex, and the effector protein occur *in the plasma membrane* and lead to activation of the second messenger. The extracellular messenger cannot enter the cell to "personally" deliver its message to the proteins that carry out the desired response. Instead, it initiates membrane events that activate an intracellular second messenger, cAMP. The second messenger then triggers a chain reaction of biochemical events *inside the cell* that leads to the cellular response.

Different types of cells have different proteins available for phosphorylation and modification by protein kinase A. Therefore, a *common second messenger such as cAMP can induce widely differing responses in different cells,* depending on what proteins are modified. Cyclic AMP can be thought of as a molecular "switch" that can "turn on" (or "turn off") different cell events, depending on the kinds of protein activity ultimately modified in the various target cells. The type of proteins altered by a second messenger depends on the unique specialization of a particular cell type. This can be likened to being able to either illuminate or cool off a room depending on whether the wall switch you flip on is wired to a device specialized to light up (a

GDP molecule bound to the α subunit. A number of different G proteins with varying α subunits have been identified. The different G proteins are activated in response to binding of various first messengers to surface receptors. When an appropriate extracellular messenger (a first messenger) binds with its receptor, the receptor attaches to the appropriate G protein, resulting in release of GDP from the G-protein complex. GTP then attaches to the α subunit, an action that activates the G protein. Once activated, the G protein breaks away from the G-protein complex and moves along the inner surface of the plasma membrane until it reaches an effector protein. An effector protein is either an enzyme or an ion channel within the membrane. The α subunit links up with the effector protein and alters its activity. Researchers have identified more than 300 different receptors that convey instructions of extracellular meschandelier) or one specialized to create air movement (a ceiling fan). In the body, the variable responsiveness once the switch is turned on results from genetically programmed differences in the sets of proteins within different cells. For example, depending on its cellular location, activating the cAMP pathway can modify heart rate in the heart, stimulate the formation of female sex hormones in the ovaries, break down stored glucose in the liver, control water conservation during urine formation in the kidneys, create simple memory traces in the brain, or cause perception of a sweet taste by a taste bud.

After the response is completed the α subunit cleaves off a phosphate, converting GTP to GDP, in essence shutting itself off, then rejoins the β and γ subunits to restore the inactive G-protein complex. Cyclic AMP and the other participating chemicals are inactivated so that the intracellular message is "erased" and the response can be terminated. For example, cAMP is quickly degraded by **phosphodiesterase,** a cytosolic enzyme that is continuously active. This action provides another highly effective means of turning off the response when it is no longer needed. Other complementary means of terminating the response are removal of the added phosphates by protein phosphatase or removal of the first messenger.

Ca^{2+} SECOND-MESSENGER PATHWAY Some cells use Ca^{2+} instead of cAMP as a second messenger. In such cases, binding of the first messenger to the surface receptor eventually leads by means of G proteins to activation of the enzyme **phospholipase C,** an effector protein bound to the inner side of the membrane (step **1** in ● Figure 4-25). This enzyme breaks down **phosphatidylinositol bisphosphate** (abbreviated **PIP_2**), a component of the tails of the phospholipid molecules within the membrane itself. The products of PIP_2 breakdown are **diacylglycerol (DAG)** and **inositol trisphosphate (IP_3)** (step **2**). Lipid-soluble DAG remains in the lipid bilayer of the plasma membrane, but water-soluble IP_3 diffuses into the cytosol. IP_3 mobilizes intracellular Ca^{2+} stored in the endoplasmic reticulum to increase cytosolic Ca^{2+} by binding with IP_3-gated receptor-channels in the ER membrane (step **3a**). Calcium then takes on the role of second messenger, ultimately bringing about the response commanded by the first messenger. Many of the Ca^{2+}-dependent cellular events are triggered by activation of **calmodulin,** an intracellular Ca^{2+}-binding protein (step **4a**). Activation of calmodulin by Ca^{2+} is similar to activation of protein kinase A by cAMP. From here the patterns of the two pathways are similar. The activated calmodulin phosphorylates the designated proteins (perhaps through multiple steps), thereby causing these proteins to change their shape and function (activating them (step **5a**). The active designated proteins bring about the ultimate desired cellular response (step **6a**). For example, this pathway is the means by which chemical messengers can activate smooth muscle contraction. Note that in this pathway, calmodulin, not protein kinase, brings about the necessary change in form and function of the designated proteins to get the job done.

Simultaneous to the IP_3 pathway, the other PIP_2 breakdown product, DAG, sets off another second-messenger pathway. (IP_3 and DAG themselves are sometimes considered to be second messengers.) DAG activates **protein kinase C (PKC)** (step **3b**),

which phosphorylates designated proteins, different from those phosphorylated by calmodulin (step **4b**). The resultant change in shape and function of these proteins activates them. These active proteins produce another cellular response (step **5b**). Although currently the subject of considerable investigation, the DAG pathway is the not yet understood as well as the other signaling pathways. IP_3 and DAG typically trigger complementary actions inside a target cell to accomplish a common goal because both of these products are formed at the same time in response to the same first messenger. For example, extracellular chemical messengers promote increased contractile activity of blood-vessel smooth muscle via the IP_3/intracellular Ca^{2+}/calmodulin pathway, and the DAG pathway enhances the sensitivity of the contractile apparatus to Ca^{2+}.

The IP_3 pathway is not the only means of increasing intracellular Ca^{2+}. Intracellular Ca^{2+} can be increased by entry from the ECF or by release from Ca^{2+} stores inside the cell via means other than the IP_3 pathway. Calcium channels in both the surface membrane and in the endoplasmic reticulum may be opened by either electrical or chemical means. For example, the opening of voltage-gated surface-membrane Ca^{2+} channels is responsible for exocytosis of neurotransmitter from the axon terminal. Alternatively, surface-membrane Ca^{2+} channels may be opened via activation of receptors that serve as channels themselves or via activation of G-protein-coupled receptors. In yet another pathway, the opening of surface-membrane Na^+ and K^+ channels by means of receptor-channels leads to electrical signals that open Ca^{2+} channels in the endoplasmic reticulum. The latter pathway is how neurotransmitter released from motor neuron terminals triggers skeletal muscle contraction. The resultant rise in intracellular Ca^{2+} turns on the contractile apparatus. The pathways get even more complex than this. Calcium entering from the ECF can serve as a second messenger to trigger an even larger release of Ca^{2+} from intracellular stores, as it does to bring about contraction in cardiac muscle. This all sounds confusing, but these examples are meant to illustrate the complexity of signaling, not to overwhelm. You will learn more about the details of these pathways when appropriate in later chapters.

Furthermore, the cAMP and Ca^{2+} pathways frequently overlap in bringing about a particular cellular activity. For example, cAMP and Ca^{2+} can influence each other. Calcium-activated calmodulin can regulate adenylyl cyclase and thus influence cAMP, whereas protein kinase A may phosphorylate and thereby change the activity of Ca^{2+} channels or carriers. Therefore, the cell-signaling pathways may communicate with one another to integrate their responses. This inter-pathway interaction is called **cross talk.**

Although the cAMP and Ca^{2+} pathways are the most prevalent second-messenger systems, they are not the only ones. For example, in a few cells **cyclic guanosine monophosphate (cyclic GMP)** serves as a second messenger in a system analogous to the cAMP system. In other cells, the second messenger is still unknown. Remember that activation of second messengers is a universal mechanism employed by a variety of extracellular messengers in addition to hydrophilic hormones. (See the boxed feature on pp. 124–125, ■ Concepts, Challenges, and Controversies, for a description of a surprising

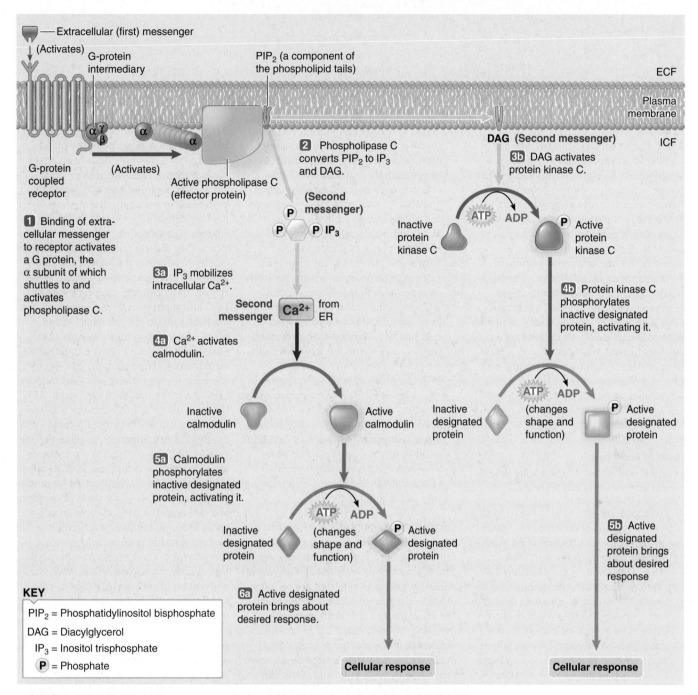

FIGURE 4-25 Mechanism of action of hydrophilic hormones via concurrent activation of the IP$_3$/Ca^{2+} second-messenger pathway and the DAG pathway.

Labels within figure:

— Extracellular (first) messenger

(Activates)

G-protein intermediary

PIP$_2$ (a component of the phospholipid tails)

ECF

Plasma membrane

α γ β, α, α

G-protein coupled receptor

(Activates)

Active phospholipase C (effector protein)

DAG (Second messenger)

ICF

2 Phospholipase C converts PIP$_2$ to IP$_3$ and DAG.

3b DAG activates protein kinase C.

1 Binding of extra-cellular messenger to receptor activates a G protein, the α subunit of which shuttles to and activates phospholipase C.

P, P, P IP$_3$ (Second messenger)

ATP ADP

Inactive protein kinase C

P Active protein kinase C

3a IP$_3$ mobilizes intracellular Ca^{2+}.

Second messenger Ca^{2+} from ER

4a Ca^{2+} activates calmodulin.

4b Protein kinase C phosphorylates inactive designated protein, activating it.

Inactive calmodulin

Active calmodulin

Inactive designated protein

ATP ADP (changes shape and function)

P Active designated protein

5a Calmodulin phosphorylates inactive designated protein, activating it.

Inactive designated protein

ATP ADP (changes shape and function)

P Active designated protein

5b Active designated protein brings about desired response

KEY

PIP$_2$ = Phosphatidylinositol bisphosphate

DAG = Diacylglycerol

IP$_3$ = Inositol trisphosphate

P = Phosphate

6a Active designated protein brings about desired response.

Cellular response

Cellular response

signal-transduction pathway—one that causes a cell to kill itself.)

AMPLIFICATION BY A SECOND-MESSENGER PATHWAY Several remaining points about receptor activation and the ensuing events merit attention. First, considering the number of steps in a second-messenger relay chain, you might wonder why so many cell types use the same complex system to accomplish such a wide range of functions. The multiple steps of a second-messenger pathway are actually advantageous because the cascading (multiplying) effect of these pathways greatly am-plifies the initial signal (● Figure 4-26, p. 126). *Amplification* means that the output of a system is much greater than the input. Using the cAMP pathway as an example, binding of one extracellular messenger molecule to a receptor activates a number of adenylyl cyclase molecules (let's arbitrarily say 10), each of which activates many (in our hypothetical example, let's say 100) cAMP molecules. Each cAMP molecule then acts on a single protein kinase A, which phosphorylates and thereby influences many (again, let's say 100) specific proteins, such as enzymes. Each enzyme, in turn, is responsible for producing many (perhaps 100) molecules of a particular

Programmed Cell Suicide: A Surprising Example of a Signal Transduction Pathway

In the vast majority of cases, the signal transduction pathways triggered by the binding of an extracellular chemical messenger to a cell's surface membrane receptor are aimed at promoting proper functioning, growth, survival, or reproduction of the cell. But every cell also has an unusual built-in pathway that, if triggered, causes the cell to commit suicide by activating intracellular protein-snipping enzymes, which slice the cell into small, disposable pieces. Such intentional, programmed cell death is called **apoptosis.** (This term means "dropping off," in reference to the dropping off of cells that are no longer useful, much as autumn leaves drop off trees.) Apoptosis is a normal part of life—individual cells that have become superfluous or disordered are triggered to self-destruct for the greater good of maintaining the whole body's health.

Roles of Apoptosis

Here are examples of the vital roles played by this intrinsic sacrificial program:

- *Predictable self-elimination of selected cells is a normal part of development.* Certain unwanted cells produced during development are programmed to kill themselves as the body is sculpted into its final form. During the development of a female, for example, apoptosis deliberately removes the embryonic ducts capable of forming a male reproductive tract. Likewise, apoptosis carves fingers from a mitten-shaped developing hand by eliminating the weblike membranes between them.

- *Apoptosis is important in tissue turnover in the adult body.* Optimal functioning of most tissues depends on a balance between controlled production of new cells and regulated cell self-destruction. This balance maintains the proper number of cells in a given tissue while ensuring a controlled supply of fresh cells that are at their peak of performance.

- *Programmed cell death plays an important role in the immune system.* Apoptosis provides a means to remove cells infected with harmful viruses. Furthermore, infection-fighting white blood cells that have finished their prescribed function and are no longer needed execute themselves.

- *Undesirable cells that threaten homeostasis are typically culled from the body by apoptosis.* Included in this hit list are aged cells, cells that have suffered irreparable damage by exposure to radiation or other poisons, and cells that have somehow gone awry. Many mutated cells are eliminated by this means before they become fully cancerous.

Comparison of Apoptosis and Necrosis

Apoptosis is not the only means by which a cell can die, but it is the neatest way. Apoptosis is a controlled, intentional, tidy way of removing individual cells that are no longer needed or that pose a threat to the body. The other form of cell death, **necrosis** (meaning "making dead"), is uncontrolled, accidental, messy murder of useful cells that have been severely injured by an agent external to the cell, as by a physical blow, O_2 deprivation, or disease. For example,

heart muscle cells deprived of their O_2 supply by complete blockage of the blood vessels supplying them during a heart attack die as a result of necrosis (see p. 335).

Even though necrosis and apoptosis both result in cell death, the steps involved are very different. In necrosis the dying cells are passive victims, whereas in apoptosis the cells actively participate in their own deaths. In necrosis, the injured cell cannot pump out Na^+ as usual. As a result, water streams in by osmosis, causing the cell to swell and rupture. Typically, in necrosis the insult that prompted cell death injures many cells in the vicinity, so many neighboring cells swell and rupture together. Release of intracellular contents into the surrounding tissues initiates an inflammatory response at the damaged site (see p. 420). Unfortunately, this inflammatory response may harm healthy neighboring cells.

By contrast, apoptosis targets individual cells for destruction, leaving the surrounding cells intact. A condemned cell signaled to commit suicide detaches itself from its neighbors, then shrinks instead of swelling and bursting. The cell's mitochondria become leaky, permitting cytochrome *c* to leak out into the cytosol. Cytochrome *c*, a component of the electron transport system, usually participates in oxidative phosphorylation to produce ATP (see p. 35). Outside its typical mitochondrial environment, however, cytochrome *c* activates a cascade of normally inactive intracellular protein-cutting enzymes, the **caspases,** which kill the cell from within. The unleashed caspases act like molecular scissors to systematically dismantle the cell. Snip-

---

product, such as a secretory product. The result of this cascade, with one event triggering the next in sequence, is a tremendous amplification of the initial signal. In our hypothetical example, one extracellular messenger molecule has been responsible for inducing a yield of 10 million molecules of a secretory product. In this way, very low concentrations of hormones and other chemical messengers can trigger pronounced cell responses.

REGULATION OF RECEPTORS Although membrane receptors serve as links between extracellular first messengers and intracellular second messengers in the regulation of specific cellular

activities, the receptors themselves are also frequently subject to regulation. In many instances, receptor number and affinity (attraction of a receptor for its chemical messenger) can be altered, depending on the circumstances. For example, a chronic elevation of insulin in the blood leads to a reduction in the number of insulin receptors. Later, when we cover the specific endocrine glands in detail, you will learn more about this mechanism to regulate the responsiveness of a target cell to its hormone.

 Many diseases can be linked to malfunctioning receptors or to defects in the ensuing signal-transduction pathways. For example, defective receptors are responping protein after protein, they chop up the nucleus, disassembling its life-essential DNA, then break down the internal shape-holding cytoskeleton, and finally fragment the cell itself into disposable membrane-enclosed packets (see accompanying photo). Importantly, the contents of the dying cell remain wrapped by plasma membrane throughout the entire self-execution process, thus avoiding the spewing of potentially harmful intracellular contents characteristic of necrosis. No inflammatory response is triggered, so no neighboring cells are harmed. Instead, cells in the vicinity swiftly engulf and destroy the apoptotic cell fragments by phagocytosis. The breakdown products are recycled for other purposes as needed. The tissue as a whole has continued to function normally while the targeted cell has unobtrusively killed itself.

Control of Apoptosis

If every cell contains caspases, what normally keeps these powerful self-destructive enzymes under control (that is, in inactive form) in cells that are useful to the body and deserve to live? Likewise, what activates the death-wielding caspase cascade in unwanted cells destined to eliminate themselves? Given the importance of these life-or-death decisions, it is not surprising that multiple pathways tightly control whether a cell is "to be or not to be." A cell normally receives a constant stream of "survival signals," which reassure the cell that it is useful to the body, that all is right in the internal environment surrounding the cell, and that everything is in good working order

within the cell. These signals include tissue-specific growth factors, certain hormones, and appropriate contact with neighboring cells and the extracellular matrix. These extracellular survival signals trigger intracellular pathways that block activation of the caspase cascade, thus restraining the cell's death machinery. Most cells are programmed to commit suicide if they do not receive their normal reassuring survival signals. With the usual safeguards removed, the lethal protein-snipping enzymes are unleashed. For example, withdrawal of growth factors or detachment from the extracellular matrix causes a cell to promptly execute itself.

Furthermore, cells display "death receptors" in their plasma membrane that receive specific extracellular "death signals," such as a particular hormone or a specific chemical messenger from white blood cells. Activation of death pathways by these signals can override the life-saving pathways triggered by the survival signals. The death-signal transduction pathway swiftly ignites the internal apoptotic machinery, driving the cell to its own demise. Likewise, the self-execution machinery is set in motion when a cell suffers irreparable intracellular damage. Thus, some signals block apoptosis, and others promote it. Whether a cell lives or dies depends on which of these competing signals dominates at any given time. Although all cells have the same death machinery, they vary in the specific signals that induce them to commit suicide.

Considering that every cell's life hangs in delicate balance at all times, it is not sur-

prising that faulty control of apoptosis—resulting in either too much or too little cell suicide—appears to participate in many major diseases. Excessive apoptotic activity is believed to contribute to the brain cell death associated with Alzheimer's disease, Parkinson's disease, and stroke, as well as to the premature demise of important infection-fighting cells in AIDS. Conversely, too little apoptosis most likely plays a role in cancer. Evidence suggests that cancer cells fail to respond to the normal extracellular signals that promote cell death. Because these cells neglect to die on cue, they grow in unchecked fashion, forming a chaotic, out-of-control mass.

Apoptosis is currently one of the hottest topics of investigation in the field. Researchers are scrambling to sort out the multiple factors involved in the signal transduction pathways controlling this process. Their hope is to find ways to tinker with the apoptotic machinery to find badly needed new therapies for treating a variety of big killers.

A normal cell (left) and a cell undergoing apoptosis (right)

sible for **Laron dwarfism.** In this condition, the person is abnormally short, despite having normal levels of growth hormone, because the tissues cannot respond normally to growth hormone. This is in contrast to the more usual type of dwarfism in which the person is abnormally short because of growth hormone deficiency. As another example, the toxins released by some infecting bacteria, such as those that cause cholera and whooping cough, keep second-messenger pathways "turned on" at a high level. **Cholera toxin** prevents the involved G protein from converting GTP to GDP, thus keeping the G protein in its active state. **Pertussis (whooping cough) toxin** blocks the inhibition of adenylyl cyclase, thereby

keeping the ensuing second-messenger pathway continuously active.

Having examined the means by which hydrophilic hormones alter their target cells, we now focus on the mechanism of lipophilic hormone action.

By stimulating genes, lipophilic hormones promote synthesis of new proteins.

All lipophilic hormones (steroids and thyroid hormone) bind with intracellular receptors and primarily produce effects in their target cells by activating specific genes that

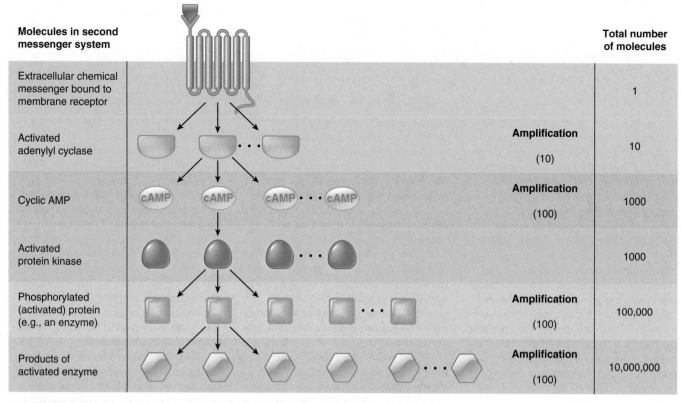

Molecules in second messenger system		Total number of molecules
Extracellular chemical messenger bound to membrane receptor		1
Activated adenylyl cyclase	**Amplification** (10)	10
Cyclic AMP	**Amplification** (100)	1000
Activated protein kinase		1000
Phosphorylated (activated) protein (e.g., an enzyme)	**Amplification** (100)	100,000
Products of activated enzyme	**Amplification** (100)	10,000,000

● **FIGURE 4-26 Amplification of the initial signal by a second-messenger pathway.** Through amplification, very low concentrations of extracellular chemical messengers, such as hormones, can trigger pronounced cellular responses.

cause the synthesis of new proteins, as summarized in ● Figure 4-27.

Free lipophilic hormone (hormone not bound with its plasma-protein carrier) diffuses through the plasma membrane of the target cell (step **1** in ● Figure 4-27) and binds with its specific receptor inside the cell, either in the cytoplasm or in the nucleus (step **2**). Each receptor has a specific region for binding with its hormone and another region for binding with DNA. The receptor cannot bind with DNA unless it first binds with the hormone. Once the hormone is bound to the receptor, the hormone receptor complex binds with DNA at a specific attachment site on the DNA known as the **hormone response element (HRE)** (step **3**). Different steroid hormones and thyroid hormone, once bound with their respective receptors, attach at different HREs on DNA. For example, the estrogen receptor complex binds at DNA's estrogen response element.

Binding of the hormone receptor complex with DNA "turns on" or activates a specific gene within the target cell (step **4**). This gene contains a code for synthesizing a given protein. The code of the activated gene is transcribed into complementary messenger RNA (mRNA) (step **5**). The new mRNA leaves the nucleus and enters the cytoplasm (step **6**), where it binds to a ribosome, the "workbench" that mediates the assembly of new proteins. Here, mRNA directs the synthesis of the designated new proteins according to the DNA code in the activated genes (step **7**). The newly synthesized protein, either enzymatic or structural, is released from the ribosome (step **8**) and produces the target cell's ultimate response to the hormone (step **9**). By

means of this mechanism, different genes are activated by different lipophilic hormones, resulting in different biological effects.

Even though most steroid actions are accomplished by hormonal binding with intracellular receptors that leads to gene activation, recent studies have unveiled another mechanism by which steroid hormones induce effects that occur too rapidly to be mediated by gene transcription. Some steroid hormones, most notably some of the sex hormones, bind with unique steroid receptors in the plasma membrane, in addition to binding with the traditional steroid receptors in the nucleus. This membrane binding leads to *nongenomic steroid receptor actions,* that is, actions accomplished by something other than altering gene activity, such as by inducing changes in ionic flux across the membrane or by altering activity of cellular enzymes.

Next we're going to compare the similarities and differences between neural and hormonal responses at the system level.

Comparison of the Nervous and Endocrine Systems

The nervous and endocrine systems are the two main regulatory systems of the body. The *nervous system* swiftly transmits electrical impulses to the skeletal muscles and the exocrine glands that it innervates. The *endocrine system* secretes hormones into the blood for delivery to distant sites of action. Although these two systems differ in many respects, they have much in common (▲ Table 4-5). They both ultimately alter

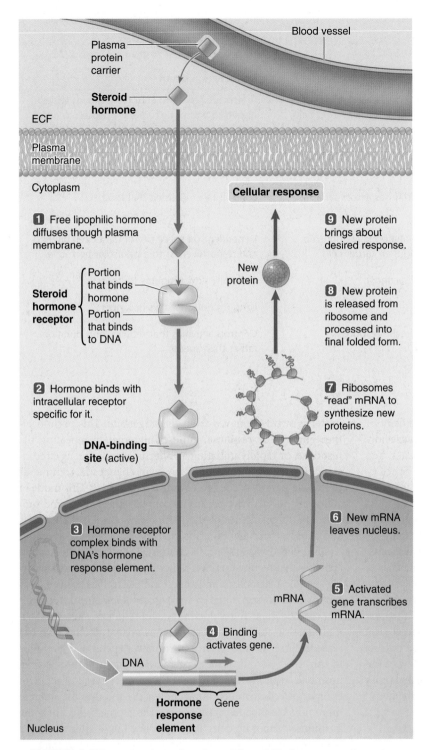

1 Free lipophilic hormone diffuses though plasma membrane.

Steroid hormone receptor { Portion that binds hormone / Portion that binds to DNA

2 Hormone binds with intracellular receptor specific for it.

DNA-binding site (active)

3 Hormone receptor complex binds with DNA's hormone response element.

4 Binding activates gene.

DNA

Hormone response element   Gene

Nucleus

Plasma protein carrier

Steroid hormone

ECF

Plasma membrane

Cytoplasm

Cellular response

Blood vessel

9 New protein brings about desired response.

New protein

8 New protein is released from ribosome and processed into final folded form.

7 Ribosomes "read" mRNA to synthesize new proteins.

6 New mRNA leaves nucleus.

mRNA

5 Activated gene transcribes mRNA.

● **FIGURE 4-27 Mechanism of action of lipophilic hormones via activation of genes.**

their target cells (their sites of action) by releasing chemical messengers (neurotransmitters in the case of neurons, hormones in the case of endocrine cells) that bind with specific receptors of the target cells. This binding triggers the cellular response dictated by the regulatory system.

Now let us examine the anatomic distinctions between these two systems and the different ways in which they accomplish specificity of action.

The nervous system is "wired," and the endocrine system is "wireless."

Anatomically, the nervous and endocrine systems are quite different. In the nervous system, each neuron terminates directly on its specific target cells; that is, the nervous system is "wired" into highly organized, distinct anatomic pathways for transmission of signals from one part of the body to another. Information is carried along chains of neurons to the desired destination through action potential propagation coupled with synaptic transmission. In contrast, the endocrine system is a "wireless" system in that the endocrine glands are not anatomically linked with their target cells. Instead, the endocrine chemical messengers are secreted into the blood and delivered to distant target sites. In fact, the components of the endocrine system itself are not anatomically interconnected; the endocrine glands are scattered throughout the body (see ● Figure 18-1, p. 662). These glands constitute a system in a functional sense, however, because they all secrete hormones and many interactions take place between various endocrine glands.

Neural specificity is due to anatomic proximity; and endocrine specificity, to receptor specialization.

As a result of their anatomic differences, the nervous and endocrine systems accomplish specificity of action by distinctly different means. Specificity of neural communication depends on neurons having a close anatomic relationship with their target cells, so each neuron has a very narrow range of influence. A neurotransmitter is released only to specific adjacent target cells, then is swiftly inactivated or removed before it can enter the blood. The target cells for a particular neuron have receptors for the neurotransmitter, but so do many other cells in other locations, and they could respond to this same mediator if it were delivered to them. For example, the entire system of motor neurons supplying your skeletal muscles uses the same neurotransmitter, *acetylcholine (ACh)*, and all your skeletal muscles bear complementary ACh receptors (see Chapter 8). Yet you can wiggle your big toe without influencing any of your other muscles because ACh can be discretely released from the motor neurons specifically wired to the muscles controlling your toe. If ACh were indiscriminately released into the blood, as are hormones, all the skeletal muscles would simultaneously respond by contracting, because they all have identical receptors for ACh. This does not happen, of course, because of the precise wiring patterns that provide direct lines of communication between motor neurons and their target cells.

▲ TABLE 4-5 — Comparison of the Nervous System and the Endocrine System

Property	Nervous System	Endocrine System
Anatomic Arrangement	A "wired" system: specific structural arrangement between neurons and their target cells, with structural continuity in the system	A "wireless" system: endocrine glands widely dispersed and not structurally related to one another or to their target cells
Type of Chemical Messenger	Neurotransmitters released into synaptic cleft	Hormones released into blood
Distance of Action of Chemical Messenger	Very short distance (diffuses across synaptic cleft)	Long distance (carried by blood)
Specificity of Action on Target Cell	Dependent on close anatomic relationship between neurons and their target cells	Dependent on specificity of target cell binding and responsiveness to a particular hormone
Speed of Response	Generally rapid (milliseconds)	Generally slow (minutes to hours)
Duration of Action	Brief (milliseconds)	Long (minutes to days or longer)
Major Functions	Coordinates rapid, precise responses	Controls activities that require long duration rather than speed

This specificity sharply contrasts to the way specificity of communication is built into the endocrine system. Because hormones travel in the blood, they reach virtually all tissues. Yet, only specific target cells can respond to each hormone. Specificity of hormonal action depends on specialization of target cell receptors. For a hormone to exert its effect, the hormone must first bind with receptors specific for it that are located only on or in the hormone's target cells. Target cell receptors are highly selective in their binding function. A receptor recognizes a specific hormone because a portion of its conformation matches a unique portion of its binding hormone in "lock-and-key" fashion. Binding of a hormone with target cell receptors initiates a reaction that culminates in the hormone's final effect. The hormone cannot influence any other cells because nontarget cells lack the right binding receptors. Likewise, a given target cell has receptors that are "tuned" to recognize only one or a few of the many hormones that circulate in its vicinity. Other signals pass by without effect because the cell has no receptors for them.

The nervous and endocrine systems have their own realms of authority but interact functionally.

The nervous and endocrine systems are specialized for controlling different types of activities. In general, the nervous system governs the coordination of rapid, precise responses. It is especially important in the body's interactions with the external environment. Neural signals in the form of action potentials are rapidly propagated along neuronal fibers, resulting in the release at the axon terminal of a neurotransmitter that must diffuse only a microscopic distance to its target cell before a response is effected. A neurally mediated response is not only rapid but brief; the action is quickly halted as the neurotrans-

mitter is swiftly removed from the target site. This permits either ending the response, almost immediately repeating the response, or rapidly initiating an alternate response, as circumstances demand (for example, the swift changes in commands to muscle groups needed to coordinate walking). This mode of action makes neural communication extremely rapid and precise. The target tissues of the nervous system are the muscles and glands, especially exocrine glands, of the body.

The endocrine system, in contrast, is specialized to control activities that require duration rather than speed, such as regulating organic metabolism and water and electrolyte balance; promoting smooth, sequential growth and development; and controlling reproduction. The endocrine system responds more slowly to its triggering stimuli than the nervous system does, for several reasons. First, the endocrine system must depend on blood flow to convey its hormonal messengers over long distances. Second, hormones typically have a more complex mechanism of action at their target cells than neurotransmitters do, and thus they require more time before a response occurs. The ultimate effect of some hormones cannot be detected until a few hours after they bind with target cell receptors. Also, because of the receptors' high affinity for their respective hormone, hormones often remain bound to receptors for some time, thus prolonging their biological effectiveness. Furthermore, unlike the brief, neurally induced responses that stop almost immediately after the neurotransmitter is removed, endocrine effects usually last for some time after the hormone's withdrawal. Neural responses to a single burst of neurotransmitter release usually last only milliseconds to seconds, whereas the alterations that hormones induce in target cells range from minutes to days or, in the case of growth-promoting effects, even a lifetime. Thus, hormonal action is relatively slow and prolonged, making endocrine control particularly suitable for regulating metabolic activities that require long-term stability.

Although the endocrine and nervous systems have their own areas of specialization, they are intimately interconnected functionally. Some neurons do not release neurotransmitters at synapses but instead end at blood vessels and release their chemical messengers (neurohormones) into the blood, where these chemicals act as hormones. A given messenger may even be a neurotransmitter when released from a nerve ending and a hormone when secreted by an endocrine cell. An example is *norepinephrine* (see p. 243). The nervous system directly or indirectly controls the secretion of many hormones (see Chapter 18). At the same time, many hormones act as neuromodulators, altering synaptic effectiveness and thereby influencing the excitability of the nervous system. The presence of certain key hormones is even essential for the proper development and maturation of the brain during fetal life. Furthermore, in many instances the nervous and endocrine systems both influence the same target cells in supplementary fashion. For example, these two major regulatory systems both help regulate the circulatory and digestive systems. Thus, many important regulatory interfaces exist between the nervous and endocrine systems. The study of these relationships is known as **neuroendocrinology.**

In the next three chapters, we will concentrate on the nervous system and will examine the endocrine system in more detail in later chapters. Throughout the text we will continue to point out the numerous ways these two regulatory systems interact so that the body is a coordinated whole, even though each system has its own realm of authority.

Chapter in Perspective: Focus on Homeostasis

To maintain homeostasis, cells must communicate so that they work together to accomplish life-sustaining activities. To bring about desired responses, the two major regulatory systems of the body, the nervous system and the endocrine system, must communicate with the target cells they are controlling. Neural and hormonal communication is therefore critical in maintaining a stable internal environment as well as in coordinating nonhomeostatic activities.

Neurons are specialized to receive, process, encode, and rapidly transmit information from one part of the body to another. The information is transmitted over intricate neuronal pathways by propagation of action potentials along the neuron's length as well as by chemical transmission of the signal from neuron to neuron at synapses and from neuron to muscles and glands through other neurotransmitter–receptor interactions at these junctions.

Collectively, the neurons make up the nervous system. Many of the activities controlled by the nervous system are geared toward maintaining homeostasis. Some neuronal electrical signals convey information about changes to which the body must rapidly respond in order to maintain homeostasis—for example, information about a fall in blood pressure. Other neuronal electrical signals swiftly convey messages to muscles and glands stimulating appropriate responses to counteract these changes—for example, adjustments in heart and blood vessel activity that restore blood pressure to normal when it starts to fall. Furthermore, the nervous system directs many activities not geared toward maintaining homeostasis, many of which are subject to voluntary control, such as playing basketball or browsing on the Internet.

The endocrine system secretes hormones into the blood, which carries these chemical messengers to distant target cells where they bring about their effect by changing the activity of enzymatic or structural proteins within these cells. Through its relatively slow-acting hormonal messengers, the endocrine system generally regulates activities that require duration rather than speed. Most of these activities are directed toward maintaining homeostasis. For example, hormones help maintain the proper concentration of nutrients in the internal environment by directing chemical reactions involved in the cellular uptake, storage, release, and use of these molecules. Also, hormones help maintain the proper water and electrolyte balance in the internal environment. Unrelated to homeostasis, hormones direct growth and control most aspects of the reproductive system.

Together the nervous and endocrine systems orchestrate a wide range of adjustments that help the body maintain homeostasis in response to stress. Likewise, these systems work in concert to control the circulatory and digestive systems, which carry out many homeostatic activities.

REVIEW EXERCISES

Objective Questions (Answers on p. A-41)

1. Conformational changes in channel proteins brought about by voltage changes are responsible for opening and closing Na^+ and K^+ gates during the generation of an action potential. *(True or false?)*

2. The $Na^+–K^+$ pump restores the membrane to resting potential after it reaches the peak of an action potential. *(True or false?)*

3. Following an action potential, there is more K^+ outside the cell than inside because of the efflux of K^+ during the falling phase. *(True or false?)*

4. Postsynaptic neurons can either excite or inhibit presynaptic neurons. *(True or false?)*

5. Second-messenger systems ultimately bring about the desired cell response by inducing a change in the shape and function of particular designated intracellular proteins. *(True or false?)*

6. Each steroidogenic organ has all the enzymes necessary to produce any steroid hormone. *(True or false?)*

7. The one-way propagation of action potentials away from the original site of activation is ensured by the _____.

8. The _____ is the site of action potential initiation in most neurons because it has the lowest threshold.

9. A junction in which electrical activity in one neuron influences the electrical activity in another neuron by means of a neurotransmitter is called a _____.

10. Summing of EPSPs occurring very close together in time as a result of repetitive firing of a single presynaptic input is known as _____.

11. Summing of EPSPs occurring simultaneously from several different presynaptic inputs is known as _____.

12. The neuronal relationship in which synapses from many presynaptic inputs act on a single postsynaptic cell is called _____, whereas the relationship in which a single presynaptic neuron synapses with and thereby influences the activity of many postsynaptic cells is known as _____.

13. A common membrane-bound intermediary between the receptor and the effector protein within the plasma membrane is the _____.

14. The three types of receptors with regard to mode of action are _____, _____, and _____.

15. Using the answer code on the right, indicate which potential is being described:
    ___ 1. behaves in all-or-none fashion
    ___ 2. magnitude of potential change varies with magnitude of triggering event
    ___ 3. decremental spread away from original site
    ___ 4. undiminishing spread throughout membrane
    ___ 5. serves as a long-distance signal
    ___ 6. serves as a short-distance signal

    (a) graded potential
    (b) action potential

16. Using the answer code on the right, indicate which characteristics apply to peptide and steroid hormones:
    ___ 1. are hydrophilic
    ___ 2. are lipophilic
    ___ 3. are synthesized by the ER
    ___ 4. are synthesized by modifying cholesterol
    ___ 5. include epinephrine from the adrenal medulla
    ___ 6. include cortisol from the adrenal cortex
    ___ 7. bind to plasma proteins
    ___ 8. bind to intracellular receptors
    ___ 9. bind to surface membrane receptors
    ___ 10. activate genes to promote synthesis of new proteins
    ___ 11. act via second messenger to alter preexisting proteins
    ___ 12. are secreted into blood by endocrine glands and carried to distant target sites

    (a) peptide hormones
    (b) steroid hormones
    (c) both peptide and steroid hormones
    (d) neither peptide nor steroid hormones

Essay Questions

1. What are the two types of excitable tissue?

2. Define the following terms: *polarization, depolarization, hyperpolarization, repolarization, resting membrane potential, threshold potential, action potential, refractory period,* and *all-or-none law.*

3. Describe the permeability changes and ion fluxes that occur during an action potential.

4. Compare contiguous conduction and saltatory conduction.

5. Compare the events that occur at excitatory and inhibitory synapses.

6. Compare the four kinds of gated channels in terms of the factor that opens or closes them.

7. Distinguish between classical neurotransmitters and neuropeptides. Explain what a neuromodulator is.

8. Discuss the possible outcomes of the GPSP brought about by interactions between EPSPs and IPSPs.

9. Distinguish between presynaptic inhibition and an inhibitory postsynaptic potential.

10. List and describe the types of intercellular communication.

11. Define the term *signal transduction.*

12. Distinguish between first and second messengers.

13. Describe the sequence of events in the cAMP second-messenger pathway.

14. Describe the sequence of events in the Ca^{2+} second-messenger pathway.

15. Explain how the cascading effect of hormonal pathways amplifies the response.

16. Compare the nervous and endocrine systems.

Quantitative Exercises (Solutions on p. A-41)

(See Appendix D, "Principles of Quantitative Reasoning.")

1. The following calculations give some insight into action potential conduction. Using the velocities given on p. 102:
   a. How long would it take for an action potential to travel 0.6 m along the axon of an unmyelinated neuron of the digestive tract?
   b. How long would it take for an action potential to travel the same distance along the axon of a large myelinated neuron innervating a skeletal muscle?
   c. Suppose there were two synapses in a 0.6 m nerve tract and the delay at each synapse is 1 msec. How long would it take an action potential/chemical signal to travel the 0.6 m now, for both the myelinated and unmyelinated neurons?
   d. What if there were five synapses?

2. Suppose point A is 1 m from point B. Compare the following situations:
   a. A single axon spans the distance from A to B, and its conduction velocity is 60 m/sec.
   b. Three neurons span the distance from A to B, all three neurons have the same conduction velocity, and the synaptic delay at both synapses (draw a picture) is 1 msec. What are the conduction velocities of the three neurons in this second situation if the total conduction time in both cases is the same?

3. One can predict what the Na^+ current produced by the Na^+–K^+ pump is with the following equation:[1]

$$p = \frac{kT}{q}\left(\frac{G_{Na^+}G_{K^+}}{G_{Na^+}G_{K^+}}\right)\log\frac{G_{K^+}[Na^+]_o}{G_{Na^+}[K^+]_i}$$

---

[1] F. C. Hoppensteadt and C. S. Peskin, *Mathematics in Medicine and the Life Sciences* (New York: Springer, 1992), equation 7.4.35, p. 178.

where p is the Na^+ pump current, G is membrane conductance to the indicated ion expressed in $\mu S/cm^2$ (S = Siemens), $[x]_o$ and $[x]_i$ are the concentrations of ion x outside and inside the cell respectively, k is Boltzmann's constant, T is the temperature in kelvins, and q is the elementary charge constant. Suppose $kT/q = 25$ mV, $G_{Na^+} = 3.3 \ \mu S/cm^2$, $G_{K^+} = 240 \ \mu S/cm^2$, $[Na^+]_o = 145$ mM, and $[K^+]_i = 4$ mM. What is the pump current for Na^+, in $\mu A/cm^2$ (A = amperes, an expression of current)?

POINTS TO PONDER

(Explanations on p. A-42)

1. Which of the following would occur if a neuron were experimentally stimulated simultaneously at both ends?
   a. The action potentials would pass in the middle and travel to the opposite ends.
   b. The action potentials would meet in the middle and then be propagated back to their starting positions.
   c. The action potentials would stop as they met in the middle.
   d. The stronger action potential would override the weaker action potential.
   e. Summation would occur when the action potentials met in the middle, resulting in a larger action potential.

2. Compare the expected changes in membrane potential of a neuron stimulated with a *subthreshold stimulus* (a stimulus not sufficient to bring a membrane to threshold), a *threshold stimulus* (a stimulus just sufficient to bring the membrane to threshold), and a *suprathreshold stimulus* (a stimulus larger than that necessary to bring the membrane to threshold).

3. Assume you touched a hot stove with your finger. Contraction of the biceps muscle causes flexion (bending) of the elbow, whereas contraction of the triceps muscle causes extension (straightening) of the elbow. What pattern of postsynaptic potentials (EPSPs and IPSPs) would you expect to be initiated as a reflex in the cell bodies of the neurons controlling these muscles to pull your hand away from the painful stimulus?

   Now assume your finger is being pricked to obtain a blood sample. The same *withdrawal reflex* would be initiated. What pattern of postsynaptic potentials would you voluntarily produce in the neurons controlling the biceps and triceps to keep your arm extended in spite of the painful stimulus?

4. Assume presynaptic excitatory neuron A terminates on a postsynaptic cell near the axon hillock and presynaptic excitatory neuron B terminates on the same postsynaptic cell on a dendrite located on the side of the cell body opposite the axon hillock. Explain why rapid firing of presynaptic neuron A could bring the postsynaptic neuron to threshold through temporal summation, thus initiating an action potential, whereas firing of presynaptic neuron B at the same frequency and the same magnitude of EPSPs may not bring the postsynaptic neuron to threshold.

5. Two classes of drugs that block different receptors are among those used to treat high blood pressure. (1) *Angiotensin receptor blockers (ARBs)* block binding of angiotensin to its receptors on the cells of the adrenal cortex that secrete a salt-conserving hormone, aldosterone. Aldosterone acts on the kidneys to conserve salt (specifically Na^+, with Cl^- following along the resulting electrical gradient) during urine formation. Salt accounts for over 90% of the osmotic (water-holding) activity of the ECF, including the plasma. (2) B_1-*adrenergic receptor blockers* block binding of epinephrine to its receptors in the heart. Epinephrine, a hormone secreted by the adrenal medulla, increases the rate and strength of contraction of the heart. Explain how each of these drugs lowers blood pressure.

6. Sometimes patients are treated for a number of years following surgical removal of a breast because of estrogen-dependent breast cancer with *selective estrogen receptor modulators (SERMs)*. Speculate how this drug might be beneficial. Indicate by what route this drug would be administered and explain why.

CLINICAL CONSIDERATION

(Explanation on p. A-42)

Becky N. was apprehensive as she sat in the dentist's chair awaiting the placement of her first silver amalgam (the "filling" in a cavity in a tooth). Before preparing the tooth for the amalgam by drilling away the decayed portion of the tooth, the dentist injected a local anesthetic in the nerve pathway supplying the region. As a result, Becky, much to her relief, did not feel any pain during the drilling and filling procedure. Local anesthetics block voltage-gated Na^+ channels. Explain how this action prevents the transmission of pain impulses to the brain.

Nervous System
(Central Nervous System)

Homeostasis
The nervous system, as one of the body's two major regulatory systems, regulates many body activities aimed at maintaining a stable internal fluid environment.

Body systems maintain homeostasis

Homeostasis is essential for survival of cells

Cells

Cells make up body systems

The **nervous system** is one of the two major regulatory systems of the body; the other is the endocrine system. The three basic functional types of neurons—afferent neurons, efferent neurons, and interneurons—form a complex interactive network of excitable cells. Ninety percent of the cells of the nervous system are glial cells, which serve as the connective tissue of the nervous system and interact extensively both structurally and functionally with neurons. The **central nervous system (CNS),** which consists of the brain and spinal cord, re-

ceives input about the external and internal environment from the afferent neurons. The CNS sorts and processes this input, then initiates appropriate directions in the efferent neurons, which carry the instructions to glands or muscles to bring about the desired response—some type of secretion or movement. Many of these neurally controlled activities are directed toward maintaining homeostasis. In general, the nervous system acts by means of its electrical signals (action potentials) to control the rapid responses of the body.

5

The Central Nervous System

CENGAGENOW Log on to CengageNOW at **http://www .cengage.com/sso/** for an opportunity to explore animations and interactive quizzes to help you learn, review, and master physiology concepts.

Organization and Cells of the Nervous System

The way humans act and react depends on complex, organized, discrete neuronal processing. Many basic life-supporting neuronal patterns, such as those controlling respiration and circulation, are similar in all individuals. However, there must be subtle differences in neuronal integration between someone who is a talented composer and someone who cannot carry a tune, or between someone who is a math wizard and someone who struggles with long division. Some differences in the nervous systems of individuals are genetically endowed. The rest, however, are due to environmental encounters and experiences. When the immature nervous system develops according to its genetic plan, an overabundance of neurons and synapses is formed. Depending on external stimuli and the extent to which these pathways are used, some are retained, firmly established, and even enhanced, whereas others are eliminated.

 A case in point is **amblyopia** (lazy eye), in which the weaker of the two eyes is not used for vision. A lazy eye that does not get appropriate visual stimulation during a critical developmental period will almost completely and permanently lose the power of vision. The functionally blind eye itself is completely normal; the defect lies in the lost neuronal connections in the brain's visual pathways. However, if the weak eye is forced to work by covering the stronger eye with a patch during the sensitive developmental period, the weaker eye will retain full vision.

The maturation of the nervous system involves many instances of "use it or lose it." Once the nervous system has matured, modifications still occur as we continue to learn from our unique set of experiences. For example, the act of reading this page is somehow altering the neuronal activity of your brain as you (it is hoped) tuck the information away in your memory.

The nervous system is organized into the central nervous system and the peripheral nervous system.

The nervous system is organized into the **central nervous system (CNS),** consisting of the brain and spinal cord, and the **peripheral nervous system (PNS),** consisting of nerve fibers that carry information between the CNS and other

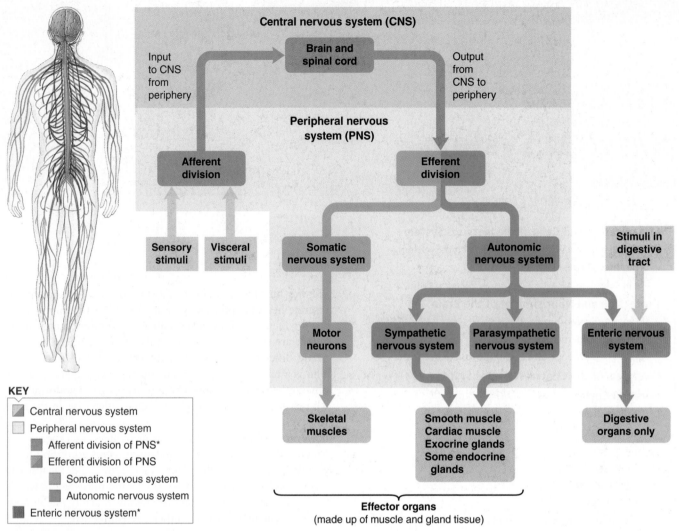

KEY

- Central nervous system
- Peripheral nervous system
- Afferent division of PNS*
- Efferent division of PNS
- Somatic nervous system
- Autonomic nervous system
- Enteric nervous system*

● **FIGURE 5-1 Organization of the nervous system.** *The afferent division of the PNS and enteric nervous system are not shown in the human figure. Afferent fibers travel within the same nerves as efferent fibers but in the opposite direction. The enteric nervous system lies entirely within the wall of the digestive tract.

parts of the body (the periphery) (● Figure 5-1). The PNS is further subdivided into afferent and efferent divisions. The **afferent division** carries information to the CNS, apprising it of the external environment and providing status reports on internal activities being regulated by the nervous system (*a* is from *ad,* meaning "toward," as in *advance; ferent* means "carrying"; thus, *afferent* means "carrying toward"). Instructions from the CNS are transmitted via the **efferent division** to **effector organs**—the muscles or glands that carry out the orders to bring about the desired effect (*e* is from *ex,* meaning "from," as in *exit;* thus, *efferent* means "carrying from"). The efferent nervous system is divided into the **somatic nervous system,** which consists of the fibers of the motor neurons that supply the skeletal muscles; and the **autonomic nervous system,** which consists of fibers that innervate smooth muscle, cardiac muscle, and glands. The latter system is further subdivided into the **sympathetic nervous**

system and the **parasympathetic nervous system,** both of which innervate most of the organs supplied by the autonomic system. In addition to the CNS and PNS, the **enteric nervous system** is an extensive nerve network in the wall of the digestive tract. Digestive activities are controlled by the autonomic nervous system and the enteric nervous system as well as by hormones. The enteric nervous system can act independently of the rest of the nervous system but is also influenced by autonomic fibers that terminate on the enteric neurons. Sometimes the enteric nervous system is considered a third component of the autonomic nervous system, one that supplies the digestive organs only.

It is important to recognize that all these "nervous systems" are really subdivisions of a single, integrated nervous system. They are arbitrary divisions based on differences in the structure, location, and functions of the various diverse parts of the whole nervous system.

The three functional classes of neurons are afferent neurons, efferent neurons, and interneurons.

Three functional classes of neurons make up the nervous system: *afferent neurons, efferent neurons,* and *interneurons.* The afferent division of the peripheral nervous system consists of **afferent neurons,** which are shaped differently from efferent neurons and interneurons (● Figure 5-2). At its peripheral ending, a typical afferent neuron has a **sensory receptor** that generates action potentials in response to a particular type of stimulus. (This stimulus-sensitive afferent neuronal receptor should not be confused with the special protein receptors that bind chemical messengers and are found in the plasma membrane of all cells.) The afferent neuron cell body, which is devoid of dendrites and presynaptic inputs, is located adjacent to the spinal cord. A long *peripheral axon,* commonly called the *afferent fiber,* extends from the receptor to the cell body, and a short *central axon* passes from the cell body into the spinal cord. Action potentials are initiated at the receptor end of the peripheral axon in response to a stimulus and are propagated along the peripheral axon and central axon toward the spinal cord. The terminals of the central axon diverge and synapse with other neurons within the spinal cord, thus disseminating information about the stimulus. Afferent neurons lie primarily within the peripheral nervous system. Only a small portion of their central axon endings projects into the spinal cord to relay signals from the periphery to the CNS.

Efferent neurons also lie primarily in the peripheral nervous system. Efferent neuron cell bodies originate in the CNS, where many centrally located presynaptic inputs converge on them to influence their outputs to the effector organs. Efferent axons (*efferent fibers*) leave the CNS to course their way to the muscles or glands they innervate, conveying their integrated output for the effector organs to put into effect. (An autonomic nerve pathway consists of a two-neuron chain between the CNS and the effector organ.)

About 99% of all neurons are **interneurons,** which lie entirely within the CNS. The human CNS is estimated to have more than 100 billion interneurons! As their name implies, interneurons lie between the afferent and efferent neurons and are important in integrating peripheral responses to peripheral information (*inter* means "between"). For example, on receiving information through afferent neurons that you are touching a hot object, appropriate interneurons signal efferent neurons that transmit to your hand and arm muscles the message, "Pull the hand away from the hot object!" The more complex the required action, the greater is the number of interneurons interposed between the afferent message and efferent response. In addition, interconnections between interneurons themselves are responsible for the abstract phenomena associated with the "mind," such as thoughts, emotions, memory, creativity, intel-

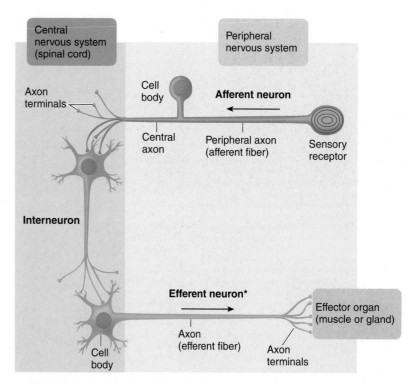

● **FIGURE 5-2 Structure and location of the three functional classes of neurons.** *Efferent autonomic nerve pathways consist of a two-neuron chain between the CNS and the effector organ.

lect, and motivation. These activities are the least understood functions of the nervous system.

Glial cells support the interneurons physically, metabolically, and functionally.

About 90% of the cells within the CNS are not neurons but **glial cells** or **neuroglia.** Despite their large numbers, glial cells occupy only about half the volume of the brain, because they do not branch as extensively as neurons do.

Unlike neurons, glial cells do not initiate or conduct nerve impulses. However, they do communicate with neurons and among themselves by means of chemical signals. For much of the time since the discovery of glial cells in the 19th century, scientists thought these cells were passive "mortar" that physically supported the functionally important neurons. In the last decade, however, the varied and important roles of these dynamic cells have become apparent. Glial cells serve as the connective tissue of the CNS and, as such, help support the neurons both physically and metabolically. They homeostatically maintain the composition of the specialized extracellular environment surrounding the neurons within the narrow limits optimal for normal neuronal function. Furthermore, they actively modulate synaptic function and are considered nearly as important as neurons to learning and memory. There are four major types of glial cells in the CNS—*astrocytes, oligodendrocytes, microglia,* and *ependymal cells*—each having specific roles (● Figure 5-3 and ▲ Table 5-1).

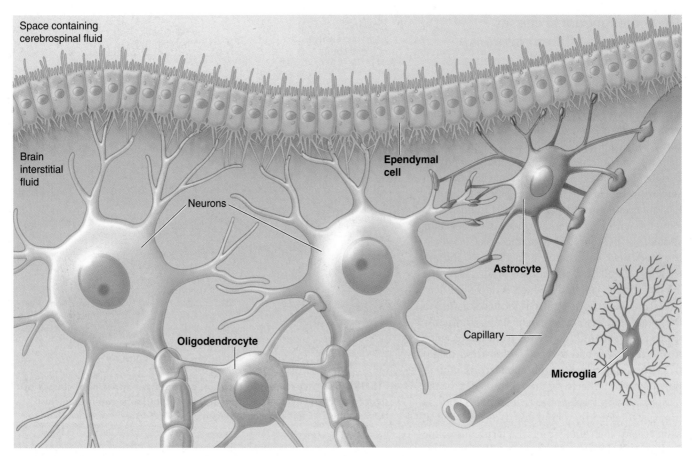

● **FIGURE 5-3 Glial cells of the central nervous system.** The glial cells include the astro-cytes, oligodendrocytes, microglia, and ependymal cells.

▲ TABLE 5-1	Functions of Glial Cells
Type of Glial Cell	**Functions**
Astrocytes	Physically support neurons in proper spatial relationships
	Serve as a scaffold during fetal brain development
	Induce formation of blood–brain barrier
	Help transfer nutrients to neurons
	Form neural scar tissue
	Take up and degrade released neurotransmitters
	Take up excess K^+ to help maintain proper brain-ECF ion concentration and normal neural excitability
	Enhance synapse formation and strengthen synaptic transmission via chemical signaling with neurons
	Communicate by chemical means with neurons and among themselves
Oligodendrocytes	Form myelin sheaths in CNS
Microglia	Play a role in defense of brain as phagocytic scavengers
	Release nerve growth factor
Ependymal Cells	Line internal cavities of brain and spinal cord
	Contribute to formation of cerebrospinal fluid
	Serve as neural stem cells with the potential to form new neurons and glial cells

ASTROCYTES Named for their starlike shape (*astro* means "star"; *cyte* means "cell") (● Figure 5-4), **astrocytes** are the most abundant glial cells. They fill a number of critical functions:

1. As the main "glue" (*glia* means "glue") of the CNS, astrocytes hold the neurons together in proper spatial relationships.

2. Astrocytes serve as a scaffold that guides neurons to their proper final destination during fetal brain development.

3. These glial cells induce the small blood vessels (capillaries) of the brain to undergo the anatomic and functional changes that establish the blood–brain barrier, a highly selective barricade between the blood and brain that we will soon describe in greater detail.

4. Through their close association with both local capillaries and neurons, astrocytes help transfer nutrients from the blood to the neurons.

5. Astrocytes are important in the repair of brain injuries and in neural scar formation.

6. They play a role in neurotransmitter activity. Astrocytes take up and degrade the excitatory neurotransmitter glutamate and the inhibitory neurotransmitter gamma-aminobutyric acid (GABA), thus bringing the actions of these chemical messengers to a halt.

7. Astrocytes take up excess K^+ from the brain ECF when high action potential activity outpaces the ability of the Na^+–K^+ pump to return the effluxed K^+ to the neurons. (Recall that K^+ leaves a neuron during the falling phase of an action potential; see p. 93.) By taking up excess K^+, astrocytes help maintain the optimal ion conditions around neurons to sustain normal neural excitability. If K^+ levels in the brain ECF were allowed to rise, the resultant lower K^+-concentration gradient between the neuronal ICF and surrounding ECF would lower the neuronal membrane closer to threshold, even at rest. This would increase the excitability of the brain. In fact, an elevation in brain-ECF K^+ concentration may be one of the factors responsible for the explosive convulsive discharge that occurs in brain cells during epileptic seizures.

8. Astrocytes along with other glial cells enhance synapse formation and modify synaptic transmission. Elongated, fine starlike extensions from an astrocyte's cell body are often wedged between presynaptic and postsynaptic portions of adjacent neurons. Recently scientists learned that an astrocyte can selectively retract these fine processes through actin activity, in a sort of reverse of pseudopod formation in amoeboid movement (see p. 48). Scientists suggest that withdrawal of these processes allows new synapses to form between neurons that were otherwise physically separated by the intervening astrocyte extension. Astrocytes also influence synapse formation and function by chemical means.

9. Astrocytes communicate with neurons and with one another by means of chemical signals in two ways. First, gap junctions (see p. 59) have been identified between astrocytes themselves and between astrocytes and neurons. Chemical signals pass directly between cells through the gap junctions without entering the ECF. Second, astrocytes have receptors for the common neurotransmitter glutamate. Furthermore, the firing of neurons in

Astrocyte

● **FIGURE 5-4 Astrocytes.** Note the starlike shape of these astrocytes, which have been grown in tissue culture.

the brain in some instances triggers the release of ATP along with the classical neurotransmitter from the axon terminal. Binding of glutamate to an astrocyte's receptors and/or detection of extracellular ATP by the astrocyte leads to calcium influx into this glial cell. The resultant rise in intracellular calcium prompts the astrocyte itself to release ATP, thereby activating adjacent glial cells. In this way, astrocytes share information about action potential activity in a nearby neuron. In addition, astrocytes and other glial cells can also release the same neurotransmitters as neurons do, as well as other chemical signals. These extracellular chemical signals from the glia can affect neuronal excitability and strengthen synaptic activity, such as by increasing neuronal release of neurotransmitter or promoting the formation of new synapses. Scientists have learned that astrocytes secrete **thrombospondin,** a large protein that can trigger synapse formation. Glial modulation of synaptic activity is likely important in memory and learning. Also, astrocytes are thought to coordinate and integrate synaptic activity among networks of neurons working together.

Scientists are trying to sort out the two-directional chatter that takes place between and among the glial cells and neurons because this dialogue plays an important role in the processing of information in the brain. In fact, some neuroscientists suggest that synapses should be considered "three-party" junctures involving the glial cells as well as the presynaptic and postsynaptic neurons. This point of view is indicative of the increasingly important role being placed on astrocytes in synapse function. Thus astrocytes have come a long way from their earlier reputation as "support staff" for neurons; these glial cells might turn out to be the "board members" commanding the neurons.

Right lateral ventricle

Left lateral ventricle

Third ventricle

Fourth ventricle

Central canal of spinal cord

(a) Lateral view of ventricles

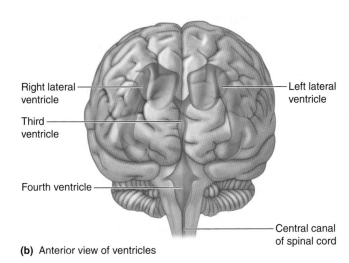

Right lateral ventricle

Left lateral ventricle

Third ventricle

Fourth ventricle

Central canal of spinal cord

(b) Anterior view of ventricles

● **FIGURE 5-5 The ventricles of the brain.**

OLIGODENDROCYTES Oligodendrocytes form the insulative myelin sheaths around axons in the CNS. An oligodendrocyte has several elongated projections, each of which is wrapped jelly-roll fashion around a section of an interneuronal axon to form a patch of myelin (see ● Figure 4-14b, p. 104; and ● Figure 5-3).

MICROGLIA Microglia are the immune defense cells of the CNS. These scavengers are "cousins" of monocytes, a type of white blood cell that leaves the blood and sets up residence as a front-line defense agent in various tissues throughout the body. Microglia are derived from the same bone-marrow tissue that gives rise to monocytes. During embryonic development, microglia migrate to the CNS, where they remain stationary until activated by an infection or injury.

In the resting state, microglia are wispy cells with many long branches that radiate outward. Resting microglia are not just waiting watchfully, however. They release low levels of growth factors, such as *nerve growth factor,* which help neurons and other glial cells survive and thrive. When trouble occurs in the CNS, microglia retract their branches, round up, and become highly mobile, moving toward the affected area to remove any foreign invaders or tissue debris by phagocytosis (see p. 31). Activated microglia also release destructive chemicals for assault against their target.

Researchers increasingly suspect that excessive release of these chemicals from overzealous microglia may damage the neurons they are meant to protect, thus contributing to the insidious neuronal damage seen in stroke, Alzheimer's disease, multiple sclerosis, the dementia (mental failing) of AIDS, and other *neurodegenerative diseases.*

EPENDYMAL CELLS Ependymal cells line the internal, fluid-filled cavities of the CNS. As the nervous system develops embryonically from a hollow neural tube, the original central cavity of this tube is maintained and modified to form the ventricles and central canal. The four **ventricles** are interconnected chambers within the brain that are continuous with the narrow, hollow **central canal** through the middle of the spinal cord (● Figure 5-5). The ependymal cells lining the ventricles help form cerebrospinal fluid, a topic to be discussed shortly. Ependymal cells are one of the few cell types with cilia (see p. 45). Beating of ependymal cilia contributes to the flow of cerebrospinal fluid through the ventricles.

Exciting new research has identified a totally different role for the ependymal cells as well: They serve as neural stem cells with the potential of forming not only other glial cells but new neurons as well (see p. 8). The traditional view has long held that new neurons are not produced in the mature brain. Then, in the late 1990s, scientists discovered that new neurons are produced in a specific part of the hippocampus, a structure important for learning and memory (see p. 163). Neurons in the rest of the brain are considered irreplaceable. But the discovery that ependymal cells are precursors for new neurons suggests that the adult brain has more potential for repairing damaged regions than previously assumed. Currently there is no evidence that the brain spontaneously repairs itself following neuron-losing insults such as head trauma, strokes, and neurodegenerative disorders. Apparently, most brain regions cannot activate this mechanism for replenishing neurons, probably because the appropriate "cocktail" of supportive chemicals is not present. Researchers hope that probing into why these ependymal cells are dormant and how they might be activated will lead to the possibility of unlocking the brain's latent capacity for self-repair.

Unlike neurons, glial cells do not lose the ability to undergo cell division, so most brain tumors of neural origin consist of glial cells (**gliomas**). Neurons themselves do not form tumors because they are unable to divide and multiply. Brain tumors of non-neural origin are of two types: (1) those that metastasize (spread) to the brain from other sites and (2) **meningiomas,** which originate from the meninges, the protective membranes covering the CNS. We next examine the meninges and other means by which the CNS is protected.

Protection and Nourishment of the Brain

Central nervous tissue is very delicate. This characteristic, coupled with the fact that damaged nerve cells cannot be replaced, makes it imperative that this fragile, irreplaceable tissue be well protected. Four major features help protect the CNS from injury:

1. It is enclosed by hard, bony structures. The *cranium (skull)* encases the brain, and the *vertebral column* surrounds the spinal cord.

2. Three protective and nourishing membranes, the *meninges,* lie between the bony covering and the nervous tissue.

3. The brain "floats" in a special cushioning fluid, the *cerebrospinal fluid (CSF)*.

4. A highly selective *blood–brain barrier* limits access of blood-borne materials into the vulnerable brain tissue.

The role of the first of these protective devices, the bony covering, is self-evident. The latter three protective mechanisms warrant further discussion.

Three meningeal membranes wrap, protect, and nourish the central nervous system.

Three membranes, the **meninges,** wrap the CNS. From the outermost to the innermost layer they are the *dura mater,* the *arachnoid mater,* and the *pia mater* (● Figure 5-6). (*Mater* means "mother," indicative of these membranes' protective and supportive role.)

The **dura mater** is a tough, inelastic covering that consists of two layers (*dura* means "tough"). Usually, these layers adhere closely, but in some regions they are separated to form blood-filled cavities, **dural sinuses,** or in the case of the larger cavities, **venous sinuses.** Venous blood draining from the brain empties into these sinuses to be returned to the heart. Cerebrospinal fluid also reenters the blood at one of these sinus sites.

The **arachnoid mater** is a delicate, richly vascularized layer with a "cobwebby" appearance (*arachnoid* means "spiderlike"). The space between the arachnoid layer and the underlying pia mater, the **subarachnoid space,** is filled with CSF. Protrusions of arachnoid tissue, the **arachnoid villi,** penetrate through gaps in the overlying dura and project into the dural sinuses. CSF is reabsorbed across the surfaces of these villi into the blood circulating within the sinuses.

The innermost meningeal layer, the **pia mater,** is the most fragile (*pia* means "gentle"). It is highly vascular and closely adheres to the surfaces of the brain and spinal cord, following every ridge and valley. In certain areas it dips deeply into the brain to bring a rich blood supply into close contact with the ependymal cells lining the ventricles. This relationship is important in the formation of CSF, a topic to which we now turn our attention.

The brain floats in its own special cerebrospinal fluid.

Cerebrospinal fluid (CSF) surrounds and cushions the brain and spinal cord. The CSF has about the same density as the brain itself, so the brain essentially floats or is suspended in this special fluid environment. The major function of CSF is to serve as a shock-absorbing fluid to prevent the brain from bumping against the interior of the hard skull when the head is subjected to sudden, jarring movements.

In addition to protecting the delicate brain from mechanical trauma, the CSF plays an important role in the exchange of materials between the neural cells and the interstitial fluid surrounding the brain. Only the brain interstitial fluid—not the blood or CSF—comes into direct contact with the neurons and glial cells. Because the brain interstitial fluid directly bathes the neural cells, its composition is critical. The composition of the brain interstitial fluid is influenced more by changes in the composition of the CSF than by alterations in the blood. Materials are exchanged fairly freely between the CSF and brain interstitial fluid, whereas only limited exchange occurs between the blood and brain interstitial fluid. Thus, the composition of the CSF must be carefully regulated.

Cerebrospinal fluid is formed primarily by the **choroid plexuses** found in particular regions of the ventricles. Choroid plexuses consist of richly vascularized, cauliflower-like masses of pia mater tissue that dip into pockets formed by ependymal cells. Cerebrospinal fluid forms as a result of selective transport mechanisms across the membranes of the choroid plexuses. The composition of CSF differs from that of blood. For example, CSF is lower in K^+ and slightly higher in Na^+, making the brain interstitial fluid an ideal environment for movement of these ions down concentration gradients, a process essential for conduction of nerve impulses (see p. 96). The biggest difference is the presence of plasma proteins in the blood but almost no proteins normally present in the CSF. Plasma proteins cannot exit the brain capillaries to leave the blood during formation of CSF.

Once CSF is formed, it flows through the four interconnected ventricles of the brain and through the spinal cord's narrow central canal, which is continuous with the last ventricle. Cerebrospinal fluid also escapes through small openings from the fourth ventricle at the base of the brain to enter the subarachnoid space and subsequently flows between the meningeal layers over the entire surface of the brain and spinal cord (● Figure 5-6). When the CSF reaches the upper regions of the brain, it is reabsorbed from the subarachnoid space into the venous blood through the arachnoid villi. Flow of CSF through this system is facilitated by ciliary beating along with circulatory and postural factors that result in a CSF pressure of about 10 mm Hg. Reduction of this pressure by removal of even a few milliliters (ml) of CSF during a spinal tap for laboratory analysis may produce severe headaches.

Clinical Note Through the ongoing processes of formation, circulation, and reabsorption, the entire CSF volume of about 125 to 150 ml is replaced more than three times a day. If any one of these processes is defective so that excess CSF accumulates, **hydrocephalus** ("water on the brain") occurs. The

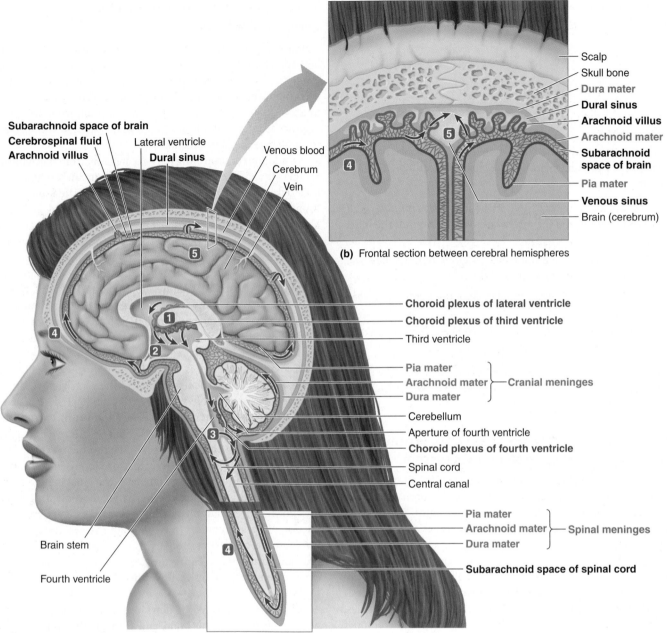

Subarachnoid space of brain
Cerebrospinal fluid
Arachnoid villus

Lateral ventricle

Dural sinus

Venous blood

Cerebrum

Vein

Scalp
Skull bone
Dura mater
Dural sinus
Arachnoid villus
Arachnoid mater
Subarachnoid space of brain
Pia mater
Venous sinus
Brain (cerebrum)

(b) Frontal section between cerebral hemispheres

Choroid plexus of lateral ventricle
Choroid plexus of third ventricle
Third ventricle

Pia mater
Arachnoid mater } **Cranial meninges**
Dura mater

Cerebellum
Aperture of fourth ventricle
Choroid plexus of fourth ventricle
Spinal cord
Central canal

Brain stem

Pia mater
Arachnoid mater } **Spinal meninges**
Dura mater

Subarachnoid space of spinal cord

Fourth ventricle

(a) Sagittal section of brain and spinal cord

Cerebrospinal fluid

1 is produced by the choroid plexuses,

2 circulates throughout the ventricles,

3 exits the fourth ventricle at the base of the brain,

4 flows in the subarachnoid space between the meningeal layers, and

5 is finally reabsorbed from the subarachnoid space into the venous blood across the arachnoid villi.

● **FIGURE 5-6 Relationship of the meninges and cerebrospinal fluid to the brain and spinal cord.** (a) Brain, spinal cord, and meninges in sagittal section. The arrows and numbered steps indicate the direction of flow of cerebrospinal fluid (in yellow). (b) Frontal section in the region between the two cerebral hemispheres of the brain, depicting the meninges in greater detail.

resulting increase in CSF pressure can lead to brain damage and mental retardation if untreated. Treatment consists of surgically shunting the excess CSF to veins elsewhere in the body.

A highly selective blood–brain barrier regulates exchanges between the blood and brain.

The brain is carefully shielded from harmful changes in the blood by a highly selective **blood–brain barrier (BBB).** Throughout the body, materials can be exchanged between the blood and interstitial fluid only across the walls of capillaries, the smallest blood vessels. Unlike the rather free exchange across capillaries elsewhere, only selected, carefully regulated exchanges can be made across the BBB. For example, even if the K^+ level in the blood is doubled, little change occurs in the K^+ concentration of the fluid bathing the central neurons. This is beneficial because alterations in interstitial fluid K^+ would be detrimental to neuronal function.

The BBB has both anatomic and physiologic features. Capillary walls throughout the body are formed by a single layer of cells. Usually, all blood plasma components (except the large plasma proteins) can be freely exchanged between the blood and the interstitial fluid through holes or pores between the cells of the capillary wall. In brain capillaries, however, the cells are joined by *tight junctions* (see p. 59), which completely seal the capillary wall so that nothing can be exchanged across the wall by passing between the cells. The only possible exchanges are through the capillary cells themselves. Lipid-soluble substances such as O_2, CO_2, alcohol, and steroid hormones penetrate these cells easily by dissolving in their lipid plasma membrane. Small water molecules also diffuse through readily, by passing between the phospholipid molecules of the plasma membrane or through aquaporins (water channels) (see p. 63). All other substances exchanged between the blood and brain interstitial fluid, including such essential materials as glucose, amino acids, and ions, are transported by highly selective membrane-bound carriers. Thus, transport across brain capillary walls *between* the wall-forming cells is prevented *anatomically* and transport *through* the cells is restricted *physiologically.* Together, these mechanisms constitute the BBB.

By strictly limiting exchange between the blood and brain, the BBB protects the delicate brain from chemical fluctuations in the blood and minimizes the possibility that potentially harmful blood-borne substances might reach the central neural tissue. It further prevents certain circulating hormones that could also act as neurotransmitters from reaching the brain, where they could produce uncontrolled nervous activity. On the negative side, the BBB limits the use of drugs for the treatment of brain and spinal cord disorders because many drugs are unable to penetrate this barrier.

Brain capillaries are surrounded by astrocyte processes, which at one time were erroneously thought to be physically responsible for the BBB. Scientists now know that astrocytes play two roles in the BBB: (1) They signal the cells forming the brain capillaries to "get tight." Capillary cells do not have an inherent ability to form tight junctions; they do so only at the command of a signal within their neural environment. (2) Astrocytes participate in the cross-cellular transport of some substances, such as K^+.

Certain areas of the brain, most notably a portion of the hypothalamus, are not subject to the BBB. Functioning of the hypothalamus depends on its "sampling" the blood and adjusting its controlling output accordingly to maintain homeostasis. Part of this output is in the form of water-soluble hormones that must enter hypothalamic capillaries to be transported to their sites of action. Appropriately, these hypothalamic capillaries are not sealed by tight junctions.

The brain depends on constant delivery of oxygen and glucose by the blood.

Even though many substances in the blood never actually come in contact with the brain tissue, the brain is more dependent than any other tissue on a constant blood supply. Unlike most tissues, which can resort to anaerobic metabolism to produce ATP in the absence of O_2 for at least short periods (see p. 37), the brain cannot produce ATP without O_2. Scientists recently discovered an O_2-binding protein, **neuroglobin,** in the brain. This molecule, which is similar to hemoglobin, the O_2-carrying protein in red blood cells (see p. 142), is thought to play a key role in O_2 handling in the brain, although its exact function remains to be determined. Also in contrast to most tissues, which can use other sources of fuel for energy production in lieu of glucose, the brain normally uses only glucose but does not store any of this nutrient. Because of its high rate of demand for ATP, under resting conditions the brain uses 20% of the O_2 and 50% of the glucose consumed in the body. Therefore, the brain depends on a continuous, adequate blood supply of O_2 and glucose. Although it constitutes only 2% of body weight, the brain receives 15% of the blood pumped out by the heart. (Instead of using glucose during starvation, the brain can resort to using ketone bodies produced by the liver, but this alternate nutrient source also must be delivered by the blood to the brain.)

 Brain damage results if this organ is deprived of its critical O_2 supply for more than 4 to 5 minutes or if its glucose supply is cut off for more than 10 to 15 minutes. The most common cause of inadequate blood supply to the brain is a stroke. (See the boxed feature on p. 142, ■ Concepts, Challenges, and Controversies, for details.)

Overview of the Central Nervous System

The CNS consists of the brain and spinal cord. The estimated 100 billion neurons in your brain are assembled into complex networks that enable you to (1) subconsciously regulate your internal environment by neural means, (2) experience emotions, (3) voluntarily control your movements, (4) perceive (be consciously aware of) your own body and your surroundings, and (5) engage in other higher cognitive processes such as thought and memory. The term **cognition** refers to the act or process of "knowing," including both awareness and judgment.

Strokes: A Deadly Domino Effect

The most common cause of brain damage is a **cerebrovascular accident (CVA or stroke).** When a cerebral (brain) blood vessel is blocked by a clot (which accounts for over 80% of strokes) or ruptures, the brain tissue supplied by that vessel loses its vital O_2 and glucose supply. The result is damage and usually death of the deprived tissue. New findings show that neural damage (and the subsequent loss of neural function) extends well beyond the blood-deprived area as a result of a neurotoxic effect that leads to the death of additional nearby cells. The initial blood-deprived cells die by necrosis (unintentional cell death), but the doomed neighbors undergo apoptosis (deliberate cell suicide; see p. 124). In a process known as **excitotoxicity,** the initial O_2-starved cells release excessive amounts of glutamate, a common excitatory neurotransmitter. The excitatory overdose of glutamate from the damaged brain cells binds with and overexcites surrounding neurons. Specifically, glutamate binds with excitatory receptors known as NMDA receptors, which function as Ca^{2+} channels. As a result of toxic activation of these receptor channels, they remain open for too long, permitting too much Ca^{2+} to rush into the affected neighboring neurons. This elevated intracellular Ca^{2+} triggers the cells to self-destruct, producing **free radicals** in the process. These highly reactive, electron-deficient particles cause further cell damage by snatching electrons from other molecules. Adding to the injury, researchers speculate that the Ca^{2+} apoptotic signal may spread from the dying cells to abutting healthy cells through gap junctions, which allow Ca^{2+} and other small ions to diffuse freely between cells. This action kills even more neurons. Thus, most neurons that die following a stroke are originally unharmed cells that commit suicide in response to the chain of reactions unleashed by the toxic release of glutamate from the initial site of O_2 deprivation.

Until the last decade, physicians could do nothing to halt the inevitable neuronal loss following a stroke, leaving patients with an unpredictable mix of neural deficits. Treatment was limited to rehabilitative therapy after the damage was already complete. In recent years, armed with new knowledge about the underlying factors in stroke-related neuronal death, the medical community has been seeking ways to halt the cell-killing domino effect. The goal, of course, is to limit the extent of neuronal damage and thus minimize or even prevent clinical symptoms such as paralysis. In the early 1990s, doctors started administering clot-dissolving drugs within the first three hours after the onset of a stroke to restore blood flow through blocked cerebral vessels. Clot busters were the first drugs used to treat strokes, but they are only the beginning of new stroke therapies. Other methods are currently under investigation to prevent adjacent neurons from succumbing to the neurotoxic release of glutamate. These include blocking the NMDA receptors that initiate the death-wielding chain of events in response to glutamate, halting the apoptosis pathway that results in self-execution, and blocking the gap junctions that permit the Ca^{2+} death messenger to spread to adjacent cells. These tactics hold much promise for treating strokes, which are the most common cause of adult disability and the third leading cause of death in the United States. However, to date no new neuroprotective drugs have been found that do not cause serious side effects.

No part of the brain acts in isolation from other brain regions, because networks of neurons are anatomically linked by synapses, and neurons throughout the brain communicate extensively with each other by electrical and chemical means. However, neurons that work together to ultimately accomplish a given function tend to be organized within a discrete location. Therefore, even though the brain operates as a whole, it is organized into different regions. The parts of the brain can be grouped in various ways based on anatomic distinctions, functional specialization, and evolutionary development. We will use the following grouping:

1. Brain stem
2. Cerebellum
3. Forebrain
   a. Diencephalon
      (1) Hypothalamus
      (2) Thalamus
   b. Cerebrum
      (1) Basal nuclei
      (2) Cerebral cortex

The order in which these components are listed generally represents both their anatomic location (from bottom to top) and their complexity and sophistication of function (from the least specialized, oldest level to the newest, most specialized level).

A primitive nervous system consists of comparatively few interneurons interspersed between afferent and efferent neurons. During evolutionary development, the interneuronal component progressively expanded, formed more complex interconnections, and became localized at the head end of the nervous system, forming the brain. Newer, more sophisticated layers of the brain were added on to the older, more primitive layers. The human brain represents the present peak of development.

The *brain stem*, the oldest region of the brain, is continuous with the spinal cord (▲ Table 5-2 and ● Figure 5-7b). It consists of the midbrain, pons, and medulla. The brain stem controls many of the life-sustaining processes, such as respiration, circulation, and digestion, that are common to all vertebrates. These processes are often referred to as *vegetative functions*, meaning functions performed unconsciously or involuntarily. With the loss of higher brain functions, these lower brain levels, in accompaniment with appropriate supportive therapy such as provision of adequate nourishment, can still sustain the functions essential for survival, but the person has no awareness or control of that life.

Attached at the top rear portion of the brain stem is the *cerebellum,* which is concerned with maintaining proper position of the body in space and subconscious coordination of motor activity (movement). The cerebellum also plays a key role in learning skilled motor tasks, such as a dance routine.

On top of the brain stem, tucked within the interior of the cerebrum, is the *diencephalon.* It houses two brain components: the *hypothalamus,* which controls many homeostatic functions important in maintaining stability of the internal environment; and the *thalamus,* which performs some primitive sensory processing.

Using an ice cream cone as an analogy, on top of this "cone" of lower brain regions is the *cerebrum,* whose "scoop" gets progressively larger and more highly convoluted (that is, has tortuous ridges delineated by deep grooves or folds) the more advanced the vertebrate species is. The cerebrum is most highly developed in humans, where it constitutes about 80% of the total brain weight. The outer layer of the cerebrum is the highly convoluted *cerebral cortex,* which caps an inner core that houses the *basal nuclei.* The myriad convolutions of the human cerebral cortex give it the appearance of a much-folded walnut (● Figure 5-7a). In many lower mammals, the cortex is smooth. Without these surface wrinkles, the human cortex would take up to three times the area it does, and, thus, would not fit like a cover over the underlying structures. The increased neural circuitry housed in the extra cerebral cortical area not found in lower species is responsible for many of our unique human abilities. The cerebral cortex plays a key role in the most sophisticated neural functions, such as voluntary initiation of movement, final sensory perception, conscious thought, language, personality traits, and other factors we associate with the mind or intellect. It is the highest, most complex, integrating area of the brain.

Each of these regions of the CNS will be discussed in turn, starting with the highest level, the cerebral cortex, and moving down to the lowest level, the spinal cord.

Cerebral Cortex

The **cerebrum,** by far the largest portion of the human brain, is divided into two halves, the right and left **cerebral hemispheres** (● Figure 5-7a). They are connected to each other by the **corpus callosum,** a thick band consisting of an estimated 300 million neuronal axons that connect the two hemispheres (● Figure 5-7b; also see ● Figure 5-14, p. 154). The corpus callosum is the body's "information superhighway." The two hemispheres communicate and cooperate with each other by means of constant information exchange through this neural connection.

The cerebral cortex is an outer shell of gray matter covering an inner core of white matter.

Each hemisphere is composed of a thin outer shell of *gray matter,* the **cerebral cortex,** covering a thick central core of *white matter* (see ● Figure 5-14). Several other masses of gray matter

that collectively constitute the basal nuclei are located deep within the white matter. Throughout the entire CNS, **gray matter** consists predominantly of densely packaged neuronal cell bodies and their dendrites as well as most glial cells. Bundles or tracts of myelinated nerve fibers (axons) constitute the **white matter;** its white appearance is due to the lipid composition of the myelin. The gray matter can be viewed as the "computers" of the CNS and the white matter as the "wires" that connect the computers to one another. Integration of neural input and initiation of neural output take place at synapses within the gray matter. The fiber tracts in the white matter transmit signals from one part of the cerebral cortex to another or between the cortex and other regions of the CNS. Such communication between different areas of the cortex and elsewhere facilitates integration of their activity. This integration is essential for even a relatively simple task such as picking a flower. Vision of the flower is received by one area of the cortex, reception of its fragrance takes place in another area, and movement is initiated by still another area. More subtle neuronal responses, such as appreciation of the flower's beauty and the urge to pick it, are poorly understood but undoubtedly involve extensive interconnection of fibers between different cortical regions.

The cerebral cortex is organized into layers and functional columns.

The cerebral cortex is organized into six well-defined layers based on varying distributions of several distinctive cell types. These layers are organized into functional vertical columns that extend perpendicularly about 2 mm from the cortical surface down through the thickness of the cortex to the underlying white matter. The neurons within a given column function as a "team," with each cell being involved in different aspects of the same specific activity—for example, perceptual processing of the same stimulus from the same location.

The functional differences between various areas of the cortex result from different layering patterns within the columns and from different input–output connections, not from the presence of unique cell types or different neuronal mechanisms. For example, those regions of the cortex responsible for sensory perception have an expanded layer 4, a layer rich in **stellate cells,** which are neurons responsible for initial processing of sensory input to the cortex. In contrast, cortical areas that control output to skeletal muscles have a thickened layer 5, which contains an abundance of large neurons known as **pyramidal cells.** These nerve cells send fibers down the spinal cord from the cortex to terminate on efferent motor neurons that innervate skeletal muscles.

The four pairs of lobes in the cerebral cortex are specialized for different activities.

We now consider the locations of the major functional areas of the cerebral cortex. Throughout this discussion, keep in mind that even though a discrete activity is ultimately attributed to a particular region of the brain, no part of the brain functions in isolation. Each part depends on complex interplay among numerous other regions for both incoming and outgoing messages.

Overview of Structures and Functions of the Major Components of the Brain

Brain Component

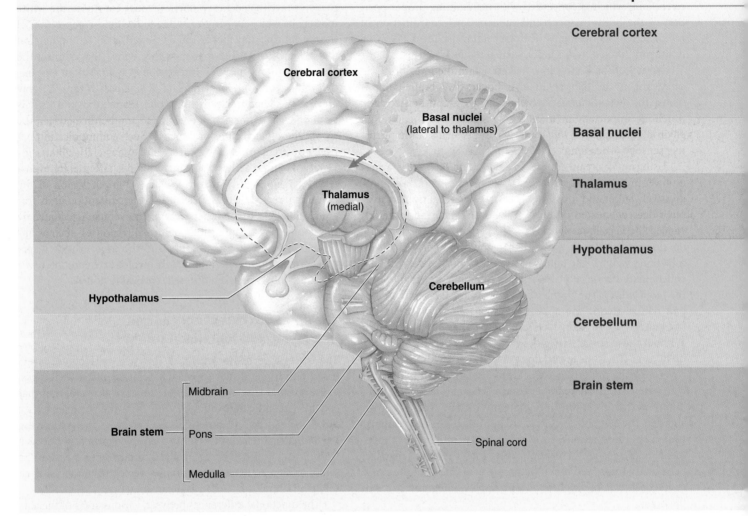

Cerebral cortex

Basal nuclei

Thalamus

Hypothalamus

Cerebellum

Brain stem

The anatomic landmarks used in cortical mapping are specific deep folds that divide each half of the cortex into four major lobes: the *occipital, temporal, parietal,* and *frontal lobes* (● Figure 5-8). Look at the basic functional map of the cortex in ● Figure 5-9a during the following discussion of the major activities attributed to various regions of these lobes.

The **occipital lobes,** located posteriorly (at the back of the head), carry out the initial processing of visual input. Auditory (sound) sensation is initially received by the **temporal lobes,** located laterally (on the sides of the head) (● Figure 5-9a and b). You will learn more about the functions of these regions in Chapter 6 when we discuss vision and hearing.

The parietal lobes and frontal lobes, located on the top of the head, are separated by a deep infolding, the **central sulcus,** which runs roughly down the middle of the lateral surface of each hemisphere. The **parietal lobes** lie to the rear of the central sulcus on each side, and the **frontal lobes** lie in front of it. The parietal lobes are primarily responsible for receiving and processing sensory input. The frontal lobes are responsible for three main functions: (1) voluntary motor activity, (2) speaking

ability, and (3) elaboration of thought. We next examine the role of the parietal lobes in sensory perception, then turn our attention to the functions of the frontal lobes in more detail.

The parietal lobes accomplish somatosensory processing.

Sensations from the surface of the body, such as touch, pressure, heat, cold, and pain, are collectively known as **somesthetic sensations** (*somesthetic* means "body feelings"). The means by which afferent neurons detect and relay information to the CNS about these sensations will be covered in Chapter 6 when we explore the afferent division of the peripheral nervous system in detail. Within the CNS, this information is *projected* (transmitted along specific neural pathways to higher brain levels) to the **somatosensory cortex.** The somatosensory cortex is located in the front portion of each parietal lobe immediately behind the central sulcus (● Figures 5-9a and 5-10a). It is the site for initial cortical processing and perception of both somesthetic and proprioceptive input. **Proprioception** is the awareness of body position.

Major Functions

1. Sensory perception
2. Voluntary control of movement
3. Language
4. Personality traits
5. Sophisticated mental events, such as thinking, memory, decision making, creativity, and self-consciousness

1. Inhibition of muscle tone
2. Coordination of slow, sustained movements
3. Suppression of useless patterns of movement

1. Relay station for all synaptic input
2. Crude awareness of sensation
3. Some degree of consciousness
4. Role in motor control

1. Regulation of many homeostatic functions, such as temperature control, thirst, urine output, and food intake
2. Important link between nervous and endocrine systems
3. Extensive involvement with emotion and basic behavioral patterns
4. Role in sleep–wake cycle

1. Maintenance of balance
2. Enhancement of muscle tone
3. Coordination and planning of skilled voluntary muscle activity

1. Origin of majority of peripheral cranial nerves
2. Cardiovascular, respiratory, and digestive control centers
3. Regulation of muscle reflexes involved with equilibrium and posture
4. Reception and integration of all synaptic input from spinal cord; arousal and activation of cerebral cortex
5. Role in sleep–wake cycle

Each region within the somatosensory cortex receives somesthetic and proprioceptive input from a specific area of the body. This distribution of cortical sensory processing is depicted in ● Figure 5-10b. Note that on this so-called **sensory homunculus** (*homunculus* means "little man"), the different parts of the body are not equally represented. The size of each body part in this homunculus indicates the relative proportion of the somatosensory cortex devoted to that area. The exaggerated size of the face, tongue, hands, and genitalia indicates the high degree of sensory perception associated with these body parts.

The somatosensory cortex on each side of the brain for the most part receives sensory input from the opposite side of the body, because most of the ascending pathways that carry sensory information up the spinal cord cross over to the opposite side before eventually terminating in the cortex (see ● Figure 5-28a, p. 175). Thus, damage to the somatosensory cortex in the left hemisphere produces sensory deficits on the right side of the body, whereas sensory losses on the left side are associated with damage to the right half of the cortex.

Simple awareness of touch, pressure, temperature, or pain is detected by the thalamus, a lower level of the brain, but the somatosensory cortex goes beyond mere recognition of sensations to fuller sensory perception. The thalamus makes you aware that something hot versus something cold is touching your body, but it does not tell you where or of what intensity. The somatosensory cortex localizes the source of sensory input and perceives the level of intensity of the stimulus. It also is capable of spatial discrimination, so it can discern shapes of objects being held and can distinguish subtle differences in similar objects that come into contact with the skin.

The somatosensory cortex, in turn, projects this sensory input via white matter fibers to adjacent higher sensory areas for even further elaboration, analysis, and integration of sensory information. These higher areas are important in perceiving complex patterns of somatosensory stimulation—for example, simultaneous appreciation of the texture, firmness, temperature, shape, position, and location of an object you are holding.

The primary motor cortex located in the frontal lobes controls the skeletal muscles.

The area in the rear portion of the frontal lobe immediately in front of the central sulcus and next to the somatosensory cortex is the **primary motor cortex** (see ● Figures 5-9a and 5-10a). It confers voluntary control over movement produced by skeletal muscles. As in sensory processing, the motor cortex on each side of the brain primarily controls muscles on the opposite side of the body. Neuronal tracts originating in the motor cortex of the left hemisphere cross over before passing down the spinal cord to terminate on efferent motor neurons that trigger skeletal muscle contraction on the right side of the body (see ● Figure 5-28b, p. 175). Accordingly, damage to the motor cortex on the left side of the brain produces paralysis on the right side of the body, and the converse is also true.

Stimulation of different areas of the primary motor cortex brings about movement in different regions of the body. Like the sensory homunculus for the somatosensory cortex, the **motor homunculus,** which depicts the location and relative amount of motor cortex devoted to output to the muscles of each body part, is distorted (● Figure 5-10c). The fingers, thumbs, and muscles important in speech, especially those of the lips and tongue, are grossly exaggerated, indicating the fine degree of motor control these body parts have. Compare this to how little brain tissue is devoted to the trunk, arms, and lower extremities, which are not capable of such complex movements. Thus, the extent of representation in the motor cortex is proportional to the precision and complexity of motor skills required of the respective part.

The higher motor areas are also important in motor control.

Even though signals from the primary motor cortex terminate on the efferent neurons that trigger voluntary skeletal muscle contraction, the motor cortex is not the only region of the brain involved with motor control. First, lower brain regions and the

Left cerebral hemisphere　Right cerebral hemisphere

Longitudinal fissure

(a) Brain, dorsal view

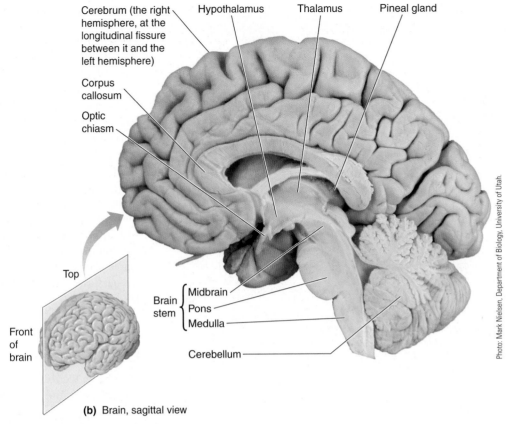

Cerebrum (the right hemisphere, at the longitudinal fissure between it and the left hemisphere)

Corpus callosum

Optic chiasm

Hypothalamus　Thalamus　Pineal gland

Brain stem { Midbrain
Pons
Medulla }

Cerebellum

Top

Front of brain

(b) Brain, sagittal view

Photo: Mark Nielsen, Department of Biology, University of Utah.

● **FIGURE 5-7 Brain of a human cadaver.** (a) Dorsal view looking down on the top of the brain. Note that the deep longitudinal fissure divides the cerebrum into the right and left cerebral hemispheres. (b) Sagittal view of the right half of the brain. All major brain regions are visible from this midline interior view. The corpus callosum serves as a neural bridge between the two cerebral hemispheres.

spinal cord control involuntary skeletal muscle activity, such as in maintaining posture. Some of these same regions also play an important role in monitoring and coordinating voluntary motor activity that the primary motor cortex has set in motion. Second, although fibers originating from the motor cortex can activate motor neurons to bring about muscle contraction, the motor cortex itself does not *initiate* voluntary movement. The motor cortex is activated by a widespread pattern of neuronal discharge, the **readiness potential,** which occurs about 750 msec before specific electrical activity is detectable in the motor cortex. Three higher motor areas of the cortex are involved in this voluntary decision-making period. These higher areas, which all command the primary motor cortex, include the *supplementary motor area,* the *premotor cortex,* and the *posterior parietal cortex* (see ● Figure 5-9a). Furthermore, a subcortical region of the brain, the *cerebellum,* plays an important role in planning, initiating, and timing certain kinds of movement by sending input to the motor areas of the cortex.

The three higher motor areas of the cortex and the cerebellum carry out different, related functions that are all important in programming and coordinating complex movements that involve simultaneous contraction of many muscles. Even though electrical stimulation of the primary motor cortex brings about

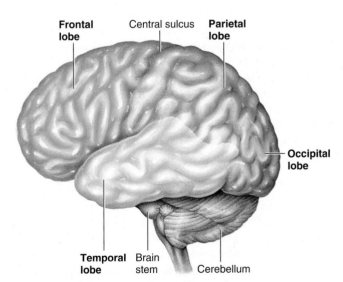

Frontal lobe　Central sulcus　Parietal lobe

Occipital lobe

Temporal lobe　Brain stem　Cerebellum

● **FIGURE 5-8 Cortical lobes.** Each half of the cerebral cortex is divided into the occipital, temporal, parietal, and frontal lobes, as depicted in this lateral view of the brain.

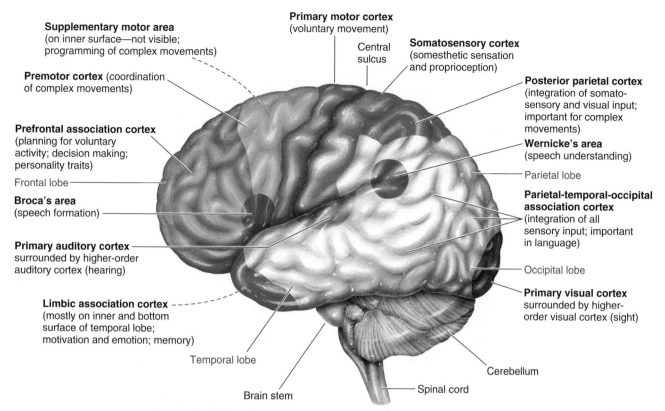

Supplementary motor area (on inner surface—not visible; programming of complex movements)

Premotor cortex (coordination of complex movements)

Prefrontal association cortex (planning for voluntary activity; decision making; personality traits)

Frontal lobe

Broca's area (speech formation)

Primary auditory cortex surrounded by higher-order auditory cortex (hearing)

Limbic association cortex (mostly on inner and bottom surface of temporal lobe; motivation and emotion; memory)

Temporal lobe

Brain stem

Primary motor cortex (voluntary movement)

Central sulcus

Somatosensory cortex (somesthetic sensation and proprioception)

Posterior parietal cortex (integration of somatosensory and visual input; important for complex movements)

Wernicke's area (speech understanding)

Parietal lobe

Parietal-temporal-occipital association cortex (integration of all sensory input; important in language)

Occipital lobe

Primary visual cortex surrounded by higher-order visual cortex (sight)

Cerebellum

Spinal cord

(a) Regions of the cerebral cortex responsible for various functions

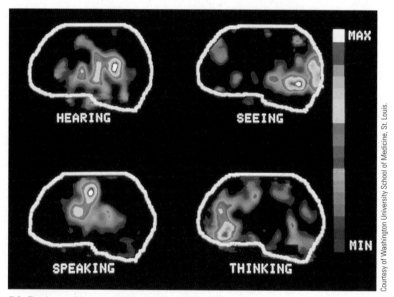

HEARING   SEEING

SPEAKING   THINKING

MAX

MIN

Courtesy of Washington University School of Medicine, St. Louis.

(b) Regions of increased blood flow during different tasks

● **FIGURE 5-9 Functional areas of the cerebral cortex.** (a) Various regions of the cerebral cortex are primarily responsible for various aspects of neural processing, as indicated in this lateral view of the brain. (b) Different areas of the brain "light up" on positron-emission tomography (PET) scans as a person performs different tasks. PET scans detect the magnitude of blood flow in various regions of the brain. Because more blood flows into a particular region of the brain when it is more active, neuroscientists can use PET scans to "take pictures" of the brain at work on various tasks.

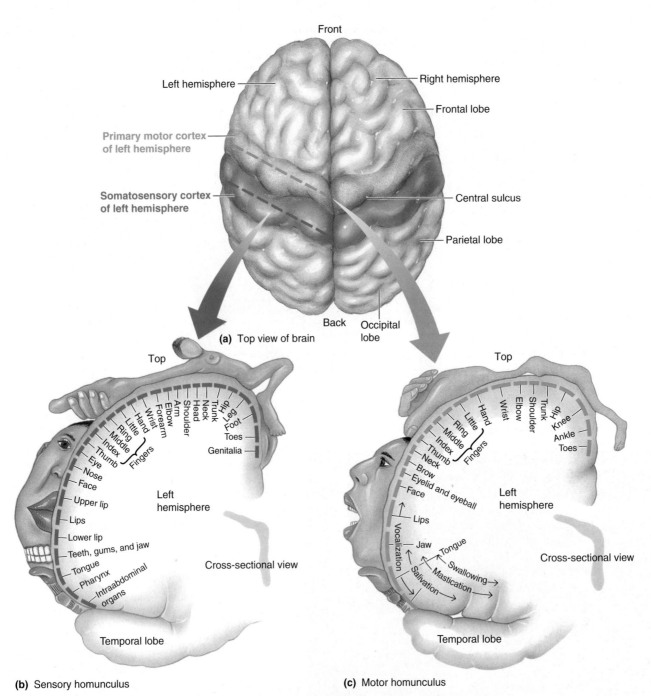

(a) Top view of brain

(b) Sensory homunculus

(c) Motor homunculus

● **FIGURE 5-10 Somatotopic maps of the somatosensory cortex and primary motor cortex.** (a) Top view of cerebral hemispheres showing somatosensory cortex and primary motor cortex. (b) Sensory homunculus showing the distribution of sensory input to the somatosensory cortex from different parts of the body. The distorted graphic representation of the body parts indicates the relative proportion of the somatosensory cortex devoted to reception of sensory input from each area. (c) Motor homunculus showing the distribution of motor output from the primary motor cortex to different parts of the body. The distorted graphic representation of the body parts indicates the relative proportion of the primary motor cortex devoted to controlling skeletal muscles in each area.

contraction of particular muscles, no purposeful coordinated movement can be elicited, just as pulling on isolated strings of a puppet does not produce any meaningful movement. A puppet displays purposeful movements only when a skilled puppeteer manipulates the strings in a coordinated manner. In the same way, these four regions (and perhaps other areas as yet undetermined) develop a **motor program** for the specific voluntary task and then "pull" the appropriate pattern of "strings" in the primary motor cortex to produce the sequenced contraction of appropriate muscles that accomplishes the desired complex movement.

The **supplementary motor area** lies on the medial (inner) surface of each hemisphere anterior to (in front of) the primary motor cortex. It plays a preparatory role in programming complex sequences of movement. Stimulation of various regions of this motor area brings about complex patterns of movement, such as opening or closing the hand. Lesions here do not result in paralysis, but they do interfere with performance of more complex, useful integrated movements.

The **premotor cortex,** located on the lateral surface of each hemisphere in front of the primary motor cortex, is important in orienting the body and arms toward a specific target. To command the primary motor cortex to produce the appropriate skeletal muscle contraction that accomplishes the desired movement, the premotor cortex must be informed of the body's momentary position in relation to the target. The premotor cortex is guided by sensory input processed by the **posterior parietal cortex,** a region that lies posterior to (in back of) the primary somatosensory cortex. These two higher motor areas have many anatomic interconnections and are closely related functionally. When either of these areas is damaged, the person cannot process complex sensory information to accomplish purposeful movement in a spatial context; for example, the person cannot successfully manipulate eating utensils.

Even though these higher motor areas command the primary motor cortex and are important in preparing for execution of deliberate, meaningful movement, researchers cannot say that voluntary movement is actually initiated by these areas. This pushes the question of how and where voluntary activity is initiated one step further. Probably no single area is responsible; undoubtedly, numerous pathways can ultimately bring about deliberate movement.

Think about the neural systems called into play, for example, during the simple act of picking up an apple to eat. Your memory tells you the fruit is in a bowl on the kitchen counter. Sensory systems, coupled with your knowledge based on past experience, enable you to distinguish the apple from the other kinds of fruit in the bowl. On receiving this integrated sensory information, motor systems issue commands to the exact muscles of the body in the proper sequence to enable you to move to the fruit bowl and pick up the targeted apple. During execution of this act, minor adjustments in the motor command are made as needed, based on continual updating provided by sensory input about the position of your body relative to the goal. Then there is the issue of motivation and behavior. Are you reaching for the apple because you are hungry (detected by a neural system in the hypothalamus) or because of a more complex behavioral scenario, such as the fact that you started to think about food because you just saw someone eating on tele-

vision? Why did you choose an apple rather than a banana when both are in the fruit bowl and you like the taste of both, and so on? Thus, initiating and executing purposeful voluntary movement actually include a complex neuronal interplay that involves output from the motor regions guided by integrated sensory information and ultimately depends on motivational systems and elaboration of thought. All this plays against a background of memory stores from which you can make meaningful decisions about desirable movements.

Somatotopic maps vary slightly between individuals and are dynamic, not static.

Although the general organizational pattern of sensory and motor somatotopic ("body representation") maps of the cortex is similar in all people, the precise distribution is unique for each individual. Just as each of us has two eyes, a nose, and a mouth and yet no two faces have these features arranged in exactly the same way, so it is with brains. Furthermore, an individual's somatotopic mapping is not "carved in stone" but is subject to constant subtle modifications based on use. The general pattern is governed by genetic and developmental processes, but the individual cortical architecture can be influenced by **use-dependent competition** for cortical space. For example, when monkeys were encouraged to use their middle fingers instead of their other fingers to press a bar for food, after only several thousand bar presses the "middle finger area" of the motor cortex was greatly expanded and encroached upon territory previously devoted to the other fingers. Similarly, modern neuroimaging techniques reveal that the left hand of a right-handed string musician is represented by a larger area of the somatosensory cortex than is the left hand of a person who does not play a string instrument. In this way, the musician's left-hand fingers develop a greater "feel" for the instrument as they skillfully manipulate the strings.

Other regions of the brain besides the somatosensory cortex and motor cortex can also be modified by experience. We now turn our attention to this plasticity of the brain.

Because of its plasticity, the brain can be remodeled in response to varying demands.

The brain displays a degree of **plasticity,** that is, an ability to change or be functionally remodeled in response to the demands placed on it. The term *plasticity* is used to describe this ability because plastics can be manipulated into any desired shape to serve a particular purpose. The ability of the brain to modify as needed is more pronounced in the early developmental years, but even adults retain some plasticity. When one area of the brain associated with a particular activity is destroyed, other areas may gradually assume some or all of the functions of the damaged region. Researchers are only beginning to unravel the molecular mechanisms responsible for the brain's plasticity. Current evidence suggests that the formation of new neural pathways (not new neurons, but new connections between existing neurons) in response to changes in experience are mediated in part by alterations in dendritic shape resulting from modifications in certain cytoskeletal elements (see p. 42). As its dendrites become more branched and elongated, a neuron becomes able to receive and integrate more signals from other neurons. Thus, the precise synaptic connections between neurons are not fixed but can be modified by experience. The gradual modification of each person's brain by a unique set of experiences provides a biological basis for individuality. Even though the particular architecture of your own rather plastic brain has been and continues to be influenced by your unique experiences, it is important to realize that what you do and do not do cannot totally shape the organization of your cortex and other parts of the brain. Some limits are genetically established, and others are developmental limits on the extent to which modeling can be influenced by patterns of use. For example, some cortical regions maintain their plasticity throughout life, especially the ability to learn and to add new memories, but other cortical regions can be modified by use for only a specified time after birth before becoming permanently fixed. The length of this critical developmental period varies for different cortical regions.

Different regions of the cortex control different aspects of language.

Language ability is an excellent example of early cortical plasticity coupled with later permanence. Unlike the sensory and motor regions of the cortex, which are present in both hemispheres, in the vast majority of people the areas of the brain responsible for language ability are found in only one hemisphere—the left hemisphere.

Clinical Note However, if a child under the age of two accidentally suffers damage to the left hemisphere, language functions are transferred to the right hemisphere with no delay in language development but at the expense of less obvious nonverbal abilities for which the right hemisphere is normally responsible. Up to about the age of 10, after damage to the left hemisphere, language ability can usually be reestablished in the right hemisphere following a temporary period of loss. If damage occurs beyond the early teens, however, language ability is permanently impaired, even though some limited restoration may be possible. The regions of the brain involved in comprehending and expressing language apparently are permanently assigned before adolescence.

Even in normal individuals, there is evidence for early plasticity and later permanence in language development. Infants can distinguish between and articulate the entire range of speech sounds, but each language uses only a portion of these sounds. As children mature, they often lose the ability to distinguish between or express speech sounds that are not important in their native language. For example, Japanese children can distinguish between the sounds of *r* and *l*, but many Japanese adults cannot perceive the difference between them.

ROLES OF BROCA'S AREA AND WERNICKE'S AREA Language is a complex form of communication in which written or spoken words symbolize objects and convey ideas. It involves the integration of two distinct capabilities—namely, *expression* (speaking ability) and *comprehension*—each of which is related to a specific area of the cortex. The primary areas of cortical specialization for language are Broca's area and Wernicke's area. **Broca's area,** which governs speaking ability, is located in the left frontal lobe in close association with the motor areas of the cortex that control the muscles necessary for articulation (see ● Figures 5-9 and 5-11). **Wernicke's area,** located in the left cortex at the juncture of the parietal, temporal, and occipital lobes, is concerned with language comprehension. It plays a critical role in understanding both spoken and written messages. Furthermore, it is responsible for formulating coherent patterns of speech that are transferred via a bundle of fibers to Broca's area, which in turn controls articulation of this speech. Wernicke's area receives input from the visual cortex in the occipital lobe, a pathway important in reading comprehension and in describing objects seen, as well as from the auditory cortex in the temporal lobe, a pathway essential for understanding spoken words. Wernicke's area also receives input from the somatosensory cortex, a pathway important in the ability to read Braille. Precise interconnecting pathways between these localized cortical areas are involved in the various aspects of speech (● Figure 5-11).

Clinical Note **LANGUAGE DISORDERS** Because various aspects of language are localized in different regions of the cortex, damage to specific regions of the brain can result in selective disturbances of language. Damage to Broca's area results in a failure of word formation, although the patient can still understand the spoken and written word. Such people know what they want to say but cannot express themselves. Even though they can move their lips and tongue, they cannot establish the proper motor command to articulate the desired words. In contrast, patients with a lesion in Wernicke's area cannot understand words they see or hear. They can speak fluently, but their perfectly articulated words make no sense. They cannot attach meaning to words or choose appropriate words to convey their thoughts. Such language disorders caused by damage to specific cortical areas are known as **aphasias,** most of which result from strokes. Aphasias should not be confused with **speech impediments,** which are caused by a defect in the mechanical aspect of speech, such as weakness or incoordination of the muscles controlling the vocal apparatus.

Dyslexia, another language disorder, is a difficulty in learning to read because of inappropriate interpretation of words. The problem arises from developmental abnormalities in connections between the visual and language areas of the cortex or within the language areas themselves; that is, the person is born with "faulty wiring" within the language-processing system. Emerging evidence suggests that dyslexia stems from a deficit in phonological processing, meaning an impaired ability to break down written words into their underlying phonetic components. Dyslexics have difficulty decoding and thus identifying and assigning meaning to words. The condition is in no way related to intellectual ability.

The association areas of the cortex are involved in many higher functions.

The motor, sensory, and language areas account for only about half of the total cerebral cortex. The remaining areas, called **association areas,** are involved in higher functions. There are three association areas: (1) the *prefrontal association cortex,* (2) the *parietal-temporal-occipital association cortex,* and (3) the *limbic*

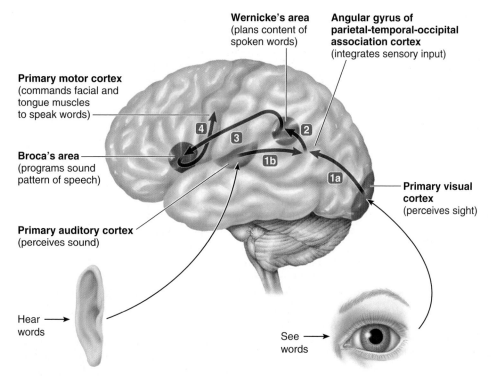

Wernicke's area
(plans content of
spoken words)

Angular gyrus of
parietal-temporal-occipital
association cortex
(integrates sensory input)

Primary motor cortex
(commands facial and
tongue muscles
to speak words)

Broca's area
(programs sound
pattern of speech)

Primary auditory cortex
(perceives sound)

Primary visual
cortex
(perceives sight)

Hear
words

See
words

1a To speak about something seen, the brain transfers the visual information from the primary visual cortex to the angular gyrus of the parietal-temporal-occipital association cortex, which integrates inputs such as sight, sound, and touch.

1b To speak about something heard, the brain transfers the auditory information from the primary auditory cortex to the angular gyrus.

2 The information is transferred to Wernicke's area, where the choice and sequence of words to be spoken are formulated.

3 This language command is then transmitted to Broca's area, which translates the message into a programmed sound pattern.

4 This sound program is conveyed to the precise areas of the primary motor cortex that activate the appropriate facial and tongue muscles for causing the desired words to be spoken.

● **FIGURE 5-11 Cortical pathway for speaking a word seen or heard.** The arrows and numbered steps describe the pathway used to speak about something seen or heard. Similarly, appropriate muscles of the hand can be commanded to write the desired words.

association cortex (see ● Figure 5-9a). At one time the association areas were called "silent" areas, because stimulation does not produce any observable motor response or sensory perception. (During brain surgery, typically the patient remains awake and only local anesthetic is used along the cut scalp. This is possible because the brain itself is insensitive to pain. Before cutting into this precious, nonregenerative tissue, the neurosurgeon explores the exposed region with a tiny stimulating electrode. The patient is asked to describe what happens with each stimulation—the flick of a finger, a prickly feeling on the bottom of the foot, nothing? In this way, the surgeon can ascertain the appropriate landmarks on the neural map before making an incision.)

The **prefrontal association cortex** is the front portion of the frontal lobe just anterior to the premotor cortex. This is the part of the brain that "brainstorms" or thinks (see ● Figure 5-9b). Specifically, the roles attributed to this region are (1) planning for voluntary activity; (2) decision making (that is, weighing consequences of future actions and choosing between different options for various social or physical situations); (3) creativity; and (4) personality traits. To carry out these highest of neural functions, the prefrontal cortex is the site of operation of *working memory,* where the brain temporarily stores and actively manipulates information used in reasoning and planning. You will learn more about working memory later.

Stimulating the prefrontal cortex does not produce any observable effects, but deficits in this area change personality and social behavior. Because damage to the prefrontal lobe was known to produce these changes, about 70 years ago prefrontal lobotomy (surgical removal) was used to

treat violent individuals or others with "bad" personality traits or social behavior in the hope that their personality change would be for the better. Of course, the other functions of the prefrontal cortex were lost in the process (fortunately, the technique was used for only a short time).

The **parietal-temporal-occipital association cortex** lies at the interface of the three lobes for which it is named. In this strategic location, it pools and integrates somatic, auditory, and visual sensations projected from these three lobes for complex perceptual processing. It enables you to "get the complete picture" of the relationship of various parts of your body with the external world. For example, it integrates visual information with proprioceptive input to let you place what you are seeing in proper perspective, such as realizing that a bottle is in an upright position despite the angle from which you view it (that is, whether you are standing up, lying down, or hanging upside down from a tree branch). This region is also involved in the language pathway connecting Wernicke's area to the visual and auditory cortices.

The **limbic association cortex** is located mostly on the bottom and adjoining inner portion of each temporal lobe. This area is concerned primarily with motivation and emotion and is extensively involved in memory.

The cortical association areas are all interconnected by bundles of fibers within the cerebral white matter. Collectively, the association areas integrate diverse information for purposeful action. An oversimplified basic sequence of linkage between the various functional areas of the cortex is schematically represented in ● Figure 5-12.

Sensory input	Relayed from afferent neuronal receptors
Primary sensory areas (somatosensory, primary visual, primary auditory cortexes)	Initial cortical processing of specific sensory input
Higher sensory areas	Further elaboration and processing of specific sensory input
Association areas	Integration, storage, and use of diverse sensory input for planning of purposeful action
Higher motor areas	Programming of sequences of movement in context of diverse information provided
Primary motor cortex	Commanding of efferent motor neurons to initiate voluntary movement
Motor output	Relayed through efferent motor neurons to appropriate skeletal muscles, which carry out desired action

● **FIGURE 5-12 Linking of various regions of the cortex.** For simplicity, a number of interconnections have been omitted.

The cerebral hemispheres have some degree of specialization.

The cortical areas described thus far appear to be equally distributed in both the right and left hemispheres, except for the language areas, which are found only on one side, usually the left. The left side is also most commonly the dominant hemisphere for fine motor control. Thus, most people are right-handed, because the left side of the brain controls the right side of the body. Furthermore, each hemisphere is somewhat specialized in the types of mental activities it carries out best. The **left cerebral hemisphere** excels in logical, analytic, sequential, and verbal tasks, such as math, language forms, and philosophy. In contrast, the **right cerebral hemisphere** excels in nonlanguage skills, especially spatial perception and artistic and musical talents. The left hemisphere tends to process information in a fine-detail, fragmentary way, while the right hemisphere views the world in a big-picture, holistic way. Normally, the two hemispheres share so much information that they complement each other, but in many individuals the skills associated with one hemisphere are more strongly developed. Left cerebral hemisphere dominance tends to be associated with "thinkers," whereas right hemispheric skills dominate in "creators."

An electroencephalogram is a record of postsynaptic activity in cortical neurons.

Extracellular current flow arising from electrical activity within the cerebral cortex can be detected by placing recording electrodes on the scalp to produce a graphic record known as an **electroencephalogram,** or **EEG.** These "brain waves" for the most part are not due to action potentials but instead represent

Eyes closed	**Eyes open**	Eyes closed
Alpha waves	**Beta waves**	Alpha waves

● **FIGURE 5-13 Replacement of an alpha rhythm on an EEG with a beta rhythm when the eyes are opened.**

the momentary collective postsynaptic potential activity (that is, EPSPs and IPSPs; see p. 106) in the cell bodies and dendrites located in the cortical layers under the recording electrode.

Electrical activity can always be recorded from the living brain, even during sleep and unconscious states, but the waveforms vary, depending on the degree of activity in the cerebral cortex. Often the waveforms appear irregular, but sometimes distinct patterns in the wave's amplitude and frequency can be observed. A dramatic example of this is illustrated in ● Figure 5-13, in which the EEG waveform recorded over the occipital (visual) cortex changes markedly in response to simply opening and closing the eyes.

The EEG has three major uses:

1. The EEG is often used as a *clinical tool in the diagnosis of cerebral dysfunction.* Diseased or damaged cortical tissue often gives rise to altered EEG patterns. One of the most common neurologic diseases accompanied by a distinctively abnormal EEG is **epilepsy.** Epileptic seizures occur when a large collection of neurons undergo abnormal, synchronous action potentials that produce stereotypical, involuntary spasms and alterations in behavior. Different underlying problems, including genetic defects and traumatic brain injuries, can lead to the neuronal hyperexcitability that characterizes epilepsy. Typically there is too little inhibitory compared to excitatory activity, as with compromised functioning of the inhibitory neurotransmitter GABA or prolonged action of the excitatory neurotransmitter glutamate. The seizures may be partial or generalized, depending on the location and extent of the abnormal neuronal discharge. Each type of seizure displays different EEG features.

2. The EEG is also used in the *legal determination of brain death.* Even though a person may have stopped breathing and the heart may have stopped pumping blood, it is often possible to restore and maintain respiratory and circulatory activity if resuscitative measures are instituted soon enough. Yet because the brain is susceptible to O_2 deprivation, irreversible brain damage may occur before lung and heart function can be reestablished, resulting in the paradoxical situation of a dead brain in a living body. The determination of whether a comatose patient being maintained by artificial respiration and other supportive measures is alive or dead has important medical, legal, and social implications. The need for viable organs for modern transplant surgery has made the timeliness of such life/death determinations of utmost importance. Physicians, lawyers, and the American public in general have accepted the notion of brain death—that is, a brain that is not functioning, with no possibility of recovery—as the determinant of death under such circumstances. Brain-dead people make good organ donors because the organs are still being supplied with circulating blood and thus are in better shape than those obtained from a person whose heart has stopped beating. The most widely accepted indication of brain death is *electrocerebral silence*—an essentially flat EEG. This must be coupled with other stringent criteria, such as absence of eye reflexes, to guard against a false terminal diagnosis in individuals with a flat EEG due to causes that can be reversed, as in certain kinds of drug intoxication.

3. The EEG is also used to *distinguish various stages of sleep*, as described later in this chapter.

Neurons in different regions of the cerebral cortex may fire in rhythmic synchrony.

Most information about the electrical activity of the brain has been gleaned not from EEG studies but from direct recordings of individual neurons in experimental animals engaged in various activities. Following surgical implantation of an extremely thin recording microelectrode into a single neuron within a specific region of the cerebral cortex, scientists have been able to observe changes in the electrical activity of the neuron as the animal engages in particular motor tasks or encounters various sensations. Through these studies, investigators have concluded that neural information is coded by changes in the frequency of action potentials in specific neurons: The larger the triggering event, the greater the firing rate of the neuron.

Single-neuron recordings cannot identify concurrent changes in electrical activity in a group of neurons working together to accomplish a particular activity, however. As an analogy, imagine trying to record a concert by using a single microphone that could pick up only the sounds produced by one musician. You would get a very limited impression of the performance by hearing only the changes in notes and tempo as played by this one individual. You would miss the richness of the melody and rhythm being performed in synchrony by the entire orchestra. Similarly, in recording from single neurons and detecting their changes in firing rates, scientists have overlooked a parallel information mechanism involving changes in the relative timing of action potential discharges among a functional group of neurons, called a *neural network* or *assembly*. Recent studies involving simultaneous recordings from multiple neurons suggest that interacting neurons may transiently fire together for fractions of a second. Many neuroscientists believe that the brain encodes information not just by changing the firing rates of individual neurons but also by changing the patterns of these brief neural synchronizations. That is, groups of neurons communicate, or send messages about what is happening, by changing their pattern of synchronous firing. Recall that astrocytes also help coordinate synaptic activity among neural networks.

Neurons within an assembly that fire together may be widely scattered. For example, when you view a bouncing ball, different visual units initially process different aspects of this object—its shape, its color, its movement, and so on. Somehow all these separate processing pathways must be integrated, or

"bound together," for you to "see" the bouncing ball as a whole unit without stopping to contemplate its many separate features. The solution to the longtime mystery of how the brain accomplishes this integration might lie in the synchronous firing of neurons in separate regions of the brain that are functionally linked by virtue of being responsive to different aspects of the same objects, such as the bouncing ball.

We now shift our attention to the **subcortical regions** of the brain, which interact extensively with the cortex in the performance of their functions (*subcortical* means "under the cortex"). These regions include the *basal nuclei*, located in the cerebrum, and the *thalamus* and *hypothalamus*, located in the diencephalon.

Basal Nuclei, Thalamus, and Hypothalamus

The **basal nuclei** (also known as **basal ganglia**) consist of several masses of gray matter located deep within the cerebral white matter (▲ Table 5-2 and ● Figure 5-14). In the CNS, a **nucleus** (plural, *nuclei*) is a functioning group of neuron cell bodies.

The basal nuclei play an important inhibitory role in motor control.

Besides having nonmotor functions that are not clearly understood, the basal nuclei play a complex role in controlling movement. In particular, they are important in (1) inhibiting muscle tone throughout the body (proper muscle tone is normally maintained by a balance of excitatory and inhibitory inputs to the neurons that innervate skeletal muscles); (2) selecting and maintaining purposeful motor activity while suppressing useless or unwanted patterns of movement; and (3) helping monitor and coordinate slow, sustained contractions, especially those related to posture and support. The basal nuclei do not directly influence the efferent motor neurons that bring about muscle contraction but act instead by modifying ongoing activity in motor pathways.

To accomplish these complex integrative roles, the basal nuclei receive and send out much information, as is indicated by the tremendous number of fibers linking them to other regions of the brain. One important pathway consists of strategic interconnections that form a complex feedback loop linking the cerebral cortex (especially its motor regions), the basal nuclei, and the thalamus. The thalamus positively reinforces voluntary motor behavior initiated by the cortex, whereas the basal nuclei modulate this activity by exerting an inhibitory effect on the thalamus to eliminate antagonistic or unnecessary movements. The basal nuclei also inhibit motor activity by acting through neurons in the brain stem.

 The importance of the basal nuclei in motor control is evident in diseases involving this region, the most common of which is **Parkinson's disease (PD)**. This condition is associated with a gradual destruction of neurons that release the neurotransmitter dopamine in the basal nuclei (see p. 112). Because the basal nuclei lack enough dopamine to exert

Right cerebral hemisphere

Left cerebral hemisphere

Cerebral cortex (gray matter)

White matter

Corpus callosum

Lateral ventricle

Thalamus

Third ventricle

Caudate nucleus

Putamen

Globus pallidus

Claustrum

Basal nuclei (gray matter)

Part of hypothalamus

Longitudinal fissure

Corpus callosum

Lateral ventricle

Third ventricle

Cerebral cortex (gray matter)

White matter

Thalamus

Basal nuclei (gray matter)

Part of hypothalamus

Photo: Mark Nielsen.

● **FIGURE 5-14 Frontal section of the brain.** The cerebral cortex, an outer shell of gray matter, surrounds an inner core of white matter. Deep within the cerebral white matter are several masses of gray matter, the basal nuclei. The ventricles are cavities in the brain through which the cerebrospinal fluid flows. The thalamus forms the walls of the third ventricle. For comparison, the colors used for these brain components are the same as those used in the lateral view depicted in Table 5-2, p. 144. Also, compare this frontal section of a cadaver brain with the sagittal section of a cadaver brain in Figure 5-7, p. 146.

their normal roles, three types of motor disturbances characterize PD: (1) increased muscle tone, or rigidity; (2) involuntary, useless, or unwanted movements, such as resting tremors (for example, hands rhythmically shaking, making it difficult or impossible to hold a cup of coffee); and (3) slowness in initiating and carrying out different motor behaviors. People with PD find it difficult to stop ongoing activities. If sitting down, they tend to remain seated, and if they get up, they do so very slowly.

The thalamus is a sensory relay station and is important in motor control.

Deep within the brain near the basal nuclei is the **diencephalon,** a midline structure that forms the walls of the third ventricular cavity, one of the spaces through which CSF flows. The diencephalon consists of two main parts, the *thalamus* and the *hypothalamus* (see ▲ Table 5-2 and ● Figures 5-7b, 5-14, and 5-15).

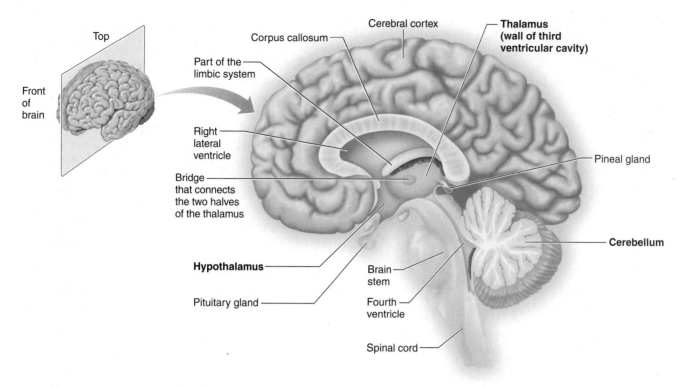

● **FIGURE 5-15 Location of the thalamus, hypothalamus, and cerebellum in sagittal section.**

The **thalamus** serves as a "relay station" for preliminary processing of sensory input. All sensory input synapses in the thalamus on its way to the cortex. The thalamus screens out insignificant signals and routes the important sensory impulses to appropriate areas of the somatosensory cortex, as well as to other regions of the brain. Along with the brain stem and cortical association areas, the thalamus helps to direct attention to stimuli of interest. For example, parents can sleep soundly through the noise of outdoor traffic but become instantly aware of their baby's slightest whimper. The thalamus is also capable of crude awareness of various sensations but cannot distinguish their location or intensity. Some degree of consciousness resides here as well. As described in the preceding section, the thalamus also plays an important role in motor control by positively reinforcing voluntary motor behavior initiated by the cortex.

The hypothalamus regulates many homeostatic functions.

The **hypothalamus** is a collection of specific nuclei and associated fibers that lie beneath the thalamus. It is an integrating center for many important homeostatic functions and serves as an important link between the autonomic nervous system and the endocrine system. Specifically, the hypothalamus (1) controls body temperature; (2) controls thirst and urine output; (3) controls food intake; (4) controls anterior pituitary hormone secretion; (5) produces posterior pituitary hormones; (6) controls uterine contractions and milk ejection; (7) serves as a major autonomic nervous system coordinating center, which in turn affects all smooth muscle, cardiac muscle, and exocrine

glands; (8) plays a role in emotional and behavioral patterns; and (9) participates in the sleep–wake cycle.

The hypothalamus is the brain area most involved in directly regulating the internal environment. For example, when the body is cold, the hypothalamus initiates internal responses to increase heat production (such as shivering) and to decrease heat loss (such as constricting the skin blood vessels to reduce the flow of warm blood to the body surface, where heat could be lost to the external environment). Other areas of the brain, such as the cerebral cortex, act more indirectly to regulate the internal environment. For example, a person who feels cold is motivated to voluntarily put on warmer clothing, close the window, turn up the thermostat, and so on. Even these voluntary behavioral activities are strongly influenced by the hypothalamus, which, as a part of the limbic system, functions together with the cortex in controlling emotions and motivated behavior. We now turn attention to the limbic system and its functional relations with the higher cortex, first in emotion and behavior, then in learning and memory.

Emotion, Behavior, and Motivation

The **limbic system** is not a separate structure but a ring of forebrain structures that surround the brain stem and are interconnected by intricate neuron pathways (● Figure 5-16). It includes portions of each of the following: the lobes of the cerebral cortex (especially the limbic association cortex), the basal nuclei, the thalamus, and the hypothalamus. This complex interacting

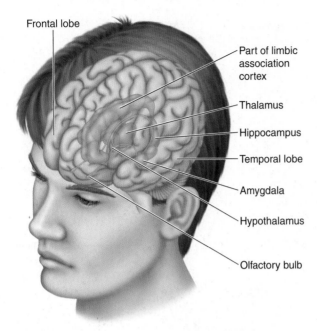

Frontal lobe

Part of limbic
association
cortex

Thalamus

Hippocampus

Temporal lobe

Amygdala

Hypothalamus

Olfactory bulb

● **FIGURE 5-16 Limbic system.** This partially transparent view of the brain reveals the structures composing the limbic system.

network is associated with emotions, basic survival and sociosexual behavioral patterns, motivation, and learning. Let us examine each of these brain functions further.

The limbic system plays a key role in emotion.

The concept of **emotion** encompasses subjective emotional feelings and moods (such as anger, fear, and happiness) plus the overt physical responses associated with these feelings. These responses include specific behavioral patterns (for example, preparing for attack or defense when angered by an adversary) and observable emotional expressions (for example, laughing, crying, or blushing). Evidence points to a central role for the limbic system in all aspects of emotion. Stimulating specific regions of the limbic system during brain surgery produces vague subjective sensations that the patient may describe as joy, satisfaction, or pleasure in one region and discouragement, fear, or anxiety in another. For example, the **amygdala,** on the interior underside of the temporal lobe (● Figure 5-16), is especially important in processing inputs that give rise to the sensation of fear. In humans and to an undetermined extent in other species, higher levels of the cortex are also crucial for conscious awareness of emotional feelings.

The limbic system and higher cortex participate in controlling basic behavioral patterns.

Basic inborn behavioral patterns controlled at least in part by the limbic system include those aimed at individual survival (attack, searching for food) and those directed toward perpetuating the species (sociosexual behaviors conducive to mating). In experimental animals, stimulating the limbic system brings about complex and even bizarre behaviors. For example, stimu-

lation in one area can elicit responses of anger and rage in a normally docile animal, whereas stimulation in another area results in placidity and tameness, even in an otherwise vicious animal. Stimulation in yet another limbic area can induce sexual behaviors such as copulatory movements.

The relationships among the hypothalamus, limbic system, and higher cortical regions regarding emotions and behavior are still not well understood. Apparently the extensive involvement of the hypothalamus in the limbic system governs the involuntary internal responses of various body systems in preparation for appropriate action to accompany a particular emotional state. For example, the hypothalamus controls the increase of heart rate and respiratory rate, elevation of blood pressure, and diversion of blood to skeletal muscles that occur in anticipation of attack or when angered. These preparatory changes in the internal state require no conscious control.

In executing complex behavioral activities such as attack, flight, or mating, the individual (animal or human) must interact with the external environment. Higher cortical mechanisms are called into play to connect the limbic system and hypothalamus with the outer world so that appropriate overt behaviors are manifested. At the simplest level, the cortex provides the neural mechanisms necessary for implementing the appropriate skeletal muscle activity required to approach or avoid an adversary, participate in sexual activity, or display emotional expression. For example, the stereotypical sequence of movement for the universal human emotional expression of smiling is apparently preprogrammed in the cortex and can be called forth by the limbic system. One can also voluntarily call forth the smile program, as when posing for a picture. Even individuals blind from birth have normal facial expressions; that is, they do not learn to smile by observation. Smiling means the same thing in every culture, despite widely differing environmental experiences. Such behavior patterns shared by all members of a species are believed to be more abundant in lower animals.

Higher cortical levels also can reinforce, modify, or suppress basic behavioral responses so that actions can be guided by planning, strategy, and judgment based on an understanding of the situation. Even if you are angry at someone and your body is internally preparing for attack, you can usually judge whether an attack is appropriate and can consciously suppress the external manifestation of this basic emotional behavior. Thus, the higher levels of the cortex, particularly the prefrontal and limbic association areas, are important in conscious learned control of innate behavioral patterns. Using fear as an example, exposure to an aversive experience calls two parallel tracks into play for processing this emotional stimulus: a fast track in which the lower-level amygdala plays a key role and a slower track mediated primarily by the higher-level prefrontal cortex. The fast track permits a rapid, rather crude, instinctive response ("gut reaction") and is essential for the "feeling" of being afraid. The slower track involving the prefrontal cortex permits a more refined response to the aversive stimulus based on a rational analysis of the current situation compared to stored past experiences. The prefrontal cortex formulates plans and guides behavior, suppressing amygdala-induced responses that may be inappropriate for the situation at hand.

Motivated behaviors are goal directed.

An individual tends to reinforce behaviors that have proved gratifying and to suppress behaviors that have been associated with unpleasant experiences. Certain regions of the limbic system have been designated as **"reward"** and **"punishment" centers,** because stimulation in these respective areas gives rise to pleasant or unpleasant sensations. When a self-stimulating device is implanted in a reward center, an experimental animal will self-deliver up to 5000 stimulations per hour and will continue self-stimulation in preference to food, even when starving. When the device is implanted in a punishment center, animals will avoid stimulation at all costs. Reward centers are found most abundantly in regions involved in mediating the highly motivated behavioral activities of eating, drinking, and sexual activity.

Motivation is the ability to direct behavior toward specific goals. Some goal-directed behaviors are aimed at satisfying specific identifiable physical needs related to homeostasis. **Homeostatic drives** represent the subjective urges associated with specific bodily needs that motivate appropriate behavior to satisfy those needs. As an example, the sensation of thirst accompanying a water deficit in the body drives an individual to drink to satisfy the homeostatic need for water. However, whether water, a soft drink, or another beverage is chosen as the thirst quencher is unrelated to homeostasis. Much human behavior does not depend on purely homeostatic drives related to simple tissue deficits such as thirst. Human behavior is influenced by experience, learning, and habit, shaped in a complex framework of unique personal gratifications blended with cultural expectations. To what extent, if any, motivational drives unrelated to homeostasis, such as the drive to pursue a particular career or win a certain race, are involved with the reinforcing effects of the reward and punishment centers is unknown. Indeed, some individuals motivated toward a particular goal may even deliberately "punish" themselves in the short term to achieve their long-range gratification (for example, the temporary pain of training in preparation for winning a competitive athletic event).

Norepinephrine, dopamine, and serotonin are neurotransmitters in pathways for emotions and behavior.

The underlying neurophysiological mechanisms responsible for the psychological observations of emotions and motivated behavior largely remain a mystery, although the neurotransmitters norepinephrine, dopamine, and serotonin all have been implicated. Norepinephrine and dopamine, both chemically classified as catecholamines (see p. 118), are known transmitters in the regions that elicit the highest rates of self-stimulation in animals equipped with do-it-yourself devices. A number of **psychoactive drugs** affect moods in humans, and some of these drugs have also been shown to influence self-stimulation in experimental animals. For example, increased self-stimulation is observed after the administration of drugs that increase catecholamine synaptic activity, such as *amphetamine,* an "upper" drug. Amphetamine stimulates the release of dopamine from dopamine-secreting neurons.

Although most psychoactive drugs are used therapeutically to treat various mental disorders, others, unfortunately, are abused. Many abused drugs act by enhancing the effectiveness of dopamine in the "pleasure" pathways, thus initially giving rise to an intense sensation of pleasure. As you have already learned, an example is *cocaine*, which blocks the reuptake of dopamine at synapses (see p. 112).

Depression is among the psychiatric disorders associated with defects in limbic system neurotransmitters. (As a distinction, *psychiatric disorders* involve abnormal activity in specific neurotransmitter pathways in the absence of detectable brain lesions, whereas *neurological disorders* are associated with specific lesions of the brain and may or may not involve abnormalities in neurotransmission. Examples of neurological disorders include Parkinson's disease and Alzheimer's disease.) A functional deficiency of serotonin or norepinephrine or both is implicated in depression, a disorder characterized by a pervasive negative mood accompanied by a generalized loss of interests, an inability to experience pleasure, and suicidal tendencies. All effective antidepressant drugs increase the available concentration of these neurotransmitters in the CNS. Prozac, the most widely prescribed drug in American psychiatry, is illustrative. It blocks the reuptake of released serotonin, thus prolonging serotonin activity at synapses (see p. 107). Serotonin and norepinephrine are synaptic messengers in the limbic regions of the brain involved in pleasure and motivation, suggesting that the pervasive sadness and lack of interest (no motivation) in depressed patients are related at least in part to deficiencies or decreased effectiveness of these neurotransmitters. Events perceived as being stressful can trigger depression, but the underlying link has not been determined.

Researchers are optimistic that as understanding of the molecular mechanisms of mental disorders is expanded in the future, many psychiatric problems can be corrected or better managed through drug intervention, a hope of great medical significance.

Learning and Memory

In addition to their involvement in emotion and basic behavioral patterns, the limbic system and higher cortex are involved in learning and memory. The cerebellum also plays a key role in some types of learning and memory, as you will see as we examine these processes.

Learning is the acquisition of knowledge as a result of experiences.

Learning is the acquisition of knowledge or skills as a consequence of experience, instruction, or both. It is widely believed that rewards and punishments are integral parts of many types of learning. If an animal is rewarded on responding in a particular way to a stimulus, the likelihood increases that the animal will respond in the same way again to the same stimulus as a consequence of this experience. Conversely, if a particular response is accompanied by punishment, the animal is less likely to repeat the same response to the same stimulus. When behavioral responses that give rise to pleasure are reinforced or

Comparison of Short-Term and Long-Term Memory

Characteristic	Short-Term Memory	Long-Term Memory
Time of Storage after Acquisition of New Information	Immediate	Later; must be transferred from short-term to long-term memory through consolidation; enhanced by practice or recycling of information through short-term mode
Duration	Lasts for seconds to hours	Retained for days to years
Capacity of Storage	Limited	Very large
Retrieval Time (remembering)	Rapid retrieval	Slower retrieval, except for thoroughly ingrained memories, which are rapidly retrieved
Inability to Retrieve (forgetting)	Permanently forgotten; memory fades quickly unless consolidated into long-term memory	Usually only transiently unable to access; relatively stable memory trace
Mechanism of Storage	Involves transient modifications in functions of preexisting synapses, such as altering amount of neurotransmitter released	Involves relatively permanent functional or structural changes between existing neurons, such as formation of new synapses; synthesis of new proteins plays a key role

those accompanied by punishment are avoided, learning has taken place. Housebreaking a puppy is an example. If the puppy is praised when it urinates outdoors but scolded when it wets the carpet, it will soon learn the acceptable place to empty its bladder. Thus, learning is a change in behavior that occurs as a result of experiences. It highly depends on the organism's interaction with its environment. The only limits to the effects that environmental influences can have on learning are the biological constraints imposed by species-specific and individual genetic endowments.

Memory is laid down in stages.

Memory is the storage of acquired knowledge for later recall. Learning and memory form the basis by which individuals adapt their behavior to their particular external circumstances. Without these mechanisms, it would be impossible for individuals to plan for successful interactions and to intentionally avoid predictably disagreeable circumstances.

The neural change responsible for retention or storage of knowledge is known as the **memory trace.** Generally, concepts, not verbatim information, are stored. As you read this page, you are storing the concept discussed, not the specific words. Later, when you retrieve the concept from memory, you will convert it into your own words. It is possible, however, to memorize bits of information word by word.

Storage of acquired information is accomplished in at least two stages: short-term memory and long-term memory (▲ Table 5-3). **Short-term memory** lasts for seconds to hours, whereas **long-term memory** is retained for days to years. The process of transferring and fixing short-term memory traces into long-term memory stores is known as **consolidation.**

A recently developed concept is that of **working memory,** or what has been called "the erasable blackboard of the mind."

Working memory temporarily holds and interrelates various pieces of information relevant to a current mental task. Through your working memory, you briefly hold and process data for immediate use—both newly acquired information and related, previously stored knowledge that is transiently brought forth into working memory—so that you can evaluate the incoming data in context. This integrative function is crucial to your ability to reason, plan, and make judgments. By comparing and manipulating new and old information within your working memory, you can comprehend what you are reading, carry on a conversation, calculate a restaurant tip in your head, find your way home, and know that you should put on warm clothing if you see snow outside. In short, working memory enables people to string thoughts together in a logical sequence and plan for future action.

Although strong evidence is still lacking, new findings hint that once an established memory is actively recalled, it becomes labile (unstable or subject to change) and must be **reconsolidated** into a restabilized, inactive state. According to this controversial proposal, new information may be incorporated into the old memory trace during reconsolidation.

COMPARISON OF SHORT-TERM AND LONG-TERM MEMORY
Newly acquired information is initially deposited in short-term memory, which has a limited capacity for storage. Information in short-term memory has one of two eventual fates. Either it is soon forgotten (for example, forgetting a telephone number after you have looked it up and finished dialing), or it is transferred into the more permanent long-term memory mode through *active practice* or *rehearsal.* The recycling of newly acquired information through short-term memory increases the likelihood that the information will be consolidated into long-term memory. (Therefore, when you cram for an exam, your long-term retention of the information is poor!) This relationship can be

likened to developing photographic film. The originally developed image (short-term memory) will rapidly fade unless it is chemically fixed (consolidated) to provide a more enduring image (long-term memory). Sometimes only parts of memories are fixed, and other parts fade away. Information of interest or importance to the individual is more likely to be recycled and fixed in long-term stores, whereas less important information is quickly erased.

The storage capacity of the long-term memory bank is much larger than the capacity of short-term memory. Different informational aspects of long-term memory traces seem to be processed and codified, then stored with other memories of the same type; for example, visual memories are stored separately from auditory memories. This organization facilitates future searching of memory stores to retrieve desired information. For example, in remembering a woman you once met, you may use various recall cues from different storage pools, such as her name, her appearance, the fragrance she wore, an incisive comment she made, or the song playing in the background.

Stored knowledge is of no use unless it can be retrieved and used to influence current or future behavior. Because long-term memory stores are larger, it often takes longer to retrieve information from long-term memory than from short-term memory. *Remembering* is the process of retrieving specific information from memory stores; *forgetting* is the inability to retrieve stored information. Information lost from short-term memory is permanently forgotten, but information in long-term storage is frequently forgotten only temporarily. For example, you may be transiently unable to remember an acquaintance's name, then have it suddenly "come to you" later.

Some forms of long-term memory involving information or skills used on a daily basis are essentially never forgotten and are rapidly accessible, such as knowing your own name or being able to write. Even though long-term memories are relatively stable, stored information may be gradually lost or modified over time unless it is thoroughly ingrained by years of practice.

AMNESIA Occasionally, individuals suffer from a lack of memory that involves whole portions of time rather than isolated bits of information. This condition, known as **amnesia**, occurs in two forms. The most common form, *retrograde* ("going backward") *amnesia*, is the inability to recall recent past events. It usually follows a traumatic event that interferes with electrical activity of the brain, such as a concussion or stroke. If a person is knocked unconscious, the content of short-term memory is essentially erased, resulting in loss of memory about activities that occurred within about the last half hour before the event. Severe trauma may interfere with access to recently acquired information in long-term stores as well.

Anterograde ("going forward") *amnesia,* conversely, is the inability to store memory in long-term storage for later retrieval. It is usually associated with lesions of the medial portions of the temporal lobes, which are generally considered critical regions for memory consolidation. People suffering from this condition may be able to recall things they learned before the onset of their problem, but they cannot establish new permanent memories. New information is lost as quickly as it fades from short-term memory. In one case study, the person could not remember where the bathroom was in his new home but still had total recall of his old home.

Short-term memory and long-term memory involve different molecular mechanisms.

Despite a vast amount of psychological data, only a few tantalizing scraps of physiologic evidence concerning the cellular basis of memory traces are available. Obviously, some change must take place within the neural circuitry of the brain to account for the altered behavior that follows learning. A single memory does not reside in a single neuron but rather in changes in the pattern of signals transmitted across synapses within a vast neuronal network.

Different mechanisms are responsible for short-term and long-term memory. Short-term memory involves transient modifications in the function of preexisting synapses, such as a temporary change in the amount of neurotransmitter released in response to stimulation or temporarily increased responsiveness of the postsynaptic cell to the neurotransmitter within affected nerve pathways. Long-term memory, in contrast, involves relatively permanent functional or structural changes between existing neurons in the brain. Let us look at each of these types of memory in more detail.

Short-term memory involves transient changes in synaptic activity.

Ingenious experiments in the sea snail *Aplysia* have shown that two forms of short-term memory—habituation and sensitization—are due to modification of different channel proteins in presynaptic terminals of specific afferent neurons involved in the pathway that mediates the behavior being modified. This modification, in turn, brings about changes in transmitter release. **Habituation** is a decreased responsiveness to repetitive presentations of an indifferent stimulus—that is, one that is neither rewarding nor punishing. **Sensitization** is increased responsiveness to mild stimuli following a strong or noxious stimulus. *Aplysia* reflexly withdraws its gill when its siphon, a breathing organ at the top of its gill, is touched. Afferent (presynaptic) neurons responding to touch of the siphon directly synapse on efferent (postsynaptic) motor neurons controlling gill withdrawal. The snail becomes habituated when its siphon is repeatedly touched; that is, it learns to ignore the stimulus and no longer withdraws its gill in response. Sensitization, a more complex form of learning, takes place in *Aplysia* when it is given a hard bang on the siphon. Subsequently, the snail withdraws its gill more vigorously in response to even mild touch. Interestingly, these different forms of learning affect the same site—the synapse between a siphon afferent and a gill efferent—in opposite ways. Habituation depresses this synaptic activity, whereas sensitization enhances it (● Figure 5-17). These transient modifications persist for as long as the time course of the memory.

MECHANISM OF HABITUATION Recall that when an action potential arrives in a presynaptic axon terminal, voltage-gated Ca^{2+} channels open, permitting the entry of Ca^{2+} that triggers

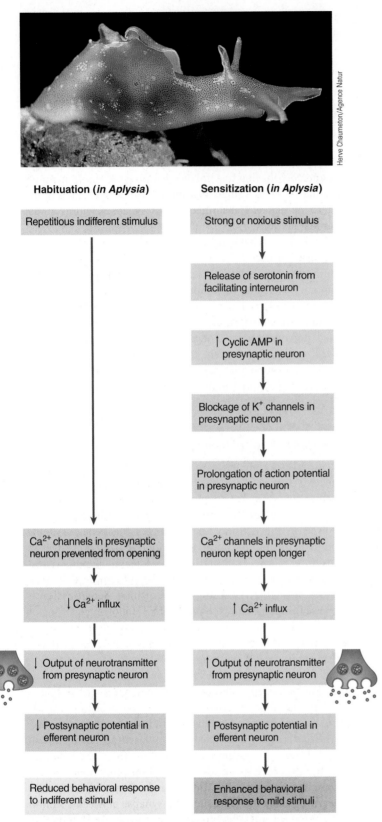

Habituation (in Aplysia)

Repetitious indifferent stimulus

↓

Ca²⁺ channels in presynaptic neuron prevented from opening

↓

↓ Ca²⁺ influx

↓

↓ Output of neurotransmitter from presynaptic neuron

↓

↓ Postsynaptic potential in efferent neuron

↓

Reduced behavioral response to indifferent stimuli

Sensitization (in Aplysia)

Strong or noxious stimulus

↓

Release of serotonin from facilitating interneuron

↓

↑ Cyclic AMP in presynaptic neuron

↓

Blockage of K⁺ channels in presynaptic neuron

↓

Prolongation of action potential in presynaptic neuron

↓

Ca²⁺ channels in presynaptic neuron kept open longer

↓

↑ Ca²⁺ influx

↓

↑ Output of neurotransmitter from presynaptic neuron

↓

↑ Postsynaptic potential in efferent neuron

↓

Enhanced behavioral response to mild stimuli

● **FIGURE 5-17 Habituation and sensitization in *Aplysia*.**
Researchers have shown that in the sea snail *Aplysia* (shown in the photo), two forms of short-term memory—habituation and sensitization—result from opposite changes in neurotransmitter release from the same presynaptic neuron, caused by different transient channel modifications.

the exocytosis of neurotransmitter (see p. 105). As a result of habituation, these Ca²⁺ channels do not open as readily when an action potential arrives, reducing Ca²⁺ entry into the presynaptic terminal, which leads to a decrease in neurotransmitter release. As a consequence, the postsynaptic potential is reduced compared to normal, resulting in a decrease or absence of the behavioral response controlled by the postsynaptic efferent neuron (gill withdrawal). Thus, the memory for habituation in *Aplysia* is stored in the form of modification of specific Ca²⁺ channels. With no further training, this reduced responsiveness lasts for several hours. A similar process is responsible for short-term habituation in other species studied. This suggests that Ca²⁺ channel modification is a general mechanism of habituation, although in higher species the involvement of intervening interneurons makes the process somewhat more complicated. Habituation is probably the most common form of learning and is believed to be the first learning process to take place in human infants. By learning to ignore indifferent stimuli, the animal or person is free to attend to other more important stimuli.

MECHANISM OF SENSITIZATION Sensitization in *Aplysia* also involves channel modification, but a different channel and mechanism are involved. In contrast to what happens in habituation, Ca²⁺ entry into the presynaptic terminal is enhanced in sensitization. The subsequent increase in neurotransmitter release produces a larger postsynaptic potential, resulting in a more vigorous gill-withdrawal response. Sensitization does not have a direct effect on presynaptic Ca²⁺ channels. Instead, it indirectly enhances Ca²⁺ entry via presynaptic facilitation (see p. 111). The neurotransmitter serotonin is released from a facilitating interneuron that synapses on the presynaptic terminal to bring about increased release of presynaptic neurotransmitter in response to an action potential. It does so by triggering activation of a cyclic AMP second-messenger pathway (see p. 121) within the presynaptic terminal, which ultimately brings about blocking of K⁺ channels. This blockage prolongs the action potential in the presynaptic terminal. Remember that K⁺ efflux through opened K⁺ channels hastens the return to resting potential (repolarization) during the falling phase of the action potential. Because the presence of a local action potential is responsible for opening of Ca²⁺ channels in the terminal, a prolonged action potential permits the greater Ca²⁺ influx associated with sensitization.

Thus, existing synaptic pathways may be functionally interrupted (habituated) or enhanced (sensitized) during simple learning. Scientists speculate that much of short-term memory is similarly a temporary modification of already existing processes. Several lines of research suggest that the cyclic AMP cascade, especially activation of protein kinase, plays an important role, at least in elementary forms of learning and memory.

Further studies have revealed that memories more complex than habituation and sensitization that involve conscious awareness are initially stored by means of more persistent changes in activity of existing synapses. Specifically, initial storage of this information appears to be accomplished by means of long-term potentiation, to which we now turn our attention.

MECHANISM OF LONG-TERM POTENTIATION With **long-term potentiation (LTP),** modifications take place as a result of increased use at a given preexisting synapse that enhance the future ability of the presynaptic neuron to excite the postsynaptic neuron. That is, this connection gets stronger the more often it is used. Such strengthening of synaptic activity results in the formation of more EPSPs in the postsynaptic neuron in response to chemical signals from this particular excitatory presynaptic input. The increased excitatory responsiveness is ultimately translated into more action potentials being sent along this postsynaptic cell to other neurons. LTP lasts for days or even weeks—long enough for this short-term memory to be consolidated into more permanent long-term memory. LTP is especially prevalent in the hippocampus, a site critical for converting short-term memories into long-term memories. Less commonly *long-term depression (LTD),* or weakening of synaptic transmission, has been demonstrated. The mechanism and function of LTD are poorly understood at this time.

Enhanced synaptic transmission could theoretically result from either changes in the postsynaptic neuron (such as increased responsiveness to the neurotransmitter) or in the presynaptic neuron (such as increased release of neurotransmitter). The underlying mechanisms for LTP are still the subject of much research and debate. Most likely, multiple mechanisms are involved in this complex phenomenon. It appears that there are several forms of LTP, some arising from changes only in the postsynaptic neuron and others also having a presynaptic component. Based on current scientific evidence, the following is a plausible mechanism for LTP involving both a postsynaptic change and a presynaptic modification (● Figure 5-18).

LTP begins when a presynaptic neuron releases the common excitatory neurotransmitter glutamate in response to an action potential. Glutamate binds to two types of receptors on the postsynaptic neuron: *AMPA receptors* and *NMDA receptors.* An **AMPA receptor** is a chemically mediated receptor-channel that opens on binding of glutamate and permits net entry of Na^+ ions, leading to formation of an EPSP at the postsynaptic neuron (see pp. 115 and 106). This is the ordinary receptor at excitatory synapses about which you already learned. An **NMDA receptor** is a receptor-channel that permits Ca^{2+} entry when it is open. This receptor-channel is unusual because it is both chemically gated and voltage dependent. It is closed by both a gate and by a magnesium ion (Mg^{2+}) that physically blocks the channel opening at resting potential. Two events must happen almost simultaneously to open an NMDA receptor-channel: presynaptic glutamate release and postsynaptic depolarization by other inputs. The gate opens on binding of glutamate, but this action alone does not permit Ca^{2+} entry. Additional depolarization of the postsynaptic neuron beyond that produced by the EPSP resulting from glutamate binding to the AMPA receptor is needed to depolarize the postsynaptic neuron enough to force Mg^{2+} out of the channel. Thus even though glutamate binds with the NMDA receptor, the channel does not open unless the postsynaptic cell is sufficiently depolarized as a result of other excitatory activity. The postsynaptic cell can be sufficiently depolarized to expel Mg^{2+} in two ways: by repeated input from this single excitatory presynaptic neuron, resulting in temporal summation of EPSPs from this source (see p. 108), or by additional excitatory input from another presynaptic neuron at about the same time. When the NMDA receptor channel opens as a result of simultaneous gate opening and Mg^{2+} expulsion, Ca^{2+} enters the postsynaptic cell. The entering Ca^{2+} activates a Ca^{2+} second-messenger pathway in this neuron. This second-messenger pathway leads to the physical insertion of additional AMPA receptors in the postsynaptic membrane. Because of the increased availability of AMPA receptors, the postsynaptic cell exhibits a greater EPSP response to subsequent release of glutamate from the presynaptic cell. This heightened sensitivity of the postsynaptic neuron to glutamate from the presynaptic cell helps maintain LTP.

Furthermore, at some synapses activation of the Ca^{2+} second-messenger pathway in the postsynaptic neuron causes this cell to release a retrograde ("going backward") paracrine that diffuses to the presynaptic neuron (see p. 114). Here, the retrograde paracrine activates a second-messenger pathway in the presynaptic neuron, ultimately enhancing the release of glutamate from the presynaptic neuron. This positive feedback strengthens the signaling process at this synapse, also helping sustain LTP. Note that in this mechanism, a chemical factor from the postsynaptic neuron influences the presynaptic neuron, just the opposite direction of neurotransmitter activity at a synapse. The retrograde paracrine is distinct from classical neurotransmitters or neuropeptides. Most investigators believe that the retrograde messenger is **nitric oxide,** a chemical that performs a bewildering array of other functions in the body. These other functions range from dilation of blood vessels in the penis during erection to destruction of foreign invaders by the immune system (see p. 356).

The modifications that take place during the development of LTP are sustained long after the activity that led to these changes has ceased. Therefore, information can be transmitted along this same synaptic pathway more efficiently when activated in the future. That is, the synapse "remembers." LTP is specific for the activated pathway. Pathways between other inactive presynaptic inputs and the same postsynaptic cell are not affected. Note that LTP develops in response to frequent activity across a synapse as a result of repetitive, intense firing of a given input (as with repeatedly going over a particular fact during studying) or to the linking of one input with another input firing at the same time. For example, when you smell a pie baking in the oven, your mouth waters in anticipation of the imminent arrival of a tasty treat you have come to associate with this aroma. The taste and feel of food in the mouth is the built-in trigger for salivation. However, through experience, neurons in the pathway that control salivation link input arising from the smell of pie with input from its delicious taste. After the smell-input pathway is strengthened through development of LTP and ultimate consolidation into long-term storage, the smell of pie alone can cause salivation.

The *ethanol* in alcoholic beverages blocks the NMDA receptors while facilitating GABA function. Ethanol's blockage of the NMDA receptors is likely the reason people have difficulty remembering what happened during a time of heavy drinking. By enhancing the actions of GABA, the major inhibitory neu-

1 Glutamate is released from activated presynaptic neuron.

2 Glutamate binds with both AMPA and NMDA receptors.

3 Binding opens AMPA receptor-channel.

4 Na^+ entry through open AMPA channel depolarizes postsynaptic neuron, producing EPSP.

5 Binding opens gate of NMDA receptor-channel but Mg^{2+} still blocks channel. Sufficient depolarization from this AMPA opening plus other EPSPs drives Mg^{2+} out.

6 Ca^{2+} entry through open NMDA channel activates Ca^{2+} second-messenger pathway.

7 Second-messenger pathway promotes insertion of additional AMPA receptors in postsynaptic membrane, increasing its sensitivity to glutamate.

8 Second-messenger pathway also triggers release of retrograde paracrine (likely nitric oxide).

9 Nitric oxide stimulates long-lasting increase in glutamate release by presynaptic neuron.

● **FIGURE 5-18 Possible pathways for long-term potentiation.**

rotransmitter in the brain, ethanol depresses the overall activity of the CNS.

Studies suggest a regulatory role for the cAMP second-messenger pathway in the development and maintenance of LTP in addition to the Ca^{2+} second-messenger pathway. Participation of cAMP may hold a key to linking short-term memory to long-term memory consolidation.

Long-term memory involves formation of new, permanent synaptic connections.

Whereas short-term memory involves transient strengthening of preexisting synapses, long-term memory storage requires the activation of specific genes that control synthesis of proteins needed for lasting structural or functional changes at specific synapses. Examples of such changes include the formation of new synaptic connections or permanent changes in pre- or postsynaptic membranes. Thus, long-term memory storage involves rather permanent physical changes in the brain.

Studies comparing the brains of experimental animals reared in a sensory-deprived environment with those raised in a sensory-rich environment demonstrate readily observable microscopic differences. The animals afforded more environmental interactions—and supposedly, therefore, more opportunity to learn—displayed greater branching and elongation of dendrites in nerve cells in regions of the brain suspected to be involved with memory storage. Greater dendritic surface area presumably provides more sites for synapses. Thus, long-term memory may be stored at least in part by a particular pattern of dendritic branching and synaptic contacts.

No one knows for sure how the transient short-term memory is converted to the permanent long-term mode, but many researchers believe that cAMP initiates an intracellular pathway that ultimately turns on genes that lead to new protein synthesis in the involved neurons. This second messenger plays a regulatory role in LTP as well as in simpler forms of short-term memory such as sensitization and has been implicated in long-term memory formation in *Aplysia*. The **immediate early genes (IEGs)** play a critical role in memory consolidation. These genes govern the synthesis of the proteins that encode long-term memory. The exact role that these critical new long-term memory proteins might play remains speculative. They may be needed for structural changes in dendrites or used for synthesis of more neurotransmitters or additional receptor sites. Alternatively, they may accomplish long-term modification of neurotransmitter release by sustaining biochemical events initially activated by short-term memory processes.

Most investigation of learning and memory has focused on changes in synaptic connections within the brain's gray matter. To complicate the issue even further, scientists now have evidence that white matter also changes during learning and memory formation as more myelin surrounds axons, especially during adolescence, speeding up transmission between connected neurons. Apparently neurons produce chemical signals such as **neuregulin** that regulate the extent to which myelin-forming cells wrap themselves around the axon. The amount of neuregulin produced is correlated with the extent of action potential propagation within the axon. Accordingly, researchers propose that conduction velocity can be increased by means of further myelination in more active pathways, and that these changes support learning and memory. In addition to a probable role of white matter, numerous hormones and neuropeptides have also been shown to affect learning and memory processes.

Memory traces are present in multiple regions of the brain.

Another question besides the "how" of memory is the "where" of memory. What parts of the brain are responsible for memory? There is no single "memory center" in the brain. Instead, the neurons involved in memory traces are widely distributed throughout the subcortical and cortical regions of the brain. The regions of the brain most extensively implicated in memory include the hippocampus and associated structures of the medial temporal lobes, the limbic system, the cerebellum, the prefrontal cortex, and other areas of the cerebral cortex.

THE HIPPOCAMPUS AND DECLARATIVE MEMORIES The **hippocampus,** the elongated, medial portion of the temporal lobe that is part of the limbic system (see ● Figure 5-16), plays a vital role in short-term memory involving the integration of various related stimuli. It is a prominent site where LTP takes place and is also crucial for consolidation into long-term memory. The hippocampus is believed to store new long-term memories only temporarily and then transfer them to other cortical sites for more permanent storage. The sites for long-term storage of various types of memories are only beginning to be identified by neuroscientists.

The hippocampus and surrounding regions play an especially important role in **declarative memories**—the "what" memories of specific people, places, objects, facts, and events that often result after only one experience and that can be declared or verbalized in a statement such as "I saw the Statue of Liberty last summer" or conjured up in a mental image. Declarative memories involve conscious recall. This memory type is sometimes subdivided into **semantic memories** (memories of facts) and **episodic memories** (memories of events in our lives).

Clinical Note People with hippocampal damage are profoundly forgetful of facts critical to daily functioning. Declarative memories typically are the first to be lost. Extensive damage in the hippocampus is evident in patients with Alzheimer's disease during autopsy. (For an expanded discussion of Alzheimer's disease, see the boxed feature on pp. 164–165, ■ Concepts, Challenges, and Controversies.)

THE CEREBELLUM AND PROCEDURAL MEMORIES The cerebellum and relevant cortical regions play an essential role in the "how to" **procedural memories** involving motor skills gained through repetitive training, such as memorizing a particular dance routine. The cortical areas important for a given procedural memory are the specific motor or sensory systems engaged in performing the routine. In contrast to declarative memories, which are consciously recollected from previous experiences, procedural memories can be brought forth without conscious effort. For example, an ice skater during a competition typically performs best by "letting the body take over" the routine instead of thinking about exactly what needs to be done next.

Clinical Note The distinct localization in different parts of the brain of declarative and procedural memory is apparent in people who have temporal/limbic lesions. They can perform a skill, such as playing a piano, but the next day they have no recollection of having done so.

THE PREFRONTAL CORTEX AND WORKING MEMORY The major orchestrator of the complex reasoning skills associated with working memory is the prefrontal association cortex. The prefrontal cortex serves as a temporary storage site for holding

Alzheimer's Disease: A Tale of Beta Amyloid Plaques, Tau Tangles, and Dementia

"I can't remember where I put my keys. I must be getting Alzheimer's." The incidence and awareness of **Alzheimer's disease (AD),** which is characterized in its early stages by loss of recent memories, have become so commonplace that people sometimes jest about having it when they can't remember something. AD is no joking matter, however.

Incidence

AD is the most common and most costly neurological disorder of the CNS. About 4.5 million Americans currently have AD, but because it is an age-related condition and the population is aging, the incidence is expected to climb. The number of affected individuals is expected to swell to 7 million as "baby boomers" age. About 0.1% of those between 60 and 65 years of age are afflicted with the disease, but the incidence rises to 30% to 47% among those over age 85. According to the *National Institute of Aging Report to Congress,* the most rapidly growing segment of the AD population by percentage is the over-85 age group.

Symptoms

AD accounts for about two thirds of the cases of *senile dementia,* which is a generalized age-related diminution of mental abilities. In the earliest stages of AD, only short-term memory is impaired, but as the disease progresses, even firmly entrenched long-term memories, such as recognition of family members, are lost. Confusion, disorientation, and personality changes characterized by irritability and emotional outbursts are common. Higher mental abilities gradually deteriorate as the patient inexorably loses the ability to read, write, and calculate. Language ability and speech are often impaired. In later stages, AD victims become childlike and are unable to feed, dress, and groom themselves. Patients usually die in a severely debilitated state 4 to 12 years after the onset of the disease.

First described nearly a century ago by Alois Alzheimer, a German neurologist, the condition can only be confirmed at autopsy on finding the characteristic brain lesions associated with the disease. Currently, AD is diagnosed prior to death by a process of elimination; that is, all other disorders that could produce dementia, such as a stroke or brain tumor, must be ruled out. A battery of cognitive tests is sometimes used to support a probable diagnosis of AD. Although scientists have been searching various avenues for a reliable test for diagnosing AD in patients, to date they have not been successful.

Characteristic Brain Lesions

The characteristic brain lesions, extracellular *neuritic (senile) plaques* and intracellular *neurofibrillary tangles,* are dispersed throughout the cerebral cortex and are especially abundant in the hippocampus. A **neuritic plaque** consists of a central core of extracellular, waxy, fibrous protein known as **beta amyloid (Aβ)** surrounded by degenerating dendritic and axonal nerve endings. **Neurofibrillary tangles** are dense bundles of abnormal, paired helical filaments that accumulate in the cell bodies of affected neurons. AD is also characterized by degeneration of the cell bodies of certain neurons in the basal forebrain. The acetylcholine-releasing axons of these neurons normally terminate in the cerebral cortex and hippocampus, so the loss of these neurons results in a deficiency of acetylcholine in these areas. Neuron death and loss of synaptic communication are responsible for the ensuing dementia.

Underlying Pathology

Much progress has been made in understanding the pathology underlying the condition in recent years. **Amyloid precursor protein (APP),** a structural component of all neuronal plasma membranes, is especially abundant in presynaptic terminal endings. APP can be cleaved at different locations to produce different products. Cleavage of APP at one site yields a product believed to exert a normal physiologic function, possibly playing a role in learning and memory. Cleavage of APP at an alternative site yields Aβ. Depending on the exact site of cleavage, two different variants of Aβ are produced and released from the neuron. Normally about 90% of the Aβ is a soluble and harmless form of this product. The other 10% is the dangerous, plaque-forming version, which forms thin, insoluble filaments that readily aggregate into Aβ plaques and also appears to be neurotoxic. Furthermore, researchers recently identified a potentially toxic soluble molecule derived from Aβ that diffuses through the brain instead of aggregating into plaques. The balance between these APP products can be shifted by mutations in APP, other genetic defects, age-related or pathological changes in the brain, or perhaps environmental factors. The end result is increased production of toxic Aβ.

Aβ formation is seen early in the course of the disease, with neurofibrillary tangles developing somewhat later. AD does not "just happen" in old age. Instead, it results from a host of gradual, insidious processes that occur over the course of years or decades. Although some pieces of the puzzle have not been identified, the following is a possible scenario based on the findings to date. The deposited Aβ is directly toxic to neurons. Furthermore, the gradual buildup of Aβ plaques attracts microglia to the plaque sites. These immune cells of the brain launch an inflammatory attack on the plaque, releasing toxic chemicals that can damage surrounding "innocent bystander" neurons in the process.

These inflammatory assaults, along with the direct toxicity of the deposited Aβ, also cause changes in the neuronal cytoskeleton that lead to formation of nerve-cell clogging neurofibrillary tangles. The protein

relevant data and is largely responsible for the so-called executive functions involving manipulation and integration of this information for planning, juggling competing priorities, problem solving, and organizing activities. The prefrontal cortex carries out these complex reasoning functions in cooperation with all the brain's sensory regions, which are linked to the prefrontal cortex through neural connections.

tau normally associates with tubulin molecules in the formation of microtubules, which serve as axonal "highways" for transport of materials back and forth between the cell body and the axon terminal (see p. 43). Tau molecules act like "railroad ties" anchoring the "rails" of tubulin molecules within the microtubule. If tau molecules become hyperphosphorylated (have too many phosphate groups attached), they cannot interact with tubulin. Research suggests that Aβ binds to receptors on the surface of nerve cells, triggering a chain of intracellular events that leads to tau hyperphosphorylation. When not bound to tubulin, the incapacitated tau molecules intertwine, forming paired helical filaments that aggregate to form neurofibrillary tangles. More important, just as train tracks start to fall apart if too many ties are missing, the microtubules start to break down as increasing numbers of tau molecules can no longer do their job. The resultant loss of the neuron's transport system can lead to the death of the cell.

Other factors also play a role in the complex story of AD, but exactly where they fit in is unclear. According to a leading proposal, Aβ causes excessive influx of Ca^{2+}, which triggers a chain of biochemical events that kills the cells. Brain cells that have an abundance of glutamate NMDA receptors, most notably the hippocampal cells involved in long-term potentiation (see p. 161), are especially vulnerable to glutamate toxicity. Loss of hippocampal memory-forming capacity is a hallmark feature of AD. Some evidence suggests that the injured neurons commit apoptosis (cell suicide; see p. 124), but the self-induced death seems to be a much slower process than typical apoptosis. Findings indicate that Aβ can activate the self-execution biochemical pathway. Other studies suggest that cell-damaging free radicals (see p. 142) are produced during the course of the disease. All these neuron-destroying pathways eventually lead to the gradual development of symptoms.

Possible Causes

The underlying trigger for abnormal Aβ formation in AD is unknown in the majority of cases. Many investigators believe the condition has many underlying causes. Both genetic and environmental factors have been implicated in an increased risk of acquiring AD. About 15% of cases are linked to specific, known genetic defects that run in families and cause early onset of the disease. Individuals with this *familial Alzheimer's disease* typically develop clinical symptoms in their 40s or 50s.

The other 85% of AD patients do not begin to manifest symptoms until later in life, somewhere between 65 and 85 years of age. Specific gene traits have also been identified that increase an individual's vulnerability of acquiring *late-onset Alzheimer's disease.* However, not everyone with genetic tendencies for AD develops the disease. Furthermore, many develop the illness with no apparent genetic predisposition. Obviously, other factors must also be at play in producing the condition. Hormonal imbalances may play a role. In particular, research findings suggest that *cortisol,* the stress hormone, increases the propensity to develop the condition, whereas *estrogen,* the female sex hormone, seems to protect against the onslaught of the disease. In addition, investigators have been searching for possible environmental triggers, but none have been found to date. Although aluminum from pots and pans and aerosol antiperspirants was implicated for a while, it is no longer believed to be a factor in the development of the disease.

Treatment

Currently two classes of drugs are specifically approved for treatment of AD. One class raises the levels of acetylcholine (the deficient neurotransmitter) in the brain. For example, Aricept, the most commonly prescribed drug for AD, inhibits the enzyme that normally clears released acetylcholine from the synapse. Such drugs transiently improve the symptoms in some patients, but they do nothing to halt or slow the deterioration caused by the disease. The second, newer class of approved drugs, an example being Nemenda, interferes with the NMDA receptors, blocking the toxic effects of excess glutamate release. Several over-the-counter agents are also used to treat AD. Antioxidants hold some promise of thwarting free-radical damage. Aspirin and other anti-inflammatory drugs may slow the course of AD by blocking the inflammatory components of the condition. Some studies hint that the spice turmeric might help clear plaque deposits.

As researchers continue to unravel the underlying factors, the likelihood of finding various means to block the gradual, relentless progression of AD increases. For example, investigators are cautiously optimistic about a newly developed vaccine targeted against Aβ plaque. Also, the search is on for new drugs that might block the cleavage of plaque-forming Aβ from APP or might inhibit the aggregation of this amyloid into dangerous plaques, thus halting AD in its tracks in the earliest stages. Almost 200 drugs targeted at different steps of the AD pathway are under development by a number of pharmaceutical companies.

The financial payoff for the drug companies that come up with successful products will be huge, and the broader impact will be even more significant. Prevention or treatment of AD cannot come too soon in view of the tragic toll the condition takes on its victims, their families, and society. The cost of custodial care for AD patients is currently estimated at $100 billion annually and will continue to rise as a greater percentage of our population ages and becomes afflicted with the condition.

Researchers have identified different storage bins in the prefrontal cortex, depending on the nature of the current relevant data. For example, working memory involving spatial cues is in a prefrontal location distinct from working memory involving verbal cues or cues about an object's appearance. One recent fascinating proposal suggests that a person's intelligence may be determined by the capacity of his or her working memory to temporarily hold and relate a variety of relevant data.

Brain stem

Cerebellum

(a) Gross structure of cerebellum

Unfolded

Regulation of
muscle tone,
coordination of
skilled voluntary
movement

Planning and
initiation of
voluntary activity,
storage of
procedural
memories

Maintenance of
balance, control
of eye movements

(b) Unfolded cerebellum, revealing its
three functionally distinct parts

Cut

Median sagittal
section of
cerebellum
and brain stem

(c) Internal structure of cerebellum

KEY
- Vestibulocerebellum
- Spinocerebellum
- Cerebrocerebellum

● **FIGURE 5-19** Cerebellum.

Cerebellum

The **cerebellum** is a highly folded, baseball-sized part of the brain that lies underneath the occipital lobe of the cortex and is attached to the back of the upper portion of the brain stem (see ▲ Table 5-2 and ● Figures 5-7b and 5-15).

The cerebellum is important in balance and in planning and executing voluntary movement.

More individual neurons are found in the cerebellum than in the entire rest of the brain, indicative of the importance of this structure. The cerebellum consists of three functionally distinct parts with different roles concerned primarily with subconscious control of motor activity (● Figure 5-19). Specifically, the different parts of the cerebellum perform the following functions:

1. The **vestibulocerebellum** is important for maintaining balance and controls eye movements.

2. The **spinocerebellum** enhances muscle tone and coordinates skilled, voluntary movements. This brain region is especially important in ensuring the accurate timing of various

muscle contractions to coordinate movements involving multiple joints. For example, the movements of your shoulder, elbow, and wrist joints must be synchronized even during the simple act of reaching for a pencil. When cortical motor areas send messages to muscles for executing a particular movement, the spinocerebellum is informed of the intended motor command. This region also receives input from peripheral receptors that inform it about the body movements and positions that are actually taking place. The spinocerebellum essentially acts as "middle management," comparing the "intentions" or "orders" of the higher centers with the "performance" of the muscles and then correcting any "errors" or deviations from the intended movement. The spinocerebellum even seems able to predict the position of a body part in the next fraction of a second during a complex movement and to make adjustments accordingly. If you are reaching for a pencil, for example, this region "puts on the brakes" soon enough to stop the forward movement of your hand at the intended location rather than letting you overshoot your target. These ongoing adjustments, which ensure smooth, precise, directed movement, are especially important for rapidly changing (phasic) activities such as typing, playing the piano, or running.

3. The **cerebrocerebellum** plays a role in planning and initiating voluntary activity by providing input to the cortical motor areas. This is also the cerebellar region that stores procedural memories.

Recent discoveries suggest that in addition to these well-established functions, the cerebellum has even broader responsibilities, such as perhaps coordinating the brain's acquisition of sensory input. Researchers are currently trying to make sense of new and surprising findings that do not fit with the cerebellum's traditional roles in motor control.

Clinical Note The following symptoms of cerebellar disease can be referred to as a loss of the cerebellum's established motor functions: poor balance; "drunken sailor" gait with wide stance and unsteady walking; nystagmus (rhythmic, oscillating eye movements); reduced muscle tone but no paralysis; inability to perform rapid alternating movements smoothly, such as being unable to swiftly slap the open palm of one hand alternately with the palm or back of the other hand; and inability to stop and start skeletal muscle action quickly. The latter gives rise to an *intention tremor* characterized by oscillating to-and-fro movements of a limb as it approaches its intended destination. As a person with cerebellar damage tries to pick up a pencil, he or she may overshoot the pencil and then rebound excessively, repeating this to-and-fro process until success is finally achieved. No tremor is observed except in performing intentional activity, in contrast to the resting tremor associated with disease of the basal nuclei, most notably Parkinson's disease.

The cerebellum and basal nuclei both monitor and adjust motor activity commanded from the motor cortex, and like the basal nuclei, the cerebellum does not directly influence the efferent motor neurons. Although they perform different roles (for example, the cerebellum enhances muscle tone, whereas the basal nuclei inhibit it), both function indirectly by modifying the output of major motor systems in the brain. The motor command for a particular voluntary activity arises from the motor cortex, but the actual execution of that activity is coordinated subconsciously by these subcortical regions. To illustrate, you can voluntarily decide you want to walk, but you do not have to consciously think about the specific sequence of movements you have to perform to accomplish this intentional act. Accordingly, much voluntary activity is actually involuntarily regulated.

You will learn more about motor control when we discuss skeletal muscle physiology in Chapter 8. For now, we are going to move on to the remaining part of the brain, the brain stem.

Brain Stem

The **brain stem** consists of the **medulla, pons,** and **midbrain** (see ▲ Table 5-2 and ● Figure 5-7b).

The brain stem is a vital link between the spinal cord and higher brain regions.

All incoming and outgoing fibers traversing between the periphery and higher brain centers must pass through the brain stem, with incoming fibers relaying sensory information to the brain and outgoing fibers carrying command signals from the brain for efferent output. A few fibers merely pass through, but most synapse within the brain stem for important processing. Thus, the brain stem is a critical connecting link between the rest of the brain and the spinal cord.

The functions of the brain stem include the following:

1. The majority of the 12 pairs of **cranial nerves** arise from the brain stem (● Figure 5-20). With one major exception, these nerves supply structures in the head and neck with both sensory and motor fibers. They are important in sight, hearing, taste, smell, sensation of the face and scalp, eye movement, chewing, swallowing, facial expressions, and salivation. The major exception is cranial nerve X, the **vagus nerve.** Instead of innervating regions in the head, most branches of the vagus nerve supply organs in the thoracic and abdominal cavities. The vagus is the major nerve of the parasympathetic nervous system.

2. Collected within the brain stem are neuronal clusters or "centers" that control heart and blood vessel function, respiration, and many digestive activities. A functional collection of neuronal cell bodies within the CNS is alternately known as a **center,** such as the *respiratory control center* in the brain stem, or as a *nucleus* (plural *nuclei*), such as the basal nuclei.

3. The brain stem plays a role in regulating muscle reflexes involved in equilibrium and posture.

4. A widespread network of interconnected neurons called the **reticular formation** runs throughout the entire brain stem and into the thalamus. This network receives and integrates all incoming sensory synaptic input. Ascending fibers originating in the reticular formation carry signals upward to arouse and activate the cerebral cortex (● Figure 5-21). These fibers compose the **reticular activating system (RAS),** which controls the overall degree of cortical alertness and is important in the ability to direct attention. In turn, fibers descending from the cortex, especially its motor areas, can activate the RAS.

5. The centers that govern sleep traditionally have been considered to be housed within the brain stem, although recent evidence suggests that the center promoting slow-wave sleep lies in the hypothalamus.

We are now going to examine sleep and the other states of consciousness in further detail.

Sleep is an active process consisting of alternating periods of slow-wave and paradoxical sleep.

The term **consciousness** refers to subjective awareness of the external world and self, including awareness of the private inner world of one's own mind—that is, awareness of thoughts, perceptions, dreams, and so on. Even though the final level of awareness resides in the cerebral cortex and a crude sense of awareness is detected by the thalamus, conscious experience depends on the integrated functioning of many parts of the nervous system. The cellular and molecular basis underlying consciousness is one of the greatest unanswered questions in neuroscience.

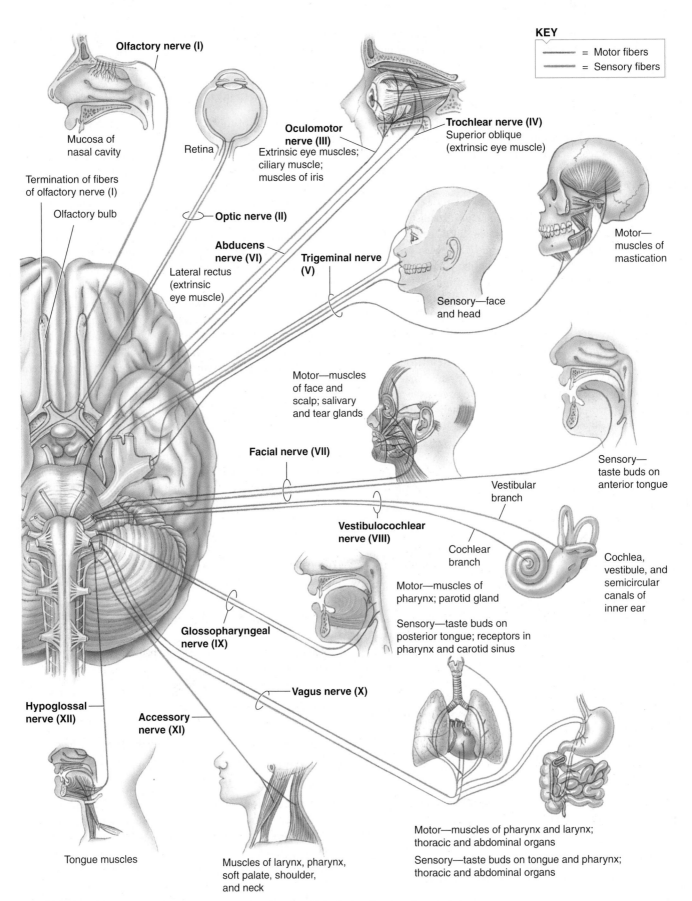

Olfactory nerve (I)

Mucosa of
nasal cavity

Termination of fibers
of olfactory nerve (I)

Olfactory bulb

Retina

Optic nerve (II)

**Abducens
nerve (VI)**

Lateral rectus
(extrinsic
eye muscle)

**Oculomotor
nerve (III)**
Extrinsic eye muscles;
ciliary muscle;
muscles of iris

Trochlear nerve (IV)
Superior oblique
(extrinsic eye muscle)

**Trigeminal nerve
(V)**

Motor—
muscles of
mastication

Sensory—face
and head

Motor—muscles
of face and
scalp; salivary
and tear glands

Facial nerve (VII)

**Vestibulocochlear
nerve (VIII)**

Vestibular
branch

Cochlear
branch

Sensory—
taste buds on
anterior tongue

Cochlea,
vestibule, and
semicircular
canals of
inner ear

Motor—muscles of
pharynx; parotid gland

Sensory—taste buds on
posterior tongue; receptors in
pharynx and carotid sinus

**Glossopharyngeal
nerve (IX)**

Vagus nerve (X)

**Hypoglossal
nerve (XII)**

**Accessory
nerve (XI)**

Tongue muscles

Muscles of larynx, pharynx,
soft palate, shoulder,
and neck

Motor—muscles of pharynx and larynx;
thoracic and abdominal organs

Sensory—taste buds on tongue and pharynx;
thoracic and abdominal organs

KEY
——— = Motor fibers
——— = Sensory fibers

● **FIGURE 5-20 Cranial nerves.** Inferior (underside) view of the brain, showing the attachments of the 12 pairs of cranial nerves to the brain and many of the structures innervated by
those nerves.

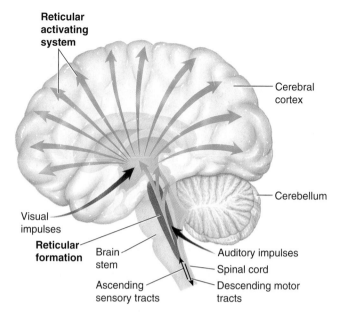

Reticular
activating
system

Cerebral cortex

Cerebellum

Visual impulses

Reticular formation

Brain stem

Auditory impulses

Spinal cord

Ascending sensory tracts

Descending motor tracts

● **FIGURE 5-21 The reticular activating system.** The reticular formation, a widespread network of neurons within the brain stem (in red), receives and integrates all synaptic input. The reticular activating system, which promotes cortical alertness and helps direct attention toward specific events, consists of ascending fibers (in blue) that originate in the reticular formation and carry signals upward to arouse and activate the cerebral cortex.

The following states of consciousness are listed in decreasing order of arousal level, based on the extent of interaction between peripheral stimuli and the brain:

- maximum alertness
- wakefulness
- sleep (several different types)
- coma

Maximum alertness depends on attention-getting sensory input that "energizes" the RAS and subsequently the activity level of the CNS as a whole. At the other extreme, coma is the total unresponsiveness of a living person to external stimuli, caused either by brain stem damage that interferes with the RAS or by widespread depression of the cerebral cortex, such as accompanies O_2 deprivation.

The **sleep–wake cycle** is a normal cyclic variation in awareness of surroundings. In contrast to being awake, sleeping people are not consciously aware of the external world, but they do have inward conscious experiences such as dreams. Furthermore, they can be aroused by external stimuli, such as an alarm going off.

Sleep is an active process, not just the absence of wakefulness. The brain's overall level of activity is not reduced during sleep. During certain stages of sleep, O_2 uptake by the brain is even increased above normal waking levels.

There are two types of sleep, characterized by different EEG patterns and different behaviors: *slow-wave sleep* and *paradoxical*, or *REM, sleep* (▲ Table 5-4).

EEG PATTERNS DURING SLEEP Slow-wave sleep occurs in four stages, each displaying progressively slower EEG waves of higher amplitude (hence, "slow-wave" sleep) (● Figure 5-22). At the onset of sleep, you move from the light sleep of stage 1 to the deep sleep of stage 4 of slow-wave sleep during a period of 30 to 45 minutes; then you reverse through the same stages in the same amount of time. A 10- to 15-minute episode of **paradoxical sleep** punctuates the end of each slow-wave sleep cycle. Paradoxically, your EEG pattern during this time abruptly becomes similar to that of a wide-awake, alert individual, even

▲ TABLE 5-4	Comparison of Slow-Wave and Paradoxical Sleep	
	TYPE OF SLEEP	
Characteristic	**Slow-Wave Sleep**	**Paradoxical Sleep**
EEG	Displays slow waves	Similar to EEG of alert, awake person
Motor Activity	Considerable muscle tone; frequent shifting	Abrupt inhibition of muscle tone; no movement
Heart Rate, Respiratory Rate, Blood Pressure	Minor reductions	Irregular
Dreaming	Rare (mental activity is extension of waking-time thoughts)	Common
Arousal	Sleeper easily awakened	Sleeper hard to arouse but apt to wake up spontaneously
Percentage of Sleeping Time	80%	20%
Other Important Characteristics	Has four stages; sleeper must pass through this type of sleep first	Rapid eye movements

Slow-wave sleep, stage 4

Paradoxical sleep

Awake, eyes open

● **FIGURE 5-22 EEG patterns during different types of sleep.** Note that the EEG pattern during paradoxical sleep is similar to that of an alert, awake person, whereas the pattern during slow-wave sleep displays distinctly different waves.

KEY

	Awake
	Slow-wave sleep
	Paradoxical (REM) sleep

● **FIGURE 5-23 Typical cyclical sleep pattern in a young adult.**

though you are still asleep (hence, "paradoxical" sleep) (● Figure 5-22). After the paradoxical episode, the stages of slow-wave sleep repeat. A person cyclically alternates between the two types of sleep throughout the night. Brief periods of wakefulness occasionally occur. Most stage 4 deep sleep occurs during the first several hours of sleep, with REM sleep occupying an increasingly greater share of sleep time as morning approaches (● Figure 5-23). Because of the resemblance of this graphic representation of the cyclical sleep pattern to a city skyline, the pattern of sleep is sometimes referred to as *sleep architecture.*

In a normal sleep cycle, you always pass through slow-wave sleep before entering paradoxical sleep. On average, paradoxical sleep occupies 20% of total sleeping time throughout adolescence and most of adulthood. Infants spend considerably more time in paradoxical sleep. In contrast, paradoxical as well as deep stage 4 slow-wave sleep declines in the elderly. People who require less total sleeping time than normal spend proportionately more time in paradoxical and deep stage 4 slow-wave sleep and less time in the lighter stages of slow-wave sleep.

BEHAVIORAL PATTERNS DURING SLEEP In addition to distinctive EEG patterns, the two types of sleep are distinguished by behavioral differences. It is difficult to pinpoint exactly when an individual drifts from drowsiness into slow-wave sleep. In this type of sleep, the person still has considerable muscle tone and frequently shifts body position. Only minor reductions in respi-

ratory rate, heart rate, and blood pressure occur. During this time the sleeper can be easily awakened and rarely dreams. The mental activity associated with slow-wave sleep is less visual than dreaming. It is more conceptual and plausible—like an extension of waking-time thoughts concerned with everyday events—and it is less likely to be recalled. The major exception is nightmares, which occur during stages 3 and 4. People who walk and talk in their sleep do so during slow-wave sleep.

The behavioral pattern accompanying paradoxical sleep is marked by abrupt inhibition of muscle tone throughout the body. The muscles are completely relaxed, with no movement taking place except in the eye muscles. Paradoxical sleep is characterized by *rapid eye movements,* hence the alternative name, REM sleep. Heart rate and respiratory rate become irregular, and blood pressure may fluctuate. Another characteristic of REM sleep is *dreaming.* The rapid eye movements are not related to "watching" the dream imagery. The eye movements are driven in a locked, oscillating pattern uninfluenced by dream content.

Brain imaging of volunteers during REM sleep shows heightened activity in the higher-level visual processing areas and limbic system (the seat of emotions), coupled with reduced activity in the prefrontal cortex (the seat of reasoning). This pattern of activity lays the groundwork for the characteristics of dreaming: internally generated visual imagery reflecting activation of the person's "emotional memory bank" with little guidance or interpretation from the complex thinking areas. As a result, dreams are often charged with intense emotions, a distorted sense of time, and bizarre content that is uncritically accepted as real, with little reflection about all the strange happenings.

The sleep–wake cycle is controlled by interactions among three neural systems.

The sleep–wake cycle as well as the various stages of sleep are due to the cyclic interplay of three different neural systems: (1) an **arousal system,** which is regulated by a group of neurons in the hypothalamus and involves the reticular activating system originating in the brain stem; (2) a **slow-wave sleep center** in the hypothalamus that contains *sleep-on neurons* that induce sleep; and (3) a **paradoxical sleep center** in the brain stem that houses *REM sleep-on neurons,* which become very active during REM sleep. The patterns of interaction among these three neural regions, which bring about the fairly predictable cyclical sequence between being awake and passing alternately between the two types of sleep, are the subject of intense investigation. A growing body of evidence suggests that a group of neurons in the hypothalamus that secrete the excitatory neurotransmitter *hypocretin* (also known as *orexin*) is at the top of the chain of command for regulating the arousal system. Surprisingly, hypocretin is better known as an appetite-enhancing signal, but it apparently plays an important role in arousal too. These hypocretin-secreting neurons fire autonomously (on their own) and continuously and keep you awake and alert by stimulating the reticular activating system. They must be inhibited to induce sleep, as perhaps by IPSPs generated by input from the sleep-on neurons or by other inhibitory inputs.

The sleep-on neurons make up the slow-wave sleep center in the hypothalamus. As their name implies, they appear to be responsible for bringing on sleep, likely by inhibiting the arousal-promoting neurons by releasing the inhibitory neurotransmitter GABA. This mechanism would explain why we enter slow-wave sleep first when we fall asleep. The sleep-on neurons are inactive when a person is awake and are maximally active only during slow-wave sleep. But what turns on the sleep-on neurons to induce sleep? Scientists do not know much about the factors that activate the sleep-on neurons.

The REM sleep-on neurons in the brain stem are believed to serve as the switch between slow-wave sleep and REM sleep. It appears that the REM sleep-on neurons can turn off the sleep-on neurons and switch the sleep pattern from slow-wave sleep to REM sleep. The underlying molecular mechanisms responsible for the cyclical interplay between the two types of sleep remain poorly understood.

The normal cycle can easily be interrupted, with the arousal system more readily overriding the sleep systems than vice versa; that is, it is easier to stay awake when you are sleepy than to fall asleep when you are wide awake. The arousal system can be activated by afferent sensory input (for example, a person has difficulty falling asleep when it is noisy) or by input descending to the brain stem from higher brain regions. Intense concentration or strong emotional states, such as anxiety or excitement, can keep a person from falling asleep, just as motor activity, such as getting up and walking around, can arouse a drowsy person.

The function of sleep is unclear.

Even though humans spend about a third of their lives sleeping, why sleep is needed largely remains a mystery. Sleep is not accompanied by a reduction in neural activity (that is, the brain cells are not "resting"), as once was suspected, but rather by a profound *change* in activity. Sleep is for the brain, not for other parts of the body. Although still speculative, recent studies suggest that slow-wave sleep and REM sleep serve different purposes.

One widely accepted proposal holds that sleep provides "catch-up" time for the brain to restore biochemical or physiological processes that have progressively degraded during wakefulness. The most direct evidence supporting this proposal is the potential role of *adenosine* as a neural sleep factor. Adenosine, the backbone of adenosine triphosphate (ATP), the body's energy currency, is generated during the awake state by metabolically active neurons and glial cells. Thus, the brain's extracellular concentration of adenosine continues to rise the longer a person has been awake. Adenosine, which acts as a neuromodulator, has been shown experimentally to inhibit the arousal center. This action can bring on slow-wave sleep, during which restoration and recovery activities are believed to take place. Injections of adenosine induce apparently normal sleep, whereas *caffeine,* which blocks adenosine receptors in the brain, revives drowsy people by removing adenosine's inhibitory influence on the arousal center. Adenosine levels diminish during sleep, presumably because the brain uses this adenosine as a raw ingredient for replenishing its limited energy stores. Thus, the body's need for sleep may stem from the brain's periodic need to replenish diminishing energy stores. Because adenosine reflects the level of brain cell activity, the concentration of this chemical in the brain may serve as a gauge of how much energy has been depleted.

Another "restoration and recovery" proposal suggests that slow-wave sleep provides time for the brain to repair the damage caused by the toxic free radicals (see p. 142) produced as by-products of the stepped-up metabolism during the waking state. Other organs can sacrifice and replace cells damaged by free radicals, but this is not an option for the nonregenerative brain.

A possible "restoration and recovery" function of REM sleep is to let some of the neural pathways regain full sensitivity. When a person is awake, the brain neurons that release the neurotransmitters norepinephrine and serotonin are maximally and continuously active. Release of these neurotransmitters ceases during REM sleep. Studies suggest that constant release of norepinephrine and serotonin can desensitize their receptors. Perhaps REM sleep is needed to restore receptor sensitivity for optimal functioning during the next period of wakefulness.

Another leading theory has nothing to do with restoration and recovery. Instead, other researchers believe that sleep is necessary to allow the brain to "shift gears" to accomplish the long-term structural and chemical adjustments necessary for learning and memory. This theory might explain why infants need so much sleep. Their highly plastic brains are rapidly undergoing profound synaptic modifications in response to environmental stimulation. In contrast, mature individuals, in whom neural changes are less dramatic, sleep less. Some evidence suggests that the different types of sleep might be involved in consolidation of different kinds of memories, with declarative memories being consolidated during slow-wave sleep and procedural memories during REM sleep. A recent memory-related theory is that sleep, especially slow-wave sleep, is a time for replaying the events of the day, not only to help consolidate memories but perhaps to make recent experiences more meaningful by catching

Spinal Cord

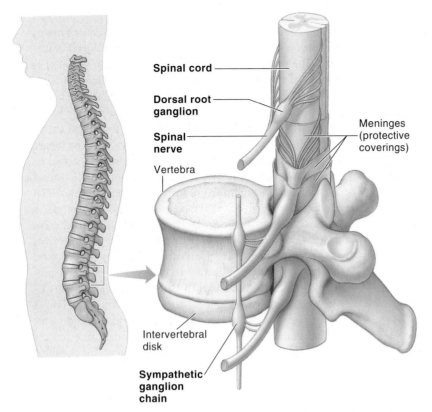

The **spinal cord** is a long, slender cylinder of nerve tissue that extends from the brain stem. It is about 45 cm long (18 in.) and 2 cm in diameter (about the size of your thumb).

The spinal cord extends through the vertebral canal and is connected to the spinal nerves.

Exiting through a large hole in the base of the skull, the spinal cord is enclosed by the protective vertebral column as it descends through the vertebral canal (● Figure 5-24). Paired **spinal nerves** emerge from the spinal cord through spaces formed between the bony, winglike arches of adjacent vertebrae. The spinal nerves are named according to the region of the vertebral column from which they emerge (● Figure 5-25): there are 8 pairs of *cervical (neck) nerves* (namely C1–C8), 12 *thoracic (chest) nerves,* 5 *lumbar (abdominal) nerves,* 5 *sacral (pelvic) nerves,* and 1 *coccygeal (tailbone) nerve.*

During development, the vertebral column grows about 25 cm longer than the spinal cord. Because of this differential growth, segments of the spinal cord that give rise to various spinal nerves are not aligned with the corresponding intervertebral spaces. Most spinal nerve roots must descend along the cord before emerging from the vertebral column at the corresponding space. The spinal cord itself extends only to the level of the first or second lumbar vertebra (about waist level), so the nerve roots of the remaining nerves are greatly elongated, to exit the vertebral column at their appropriate space. The thick bundle of elongated nerve roots within the lower vertebral canal is called the **cauda equina** ("horse's tail") because of its appearance (● Figure 5-25b).

Clinical Note Spinal taps to obtain a sample of CSF are performed by inserting a needle into the vertebral canal below the level of the second lumbar vertebra. Insertion at this site does not run the risk of penetrating the spinal cord. The needle pushes aside the nerve roots of the cauda equina so that a sample of the surrounding fluid can be withdrawn safely.

The white matter of the spinal cord is organized into tracts.

Although there are some slight regional variations, the cross-sectional anatomy of the spinal cord is generally the same throughout its length (● Figure 5-26). In contrast to the gray matter forming an outer shell capping an inner white core in the brain, the gray matter in the spinal cord forms an inner butterfly-shaped region surrounded by the outer white matter. As in the brain, the cord gray matter consists primarily of neuronal cell bodies and their dendrites, short interneurons, and glial cells. The white matter is organized into *tracts,* which are bundles of nerve fibers (axons of long interneurons) with a similar function. The bundles are grouped into columns that

● FIGURE 5-24 Location of the spinal cord relative to the vertebral column.

information missed on first pass and by "connecting the dots" between new pieces of information. This information-processing proposal could explain why people with an important decision to make sometimes say they will "sleep on it" before arriving at a conclusion.

Another theory of why we need sleep is for energy conservation, but this proposal is not as widely accepted. The various theories are not mutually exclusive. Sleep might serve multiple purposes.

Not much is known about the brain's need for cycling between the two types of sleep, although a specified amount of paradoxical sleep appears to be required. Individuals experimentally deprived of paradoxical sleep for a night or two by being aroused every time the paradoxical EEG pattern appeared suffered hallucinations and spent proportionally more time in paradoxical sleep during subsequent undisturbed nights, as if to make up for lost time.

Clinical Note An unusual sleep disturbance is **narcolepsy.** It is characterized by brief (5- to 30-minute) irresistible sleep attacks during the day. A person suffering from this condition suddenly falls asleep during any ongoing activity, often without warning. Narcoleptic patients typically enter into paradoxical sleep directly without the normal prerequisite passage through slow-wave sleep. Investigators recently learned that narcolepsy is linked to a deficiency of hypocretin.

We have now completed our discussion of the brain and are going to shift our attention to the other component of the CNS, the spinal cord.

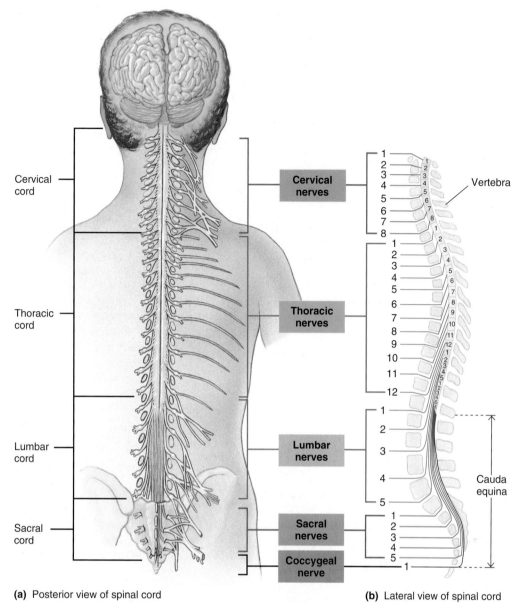

(a) Posterior view of spinal cord   **(b)** Lateral view of spinal cord

● **FIGURE 5-25 Spinal nerves.** The 31 pairs of spinal nerves are named according to the region of the vertebral column from which they emerge. Because the spinal cord is shorter than the vertebral column, spinal nerve roots must descend along the cord before emerging from the vertebral column at the corresponding intervertebral space, especially those beyond the level of the first lumbar vertebra (L1). Collectively these rootlets are called the cauda equina, literally "horse's tail." (a) Posterior view of the brain, spinal cord, and spinal nerves (on the right side only). (b) Lateral view of the spinal cord and spinal nerves emerging from the vertebral column.

extend the length of the cord. Each of these tracts begins or ends within a particular area of the brain, and each transmits a specific type of information. Some are **ascending** (cord to brain) **tracts** that transmit to the brain signals derived from afferent input. Others are **descending** (brain to cord) **tracts** that relay messages from the brain to efferent neurons (● Figure 5-27).

The tracts are generally named for their origin and termination. For example, the **ventral spinocerebellar tract** is an ascending pathway that originates in the spinal cord and runs up the ventral (toward the front) margin of the cord with several synapses along the way until it eventually terminates in the cer-

ebellum (● Figure 5-28a). This tract carries information derived from muscle stretch receptors that has been delivered to the spinal cord by afferent fibers for use by the spinocerebellum. In contrast, the **ventral corticospinal tract** is a descending pathway that originates in the motor region of the cerebral cortex, then travels down the ventral portion of the spinal cord, and terminates in the spinal cord on the cell bodies of efferent motor neurons supplying skeletal muscles (● Figure 5-28b). Because various types of signals are carried in different tracts within the spinal cord, damage to particular areas of the cord can interfere with some functions, whereas other functions remain intact.

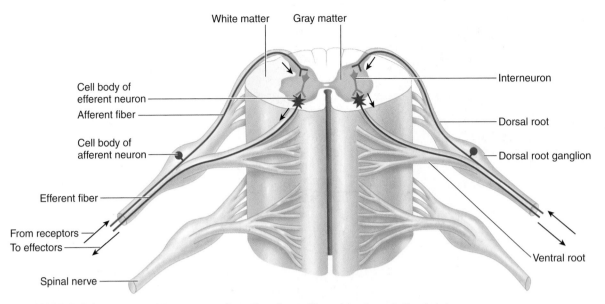

● **FIGURE 5-26 Spinal cord in cross section.** The afferent fibers enter through the dorsal root, and the efferent fibers exit through the ventral root. Afferent and efferent fibers are enclosed together within a spinal nerve.

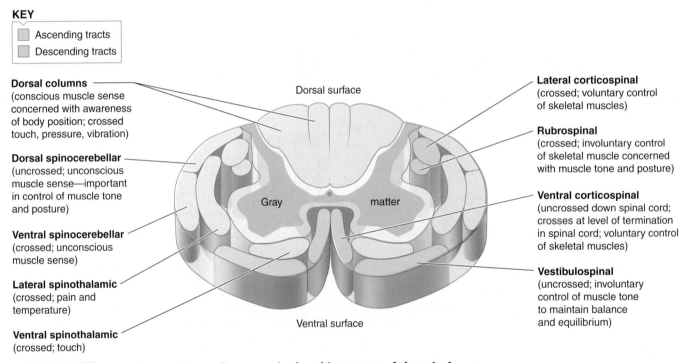

KEY
- Ascending tracts
- Descending tracts

Dorsal columns
(conscious muscle sense concerned with awareness of body position; crossed touch, pressure, vibration)

Dorsal spinocerebellar
(uncrossed; unconscious muscle sense—important in control of muscle tone and posture)

Ventral spinocerebellar
(crossed; unconscious muscle sense)

Lateral spinothalamic
(crossed; pain and temperature)

Ventral spinothalamic
(crossed; touch)

Dorsal surface

Gray   matter

Ventral surface

Lateral corticospinal
(crossed; voluntary control of skeletal muscles)

Rubrospinal
(crossed; involuntary control of skeletal muscle concerned with muscle tone and posture)

Ventral corticospinal
(uncrossed down spinal cord; crosses at level of termination in spinal cord; voluntary control of skeletal muscles)

Vestibulospinal
(uncrossed; involuntary control of muscle tone to maintain balance and equilibrium)

● **FIGURE 5-27 Ascending and descending tracts in the white matter of the spinal cord in cross section.**

Each horn of the spinal cord gray matter houses a different type of neuronal cell body.

The centrally located gray matter is also functionally organized (● Figure 5-29). The central canal, which is filled with CSF, lies in the center of the gray matter. Each half of the gray matter is arbitrarily divided into a *dorsal (posterior) horn*, a *ventral (ante-* *rior) horn*, and a *lateral horn*. The **dorsal horn** contains cell bodies of interneurons on which afferent neurons terminate. The **ventral horn** contains cell bodies of the efferent motor neurons supplying skeletal muscles. Autonomic nerve fibers supplying cardiac and smooth muscle and exocrine glands originate at cell bodies found in the **lateral horn.**

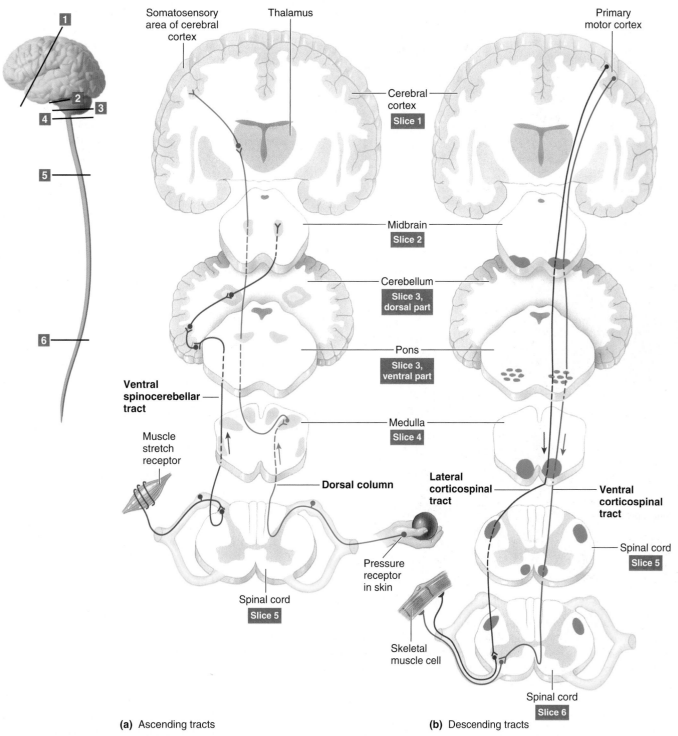

(a) Ascending tracts **(b)** Descending tracts

● **FIGURE 5-28 Examples of ascending and descending pathways in the white matter of the spinal cord.** (a) Cord-to-brain pathways of several ascending tracts (a dorsal column tract and ventral spinocerebellar tract). (b) Brain-to-cord pathways of several descending tracts (lateral corticospinal and ventral corticospinal tracts).

Spinal nerves carry both afferent and efferent fibers.

Spinal nerves connect with each side of the spinal cord by a **dorsal root** and a **ventral root** (see ● Figure 5-26). Afferent fibers carrying incoming signals from peripheral receptors enter the spinal cord through the dorsal root. The cell bodies for the afferent neurons at each level are clustered together in a **dorsal root ganglion.** (A collection of neuronal cell bodies located outside the CNS is called a **ganglion.** Remember a functional collection of cell bodies within the CNS is referred to as a *center* or a *nucleus.*) The cell bodies for the efferent neurons originate in the gray matter and send axons out through the ventral root.

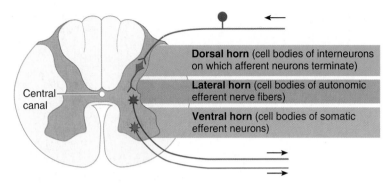

● **FIGURE 5-29 Regions of the gray matter.**

Axon

Myelin sheath

Connective tissue around the axon

Connective tissue around a fascicle

Connective tissue around the nerve

Blood vessels

Nerve fascicle (many axons bundled in connective tissue)

Nerve

Dr. R. G. Kessel and Dr. R. H. Kardon/Visuals Unlimited

● **FIGURE 5-30 Structure of a nerve.** Neuronal axons (both afferent and efferent fibers) are bundled together into connective tissue–wrapped fascicles. A nerve consists of a group of fascicles enclosed by a connective tissue covering and following the same pathway. The photograph is a scanning electron micrograph of several nerve fascicles in cross section.

Therefore, efferent fibers carrying outgoing signals to muscles and glands exit through the ventral root.

The dorsal and ventral roots at each level join to form a **spinal nerve** that emerges from the vertebral column (see ● Figure 5-26). A spinal nerve contains both afferent and efferent fibers traversing between a particular region of the body and the spinal cord. Note the relationship between a *nerve* and a *neuron*. A **nerve** is a bundle of peripheral neuronal axons, some afferent and some efferent, enclosed by a connective tissue covering and following the same pathway (● Figure 5-30). A nerve does not contain a complete nerve cell, only the axonal portions of many neurons. (By this definition, there are no nerves in the CNS! Bundles of axons in the CNS are called *tracts*.) The individual fibers within a nerve gener-

ally do not have any direct influence on one another. They travel together for convenience, just as many individual telephone lines are carried within a telephone cable, yet any particular phone connection can be private without interference or influence from other lines in the cable.

The 31 pairs of spinal nerves, along with the 12 pairs of cranial nerves that arise from the brain, constitute the *peripheral nervous system.* After they emerge, the spinal nerves progressively branch to form a vast network of peripheral nerves that supply the tissues. Each segment of the spinal cord gives rise to a pair of spinal nerves that ultimately supply a particular region of the body with both afferent and efferent fibers. Thus, the location and extent of sensory and motor deficits associated with spinal-cord injuries can be clinically important in determining the level and extent of the cord injury.

With reference to sensory input, each specific region of the body surface supplied by a particular spinal nerve is called a **dermatome.** These same spinal nerves also carry fibers that branch off to supply internal organs, and sometimes pain originating from one of these organs is "referred" to the corresponding dermatome supplied by the same spinal nerve. **Referred pain** originating in the heart, for example, may appear to come from the left shoulder and arm. The mechanism responsible for referred pain is not completely understood. Inputs arising from the heart presumably share a pathway to the brain in common with inputs from the left upper extremity. The higher perception levels, being more accustomed to receiving sensory input from the left arm than from the heart, may interpret the input from the heart as having arisen from the left arm.

The spinal cord is responsible for the integration of many basic reflexes.

The spinal cord is strategically located between the brain and afferent and efferent fibers of the peripheral nervous system; this location enables the spinal cord to fulfill its two primary functions: (1) serving as a link for transmission of information between the brain and the remainder of the body and (2) integrating reflex activity between afferent input and efferent output without involving the brain. This type of reflex activity is called a *spinal reflex.*

A **reflex** is any response that occurs automatically without conscious effort. There are two types of reflexes: (1) **simple,** or **basic, reflexes,** which are built-in, unlearned responses, such as pulling the hand away from a burning hot object; and (2) **acquired,** or **conditioned, reflexes,** which are a result of practice and learning, such as a pianist striking a particular key on seeing a given note on the music staff. The musician reads music and plays it automatically, but only after considerable conscious training effort. (For a discussion of the role of acquired reflexes in many sports skills, see the accompanying boxed feature, ■ A Closer Look at Exercise Physiology.)

Swan Dive or Belly Flop: It's a Matter of CNS Control

Sport skills must be learned. Much of the time, strong basic reflexes must be overridden in order to perform the skill. Learning to dive into water, for example, is very difficult initially. Strong head-righting reflexes controlled by sensory organs in the neck and ears initiate a straightening of the neck and head before the beginning diver enters the water, causing what is commonly known as a "belly flop." In a backward dive, the head-righting reflex causes the beginner to land on his or her back or even in a sitting position. To perform any motor skill that involves body inversions, somersaults, back flips, or other abnormal postural movements, the person must learn to consciously inhibit basic postural reflexes. This is accomplished by having the person concentrate on specific body positions during the movement. For example, to perform a somersault, the person must concentrate on keeping the chin tucked and grabbing the knees. After the skill is performed repeatedly, new synaptic patterns are formed in the CNS, and the new or conditioned response substitutes for the natural reflex responses. Sport skills must be practiced until the movement becomes automatic; then the athlete is free during competition to think about strategy or the next move to be performed in a routine.

REFLEX ARC The neural pathway involved in accomplishing reflex activity is known as a **reflex arc,** which typically includes five basic components:

1. sensory receptor
2. afferent pathway
3. integrating center
4. efferent pathway
5. effector

The **sensory receptor** (*receptor* for short) responds to a **stimulus,** which is a detectable physical or chemical change in the environment of the receptor. In response to the stimulus, the receptor produces an action potential that is relayed by the **afferent pathway** to the **integrating center** (usually the CNS) for processing. The spinal cord and brain stem integrate basic reflexes, whereas higher brain levels usually process acquired reflexes. The integrating center processes all information available to it from this receptor as well as from all other inputs, then "makes a decision" about the appropriate response. The instructions from the integrating center are transmitted via the **efferent pathway** to the **effector**—a muscle or gland—which carries out the desired response. Unlike conscious behavior, in which any one of a number of responses is possible, a reflex response is predictable, because the pathway is always the same.

STRETCH REFLEX A basic **spinal reflex** is one integrated by the spinal cord; that is, all components necessary for linking afferent input to efferent response are present within the spinal cord. The simplest reflex is the **stretch reflex,** in which an afferent neuron originating at a stretch-detecting receptor in a skeletal muscle terminates directly on the efferent neuron supplying the same skeletal muscle to cause it to contract and counteract the stretch. In this reflex the integrating center is the single synapse within the spinal cord between the afferent and efferent pathways. The output of this system (whether or not the muscle contracts in response to passive stretch) depends on the extent of summation of EPSPs at the cell body of the efferent neuron arising from the frequency of afferent input (determined by the extent of stretch detected by the receptor). Integration in this case simply involves summation of EPSPs from a single source. (You will learn more about the role of this reflex in Chapter 8.) The stretch reflex is a **monosynaptic** ("one synapse") **reflex,** because the only synapse in the reflex arc is the one between the afferent neuron and the efferent neuron. All other reflexes are **polysynaptic** ("many synapses"), because interneurons are interposed in the reflex pathway and, therefore, a number of synapses are involved. The withdrawal reflex is an example of a polysynaptic basic spinal reflex.

WITHDRAWAL REFLEX When a person touches a hot stove (or receives another painful stimulus), a **withdrawal reflex** is initiated to withdraw from the painful stimulus (● Figure 5-31). The skin has different receptors for warmth, cold, light touch, pressure, and pain. Even though all information is sent to the CNS by way of action potentials, the CNS can distinguish between various stimuli because different receptors and consequently different afferent pathways are activated by different stimuli. When a receptor is stimulated enough to reach threshold, an action potential is generated in the afferent neuron. The stronger the stimulus is, the greater the frequency of action potentials generated and propagated to the CNS. Once the afferent neuron enters the spinal cord, it diverges to synapse with the following different interneurons (the letters correspond to those in ● Figure 5-31).

(a) An excited afferent neuron stimulates excitatory interneurons that stimulate the efferent motor neurons supplying the biceps, the muscle in the arm that flexes (bends) the elbow joint, pulling the hand away from the hot stove.

(b) The afferent neuron also stimulates inhibitory interneurons that inhibit the efferent neurons supplying the triceps to prevent it from contracting. The triceps is the muscle that extends (straightens out) the elbow joint. When the biceps contracts to flex the elbow, it would be counterproductive for the triceps to contract. Therefore, built into the withdrawal reflex is inhibition of the muscle that antagonizes (opposes) the desired response. This type of connection involving stimulation of the nerve supply to one muscle and simultaneous inhibition of the nerves to its antagonistic muscle is known as **reciprocal innervation.**

(c) The afferent neuron stimulates still other interneurons that carry the signal up the spinal cord to the brain via an ascending pathway. Only when the impulse reaches the sensory area of the cortex is the person aware of the pain, its location, and the type

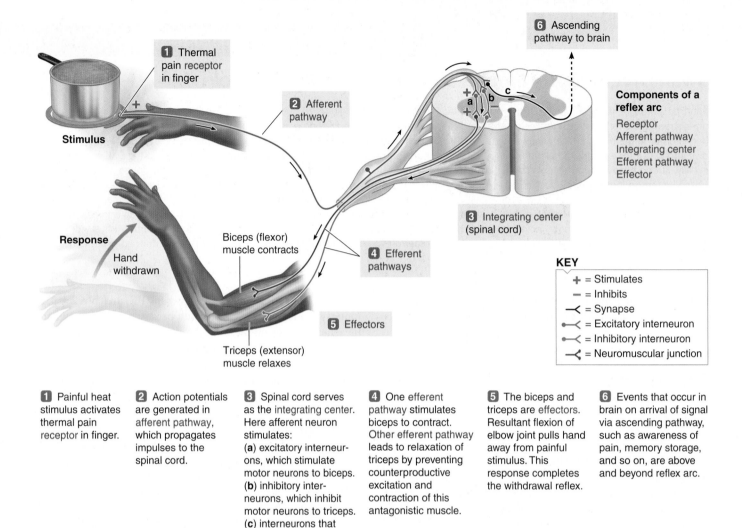

1 Painful heat stimulus activates thermal pain receptor in finger.

2 Action potentials are generated in afferent pathway, which propagates impulses to the spinal cord.

3 Spinal cord serves as the integrating center. Here afferent neuron stimulates:
(**a**) excitatory interneurons, which stimulate motor neurons to biceps.
(**b**) inhibitory interneurons, which inhibit motor neurons to triceps.
(**c**) interneurons that are part of ascending pathway to brain.

4 One efferent pathway stimulates biceps to contract. Other efferent pathway leads to relaxation of triceps by preventing counterproductive excitation and contraction of this antagonistic muscle.

5 The biceps and triceps are effectors. Resultant flexion of elbow joint pulls hand away from painful stimulus. This response completes the withdrawal reflex.

6 Events that occur in brain on arrival of signal via ascending pathway, such as awareness of pain, memory storage, and so on, are above and beyond reflex arc.

● **FIGURE 5-31 The withdrawal reflex.**

of stimulus. Also, when the impulse reaches the brain, the information can be stored as memory, and the person can start thinking about the situation—how it happened, what to do about it, and so on. All this activity at the conscious level is above and beyond the basic reflex.

As with all spinal reflexes, the brain can modify the withdrawal reflex. Impulses may be sent down descending pathways to the efferent motor neurons supplying the involved muscles to override the input from the receptors, actually preventing the biceps from contracting in spite of the painful stimulus. When your finger is being pricked to obtain a blood sample, pain receptors are stimulated, initiating the withdrawal reflex. Knowing that you must be brave and not pull your hand away, you can consciously override the reflex by sending IPSPs via descending pathways to the motor neurons supplying the biceps and EPSPs to those supplying the triceps. The activity in these efferent neurons depends on the sum of activity of all their synaptic inputs. Because the neurons supplying the biceps are now receiving more IPSPs from the brain (voluntary) than EPSPs from the afferent pain pathway (reflex), these neurons are in-

hibited and do not reach threshold. Therefore, the biceps is not stimulated to contract and withdraw the hand. Simultaneously, the neurons to the triceps are receiving more EPSPs from the brain than IPSPs via the reflex arc, so they reach threshold, fire, and consequently stimulate the triceps to contract. Thus, the arm is kept extended despite the painful stimulus. In this way, the withdrawal reflex has been voluntarily overridden.

OTHER REFLEX ACTIVITY Spinal reflex action is not necessarily limited to motor responses on the side of the body to which the stimulus is applied. Assume that a person steps on a tack instead of burning a finger. A reflex arc is initiated to withdraw the injured foot from the painful stimulus, while the opposite leg simultaneously prepares to suddenly bear all the weight so that the person does not lose balance or fall (● Figure 5-32). Unimpeded bending of the injured extremity's knee is accomplished by concurrent reflex stimulation of the muscles that flex the knee and inhibition of the muscles that extend the knee. This response is a typical withdrawal reflex. At the same time, unimpeded extension of the opposite limb's knee is accomplished by activation of pathways that cross over to the opposite

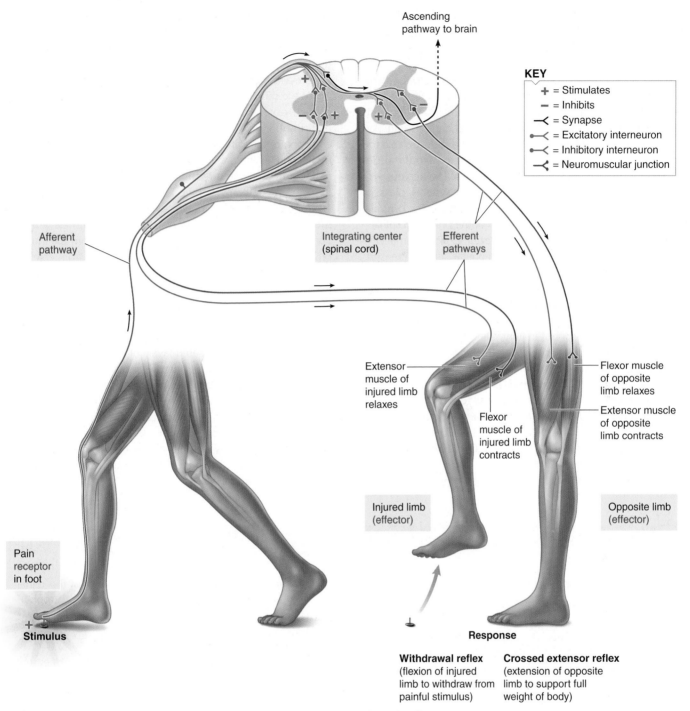

KEY
- + = Stimulates
- − = Inhibits
- ≺ = Synapse
- •≺ = Excitatory interneuron
- •≺ = Inhibitory interneuron
- ≺ = Neuromuscular junction

Ascending pathway to brain

Afferent pathway

Integrating center (spinal cord)

Efferent pathways

Extensor muscle of injured limb relaxes

Flexor muscle of injured limb contracts

Flexor muscle of opposite limb relaxes

Extensor muscle of opposite limb contracts

Injured limb (effector)

Opposite limb (effector)

Pain receptor in foot

Stimulus

Response

Withdrawal reflex
(flexion of injured limb to withdraw from painful stimulus)

Crossed extensor reflex
(extension of opposite limb to support full weight of body)

● **FIGURE 5-32 The crossed extensor reflex coupled with the withdrawal reflex.**

side of the spinal cord to reflexly stimulate this knee's extensors and inhibit its flexors. This **crossed extensor reflex** ensures that the opposite limb will be in a position to bear the weight of the body as the injured limb is withdrawn from the stimulus.

Besides protective reflexes (such as the withdrawal reflex) and simple postural reflexes (such as the crossed extensor reflex), basic spinal reflexes mediate emptying of pelvic organs (for example, urination). All spinal reflexes can be voluntarily overridden at least temporarily by higher brain centers.

Not all reflex activity involves a clear-cut reflex arc, although the basic principles of a reflex (that is, an automatic response to a detectable change) are present. Pathways for unconscious responsiveness digress from the typical reflex arc in two general ways:

1. *Responses mediated at least in part by hormones.* A particular reflex may be mediated solely by either neurons or hormones or may involve a pathway using both.

2. *Local responses that do not involve either nerves or hormones.* For example, the blood vessels in an exercising muscle dilate because of local metabolic changes, thereby increasing blood flow to match the active muscle's metabolic needs.

Chapter in Perspective: Focus on Homeostasis

To interact in appropriate ways with the external environment to sustain the body's viability, such as in acquiring food, and to make the internal adjustments necessary to maintain homeostasis, the body must be informed about any changes taking place in the external and internal environment and must be able to process this information and send messages to various muscles and glands to accomplish the desired results. The nervous system, one of the body's two major regulatory systems, plays a central role in this life-sustaining communication. The central nervous system (CNS), which consists of the brain and spinal cord, receives information about the external and internal environment by means of the afferent peripheral nerves. After sorting, processing, and integrating this input, the CNS sends directions, by means of efferent peripheral nerves, to bring about appropriate muscular contractions and glandular secretions.

With its swift electrical signaling system, the nervous system is especially important in controlling the rapid responses of the body. Many neurally controlled muscular and glandular activities are aimed toward maintaining homeostasis. The CNS is the main site of integration between afferent input and efferent output. It links the appropriate response to a particular input so that conditions compatible with life are maintained in the body. For example, when informed by the afferent nervous system that blood pressure has fallen, the CNS sends appropriate commands to the heart and blood vessels to increase the blood pressure to normal. Likewise, when informed that the body is overheated, the CNS promotes secretion of sweat by the sweat glands. Evaporation of sweat helps cool the body to normal temperature. Were it not for this processing and integrating ability of the CNS, maintaining homeostasis in an organism as complex as a human would be impossible.

At the simplest level, the spinal cord integrates many basic protective and evacuative reflexes that do not require conscious participation, such as withdrawing from a painful stimulus and emptying of the urinary bladder. In addition to serving as a more complex integrating link between afferent input and efferent output, the brain is responsible for the initiation of all voluntary movement, complex perceptual awareness of the external environment and self, language, and abstract neural phenomena such as thinking, learning, remembering, consciousness, emotions, and personality traits. All neural activity—from the most private thoughts to commands for motor activity, from enjoying a concert to retrieving memories from the distant past—is ultimately attributable to propagation of action potentials along individual nerve cells and chemical transmission between cells.

During evolutionary development, the nervous system has become progressively more complex. Newer, more complicated, and more sophisticated layers of the brain have been piled on top of older, more primitive regions. Mechanisms for governing many basic activities necessary for survival are built into the older parts of the brain. The newer, higher levels progressively modify, enhance, or nullify actions coordinated by lower levels in a hierarchy of command, and they also add new capabilities. Many of these higher neural activities are not aimed toward maintaining life, but they add immeasurably to the quality of being alive.

REVIEW EXERCISES

Objective Questions (Answers on p. A-42)

1. The major function of the CSF is to nourish the brain. *(True or false?)*

2. In emergencies when O_2 supplies are low, the brain can perform anaerobic metabolism. *(True or false?)*

3. Damage to the left cerebral hemisphere brings about paralysis and loss of sensation on the left side of the body. *(True or false?)*

4. The hands and structures associated with the mouth have a disproportionately large share of representation in both the sensory and motor cortexes. *(True or false?)*

5. The left cerebral hemisphere specializes in artistic and musical ability, whereas the right side excels in verbal and analytical skills. *(True or false?)*

6. The specific function a particular cortical region will carry out is permanently determined during embryonic development. *(True or false?)*

7. _____ is a decreased responsiveness to an indifferent stimulus that is repeatedly presented.

8. The process of transferring and fixing short-term memory traces into long-term memory stores is known as _____.

9. Afferent fibers enter through the _____ root of the spinal cord, and efferent fibers leave through the _____ root.

10. Using the answer code on the right, indicate which neurons are being described (a characteristic may apply to more than one class of neurons):
    ___ 1. have receptor at peripheral endings
    ___ 2. lie entirely within the CNS
    ___ 3. lie primarily within the peripheral nervous system
    ___ 4. innervate muscles and glands
    ___ 5. cell body is devoid of presynaptic inputs
    ___ 6. predominant type of neuron
    ___ 7. responsible for thoughts, emotions, memory, etc.
    (a) afferent neurons
    (b) efferent neurons
    (c) interneurons

11. Match the following:
    ___ 1. consists of nerves carrying information between the periphery and the CNS
    ___ 2. consists of the brain and spinal cord
    ___ 3. division of the peripheral nervous system that transmits signals to the CNS
    (a) somatic nervous system
    (b) autonomic nervous system
    (c) central nervous system
    (d) peripheral nervous system

(continued on p. 181)

___ 4. division of the peripheral nervous system that transmits signals from the CNS

___ 5. supplies skeletal muscles

___ 6. supplies smooth muscle, cardiac muscle, and glands

(e) efferent division
(f) afferent division

Essay Questions

1. Discuss the function of each of the following: astrocytes, oligodendrocytes, ependymal cells, microglia, cranium, vertebral column, meninges, cerebrospinal fluid, and blood–brain barrier.

2. Compare the composition of white and gray matter.

3. Draw and label the major functional areas of the cerebral cortex, indicating the functions attributable to each area.

4. Discuss the function of each of the following parts of the brain: thalamus, hypothalamus, basal nuclei, limbic system, cerebellum, and brain stem.

5. Define *somesthetic sensations* and *proprioception*.

6. What is an electroencephalogram?

7. Discuss the roles of Broca's area and Wernicke's area in language.

8. Compare short-term and long-term memory.

9. Discuss the difference between AMPA and NMDA glutamate receptors and their roles in long-term potentiation.

10. What is the reticular activating system?

11. Compare slow-wave and paradoxical (REM) sleep.

12. Draw and label a cross section of the spinal cord.

13. List the five components of a basic reflex arc.

14. Distinguish between a monosynaptic and a polysynaptic reflex.

POINTS TO PONDER

(Explanations on p. A-42)

1. Special studies designed to assess the specialized capacities of each cerebral hemisphere have been performed on "split-brain" patients. In these people the corpus callosum—the bundle of fibers that links the two halves of the brain—has been surgically cut to prevent the spread of epileptic seizures from one hemisphere to the other. Even though no overt changes in behavior, intellect, or personality occur in these patients, because both hemispheres individually receive the same information, deficits are observable with tests designed to restrict information to one brain hemisphere at a time. One such test involves limiting a visual stimulus to only half of the brain. Because of a crossover in the nerve pathways from the eyes to the occipital cortex, the visual information to the right of a midline point is transmitted to only the left half of the brain, whereas visual information to the left of this point is received by only the right half of the brain. A split-brain patient presented with a visual stimulus that reaches only the left hemisphere accurately describes the object seen, but when a visual stimulus is presented to only the right hemisphere, the patient denies having seen anything. The right hemisphere does receive the visual input, however,

as demonstrated by nonverbal tests. Even though a split-brain patient denies having seen anything after an object is presented to the right hemisphere, he or she can correctly match the object by picking it out from among a number of objects, usually to the patient's surprise. What is your explanation of this finding?

2. The hormone insulin enhances the carrier-mediated transport of glucose into most of the body's cells but not into brain cells. The uptake of glucose from the blood by neurons is not dependent on insulin. Knowing the brain's need for a continuous supply of blood-borne glucose, predict the effect that insulin excess would have on the brain.

3. Which of the following symptoms are most likely to occur as the result of a severe blow to the back of the head?
   a. paralysis
   b. hearing impairment
   c. visual disturbances
   d. burning sensations
   e. personality disorders

4. Give examples of conditioned reflexes you have acquired.

5. Under what circumstances might it be inadvisable to administer a clot-dissolving drug to a stroke victim?

CLINICAL CONSIDERATION

(Explanation on p. A-43)

Julio D., who had recently retired, was enjoying an afternoon of playing golf when suddenly he experienced a severe headache and dizziness. These symptoms were quickly followed by numbness and partial paralysis on the upper right side of his

body, accompanied by an inability to speak. After being rushed to the emergency room, Julio was diagnosed as having suffered a stroke. Given the observed neurologic impairment, what areas of his brain were affected?

**Nervous System
(Peripheral Nervous System)**

Body systems
maintain homeostasis

Homeostasis
The nervous system, as one of the body's two major regulatory systems, regulates many body activities aimed at maintaining a stable internal fluid environment.

Homeostasis is
essential for
survival of cells

Cells

Cells make up
body systems

The nervous system, one of the two major regulatory systems of the body, consists of the central nervous system (CNS), composed of the brain and spinal cord, and the **peripheral nervous system (PNS),** composed of the afferent and efferent fibers that relay signals between the CNS and periphery (other parts of the body).

The **afferent division** of the PNS detects, encodes, and transmits peripheral signals to the CNS, thus informing the

CNS about the internal and external environment. This afferent input to the controlling centers of the CNS is essential in maintaining homeostasis. To make appropriate adjustments in effector organs via efferent output, the CNS has to "know" what is going on. Afferent input is also used to plan for voluntary actions unrelated to homeostasis.

CHAPTER 6

The Peripheral Nervous System: Afferent Division; Special Senses

CENGAGENOW™ Log on to CengageNOW at **http://www .cengage.com/sso/** for an opportunity to explore a learning module that illustrates difficult concepts with self-study tutorials, animations, and interactive quizzes to help you learn, review, and master physiology concepts.

Receptor Physiology

The peripheral nervous system (PNS) consists of nerve fibers that carry information between the CNS and other parts of the body. The afferent division of the PNS sends information about the external and internal environment to the CNS.

A **stimulus** is a change detectable by the body. Stimuli exist in a variety of energy forms, or **modalities,** such as heat, light, sound, pressure, and chemical changes. Afferent neurons have **sensory receptors** (*receptors* for short) at their peripheral endings that respond to stimuli in both the external world and internal environment. (Though both are called *receptors,* stimulus-sensitive sensory receptors are distinctly different from the plasma-membrane protein receptors that bind with extracellular chemical messengers; see p. 57.) Because the only way afferent neurons can transmit information to the CNS about stimuli is via action potential propagation, receptors must convert these other forms of energy into electrical signals. Stimuli bring about graded potentials known as **receptor potentials** in the receptor. The conversion of stimulus energy into a receptor potential is known as **sensory transduction.** Receptor potentials in turn trigger action potentials in the afferent fiber.

Receptors have differential sensitivities to various stimuli.

Each type of receptor is specialized to respond to one type of stimulus, its **adequate stimulus.** For example, receptors in the eye are sensitive to light, receptors in the ear to sound waves, and warmth receptors in the skin to heat energy. Because of this differential sensitivity of receptors, we cannot "see" with our ears or "hear" with our eyes. Some receptors can respond weakly to stimuli other than their adequate stimulus, but even when activated by a different stimulus, a receptor still gives rise to the sensation usually detected by that receptor type. As an example, the adequate stimulus for eye receptors (photoreceptors) is light, to which they are exquisitely sensitive, but these receptors can also be activated to a lesser degree by mechanical stimulation. When hit in the eye, a person often "sees stars," because the mechanical pressure stimulates the photoreceptors. Because receptors typically are activated by their adequate

1 In sensory receptors that are specialized afferent neuron endings, stimulus opens stimulus-sensitive channels, permitting net Na^+ entry that produces receptor potential.

2 Local current flow between depolarized receptor ending and adjacent region opens voltage-gated Na^+ channels.

3 Na^+ entry initiates action potential in afferent fiber that self-propagates to CNS.

(a) Receptor potential in specialized afferent ending

1 In sensory receptors that are separate cells, stimulus opens stimulus-sensitive channels, permitting net Na^+ entry that produces receptor potential.

2 This local depolarization opens voltage-gated Ca^{2+} channels.

3 Ca^{2+} entry triggers exocytosis of neurotransmitter.

4 Neurotransmitter binding opens chemically gated receptor-channels at afferent ending, permitting net Na^+ entry.

5 Resultant depolarization opens voltage-gated Na^+ channels in adjacent region.

6 Na^+ entry initiates action potential in afferent fiber that self-propagates to CNS.

(b) Receptor potential in separate receptor cell

● **FIGURE 6-1 Conversion of receptor potential into action potentials.** (a) Specialized afferent ending as sensory receptor. Local current flow between a depolarized receptor ending undergoing a receptor potential and the adjacent region initiates an action potential in the afferent fiber by opening voltage-gated Na^+ channels. (b) Separate receptor cell as sensory receptor. The depolarized receptor cell undergoing a receptor potential releases a neurotransmitter that binds with chemically gated channels in the afferent fiber ending. This binding leads to a depolarization that opens voltage-gated Na^+ channels, initiating an action potential in the afferent fiber.

stimulus, the sensation usually corresponds to the stimulus modality.

TYPES OF RECEPTORS ACCORDING TO THEIR ADEQUATE STIMULUS Depending on the type of energy to which they ordinarily respond, receptors are categorized as follows:

- **Photoreceptors** are responsive to visible wavelengths of light.

- **Mechanoreceptors** are sensitive to mechanical energy. Examples include skeletal muscle receptors sensitive to stretch, the receptors in the ear containing fine hairs that are bent as a result of sound waves, and blood pressure–monitoring baroreceptors.

- **Thermoreceptors** are sensitive to heat and cold.

- **Osmoreceptors** detect changes in the concentration of solutes in the extracellular fluid and the resultant changes in osmotic activity (see p. 64).

- **Chemoreceptors** are sensitive to specific chemicals. Chemoreceptors include the receptors for taste and smell, as well as those located deeper within the body that detect O_2 and CO_2 concentrations in the blood or the chemical content of the digestive tract.

- **Nociceptors,** or **pain receptors,** are sensitive to tissue damage such as pinching or burning or to distortion of tissue. Intense stimulation of any receptor is also perceived as painful.

Some sensations are compound sensations in that their perception arises from central integration of several simultaneously activated primary sensory inputs. For example, the perception of wetness comes from touch, pressure, and thermal receptor input; there is no such thing as a "wetness receptor."

USES FOR INFORMATION DETECTED BY RECEPTORS The information detected by receptors is conveyed via afferent neurons to the CNS, where it is used for a variety of purposes:

- Afferent input is essential for the control of efferent output, both for regulating motor behavior in accordance with external circumstances and for coordinating internal activities directed at maintaining homeostasis. At the most basic level, afferent input provides information (of which the person may or may not be consciously aware) for the CNS to use in directing activities necessary for survival. On a broader level, we could not interact successfully with our environment or with one another without sensory input.

- Processing of sensory input by the reticular activating system in the brain stem is critical for cortical arousal and consciousness (see p. 167).

- Central processing of sensory information gives rise to our perceptions of the world around us.

- Selected information delivered to the CNS may be stored for future reference.

- Sensory stimuli can have a profound impact on our emotions. The smell of just-baked apple pie, the sensuous feel of silk, the sight of a loved one, hearing bad news—sensory input can gladden, sadden, arouse, calm, anger, frighten, or evoke any other range of emotions.

We next examine how adequate stimuli initiate action potentials that ultimately are used for these purposes.

● **FIGURE 6-2 Comparison of the initiation site of an action potential in the three types of neurons.**

A stimulus alters the receptor's permeability, leading to a graded receptor potential.

A receptor may be either (1) a specialized ending of the afferent neuron or (2) a separate receptor cell closely associated with the peripheral ending of the neuron. Stimulation of a receptor alters its membrane permeability, usually by causing nonspecific cation channels to open. The means by which this permeability change takes place is individualized for each receptor type. Because the electrochemical driving force is greater for Na^+ than for other small cations at resting potential, the predominant effect is an inward flux of Na^+, which depolarizes the receptor membrane (see p. 88). (There are exceptions; for example, photoreceptors are hyperpolarized on stimulation.) This local depolarization, the receptor potential, is a graded potential. As is true of all graded potentials, the stronger the stimulus is, the greater the permeability change and the larger the receptor potential (see p. 89). Also, receptor potentials have no refractory period, so summation in response to rapidly successive stimuli is possible. Because the receptor region has very few if any voltage-gated Na^+ channels and thus has a high threshold, action potentials do not take place at the receptor itself. For long-distance transmission, the receptor potential must be converted into action potentials that can be propagated along the afferent fiber.

Receptor potentials may initiate action potentials in the afferent neuron.

If a receptor potential is large enough, it may trigger an action potential in the afferent neuron membrane next to the receptor by promoting the opening of voltage-gated Na^+ channels in this region. In myelinated afferent fibers, this trigger zone is the node of Ranvier closest to the receptor. The means by which the

Na^+ channels are opened differ depending on whether the receptor is a specialized afferent ending or a separate cell.

- In the case of a specialized afferent ending, local current flow between the activated receptor ending undergoing a receptor potential and the cell membrane next to the receptor depolarizes this adjacent region (● Figure 6-1a). If the region is depolarized to threshold, voltage-gated Na^+ channels open here, triggering an action potential that is conducted along the afferent fiber to the CNS.

- In the case of a separate receptor cell, the receptor cell synapses with the ending of the afferent neuron (● Figure 6-1b). A receptor potential promotes the opening of voltage-gated Ca^{2+} channels in the receptor cell. The resultant Ca^{2+} entry causes the release of a neurotransmitter that diffuses across the synaptic cleft and binds with specific protein receptors on the afferent neuron membrane. This binding opens chemically gated Na^+ receptor-channels (see p. 115). If the resultant Na^+ entry depolarizes the afferent neuron ending to threshold, voltage-gated Na^+ channels open here, triggering an action potential that self-propagates to the CNS.

Note that the initiation site of action potentials in an afferent neuron differs from the site in an efferent neuron or interneuron. In the latter two types of neurons, action potentials are initiated at the axon hillock located at the start of the axon next to the cell body (see p. 109). By contrast, action potentials are initiated at the peripheral end of an afferent nerve fiber next to the receptor, a long distance from the cell body (● Figure 6-2).

The intensity of the stimulus is reflected by the magnitude of the receptor potential. The larger the receptor potential is, the greater the frequency of action potentials generated in the afferent neuron (● Figure 6-3). A larger receptor potential cannot bring about a larger action potential (because of the

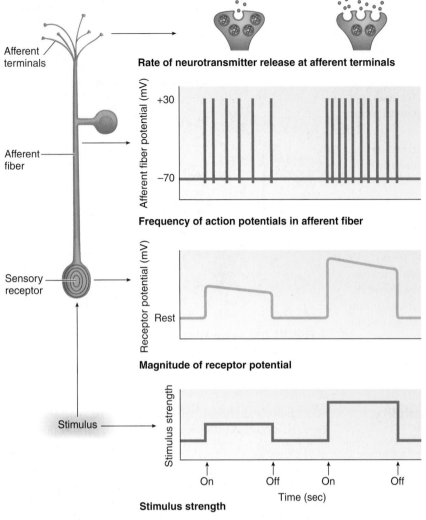

Rate of neurotransmitter release at afferent terminals

Frequency of action potentials in afferent fiber

Magnitude of receptor potential

Stimulus strength

Afferent terminals

Afferent fiber

Sensory receptor

Stimulus

● **FIGURE 6-3 Magnitude of receptor potential, frequency of action potentials in afferent fiber, and rate of neurotransmitter release at afferent terminals as a function of stimulus strength.**

(a) Tonic receptor

(b) Phasic receptor

● **FIGURE 6-4 Tonic and phasic receptors.** (a) A tonic receptor does not adapt at all or adapts slowly to a sustained stimulus and thus provides continuous information about the stimulus. (b) A phasic receptor adapts rapidly to a sustained stimulus and frequently exhibits an off response when the stimulus is removed. Thus, the receptor signals changes in stimulus intensity rather than relaying status quo information.

all-or-none law), but it can induce more rapid firing of action potentials (see p. 99). The more rapidly an afferent fiber fires, the more neurotransmitter it releases. This neurotransmitter influences the next cell in the neural pathway, passing on information about stimulus strength. Stimulus strength is also reflected by the size of the area stimulated. Stronger stimuli usually affect larger areas, so correspondingly more receptors respond. For example, a light touch does not activate as many pressure receptors in the skin as does a more forceful touch applied to the same area. Stimulus intensity is therefore distinguished both by the frequency of action potentials generated in the afferent neuron and by the number of receptors and thus afferent fibers activated within the area.

Receptors may adapt slowly or rapidly to sustained stimulation.

Stimuli of the same intensity do not always bring about receptor potentials of the same magnitude from the same receptor. Some receptors can diminish the extent of their depolarization despite sustained stimulus strength, a phenomenon called **adaptation.** Subsequently, the frequency of action potentials generated in the afferent neuron decreases. That is, the receptor "adapts" to the stimulus by no longer responding to it to the same degree.

TYPES OF RECEPTORS ACCORDING TO THEIR SPEED OF ADAPTATION There are two types of receptors—*tonic receptors* and *phasic receptors*—based on their speed of adaptation. **Tonic receptors** do not adapt at all or adapt slowly (● Figure 6-4a). These receptors are important in situations where it is valuable to maintain information about a stimulus. Examples of tonic receptors are muscle stretch receptors, which monitor muscle length, and joint proprioceptors, which measure the degree of joint flexion. To maintain posture and balance, the CNS must continually get information about the degree of muscle length and joint position. It is important, therefore, that these receptors do not adapt to a stimulus but continue to generate action potentials to relay this information to the CNS.

Phasic receptors, in contrast, are rapidly adapting receptors. The receptor rapidly adapts by no longer responding to a maintained stimulus. Some phasic receptors, most notably the *Pacinian corpuscle*, respond with a slight depolarization called the **off response** when the stimulus is removed (● Figure 6-4b). Phasic receptors are useful in situations where it is important to signal a change in stimulus intensity rather than to relay status quo information. Many of the *tactile (touch) receptors* that signal changes in pressure on the skin surface are phasic receptors. Because these receptors adapt rapidly, you are not continually conscious of wearing your watch, rings, and clothing. When you put something on, you soon become accustomed to it, because of these receptors' rapid adaptation. When you take the item off, you are aware of its removal because of the off response.

TACTILE RECEPTORS Tactile (touch) receptors in the skin are mechanoreceptors. The mechanical forces of a stimulus distort nonspecific cation channel proteins in the plasma membrane of these receptors, leading to net Na^+ entry, which causes a receptor potential that triggers an action potential in the afferent fiber. Sensory input from these receptors informs the CNS of the body's contact with objects in the external environment. Tactile receptors include the following (● Figure 6-5):

■ **Hair receptor,** which senses hair movement and very gentle touch, such as stroking the hair on your arm with a wisp of cotton, and is rapidly adapting

■ **Merkel's disc,** which detects light, sustained touch and texture, such as reading Braille, and is slowly adapting

■ **Pacinian corpuscle,** which responds to vibrations and deep pressure and is rapidly adapting

■ **Ruffini endings,** which respond to deep, sustained pressure and stretch of the skin, such as during a massage, and are slowly adapting

■ **Meissner's corpuscle,** which is sensitive to light, fluttering touch, such as tickling with a feather, and is rapidly adapting

MECHANISM OF ADAPTATION IN THE PACINIAN CORPUSCLE
The mechanism by which adaptation is accomplished varies for different receptors and is not fully understood for all receptor types. Many receptors adapt as a result of inactivation of channels that opened in response to the stimulus. Adaptation in the well-studied Pacinian corpuscle depends on the physical properties of this receptor. A Pacinian corpuscle is a specialized receptor ending that consists of concentric layers of connective tissue resembling layers of an onion wrapped around the peripheral terminal of an afferent neuron. When pressure is first applied to the Pacinian corpuscle, the underlying terminal responds with a receptor potential of a magnitude that reflects the intensity of the stimulus. As the stimulus continues, the pressure energy is dissipated because it causes the receptor layers to slip (just as steady pressure on a peeled onion causes its layers to slip). Because this physical effect filters out the steady component of the applied pressure, the underlying neuronal ending

no longer responds with a receptor potential; that is, adaptation has occurred.

Adaptation should not be confused with habituation (see p. 159). Although both these phenomena involve decreased neural responsiveness to repetitive stimuli, they operate at different points in the neural pathway. Adaptation is a receptor adjustment in the PNS, whereas habituation involves a modification in synaptic effectiveness in the CNS.

Visceral afferents carry subconscious input whereas sensory afferents carry conscious input.

Action potentials generated by receptors in afferent fibers in response to stimuli are propagated to the CNS. Afferent information about the internal environment, such as the blood pressure and the concentration of CO_2 in the body fluids, never reaches the level of conscious awareness, but this input is essential for determining the appropriate efferent output to maintain homeostasis. The incoming pathway for information derived from the internal viscera (*viscera* are organs in the body cavities, such as the abdominal cavity) is called a **visceral afferent.** Even though mostly subconscious information is transmitted via visceral

Hair receptor: hair movement and very gentle touch

Merkel's disc: light, sustained touch

Pacinian corpuscle: vibrations and deep pressure

Ruffini endings: deep pressure

Meissner's corpuscle: light, fluttering touch

● **FIGURE 6-5 Tactile receptors in the skin.**

Back Swings and Prejump Crouches: What Do They Share in Common?

Proprioception, the sense of the body's position in space, is critical to any movement and is especially important in athletic performance, whether it be a figure skater performing triple jumps on ice, a gymnast performing a difficult floor routine, or a football quarterback throwing perfectly to a spot 60 yards downfield. To control skeletal muscle contraction to achieve the desired movement, the CNS must be continuously apprised of the results of its action, through sensory feedback.

A number of receptors provide proprioceptive input. Muscle proprioceptors provide feedback information on muscle tension and length. Joint proprioceptors provide feedback on joint acceleration, angle, and direction of movement. Skin proprioceptors inform the CNS of weight-bearing pressure on the skin.

Proprioceptors in the inner ear, along with those in neck muscles, provide information about head and neck position so that the CNS can orient the head correctly. For example, neck reflexes facilitate essential trunk and limb movements during somersaults, and divers and tumblers use strong movements of the head to maintain spins.

The most complex and probably one of the most important proprioceptors is the muscle spindle (see p. 285). Muscle spindles are found throughout a muscle but tend to be concentrated in its center. Each spindle lies parallel to the muscle fibers within the muscle. The spindle is sensitive to both the muscle's rate of change in length and the final length achieved. If a muscle is stretched, each muscle spindle within the muscle is also stretched, and the afferent neuron whose peripheral axon terminates on the muscle spindle is stimulated. The afferent fiber passes into the spinal cord and synapses directly on the motor neurons that supply the same muscle. Stimulation of the stretched muscle as a result of this stretch reflex causes the muscle to contract sufficiently to relieve the stretch.

Older persons or those with weak quadriceps (thigh) muscles unknowingly take advantage of the muscle spindle by pushing on the center of the thighs when they get up from a sitting position. Contraction of the quadriceps muscle extends the knee joint, thus straightening the leg. The act of pushing on the center of the thighs when getting up slightly stretches the quadriceps muscle in both limbs, stimulating the muscle spindles. The resultant stretch reflex aids in contraction of the quadriceps muscles and helps the person assume a standing position.

In sports, people use the muscle spindle to advantage all the time. To jump high, as in basketball jump balls, an athlete starts by crouching down. This action stretches the quadriceps muscles and increases the firing rate of their spindles, thus triggering the stretch reflex that reinforces the quadriceps muscles' contractile response so that these extensor muscles of the legs gain additional power. The same is true for crouch starts in running events. The backswing in tennis, golf, and baseball similarly provides increased muscular excitation through reflex activity initiated by stretched muscle spindles.

---

afferents, people do become aware of pain signals arising from viscera. Afferent input derived from receptors located at the body surface or in the muscles or joints typically reaches the level of conscious awareness. This input is known as *sensory information,* and the incoming pathway is considered a **sensory afferent.** Sensory information is categorized as either (1) **somatic** (body sense) **sensation** arising from the body surface, including *somesthetic sensation* from the skin and *proprioception* from the muscles, joints, skin, and inner ear (see p. 144); or (2) **special senses**, including *vision, hearing, equilibrium, taste,* and *smell.* (See the accompanying boxed feature, ■ A Closer Look at Exercise Physiology, for a description of the usefulness of proprioception in athletic performance.) Final processing of sensory input by the CNS not only is essential for interaction with the environment for basic survival (for example, food procurement and defense from danger) but also adds immeasurably to the richness of life.

Each somatosensory pathway is "labeled" according to modality and location.

On reaching the spinal cord, afferent information has two possible destinies: (1) it may become part of a reflex arc, bringing about an appropriate effector response, or (2) it may be relayed upward to the brain via ascending pathways for further processing and possible conscious awareness. Pathways conveying conscious somatic sensation, the **somatosensory pathways,** consist of discrete chains of neurons, or *labeled lines,* synaptically interconnected in a particular sequence to accomplish progressively more sophisticated processing of the sensory information.

LABELED LINES The afferent neuron with its peripheral receptor that first detects the stimulus is known as a **first-order sensory neuron.** It synapses on a **second-order sensory neuron,** either in the spinal cord or the medulla, depending on which sensory pathway is involved. This neuron then synapses on a **third-order sensory neuron** in the thalamus, and so on. With each step, the input is processed further. A particular sensory modality detected by a specialized receptor type is sent over a specific afferent and ascending pathway (a neural pathway committed to that modality) to excite a defined area in the somatosensory cortex. That is, a particular sensory input is projected to a specific region of the cortex (see ● Figure 5-28a, p. 175, for an example). Thus, different types of incoming information are kept separated within specific **labeled lines** between the periphery and the cortex. In this way, even though all information is propagated to the CNS via the same type of signal (action po-

▲ TABLE 6-1	Coding of Sensory Information	
Stimulus Property	**Mechanism of Coding**	
Type of Stimulus (stimulus modality)	Distinguished by the type of receptor activated and the specific pathway over which this information is transmitted to a particular area of the cerebral cortex	
Location of Stimulus	Distinguished by the location of the activated receptive field and the pathway that is subsequently activated to transmit this information to the area of the somatosensory cortex representing that particular location	
Intensity of Stimulus (stimulus strength)	Distinguished by the frequency of action potentials initiated in an activated afferent neuron and the number of receptors (and afferent neurons) activated	

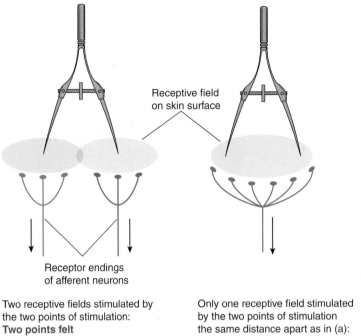

Two receptive fields stimulated by the two points of stimulation: **Two points felt**

Only one receptive field stimulated by the two points of stimulation the same distance apart as in (a): **One point felt**

(a) Region with small receptive fields

(b) Region with large receptive field

● **FIGURE 6-6 Comparison of discriminative ability of regions with small versus large receptive fields.** The relative tactile acuity of a given region can be determined by the *two-point threshold-of-discrimination test*. If the two points of a pair of calipers applied to the surface of the skin stimulate two different receptive fields, two separate points are felt. If the two points touch the same receptive field, they are perceived as only one point. By adjusting the distance between the caliper points, one can determine the minimal distance at which the two points can be recognized as two rather than one, which reflects the size of the receptive fields in the region. With this technique, it is possible to plot the discriminative ability of the body surface. The two-point threshold ranges from 2 mm in the fingertip (enabling a person to read Braille, where the raised dots are spaced 2.5 mm apart) to 48 mm in the poorly discriminative skin of the calf of the leg.

tentials), the brain can decode the type and location of the stimulus. ▲ Table 6-1 summarizes how the CNS is informed of the type (what?), location (where?), and intensity (how much?) of a stimulus.

PHANTOM PAIN Activation of a sensory pathway at any point gives rise to the same sensation that would be produced by stimulation of the receptors in the body part itself. This phenomenon is the traditional explanation for **phantom pain**—for example, pain perceived as originating in the foot by a person whose leg has been amputated at the knee. Irritation of the severed endings of the afferent pathways in the stump can trigger action potentials that, on reaching the foot region of the somatosensory cortex, are interpreted as pain in the missing foot. New evidence suggests that in addition, the sensation of phantom pain may arise from extensive remodeling of the brain region that originally handled sensations from the severed limb. This "remapping" of the "vacated" area of the brain is speculated to somehow lead to signals from elsewhere being misinterpreted as pain arising from the missing extremity.

Acuity is influenced by receptive field size and lateral inhibition.

Each somatosensory neuron responds to stimulus information only within a circumscribed region of the skin surface surrounding it; this region is called its **receptive field**. The size of a receptive field varies inversely with the density of receptors in the region; the more closely receptors of a particular type are spaced, the smaller the area of skin each monitors. The smaller the receptive field is in a region, the greater its **acuity** or **discriminative ability**. Compare the tactile discrimination in your fingertips with that in your elbow by "feeling" the same object with both. You can sense more precise information about the object with your richly innervated fingertips because the receptive fields there are small; as a result, each neuron signals information about small, discrete portions of the object's surface. An estimated 17,000 tactile mechanoreceptors are present in the fingertips and palm of each hand. In contrast, the skin over the elbow is served by relatively few sensory endings with larger receptive fields. Subtle differences within each large receptive field cannot be detected (● Figure 6-6). The distorted cortical representation of various body parts in the sensory homunculus (see p. 148) corresponds precisely with the innervation density; more cortical space is allotted for sensory reception

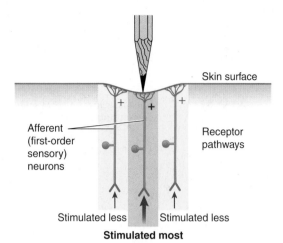

Afferent
(first-order
sensory)
neurons

Receptor
pathways

Skin surface

Stimulated less   Stimulated less

Stimulated most

Frequency
of action
potentials

Baseline
level of
activity

Location on skin

(a) Activity in afferent neurons

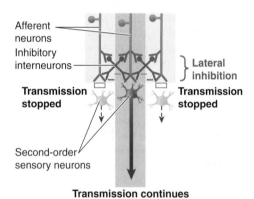

Afferent
neurons

Inhibitory
interneurons

Lateral
inhibition

**Transmission
stopped**

**Transmission
stopped**

Second-order
sensory neurons

Transmission continues

Frequency
of action
potentials

Baseline
level of
activity

Area of sensation
on skin

(b) Lateral inhibition

● **FIGURE 6-7 Lateral inhibition.** (a) The receptor at the site of most intense stimulation is activated to the greatest extent. Surrounding receptors are also stimulated but to a lesser degree. (b) The most intensely activated receptor pathway halts transmission of impulses in the less intensely stimulated pathways through lateral inhibition. This process facilitates localization of the site of stimulation.

from areas with smaller receptive fields and, accordingly, greater tactile discriminative ability.

Besides receptor density, a second factor influencing acuity is *lateral inhibition.* You can appreciate the importance of this phenomenon by slightly indenting the surface of your skin with the point of a pencil (● Figure 6-7a). The receptive field is excited immediately under the center of the pencil point where the stimulus is most intense, but the surrounding receptive fields are also stimulated, only to a lesser extent because they are less distorted. If information from these marginally excited afferent fibers in the fringe of the stimulus area were to reach the cortex, localization of the pencil point would be blurred. To facilitate localization and sharpen contrast, **lateral inhibition** occurs within the CNS (● Figure 6-7b). With lateral inhibition, each activated signal pathway inhibits the pathways next to it by stimulating inhibitory interneurons that pass laterally between ascending fibers serving neighboring receptive fields. The most strongly activated pathway originating from the center of the stimulus area inhibits the less excited pathways from the fringe areas to a greater extent than the weakly activated pathways in the fringe areas inhibit the more excited central pathway. Blockage of further transmission in the weaker inputs increases the contrast between wanted and unwanted information so that the pencil point can be precisely localized. The extent of lateral inhibitory connections within sensory pathways varies for different modalities. Those with the most lateral inhibition—touch and vision—bring about the most accurate localization.

Perception is the conscious awareness of surroundings derived from interpretation of sensory input.

Perception is our conscious interpretation of the external world as created by the brain from a pattern of nerve impulses delivered to it from sensory receptors. Is the world, as we perceive it, reality? The answer is a resounding no. Our perception is different from what is really "out there," for several reasons. First, humans have receptors that detect only a limited number of existing energy forms. We perceive sounds, colors, shapes, textures, smells, tastes, and temperature but are not informed of magnetic forces, polarized light waves, radio waves, or X-rays because we do not have receptors to respond to the latter energy forms. What is not detected by receptors, the brain will never know. Our response range is limited even for the energy forms for which we do have receptors. For example, dogs can hear a whistle whose pitch is above our level of detection. Second, the information channels to our brains are not high-fidelity recorders. During precortical processing of sensory input, some features of stimuli are accentuated and others are suppressed or ignored, as through lateral inhibition. Third, the cerebral cortex further manipulates the data, comparing the sensory input with other incoming information as well as with memories of past experiences to extract the significant features—for example, sifting out a friend's words from the hubbub of sound in a school cafeteria. In the process, the cortex often fills in or distorts the information to abstract a logical perception; that is, it "completes the picture." As a simple example, you "see" a white square in ● Figure 6-8 even though there is no white square but

● **FIGURE 6-8 Do you "see" a white square that is not really there?**

● **FIGURE 6-9 Variable perceptions from the same visual input.** Do you see two faces in profile, or a wineglass?

▲ TABLE 6-2	Characteristics of Pain
Fast Pain	**Slow Pain**
Occurs on stimulation of mechanical and thermal nociceptors	Occurs on stimulation of polymodal nociceptors
Carried by small, myelinated A-delta fibers	Carried by small, unmyelinated C fibers
Produces sharp, prickling sensation	Produces dull, aching, burning sensation
Easily localized	Poorly localized
Occurs first	Occurs second; persists for longer time; more unpleasant

merely right-angle wedges taken out of four red circles. Optical illusions illustrate how the brain interprets reality according to its own rules. Do you see two faces in profile or a wineglass in ● Figure 6-9? You can alternately see one or the other out of identical visual input. Thus, our perceptions do not replicate reality. Other species, equipped with different types of receptors and sensitivities and with different neural processing, perceive a markedly different world from what we perceive.

Having completed our general discussion of receptor physiology, we now examine one important somatic sensation in greater detail—pain.

Pain

Pain is primarily a protective mechanism meant to bring to conscious awareness the fact that tissue damage is occurring or is about to occur. Furthermore, storage of painful experiences in memory helps us avoid potentially harmful events in the future.

Stimulation of nociceptors elicits the perception of pain plus motivational and emotional responses.

Unlike other somatosensory modalities, the sensation of pain is accompanied by motivated behavioral responses (such as withdrawal or defense) as well as emotional reactions (such as crying or fear). Also, unlike other sensations, the subjective perception of pain can be influenced by other past or present experiences (for example, heightened pain perception accom-

panying fear of the dentist or lowered pain perception in an injured athlete during a competitive event).

CATEGORIES OF PAIN RECEPTORS There are three categories of pain receptors, or nociceptors: **Mechanical nociceptors** respond to mechanical damage such as cutting, crushing, or pinching; **thermal nociceptors** respond to temperature extremes, especially heat; and **polymodal** nociceptors respond equally to all kinds of damaging stimuli, including irritating chemicals released from injured tissues. Because of their value to survival, nociceptors do not adapt to sustained or repetitive stimulation.

Clinical Note All nociceptors can be sensitized by the presence of *prostaglandins,* which greatly enhance the receptor response to noxious stimuli (that is, it hurts more when prostaglandins are present). Prostaglandins are a special group of fatty acid derivatives that are cleaved from the lipid bilayer of the plasma membrane and act locally on being released (see p. 758). Tissue injury, among other things, can lead to the local release of prostaglandins. These chemicals act on the nociceptors' peripheral endings to lower their threshold for activation. Aspirin-like drugs inhibit the synthesis of prostaglandins, accounting at least in part for the analgesic (pain-relieving) properties of these drugs.

FAST AND SLOW AFFERENT PAIN FIBERS Pain impulses originating at nociceptors are transmitted to the CNS via one of two types of afferent fibers (▲ Table 6-2). Signals arising from mechanical and thermal nociceptors are transmitted over small, myelinated **A-delta fibers** at rates of up to 30 m/sec (the **fast pain pathway**). Impulses from polymodal nociceptors are carried by small, unmyelinated **C fibers** at a much slower rate of 12 m/sec (the **slow pain pathway**). Think about the last time you cut or burned your finger. You undoubtedly felt a sharp twinge of pain at first, with a more diffuse, disagreeable pain commencing shortly thereafter. Pain typically is perceived initially as a brief, sharp, prickling sensation that is easily localized; this is the fast pain pathway originating from specific mechanical or heat nociceptors. This feeling is followed by a dull, aching, poorly localized sensation that persists for a longer time and is more unpleasant; this is the slow pain pathway, which is activated by chemicals, especially **bradykinin**, a normally inactive substance that is activated by enzymes released into the ECF from damaged tissue. Bradykinin and related compounds not only provoke pain, presumably by stimulating the polymodal nociceptors, but they also contribute to the inflammatory response to tissue injury (Chapter 12). The persistence of these chemicals might explain the long-lasting, aching pain that continues after removal of the mechanical or thermal stimulus that caused the tissue damage.

Interestingly, the peripheral receptors of afferent C fibers are activated by **capsaicin**, the ingredient in hot chili peppers that gives them their fiery zing. (In addition to binding with pain receptors, capsaicin binds with the thermal receptors that are normally activated by heat—hence the burning sensation when eating hot peppers.) Ironically, local application of capsaicin can actually reduce clinical pain, most likely by overstimulating and damaging the nociceptors with which it binds.

HIGHER-LEVEL PROCESSING OF PAIN INPUT Multiple structures are involved in pain processing. The primary afferent pain fibers synapse with specific second-order excitatory interneurons in the dorsal horn of the spinal cord. In response to stimulus-induced action potentials, afferent pain fibers release neurotransmitters that influence these next neurons in line. The two best known of these pain neurotransmitters are *substance P* and *glutamate*. **Substance P** activates ascending pathways that transmit nociceptive signals to higher levels for further processing (● Figure 6-10a). Ascending pain pathways have different destinations in the *cortex*, the *thalamus*, and the *reticular formation*. Cortical somatosensory processing areas localize the pain, whereas other cortical areas participate in other conscious components of the pain experience, such as deliberation about the incident. Pain can still be perceived in the absence of the cortex, presumably at the level of the thalamus. The reticular formation increases the level of alertness associated with the noxious encounter. Interconnections from the thalamus and reticular formation to the *hypothalamus* and *limbic system* elicit the behavioral and emotional responses accompanying the painful experience. The limbic system appears to be especially important in perceiving the unpleasant aspects of pain.

Glutamate, the other neurotransmitter released from primary afferent pain terminals, is a major excitatory neurotransmitter (see p. 107). Glutamate acts on two different plasma membrane receptors on the dorsal horn excitatory interneurons, with two different outcomes (see p. 161). First, binding of glutamate with its *AMPA receptors* leads to permeability changes that ultimately result in the generation of action potentials in the dorsal horn cells. These action potentials transmit the pain message to higher centers. Second, binding of glutamate with its *NMDA receptors* leads to Ca^{2+} entry into these neurons. This pathway is not involved in the transmission of pain messages. Instead, Ca^{2+} initiates second-messenger systems that make the dorsal horn cells more excitable than usual (see p. 117). This hyperexcitability contributes in part to the exaggerated sensitivity of an injured area to subsequent exposure to painful or even normally nonpainful stimuli, such as a light touch.

Think about how exquisitely sensitive your sunburned skin is, even to clothing. Other mechanisms also contribute to supersensitivity of an injured area. For example, responsiveness of the pain-sensing peripheral receptors can be boosted so that they react more vigorously to subsequent stimuli. This exaggerated sensitivity presumably serves a useful purpose by discouraging activities that could cause further damage or interfere with healing of the injured area. Usually this hypersensitivity resolves as the injury heals.

Clinical Note Chronic pain, sometimes excruciating, occasionally occurs in the absence of tissue injury. In contrast to the pain accompanying peripheral injury, which serves as a normal protective mechanism to warn of impending or actual damage to the body, abnormal chronic pain states result from damage within the pain pathways in the peripheral nerves or the CNS. That is, pain is perceived because of abnormal signaling within the pain pathways in the absence of peripheral injury or typical painful stimuli. For example, strokes that damage ascending pathways can lead to an abnormal, persistent sensation of pain. Abnormal chronic pain is sometimes categorized as *neuropathic pain*.

The brain has a built-in analgesic system.

In addition to the chain of neurons connecting peripheral nociceptors with higher CNS structures for pain perception, the CNS also contains a built-in pain-suppressing or **analgesic system** that suppresses transmission in the pain pathways as they enter the spinal cord. Three brain-stem regions are part of this descending analgesic pathway: the *periaqueductal gray matter* (gray matter surrounding the cerebral aqueduct, a narrow canal that connects the third and fourth ventricular cavities) and specific nuclei in the *medulla* and *reticular formation*. Electrical stimulation of any of these parts of the brain produces profound analgesia.

The periaquaductal gray matter stimulates particular neurons whose cell bodies lie in the medulla and reticular formation and that terminate on inhibitory interneurons in the dorsal horn of the spinal cord (● Figure 6-10b). These inhibitory interneurons release enkephalin, which binds with **opiate receptors** at the afferent pain-fiber terminal. People have long known that **morphine**, a component of the opium poppy, is a powerful analgesic. Researchers considered it very unlikely that the body has been endowed with opiate receptors only to interact with chemicals derived from a flower! They therefore began to search for the substances that normally bind with these opiate receptors. The result was the discovery of **endogenous opiates** (morphinelike substances)—the *endorphins, enkephalins,* and *dynorphin*—which are important in the body's natural analgesic system. These endogenous opiates serve as analgesic neurotransmitters. Binding of enkephalin from the dorsal-horn inhibitory interneuron with the afferent pain-fiber terminal suppresses the release of substance P via presynaptic inhibition, thereby blocking further transmission of the pain signal (see p. 111). Morphine binds to these same opiate receptors, which accounts in large part for its analgesic properties. Furthermore, injection of morphine into the periaqueductal gray matter and medulla causes profound analgesia, suggesting that endogenous

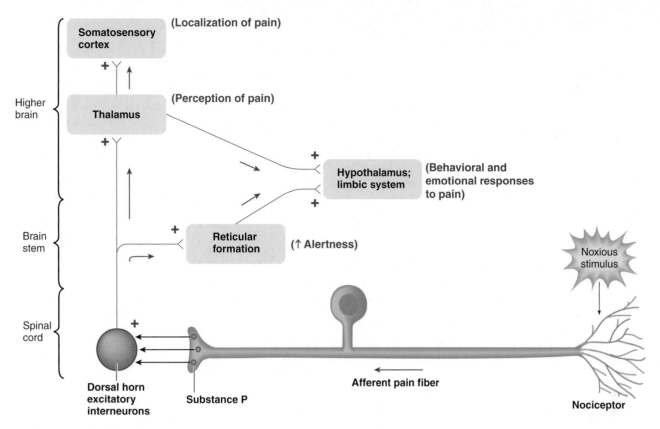

(a) Substance P pain pathway

(b) Analgesic pathway

● **FIGURE 6-10 Substance P pain pathway and analgesic pathway.** (a) When activated by a noxious stimulus, some afferent pain pathways release substance P, which activates ascending pain pathways that provide various brain regions with input for processing different aspects of the painful experience. (b) Endogenous opiates released from descending analgesic (pain-relieving) pathways bind with opiate receptors at the synaptic knob of the afferent pain fiber. This binding inhibits the release of substance P, thereby blocking transmission of pain impulses along the ascending pain pathways.

Acupuncture: Is It for Real?

It sounds like science fiction. How can a needle inserted in the hand relieve a toothache? **Acupuncture analgesia (AA),** the technique of relieving pain by inserting and manipulating threadlike needles at key points, has been practiced in China for more than 2000 years but is relatively new to Western medicine and still remains controversial in the United States.

Brief History

Traditional Chinese teaching holds that disease can occur when the normal patterns of flow of healthful energy (called *qi;* pronounced "chee") just under the skin become disrupted, with acupuncture being able to correct this imbalance and restore health. Many Western scientists have been skeptical because, until recently, the phenomenon could not be explained on the basis of any known, logical, physiologic principles, although a tremendous body of anecdotal evidence in support of the effectiveness of AA existed in China. In Western medicine, the success of acupuncture was considered a placebo effect. The term *placebo effect* refers to a chemical or technique that brings about a desired response through the power of suggestion or distraction rather than through any direct action.

Because the Chinese were content with anecdotal evidence for the success of AA, this phenomenon did not come under close scientific scrutiny until the last several decades, when European and American scientists started studying it. As a result of these efforts, an impressive body of rigorous scientific investigation supports the contention that AA really works (that is, by a physiologic rather than a placebo/psychological effect). In controlled clinical studies, 55% to 85% of patients were helped by AA. Pain relief was reported by only 30% to 35% of placebo controls (people who thought they were receiving proper AA treatment, but in whom needles were inserted in the wrong places or not deep enough). Furthermore, its mechanisms of action have become apparent. Indeed, more is known about the underlying physiologic mechanisms of AA than about those of many conventional medical techniques, such as gas anesthesia.

Mechanism of Action

The overwhelming body of evidence supports the *acupuncture endorphin hypothesis* as the primary mechanism of AA's action. According to this hypothesis, acupuncture needles activate specific afferent nerve fibers, which send impulses to the CNS. Here the incoming impulses cause analgesia by blocking pain transmission at both the spinal-cord and brain levels through use of endorphins and closely related endogenous opiates. Several other neurotransmitters, such as serotonin and norepinephrine, as well as cortisol, the major hormone released during stress, are implicated as well. (Pain relief in placebo controls is believed to occur as a result of placebo responders subconsciously activating their own built-in analgesic system.)

Acupuncture in the United States

In the United States, AA has not been used in mainstream medicine, even by physicians who have been convinced by scientific evidence that the technique is valid. AA methodology has traditionally not been taught in U.S. medical colleges, and the techniques take time to learn. Also, AA is much more time-consuming than using drugs. Western physicians who have been trained to use drugs to solve most pain problems are generally reluctant to scrap their known methods for an unfamiliar, time-consuming technique. However, acupuncture is gaining favor as an alternative treatment for relief of chronic pain, especially because analgesic drugs can have troublesome side effects. After decades of being spurned by the vast majority of the U.S. medical community, acupuncture started gaining respectability following a 1997 report issued by an expert panel convened by the National Institutes of Health (NIH). This report, based on an evaluation of published scientific studies, concluded that acupuncture is effective as an alternative or adjunct to conventional treatment for many kinds of pain and nausea. Now that acupuncture has been sanctioned by NIH, some medical insurers have taken the lead in paying for this now scientifically legitimate treatment, and some of the nation's medical schools are beginning to incorporate the technique into their curricula. There are also 40 nationally accredited acupuncture schools for nonphysicians. Forty-two states now require certification to practice acupuncture. Of the 13,000 licensed acupuncture practitioners in the United States, only 3000 are physicians. According to a recent national health survey, over 8 million adults in the United States have undergone acupuncture.

---

opiates also are released centrally to block the descending pain-suppressing pathway.

It is not clear how this natural pain-suppressing mechanism is normally activated. Factors known to modulate pain include exercise, stress, and acupuncture. Researchers believe that endorphins are released during prolonged exercise and presumably produce the "runner's high." Some types of stress also induce analgesia. It is sometimes disadvantageous for a stressed organism to display the normal reaction to pain. For example, when two male lions are fighting for dominance of the group, withdrawing, escaping, or resting when injured would mean certain defeat. (See the accompanying boxed feature, ■ Concepts, Challenges, and Controversies, for an examination of how acupuncture relieves pain.)

We have now completed our coverage of somatic sensation. As you are now aware, somatic sensation is detected by widely distributed receptors that provide information about the body's interactions with the environment in general. In contrast, each of the special senses has highly localized, extensively specialized receptors that respond to unique environmental stimuli. The special senses include **vision, hearing, equilibrium, taste,** and **smell,** to which we now turn our attention, starting with vision.

Suspensory ligament

Ciliary body

Conjunctiva

Iris

Pupil

Lens

Cornea

Aqueous humor

Vitreous humor

Extrinsic eye muscle

Choroid

Retina

Sclera

Fovea

Optic nerve

Optic disc

Blood vessels in retina

Sclera

Iris

Canal for tear drainage

Pupil

(a) External front view

(b) Internal sagittal view

● **FIGURE 6-11 Structure of the eye.**

Eye: Vision

For vision, the eyes capture the patterns of illumination in the environment as an "optical picture" on a layer of light-sensitive cells, the *retina,* much as a nondigital camera captures an image on film. Just as film can be developed into a visual likeness of the original image, the coded image on the retina is transmitted through the steps of visual processing until it is finally consciously perceived as a visual likeness of the original image. Before considering the steps involved in the process of vision, we are first going to examine how the eyes are protected from injury.

Protective mechanisms help prevent eye injuries.

Several mechanisms help protect the eyes from injury. Except for its anterior (front) portion, the eyeball is sheltered by the bony socket in which it is positioned. The **eyelids** act like shutters to protect the anterior portion of the eye from environmental insults. They close reflexly to cover the eye under threatening circumstances, such as rapidly approaching objects, dazzling light, and instances when the exposed surface of the eye or eyelashes are touched. Frequent spontaneous blinking of the eyelids helps disperse the lubricating, cleansing, bactericidal ("germ-killing") **tears.** Tears are produced continuously by the **lacrimal gland** in the upper lateral corner under the eyelid. This eye-washing fluid flows across the anterior surface of the eye and drains into tiny canals in the corner of each eye (● Figure 6-11a), eventually emptying into the back of the nasal passageway. This drainage system cannot handle the profuse tear production during crying, so the tears overflow from the eyes. The eyes are also equipped with protective **eyelashes,** which trap fine, airborne debris such as dust before it can fall into the eye.

The eye is a fluid-filled sphere enclosed by three specialized tissue layers.

Each **eye** is a spherical, fluid-filled structure enclosed by three layers. From outermost to innermost, these are (1) the *sclera/cornea;* (2) the *choroid/ciliary body/iris;* and (3) the *retina* (● Figure 6-11b). Most of the eyeball is covered by a tough outer layer of connective tissue, the **sclera,** which forms the visible white part of the eye (● Figure 6-11a).

Anteriorly, the outer layer consists of the transparent **cornea,** through which light rays pass into the interior of the eye. The middle layer underneath the sclera is the highly pigmented **choroid,** which contains many blood vessels that nourish the retina. The choroid layer becomes specialized anteriorly to form the *ciliary body* and *iris,* which will be described shortly. The innermost coat under the choroid is the **retina,** which consists of an outer pigmented layer and an inner nervous-tissue layer. The latter contains the **rods** and **cones,** the photoreceptors that convert light energy into nerve impulses. Like the black walls of a photographic studio, the pigment in the choroid and retina absorbs light after it strikes the retina to prevent reflection or scattering of light within the eye.

The interior of the eye consists of two fluid-filled cavities, separated by an elliptical **lens,** all of which are transparent to permit light to pass through the eye from the cornea to the retina. The larger posterior (rear) cavity between the lens and retina contains a semifluid, jellylike substance, the **vitreous humor.** The vitreous humor helps maintain the spherical shape of the eyeball. The anterior cavity between the cornea and lens contains a clear, watery fluid, the **aqueous humor.** The aqueous humor carries nutrients for the cornea and lens, both of which lack a blood supply. Blood vessels in these structures would impede the passage of light to the photoreceptors.

Aqueous humor is produced at a rate of about 5 ml/day by a capillary network within the **ciliary body,** a specialized anterior derivative of the choroid layer. This fluid drains into a canal (*the canal of Schlemm*) at the edge of the cornea and eventually enters the blood (● Figure 6-12).

Clinical Note If aqueous humor is not drained as rapidly as it forms (for example, because of a blockage in the drainage canal), the excess accumulates in the anterior cavity, causing the pressure to rise within the eye. This condition is known as **glaucoma.** The excess aqueous humor pushes the lens backward into the vitreous humor, which in turn pushes against the inner neural layer of the retina. This compression causes retinal and optic nerve damage that can lead to blindness if the condition is not treated.

The amount of light entering the eye is controlled by the iris.

Not all the light passing through the cornea reaches the light-sensitive photoreceptors, because of the presence of the iris, a thin, pigmented smooth muscle that forms a visible ringlike structure within the aqueous humor (see ● Figure 6-11a and b). The pigment in the iris is responsible for eye color. The varied flecks, lines, and other nuances of the iris are unique for each individual, making the iris the basis of the latest identification technology. Recognition of iris patterns by a video camera that captures iris images and translates the landmarks into a computerized code is more foolproof than fingerprinting or even DNA testing.

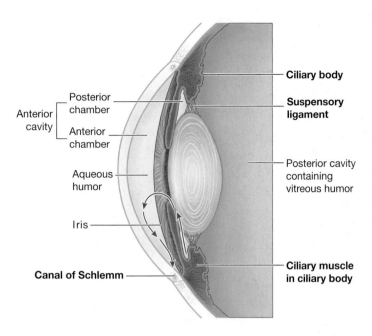

● **FIGURE 6-12 Formation and drainage of aqueous humor.** Aqueous humor is formed by a capillary network in the ciliary body, then drains into the canal of Schlemm, and eventually enters the blood.

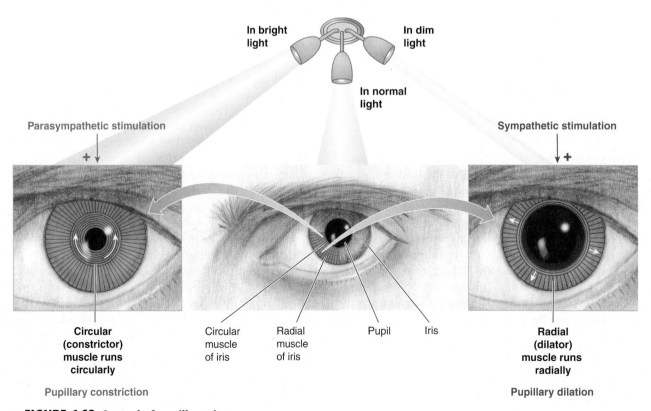

● **FIGURE 6-13 Control of pupillary size.**

The round opening in the center of the iris through which light enters the interior portions of the eye is the **pupil.** The size of this opening can be adjusted by variable contraction of the iris muscles to admit more or less light as needed, much as the diaphragm controls the amount of light entering a camera. The iris contains two sets of smooth muscle networks, one *circular* (the muscle fibers run in a ringlike fashion within the iris) and the other *radial* (the fibers project outward from the pupillary margin like bicycle spokes) (● Figure 6-13). Because muscle fibers shorten when they contract, the pupil gets smaller when the **circular** (or **constrictor**) **muscle** contracts and forms a smaller ring. This reflex pupillary constriction occurs in bright light to decrease the amount of light entering the eye. When the **radial** (or **dilator**) **muscle** shortens, the size of the pupil increases. Such pupillary dilation occurs in dim light to allow the entrance of more light. Iris muscles are controlled by the autonomic nervous system. Parasympathetic nerve fibers innervate the circular muscle (causing pupillary constriction), and sympathetic fibers supply the radial muscle (causing pupillary dilation).

The eye refracts the entering light to focus the image on the retina.

Light is a form of electromagnetic radiation composed of particle-like individual packets of energy called **photons** that travel in wavelike fashion. The distance between two wave peaks is known as the **wavelength** (● Figure 6-14). The wavelengths in the electromagnetic spectrum range from 10^{-14} m (quadrillionths of a meter, as in the extremely short cosmic rays) to 10^4 m (10 km, as in long radio waves) (● Figure 6-15). The photoreceptors in the eye are sensitive only to wavelengths between 400 and 700 nanometers (nm; billionths of a meter). Thus, **visible light** is only a small portion of the total electro-

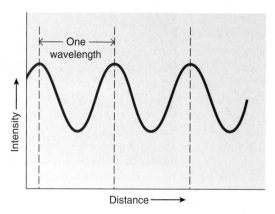

● **FIGURE 6-14 Properties of an electromagnetic wave.** A wavelength is the distance between two wave peaks. The intensity is the amplitude of the wave.

magnetic spectrum. Light of different wavelengths in this visible band is perceived as different color sensations. The shorter visible wavelengths are sensed as violet and blue; the longer wavelengths are interpreted as orange and red.

In addition to having variable wavelengths, light energy also varies in **intensity;** that is, the amplitude, or height, of the wave (● Figure 6-14). Dimming a bright red light does not change its color; it just becomes less intense or less bright.

Light waves *diverge* (radiate outward) in all directions from every point of a light source. The forward movement of a light wave in a particular direction is known as a **light ray.** Divergent light rays reaching the eye must be bent inward to be focused back into a point (the **focal point**) on the light-sensitive retina to provide an accurate image of the light source (● Figure 6-16).

● **FIGURE 6-15 Electromagnetic spectrum.** The wavelengths in the electromagnetic spectrum range from less than 10^{-14} m to 10^4 m. The visible spectrum includes wavelengths ranging from 400 to 700 nanometers (nm).

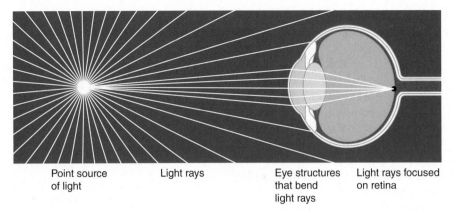

Point source of light   Light rays   Eye structures that bend light rays   Light rays focused on retina

● **FIGURE 6-16 Focusing of diverging light rays.** Diverging light rays must be bent inward to be focused.

● **FIGURE 6-17 Refraction.** A light ray is bent (refracted) when it strikes the surface of a medium of different density from the one in which it had been traveling (for example, moving from air into glass) at any angle other than perpendicular to the new medium's surface. Thus, the pencil in the glass of water appears to bend. What is happening, though, is that the light rays coming to the camera (or your eyes) are bent as they pass through the water, then the glass, and then the air. Consequently, the pencil appears distorted.

PROCESS OF REFRACTION Light travels faster through air than through other transparent media such as water and glass. When a light ray enters a medium of greater density, it is slowed down (the converse is also true). The course of direction of the ray changes if it strikes the surface of the new medium at any angle other than perpendicular (● Figure 6-17). The bending of a light ray is known as **refraction.** With a curved surface such as a lens, the greater the curvature is, the greater the degree of bending and the stronger the lens. When a light ray strikes the curved surface of any object of greater density, the direction of refraction depends on the angle of the curvature (● Figure 6-18). A **convex** surface curves outward

(like the outer surface of a ball), whereas a **concave** surface curves inward (like a cave). Convex surfaces converge light rays, bringing them closer together. Because convergence is essential for bringing an image to a focal point, refractive surfaces of the eye are convex. Concave surfaces diverge light rays (spread them farther apart). A concave lens is useful for correcting certain refractive errors of the eye, such as nearsightedness.

THE EYE'S REFRACTIVE STRUCTURES The two structures most important in the eye's refractive ability are the cornea and the lens. The curved corneal surface, the first structure light passes

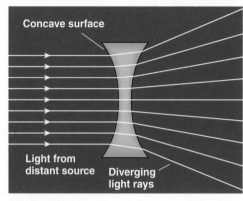

FIGURE 6-18 Refraction by convex and concave lenses. (a) A lens with a convex surface converges the rays (brings them closer together). (b) A lens with a concave surface diverges the rays (spreads them farther apart).

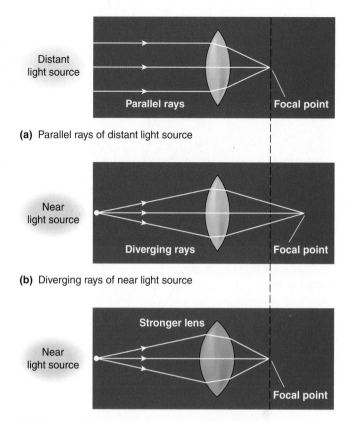

(a) Parallel rays of distant light source

(b) Diverging rays of near light source

(c) Stronger lens needed to focus near light source

● **FIGURE 6-19 Focusing of distant and near sources of light.** (a) The rays from a distant (far) light source (more than 20 feet from the eye) are parallel by the time the rays reach the eye. (b) The rays from a near light source (less than 20 feet from the eye) are still diverging when they reach the eye. A longer distance is required for a lens of a given strength to bend the diverging rays from a near light source into focus than to bend the parallel rays from a distant light source into focus. (c) To focus both a distant and a near light source in the same distance (the distance between the lens and retina), a stronger lens must be used for the near source.

through as it enters the eye, contributes most extensively to the eye's total refractive ability because the difference in density at the air–cornea interface is much greater than the differences in density between the lens and the fluids surrounding it. In **astigmatism** the curvature of the cornea is uneven, so light rays are unequally refracted. The refractive ability of a person's cornea remains constant, because the curvature of the cornea never changes. In contrast, the refractive ability of the lens can be adjusted by changing its curvature as needed for near or far vision.

Rays from light sources more than 20 feet away are considered parallel by the time they reach the eye. Light rays originating from near objects are still diverging when they reach the eye. For a given refractive ability of the eye, the diverging rays of a near source come to a focal point a greater distance behind the lens than the parallel rays of a far source come to a focal point (● Figure 6-19a and b). However, in a particular eye the distance between the lens and the retina always remains the same. Therefore, a greater distance beyond the lens is not available for bringing near objects into focus. Yet for clear vision the refractive structures of the eye must bring both near and far light sources into focus on the retina. If an image is focused before it reaches the retina or is not yet focused when it reaches the retina, it will be blurred (● Figure 6-20). To bring both near and far light sources into focus on the retina (that is, in the same

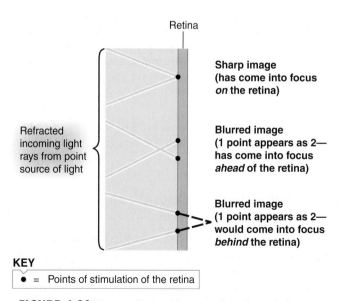

KEY

● = Points of stimulation of the retina

● **FIGURE 6-20 Comparison of images that do and do not come into focus on the retina.**

(a) Anterior view of suspensory ligaments extending from ciliary muscles to lens

(b) Sagittal and anterior view when ciliary muscle is relaxed

(c) Sagittal and anterior view during accommodation, when ciliary muscle is contracted

● **FIGURE 6-21 Mechanism of accommodation.** (a) Suspensory ligaments extend from the ciliary muscle to the outer edge of the lens. (b) When the ciliary muscle is relaxed, the suspensory ligaments are taut, putting tension on the lens so that it is flat and weak. (c) When the ciliary muscle is contracted, the suspensory ligaments become slack, reducing the tension on the lens, allowing it to assume a stronger, rounder shape because of its elasticity.

distance), a stronger lens must be used for the near source (● Figure 6-19c). Let us see how the strength of the lens can be adjusted as needed.

Accommodation increases the strength of the lens for near vision.

The ability to adjust the strength of the lens is known as **accommodation.** The strength of the lens depends on its shape, which in turn is regulated by the ciliary muscle. The **ciliary muscle** is part of the ciliary body, an anterior specialization of the choroid layer. The ciliary body has two major components: the ciliary muscle and the capillary network that produces aqueous humor

(see ● Figure 6-12, p. 196). The ciliary muscle is a circular ring of smooth muscle attached to the lens by **suspensory ligaments** (● Figure 6-21a).

When the ciliary muscle is relaxed, the suspensory ligaments are taut, and they pull the lens into a flattened, weakly refractive shape (● Figure 6-21b). As the muscle contracts, its circumference decreases, slackening the tension in the suspensory ligaments (● Figure 6-21c). When the suspensory ligaments are not pulling on the lens, it becomes more spherical, because of its inherent elasticity. The greater curvature of the more rounded lens increases its strength, further bending light rays. In the normal eye, the ciliary muscle is relaxed and the lens is flat for far vision, but the muscle contracts to let the lens become more convex and stronger for near vision. The ciliary muscle is controlled by the autonomic nervous system, with sympathetic stimulation causing its relaxation and parasympathetic stimulation causing its contraction.

Clinical Note The lens is made up of about 1000 layers of cells that destroy their nucleus and organelles during development so that the cells are perfectly transparent. Lacking DNA and protein-synthesizing machinery, mature lens cells cannot regenerate or repair themselves. Cells in the center of the lens are in double jeopardy. Not only are they oldest, but they also are farthest away from the aqueous humor, the lens's nutrient source. With advancing age, these nonrenewable central cells die and become stiff. With loss of elasticity, the lens can no longer assume the spherical shape required to accommodate for near vision. This age-related reduction in accommodative ability, **presbyopia,** affects most people by middle age (45 to 50), requiring them to resort to corrective lenses for near vision (reading).

The elastic fibers in the lens are normally transparent. These fibers occasionally become opaque so that light rays cannot pass through, a condition known as a **cataract.** The defective lens can be surgically removed and vision restored by an implanted artificial lens or by compensating eyeglasses.

Other common vision disorders are *nearsightedness (myopia)* and *farsightedness (hyperopia)*. In a normal eye (**emmetropia**) (● Figure 6-22a), a far light source is focused on the retina without accommodation, whereas the strength of the lens is increased by accommodation to bring a near source into focus. In **myopia** (● Figure 6-22b1), because the eyeball is too long or the lens is too strong, a near light source is brought into focus on the retina without accommodation (even though accommodation is normally used for near vision), whereas a far light source is focused in front of the retina and is blurry. Thus, a myopic individual has better near vision than far vision, a condition that can be corrected by a concave lens (● Figure 6-22b2). With **hyperopia** (● Figure 6-22c1), either the eyeball is too short or the lens is too weak. Far objects are focused on the retina only with accommodation, whereas near objects are focused behind the retina even with accommodation and, accordingly, are blurry. Thus, a hyperopic individual has better far vision than near vision, a condition that can be corrected by a convex lens (● Figure 6-22c2). Instead of using corrective eyeglasses or contact lenses, many people are now opting to compensate for refractive errors with laser eye surgery (such as LASIK) to permanently change the shape of the cornea.

Light must pass through several retinal layers before reaching the photoreceptors.

The major function of the eye is to focus light rays from the environment on the rods and cones, the photoreceptor cells of the retina. The photoreceptors then transform the light energy into electrical signals for transmission to the CNS.

The receptor-containing portion of the retina is actually an anatomical extension of the CNS and not a separate peripheral organ. During embryonic development, the retinal cells "back out" of the nervous system, so the retinal layers, surprisingly,

are facing backward! The neural portion of the retina consists of three layers of excitable cells (● Figure 6-23): (1) the outermost layer (closest to the choroid) containing the **rods** and **cones,** whose light-sensitive ends face the choroid (away from the incoming light); (2) a middle layer of **bipolar cells** and associated interneurons; and (3) an inner layer of **ganglion cells.** Axons of the ganglion cells join to form the **optic nerve,** which leaves the retina slightly off center. The point on the retina at which the optic nerve leaves and through which blood vessels pass is the **optic disc** (see ● Figure 6-11b). This region is often called the **blind spot;** no image can be detected in this area because it has no rods and cones (● Figure 6-24). We are normally not aware of the blind spot, because central processing somehow "fills in" the missing spot. You can discover the existence of your own blind spot by a simple demonstration (● Figure 6-25).

Light must pass through the ganglion and bipolar layers before reaching the photoreceptors in all areas of the retina except the fovea. In the **fovea,** which is a pinhead-sized depression located in the exact center of the retina (see ● Figure 6-11b), the bipolar and ganglion cell layers are pulled aside so that light strikes the photoreceptors directly. This feature, coupled with the fact that only cones (which have greater acuity or discriminative ability than the rods) are found here, makes the fovea the point of most distinct vision. In fact, the fovea has the greatest concentration of cones in the retina. Thus, we turn our eyes so that the image of the object at which we are looking is focused on the fovea. The area immediately surrounding the fovea, the **macula lutea,** also has a high concentration of cones and fairly high acuity (● Figure 6-24). Macular acuity, however, is less than that of the fovea, because of the overlying ganglion and bipolar cells in the macula.

Clinical Note **Macular degeneration** is the leading cause of blindness in the Western Hemisphere. This condition is characterized by loss of photoreceptors in the macula lutea in association with advancing age. Its victims have "doughnut" vision. They suffer a loss in the middle of their visual field, which normally has the highest acuity, and are left with only the less distinct peripheral vision.

Phototransduction by retinal cells converts light stimuli into neural signals.

Photoreceptors (rod and cone cells) consist of three parts (● Figure 6-26a):

1. An *outer segment,* which lies closest to the eye's exterior, facing the choroid. It detects the light stimulus.

2. An *inner segment,* which lies in the middle of the photoreceptor's length. It contains the metabolic machinery of the cell.

3. A *synaptic terminal,* which lies closest to the eye's interior, facing the bipolar cells. It transmits the signal generated in the photoreceptor on light stimulation to these next cells in the visual pathway.

The outer segment, which is rod shaped in rods and cone shaped in cones (● Figure 6-26a and c), consists of stacked, flattened, membranous discs containing an abundance of light-sensitive *photopigment* molecules. Each retina has over

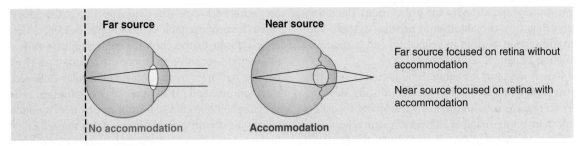

(a) Normal eye (Emmetropia)

Far source focused on retina without accommodation

Near source focused on retina with accommodation

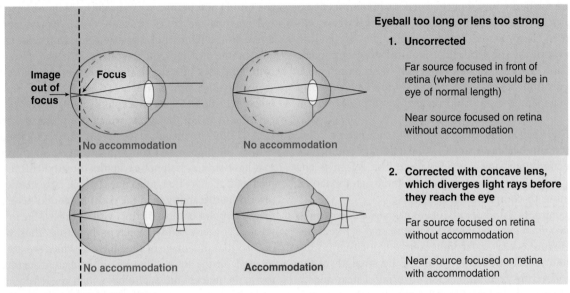

Eyeball too long or lens too strong

1. Uncorrected

Far source focused in front of retina (where retina would be in eye of normal length)

Near source focused on retina without accommodation

2. Corrected with concave lens, which diverges light rays before they reach the eye

Far source focused on retina without accommodation

Near source focused on retina with accommodation

(b) Nearsightedness (Myopia)

Eyeball too short or lens too weak

1. Uncorrected

Far source focused on retina with accommodation

Near source focused behind retina even with accommodation

2. Corrected with convex lens, which converges light rays before they reach the eye

Far source focused on retina without accommodation

Near source focused on retina with accommodation

(c) Farsightedness (Hyperopia)

● **FIGURE 6-22 Emmetropia, myopia, and hyperopia.** This figure compares far vision and near vision (a) in the normal eye with (b) nearsightedness and (c) farsightedness in both their (1) uncorrected and (2) corrected states. The vertical dashed line represents the normal distance of the retina from the cornea; that is, the site at which an image is brought into focus by the refractive structures in a normal eye.

Direction of light

Direction of retinal visual processing

Optic nerve

Retina

Pigment layer
Choroid layer
Sclera

Front of retina

Back of retina

Fibers of the optic nerve | **Ganglion cell** | Amacrine cell | **Bipolar cell** | Horizontal cell | **Cone** **Rod**
Photoreceptor cells

Retina

● **FIGURE 6-23 Retinal layers.** The retinal visual pathway extends from the photoreceptor cells (rods and cones, whose light-sensitive ends face the choroid away from the incoming light) to the bipolar cells to the ganglion cells. The horizontal and amacrine cells act locally for retinal processing of visual input.

A. L. Blum/Visuals Unlimited

Blind spot   Macula lutea

● **FIGURE 6-24 View of the retina seen through an ophthalmoscope.** With an ophthalmoscope, a lighted viewing instrument, it is possible to view the optic disc (blind spot) and macula lutea within the retina at the rear of the eye.

● **FIGURE 6-25 Demonstration of the blind spot.** Find the blind spot in your left eye by closing your right eye and holding the book about 4 inches from your face. While focusing on the cross, gradually move the book away from you until the circle vanishes from view. At this time, the image of the circle is striking the blind spot of your left eye. You can similarly locate the blind spot in your right eye by closing your left eye and focusing on the circle. The cross will disappear when its image strikes the blind spot of your right eye.

(a) Structure of rods and cones

(b) Photopigment rhodopsin in the dark and light

● **FIGURE 6-26 Photoreceptors.** (a) The three parts of the rods and cones, the eye's photoreceptors. Note in the outer segment of the rod and cone the stacked, flattened, membranous discs, which contain an abundance of photopigment molecules. (b) A photopigment, such as rhodopsin, depicted here and found in rods, consists of opsin, a plasma-membrane protein, and retinal, a vitamin-A derivative. In the dark, 11-*cis*-retinal is bound within the interior of opsin and the photopigment is inactive. In the light, retinal changes to all-*trans*-retinal, activating the photopigment. (c) Scanning electron micrograph of the outer segments of rods and cones. Note the rod shape in rods and the cone shape in cones.

(c) Outer segments of rods and cones

125 million photoreceptors, and more than a billion photopigment molecules may be packed into the outer segment of each photoreceptor.

Photopigments undergo chemical alterations when activated by light. Through a series of steps, this light-induced change and subsequent activation of the photopigment bring about a receptor potential in the photoreceptor that ultimately leads to the generation of action potentials in ganglion cells, which transmit this information to the brain for visual processing. A photopigment consists of two components: **opsin,** an integral protein in the disc plasma membrane; and **retinal,** a derivative of vitamin A. Retinal is the light-absorbing part of the photopigment.

Phototransduction, the process of converting light stimuli into electrical signals, is basically the same for all photoreceptors, but the mechanism is contrary to the usual means by which receptors respond to their adequate stimulus. Receptors typically *depolarize* when stimulated, but photoreceptors *hyperpolarize* on light absorption. Let's first examine the status of the photoreceptors in the dark, then consider what happens when they are exposed to light. We will use rods as an example, but the same events take place in cones, except that they preferentially absorb light in different parts of the visible spectrum.

PHOTORECEPTOR ACTIVITY IN THE DARK The photopigment in rods is **rhodopsin.** Retinal exists in different conformations in the dark and light. In the dark, it exists as 11-*cis* retinal, which fits into a binding site within the interior of the opsin portion of rhodopsin (● Figure 6-26b). The plasma membrane of a photoreceptor's outer segment contains chemically gated Na^+ channels. Unlike other chemically gated channels that respond to extracellular chemical messengers, these channels respond to an internal second messenger, **cyclic GMP,** or **cGMP** (cyclic guanosine monophosphate). Binding of cGMP to these Na^+ channels keeps them open. In the absence of light, the concentration of cGMP is high (● Figure 6-27a). (Light absorption leads to the breakdown of cGMP.) Therefore, the Na^+ channels of a photoreceptor, unlike most receptors, are open in the absence of stimulation, that is, in the dark. The resultant passive inward Na^+ leak, the so-called *dark current,* depolarizes the photoreceptor. The passive spread of this depolarization from the outer segment (where the Na^+ channels are located) to the synaptic terminal (where the photoreceptor's neurotransmitter is stored) keeps the synaptic terminal's voltage-gated Ca^{2+} channels open. Calcium entry triggers the release of the neurotransmitter glutamate from the synaptic terminal while in the dark.

PHOTORECEPTOR ACTIVITY IN THE LIGHT On exposure to light, the concentration of cGMP is decreased through a series of biochemical steps triggered by photopigment activation (● Figure 6-27b). When 11-*cis* retinal absorbs light, it changes to the all-*trans* retinal conformation (see ● Figure 6-26b). This is the only light-dependent step in the entire process of phototransduction. As a result of this change in shape, retinal no longer fits snugly in its binding site in opsin, causing opsin to also change its conformation, which activates the photopigment. Membrane-bound opsin is similar in shape and behavior to G-protein-coupled receptors, except that instead of being activated by binding with an extracellular chemical messenger, photopigments are activated in response to light absorption by retinal (see p. 117). Rod and cone cells contain a G protein called **transducin.** The activated photopigment activates transducin, which in turn activates the intracellular enzyme phosphodiesterase. This enzyme degrades cGMP, thus decreasing the concentration of this second messenger in the photoreceptor. During the light excitation process, the reduction in cGMP permits the chemically gated Na^+ channels to close. This channel closure stops the depolarizing Na^+ leak and causes membrane hyperpolarization. The hyperpolarization, which is the receptor potential, passively spreads from the outer segment to the synaptic terminal of the photoreceptor. Here the potential change leads to closure of the voltage-gated Ca^{2+} channels and a subsequent reduction in glutamate release from the synaptic terminal. Thus, photoreceptors are *inhibited by their adequate stimulus* (hyperpolarized by light) and *excited in the absence of stimulation* (depolarized by darkness). The hyperpolarizing potential and subsequent decrease in neurotransmitter release are graded according to the intensity of light. The brighter the light is, the greater the hyperpolarizing response and the greater the reduction in glutamate release.

The short-lived active form of the photopigment quickly dissociates into opsin and retinal. The retinal is converted back into its 11-*cis* form. In the dark, enzyme-mediated mechanisms rejoin opsin and this recycled retinal to restore the photopigment to its original inactive conformation (see ● Figure 6-26b).

FURTHER RETINAL PROCESSING OF LIGHT INPUT How does the retina signal the brain about light stimulation through such an inhibitory response? Further retinal processing involves different influences of glutamate on two parallel pathways. Each photoreceptor synapses with two side-by-side bipolar cells, one an *on-center bipolar cell* and the other an *off-center bipolar cell.* These cells, in turn, terminate respectively on *on-center ganglion cells* and *off-center ganglion* cells, whose axons collectively form the optic nerve for transmission of signals to the brain.

The receptive field of a bipolar or ganglion cell is determined by the field of light detection by the photoreceptor with which it is linked. (Of course light is not directly detected by the bipolar or ganglion cells; light stimulates the photoreceptors, which signal the bipolar cells that in turn send the message to the ganglion cells.) **On-center** and **off-center cells** respond in opposite ways, depending on the relative comparison of illumination between the center and periphery (surround) of their receptive fields. Think of the receptive field as a doughnut. An on-center cell increases its rate of firing when light is most intense at the center of its receptive field (that is, when the doughnut hole is lit up) and stops firing when its surround is most intensely illuminated. In contrast, an off-center cell increases its firing rate when light is the brightest in the periphery of its receptive field (that is, when the doughnut itself is lit up) and stops firing when light is most intense in its center (● Figure 6-28a). Thus, on-center cells are "turned on" and off-center cells are "turned off" when light shines most intensely on their centers. Both cells respond only weakly when light shines evenly on both their centers and surrounds. This response pattern is useful for enhancing the differ-

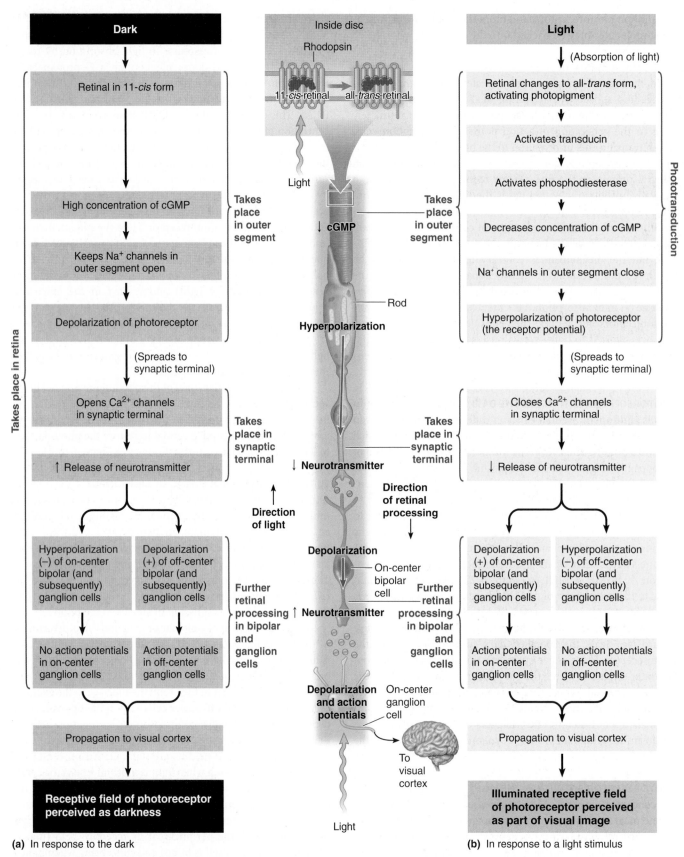

Dark

Retinal in 11-*cis* form

High concentration of cGMP

Keeps Na⁺ channels in outer segment open

Depolarization of photoreceptor

(Spreads to synaptic terminal)

Opens Ca²⁺ channels in synaptic terminal

↑ Release of neurotransmitter

Hyperpolarization (−) of on-center bipolar (and subsequently) ganglion cells

Depolarization (+) of off-center bipolar (and subsequently) ganglion cells

No action potentials in on-center ganglion cells

Action potentials in off-center ganglion cells

Propagation to visual cortex

Receptive field of photoreceptor perceived as darkness

Takes place in outer segment

Takes place in synaptic terminal

Further retinal processing in bipolar and ganglion cells

Takes place in retina

(a) In response to the dark

Inside disc

Rhodopsin

11-*cis*-retinal → all-*trans*-retinal

Light

↓ **cGMP**

Rod

Hyperpolarization

Direction of light

↑ Further retinal processing

↓ **Neurotransmitter**

Direction of retinal processing

Depolarization

On-center bipolar cell

↑ **Neurotransmitter**

Depolarization and action potentials

On-center ganglion cell

To visual cortex

Light

Light

↓ (Absorption of light)

Retinal changes to all-*trans* form, activating photopigment

Activates transducin

Activates phosphodiesterase

Decreases concentration of cGMP

Na⁺ channels in outer segment close

Hyperpolarization of photoreceptor (the receptor potential)

(Spreads to synaptic terminal)

Closes Ca²⁺ channels in synaptic terminal

↓ Release of neurotransmitter

Depolarization (+) of on-center bipolar (and subsequently) ganglion cells

Hyperpolarization (−) of off-center bipolar (and subsequently) ganglion cells

Action potentials in on-center ganglion cells

No action potentials in off-center ganglion cells

Propagation to visual cortex

Illuminated receptive field of photoreceptor perceived as part of visual image

Takes place in outer segment

Takes place in synaptic terminal

Further retinal processing in bipolar and ganglion cells

Phototransduction

(b) In response to a light stimulus

● **FIGURE 6-27 Phototransduction, further retinal processing, and initiation of action potentials in the visual pathway.** (a) Events occurring in the retina and visual pathway in response to the dark. (b) Events occurring in the retina and visual pathway in response to a light stimulus.

Receptive field of on-center cell

Receptive field of off-center cell

Off

On — Inhibited by light

Excited by light

Excited — by light

Inhibited — by light

On

Off

Both types of cells are weakly stimulated by uniform light on both center and surround.

(a) Receptive fields of on-center and off-center cells

(b) Outcome of retinal processing by on-center and off-center cells

● **FIGURE 6-28 On-center and off-center cells in retina.**
(a) On-center cells are excited and off-center cells are inhibited by bright light in the centers of their receptive fields. (b) Retinal processing by on-center and off-center ganglion cells is largely responsible for enhancing differences in relative (rather than absolute) brightness, which helps define contours. Note that the gray circle surrounded by black appears brighter than the one surrounded by white, even though the two circles are identical (same shade and size).

ence in light level between one small area at the center of a receptive field and the illumination immediately around it. By emphasizing differences in relative brightness, this mechanism helps define contours of images, but in so doing, information about absolute brightness is sacrificed (● Figure 6-28b).

Glutamate released from the photoreceptor terminal in the dark has opposite effects on the two types of bipolar cells because they have different types of receptors that lead to different channel responses on binding with this neurotransmitter. Glutamate hyperpolarizes (inhibits) on-center bipolar cells and depolarizes (excites) off-center bipolar cells. When glutamate secretion decreases on light exposure, this reduction depolarizes (stimulates) the hyperpolarized on-center bipolar cells and hyperpolarizes (inhibits) the depolarized off-center bipolar cells. The bipolar cells pass on the information about patterns of illumination to the next neurons in the processing chain, the ganglion cells, by changing their rate of neurotransmitter release in accordance with their state of polarization—increased neurotransmitter release on depolarization and decreased neurotransmitter release on hyperpolarization.

Bipolar cells, similar to the photoreceptors, display graded potentials. Action potentials do not originate until the ganglion cells, the first neurons in the chain that must propagate the visual message over long distances to the brain, are stimulated. As

the firing rates of the on-center and off-center ganglion cells change in response to the changing pattern of illumination, the brain is informed about the rapidity and extent of change in contrast within the visual image.

 Researchers have developed an ambitious microelectronic chip designed to serve as a partial substitute retina. Implanted in the eye, this "vision chip" bypasses the photoreceptor step altogether: Images received by means of a wireless video camera mounted on eyeglasses are fed to the chip, which translates the images into electrical signals detectable by the ganglion cells and transmitted on for further optical processing. The device is still in an early clinical trial phase, but preliminary results are promising. It has restored at least some sight (the ability to identify objects and detect motion) in people who are blinded by loss of photoreceptor cells but whose ganglion cells and optic pathways remain healthy. If the research is successful, the chip could benefit people with advanced macular degeneration. Another promising avenue under investigation for halting or even reversing the loss of sight in degenerative eye diseases is regenerating the retina through fetal retinal transplants or stem cell therapy (see p. 8).

Rods provide indistinct gray vision at night, whereas cones provide sharp color vision during the day.

The retina contains 20 times more rods than cones (120 million rods compared to 6 million cones per eye). Cones are most abundant in the macula lutea in the center of the retina. From this point outward, the concentration of cones decreases and the concentration of rods increases. Rods are most abundant in the periphery. We have examined the similar way in which phototransduction takes place in rods and cones. Now we will focus on the differences between these photoreceptors (▲ Table 6-3).

RODS HAVE HIGH SENSITIVITY; CONES HAVE LOWER SENSITIVITY
The outer segments are longer in rods than in cones so they contain more photopigments and thus can absorb light more readily. Also, as you will see shortly, the way in which rods con-

nect with other neurons in their processing pathway further increases the sensitivity of rod vision. Because rods have high sensitivity, they can respond to the dim light of night. Cones by contrast have lower sensitivity to light, being activated only by bright daylight. Thus rods are specialized for night vision and cones for day vision.

CONE VISION HAS HIGH ACUITY; ROD VISION HAS LOW ACUITY The pathways by which cones are "wired" to the other retinal neuronal layers confer high acuity (sharpness; ability to distinguish between two nearby points). Thus cones provide sharp vision with high resolution for fine detail. By contrast, the wiring pathways of rods provide low acuity, so you can see at night with your rods but at the expense of distinctness. Let us see how wiring patterns influence sensitivity and acuity.

Little convergence of neurons takes place in the retinal pathways for cone output (see p. 112). Each cone generally has a private line connecting it to a particular ganglion cell. In contrast, much convergence occurs in rod pathways. Each retina has about 1 million ganglion cells, which receive the output from around 125 million photoreceptors, 120 million of which are rods. Thus, output from more than 100 rods may converge via bipolar cells on a single ganglion cell. Before a ganglion cell can have an action potential, the cell must be brought to threshold through influence of the graded potentials in the photoreceptors to which it is wired. Because a single-cone ganglion cell is influenced by only one cone, only bright daylight is intense enough to induce a sufficient receptor potential in the cone to ultimately bring the ganglion cell to threshold. The abundant convergence in the rod visual pathways, in contrast, offers good opportunities for summation of subthreshold events in a rod ganglion cell (see p. 108). Whereas a small receptor potential induced by dim light in a single cone would not be sufficient to bring its ganglion cell to threshold, similar small receptor potentials induced by the same dim light in multiple rods converging on a single ganglion cell would have an additive effect to bring the rod ganglion cell to threshold. Because rods can bring about action potentials in response to small amounts of light, they are much more sensitive than cones. (Rods are also more sensitive than cones because they have more photopigment.) However, because cones have private lines into the optic nerve, each cone transmits information about an extremely small receptive field on the retinal surface. Cones are thus able to provide highly detailed vision at the expense of sensitivity. With rod vision, acuity is sacrificed for sensitivity. Because many rods share a single ganglion cell, once an action potential is initiated, it is impossible to discern which of the multiple rod inputs were activated to bring the ganglion cell to threshold. Objects appear fuzzy when rod vision is used, because of this poor ability to distinguish between two nearby points.

CONES PROVIDE COLOR VISION; RODS PROVIDE VISION IN SHADES OF GRAY There are four different photopigments, one in the rods and one in each of three types of cones—**red, green,** and **blue cones.** Each photopigment has the same retinal but a different opsin. Because each opsin binds retinal in a unique way, each of the four photopigments absorb different wavelengths of light in the visible spectrum to varying degrees. Each

● **FIGURE 6-29 Sensitivity of the three types of cones to different wavelengths.**

photopigment maximally absorbs a particular wavelength but also absorbs a range of wavelengths shorter and longer than this peak absorption. The farther a wavelength is from the peak wavelength absorbed, the less strongly the photopigment responds. Rods absorb the greatest range of wavelengths. The absorption curves for the three cone types overlap so that two or three cones may respond to a given wavelength but to a different extent (● Figure 6-29). Because the photopigments in the three types of cones each respond selectively to a different part of the visible light spectrum, the brain can compare the responses of the three cone types, making color vision in daylight possible. In contrast, the brain cannot discriminate between various wavelengths when using visual input from the rods. The rhodopsin in every rod responds in the same way to a given wavelength, so no comparison among rod inputs is possible. Therefore, rods provide vision at night only in shades of gray by detecting different intensities, not different colors. We now examine color vision in further detail.

Color vision depends on the ratios of stimulation of the three cone types.

Vision depends on stimulation of photoreceptors by light. Certain objects in the environment such as the sun, fire, and light bulbs emit light. But how do you see objects such as chairs, trees, and people, which do not emit light? The pigments in various objects selectively absorb particular wavelengths of light transmitted to them from light-emitting sources, and the unabsorbed wavelengths are reflected from the objects' surfaces. These reflected light rays enable you to see the objects. An object perceived as blue absorbs the longer red and green wavelengths of light and reflects the shorter blue wavelengths, which can be absorbed by the photopigment in the eyes' blue cones, thereby activating them.

Each cone type is most effectively activated by a particular wavelength of light in the range of color indicated by its name—blue, green, or red. However, cones also respond in varying degrees to other wavelengths (● Figure 6-29). **Color vision,** the perception of the many colors of the world, depends on the three cone types' various *ratios of stimulation* in response to different wavelengths. A wavelength perceived as blue does not stimulate red or green cones at all but excites blue cones maximally (the percentage of maximal stimulation for red, green, and blue cones, respectively, is 0:0:100). The sensation of yellow, in comparison, arises from a stimulation ratio of 83:83:0, red and green cones each being stimulated 83% of maximum, while blue cones are not excited at all. The ratio for green is 31:67:36, and so on, with various combinations giving rise to the sensation of all the different colors. White is a mixture of all wavelengths of light, whereas black is the absence of light.

The extent to which each of the cone types is excited is coded and transmitted in separate parallel pathways to the brain. A distinct color vision center in the primary visual cortex combines and processes these inputs to generate the perception of color, taking into consideration the object in comparison with its background. The concept of color is therefore in the mind of the beholder. Most of us agree on what color we see because we have the same types of cones and use similar neural pathways for comparing their output. Occasionally, however, individuals lack a particular cone type, so their color vision is a product of the differential sensitivity of only two types of cones, a condition known as **color blindness.** Color blindness can also occur when a person has all three cone types, but one of the photopigments has a slightly different absorption curve than normal. Not only do color-defective individuals perceive certain colors differently, but they are also unable to distinguish as many varieties of colors (● Figure 6-30). For example, people with certain color defects cannot distinguish between red and green. At a traffic light they can tell which light is "on" by its intensity, but they must rely on the position of the bright light to know whether to stop or go.

Although the three-cone system has been accepted as the standard model of color vision for more than two centuries, new evidence suggests that perception of color may be more complex. DNA studies showed that men with normal color vision have a variable number of genes coding for cone pigments. For example, men who have multiple genes (from two up to four) for red light detection can distinguish more subtle differences in colors in this long-wavelength range than those with single copies of red-cone genes. This finding will undoubtedly lead to a reevaluation of how the various photopigments contribute to color vision.

The sensitivity of the eyes can vary markedly through dark and light adaptation.

The eyes' sensitivity to light depends on the amount of light-responsive photopigment present in the rods and cones. When you go from bright sunlight into darkened surroundings, you cannot see anything at first, but gradually you begin to distinguish objects as a result of the process of **dark adaptation.** Breakdown of photopigments during exposure to sunlight tremendously decreases photoreceptor sensitivity. In the dark, the photopigments broken down during light exposure are gradually regenerated. As a result, the sensitivity of your eyes gradually increases so you can begin to see in the darkened surroundings. However, only the highly sensitive, rejuvenated rods are "turned on" by the dim light.

● **FIGURE 6-30 Color blindness chart.** People with red–green color blindness cannot detect the number 29 in this chart.

Conversely, when you move from the dark to the light (for example, leaving a movie theater and entering the bright sunlight), at first your eyes are very sensitive to the dazzling light. With little contrast between lighter and darker parts, the entire image appears bleached. As some of the photopigments are rapidly broken down by the intense light, the sensitivity of the eyes decreases and normal contrasts can once again be detected, a process known as **light adaptation.** The rods are so sensitive to light that enough rhodopsin is broken down in bright light to essentially "burn out" the rods; that is, after the rod photopigments have already been broken down by the bright light, they no longer can respond to the light. Therefore, only the less sensitive cones are used for day vision.

Researchers estimate that our eyes' sensitivity can change as much as 1 million times as they adjust to various levels of illumination through dark and light adaptation. These adaptive measures are also enhanced by pupillary reflexes that adjust the amount of available light permitted to enter the eye.

Clinical Note Because retinal is a derivative of vitamin A, adequate amounts of this nutrient must be available for the synthesis of photopigments. **Night blindness** occurs as a result of dietary deficiencies of vitamin A. Although photopigment concentrations in both rods and cones are reduced in this condition, there is still enough cone photopigment to respond to the intense stimulation of bright light, except in the most severe cases. However, even modest reductions in rhodopsin content can decrease the sensitivity of rods so much that they cannot respond to dim light. The person can see in the day using cones but cannot see at night because the rods are no longer functional. Thus, carrots are "good for your eyes" because they are rich in vitamin A.

Visual information is modified and separated before reaching the visual cortex.

The field of view that can be seen without moving the head is known as the **visual field.** The information that reaches the visual cortex in the occipital lobe is not a replica of the visual field for several reasons:

1. The image detected on the retina at the onset of visual processing is upside down and backward because of bending of the light rays. Once it is projected to the brain, the inverted image is interpreted as being in its correct orientation.

2. The information transmitted from the retina to the brain is not merely a point-to-point record of photoreceptor activation. Before the information reaches the brain, the retinal neuronal layers beyond the rods and cones reinforce selected information and suppress other information to enhance contrast. The differential activity of on-center and off-center cells along with the contributions of specialized retinal interneurons, the *horizontal cells* and *amacrine cells* (see ● Figure 6-23), are responsible for much of this retinal processing. For example, horizontal cells participate in lateral inhibition, by which strongly excited cone pathways suppress activity in surrounding pathways of weakly stimulated cones. This increases the dark–bright contrast to enhance the sharpness of boundaries.

3. Various aspects of visual information, such as form, movement, color, and depth, are separated and projected in parallel pathways to different regions of the cortex. Only when these separate bits of processed information are integrated by higher visual regions is a reassembled picture of the visual scene perceived. This is similar to the blobs of paint on an artist's palette versus the finished portrait; the separate pigments do not represent a portrait of a face until they are appropriately integrated on a canvas.

Clinical Note Patients with lesions in specific visual-processing regions of the brain may be unable to completely combine components of a visual impression. For example, a person may be unable to discern movement of an object but have reasonably good vision for shape, pattern, and color. Sometimes the defect can be remarkably specific, like being unable to recognize familiar faces while retaining the ability to recognize inanimate objects.

4. Because of the pattern of wiring between the eyes and the visual cortex, the left half of the cortex receives information only from the right half of the visual field as detected by both eyes, and the right half receives input only from the left half of the visual field of both eyes.

As light enters the eyes, light rays from the left half of the visual field fall on the right half of the retina of both eyes (the medial or inner half of the left retina and the lateral or outer half of the right retina) (● Figure 6-31a). Similarly, rays from the right half of the visual field reach the left half of each retina (the lateral half of the left retina and the medial half of the right retina). Each optic nerve exiting the retina carries information from both halves of the retina it serves. This information is separated as the optic nerves meet at the **optic chiasm** located underneath the hypothalamus (*chiasm* means "cross") (see ● Figure 5-7b, p. 146). Within the optic chiasm, the fibers from the medial half of each retina cross to the opposite side, but those from the lateral half remain on the original side. The reorganized bundles of fibers leaving the optic chiasm are known as **optic tracts.** Each optic tract carries information from the

lateral half of one retina and the medial half of the other retina. Therefore, this partial crossover brings together, from the two eyes, fibers that carry information from the same half of the visual field. Each optic tract, in turn, delivers to the half of the brain on its same side information about the opposite half of the visual field.

 A knowledge of these pathways can facilitate diagnosis of visual defects arising from interruption of the visual pathway at various points (● Figure 6-31b).

Before we move on to how the brain processes visual information, take a look at ▲ Table 6-4, which summarizes the functions of the various components of the eyes.

The thalamus and visual cortexes elaborate the visual message.

The first stop in the brain for information in the visual pathway is the *lateral geniculate nucleus* of the thalamus (● Figure 6-31a). It separates information received from the eyes and relays it via fiber bundles known as **optic radiations** to different zones in the cortex, each of which processes different aspects of the visual stimulus (for example, form, movement, color, and depth). This sorting process is no small task, because each optic nerve contains more than a million fibers carrying information from the photoreceptors in one retina. This is more than all the afferent fibers carrying somatosensory input from all the other regions of the body! Researchers estimate that hundreds of millions of neurons occupying about 40% of the cortex participate in visual processing, compared to 8% devoted to touch perception and 3% to hearing. Yet the connections in the visual pathways are precise. The lateral geniculate nucleus and each of the zones in the cortex that processes visual information have a topographical map representing the retina point for point. As with the somatosensory cortex, the neural maps of the retina are distorted. The fovea, the retinal region capable of greatest acuity, has much greater representation in the neural map than the more peripheral regions of the retina do.

DEPTH PERCEPTION Although each half of the visual cortex receives information simultaneously from the same part of the visual field as received by both eyes, the messages from the two eyes are not identical. Each eye views an object from a slightly different vantage point, even though the overlap is tremendous. The overlapping area seen by both eyes at the same time is known as the **binocular** ("two-eyed") **field of vision,** which is important for **depth perception.** Like other areas of the cortex, the primary visual cortex is organized into functional columns, each processing information from a small region of the retina. Independent alternating columns are devoted to information about the same point in the visual field from the right and left eyes. The brain uses the slight disparity in the information received from the two eyes to estimate distance, allowing you to perceive three-dimensional objects in spatial depth. Some depth perception is possible using only one eye, based on experience and comparison with other cues. For example, if your one-eyed view includes a car and a building and the car is much larger, you correctly interpret that the car must be closer to you than the building is.

KEY

—— = Site of lesion	✗ = Visual deficit

1 Left optic nerve

2 Optic chiasm

3 Left optic tract (or radiation)

Left eye | Right eye

Viewing brain from above with overlying structures removed

(a) Visual pathway

(b) Visual deficits with specific lesions in visual pathway

● **FIGURE 6-31 The visual pathway and visual deficits associated with lesions in the pathway.** (a) Note that the left half of the visual cortex in the occipital lobe receives information from the right half of the visual field of both eyes (in green), and the right half of the cortex receives information from the left half of the visual field of both eyes (in red). (b) Each visual deficit illustrated is associated with a lesion at the corresponding numbered point of the visual pathway in part (a).

Clinical Note Sometimes the two views are not successfully merged. This condition may occur for two reasons: (1) The eyes are not both focused on the same object simultaneously, because of defects of the external eye muscles that make fusion of the two eyes' visual fields impossible; or (2) the binocular information is improperly integrated during visual processing. The result is double vision, or **diplopia,** a condition in which the disparate views from both eyes are seen simultaneously.

HIERARCHY OF VISUAL CORTICAL PROCESSING Within the cortex, visual information is first processed in the primary visual cortex, then is sent to higher-level visual areas for even more complex processing and abstraction. The visual cortex is precisely organized both vertically and horizontally. Vertical columns extend through the thickness of the cortex from its outer surface to the white matter. Each column is made up of cells that process the same bit of visual input. There are three types of columns, based on the type of visual input they process: (1) as discussed in the preceding section, an alternating system of *ocular dominance columns* devoted to input from the left or the right eye is important for binocular interaction and depth perception; (2) *orientation columns* related to axis of orientation of visual stimuli play a key role in perceiving form and movement; and (3) shortened columns known as *blobs* process color.

The orientation columns contain a hierarchy of visual cells that respond to increasingly complex stimuli. Three types of visual cortical neurons have been identified based on the com-

plexity of stimulus requirements needed for the cell to respond; these are called **simple, complex,** and **hypercomplex cells.** All of the cells within a given orientation column process input arising from visual stimuli in the same axis of orientation, such as a slit of light oriented vertically, horizontally, or at some oblique angle. The primary visual cortex has orientation columns for every possible axis of orientation. Dissection of visual input into these various orientations is important for discerning form and movement. The visual cortex is also organized into six layers, with each layer consisting of specific cell types. For example, simple cells are found in layer 4. Simple and complex cells are stacked on top of one another in a specific way within each orientation column. Hypercomplex cells are found only in higher visual-processing areas. Horizontal connections within the layers link vertical columns that carry out similar functions. Each layer has different inputs and outputs and is specialized to perform a particular task.

Unlike a retinal cell, which responds to the amount of light, a cortical cell fires only when it receives a particular pattern of illumination for which it is programmed. These patterns are built up by converging connections that originate from closely aligned photoreceptor cells in the retina. For example, in the orientation columns some simple cells fire only when a bar is viewed vertically in a specific location, others when a bar is horizontal, and others at various oblique orientations. Movement of a critical axis of orientation becomes important for response by some of the complex cells. Hypercomplex cells add a

Functions of the Major Components of the Eye

Structures	Location	Function
(in alphabetical order)		
Aqueous Humor	Anterior cavity between cornea and lens	Clear watery fluid that is continually formed and carries nutrients to the cornea and lens
Bipolar Cells	Middle layer of nerve cells in retina	Important in retinal processing of light stimulus
Blind Spot	Point slightly off center on retina where optic nerve exits; is devoid of photoreceptors (also known as *optic disc*)	Route for passage of optic nerve and blood vessels
Choroid	Middle layer of eye	Pigmented to prevent scattering of light rays in eye; contains blood vessels that nourish retina; anteriorly specialized to form ciliary body and iris
Ciliary Body	Specialized anterior derivative of the choroid layer; forms a ring around the outer edge of the lens	Produces aqueous humor and contains ciliary muscle
Ciliary Muscle	Circular muscular component of ciliary body; attached to lens by means of suspensory ligaments	Important in accommodation
Cones	Photoreceptors in outermost layer of retina	Responsible for high acuity, color, and day vision
Cornea	Anterior clear outermost layer of eye	Contributes most extensively to eye's refractive ability
Fovea	Exact center of retina	Region with greatest acuity
Ganglion Cells	Inner layer of nerve cells in retina	Important in retinal processing of light stimulus; form optic nerve
Iris	Visible pigmented ring of muscle within aqueous humor	Varies size of pupil by variable contraction; responsible for eye color
Lens	Between aqueous humor and vitreous humor; attaches to ciliary muscle by suspensory ligaments	Provides variable refractive ability during accommodation
Macula Lutea	Area immediately surrounding the fovea	Has high acuity because of abundance of cones
Optic Disc	(see entry for blind spot)	
Optic Nerve	Leaves each eye at optic disc (blind spot)	First part of visual pathway to the brain
Pupil	Anterior round opening in middle of iris	Permits variable amounts of light to enter eye
Retina	Innermost layer of eye	Contains the photoreceptors (rods and cones)
Rods	Photoreceptors in outermost layer of retina	Responsible for high-sensitivity, black-and-white, and night vision
Sclera	Tough outer layer of eye	Protective connective tissue coat; forms visible white part of eye; anteriorly specialized to form cornea
Suspensory Ligaments	Suspended between ciliary muscle and lens	Important in accommodation
Vitreous Humor	Between lens and retina	Semifluid, jellylike substance that helps maintain spherical shape of eye

new dimension to visual processing by responding only to particular edges, corners, and curves. Each level of cortical visual neurons has increasingly greater capacity for abstraction of information built up from the increasing convergence of input from lower-level neurons. In this way, the cortex transforms the dotlike pattern of photoreceptors stimulated to varying degrees by varying light intensities in the retinal image into information about position, orientation, movement, contour, and length. Other aspects of visual information, such as depth perception and color perception, are processed simultaneously by the other vertical and horizontal organizational systems. How and where the entire image is finally put together is still unresolved. Only when these separate bits of processed information are integrated by higher visual regions is a reassembled picture of the visual scene perceived.

Visual input goes to other areas of the brain not involved in vision perception.

Not all fibers in the visual pathway terminate in the visual cortexes. Some are projected to other regions of the brain for purposes other than direct vision perception. Examples of nonsight activities dependent on input from the rods and cones include (1) contribution to cortical alertness and attention, (2) control of pupil size, and (3) control of eye movements. Each eye is equipped with a set of six **external eye muscles** that position and move the eye so that it can better locate, see, and track objects. Eye movements are among the fastest, most discretely controlled movements of the body.

Three percent of the ganglion cells are not involved in visual processing. Instead they make **melanopsin,** a light-sensitive pigment that plays a key role in setting the body's "biological clock" to march in step with the light–dark cycles (see p. 686).

Some sensory input may be detected by multiple sensory-processing areas in the brain.

Before shifting gears to another sense—hearing—we should mention a controversial new theory regarding the senses that challenges the prevailing view that the separate senses feed into distinct brain regions that handle only one sense. A growing body of evidence suggests that the brain regions devoted almost exclusively to a certain sense, such as the visual cortex for visual input and the somatosensory cortex for touch input, actually receive a variety of sensory signals. Therefore, tactile and auditory signals also arrive in the visual cortex. For example, one study using new brain-imaging techniques showed that people blind from birth use the visual cortex when they read Braille, even though they are not "seeing" anything. The tactile input from their fingers reaches the visual area of the brain as well as the somatosensory cortex. This input helps them "visualize" the patterns of the Braille bumps.

Also reinforcing the notion that central processing of different types of sensory input overlaps to some extent, scientists recently discovered *multisensory neurons*—brain cells that react to multiple sensory inputs instead of just to one. No one knows whether these cells are rare or commonplace in the brain. (See the boxed feature on p. 214, ■ Concepts, Challenges, and Controversies, for one way in which researchers are exploiting this sharing of sensory input by multiple regions of the brain.)

For the remainder of the chapter, we will concentrate on the mainstream function of the other special senses. We now shift attention from the eyes to the ears.

Ear: Hearing and Equilibrium

Each **ear** consists of three parts: the *external,* the *middle,* and the *inner ear* (● Figure 6-32). The external and middle portions of the ear transmit airborne sound waves to the fluid-filled inner ear, amplifying sound energy in the process. The inner ear houses two different sensory systems: the *cochlea,* which contains the receptors for conversion of sound waves into nerve impulses, making hearing possible; and the *vestibular apparatus,* which is necessary for the sense of equilibrium.

Sound waves consist of alternate regions of compression and rarefaction of air molecules.

Hearing is the neural perception of sound energy. Hearing involves two aspects: identification of the sounds ("what") and their localization ("where"). We will first examine the characteristics of sound waves, then how the ears and brain process sound input to accomplish hearing.

Sound waves are traveling vibrations of air that consist of regions of high pressure caused by compression of air molecules alternating with regions of low pressure caused by rarefaction of the molecules (● Figure 6-33a). Any device capable of producing such a disturbance pattern in air molecules is a source of sound. A simple example is a tuning fork. When a tuning fork is struck, its prongs vibrate. As a prong of the fork moves in one direction (● Figure 6-33b), air molecules ahead of it are pushed closer together, or compressed, increasing the pressure in this area. Simultaneously, as the prong moves forward the air molecules behind the prong spread out, or are rarefied, lowering the pressure in that region. As the prong moves in the opposite direction, an opposite wave of compression and rarefaction is created. Even though individual molecules are moved only short distances as the tuning fork vibrates, alternating waves of compression and rarefaction spread out considerable distances in a rippling fashion. Disturbed air molecules disturb other molecules in adjacent regions, setting up new regions of compression and rarefaction, and so on (● Figure 6-33c). Sound energy gradually dissipates as sound waves travel farther from the original sound source. The intensity of the sound decreases, until it finally dies out when the last sound wave is too weak to disturb the air molecules around it.

Sound waves can also travel through media other than air, such as water. They do so less efficiently, however; greater pressures are required to cause movements of fluid than movements of air because of the fluid's greater inertia (resistance to change).

Sound is characterized by its pitch (tone), intensity (loudness), and timbre (quality) (● Figure 6-34, p. 216):

■ The **pitch,** or **tone,** of a sound (for example, whether it is a C or a G note) is determined by the frequency of vibrations. The

"Seeing" with the Tongue

Although each type of sensory input is received primarily by a distinct brain region responsible for perception of that modality, the regions of the brain involved with perceptual processing receive sensory signals from a variety of sources. Thus, the visual cortex receives sensory input not only from the eyes but from the body surface and ears as well. One group of scientists is exploiting in an unusual but exciting way this sharing of sensory input by multiple regions of the brain. In this research, blind or sighted-but-blindfolded volunteers are able to crudely perceive shapes and features in space by means of a tongue display unit. When this device, which consists of a grid of electrodes, is positioned on the tongue, it translates images detected by a camera into a pattern of electrical signals that activate touch receptors on the tongue (see the accompanying figure). The pattern of "tingling" on the tongue as a result of the light-induced electrical signals corresponds with the image recorded by the camera. With practice, the visual cortex interprets this alternate sensory input as a visual image. As one of the investigators who developed this technique claims, a person sees with the visual cortex, not with the eyes. Any means of sending signals to the visual cortex can be perceived as a visual image. For example, one blind participant in the study saw the flickering of a candle flame for the first time by means of this tongue device.

The tongue is a better choice than the skin for receiving this light-turned-tactile input because the saliva is an electrically conducive fluid that readily conducts the current generated in the device by the visual

● A blindfolded subject mimics hand gestures being recorded by a video camera (white box beside laptop computer) and transmitted to an image-translating tongue display unit.

input. Furthermore, the tongue is densely populated with tactile receptors, opening up the possibility that the tongue can provide higher acuity of visual input than the skin could. This feature will be important if such a device is ever used to help the visually impaired. The researchers' goal is to improve the resolution of the device by increasing the number of in-the-mouth electrodes. Even so, the perceived image will still be crude because the acuity afforded by this device can never come close to matching that provided by the eyes' small receptive fields.

Although using the tongue as a surrogate eye could never provide anywhere

near the same vision as a normal eye, the hope is that this technique will afford people who are blind a means to make out doorways, to see objects as vague shapes, and to track motion. Even this limited visual input would enable a sightless person to get around more easily and improve the quality of his or her life. The device's developers plan to shrink the size of the unit so that it will fit inconspicuously in the user's mouth and be connected by a wireless link to a miniature camera mounted on eyeglasses. Such a trimmed-down unit would be practical to use and cosmetically acceptable.

greater the frequency of vibration, the higher the pitch is. Human ears can detect sound waves with frequencies from 20 to 20,000 cycles per second but are most sensitive to frequencies between 1000 and 4000 cycles per second.

■ The **intensity,** or **loudness,** of a sound depends on the amplitude of the sound waves, or the pressure differences between a high-pressure region of compression and a low-pressure region of rarefaction. Within the hearing range, the greater the amplitude, the louder the sound is. Human ears can detect a wide range of sound intensities, from the slightest whisper to the painfully loud takeoff of a jet. Loudness is measured in decibels (dB), which are a logarithmic measure of intensity compared with the faintest sound that can be heard—the **hearing threshold.** Because of the logarithmic relationship, every 10 dB indicates a 10-fold increase in loudness. A few examples of common sounds illustrate the magnitude of these increases (▲ Table 6-5). Note that the rustle of leaves at 10 dB is 10 times louder than hearing threshold, but the sound of a jet taking off at 150 dB is a quadrillion (a million billion) times, not 150 times, louder than the faintest audible sound. Sounds greater than 100 dB can permanently damage the sensitive sensory apparatus in the cochlea.

■ The **timbre,** or **quality,** of a sound depends on its overtones, which are additional frequencies superimposed on the funda-

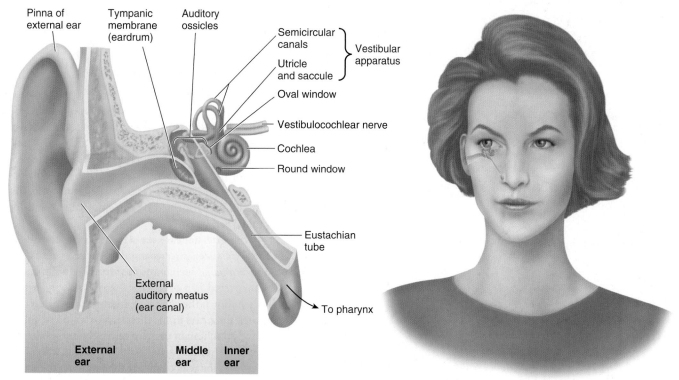

● **FIGURE 6-32 Anatomy of the ear.**

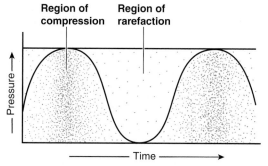

Region of compression **Region of rarefaction**

(a) Sound waves

Pressure →

← Time →

Normal density of air molecules (when tuning fork is at rest)

Region of rarefaction **Region of compression**

(b) Tuning fork setting up sound waves

● **FIGURE 6-33 Formation of sound waves.**
(a) Sound waves are alternating regions of compression and rarefaction of air molecules. (b) A vibrating tuning fork sets up sound waves as the air molecules ahead of the advancing arm of the tuning fork are compressed while the molecules behind the arm are rarefied. (c) Disturbed air molecules bump into molecules beyond them, setting up new regions of air disturbance more distant from the original source of sound. In this way, sound waves travel progressively farther from the source, even though each individual air molecule travels only a short distance when it is disturbed. The sound wave dies out when the last region of air disturbance is too weak to disturb the region beyond it.

(c) Sound waves traveling from source

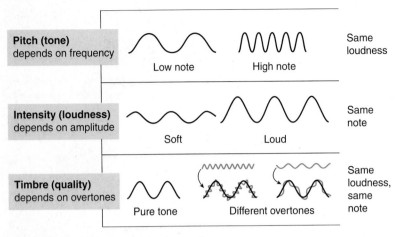

● **FIGURE 6-34 Properties of sound waves.**

▲ TABLE 6-5	Relative Magnitude of Common Sounds	
Sound	**Loudness in Decibels (Db)**	**Comparison to Faintest Audible Sound (Hearing Threshold)**
Rustle of Leaves	10 dB	10 times louder
Ticking of Watch	20 dB	100 times louder
Whispering	30 dB	1 thousand times louder
Normal Conversation	60 dB	1 million times louder
Food Blender, Lawn Mower, Hair Dryer	90 dB	1 billion times louder
Loud Rock Concert, Ambulance Siren	120 dB	1 trillion times louder
Takeoff of Jet Plane	150 dB	1 quadrillion times louder

mental pitch or tone. A tuning fork has a pure tone, but most sounds lack purity. For example, complex mixtures of overtones impart different sounds to different instruments playing the same note (a C note on a trumpet sounds different from C on a piano). Overtones are likewise responsible for characteristic differences in voices. Timbre enables the listener to distinguish the source of sound waves, because each source produces a different pattern of overtones. Thanks to timbre, you can tell whether it is your mother or girlfriend calling on the telephone before you say the wrong thing.

The external ear plays a role in sound localization.

The specialized receptor cells for sound are located in the fluid-filled inner ear. Airborne sound waves must therefore be channeled toward and transferred into the inner ear, compensating in the process for the loss in sound energy that naturally occurs as sound waves pass from air into water. This function is performed by the external ear and the middle ear.

The **external ear** (see ● Figure 6-32) consists of the *pinna* (ear), *external auditory meatus* (ear canal), and *tympanic membrane* (eardrum). The **pinna,** a prominent skin-covered flap of cartilage, collects sound waves and channels them down the ear canal. Many species (dogs, for example) can cock their ears in the direction of sound to collect more sound waves, but human ears are relatively immobile. Because of its shape, the pinna partially shields sound waves that approach the ear from the rear, changing the timbre of the sound and thus helping a person distinguish whether a sound is coming from directly in front or behind.

Sound localization for sounds approaching from the right or left is determined by two cues. First, the sound wave reaches the ear closer to the sound source slightly before it arrives at the farther ear. Second, the sound is less intense as it reaches the farther ear because the head acts as a sound barrier that partially disrupts the propagation of sound waves. The auditory cortex integrates all these cues to determine the location of the sound source. It is difficult to localize sound with only one ear. Recent evidence suggests that the auditory cortex pinpoints the location of a sound by differences in the timing of neuronal firing patterns, not by any spatially organized map such as the one projected on the visual cortex point for point from the retina that enables the location of a visual object to be identified.

The entrance to the **ear canal** is guarded by fine hairs. The skin lining the canal contains modified sweat glands that produce **cerumen** (earwax), a sticky secretion that traps fine foreign particles. Together the hairs and earwax help prevent airborne particles from reaching the inner portions of the ear canal, where they could accumulate or injure the tympanic membrane and interfere with hearing.

The tympanic membrane vibrates in unison with sound waves in the external ear.

The **tympanic membrane,** which is stretched across the entrance to the middle ear, vibrates when struck by sound waves. The alternating higher- and lower-pressure regions of a sound wave cause the exquisitely sensitive eardrum to bow inward and outward in unison with the wave's frequency.

For the membrane to be free to move as sound waves strike it, the resting air pressure on both sides of the tympanic membrane must be equal. The outside of the eardrum is exposed to atmospheric pressure that reaches it through the ear canal. The inside of the eardrum facing the middle ear cavity is also exposed to atmospheric pressure via the **eustachian (auditory) tube,** which connects the middle ear to the **pharynx** (back of the throat) (see ● Figure 6-32). The eustachian tube is normally closed, but it can be pulled open by yawning, chewing, and swallowing. Such opening permits air pressure within the middle ear to equilibrate with atmospheric pressure so that pressures on both sides of the tympanic membrane are equal. During rapid external pressure changes (for example, during air flight), the eardrum bulges painfully as the pressure outside the ear changes while the pressure in the middle ear remains unchanged. Opening the eustachian tube by yawning allows the pressure on both sides of the tympanic membrane to equalize, relieving the pressure distortion as the eardrum "pops" back into place.

Clinical Note Infections originating in the throat sometimes spread through the eustachian tube to the middle ear. The resulting fluid accumulation in the middle ear not only is painful but also interferes with sound conduction across the middle ear.

The middle ear bones convert tympanic membrane vibrations into fluid movements in the inner ear.

The **middle ear** transfers the vibratory movements of the tympanic membrane to the fluid of the inner ear. This transfer is facilitated by a movable chain of three small bones, or **ossicles** (the **malleus, incus,** and **stapes**), that extend across the middle ear (● Figure 6-35a). The first bone, the malleus, is attached to the tympanic membrane, and the last bone, the stapes, is attached to the **oval window,** the entrance into the fluid-filled cochlea. As the tympanic membrane vibrates in response to sound waves, the chain of bones is set into motion at the same frequency, transmitting this frequency of movement from the tympanic membrane to the oval window. The resulting pressure on the oval window with each vibration produces wavelike movements in the inner ear fluid at the same frequency as the original sound waves. However, as noted earlier, greater pressure is required to set fluid in motion. The ossicular system amplifies the pressure of the airborne sound waves by two mechanisms to set up fluid vibrations in the cochlea. First, because the surface area of the tympanic membrane is much larger than that of the oval window, pressure is increased as force exerted on the tympanic membrane is conveyed by the ossicles to the oval window (pressure = force/unit area). Second, the lever action of the ossicles provides an additional mechanical advantage. Together, these mechanisms increase the force exerted on the oval window by 20 times what it would be if the sound wave struck the oval window directly. This additional pressure is sufficient to set the cochlear fluid in motion.

Several tiny muscles in the middle ear contract reflexly in response to loud sounds (over 70 dB), causing the tympanic membrane to tighten and limiting movement of the ossicular chain. This reduced movement of middle ear structures diminishes the transmission of loud sound waves to the inner ear to protect the delicate sensory apparatus from damage. This reflex response is relatively slow, however, happening at least 40 msec after exposure to a loud sound. It thus provides protection only from prolonged loud sounds, not from sudden sounds like an explosion. Taking advantage of this reflex, World War II antiaircraft guns were designed to make a loud prefiring sound to protect the gunner's ears from the much louder boom of the actual firing.

The cochlea contains the organ of Corti, the sense organ for hearing.

The pea-sized, snail-shaped **cochlea,** the "hearing" portion of the inner ear, is a coiled tubular system lying deep within the temporal bone (see ● Figure 6-32) (*cochlea* means "snail"). It is easier to understand the functional components of the cochlea by "uncoiling" it, as shown in ● Figure 6-35a. The cochlea is divided throughout most of its length into three fluid-filled longitudinal compartments. A blind-ended **cochlear duct,** which is also known as the **scala media,** constitutes the middle compartment. It tunnels lengthwise through the center of the cochlea, almost but not quite reaching its end. The upper compartment, the **scala vestibuli,** follows the inner contours of the spiral, and the **scala tympani,** the lower compartment, follows the outer contours (● Figure 6-35a and b). The fluid within the scala vestibuli and scala tympani is called **perilymph.** The cochlear duct contains a slightly different fluid, the **endolymph** (● Figure 6-36a). The region beyond the tip of the cochlear duct where the fluid in the upper and lower compartments is continuous is called the **helicotrema.** The scala vestibuli is sealed from the middle ear cavity by the oval window, to which the stapes is attached. Another small membrane-covered opening, the **round window,** seals the scala tympani from the middle ear. The thin **vestibular membrane** forms the ceiling of the cochlear duct and separates it from the scala vestibuli. The **basilar membrane** forms the floor of the cochlear duct, separating it from the scala tympani. The basilar membrane is especially important because it bears the **organ of Corti,** the sense organ for hearing.

Hair cells in the organ of Corti transduce fluid movements into neural signals.

The organ of Corti, which rests on top of the basilar membrane throughout its full length, contains **auditory hair cells** that are the receptors for sound. The 15,000 hair cells within each cochlea are arranged in four parallel rows along the length of the basilar membrane: one row of **inner hair cells** and three rows of **outer hair cells** (● Figure 6-35c). Protruding from the surface of each hair cell are about 100 hairs known as **stereocilia,** which are actin-stiffened microvilli (see p. 48). Hair cells are mechanoreceptors; they generate neural signals when their surface hairs are mechanically deformed in association with fluid movements in the inner ear. These stereocilia contact the **tectorial membrane,** an awninglike projection overhanging the organ of Corti throughout its length (● Figure 6-35b and c).

The pistonlike action of the stapes against the oval window sets up pressure waves in the upper compartment. Because fluid

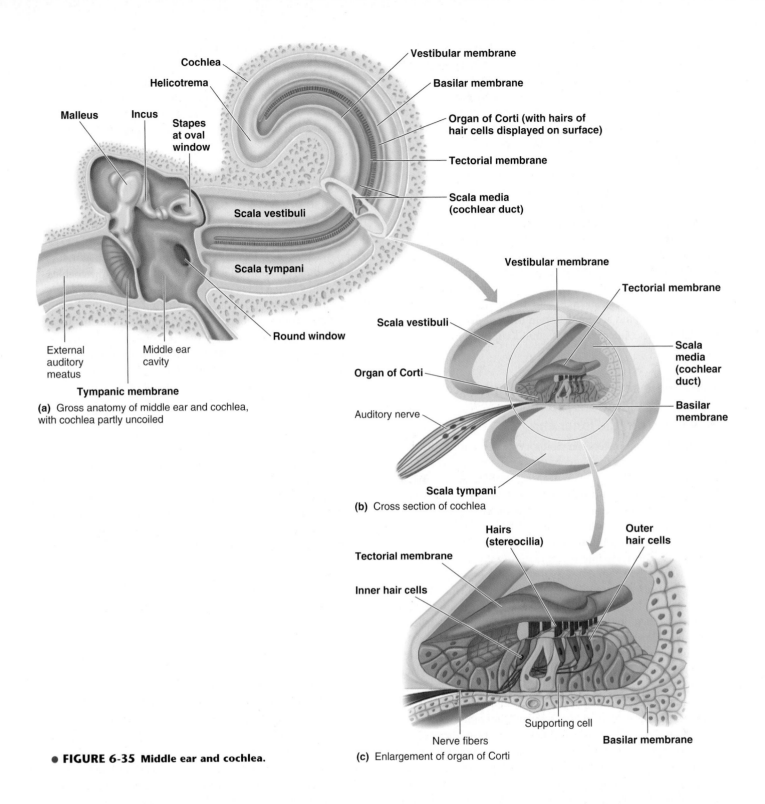

Cochlea

Helicotrema

Malleus   **Incus**

Stapes at oval window

Vestibular membrane

Basilar membrane

Organ of Corti (with hairs of hair cells displayed on surface)

Tectorial membrane

Scala vestibuli

Scala media (cochlear duct)

Scala tympani

Round window

External auditory meatus

Middle ear cavity

Tympanic membrane

(a) Gross anatomy of middle ear and cochlea, with cochlea partly uncoiled

Vestibular membrane

Tectorial membrane

Scala vestibuli

Scala media (cochlear duct)

Organ of Corti

Basilar membrane

Auditory nerve

Scala tympani

(b) Cross section of cochlea

Hairs (stereocilia)

Outer hair cells

Tectorial membrane

Inner hair cells

Supporting cell

Nerve fibers

Basilar membrane

(c) Enlargement of organ of Corti

● **FIGURE 6-35 Middle ear and cochlea.**

is incompressible, pressure is dissipated in two ways as the stapes causes the oval window to bulge inward: (1) displacement of the round window and (2) deflection of the basilar membrane (● Figure 6-36a). In the first of these pathways, the pressure wave pushes the perilymph forward in the upper compartment, then around the helicotrema, and into the lower compartment, where it causes the round window to bulge outward into the middle ear cavity to compensate for the pressure increase. As the stapes rocks backward and pulls the oval window outward toward the middle ear, the perilymph shifts in the opposite direction, displacing the round window inward. This pathway does not result in sound reception; it just dissipates pressure.

Pressure waves of frequencies associated with sound reception take a "shortcut" (Figure 6-36a). Pressure waves in the upper compartment are transferred through the thin vestibular membrane, into the cochlear duct, and then through the basilar membrane into the lower compartment, where they cause the round window to alternately bulge outward and inward. The main difference in this pathway is that transmission of pressure

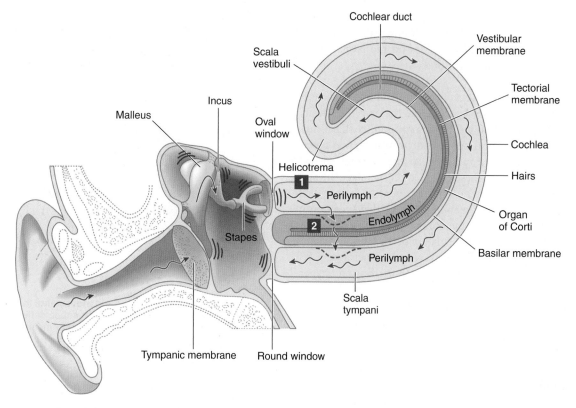

Fluid movement within the perilymph set up by vibration of the oval window follows two pathways:

Pathway 1: Through the scala vestibuli, around the heliocotrema, and through the scala tympani, causing the round window to vibrate. This pathway just dissipates sound energy.

Pathway 2: A "shortcut" from the scala vestibuli through the basilar membrane to the scala tympani. This pathway triggers activation of the receptors for sound by bending the hairs of hair cells as the organ of Corti on top of the vibrating basilar membrane is displaced in relation to the overlying tectorial membrane.

(a) Fluid movement in cochlea

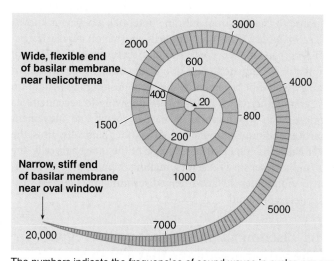

The numbers indicate the frequencies of sound waves in cycles per second with which different regions of the basilar membrane maximally vibrate.

(b) Basilar membrane, partly uncoiled

(c) Basilar membrane, completely uncoiled

● **FIGURE 6-36 Transmission of sound waves.** (a) Fluid movement within the cochlea set up by vibration of the oval window follows two pathways, one dissipating sound energy and the other initiating the receptor potential. (b) Different regions of the basilar membrane vibrate maximally at different frequencies. (c) The narrow, stiff end of the basilar membrane nearest the oval window vibrates best with high-frequency pitches. The wide, flexible end of the basilar membrane near the helicotrema vibrates best with low-frequency pitches.

waves through the basilar membrane causes this membrane to move up and down, or vibrate, in synchrony with the pressure wave. Because the organ of Corti rides on the basilar membrane, the hair cells also move up and down.

ROLE OF THE INNER HAIR CELLS The inner and outer hair cells differ in function. The inner hair cells are the ones that "hear": They transform the mechanical forces of sound (cochlear fluid vibration) into the electrical impulses of hearing (action potentials propagating auditory messages to the cerebral cortex). Because the stereocilia of these receptor cells contact the stiff, stationary tectorial membrane, they are bent back and forth when the oscillating basilar membrane shifts their position in relationship to the tectorial membrane (● Figure 6-37). This back-and-forth mechanical deformation of the hairs alternately opens and closes mechanically gated cation channels (see p. 88) in the hair cell, resulting in alternating depolarizing and hyperpolarizing potential changes—the receptor potential—at the same frequency as the original sound stimulus.

The stereocilia of each hair cell are organized into rows of graded heights ranging from short to tall in a precise staircase pattern (● Figure 6-38a). **Tip links,** which are CAMs (cell adhesion molecules; see p. 57), link the tips of stereocilia in adjacent rows. When the basilar membrane moves upward, the bundle of stereocilia bends toward its tallest membrane, stretching the tip links. Stretched tip links tug open the cation channels to which they are attached (● Figure 6-38b). The resultant ion movement is unusual because of the unique composition of the endolymph that bathes the stereocilia. In sharp contrast to ECF elsewhere, endolymph has a higher concentration of K^+ than found inside the hair cell. Some cation channels are open in a resting hair cell, allowing low-level K^+ entry down its concentration gradient. When more cation channels are pulled open, more K^+ enters the hair cell. This additional entering K^+ depolarizes (excites) the hair cell (● Figure 6-38c). When the basilar membrane moves in the opposite direction, the hair bundle bends away from the tallest stereocilium, slackening the tip links and closing all the channels. As a result, K^+ entry ceases, hyperpolarizing the hair cell.

Like photoreceptors, hair cells do not undergo action potentials. The inner hair cells communicate via a chemical synapse with the terminals of afferent nerve fibers making up the **auditory (cochlear) nerve.** Because of low-level K^+ entry, the inner hair cells spontaneously release some neurotransmitter via Ca^{2+}-induced exocytosis in the absence of stimulation. Depolarization of these hair cells opens more voltage-gated Ca^{2+} channels. The resultant additional Ca^{2+} entry increases their rate of neurotransmitter secretion, which steps up the rate of firing in the afferent fibers with which the inner hair cells synapse. Conversely, the firing rate decreases below resting level as these hair cells release less neurotransmitter when they are hyperpolarized on displacement in the opposite direction.

To summarize, the ear converts sound waves in the air into oscillating movements of the basilar membrane that bends the hairs of the receptor cells back and forth. This shifting mechanical deformation of the hairs alternately opens and closes the receptor cells' channels, bringing about graded potential changes in the receptor that lead to changes in the rate of action

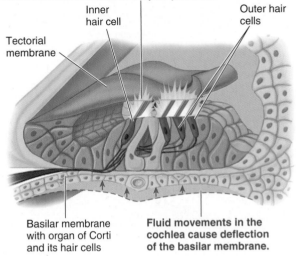

The stereocilia (hairs) from the hair cells of the basilar membrane contact the overlying tectorial membrane. These hairs are bent when the basilar membrane is deflected in relation to the stationary tectorial membrane. This bending of the inner hair cells' hairs opens mechanically gated channels, leading to ion movements that result in a receptor potential.

Inner hair cell   Outer hair cells

Tectorial membrane

Basilar membrane with organ of Corti and its hair cells

Fluid movements in the cochlea cause deflection of the basilar membrane.

● **FIGURE 6-37** Bending of hairs on deflection of the basilar membrane.

potentials propagated to the brain. These neural signals can be perceived by the brain as sound sensations (● Figure 6-39).

ROLE OF THE OUTER HAIR CELLS Whereas the inner hair cells send auditory signals to the brain over afferent fibers, the outer hair cells do not signal the brain about incoming sounds. Instead, the outer hair cells actively and rapidly change length in response to changes in membrane potential, a behavior known as *electromotility*. The outer hair cells shorten on depolarization and lengthen on hyperpolarization. These changes in length amplify or accentuate the motion of the basilar membrane. An analogy would be a person deliberately pushing the pendulum of a grandfather clock in time with its swing to accentuate its motion. Such modification of basilar membrane movement improves and tunes stimulation of the inner hair cells. Thus, the outer hair cells enhance the response of the inner hair cells, the real auditory sensory receptors, making them exquisitely sensitive to sound intensity and highly discriminatory between various pitches of sound.

Pitch discrimination depends on the region of the basilar membrane that vibrates.

Pitch discrimination (that is, the ability to distinguish between various frequencies of incoming sound waves) depends on the shape and properties of the basilar membrane, which is narrow and stiff at its oval window end and wide and flexible at its helicotrema end (see ● Figure 6-36b). Different regions of the basilar membrane naturally vibrate maximally at different frequencies; that is, each frequency displays peak vibration at a different position along the membrane. The narrow end nearest the oval window vibrates best with high-frequency pitches,

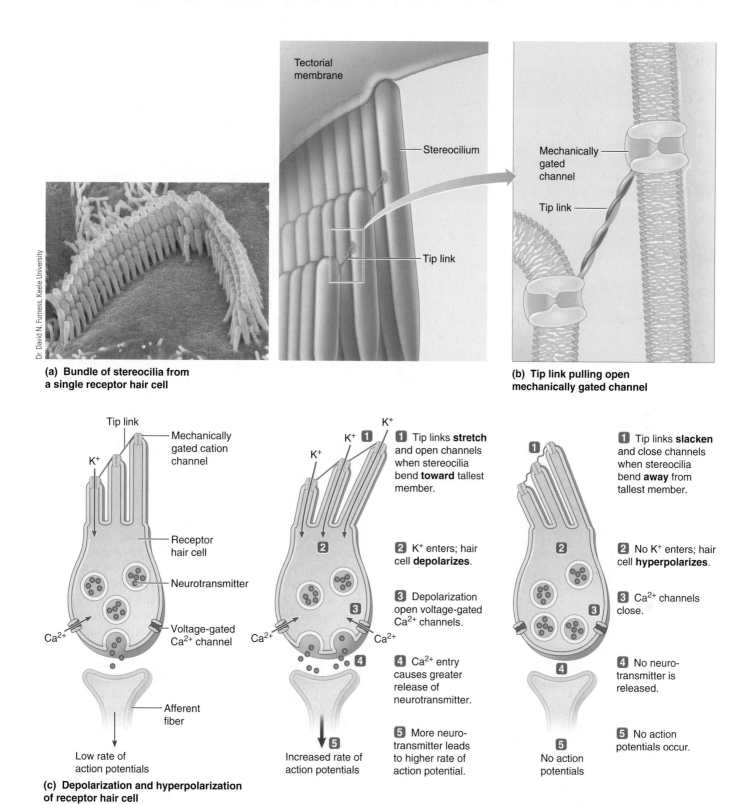

(a) Bundle of stereocilia from a single receptor hair cell

Tectorial membrane

Stereocilium

Tip link

Mechanically gated channel

Tip link

(b) Tip link pulling open mechanically gated channel

Tip link

Mechanically gated cation channel

K⁺

Receptor hair cell

Neurotransmitter

Ca²⁺

Voltage-gated Ca²⁺ channel

Afferent fiber

Low rate of action potentials

K⁺

K⁺

K⁺

1 Tip links **stretch** and open channels when stereocilia bend **toward** tallest member.

2 K⁺ enters; hair cell **depolarizes**.

3 Depolarization open voltage-gated Ca²⁺ channels.

Ca²⁺

Ca²⁺

4 Ca²⁺ entry causes greater release of neurotransmitter.

Increased rate of action potentials

5 More neurotransmitter leads to higher rate of action potential.

1 Tip links **slacken** and close channels when stereocilia bend **away** from tallest member.

2 No K⁺ enters; hair cell **hyperpolarizes**.

3 Ca²⁺ channels close.

4 No neurotransmitter is released.

No action potentials

5 No action potentials occur.

(c) Depolarization and hyperpolarization of receptor hair cell

● **FIGURE 6-38 The role of stereocilia in sound transduction.**

whereas the wide end nearest the helicotrema vibrates maximally with low-frequency tones (see ● Figure 6-36c). The pitches in between are sorted out precisely along the length of the membrane from higher to lower frequency. As a sound wave of a particular frequency is set up in the cochlea ᵥy oscillation of

the stapes, the wave travels to the region of the basilar membrane that naturally responds maximally to that frequency. The energy of the pressure wave is dissipated with this vigorous membrane oscillation, so the wave dies out at the region of maximal displacement.

● **FIGURE 6-39 Pathway for sound transduction.**

Scientists recently learned that the spiral shape of the cochlea steers the low-frequency sound waves toward the tightest turn at its center, where the region of the basilar membrane that responds maximally to these bass sounds is located. Therefore, the cochlea's coil is not just a means of packaging a lot of membrane into a small space; this curvature also boosts detection of low notes.

The hair cells in the region of peak vibration of the basilar membrane undergo the most mechanical deformation and accordingly are the most excited. You can think of the organ of Corti as a piano with 15,000 keys (represented by the 15,000 hair cells) rather than the usual 88 keys. Each hair cell is "tuned" to an optimal sound frequency, determined by its location on the organ of Corti. Different sound waves promote maximal movement of different regions of the basilar membrane, and thus activate differently tuned hair cells (that is, different sound waves "strike" different "piano keys"). This information is propagated to the CNS, which interprets the pattern of hair cell stimulation as a sound of a particular frequency. Modern techniques have determined that the basilar membrane is so fine-tuned that the peak membrane response to a single pitch probably extends no more than the width of a few hair cells.

Overtones of varying frequencies cause many points along the basilar membrane to vibrate simultaneously but less intensely than the fundamental tone, enabling the CNS to distinguish the timbre of the sound (**timbre discrimination**).

Loudness discrimination depends on the amplitude of vibration.

Intensity (loudness) discrimination depends on the amplitude of vibration. As sound waves originating from louder sound sources strike the eardrum, they cause it to vibrate more vigorously (that is, bulge in and out to a greater extent) but at the same frequency as a softer sound of the same pitch. The greater tympanic membrane deflection is converted into a greater amplitude of basilar membrane movement in the region of peak responsiveness, causing greater bending of the stereocilia in this region. The CNS interprets this greater basilar membrane oscillation and hair bending as a louder sound. Thus pitch discrimination depends on *where* the basilar membrane maximally vibrates and loudness discrimination depends on *how much* this place vibrates.

The auditory system is so sensitive and can detect sounds so faint that the distance of basilar membrane deflection is comparable to only a fraction of the diameter of a hydrogen atom, the smallest of atoms. No wonder very loud sounds, which cannot be sufficiently attenuated by protective middle ear reflexes (for example, the sounds of a typical rock concert), can set up such violent vibrations of the basilar membrane that irreplaceable hair cells are actually sheared off or permanently distorted, leading to partial hearing loss (● Figure 6-40).

The auditory cortex is mapped according to tone.

Just as various regions of the basilar membrane are associated with particular tones, the **primary auditory cortex** in the temporal lobe is also *tonotopically* organized. Each region of the basilar membrane is linked to a specific region of the primary auditory cortex. Accordingly, specific cortical neurons are acti-

(a) Normal hair cells

R.S. Preston and J. E. Hawkins, University of Michigan, Kresge Hearing Institute

(b) Damaged hair cells

● **FIGURE 6-40 Loss of hair cells caused by loud noises.** The scanning electron micrographs show portions of the organ of Corti, with its three rows of outer hair cells and one row of inner hair cells, from the inner ear of (a) a normal guinea pig and (b) a guinea pig after a 24-hour exposure to noise at 120 decibels SPL (sound pressure level), a level approached by loud rock music.

vated only by particular tones; that is, each region of the auditory cortex becomes excited only in response to a specific tone detected by a selected portion of the basilar membrane.

The afferent neurons that pick up the auditory signals from the inner hair cells exit the cochlea via the auditory nerve. The neural pathway between the organ of Corti and the auditory cortex involves several synapses en route, the most notable of which are in the brain stem and *medial geniculate nucleus* of the thalamus. The brain stem uses the auditory input for alertness and arousal. The thalamus sorts and relays the signals upward. Unlike signals in the visual pathways, auditory signals from each ear are transmitted to both temporal lobes because the fibers partially cross over in the brain stem. For this reason, a disruption of the auditory pathways on one side beyond the brain stem does not affect hearing in either ear to any extent.

The primary auditory cortex appears to perceive discrete sounds, whereas the surrounding higher-order auditory cortex integrates the separate sounds into a coherent, meaningful pattern. Think about the complexity of the task accomplished by your auditory system. When you are at a concert, your organ of Corti responds to the simultaneous mixture of the instruments, the applause and hushed talking of the audience, and the background noises in the theater. You can distinguish these separate parts of the many sound waves reaching your ears and can pay attention to those of importance to you.

Deafness is caused by defects in either conduction or neural processing of sound waves.

 Clinical Note Loss of hearing, or **deafness,** may be temporary or permanent, partial or complete. Hearing loss, which affects about 10% of all Americans, is the second most com-

mon physical disability in the United States. Deafness is classified into two types—*conductive deafness* and *sensorineural deafness*—depending on the part of the hearing mechanism that fails to function adequately. **Conductive deafness** occurs when sound waves are not adequately conducted through the external and middle portions of the ear to set the fluids in the inner ear in motion. Possible causes include physical blockage of the ear canal with earwax, rupture of the eardrum, middle ear infections with accompanying fluid accumulation, or restriction of ossicular movement because of bony adhesions between the stapes and oval window. In **sensorineural deafness,** sound waves are transmitted to the inner ear, but they are not translated into nerve signals that are interpreted by the brain as sound sensations. The defect can lie in the organ of Corti, in the auditory nerves or, rarely, in the ascending auditory pathways or auditory cortex.

One of the most common causes of partial hearing loss, **neural presbycusis,** is a degenerative, age-related process that occurs as hair cells "wear out" with use. Over time, exposure to even ordinary modern-day sounds eventually damages hair cells so that, on average, adults have lost more than 40% of their cochlear hair cells by age 65. Unfortunately, partial hearing loss caused by excessive exposure to loud noises is affecting people at younger ages than in the past because we live in an increasingly noisy environment. Currently more than 28 million Americans have some degree of hearing loss, and this number is expected to climb to 78 million by 2030. According to one study, an estimated 5 million children between 6 and 19 years of age in the United States already have some hearing damage resulting from amplified music and other noise pollution. Hair cells that process high-frequency sounds are the most vulnerable to destruction.

Hearing aids are helpful in conductive deafness but are less beneficial for sensorineural deafness. These devices increase the intensity of airborne sounds and may modify the sound spectrum and tailor it to the person's particular pattern of hearing loss at higher or lower frequencies. For the sound to be perceived, however, the receptor cell–neural pathway system must still be intact.

In recent years, **cochlear implants** have become available. These electronic devices, which are surgically implanted, transduce sound signals into electrical signals that can directly stimulate the auditory nerve, thus bypassing a defective cochlear system. Cochlear implants cannot restore normal hearing, but they do permit recipients to recognize sounds. Success ranges from an ability to "hear" a phone ringing to being able to carry on a conversation over the telephone.

Exciting new findings suggest that in the future it may be possible to restore hearing by stimulating an injured inner ear to repair itself. Scientists have long considered the hair cells of the inner ear irreplaceable. Thus, hearing loss resulting from hair cell damage due to the aging process or exposure to loud noises is considered permanent. Encouraging new studies suggest, to the contrary, that hair cells in the inner ear have the latent ability to regenerate in response to an appropriate chemical signal. Researchers are currently trying to develop a drug that will spur regrowth of hair cells, thus repairing inner ear damage and hopefully restoring hearing. Other investigators are using neural growth factors to coax auditory nerve cell endings to resprout in the hopes of reestablishing lost neural pathways.

The vestibular apparatus is important for equilibrium by detecting position and motion of the head.

In addition to its cochlear-dependent role in hearing, the inner ear has another specialized component, the **vestibular apparatus,** which provides information essential for the sense of equilibrium and for coordinating head movements with eye and postural movements (● Figure 6-41). The vestibular apparatus consists of two sets of structures lying within a tunneled-out region of the temporal bone near the cochlea—the *semicircular canals* and the *otolith organs.*

The vestibular apparatus detects changes in position and motion of the head. As in the cochlea, all components of the vestibular apparatus contain endolymph and are surrounded by perilymph. Also, similar to the organ of Corti, the vestibular components each contain hair cells that respond to mechanical deformation triggered by specific movements of the endolymph. And like the auditory hair cells, the vestibular receptors may be either depolarized or hyperpolarized, depending on the direction of the fluid movement. Unlike information from the auditory system, much of the information provided by the vestibular apparatus does not reach the level of conscious awareness.

ROLE OF THE SEMICIRCULAR CANALS The **semicircular canals** detect rotational or angular acceleration or deceleration of the head, such as when starting or stopping spinning, somersaulting, or turning the head. Each ear contains three semicircular canals arranged three-dimensionally in planes that lie at right angles to each other. The receptor hair cells of each semicircular canal are situated on top of a ridge located in the **ampulla,** a swelling at the base of the canal (● Figure 6-41a and b). The hairs are embedded in an overlying, caplike, gelatinous layer, the **cupula,** which protrudes into the endolymph within the ampulla. The cupula sways in the direction of fluid movement, much like seaweed leaning in the direction of the prevailing tide.

Acceleration or deceleration during rotation of the head in any direction causes endolymph movement in at least one of the semicircular canals, because of their three-dimensional arrangement. As you start to move your head, the bony canal and the ridge of hair cells embedded in the cupula move with your head. Initially, however, the fluid within the canal, not being attached to your skull, does not move in the direction of the rotation but lags behind because of its inertia. (Because of inertia, a resting object remains at rest, and a moving object contin-

ues to move in the same direction unless the object is acted on by some external force that induces change.) When the endolymph is left behind as you start to rotate your head, the fluid in the same plane as the head movement is in effect shifted in the opposite direction from the movement (similar to your body tilting to the right as the car in which you are riding suddenly turns to the left) (● Figure 6-41c). This fluid movement causes the cupula to lean in the opposite direction from the head movement, bending the sensory hairs embedded in it. If your head movement continues at the same rate in the same direction, the endolymph catches up and moves in unison with your head so that the hairs return to their unbent position. When your head slows down and stops, the reverse situation occurs. The endolymph briefly continues to move in the direction of the rotation while your head decelerates to a stop. As a result, the cupula and its hairs are transiently bent in the direction of the preceding spin, which is opposite to the way they were bent during acceleration.

The hairs of a **vestibular hair cell** consist of one cilium, the **kinocilium,** along with a tuft of 20 to 50 microvilli—the **stereocilia**—arranged in rows of decreasing height from the taller kinocilium (● Figure 6-41d) (see p. 45). As in the auditory hair cell, the stereocilia are linked by tip links. When the stereocilia are deflected by endolymph movement, the resultant tension on the tip links pulls on mechanically gated ion channels in the hair cell. Depending on whether the ion channels are mechanically opened or closed by hair bundle displacement, the hair cell either depolarizes or hyperpolarizes. Each hair cell is oriented so that it depolarizes when its stereocilia are bent toward the kinocilium; bending in the opposite direction hyperpolarizes the cell. The hair cells form a chemically mediated synapse with terminal endings of afferent neurons whose axons join with those of the other vestibular structures to form the **vestibular nerve.** This nerve unites with the auditory nerve from the cochlea to form the **vestibulocochlear nerve.** Depolarization increases the release of neurotransmitter from the hair cells, thereby bringing about an increased rate of firing in the afferent fibers; conversely, hyperpolarization reduces neurotransmitter release from the hair cells, in turn decreasing the frequency of action potentials in the afferent fibers. When the fluid gradually comes to a halt, the hairs straighten again. Thus, the semicircular canals detect changes in the rate of rotational movement (rotational acceleration or deceleration) of your head. They do not respond when your head is motionless or when it is moving in a circle at a constant speed.

ROLE OF THE OTOLITH ORGANS The **otolith organs** provide information about the position of the head relative to gravity (that is, static head tilt) and also detect changes in the rate of linear motion (moving in a straight line regardless of direction). The otolith organs, the **utricle** and the **saccule,** are saclike structures housed within a bony chamber situated between the semicircular canals and the cochlea (● Figure 6-41a). The hairs (kinocilium and stereocilia) of the receptor hair cells in these sense organs also protrude into an overlying gelatinous sheet, whose movement displaces the hairs and results in changes in hair cell potential. Many tiny crystals of calcium carbonate—the **otoliths** ("ear stones")—are suspended within the gelatinous

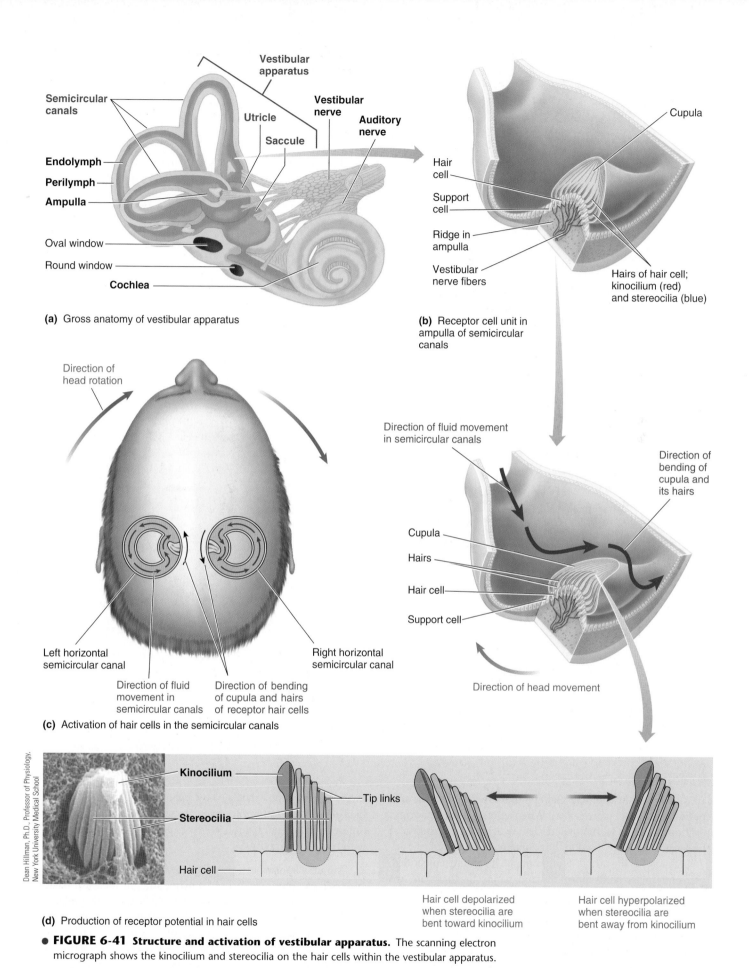

(a) Gross anatomy of vestibular apparatus

(b) Receptor cell unit in ampulla of semicircular canals

(c) Activation of hair cells in the semicircular canals

(d) Production of receptor potential in hair cells

Hair cell depolarized when stereocilia are bent toward kinocilium

Hair cell hyperpolarized when stereocilia are bent away from kinocilium

● **FIGURE 6-41 Structure and activation of vestibular apparatus.** The scanning electron micrograph shows the kinocilium and stereocilia on the hair cells within the vestibular apparatus.

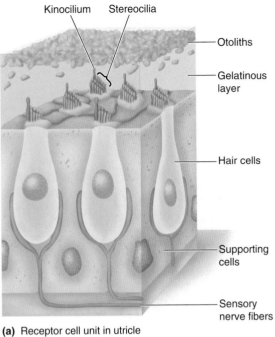

Kinocilium   Stereocilia

Otoliths

Gelatinous layer

Hair cells

Supporting cells

Sensory nerve fibers

(a) Receptor cell unit in utricle

Gravitational force

(b) Activation of utricle by change in head position

(c) Activation of utricle by horizontal linear acceleration

layer, making it heavier and giving it more inertia than the surrounding fluid (● Figure 6-42a). When a person is in an upright position, the hairs within the utricles are oriented vertically and the saccule hairs are lined up horizontally.

Let's look at the utricle as an example. Its otolith-embedded, gelatinous mass shifts positions and bends the hairs in two ways:

1. When you tilt your head in any direction other than vertical (that is, other than straight up and down), the hairs are bent in the direction of the tilt because of the gravitational force exerted on the top-heavy gelatinous layer (● Figure 6-42b). This bend-

ing produces depolarizing or hyperpolarizing receptor potentials depending on the tilt of your head. The CNS thus receives different patterns of neural activity depending on head position with respect to gravity.

2. The utricle hairs are also displaced by any change in horizontal linear motion (such as moving straight forward, backward, or to the side). As you start to walk forward (● Figure 6-42c), the top-heavy otolith membrane at first lags behind the endolymph and hair cells because of its greater inertia. The hairs are thus bent to the rear, in the opposite direction of the forward movement of your head. If you maintain your

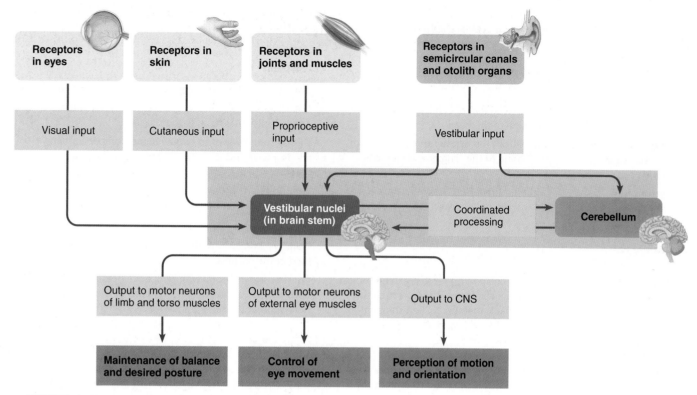

● **FIGURE 6-43 Input and output of the vestibular nuclei.**

walking pace, the gelatinous layer soon catches up and moves at the same rate as your head so that the hairs are no longer bent. When you stop walking, the otolith sheet continues to move forward briefly as your head slows and stops, bending the hairs toward the front. Thus, the hair cells of the utricle detect horizontally directed linear acceleration and deceleration, but they do not provide information about movement in a straight line at constant speed.

The saccule functions similarly to the utricle, except that it responds selectively to tilting of the head away from a horizontal position (such as getting up from bed) and to vertically directed linear acceleration and deceleration (such as jumping up and down or riding in an elevator).

Signals arising from the various components of the vestibular apparatus are carried through the vestibulocochlear nerve to the **vestibular nuclei,** a cluster of neuronal cell bodies in the brain stem, and to the cerebellum. Here the vestibular information is integrated with input from the eyes, skin surface, joints, and muscles for (1) maintaining balance and desired posture; (2) controlling the external eye muscles so that the eyes remain fixed on the same point, despite movement of the head; and (3) perceiving motion and orientation (● Figure 6-43).

Some people, for poorly understood reasons, are especially sensitive to particular motions that activate the vestibular apparatus and cause symptoms of dizziness and nausea; this sensitivity is called **motion sickness.** Occasionally, fluid imbalances within the inner ear lead to **Ménière's disease.** Not surprisingly, because both the vestibular apparatus

and cochlea contain the same inner ear fluids, both vestibular and auditory symptoms occur with this condition. An afflicted individual suffers transient attacks of severe vertigo (dizziness) accompanied by pronounced ringing in the ears and some loss of hearing. During these episodes, the person cannot stand upright and reports feeling as though self or surrounding objects in the room are spinning around.

▲ Table 6-6 summarizes the functions of the major components of the ear.

Chemical Senses: Taste and Smell

Unlike the eyes' photoreceptors and the ears' mechanoreceptors, the receptors for taste and smell are chemoreceptors, which generate neural signals on binding with particular chemicals in their environment. The sensations of taste and smell in association with food intake influence the flow of digestive juices and affect appetite. Furthermore, stimulation of taste or smell receptors induces pleasurable or objectionable sensations and signals the presence of something to seek (a nutritionally useful, good-tasting food) or to avoid (a potentially toxic, bad-tasting substance). Thus, the chemical senses provide a "quality-control" checkpoint for substances available for ingestion. In lower animals, smell also plays a major role in finding direction, in seeking prey or avoiding predators, and in sexual attraction to a mate. The sense of smell is less sensitive in humans and much less important in influencing our behavior (although millions of dollars are spent annually on per-

Functions of the Major Components of the Ear

Structure	Location	Function
External Ear		Collects and transfers sound waves to the middle ear
Pinna (ear)	Skin-covered flap of cartilage located on each side of the head	Collects sound waves and channels them down the ear canal; contributes to sound localization
External auditory meatus (ear canal)	Tunnel from the exterior through the temporal bone to the tympanic membrane	Directs sound waves to tympanic membrane; contains filtering hairs and secretes earwax to trap foreign particles
Tympanic membrane (eardrum)	Thin membrane that separates the external ear and the middle ear	Vibrates in synchrony with sound waves that strike it, setting middle ear bones in motion
Middle Ear		Transfers vibrations of the tympanic membrane to the fluid in the cochlea
Malleus, incus, stapes	Movable chain of bones that extends across the middle ear cavity; malleus attaches to the tympanic membrane, and stapes attaches to the oval window	Oscillate in synchrony with tympanic membrane vibrations and set up wavelike movements in the cochlear perilymph at the same frequency
Inner Ear: Cochlea		Houses sensory system for hearing
Oval window	Thin membrane at the entrance to the cochlea; separates the middle ear from the scala vestibuli	Vibrates in unison with movement of stapes, to which it is attached; oval window movement sets cochlear perilymph in motion
Scala vestibuli	Upper compartment of the cochlea, a snail-shaped tubular system that lies deep within the temporal bone	Contains perilymph that is set in motion by oval window movement driven by oscillation of middle ear bones
Scala tympani	Lower compartment of the cochlea	Contains perilymph that is continuous with the scala vestibuli
Cochlear duct (scala media)	Middle compartment of the cochlea; a blind-ended tubular compartment that tunnels through the center of the cochlea	Contains endolymph; houses the basilar membrane
Basilar membrane	Forms the floor of the cochlear duct	Vibrates in unison with perilymph movements; bears the organ of Corti, the sense organ for hearing
Organ of Corti	Rests on top of the basilar membrane throughout its length	Contains hair cells, the receptors for sound; inner hair cells undergo receptor potentials when their hairs are bent as a result of fluid movement in cochlea
Tectorial membrane	Stationary membrane that overhangs the organ of Corti and contacts the surface hairs of the receptor hair cells	Serves as the stationary site against which the hairs of the receptor cells are bent and undergo receptor potentials as the vibrating basilar membrane moves in relation to this overhanging membrane
Round window	Thin membrane that separates the scala tympani from the middle ear	Vibrates in unison with fluid movements in perilymph to dissipate pressure in cochlea; does not contribute to sound reception
Inner Ear: Vestibular Apparatus		Houses sensory systems for equilibrium and provides input essential for maintenance of posture and balance
Semicircular canals	Three semicircular canals arranged three-dimensionally in planes at right angles to each other near the cochlea	Detect rotational or angular acceleration or deceleration
Utricle	Saclike structure in a bony chamber between the cochlea and semicircular canals	Detects (1) changes in head position away from vertical and (2) horizontally directed linear acceleration and deceleration
Saccule	Lies next to the utricle	Detects (1) changes in head position away from horizontal and (2) vertically directed linear acceleration and deceleration

fumes and deodorants to make us smell better and thereby be more socially attractive). We first examine the mechanism of taste (**gustation**) and then turn our attention to smell (**olfaction**).

Taste receptor cells are located primarily within tongue taste buds.

The chemoreceptors for taste sensation are packaged in taste buds, about 10,000 of which are present in the oral cavity and throat, with the greatest percentage on the upper surface of the tongue (● Figure 6-44). A taste bud consists of about 50 long, spindle-shaped *taste receptor cells* packaged with *supporting cells* in an arrangement like slices of an orange. Each taste bud has a small opening, the **taste pore**, through which fluids in the mouth come into contact with the surface of its receptor cells. **Taste receptor cells** are modified epithelial cells with many surface folds, or microvilli, that protrude slightly through the taste pore, greatly increasing the surface area exposed to the oral contents (see p. 48). The plasma membrane of the microvilli contains receptor sites that bind selectively with chemical molecules in the environment. Only chemicals in solution—either ingested liquids or solids that have been dissolved in saliva—can attach to receptor cells and evoke the sensation of taste. Binding of a taste-provoking chemical, a **tastant,** with a receptor cell alters the cell's ionic channels to produce a depolarizing receptor potential. Like the other special sense receptors, a depolarizing receptor potential opens voltage-gated Ca^{2+} channels, leading to the entry of Ca^{2+}, which promotes release of neurotransmitter. This neurotransmitter, in turn, initiates action potentials within terminal endings of afferent nerve fibers with which the receptor cell synapses.

Most receptors are carefully sheltered from direct exposure to the environment, but the taste receptor cells, by virtue of their task, frequently come into contact with potent chemicals. Unlike the eye or ear receptors, which are irreplaceable, taste receptors have a life span of about 10 days. Epithelial cells surrounding the taste bud differentiate first into supporting cells and then into receptor cells to constantly renew the taste bud components.

Terminal afferent endings of several cranial nerves synapse with taste buds in various regions of the mouth. Signals in these sensory inputs are conveyed via synaptic stops in the brain stem and thalamus to the **cortical gustatory area,** a region in the parietal lobe adjacent to the "tongue" area of the somatosensory cortex. Unlike most sensory input, the gustatory pathways are primarily uncrossed. The brain stem also projects fibers to the hypothalamus and limbic system to add affective dimensions, such as whether the taste is pleasant or unpleasant, and to process behavioral aspects associated with taste and smell.

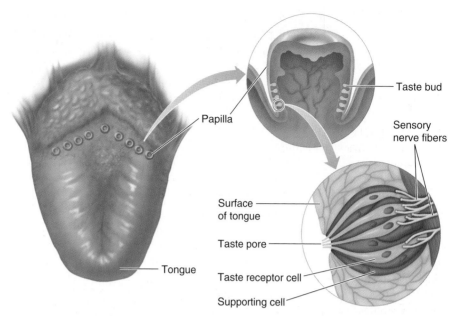

● **FIGURE 6-44 Location and structure of the taste buds.** Taste buds are located primarily along the edges of moundlike papillae on the upper surface of the tongue. The receptor cells and supporting cells of a taste bud are arranged like slices of an orange.

Taste discrimination is coded by patterns of activity in various taste bud receptors.

We can discriminate among thousands of different taste sensations, yet all tastes are varying combinations of five **primary tastes:** *salty, sour, sweet, bitter,* and *umami.* Umami, a meaty or savory taste, has recently been added to the list of primary tastes. Another new taste sensation has also been proposed—*fatty.* Scientists have identified a potential sensor in the mouth for long-chain fatty acids.

Each receptor cell responds in varying degrees to all five (maybe six) primary tastes but generally responds preferentially to one of the taste modalities. Fine taste discrimination beyond the primary tastes depends on subtle differences in the stimulation patterns of all the taste buds in response to various substances, similar to the variable stimulation of the three cone types that gives rise to the range of color sensations. Receptor cells use different pathways to bring about a depolarizing receptor potential in response to each tastant category:

■ **Salty taste** is stimulated by chemical salts, especially NaCl (table salt). Direct entry of positively charged Na^+ ions through specialized Na^+ channels in the receptor cell membrane, a movement that reduces the cell's internal negativity, is responsible for receptor depolarization in response to salt.

■ **Sour taste** is caused by acids, which contain a free hydrogen ion, H^+. The citric acid content of lemons, for example, accounts for their distinctly sour taste. Depolarization of the receptor cell by sour tastants occurs because H^+ blocks K^+ channels in the receptor cell membrane. The resultant decrease in the passive movement of positively charged K^+ ions out of the cell reduces the internal negativity, producing a depolarizing receptor potential.

■ **Sweet taste** is evoked by the particular configuration of glucose. From an evolutionary perspective, we crave sweet foods because they supply necessary calories in a readily usable form. However, other organic molecules with similar structures but no calories, such as saccharin, aspartame, sucralose, and other artificial sweeteners, can also interact with "sweet" receptor binding sites. Binding of glucose or another chemical with the taste cell receptor activates a G protein, which turns on the cAMP second-messenger pathway in the taste cell (see p. 121). The second-messenger pathway ultimately results in phosphorylation and blockage of K^+ channels in the receptor cell membrane, leading to a depolarizing receptor potential.

■ **Bitter taste** is elicited by a more chemically diverse group of tastants than the other taste sensations. For example, alkaloids (such as caffeine, nicotine, strychnine, morphine, and other toxic plant derivatives), as well as poisonous substances, all taste bitter, presumably as a protective mechanism to discourage ingestion of these potentially dangerous compounds. Taste cells that detect bitter flavors possess 50 to 100 bitter receptors, each of which responds to a different bitter flavor. Because each receptor cell has this diverse family of bitter receptors, a wide variety of unrelated chemicals all taste bitter despite their diverse structures. This mechanism expands the ability of the taste receptor to detect a wide range of potentially harmful chemicals. The first G protein in taste—**gustducin**—was identified in one of the bitter signaling pathways. This G protein, which sets off a second-messenger pathway in the taste cell, is very similar to the visual G protein, transducin.

■ **Umami taste,** which was first identified and named by a Japanese researcher, is triggered by amino acids, especially glutamate. The presence of amino acids, as found in meat for example, serves as a marker for a desirable, nutritionally protein-rich food. Glutamate binds to a G protein–coupled receptor and activates a second-messenger pathway. In addition to giving us our sense of meaty flavors, this pathway is responsible for the distinctive taste of the flavor additive monosodium glutamate (MSG), which is especially popular in Asian dishes.

In addition to input from the taste receptors, taste perception is also influenced by information derived from other receptors, especially odor. When you temporarily lose your sense of smell because of swollen nasal passageways during a cold, your sense of taste is also markedly reduced, even though your taste receptors are unaffected by the cold. Other factors affecting taste include temperature and texture of the food as well as psychological factors associated with past experiences with the food. How the gustatory cortex accomplishes the complex perceptual processing of taste sensation is currently not known.

The olfactory receptors in the nose are specialized endings of renewable afferent neurons.

The **olfactory** (smell) **mucosa,** a 3 cm^2 patch of mucosa in the ceiling of the nasal cavity, contains three cell types: *olfactory receptor cells, supporting cells,* and *basal cells* (● Figure 6-45). The supporting cells secrete mucus, which coats the nasal passages. The basal cells are precursors for new olfactory receptor

cells, which are replaced about every two months. An **olfactory receptor cell** is an afferent neuron whose receptor portion lies in the olfactory mucosa in the nose and whose afferent axon traverses into the brain. The axons of the olfactory receptor cells collectively form the **olfactory nerve.**

The receptor portion of an olfactory receptor cell consists of an enlarged knob bearing several long cilia that extend like a tassel to the surface of the mucosa. These cilia contain the binding sites for attachment of **odorants,** molecules that can be smelled. During quiet breathing, odorants typically reach the sensitive receptors only by diffusion because the olfactory mucosa is above the normal path of airflow. The act of sniffing enhances this process by drawing the air currents upward within the nasal cavity so that a greater percentage of the odoriferous molecules in the air come into contact with the olfactory mucosa. Odorants also reach the olfactory mucosa during eating by wafting up to the nose from the mouth through the pharynx (back of the throat).

To be smelled, a substance must be (1) sufficiently volatile (easily vaporized) that some of its molecules can enter the nose in the inspired air and (2) sufficiently water soluble that it can dissolve in the mucus coating the olfactory mucosa. As with taste receptors, molecules must be dissolved to be detected by olfactory receptors.

Various parts of an odor are detected by different olfactory receptors and sorted into "smell files."

The human nose contains 5 million olfactory receptors, of which there are 1000 different types. During smell detection, an odor is "dissected" into various components. Each receptor responds to only one discrete component of an odor rather than to the whole odorant molecule. Accordingly, each of the various parts of an odor is detected by one of the thousand different receptors, and a given receptor can respond to a particular odor component shared in common by different scents. Compare this to the three cone types for coding color vision and the taste buds that respond differentially to only five (maybe six) primary tastes to accomplish coding for taste discrimination.

Binding of an appropriate scent signal to an olfactory receptor activates a G protein, triggering a cascade of cAMP-dependent intracellular reactions that leads to opening of cAMP-gated nonspecific cation channels. The resultant net Na^+ entry causes a depolarizing receptor potential that generates action potentials in the afferent fiber. The frequency of the action potentials depends on the concentration of the stimulating chemical molecules.

The afferent fibers arising from the receptor endings in the nose pass through tiny holes in the flat bone plate separating the olfactory mucosa from the overlying brain tissue (● Figure 6-45). They immediately synapse in the **olfactory bulb,** a complex neural structure containing several different layers of cells that are functionally similar to the retinal layers of the eye. The twin olfactory bulbs, one on each side, are about the size of small grapes (see ● Figure 5-16, p. 156). Each olfactory bulb is lined by small ball-like neural junctions known as **glomeruli** ("little balls") (● Figure 6-46). Within each glomerulus, the terminals of

● **FIGURE 6-45 Location and structure of the olfactory receptor cells.**

receptor cells carrying information about a particular scent component synapse with the next cells in the olfactory pathway, the **mitral cells.** Because each glomerulus receives signals only from receptors that detect a particular odor component, the glomeruli serve as "smell files." The separate components of an odor are sorted into different glomeruli, one component per file. Thus, the glomeruli, which serve as the first relay station in the brain for processing olfactory information, play a key role in organizing scent perception.

The mitral cells on which the olfactory receptors terminate in the glomeruli refine the smell signals and relay them to the brain for further processing. Fibers leaving the olfactory bulb travel in two different routes:

1. A subcortical route going primarily to regions of the limbic system, especially the lower medial sides of the temporal lobes (considered the **primary olfactory cortex**). This route, which includes hypothalamic involvement, permits close coordination between smell and behavioral reactions associated with feeding, mating, and direction orienting.

2. A route through the thalamus to the cortex. As with other senses, the cortical route is important for conscious perception and fine discrimination of smell.

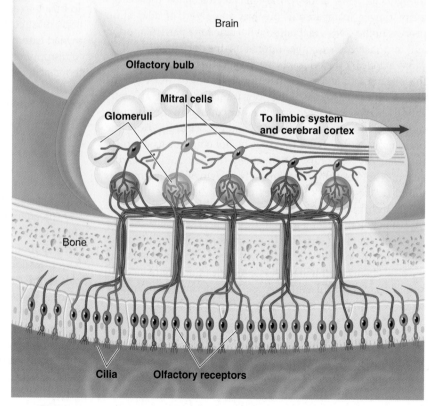

● **FIGURE 6-46 Processing of scents in the olfactory bulb.** Each of the glomeruli lining the olfactory bulb receives synaptic input from only one type of olfactory receptor, which, in turn, responds to only one discrete component of an odorant. Thus, the glomeruli sort and file the various components of an odoriferous molecule before relaying the smell signal to the mitral cells and higher brain levels for further processing.

Odor discrimination is coded by patterns of activity in the olfactory bulb glomeruli.

Because each given odorant activates multiple receptors and glomeruli in response to its various odor components, odor discrimination is based on different patterns of glomeruli activated by various scents. In this way, the cortex can distinguish more than 10,000 different scents. This mechanism for sorting out and distinguishing different odors is very effective. A noteworthy example is our ability to detect methyl mercaptan (garlic odor) at a concentration of 1 molecule per 50 billion molecules in the air! This substance is added to odorless natural gas to enable us to detect potentially lethal gas leaks. Despite this impressive sensitivity, humans have a poor sense of smell compared to other species. By comparison, dogs' sense of smell is hundreds of times more sensitive than that of humans. Bloodhounds, for example, have about 4 billion olfactory receptor cells compared to our 5 million such cells, accounting for bloodhounds' superior scent-sniffing ability.

The olfactory system adapts quickly, and odorants are rapidly cleared.

Although the olfactory system is sensitive and highly discriminating, it is also quickly adaptive. Sensitivity to a new odor diminishes rapidly after a short period of exposure to it, even though the odor source continues to be present. This reduced sensitivity does not involve receptor adaptation, as researchers thought for years; actually, the olfactory receptors themselves adapt slowly. It apparently involves some sort of adaptation process in the CNS. Adaptation is specific for a particular odor, and responsiveness to other odors remains unchanged.

What clears the odorants away from their binding sites on the olfactory receptors so that the sensation of smell doesn't "linger" after the source of the odor is removed? Several "odor-eating" enzymes have recently been discovered in the olfactory mucosa that serve as molecular janitors, clearing away the odoriferous molecules so that they do not continue to stimulate the olfactory receptors. Interestingly, these odorant-clearing enzymes are very similar chemically to detoxification enzymes found in the liver. (These liver enzymes inactivate potential toxins absorbed from the digestive tract; see p. 27.) This resemblance may not be coincidental. Researchers speculate that the nasal enzymes may serve the dual purpose of clearing the olfactory mucosa of old odorants and transforming potentially harmful chemicals into harmless molecules. Such detoxification would serve a very useful purpose, considering the open passageway between the olfactory mucosa and the brain.

The vomeronasal organ detects pheromones.

In addition to the olfactory mucosa, the nose contains another sense organ, the **vomeronasal organ (VNO),** which is common in mammals but until recently was thought nonexistent in humans. The VNO is located about half an inch inside the human nose next to the vomer bone, hence its name. It detects **pheromones,** nonvolatile chemical signals passed subconsciously from one individual to another. In animals, binding of a pheromone to its receptor on the surface of a neuron in the VNO triggers an action potential that travels through nonolfactory pathways to the limbic system, the brain region that governs emotional responses and sociosexual behaviors. These signals never reach the higher levels of conscious awareness. In animals, the VNO is known as the "sexual nose" for its role in governing reproductive and social behaviors, such as identifying and attracting a mate and communicating social status.

Some scientists now claim the existence of pheromones in humans, although many skeptics doubt these findings. Although the role of the VNO in human behavior has not been validated, some researchers suspect that it is responsible for spontaneous "feelings" between people, either "good chemistry," such as "love at first sight," or "bad chemistry," such as "getting bad vibes" from someone you just met. They speculate that pheromones in humans subtly influence sexual activity, compatibility with others, or group behavior, similar to the role they play in other mammals, although this messenger system is nowhere as powerful or important in humans as in animals. Because messages conveyed by the VNO seem to bypass cortical consciousness, the response to the largely odorless pheromones is not a distinct, discrete perception, such as smelling a favorite fragrance, but more like an inexplicable impression.

Chapter in Perspective: Focus on Homeostasis

To maintain a life-sustaining stable internal environment, the body must constantly make adjustments to compensate for myriad external and internal factors that continuously threaten to disrupt homeostasis, such as external exposure to cold or internal acid production. Many of these adjustments are directed by the nervous system, one of the body's two major regulatory systems. The central nervous system (CNS), the integrating and decision-making component of the nervous system, must continuously be informed of "what's happening" in both the internal and the external environment so that it can command appropriate responses in the organ systems to maintain the body's viability. In other words, the CNS must know what changes are taking place before it can respond to these changes.

The afferent division of the peripheral nervous system (PNS) is the communication link by which the CNS is informed about the internal and external environment. The afferent division detects, encodes, and transmits peripheral signals to the CNS for processing. Afferent input is necessary for arousal, perception, and determination of efferent output.

Afferent information about the internal environment, such as the CO_2 level in the blood, never reaches the level of conscious awareness, but this input to the controlling centers of the CNS is essential for maintaining homeostasis. Afferent input that reaches the level of conscious awareness, called sensory information, includes somesthetic and proprioceptive sensation (body sense) and special senses (vision, hearing, equilibrium, taste, and smell).

The body sense receptors are distributed over the entire body surface as well as throughout the joints and muscles. Afferent signals from these receptors provide information about what's happening directly to each specific body part in relation to the external environment (that is, the "what," "where," and "how much" of stimulatory inputs to the body's surface and the momentary position of the body in space). In contrast, each special sense organ is restricted to a single site in the body. Rather than provide information about a specific body part, a special sense organ provides a specific type of information about the external environment that is useful to the body as a whole. For example, through their ability to detect, extensively analyze, and integrate patterns of illumination in the external environment, the eyes and visual processing system enable you to "see" your surroundings. The same integrative effect could not be achieved if photoreceptors were scattered over your entire body surface, as are touch receptors.

Sensory input (both body sense and special senses) enables a complex multicellular organism such as a human to interact in meaningful ways with the external environment in procuring food, defending against danger, and engaging in other behavioral actions geared toward maintaining homeostasis. In addition to providing information essential for interactions with the external environment for basic survival, the perceptual processing of sensory input adds immeasurably to the richness of life, such as enjoyment of a good book, concert, or meal.

REVIEW EXERCISES

Objective Questions (Answers on p. A-43)

1. Conversion of the energy forms of stimuli into electrical energy by the receptors is known as _____.

2. The type of stimulus to which a particular receptor is most responsive is called its _____.

3. All afferent information is sensory information. (True or false?)

4. Off-center ganglion cells increase their rate of firing when a beam of light strikes the periphery of their receptive field. (True or false?)

5. During dark adaptation, rhodopsin is gradually regenerated to increase the sensitivity of the eyes. (True or false?)

6. An optic nerve carries information from the lateral and medial halves of the same eye, whereas an optic tract carries information from the lateral half of one eye and the medial half of the other. (True or false?)

7. Displacement of the round window generates neural impulses perceived as sound sensations. (True or false?)

8. Stereocilia of the inner hair cells hyperpolarize when they bend toward and depolarize when they bend away from their tallest member. (True or false?)

9. Hair cells in different regions of the organ of Corti and neurons in different regions of the auditory cortex are activated by different tones. (True or false?)

10. Each taste receptor responds to just one of the five primary tastes. (True or false?)

11. Rapid adaptation to odors results from adaptation of the olfactory receptors. (True or false?)

12. Match the following (see the next column, too):
    ___ 1. layer that contains photoreceptors
    ___ 2. point from which optic nerve leaves retina
    ___ 3. forms white part of eye
    ___ 4. thalamic structure that processes visual input
    ___ 5. colored diaphragm of muscle that controls amount of light entering eye

    (a) choroid
    (b) aqueous humor
    (c) fovea
    (d) lateral geniculate nucleus
    (e) cornea
    (f) retina
    (g) lens
    (h) optic disc; blind spot
    (i) iris

    ___ 6. contributes most to eye's refractive ability
    ___ 7. supplies nutrients to lens and cornea
    ___ 8. produces aqueous humor
    ___ 9. contains vascular supply for retina and a pigment that minimizes scattering of light within eye
    ___ 10. has adjustable refractive ability
    ___ 11. portion of retina with greatest acuity
    ___ 12. point at which fibers from medial half of each retina cross to opposite side

    (j) ciliary body
    (k) optic chiasm
    (l) sclera

13. Using the answer code on the right, indicate which properties apply to taste and/or smell:
    ___ 1. Receptors are separate cells that synapse with terminal endings of afferent neurons.
    ___ 2. Receptors are specialized endings of afferent neurons.
    ___ 3. Receptors are regularly replaced.
    ___ 4. Specific chemicals in the environment attach to special binding sites on the receptor surface, leading to a depolarizing receptor potential.
    ___ 5. There are two processing pathways: a limbic system route and a thalamic-cortical route.
    ___ 6. The discriminative ability is based on patterns of receptor stimulation by five (maybe six) different modalities.
    ___ 7. A thousand different receptor types are used.
    ___ 8. Information from receptor cells is filed and sorted by neural junctions called glomeruli.

    (a) applies to taste
    (b) applies to smell
    (c) applies to both taste and smell

Essay Questions

1. List and describe the receptor types according to their adequate stimulus.
2. Compare tonic and phasic receptors.
3. Explain how acuity is influenced by receptive field size and by lateral inhibition.
4. Compare the fast and slow pain pathways.
5. Describe the built-in analgesic system of the brain.
6. Describe the steps involved in phototransduction by photoreceptors and further retinal processing by bipolar and ganglion cells.
7. Compare the functional characteristics of rods and cones.
8. What are sound waves? What is responsible for the pitch, intensity, and timbre of a sound?
9. Describe the function of each of the following parts of the ear: pinna, ear canal, tympanic membrane, ossicles, oval window, and the various parts of the cochlea. Include a discussion of how sound waves are transduced into action potentials.
10. Discuss the functions of the semicircular canals, the utricle, and the saccule.
11. Describe the location, structure, and activation of the receptors for taste and smell.
12. Compare the processes of color vision, hearing, taste, and smell discrimination.

Quantitative Exercises (Solutions on p. A-43)

1. Calculate the difference in the time it takes for an action potential to travel 1.3 m between the slow (12 m/sec) and fast (30 m/sec) pain pathways.
2. Have you ever noticed that humans have circular pupils, whereas cats' pupils are more elongated from top to bottom? For simplicity in calculation, assume the cat pupil is rectangular. The following calculations will help you understand the implication of this difference. For simplicity, assume a constant intensity of light.
   a. If the diameter of a human's circular pupil were decreased by half on contraction of the constrictor muscle of the iris, by what percentage would the amount of light allowed into the eye be decreased?
   b. If a cat's rectangular pupil were decreased by half along one axis only, by what percentage would the amount of light allowed into the eye be decreased?
   c. Comparing these calculations, do humans or cats have more precise control over the amount of light falling on the retina?
3. A decibel is the unit of sound level, β, defined as follows:

$$\beta = (10 \text{ dB}) \log_{10}(I/I_0),$$

where I is *sound intensity*, or the rate at which sound waves transmit energy per unit area. The units of I are watts per square meter (W/m^2). I_0 is a constant intensity close to the human hearing threshold, namely, 10^{-12} W/m^2.
   a. For the following sound levels, calculate the corresponding sound intensities:
      (1) 20 dB (a whisper)
      (2) 70 dB (a car horn)
      (3) 120 dB (a low-flying jet)
      (4) 170 dB (a space shuttle launch)
   b. Explain why the sound levels of these sounds increase by the same increment (that is, each sound is 50 dB higher than the one preceding it), yet the incremental increases in sound intensities you calculated are so different. What implications does this have for performance of the human ear?

POINTS TO PONDER

(Explanations on p. A-43)

1. Patients with certain nerve disorders are unable to feel pain. Why is this disadvantageous?
2. Ophthalmologists frequently instill eye drops in their patients' eyes to bring about pupillary dilation, which makes it easier for the physician to view the eye's interior. In what way would the drug in the eye drops affect autonomic nervous system activity in the eye to cause the pupils to dilate?
3. A patient complains of not being able to see the right half of the visual field with either eye. At what point in the patient's visual pathway does the defect lie?
4. Explain how middle ear infections interfere with hearing. Of what value are the "tubes" that are sometimes surgically placed in the eardrums of patients with a history of repeated middle ear infections accompanied by chronic fluid accumulation?
5. Explain why your sense of smell is reduced when you have a cold, even though the cold virus does not directly adversely affect the olfactory receptor cells.

(Explanation on p. A-44)

Suzanne J. complained to her physician of bouts of dizziness. The physician asked her whether by "dizziness" she meant a feeling of lightheadedness, as if she were going to faint (a condition known as **syncope**), or a feeling that she or surrounding objects in the room were spinning around (a condition known as **vertigo**). Why is this distinction important in the differential diagnosis of her condition? What are some possible causes of each of these symptoms?

**Nervous System
(Peripheral Nervous System)**

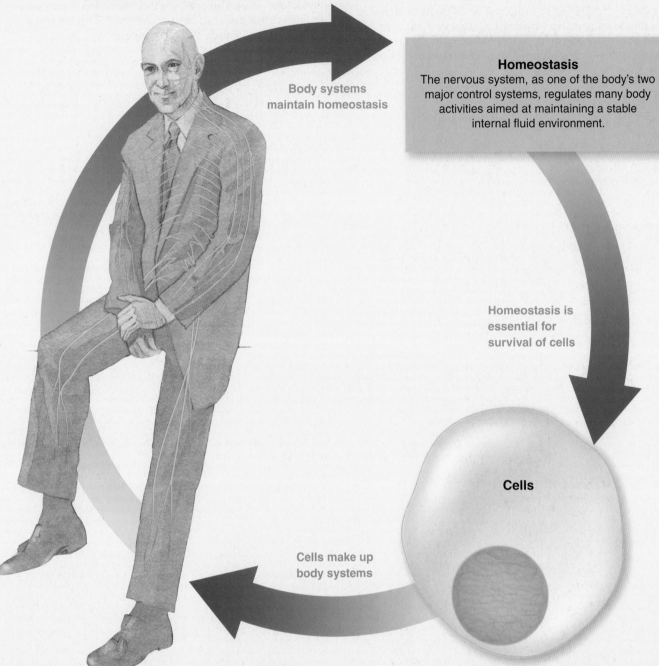

Body systems
maintain homeostasis

Homeostasis
The nervous system, as one of the body's two major control systems, regulates many body activities aimed at maintaining a stable internal fluid environment.

Homeostasis is
essential for
survival of cells

Cells

Cells make up
body systems

The nervous system, one of the two major regulatory systems of the body, consists of the central nervous system (CNS), composed of the brain and spinal cord, and the **peripheral nervous system (PNS),** composed of the afferent and efferent fibers that relay signals between the CNS and periphery (other parts of the body).

Once informed by the afferent division of the PNS that a change in the internal or external environment is threatening homeostasis, the CNS makes appropriate adjustments to maintain homeostasis. The CNS makes these adjustments by controlling the activities of effectors (muscles and glands) by transmitting signals from the CNS to these organs through the **efferent division** of the PNS.

The Peripheral Nervous System: Efferent Division

Autonomic Nervous System

The efferent division of the peripheral nervous system (PNS) is the communication link by which the central nervous system (CNS) controls the activities of muscles and glands, the effector organs that carry out the intended effects, or actions. The CNS regulates these effectors by initiating action potentials in the cell bodies of efferent neurons whose axons terminate on these organs. Cardiac muscle, smooth muscle, most exocrine glands, some endocrine glands, and adipose tissue (fat) are innervated by the **autonomic nervous system,** the involuntary branch of the peripheral efferent division. Skeletal muscle is innervated by the **somatic nervous system,** the branch of the efferent division subject to voluntary control. Efferent output typically influences either movement or secretion, as illustrated in ▲ Table 7-1, which provides examples of the effects of neural control on various effectors composed of different types of muscle and gland tissue. Much of this efferent output is directed toward maintaining homeostasis. The efferent output to skeletal muscles is also directed toward voluntarily controlled nonhomeostatic activities, such as riding a bicycle. (It is important to realize that many effector organs are also subject to hormonal control and/or intrinsic control mechanisms; see p. 14.)

How many different neurotransmitters would you guess are released from the various efferent neuronal terminals to elicit essentially all the neurally controlled effector organ responses? Only two—acetylcholine and norepinephrine! Acting independently, these neurotransmitters bring about such diverse effects as salivary secretion, bladder contraction, and voluntary motor movements. These effects are a prime example of how the same chemical messenger may cause different responses in various tissues, depending on specialization of the effector organs.

An autonomic nerve pathway consists of a two-neuron chain.

Each autonomic nerve pathway extending from the CNS to an innervated organ is a two-neuron chain (● Figure 7-1) (except to the adrenal medulla; you will learn more about this exception later). The cell body of the first neuron in the series is located in the CNS. Its axon, the **preganglionic fiber,** synapses with the cell body of the second neuron, which lies within a ganglion. (Recall that a ganglion is a cluster of neuronal cell bodies out-

Examples of the Influence of Efferent Output on Movement and Secretion by Effector Organs

Category of Influence	Examples of Effectors with Different Types of Tissues	Sample Outcome in Response to Efferent Output
Influence on Movement	Heart (cardiac muscle)	Increased pumping of blood when the blood pressure falls too low
	Stomach (smooth muscle)	Delayed emptying of the stomach until the intestine is ready to process the food
	Diaphragm—a respiratory muscle (skeletal muscle)	Augmented breathing in response to exercise
Influence on Secretion	Sweat glands (exocrine glands)	Initiation of sweating on exposure to a hot environment
	Endocrine pancreas (endocrine gland)	Increased secretion of insulin, a hormone that puts excess nutrients in storage following a meal

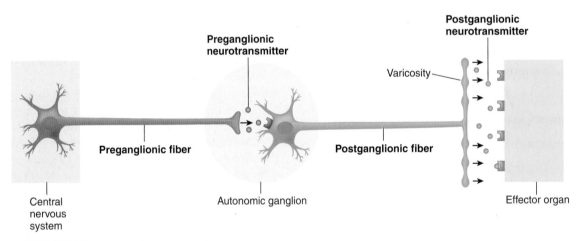

● **FIGURE 7-1 Autonomic nerve pathway.**

side the CNS.) The axon of the second neuron, the **postganglionic fiber,** innervates the effector organ.

The autonomic nervous system has two subdivisions—the **sympathetic** and the **parasympathetic nervous systems** (● Figure 7-2). Sympathetic nerve fibers originate in the thoracic and lumbar regions of the spinal cord (see p. 43). Most sympathetic preganglionic fibers are very short, synapsing with cell bodies of postganglionic neurons within ganglia that lie in a **sympathetic ganglion chain** (also called the **sympathetic trunk**) located along either side of the spinal cord (see ● Figure 5-24, p. 42). Long postganglionic fibers originate in the ganglion chain and end on the effector organs. Some preganglionic fibers pass through the ganglion chain without synapsing. Instead, they end later in sympathetic **collateral ganglia** about halfway between the CNS and the innervated organs, with postganglionic fibers traveling the rest of the distance.

Parasympathetic preganglionic fibers arise from the cranial (brain) and sacral (lower spinal cord) areas of the CNS. These fibers are longer than sympathetic preganglionic fibers because they do not end until they reach **terminal ganglia** that lie in or near the effector organs. Very short postganglionic fibers end on the cells of an organ itself.

Parasympathetic postganglionic fibers release acetylcholine; sympathetic ones release norepinephrine.

Sympathetic and parasympathetic preganglionic fibers release the same neurotransmitter, **acetylcholine (ACh),** but the postganglionic endings of these two systems release different neurotransmitters (the neurotransmitters that influence the autonomic effectors). Parasympathetic postganglionic fibers release acetylcholine. Accordingly, they, along with all autonomic preganglionic fibers, are called **cholinergic fibers.** Most sympathetic postganglionic fibers, in contrast, are called **adrenergic fibers** because they release **noradrenaline,** com-

KEY

⟨	Parasympathetic preganglionic fiber
⟨	Parasympathetic postganglionic fiber
⟨	Sympathetic preganglionic fiber
⟨	Sympathetic postganglionic fiber
ACh ●	Acetylcholine
NE ■	Norepinephrine
E ▲	Epinephrine

● **FIGURE 7-2 Autonomic nervous system.** The sympathetic nervous system, which originates in the thoracolumbar regions of the spinal cord, has short cholinergic (acetylcholine-releasing) preganglionic fibers and long adrenergic (norepinephrine-releasing) postganglionic fibers. The parasympathetic nervous system, which originates in the brain and sacral region of the spinal cord, has long cholinergic preganglionic fibers and short cholinergic postganglionic fibers. In most instances, sympathetic and parasympathetic postganglionic fibers both innervate the same effector organs. The adrenal medulla is a modified sympathetic ganglion, which releases epinephrine and norepinephrine into the blood. Nicotinic cholinergic receptors are located in the autonomic ganglia and adrenal medulla and respond to ACh released by all autonomic preganglionic fibers. Muscarinic cholinergic receptors are located at the autonomic effectors and respond to ACh released by parasympathetic postganglionic fibers. α_1, α_2, β_1, β_2 adrenergic receptors are variably located at the autonomic effectors and differentially respond to norepinephrine released by sympathetic postganglionic fibers and to epinephrine released by the adrenal medulla.

▲ TABLE 7-2	Sites of Release for Acetylcholine and Norepinephrine
Acetylcholine	**Norepinephrine**
All preganglionic terminals of the autonomic nervous system	Most sympathetic postganglionic terminals
All parasympathetic postganglionic terminals	Adrenal medulla
Sympathetic postganglionic terminals at sweat glands and some blood vessels in skeletal muscle	Central nervous system
Terminals of efferent neurons supplying skeletal muscle (motor neurons)	
Central nervous system	

monly known as **norepinephrine.**[1] Both acetylcholine and norepinephrine also serve as chemical messengers elsewhere in the body (▲ Table 7-2).

Postganglionic autonomic fibers do not end in a single terminal swelling like a synaptic knob. Instead, the terminal branches of autonomic fibers have numerous swellings, or **varicosities,** that simultaneously release neurotransmitter over a large area of the innervated organ rather than on single cells (see ● Figures 7-1 and 8-32, p. 297). This diffuse release of neurotransmitter, coupled with the fact that any resulting change in electrical activity is spread throughout a smooth or cardiac muscle mass via gap junctions (see p. 59), means that autonomic activity typically influences whole organs instead of discrete cells.

The sympathetic and parasympathetic nervous systems dually innervate most visceral organs.

Afferent information coming from the viscera (internal organs) usually does not reach conscious level (see p. 187). Examples of visceral afferent information include input from the baroreceptors that monitor blood pressure and input from the chemoreceptors that monitor the protein or fat content of ingested food. This input is used to direct the activity of the autonomic efferent neurons. Autonomic efferent output regulates visceral activities

---

[1] *Noradrenaline (norepinephrine)* is chemically very similar to *adrenaline (epinephrine),* the primary hormone product secreted by the adrenal medulla gland. Because a U.S. pharmaceutical company marketed this product for use as a drug under the trade name Adrenalin, the scientific community in this country prefers the alternative name "epinephrine" as a generic term for this chemical messenger, and accordingly, "noradrenaline" is known as "norepinephrine." In most other English-speaking countries, however, "adrenaline" and "noradrenaline" are the terms of choice.

such as circulation, digestion, sweating, and pupillary size. Like visceral afferent input, autonomic efferent output operates outside the realm of consciousness and voluntary control.

Most visceral organs are innervated by both sympathetic and parasympathetic nerve fibers (● Figure 7-3). Innervation of a single organ by both branches of the autonomic nervous system is known as **dual innervation** (*dual* means "pertaining to two"). ▲ Table 7-3 summarizes the major effects of these autonomic branches. Although the details of this wide array of autonomic responses are described more fully in later chapters that discuss the individual organs involved, you can consider several general concepts now. As you can see from the table, the sympathetic and parasympathetic nervous systems generally exert opposite effects in a particular organ. Sympathetic stimulation increases the heart rate, whereas parasympathetic stimulation decreases it; sympathetic stimulation slows down movement within the digestive tract, whereas parasympathetic stimulation enhances digestive motility. Note that both systems increase the activity of some organs and reduce the activity of others.

Rather than memorize a list such as in ▲ Table 7-3, it is better to logically deduce the actions of the two systems by first understanding the circumstances under which each system dominates. Usually both systems are partially active; that is, normally some level of action potential activity exists in both the sympathetic and the parasympathetic fibers supplying a particular organ. This ongoing activity is called **sympathetic or parasympathetic tone** or **tonic activity.** Under given circumstances, activity of one division can dominate the other. *Sympathetic dominance* to a particular organ exists when the sympathetic fibers' rate of firing to that organ increases above tonic level, coupled with a simultaneous decrease below tonic level in the parasympathetic fibers' frequency of action potentials to the same organ. The reverse situation is true for *parasympathetic dominance.* The balance between sympathetic and parasympathetic activity can be shifted separately for individual organs to meet specific demands (for example, sympathetically induced dilation of the pupil in dim light; see p. 197), or a more generalized, widespread discharge of one autonomic system in favor of the other can be elicited to control bodywide functions. Massive widespread discharges take place more frequently in the sympathetic system. The value of massive sympathetic discharge is clear, considering the circumstances during which this system usually dominates.

TIMES OF SYMPATHETIC DOMINANCE The sympathetic system promotes responses that prepare the body for strenuous physical activity in emergency or stressful situations, such as a physical threat from the outside. This response is typically referred to as a **fight-or-flight response** (some physiologists even throw in **fright,** too)**,** because the sympathetic system readies the body to fight against or flee from (and be frightened by) the threat. Think about the body resources needed in such circumstances. The heart beats more rapidly and more forcefully, blood pressure is elevated by generalized constriction (narrowing) of the blood vessels, respiratory airways open wide to permit maximal airflow, glycogen

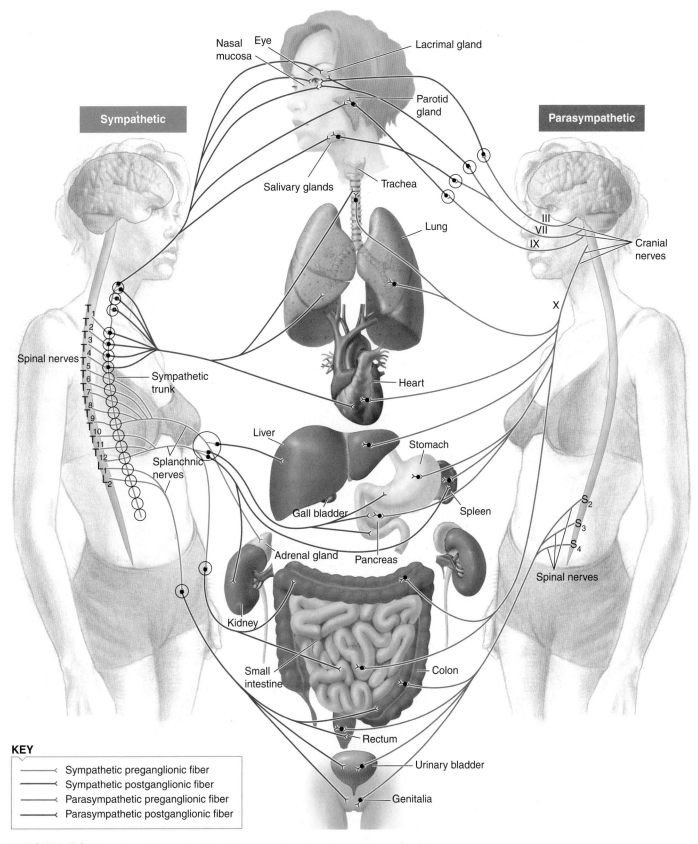

Sympathetic

Parasympathetic

Nasal mucosa
Eye
Lacrimal gland
Parotid gland
Salivary glands
Trachea
Lung
III
VII
IX
X
Cranial nerves
Heart
Spinal nerves
T₁
T₂
T₃
T₄
T₅
T₆
T₇
T₈
T₉
T₁₀
T₁₁
T₁₂
L₁
L₂
Sympathetic trunk
Liver
Stomach
Splanchnic nerves
Spleen
Gall bladder
Pancreas
S₂
S₃
S₄
Spinal nerves
Adrenal gland
Kidney
Small intestine
Colon
Rectum
Urinary bladder
Genitalia

KEY

Sympathetic preganglionic fiber
Sympathetic postganglionic fiber
Parasympathetic preganglionic fiber
Parasympathetic postganglionic fiber

● **FIGURE 7-3 Structures innervated by the sympathetic and parasympathetic nervous systems.**

Effects of Autonomic Nervous System on Various Organs

Organ	Effect of Sympathetic Stimulation (and Types of Adrenergic Receptors)	Effect of Parasympathetic Stimulation
Heart	Increased rate, increased force of contraction (of whole heart) (β_1)	Decreased rate, decreased force of contraction (of atria only)
Most Innervated Blood Vessels	Constriction (α_1)	Dilation of vessels supplying the penis and clitoris only
Lungs	Dilation of bronchioles (airways) (β_2)	Constriction of bronchioles
	Inhibition of mucus secretion (α)	Stimulation of mucus secretion
Digestive Tract	Decreased motility (movement) (α_2, β_2)	Increased motility
	Contraction of sphincters (to prevent forward movement of contents) (α_1)	Relaxation of sphincters (to permit forward movement of contents)
	Inhibition of digestive secretions (α_2)	Stimulation of digestive secretions
Urinary Bladder	Relaxation (β_2)	Contraction (emptying)
Eye	Dilation of pupil (contraction of radial muscle) (α_1)	Constriction of pupil (contraction of circular muscle)
	Adjustment of eye for far vision (β_2)	Adjustment of eye for near vision
Liver (glycogen stores)	Glycogenolysis (glucose released) (β_2)	None
Adipose Cells (fat stores)	Lipolysis (fatty acids released) (β_2)	None
Exocrine Glands		
Exocrine pancreas	Inhibition of pancreatic exocrine secretion (α_2)	Stimulation of pancreatic exocrine secretion (important for digestion)
Sweat glands	Stimulation of secretion by most sweat glands (α_1; most are cholinergic)	Stimulation of secretion by some sweat glands
Salivary glands	Stimulation of small volume of thick saliva rich in mucus (α_1)	Stimulation of large volume of watery saliva rich in enzymes
Endocrine Glands		
Adrenal medulla	Stimulation of epinephrine and norepinephrine secretion (cholinergic)	None
Endocrine pancreas	Inhibition of insulin secretion; stimulation of glucagon secretion (α_2)	Stimulation of insulin and glucagon secretion
Genitals	Ejaculation and orgasmic contractions (males); orgasmic contractions (females) (α_1)	Erection (caused by dilation of blood vessels in penis [male] and clitoris [female])
Brain Activity	Increased alertness (receptors unknown)	None

(stored sugar) and fat stores are broken down to release extra fuel into the blood, and blood vessels supplying skeletal muscles dilate (open more widely). All these responses are aimed at providing increased flow of oxygenated, nutrient-rich blood to the skeletal muscles in anticipation of strenuous physical activity. Furthermore, the pupils dilate and the eyes adjust for far vision, letting the person visually assess the entire threatening scene. Sweating is promoted in anticipation of excess heat production by the physical exertion. Because digestive and urinary activities are not essential in meeting the threat, the sympathetic system inhibits these activities.

TIMES OF PARASYMPATHETIC DOMINANCE The parasympathetic system dominates in quiet, relaxed situations. Under such nonthreatening circumstances, the body can be concerned with its own "general housekeeping" activities, such as digestion. The parasympathetic system promotes these **"rest-and-digest"** types of bodily functions while slowing down those activities that are enhanced by the sympathetic system. There is no need, for example, to have the heart beating rapidly and forcefully when the person is in a tranquil setting.

ADVANTAGE OF DUAL AUTONOMIC INNERVATION What is the advantage of dual innervation of organs with nerve fibers whose actions oppose each other? It enables precise control over an organ's activity, like having both an accelerator and a brake to control the speed of a car. If an animal suddenly darts across the road as you are driving, you could eventually stop if you just took your foot off the accelerator, but you might stop too slowly to avoid hitting the animal. If you simultaneously apply the brake as you lift up on the accelerator, however, you can come to a more rapid, controlled stop. In a similar manner, a sympathetically accelerated heart rate could gradually be reduced to normal following a stressful situation by decreasing the firing rate in the cardiac sympathetic nerve (letting up on the accelerator). However, the heart rate can be reduced more rapidly by simultaneously increasing activity in the parasympathetic supply to the heart (applying the brake). Indeed, the two divisions of the autonomic nervous system are usually reciprocally controlled; increased activity in one division is accompanied by a corresponding decrease in the other.

There are several exceptions to the general rule of dual reciprocal innervation by the two branches of the autonomic nervous system; the most notable are the following:

■ *Innervated blood vessels* (most arterioles and veins are innervated; arteries and capillaries are not) receive only sympathetic nerve fibers. Regulation is accomplished by increasing or decreasing the firing rate above or below the tonic level in these sympathetic fibers. The only blood vessels to receive both sympathetic and parasympathetic fibers are those supplying the penis and clitoris. The precise vascular control this dual innervation affords these organs is important in accomplishing erection.

■ Most *sweat glands* are innervated only by sympathetic nerves. The postganglionic fibers of these nerves are unusual because they secrete acetylcholine rather than norepinephrine.

■ *Salivary glands* are innervated by both autonomic divisions, but unlike elsewhere, sympathetic and parasympathetic activity is not antagonistic. Both stimulate salivary secretion, but the saliva's volume and composition differ, depending on which autonomic branch is dominant.

You will learn more about these exceptions in later chapters. We are now going to turn our attention to the adrenal medulla, a unique endocrine component of the sympathetic nervous system.

The adrenal medulla is a modified part of the sympathetic nervous system.

There are two *adrenal glands,* one lying above the kidney on each side (*ad* means "next to"; *renal* means "kidney"). The adrenal glands are endocrine glands, each with an outer portion, the *adrenal cortex,* and an inner portion, the *adrenal medulla* (see pp. 698 and 705–707). The **adrenal medulla** is a modified sympathetic ganglion that does not give rise to postganglionic fibers. Instead, on stimulation by the preganglionic fiber that originates in the CNS, it secretes hormones into the blood (see ● Figure 7-2). Not surprisingly, the hormones are identical or similar to postganglionic sympathetic neurotransmitters. About 20% of the adrenal medullary hormone output is norepinephrine, and the remaining 80% is the closely related **epinephrine (adrenaline)** (see footnote 1, p. 240). These hormones, in general, reinforce activity of the sympathetic nervous system.

Several different receptor types are available for each autonomic neurotransmitter.

Because each autonomic neurotransmitter and medullary hormone stimulates activity in some tissues but inhibits activity in others, the particular responses must depend on specialization of the tissue cells rather than on properties of the chemicals themselves. Responsive tissue cells have one or more of several different types of plasma membrane receptor proteins for these chemical messengers. Binding of a neurotransmitter to a receptor induces the tissue-specific response.

CHOLINERGIC RECEPTORS Researchers have identified two types of acetylcholine (cholinergic) receptors—*nicotinic* and *muscarinic receptors*—on the basis of their response to particular drugs. **Nicotinic receptors** are activated by the tobacco plant derivative nicotine, whereas **muscarinic receptors** are activated by the mushroom poison muscarine (▲ Table 7-4).

Nicotinic receptors are found on the postganglionic cell bodies in all autonomic ganglia. These receptors respond to acetylcholine released from both sympathetic and parasympathetic preganglionic fibers. Binding of acetylcholine to these receptors brings about opening of nonspecific cation channels in the postganglionic cell that permit passage of both Na^+ and K^+. Because of the greater electrochemical gradient for Na^+ than for K^+, more Na^+ enters the cell than K^+ leaves, bringing about a depolarization that leads to initiation of an action potential in the postganglionic cell.

Muscarinic receptors are found on effector cell membranes (smooth muscle, cardiac muscle, and glands). They bind with acetylcholine released from parasympathetic postganglionic fibers. There are five subtypes of muscarinic receptors, all of which are linked to G proteins that activate second-messenger pathways leading to the target cell response (see p. 117).

ADRENERGIC RECEPTORS The two major classes of adrenergic receptors for norepinephrine and epinephrine are **alpha (α)** and **beta (β) receptors,** which are further subclassified into α_1 and α_2 and β_1 and β_2 **receptors** (▲ Table 7-4). These various receptor types are distinctly distributed among the effector

Properties of Autonomic Receptor Types

Receptor Type	Neurotransmitter Affinity	Effector(s) with Receptor Type	Mechanism of Action at Effector	Effect on Effector
Nicotinic	Acetylcholine from autonomic preganglionic fibers	All autonomic postganglionic cell bodies; adrenal medulla	Opens nonspecific cation receptor-channels	Excitatory
	Acetylcholine from motor neurons	Motor end plates of skeletal muscle fibers	Opens nonspecific cation receptor-channels	Excitatory
Muscarinic	Acetylcholine from parasympathetic postganglionic fibers	Cardiac muscle, smooth muscle, most exocrine and some endocrine glands	Activates various G-protein-coupled receptor pathways, depending on effector	Excitatory or inhibitory, depending on effector
α_1	Norepinephrine from sympathetic postganglionic fibers; epinephrine from adrenal medulla: Norepinephrine > epinephrine	Most sympathetic target tissues	Activates IP_3/Ca^{2+} second-messenger pathway	Excitatory
α_2	Norepinephrine > epinephrine	Digestive organs	Inhibits cAMP	Inhibitory
β_1	Norepinephrine = epinephrine	Heart	Activates cAMP	Excitatory
β_2	Epinephrine only	Smooth muscles of arterioles and bronchioles	Activates cAMP	Inhibitory

organs. Receptors of the β_2 type bind almost exclusively with epinephrine, whereas β_1 receptors have about equal affinities for norepinephrine and epinephrine, and α receptors of both subtypes have a greater sensitivity to norepinephrine than to epinephrine (see ● Figure 7-2).

All adrenergic receptors are coupled to G proteins, but the ensuing pathway differs for the various receptor types. Activation of both β_1 and β_2 receptors brings about the target cell response by activating the cAMP second-messenger pathway (see p. 121). Stimulation of α_1 receptors elicits the desired response via the IP_3/Ca^{2+} second-messenger system (see p. 122). By contrast, binding of a neurotransmitter to an α_2 receptor inhibits cAMP production in the target cell.

Activation of α_1 receptors usually brings about an excitatory response in the effector—for example, arteriolar constriction caused by increased contraction of smooth muscle in the walls of these blood vessels. The α_1 receptors are present in most sympathetic target tissues. Activation of α_2 receptors, in contrast, brings an inhibitory response in the effector, such as decreased smooth muscle contraction in the digestive tract. Stimulation of β_1 receptors, which are found primarily in the heart, causes an excitatory response, namely, increased rate and force of cardiac contraction. The response to β_2 receptor activation is generally inhibitory, such as arteriolar or bronchiolar (respiratory airway) dilation caused by relaxation of the smooth muscle in the walls of these tubular structures. As a quick rule of thumb, activation of the subscript "1" versions of adrenergic receptors lead to excitatory

responses and activation of the subscript "2" versions lead to inhibitory responses.

 AUTONOMIC AGONISTS AND ANTAGONISTS Drugs are available that selectively alter autonomic responses at each of the receptor types. An **agonist** binds to the neurotransmitter's receptor and causes the same response as the neurotransmitter would. An **antagonist,** by contrast, binds with the receptor, preventing the neurotransmitter from binding and causing a response, yet the antagonist itself produces no response. Thus an agonist mimics the neurotransmitter's response and an antagonist blocks the neurotransmitter's response. Some of these drugs are only of experimental interest, but others are very important therapeutically. For example, *atropine* blocks the effect of acetylcholine at muscarinic receptors but does not affect nicotinic receptors. Because acetylcholine released at both parasympathetic and sympathetic preganglionic fibers combines with nicotinic receptors, blockage at nicotinic synapses would knock out both these autonomic branches. By acting selectively to interfere with acetylcholine action only at muscarinic junctions, which are the sites of parasympathetic postganglionic action, atropine effectively blocks parasympathetic effects but does not influence sympathetic activity at all. Doctors use this principle to suppress salivary and bronchial secretions before surgery, to reduce the risk of a patient inhaling these secretions into the lungs.

Likewise, drugs that act selectively at α and β adrenergic receptor sites to either activate or block specific sympathetic effects are widely used. Following are several examples. *Salbuta-*

Distinguishing Features of Sympathetic Nervous System and Parasympathetic Nervous System

Feature	Sympathetic System	Parasympathetic System
Origin of Preganglionic Fiber	Thoracic and lumbar regions of spinal cord	Brain and sacral region of spinal cord
Origin of Postganglionic Fiber (location of ganglion)	Sympathetic ganglion chain (near spinal cord) or collateral ganglia (about halfway between spinal cord and effector organs)	Terminal ganglia (in or near effector organs)
Length and Type of Fiber	Short cholinergic preganglionic fibers	Long cholinergic preganglionic fibers
	Long adrenergic postganglionic fibers	Short cholinergic postganglionic fibers
Effector Organs Innervated	Cardiac muscle, almost all smooth muscle, most exocrine glands, and some endocrine glands	Cardiac muscle, most smooth muscle, most exocrine glands, and some endocrine glands
Types of Receptors for Neurotransmitters	For preganglionic neurotransmitter: nicotinic	For preganglionic neurotransmitter: nicotinic
	For postganglionic neurotransmitter: α_1, α_2, β_1, β_2	For postganglionic neurotransmitter: muscarinic
Dominance	Dominates in emergency "fight-or-flight" situations; prepares body for strenuous physical activity	Dominates in quiet, relaxed situations; promotes "general housekeeping" activities such as digestion

mol selectively activates β_2 adrenergic receptors at low doses, making it possible to dilate the bronchioles in the treatment of asthma without undesirably stimulating the heart (the heart has mostly β_1 receptors). By contrast, *metoprolol* selectively blocks β_1 adrenergic receptors and is prescribed to treat high blood pressure because it decreases the amount of blood the heart pumps into the blood vessels. Metoprolol does not affect β_2 receptors and so has no effect on the bronchioles. The less specific drug *propanolol* also lowers blood pressure by blocking β_1 receptors, but it also blocks β_2 receptors located in the bronchioles, leading to narrowing of these airways and asthma as an undesirable side effect in susceptible people.

Many regions of the central nervous system are involved in the control of autonomic activities.

Messages from the CNS are delivered to cardiac muscle, smooth muscle, and glands via the autonomic nerves, but which regions of the CNS regulate autonomic output? Autonomic control of these effectors is mediated by reflexes and through centrally located control centers. Going back one step further, ultimately information carried to the CNS via the visceral afferents is used to determine the appropriate output via the autonomic efferents to the effectors to maintain homeostasis. Indeed, some physiologists regard the visceral afferents as part of the autonomic nervous system, while others consider the sympathetic and parasympathetic efferents as being the only components of the autonomic nervous system. Whatever way you classify these

parts of the nervous system, the point remains that visceral afferent input is critical for determining sympathetic and parasympathetic output.

■ Some autonomic reflexes, such as urination, defecation, and erection, are integrated at the spinal-cord level, but all these spinal reflexes are subject to control by higher levels of consciousness.

■ The medulla within the brain stem is the region most directly responsible for autonomic output. Centers for controlling cardiovascular, respiratory, and digestive activity via the autonomic system are located there.

■ The hypothalamus plays an important role in integrating the autonomic, somatic, and endocrine responses that automatically accompany various emotional and behavioral states. For example, the increased heart rate, blood pressure, and respiratory activity associated with anger or fear are brought about by the hypothalamus acting through the medulla.

■ Autonomic activity can also be influenced by the prefrontal association cortex through its involvement with emotional expression characteristic of the individual's personality. An example is blushing when embarrassed, which is caused by dilation of blood vessels supplying the skin of the cheeks. Such responses are mediated through hypothalamic-medullary pathways.

▲ Table 7-5 summarizes the main distinguishing features of the sympathetic and parasympathetic nervous systems.

Somatic Nervous System

Motor neurons supply skeletal muscle.

Skeletal muscle is innervated by **motor neurons,** the axons of which constitute the somatic nervous system. (Sometimes all efferent neurons are referred to as *motor neurons,* but we reserve this term for the efferent somatic fibers that supply skeletal muscle and bring about body movement; *motor* means "movement.") The cell bodies of almost all motor neurons are within the ventral horn of the spinal cord. The only exception is that the cell bodies of motor neurons supplying muscles in the head are in the brain stem. Unlike the two-neuron chain of autonomic nerve fibers, the axon of a motor neuron is continuous from its origin in the CNS to its ending on skeletal muscle. Motor neuron axon terminals release acetylcholine, which brings about excitation and contraction of the innervated muscle cells. Motor neurons can only stimulate skeletal muscles, in contrast to autonomic fibers, which can either stimulate or inhibit their effector organs. Inhibition of skeletal muscle activity can be accomplished only within the CNS through inhibitory synaptic input to the dendrites and cell bodies of the motor neurons supplying that particular muscle.

Motor neurons are the final common pathway.

Motor neuron dendrites and cell bodies are influenced by many converging presynaptic inputs, both excitatory and inhibitory. Some of these inputs are part of spinal reflex pathways originating with peripheral sensory receptors. Others are part of descending pathways originating within the brain. Areas of the brain that exert control over skeletal muscle movements include the motor regions of the cortex, the basal nuclei, the cerebellum, and the brain stem (see pp. 145–149, 153, and 166–167; also see ● Figure 8-23, p. 286, for a summary of motor control and ● Figure 5-28b, p. 175, for specific examples of these descending motor pathways).

Motor neurons are considered the **final common pathway** because the only way any other parts of the nervous system can influence skeletal muscle activity is by acting on these motor neurons. The level of activity in a motor neuron and its subsequent output to the skeletal muscle fibers it innervates depend on the relative balance of EPSPs and IPSPs (see p. 106) brought about by its presynaptic inputs originating from these diverse sites in the brain.

The somatic system is under voluntary control, but much of the skeletal muscle activity involving posture, balance, and stereotypical movements is subconsciously controlled. You may decide you want to start walking, but you do not have to consciously bring about the alternate contraction and relaxation of the involved muscles because these movements are involuntarily coordinated by lower brain centers.

 The cell bodies of the crucial motor neurons may be selectively destroyed by **poliovirus.** The result is paralysis of the muscles innervated by the affected neurons. **Amyotrophic lateral sclerosis (ALS),** also known as **Lou Gehrig's disease,** is the most common motor-neuron disease. This incurable condition is characterized by the degeneration and eventual death of motor neurons. The result is gradual loss of motor control, progressive paralysis, and finally, death within three to five years of onset. The exact cause is uncertain, although researchers are investigating a variety of potential underlying problems. Among these are pathological changes in neurofilaments that block axonal transport of crucial materials (see p. 49), extracellular accumulation of toxic levels of the excitatory neurotransmitter glutamate, aggregation of misfolded intracellular proteins, mitochondrial dysfunction leading to reduced energy production, and activation of protein-cutting enzymes (caspases, the ones involved in apoptosis; see p. 124) that selectively cut up the neuronal cell body and nucleus. Scientists are working on potential therapies aimed at slowing the progress of this devastating disease.

Before turning our attention to the junction between a motor neuron and the muscle cells it innervates, we are going to pull together in tabular form two groups of information we have been examining. ▲ Table 7-6 summarizes the features of the two branches of the efferent division of the PNS—the autonomic nervous system and the somatic nervous system. ▲ Table 7-7 compares the three functional types of neurons—afferent neurons, efferent neurons, and interneurons.

Neuromuscular Junction

Motor neurons and skeletal muscle fibers are chemically linked at neuromuscular junctions.

An action potential in a motor neuron is rapidly propagated from the cell body within the CNS to the skeletal muscle along the large myelinated axon (efferent fiber) of the neuron. As the axon approaches a muscle, it divides and loses its myelin sheath. Each of these axon terminals forms a special junction, a **neuromuscular junction,**[2] with one of the many muscle cells that compose the whole muscle (● Figure 7-4). Each branch innervates only one muscle cell; therefore, each muscle cell has only one neuromuscular junction. A single muscle cell, called a **muscle fiber,** is long and cylindrical. Within a neuromuscular junction, the axon terminal splits into multiple fine branches each of which ends in an enlarged knoblike structure called the **terminal button,** or **bouton.** The entire axon terminal ending (all the fine branches with terminal buttons) fits into a shallow depression, or groove, in the underlying muscle fiber. This spe-

---

[2]Many scientists refer to a **synapse** as any junction between two cells that handle information electrically. According to this broad point of view, *chemical synapses* include junctions between two neurons as well as those between a neuron and an effector cell (such as muscle cells of any type or gland cells), and *electrical synapses* include gap junctions between smooth muscle cells, between cardiac muscle cells, or between some neurons. We narrowly reserve the term *synapse* specifically for neuron-to-neuron junctions and use different terms for other types of junctions, such as the term *neuromuscular junction* for a junction between a motor neuron and a skeletal muscle cell.

Comparison of the Autonomic Nervous System and the Somatic Nervous System

Feature	Autonomic Nervous System	Somatic Nervous System
Site of Origin	Sympathetic: lateral horn of thoracic and lumbar spinal cord Parasympathetic: brain and sacral spinal cord	Ventral horn of spinal cord for most; those supplying muscles in head originate in brain
Number of Neurons from Origin in CNS to Effector Organ	Two-neuron chain (preganglionic and postganglionic)	Single neuron (motor neuron)
Organs Innervated	Cardiac muscle, smooth muscle, exocrine and some endocrine glands	Skeletal muscle
Type of Innervation	Most effector organs dually innervated by the two antagonistic branches of this system (sympathetic and parasympathetic)	Effector organs innervated only by motor neurons
Neurotransmitter at Effector Organs	May be acetylcholine (parasympathetic terminals) or norepinephrine (sympathetic terminals)	Only acetylcholine
Effects on Effector Organs	Either stimulation or inhibition (antagonistic actions of two branches)	Stimulation only (inhibition possible only centrally through IPSPs on dendrites and cell body of motor neuron)
Types of Control	Under involuntary control	Subject to voluntary control; much activity subconsciously coordinated
Higher Centers Involved in Control	Spinal cord, medulla, hypothalamus, prefrontal association cortex	Spinal cord, motor cortex, basal nuclei, cerebellum, brain stem

cialized underlying portion of the muscle cell membrane is called the **motor end plate** (● Figure 7-5, p. 250).

Acetylcholine is the neuromuscular junction neurotransmitter.

Nerve and muscle cells do not actually come into direct contact at a neuromuscular junction. The space, or cleft, between these two structures is too large to permit electrical transmission of an impulse between them (that is, the action potential cannot "jump" that far). Therefore, just as at a neuronal chemical synapse (see p. 104), a chemical messenger carries the signal between a terminal button and the muscle fiber. This neurotransmitter is acetylcholine (ACh).

RELEASE OF ACh AT THE NEUROMUSCULAR JUNCTION Each terminal button contains thousands of vesicles that store ACh. Propagation of an action potential to the axon terminal (● Figure 7-5, step **1**) triggers the opening of voltage-gated Ca^{2+} channels in all of its terminal buttons (see p. 88). We will focus on one terminal button, but the same events take place concurrently at all the terminal buttons of a given neuromuscular junction. Opening of Ca^{2+} channels permits Ca^{2+} to diffuse into the terminal button from its higher extracellular concentration (step **2**), which in turn causes the release of ACh by exocytosis from several hundred of the vesicles into the cleft (step **3**).

FORMATION OF AN END-PLATE POTENTIAL The released ACh diffuses across the cleft and binds with specific receptor-channels, which are specialized membrane proteins unique to the motor end-plate portion of the muscle fiber membrane (step **4**). (These cholinergic receptors are of the nicotinic type.) Binding of ACh with these chemically gated receptor-channels causes them to open. They are nonspecific cation channels that permit a small amount of cation traffic through them (both Na^+ and K^+) but no anions (step **5**). Because the permeability of the end-plate membrane to Na^+ and K^+ on opening of these channels is essentially equal, the relative movement of these ions through the channels depends on their electrochemical driving forces. Recall that at resting potential, the net driving force for Na^+ is much greater than that for K^+ because the rest-

Comparison of Types of Neurons

FEATURE	AFFERENT NEURON	EFFERENT NEURON		INTERNEURON
		Autonomic Nervous System	Somatic Nervous System	
Origin, Structure, Location	Receptor at peripheral ending; elongated peripheral axon, which travels in peripheral nerve; cell body located in dorsal root ganglion; short central axon entering spinal cord	Two-neuron chain; first neuron (preganglionic fiber) originating in CNS and terminating on a ganglion; second neuron (postganglionic fiber) originating in the ganglion and terminating on the effector organ	Cell body of motor neuron lying in spinal cord; long axon traveling in peripheral nerve and terminating on the effector organ	Various shapes; lying entirely within CNS; some cell bodies originating in brain, with long axons traveling down the spinal cord in descending pathways; some originating in spinal cord, with long axons traveling up the cord to the brain in ascending pathways; others forming short local connections
Termination	Interneurons*	Effector organs (cardiac muscle, smooth muscle, exocrine and some endocrine glands)	Effector organs (skeletal muscle)	Other interneurons and efferent neurons
Function	Carries information about the external and internal environment to CNS	Carries instructions from CNS to effector organs	Carries instructions from CNS to effector organs	Processes and integrates afferent input; initiates and coordinates efferent output; responsible for thought and other higher mental functions
Convergence of Input on Cell Body	No (only input is through receptor)	Yes	Yes	Yes
Effect of Input to Neuron	Can only be excited (through receptor potential induced by stimulus; must reach threshold for action potential)	Can be excited or inhibited (through EPSPs and IPSPs at first neuron; must reach threshold for action potential)	Can be excited or inhibited (through EPSPs and IPSPs; must reach threshold for action potential)	Can be excited or inhibited (through EPSPs and IPSPs; must reach threshold for action potential)
Site of Action Potential Initiation	First excitable portion of membrane adjacent to receptor	Axon hillock	Axon hillock	Axon hillock
Divergence of Output	Yes	Yes	Yes	Yes
Effect of Output on Structure on Which It Terminates	Only excites	Postganglionic fiber either excites or inhibits	Only excites	Either excites or inhibits

*Except in stretch reflex where afferent neuron terminates directly on efferent neuron; see p. 177

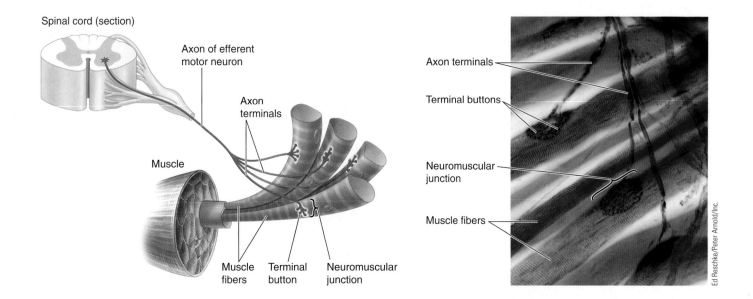

Spinal cord (section)

Axon of efferent motor neuron

Axon terminals

Muscle

Muscle fibers

Terminal button

Neuromuscular junction

Axon terminals

Terminal buttons

Neuromuscular junction

Muscle fibers

Ed Reschke/Peter Arnold/Inc.

● **FIGURE 7-4 Motor neuron innervating skeletal muscle cells.** The cell body of a motor neuron originates in the ventral horn of the spinal cord. The axon (somatic efferent fiber) exits through the ventral root and travels through a spinal nerve to the skeletal muscle it innervates. When the axon reaches a skeletal muscle, it divides into many axon terminals, each of which forms a neuromuscular junction with a single muscle cell (muscle fiber). The axon terminal within a neuromuscular junction further divides into fine branches, each of which ends in an enlarged terminal button. Note that the muscle fibers innervated by a single axon terminal are dispersed throughout the muscle, but for simplicity they are grouped together in this figure.

ing potential is much closer to the K^+ than to the Na^+ equilibrium potential. Both the concentration and electrical gradients for Na^+ are inward, whereas the outward concentration gradient for K^+ is almost, but not quite, balanced by the opposing inward electrical gradient. As a result, when ACh triggers the opening of these channels, considerably more Na^+ moves inward than K^+ outward, depolarizing the motor end plate. This potential change is called the **end-plate potential (EPP).** It is a graded potential similar to an EPSP (excitatory postsynaptic potential; see p. 106), except that an EPP is much larger for the following reasons: (1) a neuromuscular junction consists of multiple terminal buttons, each of which simultaneously releases ACh on activation of the axon terminal; (2) more neurotransmitter is released from a terminal button than from a presynaptic knob in response to an action potential; (3) the motor end plate has a larger surface area and a higher density of neurotransmitter receptor-channels and thus has more sites for binding with neurotransmitter than a subsynaptic membrane has; and (4) accordingly, many more receptor-channels are opened in response to neurotransmitter release at a neuromuscular junction than at a synapse. This permits a greater net influx of positive ions and a larger depolarization (an EPP) at a neuromuscular junction. As with an EPSP, an EPP is a graded potential, whose magnitude depends on the amount and duration of ACh at the end plate.

INITIATION OF AN ACTION POTENTIAL The motor end-plate region itself does not have a threshold potential, so an action potential cannot be initiated at this site. However, an EPP brings about an action potential in the rest of the muscle fiber, as follows. The neuromuscular junction is usually in the middle of the long, cylindrical muscle fiber. When an EPP takes place, local current flow occurs between the depolarized end plate and the adjacent, resting cell membrane in both directions (step **6**), opening voltage-gated Na^+ channels and thus reducing the potential to threshold in the adjacent areas (step **7**). The subsequent action potential initiated at these sites propagates throughout the muscle fiber membrane by contiguous conduction (step **8**) (see p. 96). The spread runs in both directions, away from the motor end plate toward both ends of the fiber. This electrical activity triggers contraction of the muscle fiber. Thus, by means of ACh, an action potential in a motor neuron brings about an action potential and subsequent contraction in the muscle fiber. (See the boxed feature on p. 251, ■ A Closer Look at Exercise Physiology, to examine the importance of motor neuron stimulation in maintaining the integrity of skeletal muscles.)

Unlike synaptic transmission, the magnitude of an EPP is normally enough to cause an action potential in the muscle cell. Therefore, one-to-one transmission of an action potential typically occurs at a neuromuscular junction; one action potential in a nerve cell triggers one action potential in a muscle cell that

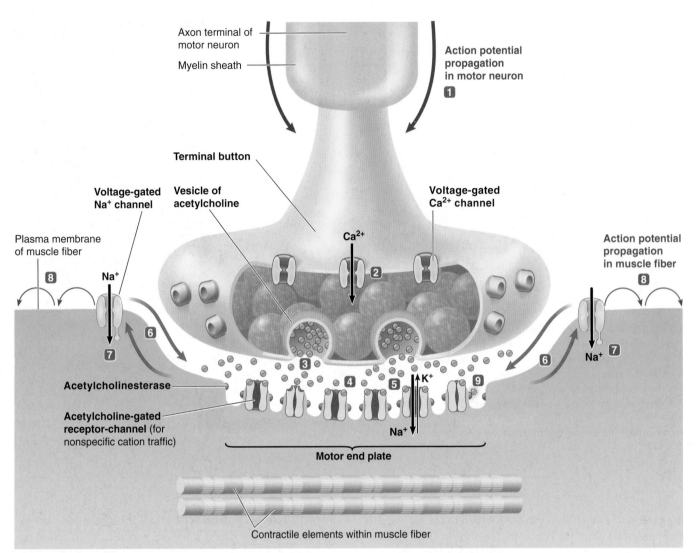

Axon terminal of motor neuron

Myelin sheath

Action potential propagation in motor neuron [1]

Terminal button

Voltage-gated Na⁺ channel

Vesicle of acetylcholine

Voltage-gated Ca²⁺ channel

Plasma membrane of muscle fiber

Ca^{2+}

[2]

[8]

Na^+

Action potential propagation in muscle fiber [8]

[6]

[7]

[6]

Na^+ [7]

Acetylcholinesterase

[3]

[4]

[5]

↑K^+

[9]

Acetylcholine-gated receptor-channel (for nonspecific cation traffic)

Na^+

Motor end plate

Contractile elements within muscle fiber

[1] An action potential in a motor neuron is propagated to the axon terminal (terminal button).

[2] This local action potential triggers the opening of voltage-gated Ca^{2+} channels and the subsequent entry of Ca^{2+} into the terminal button.

[3] Ca^{2+} triggers the release of acetylcholine (ACh) by exocytosis from a portion of the vesicles.

[4] ACh diffuses across the space separating the nerve and muscle cells and binds with receptor-channels specific for it on the motor end plate of the muscle cell membrane.

[5] This binding brings about the opening of these nonspecific cation channels, leading to a relatively large movement of Na^+ into the muscle cell compared to a smaller movement of K^+ outward.

[6] The result is an end-plate potential. Local current flow occurs between the depolarized end plate and the adjacent membrane.

[7] This local current flow opens voltage-gated Na^+ channels in the adjacent membrane.

[8] The resultant Na^+ entry reduces the potential to threshold, initiating an action potential, which is propagated throughout the muscle fiber.

[9] ACh is subsequently destroyed by acetylcholinesterase, an enzyme located on the motor end-plate membrane, terminating the muscle cell's response.

● **FIGURE 7-5 Events at a neuromuscular junction.**

it innervates. Other comparisons of neuromuscular junctions with synapses can be found in ▲ Table 7-8.

Acetylcholinesterase ends acetylcholine activity at the neuromuscular junction.

To ensure purposeful movement, a muscle cell's electrical and resultant contractile response to stimulation by its motor neu-

ron must be switched off promptly when there is no longer a signal from the motor neuron. The muscle cell's electrical response is turned off by an enzyme in the motor end-plate membrane, **acetylcholinesterase (AChE),** which inactivates ACh.

As a result of diffusion, many of the released ACh molecules come into contact with and bind to receptor-channels on the surface of the motor end-plate membrane. However, some of the ACh molecules bind with AChE, which is also at the end-

Loss of Muscle Mass: A Plight of Spaceflight

Skeletal muscles are a case of "use it or lose it." Stimulation of skeletal muscles by motor neurons is essential not only to cause the muscles to contract but also to maintain their size and strength. Muscles not routinely stimulated gradually atrophy, or diminish in size and strength.

Our skeletal muscles are important in supporting our upright posture in the face of gravitational forces in addition to moving body parts. When humans entered the weightlessness of space, it became apparent early that the muscular system required the stress of work or gravity to maintain its size and strength. In 1991, the space shuttle *Columbia* was launched for a nine-day mission dedicated among other things to comprehensive research on physiologic changes brought on by weightlessness. The three female and four male astronauts aboard suffered a dramatic and significant 25% reduction of mass in their weight-bearing muscles. Remarkably less effort is required to move the body in space than on Earth, and there is no need for active muscular opposition to gravity. Furthermore, the muscles used to move around the confines of a space capsule differ from those used for walking down the street. As a result, some muscles rapidly undergo what is known as *functional atrophy.*

The muscles most affected are those in the lower extremities, the gluteal (buttocks) muscles, the extensor muscles of the neck and back, and the muscles of the trunk—

that is, the muscles used for antigravity support on the ground. Changes include a decrease in muscle volume and mass, decrease in strength and endurance, increased breakdown of muscle protein, and loss of muscle nitrogen (an important component of muscle protein). The exact biological mechanisms that induce muscle atrophy are unknown, but a majority of scientists believe that the lack of customary forcefulness of contraction plays a major role. This atrophy presents no problem while within the space capsule, but such loss of muscle mass must be restricted if astronauts are to perform heavy work during space walks and are to resume normal activities on their return to Earth.

Space programs in the United States and the former Soviet Union have employed intervention techniques that emphasize both diet and exercise in an attempt to prevent muscle atrophy. Faithful performance of vigorous, carefully designed physical exercise for several hours daily has helped reduce the severity of functional atrophy. Studies of nitrogen and mineral balances, however, suggest that muscle atrophy continues to progress during exposure to weightlessness despite efforts to prevent it. Furthermore, only half of the muscle mass was restored in the *Columbia* crew after the astronauts had been back on the ground a length of time equal to their flight. These and other findings suggested

that further muscle-preserving interventions would be necessary for extended stays in space.

The International Space Station, which became operational in 2000, has nearly five times more cubic feet of workspace than the Mir or Skylab stations. This additional space includes more exercise capabilities plus sophisticated laboratory equipment for further studies on the effect of weightlessness on the body, not only on muscle but on other systems as well.

Despite these efforts, the effect of weightlessness on muscle especially continues to be a problem. The longest stay in the International Space Station has been 215 days. On return, the astronauts have difficulty walking because of significant muscle atrophy. One of the concerns with the much-talked-about mission to Mars is the expected magnitude of loss of muscle mass by the time the astronauts arrive at Mars, leaving them too weak to walk, let alone carry out their mission or make repairs.

Multiple research groups are currently exploring pharmaceutical avenues for slowing muscle breakdown. These so-called antiwasting drugs are of interest not only for space application but for medical application as well. Muscle wasting is a problem for people confined to bed for long periods of time and for those suffering from certain chronic diseases such as cancer and AIDS.

---

plate surface. Being quickly inactivated, this ACh never contributes to the end-plate potential. The acetylcholine that does bind with receptor-channels does so very briefly (for about 1 millionth of a second), then detaches. Some of the detached ACh molecules quickly rebind with receptor-channels, keeping these end-plate channels open, but some randomly contact AChE instead this time around and are inactivated (step **9**). As this process repeats, more and more ACh is inactivated until it has been virtually removed from the cleft within a few milliseconds after its release. ACh removal ends the EPP, so the remainder of the muscle cell membrane returns to resting potential. Now the muscle cell can relax. Or, if sustained contraction is essential for the desired movement, another motor neuron action potential leads to the release of more ACh, which keeps the contractile process going. By removing contraction-inducing ACh from the motor end plate, AChE permits the choice of al-

lowing relaxation to take place (no more ACh released) or keeping the contraction going (more ACh released), depending on the body's momentary needs.

The neuromuscular junction is vulnerable to several chemical agents and diseases.

Clinical Note Several chemical agents and diseases affect the neuromuscular junction by acting at different sites in the transmission process (▲ Table 7-9). Two well-known toxins—*black widow spider venom* and *botulinum toxin*—alter the release of ACh, but in opposite directions.

BLACK WIDOW SPIDER VENOM CAUSES EXPLOSIVE RELEASE OF ACh The venom of black widow spiders exerts its deadly effect by triggering explosive release of ACh from the storage vesicles,

TABLE 7-8 Comparison of a Synapse and a Neuromuscular Junction

Similarities	Differences
Both consist of two excitable cells separated by a narrow cleft that prevents direct transmission of electrical activity between them.	A synapse is a junction between two neurons. A neuromuscular junction exists between a motor neuron and a skeletal muscle fiber.
Axon terminals of both store chemical messengers (neurotransmitters) that are released by the Ca^{2+}-induced exocytosis of storage vesicles when an action potential reaches the terminal.	One-to-one transmission of action potentials occurs at a neuromuscular junction, whereas one action potential in a presynaptic neuron usually cannot by itself bring about an action potential in a postsynaptic neuron. An action potential in a postsynaptic neuron typically occurs only when summation of EPSPs brings the membrane to threshold.
In both, binding of the neurotransmitter with receptor-channels in the membrane of the cell underlying the axon terminal opens the channels, permitting ionic movements that alter the membrane potential of the cell.	A neuromuscular junction is always excitatory (an EPP); a synapse may be either excitatory (an EPSP) or inhibitory (an IPSP).
The resultant change in membrane potential in both cases is a graded potential.	Inhibition of skeletal muscles cannot be accomplished at the neuromuscular junction; it can occur only in the CNS through IPSPs at the dendrites and cell body of the motor neuron.

TABLE 7-9 Examples of Chemical Agents and Diseases That Affect the Neuromuscular Junction

Mechanism	Chemical Agent or Disease
Alters Release of Acetylcholine	
Causes explosive release of acetylcholine	Black widow spider venom
Blocks release of acetylcholine	*Clostridium botulinum* toxin
Blocks Acetylcholine Receptor-Channels	
Reversibly binds with acetylcholine receptor-channels	Curare
Autoimmune (self-produced) antibodies inactivate acetylcholine receptor-channels	Myasthenia gravis
Prevents Inactivation of Acetylcholine	
Irreversibly inhibits acetylcholinesterase	Organophosphates (certain pesticides and military nerve gases)
Temporarily inhibits acetylcholinesterase	Neostigmine

not only at neuromuscular junctions but at all cholinergic sites. All cholinergic sites undergo prolonged depolarization, the most harmful result of which is respiratory failure. Breathing is accomplished by alternate contraction and relaxation of skeletal muscles, particularly the diaphragm. Respiratory paralysis occurs as a result of prolonged depolarization of the diaphragm. During this so-called **depolarization block,** the voltage-gated Na^+ channels are trapped in their inactivated state (that is, their activation gate remains open and their inactivation gate remains closed; they are in their closed and not capable of opening conformation; see p. 92). This depolarization block prohibits the initiation of new action potentials and resultant contraction of the diaphragm. As a consequence, the victim cannot breathe.

BOTULINUM TOXIN BLOCKS RELEASE OF ACh Botulinum toxin, in contrast, exerts its lethal blow by blocking the release of ACh from the terminal button in response to a motor neuron action potential. *Clostridium botulinum* toxin causes **botulism,** a form of food poisoning. It most frequently results from improperly canned foods contaminated with clostridial bacteria that survive and multiply, producing their toxin in the process. When this toxin is consumed, it prevents muscles from responding to nerve impulses. Death is due to respiratory failure caused by inability to contract the diaphragm. Botulinum toxin is one of the most lethal poisons known; ingesting less than 0.0001 mg can kill an adult human. (See the accompanying boxed feature, ■ Concepts, Challenges, and Controversies, to learn about a surprising wrinkle in the botulinum toxin story.)

CURARE BLOCKS ACTION OF ACh AT RECEPTOR-CHANNELS Other chemicals interfere with neuromuscular junction activity by blocking the effect of released ACh. The best-known example

Botulinum Toxin's Reputation Gets a Facelift

The powerful toxin produced by *Clostridium botulinum* causes the deadly food poisoning botulism. Yet this dreaded, highly lethal poison has been put to use as a treatment for alleviating specific movement disorders and, more recently, has been added to the list of tools that cosmetic surgeons use to fight wrinkles.

During the last decade, botulinum toxin, marketed in therapeutic doses as *Botox*, has offered welcome relief to people with a number of painful, disruptive neuromuscular diseases known categorically as **dystonias.** These conditions are characterized by spasms (excessive, sustained, involuntarily produced muscle contractions) that result in involuntary twisting or abnormal postures, depending on the body part affected. For example, painful neck spasms that twist the head to one side result from *spasmodic torticollis* (*tortus* means "twisted"; *collum* means "neck"), the most common dystonia. The problem is believed to arise from too little inhibitory compared to excitatory input to the motor neurons that supply the affected muscle. The reasons for this imbalance in motor neuron input are unknown. The end result of excessive motor neuron activation is sustained, disabling contraction of the muscle supplied by the overactive motor neurons. Fortunately, injecting minuscule amounts of botulinum toxin into the affected muscle causes a reversible, partial paralysis of the muscle. Botulinum toxin interferes with the release of muscle-contraction-causing acetylcholine from the overactive motor neurons at the neuromuscular junctions in the treated muscle. The goal is to inject just enough botulinum toxin to alleviate the troublesome spasmodic contractions but not enough to eliminate the normal contractions needed for ordinary movements. The therapeutic dose is considerably less than the amount of toxin needed to induce even mild symptoms of botulinum poisoning. Botulinum toxin is eventually cleared away, so its muscle-relaxing effects wear off after 3 to 6 months, at which time the treatment must be repeated.

The first dystonia for which Botox was approved as a treatment by the Food and Drug Administration (FDA) was *blepharospasm* (*blepharo* means "eyelid"). In this condition, sustained and involuntary contractions of the muscles around the eye nearly permanently close the eyelids.

Botulinum toxin's potential as a treatment option for cosmetic surgeons was accidentally discovered when physicians noted that injections used to counter abnormal eye muscle contractions also smoothed the appearance of wrinkles in the treated areas. It turns out that frown lines, crow's feet, and furrowed brows are caused by facial muscles that have become overactivated, or permanently contracted, as a result of years of performing certain repetitive facial expressions. By relaxing these muscles, botulinum toxin temporarily smoothes out these age-related wrinkles. Botox now has FDA approval as an anti-wrinkle treatment. The agent is considered an excellent alternative to facelift surgery for combating lines and creases. This treatment is among the most rapidly growing cosmetic procedures in the United States, especially in the entertainment industry and in high-fashion circles. However, as with its therapeutic use to treat dystonias, the costly injections of botulinum toxin must be repeated every 3 to 6 months to maintain the desired effect in appearance. Furthermore, Botox does not work against the fine, crinkly wrinkles associated with years of excessive sun exposure, because these wrinkles are caused by skin damage, not by contracted muscles.

is the antagonist **curare,** which reversibly binds to the ACh receptor-channels on the motor end plate. Unlike ACh, however, curare does not alter membrane permeability, nor is it inactivated by AChE. When curare occupies ACh receptors, ACh cannot combine with these sites to open the channels that would permit the ionic movement responsible for an EPP. Consequently, because muscle action potentials cannot occur in response to nerve impulses to these muscles, paralysis ensues. When enough curare is present to block a significant number of ACh receptor-channels, the person dies from respiratory paralysis caused by inability to contract the diaphragm. In the past some peoples used curare as a deadly arrowhead poison.

ORGANOPHOSPHATES PREVENT INACTIVATION OF ACh Organophosphates are a group of chemicals that modify neuromuscular junction activity in yet another way—namely, by irreversibly inhibiting AChE. Inhibition of AChE prevents the inactivation of released ACh. Death from organophosphates is also due to respiratory failure because the diaphragm cannot repolarize and return to resting conditions, then contract again to bring in a fresh breath of air. These toxic agents are used in some pesticides and military nerve gases.

MYASTHENIA GRAVIS INACTIVATES ACh RECEPTOR-CHANNELS Myasthenia gravis, a disease involving the neuromuscular junction, is characterized by extreme muscular weakness (*myasthenia* means "muscular weakness"; *gravis* means "severe"). It is an autoimmune condition (*autoimmune* means "immunity against self") in which the body erroneously produces antibodies against its own motor end-plate ACh receptor-channels. Thus, not all the released ACh molecules can find a functioning receptor with which to bind. As a result, AChE destroys much of the ACh before it ever has a chance to interact with a receptor and contribute to the EPP. Treatment consists of administering a drug such as *neostigmine* that inhibits AChE temporarily (in contrast to the toxic organophosphates, which irreversibly block this enzyme). This drug prolongs the action of ACh at the neuromuscular junction by permitting it to build up for the short term. The resultant EPP is of sufficient magnitude to initiate an action potential and subsequent contraction in the muscle fiber, as it normally would.

Chapter in Perspective: Focus on Homeostasis

The nervous system, along with the other major regulatory system, the endocrine system, controls most of the body's muscular and glandular activities. Whereas the afferent division of the PNS detects and carries information to the CNS for processing and decision making, the efferent division of the PNS carries directives from the CNS to the effector organs (muscles and glands), which carry out the intended response. Much of this efferent output is directed toward maintaining homeostasis.

The autonomic nervous system, which is the efferent branch that innervates smooth muscle, cardiac muscle, and glands, plays a major role in the following range of homeostatic activities:

- regulating blood pressure

- controlling digestive juice secretion and digestive tract contractions that mix ingested food with the digestive juices
- controlling sweating to help maintain body temperature

The somatic nervous system, the efferent branch that innervates skeletal muscle, contributes to homeostasis by stimulating the following activities:

- Skeletal muscle contractions that enable the body to move in relation to the external environment contribute to homeostasis by moving the body toward food or away from harm.
- Skeletal muscle contraction also accomplishes breathing to maintain appropriate levels of O_2 and CO_2 in the body.
- Shivering is a skeletal muscle activity important in maintaining body temperature.

In addition, efferent output to skeletal muscles accomplishes many movements that are not aimed at maintaining a stable internal environment but nevertheless enrich our lives and enable us to engage in activities that contribute to society, such as dancing, building bridges, or performing surgery.

REVIEW EXERCISES

Objective Questions (Answers on p. A-44)

1. Sympathetic preganglionic fibers begin in the thoracic and lumbar segments of the spinal cord. *(True or false?)*

2. Action potentials are transmitted on a one-to-one basis at both a neuromuscular junction and a synapse. *(True or false?)*

3. The sympathetic nervous system
   a. is always excitatory
   b. innervates only tissues concerned with protecting the body against challenges from the outside environment
   c. has short preganglionic and long postganglionic fibers
   d. is part of the afferent division of the PNS
   e. is part of the somatic nervous system

4. Acetylcholinesterase
   a. is stored in vesicles in the terminal button
   b. combines with receptor-channels on the motor end plate to bring about an end-plate potential
   c. is inhibited by organophosphates
   d. is the chemical transmitter at the neuromuscular junction
   e. paralyzes skeletal muscle by strongly binding with acetylcholine receptor-channels

5. The two divisions of the autonomic nervous system are the _____ nervous system, which dominates in "fight-or-flight" situations, and the _____ nervous system, which dominates in "rest-and-digest" situations.

6. The _____ is a modified sympathetic ganglion that does not give rise to postganglionic fibers but instead secretes hormones similar or identical to sympathetic postganglionic neurotransmitters into the blood.

7. Using the answer code on the right, identify the autonomic neurotransmitter being described:
   ___ 1. secreted by all preganglionic fibers
   ___ 2. secreted by sympathetic postganglionic fibers
   ___ 3. secreted by parasympathetic postganglionic fibers
   ___ 4. secreted by the adrenal medulla
   ___ 5. secreted by somatic motor neurons
   ___ 6. binds to muscarinic or nicotinic receptors
   ___ 7. binds to α or β receptors

   (a) acetylcholine
   (b) norepinephrine

8. Using the answer code on the right, indicate which type of efferent output is being described:
   ___ 1. composed of two-neuron chains
   ___ 2. innervates cardiac muscle, smooth muscle, and glands
   ___ 3. innervates skeletal muscle
   ___ 4. consists of the axons of motor neurons
   ___ 5. exerts either an excitatory or an inhibitory effect on its effector organs
   ___ 6. dually innervates its effector organs
   ___ 7. exerts only an excitatory effect on its effector organs

   (a) characteristic of the somatic nervous system
   (b) characteristic of the autonomic nervous system

9. Using the answer code on the right, indicate what types of receptors are present for each of the organs listed (more than one answer may apply).

  ___ 1. heart   (a) α_1
  ___ 2. arteriolar smooth muscle   (b) α_2
  ___ 3. bronchiolar smooth muscle   (c) β_1
  ___ 4. skeletal muscle fibers   (d) β_2
  ___ 5. adrenal medulla   (e) nicotinic
  ___ 6. digestive glands   (f) muscarinic

Essay Questions

1. Distinguish between preganglionic and postganglionic fibers.

2. Compare the origin, preganglionic and postganglionic fiber length, and neurotransmitters of the sympathetic and parasympathetic nervous systems.

3. What is the advantage of dual innervation of many organs by both branches of the autonomic nervous system?

4. Distinguish among the following types of receptors: nicotinic receptors, muscarinic receptors, α_1 receptors, α_2 receptors, β_1 receptors, and β_2 receptors.

5. What regions of the CNS regulate autonomic output?

6. Why are somatic motor neurons called the "final common pathway"?

7. Describe the sequence of events that occurs at a neuromuscular junction.

8. Discuss the effect each of the following has at the neuromuscular junction: black widow spider venom, botulinum toxin, curare, myasthenia gravis, organophosphates and neostigmine.

Quantitative Exercises (Solutions on p. A-44)

1. When a muscle fiber is activated at the neuromuscular junction, tension does not begin to rise until about 1 msec after initiation of the action potential in the muscle fiber. Many things occur during this delay, one time-consuming event being diffusion of ACh across the neuromuscular junction. The following equation can be used to calculate how long this diffusion takes:

$$t = x^2/2D$$

In this equation, x is the distance covered, D is the diffusion coefficient, and t is the time it takes for the substance to diffuse across the distance x. In this example, x is the width of the cleft between the neuronal axon terminal and the muscle fiber at the neuromuscular junction (assume 200 nm), and D is the diffusion coefficient of ACh (assume 1×10^{-5} cm^2/sec). How long does it take ACh to diffuse across the neuromuscular junction?

POINTS TO PONDER

(Explanations on p. A-44)

1. Explain why epinephrine, which causes arteriolar constriction (narrowing) in most tissues, is frequently administered in conjunction with local anesthetics.

2. Would skeletal muscle activity be affected by atropine? Why or why not?

3. Considering that you can voluntarily control the emptying of your urinary bladder by contracting (preventing emptying) or relaxing (permitting emptying) your external urethral sphincter, a ring of muscle that guards the exit from the bladder, of what type of muscle is this sphincter composed and what branch of the nervous system supplies it?

4. The venom of certain poisonous snakes contains α bungarotoxin, which binds tenaciously to ACh receptor sites on the motor end-plate membrane. What would the resultant symptoms be?

5. Explain how destruction of motor neurons by poliovirus or amyotrophic lateral sclerosis can be fatal.

CLINICAL CONSIDERATION

(Explanation on p. A-44)

Christopher K. experienced chest pains when he climbed the stairs to his fourth-floor office or played tennis but had no symptoms when not physically exerting himself. His condition was diagnosed as **angina pectoris** (*angina* means "pain"; *pectoris* means "chest"), heart pain that occurs whenever the blood supply to the heart muscle cannot meet the muscle's need for oxygen delivery. This condition usually is caused by narrowing of the blood vessels supplying the heart by cholesterol-containing deposits. Most people with this condition do not have any pain at rest but experience bouts of pain whenever the heart's need for oxygen increases, such as during exercise or emotionally stressful situations that increase sympathetic nervous activity. Christopher obtains immediate relief of angina attacks by promptly taking a vasodilator drug such as *nitroglycerin*, which relaxes the smooth muscle in the walls of his narrowed heart vessels. Consequently, the vessels open more widely and more blood can flow through them. For prolonged treatment, his doctor has indicated that Christopher will experience fewer and less severe angina attacks if he takes a β_1-blocker drug, such as metoprolol, on a regular basis. Explain why.

Muscular System

Homeostasis
Skeletal muscles contribute to homeostasis by playing a major role in the procurement of food, breathing, heat generation for maintenance of body temperature, and movement away from harm.

Body systems maintain homeostasis

Homeostasis is essential for survival of cells

Cells

Cells make up body systems

Muscles are the contraction specialists of the body. **Skeletal muscle** attaches to the skeleton. Contraction of skeletal muscles moves bones to which they are attached, allowing the body to perform a variety of motor activities. Skeletal muscles that support homeostasis include those important in acquiring, chewing, and swallowing food and those essential for breathing. Also, heat-generating muscle contractions are important in regulating temperature. Skeletal muscles are further used to move the body away from harm. Skeletal muscle contractions are also important for nonhomeostatic activities, such as dancing or operating a computer. **Smooth muscle** is found in the walls of hollow organs and tubes. Controlled contraction of smooth muscle regulates movement of blood through blood vessels, food through the digestive tract, air through respiratory airways, and urine to the exterior. **Cardiac muscle** is found only in the walls of the heart, whose contraction pumps life-sustaining blood throughout the body.

Muscle Physiology

CENGAGENOW™ Log on to CengageNOW at **http://www.cengage.com/sso/** for an opportunity to explore a learning module that illustrates difficult concepts with self-study tutorials, animations, and interactive quizzes to help you learn, review, and master physiology concepts.

Structure of Skeletal Muscle

By moving specialized intracellular components, muscle cells can develop tension and shorten, that is, contract. Recall that the three types of muscle are *skeletal muscle, cardiac muscle,* and *smooth muscle* (see p. 4). Through their highly developed ability to contract, groups of muscle cells working together within a muscle can produce movement and do work. Controlled contraction of muscles allows (1) purposeful movement of the whole body or parts of the body (such as walking or waving your hand), (2) manipulation of external objects (such as driving a car or moving a piece of furniture), (3) propulsion of contents through various hollow internal organs (such as circulation of blood or movement of a meal through the digestive tract), and (4) emptying the contents of certain organs to the external environment (such as urination or giving birth).

Muscle comprises the largest group of tissues in the body, accounting for approximately half of the body's weight. Skeletal muscle alone makes up about 40% of body weight in men and 32% in women, with smooth and cardiac muscle making up another 10% of the total weight. Although the three muscle types are structurally and functionally distinct, they can be classified in two different ways according to their common characteristics (● Figure 8-1). First, muscles are categorized as *striated* (skeletal and cardiac muscle) or *unstriated* (smooth muscle), depending on whether alternating dark and light bands, or striations (stripes), can be seen when the muscle is viewed under a light microscope. Second, muscles are categorized as *voluntary* (skeletal muscle) or *involuntary* (cardiac and smooth muscle), depending, respectively, on whether they are innervated by the somatic nervous system and are subject to voluntary control, or are innervated by the autonomic nervous system and are not subject to voluntary control (see p. 237). Although skeletal muscle is categorized as voluntary, because it can be consciously controlled, much skeletal muscle activity is also subject to subconscious, involuntary regulation, such as that related to posture, balance, and stereotypical movements like walking.

Most of this chapter is a detailed examination of the most abundant and best understood muscle, skeletal muscle. Skeletal muscles make up the muscular system. We will begin with a discussion of skeletal muscle structure, then examine how it works from the molecular level through the cell level and finally to the whole muscle. The chapter concludes with a discussion of the unique properties of smooth and cardiac muscle in comparison to skeletal muscle. Smooth muscle appears throughout the body systems as components of hollow organs and tubes. Cardiac muscle is found only in the heart.

Skeletal muscle fibers are striated by a highly organized internal arrangement.

A single skeletal muscle cell, known as a **muscle fiber,** is relatively large, elongated, and cylinder shaped, measuring from 10 to 100 micrometers (μm) in diameter and up to 750,000 μm, or 2.5 feet, in length (1 μm = 1 millionth of a meter). A skeletal muscle consists of a number of muscle fibers lying parallel to one another and bundled together by connective tissue ($\bullet$ Figure 8-2a). The fibers usually extend the entire length of the muscle. During embryonic development, the huge skeletal muscle fibers are formed by the fusion of many smaller cells called **myoblasts** (*myo* means "muscle"; *blast* refers to a primitive cell that forms more specialized cells); thus, one striking feature is the presence of multiple nuclei in a single muscle cell. Another feature is the abundance of mitochondria, the energy-generating organelles, as would be expected with the high energy demands of a tissue as active as skeletal muscle.

The predominant structural feature of a skeletal muscle fiber is numerous **myofibrils.** These specialized contractile elements, which constitute 80% of the volume of the muscle fiber, are cylindrical intracellular structures 1 mm in diameter that extend the entire length of the muscle fiber (Figure 8-2b). Each myofibril consists of a regular arrangement of highly organized cytoskeletal elements—the thick and the thin filaments ($\bullet$ Figure 8-2c). The **thick filaments,** which are 12 to 18 nm in diameter and 1.6 μm in length, are special assemblies of the protein *myosin,* whereas the **thin filaments,** which are 5 to 8 nm in diameter and 1.0 μm long, are made up primarily of the protein *actin* ($\bullet$ Figure 8-2d). The levels of organization in a skeletal muscle can be summarized as follows:

$$\text{Whole muscle} \rightarrow \text{muscle fiber} \rightarrow \text{myofibril} \rightarrow \begin{array}{c}\text{thick and thin filaments}\end{array} \rightarrow \begin{array}{c}\text{myosin and actin}\end{array}$$

(an organ) (a cell) (a specialized intracellular structure) (cytoskeletal elements) (protein molecules)

A AND I BANDS Viewed with an electron microscope, a myofibril displays alternating dark bands (the *A bands*) and light bands (the *I bands*) ($\bullet$ Figure 8-3a). The bands of all the myofibrils lined up parallel to one another collectively produce the striated or striped appearance of a skeletal muscle fiber visible under a light microscope ($\bullet$ Figure 8-3b). Alternate stacked sets of thick and thin filaments that slightly overlap one another are responsible for the A and I bands (see $\bullet$ Figure 8-2c). This precise filament geometry is maintained by several cytoskeletal proteins.

An **A band** is made up of a stacked set of thick filaments along with the portions of the thin filaments that overlap on both ends of the thick filaments. The thick filaments lie only within the A band and extend its entire width; that is, the two ends of the thick filaments within a stack define the outer limits of a given A band. The lighter area within the middle of the A band, where the thin filaments do not reach, is the **H zone.** Only the central portions of the thick filaments are found in this region. A system of supporting proteins holds the thick filaments together vertically within each stack. These proteins can be seen as the **M line,** which extends vertically down the middle of the A band within the center of the H zone.

An **I band** consists of the remaining portion of the thin filaments that do not project into the A band. Visible in the middle of each I band is a dense, vertical **Z line.** The area between two Z lines is called a **sarcomere,** which is the functional unit of skeletal muscle. A **functional unit** of any organ is the smallest component that can perform all the functions of that organ. Accordingly, a sarcomere is the smallest component of a muscle fiber that can contract. The Z line is a flat, cytoskeletal disc that connects the thin filaments of two adjoining sarcomeres. Each relaxed sarcomere is about 2.5 μm in width and consists of one whole A band and half of each of the two I bands located on either side. An I band contains only thin filaments from two adjacent sarcomeres but not the entire length of these filaments. During growth, a muscle increases in length by adding new sarcomeres on the ends of the myofibrils, not by increasing the size of each sarcomere.

Not shown in the figure, single strands of a giant, highly elastic protein known as **titin** extend in both directions from the M line along the length of the thick filament to the Z lines at opposite ends of the sarcomere. Titin is the largest protein in the body, being made up of nearly 30,000 amino acids. It serves two important roles: (1) along with the M-line proteins, titin helps stabilize the position of the thick filaments in relation to the thin filaments; and (2) by acting like a spring, it greatly augments a muscle's elasticity. That is, titin helps a muscle stretched by an external force passively recoil or

Striated muscle

Unstriated muscle

Skeletal muscle | Cardiac muscle | Smooth muscle

Ed Reschke

Ed Reschke

Biophoto/Photo Researchers, Inc.

Voluntary muscle

Involuntary muscle

$\bullet$ **FIGURE 8-1 Categorization of muscle.**

(a) Relationship of a whole muscle and a muscle fiber

(b) Relationship of a muscle fiber and a myofibril

(c) Cytoskeletal components of a myofibril

(d) Protein components of thick and thin filaments

● **FIGURE 8-2 Levels of organization in a skeletal muscle.** Note in the cross section of a myofibril in part (c) that each thick filament is surrounded by six thin filaments, and each thin filament is surrounded by three thick filaments.

spring back to its resting length when the stretching force is removed, much like a stretched spring.

CROSS BRIDGES With an electron microscope, fine **cross bridges** can be seen extending from each thick filament toward the surrounding thin filaments in the areas where the thick and thin filaments overlap (see the longitudinal view in ● Figure 8-2c). Three-dimensionally, the thin filaments are arranged hexagonally around the thick filaments. Cross bridges project from each thick filament in all six directions toward the sur-

Z line   Z line   M line

|←——— Sarcomere ———→|

H zone

|→|←|←— I band —→|←— A band —→|

(a) Electron micrograph of a myofibril

Sydney Schochet Jr., M.D.

M. Abbey/Science Source/Photo Researchers, Inc.

(b) Low-power light micrograph of skeletal muscle fibers

● **FIGURE 8-3 Microscope view of skeletal muscle components.** (a) Note the A and I bands. (b) Note striated appearance.

Source: Reprinted with permission from Sydney Schochet Jr., M.D., Professor, Department of Pathology, School of Medicine, West Virginia University, *Diagnostic Pathology of Skeletal Muscle and Nerve,* ● Fig. 1-13 (Stamford, CT: Appleton & Lange, 1986).

rounding thin filaments. Each thin filament, in turn, is surrounded by three thick filaments (see the cross section view in ● Figure 8-2c). To give you an idea of the magnitude of these filaments, a single muscle fiber may contain an estimated 16 billion thick and 32 billion thin filaments, all arranged in this very precise pattern within the myofibrils.

Myosin forms the thick filaments.

Each thick filament has several hundred myosin molecules packed together in a specific arrangement. A **myosin** molecule is a protein consisting of two identical subunits, each

shaped somewhat like a golf club (● Figure 8-4a). The protein's tail ends are intertwined around each other like golf-club shafts twisted together, with the two globular heads projecting out at one end. The two halves of each thick filament are mirror images made up of myosin molecules lying lengthwise in a regular, staggered array, with their tails oriented toward the center of the filament and their globular heads protruding outward at regular intervals (● Figure 8-4b). These heads form the cross bridges between the thick and thin filaments. Each cross bridge has two important sites crucial to the contractile process: (1) an actin-binding site and (2) a myosin ATPase (ATP-splitting) site.

● **FIGURE 8-4 Structure of myosin molecules and their organization within a thick filament.** (a) Each myosin molecule consists of two identical, golf-club-shaped subunits with their tails intertwined and their globular heads, each of which contains an actin-binding site and a myosin ATPase site, projecting out at one end. (b) A thick filament is made up of myosin molecules lying lengthwise parallel to one another. Half are oriented in one direction and half in the opposite direction. The globular heads, which protrude at regular intervals along the thick filament, form the cross bridges.

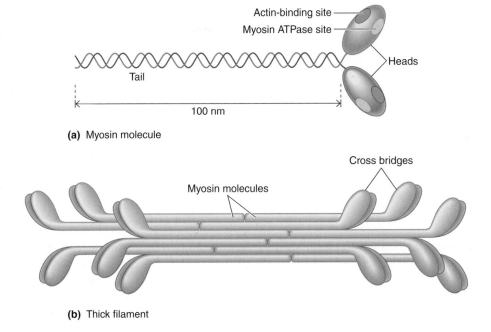

Actin-binding site

Myosin ATPase site

Heads

Tail

100 nm

(a) Myosin molecule

Cross bridges

Myosin molecules

(b) Thick filament

Actin is the main structural component of the thin filaments.

Thin filaments consist of three proteins: *actin, tropomyosin,* and *troponin* (● Figure 8-5). **Actin** molecules, the primary structural proteins of the thin filament, are spherical. The backbone of a thin filament is formed by actin molecules joined into two strands and twisted together, like two intertwined strings of pearls. Each actin molecule has a special binding site for attachment with a myosin cross bridge. By a mechanism to be described shortly, binding of myosin and actin molecules at the cross bridges results in energy-consuming contraction of the muscle fiber. Accordingly, myosin and actin are often called **contractile proteins,** even though, as you will see, neither myosin nor actin actually contracts. Myosin and actin are not unique to muscle cells, but these proteins are more abundant and more highly organized in muscle cells (see p. 47).

In a relaxed muscle fiber, contraction does not take place; actin cannot bind with cross bridges, because of the way the two other types of protein—tropomyosin and troponin—are positioned within the thin filament. **Tropomyosin** molecules are threadlike proteins that lie end to end alongside the groove of the actin spiral. In this position, tropomyosin covers the actin sites that bind with the cross bridges, blocking the interaction that leads to muscle contraction. The other thin filament component, **troponin,** is a protein complex made of three polypeptide units: one binds to tropomyosin, one binds to actin, and a third can bind with Ca^{2+}.

When troponin is not bound to Ca^{2+}, this protein stabilizes tropomyosin in its blocking position over actin's cross-bridge binding sites (● Figure 8-6a). When Ca^{2+} binds to troponin, the shape of this protein is changed in such a way that tropomyosin slips away from its blocking position (● Figure 8-6b). With tropomyosin out of the way, actin and myosin can bind and interact at the cross bridges, resulting in muscle contraction. Tropomyosin and troponin are often called **regulatory proteins** because of their role in covering (preventing contraction) or exposing (permitting contraction) the binding sites for cross-bridge interaction between actin and myosin.

Molecular Basis of Skeletal Muscle Contraction

Several important links in the contractile process remain to be discussed. How does cross-bridge interaction between actin and myosin bring about muscle contraction? How does a muscle action potential trigger this contractile process? What is the source of the Ca^{2+} that physically repositions troponin and tropomyo-

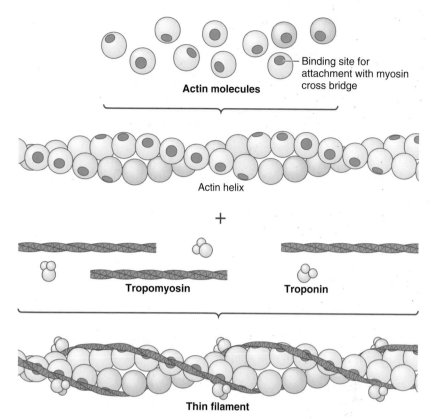

● **FIGURE 8-5 Composition of a thin filament.** The main structural component of a thin filament is two chains of spherical actin molecules that are twisted together. Troponin molecules (which consist of three small, spherical subunits) and threadlike tropomyosin molecules are arranged to form a ribbon that lies alongside the groove of the actin helix and physically covers the binding sites on actin molecules for attachment with myosin cross bridges. (The thin filaments shown here are not drawn in proportion to the thick filaments in ● Figure 8-4. Thick filaments are two to three times larger in diameter than thin filaments.)

sin to permit cross-bridge binding? We will turn our attention to these topics in this section.

During contraction, cycles of cross-bridge binding and bending pull the thin filaments inward.

Cross-bridge interaction between actin and myosin brings about muscle contraction by means of the sliding filament mechanism.

SLIDING FILAMENT MECHANISM The thin filaments on each side of a sarcomere slide inward over the stationary thick filaments toward the A band's center during contraction (● Figure 8-7). As they slide inward, the thin filaments pull the Z lines to which they are attached closer together, so the sarcomere shortens. As all the sarcomeres throughout the muscle fiber's length shorten simultaneously, the entire fiber shortens. This is the **sliding filament mechanism** of muscle contraction. The H zone, in the center of the A band where the thin filaments do not reach, becomes smaller as the thin filaments approach each

Longitudinal view

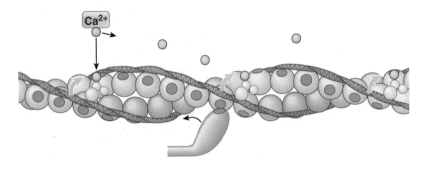

Tropomyosin

Troponin

Actin

Myosin cross-bridge binding sites

Actin-binding site

Myosin cross bridge

Cross-sectional view

Actin

Myosin cross-bridge binding site

Troponin

Tropomyosin

Actin-binding site

Myosin cross bridge

(a) Relaxed

1 No excitation.

2 No cross-bridge binding because cross-bridge binding site on actin is physically covered by troponin-tropomyosin complex.

3 Muscle fiber is relaxed.

Ca^{2+}

Ca^{2+}

(b) Excited

1 Muscle fiber is excited and Ca^{2+} is released.

2 Released Ca^{2+} binds with troponin, pulling troponin-tropomyosin complex aside to expose cross-bridge binding site.

3 Cross-bridge binding occurs.

4 Binding of actin and myosin cross bridge triggers power stroke that pulls thin filament inward during contraction.

● **FIGURE 8-6 Role of calcium in turning on cross bridges.**

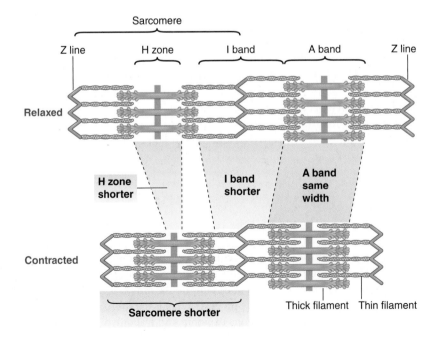

Sarcomere

Z line   H zone   I band   A band   Z line

Relaxed

H zone shorter

I band shorter

A band same width

Contracted

Sarcomere shorter

Thick filament   Thin filament

● **FIGURE 8-7 Changes in banding pattern during shortening.** During muscle contraction, each sarcomere shortens as the thin filaments slide closer together between the thick filaments so that the Z lines are pulled closer together. The width of the A bands does not change as a muscle fiber shortens, but the I bands and H zones become shorter.

other when they slide more deeply inward. The I band, which consists of the portions of the thin filaments that do not overlap with the thick filaments, narrows as the thin filaments further overlap the thick filaments during their inward slide. The thin filaments themselves do not change length during muscle fiber shortening. The width of the A band remains unchanged during contraction because its width is determined by the length of the thick filaments, and the thick filaments do not change length during the shortening process. Note that neither the thick nor the thin filaments decrease in length to shorten the sarcomere. Instead, contraction is accomplished by the thin filaments from the opposite sides of each sarcomere sliding closer together between the thick filaments.

POWER STROKE During contraction, with the tropomyosin and troponin "chaperones" pulled out of the way by Ca^{2+}, the myosin cross bridges from a thick filament can bind with the actin molecules in the surrounding thin filaments. Myosin is a molecular motor, similar to kinesin and dynein. Recall that kinesin and dynein have little feet that "walk" along microtubules to transport specific products from one part of the cell to another (as in protein transport within a neuronal axon; see p. 45) and to move microtubules in relation to one another (as to accomplish beating of cilia or flagella; see p. 47). In the same way, the myosin heads or cross bridges "walk" along an actin filament to pull it inward relative to the stationary thick filament. Let's concentrate on a single cross-bridge interaction (● Figure 8-8a). The two myosin heads of each myosin molecule act independently, with only one head attaching to actin at a given time. When myosin and actin make contact at a cross bridge, the bridge changes shape, bending 45° inward as if it were on a hinge, "stroking" toward the center of the sarcomere, like the stroking of a boat oar. This so-called **power stroke** of a cross bridge pulls inward the thin filament to which it is attached. A single power stroke pulls the thin filament inward only a small percentage of the total shortening distance. Repeated cycles of cross-bridge binding and bending complete the shortening.

At the end of one cross-bridge cycle, the link between the myosin cross bridge and actin molecule breaks. The cross bridge returns to its original shape and binds to the next actin molecule behind its previous actin partner. The cross bridge bends once again to pull the thin filament in farther, then detaches and repeats the cycle. Repeated cycles of cross-bridge power strokes successively pull in the thin filaments, much like pulling in a rope hand over hand.

Because of the way myosin molecules are oriented within a thick filament (● Figure 8-8b), all the cross bridges stroke toward the center of the sarcomere so that all six of the surrounding thin filaments on each end of the sarcomere are pulled inward simul-

1 Binding: Myosin cross bridge binds to actin molecule.

2 Power stroke: Cross bridge bends, pulling thin myofilament inward.

3 Detachment: Cross bridge detaches at end of power stroke and returns to original conformation.

4 Binding: Cross bridge binds to more distal actin molecule; cycle repeats.

(a) Single cross-bridge cycle

(b) All cross-bridge stroking directed toward center of thick filament

Thin myofilament   Thick myofilament

(c) Simultaneous pulling inward of all six thin filaments surrounding a thick filament

● **FIGURE 8-8 Cross-bridge activity.** (a) During each cross-bridge cycle, the cross bridge binds with an actin molecule, bends to pull the thin filament inward during the power stroke, then detaches and returns to its resting conformation, ready to repeat the cycle. (b) The power strokes of all cross bridges extending from a thick filament are directed toward the center of the thick filament. (c) All six thin filaments surrounding each thick filament are pulled inward simultaneously through cross-bridge cycling during muscle contraction.

taneously (● Figure 8-8c). The cross bridges aligned with given thin filaments do not all stroke in unison, however. At any time during contraction, part of the cross bridges are attached to the thin filaments and are stroking, while others are returning to their original conformation in preparation for binding with another actin molecule. Thus, some cross bridges are "holding on" to the thin filaments, whereas others "let go" to bind with new actin. Were it not for this asynchronous cycling of the cross bridges, the thin filaments would slip back toward their resting position between strokes.

How does muscle excitation switch on this cross-bridge cycling? The term **excitation–contraction coupling** refers to the series of events linking muscle excitation (the presence of an action potential in a muscle fiber) to muscle contraction (cross-bridge activity that causes the thin filaments to slide closer together to produce sarcomere shortening). We now turn attention to excitation–contraction coupling.

Calcium is the link between excitation and contraction.

Skeletal muscles are stimulated to contract by release of acetylcholine (ACh) at neuromuscular junctions between motor-neuron terminal buttons and muscle fibers. Recall that binding of ACh with the motor end plate of a muscle fiber brings about permeability changes in the muscle fiber, resulting in an action potential that is conducted over the entire surface of the muscle cell membrane (see p. 247). Two membranous structures within the muscle fiber play an important role in linking this excitation to contraction—*transverse tubules* and the *sarcoplasmic reticulum*. Let's examine the structure and function of each.

SPREAD OF THE ACTION POTENTIAL DOWN THE T TUBULES At each junction of an A band and I band, the surface membrane dips into the muscle fiber to form a **transverse tubule (T tubule),** which runs perpendicularly from the surface of the muscle cell membrane into the central portions of the muscle fiber (● Figure 8-9). Because the T tubule membrane is continuous with the surface membrane, an action potential on the surface membrane also spreads down into the T tubule, rapidly transmitting the surface electrical activity into the central portions of the fiber. The presence of a local action potential in the T tubules leads to permeability changes in a separate membranous network within the muscle fiber, the sarcoplasmic reticulum.

RELEASE OF CALCIUM FROM THE SARCOPLASMIC RETICULUM The **sarcoplasmic reticulum** is a modified endoplasmic reticulum (see p. 24) that consists of a fine network of interconnected membrane-enclosed compartments surrounding each myofibril like a mesh sleeve (● Figure 8-9). This membranous network encircles the myofibril throughout its length but is not continuous. Separate segments of sarcoplasmic reticulum are wrapped around each A band and each I band. The ends of each segment expand to form saclike regions, the **lateral sacs** (alternatively known as **terminal cisternae**), which are separated from the adjacent T tubules by a slight gap. The sarcoplasmic reticulum's lateral sacs store Ca^{2+}. Spread of an action potential down a T tubule triggers release of Ca^{2+} from the sarcoplasmic reticulum into the cytosol.

How is a change in T tubule potential linked with the release of Ca^{2+} from the lateral sacs? An orderly arrangement of **foot proteins** extends from the sarcoplasmic reticulum and spans the gap between the lateral sac and T tubule. Each foot protein contains four subunits arranged in a specific pattern (● Figure 8-10a). These foot proteins not only bridge the gap but also serve as *Ca^{2+}-release channels.* These foot-protein Ca^{2+} channels are also known as **ryanodine receptors** because they are locked in the open position by the plant chemical ryanodine.

Half of the sarcoplasmic reticulum's foot proteins are "zipped together" with complementary receptors on the T tubule side of the junction. These T tubule receptors, which are made up of four subunits in exactly the same pattern as the foot proteins, are located like mirror images in contact with every other foot protein protruding from the sarcoplasmic reticulum (● Figure 8-10b and c). These T tubule receptors are known as **dihydropyridine receptors** because they are blocked by the drug dihydropyridine. The dihydropyridine receptors are voltage-gated sensors. When an action potential is propagated down the

Surface membrane of muscle fiber

Myofibrils

Segments of sarcoplasmic reticulum

Lateral sacs

Transverse (T) tubule

|←—I band —→| |————— A band —————| |←—I band —→|

● **FIGURE 8-9 The T tubules and sarcoplasmic reticulum in relationship to the myofibrils.** The transverse (T) tubules are membranous, perpendicular extensions of the surface membrane that dip deep into the muscle fiber at the junctions between the A and I bands of the myofibrils. The sarcoplasmic reticulum is a fine, membranous network that runs longitudinally and surrounds each myofibril, with separate segments encircling each A band and I band. The ends of each segment are expanded to form lateral sacs that lie next to the adjacent T tubules.

(a) Receptors in sarcoplasmic reticulum

(b) Receptors in T tubule

(c) Ca²⁺ release from sarcoplasmic reticulum on opening of this structure's Ca²⁺-release channels (foot proteins/ryanodine receptors) in response to voltage-gated activation of the T tubule's dihydropyridine receptors

• **FIGURE 8-10 Relationship between dihydropyridine receptors on the T tubule and ryanodine receptors (Ca²⁺-release channels) on the adjacent lateral sacs of the sarcoplasmic reticulum.**

T tubule, the local depolarization activates the voltage-gated dihydropyridine receptors. The activated T tubule receptors, in turn, trigger the opening of the directly abutting Ca²⁺-release channels (alias ryanodine receptors alias foot proteins) in the adjacent lateral sacs of the sarcoplasmic reticulum. Opening of the half of the Ca²⁺-release channels in direct contact with the dihydropyridine receptors triggers the opening of the other half of the Ca²⁺-release channels that are not directly associated with the T tubule receptors.

Calcium is released into the cytosol from the lateral sacs through all these open Ca²⁺-release channels. By slightly repositioning the troponin and tropomyosin molecules, this released Ca²⁺ exposes the binding sites on the actin molecules so they can link with the myosin cross bridges at their complementary binding sites. Excitation–contraction coupling is summarized in • Figure 8-11.

ATP-POWERED CROSS-BRIDGE CYCLING Recall that a myosin cross bridge has two special sites, an actin-binding site and an ATPase site (see • Figure 8-4a). The latter is an enzymatic site that can bind the energy carrier *adenosine triphosphate (ATP)*

and split it into *adenosine diphosphate (ADP)* and *inorganic phosphate (Pᵢ)*, yielding energy in the process. The breakdown of ATP occurs on the myosin cross bridge before the bridge ever links with an actin molecule (step **1** in • Figure 8-12). The ADP and Pᵢ remain tightly bound to the myosin, and the generated energy is stored within the cross bridge to produce a high-energy form of myosin. To use an analogy, the cross bridge is "cocked" like a gun, ready to be fired when the trigger is pulled. When the muscle fiber is excited, Ca²⁺ pulls the troponin–tropomyosin complex out of its blocking position so that the energized (cocked) myosin cross bridge can bind with an actin molecule (step **2a**). This contact between myosin and actin "pulls the trigger," causing the cross-bridge bending that produces the power stroke (step **3**). Researchers have not found the mechanism by which the chemical energy released from ATP is stored within the myosin cross bridge and then translated into the mechanical energy of the power stroke. Inorganic phosphate is released from the cross bridge during the power stroke. After the power stroke is complete, ADP is released.

When the muscle is not excited and Ca²⁺ is not released, troponin and tropomyosin remain in their blocking position so

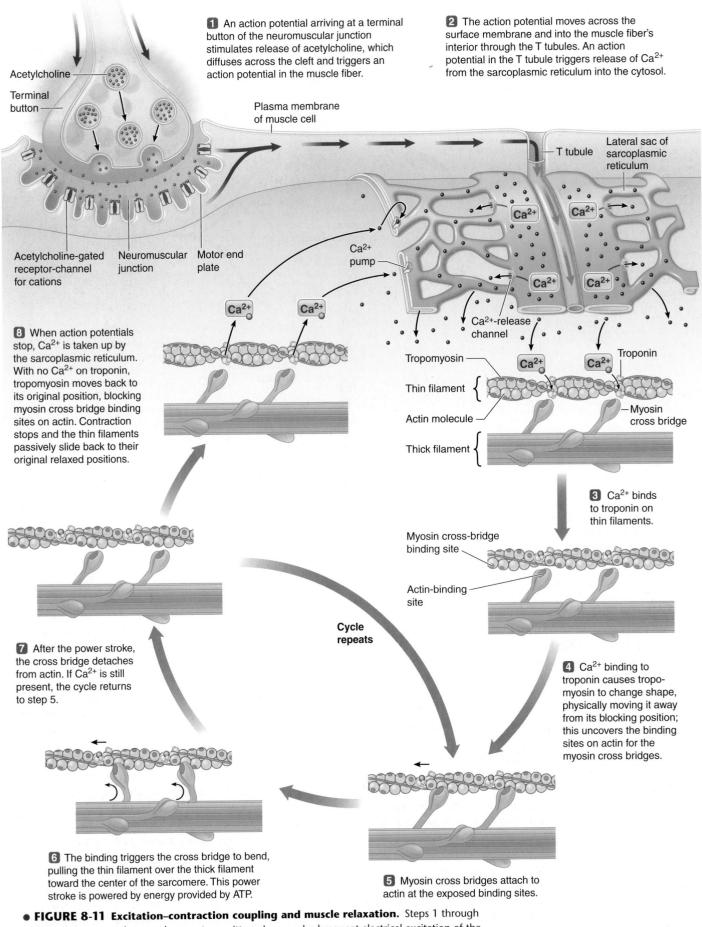

1 An action potential arriving at a terminal button of the neuromuscular junction stimulates release of acetylcholine, which diffuses across the cleft and triggers an action potential in the muscle fiber.

2 The action potential moves across the surface membrane and into the muscle fiber's interior through the T tubules. An action potential in the T tubule triggers release of Ca^{2+} from the sarcoplasmic reticulum into the cytosol.

Acetylcholine

Terminal button

Plasma membrane of muscle cell

T tubule

Lateral sac of sarcoplasmic reticulum

Acetylcholine-gated receptor-channel for cations

Neuromuscular junction

Motor end plate

Ca^{2+} pump

Ca^{2+}

Ca^{2+}-release channel

Tropomyosin

Thin filament

Actin molecule

Thick filament

Troponin

Myosin cross bridge

8 When action potentials stop, Ca^{2+} is taken up by the sarcoplasmic reticulum. With no Ca^{2+} on troponin, tropomyosin moves back to its original position, blocking myosin cross bridge binding sites on actin. Contraction stops and the thin filaments passively slide back to their original relaxed positions.

3 Ca^{2+} binds to troponin on thin filaments.

Myosin cross-bridge binding site

Actin-binding site

Cycle repeats

7 After the power stroke, the cross bridge detaches from actin. If Ca^{2+} is still present, the cycle returns to step 5.

4 Ca^{2+} binding to troponin causes tropomyosin to change shape, physically moving it away from its blocking position; this uncovers the binding sites on actin for the myosin cross bridges.

6 The binding triggers the cross bridge to bend, pulling the thin filament over the thick filament toward the center of the sarcomere. This power stroke is powered by energy provided by ATP.

5 Myosin cross bridges attach to actin at the exposed binding sites.

● **FIGURE 8-11 Excitation–contraction coupling and muscle relaxation.** Steps 1 through 7 show the events that couple neurotransmitter release and subsequent electrical excitation of the muscle cell with muscle contraction. At step 7, if Ca^{2+} is still present, the cross-bridge cycle returns to step 5 for another power stroke. If Ca^{2+} is no longer present as a consequence of step 8, relaxation occurs.

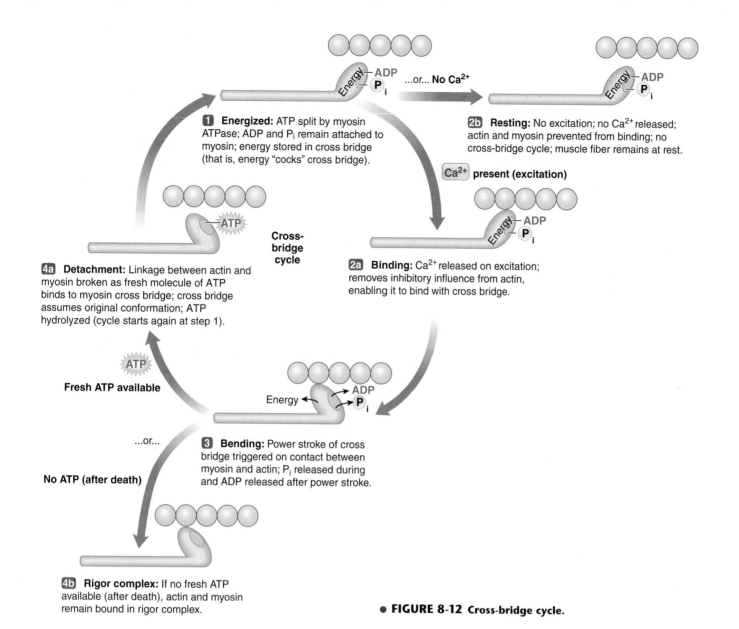

1. **Energized:** ATP split by myosin ATPase; ADP and P_i remain attached to myosin; energy stored in cross bridge (that is, energy "cocks" cross bridge).

...or... **No Ca²⁺**

2b. **Resting:** No excitation; no Ca²⁺ released; actin and myosin prevented from binding; no cross-bridge cycle; muscle fiber remains at rest.

Ca²⁺ present (excitation)

Cross-bridge cycle

4a. **Detachment:** Linkage between actin and myosin broken as fresh molecule of ATP binds to myosin cross bridge; cross bridge assumes original conformation; ATP hydrolyzed (cycle starts again at step 1).

2a. **Binding:** Ca²⁺ released on excitation; removes inhibitory influence from actin, enabling it to bind with cross bridge.

Fresh ATP available

...or...

No ATP (after death)

3. **Bending:** Power stroke of cross bridge triggered on contact between myosin and actin; P_i released during and ADP released after power stroke.

4b. **Rigor complex:** If no fresh ATP available (after death), actin and myosin remain bound in rigor complex.

● **FIGURE 8-12 Cross-bridge cycle.**

that actin and the myosin cross bridges do not bind and no power stroking takes place (step 2b).

When P_i and ADP are released from myosin following contact with actin and the subsequent power stroke, the myosin ATPase site is free for attachment of another ATP molecule. The actin and myosin remain linked at the cross bridge until a fresh molecule of ATP attaches to myosin at the end of the power stroke. Attachment of the new ATP molecule permits detachment of the cross bridge, which returns to its unbent form, ready to start another cycle (step 4a). The newly attached ATP is then split by myosin ATPase, energizing the myosin cross bridge once again (step 1). On binding with another actin molecule, the energized cross bridge again bends, and so on, successively pulling the thin filament inward to accomplish contraction.

 RIGOR MORTIS Note that fresh ATP must attach to myosin to permit the cross-bridge link between myosin and actin to break at the end of a cycle, even though the ATP

is not split during this dissociation process. The need for ATP in separating myosin and actin is amply shown in **rigor mortis.** This "stiffness of death" is a generalized locking in place of the skeletal muscles that begins 3 to 4 hours after death and completes in about 12 hours. Following death, the cytosolic concentration of Ca²⁺ begins to rise, most likely because the inactive muscle cell membrane cannot keep out extracellular Ca²⁺ and perhaps also because Ca²⁺ leaks out of the lateral sacs. This Ca²⁺ moves the regulatory proteins aside, letting actin bind with the myosin cross bridges, which were already charged with ATP before death. Dead cells cannot produce any more ATP, so actin and myosin, once bound, cannot detach, because they lack fresh ATP. The thick and thin filaments thus stay linked by the immobilized cross bridges, leaving dead muscles stiff (step 4b). During the next several days, rigor mortis gradually subsides as the proteins involved in the rigor complex begin to degrade.

RELAXATION How is **relaxation** normally accomplished in a living muscle? Just as an action potential in a muscle fiber turns

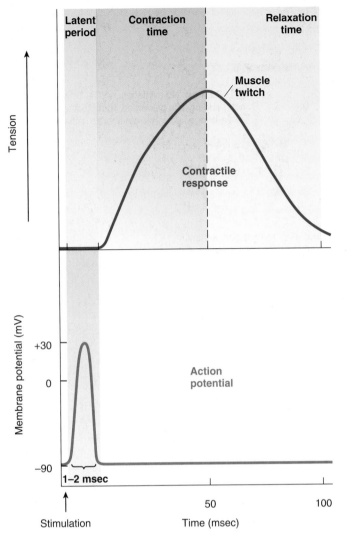

Latent period | Contraction time | Relaxation time

Tension

Muscle twitch

Contractile response

Membrane potential (mV)

+30

0

−90

Action potential

1–2 msec

Stimulation

Time (msec)

50   100

● **FIGURE 8-13 Relationship of an action potential to the resultant muscle twitch.** The duration of the action potential is not drawn to scale but is exaggerated. Note that the resting potential of a skeletal muscle fiber is −90 mV, compared to a resting potential of −70 mV in a neuron.

How long does the contractile activity initiated by an action potential last in response to a single action potential before relaxation occurs?

Contractile activity far outlasts the electrical activity that initiated it.

A single action potential in a skeletal muscle fiber lasts only 1 to 2 msec. The onset of the resulting contractile response lags behind the action potential because the entire excitation–contraction coupling must occur before cross-bridge activity begins. In fact, the action potential is completed before the contractile apparatus even becomes operational. This time delay of a few milliseconds between stimulation and the onset of contraction is called the **latent period** (● Figure 8-13).

Time is also required for generating tension within the muscle fiber, produced by the sliding interactions between the thick and thin filaments through cross-bridge activity. The time from contraction onset until peak tension develops—**contraction time**—averages about 50 msec, although this time varies depending on the type of muscle fiber. The contractile response does not end until the lateral sacs have taken up all the Ca^{2+} released in response to the action potential. This reuptake of Ca^{2+} is also time-consuming. Even after Ca^{2+} is removed, it takes time for the filaments to return to their resting positions. The time from peak tension until relaxation is complete—the **relaxation time**—usually lasts another 50 msec or more. Consequently, the entire contractile response to a single action potential may last up to 100 msec or more; this is much longer than the duration of the action potential that initiated it (100 msec compared to 1 to 2 msec). This fact is important in the body's ability to produce muscle contractions of variable strength, as you will discover in the next section.

Skeletal Muscle Mechanics

Thus far we have described the contractile response in a single muscle fiber. In the body, groups of muscle fibers are organized into whole muscles. We now turn attention to the contraction of whole muscles.

Whole muscles are groups of muscle fibers bundled together and attached to bones.

Each person has about 600 skeletal muscles, which range in size from the delicate external eye muscles that control eye movements and contain only a few hundred fibers, to the large, powerful leg muscles that contain several hundred thousand fibers.

Each muscle is sheathed by connective tissue that penetrates from the surface into the muscle to envelop each individual fiber and divide the muscle into columns or bundles. The connective tissue extends beyond the ends of the muscle to form tough, collagenous **tendons** that attach the muscle to bones. A tendon may be quite long, attaching to a bone some distance from the fleshy part of the muscle. For example, some of the muscles involved in finger movement are in the forearm, with long tendons extending down to attach to the bones of the fingers. (You can readily see these tendons move on the top of your

on the contractile process by triggering release of Ca^{2+} from the lateral sacs into the cytosol, the contractile process is turned off when Ca^{2+} is returned to the lateral sacs when local electrical activity stops. The sarcoplasmic reticulum has an energy-consuming carrier, a *Ca^{2+}–ATPase pump*, which actively transports Ca^{2+} from the cytosol and concentrates it in the lateral sacs (see ● Figure 8-11). Recall that the end-plate potential and resultant muscle-fiber action potential stop when the membrane-bound enzyme acetylcholinesterase removes ACh from the neuromuscular junction (see p. 250). When a local action potential is no longer in the T tubules to trigger release of Ca^{2+}, the ongoing activity of the sarcoplasmic reticulum's Ca^{2+} pump returns released Ca^{2+} back into its lateral sacs. Removing cytosolic Ca^{2+} lets the troponin–tropomyosin complex slip back into its blocking position, so actin and myosin can no longer bind at the cross bridges. The thin filaments, freed from cycles of cross-bridge attachment and pulling, return passively to their resting position. The muscle fiber has relaxed.

Contractile
component
(sarcomeres)

Series-elastic
component
(intracellular
titin/connective
tissue/tendon)

Load

Load

● **FIGURE 8-14 Relationship between the contractile component and the series-elastic component in transmitting muscle tension to bone.** Muscle tension is transmitted to the bone by means of the stretching and tightening of the muscle's elastic connective tissue and tendon as a result of sarcomere shortening brought about by cross-bridge cycling.

hand when you wiggle your fingers.) This arrangement permits greater dexterity; the fingers would be much thicker and more awkward if all the muscles used in finger movement were actually in the fingers.

Muscle tension is transmitted to bone as the contractile component tightens the series-elastic component.

Tension is produced internally within the sarcomeres, considered the **contractile component** of the muscle, as a result of cross-bridge activity and the resulting sliding of filaments. However, the sarcomeres are not attached directly to the bones. Instead, the tension generated by these contractile elements must be transmitted to the bone via the connective tissue and tendons before the bone can be moved. Connective tissue and tendon, as well as other components of the muscle, such as the intracellular titin, have a certain degree of passive elasticity. These noncontractile tissues are called

the **series-elastic component** of the muscle; they behave like a stretchy spring placed between the internal tension-generating elements and the bone that is to be moved against an external load (● Figure 8-14). Shortening of the sarcomeres stretches the series-elastic component. Muscle tension is transmitted to the bone by this tightening of the series-elastic component. This force applied to the bone moves the bone against a load.

A muscle is typically attached to at least two different bones across a joint by means of tendons that extend from each end of the muscle (● Figure 8-15). When the muscle shortens during contraction, the position of the joint is changed as one bone is moved in relation to the other—for example, *flexion* (bending) of the elbow joint by contraction of the biceps muscle and *extension* (straightening) of the elbow by contraction of the triceps. The end of the muscle attached to the more stationary part of the skeleton is called the **origin,** and the end attached to the skeletal part that moves is the **insertion.**

Contractions of a whole muscle can be of varying strength.

A single action potential in a muscle fiber produces a brief, weak contraction called a **twitch,** which is too short and too weak to be useful and normally does not take place in the body. Muscle fibers are arranged into whole muscles, where they function cooperatively to produce contractions of variable grades of strength stronger than a twitch. In other words, you can vary the force you exert by the same muscle, depending on whether you are picking up a piece of paper, a book, or a 50-pound weight. Two primary factors can be adjusted to accomplish gradation of whole-muscle tension: (1) *the number of muscle fibers contracting within a muscle* and (2) *the tension developed by each contracting fiber.* We will discuss each of these factors in turn.

The number of fibers contracting within a muscle depends on the extent of motor unit recruitment.

The greater the number of fibers contracting, the greater the total muscle tension is. Therefore, larger muscles consisting of more muscle fibers obviously can generate more tension than smaller muscles with fewer fibers can.

Flexion

Origin of biceps

Biceps contracts

Insertion of biceps

Extension

Origin of triceps

Triceps contracts

Insertion of triceps

● **FIGURE 8-15 Flexion and extension of the elbow joint.**

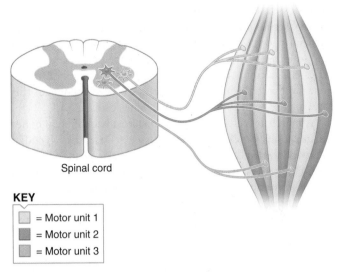

Spinal cord

KEY

☐ = Motor unit 1
☐ = Motor unit 2
☐ = Motor unit 3

● **FIGURE 8-16 Motor units in a skeletal muscle.**

Each whole muscle is innervated by a number of different motor neurons. When a motor neuron enters a muscle, it branches, with each axon terminal supplying a single muscle fiber (● Figure 8-16). One motor neuron innervates a number of muscle fibers, but each muscle fiber is supplied by only one motor neuron. When a motor neuron is activated, all the muscle fibers it supplies are stimulated to contract simultaneously. This team of concurrently activated components—one motor neuron plus all the muscle fibers it innervates—is called a **motor unit.** The muscle fibers that compose a motor unit are dispersed throughout the whole muscle; thus, their simultaneous contraction results in an evenly distributed, although weak, contraction of the whole muscle. Each muscle consists of a number of intermingled motor units. For a weak contraction of the whole muscle, only one or a few of its motor units are activated. For stronger and stronger contractions, more and more motor units are recruited, or stimulated to contract simultaneously, a phenomenon known as **motor unit recruitment.**

How much stronger the contraction will be with the recruitment of each additional motor unit depends on the size of the motor units (that is, the number of muscle fibers controlled by a single motor neuron). The number of muscle fibers per motor unit and the number of motor units per muscle vary widely, depending on the specific function of the muscle. For muscles that produce precise, delicate movements, such as external eye muscles and hand muscles, a single motor unit may contain as few as a dozen muscle fibers. Because so few muscle fibers are involved with each motor unit, recruitment of each additional motor unit adds only a small increment to the whole muscle's strength of contraction (● Figure 8-17a). These small motor units allow very fine control over muscle tension. In contrast, in muscles designed for powerful, coarsely controlled movement, such as those of the legs, a single motor unit may

contain 1500 to 2000 muscle fibers. Recruitment of motor units in these muscles results in large incremental increases in whole-muscle tension (● Figure 8-17b). More powerful contractions occur at the expense of less precisely controlled gradations. Thus, the number of muscle fibers participating in the whole muscle's total contractile effort depends on the number of motor units recruited and the number of muscle fibers per motor unit in that muscle.

To delay or prevent **fatigue** (inability to maintain muscle tension at a given level) during a sustained contraction involving only a portion of a muscle's motor units, as is necessary in muscles supporting the weight of the body against the force of gravity, **asynchronous recruitment of motor units** takes place. The body alternates motor unit activity, like shifts at a factory, to give motor units that have been active an opportunity to rest while others take over. Changing of the shifts is carefully coordinated, so the sustained contraction is smooth rather than jerky. Asynchronous motor unit recruitment is possible only for submaximal contractions, during which only some of the motor units must maintain the desired level of tension. During maximal contractions, when all the muscle fibers must participate, it is impossible to alternate motor unit activity to prevent fatigue. This is one reason why you cannot support a heavy object as long as a light one.

Furthermore, the type of muscle fiber that is activated varies with the extent of gradation. Most muscles consist of a mixture of fiber types that differ metabolically, some being more resistant to fatigue than others. During weak or moderate endurance-type activities (aerobic exercise), the motor units most resistant to fatigue are recruited first. The last fibers to be called into play in the face of demands for further increases in tension are those that fatigue rapidly. An individual can therefore engage in endurance activities for prolonged periods of time but can only briefly maintain bursts of all-out, powerful effort. Of course,

(a) Recruitment of small motor units

(b) Recruitment of large motor units

● **FIGURE 8-17 Comparison of motor unit recruitment in skeletal muscles with small motor units and muscles with large motor units.** (a) Small incremental increases in strength of contraction occur during motor unit recruitment in muscles with small motor units because only a few additional fibers are called into play as each motor unit is recruited. (b) Large incremental increases in strength of contraction occur during motor unit recruitment in muscles with large motor units, because so many additional fibers are stimulated with recruitment of each additional motor unit.

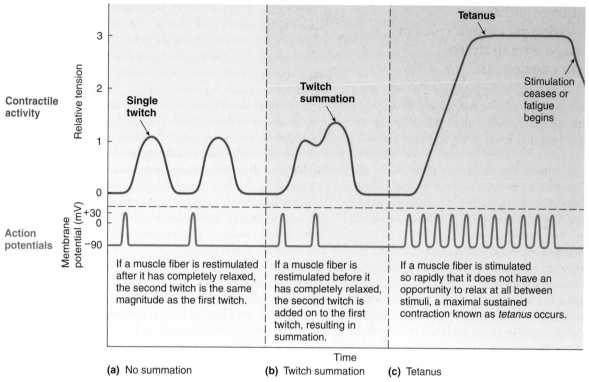

Contractile activity

Action potentials

Single twitch

Twitch summation

Tetanus

Stimulation ceases or fatigue begins

+30
0
−90

If a muscle fiber is restimulated after it has completely relaxed, the second twitch is the same magnitude as the first twitch.

If a muscle fiber is restimulated before it has completely relaxed, the second twitch is added on to the first twitch, resulting in summation.

If a muscle fiber is stimulated so rapidly that it does not have an opportunity to relax at all between stimuli, a maximal sustained contraction known as *tetanus* occurs.

Time

(a) No summation **(b)** Twitch summation **(c)** Tetanus

● **FIGURE 8-18 Twitch summation and tetanus.**

even the muscle fibers most resistant to fatigue will eventually fatigue if required to maintain a certain level of sustained tension.

The frequency of stimulation can influence the tension developed by each muscle fiber.

Whole-muscle tension depends not only on the number of muscle fibers contracting but also on the tension developed by each contracting fiber. Various factors influence the extent to which tension can be developed. These factors include the following:

1. Frequency of stimulation
2. Length of the fiber at the onset of contraction
3. Extent of fatigue
4. Thickness of the fiber

We now examine the effect of frequency of stimulation. (The other factors are discussed in later sections.)

TWITCH SUMMATION AND TETANUS Even though a single action potential in a muscle fiber produces only a twitch, contractions with longer duration and greater tension can be achieved by repeated stimulation of the fiber. Let us see what happens when a second action potential occurs in a muscle fiber. If the muscle fiber has completely relaxed before the next action potential takes place, a second twitch of the same magnitude as the first occurs (● Figure 8-18a). The same excitation–contraction

events take place each time, resulting in identical twitch responses. If, however, the muscle fiber is stimulated a second time before it has completely relaxed from the first twitch, a second action potential causes a second contractile response, which is added "piggyback" on top of the first twitch (● Figure 8-18b). The two twitches from the two action potentials add together, or sum, to produce greater tension in the fiber than that produced by a single action potential. This **twitch summation** is similar to temporal summation of EPSPs at the postsynaptic neuron (see p. 108).

Twitch summation is possible only because the duration of the action potential (1 to 2 msec) is much shorter than the duration of the resulting twitch (100 msec). Once an action potential has been initiated, a brief refractory period occurs during which another action potential cannot be initiated (see p. 96). It is therefore impossible to achieve summation of action potentials. The membrane must return to resting potential and recover from its refractory period before another action potential can occur. However, because the action potential and refractory period are over long before the resulting muscle twitch is completed, the muscle fiber may be restimulated while some contractile activity still exists, to produce summation of the mechanical response.

If the muscle fiber is stimulated so rapidly that it does not have a chance to relax at all between stimuli, a smooth, sustained contraction of maximal strength known as **tetanus** occurs (● Figure 8-18c). A tetanic contraction is usually three to four times stronger than a single twitch. (Don't confuse this normal physiologic tetanus with the disease tetanus; see p. 112.)

Twitch summation results from a sustained elevation in cytosolic calcium and from increased time to stretch the series-elastic component.

What is the mechanism of twitch summation and tetanus at the cell level? The tension produced by a contracting muscle fiber increases as a result of greater cross-bridge cycling. The series-elastic component (connective tissue, titin, and tendons) must be stretched to transmit the tension generated in the muscle fiber to the bone, and it takes time to stretch these elastic elements. Accordingly, two factors contribute to twitch summation: (1) sustained elevation in cytosolic Ca^{2+} permitting greater cross-bridge cycling, and (2) more time to stretch the series-elastic component.

The most important factor in the development of twitch summation is sustained elevation in cytosolic Ca^{2+} as the frequency of action potentials increases. Enough Ca^{2+} is released in response to a single action potential to interact with all the troponin within the cell. As a result, all the cross bridges are free to participate in the contractile response. How, then, can repetitive action potentials bring about a greater contractile response? The difference depends on how long enough Ca^{2+} is available. The cross bridges remain active and continue to cycle as long as enough Ca^{2+} is present to keep the troponin–tropomyosin complex away from the cross-bridge binding sites on actin. Each troponin–tropomyosin complex spans a distance of seven actin molecules. Thus, binding of Ca^{2+} to one troponin molecule leads to the uncovering of only seven cross-bridge binding sites on the thin filament.

As soon as Ca^{2+} is released in response to an action potential, the sarcoplasmic reticulum starts pumping Ca^{2+} back into the lateral sacs. As the cytosolic Ca^{2+} concentration declines with the reuptake of Ca^{2+} by the lateral sacs, less Ca^{2+} is present to bind with troponin, so some of the troponin–tropomyosin complexes slip back into their blocking positions. Consequently, not all the cross-bridge binding sites remain available to participate in the cycling process during a single twitch induced by a single action potential. Because not all the cross bridges find a binding site, the resulting contraction during a single twitch is not of maximal strength.

If action potentials and twitches occur far enough apart in time for all the released Ca^{2+} from the first contractile response to be pumped back into the lateral sacs between the action potentials, an identical twitch response will occur as a result of the second action potential. If, however, a second action potential occurs and more Ca^{2+} is released while the Ca^{2+} that was released in response to the first action potential is being taken back up, the cytosolic Ca^{2+} concentration remains high and might even be elevated further. This prolonged availability of Ca^{2+} in the cytosol permits more of the cross bridges to continue participating in the cycling process for a longer time. As a result, tension development increases correspondingly. As the frequency of action potentials increases, the duration of elevated cytosolic Ca^{2+} concentration increases, and contractile activity likewise increases until a maximum tetanic contraction is reached. With tetanus, the maximum number of cross-bridge binding sites remain uncovered so that cross-bridge cycling, and consequently tension development, is at its peak.

The second factor contributing to twitch summation is related to the elastic structures of the muscle fiber. During a single twitch, the contraction does not last long enough to completely stretch the series-elastic component and allow the full sarcomere-generated tension to be transmitted to the bone. At the end of the twitch, the elastic elements slowly relax, or recoil to their initial nonstretched state. If another twitch occurs before the elastic elements have completely relaxed, the tension from the second twitch adds to the residual tension in the series-elastic component remaining from the first twitch. With greater frequencies of action potentials and more frequent twitches, less time is available for the elastic elements to recoil between twitches. Consequently, as the frequency of action potentials increases, the tension in the series-elastic component transmitted to the bone progressively increases until it reaches its maximum during tetanus.

Because skeletal muscle must be stimulated by motor neurons to contract, the nervous system plays a key role in regulating contraction strength. The two main factors subject to control to accomplish gradation of contraction are the *number of motor units stimulated* and the *frequency of their stimulation*. The areas of the brain that direct motor activity combine tetanic contractions and precisely timed shifts of asynchronous motor unit recruitment to execute smooth rather than jerky contractions.

Additional factors not directly under nervous control also influence the tension developed during contraction. Among these is the length of the fiber at the onset of contraction, to which we now turn attention.

There is an optimal muscle length at which maximal tension can be developed.

A relationship exists between the length of the muscle before the onset of contraction and the tetanic tension that each contracting fiber can subsequently develop at that length. For every muscle there is an **optimal length (l_o)** at which maximal force can be achieved on a subsequent tetanic contraction. More tension can be achieved during tetanus when beginning at the optimal muscle length than can be achieved when the contraction begins with the muscle less than or greater than its optimal length. This **length–tension relationship** can be explained by the sliding filament mechanism of muscle contraction.

AT OPTIMAL LENGTH (l_o) At l_o, when maximum tension can be developed (point **A** in ● Figure 8-19), the thin filaments optimally overlap the regions of the thick filaments from which the cross bridges project. At this length, a maximal number of cross bridges and actin molecules are accessible to each other for cycles of binding and bending. The central region of thick filaments, where the thin filaments do not overlap at l_o, lacks cross bridges; only myosin tails are found here.

AT LENGTHS GREATER THAN l_o At greater lengths, as when a muscle is passively stretched (point **B**), the thin filaments are pulled out from between the thick filaments, decreasing the number of actin sites available for cross-bridge binding; that is, some of the actin sites and cross bridges no longer "match up," so they "go unused." When less cross-bridge activity can occur,

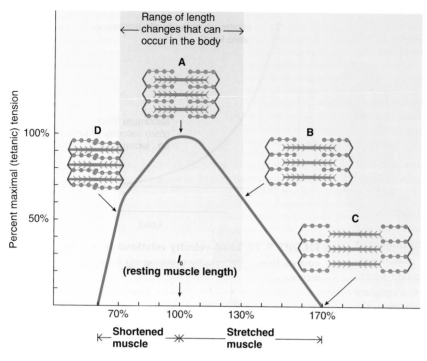

Range of length changes that can occur in the body

l_o (resting muscle length)

Shortened muscle — Stretched muscle

Muscle fiber length compared with resting length

- **FIGURE 8-19 Length–tension relationship.** Maximal tetanic contraction can be achieved when a muscle fiber is at its optimal length (l_o) before the onset of contraction, because this is the point of optimal overlap of thick-filament cross bridges and thin-filament cross-bridge binding sites (point A). The percentage of maximal tetanic contraction that can be achieved decreases when the muscle fiber is longer or shorter than l_o before contraction. When it is longer, fewer thin-filament binding sites are accessible for binding with thick-filament cross bridges, because the thin filaments are pulled out from between the thick filaments (points B and C). When the fiber is shorter, fewer thin-filament binding sites are exposed to thick-filament cross bridges because the thin filaments overlap (point D). Also, further shortening and tension development are impeded as the thick filaments become forced against the Z lines (point D). In the body, the resting muscle length is at l_o. Furthermore, because of restrictions imposed by skeletal attachments, muscles cannot vary beyond 30% of their l_o in either direction (the range screened in light green). At the outer limits of this range, muscles still can achieve about 50% of their maximal tetanic contraction.

less tension can develop. In fact, when the muscle is stretched to about 70% longer than its l_o (point **C**) the thin filaments are completely pulled out from between the thick filaments, preventing cross-bridge activity, and consequently no contraction can occur.

AT LENGTHS LESS THAN l_o If a muscle is shorter than l_o before contraction (point **D**), less tension can be developed for three reasons:

1. The thin filaments from the opposite sides of the sarcomere become overlapped, which limits the opportunity for the cross bridges to interact with actin.

2. The ends of the thick filaments become forced against the Z lines, so further shortening is impeded.

3. Besides these two mechanical factors, at muscle lengths less than 80% of l_o, not as much Ca^{2+} is released during excitation–contraction coupling for reasons unknown. Furthermore, by an

unknown mechanism the ability of Ca^{2+} to bind to troponin and pull the troponin–tropomyosin complex aside is reduced at shorter muscle lengths. Consequently, fewer actin sites are uncovered for participation in cross-bridge activity.

LIMITATIONS ON MUSCLE LENGTH The extremes in muscle length that prevent development of tension occur only under experimental conditions, when a muscle is removed and stimulated at various lengths. In the body the muscles are positioned so that their relaxed length is approximately their optimal length; thus, they can achieve near-maximal tetanic contraction most of the time. Because attachment to the skeleton imposes limitations, a muscle cannot be stretched or shortened more than 30% of its resting optimal length, and usually it deviates much less than 30% from normal length. Even at the outer limits (130% and 70% of l_o), the muscles still can generate half their maximum tension.

The factors we have discussed thus far that influence how much tension a contracting muscle fiber can develop—the frequency of stimulation and the muscle length at the onset of contraction—can vary from contraction to contraction. Other determinants of muscle fiber tension—the metabolic capability of the fiber relative to resistance to fatigue and the thickness of the fiber—do not vary from contraction to contraction but depend on the fiber type and can be modified over a period of time. After we complete our discussion of skeletal muscle mechanics, we will consider these other factors in the next section, on skeletal muscle metabolism and fiber types.

The two primary types of contraction are isotonic and isometric.

Not all muscle contractions shorten muscles and move bones. For a muscle to shorten during contraction, the tension developed in the muscle must exceed the forces that oppose movement of the bone to which the muscle's insertion is attached. In the case of elbow flexion, the opposing force, or **load,** is the weight of an object being lifted. When you flex your elbow without lifting any external object, there is still a load, albeit a minimal one—the weight of your forearm being moved against the force of gravity.

There are two primary types of contraction, depending on whether the muscle changes length during contraction. In an **isotonic contraction,** muscle tension remains constant as the muscle changes length. In an **isometric contraction,** the muscle is prevented from shortening, so tension develops at constant muscle length. The same internal events occur in both isotonic and isometric contractions: Muscle excitation turns on the

The disadvantage of this lever system is that at its insertion the muscle must exert a force seven times greater than the load. The product of the length of the power arm times the upward force applied must equal the product of the length of the load arm times the downward force exerted by the load. Because the load arm times the downward force is 35 cm × 5 kg, the power arm times the upward force must be 5 cm × 35 kg (the force that must be exerted by the muscle to be in mechanical equilibrium). Thus, skeletal muscles typically work at a mechanical disadvantage in that they must exert a considerably greater force than the actual load to be moved. Nevertheless, the amplification of velocity and distance afforded by the lever arrangement enables muscles to move loads faster over greater distances than would otherwise be possible. This amplification provides valuable maneuverability and speed.

Now let's shift attention from muscle mechanics to the metabolic means by which muscles power these movements.

Skeletal Muscle Metabolism and Fiber Types

Three different steps in the contraction–relaxation process require ATP:

1. Splitting of ATP by myosin ATPase provides the energy for the power stroke of the cross bridge.

2. Binding (but not splitting) of a fresh molecule of ATP to myosin lets the bridge detach from the actin filament at the end of a power stroke so that the cycle can be repeated. This ATP is later split to provide energy for the next stroke of the cross bridge.

3. Active transport of Ca^{2+} back into the sarcoplasmic reticulum during relaxation depends on energy derived from the breakdown of ATP.

Muscle fibers have alternate pathways for forming ATP.

Because ATP is the only energy source that can be directly used for these activities, for contractile activity to continue, ATP must constantly be supplied. Only limited stores of ATP are immediately available in muscle tissue, but three pathways supply additional ATP as needed during muscle contraction: (1) transfer of a high-energy phosphate from creatine phosphate to ADP, (2) oxidative phosphorylation (the electron transport system and chemiosmosis), and (3) glycolysis.

CREATINE PHOSPHATE Creatine phosphate is the first energy storehouse tapped at the onset of contractile activity (● Figure 8-22, step **3a**). Like ATP, creatine phosphate contains a high-energy phosphate group, which can be donated directly to ADP to form ATP. Just as energy is released when the terminal phosphate bond in ATP is split, similarly energy is released when the bond between phosphate and creatine is broken. The energy released from the hydrolysis of creatine phosphate, along with the phosphate, can be donated directly to ADP to form ATP. This reaction, which is catalyzed by the muscle cell enzyme

creatine kinase, is reversible; energy and phosphate from ATP can be transferred to creatine to form creatine phosphate:

$$\text{Creatine phosphate} + \text{ADP} \underset{}{\overset{\text{creatine kinase}}{\rightleftharpoons}} \text{creatine} + \text{ATP}$$

In accordance with the law of mass action (see p. 491), as energy reserves are built up in a resting muscle, the increased concentration of ATP favors transfer of the high-energy phosphate group from ATP to form creatine phosphate. By contrast, at the onset of contraction when myosin ATPase splits the meager reserves of ATP, the resultant fall in ATP favors transfer of the high-energy phosphate group from stored creatine phosphate to form more ATP. A rested muscle contains about five times as much creatine phosphate as ATP. Thus, most energy is stored in muscle in creatine phosphate pools. Because only one enzymatic reaction is involved in this energy transfer, ATP can be formed rapidly (within a fraction of a second) by using creatine phosphate.

Thus, creatine phosphate is the first source for supplying additional ATP when exercise begins. Muscle ATP levels actually remain fairly constant early in contraction, but creatine phosphate stores become depleted. In fact, short bursts of high-intensity contractile effort, such as high jumps, sprints, or weight lifting, are supported primarily by ATP derived at the expense of creatine phosphate. Other energy systems do not have a chance to become operable before the activity is over. Creatine phosphate stores typically power the first minute or less of exercise.

Some athletes hoping to gain a competitive edge take oral creatine supplements to boost their performance in short-term, high-intensity activities lasting less than a minute. (We naturally get creatine in our diets, especially in meat.) Loading the muscles with extra creatine means larger creatine phosphate stores—that is, larger energy stores that can translate into a small edge in performance of activities requiring short, explosive bursts of energy. Yet creatine supplements should be used with caution, because the long-term health effects are unknown. Also, extra creatine stores are of no use in activities of longer duration that rely on more long-term energy-supplying mechanisms.

OXIDATIVE PHOSPHORYLATION If the energy-dependent contractile activity is to continue, the muscle shifts to the alternate pathways of oxidative phosphorylation and glycolysis to form ATP. These multistepped pathways require time to pick up their rates of ATP formation to match the increased demands for energy, time provided by the immediate supply of energy from the one-step creatine phosphate system.

Oxidative phosphorylation takes place within the muscle mitochondria if sufficient O_2 is present. Oxygen is required to support the mitochondrial electron-transport system, which, together with chemiosmosis by ATP synthase, efficiently harnesses energy captured from the breakdown of nutrient molecules and uses it to generate ATP (see p. 35). This pathway is fueled by glucose or fatty acids, depending on the intensity and duration of the activity (● Figure 8-22, step **3b**). Although it provides a rich yield of 32 ATP molecules for each glucose molecule processed, oxidative phosphorylation is relatively slow because of the number of steps involved.

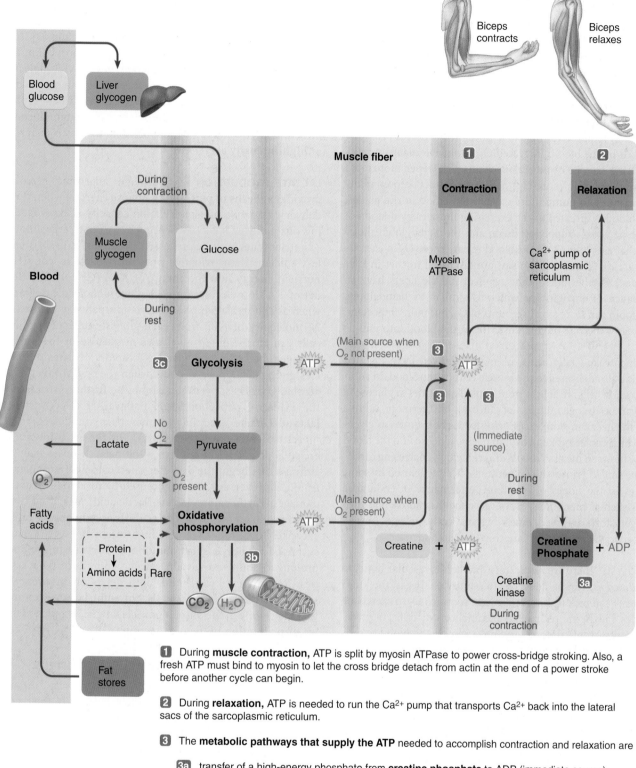

■ **FIGURE 8-22 Metabolic pathways producing ATP used during muscle contraction and relaxation.**

1 During **muscle contraction,** ATP is split by myosin ATPase to power cross-bridge stroking. Also, a fresh ATP must bind to myosin to let the cross bridge detach from actin at the end of a power stroke before another cycle can begin.

2 During **relaxation,** ATP is needed to run the Ca^{2+} pump that transports Ca^{2+} back into the lateral sacs of the sarcoplasmic reticulum.

3 The **metabolic pathways that supply the ATP** needed to accomplish contraction and relaxation are

3a transfer of a high-energy phosphate from **creatine phosphate** to ADP (immediate source);

3b **oxidative phosphorylation** (the main source when O_2 is present), fueled by glucose derived from muscle glycogen stores or by glucose and fatty acids delivered by the blood; and

3c **glycolysis** (the main source when O_2 is not present). Pyruvate, the end product of glycolysis, is converted to lactate when lack of O_2 prevents the pyruvate from being further processed by the oxidative phosphorylation pathway.

During light exercise (such as walking) to moderate exercise (such as jogging or swimming), muscle cells can form enough ATP through oxidative phosphorylation to keep pace with the modest energy demands of the contractile machinery for prolonged periods of time. To sustain ongoing oxidative phosphorylation, the exercising muscles depend on delivery of adequate O_2 and nutrients to maintain their activity. Activity that can be supported in this way is **aerobic** ("with O_2") or **endurance-type exercise.**

The O_2 required for oxidative phosphorylation is primarily delivered by the blood. Increased O_2 is made available to muscles during exercise by several mechanisms: Deeper, more rapid breathing brings in more O_2; the heart contracts more rapidly and forcefully to pump more oxygenated blood to the tissues; more blood is diverted to the exercising muscles by dilation of the blood vessels supplying them; and the hemoglobin molecules that carry the O_2 in the blood release more O_2 in exercising muscles. (These mechanisms are discussed further in later chapters.) Furthermore, some types of muscle fibers have an abundance of **myoglobin,** which is similar to hemoglobin. Myoglobin can store small amounts of O_2, but more important, it increases the rate of O_2 transfer from the blood into muscle fibers.

Glucose and fatty acids, ultimately derived from ingested food, are also delivered to the muscle cells by the blood. In addition, muscle cells are able to store limited quantities of glucose in the form of glycogen (chains of glucose). Furthermore, up to a point the liver can store excess ingested carbohydrates as glycogen, which can be broken down to release glucose into the blood for use between meals. *Carbohydrate loading*—increasing carbohydrate intake prior to a competition—is a tactic used by some athletes in the hopes of boosting performance in endurance events such as marathons. However, once the muscle and liver glycogen stores are filled, excess ingested carbohydrates (or any other energy-rich nutrient) are converted to body fat.

GLYCOLYSIS There are respiratory and cardiovascular limits to how much O_2 can be delivered to a muscle. That is, the lungs and heart can pick up and deliver just so much O_2 to exercising muscles. Furthermore, in near-maximal contractions, the powerful contraction compresses almost closed the blood vessels that course through the muscle, severely limiting the O_2 available to the muscle fibers. Even when O_2 is available, the relatively slow oxidative-phosphorylation system may not be able to produce ATP rapidly enough to meet the muscle's needs during intense activity. A skeletal muscle's energy consumption may increase up to 100-fold when going from rest to high-intensity exercise. When O_2 delivery or oxidative phosphorylation cannot keep pace with the demand for ATP formation as the intensity of exercise increases, the muscle fibers rely increasingly on glycolysis to generate ATP (● Figure 8-22, step **3c**) (see p. 33). The chemical reactions of **glycolysis** yield products for ultimate entry into the oxidative phosphorylation pathway, but glycolysis can also proceed alone in the absence of further processing of its products by oxidative phosphorylation. During glycolysis, a glucose molecule is broken down into two **pyruvate** molecules, yielding two ATP molecules in the process. Pyruvate can be further degraded by oxidative

phosphorylation to extract more energy. However, glycolysis alone has two advantages over the oxidative phosphorylation pathway: (1) glycolysis can form ATP in the absence of O_2 (operating *anaerobically,* that is, "without O_2"), and (2) it can proceed more rapidly than oxidative phosphorylation. Although glycolysis extracts considerably fewer ATP molecules from each nutrient molecule processed, it can proceed so much more rapidly that it can outproduce oxidative phosphorylation over a given period of time if enough glucose is present. Activity that can be supported in this way is **anaerobic** or **high-intensity exercise.**

LACTATE PRODUCTION Even though anaerobic glycolysis provides a means of performing intense exercise when the O_2 delivery/oxidative phosphorylation capacity is exceeded, using this pathway has two consequences. First, large amounts of nutrient fuel must be processed, because glycolysis is much less efficient than oxidative phosphorylation in converting nutrient energy into the energy of ATP. (Glycolysis yields a net of 2 ATP molecules for each glucose molecule degraded, whereas the oxidative phosphorylation pathway can extract 32 molecules of ATP from each glucose molecule.) Muscle cells can store limited quantities of glucose in the form of glycogen, but anaerobic glycolysis rapidly depletes the muscle's glycogen supplies. Second, when the end product of anaerobic glycolysis, pyruvate, cannot be further processed by the oxidative phosphorylation pathway, it is converted to **lactate.** Lactate accumulation has been implicated in the muscle soreness that occurs during the time that intense exercise is actually taking place. (The delayed-onset pain and stiffness that begin the day after unaccustomed muscular exertion, however, are probably caused by reversible structural damage.) Furthermore, lactate (lactic acid) picked up by the blood produces the metabolic acidosis accompanying intense exercise. Therefore, anaerobic high-intensity exercise can be sustained for only a short duration, in contrast to the body's prolonged ability to sustain aerobic endurance-type activities. Researchers believe that both depletion of energy reserves and the fall in muscle pH caused by lactate accumulation play a role in the onset of muscle fatigue, a topic to which we turn attention in the next section.

Fatigue may be of muscle or central origin.

Contractile activity in a particular skeletal muscle cannot be maintained at a given level indefinitely. Eventually, the tension in the muscle declines as fatigue sets in. There are two different types of fatigue: muscle fatigue and central fatigue.

Muscle fatigue occurs when an exercising muscle can no longer respond to stimulation with the same degree of contractile activity. Muscle fatigue is a defense mechanism that protects a muscle from reaching a point at which it can no longer produce ATP. An inability to produce ATP would result in rigor mortis (obviously not an acceptable outcome of exercise). The underlying causes of muscle fatigue are unclear. The primary implicated factors include the following:

■ The *local increase in ADP and inorganic phosphate* from ATP breakdown may directly interfere with cross-bridge cycling and/or block Ca^{2+} release and uptake by the sarcoplasmic reticulum.

- *Accumulation of lactate* may inhibit key enzymes in the energy-producing pathways and/or excitation–contraction coupling process.

- The *accumulation of extracellular K^+* that occurs in the muscle when the Na^+–K^+ pump cannot actively transport K^+ back into the muscle cells as rapidly as this ion leaves during the falling phase of repeated action potentials (see p. 93) causes a local reduction in membrane potential. This altered potential may decrease the release of Ca^{2+} intracellularly by impairing the coupling of voltage-gated dihydropyridine receptors in the T tubules and Ca^{2+}-release channels in the sarcoplasmic reticulum.

- *Depletion of glycogen energy reserves* may lead to muscle fatigue in exhausting exercise.

The time of onset of fatigue varies with the type of muscle fiber, some fibers being more resistant to fatigue than others, and with the intensity of the exercise, more rapid onset of fatigue being associated with high-intensity activities.

Central fatigue occurs when the CNS no longer adequately activates the motor neurons supplying the working muscles. The person slows down or stops exercising even though the muscles are still able to perform. Central fatigue often is psychologically based. During strenuous exercise, central fatigue may stem from discomfort associated with the activity; it takes strong motivation (a will to win) to deliberately persevere when in pain. In less strenuous activities, central fatigue may reduce physical performance in association with boredom and monotony (such as assembly-line work) or tiredness (lack of sleep). The mechanisms involved in central fatigue are poorly understood. In some cases, central fatigue may stem from biochemical insufficiencies within the brain.

Neuromuscular fatigue in exercise—an inability of active motor neurons to synthesize acetylcholine rapidly enough to sustain chemical transmission of action potentials from the motor neurons to the muscles—can be produced experimentally but does not occur under normal physiologic conditions.

Increased oxygen consumption is necessary to recover from exercise.

A person continues to breathe deeply and rapidly for a period of time after exercising. The need for elevated O_2 uptake during recovery from exercise (**excess postexercise oxygen consumption,** or **EPOC**) is due to a variety of factors. The best known is repayment of an **oxygen deficit** incurred during exercise, when contractile activity was being supported by ATP derived from nonoxidative sources such as creatine phosphate and anaerobic glycolysis. During exercise, the creatine phosphate stores of active muscles are reduced, lactate may accumulate, and glycogen stores may be tapped; the extent of these effects depends on the intensity and duration of the activity. Oxygen is needed for recovery of the energy systems. During the recovery period, fresh supplies of ATP are formed by oxidative phosphorylation using the newly acquired O_2, which is provided by the sustained increase in breathing after exercise has stopped. Most of this ATP

is used to resynthesize creatine phosphate to restore its reserves. This can be accomplished in a matter of a few minutes. Any accumulated lactate is converted back into pyruvate, part of which is used by the oxidative phosphorylation system for ATP production. The remainder of the pyruvate is converted back into glucose by the liver. Most of this glucose in turn is used to replenish the glycogen stores drained from the muscles and liver during exercise. These biochemical reactions involving pyruvate require O_2 and take several hours for completion. Thus, EPOC provides the O_2 needed to restore the creatine phosphate system, remove lactate, and at least partially replenish glycogen stores.

Unrelated to increased O_2 uptake is the need to restore nutrients after grueling exercise, such as marathon races, in which glycogen stores are severely depleted. In such cases, long-term recovery can take a day or more, because the exhausted energy stores require nutrient intake for full replenishment. Therefore, depending on the type and duration of activity, recovery can be complete within a few minutes or can require more than a day.

Part of the EPOC is not directly related to repayment of energy stores but instead results from a general metabolic disturbance following exercise. For example, the local increase in muscle temperature arising from heat-generating contractile activity speeds up the rate of all chemical reactions in the muscle tissue, including those dependent on O_2. Likewise, the secretion of epinephrine, a hormone that increases O_2 consumption by the body, is elevated during exercise. Until the circulating level of epinephrine returns to its preexercise state, O_2 uptake is increased above normal. Furthermore, during exercise body temperature rises several degrees Fahrenheit. A rise in temperature speeds up O_2-consuming chemical reactions. Until body temperature returns to preexercise levels, the increased speed of these chemical reactions accounts in part for the EPOC.

We have been looking at the contractile and metabolic activities of skeletal muscle fibers in general. Yet not all skeletal muscle fibers use these mechanisms to the same extent. We next examine the different types of muscle fibers based on their speed of contraction and how they are metabolically equipped to generate ATP.

There are three types of skeletal muscle fibers, based on differences in ATP hydrolysis and synthesis.

Classified by their biochemical capacities, there are three major types of muscle fibers (▲ Table 8-1):

1. Slow-oxidative (type I) fibers

2. Fast-oxidative (type IIa) fibers

3. Fast-glycolytic (type IIx) fibers

As their names imply, the two main differences among these fiber types are their speed of contraction (slow or fast) and the type of enzymatic machinery they primarily use for ATP formation (oxidative or glycolytic).

FAST VERSUS SLOW FIBERS Fast fibers have higher myosin ATPase (ATP-splitting) activity than slow fibers do. The higher the ATPase activity, the more rapidly ATP is split and the faster

Characteristics of Skeletal Muscle Fibers

Characteristic	TYPE OF FIBER		
	Slow-Oxidative (Type I)	Fast-Oxidative (Type IIa)	Fast-Glycolytic (Type IIx)
Myosin–ATPase Activity	Low	High	High
Speed of Contraction	Slow	Fast	Fast
Resistance to Fatigue	High	Intermediate	Low
Oxidative Phosphorylation Capacity	High	High	Low
Enzymes for Anaerobic Glycolysis	Low	Intermediate	High
Mitochondria	Many	Many	Few
Capillaries	Many	Many	Few
Myoglobin Content	High	High	Low
Color of Fiber	Red	Red	White
Glycogen Content	Low	Intermediate	High

the rate at which energy is made available for cross-bridge cycling. The result is a fast twitch, compared to the slower twitches of those fibers that split ATP more slowly. On average, the time to peak twitch tension for fast fibers is 20–40 msec compared to 60–100 msec for slow fibers. Thus, two factors determine the speed with which a muscle contracts: the load (load–velocity relationship) and the myosin ATPase activity of the contracting fibers (fast or slow twitch).

OXIDATIVE VERSUS GLYCOLYTIC FIBERS Fiber types also differ in ATP-synthesizing ability. Those with a greater capacity to form ATP are more resistant to fatigue. Some fibers are better equipped for oxidative phosphorylation, whereas others rely primarily on anaerobic glycolysis for synthesizing ATP. Because oxidative phosphorylation yields considerably more ATP from each nutrient molecule processed, it does not readily deplete energy stores. Furthermore, it does not result in lactate accumulation. Oxidative types of muscle fibers are therefore more resistant to fatigue than glycolytic fibers are.

Other related characteristics distinguishing these three fiber types are summarized in ▲ Table 8-1. As you would expect, the oxidative fibers, both slow and fast, contain an abundance of mitochondria, the organelles that house the enzymes involved in oxidative phosphorylation. Because adequate oxygenation is essential to support this pathway, these fibers are richly supplied with capillaries. Oxidative fibers also have a high myoglobin content. Myoglobin not only helps support oxidative fibers' O_2 dependency but also gives them a red color, just as oxygenated hemoglobin produces the red color of arterial blood. Accord-

ingly, these muscle fibers are called **red fibers.**

In contrast, the fast fibers specialized for glycolysis contain few mitochondria but have a high content of glycolytic enzymes instead. Also, to supply the large amounts of glucose needed for glycolysis, they contain a lot of stored glycogen. Because the glycolytic fibers need relatively less O_2 to function, they have only a meager capillary supply compared with the oxidative fibers. The glycolytic fibers contain very little myoglobin and therefore are pale in color, so they are sometimes called **white fibers.** (The most readily observable comparison between red and white fibers is the dark and white meat in poultry.)

GENETIC ENDOWMENT OF MUSCLE FIBER TYPES In humans, most muscles contain a mixture of all three fiber types; the percentage of each type is largely determined by the type of activity for which the muscle is specialized. Accordingly, a high proportion of slow-oxidative fibers are found in muscles specialized for maintaining low-intensity contractions for long periods of time without fatigue, such as the muscles of the back and legs that support the body's weight against the force of gravity. A preponderance of fast-glycolytic fibers are found in the arm muscles, which are adapted for performing rapid, forceful movements such as lifting heavy objects.

The percentage of these various fibers not only differs between muscles within an individual but also varies considerably among individuals. Athletes genetically endowed with a higher percentage of the fast-glycolytic fibers are good candidates for power and sprint events, whereas those with a greater proportion of slow-oxidative fibers are more likely to succeed in endurance activities such as marathon races.

Of course, success in any event depends on many factors other than genetic endowment, such as the extent and type of training and the level of dedication. Indeed, the mechanical and metabolic capabilities of muscle fibers can change a lot in response to the patterns of demands placed on them. Let us see how.

Muscle fibers adapt considerably in response to the demands placed on them.

Different types of exercise produce different patterns of neuronal discharge to the muscle involved. Depending on the pattern of neural activity, long-term adaptive changes occur in the muscle fibers, enabling them to respond most efficiently to the types of demands placed on the muscle. Therefore, skeletal muscle has a high degree of *plasticity* (see p. 149). Two types of changes can be induced in muscle fibers: changes in their ATP synthesizing capacity and changes in their diameter.

IMPROVEMENT IN OXIDATIVE CAPACITY Regular aerobic endurance exercise, such as long-distance jogging or swimming, promotes metabolic changes within the oxidative fibers, which are the ones primarily recruited during aerobic exercise. For example, the number of mitochondria and the number of capillaries supplying blood to these fibers both increase. Muscles so adapted can use O_2 more efficiently and therefore can better endure prolonged activity without fatiguing. However, they do not change in size.

MUSCLE HYPERTROPHY The actual size of the muscles can be increased by regular bouts of anaerobic, short-duration, high-intensity resistance training, such as weight lifting. The resulting muscle enlargement comes primarily from an increase in diameter (**hypertrophy**) of the fast-glycolytic fibers called into play during such powerful contractions. Most of the fiber thickening results from increased synthesis of myosin and actin filaments, which permits a greater opportunity for cross-bridge interaction and consequently increases the muscle's contractile strength. The mechanical stress that resistance training exerts on a muscle fiber triggers signaling proteins, which turn on genes that direct the synthesis of more of these contractile proteins. Vigorous weight training can double or triple a muscle's size. The resultant bulging muscles are better adapted to activities that require intense strength for brief periods, but endurance has not been improved.

INFLUENCE OF TESTOSTERONE Men's muscle fibers are thicker, and accordingly, their muscles are larger and stronger than those of women, even without weight training, because of the actions of testosterone, a steroid hormone secreted primarily in males. Testosterone promotes the synthesis and assembly of myosin and actin. This fact has led some athletes, both males and females, to the dangerous practice of taking this or closely related steroids to increase their athletic performance. (To explore this topic further, see the boxed feature on p. 282, ■ A Closer Look at Exercise Physiology.)

INTERCONVERSION BETWEEN FAST MUSCLE TYPES All the muscle fibers within a single motor unit are of the same fiber type. This pattern usually is established early in life, but the two types of fast-twitch fibers are interconvertible, depending on training efforts. That is, fast-glycolytic fibers can be converted to fast-oxidative fibers, and vice versa, depending on the types of demands repetitively placed on them. Adaptive changes in skeletal muscle gradually reverse to their original state over a period of months if the regular exercise program that induced these changes is discontinued.

Slow and fast fibers are not interconvertible, however. Although training can induce changes in muscle fibers' metabolic support systems, whether a fiber is fast or slow twitch depends on the fiber's nerve supply. Slow-twitch fibers are supplied by motor neurons that exhibit a low-frequency pattern of electrical activity, whereas fast-twitch fibers are innervated by motor neurons that display intermittent rapid bursts of electrical activity. Experimental switching of motor neurons supplying slow muscle fibers with those supplying fast fibers gradually reverses the speed at which these fibers contract.

MUSCLE ATROPHY At the other extreme, if a muscle is not used, its actin and myosin content decreases, its fibers become smaller, and the muscle accordingly **atrophies** (decreases in mass) and becomes weaker. Muscle atrophy can take place in two ways. **Disuse atrophy** occurs when a muscle is not used for a long period of time even though the nerve supply is intact, as when a cast or brace must be worn or during prolonged bed confinement. **Denervation atrophy** occurs after the nerve supply to a muscle is lost. If the muscle is stimulated electrically until innervation can be reestablished, such as during regeneration of a severed peripheral nerve, atrophy can be diminished but not entirely prevented. Contractile activity itself obviously plays an important role in preventing atrophy; however, poorly understood factors released from active nerve endings, perhaps packaged with the ACh vesicles, apparently contribute to the integrity and growth of muscle tissue.

LIMITED REPAIR OF MUSCLE When a muscle is damaged, limited repair is possible, even though muscle cells cannot divide mitotically to replace lost cells. A small population of inactive muscle-specific stem cells called **satellite cells** are located close to the muscle surface (see p. 8). When a muscle fiber is damaged, locally released factors activate the satellite cells, which divide to give rise to myoblasts, the same undifferentiated cells that formed the muscle during embryonic development. A group of myoblasts fuse to form a large, multinucleated cell, which immediately begins to synthesize and assemble the intracellular machinery characteristic of the muscle, ultimately differentiating completely into a mature muscle fiber. With extensive injury, this limited mechanism is not adequate to completely replace all the lost fibers. In that case, the remaining fibers often hypertrophy to compensate.

Transplantation of satellite cells or myoblasts provides one of several glimmers of hope for victims of **muscular dystrophy**, a hereditary pathological condition characterized by progressive degeneration of contractile elements, which are ultimately replaced by fibrous tissue. (See the boxed feature on p. 284, ■ Concepts, Challenges, and Controversies, for further information on this devastating condition.)

We have now completed our discussion of all the determinants of whole-muscle tension in a skeletal muscle, which are summarized in ▲ Table 8-2. Next, we examine the central and local mechanisms involved in regulating the motor activity performed by these muscles.

Control of Motor Movement

Particular patterns of motor unit output govern motor activity, ranging from maintenance of posture and balance to stereotypical locomotor movements, such as walking, to individual, highly skilled motor activity, such as gymnastics. Control of any

Are Athletes Who Use Steroids to Gain Competitive Advantage Really Winners or Losers?

The testing of athletes for drugs, and the much publicized exclusion from competition of those found to be using substances outlawed by sports federations, have stirred considerable controversy. One such group of drugs is **anabolic androgenic steroids** (*anabolic* means "buildup of tissues," *androgenic* means "male producing," and *steroids* are a class of hormone). These agents are closely related to testosterone, the natural male sex hormone, which is responsible for promoting the increased muscle mass characteristic of males.

Although their use is outlawed (possessing anabolic steroids without a prescription became a federal offense in 1991), these agents are taken by many athletes who specialize in power events such as weight lifting and sprinting in the hopes of increasing muscle mass and, accordingly, muscle strength. Both male and female athletes have resorted to using these substances in an attempt to gain a competitive edge. Bodybuilders also take anabolic steroids. Furthermore, although most players deny using these agents, experts believe these performance enhancers are widely used in professional sports such as baseball, football, basketball, competitive cycling, and hockey. There are an estimated 1 million anabolic steroid abusers in the United States. Compounding the problem, underground chemists recently created new synthetic performance-enhancing steroids undetectable by standard drug tests. Unfortunately, use of anabolic steroids has spread into our nation's high schools and even younger age groups. Recent studies indicate that 10% of male and 3% of female high school athletes use banned steroids. The manager of the steroid abuse hotline of the National Steroid Research Center reports having calls for help from abusers as young as 12 years old.

Studies have confirmed that steroids can increase muscle mass when used in large amounts and coupled with heavy exercise. One reputable study demonstrated an average 8.9-pound gain of lean muscle in bodybuilders who used steroids during a 10-week period. Anecdotal evidence suggests that some steroid users have added as much as 40 pounds of muscle in a year.

The adverse effects of these drugs, however, outweigh any benefits derived. In females, who normally lack potent androgenic hormones, anabolic steroid drugs not only promote "male-type" muscle mass and strength but also "masculinize" the users in other ways, such as by inducing growth of facial hair and by lowering the voice. More important, in both males and females, these agents adversely affect the reproductive and cardiovascular systems and the liver, may have an impact on behavior, and may be addictive.

Adverse Effects on the Reproductive System

In males, testosterone secretion and sperm production by the testes are normally controlled by hormones from the anterior pituitary gland. In negative-feedback fashion, testosterone inhibits secretion of these controlling hormones so that a constant circulating level of testosterone is maintained. The anterior pituitary is similarly inhibited by androgenic steroids taken as a drug. As a result, because the testes do not receive their normal stimulatory input from the anterior pituitary, testosterone secretion and sperm production decrease and the testes shrink. This hormone abuse also may set the stage for cancer of the testes and prostate gland.

In females, inhibition of the anterior pituitary by androgenic drugs suppresses the hormonal output that controls ovarian function. The result is failure to ovulate, menstrual irregularities, and decreased secretion of "feminizing" female sex hormones. Their decline diminishes breast size and other female characteristics.

Adverse Effects on the Cardiovascular System

Use of anabolic steroids induces several cardiovascular changes that increase the risk of developing atherosclerosis, which in turn is associated with an increased incidence of heart attacks and strokes (see p. 333). Among these adverse cardiovascular effects are (1) a reduction in high-density lipoproteins (HDL), the "good" cholesterol carriers that help remove cholesterol from the body; and (2) elevated blood pressure. Animal studies have also demonstrated damage to the heart muscle itself.

Adverse Effects on the Liver

Liver dysfunction is common with high steroid intake because the liver, which normally inactivates steroid hormones and prepares them for urinary excretion, is overloaded by the excess steroid intake. The incidence of liver cancer is also increased.

Adverse Effects on Behavior

Although the evidence is still controversial, anabolic steroid use appears to promote aggressive, even hostile behavior—the so-called 'roid rages.

Addictive Effects

A troubling new concern is the addiction to anabolic steroids of some who abuse these drugs. On the basis of their responses in one study involving face-to-face interviews, 14% of steroid users were judged to be addicted. In another survey, using anonymous, self-administered questionnaires, 57% of steroid users qualified as being addicted. This apparent tendency to become chemically dependent on steroids is alarming because the potential for adverse effects on health increases with long-term, heavy use, the kind of use that would be expected from someone hooked on the drug.

Thus, for health reasons, without even taking into account the legal and ethical issues, people should not use anabolic steroids. However, the problem appears to be worsening. Currently, the international black market for anabolic steroids is estimated at $1 billion per year.

Other Cheating Ways to Build Muscle Mass

Athletes seeking an artificial competitive edge have resorted to other illicit measures besides taking anabolic steroids, such as using human growth hormone or related compounds in the hopes of spurring muscle buildup. More worrisome, scientists predict the next illicit frontier will be gene doping. **Gene doping** refers to gene therapy aimed at improving athletic performance, such as by promoting production of naturally occurring muscle-building chemicals (for example, *insulin-like growth factor-I*, or *IGF-I*) or by blocking production of *myostatin*, a natural body chemical that puts the brakes on muscle growth. Because these chemicals occur naturally in the body, detection of gene doping will be a challenge.

Determinants of Whole-Muscle Tension in Skeletal Muscle

Number of Fibers Contracting

Number of motor units recruited*

Number of muscle fibers per motor unit

Number of muscle fibers available to contract (size of muscle)

Tension Developed by Each Contracting Fiber

Frequency of stimulation (twitch summation and tetanus)*

Length of fiber at onset of contraction (length–tension relationship)

Extent of fatigue
   Duration of activity
   Type of fiber (fatigue-resistant oxidative or fatigue-prone glycolytic)

Thickness of fiber
   Pattern of neural activity (hypertrophy, atrophy)
   Amount of testosterone (larger fibers in males than females)

*Factors controlled to accomplish gradation of contraction.

---

motor movement, no matter what its level of complexity, depends on converging input to the motor neurons of specific motor units. The motor neurons in turn trigger contraction of the muscle fibers within their respective motor units by means of the events that occur at the neuromuscular junction.

Multiple neural inputs influence motor unit output.

Three levels of input to the motor neurons control these neurons' output to the muscle fibers they innervate:

1. *Input from afferent neurons* (● Figure 8-23, **2a**), usually through intervening interneurons, at the level of the spinal cord—that is, spinal reflexes (see p. 177).

2. *Input from the primary motor cortex* **2b**. Fibers originating from neuronal cell bodies known as **pyramidal cells** within the primary motor cortex (see p. 143) descend directly without synaptic interruption to terminate on motor neurons (or on local interneurons that terminate on motor neurons) in the spinal cord. These fibers make up the **corticospinal (or pyramidal) motor system.**

3. *Input from the brain stem* **2c** as part of the multineuronal motor system. The pathways composing the **multineuronal (or extrapyramidal) motor system** include a number of synapses that involve many regions of the brain (*extra* means "outside of"; *pyramidal* refers to the pyramidal system). The final link in multineuronal pathways is the brain stem, especially the reticular formation (see p. 167), which in turn is influenced by motor regions of the cortex, the cerebellum, and the basal nuclei. In

addition, the motor cortex itself is interconnected with the thalamus as well as with premotor and supplementary motor areas, all part of the multineuronal system.

The only brain regions that directly influence motor neurons are the primary motor cortex and brain stem; the other involved brain regions indirectly regulate motor activity by adjusting motor output from the motor cortex and brain stem. A number of complex interactions take place between these various brain regions; the most important are represented in ● Figure 8-23. (See Chapter 5 for further discussion of the specific roles and interactions of these brain regions.)

Spinal reflexes involving afferent neurons are important in maintaining posture and in executing basic protective movements, such as the withdrawal reflex. The corticospinal system primarily mediates performance of fine, discrete, voluntary movements of the hands and fingers, such as those required for doing intricate needlework. Premotor and supplementary motor areas, with input from the cerebrocerebellum, plan the voluntary motor command that is issued to the appropriate motor neurons by the primary motor cortex through this descending system. The multineuronal system, in contrast, primarily regulates overall body posture involving involuntary movements of large muscle groups of the trunk and limbs. The corticospinal and multineuronal systems show considerable complex interaction and overlapping of function. To voluntarily manipulate your fingers to do needlework, for example, you subconsciously assume a particular posture of your arms that lets you hold your work.

Clinical Note Some of the inputs converging on motor neurons are excitatory, whereas others are inhibitory. Coordinated movement depends on an appropriate balance of activity in these inputs. The following types of motor abnormalities result from defective motor control:

■ If an inhibitory system originating in the brain stem is disrupted, muscles become hyperactive because of the unopposed activity in excitatory inputs to motor neurons. This condition, characterized by increased muscle tone and augmented limb reflexes, is known as **spastic paralysis.**

■ In contrast, loss of excitatory input, such as that accompanying destruction of descending excitatory pathways exiting the primary motor cortex, brings about **flaccid paralysis.** In this condition, the muscles are relaxed and the person cannot voluntarily contract muscles, although spinal reflex activity is still present. Damage to the primary motor cortex on one side of the brain, as with a stroke, leads to flaccid paralysis on the opposite half of the body (**hemiplegia,** or paralysis of one side of the body). Disruption of all descending pathways, as in traumatic severance of the spinal cord, produces flaccid paralysis below the level of the damaged region—**quadriplegia** (paralysis of all four limbs) in upper spinal cord damage and **paraplegia** (paralysis of the legs) in lower spinal cord injury.

■ Destruction of motor neurons—either their cell bodies or efferent fibers—causes flaccid paralysis and lack of reflex responsiveness in the affected muscles.

■ Damage to the cerebellum or basal nuclei does not result in paralysis but instead in uncoordinated, clumsy activity and in-

Muscular Dystrophy: When One Small Step Is a Big Deal

Hope of treatment is on the horizon for **muscular dystrophy (MD)**, a fatal muscle-wasting disease that primarily strikes boys and relentlessly leads to their death before age 20.

Symptoms

Muscular dystrophy encompasses more than 30 distinct hereditary pathological conditions, which have in common a progressive degeneration of contractile elements and their replacement by fibrous tissue. The gradual muscle wasting is characterized by progressive weakness over a period of years. Typically, an MD patient begins to show symptoms of muscle weakness at about 2 to 3 years of age, becomes wheelchair bound when he is 10 to 12 years old, and dies within the next 10 years, either from respiratory failure when his respiratory muscles become too weak or from heart failure when his heart becomes too weak.

Cause

The disease is caused by a recessive genetic defect on the X sex chromosome, of which males have only one copy. (Males have XY sex chromosomes; females have XX sex chromosomes.) If a male inherits from his mother an X chromosome bearing the defective dystrophic gene, he is destined to develop the disease, which affects one out of every 3500 boys worldwide. To acquire the condition, females must inherit a dystrophic-carrying X gene from both parents, a much rarer occurrence.

The defective gene responsible for *Duchenne muscular dystrophy (DMD),* the most common and most devastating form of the disease, was pinpointed in 1986. The gene normally produces **dystrophin,** a large protein that provides structural stability to the muscle cell's plasma membrane. Dystrophin is part of a complex of membrane-associated proteins that form a mechanical link between actin, a major component of the muscle cell's internal cytoskeleton, and the extracellular matrix, an external support network (see p. 58). This mechanical reinforcement of the plasma membrane enables the muscle cell to withstand the stresses and strains encountered during repeated cycles of contraction and stretching.

Dystrophic muscles are characterized by a lack of dystrophin. Even though this protein represents only 0.002% of the total amount of skeletal muscle protein, its presence is crucial in maintaining the integrity of the muscle cell membrane. The absence of dystrophin permits a constant leakage of Ca^{2+} into the muscle cells. This Ca^{2+} acti-vates proteases, protein-snipping enzymes that harm the muscle fibers. The resultant damage leads to the muscle wasting and ultimate fibrosis that characterize the disorder.

With the discovery of the dystrophin gene and its deficiency in DMD came the hope that scientists could somehow replenish this missing protein in the muscles of the disease's young victims. Although the disease is still considered untreatable and fatal, several lines of research are being pursued vigorously to intervene in the relentless muscle loss.

Gene-Therapy Approach

One approach is a possible "gene fix." With gene therapy, healthy genes are usually delivered to the defective cells by means of viruses. Viruses operate by invading a body cell and micromanaging the cell's genetic machinery. In this way, the virus directs the host cell to synthesize the proteins needed for viral replication. With gene therapy, the desired gene is inserted into an incapacitated virus that cannot cause disease but can still enter the target cell and take over genetic commands.

One of the big challenges for gene therapy for DMD is the enormous size of the dystrophin gene. This gene, being more than 3 million base pairs long, is the largest

---

appropriate patterns of movement. These regions normally smooth out activity initiated voluntarily.

■ Damage to higher cortical regions involved in planning motor activity results in the inability to establish appropriate motor commands to accomplish desired goals.

Muscle receptors provide afferent information needed to control skeletal muscle activity.

Coordinated, purposeful skeletal muscle activity depends on afferent input from a variety of sources. At a simple level, afferent signals indicating that your finger is touching a hot stove trigger reflex contractile activity in appropriate arm muscles to withdraw the hand from the injurious stimulus. At a more complex level, if you are going to catch a ball, the motor systems of your brain must program sequential motor commands that will move and position your body correctly for the catch, using predictions of the ball's direction and rate of movement provided by visual input. Many muscles acting simultaneously or alternately at different joints are called into play to shift your body's location and position rapidly, while maintaining your balance in the process. It is critical to have ongoing input about your body position with respect to the surrounding environment, as well as the position of your various body parts in relationship to one another. This information is necessary for establishing a neuronal pattern of activity to perform the desired movement. To appropriately program muscle activity, your CNS must know the starting position of your body. Further, it must be constantly informed about the progression of movement it has initiated, so that it can make adjustments as needed. Your brain receives this information, which is known as proprioceptive input (see p. 144), from receptors in your eyes, joints, vestibular apparatus, and skin, as well as from the muscles themselves.

You can demonstrate your joint and muscle proprioceptive receptors in action by closing your eyes and bringing the tips of your right and left index fingers together at any point in space.

gene ever found. It will not fit inside the viruses usually used to deliver genes to cells—they only have enough space for a gene one thousandth the size of the dystrophin gene. Therefore, researchers have created a minigene that is one thousandth of the size of the dystrophin gene but still contains the essential components for directing the synthesis of dystrophin. This stripped-down minigene can fit inside the viral carrier. Injection of these agents has stopped and even reversed the progression of MD in experimental animals. Gene-therapy clinical trials in humans have not been completed.

Cell-Transplant Approach
Another approach involves injecting cells that can functionally rescue the dystrophic muscle tissue. *Myoblasts* are undifferentiated cells that fuse to form the large, multinucleated skeletal muscle cells during embryonic development. After development, a small group of stem cells known as *satellite cells* remain close to the muscle surface. Satellite cells can be activated to form myoblasts, which can fuse together to form a new skeletal muscle cell to replace a damaged cell. When the loss of muscle cells is extensive, however, as in MD, this limited mechanism is not adequate to replace all the lost fibers.

One therapeutic approach for MD under study involves the transplantation of dystrophin-producing myoblasts harvested from muscle biopsies of healthy donors into the patient's dwindling muscles. Other researchers are pinning their hopes on delivery of satellite cells or partially differentiated adult stem cells that can be converted into healthy muscle cells (see p. 8).

Utrophin Approach
An alternative strategy that holds considerable promise for treating MD is upregulation of **utrophin,** a naturally occurring protein in muscle that is closely related to dystrophin. Eighty percent of the amino acid sequence for dystrophin and utrophin is identical, but these two proteins normally have different functions. Whereas dystrophin is dispersed throughout the muscle cell's surface membrane, where it contributes to the membrane's structural stability, utrophin is concentrated at the motor end plate. Here, utrophin plays a role in anchoring the acetylcholine receptors.

When researchers genetically engineered dystrophin-deficient mice that produced extra amounts of utrophin, this utrophin upregulation compensated in large part for the absent dystrophin; that is, the additional utrophin dispersed throughout the muscle cell membrane, where it assumed dystrophin's responsibilities. The result was improved intracellular Ca^{2+} homeostasis, enhanced muscle strength, and a marked reduction in the microscopic signs of muscle degeneration. Researchers are now scrambling to find a drug that will entice muscle cells to overproduce utrophin in humans, in the hopes of preventing or even repairing the muscle wasting that characterizes this devastating condition.

Anti-Myostatin Approach
Still other groups are exploring different tactics, such as interventions with newly designed drugs that increase the size of dwindling muscle fibers to counter the functional decline of dystrophic muscles. As an example, scientists have learned that **myostatin,** a protein produced in muscle cells, normally inhibits skeletal muscle growth in a check-and-balance fashion. They are working on ways to inhibit this inhibitor in patients with MD, thereby stimulating muscle growth.

These steps toward an eventual treatment mean that hopefully one day the afflicted boys will be able to take steps on their own instead of being destined to wheelchairs and early death.

You can do so without seeing where your hands are, because your brain is informed of the position of your hands and other body parts at all times by afferent input from the joint and muscle receptors.

Two types of muscle receptors—*muscle spindles* and *Golgi tendon organs*—monitor changes in muscle length and tension. Muscle length is monitored by muscle spindles; changes in muscle tension are detected by Golgi tendon organs. Both these receptor types are activated by muscle stretch, but they convey different types of information. Let us see how.

MUSCLE SPINDLE STRUCTURE Muscle spindles, which are distributed throughout the fleshy part of a skeletal muscle, consist of collections of specialized muscle fibers known as **intrafusal fibers,** which lie within spindle-shaped connective tissue capsules parallel to the "ordinary" **extrafusal fibers** (*fusus* means "spindle") (● Figure 8-24a). Unlike an ordinary extrafusal skeletal muscle fiber, which contains contractile elements (myofibrils) throughout its entire length, an intrafusal fiber has a noncontractile central portion, with the contractile elements being limited to both ends.

Each muscle spindle has its own private efferent and afferent nerve supply. The efferent neuron that innervates a muscle spindle's intrafusal fibers is known as a **gamma motor neuron,** whereas the motor neurons that supply the extrafusal fibers are called **alpha motor neurons.** Two types of afferent sensory endings terminate on the intrafusal fibers and serve as muscle spindle receptors, both of which are activated by stretch. The **primary (annulospiral) endings** are wrapped around the central portion of the intrafusal fibers; they detect changes in the length of the fibers during stretching as well as the speed with which it occurs. The **secondary (flower-spray) endings,** which are clustered at the end segments of many of the intrafusal fibers, are sensitive only to changes in length. Muscle spindles play a key role in the stretch reflex.

STRETCH REFLEX Whenever a whole muscle is passively stretched, its muscle spindle intrafusal fibers are likewise stretched, increas-

1 The CNS is constantly apprised of muscle length and tension and other peripheral events via pathways conveying afferent input, so it can program coordinated, purposeful skeletal muscle activity.

2 Motor movement is controlled by input to the motor neurons from

2a afferent neuron terminals at the level of the spinal cord,

2b the primary motor cortex, via the corticospinal motor system, and

2c brain stem nuclei, which serve as the final link in the complex, multineuronal motor system involving many regions of the brain.

KEY

⟵ = Pathways conveying afferent input   ⟵ = Corticospinal motor system   ⟵ = Multineuronal motor system

● **FIGURE 8-23 Motor control.** Arrows imply influence, whether excitatory or inhibitory; connections are not necessarily direct but may involve interneurons.

ing the firing rate in the afferent nerve fibers whose sensory endings terminate on the stretched spindle fibers. The afferent neuron directly synapses on the alpha motor neuron that innervates the extrafusal fibers of the same muscle, resulting in contraction of that muscle (● Figure 8-25, **1** and **2**). This monosynaptic **stretch reflex** (see p. 177) serves as a local negative-feedback mechanism to resist any passive changes in muscle length so that optimal resting length can be maintained.

The classic example of the stretch reflex is the **patellar tendon,** or **knee-jerk, reflex** (● Figure 8-26). The extensor muscle of the knee is the *quadriceps femoris,* which forms the anterior (front) portion of the thigh and is attached just below the knee to the tibia (shinbone) by the *patellar tendon.* Tapping this tendon with a rubber mallet passively stretches the quadriceps muscle, activating its spindle receptors. The resulting stretch reflex brings about contraction of this extensor muscle, causing

Alpha motor neuron axon

Gamma motor neuron axon

Afferent neuron axons

Primary (annulospiral) endings of afferent fibers

Secondary (flower-spray) endings of afferent fibers

Extrafusal ("ordinary") muscle fibers

Capsule

Intrafusal (spindle) muscle fibers

Contractile end portion of intrafusal fiber

Noncontractile central portion of intrafusal fiber

(a) Muscle spindle

Skeletal muscle

Afferent fiber

Golgi tendon organ

Collagen

Tendon

Bone

(b) Golgi tendon organ

● **FIGURE 8-24 Muscle receptors.** (a) A muscle spindle consists of a collection of specialized intrafusal fibers that lie within a connective tissue capsule parallel to the ordinary extrafusal skeletal muscle fibers. The muscle spindle is innervated by its own gamma motor neuron and is supplied by two types of afferent sensory terminals, the primary (annulospiral) endings and the secondary (flower-spray) endings, both of which are activated by stretch. (b) The Golgi tendon organ is entwined with the collagen fibers in a tendon and monitors changes in muscle tension transmitted to the tendon.

the knee to extend and raise the foreleg in the well-known knee-jerk fashion.

Clinical Note This test is routinely done as a preliminary assessment of nervous system function. A normal knee jerk indicates that a number of neural and muscular components—muscle spindle, afferent input, motor neurons, efferent output, neuromuscular junctions, and the muscles themselves—are functioning normally. It also indicates an appropriate balance of excitatory and inhibitory input to the motor neurons from higher brain levels. Muscle jerks may be absent or depressed with loss of higher-level excitatory inputs, or may be greatly exaggerated with loss of inhibitory input to the motor neurons from higher brain levels.

The primary purpose of the stretch reflex is to resist the tendency for the passive stretch of extensor muscles by gravita-

tional forces when you are standing upright. Whenever your knee joint tends to buckle because of gravity, your quadriceps muscle is stretched. The resulting enhanced contraction of this extensor muscle brought about by the stretch reflex quickly straightens out the knee, keeping your leg extended so that you remain standing. The stretch reflex also normally comes into play when a load gets heavier as you are holding it, as for example if you are filling a glass of water. The resultant stretch of your biceps muscle as the glass gets heavier initiates a stretch reflex in this muscle that helps you keep the glass in position instead of lowering it and perhaps spilling the water.

COACTIVATION OF GAMMA AND ALPHA MOTOR NEURONS
Gamma motor neurons initiate contraction of the muscular end regions of intrafusal fibers (● Figure 8-25, **3**). This contractile

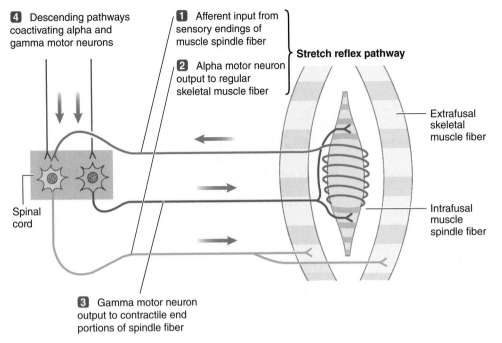

4 Descending pathways coactivating alpha and gamma motor neurons

1 Afferent input from sensory endings of muscle spindle fiber

2 Alpha motor neuron output to regular skeletal muscle fiber

Stretch reflex pathway

Spinal cord

Extrafusal skeletal muscle fiber

Intrafusal muscle spindle fiber

3 Gamma motor neuron output to contractile end portions of spindle fiber

(a) Pathways involved in monosynaptic stretch reflex and coactivation of alpha and gamma motor neurons

Relaxed muscle; spindle fiber sensitive to stretch of muscle

(b) Relaxed muscle

Contracted muscle in hypothetical situation of no spindle coactivation; slackened spindle fiber not sensitive to stretch of muscle

(c) Contracted muscle with no spindle coactivation

Contracted muscle in normal situation of spindle coactivation; contracted spindle fiber sensitive to stretch of muscle

(d) Contracted muscle with spindle coactivation

● **FIGURE 8-25 Muscle spindle function.**

response is too weak to have any influence on whole-muscle tension, but it does have an important localized effect on the muscle spindle itself. If there were no compensating mechanisms, shortening of the whole muscle by alpha motor-neuron stimulation of extrafusal fibers would slacken the spindle fibers so that they would be less sensitive to stretch and therefore not as effective as muscle length detectors (● Figure 8-25b and c). **Coactivation** of the gamma motor-neuron system along with the alpha motor-neuron system during reflex and voluntary contractions (● Figure 8-25, **4**) takes the slack out of the spindle fibers as the whole muscle shortens, letting these receptor structures maintain their high sensitivity to stretch over a wide range of muscle lengths. When gamma motor-neuron stimulation triggers simultaneous contraction of both end muscular portions of an intrafusal fiber, the noncontractile central portion is pulled in opposite direc-

tions, tightening this region and taking out the slack (● Figure 8-25d). Whereas the extent of alpha motor-neuron activation depends on the intended strength of the motor response, the extent of simultaneous gamma motor-neuron activity to the same muscle depends on the anticipated distance of shortening.

GOLGI TENDON ORGANS In contrast to muscle spindles, which lie within the belly of the muscle, **Golgi tendon organs** are in the tendons of the muscle, where they can respond to changes in the muscle's tension rather than to changes in its length. Because a number of factors determine the tension developed in the whole muscle during contraction (for example, frequency of stimulation or length of the muscle at the onset of contraction), it is essential that motor control systems be apprised of the tension actually achieved so that adjustments can be made if necessary.

Golgi tendon organs consist of endings of afferent fibers entwined within bundles of connective tissue (collagen) fibers that make up the tendon (● Figure 8-24b). When the extrafusal muscle fibers contract, the resulting pull on the tendon tightens the connective tissue bundles, which in turn increase the tension exerted on the bone to which the tendon is attached. In the process, the entwined Golgi organ afferent receptor endings are stretched, causing the afferent fibers to fire; the frequency of firing is directly related to the tension developed. This afferent information is sent to the brain for processing. Much of this information is used subconsciously for smoothly executing motor activity, but unlike afferent information from the muscle spindles, afferent information from the Golgi tendon organ reaches the level of conscious awareness. You are aware of the tension within a muscle but not of its length.

Scientists once thought the Golgi tendon organ triggered a protective spinal reflex that halted further contraction and brought about sudden reflex relaxation when the muscle tension became great enough, thus helping prevent damage to the muscle or tendon from excessive, tension-developing muscle contractions. Scientists now believe, however, that this receptor

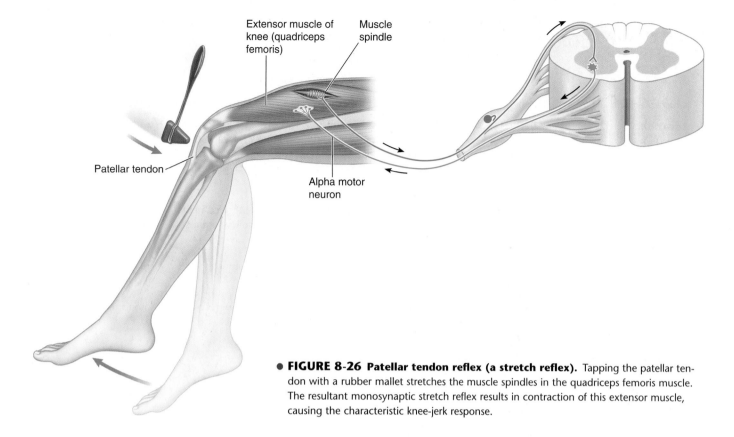

● **FIGURE 8-26 Patellar tendon reflex (a stretch reflex).** Tapping the patellar tendon with a rubber mallet stretches the muscle spindles in the quadriceps femoris muscle. The resultant monosynaptic stretch reflex results in contraction of this extensor muscle, causing the characteristic knee-jerk response.

Labels in figure: Extensor muscle of knee (quadriceps femoris); Muscle spindle; Patellar tendon; Alpha motor neuron

is a pure sensor and does not initiate any reflexes. Other unknown mechanisms are apparently involved in inhibiting further contraction to prevent tension-induced damage.

Having completed our discussion of skeletal muscle, we now examine smooth and cardiac muscle.

Smooth and Cardiac Muscle

The two other types of muscle—smooth muscle and cardiac muscle—share some basic properties with skeletal muscle, but each also displays unique characteristics (▲ Table 8-3). The three muscle types all have a specialized contractile apparatus made up of thin actin filaments that slide relative to stationary thick myosin filaments in response to a rise in cytosolic Ca^{2+} to accomplish contraction. Also, they all directly use ATP as the energy source for cross-bridge cycling. However, the structure and organization of fibers within these different muscle types vary, as do their mechanisms of excitation and the means by which excitation and contraction are coupled. Furthermore, there are important distinctions in the contractile response itself. We will spend the rest of this chapter highlighting unique features of smooth and cardiac muscle as compared with skeletal muscle, saving more detailed discussion of their function for chapters on organs containing these muscle types.

Smooth muscle cells are small and unstriated.

Most smooth muscle cells are found in the walls of hollow organs and tubes. Their contraction exerts pressure on and regulates the forward movement of the contents of these structures.

Both smooth and skeletal muscle cells are elongated, but in contrast to their large, cylindrical skeletal muscle counterparts, smooth muscle cells are spindle shaped, have a single nucleus, and are considerably smaller (2 to 10 μm in diameter and 50 to 400 μm long). Also unlike skeletal muscle cells, a single smooth muscle cell does not extend the full length of a muscle. Instead, groups of smooth muscle cells are typically arranged in sheets (● Figure 8-27a, p. 292).

A smooth muscle cell has three types of filaments: (1) thick myosin filaments, which are longer than those in skeletal muscle; (2) thin actin filaments, which contain tropomyosin but lack the regulatory protein troponin; and (3) filaments of intermediate size, which do not directly participate in contraction but are part of the cytoskeletal framework that supports the cell shape. Smooth muscle filaments do not form myofibrils and are not arranged in the sarcomere pattern found in skeletal muscle. Thus, smooth muscle cells do not show the banding or striation of skeletal muscle, hence the term "smooth" for this muscle type.

Lacking sarcomeres, smooth muscle does not have Z lines as such but has **dense bodies** containing the same protein constituent found in Z lines (● Figure 8-27b). Dense bodies are positioned throughout the smooth muscle cell as well as attached to the internal surface of the plasma membrane. Dense bodies are held in place by a scaffold of intermediate filaments. The actin filaments are anchored to the dense bodies. Considerably more actin is present in smooth muscle cells than in skeletal muscle cells, with 10 to 15 thin filaments for each thick myosin filament in smooth muscle compared to 2 thin filaments for each thick filament in skeletal muscle.

Comparison of Muscle Types

	TYPE OF MUSCLE			
Characteristic	**Skeletal**	**Multiunit Smooth**	**Single-Unit Smooth**	**Cardiac**
Location	Attached to skeleton	Large blood vessels, small airways, eye, and hair follicles	Walls of hollow organs in digestive, reproductive, and urinary tracts and in small blood vessels	Heart only
Function	Movement of body in relation to external environment	Varies with structure involved	Movement of contents within hollow organs	Pumps blood out of heart
Mechanism of Contraction	Sliding filament mechanism	Sliding filament mechanism	Sliding filament mechanism	Sliding filament mechanism
Innervation	Somatic nervous system (alpha motor neurons)	Autonomic nervous system	Autonomic nervous system	Autonomic nervous system
Level of Control	Under voluntary control; also subject to subconscious regulation	Under involuntary control	Under involuntary control	Under involuntary control
Initiation of Contraction	Neurogenic	Neurogenic	Myogenic (pacemaker potentials and slow-wave potentials)	Myogenic (pacemaker potentials)
Role of Nervous Stimulation	Initiates contraction; accomplishes gradation	Initiates contraction; contributes to gradation	Modifies contraction; can excite or inhibit; contributes to gradation	Modifies contraction; can excite or inhibit; contributes to gradation
Modifying Effect of Hormones	No	Yes	Yes	Yes
Presence of Thick Myosin and Thin Actin Filaments	Yes	Yes	Yes	Yes
Striated by Orderly Longitudinal Arrangement of Filaments	Yes	No	No	Yes
Presence of Troponin and Tropomyosin	Yes	Tropomyosin only	Tropomyosin only	Yes
Presence of T Tubules	Yes	No	No	Yes
Level of Development of Sarcoplasmic Reticulum	Well developed	Poorly developed	Poorly developed	Moderately developed
Cross Bridges Turned on by Ca^{2+}	Yes	Yes	Yes	Yes
Source of Increased Cytosolic Ca^{2+}	Sarcoplasmic reticulum	ECF and sarcoplasmic reticulum	ECF and sarcoplasmic reticulum	ECF and sarcoplasmic reticulum

Characteristic	Skeletal	Multiunit Smooth	Single-Unit Smooth	Cardiac
TYPE OF MUSCLE				
Site of Ca²⁺ Regulation	Troponin in thin filaments	Myosin in thick filaments	Myosin in thick filaments	Troponin in thin filaments
Mechanism of Ca²⁺ Action	Physically repositions troponin–tropomyosin complex to uncover actin cross-bridge binding sites	Chemically brings about phosphorylation of myosin cross bridges so they can bind with actin	Chemically brings about phosphorylation of myosin cross bridges so they can bind with actin	Physically repositions troponin–tropomyosin complex
Presence of Gap Junctions	No	Yes (very few)	Yes	Yes
ATP Used Directly by Contractile Apparatus	Yes	Yes	Yes	Yes
Myosin ATPase Activity; Speed of Contraction	Fast or slow, depending on type of fiber	Very slow	Very slow	Slow
Means by Which Gradation Accomplished	Varying number of motor units contracting (motor unit recruitment) and frequency at which they are stimulated (twitch summation)	Varying number of muscle fibers contracting and varying cytosolic Ca²⁺ concentration in each fiber by autonomic and hormonal influences	Varying cytosolic Ca²⁺ concentration through myogenic activity and influences of the autonomic nervous system, hormones, mechanical stretch, and local metabolites	Varying length of fiber (depending on extent of filling of the heart chambers) and varying cytosolic Ca²⁺ concentration through autonomic, hormonal, and local metabolite influence
Presence of Tone in Absence of External Stimulation	No	No	Yes	No
Clear-Cut Length–Tension Relationship	Yes	No	No	Yes

Smooth muscle cells   Nucleus

Brian Eyden/Science Source/Photo Researchers, Inc.

(a) Low-power light micrograph of smooth muscle cells

Smooth muscle cells   Dense bodies

Brenda Russell, University of Illinois

(b) Electron micrograph of smooth muscle cells

● **FIGURE 8-27 Microscopic view of smooth muscle cells.**
(a) Note the spindle shape and single, centrally located nucleus.
(b) Note the presence of dense bodies and lack of banding.

The thick- and thin-filament contractile units are oriented slightly diagonally from side to side within the smooth muscle cell in an elongated, diamond-shaped lattice, rather than running parallel with the long axis as myofibrils do in skeletal muscle (● Figure 8-28a). Relative sliding of the thin filaments past the thick filaments during contraction causes the filament lattice to shorten and expand from side to side. As a result, the whole cell shortens and bulges out between the points where the thin filaments are attached to the inner surface of the plasma membrane (● Figure 8-28b).

Unlike in skeletal muscle, myosin molecules are arranged in a smooth-muscle thick filament so that cross bridges are present along the entire filament length (that is, there is no bare portion in the center of a smooth-muscle thick filament). As a result, the surrounding thin filaments can be pulled along the thick filaments for longer distances than in skeletal muscle. Also dissimilar to skeletal muscle, the myosin proteins in smooth-muscle

thick filaments are organized so that half of the surrounding thin filaments are pulled in one direction and the other half are pulled in the opposite direction (● Figure 8-28b).

Smooth muscle cells are turned on by Ca²⁺-dependent phosphorylation of myosin.

The thin filaments of smooth muscle cells do not contain troponin, and tropomyosin does not block actin's cross-bridge binding sites. What, then, prevents actin and myosin from binding at the cross bridges in the resting state, and how is cross-bridge activity switched on in the excited state? Lightweight chains of proteins are attached to the heads of myosin molecules, near the "neck" region. These so-called **light chains** are only of secondary importance in skeletal muscle, but they have a crucial regulatory function in smooth muscle. The smooth muscle myosin heads can interact with actin only when the myosin light chain is *phosphorylated* (that is, has an inorganic phosphate from ATP attached to it). During excitation, the increased cytosolic Ca²⁺ acts as an intracellular messenger, initiating a chain of biochemical events that results in phosphorylation of the myosin light chain (● Figure 8-29). Smooth muscle Ca²⁺ binds with **calmodulin,** an intracellular protein found in most cells that is structurally similar to troponin (see p. 122). This Ca²⁺-calmodulin complex binds to and activates another protein, **myosin light chain kinase (MLC kinase),** which in turn phosphorylates the myosin light chain. Note that this inorganic phosphate on the myosin light chain is in addition to the inorganic phosphate accompanying ADP on the myosin cross-bridge ATPase site. The P_i on the ATPase site is part of the energy-supplying cycle that powers cross-bridge bending. The P_i on the light chain permits the myosin cross bridge to bind with actin so that cross-bridge cycling can begin. Therefore, smooth muscle is triggered to contract by a rise in cytosolic Ca²⁺ similar to what happens in skeletal muscle, but in smooth muscle, Ca²⁺ ultimately turns on the cross bridges by inducing a *chemical* change in myosin in the *thick* filaments, whereas in skeletal muscle it exerts its effects by invoking a *physical* change at the *thin* filaments (● Figure 8-30). Recall that in skeletal muscle, Ca²⁺ moves troponin and tropomyosin from their blocking position, so actin and myosin are free to bind with each other.

Phasic smooth muscle contracts in bursts of activity; tonic smooth muscle maintains an ongoing level of contraction.

The means by which cytosolic Ca²⁺ concentration increases in smooth muscle cells to turn on the cross bridges also differs from that for skeletal muscle and even varies among the highly diversified smooth muscle found in different organs. Smooth muscle can be classified in various ways, depending on the timing and means of increasing cytosolic Ca²⁺: phasic or tonic, multiunit or single-unit, and neurogenic or myogenic smooth muscle. Each smooth muscle belongs to one class of each of these three categories. That is, the smooth muscle in one organ might be phasic, multiunit, and neurogenic, and the smooth muscle in another organ might be tonic, single-unit, and myogenic. Let's examine each of these categories.

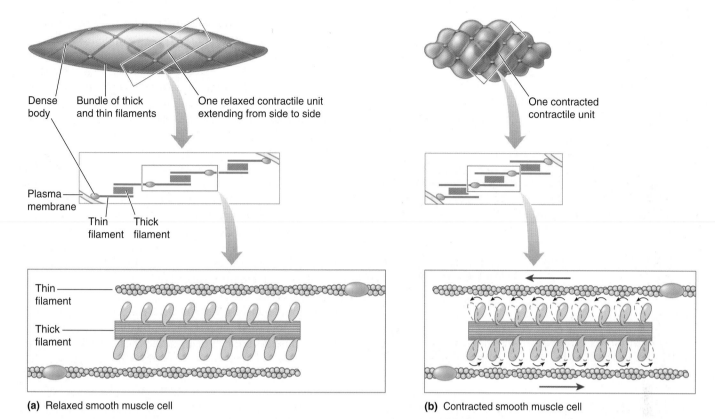

Dense
body

Bundle of thick
and thin filaments

One relaxed contractile unit
extending from side to side

One contracted
contractile unit

Plasma
membrane

Thin
filament

Thick
filament

Thin
filament

Thick
filament

(a) Relaxed smooth muscle cell

(b) Contracted smooth muscle cell

● **FIGURE 8-28 Arrangement of thick and thin filaments in a smooth muscle cell in relaxed and contracted states.**

Calmodulin

Ca^{2+}

Ca^{2+}-calmodulin

Inactive myosin
light chain kinase

Active myosin
light chain kinase

ATP

P_i

ADP

Inactive myosin
cross bridge

Phosphorylated
myosin cross bridge
(can bind with actin)

Permits binding
with actin

P_i

ADP
P_i

Part of
cross-bridge
energy cycle

Myosin light chain

● **FIGURE 8-29 Calcium activation of myosin cross bridge in smooth muscle.**

Smooth muscle can be grouped into two categories depending on how its cytosolic Ca^{2+} concentration increases and its level of ongoing contractile activity: *phasic smooth muscle* and *tonic smooth muscle*. **Phasic smooth muscle** contracts in bursts, triggered by action potentials that lead to increased cytosolic Ca^{2+}. These bursts in contraction are characterized by pro-

nounced increases in contractile activity. Phasic smooth muscle is most abundant in the walls of hollow organs that push contents through them, such as the digestive organs. Phasic digestive contractions mix the food with digestive juices and propel the mass forward for further processing. **Tonic smooth muscle** is usually partially contracted at all times. This state of partial

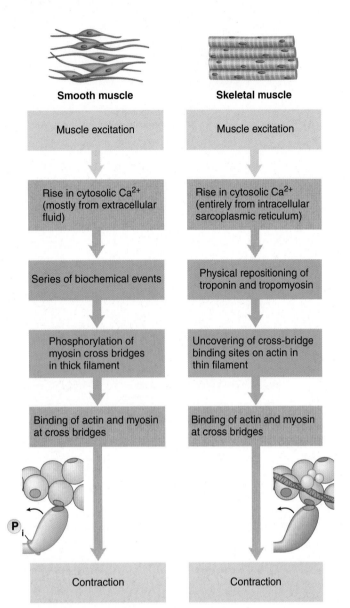

Smooth muscle	Skeletal muscle
Muscle excitation	Muscle excitation
Rise in cytosolic Ca^{2+} (mostly from extracellular fluid)	Rise in cytosolic Ca^{2+} (entirely from intracellular sarcoplasmic reticulum)
Series of biochemical events	Physical repositioning of troponin and tropomyosin
Phosphorylation of myosin cross bridges in thick filament	Uncovering of cross-bridge binding sites on actin in thin filament
Binding of actin and myosin at cross bridges	Binding of actin and myosin at cross bridges
Contraction	Contraction

● **FIGURE 8-30 Comparison of the role of calcium in bringing about contraction in smooth muscle and skeletal muscle.**

A smooth muscle cell has no T tubules and a poorly developed sarcoplasmic reticulum. In phasic smooth muscle, the increased cytosolic Ca^{2+} that triggers contraction comes from two sources: Most Ca^{2+} enters from the ECF, but some is released intracellularly from the sparse sarcoplasmic reticulum stores. Unlike their role in skeletal muscle cells, voltage-gated dihydropyridine receptors in the plasma membrane of smooth muscle cells function as Ca^{2+} channels. When these surface-membrane channels are opened in response to an action potential, Ca^{2+} enters down its concentration gradient from the ECF. The entering Ca^{2+} triggers the opening of Ca^{2+} channels in the sarcoplasmic reticulum so that small additional amounts of Ca^{2+} are released intracellularly from this meager source. Because smooth muscle cells are so much smaller in diameter than skeletal muscle fibers, most Ca^{2+} entering from the ECF can influence cross-bridge activity, even in the central portions of the cell, without requiring an elaborate T tubule–sarcoplasmic reticulum mechanism.

One of the major means of increasing cytosolic Ca^{2+} concentration and thus increasing contractile activity in tonic smooth muscle is binding of an extracellular chemical messenger, such as norepinephrine or various hormones, to a G-protein-coupled receptor, which activates the IP_3/Ca^{2+} second-messenger pathway (see p. 122). The membrane of the sarcoplasmic reticulum in tonic smooth muscle has IP_3 receptors, which like ryanodine receptors, are Ca^{2+}-release channels. IP_3 binding leads to release of contractile-inducing Ca^{2+} from this intracellular store into the cytosol. This is how norepinephrine released from the sympathetic nerve endings acts on arterioles to increase blood pressure.

Relaxation in smooth muscle is accomplished by removal of Ca^{2+} as it is actively transported out across the plasma membrane and/or back into the sarcoplasmic reticulum, depending on its source. When Ca^{2+} is removed, myosin is dephosphorylated (the phosphate is removed) and can no longer interact with actin, so the muscle relaxes.

We still have not addressed the question of what brings about action potentials in smooth muscle. Smooth muscle is grouped into two categories—*multiunit* and *single-unit smooth muscle*—based on differences in how the muscle fibers become excited. Let us compare these two types of smooth muscle.

Multiunit smooth muscle is neurogenic.

Multiunit smooth muscle exhibits properties partway between those of skeletal muscle and single-unit smooth muscle. As the name implies, a multiunit smooth muscle consists of multiple discrete units that function independently of one another and must be separately stimulated by nerves to undergo action potentials and contract, similar to skeletal muscle motor units. Thus, contractile activity in both skeletal muscle and multiunit smooth muscle is **neurogenic** ("nerve produced"). That is, contraction in these muscle types is initiated only in response to stimulation by the nerves supplying the muscle. All multiunit smooth muscle is phasic, contracting only when neurally stimulated. Whereas skeletal muscle is innervated by the voluntary somatic nervous system (motor neurons), multiunit (as well as single-unit) smooth muscle is supplied by the involuntary autonomic nervous system.

contraction is called **tone.** Tone exists because this type of smooth muscle has a relatively low resting potential of -55 mV to -40 mV. Some surface-membrane voltage-gated Ca^{2+} channels are open at these potentials. The resultant Ca^{2+} entry maintains a state of partial contraction. Therefore, maintenance of tone in tonic smooth muscle does not depend on action potentials. Tonic smooth muscle does not display bursts of contractile activity but instead incrementally varies its extent of contraction above or below this tonic level in response to regulatory factors, which alter the cytosolic Ca^{2+} concentration. The smooth muscle in walls of arterioles is an example of tonic smooth muscle. The ongoing tonic contraction in these small blood vessels squeezes down on the blood flowing through them and is one of the major contributing factors to maintenance of blood pressure.

Multiunit smooth muscle is found (1) in the walls of large blood vessels; (2) in small airways to the lungs; (3) in the muscle of the eye that adjusts the lens for near or far vision; (4) in the iris of the eye, which alters the pupil size to adjust the amount of light entering the eye; and (5) at the base of hair follicles, contraction of which causes "goose bumps."

Single-unit smooth muscle cells form functional syncytia.

Most smooth muscle is **single-unit smooth muscle,** alternatively called **visceral smooth muscle,** because it is found in the walls of the hollow organs or viscera (for example, the digestive, reproductive, and urinary tracts and small blood vessels). The term "single-unit smooth muscle" derives from the fact that the muscle fibers that make up this type of muscle become excited and contract as a single unit. The muscle fibers in single-unit smooth muscle are electrically linked by gap junctions (see p. 59). When an action potential occurs anywhere within a sheet of single-unit smooth muscle, it is quickly propagated via these special points of electrical contact throughout the entire group of interconnected cells, which then contract as a single, coordinated unit. Such a group of interconnected muscle cells that function electrically and mechanically as a unit is known as a **functional syncytium** (plural, *syncytia; syn* means "together"; *cyt* means "cell").

Thinking about the role of the uterus during labor can help you appreciate the significance of this arrangement. Muscle cells composing the uterine wall act as a functional syncytium. They repetitively become excited and contract as a unit during labor, exerting a series of coordinated "pushes" that eventually deliver the baby. Independent, uncoordinated contractions of individual muscle cells in the uterine wall could not exert the uniformly applied pressure needed to expel the baby. Single-unit smooth muscle elsewhere in the body is arranged in similar functional syncytia.

Single-unit smooth muscle is myogenic.

Single-unit smooth muscle is **self-excitable,** so it does not require nervous stimulation for contraction. Single-unit smooth muscle may be of the phasic or tonic type. In phasic single-unit smooth muscle, clusters of specialized smooth muscle cells within a functional syncytium display spontaneous electrical activity; that is, they can undergo action potentials without any external stimulation. In contrast to the other excitable cells we have been discussing (such as neurons, skeletal muscle fibers, and multiunit smooth muscle), the self-excitable cells of single-unit smooth muscle do not maintain a constant resting potential. Instead, their membrane potential inherently fluctuates without any influence by factors external to the cell. Two major types of spontaneous depolarizations displayed by self-excitable cells are *pacemaker potentials* and *slow-wave potentials.*

PACEMAKER POTENTIALS With **pacemaker potentials,** the membrane potential gradually depolarizes on its own because of shifts in passive ionic fluxes accompanying automatic changes in channel permeability (● Figure 8-31a). When the membrane has depolarized to threshold, an action potential is initiated.

(a) Pacemaker potential

(b) Slow-wave potential

● **FIGURE 8-31 Self-generated electrical activity in smooth muscle.** (a) With pacemaker potentials, the membrane gradually depolarizes to threshold on a regular periodic basis without any nervous stimulation. These regular depolarizations cyclically trigger self-induced action potentials. (b) In slow-wave potentials, the membrane gradually undergoes self-induced hyperpolarizing and depolarizing swings in potential. A burst of action potentials occurs if a depolarizing swing brings the membrane to threshold.

After repolarizing, the membrane potential once again depolarizes to threshold, cyclically continuing in this manner to repetitively self-generate action potentials.

SLOW-WAVE POTENTIALS Slow-wave potentials are spontaneous, gradually alternating hyperpolarizing and depolarizing swings in potential (● Figure 8-31b). The potential is moved farther from threshold during each hyperpolarizing swing and closer to threshold during each depolarizing swing. If threshold is reached, a burst of action potentials occurs at the peak of a depolarizing swing. Threshold is not always reached, however, so the oscillating slow-wave potentials can continue without generating action potentials. Whether threshold is reached depends on the starting point of the membrane potential at the onset of its depolarizing swing. The starting point, in turn, is influenced by neural and local factors. Slow-wave potentials occur only in smooth muscle of the digestive tract. (We have now discussed all the means by which excitable tissues can be brought to threshold.

Various Means of Initiating Action Potentials in Excitable Tissues

Method of Depolarizing the Membrane to Threshold Potential	Type of Excitable Tissue Involved	Description of This Triggering Event
Summation of excitatory post-synaptic potentials (EPSPs) (see p. 108)	Efferent neurons, interneurons	Temporal or spatial summation of slight depolarizations (EPSPs) of the dendrite/cell body end of the neuron brought about by changes in channel permeability in response to binding of excitatory neurotransmitter with surface membrane receptors
Receptor potential (see p. 185)	Afferent neurons	Typically a depolarization of the afferent neuron's receptor initiated by changes in channel permeability in response to the neuron's adequate stimulus
End-plate potential (see p. 249)	Skeletal muscle	Depolarization of the motor end plate brought about by changes in channel permeability in response to binding of the neurotransmitter acetylcholine with receptors on the end-plate membrane
Pacemaker potential (see p. 295)	Smooth muscle, cardiac muscle	Gradual depolarization of the membrane on its own because of shifts in passive ionic fluxes accompanying automatic changes in channel permeability
Slow-wave potential (see p. 295)	Smooth muscle (in digestive tract only)	Gradual alternating hyperpolarizing and depolarizing swings in potential brought about by unknown mechanisms; depolarizing swing may or may not reach threshold

▲ Table 8-4 summarizes the different triggering events that can initiate action potentials in various excitable tissues.)

MYOGENIC ACTIVITY Self-excitable smooth muscle cells are specialized to initiate action potentials, but they are not equipped to contract. Only a very few of all the cells in a functional syncytium are noncontractile, pacemaker cells. Pacemaker cells are typically clustered in a specific location. The vast majority of smooth muscle cells in a functional syncytium are specialized to contract but cannot self-initiate action potentials. However, once an action potential is initiated by a self-excitable pacemaker cell, it is conducted to the remaining contractile, nonpacemaker cells of the functional syncytium via gap junctions, so the entire group of connected cells contracts as a unit without any nervous input. Such nerve-independent contractile activity initiated by the muscle itself is called **myogenic** ("muscle-produced") activity, in contrast to the neurogenic activity of skeletal muscle and multiunit smooth muscle. Recall that tonic, single-unit smooth muscle cells have sufficient levels of cytosolic Ca^{2+} to maintain a low level of tension even in the absence of action potentials, so they too are myogenic. (Thus multiunit smooth muscles are all neurogenic and phasic; single-unit smooth muscles are all myogenic and may be phasic or tonic.)

Gradation of single-unit smooth muscle contraction differs from that of skeletal muscle.

Single-unit smooth muscle differs from skeletal muscle in the way contraction is graded. Gradation of skeletal muscle contraction is entirely under neural control, primarily involving motor unit recruitment and twitch summation. In single-unit smooth muscle, gap junctions ensure that an entire smooth muscle mass contracts as a single unit, making it impossible to vary the number of muscle fibers contracting. Only the tension of the fibers can be modified to achieve varying strengths of contraction of the whole organ. The portion of cross bridges activated and the tension subsequently developed in single-unit smooth muscle can be graded by varying the cytosolic Ca^{2+} concentration. A single excitation in smooth muscle does not cause all the cross bridges to switch on, in contrast to skeletal muscles, where a single action potential triggers release of enough Ca^{2+} to permit all cross bridges to cycle. As Ca^{2+} concentration increases in smooth muscle, more cross bridges are brought into play, and greater tension develops.

MODIFICATION OF SMOOTH MUSCLE ACTIVITY BY THE AUTONOMIC NERVOUS SYSTEM

Smooth muscle is typically innervated by both branches of the autonomic nervous system. In single-unit smooth muscle (both phasic and tonic), this nerve supply does not *initiate* contraction, but it can *modify* the rate and strength of contraction, either enhancing or retarding the inherent contractile activity of a given organ. Recall that the isolated motor end-plate region of a skeletal muscle fiber interacts with ACh released from a single axon terminal of a motor neuron. In contrast, the receptor proteins that bind with autonomic neurotransmitters are dispersed throughout the entire surface membrane of a smooth muscle cell. Smooth muscle cells are sensitive to varying degrees and in varying ways to autonomic neurotransmitters, depending on the cells' distribution of cholinergic and adrenergic receptors (see pp. 243–244).

Each terminal branch of a postganglionic autonomic fiber travels across the surface of one or more smooth muscle cells, releasing neurotransmitter from the vesicles within its multiple **varicosities** (bulges) as an action potential passes along the terminal (● Figure 8-32). The neurotransmitter diffuses to the many receptors specific for it on the cells underlying the terminal. Thus, in contrast to the discrete one-to-one relationship at motor end plates, a given smooth muscle cell can be influenced by more than one type of neurotransmitter, and each autonomic terminal can influence more than one smooth muscle cell.

OTHER FACTORS INFLUENCING SMOOTH MUSCLE ACTIVITY

Other factors (besides autonomic neurotransmitters) can influence the rate and strength of both multiunit and single-unit smooth muscle contraction, including mechanical stretch, certain hormones, local metabolites, and specific drugs. The smooth muscle of digestive organs is also influenced by the enteric nervous system, which is a specialized network of nerve fibers built into the wall of the digestive tract (see p. 134 and p. 595). Some smooth muscle is poorly innervated, an example being the uterus, in which case the rate and strength of contraction is regulated entirely by circulating and locally released chemical messengers, which vary with the stage of the menstrual cycle and with the stage of pregnancy. All these factors ultimately act by modifying the permeability of Ca^{2+} channels in the plasma membrane, the sarcoplasmic reticulum, or both, through a variety of mechanisms. Thus, smooth muscle is sub-

● **FIGURE 8-32 Innervation of smooth muscle by autonomic postganglionic nerve terminals.**

ject to more external influences than skeletal muscle is, even though smooth muscle can contract on its own and skeletal muscle cannot.

For now, as we look at the length–tension relationship in smooth muscle, we are going to discuss in more detail only the effect of mechanical stretch (as occurs during filling of a hollow organ) on smooth muscle contractility. We will examine the extracellular chemical influences on smooth muscle contractility in later chapters when we discuss regulation of the various organs that contain smooth muscle.

Smooth muscle can still develop tension yet inherently relaxes when stretched.

The relationship between the length of the muscle fibers before contraction and the tension that can be developed on a subsequent contraction is less closely linked in smooth muscle than in skeletal muscle. The range of lengths over which a smooth muscle fiber can develop near-maximal tension is much greater than for skeletal muscle. Smooth muscle can still develop considerable tension even when stretched up to 2.5 times its resting length, for two likely reasons. First, in contrast to skeletal muscle, in which the resting length is at l_o, in smooth muscle the resting (nonstretched) length is much shorter than at l_o. Therefore, smooth muscle can be stretched considerably before reaching its optimal length. Second, the thin filaments still overlap the much longer thick filaments even in the stretched-out position, so cross-bridge interaction and tension development can still take place. In contrast, when skeletal muscle is stretched only three fourths longer than its resting length, the thick and thin filaments are completely pulled apart and can no longer interact (see ● Figure 8-19, p. 273).

The ability of a considerably stretched smooth muscle fiber to still develop tension is important, because the smooth muscle fibers within the wall of a hollow organ are progressively stretched as the volume of the organ's contents expands. Consider the urinary bladder as an example. Even though the muscle fibers in the urinary bladder are stretched as the bladder gradually fills with urine, they still maintain their tone and can even develop further tension in response to inputs that regulate bladder emptying. If considerable stretching prevented tension development, as in skeletal muscle, a filled bladder would not be capable of contracting to empty.

STRESS RELAXATION RESPONSE When a smooth muscle is suddenly stretched, it initially increases its tension, much like the tension created in a stretched rubber band. The muscle quickly adjusts to this new length, however, and inherently relaxes to the tension level prior to the stretch, probably as a consequence of rearrangement of cross-bridge attachments. Smooth muscle cross bridges detach comparatively slowly. On sudden stretching, it is speculated that any attached cross bridges would strain against the stretch, contributing to a passive (not actively generated) increase in tension. As these cross bridges detach, the filaments would be permitted to slide into an unstrained stretched position, restoring the tension to its original level. This inherent property of smooth muscle is called the **stress relaxation response.**

ADVANTAGES OF THE SMOOTH MUSCLE LENGTH–TENSION RELATIONSHIP These two responses of smooth muscle to being stretched—being able to develop tension even when considerably stretched and inherently relaxing when stretched—are highly advantageous. They enable smooth muscle to exist at a variety of lengths with little change in tension. As a result, a hollow organ enclosed by smooth muscle can accommodate variable volumes of contents with little change in the pressure exerted on the contents except when the contents are to be pushed out of the organ. At that time, the tension is deliberately increased by fiber shortening. It is possible for smooth muscle fibers to contract to half their normal length, enabling hollow organs to dramatically empty their contents on increased contractile activity; thus, smooth-muscled viscera can easily accommodate large volumes but can empty to practically zero volume. This length range in which smooth muscle normally functions (anywhere from 0.5 to 2.5 times the normal length) is much greater than the limited length range within which skeletal muscle remains functional.

Smooth muscle contains a lot of connective tissue, which resists being stretched. Unlike skeletal muscle, in which the skeletal attachments restrict how far the muscle can be stretched, this connective tissue prevents smooth muscle from being overstretched and thus puts an upper limit on how much a smooth-muscled hollow organ can hold.

Smooth muscle is slow and economical.

A smooth muscle contractile response proceeds more slowly than a skeletal muscle twitch. ATP splitting by myosin ATPase is much slower in smooth muscle, so cross-bridge activity and filament sliding occur about 10 times more slowly in smooth muscle than in skeletal muscle. A single smooth muscle contraction may

last as long as 3 seconds (3000 msec), compared to the maximum of 100 msec required for a single contractile response in skeletal muscle. Smooth muscle also relaxes more slowly because of slower Ca^{2+} removal. Slowness should not be equated with weakness, however. Smooth muscle can generate the same contractile tension per unit of cross-sectional area as skeletal muscle, but it does so more slowly and at considerably less energy expense. Because of slow cross-bridge cycling during smooth muscle contraction, cross bridges stay attached for more time during each cycle, compared with skeletal muscle; that is, the cross bridges "latch onto" the thin filaments for a longer time each cycle. This so-called **latch phenomenon** enables smooth muscle to maintain tension with comparatively less ATP consumption because each cross-bridge cycle uses up one molecule of ATP. The duration of force maintained by a single cross-bridge interaction lasts about eight times longer in smooth muscle than in skeletal muscle. Smooth muscle is therefore an economical contractile tissue, making it well suited for long-term sustained contractions with little energy consumption and without fatigue. In contrast to the rapidly changing demands placed on your skeletal muscles as you maneuver through and manipulate your external environment, your smooth muscle activities are geared for long-term duration and slower adjustments to change. Because of its slowness and the less ordered arrangement of its filaments, smooth muscle has often been mistakenly viewed as a poorly developed version of skeletal muscle. Actually, smooth muscle is just as highly specialized for the demands placed on it. It is an extremely adaptive, efficient tissue.

Nutrient and O_2 delivery are generally adequate to support the smooth muscle contractile process. Smooth muscle can use a wide variety of nutrient molecules for ATP production. There are no energy storage pools comparable to creatine phosphate in smooth muscle; they are not necessary. Oxygen delivery is usually adequate to keep pace with the low rate of oxidative phosphorylation needed to provide ATP for the energy-efficient smooth muscle. If necessary, anaerobic glycolysis can sustain adequate ATP production if O_2 supplies are diminished.

Cardiac muscle blends features of both skeletal and smooth muscle.

Cardiac muscle, found only in the heart, shares structural and functional characteristics with both skeletal and single-unit smooth muscle. Like skeletal muscle, cardiac muscle is striated, with its thick and thin filaments highly organized into a regular banding pattern. Cardiac thin filaments contain troponin and tropomyosin, which constitute the site of Ca^{2+} action in switching on cross-bridge activity, as in skeletal muscle. Also like skeletal muscle, cardiac muscle has a clear length–tension relationship. Like the oxidative skeletal muscle fibers, cardiac muscle cells have lots of mitochondria and myoglobin. They also have T tubules and a moderately well-developed sarcoplasmic reticulum.

As in smooth muscle, Ca^{2+} enters the cytosol from both the ECF and the sarcoplasmic reticulum during cardiac excitation. Ca^{2+} entry from the ECF occurs through voltage-gated dihydropyridine receptors, which also act as Ca^{2+} channels in the T tubule membrane. This Ca^{2+} entry from the ECF triggers release of Ca^{2+} intracellularly from the sarcoplasmic reticulum. Like single-unit smooth muscle, the heart displays pacemaker (but not slow-wave) activity, initiating its own action potentials without any external influence. Cardiac cells are interconnected by gap junctions that enhance the spread of action potentials throughout the heart, just as in single-unit smooth muscle. Also similarly, the heart is innervated by the autonomic nervous system, which, along with certain hormones and local factors, can modify the rate and strength of contraction.

Unique to cardiac muscle, the cardiac fibers are joined in a branching network, and the action potentials of cardiac muscle last much longer before repolarizing. Further details and the importance of cardiac muscle's features are addressed in the next chapter.

Chapter in Perspective: Focus on Homeostasis

Skeletal muscles comprise the muscular system itself. Cardiac and smooth muscle are part of the organs that make up other body systems. Cardiac muscle is found only in the heart, which is part of the circulatory system. Smooth muscle is found in the walls of hollow organs and tubes, including the blood vessels in the circulatory system, airways in the respiratory system, bladder in the urinary system, stomach and intestines in the digestive system, and uterus and ductus deferens (the duct through which sperm leave the testes) in the reproductive system.

Contraction of skeletal muscles accomplishes movement of body parts in relation to one another and movement of the whole body in relation to the external environment. Thus, these muscles permit you to move through and manipulate your external environment. At a very general level, some of these movements are aimed at maintaining homeostasis, such as moving the body toward food or away from harm. Examples of more specific homeostatic functions accomplished by skeletal muscles include chewing and swallowing food for further breakdown in the digestive system into usable energy-producing nutrient molecules (the mouth and throat muscles are all skeletal muscles), and breathing to obtain O_2 and get rid of CO_2 (the respiratory muscles are all skeletal muscles). Heat generated by contracting skeletal muscles also is the major source of heat production in maintaining body temperature. The skeletal muscles further accomplish many nonhomeostatic activities that enable us to work and play—for example, operating a computer or riding a bicycle—so that we can contribute to society and enjoy ourselves.

All the other systems of the body, except the immune (defense) system, depend on their nonskeletal muscle components to enable them to accomplish their homeostatic functions. For example, contraction of cardiac muscle in the heart pushes life-sustaining blood forward into the blood vessels, and contraction of smooth muscle in the stomach and intestines pushes the ingested food through the digestive tract at a rate appropriate for the digestive juices secreted along the route to break down the food into usable units.

REVIEW EXERCISES

Objective Questions (Answers on p. A-44)

1. On completion of an action potential in a muscle fiber, the contractile activity initiated by the action potential ceases. *(True or false?)*

2. The velocity at which a muscle shortens depends entirely on the ATPase activity of its fibers. *(True or false?)*

3. When a skeletal muscle is maximally stretched, it can develop maximal tension on contraction, because the actin filaments can slide in a maximal distance. *(True or false?)*

4. A pacemaker potential always initiates an action potential. *(True or false?)*

5. A slow-wave potential always initiates an action potential. *(True or false?)*

6. Smooth muscle can develop tension even when considerably stretched, because the thin filaments still overlap with the long thick filaments. *(True or false?)*

7. A(n) _____ contraction is an isotonic contraction in which the muscle shortens, whereas the muscle lengthens in a(n) _____ isotonic contraction.

8. _____ motor neurons supply extrafusal muscle fibers, whereas intrafusal fibers are innervated by _____ motor neurons.

9. The two types of atrophy are _____ and _____.

10. Which of the following provide(s) direct input to alpha motor neurons? (Indicate all correct answers.)
    a. primary motor cortex
    b. brain stem
    c. cerebellum
    d. basal nuclei
    e. spinal reflex pathways

11. Which of the following is *not* involved in bringing about muscle relaxation?
    a. reuptake of Ca^{2+} by the sarcoplasmic reticulum
    b. no more ATP
    c. no more action potential
    d. removal of ACh at the end plate by acetylcholinesterase
    e. filaments sliding back to their resting position

12. Match the following (with reference to skeletal muscle):
___ 1. Ca^{2+}
___ 2. T tubule
___ 3. ATP
___ 4. lateral sac of the sarcoplasmic reticulum
___ 5. myosin
___ 6. troponin–tropomyosin complex
___ 7. actin

(a) cyclically binds with the myosin cross bridges during contraction
(b) has ATPase activity
(c) supplies energy for the power stroke of a cross bridge
(d) rapidly transmits the action potential to the central portion of the muscle fiber
(e) stores Ca^{2+}
(f) pulls the troponin–tropomyosin complex out of its blocking position
(g) prevents actin from interacting with myosin when the muscle fiber is not excited

13. Using the answer code at the right, indicate what happens in the banding pattern during contraction:
___ 1. thick myofilament
___ 2. thin myofilament
___ 3. A band
___ 4. I band
___ 5. H zone
___ 6. sarcomere

(a) remains the same size during contraction
(b) decreases in length (shortens) during contraction

Essay Questions

1. Describe the levels of organization in a skeletal muscle.

2. What produces the striated appearance of skeletal muscles? Describe or draw the arrangement of thick and thin filaments that gives rise to the banding pattern.

3. What is the functional unit of skeletal muscle?

4. Describe the composition of thick and thin filaments.

5. Describe the sliding filament mechanism of muscle contraction. How do cross-bridge power strokes bring about shortening of the muscle fiber?

6. Compare the excitation–contraction coupling process in skeletal muscle with that in smooth muscle.

7. How can gradation of skeletal muscle contraction be accomplished?

8. What is a motor unit? Compare the size of motor units in finely controlled muscles with those specialized for coarse, powerful contractions. Describe motor unit recruitment.

9. Explain twitch summation and tetanus.

10. How does a skeletal muscle fiber's length at the onset of contraction affect the strength of the subsequent contraction?

11. Compare isotonic and isometric contractions.

12. Describe the role of each of the following in powering skeletal muscle contraction: ATP, creatine phosphate, oxidative phosphorylation, and glycolysis. Distinguish between aerobically and anaerobically supported exercise.

13. Compare the three types of skeletal muscle fibers.

14. What are the roles of the corticospinal system and multineuronal system in controlling motor movement?

15. Describe the structure and function of muscle spindles and Golgi tendon organs.

16. Distinguish between phasic and tonic smooth muscle.

17. Distinguish between multiunit and single-unit smooth muscle.

18. Differentiate between neurogenic and myogenic muscle activity.

19. How can smooth muscle contraction be graded?

20. Compare the contractile speed and relative energy expenditure of skeletal muscle with that of smooth muscle.

21. In what ways is cardiac muscle functionally similar to skeletal muscle and to single-unit smooth muscle?

Quantitative Exercises (Solutions on p. A-44)

1. Consider two individuals each throwing a baseball, one a weekend athlete and the other a professional pitcher.
   a. Given the following information, calculate the velocity of the ball as it leaves the amateur's hand:
      • The distance from his shoulder socket (humeral head) to the ball is 70 cm.
      • The distance from his humeral head to the points of insertion of the muscles moving his arm forward (we must simplify here, because the shoulder is such a complex joint) is 9 cm.
      • The velocity of muscle shortening is 2.6 m/sec.
   b. The professional pitcher throws the ball 85 miles per hour. If his points of insertion are also 9 cm from the humeral head and the distance from his humeral head to the ball is 90 cm, how much faster did the professional pitcher's muscles shorten compared to the amateur's?

2. The velocity at which a muscle shortens is related to the force that it can generate in the following way:[1]

$$v = b(F_0 - F)/(F + a)$$

where v is the velocity of shortening, and F_0 can be thought of as an "upper load limit," or the maximum force a muscle can generate against a resistance. The parameter a is inversely proportional to the cross-bridge cycling rate, and b is proportional to the number of sarcomeres in line in a muscle. Draw the resistance (load)–velocity curve predicted by this equation by plotting the points $F = 0$ and $F = F_0$. Values of v are on the vertical axis; values of F are on the horizontal axis; a, b, and F_0 are constants.
   a. Notice that the curve generated from this equation is the same as that in ● Figure 8-20, p. 274. Why does the curve have this shape? That is, what does the shape of the curve tell you about muscle performance in general?
   b. What happens to the resistance (load)–velocity curve when F_0 is increased? When the cross-bridge cycling rate is increased? When the size of the muscle is increased? How will each of these changes affect the performance of the muscle?

---

[1]F. C. Hoppensteadt and C. S. Peskin, *Mathematics in Medicine and the Life Sciences* (New York: Springer, 1992), equation 9.1.1, p. 199.

(Explanations on p. A-45)

1. Why does regular aerobic exercise provide more cardio-vascular benefit than weight training does? (*Hint:* The heart responds to the demands placed on it in a way similar to that of skeletal muscle.)

2. If the biceps muscle of a child inserts 4 cm from the elbow and the length of the arm from the elbow to the hand is 28 cm, how much force must the biceps generate in order for the child to lift an 8 kg stack of books with one hand?

3. Put yourself in the position of the scientists who discovered the sliding filament mechanism of muscle contraction, by considering what molecular changes must be involved to account for the observed alterations in the banding pattern during contraction. If you were comparing a relaxed and contracted muscle fiber under an electron microscope (see ● Figure 8-3a, p. 260), how could you determine that the thin filaments do not change in length during muscle contraction? You cannot see or measure a single thin filament at this magnification. (*Hint:* What landmark in the banding pattern represents each end of the thin filament? If these landmarks are the same distance apart in a relaxed and contracted fiber, then the thin filaments must not change in length.)

4. What type of off-the-snow training would you recommend for a competitive downhill skier versus a competitive cross-country skier? What adaptive skeletal muscle changes would you hope to accomplish in the athletes in each case?

5. When the bladder is filled and the micturition (urination) reflex is initiated, the nervous supply to the bladder promotes contraction of the bladder and relaxation of the external urethral sphincter, a ring of muscle that guards the exit from the bladder. If the time is inopportune for bladder emptying when the micturition reflex is initiated, the external urethral sphincter can be voluntarily tightened to prevent urination even though the bladder is contracting. Using your knowledge of the muscle types and their innervation, of what types of muscle are the bladder and the external urethral sphincter composed, and what branch of the efferent division of the peripheral nervous system supplies each of these muscles?

(Explanation on p. A-45)

Jason W. is waiting impatiently for the doctor to finish removing the cast from his leg, which Jason broke the last day of school six weeks ago. Summer vacation is half over, and he hasn't been able to swim, play baseball, or participate in any of his favorite sports. When the cast is finally off, Jason's excitement gives way to concern when he sees that the injured limb is noticeably smaller in diameter than his normal leg. What explains this reduction in size? How can the leg be restored to its normal size and functional ability?

Circulatory System (Heart)

Body systems maintain homeostasis

Homeostasis
The circulatory system contributes to homeostasis by transporting O_2, CO_2, wastes, electrolytes, and hormones from one part of the body to another.

Homeostasis is essential for survival of cells

Cells
Cells need a constant supply of O_2 and nutrients delivered to them by the circulatory system, which also carries away CO_2 and other wastes, in order to power life-sustaining cellular activities by the chemical reaction:

$$Food + O_2 \rightarrow CO_2 + H_2O + Energy$$

Cells make up body systems

To maintain homeostasis, essential materials such as O_2 and nutrients must continually be picked up from the external environment and delivered to the cells and waste products must continually be removed. Furthermore, excess heat generated by muscles must be transported to the skin where it can be lost from the body surface to help maintain body temperature. Homeostasis also depends on the transfer of hormones, which are important regulatory chemical messengers, from their site of production to their site of action. The circulatory system, which contributes to homeostasis by serving as the body's transport system, consists of the heart, blood vessels, and blood.

All body tissues constantly depend on the life-supporting blood flow the heart provides them by contracting or beating. The heart drives blood through the blood vessels for delivery to the tissues in sufficient amounts, whether the body is at rest or engaging in vigorous exercise.

Cardiac Physiology

CONTENTS AT A GLANCE

CENGAGENOW™ Log on to CengageNOW at **http://www .cengage.com/sso/** for an opportunity to explore a learning module that illustrates difficult concepts with self-study tutorials, animations, and interactive quizzes to help you learn, review, and master physiology concepts.

Anatomy of the Heart

From just a matter of days following conception until death, the beat goes on. In fact, throughout an average human life span, the heart contracts about 3 billion times, never stopping except for a fraction of a second to fill between beats. Within about three weeks after conception, the heart of the developing embryo starts to function. It is the first organ to become functional. At this time the human embryo is only a few millimeters long, about the size of a capital letter on this page.

Why does the heart develop so early, and why is it so crucial throughout life? It is this important because the circulatory system is the body's transport system. A human embryo, having very little yolk available as food, depends on promptly establishing a circulatory system that can interact with the mother's circulation to pick up and distribute to the developing tissues the supplies so critical for survival and growth. Thus begins the story of the circulatory system, which continues throughout life to be a vital pipeline for transporting materials on which the cells of the body absolutely depend.

The **circulatory system** has three basic components:

1. The **heart** serves as the pump that imparts pressure to the blood to establish the pressure gradient needed for blood to flow to the tissues. Like all liquids, blood flows down a pressure gradient from an area of higher pressure to an area of lower pressure. This chapter focuses on cardiac physiology (*cardia* means "heart").

2. The **blood vessels** serve as the passageways through which blood is directed and distributed from the heart to all parts of the body and subsequently returned to the heart. The smallest of the blood vessels are designed for rapid exchange of materials between the surrounding tissues and the blood within the vessels (see Chapter 10).

3. **Blood** is the transport medium within which materials being transported long distances in the body (such as O_2, CO_2, nutrients, wastes, electrolytes, and hormones) are dissolved or suspended (see Chapter 11).

Blood travels continuously through the circulatory system to and from the heart through two separate vascular (blood vessel) loops, both originating and terminating at the heart (● Figure 9-1). The **pulmonary circulation** consists of a closed loop of vessels carrying blood between the heart and lungs

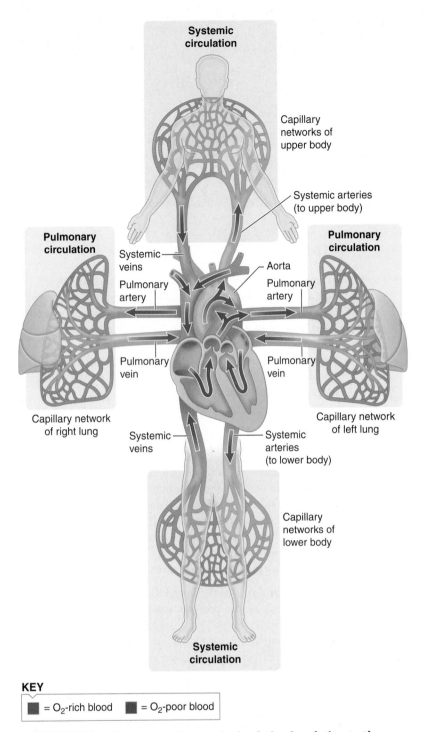

Systemic circulation

Capillary networks of upper body

Systemic arteries (to upper body)

Pulmonary circulation

Systemic veins

Pulmonary artery

Aorta

Pulmonary artery

Pulmonary circulation

Capillary network of right lung

Pulmonary vein

Pulmonary vein

Capillary network of left lung

Systemic veins

Systemic arteries (to lower body)

Capillary networks of lower body

Systemic circulation

KEY

■ = O₂-rich blood   ■ = O₂-poor blood

● **FIGURE 9-1 Pulmonary and systemic circulation in relation to the heart.** The circulatory system consists of two separate vascular loops: the pulmonary circulation, which carries blood between the heart and lungs; and the systemic circulation, which carries blood between the heart and organ systems. Each of these loops forms a figure "8," with the pulmonary circulation simultaneously supplying the right and left lungs, and the systemic circulation simultaneously supplying the upper body and lower body.

(*pulmo* means "lung"). The **systemic circulation** is a circuit of vessels carrying blood between the heart and other body systems. Each of these vascular loops forms a figure "8." The pulmonary circulation simultaneously loops through the right lung and the left lung; the systemic circulation simultaneously loops through the upper half and the lower half of the body.

The heart is positioned in the middle of the thoracic cavity.

The heart is a hollow, muscular organ about the size of a clenched fist. It lies in the **thoracic (chest) cavity** about midline between the **sternum** (breastbone) anteriorly and the **vertebrae** (backbone) posteriorly. Place your hand over your heart. People usually put their hand on the left side of the chest, even though the heart is actually in the middle of the chest. The heart has a broad **base** at the top and tapers to a pointed tip, the **apex,** at the bottom. It is situated at an angle under the sternum so that its base lies predominantly to the right and the apex to the left of the sternum. When the heart beats forcefully, the apex actually thumps against the inside of the chest wall on the left side. Because we become aware of the beating heart through the apex beat on the left side of the chest, we tend to think—erroneously—that the entire heart is on the left.

Clinical Note The heart's position between two bony structures, the sternum and vertebrae, makes it possible to manually drive blood out of the heart when it is not pumping effectively, by rhythmically depressing the sternum. This maneuver compresses the heart between the sternum and vertebrae so that blood is squeezed out into the blood vessels, maintaining blood flow to the tissues. Often this *external cardiac compression,* which is part of **cardiopulmonary resuscitation (CPR),** serves as a lifesaving measure until appropriate therapy can restore the heart to normal function.

The heart is a dual pump.

Even though anatomically the heart is a single organ, the right and left sides of the heart function as two separate pumps. The heart is divided into right and left halves and has four chambers, an upper and a lower chamber within each half (● Figure 9-2a). The upper chambers, the **atria** (singular, **atrium**), receive blood returning to the heart and transfer it to the lower chambers, the **ventricles,** which pump blood from the heart. The vessels that return blood from the tissues to the atria are **veins,** and those that carry blood away from the ventricles to the tissues are **arteries.** The two halves of the heart are separated by the **septum,** a continuous muscular partition that prevents mixture of blood from the two sides of the heart. This separation is extremely important, because the right side of the heart is receiving and pumping O₂-poor blood, whereas the left side of the heart receives and pumps O₂-rich blood.

(a) Blood flow through the heart

(c) Thickness of right and left ventricles

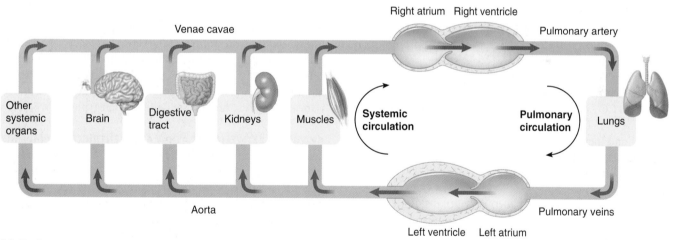

(b) Dual pump action of the heart

● **FIGURE 9-2 Blood flow through and pump action of the heart.** (a) The arrows indicate the direction of blood flow. To illustrate the direction of blood flow through the heart, all of the heart valves are shown open, which is never the case. The right side of the heart receives O_2-poor blood from the systemic circulation and pumps it into the pulmonary circulation. The left side of the heart receives O_2-rich blood from the pulmonary circulation and pumps it into the systemic circulation. (b) Note the parallel pathways of blood flow through the systemic organs. (The relative volume of blood flowing through each organ is not drawn to scale.) (c) Note that the left ventricular wall is much thicker than the right wall.

THE COMPLETE CIRCUIT OF BLOOD FLOW Let us look at how the heart functions as a dual pump by tracing a drop of blood through one complete circuit (● Figure 9-2a and b). Blood returning from the systemic circulation enters the right atrium via two large veins, the **venae cavae**, one returning blood from above and the other returning blood from below heart level. The drop of blood entering the right atrium has returned from the body tissues, where O_2 has been taken from it and CO_2 has been added to it. This partially deoxygenated blood flows from the right atrium into the right ventricle, which pumps it out through the **pulmonary artery**, which immediately forms two branches, one going to each of the two lungs. Thus, *the right side of the heart receives blood from the systemic circulation and pumps it into the pulmonary circulation.*

Within the lungs, the drop of blood loses its extra CO_2 and picks up a fresh supply of O_2 before being returned to the left atrium via the **pulmonary veins** coming from both lungs. This O_2-rich blood returning to the left atrium subsequently flows into the left ventricle, the pumping chamber that propels the blood to all body systems except the lungs; that is, *the left side of the heart receives blood from the pulmonary circulation and pumps it into the systemic circulation.* The single large artery carrying blood away from the left ventricle is the **aorta**. Major arteries branch from the aorta to supply the various organs of the body.

In contrast to the pulmonary circulation, in which all the blood flows through the lungs, the systemic circulation may be viewed as a series of parallel pathways. Part of the blood pumped out by the left ventricle goes to the muscles, part to the kidneys, part to the brain, and so on (● Figure 9-2b). Thus, the output of the left ventricle is distributed so that each part of the body receives a fresh blood supply; the same arterial blood does not pass from organ to organ. Accordingly, the drop of blood we are tracing goes to only one of the systemic organs. Tissue cells within the organ take O_2 from the blood and use it to oxidize nutrients for energy production; in the process, the tissue cells form CO_2 as a waste product that is added to the blood (see pp. 2 and 37). The drop of blood, now partially depleted of O_2 content and increased in CO_2 content, returns to the right side of the heart, which once again will pump it to the lungs. One circuit is complete.

COMPARISON OF THE RIGHT AND LEFT PUMPS Both sides of the heart simultaneously pump equal amounts of blood. The volume of O_2-poor blood being pumped to the lungs by the right side of the heart soon becomes the same volume of O_2-rich blood being delivered to the tissues by the left side of the heart. The pulmonary circulation is a low-pressure, low-resistance system, whereas the systemic circulation is a high-pressure, high-resistance system. Pressure is the force exerted on the vessel walls by the blood pumped into the vessels by the heart. Resistance is the opposition to blood flow, largely caused by friction between the flowing blood and the vessel wall. Even though the right and left sides of the heart pump the same amount of blood, the left side performs more work, because it pumps an equal volume of blood at a higher pressure into a higher-resistance and longer system. Accordingly, the heart muscle on the left side is much thicker than the

When pressure is greater behind the valve, it opens.

Valve opened

When pressure is greater in front of the valve, it closes. Note that when pressure is greater in front of the valve, it does not open in the opposite direction; that is, it is a one-way valve.

Valve closed; does not open in opposite direction

● **FIGURE 9-3 Mechanism of valve action.**

muscle on the right side, making the left side a stronger pump (● Figure 9-2c).

Pressure-operated heart valves ensure that blood flows in the right direction through the heart.

Blood flows through the heart in one fixed direction from veins to atria to ventricles to arteries. The presence of four one-way heart valves ensures this unidirectional flow of blood. The valves are positioned so that they open and close passively because of pressure differences, similar to a one-way door (● Figure 9-3). A forward pressure gradient (that is, a greater pressure behind the valve) forces the valve open, much as you open a door by pushing on one side of it, whereas a backward pressure gradient (that is, a greater pressure in front of the valve) forces the valve closed, just as you apply pressure to the opposite side of the door to close it. Note that a backward gradient can force the valve closed but cannot force it to swing open in the opposite direction; that is, heart valves are not like swinging, saloon-type doors.

AV VALVES BETWEEN THE ATRIA AND VENTRICLES Two of the heart valves, the **right** and **left atrioventricular (AV) valves**, are positioned between the atrium and the ventricle on the right and left sides, respectively (● Figure 9-4a). These valves let blood flow from the atria into the ventricles during ventricular filling (when atrial pressure exceeds ventricular pressure) but prevent the backflow of blood from the ventricles into the atria during ventricular emptying (when ventricular pressure greatly exceeds atrial pressure). If the rising ventricular pressure did not force the AV valves to close as the ventricles contracted to empty, much of the blood would inefficiently be forced back into the atria and veins instead of being pumped into the arteries. The right AV valve is also called the **tricuspid valve** (*tri* means "three") because it consists of three cusps or leaflets (● Figure 9-4b). Likewise, the left AV valve, which has two cusps, is often called the **bicuspid valve** (*bi* means "two") or, alternatively, the **mitral valve** (because of its physical resemblance to a mitre, or bishop's traditional hat).

The edges of the AV valve leaflets are fastened by tough, thin, fibrous cords of tendinous-type tissue, the **chordae ten-**

dineae, which prevent the valves from being everted. That is, the chordae tendineae prevent the AV valve from being forced by the high ventricular pressure to open in the opposite direction into the atria. These cords extend from the edges of each cusp and attach to small, nipple-shaped **papillary muscles,** which protrude from the inner surface of the ventricular walls (*papilla* means "nipple"). When the ventricles contract, the papillary muscles also contract, pulling downward on the chordae tendineae. This pulling exerts tension on the closed AV valve cusps to hold them in position, much like tethering ropes hold down a hot-air balloon. This action helps keep the valve tightly sealed in the face of a strong backward pressure gradient (● Figure 9-4c).

SEMILUNAR VALVES BETWEEN THE VENTRICLES AND MAJOR ARTERIES The two remaining heart valves, the **aortic** and **pulmonary valves,** lie at the juncture where the major arteries leave the ventricles (● Figure 9-4a). They are known as **semilunar valves** because they have three cusps, each resembling a shallow half-moon-shaped pocket (*semi* means "half"; *lunar* means "moon") (● Figure 9-4b). These valves are forced open when the left and right ventricular pressures exceed the pressure in the aorta and pulmonary artery, respectively, during ventricular contraction and emptying. Closure results when the ventricles relax and ventricular pressures fall below the aortic and pulmonary artery pressures. The closed valves prevent blood from flowing from the arteries back into the ventricles from which it has just been pumped.

The semilunar valves are prevented from everting by the anatomic structure and positioning of the cusps. When on ventricular relaxation a backward pressure gradient is created, the back surge of blood fills the pocketlike cusps and sweeps them into a closed position, with their unattached upturned edges fitting together in a deep, leakproof seam (● Figure 9-4d).

NO VALVES BETWEEN THE ATRIA AND VEINS Even though there are no valves between the atria and veins, backflow of blood from

(a) Location of the heart valves in a longitudinal section of the heart

Right AV valve   Left AV valve   Aortic or pulmonary valve

(b) Heart valves in closed position, viewed from above

(c) Prevention of eversion of AV valves   **(d)** Prevention of eversion of semilunar valves

● **FIGURE 9-4 Heart valves.** Eversion of the AV valves is prevented by tension on the valve leaflets exerted by the chordae tendineae when the papillary muscles contract. When the semilunar valves are swept closed, their upturned edges fit together in a deep, leakproof seam that prevents valve eversion.

the atria into the veins usually is not a significant problem, for two reasons: (1) atrial pressures usually are not much higher than venous pressures, and (2) the sites where the venae cavae enter the atria are partially compressed during atrial contraction.

FIBROUS SKELETON OF THE VALVES Four interconnecting rings of dense connective tissue provide a firm base for attachment of the four heart valves (● Figure 9-5). This **fibrous**

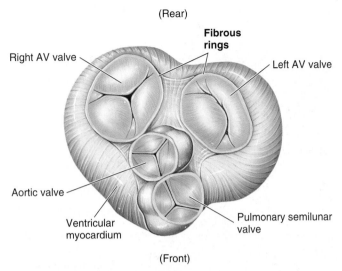

(Rear)

Fibrous rings

Right AV valve

Left AV valve

Aortic valve

Ventricular myocardium

Pulmonary semilunar valve

(Front)

● **FIGURE 9-5 Fibrous skeleton of the heart.** A view of the heart from above, with the atria and major vessels removed to show the heart valves and fibrous rings. Note that the inlet and outlet valves to the ventricle all lie on the same plane through the heart.

skeleton, which separates the atria from the ventricles, also provides a fairly rigid structure for attachment of the cardiac muscle. The atrial muscle mass is anchored above the rings, and the ventricular muscle mass is attached to the bottom of the rings.

It might seem rather surprising that the inlet valves to the ventricles (the AV valves) and the outlet valves from the ventricles (the semilunar valves) all lie on the same plane through the heart, as delineated by the fibrous skeleton. This relationship comes about because the heart forms from a single tube that bends on itself and twists on its axis during embryonic development. Although this turning and twisting make studying the structural relationships of the heart more difficult, the twisted structure has functional importance in that it helps the heart pump more efficiently. We will see how by turning our attention to the portion of the heart that actually generates the forces responsible for blood flow, the cardiac muscle within the heart walls.

The heart walls are composed primarily of spirally arranged cardiac muscle fibers.

The heart wall has three distinct layers:

- A thin inner layer, the **endothelium,** a unique type of epithelial tissue that lines the entire circulatory system

- A middle layer, the **myocardium,** which is composed of cardiac muscle and constitutes the bulk of the heart wall (*myo* means "muscle"; *cardia* means "heart")

- A thin external layer, the **epicardium,** that covers the heart (*epi* means "on")

The myocardium consists of interlacing bundles of cardiac muscle fibers arranged spirally around the circumference of the heart (● Figure 9-6a). The spiral arrangement is the result of the heart's complex twisting during development. As a result of this arrangement, when the ventricular muscle contracts and shortens, the diameter of the ventricular chambers is reduced while the apex is simultaneously pulled upward toward the top of the heart in a rotating manner. This exerts a "wringing" effect, efficiently exerting pressure on the blood within the enclosed chambers and directing it upward toward the openings of the major arteries that exit at the base of the ventricles.

To support their rhythmic, lifelong contractile activity (remember heart muscle cells cannot divide, so at least most of them live as long as you do), they have an abundance of energy-generating mitochondria, and they receive a rich blood supply, about one capillary for each myocardial fiber.

Cardiac muscle fibers are interconnected by intercalated discs and form functional syncytia.

The individual cardiac muscle cells are interconnected to form branching fibers, with adjacent cells joined end to end at specialized structures known as **intercalated discs.** Two types of membrane junctions are present within an intercalated disc: desmosomes and gap junctions (● Figure 9-6b and c). A *desmosome,* a type of adhering junction that mechanically holds cells together, is particularly abundant in tissues such as the heart that are subject to considerable mechanical stress (see p. 58). At intervals along the intercalated disc, the opposing membranes approach each other very closely to form *gap junctions,* which are areas of low electrical resistance that allow action potentials to spread from one cardiac cell to adjacent cells (see p. 59). Some cardiac muscle cells can generate action potentials without any nervous stimulation. When one of the cardiac cells spontaneously undergoes an action potential, the electrical impulse spreads to all the other cells that are joined by gap junctions in the surrounding muscle mass, so that they become excited and contract as a single, *functional syncytium* (see p. 295). The atria and the ventricles each form a functional syncytium and contract as separate units. The synchronous contraction of the muscle cells that make up the walls of each of these chambers produces the force needed to eject the enclosed blood.

No gap junctions join the atrial and ventricular contractile cells, and furthermore, the atria and ventricles are separated by the electrically nonconductive fibrous skeleton that surrounds and supports the valves. However, an important, specialized conduction system facilitates and coordinates transmission of electrical excitation from the atria to the ventricles to ensure synchronization between atrial and ventricular pumping.

Because of both the syncytial nature of cardiac muscle and the conduction system between the atria and ventricles, an impulse spontaneously generated in one part of the heart spreads throughout the entire heart. Therefore, unlike skeletal muscle, where graded contractions can be produced by varying the number of muscle cells contracting within the muscle (recruitment of motor units), either all the cardiac muscle fibers contract or none do. A "halfhearted" contraction is not possible. Cardiac contraction is graded by varying the strength of contraction of all the cardiac muscle cells. You will learn more about this process in a later section.

The heart is enclosed by the pericardial sac.

The heart is enclosed in the double-walled, membranous **pericardial sac** (*peri* means "around"). The sac consists of two layers—a tough, fibrous covering and a secretory lining. The outer fibrous covering of the sac attaches to the connective tissue partition that separates the lungs. This attachment anchors the heart so that it remains properly positioned within the chest. The sac's secretory lining secretes a thin **pericardial fluid,** which provides lubrication to prevent friction between the pericardial layers as they glide over each other with every beat of the heart.

Clinical Note **Pericarditis,** an inflammation of the pericardial sac that results in a painful friction rub between the two pericardial layers, occurs occasionally because of viral or bacterial infection.

Building on this foundation of heart structure, we now explain how action potentials are initiated and spread throughout the heart, followed by a discussion of how this electrical activity brings about coordinated pumping of the heart.

Electrical Activity of the Heart

Contraction of cardiac muscle cells to eject blood is triggered by action potentials sweeping across the muscle cell membranes. The heart contracts, or beats, rhythmically as a result of action potentials that it generates by itself, a property called **autorhythmicity** (*auto* means "self"). There are two specialized types of cardiac muscle cells:

1. **Contractile cells,** which are 99% of the cardiac muscle cells, do the mechanical work of pumping. These working cells normally do not initiate their own action potentials.

2. In contrast, the small but extremely important remainder of the cardiac cells, the **autorhythmic cells,** do not contract but instead are specialized for initiating and conducting the action potentials responsible for contraction of the working cells.

Cardiac autorhythmic cells display pacemaker activity.

In contrast to nerve and skeletal muscle cells, in which the membrane remains at constant resting potential unless the cell is stimulated, the cardiac autorhythmic cells do not have a resting potential. Instead, they display *pacemaker activity*; that is, their membrane potential slowly depolarizes, or drifts, between action potentials until threshold is reached, at which time the membrane fires or has an action potential. An autorhythmic cell membrane's slow drift to threshold is called the **pacemaker potential** (● Figure 9-7; see also p. 295). Through repeated cycles of drift and fire, these autorhythmic cells cyclically initiate action potentials, which then spread throughout the heart to trigger rhythmic beating without any nervous stimulation.

PACEMAKER POTENTIAL AND ACTION POTENTIAL IN AUTO-RHYTHMIC CELLS Complex interactions of several different ionic mechanisms are responsible for the pacemaker potential. The most important changes in ion movement that give rise to

(a) Bundles of cardiac muscle are arranged spirally around the ventricle. When they contract, they "wring" blood from the apex to the base where the major arteries exit.

© Dr. John Cunningham/Visuals Unlimited

(b) Cardiac muscle fibers branch and are interconnected by intercalated discs.

(c) Intercalated discs contain two types of membrane junctions: mechanically important desmosomes and electrically important gap junctions.

● **FIGURE 9-6 Organization of cardiac muscle fibers.** Bundles of cardiac muscle fibers are arranged spirally around the ventricle. Adjacent cardiac muscle cells are joined end to end by intercalated discs, which contain two types of specialized junctions: desmosomes, which act as spot rivets mechanically holding the cells together; and gap junctions, which permit action potentials to spread from one cell to adjacent cells.

KEY

f = Funny channels
T = Transient-type channels
L = Long-lasting channels

● **FIGURE 9-7 Pacemaker activity of cardiac autorhythmic cells.** The first half of the pacemaker potential is the result of simultaneous opening of unique funny channels, which permits inward Na^+ current, and closure of K^+ channels, which reduces outward K^+ current. The second half of the pacemaker potential is the result of opening of T-type Ca^{2+} channels. Once threshold is reached, the rising phase of the action potential is the result of opening of L-type Ca^{2+} channels, whereas the falling phase is the result of opening of K^+ channels.

the pacemaker potential are (1) an increased inward Na^+ current, (2) a decreased outward K^+ current, and (3) an increased inward Ca^{2+} current.

To elaborate, the initial phase of the slow depolarization to threshold is caused by net Na^+ entry through a type of voltage-gated channels found only in cardiac pacemaker cells. Typically voltage-gated channels open when the membrane becomes less negative (depolarizes), but these unique channels open when the potential becomes more negative (hyperpolarizes) at the end of repolarization from the previous action potential. Because of their unusual behavior, they are called **funny,** or I_f, channels. When one action potential ends and the I_f channels open, the resultant depolarizing net inward Na^+ current through these open channels starts immediately moving the pacemaker cell's membrane potential toward threshold once again.

The second ionic mechanism that contributes to this pacemaker potential is a progressive reduction in the passive outward flux of K^+. In cardiac autorhythmic cells, permeability to K^+ does not remain constant between action potentials as it does in nerve and skeletal muscle cells. The K^+ channels that opened during the falling phase of the preceding action potential slowly close at negative potentials. This slow closure gradually diminishes the outflow of positive potassium ions down their concentration gradient. As a result, the rate of K^+ efflux slowly declines at the same time the slow inward leak of Na^+

occurs through the open I_f channels, further contributing to the drift toward threshold.

The third ionic contribution to pacemaker potential is increased Ca^{2+} entry. In the second half of the pacemaker potential, the I_f channels close and transient Ca^{2+} channels (**T-type Ca^{2+} channels**), one of two types of voltage-gated Ca^{2+} channels, open before the membrane reaches threshold. ("T" stands for *transient*.) The resultant brief influx of Ca^{2+} further depolarizes the membrane, bringing it to threshold, at which time the transient Ca^{2+} channels close.

Once threshold is reached, the rising phase of the action potential occurs in response to activation of a long-lasting, voltage-gated Ca^{2+} channel (**L-type Ca^{2+} channel;** "L" standing for *long-lasting*) and a subsequently large influx of Ca^{2+}. The Ca^{2+}-induced rising phase of a cardiac pacemaker cell differs from that in nerve and skeletal muscle cells, where Na^+ influx rather than Ca^{2+} influx swings the potential in the positive direction.

The falling phase is the result, as usual, of the K^+ efflux that occurs when K^+ permeability increases on activation of voltage-gated K^+ channels, coupled with closure of the L-type Ca^{2+} channels. After the action potential is over, slow closure of these K^+ channels contributes to the next slow depolarization to threshold.

The sinoatrial node is the normal pacemaker of the heart.

The specialized noncontractile cardiac cells capable of autorhythmicity lie in the following specific sites (● Figure 9-8):

1. The **sinoatrial node (SA node),** a small, specialized region in the right atrial wall near the opening of the superior vena cava.

2. The **atrioventricular node (AV node),** a small bundle of specialized cardiac muscle cells located at the base of the right atrium near the septum, just above the junction of the atria and ventricles.

3. The **bundle of His (atrioventricular bundle),** a tract of specialized cells that originates at the AV node and enters the interventricular septum. Here, it divides to form the right and left bundle branches that travel down the septum, curve around the tip of the ventricular chambers, and travel back toward the atria along the outer walls.

4. **Purkinje fibers,** small terminal fibers that extend from the bundle of His and spread throughout the ventricular myocardium much like small twigs of a tree branch.

NORMAL PACEMAKER ACTIVITY Because these various autorhythmic cells have different rates of slow depolarization to threshold, the rates at which they are normally capable of generating action potentials also differ (▲ Table 9-1). The heart cells with the fastest rate of action potential initiation are localized in the SA node. Once an action potential occurs in any cardiac muscle cell, it is propagated throughout the rest of the myocardium via gap junctions and the specialized conduction system. Therefore, the SA node, which normally has the fastest rate of autorhythmicity, at 70 to 80 action potentials per minute, drives the rest of the heart at this rate and thus is known as the

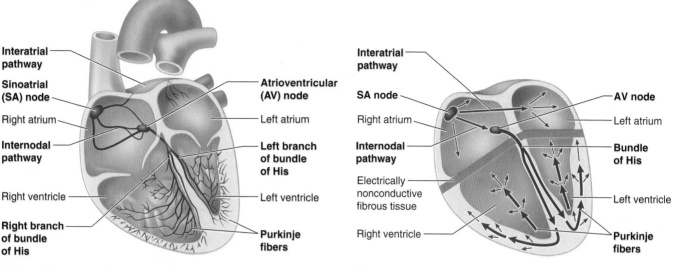

(a) Specialized conduction system of the heart   **(b)** Spread of cardiac excitation

● **FIGURE 9-8 Specialized conduction system of the heart and spread of cardiac excitation.** An action potential initiated at the SA node first spreads throughout both atria. Its spread is facilitated by two specialized atrial conduction pathways, the interatrial and internodal pathways. The AV node is the only point where an action potential can spread from the atria to the ventricles. From the AV node, the action potential spreads rapidly throughout the ventricles, hastened by a specialized ventricular conduction system consisting of the bundle of His and Purkinje fibers.

pacemaker of the heart. That is, the entire heart becomes excited, triggering the contractile cells to contract and the heart to beat at the pace or rate set by SA node autorhythmicity, normally at 70 to 80 beats per minute. The other autorhythmic tissues cannot assume their own naturally slower rates, because they are activated by action potentials originating in the SA node before they can reach threshold at their own, slower rhythm.

The following analogy shows how the SA node drives the rest of the heart at its own pace. Suppose a train has 100 cars, 3 of which are engines capable of moving on their own; the other 97 cars must be pulled (● Figure 9-9a). One engine (the SA node) can travel at 70 miles/hour (mph) on its own, another engine (the AV node) at 50 mph, and the last engine (the Purkinje fibers) at 30 mph. If all these cars are joined, the engine that can travel at 70 mph will pull the rest of the cars at that speed. The engines that can travel at lower speeds on their own will be pulled at a faster speed by the fastest engine and therefore cannot assume their own slower rate as long as they are being driven by a faster engine. The other 97 cars (nonautorhythmic, contractile cells), being unable to move on their own, will likewise travel at whatever speed the fastest engine pulls them.

ABNORMAL PACEMAKER ACTIVITY If for some reason the fastest engine breaks down (SA node damage), the next-fastest engine (AV node) takes over and the entire train travels at 50 mph; that is, if the SA node becomes nonfunctional, the AV node assumes pacemaker activity (● Figure 9-9b). The non–SA nodal autorhythmic tissues are **latent pacemakers** that can take over, although at a lower rate, if the normal pacemaker fails.

If impulse conduction becomes blocked between the atria and the ventricles, the atria continue at the typical rate of 70 beats per minute, and the ventricular tissue, not being driven by the faster SA nodal rate, assumes its own much slower autorhythmic rate of about 30 beats per minute, initiated by the ventricular autorhythmic cells (Purkinje fibers). This situation is like a breakdown of the second engine (AV node) so that the lead engine (SA node) becomes disconnected from the slow third engine (Purkinje fibers) and the rest of the cars (● Figure 9-9c). The lead engine (and cars connected directly to it; that is, the atrial cells) continues at 70 mph while the rest of the train proceeds at 30 mph. This **complete heart block** occurs when the conducting tissue between the atria and ventricles is damaged, as for example during a heart attack, and becomes non-

▲ TABLE 9-1	Normal Rate of Action Potential Discharge in Autorhythmic Tissues of the Heart

Tissue	Action Potentials per Minute*
SA node (normal pacemaker)	70–80
AV node	40–60
Bundle of His and Purkinje fibers	20–40

*In the presence of parasympathetic tone; see p. 240 and p. 327.

(a) Normal pacemaker activity: Whole train will go **70 mph** (heart rate set by SA node, the fastest autorhythmic tissue).

(b) Takeover of pacemaker activity by AV node when the SA node is nonfunctional: Train will go **50 mph** (the next fastest autorhythmic tissue, the AV node, will set the heart rate).

(c) Takeover of ventricular rate by the slower ventricular autorhythmic tissue in complete heart block: First part of train will go **70 mph**; last part will go **30 mph** (atria will be driven by SA node; ventricles will assume own, much slower rhythm).

(d) Takeover of pacemaker activity by an ectopic focus: Train will be driven by ectopic focus, which is now going faster than the SA node (the whole heart will be driven more rapidly by an abnormal pacemaker).

● **FIGURE 9-9 Analogy of pacemaker activity.** In complete heart block (c), when ventricular rate is taken over by the slower ventricular autorhythmic tissue, the atrial rate (not shown) is still driven by the SA node.

functional. A ventricular rate of 30 beats per minute will support only a very sedentary existence; in fact, the patient usually becomes comatose.

When a person has an abnormally low heart rate, as in SA node failure or heart block, an **artificial pacemaker** can be used. Such an implanted device rhythmically generates impulses that spread throughout the heart to drive both the atria and ventricles at the typical rate of 70 beats per minute.

Occasionally an area of the heart, such as a Purkinje fiber, becomes overly excitable and depolarizes more rapidly than the SA node. (The slow engine suddenly goes faster than the lead engine; see ● Figure 9-9d). This abnormally excitable area, an **ectopic focus,** initiates a premature action potential

that spreads throughout the rest of the heart before the SA node can initiate a normal action potential (*ectopic* means "out of place"). An occasional abnormal impulse from a ventricular ectopic focus produces a **premature ventricular contraction (PVC).** If the ectopic focus continues to discharge at its more rapid rate, pacemaker activity shifts from the SA node to the ectopic focus. The heart rate abruptly becomes greatly accelerated and continues this rapid rate for a variable time period until the ectopic focus returns to normal. Such overly irritable areas may be associated with organic heart disease, but more frequently they occur in response to anxiety, lack of sleep, or excess caffeine, nicotine, or alcohol consumption.

We now turn attention to how an action potential, once initiated, is conducted throughout the heart.

The spread of cardiac excitation is coordinated to ensure efficient pumping.

Once initiated in the SA node, an action potential spreads throughout the rest of the heart. For efficient cardiac function, the spread of excitation should satisfy three criteria:

1. *Atrial excitation and contraction should be complete before the onset of ventricular contraction.* Complete ventricular filling requires that atrial contraction precede ventricular contraction. During cardiac relaxation, the AV valves are open, so venous blood entering the atria continues to flow directly into the ventricles. Almost 80% of ventricular filling occurs by this means before atrial contraction. When the atria do contract, more blood is squeezed into the ventricles to complete ventricular filling. Ventricular contraction then occurs to eject blood from the heart into the arteries.

If the atria and ventricles were to contract simultaneously, the AV valves would close immediately because ventricular pressures would greatly exceed atrial pressures. The ventricles have much thicker walls and, accordingly, can generate more pressure. Atrial contraction would be unproductive because the atria could not squeeze blood into the ventricles through closed valves. Therefore, to ensure complete filling of the ventricles—to obtain the remaining 20% of ventricular filling that occurs during atrial contraction—the atria must become excited and contract before ventricular excitation and contraction. During a normal heartbeat, atrial contraction occurs about 160 msec before ventricular contraction.

2. *Excitation of cardiac muscle fibers should be coordinated to ensure that each heart chamber contracts as a unit to pump efficiently.* If the muscle fibers in a heart chamber became excited and contracted randomly rather than contracting simultaneously in a coordinated fashion, they would be unable to eject blood. A smooth, uniform ventricular contraction is essential to squeeze out the blood. As an analogy, assume you have a basting syringe full of water. If you merely poke a finger here or there into the rubber bulb of the syringe, you will not eject much water. However, if you compress the bulb in a smooth, coordinated fashion, you can squeeze out the water.

Clinical Note In a similar manner, contraction of isolated cardiac muscle fibers is not successful in pumping blood. Such random, uncoordinated excitation and contraction of the cardiac cells is known as **fibrillation.** Fibrillation of the ventricles is much more serious than atrial fibrillation. Ventricular fibrillation rapidly causes death, because the heart cannot pump blood into the arteries. This condition can often be corrected by **electrical defibrillation,** in which a very strong electrical current is applied on the chest wall. When this current reaches the heart, it stimulates (depolarizes) all parts of the heart simultaneously. Usually the first part of the heart to recover is the SA node, which takes over pacemaker activity, once again initiating impulses that trigger the synchronized contraction of the rest of the heart.

3. *The pair of atria and pair of ventricles should be functionally coordinated so that both members of the pair contract simultaneously.* This coordination permits synchronized pumping of blood into the pulmonary and systemic circulation.

The normal spread of cardiac excitation is carefully orchestrated to ensure that these criteria are met and the heart functions efficiently, as follows (see ● Figure 9-8b).

ATRIAL EXCITATION An action potential originating in the SA node first spreads throughout both atria, primarily from cell to cell via gap junctions. In addition, several poorly delineated, specialized conduction pathways speed up conduction of the impulse through the atria.

■ The **interatrial pathway** extends from the SA node within the right atrium to the left atrium. Because this pathway rapidly transmits the action potential from the SA node to the pathway's termination in the left atrium, a wave of excitation can spread across the gap junctions throughout the left atrium at the same time as excitation is similarly spreading throughout the right atrium. This ensures that both atria become depolarized to contract simultaneously.

■ The **internodal pathway** extends from the SA node to the AV node. The AV node is the only point of electrical contact between the atria and ventricles; in other words, because the atria and ventricles are structurally connected by electrically nonconductive fibrous tissue, the only way an action potential in the atria can spread to the ventricles is by passing through the AV node. The internodal conduction pathway directs the spread of an action potential originating at the SA node to the AV node to ensure sequential contraction of the ventricles following atrial contraction. Hastened by this pathway, the action potential arrives at the AV node within 30 sec of SA node firing.

CONDUCTION BETWEEN THE ATRIA AND THE VENTRICLES The action potential is conducted relatively slowly through the AV node. This slowness is advantageous because it allows time for complete ventricular filling. The impulse is delayed about 100 msec (the **AV nodal delay**), which enables the atria to become completely depolarized and to contract, emptying their contents into the ventricles, before ventricular depolarization and contraction occur.

VENTRICULAR EXCITATION After the AV nodal delay, the impulse travels rapidly down the septum via the right and left branches of the bundle of His and throughout the ventricular myocardium via the Purkinje fibers. The network of fibers in this ventricular conduction system is specialized for rapid propagation of action potentials. Its presence hastens and coordinates the spread of ventricular excitation to ensure that the ventricles contract as a unit. The action potential is transmitted through the entire Purkinje fiber system within 30 msec.

Although this system carries the action potential rapidly to a large number of cardiac muscle cells, it does not terminate on every cell. The impulse quickly spreads from the excited cells to the rest of the ventricular muscle cells by means of gap junctions.

The ventricular conduction system is more highly organized and more important than the interatrial and internodal conduction pathways. Because the ventricular mass is so much larger than the atrial mass, a rapid conduction system is crucial to hasten the spread of excitation in the ventricles. Purkinje fibers can transmit an action potential six times faster than the ventricular syncytium of contractile cells could. If the entire ventricular depolarization process depended on cell-to-cell spread of the impulse via gap junctions, the ventricular tissue immediately next to the AV node would become excited and contract before the impulse had even passed to the heart apex. This, of course, would not allow efficient pumping. Rapid conduction of the action potential down the bundle of His and its swift, diffuse distribution throughout the Purkinje network lead to almost simultaneous activation of the ventricular myocardial cells in both ventricular chambers, which ensures a single, smooth, coordinated contraction that can efficiently eject blood into both the systemic and pulmonary circulations at the same time.

The action potential of cardiac contractile cells shows a characteristic plateau.

The action potential in cardiac contractile cells, although initiated by the nodal pacemaker cells, varies considerably in ionic mechanisms and shape from the SA node potential (compare ● Figures 9-7 and 9-10). Unlike the membrane of autorhythmic cells, the membrane of contractile cells remains essentially at rest at about –90 mV until excited by electrical activity propagated from the pacemaker. Myocardial contractile cells have several subclasses of K^+ channels. At resting potential, the type of K^+ channel that is open is especially leaky.[1] The resultant outward movement of K^+ through these channels keeps the resting potential close to the K^+ equilibrium potential at -90 mV. Once the membrane of a ventricular myocardial contractile cell is depolarized to threshold via current flow through gap junctions, an action potential is generated by a complicated interplay of permeability changes and membrane potential changes as follows (● Figure 9-10):

1. During the rising phase of the action potential, the membrane potential rapidly reverses to a positive value of about +20 mV to +30 mV (depending on the myocardial cell) as a result of activation of voltage-gated Na^+ channels and Na^+ subsequently rapidly entering the cell, as it does in other excitable cells undergoing an action potential (see p. 92). These are the same type of double-gated Na^+ channels found in nerve and skeletal muscle cells. At peak potential, the Na^+ permeability then rapidly plummets to its low resting value.

2. At peak potential, another subclass of K^+ channels transiently opens. The resultant limited efflux of K^+ through these transient channels brings about a brief, small repolarization as the membrane becomes slightly less positive.

● **FIGURE 9-10 Action potential in contractile cardiac muscle cells.** The action potential in cardiac contractile cells differs considerably from the action potential in cardiac autorhythmic cells (compare with Figure 9-7). The rapid rising phase of the action potential in contractile cells is the result of Na^+ entry on opening of fast Na^+ channels at threshold. The early, brief repolarization after the potential reaches its peak is because of limited K^+ efflux on opening of transient K^+ channels, coupled with inactivation of the Na^+ channels. The prolonged plateau phase is the result of slow Ca^{2+} entry on opening of L-type Ca^{2+} channels, coupled with reduced K^+ efflux on closure of several types of K^+ channels. The rapid falling phase is the result of K^+ efflux on opening of ordinary voltage-gated K^+ channels, as in other excitable cells. Resting potential is maintained by opening of leaky K^+ channels.

3. Unique to these cardiac muscle cells, the membrane potential is maintained at this near peak positive level for several hundred milliseconds, producing a *plateau phase* of the action potential. In contrast, the short action potential of neurons and skeletal muscle cells lasts 1 to 2 msec. Whereas the rising phase of the action potential is brought about by activation of comparatively "fast" Na^+ channels and the early, brief repolarization arises from activation of fast, transient K^+ channels, this plateau is maintained by two voltage-dependent permeability changes: activation of "slow" L-type Ca^{2+} channels and a marked decrease in K^+ permeability in the cardiac contractile cell membrane. These permeability changes occur in response to the sudden change in voltage during the rising phase of the action potential. Opening of the L-type Ca^{2+} channels results in a slow, inward diffusion of Ca^{2+} because Ca^{2+} is in greater concentration in the ECF. This continued influx of positively charged Ca^{2+} prolongs the positivity inside the cell and is primarily responsible for the plateau portion of the action potential. This effect is enhanced by the concomitant decrease in K^+ permeability on closure of both the briefly opened transient K^+ channels and the leaky K^+ channels open at resting potential. The resultant reduction in outward movement of positively charged

---

[1]For reasons beyond the scope of this book, this channel type is officially known as an *inward rectifier K^+ channel,* and the "ordinary" K^+ channel discussed shortly is called a *delayed rectifier K^+ channel.*

K^+ prevents rapid repolarization of the membrane and thus contributes to prolongation of the plateau phase.

4. The rapid falling phase of the action potential results from inactivation of the Ca^{2+} channels and delayed activation of "ordinary" voltage-gated K^+ channels, yet another subclass of K^+ channels identical to the ones responsible for repolarization in neurons and skeletal muscle cells. The decrease in Ca^{2+} permeability diminishes the slow, inward movement of positive Ca^{2+}, whereas the sudden increase in K^+ permeability simultaneously promotes rapid outward diffusion of positive K^+. Thus, as in other excitable cells, the cell returns to resting potential as K^+ leaves the cell. At resting potential, the ordinary voltage-gated K^+ channels close and the leaky K^+ channels open once again.

Let's now see how this action potential brings about contraction.

Ca^{2+} entry from the ECF induces a much larger Ca^{2+} release from the sarcoplasmic reticulum.

In cardiac contractile cells, the L-type Ca^{2+} channels lie primarily in the T tubules. As you just learned, these voltage-gated channels open during a local action potential. Thus, unlike in skeletal muscle, Ca^{2+} diffuses into the cytosol from the ECF across the T tubule membrane during a cardiac action potential. This entering Ca^{2+} triggers the opening of nearby ryanodine Ca^{2+}-release channels in the adjacent lateral sacs of the sarcoplasmic reticulum (see p. 264). By means of this so-called **Ca^{2+}-induced Ca^{2+} release,** Ca^{2+} entering the cytosol from the ECF induces a much larger release of Ca^{2+} into the cytosol from the intracellular stores (● Figure 9-11). The resultant local bursts of Ca^{2+} release, known as Ca^{2+} *sparks,* from the sarcoplasmic reticulum collectively increase the cytosolic Ca^{2+} pool sufficiently to turn on the contractile machinery. Ninety percent of the Ca^{2+} needed for muscle contraction comes from the sarcoplasmic reticulum. This extra supply of Ca^{2+}, coupled with slow Ca^{2+} removal processes, is responsible for the long period of cardiac contraction, which lasts about three times longer than the contraction of a single skeletal muscle fiber (300 msec compared to 100 msec). This increased contractile time ensures adequate time to eject the blood.

ROLE OF CYTOSOLIC Ca^{2+} IN EXCITATION–CONTRACTION COUPLING As in skeletal muscle, the role of Ca^{2+} within the cytosol is to bind with the troponin–tropomyosin complex and physically pull it aside to allow cross-bridge cycling and contraction (● Figure 9-11) (see p. 260). However, unlike skeletal muscle, in which sufficient Ca^{2+} is always released to turn on all the cross bridges, in cardiac muscle the extent of cross-bridge activity varies with the amount of cytosolic Ca^{2+}. As we will show, various regulatory factors can alter the amount of cytosolic Ca^{2+}.

Removal of Ca^{2+} from the cytosol by energy-dependent mechanisms in both the plasma membrane and sarcoplasmic reticulum restores the blocking action of troponin and tropomyosin, so contraction ceases and the heart muscle relaxes.

 INFLUENCE OF ALTERED ECF K^+ AND Ca^{2+} CONCENTRATIONS Not surprisingly, changes in the ECF concentration of K^+ and Ca^{2+} can have profound effects on the

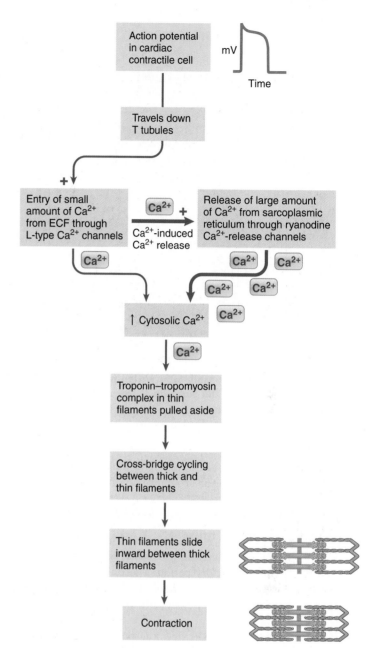

● **FIGURE 9-11 Excitation–contraction coupling in cardiac contractile cells.**

heart. Abnormal levels of K^+ are most important clinically, followed to a lesser extent by Ca^{2+} imbalances. By changing the concentration gradient between the ICF and ECF, changes in the ECF K^+ concentration (either above or below normal) can lead to abnormalities in cardiac excitation by changing the resting potential (see p. 535).

An elevated ECF Ca^{2+} concentration augments the strength of cardiac contraction by prolonging the plateau phase of the action potential and by increasing the cytosolic concentration of Ca^{2+}. Contractions tend to be of longer duration, with little time to rest between contractions. Some drugs alter cardiac function by influencing Ca^{2+} movement across the myocardial cell membranes. For example, Ca^{2+}-channel blocking agents, such as *verapamil,* block Ca^{2+} influx through the L-type Ca^{2+} channels during an action potential, reducing the force of car-

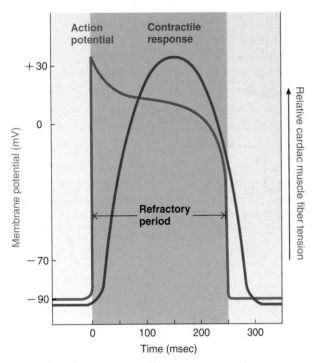

● **FIGURE 9-12 Relationship of an action potential and the refractory period to the duration of the contractile response in cardiac muscle.**

diac contraction. Other drugs, such as *digitalis,* increase cardiac contractility by inducing an accumulation of cytosolic Ca^{2+}.

A long refractory period prevents tetanus of cardiac muscle.

Like other excitable tissues, cardiac muscle has a refractory period. During the refractory period, a second action potential cannot be triggered until an excitable membrane has recovered from the preceding action potential. In skeletal muscle, the refractory period is very short compared with the duration of the resulting contraction, so the fiber can be restimulated before the first contraction is complete to produce summation of contractions. Rapidly repetitive stimulation that does not let the muscle fiber relax between stimulations results in a sustained, maximal contraction known as *tetanus* (see ● Figure 8-18, p. 271).

In contrast, cardiac muscle has a long refractory period that lasts about 250 msec because of the prolonged plateau phase of the action potential. This is almost as long as the period of contraction initiated by the action potential; a cardiac muscle fiber contraction averages about 300 msec (● Figure 9-12). Consequently, cardiac muscle cannot be restimulated until contraction is almost over, precluding summation of contractions and tetanus of cardiac muscle. This is a valuable protective mechanism, because the pumping of blood requires alternate periods of contraction (emptying) and relaxation (filling). A prolonged tetanic contraction would prove fatal. The heart chambers could not be filled and emptied again.

The chief factor responsible for the long refractory period is inactivation, during the prolonged plateau phase, of the Na^+ channels that were activated during the initial Na^+ influx of the rising phase; that is, the double-gated Na^+ channels are in their closed and not capable of opening conformation (see p. 92). Not until the membrane recovers from this inactivation process (when the membrane has already repolarized to resting), can the Na^+ channels be activated once again to begin another action potential.

The ECG is a record of the overall spread of electrical activity through the heart.

The electrical currents generated by cardiac muscle during depolarization and repolarization spread into the tissues around the heart and are conducted through the body fluids. A small part of this electrical activity reaches the body surface, where it can be detected using recording electrodes. The record produced is an **electrocardiogram,** or **ECG.** (Alternatively, the abbreviation **EKG** is often used, from the ancient Greek word "*kardia*" instead of the Latin "*cardia*" for "heart.")

Remember three important points when considering what an ECG represents:

1. An ECG is a recording of that part of the electrical activity present in body fluids from the cardiac impulse that reaches the body surface, not a direct recording of the actual electrical activity of the heart.

2. The ECG is a complex recording representing the *overall* spread of activity throughout the heart during depolarization and repolarization. It is not a recording of a *single* action potential in a single cell at a single point in time. The record at any given time represents the sum of electrical activity in all the cardiac muscle cells, some of which may be undergoing action potentials while others may not yet be activated. For example, immediately after the SA node fires, the atrial cells are undergoing action potentials while the ventricular cells are still at resting potential. At a later point, the electrical activity will have spread to the ventricular cells while the atrial cells will be repolarizing. Therefore, the overall pattern of cardiac electrical activity varies with time as the impulse passes throughout the heart.

3. The recording represents comparisons in voltage detected by electrodes at two different points on the body surface, not the actual potential. For example, the ECG does not record a potential at all when the ventricular muscle is either completely depolarized or completely repolarized; both electrodes are "viewing" the same potential, so no difference in potential between the two electrodes is recorded.

The exact pattern of electrical activity recorded from the body surface depends on the orientation of the recording electrodes. Electrodes may be loosely thought of as "eyes" that "see" electrical activity and translate it into a visible recording, the ECG record. Whether an upward deflection or downward deflection is recorded is determined by the way the electrodes are oriented with respect to the current flow in the heart. For example, the spread of excitation across the heart is "seen" differently from the right arm, from the left leg, or from a recording directly over the heart. Even though the same electrical events are occurring in the heart, different

(a) Limb leads

(b) Chest leads

● **FIGURE 9-13 Electrocardiogram leads.** (a) The six limb leads include leads I, II, III, aVR, aVL, and aVF. Leads I, II, and III are bipolar leads because two recording electrodes are used. The tracing records the *difference* in potential between the two electrodes. For example, lead I records the difference in potential detected at the right arm and left arm. The electrode placed on the right leg serves as a ground and is not a recording electrode. The aVR, aVL, and aVF leads are unipolar leads. Even though two electrodes are used, only the actual potential under one electrode, the exploring electrode, is recorded. The other electrode is set at zero potential and serves as a neutral reference point. For example, the aVR lead records the potential reaching the right arm in comparison to the rest of the body. (b) The six chest leads, V_1 through V_6, are also unipolar leads. The exploring electrode mainly records the electrical potential of the cardiac musculature immediately beneath the electrode in six different locations surrounding the heart.

waveforms representing the same electrical activity result when this activity is recorded by electrodes at different points on the body.

To provide standard comparisons, ECG records routinely consist of 12 conventional electrode systems, or leads. When an electrocardiograph machine is connected between recording electrodes at two points on the body, the specific arrangement of each pair of connections is called a **lead.** The 12 different leads each record electrical activity in the heart from different locations—six different electrical arrangements from the limbs and six chest leads at various sites around the heart. To provide a common basis for comparison and for recognizing deviations from normal, the same 12 leads are routinely used in all ECG recordings (● Figure 9-13).

Different parts of the ECG record can be correlated to specific cardiac events.

Interpretation of the wave configurations recorded from each lead depends on a thorough knowledge about the sequence of cardiac excitation spread and about the position of the heart relative to electrode placement. A normal ECG has three distinct waveforms: the P wave, the QRS complex, and the T wave (● Figure 9-14). (The letters only indicate the orderly sequence of the waves. The inventor of the technique just started in mid-alphabet when naming the waves.)

- The **P wave** represents atrial depolarization.
- The **QRS complex** represents ventricular depolarization.
- The **T wave** represents ventricular repolarization.

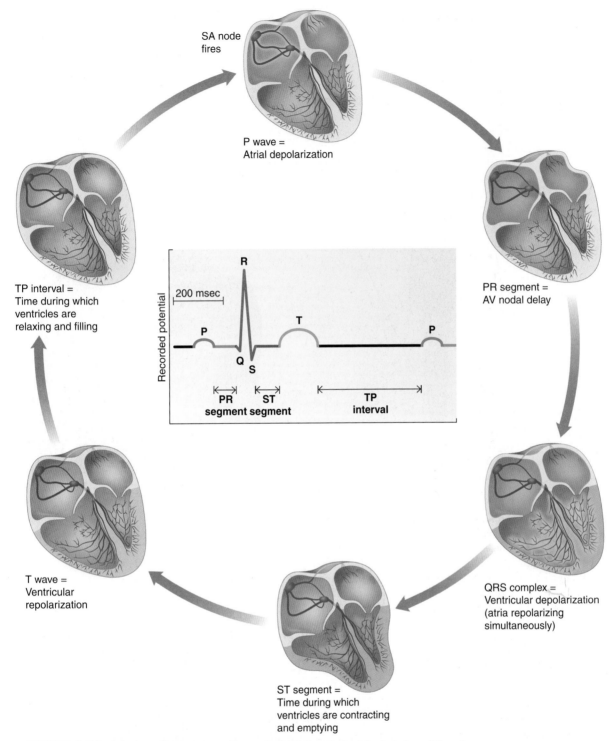

SA node
fires

P wave =
Atrial depolarization

PR segment =
AV nodal delay

QRS complex =
Ventricular depolarization
(atria repolarizing
simultaneously)

ST segment =
Time during which
ventricles are contracting
and emptying

T wave =
Ventricular
repolarization

TP interval =
Time during which
ventricles are
relaxing and filling

Recorded potential

200 msec

R

P

Q S

T

P

PR
segment

ST
segment

TP
interval

● **FIGURE 9-14 Electrocardiogram waveforms in lead II and electrical status of heart associated with each waveform.**

Because these shifting waves of depolarization and repolarization bring about the alternating contraction and relaxation of the heart, respectively, the cyclic mechanical events of the heart lag slightly behind the rhythmic changes in electrical activity.

The following points about the ECG record should also be noted:

1. Firing of the SA node does not generate enough electrical activity to reach the body surface, so no wave is recorded for SA nodal depolarization. Therefore, the first recorded wave, the P wave, occurs when the impulse or wave of depolarization spreads across the atria.

2. In a normal ECG, no separate wave for atrial repolarization is visible. The electrical activity associated with atrial repolarization normally occurs simultaneously with ventricular depolarization and is masked by the QRS complex.

3. The P wave is much smaller than the QRS complex, because the atria have a much smaller muscle mass than the ventricles and consequently generate less electrical activity.

4. At the following three points in time no net current flow is taking place in the heart musculature, so the ECG remains at baseline:

a. *During the AV nodal delay.* This delay is represented by the interval of time between the end of P and the onset of QRS; this segment of the ECG is known as the **PR segment.** (It is called the "PR segment" rather than the "PQ segment" because the Q deflection is small and sometimes absent, whereas the R deflection is the dominant wave of the complex.) Current is flowing through the AV node, but the magnitude is too small for the ECG electrodes to detect.

b. *When the ventricles are completely depolarized and the cardiac contractile cells are undergoing the plateau phase of their action potential before they repolarize,* represented by the **ST segment.** This segment lies between QRS and T; it coincides with the time during which ventricular activation is complete and the ventricles are contracting and emptying. Note that the ST segment is *not* a record of cardiac contractile activity. The ECG is a measure of the electrical activity that triggers the subsequent mechanical activity.

c. *When the heart muscle is completely repolarized and at rest and ventricular filling is taking place,* after the T wave and before the next P wave. This period is called the **TP interval.**

The ECG can be used to diagnose abnormal heart rates, arrhythmias, and damage of heart muscle.

Clinical Note Because electrical activity triggers mechanical activity, abnormal electrical patterns are usually accompanied by abnormal contractile activity of the heart. Thus, evaluation of ECG patterns can provide useful information about the status of the heart. The main deviations from normal that can be found through electrocardiography are (1) abnormalities in rate, (2) abnormalities in rhythm, and (3) cardiac myopathies (● Figure 9-15).

ABNORMALITIES IN RATE The heart rate can be determined from the distance between two consecutive QRS complexes on the calibrated paper used to record an ECG. A rapid heart rate of more than 100 beats per minute is called **tachycardia** (*tachy* means "fast"), whereas a slow heart rate of fewer than 60 beats per minute is called **bradycardia** (*brady* means "slow").

ABNORMALITIES IN RHYTHM *Rhythm* refers to the regularity or spacing of the ECG waves. Any variation from the normal rhythm and sequence of excitation of the heart is termed an **arrhythmia.** It may result from ectopic foci, alterations in SA node pacemaker activity, or interference with conduction. Heart rate is also often altered. *Extrasystoles,* or *premature ventricular contractions,* originating from an ectopic focus are common deviations from normal rhythm. Other abnormalities in rhythm

● FIGURE 9-15 Representative heart conditions detectable through electrocardiography.

easily detected on an ECG include atrial flutter, atrial fibrillation, ventricular fibrillation, and heart block.

Atrial flutter is characterized by a rapid but regular sequence of atrial depolarizations at rates between 200 and 380 beats per minute. The ventricles rarely keep pace with the racing atria. Because the conducting tissue's refractory period is longer than that of the atrial muscle, the AV node is unable to respond to every impulse that converges on it from the atria. Maybe only one out of every two or three atrial impulses successfully passes through the AV node to the ventricles. Such a situation is referred to as a *2:1* or *3:1 rhythm.* The fact that not every atrial impulse reaches the ventricle in atrial flutter is important, because it precludes a rapid ventricular rate of more than 200 beats per minute. Such a high rate would not allow adequate time for ventricular filling between beats. In such a

The What, Who, and When of Stress Testing

Stress tests, or **graded exercise tests,** are conducted primarily to aid in diagnosing or quantifying heart or lung disease and to evaluate the functional capacity of asymptomatic individuals. The tests are usually given on motorized treadmills or bicycle ergometers (stationary, variable-resistance bicycles). Workload intensity (how hard the subject is working) is adjusted by progressively increasing the speed and incline of the treadmill or by progressively increasing the pedaling frequency and resistance on the bicycle. The test starts at a low intensity and continues until a prespecified workload is achieved, physiologic symptoms occur, or the subject is too fatigued to continue.

During diagnostic testing, the patient is monitored with an ECG, and blood pressure is taken each minute. A test is considered positive if ECG abnormalities occur (such as ST segment depression, inverted T waves, or dangerous arrhythmias) or if physical symptoms such as chest pain develop. A test that is interpreted as positive in a person who does not have heart disease is called a *false positive test.* In men, false positives occur only about 10% to 20% of the time, so the diagnostic stress test for men has a *specificity* of 80% to 90%. Women have a greater frequency of false positive tests, with a corresponding lower specificity of about 70%.

The *sensitivity* of a test means that people with disease are correctly identified and there are few false negatives. The sensitivity of the stress test is reported to be 60% to 80%; that is, if 100 people with heart disease were tested, 60 to 80 would be correctly identified, but 20 to 40 would have a false negative test. Although stress testing is now an important diagnostic tool, it is just one of several tests used to determine the presence of coronary artery disease.

Stress tests are also conducted on people not suspected of having heart or lung disease to determine their present functional capacity. These functional tests are administered in the same way as diagnostic tests, but they are conducted by exercise physiologists and a physician need not be present. These tests are used to establish safe exercise prescriptions, to aid athletes in establishing optimal training programs, and to serve as research tools to evaluate the effectiveness of a particular training regimen. Functional stress testing is becoming more prevalent as more people are joining hospital- or community-based wellness programs for disease prevention.

---

case, the output of the heart would be reduced to the extent that loss of consciousness or even death could result because of decreased blood flow to the brain.

Atrial fibrillation is characterized by rapid, irregular, uncoordinated depolarizations of the atria with no definite P waves. Accordingly, atrial contractions are chaotic and asynchronized. Because impulses reach the AV node erratically, the ventricular rhythm is also very irregular. The QRS complexes are normal in shape but occur sporadically. Variable lengths of time between ventricular beats are available for ventricular filling. Some ventricular beats come so close together that little filling can occur between beats. When less filling occurs, the subsequent contraction is weaker. In fact, some of the ventricular contractions may be too weak to eject enough blood to produce a palpable wrist pulse. In this situation, if the heart rate is determined directly, either by the apex beat or via the ECG, and the pulse rate is taken concurrently at the wrist, the heart rate will exceed the pulse rate. Such a difference in heart rate and pulse rate is known as a **pulse deficit.** Normally, the heart rate coincides with the pulse rate, because each cardiac contraction initiates a pulse wave as it ejects blood into the arteries.

Ventricular fibrillation is a very serious rhythmic abnormality in which the ventricular musculature exhibits uncoordinated, chaotic contractions. Multiple impulses travel erratically in all directions around the ventricles. The ECG tracing in ventricular fibrillations is very irregular with no detectable pattern or rhythm. When contractions are so disorganized, the ventricles are ineffectual as pumps. If circulation is not restored in less than four minutes through external cardiac compression or electrical defibrillation, irreversible brain damage occurs, and death is imminent.

Another type of arrhythmia, **heart block,** arises from defects in the cardiac conducting system. The atria still beat regularly, but the ventricles occasionally fail to be stimulated and thus do not contract following atrial contraction. Impulses between the atria and ventricles can be blocked to varying degrees. In some forms of heart block, only every second or third atrial impulse is passed to the ventricles. This is known as *2:1 or 3:1 block,* which can be distinguished from the 2:1 or 3:1 rhythm associated with atrial flutter by the rates involved. In heart block, the atrial rate is normal but the ventricular rate is considerably below normal, whereas in atrial flutter the atrial rate is very high, in accompaniment with a normal or above-normal ventricular rate. *Complete heart block* is characterized by complete dissociation between atrial and ventricular activity, with impulses from the atria not being conducted to the ventricles at all. The SA node continues to govern atrial depolarization, but the ventricles generate their own impulses at a rate much slower than that of the atria. On the ECG, the P waves exhibit a normal rhythm. The QRS and T waves also occur regularly but much more slowly than the P waves and are completely independent of P wave rhythm. Because atrial and ventricular activity is not synchronized, waves for atrial repolarization may appear, no longer masked by the QRS complex.

CARDIAC MYOPATHIES Abnormal ECG waves are also important in recognizing and assessing **cardiac myopathies** (damage of the heart muscle). **Myocardial ischemia** is inadequate delivery of oxygenated blood to the heart tissue. Actual death, or **necrosis,** of heart muscle cells occurs when a blood vessel supplying that area of the heart becomes blocked or ruptured. This condition is **acute myocardial infarction,** commonly called a

heart attack. Abnormal QRS waveforms appear when part of the heart muscle becomes necrotic. In addition to ECG changes, because damaged heart muscle cells release characteristic enzymes into the blood, the level of these enzymes in the blood provides a further index of the extent of myocardial damage.

Interpretation of an ECG is a complex task requiring extensive knowledge and training. The foregoing discussion is not by any means intended to make you an ECG expert but seeks to give you an appreciation of the ways in which the ECG can be used as a diagnostic tool, as well as to present an overview of some of the more common abnormalities of heart function. (For a further use of the ECG, see the accompanying boxed feature, ■ A Closer Look at Exercise Physiology.)

Mechanical Events of the Cardiac Cycle

The mechanical events of the cardiac cycle—contraction, relaxation, and the resultant changes in blood flow through the heart—are brought about by the rhythmic changes in cardiac electrical activity.

The heart alternately contracts to empty and relaxes to fill.

The cardiac cycle consists of alternate periods of **systole** (contraction and emptying) and **diastole** (relaxation and filling). Contraction results from the spread of excitation across the heart, whereas relaxation follows the subsequent repolarization of the cardiac musculature. The atria and ventricles go through separate cycles of systole and diastole. Unless qualified, the terms *systole* and *diastole* refer to what is happening with the ventricles.

The following discussion and corresponding ● Figure 9-16 correlate various events that occur concurrently during the cardiac cycle, including ECG features, pressure changes, volume changes, valve activity, and heart sounds. Only the events on the left side of the heart are described, but keep in mind that identical events are occurring on the right side of the heart, except that the pressures are lower. To complete one full cardiac cycle, our discussion will begin and end with ventricular diastole.

MID VENTRICULAR DIASTOLE During most of ventricular diastole, the atrium is still also in diastole. This stage corresponds to the TP interval on the ECG—the interval after ventricular repolarization and before another atrial depolarization. Because of the continuous inflow of blood from the venous system into the atrium, atrial pressure slightly exceeds ventricular pressure even though both chambers are relaxed (point **1** in ● Figure 9-16). Because of this pressure differential, the AV valve is open, and blood flows directly from the atrium into the ventricle throughout ventricular diastole (heart a in ● Figure 9-16). As a result of this passive filling, the ventricular volume slowly continues to rise even before atrial contraction takes place (point **2**).

LATE VENTRICULAR DIASTOLE Late in ventricular diastole, the SA node reaches threshold and fires. The impulse spreads throughout the atria, which appears on the ECG as the P wave (point **3**). Atrial depolarization brings about atrial contraction, raising the atrial pressure curve (point **4**) and squeezing more blood into the ventricle. The excitation–contraction coupling process takes place during the short delay between the P wave and the rise in atrial pressure. The corresponding rise in ventricular pressure (point **5**) that occurs simultaneously with the rise in atrial pressure results from the additional volume of blood added to the ventricle by atrial contraction (point **6** and heart b). Throughout atrial contraction, atrial pressure still slightly exceeds ventricular pressure, so the AV valve remains open.

END OF VENTRICULAR DIASTOLE Ventricular diastole ends at the onset of ventricular contraction. By this time, atrial contraction and ventricular filling are completed. The volume of blood in the ventricle at the end of diastole (point **7**) is known as the **end-diastolic volume (EDV),** which averages about 135 ml. No more blood will be added to the ventricle during this cycle. Therefore, the end-diastolic volume is the maximum amount of blood that the ventricle will contain during this cycle.

VENTRICULAR EXCITATION AND ONSET OF VENTRICULAR SYSTOLE After atrial excitation, the impulse travels through the AV node and specialized conduction system to excite the ventricle. Simultaneously, the atria are contracting. By the time ventricular activation is complete, atrial contraction is already over. The QRS complex represents this ventricular excitation (point **8**), which induces ventricular contraction. The ventricular pressure curve sharply increases shortly after the QRS complex, signaling the onset of ventricular systole (point **9**). The slight delay between the QRS complex and the actual onset of ventricular systole is the time required for the excitation–contraction coupling process to occur. As ventricular contraction begins, ventricular pressure immediately exceeds atrial pressure. This backward pressure differential forces the AV valve closed (point **9**).

ISOVOLUMETRIC VENTRICULAR CONTRACTION After ventricular pressure exceeds atrial pressure and the AV valve has closed, to open the aortic valve, the ventricular pressure must continue to increase until it exceeds aortic pressure. Therefore, after closing of the AV valve and before opening of the aortic valve is a brief period of time when the ventricle remains a closed chamber (point **10**). Because all valves are closed, no blood can enter or leave the ventricle during this time. This interval is termed the period of **isovolumetric ventricular contraction** (*isovolumetric* means "constant volume and length") (heart c). Because no blood enters or leaves the ventricle, the ventricular chamber stays at constant volume, and the muscle fibers stay at constant length. This isovolumetric condition is similar to an isometric contraction in skeletal muscle. During isovolumetric ventricular contraction, ventricular pressure continues to increase as the volume remains constant (point **11**).

VENTRICULAR EJECTION When ventricular pressure exceeds aortic pressure (point **12**), the aortic valve is forced open and ejection of blood begins (heart d). The amount of blood pumped out of each ventricle with each contraction is called the

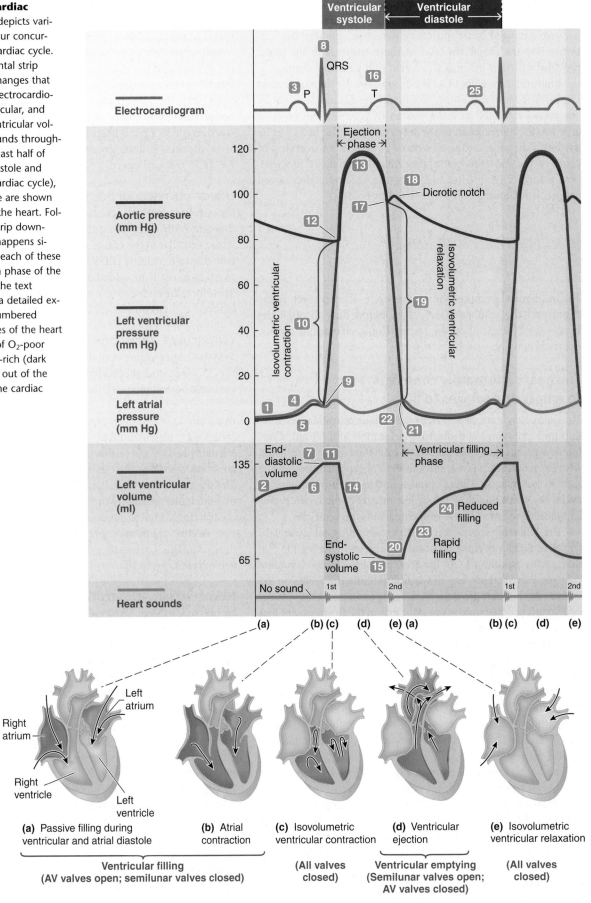

● **FIGURE 9-16 Cardiac cycle.** This graph depicts various events that occur concurrently during the cardiac cycle. Follow each horizontal strip across to see the changes that take place in the electrocardiogram; aortic, ventricular, and atrial pressures; ventricular volume; and heart sounds throughout the cycle. The last half of diastole, one full systole and diastole (one full cardiac cycle), and another systole are shown for the left side of the heart. Follow each vertical strip downward to see what happens simultaneously with each of these factors during each phase of the cardiac cycle. See the text (pp. 321–324) for a detailed explanation of the numbered points. The sketches of the heart illustrate the flow of O_2-poor (dark blue) and O_2-rich (dark pink) blood in and out of the ventricles during the cardiac cycle.

Ventricular systole

Ventricular diastole

QRS

P

T

Electrocardiogram

Ejection phase

120

Dicrotic notch

100

Aortic pressure (mm Hg)

80

Isovolumetric ventricular relaxation

60

Isovolumetric ventricular contraction

Left ventricular pressure (mm Hg)

40

20

Left atrial pressure (mm Hg)

0

End-diastolic volume

Ventricular filling phase

135

Left ventricular volume (ml)

Reduced filling

Rapid filling

End-systolic volume

65

No sound   1st   2nd   1st   2nd

Heart sounds

(a)   (b) (c)   (d)   (e) (a)   (b) (c)   (d)   (e)

Left atrium

Right atrium

Right ventricle

Left ventricle

(a) Passive filling during ventricular and atrial diastole

(b) Atrial contraction

(c) Isovolumetric ventricular contraction

(d) Ventricular ejection

(e) Isovolumetric ventricular relaxation

Ventricular filling (AV valves open; semilunar valves closed)

(All valves closed)

Ventricular emptying (Semilunar valves open; AV valves closed)

(All valves closed)

stroke volume (SV). The aortic pressure curve rises as blood is forced into the aorta from the ventricle faster than blood is draining off into the smaller vessels at the other end (point **13**). The ventricular volume decreases substantially as blood is rapidly pumped out (point **14**). Ventricular systole includes both the period of isovolumetric contraction and the ventricular ejection phase.

END OF VENTRICULAR SYSTOLE The ventricle does not empty completely during ejection. Normally, only about half the blood within the ventricle at the end of diastole is pumped out during the subsequent systole. The amount of blood left in the ventricle at the end of systole when ejection is complete is the **end-systolic volume (ESV),** which averages about 65 ml (point **15**). This is the least amount of blood that the ventricle will contain during this cycle.

The difference between the volume of blood in the ventricle before contraction and the volume after contraction is the amount of blood ejected during the contraction; that is, EDV − ESV = SV. In our example, the end-diastolic volume is 135 ml, the end-systolic volume is 65 ml, and the stroke volume is 70 ml.

VENTRICULAR REPOLARIZATION AND ONSET OF VENTRICULAR DIASTOLE The T wave signifies ventricular repolarization at the end of ventricular systole (point **16**). As the ventricle starts to relax, on repolarization, ventricular pressure falls below aortic pressure and the aortic valve closes (point **17**). Closure of the aortic valve produces a disturbance or notch on the aortic pressure curve, the **dicrotic notch** (point **18**). No more blood leaves the ventricle during this cycle, because the aortic valve has closed.

ISOVOLUMETRIC VENTRICULAR RELAXATION When the aortic valve closes, the AV valve is not yet open, because ventricular pressure still exceeds atrial pressure, so no blood can enter the ventricle from the atrium. Therefore, all valves are once again closed for a brief period of time known as **isovolumetric ventricular relaxation** (point **19** and heart e). The muscle fiber length and chamber volume (point **20**) remain constant. No blood leaves or enters as the ventricle continues to relax and the pressure steadily falls.

VENTRICULAR FILLING When ventricular pressure falls below atrial pressure, the AV valve opens (point **21**), and ventricular filling occurs again. Ventricular diastole includes both the period of isovolumetric ventricular relaxation and the ventricular filling phase.

Atrial repolarization and ventricular depolarization occur simultaneously, so the atria are in diastole throughout ventricular systole. Blood continues to flow from the pulmonary veins into the left atrium. As this incoming blood pools in the atrium, atrial pressure rises continuously (point **22**). When the AV valve opens at the end of ventricular systole, blood that accumulated in the atrium during ventricular systole pours rapidly into the ventricle (heart a again). Ventricular filling thus occurs rapidly at first (point **23**) because of the increased atrial pressure resulting from the accumulation of blood in the atria. Then

● **FIGURE 9-17 Ventricular filling profiles during normal and rapid heart rates.** Because much of ventricular filling occurs early in diastole during the rapid-filling phase, filling is not seriously impaired when diastolic time is reduced as a result of an increase in heart rate.

ventricular filling slows down (point **24**) as the accumulated blood has already been delivered to the ventricle, and atrial pressure starts to fall. During this period of reduced filling, blood continues to flow from the pulmonary veins into the left atrium and through the open AV valve into the left ventricle. During late ventricular diastole, when the ventricle is filling slowly, the SA node fires again, and the cardiac cycle starts over (point **25**).

When the body is at rest, one complete cardiac cycle lasts 800 msec, with 300 msec devoted to ventricular systole and 500 msec taken up by ventricular diastole. Significantly, much of ventricular filling occurs early in diastole during the rapid-filling phase. During times of rapid heart rate, diastole length is shortened much more than systole length is. For example, if the heart rate increases from 75 to 180 beats per minute, the duration of diastole decreases about 75%, from 500 msec to 125 msec. This greatly reduces the time available for ventricular relaxation and filling. However, because much ventricular filling is accomplished during early diastole, filling is not seriously impaired during periods of increased heart rate, such as during exercise (● Figure 9-17). There is a limit, however, to how rapidly the heart can beat without decreasing the period of diastole to the point that ventricular filling is severely impaired. At heart rates greater than 200 beats per minute, diastolic time is too short to allow adequate ventricular filling. With inadequate filling, the resultant cardiac output is deficient. Normally, ventricular rates do not exceed 200 beats per minute, because the relatively long refractory period of the AV node will not allow impulses to be conducted to the ventricles more frequently than this.

The two heart sounds are associated with valve closures.

Two major heart sounds normally can be heard with a stethoscope during the cardiac cycle. The **first heart sound** is low-pitched, soft, and relatively long—often said to sound like "lub." The **second heart sound** has a higher pitch and is shorter and sharper—often said to sound like "dup." Thus, one normally hears "lub-dup-lub-dup-lub-dup . . . ." The first heart sound is associated with closure of the AV valves, whereas the second sound is associated with closure of the semilunar valves (see "Heart sounds" in ● Figure 9-16). Opening of valves does not produce any sound.

The sounds are caused by vibrations set up within the walls of the ventricles and major arteries during valve closure, not by the valves snapping shut. Because the AV valves close at the onset of ventricular contraction, when ventricular pressure first exceeds atrial pressure, the first heart sound signals the onset of ventricular systole (point **9** in ● Figure 9-16). The semilunar valves close at the onset of ventricular relaxation, as the left and right ventricular pressures fall below the aortic and pulmonary artery pressures, respectively. The second heart sound, therefore, signals the onset of ventricular diastole (point **17**).

Turbulent blood flow produces heart murmurs.

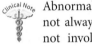 Abnormal heart sounds, or **murmurs,** are usually (but not always) associated with cardiac disease. Murmurs not involving heart pathology, so-called **functional murmurs,** are more common in young people.

Blood normally flows in a *laminar* fashion; that is, layers of the fluid slide smoothly over each other (*lamina* means "layer"). Laminar flow does not produce any sound. When blood flow becomes turbulent, however, a sound can be heard (● Figure 9-18). Such an abnormal sound is the result of vibrations that the turbulent flow creates in the surrounding structures.

STENOTIC AND INSUFFICIENT VALVES The most common cause of turbulence is valve malfunction, either a stenotic or an insufficient valve. A **stenotic valve** is a stiff, narrowed valve that does not open completely. Blood must be forced through the constricted opening at tremendous velocity, resulting in turbulence that produces an abnormal whistling sound similar to the sound produced when you force air rapidly through narrowed lips to whistle.

An **insufficient,** or **incompetent, valve** is one that cannot close completely, usually because the valve edges are scarred and do not fit together properly. Turbulence is produced when blood flows backward through the insufficient valve and collides with blood moving in the opposite direction, creating a swishing or gurgling murmur. Such backflow of blood is known as **regurgitation.** An insufficient heart valve is often called a **leaky valve,** because it lets blood leak back through at a time when the valve should be closed.

Most often, both valvular stenosis and insufficiency are caused by **rheumatic fever,** an autoimmune ("immunity against self") disease triggered by a streptococcus bacterial infection. Antibodies formed against toxins produced by these bacteria

(a) Laminar flow (does not create any sound)

(b) Turbulent flow (can be heard)

● **FIGURE 9-18 Comparison of laminar and turbulent flow.**

interact with many of the body's own tissues, resulting in immunological damage. The heart valves are among the most susceptible tissues in this regard. Large, hemorrhagic, fibrous lesions form along the inflamed edges of an affected heart valve, causing the valve to become thickened, stiff, and scarred. Sometimes the leaflet edges permanently adhere to each other. Depending on the extent and specific nature of the lesions, the valve may become either stenotic or insufficient or some degree of both. On occasion children are born with malfunctioning valves.

TIMING OF MURMURS The valve involved and the type of defect can usually be detected by the *location* and *timing* of the murmur. Each heart valve can be heard best at a specific location on the chest. Noting where a murmur is loudest helps the diagnostician tell which valve is involved.

The "timing" of the murmur refers to the part of the cardiac cycle during which the murmur is heard. Recall that the first heart sound signals the onset of ventricular systole, and the second heart sound signals the onset of ventricular diastole. Thus, a murmur between the first and second heart sounds (lub-murmur-dup, lub-murmur-dup) is a **systolic murmur.** A **diastolic murmur,** in contrast, occurs between the second and first heart sound (lub-dup-murmur, lub-dup-murmur). The sound of the murmur characterizes it as either a stenotic (whistling) murmur or an insufficient (swishy) murmur. Armed with these facts, one can determine the cause of a valvular murmur (▲ Table 9-2). As an example, a whistling murmur (denoting a stenotic valve) occurring between the first and second heart sounds (denoting a systolic murmur) indicates stenosis in a valve that should be open during systole. It could be either the aortic or the pulmonary semilunar valve through which blood is being ejected. Identifying which of these valves is stenotic is accomplished by finding where the murmur is best heard.

The main concern with heart murmurs, of course, is not the murmur itself but the harmful circulatory results of the defect.

TABLE 9-2 Timing and Type of Murmur Associated with Various Heart Valve Disorders

Pattern Heard on Auscultation	Type of Valve Defect	Timing of Murmur	Valve Disorder	Comment
Lub-Whistle-Dup	Stenotic	Systolic	Stenotic semilunar valve	A whistling systolic murmur signifies that a valve that should be open during systole (a semilunar valve) does not open completely.
Lub-Dup-Whistle	Stenotic	Diastolic	Stenotic AV valve	A whistling diastolic murmur signifies that a valve that should be open during diastole (an AV valve) does not open completely.
Lub-Swish-Dup	Insufficient	Systolic	Insufficient AV valve	A swishy systolic murmur signifies that a valve that should be closed during systole (an AV valve) does not close completely.
Lub-Dup-Swish	Insufficient	Diastolic	Insufficient semilunar valve	A swishy diastolic murmur signifies that a valve that should be closed during diastole (a semilunar valve) does not close completely.

Cardiac Output and Its Control

Cardiac output (CO) is the volume of blood pumped by *each ventricle* per minute (not the total amount of blood pumped by the heart). During any period of time, the volume of blood flowing through the pulmonary circulation is the same as the volume flowing through the systemic circulation. Therefore, the cardiac output from each ventricle normally is the same, although on a beat-to-beat basis, minor variations may occur.

Cardiac output depends on the heart rate and the stroke volume.

The two determinants of cardiac output are *heart rate* (beats per minute) and *stroke volume* (volume of blood pumped per beat or stroke). The average resting heart rate is 70 beats per minute, established by SA node rhythmicity; the average resting stroke volume is 70 ml per beat, producing an average cardiac output of 4900 ml/min, or close to 5 liters/min:

$$\text{Cardiac output (CO)} = \text{heart rate} \times \text{stroke volume}$$
$$= 70 \text{ beats/min} \times 70 \text{ ml/beat}$$
$$= 4900 \text{ ml/min} \approx 5 \text{ liters/min}$$

Because the body's total blood volume averages 5 to 5.5 liters, each half of the heart pumps the equivalent of the entire blood volume each minute. In other words, each minute the right ventricle normally pumps 5 liters of blood through the lungs, and the left ventricle pumps 5 liters through the systemic circulation. At this rate, each half of the heart would pump about 2.5 million liters of blood in just one year. Yet this is only the resting cardiac output! During exercise, cardiac output can increase to 20 to 25 liters per minute, and outputs as high as 40 liters per minute have been recorded in trained athletes during heavy endurance-type exercise. The difference between the cardiac output at rest and the maximum volume of blood the heart can pump per minute is called the **cardiac reserve.**

How can cardiac output vary so tremendously, depending on the demands of the body? You can readily answer this question by thinking about how your own heart pounds rapidly (increased heart rate) and forcefully (increased stroke volume) when you engage in strenuous physical activities (need for increased cardiac output). Thus, regulation of cardiac output depends on the control of both heart rate and stroke volume, topics that we discuss next.

Heart rate is determined primarily by autonomic influences on the SA node.

The SA node is normally the pacemaker of the heart because it has the fastest spontaneous rate of depolarization to threshold. Recall that this automatic gradual reduction of membrane potential between beats is the result of a complex interplay of ion movements involving an increase in Na^+ permeability, a reduction in K^+ permeability, and an increase in Ca^{2+} permeability. When the SA node reaches threshold, an action potential is initiated that spreads throughout the heart, inducing the heart to contract, or have a "heartbeat." This happens about 70 times per minute, setting the average heart rate at 70 beats per minute.

The heart is innervated by both divisions of the autonomic nervous system, which can modify the rate (as well as the strength) of contraction, even though nervous stimulation is not required to initiate contraction. The parasympathetic nerve to the heart, the **vagus nerve,** primarily supplies the atrium, especially the SA and AV nodes. Parasympathetic innervation of the ventricles is sparse. The cardiac sympathetic nerves also supply the atria, including the SA and AV nodes, and richly innervate the ventricles as well.

Effects of the Autonomic Nervous System on the Heart and Structures That Influence the Heart

Area Affected	Effect of Parasympathetic Stimulation	Effect of Sympathetic Stimulation
SA Node	Decreases the rate of depolarization to threshold; decreases the heart rate	Increases the rate of depolarization to threshold; increases the heart rate
AV Node	Decreases excitability; increases the AV nodal delay	Increases excitability; decreases the AV nodal delay
Ventricular Conduction Pathway	No effect	Increases excitability; hastens conduction through the bundle of His and Purkinje cells
Atrial Muscle	Decreases contractility; weakens contraction	Increases contractility; strengthens contraction
Ventricular Muscle	No effect	Increases contractility; strengthens contraction
Adrenal Medulla (an Endocrine Gland)	No effect	Promotes adrenomedullary secretion of epinephrine, a hormone that augments the sympathetic nervous system's actions on the heart
Veins	No effect	Increases venous return, which increases the strength of cardiac contraction through the Frank–Starling mechanism

Both the parasympathetic and sympathetic nervous system bring about their effects on the heart primarily by altering the activity of the cyclic AMP second-messenger system in the innervated cardiac cells. Acetylcholine released from the vagus nerve binds to a muscarinic receptor and is coupled to an inhibitory G protein that reduces activity of the cyclic AMP pathway (see pp. 243, 117, and 121). By contrast, the sympathetic neurotransmitter norepinephrine binds with a β_1 adrenergic receptor and is coupled to a stimulatory G protein that accelerates the cyclic AMP pathway in the target cells (see p. 244). The cAMP pathway leads to phosphorylation and altered activity of various proteins within cardiac muscle, for example, keeping channels open longer.

EFFECT OF PARASYMPATHETIC STIMULATION ON THE HEART
Let's examine the specific effects that parasympathetic and sympathetic stimulation have on the heart (▲ Table 9-3).

■ The parasympathetic nervous system's influence on the SA node is to decrease the heart rate (● Figure 9-19). In a different mechanism than its usual reduction in cAMP activity, ACh slows heart rate primarily by increasing K^+ permeability of the pacemaker cells in the SA node by binding with muscarinic cholinergic receptors that are coupled directly to ACh-regulated K^+ channels by a G protein. As a result, the rate at which spontaneous action potentials are initiated is reduced through a twofold effect:

1. Enhanced K^+ permeability hyperpolarizes the SA node membrane because more positive potassium ions leave than normal, making the inside even more negative. Because the "resting" potential starts even farther away from threshold, it takes longer to reach threshold.

2. The enhanced K^+ permeability induced by vagal stimulation also opposes the automatic reduction in K^+ permeability that contributes to development of the pacemaker potential. This countering effect decreases the rate of spontaneous depolarization, prolonging the time required to drift to threshold. Acetylcholine, by inhibiting the cAMP pathway, also depresses both the inward movement of Na^+ and Ca^{2+} through the I_f and T-type channels, respectively, further slowing the depolarization to threshold. Therefore, the SA node reaches threshold and fires less frequently, decreasing the heart rate.

■ Parasympathetic influence on the AV node decreases the node's excitability, prolonging transmission of impulses to the ventricles even longer than the usual AV nodal delay. This effect is brought about by increasing K^+ permeability, which hyperpolarizes the membrane, thereby retarding the initiation of excitation in the AV node.

■ Parasympathetic stimulation of the atrial contractile cells shortens the action potential, reducing the slow inward current carried by Ca^{2+}; that is, the plateau phase is shortened. As a result, atrial contraction is weakened.

■ The parasympathetic system has little effect on ventricular contraction, because of the sparseness of parasympathetic innervation to the ventricles. Thus, the heart is more "leisurely" under parasympathetic influence—it beats less rapidly, the time between atrial and ventricular contraction is stretched out, and atrial contraction is weaker. These actions are appropriate, considering that the parasympathetic system controls heart action in quiet, relaxed situations when the body is not demanding an enhanced cardiac output.

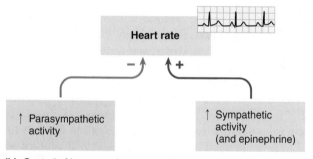

(a) Autonomic influence on SA node potential

(b) Control of heart rate by autonomic nervous system

● **FIGURE 9-19 Autonomic control of SA node activity and heart rate.** (a) Parasympathetic stimulation decreases the rate of SA nodal depolarization so that the membrane reaches threshold more slowly and has fewer action potentials, whereas sympathetic stimulation increases the rate of depolarization of the SA node so that the membrane reaches threshold more rapidly and has more frequent action potentials. (b) Because each SA node action potential ultimately leads to a heartbeat, increased parasympathetic activity decreases the heart rate, whereas increased sympathetic activity increases the heart rate.

EFFECT OF SYMPATHETIC STIMULATION ON THE HEART In contrast, the sympathetic nervous system, which controls heart action in emergency or exercise situations, when there is a need for greater blood flow, speeds up the heart rate through its effect on the pacemaker tissue.

■ The main effect of sympathetic stimulation on the SA node is to speed up depolarization so that threshold is reached more rapidly. In pacemaker cells, the rate of depolarization increases as a result of the greater inward movement of Na^+ and Ca^{2+} through the augmented I_f and T-type Ca^{2+} channels. This swifter drift to threshold under sympathetic influence permits more frequent action potentials and a correspondingly faster heart rate (● Figure 9-19 and ▲ Table 9-3).

■ Sympathetic stimulation of the AV node reduces the AV nodal delay by increasing conduction velocity, presumably by enhancing the slow, inward Ca^{2+} current.

■ Similarly, sympathetic stimulation speeds up spread of the action potential throughout the specialized conduction pathway.

■ In the atrial and ventricular contractile cells, both of which have many sympathetic nerve endings, sympathetic stimulation increases contractile strength so the heart beats more forcefully and squeezes out more blood. This effect is produced by increasing inward Ca^{2+} movement through prolonged opening of L-type Ca^{2+} channels, which enhances the slow Ca^{2+} influx and intensifies Ca^{2+} participation in excitation–contraction coupling.

■ Sympathetic stimulation not only increases the speed of contraction by allowing greater influx of Ca^{2+} into the cell through L-type Ca^{2+} channels but it also speeds up relaxation by enhancing the active Ca^{2+} pump in the sarcoplasmic reticulum that removes Ca^{2+} from the cytosol (see p. 268).

The overall effect of sympathetic stimulation on the heart, therefore, is to improve its effectiveness as a pump by increasing heart rate, decreasing the delay between atrial and ventricular contraction, decreasing conduction time throughout the heart, increasing the force of contraction, and speeding up the relaxation process so that more time is available for filling; that is, sympathetic stimulation "revs up" the heart.

CONTROL OF HEART RATE Thus, as is typical of the autonomic nervous system, parasympathetic and sympathetic effects on heart rate are antagonistic (oppose each other). At any given moment, heart rate is determined largely by the balance between inhibition of the SA node by the vagus nerve and stimulation by the cardiac sympathetic nerves. Under resting conditions, parasympathetic discharge dominates because acetylcholine suppresses sympathetic activity by inhibiting the release of norepinephrine from neighboring sympathetic nerve endings. In fact, if all autonomic nerves to the heart were blocked, the resting heart rate would increase from its average value of 70 beats per minute to about 100 beats per minute, which is the inherent rate of the SA node's spontaneous discharge when not subjected to any nervous influence. (We use 70 beats per minute as the normal rate of SA node discharge because this is the average rate under normal resting conditions when parasympathetic activity dominates.) The heart rate can be altered beyond this resting level in either direction by shifting the balance of autonomic nervous stimulation. Heart rate is speeded up by simultaneously increasing sympathetic and decreasing parasympathetic activity; heart rate is slowed by a concurrent rise in parasympathetic activity and decline in sympathetic activity. The relative level of activity in these two autonomic branches to the heart in turn is primarily coordinated by the *cardiovascular control center* in the brain stem.

● FIGURE 9-20 Intrinsic and extrinsic control of stroke volume.

Although autonomic innervation is the primary means by which heart rate is regulated, other factors affect it as well. The most important is epinephrine, a hormone secreted into the blood from the adrenal medulla on sympathetic stimulation. Epinephrine acts in a manner similar to norepinephrine to increase heart rate, thus reinforcing the direct effect that the sympathetic nervous system has on the heart.

Stroke volume is determined by the extent of venous return and by sympathetic activity.

The other component besides heart rate that determines cardiac output is stroke volume, the amount of blood pumped out by each ventricle during each beat. Two types of controls influence stroke volume: (1) *intrinsic control* related to the extent of venous return and (2) *extrinsic control* related to the extent of sympathetic stimulation of the heart. Both factors increase stroke volume by increasing the strength of heart contraction (● Figure 9-20). Let us examine each of these mechanisms in detail.

Increased end-diastolic volume results in increased stroke volume.

Intrinsic control of stroke volume, which refers to the heart's inherent ability to vary stroke volume, depends on the direct correlation between end-diastolic volume and stroke volume. As more blood returns to the heart, the heart pumps out more blood, but the relationship is not quite as simple as might seem, because the heart does not eject all the blood it contains. This intrinsic control depends on the length–tension relationship of cardiac muscle, which is similar to that of skeletal muscle. For skeletal muscle, the resting muscle length is approximately the optimal length (l_o) at which maximal tension can be developed during a subsequent contraction. When the skeletal muscle is

longer or shorter than l_o, the subsequent contraction is weaker (see ● Figure 8-19, p. 273). For cardiac muscle, the resting cardiac muscle fiber length is less than l_o. Therefore, the length of cardiac muscle fibers normally varies along the ascending limb of the length–tension curve. An increase in cardiac muscle fiber length, by moving closer to l_o, increases the contractile tension of the heart on the following systole (● Figure 9-21).

Unlike in skeletal muscle, the length–tension curve of cardiac muscle normally does not operate at lengths that fall within the region of the descending limb. That is, within physiologic limits cardiac muscle does not get stretched beyond its optimal length to the point that contractile strength diminishes with further stretching.

FRANK–STARLING LAW OF THE HEART What causes cardiac muscle fibers to vary in length before contraction? Skeletal muscle length can vary before contraction because of the positioning of the skeletal parts to which the muscle is attached, but cardiac muscle is not attached to any bones. The main determinant of cardiac muscle fiber length is the degree of diastolic filling. An analogy is a balloon filled with water—the more water you put in, the larger the balloon becomes, and the more it is stretched. Likewise, the greater the diastolic filling, the larger the end-diastolic volume (EDV), and the more the heart is stretched. The more the heart is stretched, the longer the initial cardiac fiber length before contraction. The increased length results in a greater force on the subsequent cardiac contraction and thus in a greater stroke volume. This intrinsic relationship between EDV and stroke volume is known as the **Frank–Starling law of the heart.** Stated simply, the law says that the heart normally pumps out during systole the volume of blood returned to it during diastole; increased venous return results in increased stroke volume. In ● Figure 9-21, assume that EDV increases from point A to point B. You can see that this increase in EDV is accompanied by a corresponding increase in stroke volume from point A[1] to point B[1]. The extent of filling is referred to as the **preload** because it is the workload imposed on the heart before contraction begins.

ADVANTAGES OF THE CARDIAC LENGTH–TENSION RELATIONSHIP The built-in relationship matching stroke volume with venous return has two important advantages. First, one of the most important functions of this intrinsic mechanism is equalizing output between the right and left sides of the heart so that blood pumped out by the heart is equally distributed between the pulmonary and systemic circulation. If, for example, the right side of the heart ejects a larger stroke volume, more blood enters the pulmonary circulation, so venous return to the left side of the heart increases accordingly. The increased EDV of the left side of the heart causes it to contract more forcefully, so it too pumps out a larger stroke volume. In this way, output of the two ventricular chambers is kept equal. If such equalization did not happen, too much blood would be dammed up in the venous system before the ventricle with the lower output.

Second, when a larger cardiac output is needed, such as during exercise, venous return is increased through action of the sympathetic nervous system and other mechanisms to be described in the next chapter. The resulting increase in EDV automatically increases stroke volume correspondingly. Because exercise also increases heart rate, these two factors act together to increase the cardiac output so that more blood can be delivered to the exercising muscles.

MECHANISM OF THE CARDIAC LENGTH–TENSION RELATIONSHIP Although the length–tension relationship in cardiac muscle fibers depends to a degree on the extent of overlap of thick and thin filaments, similar to the length–tension relationship in skeletal muscle, the key factor relating cardiac muscle fiber length to tension development is the dependence of myofilament Ca^{2+} sensitivity on the fiber's length. Specifically, as a cardiac muscle fiber's length increases along the ascending limb of the length–tension curve, the lateral spacing between adjacent thick and thin filaments is reduced. Stated differently, as a cardiac muscle fiber is stretched as a result of greater ventricular filling, its myofilaments are pulled closer together. As a result of this reduction in distance between the thick and thin filaments, more cross-bridge interactions between myosin and actin can take place when Ca^{2+} pulls the troponin–tropomyosin complex away from actin's cross-bridge binding sites—that is, myofilament Ca^{2+} sensitivity increases. Thus, the length–tension relationship in cardiac muscle depends not on muscle fiber length per se but on the resultant variations in the lateral spacing between the myosin and actin filaments.

We now shift attention from intrinsic control to extrinsic control of stroke volume.

Sympathetic stimulation increases the contractility of the heart.

In addition to intrinsic control, stroke volume is also subject to **extrinsic control** by factors originating outside the heart, the most important of which are actions of the cardiac sympathetic nerves and epinephrine (see ▲ Table 9-3). Sympathetic stimulation and epinephrine enhance the heart's **contractility,** which is the strength of contraction at any given EDV. In other words, on sympathetic stimulation the heart contracts more forcefully and squeezes out a greater percentage of the blood it contains, leading to more complete ejection. This increased contractility results from the increased Ca^{2+} influx triggered by norepinephrine and epinephrine. The extra cytosolic Ca^{2+} lets the myocardial fibers generate more force through greater cross-bridge cycling than they would without sympathetic influence. Normally, the EDV is 135 ml and the end-systolic volume (ESV) is 65 ml for a stroke volume of 70 ml (● Figure 9-22a). Under sympathetic influence, for the same EDV of 135 ml, the ESV might be 35 ml

● **FIGURE 9-21 Intrinsic control of stroke volume (Frank–Starling curve).** The cardiac muscle fiber's length, which is determined by the extent of venous filling, is normally less than the optimal length for developing maximal tension. Therefore, an increase in end-diastolic volume (that is, an increase in venous return), by moving the cardiac muscle fiber length closer to optimal length, increases the contractile tension of the fibers on the next systole. A stronger contraction squeezes out more blood. Thus, as more blood is returned to the heart and the end-diastolic volume increases, the heart automatically pumps out a correspondingly larger stroke volume.

● **FIGURE 9-22 Effect of sympathetic stimulation on stroke volume.**

and the stroke volume 100 ml (● Figure 9-22b). In effect, sympathetic stimulation shifts the Frank–Starling curve to the left (● Figure 9-23). Depending on the extent of sympathetic stimulation, the curve can be shifted to varying degrees, up to a maximal increase in contractile strength of about 100% greater than normal.

 The **ejection fraction** is the ratio of stroke volume to end-diastolic volume (ejection fraction = SV/EDV); that is, it is the proportion of the blood in the ventricle that is pumped out. The ejection fraction is often used clinically

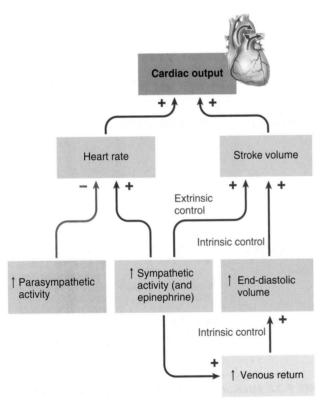

FIGURE 9-23 Shift of the Frank–Starling curve to the left by sympathetic stimulation. For the same end-diastolic volume (point A), there is a larger stroke volume (from point B to point C) on sympathetic stimulation as a result of increased contractility of the heart. The Frank–Starling curve is shifted to the left by variable degrees, depending on the extent of sympathetic stimulation.

FIGURE 9-24 Control of cardiac output. Because cardiac output equals heart rate times stroke volume, this figure is a composite of Figure 9-19b (control of heart rate) and Figure 9-20 (control of stroke volume).

as an indication of contractility. A healthy heart normally has an ejection fraction of 50% to 75% under resting conditions and may go as high as 90% during strenuous exercise, but a failing heart may pump out 30% or less.

Sympathetic stimulation increases stroke volume not only by strengthening cardiac contractility but also by enhancing

venous return (see ● Figure 9-22c). Sympathetic stimulation constricts the veins, which squeezes more blood forward from the veins to the heart, increasing the EDV and subsequently increasing stroke volume even further.

SUMMARY OF FACTORS AFFECTING STROKE VOLUME AND CARDIAC OUTPUT The strength of cardiac muscle contraction and accordingly the stroke volume can thus be graded by (1) varying the initial length of the muscle fibers, which in turn depends on the degree of ventricular filling before contraction (intrinsic control); and (2) varying the extent of sympathetic stimulation (extrinsic control) (see ● Figure 9-20). This is in contrast to gradation of skeletal muscle, in which twitch summation and recruitment of motor units produce variable strength of muscle contraction. These mechanisms do not apply to cardiac muscle. Twitch summation is impossible because of the long refractory period. Recruitment of motor units is not possible, because the heart muscle cells are arranged into functional syncytia where all contractile cells become excited and contract with every beat, instead of into distinct motor units that can be discretely activated.

All the factors that determine cardiac output by influencing heart rate or stroke volume are summarized in ● Figure 9-24. Note that sympathetic stimulation increases cardiac output by increasing both heart rate and stroke volume. Sympathetic activity to the heart increases, for example, during exercise when the working skeletal muscles need increased delivery of O_2-laden blood to support their high rate of ATP consumption.

We next examine how the afterload influences the ability of the heart to pump out blood, then how a failing heart cannot pump out enough blood, before we turn to the final section of this chapter, on how the heart muscle is nourished.

High blood pressure increases the workload of the heart.

When the ventricles contract, to force open the semilunar valves they must generate sufficient pressure to exceed the blood pressure in the major arteries. The arterial blood pressure is called the **afterload** because it is the workload imposed on the heart after the contraction has begun. If the arterial blood pressure is chronically elevated (high blood pressure) or if the exit valve is stenotic, the ventricle must generate more pressure to eject blood. For example, instead of generating the normal pressure of 120 mm Hg, the ventricular pressure may need to rise as high as 400 mm Hg to force blood through a narrowed aortic valve.

The heart may be able to compensate for a sustained increase in afterload by enlarging (through hypertrophy or enlargement of the cardiac muscle fibers; see p. 281). This enables it to contract more forcefully and maintain a normal stroke volume despite an abnormal impediment to ejection. A diseased heart or a heart weakened with age may not be able to compensate completely, however; in that case, heart failure ensues. Even if the heart is initially able to compensate for a chronic increase in afterload, the sustained extra workload placed on the heart can eventually cause pathological changes in the heart that lead to heart failure. In fact, a chronically elevated afterload is one of the two major factors that cause heart failure.

In heart failure the contractility of the heart decreases.

Clinical Note **Heart failure** is the inability of the cardiac output to keep pace with the body's demands for supplies and removal of wastes. Either one or both ventricles may progressively weaken and fail. When a failing ventricle cannot pump out all the blood returned to it, the veins behind the failing ventricle become congested with blood. Heart failure may occur for a variety of reasons, but the two most common are (1) damage to the heart muscle as a result of a heart attack or impaired circulation to the cardiac muscle and (2) prolonged pumping against a chronically increased afterload, as with a stenotic semilunar valve or a sustained elevation in blood pressure. Heart failure presently affects almost 5 million Americans, nearly 50% of whom will die within five years of diagnosis. About 500,000 new cases are diagnosed annually, with these numbers expected to rise as the population ages.

PRIME DEFECT IN HEART FAILURE The prime defect in heart failure is a decrease in cardiac contractility; that is, weakened cardiac muscle cells contract less effectively. The intrinsic ability of the heart to develop pressure and eject a stroke volume is reduced so that the heart operates on a lower length–tension curve (● Figure 9-25a). The Frank–Starling curve shifts downward and to the right such that for a given EDV, a failing heart pumps out a smaller stroke volume than a normal healthy heart.

COMPENSATORY MEASURES FOR HEART FAILURE In the early stages of heart failure, two major compensatory measures help restore stroke volume to normal. First, sympathetic activity to the heart is reflexly increased, which increases heart contractility toward normal (● Figure 9-25b). Sympathetic stimulation can help compensate only for a limited period of time, however, because the heart becomes less responsive to norepinephrine after prolonged exposure, and furthermore, norepinephrine stores in the heart's sympathetic nerve terminals become depleted. Second, when cardiac output is reduced, the kidneys, in a compensatory attempt to improve their reduced blood flow, retain extra salt and water in the body during urine formation, to expand the blood volume. The increase in circulating blood volume increases the EDV. The resultant stretching of the car-

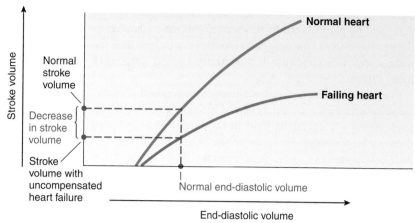

(a) Reduced contractility in a failing heart

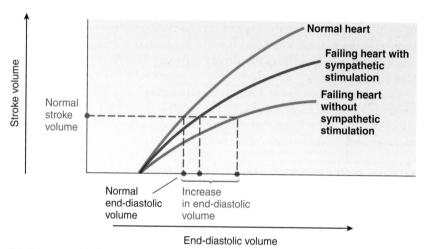

(b) Compensation for heart failure

● **FIGURE 9-25 Compensated heart failure.** (a) The Frank–Starling curve shifts downward and to the right in a failing heart. Because its contractility is decreased, the failing heart pumps out a smaller stroke volume at the same end-diastolic volume than a normal heart does. (b) During compensation for heart failure, reflex sympathetic stimulation shifts the Frank–Starling curve of a failing heart to the left, increasing the contractility of the heart toward normal. A compensatory increase in end-diastolic volume as a result of blood volume expansion further increases the strength of contraction of the failing heart. Operating at a longer cardiac muscle fiber length, a compensated failing heart is able to eject a normal stroke volume.

diac muscle fibers enables the weakened heart to pump out a normal stroke volume (● Figure 9-25b). The heart is now pumping out the blood returned to it but is operating at a greater cardiac muscle fiber length.

DECOMPENSATED HEART FAILURE As the disease progresses and heart contractility deteriorates further, the heart reaches a point at which it can no longer pump out a normal stroke volume (that is, cannot pump out all the blood returned to it) despite compensatory measures. At this point, the heart slips from compensated heart failure into a state of decompensated heart failure. Now the cardiac muscle fibers are stretched to the point that they are operating in the descending limb of the length–tension curve. *Forward failure* occurs as the heart fails to pump an adequate amount of blood forward to the tissues because the stroke volume becomes progressively smaller.

Backward failure occurs simultaneously as blood that cannot enter and be pumped out by the heart continues to dam up in the venous system. The congestion in the venous system is the reason why this condition is sometimes termed **congestive heart failure.**

Left-sided failure has more serious consequences than right-sided failure. Backward failure of the left side leads to pulmonary edema (excess tissue fluid in the lungs) because blood dams up in the lungs. This fluid accumulation in the lungs reduces exchange of O_2 and CO_2 between the air and blood in the lungs, reducing arterial oxygenation and elevating levels of acid-forming CO_2 in the blood. In addition, one of the more serious consequences of left-sided forward failure is an inadequate blood flow to the kidneys, which causes a twofold problem. First, vital kidney function is depressed; and second, the kidneys retain even more salt and water in the body during urine formation as they try to expand the plasma volume even further to improve their reduced blood flow. Excessive fluid retention further exacerbates the already existing problems of venous congestion.

Treatment of congestive heart failure therefore includes measures that reduce salt and water retention and increase urinary output as well as drugs that enhance the contractile ability of the weakened heart—digitalis, for example.

SYSTOLIC VERSUS DIASTOLIC HEART FAILURE Increasingly physicians categorize heart failure as either *systolic failure* characterized by a decrease in cardiac contractility as just described or *diastolic failure* in which the heart has trouble filling. Diastolic failure is a recently recognized problem. With diastolic failure, the ventricles do not fill normally either because the heart muscle does not adequately relax between beats or because the heart muscle stiffens and cannot expand as much as usual. Because of impeded filling, a diastolic failing heart pumps out less blood than it should with each contraction. No drugs are available yet that reliably help the heart relax, so treatment is aimed at relieving symptoms or halting underlying causes of diastolic disease.

Nourishing the Heart Muscle

Cardiac muscle cells contain an abundance of mitochondria, the O_2-dependent energy organelles. In fact, up to 40% of the cell volume of cardiac muscle cells is occupied by mitochondria, indicative of how much the heart depends on O_2 delivery and aerobic metabolism to generate the energy necessary for contraction (see p. 37). Cardiac muscle also has an abundance of myoglobin, which stores limited amounts of O_2 within the heart for immediate use (see p. 278).

The heart receives most of its own blood supply through the coronary circulation during diastole.

Although all the blood passes through the heart, the heart muscle cannot extract O_2 or nutrients from the blood within its chambers for two reasons. First, the watertight endothelial lin-

● **FIGURE 9-26 Coronary blood flow.** Most coronary blood flow occurs during diastole because the coronary vessels are compressed almost completely closed during systole.

ing does not permit blood to pass from the chamber into the myocardium. Second, the heart walls are too thick to permit diffusion of O_2 and other supplies from the blood in the chamber to the individual cardiac cells. Therefore, like other tissues of the body, heart muscle must receive blood through blood vessels, specifically via the **coronary circulation.** The coronary arteries branch from the aorta just beyond the aortic valve (see ● Figure 9-30, p. 335), and the coronary veins empty into the right atrium.

The heart muscle receives most of its blood supply during diastole. Blood flow to the heart muscle cells is substantially reduced during systole for two reasons. First, the contracting myocardium, especially in the powerful left ventricle, compresses the major branches of the coronary arteries; and second, the open aortic valve partially blocks the entrance to the coronary vessels. Thus, most coronary arterial flow (about 70%) occurs during diastole, driven by the aortic blood pressure, with flow declining as aortic pressure drops. Only about 30% of coronary arterial flow occurs during systole (● Figure 9-26).

This limited time for coronary blood flow becomes especially important during rapid heart rates, when diastolic time is much reduced. Just when increased demands are placed on the heart to pump more rapidly, it has less time to provide O_2 and nourishment to its own musculature to accomplish the increased workload.

MATCHING OF CORONARY BLOOD FLOW TO HEART MUSCLE'S O_2 NEEDS Nevertheless, under normal circumstances the heart muscle does receive adequate blood flow to support its activities, even during exercise, when the rate of coronary blood flow increases up to five times its resting rate. Let us see how.

Extra blood is delivered to the cardiac cells primarily by vasodilation, or enlargement, of the coronary vessels, which lets more blood flow through them, especially during diastole. The increased coronary blood flow is necessary to meet the heart's increased O_2 requirements because the heart, unlike most other tissues, is unable to remove much additional O_2 from the blood passing through its vessels to support increased metabolic activities. Most other tissues under resting conditions extract only about 25% of the O_2 available from the blood flowing through them, leaving a considerable O_2 reserve that can be drawn on when a tissue has increased O_2 needs; that is, the tissue can immediately increase the O_2 available to it by removing a greater percentage of O_2 from the blood passing through it. In contrast, the heart, even under resting conditions, removes up to 65% of the O_2 available in the coronary vessels, far more than is withdrawn by other tissues. This leaves little O_2 in reserve in the coronary blood should cardiac O_2 demands increase. Therefore, the primary means by which more O_2 can be made available to the heart muscle is by increasing coronary blood flow.

Coronary blood flow is adjusted primarily in response to changes in the heart's O_2 requirements. Among the proposed links between blood flow and O_2 needs is *adenosine,* which is formed from adenosine triphosphate (ATP) during cardiac metabolic activity. Cardiac cells form and release more adenosine when cardiac activity increases, and the heart accordingly needs more O_2 and is using more ATP as an energy source. The released adenosine, acting as a paracrine (see p. 114), induces dilation of the coronary blood vessels, allowing more O_2-rich blood to flow to the more active cardiac cells to meet their increased O_2 demand (● Figure 9-27). Matching O_2 delivery to O_2 needs is crucial because heart muscle depends on oxidative processes to generate energy. The heart cannot get enough ATP through anaerobic metabolism.

NUTRIENT SUPPLY TO THE HEART Although the heart has little ability to support its energy needs by means of anaerobic metabolism and must rely heavily on its O_2 supply, it can tolerate wide variations in its nutrient supply. As fuel sources, the heart primarily uses free fatty acids and, to a lesser extent, glucose and lactate, depending on their availability. Because cardiac muscle is remarkably adaptable and can shift metabolic pathways to use whatever nutrient is available, the primary danger of insufficient coronary blood flow is not fuel shortage but O_2 deficiency.

Atherosclerotic coronary artery disease can deprive the heart of essential oxygen.

Clinical Note Adequacy of coronary blood flow is relative to the heart's O_2 demands at any given moment. In the normal heart, coronary blood flow increases correspondingly as O_2 demands rise. With coronary artery disease, coronary blood flow may not be able to keep pace with rising O_2 needs. The term **coronary artery disease (CAD)** refers to pathological changes within the coronary artery walls that diminish blood flow through these vessels. A given rate of coronary blood flow may be adequate at rest but insufficient in physical exertion or other stressful situations.

Complications of CAD, including heart attacks, make it the single leading cause of death in the United States. CAD is the underlying cause of about 50% of all deaths in this country. CAD can cause myocardial ischemia and possibly lead to acute myocardial infarction by three mechanisms: (1) profound vascular spasm of the coronary arteries, (2) formation of atherosclerotic plaques, and (3) thromboembolism. We will discuss each in turn.

● **FIGURE 9-27 Matching of coronary blood flow to the O_2 need of cardiac muscle cells.**

VASCULAR SPASM **Vascular spasm** is an abnormal spastic constriction that transiently narrows the coronary vessels. Vascular spasms are associated with the early stages of CAD and are most often triggered by exposure to cold, physical exertion, or anxiety. The condition is reversible and usually does not last long enough to damage the cardiac muscle.

When too little O_2 is available in the coronary vessels, the endothelium (blood vessel lining) releases *platelet-activating factor (PAF)*. PAF, which exerts a variety of actions, was named for its first discovered effect, activating platelets. Among its other effects, PAF, once released from the endothelium, diffuses to the underlying vascular smooth muscle and causes it to contract, bringing about vascular spasm.

DEVELOPMENT OF ATHEROSCLEROSIS **Atherosclerosis** is a progressive, degenerative arterial disease that leads to occlusion (gradual blockage) of affected vessels, reducing blood flow through them. Atherosclerosis is characterized by plaques forming beneath the vessel lining within arterial walls. An **atherosclerotic plaque** consists of a lipid-rich core covered by an abnormal overgrowth of smooth muscle cells, topped off by a collagen-rich connective tissue cap. As the plaque forms, it bulges into the vessel lumen (● Figure 9-28).

Although not all the contributing factors have been identified, in recent years investigators have sorted out the following complex sequence of events in the gradual development of atherosclerosis:

1. Atherosclerosis starts with injury to the blood vessel wall, which triggers an *inflammatory response* that sets the stage for plaque buildup. Normally inflammation is a protective response that fights infection and promotes repair of damaged tissue (see p. 420). However, when the cause of the injury persists within the vessel wall, the sustained, low-grade inflammatory response

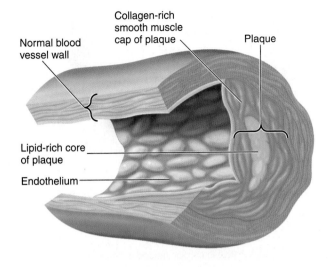

Normal blood vessel wall

Collagen-rich smooth muscle cap of plaque

Plaque

Lipid-rich core of plaque

Endothelium

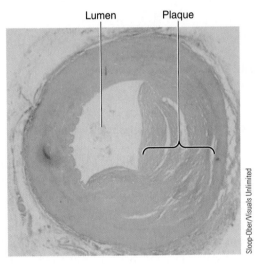

Lumen

Plaque

Sloop-Ober/Visuals Unlimited

● **FIGURE 9-28 Atherosclerotic plaque in a coronary vessel.**

over a course of decades can insidiously lead to arterial plaque formation and heart disease. Plaque formation likely has many causes. Suspected artery-abusing agents that may set off the vascular inflammatory response include oxidized cholesterol, free radicals, high blood pressure, homocysteine, chemicals released from fat cells, or even bacteria and viruses that damage blood vessel walls. The most common triggering agent appears to be oxidized cholesterol. (For a further discussion of the role of cholesterol and other factors in the development of atherosclerosis, see the boxed feature on pages 336–337, ■ Concepts, Challenges, and Controversies.)

2. Typically, the initial stage of atherosclerosis is characterized by the accumulation beneath the endothelium of excessive amounts of *low-density lipoprotein (LDL),* the so-called bad cholesterol, in combination with a protein carrier. As LDL accumulates within the vessel wall, this cholesterol product becomes oxidized, primarily by oxidative wastes produced by the blood vessel cells. These wastes are *free radicals,* very unstable electron-deficient particles that are highly reactive. Antioxidant vitamins that prevent LDL oxidation, such as *vitamin E,*

vitamin C, and *beta-carotene,* have been shown to slow plaque deposition.

3. In response to the presence of oxidized LDL and/or other irritants, the endothelial cells produce chemicals that attract *monocytes,* a type of white blood cell, to the site. These immune cells trigger a local inflammatory response.

4. Once they leave the blood and enter the vessel wall, monocytes settle down permanently, enlarge, and become large phagocytic cells called *macrophages.* Macrophages voraciously phagocytize (see p. 31) the oxidized LDL until these cells become so packed with fatty droplets that they appear foamy under a microscope. Now called *foam cells,* these greatly engorged macrophages accumulate beneath the vessel lining and form a visible *fatty streak,* the earliest form of an atherosclerotic plaque.

5. Thus, the earliest stage of plaque formation is characterized by accumulation beneath the endothelium of a cholesterol-rich deposit. The disease progresses as smooth muscle cells within the blood vessel wall migrate from the muscular layer of the blood vessel to a position on top of the lipid accumulation, just beneath the endothelium. This migration is triggered by chemicals released at the inflammatory site. At their new location, the smooth muscle cells continue to divide and enlarge, producing *atheromas,* which are benign (noncancerous) tumors of smooth muscle cells within the blood vessel walls. Together the lipid-rich core and overlying smooth muscle form a maturing plaque.

6. As it continues to develop, the plaque progressively bulges into the lumen of the vessel. The protruding plaque narrows the opening through which blood can flow.

7. Further contributing to vessel narrowing, oxidized LDL inhibits release of *nitric oxide* from the endothelial cells. Nitric oxide is a local chemical messenger that relaxes the underlying layer of normal smooth muscle cells within the vessel wall. Relaxation of these smooth muscle cells dilates the vessel. Because of reduced nitric oxide release, vessels damaged by developing plaques cannot dilate as readily as normal.

8. A thickening plaque also interferes with nutrient exchange for cells located within the involved arterial wall, leading to degeneration of the wall in the vicinity of the plaque. The damaged area is invaded by *fibroblasts* (scar-forming cells), which form a collagen-rich connective tissue cap over the plaque. (The term *sclerosis* means "excessive growth of fibrous connective tissue," hence the term *atherosclerosis* for this condition characterized by atheromas and sclerosis, along with abnormal lipid accumulation.)

9. In the later stages of the disease, Ca²⁺ often precipitates in the plaque. A vessel so afflicted becomes hard and cannot distend easily.

THROMBOEMBOLISM AND OTHER COMPLICATIONS OF ATHEROSCLEROSIS Atherosclerosis attacks arteries throughout the body, but the most serious consequences involve damage to the vessels of the brain and heart. In the brain, atherosclerosis is the prime cause of strokes, whereas in the heart it brings about myocardial ischemia and its complications. The following are potential complications of coronary atherosclerosis:

- *Angina pectoris.* Gradual enlargement of protruding plaque continues to narrow the vessel lumen and progressively diminishes coronary blood flow, triggering increasingly frequent bouts of transient myocardial ischemia as the ability to match blood flow with cardiac O_2 needs becomes more limited. Although the heart cannot normally be "felt," pain is associated with myocardial ischemia. Such cardiac pain, known as **angina pectoris** ("pain of the chest"), can be felt beneath the sternum and is often referred to (appears to come from) the left shoulder and down the left arm (see p. 176). The symptoms of angina pectoris recur whenever cardiac O_2 demands become too great in relation to the coronary blood flow—for example, during exertion or emotional stress. The pain is thought to result from stimulation of cardiac nerve endings by the accumulation of lactate when the heart shifts to its limited ability to perform anaerobic metabolism (see p. 278). The ischemia associated with the characteristically brief angina attacks is usually temporary and reversible and can be relieved by rest, taking vasodilator drugs such as *nitroglycerin,* or both. Nitroglycerin brings about coronary vasodilation by being metabolically converted to nitric oxide, which in turn relaxes the vascular smooth muscle.

- *Thromboembolism.* The enlarging atherosclerotic plaque can break through the weakened endothelial lining that covers it, exposing blood to the underlying collagen in the collagen-rich connective tissue cap of the plaque. Foam cells release chemicals that can weaken the fibrous cap of a plaque by breaking down the connective tissue fibers. Plaques with thick fibrous caps are considered stable because they are not likely to rupture. However, plaques with thinner fibrous caps are unstable because they are likely to rupture and trigger clot formation.

Blood platelets (formed elements of the blood involved in plugging vessel defects and in clot formation) normally do not adhere to smooth, healthy vessel linings. However, when platelets contact collagen at the site of vessel damage, they stick to the site and help promote the formation of a blood clot. Furthermore, foam cells produce a potent clot promoter. Such an abnormal clot attached to a vessel wall is called a **thrombus.** The thrombus may enlarge gradually until it completely blocks the vessel at that site, or the continued flow of blood past the thrombus may break it loose. As it heads downstream, such a freely floating clot, or **embolus,** may completely plug a smaller vessel (● Figure 9-29). Thus, through **thromboembolism** atherosclerosis can result in a gradual or sudden occlusion of a coronary vessel (or any other vessel).

- *Heart attack.* When a coronary vessel is completely plugged, the cardiac tissue served by the vessel soon dies from O_2 deprivation, and a heart attack occurs, unless the area can be supplied with blood from nearby vessels.

(a) Thrombus   (b) Embolus   (c) Thromboembolic lesion

● **FIGURE 9-29 Consequences of thromboembolism.** (a) A thrombus may enlarge gradually until it completely occludes the vessel at that site. (b) A thrombus may break loose from its attachment, forming an embolus that may completely occlude a smaller vessel downstream. (c) Scanning electron micrograph of a vessel completely occluded by a thromboembolic lesion.

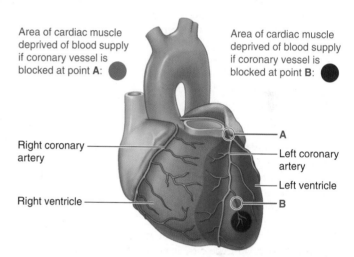

● **FIGURE 9-30 Extent of myocardial damage as a function of the size of the occluded vessel.**

Sometimes a deprived area is lucky enough to receive blood from more than one pathway. **Collateral circulation** exists when small terminal branches from adjacent blood vessels nourish the same area. These accessory vessels cannot develop suddenly after an abrupt blockage but may be lifesaving if already developed. Such alternate vascular pathways often develop over a period of time when an atherosclerotic constriction progresses slowly, or they may be induced by sustained demands on the heart through regular aerobic exercise.

In the absence of collateral circulation, the extent of the damaged area during a heart attack depends on the size of the blocked vessel: The larger the vessel occluded, the greater the area deprived of blood supply. As ● Figure 9-30 illustrates, a blockage at point A in the coronary circulation would cause more extensive damage than a blockage at point B would. Because there are only two major coronary arteries, complete blockage of either one of these main branches results in extensive myocardial damage. Left coronary artery blockage is most devastating because this vessel supplies blood to 85% of the cardiac tissue.

Atherosclerosis: Cholesterol and Beyond

The cause of atherosclerosis is still not entirely clear. Certain high-risk factors have been associated with an increased incidence of atherosclerosis and coronary heart disease. Included among them are genetic predisposition, obesity, advanced age, smoking, hypertension, lack of exercise, high blood concentrations of C-reactive protein, elevated levels of homocysteine, infectious agents, and, most notoriously, elevated cholesterol levels in the blood.

Sources of Cholesterol

There are two sources of cholesterol for the body: (1) dietary intake of cholesterol, with animal products such as egg yolk, red meats, and butter being especially rich in this lipid (animal fats contain cholesterol, whereas plant fats typically do not); and (2) manufacture of cholesterol by cells, especially liver cells.

"Good" versus "Bad" Cholesterol

Actually, it is not the total blood cholesterol level but the amount of cholesterol bound to various plasma protein carriers that is most important with regard to the risk of developing atherosclerotic heart disease. Because cholesterol is a lipid, it is not very soluble in blood. Most cholesterol in the blood is attached to specific plasma protein carriers in the form of lipoprotein complexes, which are soluble in blood. The three major lipoproteins are named for their density of protein as compared to lipid: (1) **high-density lipoproteins (HDL),** which contain the most protein and least cholesterol; (2) **low-density lipoproteins (LDL),** which have less protein and more cholesterol; and (3) **very-low-density lipoproteins (VLDL),** which have the least protein and most lipid, but the lipid they carry is neutral fat, not cholesterol.

Cholesterol carried in LDL complexes has been termed "bad" cholesterol, because cholesterol is transported to the cells, including those lining the blood vessel walls, by LDL. The propensity toward developing atherosclerosis substantially increases with elevated levels of LDL. The presence of oxidized LDL within an arterial wall is a major trigger for the inflammatory process that leads to the development of atherosclerotic plaques (see p. 333).

In contrast, cholesterol carried in HDL complexes has been dubbed "good" cholesterol, because HDL removes cholesterol from the cells and transports it to the liver for partial elimination from the body. Not only does HDL help remove excess cholesterol from the tissues but, in addition, it protects by inhibiting oxidation of LDL. Furthermore, recent research suggests that HDL also has anti-inflammatory action, helps stabilize atherosclerotic plaques so they are less prone to rupture, and reduces clot formation, all actions that counter the progressive development of atherosclerosis. The risk of atherosclerosis is inversely related to the concentration of HDL in the blood; that is, elevated levels of HDL are associated with a low incidence of atherosclerotic heart disease.

Some other factors known to influence atherosclerotic risk can be related to HDL levels; for example, cigarette smoking lowers HDL, whereas regular exercise raises HDL.

Cholesterol Uptake by Cells

Unlike most lipids, cholesterol is not used as metabolic fuel by cells. Instead, it serves as an essential component of plasma membranes. In addition, a few special cell types use cholesterol as a precursor for the synthesis of secretory products, such as steroid hormones and bile salts. Although most cells can synthesize some of the cholesterol needed for their own plasma membranes, they cannot manufacture sufficient amounts and therefore must rely on supplemental cholesterol being delivered by the blood.

Cells accomplish cholesterol uptake from the blood by synthesizing receptor proteins specifically capable of binding LDL and inserting these receptors into the plasma membrane. When an LDL particle binds to one of the membrane receptors, the cell engulfs the particle by receptor-mediated endocytosis, receptor and all (see p. 31). Within the cell, lysosomal enzymes break down the LDL to free the cholesterol, making it available to the cell for synthesis of new cellular membrane. The LDL receptor, which is also freed within the cell, is recycled back to the surface membrane.

If too much free cholesterol accumulates in the cell, both the synthesis of LDL receptor proteins (so that less cholesterol is taken up) and the cell's own cholesterol synthesis (so that less new cholesterol is made) are shut down. Faced with a cholesterol shortage, in contrast, the cell makes more LDL receptors so that it can engulf more cholesterol from the blood.

Maintenance of Blood Cholesterol Level and Cholesterol Metabolism

The maintenance of a blood-borne cholesterol supply to the cells involves an interaction between dietary cholesterol and the synthesis of cholesterol by the liver. When the amount of dietary cholesterol is increased, hepatic (liver) synthesis of cholesterol is turned off because cholesterol in the blood directly inhibits a hepatic enzyme essential for cholesterol synthesis. Thus, as more cholesterol is ingested, less is produced by the liver. Conversely, when cholesterol intake from food is reduced, the liver synthesizes more of this lipid because the inhibitory effect of cholesterol on the crucial hepatic enzyme is removed. In this way, the blood concentration of cholesterol is maintained at a fairly constant level despite changes in cholesterol intake; thus, it is difficult to significantly reduce cholesterol levels in the blood by decreasing cholesterol intake.

HDL transports cholesterol to the liver. The liver secretes cholesterol as well as cholesterol-derived bile salts into the bile. Bile enters the intestinal tract, where bile salts participate in the digestive process. Most of the secreted cholesterol and bile salts are subsequently reabsorbed from the intestine into the blood to be recycled to the liver. However, the cholesterol and bile salts not reclaimed by absorption are eliminated in the feces and lost from the body.

Thus, the liver has a primary role in determining total blood cholesterol levels, and the interplay between LDL and HDL determines the traffic flow of cholesterol between the liver and the other cells. Whenever these mechanisms are altered, blood cholesterol levels may be affected in such a way as to influence the individual's predisposition to atherosclerosis.

Varying the intake of dietary fatty acids may alter total blood cholesterol levels by influencing one or more of the mechanisms involving cholesterol balance. The blood cholesterol level tends to be raised by ingesting saturated fatty acids found predominantly in animal fats because these fatty acids stimulate cholesterol synthesis and inhibit its conversion to bile salts. In contrast, ingesting polyunsaturated fatty acids, the predominant fatty acids of most plants, tends to reduce blood cholesterol levels by enhancing elimination of both cholesterol and cholesterol-derived bile salts in the feces.

Other Risk Factors besides Cholesterol

Despite the strong links between cholesterol and heart disease, over half of all patients with heart attacks have a normal cholesterol profile and no other well-established risk factors. Clearly, other factors are involved in the development of coronary artery disease in these people. These same factors may also contribute to development of atherosclerosis in people with unfavorable cholesterol levels. Following are among the leading other possible risk factors:

- Elevated blood levels of the amino acid **homocysteine** have recently been implicated as a strong predictor for heart disease, independent of the person's cholesterol/lipid profile. Homocysteine is formed as an intermediate product during metabolism of the essential dietary amino acid *methionine*. Investigators believe homocysteine contributes to atherosclerosis by promoting proliferation of vascular smooth muscle cells, an early step in development of this artery-clogging condition. Furthermore, homocysteine appears to damage endothelial cells and may cause oxidation of LDL, both of which can contribute to plaque formation. Three B vitamins—*folic acid, vitamin B$_{12}$,* and *vitamin B$_6$*—all play key roles in pathways that clear homocysteine from the blood. Therefore, these B vitamins are all needed to keep blood homocysteine at safe levels.

- An indication of the role of inflammation in forming atherosclerotic plaques is that people with elevated levels of **C-reactive protein,** a blood-borne marker of inflammation, have a higher risk for developing coronary artery disease. In one study, people with a high level of C-reactive protein in their blood were three times more likely to have a heart attack over the next 10 years than those with a low level of this inflammatory protein. Because inflammation plays a crucial role in the development of atherosclerosis, anti-inflammatory drugs, such as aspirin, help prevent heart attacks. Furthermore, aspirin protects against heart attacks through its role in inhibiting clot formation.

- Accumulating data suggest that an infectious agent may be the underlying culprit in a significant number of cases of atherosclerotic disease. Among the leading suspects are respiratory-infection-causing *Chlamydia pneumoniae,* cold-sore-causing *herpes virus,* and gum-disease-causing bacteria. Importantly, if a link between infections and coronary artery disease can be confirmed, antibiotics may be added to the regimen of heart disease prevention strategies.

As you can see, the relationship between atherosclerosis, cholesterol, and other factors is far from clear. Much research on this complex disease is currently in progress because the incidence of atherosclerosis is so high and its consequences are potentially fatal.

A heart attack has four possible outcomes: immediate death, delayed death from complications, full functional recovery, or recovery with impaired function (▲ Table 9-4).

The discovery in 2006 of cardiac stem cells, coupled with earlier studies demonstrating that injecting damaged hearts with stem cells derived from other sources improves cardiac function, has generated hopes of future regenerative therapies. Even though the heart has the potential to produce new myocardial cells on its own through activity of its stem cells, this regenerative ability is apparently only minimally functional or nonfunctional, because

▲ TABLE 9-4	Possible Outcomes of Acute Myocardial Infarction (Heart Attack)		
Immediate Death	**Delayed Death from Complications**	**Full Functional Recovery**	**Recovery with Impaired Function**
Acute cardiac failure occurring because the heart is too weakened to pump effectively to support the body tissues	Fatal rupture of the dead, degenerating area of the heart wall	Replacement of the damaged area with a strong scar, accompanied by enlargement of the remaining normal contractile tissue to compensate for the lost cardiac musculature	Persistence of permanent functional defects, such as bradycardia or conduction blocks, caused by destruction of irreplaceable autorhythmic or conductive tissues
Fatal ventricular fibrillation brought about by damage to the specialized conducting tissue or induced by O$_2$ deprivation	Slowly progressing congestive heart failure occurring because the weakened heart is unable to pump out all the blood returned to it		

damaged heart muscle cells are replaced with scar tissue, not new muscle cells. Scientists hope to find a way to spur the latent stem cells into action.

Chapter in Perspective: Focus on Homeostasis

Survival depends on continual delivery of needed supplies to all body cells and on ongoing removal of wastes generated by the cells. Furthermore, regulatory chemical messengers, such as hormones, must be transported from their production site to their action site, where they control a variety of activities, most of which are directed toward maintaining a stable internal environment. Finally, to maintain normal body temperature, excess heat produced during muscle contraction must be carried to the skin where the heat can be lost from the body surface.

The circulatory system contributes to homeostasis by serving as the body's transport system. It provides a way to rapidly move materials from one part of the body to another. Without the circulatory system, materials would not get where they need to go to support life-sustaining activities nearly rapidly enough. For example, O_2 would take months to years to diffuse from the body surface to internal organs, yet through the heart's swift pumping action the blood can pick up and deliver O_2 and other substances to all the cells in a few seconds.

The heart serves as a dual pump to continuously circulate blood between the lungs, where O_2 is picked up, and the other body tissues, which use O_2 to support their energy-generating chemical reactions. As blood is pumped through the various tissues, other substances besides O_2 are also exchanged between the blood and tissues. For example, the blood picks up nutrients as it flows through the digestive organs, and other tissues remove nutrients from the blood as it flows through them. Even excess heat is transported by the blood from exercising muscles to the skin surface, where it is lost to the external environment.

Although all the body tissues constantly depend on the life-supporting blood flow provided to them by the heart, the heart itself is quite an independent organ. It can take care of many of its own needs without any outside influence. Contraction of this magnificent muscle is self-generated through a carefully orchestrated interplay of changing ionic permeabilities. Local mechanisms within the heart ensure that blood flow to the cardiac muscle normally meets the heart's need for O_2. In addition, the heart has built-in capabilities to vary its strength of contraction, depending on the amount of blood returned to it. The heart does not act entirely autonomously, however. It is innervated by the autonomic nervous system and is influenced by the hormone epinephrine, both of which can vary heart rate and contractility, depending on the body's needs for blood delivery. Furthermore, as with all tissues, the cells that make up the heart depend on the other body systems to maintain a stable internal environment in which they can survive and function.

REVIEW EXERCISES

Objective Questions (Answers on p. A-45)

1. Adjacent cardiac muscle cells are joined end to end at specialized structures known as _____, which contain two types of membrane junctions: _____ and _____.

2. _____ is an abnormally slow heart rate, whereas _____ is a rapid heart rate.

3. The link that coordinates coronary blood flow with myocardial oxygen needs is _____.

4. The left ventricle is a stronger pump than the right ventricle because more blood is needed to supply the body tissues than to supply the lungs. (True or false?)

5. The heart lies in the left half of the thoracic cavity. (True or false?)

6. The only point of electrical contact between the atria and ventricles is the fibrous skeletal rings. (True or false?)

7. Entrance of Ca^{2+} through funny channels is responsible for the unique plateau phase of the action potential in cardiac contractile cells. (True or false?)

8. The atria and ventricles each act as a functional syncytium. (True or false?)

9. Which of the following is the proper sequence of cardiac excitation?
   a. SA node → AV node → atrial myocardium → bundle of His → Purkinje fibers → ventricular myocardium.
   b. SA node → atrial myocardium → AV node → bundle of His → ventricular myocardium → Purkinje fibers.
   c. SA node → atrial myocardium → ventricular myocardium → AV node → bundle of His → Purkinje fibers.
   d. SA node → atrial myocardium → AV node → bundle of His → Purkinje fibers → ventricular myocardium.

10. What percentage of ventricular filling is normally accomplished before atrial contraction begins?
    a. 0%
    b. 20%
    c. 50%
    d. 80%
    e. 100%

11. Sympathetic stimulation of the heart
    a. increases the heart rate
    b. increases the contractility of the heart muscle
    c. shifts the Frank–Starling curve to the left
    d. both (a) and (b)
    e. all of the above

12. Match the following:
    ___ 1. receives O$_2$-poor blood from the venae cavae
    ___ 2. prevent backflow of blood from the ventricles to the atria
    ___ 3. pumps O$_2$-rich blood into the aorta
    ___ 4. prevent backflow of blood from the arteries into the ventricles
    ___ 5. pumps O$_2$-poor blood into the pulmonary artery
    ___ 6. receives O$_2$-rich blood from the pulmonary veins

    (a) AV valves
    (b) semilunar valves
    (c) left atrium
    (d) left ventricle
    (e) right atrium
    (f) right ventricle

13. Circle the correct choice in each instance to complete the statement: The first heart sound is associated with closing of the (AV/semilunar) valves and signals the onset of (systole/diastole), whereas the second heart sound is associated with closing of the (AV/semilunar) valves and signals the onset of (systole/diastole).

14. Circle the correct choice in each instance to complete the statements: During ventricular filling, ventricular pressure must be (greater than/less than) atrial pressure, whereas during ventricular ejection ventricular pressure must be (greater than/less than) aortic pressure. Atrial pressure is always (greater than/less than) aortic pressure. During isovolumetric ventricular contraction and relaxation, ventricular pressure is (greater than/less than) atrial pressure and (greater than/less than) aortic pressure.

15. Use the following answer code to compare the relative magnitudes of the pair of items in question:
    (a) = Item A is greater than item B
    (b) = Item B is greater than item A
    (c) = Item A and item B are approximately equal
    ___ 1. A. Resistance and pressure in pulmonary circulation
          B. Resistance and pressure in systemic circulation
    ___ 2. A. Volume of blood pumped out by left side of heart
          B. Volume of blood pumped out by right side of heart
    ___ 3. A. Spontaneous rate of depolarization to threshold in SA node
          B. Spontaneous rate of depolarization to threshold in ventricular Purkinje fibers
    ___ 4. A. Velocity of impulse conduction through the AV node
          B. Velocity of impulse conduction through the bundle of His and Purkinje fibers
    ___ 5. A. Rate of ventricular filling in early diastole
          B. Rate of ventricular filling in late diastole
    ___ 6. A. Stroke volume when EDV equals 130 ml
          B. Stroke volume when EDV equals 160 ml
    ___ 7. A. Normal stroke volume
          B. Stroke volume on sympathetic stimulation
    ___ 8. A. Normal stroke volume
          B. Stroke volume on parasympathetic stimulation
    ___ 9. A. Volume of blood in ventricles at the onset of isovolumetric ventricular contraction
          B. Volume of blood in ventricles at the end of isovolumetric ventricular contraction
    ___ 10. A. Volume of blood in left ventricle at the time aortic valve opens
           B. Volume of blood in left ventricle at the time aortic valve closes
    ___ 11. A. Volume of blood in left ventricle at the time left AV valve opens
           B. Volume of blood in left ventricle at the time left AV valve closes
    ___ 12. A. Duration of refractory period in cardiac muscle
           B. Duration of contraction in cardiac muscle

Essay Questions

1. What are the three basic components of the circulatory system?

2. Trace a drop of blood through one complete circuit of the circulatory system.

3. Describe the location and function of the four heart valves. What keeps each of these valves from everting?

4. What are the three layers of the heart wall? Describe the distinguishing features of the structure and arrangement of cardiac muscle cells. What are the two specialized types of cardiac muscle cells?

5. Why is the SA node the pacemaker of the heart?

6. Describe the normal spread of cardiac excitation. What is the significance of the AV nodal delay? Why is the ventricular conduction system important?

7. Compare the changes in membrane potential associated with an action potential in a nodal pacemaker cell with those in a myocardial contractile cell. Describe the ionic mechanisms responsible for these changes in potential in both types of cardiac muscle cell.

8. Why is tetanus of cardiac muscle impossible? Why is this inability advantageous?

9. Draw and label the waveforms of a normal ECG. What electrical event does each component of the ECG represent?

10. Describe the mechanical events (that is, pressure changes, volume changes, valve activity, and heart sounds) of the cardiac cycle. Correlate the mechanical events of the cardiac cycle with the changes in electrical activity.

11. Distinguish between a stenotic and an insufficient valve.

12. Define the following: *end-diastolic volume, end-systolic volume, stroke volume, heart rate, cardiac output, ejection fraction,* and *cardiac reserve.*

13. Discuss autonomic nervous system control of heart rate.

14. Describe the intrinsic and extrinsic control of stroke volume.

15. What are the pathological changes and consequences of coronary artery disease?

16. Discuss the sources, transport, and elimination of cholesterol in the body. Distinguish between "good" cholesterol and "bad" cholesterol.

1. During heavy exercise, the cardiac output of a trained athlete may increase to 40 liters per minute. If stroke volume could not increase above the normal value of 70 ml, what heart rate would be necessary to achieve this cardiac output? Is such a heart rate physiologically possible?

2. How much blood remains in the heart after systole if the stroke volume is 85 ml and the end-diastolic volume is 125 ml?

3. Calculate the ejection fraction in each of the three circumstances illustrated in Figure 9-22, p. 329.

POINTS TO PONDER

(Explanations on p. A-45)

1. The stroke volume ejected on the next heartbeat after a premature ventricular contraction (PVC) is usually larger than normal. Can you explain why? (*Hint:* At a given heart rate, the interval between a PVC and the next normal beat is longer than the interval between two normal beats.)

2. Trained athletes usually have lower resting heart rates than normal (for example, 50 beats/min in an athlete compared to 70 beats/min in a sedentary individual). Considering that the resting cardiac output is 5000 ml/min in both trained athletes and sedentary people, what is responsible for the bradycardia of trained athletes?

3. During fetal life, because of the tremendous resistance offered by the collapsed, nonfunctioning lungs, the pressures in the right half of the heart and pulmonary circulation are higher than in the left half of the heart and systemic circulation, a situation that reverses after birth. Also in the fetus, a vessel called the **ductus arteriosus** connects the pulmonary artery and aorta as these major vessels both leave the heart. The blood pumped out by the heart into the pulmonary circulation is shunted from the pulmonary artery into the aorta through the ductus arteriosus, bypassing the nonfunctional lungs. What force is driving blood to flow in this direction through the ductus arteriosus?

At birth, the ductus arteriosus normally collapses and eventually degenerates into a thin, ligamentous strand. On occasion, this fetal bypass fails to close properly at birth, leading to a *patent* (open) *ductus arteriosus.* In what direction would blood flow through a patent ductus arteriosus? What possible outcomes would you predict might occur as a result of this blood flow?

4. Through what regulatory mechanisms can a transplanted heart, which does not have any innervation, adjust cardiac output to meet the body's changing needs?

5. There are two branches of the bundle of His, the right and left bundle branches, each of which travels down its respective side of the ventricular septum (see Figure 9-8, p. 311). Occasionally, conduction through one of these branches becomes blocked (so-called bundle-branch block). In this case, the wave of excitation spreads out from the terminals of the intact branch and eventually depolarizes the whole ventricle, but the normally stimulated ventricle completely depolarizes a considerable time before the ventricle on the side of the defective bundle branch. For example, if the left bundle branch is blocked, the right ventricle will be completely depolarized two to three times more rapidly than the left ventricle. How would this defect affect the heart sounds?

(Explanation on p. A-46)

In a physical exam, Rachel B.'s heart rate was rapid and very irregular. Furthermore, her heart rate, determined directly by listening to her heart with a stethoscope, exceeded the pulse rate taken concurrently at her wrist. No definite P waves could be detected on Rachel's ECG. The QRS complexes were normal in shape but occurred sporadically. Given these findings, what is the most likely diagnosis of Rachel's condition? Explain why the condition is characterized by a rapid, irregular heartbeat. Would cardiac output be seriously impaired by this condition? Why or why not? What accounts for the pulse deficit?

**Cardiovascular System
(Blood Vessels)**

**Body systems
maintain homeostasis**

Homeostasis
The circulatory system contributes to
homeostasis by transporting O_2, CO_2,
wastes, electrolytes, and hormones from
one part of the body to another.

**Homeostasis is
essential for
survival of cells**

Cells
Cells need to have O_2 and nutrients
continuously delivered and CO_2
constantly removed by the circulatory
system in order to generate the energy
needed to power life-sustaining cellular
activities by the following chemical reaction:

$$Food + O_2 \rightarrow CO_2 + H_2O + Energy$$

**Cells make up
body systems**

The **circulatory system** contributes to homeostasis by serving as the body's transport system. The blood vessels transport
and distribute blood pumped through them by the heart to
meet the body's needs for O_2 and nutrient delivery, waste removal, and hormonal signaling. The highly elastic **arteries**
transport blood from the heart to the organs and serve as a
pressure reservoir to continue driving blood forward when the
heart is relaxing and filling. The **mean arterial blood pressure** is closely regulated to ensure adequate blood delivery to
the organs. The amount of blood that flows through a given
organ depends on the caliber (internal diameter) of the highly
muscular **arterioles** that supply the organ. Arteriolar caliber is
subject to control so that cardiac output can be constantly readjusted to best serve the body's needs at the moment. The
thin-walled, pore-lined **capillaries** are the actual site of exchange between blood and the surrounding tissue cells. The
highly distensible **veins** return blood from the organs to the
heart and also serve as a blood reservoir.

The Blood Vessels and Blood Pressure

Patterns and Physics of Blood Flow

Most body cells are not in direct contact with the external environment, yet these cells must make exchanges with this environment, such as picking up O_2 and nutrients and eliminating wastes. Furthermore, chemical messengers must be transported between cells to accomplish integrated activity. To achieve these long-distance exchanges, cells are linked with one another and with the external environment by vascular (blood vessel) highways. Blood is transported to all parts of the body through a system of vessels that brings fresh supplies to the vicinity of all cells while removing their wastes.

To review, all blood pumped by the right side of the heart passes through the pulmonary circulation to the lungs for O_2 pickup and CO_2 removal. The blood pumped by the left side of the heart into the systemic circulation is distributed in various proportions to the systemic organs through a parallel arrangement of vessels that branch from the aorta (● Figure 10-1) (also see p. 305). This arrangement ensures that all organs receive blood of the same composition; that is, one organ does not receive "leftover" blood that has passed through another organ. Because of this parallel arrangement, blood flow through each systemic organ can be independently adjusted as needed.

In this chapter, we will first examine some general principles regarding blood flow patterns and the physics of blood flow. Then we will turn our attention to the roles of the various types of blood vessels through which blood flows. We will end by discussing how blood pressure is regulated to ensure adequate delivery of blood to the tissues.

To maintain homeostasis, reconditioning organs receive blood flow in excess of their own needs.

Blood is constantly "reconditioned" so that its composition remains relatively constant despite an ongoing drain of supplies to support metabolic activities and despite the continual addition of wastes from the tissues. Organs that recondition the blood normally receive much more blood than is necessary to meet their basic metabolic needs, so they can adjust the extra blood to achieve homeostasis. For example, large percentages of the cardiac output are distributed to the digestive tract (to pick up nutrient supplies), to the kidneys (to eliminate metabolic wastes

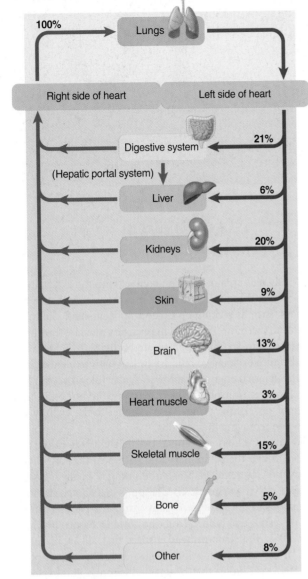

● **FIGURE 10-1 Distribution of cardiac output at rest.** The lungs receive all the blood pumped out by the right side of the heart, whereas the systemic organs each receive some of the blood pumped out by the left side of the heart. The percentage of pumped blood received by the various organs under resting conditions is indicated. This distribution of cardiac output can be adjusted as needed.

and adjust water and electrolyte composition), and to the skin (to eliminate heat). Blood flow to the other organs—heart, skeletal muscles, and so on—is solely for filling these organs' metabolic needs and can be adjusted according to their level of activity. For example, during exercise additional blood is delivered to the active muscles to meet their increased metabolic needs.

Because reconditioning organs—digestive organs, kidneys, and skin—receive blood flow in excess of their own needs, they can withstand temporary reductions in blood flow much better than other organs can that do not have this extra margin of blood supply. The brain in particular suffers irreparable damage when transiently deprived of blood supply. After only four min-

utes without O_2, permanent brain damage occurs. Therefore, constant delivery of adequate blood to the brain, which can least tolerate disrupted blood supply, is a high priority in the overall operation of the circulatory system.

In contrast, the reconditioning organs can tolerate significant reductions in blood flow for quite a long time, and often do. For example, during exercise some of the blood that normally flows through the digestive organs and kidneys is diverted to the skeletal muscles. Likewise, to conserve body heat, blood flow through the skin is markedly restricted during exposure to cold.

Later in the chapter, you will see how distribution of cardiac output is adjusted according to the body's current needs. For now, we concentrate on the factors that influence blood flow through a given blood vessel.

Blood flow through a vessel depends on the pressure gradient and vascular resistance.

The **flow rate** of blood through a vessel (that is, the volume of blood passing through per unit of time) is directly proportional to the pressure gradient and inversely proportional to vascular resistance:

$$F = \Delta P/R$$

where

F = flow rate of blood through a vessel

ΔP = pressure gradient

R = resistance of blood vessel

PRESSURE GRADIENT The **pressure gradient** is the difference in pressure between the beginning and end of a vessel. Blood flows from an area of higher pressure to an area of lower pressure down a pressure gradient. Contraction of the heart imparts pressure to the blood, which is the main driving force for flow through a vessel. Because of frictional losses (resistance), the pressure drops as blood flows throughout the vessel's length. Accordingly, pressure is higher at the beginning than at the end of the vessel, establishing a pressure gradient for forward flow of blood through the vessel. The greater the pressure gradient forcing blood through a vessel, the greater the flow rate is through that vessel (● Figure 10-2a). Think of a garden hose attached to a faucet. If you turn on the faucet slightly, a small stream of water will flow out of the end of the hose, because the pressure is slightly greater at the beginning than at the end of the hose. If you open the faucet all the way, the pressure gradient increases tremendously, so that water flows through the hose much faster and spurts from the end of the hose. Note that the *difference* in pressure between the two ends of a vessel, not the absolute pressures within the vessel, determines flow rate (● Figure 10-2b).

RESISTANCE The other factor influencing flow rate through a vessel is **resistance,** which is a measure of the hindrance or opposition to blood flow through the vessel, caused by friction between the moving fluid and the stationary vascular walls. As resistance to flow increases, it is more difficult for blood to pass

50 mm Hg pressure   10 mm Hg pressure

$\Delta P = 40$ mm Hg
Vessel 1

90 mm Hg pressure   10 mm Hg pressure

$\Delta P = 80$ mm Hg
Vessel 2

ΔP in vessel 2 = 2 times that of vessel 1

Flow in vessel 2 = 2 times that of vessel 1

Flow $\propto \Delta P$

(a) Comparison of flow rate in vessels with a different ΔP

90 mm Hg pressure   10 mm Hg pressure

$\Delta P = 80$ mm Hg
Vessel 2

180 mm Hg pressure   100 mm Hg pressure

$\Delta P = 80$ mm Hg
Vessel 3

ΔP in vessel 3 = the same as that of vessel 2, despite the larger absolute values

Flow in vessel 3 = the same as that of vessel 2

Flow $\propto \Delta P$

(b) Comparison of flow rate in vessels with the same ΔP

● **FIGURE 10-2 Relationship of flow to the pressure gradient in a vessel.** (a) As the difference in pressure (ΔP) between the two ends of a vessel increases, the flow rate increases proportionately. (b) Flow rate is determined by the *difference* in pressure between the two ends of a vessel, not the magnitude of the pressures at each end.

10 ml

10 ml

(a) Comparison of contact of a given volume of blood with the surface area of a small-radius vessel and a large-radius vessel

Same pressure gradient

Vessel 1

Vessel 2

Radius in vessel 2 = 2 times that of vessel 1

Resistance in vessel 2 = 1/16 that of vessel 1

Flow in vessel 2 = 16 times that of vessel 1

Resistance $\propto 1/r^4$
Flow $\propto r^4$

(b) Influence of vessel radius on resistance and flow

● **FIGURE 10-3 Relationship of resistance and flow to the vessel radius.** (a) The smaller-radius vessel offers more resistance to blood flow, because the blood "rubs" against a larger surface area. (b) Doubling the radius decreases the resistance to 1/16 and increases the flow 16 times, because the resistance is inversely proportional to the fourth power of the radius.

through the vessel, so flow rate decreases (as long as the pressure gradient remains unchanged). When resistance increases, the pressure gradient must increase correspondingly to maintain the same flow rate. Accordingly, when the vessels offer more resistance to flow, the heart must work harder to maintain adequate circulation.

Resistance to blood flow depends on three factors: (1) viscosity of the blood, (2) vessel length, and (3) vessel radius, which is by far the most important. The term **viscosity** (designated as η) refers to the friction developed between the molecules of a fluid as they slide over each other during flow of the fluid. The greater the viscosity, the greater the resistance to flow is. In general, the thicker a liquid, the more viscous it is. For example, molasses flows more slowly than water because molasses has greater viscosity. Blood viscosity is determined primarily by the number of circulating red blood cells. Normally, this factor is relatively constant and not important in controlling resistance. Occasionally, however, blood viscosity and resistance to flow are altered by an abnormal number of red blood cells. When excessive red blood cells are present, blood flow is more sluggish than normal.

Because blood "rubs" against the lining of the vessels as it flows past, the greater the vessel surface area in contact with the blood, the greater the resistance to flow is. Surface area is determined by both the length (L) and radius (r) of the vessel. At a constant radius, the longer the vessel is, the greater the surface area and the greater the resistance to flow. Because vessel length remains constant in the body, it is not a variable factor in the control of vascular resistance.

Therefore, the major determinant of resistance to flow is the vessel's radius. Fluid passes more readily through a large vessel than through a smaller vessel. The reason is that a given volume of blood comes into contact with much more of the surface area of a small-radius vessel than of a larger-radius vessel, resulting in greater resistance (● Figure 10-3a).

From Humors to Harvey: Historical Highlights in Circulation

Today even grade-school children know that blood is pumped by the heart and continually circulates throughout the body in a system of blood vessels. Furthermore, people accept without question that blood picks up O_2 in the lungs from the air we breathe and delivers it to the various organs. This common knowledge was unknown for most of human history, however. Even though the function of blood was described as early as the fifth century BC, our modern concept of circulation did not develop until 1628, more than 2000 years later, when William Harvey published his now classical study on the circulatory system.

Ancient Greeks believed everything material in the universe consisted of just four elements: earth, air, fire, and water. Extending this view to the human body, they thought these four elements took the form of four "humors": *black bile* (representing earth), *blood* (representing air), *yellow bile* (representing fire), and *phlegm* (representing water). According to the Greeks, disease resulted when one humor was out of normal balance with the rest. The "cure" was logical: To restore normal balance, drain off whichever humor was in excess. Because the easiest humor to drain off was the blood, bloodletting became standard procedure for treating many illnesses—a practice that persisted well into the Renaissance (which began in the 1300s and extended into the 1600s).

Although the ancient Greeks' notion of the four humors was erroneous, their concept of the necessity of balance within the body was remarkably accurate. As we now know, life depends on homeostasis, maintenance of the proper balance among all elements of the internal environment.

Aristotle (384–322 BC), a biologist as well as a philosopher, was among the first to correctly describe the heart at the center of a system of blood vessels. However, he thought the heart was both the seat of intellect (the brain was not identified as the seat of intellect until more than a century later) and a furnace that heated the blood. He considered this warmth the vital force of life, because the body cools quickly at death. Aristotle also erroneously theorized that breathing ventilated the "furnace," with air serving as a cooling agent. Aristotle

could observe with his eyes the arteries and veins in cadavers but did not have a microscope with which to observe capillaries. (The microscope was not invented until the 17th century.) Thus, he did not think arteries and veins were directly connected.

In the third century BC, Erasistratus, a Greek many consider the first "physiologist," proposed that the liver used food to make blood, which the veins delivered to the other organs. He believed the arteries contained air, not blood. According to his view, *pneuma* ("air"), a living force, was taken in by the lungs, which transferred it to the heart. The heart transformed the air into a "vital spirit" that the arteries carried to the other organs.

Galen (AD 130–206), a prolific, outspoken, dogmatic Roman physician, philosopher, and scholar, expanded on the work of Erasistratus and others who had preceded him. Galen further elaborated on the pneumatic theory. He proposed three fundamental members in the body, from lowest to highest: liver, heart, and brain. Each was dominated by a special *pneuma,* or "spirit." (In Greek, *pneuma* encompassed the related ideas of "air," "breath," and "spirit.") Like Erasistratus, Galen believed that the liver made blood from food, taking on a "natural" or "physical" spirit (*pneuma physicon*) in the process. The newly formed blood then proceeded through veins to organs. The natural spirit, which Galen considered a vapor rising from the blood, controlled the functions of nutrition, growth, and reproduction. Once its spirit supply was depleted, the blood moved in the opposite direction through the same venous pathways, returning to the liver to be replenished. When the natural spirit was carried in the venous blood to the heart, it mixed with air that was breathed in and transferred from the lungs to the heart. Contact with air in the heart transformed the natural spirit into a higher-level spirit, the "vital" spirit (*pneuma zotikon*). The vital spirit, which was carried by the arteries, conveyed heat and life throughout the body. The vital spirit was transformed further into a yet higher "animal" or "psychical" spirit (*pneuma psychikon*) in the brain. This ultimate spirit regulated the brain, nerves, feelings, and so on. Thus, according

to Galenic theory, the veins and arteries were conduits for carrying different levels of pneuma, and there was no direct connection between the veins and arteries. The heart was not involved in moving blood but instead was the site where blood and air mixed. (We now know that blood and air meet in the lungs for the exchange of O_2 and CO_2.)

Galen was one of the first to understand the need for experimentation, but unfortunately, his impatience and his craving for philosophical and literary fame led him to expound comprehensive theories that were not always based on the time-consuming collection of evidence. Even though his assumptions about bodily structure and functions often were incorrect, his theories were convincing because they seemed a logical way of pulling together what was known at the time. Furthermore, the sheer quantity of his writings helped establish him as an authority. In fact, his writings remained the anatomic and physiologic "truth" for nearly 15 centuries, throughout the Middle Ages and well into the Renaissance. So firmly entrenched was Galenic doctrine that people who challenged its accuracy risked their lives by being declared secular heretics.

Not until the Renaissance and the revival of classical learning did independent-minded European investigators begin to challenge Galen's theories. Most notably, the English physician William Harvey (1578–1657) revolutionized the view of the roles played by the heart, blood vessels, and blood. Through careful observations, experimentation, and deductive reasoning, Harvey was the first to correctly identify the heart as a pump that repeatedly moves a small volume of blood forward in one fixed direction in a circular path through a closed system of blood vessels (the *circulatory system*). He also correctly proposed that blood travels to the lungs to mix with air (instead of air traveling to the heart to mix with blood). Even though he could not see physical connections between arteries and veins, he speculated on their existence. Not until the discovery of the microscope later in the century was the existence of these connections, capillaries, confirmed, by Marcello Malpighi (1628–1694).

Furthermore, a slight change in the radius of a vessel brings about a notable change in flow, because the resistance is inversely proportional to the fourth power of the radius (multiplying the radius by itself four times):

$$R \propto 1/r^4$$

Thus, doubling the radius reduces the resistance to 1/16 its original value ($r^4 = 2 \times 2 \times 2 \times 2 = 16$; $R \propto 1/16$) and increases flow through the vessel 16-fold (at the same pressure gradient) (● Figure 10-3b). The converse is also true. Only 1/16th as much blood flows through a vessel at the same driving pressure when its radius is halved. Importantly, the radius of arterioles can be regulated and is the most important factor in controlling resistance to blood flow throughout the vascular circuit.

POISEUILLE'S LAW The factors that affect the flow rate through a vessel are integrated in **Poiseuille's law** as follows:

$$\text{Flow rate} = \frac{\pi \Delta P r^4}{8 \eta L}$$

The significance of the relationships among flow, pressure, and resistance, as largely determined by vessel radius, will become even more apparent as we embark on a voyage through the vessels in the next section.

The vascular tree consists of arteries, arterioles, capillaries, venules, and veins.

The systemic and pulmonary circulations each consist of a closed system of vessels (● Figure 10-4). (For the history leading up to the conclusion that blood vessels form a closed system, see the accompanying boxed feature, Concepts, Challenges, and Controversies.) These vascular loops each consist of a continuum of different blood vessel types that begins and ends with the heart, as follows. Looking specifically at the systemic circulation, **arteries,** which carry blood from the heart to the organs, branch or diverge into a "tree" of progressively smaller vessels, with the various branches delivering blood to different regions of the body. When a small artery reaches the organ it is supplying, it branches into numerous **arterioles.** The volume of blood flowing through an organ can be adjusted by regulating the caliber (internal diameter) of the organ's arterioles. Arterioles branch further within the organs into **capillaries,** the smallest of vessels, across which all exchanges are made with surrounding cells. Capillary exchange is the entire purpose of the circulatory system; all other activities of the system are directed toward ensuring an adequate distribution of replenished blood to capillaries for exchange with all cells. Capillaries rejoin to form small **venules,** which further merge to form small **veins** that leave the organs. The small veins progressively unite or converge to form larger veins that eventually empty into the heart. The arterioles, capillaries, and venules are collectively referred to as the **microcirculation,** because they are visible only through a microscope. The microcirculatory vessels are all located within the organs. The pulmonary circulation consists of the same vessel types, except that all the blood in this loop goes between the heart and lungs. If all of the vessels in the body were strung end to end, they could circle the circumference of Earth twice!

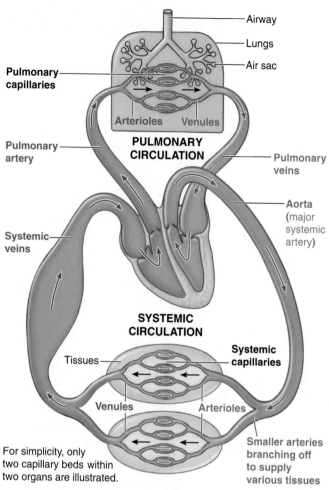

● **FIGURE 10-4 Basic organization of the cardiovascular system.** Arteries progressively branch as they carry blood from the heart to the organs. A separate small arterial branch delivers blood to each of the various organs. As a small artery enters the organ it is supplying, it branches into arterioles, which further branch into an extensive network of capillaries. The capillaries rejoin to form venules, which further unite to form small veins that leave the organ. The small veins progressively merge as they carry blood back to the heart.

In discussing the vessel types in this chapter, we will refer to their roles in the systemic circulation, starting with systemic arteries.

Arteries

The consecutive segments of the vascular tree are specialized to perform specific tasks (▲ Table 10-1).

Arteries serve as rapid-transit passageways to the organs and as a pressure reservoir.

Arteries are specialized (1) to serve as rapid-transit passageways for blood from the heart to the organs (because of their large radius, arteries offer little resistance to blood flow) and (2) to act as a **pressure reservoir** to provide the driving force for blood when the heart is relaxing.

Feature	VESSEL TYPE			
	Arteries	**Arterioles**	**Capillaries**	**Veins**
Number	Several hundred*	Half a million	Ten billion	Several hundred*
Special Features	Thick, highly elastic, walls; large radii*	Highly muscular, well-innervated walls; small radii	Very thin walled; large total cross-sectional area	Thin walled compared to arteries; highly distensible; large radii*
Functions	Passageway from heart to organs; serve as pressure reservoir	Primary resistance vessels; determine distribution of cardiac output	Site of exchange; determine distribution of extracellular fluid between plasma and interstitial fluid	Passageway to heart from organs; serve as blood reservoir

Structure

Endothelium — Basement membrane — Elastic fibers — Smooth muscle — Elastic fibers — Connective tisssue coat (mostly collagen fibers)

Large artery

Arteriole

Endothelium — Basement membrane

Capillary

Venous valve — Endothelium — Basement membrane — Smooth muscle; elastic fibers — Connective tisssue coat (mostly collagen fibers)

Large vein

Relative thickness of layers in wall

Endotheluium

Elastic fibers

Smooth muscle

Collagen fibers

*These numbers and special features refer to the large arteries and veins, not to the smaller arterial branches or venules.

Let us expand on the role of the arteries as a pressure reservoir. The heart alternately contracts to pump blood into the arteries and then relaxes to refill from the veins. When the heart is relaxing and refilling, no blood is pumped out. However, capillary flow does not fluctuate between cardiac systole and diastole; that is, blood flow is continuous through the capillaries supplying the organs. The driving force for the continued flow of blood to the organs during cardiac relaxation is provided by the elastic properties of the arterial walls.

All vessels are lined with a thin layer of smooth, flat endothelial cells that are continuous with the endothelial lining of the heart. A thick wall made up of smooth muscle and connective tissue surrounds the arteries' endothelial lining (▲ Table 10-1). Arterial connective tissue contains an abundance of two types of

Elastin fibers

© Triarch/Visuals Unlimited

● **FIGURE 10-5 Elastin fibers in an artery.** Light micrograph of a portion of the aorta wall in cross section, showing numerous wavy elastin fibers, common to all arteries.

connective tissue fibers: *collagen fibers,* which provide tensile strength against the high driving pressure of blood ejected from the heart, and *elastin fibers,* which give the arterial walls elasticity so that they behave much like a balloon (● Figure 10-5).

As the heart pumps blood into the arteries during ventricular systole, a greater volume of blood enters the arteries from the heart than leaves them to flow into smaller vessels downstream because the smaller vessels have a greater resistance to flow than the arteries do. The highly elastic arteries expand to temporarily hold this excess volume of ejected blood, storing some of the pressure energy imparted by cardiac contraction in their stretched walls—just as a balloon expands to accommodate the extra volume of air you blow into it (● Figure 10-6a). When the heart relaxes and temporarily stops pumping blood into the arteries, the stretched arterial walls passively recoil, like an inflated balloon that is released. This recoil pushes the excess blood contained in the arteries into the vessels downstream, ensuring continued blood flow to the organs when the heart is relaxing and not pumping blood into the system (● Figure 10-6b).

Arterial pressure fluctuates in relation to ventricular systole and diastole.

Blood pressure, the force exerted by the blood against a vessel wall, depends on the volume of blood contained within the vessel and the **compliance,** or **distensibility,** of the vessel walls (how easily they can be stretched). If the volume of blood entering the arteries were equal to the volume of blood leaving the arteries during the same period, arterial blood pressure would remain constant. This is not the case, however. During ventricular systole, a stroke volume of blood enters the arteries from the ventricle, while only about one third as much blood leaves the arteries to enter the arterioles. During diastole, no blood enters the arteries, while blood continues to leave, driven by elastic recoil. The maximum pressure exerted in the arteries when blood is ejected into them during systole, the **systolic pressure,** averages 120 mm Hg. The minimum pressure within the arteries when blood is draining off into the rest of the vessels during diastole, the **diastolic pressure,** averages 80 mm Hg. Although ventricular pressure falls to 0 mm Hg during diastole, arterial pressure does not fall to 0 mm Hg, because the next cardiac contraction occurs and refills the arteries before all the blood drains off (● Figure 10-7; also see ● Figure 9-16, p. 322).

Blood pressure can be measured indirectly by using a sphygmomanometer.

The changes in arterial pressure throughout the cardiac cycle can be measured directly by connecting a pressure-measuring device to a needle inserted in an artery. However, it is more convenient and reasonably accurate to measure the pressure indirectly with a **sphygmomanometer,** an externally applied inflatable cuff attached to a pressure gauge. When the cuff is wrapped around the upper arm and then inflated with air, the pressure of the cuff is transmitted through the tissues to the underlying brachial artery, the main vessel carrying blood to the forearm (● Figure 10-8a). The technique involves balancing the pressure in the cuff against

Arteries

Arterioles

From veins

To capillaries

(a) Heart contracting and emptying

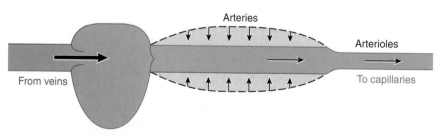

Arteries

Arterioles

From veins

To capillaries

(b) Heart relaxing and filling

● **FIGURE 10-6 Arteries as a pressure reservoir.** Because of their elasticity, arteries act as a pressure reservoir. (a) The elastic arteries distend during cardiac systole as more blood is ejected into them than drains off into the narrow, high-resistance arterioles downstream. (b) The elastic recoil of arteries during cardiac diastole continues driving the blood forward when the heart is not pumping.

● **FIGURE 10-7 Arterial blood pressure.** The systolic pressure is the peak pressure exerted in the arteries when blood is pumped into them during ventricular systole. The diastolic pressure is the lowest pressure exerted in the arteries when blood is draining off into the vessels downstream during ventricular diastole. The pulse pressure is the difference between systolic and diastolic pressure. The mean pressure is the average pressure throughout the cardiac cycle.

the pressure in the artery. When cuff pressure is greater than the pressure in the vessel, the vessel is pinched closed so that no blood flows through it. When blood pressure is greater than cuff pressure, the vessel is open and blood flows through.

DETERMINATION OF SYSTOLIC AND DIASTOLIC PRESSURE During the determination of blood pressure, a stethoscope is placed over the brachial artery at the inside bend of the elbow just below the cuff. No sound can be detected either when blood is not flowing through the vessel or when blood is flowing in the normal, smooth laminar flow (see p. 324). Turbulent blood flow, in contrast, creates vibrations that can be heard. The sounds heard when determining blood pressure, known as **Korotkoff sounds,** are distinct from the heart sounds associated with valve closure heard when listening to the heart with a stethoscope.

At the onset of a blood pressure determination, the cuff is inflated to a pressure greater than systolic blood pressure so that the brachial artery collapses. Because the externally applied pressure is greater than the peak internal pressure, the artery remains completely pinched closed throughout the entire cardiac cycle; no sound can be heard because no blood is passing through (point **1** in ● Figure 10-8b). As air in the cuff is slowly released, the pressure in the cuff is gradually reduced. When the cuff pressure falls to just below the peak systolic pressure, the artery transiently opens a bit when the blood pressure reaches this peak. Blood escapes through the partially occluded artery for a brief interval before the arterial pressure falls below the cuff pressure and the artery collapses once again. This spurt of blood is turbulent, so it can be heard. Thus, the highest cuff pressure at which the *first sound* can be heard indicates the *systolic pressure* (point **2**). As the cuff pressure continues to fall, blood intermittently spurts through the artery and produces a sound with each subsequent cardiac cycle whenever the arterial pressure exceeds the cuff pressure (point **3**).

When the cuff pressure finally falls below diastolic pressure, the brachial artery is no longer pinched closed during any part of the cardiac cycle, and blood can flow uninterrupted through the vessel (point **5**). With the return of nonturbulent blood

flow, no further sounds can be heard. Therefore, the highest cuff pressure at which the *last sound* can be detected indicates the *diastolic pressure* (point **4**).

In clinical practice, arterial blood pressure is expressed as systolic pressure over diastolic pressure, with the cutoff for desirable blood pressure being less than 120/80 (120 over 80) mm Hg.

PULSE PRESSURE The pulse that can be felt in an artery lying close to the surface of the skin is the result of the difference between systolic and diastolic pressures. This pressure difference is known as the **pulse pressure.** When blood pressure is 120/80, pulse pressure is 40 mm Hg (120 mm Hg − 80 mm Hg).

Mean arterial pressure is the main driving force for blood flow.

The **mean arterial pressure** is the *average pressure* driving blood forward into the tissues throughout the cardiac cycle. Contrary to what you might expect, mean arterial pressure is not the halfway value between systolic and diastolic pressure (for example, with a blood pressure of 120/80, mean pressure is not 100 mm Hg). The reason is that arterial pressure remains closer to diastolic than to systolic pressure for a longer portion of each cardiac cycle. At resting heart rate, about two thirds of the cardiac cycle is spent in diastole and only one third in systole. As an analogy, if a race car traveled 80 miles per hour (mph) for 40 minutes and 120 mph for 20 minutes, its average speed would be 93 mph, not the halfway value of 100 mph.

Similarly, a good approximation of the mean arterial pressure can be determined using the following formula:

Mean arterial pressure =
diastolic pressure + 1/3 pulse pressure

At 120/80, mean arterial pressure =
80 mm Hg + (1/3) 40 mm Hg = 93 mm Hg

The mean arterial pressure, not the systolic or diastolic pressure, is monitored and regulated by blood pressure reflexes described later in the chapter.

Because arteries offer little resistance to flow, only a negligible amount of pressure energy is lost in them because of friction. Therefore, arterial pressure—systolic, diastolic, pulse, or mean—is essentially the same throughout the arterial tree (● Figure 10-9).

Blood pressure exists throughout the entire vascular tree, but when discussing a person's "blood pressure" without qualifying which blood vessel type is being referred to, the term is tacitly understood to mean the pressure in the arteries.

Arterioles

When an artery reaches the organ it is supplying, it branches into numerous arterioles within the organ.

Arterioles are the major resistance vessels.

Arterioles are the major resistance vessels in the vascular tree because their radius is small enough to offer considerable resistance to flow. (Even though capillaries have a smaller radius

(a) Use of a sphygmomanometer in determining blood pressure

When blood pressure is 120/80:

When cuff pressure is greater than 120 mm Hg and exceeds blood pressure throughout the cardiac cycle:

No blood flows through the vessel.

1 No sound is heard because no blood is flowing.

When cuff pressure is between 120 and 80 mm Hg:

Blood flow through the vessel is turbulent whenever blood pressure exceeds cuff pressure.

2 The first sound is heard at peak systolic pressure.

3 Intermittent sounds are produced by turbulent spurts of flow as blood pressure cyclically exceeds cuff pressure.

When cuff pressure is less than 80 mm Hg and is below blood pressure throughout the cardiac cycle:

Blood flows through the vessel in smooth, laminar fashion.

4 The last sound is heard at minimum diastolic pressure.

5 No sound is heard thereafter because of uninterrupted, smooth, laminar flow.

● **FIGURE 10-8 Sphygmomanometry.** (a) The pressure in the sphygmomanometer (inflatable cuff) can be varied to prevent or permit blood flow in the underlying brachial artery. Turbulent blood flow can be detected with a stethoscope, whereas smooth laminar flow and no flow are inaudible. (b) The red shaded areas in the graph are the times during which blood is flowing in the brachial artery.

(b) Blood flow through the brachial artery in relation to cuff pressure and sounds

than arterioles, you will see later how collectively the capillaries do not offer as much resistance to flow as the arterioles do.) In contrast to the low resistance of the arteries, the high degree of arteriolar resistance causes a marked drop in mean pressure as blood flows through these small vessels. On average, the pressure falls from 93 mm Hg, the mean arterial pressure (the pres-

sure of the blood entering the arterioles), to 37 mm Hg, the pressure of the blood leaving the arterioles and entering the capillaries (● Figure 10-9). This decline in pressure helps establish the pressure differential that encourages the flow of blood from the heart to the various organs downstream. Arteriolar resistance also converts the pulsatile systolic-to-diastolic pres-

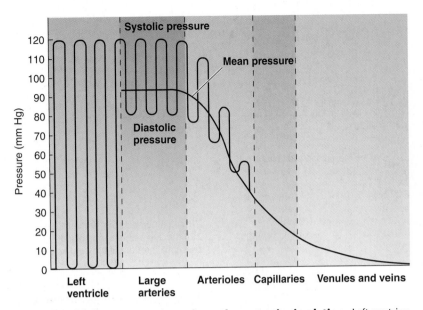

● **FIGURE 10-9 Pressures throughout the systemic circulation.** Left ventricular pressure swings between a low pressure of 0 mm Hg during diastole to a high pressure of 120 mm Hg during systole. Arterial blood pressure, which fluctuates between a peak systolic pressure of 120 mm Hg and a low diastolic pressure of 80 mm Hg each cardiac cycle, is of the same magnitude throughout the large arteries. Because of the arterioles' high resistance, the pressure drops precipitously and the systolic-to-diastolic swings in pressure are converted to a nonpulsatile pressure when blood flows through the arterioles. The pressure continues to decline but at a slower rate as blood flows through the capillaries and venous system.

sure swings in the arteries into the nonfluctuating pressure present in the capillaries.

The radius (and, accordingly, the resistances) of arterioles supplying individual organs can be adjusted independently to accomplish two functions: (1) to variably distribute the cardiac output among the systemic organs, depending on the body's momentary needs; and (2) to help regulate arterial blood pressure. Before considering how such adjustments are important in accomplishing these two functions, we discuss the mechanisms involved in adjusting arteriolar resistance.

VASOCONSTRICTION AND VASODILATION Unlike arteries, arteriolar walls contain very little elastic connective tissue. However, they do have a thick layer of smooth muscle that is richly innervated by sympathetic nerve fibers (see ▲ Table 10-1). The smooth muscle is also sensitive to many local chemical changes, to a few circulating hormones, and to mechanical factors such as stretch. The smooth muscle layer runs circularly around the arteriole (● Figure 10-10a); so when the smooth muscle layer contracts, the vessel's circumference (and its radius) becomes smaller, increasing resistance and decreasing flow through that vessel. **Vasoconstriction** is the term applied to such narrowing of a vessel (● Figure 10-10c). The term **vasodilation** refers to enlargement in the circumference and radius of a vessel as a result of its smooth muscle layer relaxing (● Figure 10-10d). Vasodilation leads to decreased resistance and increased flow through that vessel.

VASCULAR TONE The extent of contraction of arteriolar smooth muscle depends on the cytosolic concentration of Ca^{2+}. Arteriolar smooth muscle normally displays a state of partial constriction known as **vascular tone,** which establishes a baseline of arteriolar resistance (● Figure 10-10b) (also see p. 293). Two factors are responsible for vascular tone. First, arteriolar smooth muscle is tonic smooth muscle that has sufficient surface-membrane Ca^{2+} channels open even at resting potential to trigger partial contraction (see p. 294). This myogenic activity is independent of any neural or hormonal influences, leading to self-induced contractile activity (see p. 296). Second, the sympathetic fibers supplying most arterioles continually release norepinephrine, which further enhances vascular tone.

This ongoing tonic activity makes it possible to either increase or decrease the level of contractile activity to accomplish vasoconstriction or vasodilation, respectively. Were it not for tone, it would be impossible to reduce the tension in an arteriolar wall to accomplish vasodilation; only varying degrees of vasoconstriction would be possible.

A variety of factors can influence the level of contractile activity in arteriolar smooth muscle, thereby substantially changing resistance to flow in these vessels. Unlike skeletal and cardiac muscle in which action potentials trigger muscle contraction, vascular smooth muscle can undergo graded changes in force in response to chemical, physical, and neural factors with little or no accompanying change in membrane potential. These agents largely act via second-messenger systems, such as the IP_3/Ca^{2+} pathway (see p. 122). The factors that cause arteriolar vasoconstriction or vasodilation fall into two categories: local (intrinsic) controls, which are important in determining the distribution of cardiac output; and extrinsic controls, which are important in blood pressure regulation. We will look at each of these controls in turn.

Local control of arteriolar radius is important in determining the distribution of cardiac output.

The fraction of the total cardiac output delivered to each organ is not always constant; it varies, depending on the demands for blood at the time. The amount of the cardiac output received by each organ is determined by the number and caliber of the arterioles supplying that area. Recall that $F = \Delta P/R$. Because blood is delivered to all organs at the same mean arterial pressure, the driving force for flow is identical for each organ. Therefore, differences in flow to various organs are completely determined by differences in the extent of vascularization and by differences in resistance offered by the arterioles supplying each organ. On a moment-to-moment basis, the distribution of cardiac output can be varied by differentially adjusting arteriolar resistance in the various vascular beds.

Smooth
muscle cells

Cross section
of arteriole

(b) Normal arteriolar tone

Fawcett-Uehara-Suyama/Science Source/Photo Researchers, Inc.

(a) Scanning electron micrograph of an arteriole showing how
the smooth muscle cells run circularly around the vessel wall

Caused by:
↑ Myogenic activity
↑ Oxygen (O_2)
↓ Carbon dioxide (CO_2)
  and other metabolites
↑ Endothelin
↑ Sympathetic stimulation
  Vasopressin; angiotensin II
  Cold

(c) Vasoconstriction (increased contraction of circular
smooth muscle in the arteriolar wall, which leads to
increased resistance and decreased flow through the vessel)

Caused by:
↓ Myogenic activity
↓ O_2
↑ CO_2 and other metabolites
↑ Nitric oxide
↓ Sympathetic stimulation
  Histamine release
  Heat

(d) Vasodilation (decreased contraction of circular smooth
muscle in the arteriolar wall, which leads to decreased
resistance and increased flow through the vessel)

● **FIGURE 10-10 Arteriolar vasoconstriction and
vasodilation.**

As an analogy, consider a pipe carrying water, with a number of adjustable valves located throughout its length (● Figure 10-11). Assuming that water pressure in the pipe is constant, differences in the amount of water flowing into a beaker under each valve depend entirely on which valves are open and to what extent. No water enters beakers under closed valves (high resistance), and more water flows into beakers under valves that are opened completely (low resistance) than into beakers under valves that are only partially opened (moderate resistance).

Similarly, more blood flows to areas whose arterioles offer the least resistance to its passage. During exercise, for example, not only is cardiac output increased, but also, because of vasodilation in skeletal muscle and in the heart, a greater percentage of the pumped blood is diverted to these organs to support their increased metabolic activity. Simultaneously, blood flow to the digestive tract and kidneys is reduced as a result of arteriolar vasoconstriction in these organs (● Figure 10-12). Only the blood supply to the brain remains remarkably constant no matter what the person is doing, be it vigorous physical activity, intense mental concentration, or sleep. Although total blood flow

to the brain remains constant, new imaging techniques demonstrate that regional blood flow varies within the brain in close correlation with local neural activity patterns (see p. 147).

Local (intrinsic) controls are changes within an organ that alter the radius of the vessels and hence adjust blood flow through the organ by directly affecting the smooth muscle of the organ's arterioles. Local influences may be either chemical or physical. Local chemical influences on arteriolar radius include (1) local metabolic changes and (2) histamine release. Local physical influences include (1) local application of heat or cold, (2) chemical response to shear stress, and (3) myogenic response to stretch. Let's examine the role and mechanism of each of these local influences.

Local metabolic influences on arteriolar radius help match blood flow with the organs' needs.

The most important local chemical influences on arteriolar smooth muscle are related to metabolic changes within a given organ. The influence of these local changes on arteriolar radius

Constant pressure in pipe
(mean arterial pressure)

From pump
(heart)

High resistance — No flow

Moderate resistance — Moderate flow

Low resistance — Large flow

KEY
Control valves = Arterioles

● **FIGURE 10-11 Flow rate as a function of resistance.**

is important in matching blood flow through an organ with the organ's metabolic needs. Local metabolic controls are especially important in skeletal muscle and in the heart, the organs whose metabolic activity and need for blood supply normally vary most extensively, and in the brain, whose overall metabolic activity and need for blood supply remain constant. Local controls help maintain the constancy of blood flow to the brain.

ACTIVE HYPEREMIA Arterioles lie within the organ they are supplying and can be acted on by local factors within the organ. During increased metabolic activity, such as when a skeletal muscle is contracting during exercise, local concentrations of a number of the organ's chemicals change. For example, the local O_2 concentration decreases as the actively metabolizing cells use up more O_2 to support oxidative phosphorylation for ATP production (see p. 35). This and other local chemical changes produce local arteriolar dilation by triggering relaxation of the arteriolar smooth muscle in the vicinity. Local arteriolar vasodilation then increases blood flow to that particular area. This increased blood flow in response to enhanced tissue activity is called **active hyperemia** (*hyper* means "above normal"; *emia* means "blood"). When cells are more active metabolically, they need more blood to bring in O_2 and nutrients and to remove metabolic wastes. The increased blood flow meets these increased local needs.

Conversely, when an organ, such as a relaxed muscle, is less active metabolically and thus has reduced needs for blood delivery, the resultant local chemical changes (for example, increased local O_2 concentration) bring about local arteriolar vasoconstriction and a subsequent reduction in blood flow to the area. Local metabolic changes can thus adjust blood flow as needed without involving nerves or hormones.

LOCAL METABOLIC CHANGES THAT INFLUENCE ARTERIOLAR RADIUS A variety of local chemical changes act together in a cooperative, redundant manner to bring about these "selfish" local adjustments in arteriolar caliber that match a tissue's blood flow with its needs. Specifically, the following local chemical factors produce relaxation of arteriolar smooth muscle:

■ *Decreased O_2.*

■ *Increased CO_2.* More CO_2 is generated as a by-product during the stepped-up pace of oxidative phosphorylation that accompanies increased activity.

■ *Increased acid.* More carbonic acid is generated from the increased CO_2 produced as the metabolic activity of a cell increases. Also, lactate (lactic acid) accumulates if the glycolytic pathway is used for ATP production (see pp. 37 and 278).

■ *Increased K^+.* Repeated action potentials that outpace the ability of the Na^+–K^+ pump to restore the resting concentration gradients (see p. 93) result in an increase in K^+ in the tissue fluid of an actively contracting muscle or a more active region of the brain.

■ *Increased osmolarity.* Osmolarity (the concentration of osmotically active solutes) increases during elevated cellular metabolism because of increased formation of osmotically active particles.

■ *Adenosine release.* Especially in cardiac muscle, adenosine is released in response to increased metabolic activity or O_2 deprivation (see p. 333).

■ *Prostaglandin release.* Prostaglandins are local chemical messengers derived from fatty acid chains within the plasma membrane (see p. 758).

The relative contributions of these various chemical changes (and possibly others) in the local metabolic control of blood flow in systemic arterioles are still being investigated.

LOCAL VASOACTIVE MEDIATORS These local chemical changes do not act directly on vascular smooth muscle to change its contractile state. Instead, **endothelial cells,** the single layer of specialized epithelial cells that line the lumen of all blood vessels, release chemical mediators that play a key role in locally regulating arteriolar caliber. Until recently, scientists regarded endothelial cells as little more than a passive barrier between the blood and the rest of the vessel wall. It is now known that endothelial cells are active participants in a variety of vessel-

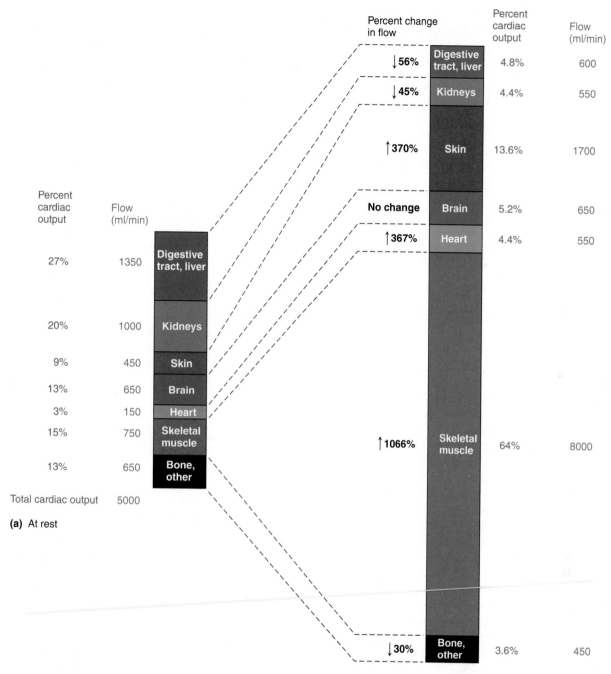

Percent change in flow

Percent change in flow		Percent cardiac output	Flow (ml/min)
↓56%	Digestive tract, liver	4.8%	600
↓45%	Kidneys	4.4%	550
↑370%	Skin	13.6%	1700
No change	Brain	5.2%	650
↑367%	Heart	4.4%	550
↑1066%	Skeletal muscle	64%	8000
↓30%	Bone, other	3.6%	450

Total cardiac output 12,500

Percent cardiac output	Flow (ml/min)		
27%	1350	Digestive tract, liver	
20%	1000	Kidneys	
9%	450	Skin	
13%	650	Brain	
3%	150	Heart	
15%	750	Skeletal muscle	
13%	650	Bone, other	

Total cardiac output 5000

(a) At rest

(b) During moderate exercise

● **FIGURE 10-12 Magnitude and distribution of the cardiac output at rest and during moderate exercise.** Not only does cardiac output increase during exercise but the distribution of cardiac output is adjusted to support the heightened physical activity. The percentage of cardiac output going to the skeletal muscles and heart rises, thereby delivering extra O$_2$ and nutrients needed to support these muscles' stepped-up rate of ATP consumption. The percentage going to the skin increases as a way to eliminate from the body surface the extra heat generated by the exercising muscles. The percentage going to most other organs shrinks. Only the magnitude of blood flow to the brain remains unchanged as the distribution of cardiac output is readjusted during exercise.

▲ TABLE 10-2	Functions of Endothelial Cells

- Line the blood vessels and heart chambers; serve as a physical barrier between the blood and the remainder of the vessel wall
- Secrete vasoactive substances in response to local chemical and physical changes; these substances cause relaxation (vasodilation) or contraction (vasoconstriction) of the underlying smooth muscle
- Secrete substances that stimulate new vessel growth and proliferation of smooth muscle cells in vessel walls
- Participate in the exchange of materials between the blood and surrounding tissue cells across capillaries through vesicular transport (see p. 25)
- Influence formation of platelet plugs, clotting, and clot dissolution (see Chapter 11)
- Participate in the determination of capillary permeability by contracting to vary the size of the pores between adjacent endothelial cells

▲ TABLE 10-3	Functions of Nitric Oxide (NO)

- Causes relaxation of arteriolar smooth muscle. By means of this action, NO plays an important role in controlling blood flow through the tissues and in maintaining mean arterial blood pressure.
- Dilates the arterioles of the penis and clitoris, thus serving as the direct mediator of erection of these reproductive organs. Erection is accomplished by rapid engorgement of these organs with blood.
- Used as chemical warfare against bacteria and cancer cells by macrophages, large phagocytic cells of the immune system.
- Interferes with platelet function and blood clotting at sites of vessel damage.
- Serves as a novel type of neurotransmitter in the brain and elsewhere.
- Plays a role in the changes underlying memory.
- By promoting relaxation of digestive-tract smooth muscle, helps regulate peristalsis, a type of contraction that pushes digestive tract contents forward.
- Relaxes the smooth muscle cells in the airways of the lungs, helping keep these passages open to facilitate movement of air in and out of the lungs.
- Modulates the filtering process involved in urine formation.
- Directs blood flow to O_2-starved tissues.
- May play a role in relaxation of skeletal muscle.

related activities, some of which will be described elsewhere (▲ Table 10-2). Among these functions, endothelial cells release locally acting chemical messengers in response to chemical changes in their environment (such as a reduction in O_2) or physical changes (such as stretching of the vessel wall). These local vasoactive ("acting on vessels") mediators act on the underlying smooth muscle to alter its state of contraction.

Among the best studied of these local vasoactive mediators is **nitric oxide (NO),** which causes local arteriolar vasodilation by inducing relaxation of arteriolar smooth muscle in the vicinity. It does so by increasing the concentration of the intracellular second messenger cyclic GMP, which leads to activation of an enzyme that reduces phosphorylation of myosin. Remember that smooth-muscle myosin can bind with actin and promote filament sliding through cycles of power strokes only when myosin is phosphorylated (see p. 292). NO is a small, highly reactive, short-lived gas molecule that once was known primarily as a toxic air pollutant. Yet studies have revealed an astonishing number of biological roles for NO, which is produced in many other tissues besides endothelial cells. In fact, NO serves as one of the body's most important messenger molecules, as shown by the range of functions identified for this chemical and listed in ▲ Table 10-3. As you can see, most areas of the body are influenced by this versatile intercellular messenger molecule.

Endothelial cells release other important chemicals besides NO. **Endothelin,** another endothelial vasoactive substance, causes arteriolar smooth muscle contraction and is one of the most potent vasoconstrictors yet identified. Still other chemicals, released from the endothelium in response to chronic changes in blood flow to an organ, trigger long-term vascular changes that permanently influence blood flow to a region. Some chemicals, for example, stimulate new vessel growth, a process known as **angiogenesis.**

Local histamine release pathologically dilates arterioles.

Histamine is another local chemical mediator that influences arteriolar smooth muscle, but it is not released in response to local metabolic changes and is not derived from endothelial cells. Although histamine normally does not participate in controlling blood flow, it is important in certain pathological conditions.

 Histamine is synthesized and stored within special connective tissue cells in many organs and in certain types of circulating white blood cells. When organs are injured or during allergic reactions, histamine is released and acts as a paracrine in the damaged region (see p. 114). By promoting relaxation of arteriolar smooth muscle, histamine is the major cause of vasodilation in an injured area. The resultant increase in blood flow into the area produces the redness and contributes to the swelling seen with inflammatory responses (see Chapter 12 for further details).

Local physical influences on arteriolar radius include temperature changes, shear stress, and stretch.

Among the physical influences on arteriolar smooth muscle, the effect of temperature changes is exploited clinically, but the chemically mediated response to shear stress and the myogenic response to stretch are most important physiologically. Let us examine each of these effects.

LOCAL HEAT OR COLD APPLICATION Heat application, by causing localized arteriolar vasodilation, is a useful therapeutic agent for promoting increased blood flow to an area. Conversely, applying ice packs to an inflamed area produces vasoconstriction, which reduces swelling by counteracting histamine-induced vasodilation.

CHEMICAL RESPONSE TO CHANGES IN SHEAR STRESS Due to friction, blood flowing over the surface of the vessel lining creates a longitudinal force known as **shear stress** on the endothelial cells. An increase in shear stress causes the endothelial cells to release NO, which diffuses to the underlying smooth muscle and promotes vasodilation. The resultant increase in arteriolar diameter reduces shear stress in the vessel. In response to shear stress on a long-term basis, endothelial cells orient themselves parallel to the direction of blood flow.

MYOGENIC RESPONSE TO STRETCH Arteriolar smooth muscle responds to being passively stretched by myogenically increasing its tone via vasoconstriction, thereby acting to resist the initial passive stretch. Conversely, a decrease in arteriolar stretching induces a reduction in myogenic vessel tone by promoting vasodilation. Endothelial-derived vasoactive substances may also contribute to these mechanically induced responses. The extent of passive stretch varies with the volume of blood delivered to the arterioles from the arteries. An increase in mean arterial pressure drives more blood forward into the arterioles and stretches them further, whereas arterial occlusion blocks blood flow into the arterioles and reduces arteriolar stretch.

Myogenic responses, coupled with metabolically induced responses, are important in reactive hyperemia and autoregulation, topics to which we now turn our attention.

REACTIVE HYPEREMIA When the blood supply to a region is completely occluded, arterioles in the region dilate because of (1) myogenic relaxation, which occurs in response to the diminished stretch accompanying no blood flow, and (2) changes in local chemical composition. Many of the same chemical changes occur in a blood-deprived tissue that occur during metabolically induced active hyperemia. When a tissue's blood supply is blocked, O_2 levels decrease in the deprived tissue; the tissue continues to consume O_2, but no fresh supplies are being delivered. Meanwhile, the concentrations of CO_2, acid, and other metabolites rise. Even though their production does not increase as it does when a tissue is more active metabolically, these substances accumulate in the tissue when the normal amounts produced are not "washed away" by blood.

After the occlusion is removed, blood flow to the previously deprived tissue is transiently much higher than normal because the arterioles are widely dilated. This post-occlusion increase in blood flow, called **reactive hyperemia,** can take place in any tissue. Such a response is beneficial for rapidly restoring the lo-

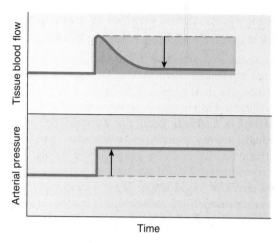

● **FIGURE 10-13 Autoregulation of tissue blood flow.** Even though blood flow through a tissue immediately increases in response to a rise in arterial pressure, the tissue blood flow is reduced gradually as a result of autoregulation within the tissue, despite a sustained increase in arterial pressure.

cal chemical composition to normal. Of course, prolonged blockage of blood flow leads to irreversible tissue damage.

AUTOREGULATION When mean arterial pressure falls (for example, because of hemorrhage or a weakened heart), the driving force is reduced, so blood flow to organs decreases. This is a milder version of what happens during vessel occlusion. The resultant changes in local metabolites and the reduced stretch in the arterioles collectively bring about arteriolar dilation to help restore tissue blood flow to normal despite the reduced driving pressure. On the negative side, widespread arteriolar dilation reduces the mean arterial pressure still further, which aggravates the problem. Conversely, in the presence of sustained elevations in mean arterial pressure (hypertension), local chemical and myogenic influences triggered by the initial increased flow of blood to tissues bring about an increase in arteriolar tone and resistance. This greater degree of vasoconstriction subsequently reduces tissue blood flow toward normal despite the elevated blood pressure (● Figure 10-13). **Autoregulation** ("self-regulation") is the term for these local arteriolar mechanisms that keep tissue blood flow fairly constant despite rather wide deviations in mean arterial driving pressure. Not all organs autoregulate equally. As examples, the brain autoregulates best, the kidneys are good at autoregulation, and skeletal muscle has poor autoregulatory abilities.

Active hyperemia, reactive hyperemia, and histamine release all deliberately increase blood flow to a particular tissue for a specific purpose by inducing local arteriolar vasodilation. In contrast, autoregulation is a means by which each tissue resists alterations in its own blood flow secondary to changes in mean arterial pressure by making appropriate adjustments in arteriolar radius.

This completes our discussion of the local control of arteriolar radius. Now let us shift attention to extrinsic control of arteriolar radius.

Extrinsic sympathetic control of arteriolar radius is important in regulating blood pressure.

Extrinsic control of arteriolar radius includes both neural and hormonal influences, the effects of the sympathetic nervous system being the most important. Sympathetic nerve fibers supply arteriolar smooth muscle everywhere in the systemic circulation except in the brain. Recall that a certain level of ongoing sympathetic activity contributes to vascular tone. Increased sympathetic activity produces generalized arteriolar vasoconstriction, whereas decreased sympathetic activity leads to generalized arteriolar vasodilation. These widespread changes in arteriolar resistance bring about changes in mean arterial pressure because of their influence on total peripheral resistance, as follows.

INFLUENCE OF TOTAL PERIPHERAL RESISTANCE ON MEAN ARTERIAL PRESSURE To find the effect of changes in arteriolar resistance on mean arterial pressure, the formula $F = \Delta P/R$ applies to the entire circulation as well as to a single vessel:

- F: Looking at the circulatory system as a whole, flow (F) through all the vessels in either the systemic or pulmonary circulation is equal to the *cardiac output*.

- ΔP: The pressure gradient (ΔP) for the entire systemic circulation is the *mean arterial pressure*. (ΔP equals the difference in pressure between the beginning and the end of the systemic circulatory system. The beginning pressure is the mean arterial pressure as the blood leaves the left ventricle at an average of 93 mm Hg. The end pressure in the right atrium is 0 mm Hg. Therefore, $\Delta P = 93$ mm Hg $- 0$ mm Hg $= 93$ mm Hg, which is equivalent to the mean arterial pressure.)

- R: The total resistance (R) offered by all the systemic peripheral vessels together is the **total peripheral resistance.** By far the greatest percentage of the total peripheral resistance is due to arteriolar resistance, because arterioles are the primary resistance vessels.

Therefore, for the entire systemic circulation, rearranging

$$F = \Delta P/R$$

to

$$\Delta P = F \times R$$

gives us the equation

Mean arterial pressure (MAP) =
cardiac output (CO) × total peripheral resistance (TPR)

Thus, the extent of total peripheral resistance offered collectively by all the systemic arterioles influences the mean arterial pressure immensely. A dam provides an analogy to this relationship. At the same time a dam restricts the flow of water downstream, it increases the pressure upstream by elevating the water level in the reservoir behind the dam. Similarly, generalized, sympathetically induced vasoconstriction reflexly reduces blood flow downstream to the organs while elevating the upstream mean arterial pressure, thereby increasing the main driving force for blood flow to all the organs.

These effects seem counterproductive. Why increase the driving force for flow to the organs by increasing arterial blood pressure while reducing flow to the organs by narrowing the vessels supplying them? In effect, the sympathetically induced arteriolar responses help maintain the appropriate driving pressure head (that is, the mean arterial pressure) to all organs. The extent to which each organ actually receives blood flow is determined by local arteriolar adjustments that override the sympathetic constrictor effect. If all arterioles were dilated, blood pressure would fall substantially, so there would not be an adequate driving force for blood flow. An analogy is the pressure head for water in the pipes in your home. If the water pressure is adequate, you can selectively obtain satisfactory water flow at any of the faucets by turning the appropriate handle to the open position. If the water pressure in the pipes is too low, however, you cannot obtain satisfactory flow at any faucet, even if you turn the handle to the maximally open position. Tonic sympathetic activity thus constricts most vessels (with the exception of those in the brain) to help maintain a pressure head on which organs can draw as needed through local mechanisms that control arteriolar radius.

NOREPINEPHRINE'S INFLUENCE ON ARTERIOLAR SMOOTH MUSCLE The norepinephrine released from sympathetic nerve endings combines with α_1-adrenergic receptors on arteriolar smooth muscle to bring about vasoconstriction (see p. 243). Cerebral (brain) arterioles are the only ones that do not have α_1 receptors, so no vasoconstriction occurs in the brain. It is important that cerebral arterioles are not reflexly constricted by neural influences, because brain blood flow must remain constant to meet the brain's continuous need for O_2, no matter what is going on elsewhere in the body. Cerebral vessels are almost entirely controlled by local mechanisms that maintain a constant blood flow to support a constant level of brain metabolic activity. In fact, reflex vasoconstrictor activity in the remainder of the cardiovascular system is aimed at maintaining an adequate pressure head for blood flow to the vital brain.

Thus, sympathetic activity contributes in an important way to maintaining mean arterial pressure, assuring an adequate driving force for blood flow to the brain at the expense of organs that can better withstand reduced blood flow. Other organs that really need additional blood, such as active muscles (including active heart muscle), obtain it through local controls that override the sympathetic effect.

LOCAL CONTROLS OVERRIDING SYMPATHETIC VASOCONSTRICTION Skeletal and cardiac muscles have the most powerful local control mechanisms with which to override generalized sympathetic vasoconstriction. For example, if you are pedaling a bicycle, the increased activity in the skeletal muscles of your legs brings about an overriding local, metabolically induced vasodilation in those particular muscles, despite the generalized sympathetic vasoconstriction that accompanies exercise. As a result, more blood flows through your leg muscles but not through your inactive arm muscles.

NO PARASYMPATHETIC INNERVATION TO ARTERIOLES Arterioles have no significant parasympathetic innervation, with the exception of the abundant parasympathetic vasodilator supply to the arterioles of the penis and clitoris. The rapid, profuse vasodilation induced by parasympathetic stimulation in these organs (by means of promoting release of NO) is largely responsible for accomplishing erection. Vasodilation elsewhere is produced by decreasing sympathetic vasoconstrictor activity below its tonic level.[1] When mean arterial pressure rises above normal, reflex reduction in sympathetic vasoconstrictor activity accomplishes generalized arteriolar vasodilation to help bring the driving pressure down toward normal.

The medullary cardiovascular control center and several hormones regulate blood pressure.

The main region of the brain that adjusts sympathetic output to the arterioles is the **cardiovascular control center** in the medulla of the brain stem. This is the integrating center for blood pressure regulation (described in further detail later in this chapter). Several other brain regions also influence blood distribution, the most notable being the hypothalamus, which, as part of its temperature-regulating function, controls blood flow to the skin to adjust heat loss to the environment.

In addition to neural reflex activity, several hormones also extrinsically influence arteriolar radius. These hormones include the adrenal medullary hormones *epinephrine* and *norepinephrine,* which generally reinforce the sympathetic nervous system in most organs, as well as *vasopressin* and *angiotensin II,* which are important in controlling fluid balance.

INFLUENCE OF EPINEPHRINE AND NOREPINEPHRINE Sympathetic stimulation of the adrenal medulla causes this endocrine gland to release epinephrine and norepinephrine. Adrenal medullary norepinephrine combines with the same α_1 receptors as sympathetically released norepinephrine to produce generalized vasoconstriction. However, epinephrine, the more abundant of the adrenal medullary hormones, combines with both β_2 and α_1 receptors but has a much greater affinity for the β_2 receptors (▲ Table 10-4). Activation of β_2 receptors produces vasodilation, but not all tissues have β_2 receptors; they are most abundant in the arterioles of the heart and skeletal muscles. During sympathetic discharge, the released epinephrine combines with the β_2 receptors in the heart and skeletal muscle to reinforce local vasodilatory mechanisms in these tissues. Arterioles in digestive organs and kidneys, in contrast, are equipped only with α_1 receptors. Therefore, the arterioles of these organs undergo more profound vasoconstriction during generalized sympathetic discharge than those in the heart and skeletal muscle do. Lacking β_2 receptors, the digestive organs and kidneys do not experience an overriding vasodilatory response on top of the α_1 receptor–induced vasoconstriction.

---

[1]A portion of the skeletal muscle fibers in some species is supplied by sympathetic cholinergic (ACh-releasing) fibers that bring about vasodilation in anticipation of exercise. However, the existence of such sympathetic vasodilator fibers in humans remains questionable.

	RECEPTOR TYPE	
Characteristic	α_1	β_2
Location of the Receptor	All arteriolar smooth muscle except in the brain	Arteriolar smooth muscle in the heart and skeletal muscles
Chemical Mediator	Norepineprine from sympathetic fibers and the adrenal medulla — Epinephrine from the adrenal medulla (less affinity for this receptor)	Epinephrine from the adrenal medulla (greater affinity for this receptor)
Arteriolar Smooth Muscle Response	Vasoconstriction	Vasodilation

▲ TABLE 10-4 **Arteriolar Smooth Muscle Adrenergic Receptors**

INFLUENCE OF VASOPRESSIN AND ANGIOTENSIN II The two other hormones that extrinsically influence arteriolar tone are vasopressin and angiotensin II. Vasopressin is primarily involved in maintaining water balance by regulating the amount of water the kidneys retain for the body during urine formation (see p. 527). Angiotensin II is part of a hormonal pathway, the *renin–angiotensin–aldosterone system,* which is important in regulating the body's salt balance. This pathway promotes salt conservation during urine formation and also leads to water retention because salt exerts a water-holding osmotic effect in the ECF (see p. 561). Thus, both these hormones play important roles in maintaining the body's fluid balance, which in turn is an important determinant of plasma volume and blood pressure.

In addition, both vasopressin and angiotensin II are potent vasoconstrictors. Their role in this regard is especially crucial during hemorrhage. A sudden loss of blood reduces the plasma volume, which triggers increased secretion of both these hormones to help restore plasma volume. Their vasoconstrictor effect also helps maintain blood pressure despite abrupt loss of plasma volume. (The functions and control of these hormones are discussed more thoroughly in later chapters.)

This completes our discussion of the various factors that affect total peripheral resistance, the most important of which are controlled adjustments in arteriolar radius. These factors are summarized in ● Figure 10-14.

We now turn attention to the next vessels in the vascular tree, the capillaries.

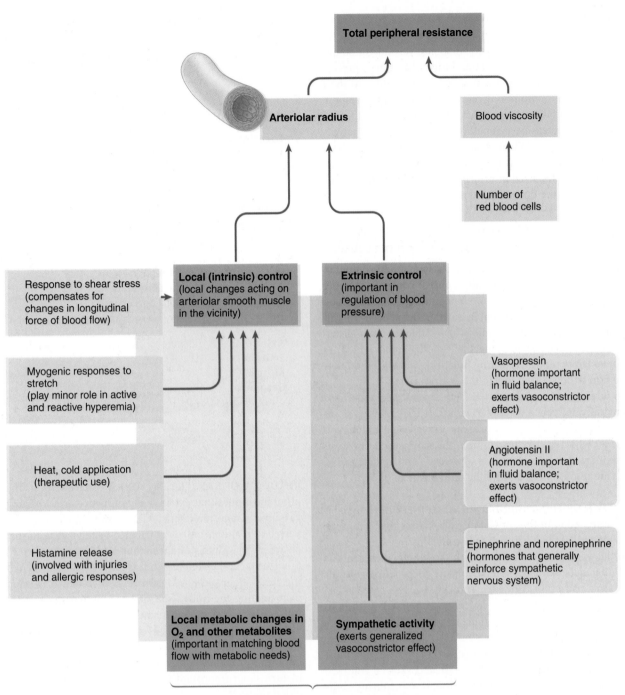

Major factors affecting arteriolar radius

● **FIGURE 10-14 Factors affecting total peripheral resistance.** The primary determinant of total peripheral resistance is the adjustable arteriolar radius. Two major categories of factors influence arteriolar radius: (1) local (intrinsic) control, which is primarily important in matching blood flow through a tissue with the tissue's metabolic needs and is mediated by local factors acting on the arteriolar smooth muscle; and (2) extrinsic control, which is important in regulating blood pressure and is mediated primarily by sympathetic influence on arteriolar smooth muscle.

Capillaries

Capillaries, the sites for exchange of materials between blood and tissue cells,[2] branch extensively to bring blood within the reach of every cell.

Capillaries are ideally suited to serve as sites of exchange.

There are no carrier-mediated transport systems across capillaries, with the exception of those in the brain that play a role in the blood–brain barrier (see p. 141). Materials are exchanged across capillary walls mainly by diffusion.

FACTORS THAT ENHANCE DIFFUSION ACROSS CAPILLARIES
Capillaries are ideally suited to enhance diffusion, in accordance with Fick's law of diffusion (see p. 62). They minimize diffusion distances while maximizing surface area and time available for exchange, as follows:

1. Diffusing molecules have only a short distance to travel between blood and surrounding cells because of the thin capillary wall and small capillary diameter, coupled with the close proximity of every cell to a capillary. This short distance is important because the rate of diffusion slows down as the diffusion distance increases.

   a. Capillary walls are very thin (1 μm in thickness; in contrast, the diameter of a human hair is 100 μm). Capillaries consist of only a single layer of flat endothelial cells—essentially the lining of the other vessel types. No smooth muscle or connective tissue is present (●Figure 10-15a; also see ▲ Table 10-1, p. 348). The endothelial cells are supported by a thin *basement membrane,* a surrounding acellular (lacking cells) layer of extracellular matrix consisting of glycoproteins and collagen. Materials entering or leaving the capillaries pass freely through the basement membrane. Capillaries also have pores through which water-soluble materials can pass. The size and number of capillary pores vary, depending on the tissue.

   b. Each capillary is so narrow (7 μm average diameter) that red blood cells (8 μm diameter) have to squeeze through in single file (●Figure 10-15b). Consequently, plasma contents are either in direct contact with the inside of the capillary wall or are only a short diffusing distance from it.

   c. Researchers estimate that because of extensive capillary branching, no cell is farther than 0.01 cm (4/1000 in.) from a capillary.

[2]Actually, some exchange takes place across the other microcirculatory vessels, especially the postcapillary venules. The entire vasculature is a continuum and does not abruptly change from one vascular type to another. When the term *capillary exchange* is used, it tacitly refers to all exchange at the microcirculatory level, the majority of which occurs across the capillaries.

(a) Cross section of a capillary

Endothelial cell nucleus   Capillary lumen

R. J. Bolander, Don W. Fawcett/Visuals Unlimited

Capillary   Red blood cell

(b) Capillary bed

Lennart Nilsson/Bonnier Alba AB

● **FIGURE 10-15 Capillary anatomy.** (a) Electron micrograph showing that the capillary wall consists of a single layer of endothelial cells. The nucleus of one of these cells is shown. (b) The capillaries are so narrow that red blood cells must pass through the capillary bed in single file.

2. Because capillaries are distributed in such incredible numbers (estimates range from 10 to 40 billion capillaries), a tremendous total surface area is available for exchange (an estimated 600 m²). Despite this large number of capillaries, at any point in time they contain only 5% of the total blood volume (250 ml out of a total of 5000 ml). As a result, a small volume of blood is exposed to an extensive surface area. If all the capillary surfaces were stretched out in a flat sheet and the volume of blood contained within the capillaries were spread over the top, this would be roughly equivalent to spreading a half pint of paint over the floor of a high school gymnasium. Imagine how thin the paint layer would be!

3. Blood flows more slowly in the capillaries than elsewhere in the circulatory system. The extensive capillary branching is re-

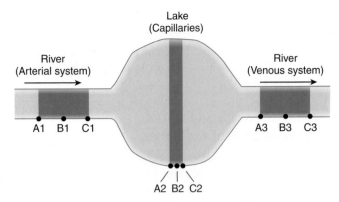

FIGURE 10-17 Relationship between total cross-sectional area and velocity of flow. The three dark blue areas represent equal volumes of water. During one minute, this volume of water moves forward from points A to points C. Therefore, an identical volume of water flows past points B1, B2, and B3 during this minute; that is, the flow rate is the same at all points along the length of this body of water. However, during that minute the identical volume of water moves forward a much shorter distance in the wide lake (A2 to C2) than in the much narrower river (A1 to C1 and A3 to C3). Thus, velocity of flow is much slower in the lake than in the river. Similarly, velocity of flow is much slower in the capillaries than in the arterial and venous systems.

FIGURE 10-16 Comparison of blood flow rate and velocity of flow in relation to total cross-sectional area. The blood flow rate (red curve) is identical through all levels of the circulatory system and is equal to the cardiac output (5 liters/min at rest). The velocity of flow (purple curve) varies throughout the vascular tree and is inversely proportional to the total cross-sectional area (green curve) of all the vessels at a given level. Note that the velocity of flow is slowest in the capillaries, which have the largest total cross-sectional area.

tance). The *velocity of flow* is the linear *speed*, or distance per unit of time, with which blood flows forward through a given segment of the circulatory system. Because the circulatory system is a closed system, the volume of blood flowing through any level of the system must equal the cardiac output. For example, if the heart pumps out 5 liters of blood per minute, and 5 liters/min return to the heart, then 5 liters/min must flow through the arteries, arterioles, capillaries, and veins. Therefore, the flow rate is the same at all levels of the circulatory system.

However, the velocity with which blood flows through the different segments of the vascular tree varies because velocity of flow is inversely proportional to the total cross-sectional area of all the vessels at any given level of the circulatory system. Even though the cross-sectional area of each capillary is extremely small compared to that of the large aorta, the total cross-sectional area of all the capillaries added together is about 750 times greater than the cross-sectional area of the aorta because there are so many capillaries. Accordingly, blood slows considerably as it passes through the capillaries (● Figure 10-16). This slow velocity allows adequate time for exchange of nutrients and metabolic end products between blood and tissue cells, which is the sole purpose of the entire circulatory system. As the capillaries rejoin to form veins, the total cross-sectional area is once again reduced, and the velocity of blood flow increases as blood returns to the heart.

As an analogy, consider a river (the arterial system) that widens into a lake (the capillaries), then narrows into a river again (the venous system) (● Figure 10-17). The flow rate is the same throughout the length of this body of water; that is, identical volumes of water are flowing past all the points along the bank of the river and lake. However, the velocity of flow is slower in the wide lake than in the narrow river because the

sponsible for this slow velocity of blood flow through the capillaries. Let us see why blood slows down in the capillaries.

SLOW VELOCITY OF FLOW THROUGH CAPILLARIES First, we need to clarify a potentially confusing point. The term *flow* can be used in two different contexts—flow rate and velocity of flow. The *flow rate* refers to the *volume* of blood per unit of time flowing through a given segment of the circulatory system (this is the flow we have been talking about in relation to the pressure gradient and resis-

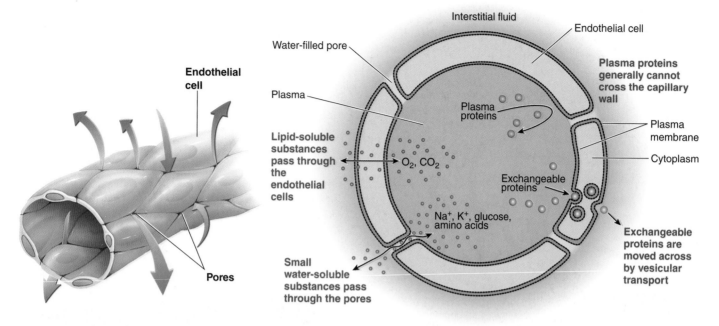

Endothelial cell

Pores

(a) Continuous capillary

Interstitial fluid

Endothelial cell

Water-filled pore

Plasma

Plasma proteins generally cannot cross the capillary wall

Plasma membrane

Cytoplasm

Plasma proteins

Lipid-soluble substances pass through the endothelial cells

O_2, CO_2

Exchangeable proteins

Na^+, K^+, glucose, amino acids

Small water-soluble substances pass through the pores

Exchangeable proteins are moved across by vesicular transport

(b) Transport across a continuous capillary wall

● **FIGURE 10-18 Exchanges across a continuous capillary wall, the most common type of capillary.** (a) Slitlike gaps between adjacent endothelial cells form pores within the capillary wall. (b) As depicted in this cross section of a capillary wall, small water-soluble substances are exchanged between the plasma and the interstitial fluid by passing through the water-filled pores, whereas lipid-soluble substances are exchanged across the capillary wall by passing through the endothelial cells. Proteins to be moved across are exchanged by vesicular transport. Plasma proteins generally cannot escape from the plasma across the capillary wall.

identical volume of water, now spread out over a larger cross-sectional area, moves forward a much shorter distance in the wide lake than in the narrow river during a given period of time. You could readily observe the forward movement of water in the swift-flowing river, but the forward motion of water in the lake would be unnoticeable.

Also, because of the capillaries' tremendous total cross-sectional area, the resistance offered by all the capillaries is much lower than that offered by all the arterioles, even though each capillary has a smaller radius than each arteriole. For this reason, the arterioles contribute more to total peripheral resistance. Furthermore, arteriolar caliber (and, accordingly resistance) is subject to control, whereas capillary caliber cannot be adjusted.

Water-filled capillary pores permit passage of small, water-soluble substances.

Diffusion across capillary walls also depends on the walls' permeability to the materials being exchanged. The endothelial cells forming the capillary walls fit together in jigsaw-puzzle fashion, but the closeness of the fit varies considerably between organs. In most capillaries, the endothelial cells are *continuous,* or closely joined, with only narrow, water-filled clefts, or **pores,** at the junctions between the cells (●Figure 10-18).

The size of the capillary pores varies from organ to organ. Following is an examination of capillaries of different porosity, from the tightest to the leakiest of these exchange vessels.

■ At one extreme, the endothelial cells in brain capillaries are joined by tight junctions so that pores are nonexistent. These junctions prevent transcapillary ("across capillary") passage of materials between the cells and thus constitute part of the protective blood–brain barrier (see p. 59).

■ In most tissues (for example, skeletal and cardiac muscle, lungs, and adipose tissue), small, water-soluble substances such as ions, glucose, and amino acids can readily pass through the water-filled pores, which are about 4 nm wide, but large, water-soluble materials that cannot fit through the pores, such as plasma proteins, are kept from passing. Lipid-soluble substances, such as O_2 and CO_2, can readily pass through the endothelial cells themselves by dissolving in the lipid bilayer barrier.

■ In addition to having the narrow clefts between endothelial cells, the more porous capillaries of the kidneys and intestines also have larger 20- to 100-nm holes or **fenestrations** (*fenestra* means "window") that extend through the thickness of the endothelial cells themselves. Fenestrations are important in the rapid movement of fluid across the capillaries in these organs during the formation of urine and during the absorption of a digested meal, respectively.

■ At the other extreme, the endothelial cells of liver cells are *discontinuous;* that is, they are not in such close contact as in continuous capillaries. The gaps between adjacent cells in discontinuous capillaries range from 10 nm to 1000 nm, creating very large pores compared to the 4 nm clefts found in continu-

Arteriole

Smooth muscle

Precapillary sphincter

Metarteriole (thoroughfare channel)

Capillary

Venule

(a) Sphincters relaxed

(b) Sphincters contracted

● **FIGURE 10-19 Capillary bed.** Capillaries branch either directly from an arteriole or from a metarteriole, a thoroughfare channel between an arteriole and venule. Capillaries rejoin at either a venule or a metarteriole. Smooth muscle cells form precapillary sphincters that encircle capillaries as they arise from a metarteriole. (a) When the precapillary sphincters are relaxed, blood flows through the entire capillary bed. (b) When the precapillary sphincters are contracted, blood flows only through the metarteriole, bypassing the capillary bed.

ous capillaries. Discontinuous capillaries form large channels known as *sinusoids* that are five times wider than traditional capillaries. Liver sinusoids have fenestrations and such large intercellular pores that even proteins pass through readily. This is adaptive, because the liver's functions include synthesis of plasma proteins and the metabolism of protein-bound substances such as cholesterol. These proteins must all pass through the liver's capillary (sinusoid) walls.

The leakiness of various capillary beds is therefore a function of how tightly the endothelial cells are joined (how wide the intercellular spaces are) and whether or not fenestrations are present, which varies according to the different organs' needs. For convenience, in the future we lump the intercellular spaces and intracellular fenestrations into the single category of "capillary pores."

Scientists traditionally considered the capillary wall a passive sieve, like a brick wall with permanent gaps in the mortar acting as pores. Recent studies, however, suggest that endothelial cells can actively change to regulate capillary permeability; that is, in response to appropriate signals, the "bricks" can readjust themselves to vary the size of the holes. Thus, the degree of leakiness does not necessarily remain constant for a given capillary bed. For example, histamine increases capillary permeability by triggering contractile responses in endothelial cells to widen the intercellular gaps. This is not a muscular contraction,

because no smooth muscle cells are present in capillaries. It is the result of an actin–myosin contractile apparatus in the nonmuscular capillary endothelial cells. Because of these enlarged pores, the affected capillary wall is leakier. As a result, normally retained plasma proteins escape into the surrounding tissue, where they exert an osmotic effect. Along with histamine-induced vasodilation, the resulting additional local fluid retention contributes to inflammatory swelling.

Vesicular transport also plays a limited role in the passage of materials across the capillary wall. Large non-lipid-soluble molecules such as protein hormones that must be exchanged between blood and surrounding tissues are transported from one side of the capillary wall to the other in endocytic-exocytic vesicles (see p. 75).

Many capillaries are not open under resting conditions.

The branching and reconverging arrangement within capillary beds varies somewhat, depending on the tissue. Capillaries typically branch either directly from an arteriole or from a thoroughfare channel known as a **metarteriole,** which runs between an arteriole and a venule. Likewise, capillaries may rejoin at either a venule or a metarteriole (● Figure 10-19a).

Unlike the true capillaries within a capillary bed, metarterioles are sparsely surrounded by wisps of spiraling smooth muscle cells. These cells also form **precapillary sphincters,** each of which consists of a ring of smooth muscle around the entrance to a capillary as it arises from a metarteriole.[3]

ROLE OF PRECAPILLARY SPHINCTERS Precapillary sphincters are not innervated, but they have a high degree of myogenic tone and are sensitive to local metabolic changes. They act as stopcocks to control blood flow through the particular capillary that each one guards. Arterioles perform a similar function for a small group of capillaries. Capillaries themselves have no smooth muscle, so they cannot actively participate in regulating their own blood flow.

Generally, tissues that are more metabolically active have a greater density of capillaries. Muscles, for example, have relatively more capillaries than their tendinous attachments. Only about 10% or less of the precapillary sphincters in a resting muscle are open at any moment, however, so blood is flowing through only about 10% of the muscle's capillaries or flowing directly through the metarteriole without entering the remaining capillary bed (● Figure 10-19b). As chemical concentrations start to change in a region of the muscle tissue supplied by closed-down capillaries, the precapillary sphincters and arterioles in the region relax. Restoration of the chemical concentrations to normal as a result of increased blood flow to that region removes the impetus for vasodilation, so the precapillary sphincters close once again and the arterioles return to normal tone. In this way, blood flow through any given capillary is often intermittent, as a result of arteriolar and precapillary sphincter action working in concert.

When the muscle as a whole becomes more active, a greater percentage of the precapillary sphincters relax, simultaneously opening up more capillary beds, while concurrent arteriolar vasodilation increases total flow to the organ. As a result of more blood flowing through more open capillaries, the total volume and surface area available for exchange increase, and the diffusion distance between the cells and an open capillary decreases (● Figure 10-20). Thus, blood flow through a particular tissue (assuming a constant blood pressure) is regulated by (1) the degree of resistance offered by the arterioles in the organ, controlled by sympathetic activity and local factors; and (2) the number of open capillaries, controlled by action of the same local metabolic factors on precapillary sphincters.

---

[3]Although generally accepted, the existence of precapillary sphincters in humans has not been conclusively established.

● **FIGURE 10-20 Complementary action of precapillary sphincters and arterioles in adjusting blood flow through a tissue in response to changing metabolic needs.**

Interstitial fluid is a passive intermediary between blood and cells.

Exchanges between blood and tissue cells are not made directly. Interstitial fluid, the true internal environment in immediate contact with the cells, acts as the go-between. Only 20% of the ECF circulates as plasma. The remaining 80% consists of interstitial fluid, which bathes all the cells in the body. Cells exchange materials directly with interstitial fluid, with the type and extent of exchange being governed by the properties of cellular plasma membranes. Movement across the plasma membrane may be either passive (that is, by diffusion down electrochemical gradients or by facilitated diffusion) or active (that is, by active carrier-mediated transport or by vesicular transport) (see ▲ Table 3-2, p. 76).

In contrast, exchanges across the capillary wall between plasma and interstitial fluid are largely passive. The only transport across this barrier that requires energy is the limited vesicular transport. Because capillary walls are highly permeable, exchange is so thorough that the interstitial fluid takes on essentially the same composition as incoming arterial blood, with the exception of the large plasma proteins that usually do not escape from the blood. Therefore, when we speak of exchanges between blood and tissue cells, we tacitly include interstitial fluid as a passive intermediary.

Exchanges between blood and surrounding tissues across the capillary walls are accomplished in two ways: (1) passive diffusion down concentration gradients, the primary mechanism for exchanging individual solutes; and (2) bulk flow, a process that fills the totally different function of determining the distribution of the ECF volume between the vascular and interstitial fluid compartments. We now examine each of these mechanisms in more detail, starting with diffusion.

Diffusion across the capillary walls is important in solute exchange.

Because there are no carrier-mediated transport systems in most capillary walls, solutes cross primarily by diffusion down concentration gradients. The chemical composition of arterial blood is carefully regulated to maintain the concentrations of individual solutes at levels that will promote each solute's movement in the appropriate direction across the capillary walls. The reconditioning organs continuously add nutrients and O_2 and remove CO_2 and other wastes as blood passes through them. Meanwhile, cells constantly use up supplies and generate metabolic wastes. As cells use up O_2 and glucose, the blood constantly brings in fresh supplies of these vital materials, maintaining concentration gradients that favor the net diffusion of these substances from blood to cells. Simultaneously, ongoing net diffusion of CO_2 and other metabolic wastes from cells to blood is maintained by the continual production of these wastes at the cell level and by their constant removal by the circulating blood (● Figure 10-21).

Because the capillary wall does not limit the passage of any constituent except plasma proteins, the extent of exchanges for each solute is independently determined by the magnitude of its concentration gradient between blood and surrounding cells. As cells increase their level of activity, they use up more O_2 and produce more CO_2, among other things. This creates larger concentration gradients for O_2 and CO_2 between these cells and blood, so more O_2 diffuses out of the blood into the cells and more CO_2 proceeds in the opposite direction, to help support the increased metabolic activity.

Bulk flow across the capillary walls is important in extracellular fluid distribution.

The second means by which exchange is accomplished across capillary walls is bulk flow. A volume of protein-free plasma actually filters out of the capillary, mixes with the surrounding interstitial fluid, and then is reabsorbed. This process is called **bulk flow** because the various constituents of the fluid are moving together in bulk, or as a unit, in contrast to the discrete diffusion of individual solutes down concentration gradients.

The capillary wall acts like a sieve, with fluid moving through its water-filled pores. When pressure inside the capillary exceeds pressure on the outside, fluid is pushed out through the pores in a process known as **ultrafiltration.** Most plasma proteins are retained on the inside during this process because of the pores' filtering effect, although a few do escape. Because

● **FIGURE 10-21 Independent exchange of individual solutes down their own concentration gradients across the capillary wall.**

all other constituents in plasma are dragged along as a unit with the volume of fluid leaving the capillary, the filtrate is essentially a protein-free plasma. When inward-driving pressures exceed outward pressures across the capillary wall, net inward movement of fluid from the interstitial fluid into the capillaries takes place through the pores, a process known as **reabsorption.**

FORCES INFLUENCING BULK FLOW Bulk flow occurs because of differences in the hydrostatic and colloid osmotic pressures between plasma and interstitial fluid. Even though pressure differences exist between plasma and surrounding fluid elsewhere in the circulatory system, only the capillaries have pores that let fluids pass through. Four forces influence fluid movement across the capillary wall (● Figure 10-22):

1. **Capillary blood pressure (P_C)** is the fluid or hydrostatic pressure exerted on the inside of the capillary walls by blood. This pressure tends to force fluid *out of* the capillaries into the interstitial fluid. By the level of the capillaries, blood pressure has dropped substantially because of frictional losses in pressure in the high-resistance arterioles upstream. On average, the hydrostatic pressure is 37 mm Hg at the arteriolar end of a tissue capillary (compared to a mean arterial pressure of 93 mm Hg). It declines even further, to 17 mm Hg, at the capillary's venular end because of further frictional loss coupled with the exit of fluid through ultrafiltration along the capillary's length (see ● Figure 10-9, p. 352).

2. **Plasma-colloid osmotic pressure (π_P),** also known as *oncotic pressure,* is a force caused by colloidal dispersion of plasma proteins (see p. A-10); it encourages fluid movement *into* the capillaries. Because plasma proteins remain in the plasma rather than entering the interstitial fluid, a protein concentration difference exists between plasma and interstitial fluid. Ac-

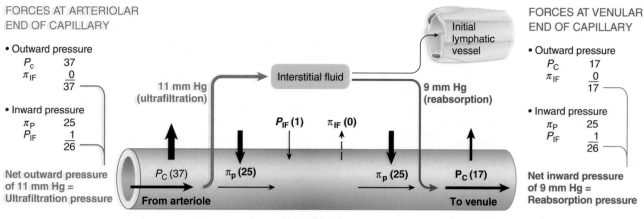

FORCES AT ARTERIOLAR END OF CAPILLARY

- Outward pressure

  P_C 37

  π_{IF} 0

  37

- Inward pressure

  π_P 25

  P_{IF} 1

  26

Net outward pressure of 11 mm Hg = Ultrafiltration pressure

All values are given in mm Hg.

Initial lymphatic vessel

Interstitial fluid

11 mm Hg (ultrafiltration)

9 mm Hg (reabsorption)

P_{IF} (1) π_{IF} (0)

P_C (37) π_p (25) π_p (25) P_C (17)

From arteriole   To venule

Blood capillary

FORCES AT VENULAR END OF CAPILLARY

- Outward pressure

  P_C 17

  π_{IF} 0

  17

- Inward pressure

  π_P 25

  P_{IF} 1

  26

Net inward pressure of 9 mm Hg = Reabsorption pressure

● **FIGURE 10-22 Bulk flow across the capillary wall.** Ultrafiltration occurs at the arteriolar end and reabsorption occurs at the venule end of the capillary as a result of imbalances in the physical forces acting across the capillary wall.

cordingly, there is also a water concentration difference between these two regions. Plasma has a higher protein concentration and a lower water concentration than interstitial fluid does. This difference exerts an osmotic effect that tends to move water from the area of higher water concentration in interstitial fluid to the area of lower water concentration (or higher protein concentration) in plasma (see p. 63). The other plasma constituents do not exert an osmotic effect, because they readily pass through the capillary wall, so their concentrations are equal in plasma and interstitial fluid. Plasma-colloid osmotic pressure averages 25 mm Hg.

3. **Interstitial fluid hydrostatic pressure** (P_{IF}) is the fluid pressure exerted on the outside of the capillary wall by interstitial fluid. This pressure tends to force fluid *into* the capillaries. Because of difficulties encountered in measuring interstitial fluid hydrostatic pressure, the actual value of the pressure is a controversial issue. It is either at, slightly above, or slightly below atmospheric pressure. For purposes of illustration, we will say it is 1 mm Hg above atmospheric pressure.

4. **Interstitial fluid–colloid osmotic pressure** (π_{IF}) is another force that does not normally contribute significantly to bulk flow. The small fraction of plasma proteins that leak across the capillary walls into the interstitial spaces are normally returned to the blood by the lymphatic system. Therefore, the protein concentration in the interstitial fluid is extremely low, and the interstitial fluid–colloid osmotic pressure is very close to zero. If plasma proteins pathologically leak into the interstitial fluid, however, as they do when histamine widens the capillary pores during tissue injury, the leaked proteins exert an osmotic effect that tends to promote movement of fluid *out of* the capillaries into the interstitial fluid.

Therefore, the two pressures that tend to force fluid out of the capillary are capillary blood pressure and interstitial fluid–colloid osmotic pressure. The two opposing pressures that tend to force fluid into the capillary are plasma-colloid osmotic pressure and interstitial fluid hydrostatic pressure. Now let us analyze the fluid movement that occurs across a capillary wall because of imbalances in these opposing physical forces (● Figure 10-22).

NET EXCHANGE OF FLUID ACROSS THE CAPILLARY WALL Net exchange at a given point across the capillary wall can be calculated using the following equation:

$$\text{Net exchange pressure} = \underbrace{(P_C + \pi_{IF})}_{\text{(outward pressure)}} - \underbrace{(\pi_P + P_{IF})}_{\text{(inward pressure)}}$$

A positive net exchange pressure (when the outward pressure exceeds the inward pressure) represents an ultrafiltration pressure. A negative net exchange pressure (when the inward pressure exceeds the outward pressure) represents a reabsorption pressure.

At the arteriolar end of the capillary, the outward pressure totals 37 mm Hg, whereas the inward pressure totals 26 mm Hg, for a net outward pressure of 11 mm Hg. Ultrafiltration takes place at the beginning of the capillary as this outward pressure gradient forces a protein-free filtrate through the capillary pores.

By the time the venular end of the capillary is reached, the capillary blood pressure has dropped, but the other pressures have remained essentially constant. At this point the outward pressure has fallen to a total of 17 mm Hg, whereas the total inward pressure is still 26 mm Hg, for a net inward pressure of 9 mm Hg. Reabsorption of fluid takes place as this inward pressure gradient forces fluid back into the capillary at its venular end.

Ultrafiltration and reabsorption, collectively known as *bulk flow,* are thus the result of a shift in the balance between the passive physical forces acting across the capillary wall. No active forces or local energy expenditures are involved in the bulk exchange of fluid between the plasma and surrounding interstitial fluid. With only minor contributions from the interstitial fluid forces, ultrafiltration occurs at the beginning of the capillary because capillary blood pressure exceeds plasma-colloid osmotic pressure, whereas by the end of the capillary, reabsorption takes place because blood pressure has fallen below osmotic pressure.

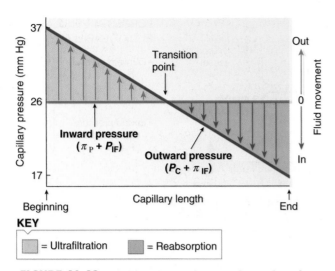

FIGURE 10-23 Net filtration and net reabsorption along the vessel length. The inward pressure ($\pi_P + P_{IF}$) remains constant throughout the length of the capillary, whereas the outward pressure ($P_C + \pi_{IF}$) progressively declines throughout the capillary's length. In the first half of the vessel, where the declining outward pressure still exceeds the constant inward pressure, progressively diminishing quantities of fluid are filtered out *(upward red arrows)*. In the last half of the vessel, progressively increasing quantities of fluid are reabsorbed *(downward blue arrows)* as the declining outward pressure falls farther below the constant inward pressure.

It is important to realize that we have taken "snapshots" at two points—at the beginning and at the end—in a hypothetical capillary. Actually, blood pressure gradually diminishes along the length of the capillary, so that progressively diminishing quantities of fluid are filtered out in the first half of the vessel and progressively increasing quantities of fluid are reabsorbed in the last half (● Figure 10-23). Even this situation is idealized. The pressures used in this figure are average values and controversial at that. Some capillaries have such high blood pressure that filtration actually occurs throughout their entire length, whereas others have such low hydrostatic pressure that reabsorption takes place throughout their length.

In fact, a recent theory that has received considerable attention is that net filtration occurs throughout the length of all *open* capillaries, whereas net reabsorption occurs throughout the length of all nearly *closed* capillaries. According to this theory, when the precapillary sphincter is relaxed, capillary blood pressure exceeds the plasma osmotic pressure even at the venular end of the capillary, promoting filtration throughout the length. When the precapillary sphincter is almost completely closed, the reduction in blood flow through the capillary diminishes capillary blood pressure below the plasma osmotic pressure even at the beginning of the capillary, so reabsorption takes place all along the capillary. Whichever mechanism is involved, the net effect is the same. A protein-free filtrate exits the capillaries and is ultimately reabsorbed.

ROLE OF BULK FLOW Bulk flow does not play an important role in the exchange of individual solutes between blood and tissues, because the quantity of solutes moved across the capillary wall by bulk flow is extremely small compared to the much larger transfer of solutes by diffusion. Thus, ultrafiltration and reabsorption are not important in the exchange of nutrients and wastes. Bulk flow is extremely important, however, in regulating the distribution of ECF between the plasma and interstitial fluid. Maintenance of proper arterial blood pressure depends in part on an appropriate volume of circulating blood. If plasma volume is reduced (for example, by hemorrhage), blood pressure falls. The resultant lowering of capillary blood pressure alters the balance of forces across the capillary walls. Because the net outward pressure is decreased while the net inward pressure remains unchanged, extra fluid is shifted from the interstitial compartment into the plasma as a result of reduced filtration and increased reabsorption. The extra fluid soaked up from the interstitial fluid provides additional fluid for the plasma, temporarily compensating for the loss of blood. Meanwhile, reflex mechanisms acting on the heart and blood vessels (described later) also come into play to help maintain blood pressure until long-term mechanisms, such as thirst (and its satisfaction) and reduction of urinary output, can restore the fluid volume to completely compensate for the loss.

Conversely, if the plasma volume becomes overexpanded, as with excessive fluid intake, the resulting rise in capillary blood pressure forces extra fluid from the capillaries into the interstitial fluid, temporarily relieving the expanded plasma volume until the excess fluid can be eliminated from the body by long-term measures, such as increased urinary output.

These internal fluid shifts between the two ECF compartments occur automatically and immediately whenever the balance of forces acting across the capillary walls is changed; they provide a temporary mechanism to help keep plasma volume fairly constant. In the process of restoring plasma volume to an appropriate level, interstitial fluid volume fluctuates, but it is much more important that plasma volume be kept constant, to ensure that the circulatory system functions effectively.

The lymphatic system is an accessory route by which interstitial fluid can be returned to the blood.

Even under normal circumstances, slightly more fluid is filtered out of the capillaries into the interstitial fluid than is reabsorbed from the interstitial fluid back into the plasma. On average, the net ultrafiltration pressure starts at 11 mm Hg at the beginning of the capillary, whereas the net reabsorption pressure only reaches 9 mm Hg by the vessel's end (see ● Figure 10-22). Because of this pressure differential, on average more fluid is filtered out of the first half of the capillary than is reabsorbed in its last half. The extra fluid filtered out as a result of this filtration–reabsorption imbalance is picked up by the **lymphatic system.** This extensive network of one-way vessels provides an accessory route by which fluid can be returned from the interstitial fluid to the blood. The lymphatic system functions much like a storm sewer that picks up and carries away excess rainwater so that it does not accumulate and flood an area.

PICKUP AND FLOW OF LYMPH Small, blind-ended terminal lymph vessels known as **initial lymphatics** permeate almost

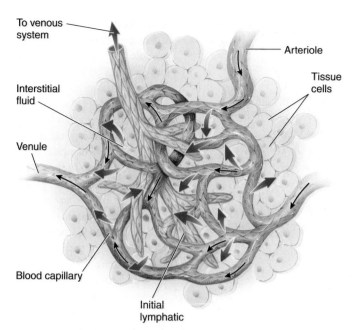

To venous system

Arteriole

Interstitial fluid

Tissue cells

Venule

Blood capillary

Initial lymphatic

(a) Relationship between initial lymphatics and blood capillaries

Fluid pressure on the outside of the vessel pushes the endothelial cell's free edge inward, permitting entrance of interstitial fluid (now lymph).

Interstitial fluid

Lymph

Overlapping endothelial cells

Fluid pressure on the inside of the vessel forces the overlapping edges together so that lymph cannot escape.

(b) Arrangement of endothelial cells in an initial lymphatic

● **FIGURE 10-24 Initial lymphatics.** (a) Blind-ended initial lymphatics pick up excess fluid filtered by blood capillaries and return it to the venous system in the chest. (b) Note that the overlapping edges of the endothelial cells create valvelike openings in the vessel wall.

every tissue of the body (● Figure 10-24a). The endothelial cells forming the walls of initial lymphatics slightly overlap like shingles on a roof, with their overlapping edges being free instead of attached to the surrounding cells. This arrangement creates one-way, valvelike openings in the vessel wall. Fluid pressure on the outside of the vessel pushes the innermost edge of a pair of overlapping edges inward, creating a gap between the edges (that is, opening the valve). This opening permits interstitial fluid to enter (● Figure 10-24b). Once interstitial fluid enters a lymphatic vessel, it is called **lymph.** Fluid pressure on the inside forces the overlapping edges together, closing the valves so that lymph does not escape. These lymphatic valvelike openings are much larger than the pores in blood capillaries.

Consequently, large particles in the interstitial fluid, such as escaped plasma proteins and bacteria, can gain access to initial lymphatics but are excluded from blood capillaries.

Initial lymphatics converge to form larger and larger **lymph vessels,** which eventually empty into the venous system near where the blood enters the right atrium (● Figure 10-25a). Because there is no "lymphatic heart" to provide driving pressure, you may wonder how lymph is directed from the tissues toward the venous system in the thoracic cavity. Lymph flow is accomplished by two mechanisms. First, lymph vessels beyond the initial lymphatics are surrounded by smooth muscle, which contracts rhythmically as a result of myogenic activity. When this muscle is stretched because the vessel is distended with lymph, the muscle inherently contracts more forcefully, pushing the lymph through the vessel. This intrinsic "lymph pump" is the major force for propelling lymph. Stimulation of lymphatic smooth muscle by the sympathetic nervous system further increases the pumping activity of the lymph vessels. Second, because lymph vessels lie between skeletal muscles, contraction of these muscles squeezes the lymph out of the vessels. One-way valves spaced at intervals within the lymph vessels direct the flow of lymph toward its venous outlet in the chest.

FUNCTIONS OF THE LYMPHATIC SYSTEM Here are the most important functions of the lymphatic system:

■ *Return of excess filtered fluid.* Normally, capillary filtration exceeds reabsorption by about 3 liters per day (20 liters filtered, 17 liters reabsorbed) (● Figure 10-25b). Yet the entire blood volume is only 5 liters, and only 2.75 liters of that is plasma. (Blood cells make up the rest of the blood volume.) With an average cardiac output, 7200 liters of blood pass through the capillaries daily under resting conditions (more when cardiac output increases). Even though only a small fraction of the filtered fluid is not reabsorbed by the blood capillaries, the cumulative effect of this process being repeated with every heartbeat results in the equivalent of more than the entire plasma volume being left behind in the interstitial fluid each day. Obviously, this fluid must be returned to the circulating plasma, and this task is accomplished by the lymph vessels. The average rate of flow through the lymph vessels is 3 liters per day, compared with 7200 liters per day through the circulatory system.

■ *Defense against disease.* The lymph percolates through **lymph nodes** located en route within the lymphatic system. Passage of this fluid through the lymph nodes is an important aspect of the body's defense mechanism against disease. For example, bacteria picked up from the interstitial fluid are destroyed by special phagocytes within the lymph nodes (see Chapter 12).

■ *Transport of absorbed fat.* The lymphatic system is important in the absorption of fat from the digestive tract. The end products of the digestion of dietary fats are packaged by cells lining the digestive tract into fatty particles that are too large to gain access to the blood capillaries but can easily enter the initial lymphatics (see Chapter 16).

■ *Return of filtered protein.* Most capillaries permit leakage of some plasma proteins during filtration. These proteins cannot

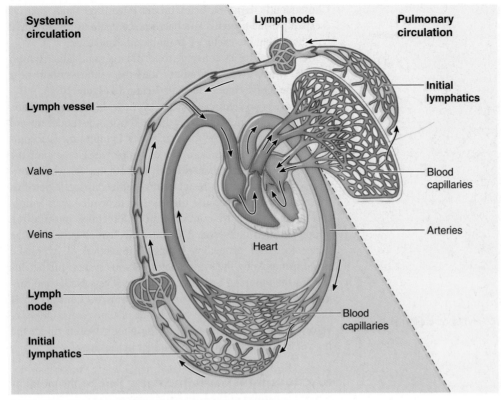

(a) Relationship of lymphatic system to circulatory system

(b) Comparison of blood flow and lymph flow per day

● **FIGURE 10-25 Lymphatic system.** (a) Lymph empties into the venous system near its entrance to the right atrium. (b) Lymph flow averages 3 liters per day, whereas blood flow averages 7200 liters per day.

readily be reabsorbed back into the blood capillaries but can easily gain access to the initial lymphatics. If the proteins were allowed to accumulate in the interstitial fluid rather than being returned to the circulation via the lymphatics, the interstitial fluid-colloid osmotic pressure (an outward pressure) would progressively increase while the plasma-colloid osmotic pressure (an inward pressure) would progressively fall. As a result, filtration forces would gradually increase and reabsorption forces would gradually decrease, resulting in progressive accumulation of fluid in the interstitial spaces at the expense of loss of plasma volume.

Edema occurs when too much interstitial fluid accumulates.

Occasionally, excessive interstitial fluid does accumulate when one of the physical forces acting across the capillary walls becomes abnormal for some reason. Swelling of the tissues because of excess interstitial fluid is known as **edema.** The causes of edema can be grouped into four general categories:

1. *A reduced concentration of plasma proteins* decreases plasma-colloid osmotic pressure. Such a drop in the major inward pressure lets excess fluid filter out, whereas less-than-normal amounts of fluid are reabsorbed; hence, extra fluid remains in the interstitial spaces. Edema can be caused by a decreased concentration of plasma proteins in several different ways: excessive loss of plasma proteins in the urine, from kidney disease; reduced synthesis of plasma proteins, from liver disease (the liver synthesizes almost all plasma proteins); a diet deficient in protein; or significant loss of plasma proteins from large burned surfaces.

2. *Increased permeability of the capillary walls* allows more plasma proteins than usual to pass from the plasma into the surrounding interstitial fluid— for example, via histamine-induced widening of the capillary pores during tissue injury or allergic reactions. The resultant fall in plasma-colloid osmotic pressure decreases the effective inward pressure, whereas the resultant rise in interstitial fluid-colloid osmotic pressure caused by excess protein in the interstitial fluid increases the effective outward force. This imbalance contributes in part to the localized edema associated with injuries (for example, blisters) and allergic responses (for example, hives).

3. *Increased venous pressure,* as when blood dams up in the veins, is accompanied by an increased capillary blood pressure, because the capillaries drain into the veins. This elevation in outward pressure across the capillary walls is largely responsible for the edema seen with congestive heart failure (see p. 331). Regional edema can also occur because of localized restriction of venous return. An example is the swelling often occurring in the legs and feet during pregnancy. The enlarged uterus compresses the major veins that drain the lower extremities as these vessels enter the abdominal cavity. The resultant damming of blood in these veins raises blood pressure in the capillaries of the legs and feet, which promotes regional edema of the lower extremities.

● **FIGURE 10-26 Elephantiasis.** This tropical condition is caused by a mosquito-borne parasitic worm that invades the lymph vessels. As a result of the interference with lymph drainage, the affected body parts, usually the extremities, become grossly edematous, appearing elephant-like.

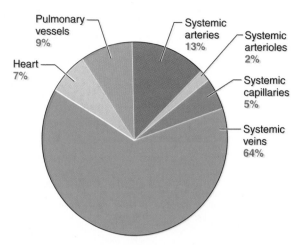

● **FIGURE 10-27 Percentage of total blood volume in different parts of the circulatory system.**

4. *Blockage of lymph vessels* produces edema because the excess filtered fluid is retained in the interstitial fluid rather than returned to the blood through the lymphatics. Protein accumulation in the interstitial fluid compounds the problem through its osmotic effect. Local lymph blockage can occur, for example, in the arms of women whose major lymphatic drainage channels from the arm have been blocked as a result of lymph node removal during surgery for breast cancer. More widespread lymph blockage occurs with *filariasis,* a mosquito-borne parasitic disease found predominantly in tropical coastal regions. In this condition, small, threadlike filaria worms infect the lymph vessels, where their presence prevents proper lymph drainage. The affected body parts, particularly the scrotum and extremities, become grossly edematous. The condition is often called *elephantiasis* because of the elephant-like appearance of the swollen extremities (● Figure 10-26).

Whatever the cause of edema, an important consequence is reduced exchange of materials between the blood and cells. As excess interstitial fluid accumulates, the distance between the blood and cells across which nutrients, O_2, and wastes must diffuse increases, so the rate of diffusion decreases. Therefore, cells within edematous tissues may not be adequately supplied.

Veins

The venous system completes the circulatory circuit. Blood leaving the capillary beds enters the venous system for transport back to the heart.

Venules communicate chemically with nearby arterioles.

At the microcirculatory level, capillaries drain into **venules,** which progressively converge to form small veins that exit the organ. In contrast to arterioles, venules have little tone and resistance. Extensive communication takes place via chemical signals between venules and nearby arterioles. This venuloarteriolar signaling is vital to matching capillary inflow and outflow within an organ.

Veins serve as a blood reservoir as well as passageways back to the heart.

Veins have a large radius, so they offer little resistance to flow. Furthermore, because the total cross-sectional area of the venous system gradually decreases as smaller veins converge into progressively fewer but larger vessels, blood flow speeds up as blood approaches the heart.

In addition to serving as low-resistance passageways to return blood from the tissues to the heart, systemic veins also serve as a *blood reservoir.* Because of their storage capacity, veins are often called **capacitance vessels.** Veins have much thinner walls with less smooth muscle than arteries do. Also, in contrast to arteries, veins have very little elasticity, because venous connective tissue contains considerably more collagen fibers than elastin fibers (see ▲ Table 10-1, p. 348). Unlike arteriolar smooth muscle, venous smooth muscle has little inherent myogenic tone. Because of these features, veins are highly distensible, or stretchable, and have little elastic recoil. They easily distend to accommodate additional volumes of blood with only a small increase in venous pressure. Arteries stretched by an excess volume of blood recoil because of the elastic fibers in their walls, driving the blood forward. Veins containing an extra volume of blood simply stretch to accommodate the additional blood without tending to recoil. In this way veins serve as a **blood reservoir;** that is, when demands for blood are low, the veins can store extra blood in reserve because of their passive distensibility. Under resting conditions, the veins contain more than 60% of the total blood volume (● Figure 10-27).

● **FIGURE 10-28 Factors that facilitate venous return.**

Let us clarify a possible point of confusion. Contrary to a common misconception, blood stored in the veins is not being held in a stagnant holding tank. Normally all the blood is circulating all the time. When the body is at rest and many of the capillary beds are closed, the capacity of the venous reservoir is increased as extra blood bypasses the capillaries and enters the veins. When this extra volume of blood stretches the veins, the blood moves forward through the veins more slowly because the total cross-sectional area of the veins has been increased as a result of the stretching. Therefore, the blood spends more time in the veins. As a result of this slower transit time through the veins, the veins are essentially storing the extra volume of blood because it is not moving forward as quickly to the heart to be pumped out again.

When the stored blood is needed, such as during exercise, extrinsic factors (soon to be described) reduce the capacity of the venous reservoir and drive the extra blood from the veins to the heart so it can be pumped to the tissues. Increased venous return leads to an increased cardiac stroke volume, in accordance with the Frank–Starling law of the heart (see p. 328). In contrast, if too much blood pools in the veins instead of being returned to the heart, cardiac output is abnormally diminished. Thus, a delicate balance exists between the capacity of the veins, the extent of venous return, and the cardiac output. We now turn attention to the factors that affect venous capacity and contribute to venous return.

Venous return is enhanced by a number of extrinsic factors.

Venous capacity (the volume of blood that the veins can accommodate) depends on the distensibility of the vein walls (how much they can stretch to hold blood) and the influence of any externally applied pressure squeezing inwardly on the veins. At a constant blood volume, as venous capacity increases, more blood remains in the veins instead of being returned to the heart. Such venous storage decreases the effective circulating blood volume, the volume of blood being returned to and pumped out of the heart. Conversely, when venous capacity decreases, more blood is returned to the heart and is subsequently pumped out. Thus, changes in venous capacity directly influence the magnitude of venous return, which in turn is an important (although not the only) determinant of effective circulating blood volume. The effective circulating blood volume is also influenced on a short-term basis by passive shifts in bulk flow between the vascular and interstitial fluid compartments and on a long-term basis by factors that control total ECF volume, such as salt and water balance.

The term **venous return** refers to the volume of blood entering each atrium per minute from the veins. Recall that the magnitude of flow through a vessel is directly proportional to the pressure gradient. Much of the driving pressure imparted to the blood by cardiac contraction has been lost by the time the blood reaches the venous system because of frictional losses along the way, especially during passage through the high-resistance arterioles. By the time the blood enters the venous system, blood pressure averages only 17 mm Hg (see ● Figure 10-9, p. 352). However, because atrial pressure is near 0 mm Hg, a small but adequate driving pressure still exists to promote the flow of blood through the large-radius, low-resistance veins.

Clinical Note If atrial pressure becomes pathologically elevated, as in the presence of a leaky AV valve, the venous-to-atrial pressure gradient is decreased, reducing venous return and causing blood to dam up in the venous system. Elevated atrial pressure is thus one cause of congestive heart failure (see p. 331).

In addition to the driving pressure imparted by cardiac contraction, five other factors enhance venous return: sympathetically induced venous vasoconstriction, skeletal muscle activity, the effect of venous valves, respiratory activity, and the effect of cardiac suction (●Figure 10-28). Most of these secondary factors affect venous return by influencing the pressure gradient between the veins and the heart. We will examine each in turn.

EFFECT OF SYMPATHETIC ACTIVITY ON VENOUS RETURN Veins are not very muscular and have little inherent tone, but venous smooth muscle is abundantly supplied with sympathetic nerve fibers. Sympathetic stimulation produces venous vasoconstriction, which modestly elevates venous pressure; this, in turn, increases the pressure gradient to drive more of the stored blood from the veins into the right atrium, thus enhancing venous return. Veins normally have such a large radius that the moderate vasoconstriction from sympathetic stimulation has little effect on resistance to flow. Even when constricted, veins still have a relatively large radius and are still low-resistance vessels.

In addition to mobilizing the stored blood, venous vasoconstriction enhances venous return by decreasing venous capacity. With the filling capacity of the veins reduced, less blood draining from the capillaries remains in the veins but continues to flow instead toward the heart. The increased venous return initiated by sympathetic stimulation leads to increased cardiac output because of the increase in end-diastolic volume. Sympathetic stimulation of the heart also increases cardiac output by increasing the heart rate and increasing the heart's contractility (see pp. 327 and 329). As long as sympathetic activity remains elevated, as during exercise, the increased cardiac output, in turn, helps sustain the increased venous return initiated in the first place by sympathetically induced venous vasoconstriction. More blood being pumped out by the heart means greater return of blood to the heart, because the reduced-capacity veins do not stretch to store any of the extra blood being pumped into the vascular system.

It is important to recognize the different outcomes of vasoconstriction in arterioles and veins. Arteriolar vasoconstriction immediately *reduces* flow through these vessels because of their

● **FIGURE 10-29 Skeletal muscle pump enhancing venous return.**

increased resistance (less blood can enter and flow through a narrowed arteriole), whereas venous vasoconstriction immediately *increases* flow through these vessels because of their decreased capacity (narrowing of veins squeezes out more of the blood already in the veins, increasing blood flow through these vessels).

EFFECT OF SKELETAL MUSCLE ACTIVITY ON VENOUS RETURN Many of the large veins in the extremities lie between skeletal muscles, so muscle contraction compresses the veins. This external venous compression decreases venous capacity and increases venous pressure, in effect squeezing fluid in the veins forward toward the heart (●Figure 10-29). This pumping action, known as the **skeletal muscle pump,** is one way extra blood stored in the veins is returned to the heart during exercise. Increased muscular activity pushes more blood out of the veins and into the heart. Increased sympathetic activity and the resultant venous vasoconstriction also accompany exercise, further enhancing venous return.

The skeletal muscle pump also counters the effect of gravity on the venous system. Let us see how.

COUNTERING THE EFFECTS OF GRAVITY ON THE VENOUS SYSTEM The average pressures mentioned thus far for various regions of the vascular tree are for a person in the horizontal position. When a person is lying down, the force of gravity is uniformly applied, so it need not be considered. When a person stands up, however, gravitational effects are not uniform. In addition to the usual pressure from cardiac contraction, vessels below heart level are subject to pressure from the weight of the column of blood extending from the heart to the level of the vessel (● Figure 10-30).

This increased pressure has two important consequences. First, the distensible veins yield under the increased hydrostatic pressure, further expanding so their capacity is increased. Even though the arteries are subject to the same gravitational effects, they are not nearly as distensible and do not expand like the veins. Much of the blood entering from the capillaries tends to pool in the expanded lower-leg veins instead of returning to the heart. Because venous return is reduced, cardiac output decreases and the effective circulating volume shrinks. Second, the marked increase in capillary blood pressure resulting from the effect of gravity causes excessive fluid to filter out of capillary beds in the lower extremities, producing localized edema (that is, swollen feet and ankles).

Two compensatory measures normally counteract these gravitational effects. First, the resultant fall in mean arterial pressure that occurs when a person moves from a lying-down to an upright position triggers sympathetically induced venous vasoconstriction, which drives some of the pooled blood forward. Second, the skeletal muscle pump "interrupts" the column of blood by completely emptying given vein segments intermittently so that a particular portion of a vein is not subjected to the weight of the entire venous column from the heart to that portion's level (● Figure 10-31; also see ● Figure 10-29). Reflex venous vasoconstriction cannot completely compensate for gravitational effects without skeletal muscle activity. When a person stands still for a long time, therefore, blood flow to the brain is reduced because of the decline in effective circulating volume, despite reflexes aimed at maintaining mean arterial pressure. Reduced flow of blood to the brain, in turn, leads to fainting, which returns the person to a horizontal position, eliminating the gravitational effects on the vascular system and restoring effective circulation. For this reason, it is counterproductive to try to hold upright someone who has fainted. Fainting is a remedy to the problem, not the problem itself.

Because the skeletal muscle pump facilitates venous return and helps counteract the detrimental effects of gravity on the circulatory system, when you are working at a desk, it's a good idea to get up periodically and when you are on your feet, to move around. The mild muscular activity "gets the blood mov-

Pressure = 0 mm Hg

1.5 m

Pressure = 90 mm Hg

Pressure = 100 mm Hg

90 mm Hg caused by gravitational effect
10 mm Hg caused by pressure imparted by cardiac contraction

● **FIGURE 10-30 Effect of gravity on venous pressure.** In an upright adult, the blood in the vessels extending between the heart and the foot is equivalent to a 1.5 m column of blood. The pressure exerted by this column of blood as a result of the effect of gravity is 90 mm Hg. The pressure imparted to the blood by the heart has declined to about 10 mm Hg in the lower-leg veins because of frictional losses in preceding vessels. Together these pressures produce a venous pressure of 100 mm Hg in the ankle and foot veins. The capillaries in the region are subjected to these same gravitational effects.

ing." It is further recommended that people who must be on their feet for long periods of time use elastic stockings that apply a continuous gentle external compression, similar to the effect of skeletal muscle contraction, to further counter the effect of gravitational pooling of blood in the leg veins.

Standing | **Walking**

Heart

Thigh

150 cm

Calf

34 cm

Foot

| 100 mm Hg | Venous pressure in foot | 27 mm Hg |

Foot vein supporting column of blood 1.5 m (150 cm) in height

Foot vein supporting column of blood 34 cm in height

● **FIGURE 10-31 Effect of contraction of the skeletal muscles of the legs in counteracting the effects of gravity.** Contraction of skeletal muscles (as in walking) completely empties given vein segments, interrupting the column of blood that the lower veins must support.

(*Source:* Adapted from *Physiology of the Heart and Circulation,* 4th ed., by R. C. Little and W. C. Little. Copyright © 1989 Year Book Medical Publishers, Inc., with permission from Elsevier.)

EFFECT OF VENOUS VALVES ON VENOUS RETURN Venous vasoconstriction and external venous compression both drive blood toward the heart. Yet if you squeeze a fluid-filled tube in the middle, fluid is pushed in both directions from the point of constriction (●Figure 10-32a). Then why isn't blood driven backward as well as forward by venous vasoconstriction and the skeletal muscle pump? Blood can only be driven forward because the large veins are equipped with one-way valves spaced at 2 to 4 cm intervals; these valves let blood move forward toward the heart but keep it from moving back toward the tissues (●Figure 10-32b). These venous valves also play a role in counteracting the gravitational effects of upright posture by helping minimize the backflow of blood that tends to occur when a person stands up and by temporarily supporting portions of the column of blood when the skeletal muscles are relaxed.

Clinical Note **Varicose veins** occur when the venous valves become incompetent and can no longer support the column of blood above them. People predisposed to this condition usually have inherited an overdistensibility and weakness of their vein walls. Aggravated by frequent, prolonged standing, the veins become so distended as blood pools in them that the edges of the valves can no longer meet to form a seal. Varicosed

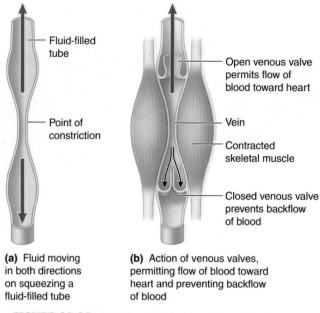

Fluid-filled tube

Point of constriction

Open venous valve permits flow of blood toward heart

Vein

Contracted skeletal muscle

Closed venous valve prevents backflow of blood

(a) Fluid moving in both directions on squeezing a fluid-filled tube

(b) Action of venous valves, permitting flow of blood toward heart and preventing backflow of blood

● **FIGURE 10-32 Function of venous valves.** (a) When a tube is squeezed in the middle, fluid is pushed in both directions. (b) Venous valves permit the flow of blood only toward the heart.

superficial leg veins become visibly overdistended and tortuous. Contrary to what might be expected, chronic pooling of blood in the pathologically distended veins does not reduce cardiac output, because there is a compensatory increase in total circulating blood volume. Instead, the most serious consequence of varicose veins is the possibility of abnormal clot formation in the sluggish, pooled blood. Particularly dangerous is the risk that these clots may break loose and block small vessels elsewhere, especially the pulmonary capillaries.

EFFECT OF RESPIRATORY ACTIVITY ON VENOUS RETURN As a result of respiratory activity, the pressure within the chest cavity averages 5 mm Hg less than atmospheric pressure. As the venous system returns blood to the heart from the lower regions of the body, it travels through the chest cavity, where it is exposed to this subatmospheric pressure. Because the venous system in the limbs and abdomen is subject to normal atmospheric pressure, an externally applied pressure gradient exists between the lower veins (at atmospheric pressure) and the chest veins (at 5 mm Hg less than atmospheric pressure). This pressure difference squeezes blood from the lower veins to the chest veins, promoting increased venous return (●Figure 10-33). This mechanism of facilitating venous return is called the **respiratory pump** because it results from respiratory activity. Increased respiratory activity as well as the effects of the skeletal muscle pump and venous vasoconstriction all enhance venous return during exercise.

EFFECT OF CARDIAC SUCTION ON VENOUS RETURN The extent of cardiac filling does not depend entirely on factors affecting the veins. The heart plays a role in its own filling. During ventricular contraction, the AV valves are drawn downward, enlarging the atrial cavities. As a result, the atrial pressure tran-

5 mm Hg less than atmospheric pressure

Atmospheric pressure

5 mm Hg less than atmospheric pressure

Atmospheric pressure

• **FIGURE 10-33 Respiratory pump enhancing venous return.** As a result of respiratory activity, the pressure surrounding the chest veins is lower than the pressure surrounding the veins in the extremities and abdomen. This establishes an externally applied pressure gradient on the veins, which drives blood toward the heart.

siently drops below 0 mm Hg, thus increasing the vein-to-atria pressure gradient so that venous return is enhanced. In addition, the rapid expansion of the ventricular chambers during ventricular relaxation creates a transient negative pressure in the ventricles so that blood is "sucked in" from the atria and veins; that is, the negative ventricular pressure increases the vein-to-atria-to-ventricle pressure gradient, further enhancing venous return. Thus, the heart functions as a "suction pump" to facilitate cardiac filling.

Blood Pressure

Mean arterial pressure is the blood pressure that is monitored and regulated in the body, not the arterial systolic or diastolic or pulse pressure nor the pressure in any other part of the vascular tree. Routine blood pressure measurements record the arterial systolic and diastolic pressures, which can be used as a yardstick for assessing mean arterial pressure. The cutoff for normal blood pressure has been designated by the National Institutes of Health (NIH) as being less than 120/80 mm Hg.

Blood pressure is regulated by controlling cardiac output, total peripheral resistance, and blood volume.

Mean arterial pressure is the main driving force for propelling blood to the tissues. This pressure must be closely regulated for two reasons. First, it must be high enough to ensure sufficient driving pressure; without this pressure, the brain and other organs will not receive adequate flow, no matter what local adjustments are made in the resistance of the arterioles supplying them. Second, the pressure must not be so high that it creates extra work for the heart and increases the risk of vascular damage and possible rupture of small blood vessels.

DETERMINANTS OF MEAN ARTERIAL PRESSURE Elaborate mechanisms involving the integrated action of the various components of the circulatory system and other body systems are vital in regulating this all-important mean arterial pressure (• Figure

10-34). Remember that the two determinants of mean arterial pressure are cardiac output and total peripheral resistance:

$$\text{Mean arterial pressure} = \text{cardiac output} \times \text{total peripheral resistance}$$

(Do not confuse this equation, which indicates the factors that *determine* the mean arterial pressure, namely, the magnitude of both the cardiac output and total peripheral resistance, with the equation used to *calculate* mean arterial pressure, namely, mean arterial pressure = diastolic pressure + 1/3 pulse pressure.)

Recall that a number of factors, in turn, determine cardiac output (see • Figure 9-24, p. 330) and total peripheral resistance (see • Figure 10-14, p. 360). Thus, you can quickly appreciate the complexity of blood pressure regulation. Let us work through • Figure 10-34, reviewing all the factors that affect mean arterial pressure. Even though we have covered all these factors before, it is useful to pull them all together. The numbers in the text correspond to the numbers in the figure.

- Mean arterial pressure depends on cardiac output and total peripheral resistance (**1** on • Figure 10-34).
- Cardiac output depends on heart rate and stroke volume **2**.
- Heart rate depends on the relative balance of parasympathetic activity **3**, which decreases heart rate, and sympathetic activity (including epinephrine throughout this discussion) **4**, which increases heart rate.
- Stroke volume increases in response to sympathetic activity **5** (extrinsic control of stroke volume).
- Stroke volume also increases as venous return increases **6** (intrinsic control of stroke volume according to the Frank–Starling law of the heart).
- Venous return is enhanced by sympathetically induced venous vasoconstriction **7**, the skeletal muscle pump **8**, the respiratory pump **9**, and cardiac suction **10**.
- The effective circulating blood volume also influences how much blood is returned to the heart **11**. The blood volume depends in the short term on the size of passive bulk-flow fluid shifts between plasma and interstitial fluid across the capillary

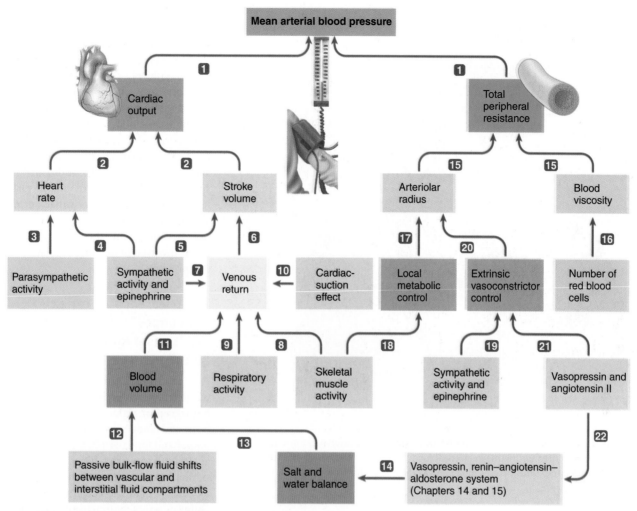

● **FIGURE 10-34 Determinants of mean arterial blood pressure.** Note that this figure is basically a composite of ● Figure 9-24, p. 330, "Control of cardiac output"; ● Figure 10-14, p. 360, "Factors affecting total peripheral resistance"; and ● Figure 10-28, p. 372, "Factors that facilitate venous return." See the text for a discussion of the numbers.

walls **12**. In the long term, the blood volume depends on salt and water balance **13**, which are hormonally controlled by the renin–angiotensin–aldosterone system and vasopressin, respectively **14**.

■ The other major determinant of mean arterial blood pressure, total peripheral resistance, depends on the radius of all arterioles as well as blood viscosity **15**. The main factor determining blood viscosity is the number of red blood cells **16**. However, arteriolar radius is the more important factor determining total peripheral resistance.

■ Arteriolar radius is influenced by local (intrinsic) metabolic controls that match blood flow with metabolic needs **17**. For example, local changes that take place in active skeletal muscles cause local arteriolar vasodilation and increased blood flow to these muscles **18**.

■ Arteriolar radius is also influenced by sympathetic activity **19**, an extrinsic control mechanism that causes arteriolar vasoconstriction **20** to increase total peripheral resistance and mean arterial blood pressure.

■ Arteriolar radius is also extrinsically controlled by the hormones vasopressin and angiotensin II, which are potent vasoconstrictors **21** as well as being important in salt and water balance **22**.

Altering any of the pertinent factors that influence blood pressure will change blood pressure, unless a compensatory change in another variable keeps the blood pressure constant. Blood flow to any given organ depends on the driving force of the mean arterial pressure and on the degree of vasoconstriction of the organ's arterioles. Because mean arterial pressure depends on the cardiac output and the degree of arteriolar vasoconstriction, if the arterioles in one organ dilate, the arterioles in other organs must constrict to maintain an adequate arterial blood pressure. An adequate pressure is needed to provide a driving force to push blood not only to the vasodilated organ but also to the brain, which depends on a constant blood supply. Thus, the cardiovascular variables must be continuously juggled to maintain a constant blood pressure despite organs' varying needs for blood.

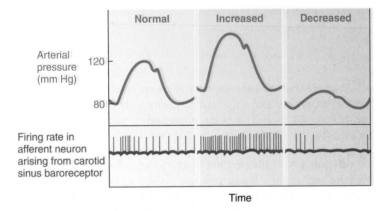

term (within seconds) adjustments are made by alterations in cardiac output and total peripheral resistance, mediated by means of autonomic nervous system influences on the heart, veins, and arterioles. *Long-term* (requiring minutes to days) control involves adjusting total blood volume by restoring normal salt and water balance through mechanisms that regulate urine output and thirst (Chapters 14 and 15). The size of the total blood volume, in turn, has a profound effect on cardiac output and mean arterial pressure. Let us now turn attention to the short-term mechanisms involved in ongoing regulation of this pressure.

The baroreceptor reflex is an important short-term mechanism for regulating blood pressure.

Any change in mean arterial pressure triggers an autonomically mediated **baroreceptor reflex** that influences the heart and blood vessels to adjust cardiac output and total peripheral resistance in an attempt to restore blood pressure to normal. Like any reflex, the baroreceptor reflex includes a receptor, an afferent pathway, an integrating center, an efferent pathway, and effector organs.

The most important receptors involved in the moment-to-moment regulation of blood pressure, the **carotid sinus** and **aortic arch baroreceptors,** are mechanoreceptors sensitive to changes in both mean arterial pressure and pulse pressure. Their responsiveness to fluctuations in pulse pressure enhances their sensitivity as pressure sensors, because small changes in systolic or diastolic pressure may alter the pulse pressure without changing the mean pressure. These baroreceptors are strategically located (●Figure 10-35) to provide critical information about arterial blood pressure in the vessels leading to the brain (the carotid sinus baroreceptor) and in the major arterial trunk before it gives off branches that supply the rest of the body (the aortic arch baroreceptor).

The baroreceptors constantly provide information about mean arterial pressure; in other words, they continuously generate action potentials in response to the ongoing pressure within the arteries. When arterial pressure (either mean or pulse pressure) increases, the receptor potential of these baroreceptors increases, thus increasing the rate of firing in the corresponding afferent neurons. Conversely, a decrease in the mean arterial pressure slows the rate of firing generated in the afferent neurons by the baroreceptors (●Figure 10-36).

The integrating center that receives the afferent impulses about the state of mean arterial pressure is the **cardiovascular**

● FIGURE 10-35 Location of the arterial baroreceptors. The arterial baroreceptors are strategically located to monitor the mean arterial blood pressure in the arteries that supply blood to the brain (cartoid sinus baroreceptor) and to the rest of the body (aortic arch baroreceptor).

● FIGURE 10-36 Firing rate in the afferent neuron from the carotid sinus baroreceptor in relation to the magnitude of mean arterial pressure.

SHORT-TERM AND LONG-TERM CONTROL MEASURES Mean arterial pressure is constantly monitored by **baroreceptors** (pressure sensors) within the circulatory system. When deviations from normal are detected, multiple reflex responses are initiated to return mean arterial pressure to its normal value. *Short-*

● **FIGURE 10-37 Summary of the effects of the parasympathetic and sympathetic nervous systems on factors that influence mean arterial blood pressure.**

control center,[4] located in the medulla within the brain stem. The efferent pathway is the autonomic nervous system. The cardiovascular control center alters the ratio between sympathetic and parasympathetic activity to the effector organs (the heart and blood vessels). To review how autonomic changes alter arterial blood pressure, study ● Figure 10-37, which summarizes the major effects of parasympathetic and sympathetic stimulation on the heart and blood vessels.

Let us fit all the pieces of the baroreceptor reflex together now by tracing the reflex activity that compensates for an elevation or fall in blood pressure. If for any reason mean arterial pressure rises above normal (● Figure 10-38a), the carotid sinus and aortic arch baroreceptors increase the rate of firing in their respective afferent neurons. On being informed by increased afferent firing that the blood pressure has become too high, the cardiovascular control center responds by decreasing sympathetic and increasing parasympathetic activity to the cardiovascular system. These efferent signals decrease heart rate, decrease stroke volume, and produce arteriolar and venous vasodilation, which in turn lead to a decrease in cardiac output and a decrease in total peripheral resistance, with a subsequent fall in blood pressure back toward normal.

Conversely, when blood pressure falls below normal (● Figure 10-38b), baroreceptor activity decreases, inducing the cardiovascular center to increase sympathetic cardiac and vasoconstrictor nerve activity while decreasing its parasympathetic output. This efferent pattern of activity leads to an increase in heart rate and stroke volume, coupled with arteriolar and venous vasoconstriction. These changes increase both cardiac output and total peripheral resistance, raising blood pressure back toward normal.

Other reflexes and responses influence blood pressure.

Besides the baroreceptor reflex, whose sole function is blood pressure regulation, several other reflexes and responses influence the cardiovascular system even though they primarily regulate other body functions. Some of these other influences deliberately move arterial pressure away from its normal value temporarily, overriding the baroreceptor reflex to accomplish a particular goal. These factors include the following:

1. Left atrial volume receptors and hypothalamic osmoreceptors are primarily important in water and salt balance in the body; they thus affect the long-term regulation of blood pressure by controlling the plasma volume.

2. Chemoreceptors located in the carotid and aortic arteries, in close association with but distinct from the baroreceptors, are sensitive to low O_2 or high acid levels in the blood. These chemoreceptors' main function is to reflexly increase respiratory activity to bring in more O_2 or to blow off more acid-

---

[4]The cardiovascular control center is sometimes divided into cardiac and vasomotor centers, which are occasionally further classified into smaller subdivisions, such as cardioacceleratory and cardioinhibitory centers and vasoconstrictor and vasodilator areas. Because these regions are highly interconnected and functionally interrelated, we refer to them collectively as the *cardiovascular control center.*

(a) Baroreceptor reflex in response to an elevation in blood pressure

(b) Baroreceptor reflex in response to a fall in blood pressure

● **FIGURE 10-38 Baroreceptor reflexes to restore blood pressure to normal.**

forming CO_2, but they also reflexly increase blood pressure by sending excitatory impulses to the cardiovascular center.

3. Cardiovascular responses associated with certain behaviors and emotions are mediated through the cerebral cortex–hypothalamic pathway and appear preprogrammed. These responses include the widespread changes in cardiovascular activity accompanying the generalized sympathetic fight-or-flight response, the characteristic marked increase in heart rate and blood pressure associated with sexual orgasm, and the localized cutaneous vasodilation characteristic of blushing.

4. Pronounced cardiovascular changes accompany exercise, including a substantial increase in skeletal muscle blood flow (see ●Figure 10-12, p. 355), a significant increase in cardiac output, a fall in total peripheral resistance (because of widespread vasodilation in skeletal muscles despite generalized arteriolar vasoconstriction in most organs), and a modest increase in mean arterial pressure (▲ Table 10-5). Evidence suggests that discrete exercise centers yet to be identified in the brain induce the appropriate cardiac and vascular changes at the onset of exercise or even in anticipation of exercise. These effects are then reinforced by afferent inputs to the medullary cardiovascular center from chemoreceptors in exercising muscles as well as by local mechanisms important in maintaining vasodilation in active muscles. The baroreceptor reflex further modulates these cardiovascular responses.

5. Hypothalamic control over cutaneous (skin) arterioles for the purpose of temperature regulation takes precedence over control that the cardiovascular center has over these same vessels for the purpose of blood pressure regulation. As a result, blood pressure can fall when the skin vessels are widely dilated to eliminate excess heat from the body, even though the baroreceptor responses are calling for cutaneous vasoconstriction to help maintain adequate total peripheral resistance.

6. Vasoactive substances released from the endothelial cells play a role in regulating blood pressure. For example, NO normally exerts an ongoing vasodilatory effect.

Despite these control measures, sometimes blood pressure is not maintained at the appropriate level. We next examine blood pressure abnormalities.

Cardiovascular Changes during Exercise

Cardiovascular Variable	Change	Comment
Heart Rate	Increases	Occurs as a result of increased sympathetic and decreased parasympathetic activity to the SA node
Venous Return	Increases	Occurs as a result of sympathetically induced venous vasoconstriction and increased activity of the skeletal muscle pump and respiratory pump
Stroke Volume	Increases	Occurs both as a result of increased venous return by means of the Frank–Starling mechanism (unless diastolic filling time is significantly reduced by a high heart rate) and as a result of a sympathetically induced increase in myocardial contractility
Cardiac Output	Increases	Occurs as a result of increases in both heart rate and stroke volume
Blood Flow to Active Skeletal Muscles and Heart Muscle	Increases	Occurs as a result of locally controlled arteriolar vasodilation, which is reinforced by the vasodilatory effects of epinephrine and overpowers the weaker sympathetic vasoconstrictor effect
Blood Flow to the Brain	Unchanged	Occurs because sympathetic stimulation has no effect on brain arterioles; local control mechanisms maintain constant cerebral blood flow whatever the circumstances
Blood Flow to the Skin	Increases	Occurs because the hypothalamic temperature control center induces vasodilation of skin arterioles; increased skin blood flow brings heat produced by exercising muscles to the body surface where the heat can be lost to the external environment
Blood Flow to the Digestive System, Kidneys, and Other Organs	Decreases	Occurs as a result of generalized sympathetically induced arteriolar vasoconstriction
Total Peripheral Resistance	Decreases	Occurs because resistance in the skeletal muscles, heart, and skin decreases to a greater extent than resistance in the other organs increases
Mean Arterial Blood Pressure	Increases (modest)	Occurs because cardiac output increases more than total peripheral resistance decreases

Hypertension is a serious national public-health problem, but its causes are largely unknown.

Clinical Note Sometimes blood pressure control mechanisms do not function properly or are unable to completely compensate for changes that have taken place. Blood pressure may be too high (**hypertension** if above 140/90 mm Hg) or too low (**hypotension** if below 100/60 mm Hg). Hypotension in its extreme form is *circulatory shock*. We will first examine hypertension, which is by far the most common of blood pressure abnormalities, and then conclude this chapter with a discussion of hypotension and shock.

There are two broad classes of hypertension, secondary hypertension and primary hypertension, depending on the cause.

SECONDARY HYPERTENSION A definite cause for hypertension can be established in only 10% of cases. Hypertension that occurs secondary to another known primary problem is called **secondary hypertension.** Here are some examples of secondary hypertension:

1. *Renal hypertension.* For example, atherosclerotic lesions protruding into the lumen of a renal artery (see p. 334) or external compression of the vessel by a tumor may reduce blood flow through the kidney. The kidney responds by initiating the hormonal pathway involving angiotensin II. This pathway promotes salt and water retention during urine formation, thus increasing the blood volume to compensate for the reduced renal blood flow. Recall that angiotensin II is also a powerful

vasoconstrictor. Although these two effects (increased blood volume and angiotensin-induced vasoconstriction) are compensatory mechanisms to improve blood flow through the narrowed renal artery, they also are responsible for elevating the arterial pressure as a whole.

2. *Endocrine hypertension.* For example, a *pheochromocytoma* is an adrenal medullary tumor that secretes excessive epinephrine and norepinephrine. Abnormally elevated levels of these hormones bring about a high cardiac output and generalized peripheral vasoconstriction, both of which contribute to the hypertension characteristic of this disorder.

3. *Neurogenic hypertension.* An example is the hypertension caused by erroneous blood pressure control arising from a defect in the cardiovascular control center.

PRIMARY HYPERTENSION The underlying cause is unknown in the remaining 90% of hypertension cases. Such hypertension is known as **primary (essential** or **idiopathic) hypertension.** Primary hypertension is a catchall category for blood pressure elevated by a variety of unknown causes rather than by a single disease entity. People show a strong genetic tendency to develop primary hypertension, which can be hastened or worsened by contributing factors such as obesity, stress, smoking, or dietary habits. Consider the following range of potential causes for primary hypertension currently being investigated.

■ *Defects in salt management by the kidneys.* Disturbances in kidney function too minor to produce outward signs of renal disease could nevertheless insidiously lead to gradual accumulation of salt and water in the body, resulting in progressive elevation of arterial pressure.

■ *Excessive salt intake.* Because salt osmotically retains water, thus expanding the plasma volume and contributing to the long-term control of blood pressure, excessive ingestion of salt can contribute to hypertension. Therefore, salt-sensitive, hypertensive individuals should limit salt intake. Yet controversy continues over whether or not restricting salt intake should be recommended as a means of preventing high blood pressure. The research data to date have been inconclusive and subject to conflicting interpretations.

■ *Diets low in fruits, vegetables, and dairy products (that is, low in K^+ and Ca^{2+}).* Dietary factors other than salt have been shown to markedly affect blood pressure. The DASH (Dietary Approaches to Stop Hypertension) studies found that a low-fat diet rich in fruits, vegetables, and dairy products could lower blood pressure in people with mild hypertension as much as any single drug treatment. Research indicates that the high K^+ intake associated with eating abundant fruits and vegetables may lower blood pressure by relaxing arteries. Furthermore, inadequate Ca^{2+} intake from dairy products has been identified as the most prevalent dietary pattern among individuals with untreated hypertension, although the role of Ca^{2+} in regulating blood pressure is unclear.

■ *Plasma membrane abnormalities such as defective $Na^+–K^+$ pumps.* Such defects, by altering the electrochemical gradient across plasma membranes, could change the excitability and contractility of the heart and the smooth muscle in blood vessel walls

in such a way as to lead to high blood pressure. In addition, the $Na^+–K^+$ pump is crucial to salt management by the kidneys. A genetic defect in the $Na^+–K^+$ pump of hypertensive-prone laboratory rats was the first gene–hypertension link to be discovered.

■ *Variation in the gene that encodes for angiotensinogen.* Angiotensinogen is part of the hormonal pathway that produces the potent vasoconstrictor angiotensin II and promotes salt and water retention. One variant of the gene in humans appears to be associated with a higher incidence of hypertension. Researchers speculate that the suspect version of the gene leads to a slight excess production of angiotensinogen, thus increasing activity of this blood pressure–raising pathway. This is the first gene–hypertension link discovered in humans.

■ *Endogenous digitalis-like substances.* Such substances act similarly to the drug digitalis (see p. 316) to increase cardiac contractility as well as to constrict blood vessels and reduce salt elimination in the urine, all of which could cause chronic hypertension.

■ *Abnormalities in NO, endothelin, or other locally acting vasoactive chemicals.* For example, a shortage of NO has been discovered in the blood vessel walls of some hypertensive patients, leading to an impaired ability to accomplish blood-pressure-lowering vasodilation. Furthermore, an underlying abnormality in the gene that codes for endothelin, a locally acting vasoconstrictor, has been strongly implicated as a possible cause of hypertension, especially among African Americans.

■ *Excess vasopressin.* Recent experimental evidence suggests that hypertension may result from a malfunction of the vasopressin-secreting cells of the hypothalamus. Vasopressin is a potent vasoconstrictor and also promotes water retention.

Whatever the underlying defect, once initiated, hypertension appears to be self-perpetuating. Constant exposure to elevated blood pressure predisposes vessel walls to the development of atherosclerosis, which further raises blood pressure.

ADAPTATION OF BARORECEPTORS DURING HYPERTENSION The baroreceptors do not respond to bring the blood pressure back to normal during hypertension because they adapt, or are "reset," to operate at a higher level. In the presence of chronically elevated blood pressure, the baroreceptors still function to regulate blood pressure, but they maintain it at a higher mean pressure.

COMPLICATIONS OF HYPERTENSION Hypertension imposes stresses on both the heart and blood vessels. The heart has an increased workload because it is pumping against an increased total peripheral resistance, whereas blood vessels may be damaged by the high internal pressure, particularly when the vessel wall is weakened by the degenerative process of atherosclerosis. Complications of hypertension include congestive heart failure caused by the heart's inability to pump continuously against a sustained elevation in arterial pressure, strokes caused by rupture of brain vessels, and heart attacks caused by rupture of coronary vessels. Spontaneous hemorrhage caused by bursting of small vessels elsewhere in the body may also occur but with less serious consequences; an example is the rupture of blood vessels in the nose, resulting in nosebleeds. Another serious complication of

The Ups and Downs of Hypertension and Exercise

When blood pressure is up, one way to bring it down is to increase the level of physical activity. Research studies have demonstrated that participation in aerobic activities protects against the development of hypertension. Furthermore, exercise can be used as a therapy to reduce hypertension once it has already developed.

Antihypertensive medication is available to lower blood pressure in severely hypertensive patients, but sometimes undesirable side effects occur. The side effects of diuretics include electrolyte imbalances, inability to handle glucose normally, and raised blood cholesterol levels. The side effects of drugs that manipulate total peripheral resistance include increased blood triglyceride levels, lower HDL cholesterol levels (the

"good" form of cholesterol), weight gain, sexual dysfunction, and depression.

Patients with mild hypertension pose a dilemma for physicians. The risks of taking the drugs may outweigh the benefits gained from lowering the blood pressure. Because of the drug therapy's possible side effects, nondrug treatment of mild hypertension may be most beneficial. The most common nondrug therapies are weight reduction, salt restriction, and exercise.

Although losing weight will almost always reduce blood pressure, research has shown that weight-reduction programs usually result in the loss of only 12 pounds, and the overall long-term success in keeping the weight off is only about 20%. Salt

restriction is beneficial for many hypertensives, but adherence to a low-salt diet is difficult for many people because fast foods and foods prepared in restaurants usually contain high amounts of salt.

The preponderance of evidence in the literature suggests that moderate aerobic exercise performed for 15 to 60 minutes three times per week is a beneficial therapy in most cases of mild to moderate hypertension. It is wise, therefore, to include a regular aerobic exercise program along with other therapeutic measures to optimally reduce high blood pressure. If more convenient, the total exercise time on a given day may even be split up into smaller sessions and still provide the same benefits.

---

hypertension is renal failure caused by progressive impairment of blood flow through damaged renal blood vessels. Furthermore, retinal damage from changes in the blood vessels supplying the eyes may result in progressive loss of vision.

Until complications occur, hypertension is symptomless, because the tissues are adequately supplied with blood. Therefore, unless blood pressure measurements are made on a routine basis, the condition can go undetected until a precipitous complicating event. When you become aware of these potential complications of hypertension and consider that one third of all adults in America are estimated to have chronic elevated blood pressure, you can appreciate the magnitude of this national health problem. Furthermore, nearly 1 billion people worldwide have high blood pressure.

TREATMENT OF HYPERTENSION Once hypertension is detected, therapeutic intervention can reduce the course and severity of the problem. Dietary management, including weight loss, along with a variety of drugs that manipulate salt and water management or autonomic activity on the cardiovascular system can be used to treat hypertension. No matter what the original cause, agents that reduce plasma volume or total peripheral resistance (or both) will decrease blood pressure toward normal. Furthermore, a regular aerobic exercise program can be employed to help reduce high blood pressure. (For details, see the accompanying boxed feature, ■A Closer Look at Exercise Physiology.)

PREHYPERTENSION In its recent guidelines, NIH identified **prehypertension** as a new category for blood pressures in the range between normal and hypertension (between 120/80 and 139/89). Blood pressures in the prehypertension range can usu-

ally be reduced by appropriate dietary and exercise measures, whereas those in the hypertension range typically must be treated with blood pressure medication in addition to changing health habits. The goal in managing blood pressures in the prehypertension range is to take action before the pressure climbs into the hypertension range, where serious complications may develop.

We now examine the other extreme, hypotension, looking first at transient orthostatic hypotension, then at the (more serious) circulatory shock.

Orthostatic hypotension results from transient inadequate sympathetic activity.

Hypotension, or low blood pressure, occurs either when there is a disproportion between vascular capacity and blood volume (in essence, too little blood to fill the vessels) or when the heart is too weak to drive the blood.

The most common situation in which hypotension occurs transiently is orthostatic hypotension. **Orthostatic (postural) hypotension** is a transient hypotensive condition resulting from insufficient compensatory responses to the gravitational shifts in blood when a person moves from a horizontal to a vertical position, especially after prolonged bed rest. When a person moves from lying down to standing up, pooling of blood in the leg veins from gravity reduces venous return, decreasing stroke volume and thus lowering cardiac output and blood pressure. This fall in blood pressure is normally detected by the baroreceptors, which initiate immediate compensatory responses to restore blood pressure to its proper level. When a long-bedridden patient first starts to rise, however, these reflex compensatory adjustments are temporarily lost or reduced be-

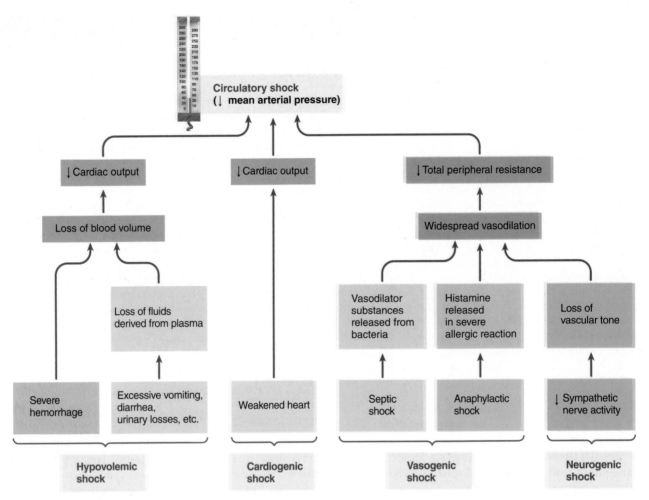

FIGURE 10-39 Causes of circulatory shock. Circulatory shock, which occurs when mean arterial blood pressure falls so low that adequate blood flow to the tissues can no longer be maintained, may result from (1) extensive loss of blood volume (hypovolemic shock), (2) failure of the heart to pump blood adequately (cardiogenic shock), (3) widespread arteriolar vasodilation (vasogenic shock), or (4) neurally defective vasoconstrictor tone (neurogenic shock).

cause of disuse. Sympathetic control of the leg veins is inadequate, so when the patient first stands up, blood pools in the lower extremities. The condition is further aggravated by the decrease in blood volume that typically accompanies prolonged bed rest. The resultant orthostatic hypotension and decrease in blood flow to the brain cause dizziness or actual fainting. Because postural compensatory mechanisms are depressed during prolonged bed confinement, patients sometimes are put on a tilt table so that they can be moved gradually from a horizontal to an upright position. This allows the body to adjust slowly to the gravitational shifts in blood.

Circulatory shock can become irreversible.

When blood pressure falls so low that adequate blood flow to the tissues can no longer be maintained, the condition known as **circulatory shock** occurs. Circulatory shock is categorized into four main types (● Figure 10-39):

1. *Hypovolemic* ("low volume") *shock* is caused by a fall in blood volume, which occurs either directly through severe hemorrhage or indirectly through loss of fluids derived from the plasma (for example, severe diarrhea, excessive urinary losses, or extensive sweating).

2. *Cardiogenic* ("heart produced") *shock* is due to a weakened heart's failure to pump blood adequately.

3. *Vasogenic* ("vessel produced") *shock* is caused by widespread vasodilation triggered by the presence of vasodilator substances. There are two types of vasogenic shock: septic and anaphylactic. *Septic shock,* which may accompany massive infections, is the result of vasodilator substances released from the infective agents. Similarly, extensive histamine release in severe allergic reactions can cause widespread vasodilation in *anaphylactic shock.*

4. *Neurogenic* ("nerve produced") *shock* also involves generalized vasodilation but not by means of the release of vasodilator substances. In this case, loss of sympathetic vascular tone leads to generalized vasodilation. This undoubtedly is responsible for the shock accompanying crushing injuries when blood loss has not been sufficient to cause hypovolemic shock. Deep, excruciating pain apparently inhibits sympathetic vasoconstrictor activity.

We now examine the consequences of and compensations for shock, using hemorrhage as an example (● Figure 10-40). This figure may look intimidating, but we will work through it step by step. It is an important example that pulls together many of the principles discussed in this chapter. As before, the numbers in the text correspond to the numbers in the figure.

CONSEQUENCES AND COMPENSATIONS OF SHOCK

■ Following severe loss of blood, the resultant reduction in circulating blood volume leads to a decrease in venous return **1** and a subsequent fall in cardiac output and arterial blood pressure. (Note the blue boxes, which indicate consequences of hemorrhage.)

■ Compensatory measures immediately attempt to maintain adequate blood flow to the brain. (Note the pink boxes, which indicate compensations for hemorrhage.)

■ The baroreceptor reflex response to the fall in blood pressure brings about increased sympathetic and decreased parasympathetic activity to the heart **2**. The result is an increase in heart rate **3** to offset the reduced stroke volume **4** brought about by the loss of blood volume. With severe fluid loss, the pulse is weak because of the reduced stroke volume but rapid because of the increased heart rate.

■ Increased sympathetic activity to the veins produces generalized venous vasoconstriction **5**, increasing venous return by means of the Frank–Starling mechanism **6**.

■ Simultaneously, sympathetic stimulation of the heart increases the heart's contractility **7** so that it beats more forcefully and ejects a greater volume of blood, likewise increasing the stroke volume.

■ The increase in heart rate and in stroke volume collectively increase cardiac output **8**.

■ Sympathetically induced generalized arteriolar vasoconstriction **9** leads to an increase in total peripheral resistance **10**.

■ Together, the increase in cardiac output and total peripheral resistance bring about a compensatory increase in arterial pressure **11**.

■ The original fall in arterial pressure is also accompanied by a fall in capillary blood pressure **12**, which results in fluid shifts from the interstitial fluid into the capillaries to expand the plasma volume **13**. This response is sometimes termed *autotransfusion*, because it restores the plasma volume as a transfusion does.

■ This ECF fluid shift is enhanced by plasma protein synthesis by the liver during the next few days following hemorrhage **14**. The plasma proteins exert a colloid osmotic pressure that helps retain extra fluid in the plasma.

■ Urinary output is reduced, thereby conserving water that normally would have been lost from the body **15**. This additional fluid retention helps expand the reduced plasma volume **16**. Expansion of plasma volume further augments the increase in cardiac output brought about by the baroreceptor reflex **17**. Reduction in urinary output results from decreased renal blood flow caused by compensatory renal arteriolar vasoconstriction **18**. The reduced plasma volume also triggers increased secretion of the hormone vasopressin and activation of the salt- and water-conserving renin–angiotensin–aldosterone hormonal pathway, which further reduces urinary output **19**.

■ Increased thirst is also stimulated by a fall in plasma volume **20**. The resultant increased fluid intake helps restore plasma volume.

■ Over a longer course of time (a week or more), lost red blood cells are replaced through increased red blood cell production triggered by a reduction in O_2 delivery to the kidneys **21**.

IRREVERSIBLE SHOCK These compensatory mechanisms are often not enough to counteract substantial fluid loss. Even if they can maintain an adequate blood pressure level, the short-term measures cannot continue indefinitely. Ultimately, fluid volume must be replaced from the outside through drinking, transfusion, or a combination of both. Blood supply to the kidneys, digestive tract, skin, and other organs can be compromised to maintain blood flow to the brain only so long before organ damage begins to occur. A point may be reached at which blood pressure continues to drop rapidly because of tissue damage, despite vigorous therapy. This condition is frequently termed *irreversible shock,* in contrast to *reversible shock,* which can be corrected by compensatory mechanisms and effective therapy.

Although the exact mechanism underlying irreversibility is not currently known, many logical possibilities could contribute to the unrelenting, progressive circulatory deterioration that characterizes irreversible shock. Metabolic acidosis arises when lactate (lactic acid) production increases as blood-deprived tissues resort to anaerobic metabolism. Acidosis deranges the enzymatic systems responsible for energy production, limiting the capability of the heart and other tissues to produce ATP. Prolonged depression of kidney function results in electrolyte imbalances that may lead to cardiac arrhythmias. The blood-deprived pancreas releases a chemical that is toxic to the heart (**myocardial toxic factor**), further weakening the heart. Vasodilator substances build up within ischemic organs, inducing local vasodilation that overrides the generalized reflex vasoconstriction. As cardiac output progressively declines because of the heart's diminishing effectiveness as a pump and total peripheral resistance continues to fall, hypotension becomes increasingly severe. This causes further cardiovascular failure, which leads to a further decline in blood pressure. Thus, when shock progresses to the point that the cardiovascular system itself starts to fail, a vicious positive-feedback cycle ensues that ultimately results in death.

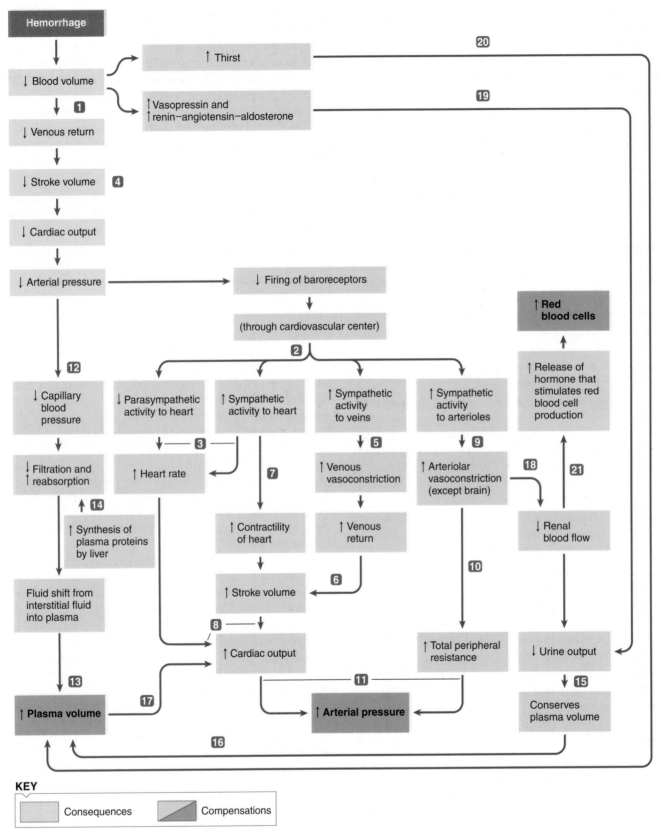

● FIGURE 10-40 Consequences and compensations of hemorrhage. The reduction in blood volume resulting from hemorrhage leads to a fall in arterial pressure. (Note the *blue boxes*, representing consequences of hemorrhage.) A series of compensations ensue *(light pink boxes)* that ultimately restore plasma volume, arterial pressure, and the number of red blood cells toward normal *(dark pink boxes)*. See the text (p. 385) for an explanation of the numbers and a detailed discussion of the compensations.

Chapter in Perspective: Focus on Homeostasis

Homeostatically, the blood vessels serve as passageways to transport blood to and from the cells for O_2 and nutrient delivery, waste removal, distribution of fluid and electrolytes, elimination of excess heat, and hormonal signaling, among other things. Cells soon die if deprived of their blood supply; brain cells succumb within four minutes. Blood is constantly recycled and reconditioned as it travels through the various organs via the vascular highways. Hence the body needs only a very small volume of blood to maintain the appropriate chemical composition of the entire internal fluid environment on which the cells depend for their survival. For example, O_2 is continually picked up by blood in the lungs and constantly delivered to all the body cells.

The smallest blood vessels, the capillaries, are the actual site of exchange between the blood and surrounding cells. Capillaries bring homeostatically maintained blood within

0.01 cm of every cell in the body; this proximity is critical, because beyond a few centimeters materials cannot diffuse rapidly enough to support life-sustaining activities. Oxygen that would take months to years to diffuse from the lungs to all the cells of the body is continuously delivered at the "doorstep" of every cell, where diffusion can efficiently accomplish short local exchanges between the capillaries and surrounding cells. Likewise, hormones must be rapidly transported through the circulatory system from their sites of production in endocrine glands to their sites of action in other parts of the body. These chemical messengers could not diffuse nearly rapidly enough to their target organs to effectively exert their controlling effects, many of which are aimed toward maintaining homeostasis.

The rest of the circulatory system is designed to transport blood to and from the capillaries. The arteries and arterioles distribute blood pumped by the heart to the capillaries for life-sustaining exchanges to take place, and the venules and veins collect blood from the capillaries and return it to the heart, where the process is repeated.

REVIEW EXERCISES

Objective Questions (Answers on p. A-46)

1. In general, the parallel arrangement of the vascular system enables each organ to receive its own separate arterial blood supply. (True or false?)

2. More blood flows through the capillaries during cardiac systole than during diastole. (True or false?)

3. The capillaries contain only 5% of the total blood volume at any point in time. (True or false?)

4. The same volume of blood passes through the capillaries in a minute as passes through the aorta, even though blood flow is much slower in the capillaries. (True or false?)

5. Because capillary walls have no carrier transport systems, all capillaries are equally permeable. (True or false?)

6. Because of gravitational effects, venous pressure in the lower extremities is greater when a person is standing up than when the person is lying down. (True or false?)

7. Which of the following functions is (are) attributable to arterioles? (Indicate all correct answers.)
   a. produce a significant decline in mean pressure, which helps establish the driving pressure gradient between the heart and organs
   b. serve as site of exchange of materials between blood and surrounding tissue cells
   c. act as main determinant of total peripheral resistance
   d. determine the pattern of distribution of cardiac output
   e. help regulate mean arterial blood pressure
   f. convert the pulsatile nature of arterial blood pressure into a smooth, nonfluctuating pressure in the vessels farther downstream
   g. act as a pressure reservoir

8. Using the answer code on the right, indicate whether the following factors increase or decrease venous return:
   ___ 1. sympathetically induced venous vasoconstriction
   ___ 2. skeletal muscle activity
   ___ 3. gravitational effects on the venous system
   ___ 4. respiratory activity
   ___ 5. increased atrial pressure associated with a leaky AV valve
   ___ 6. ventricular pressure change associated with diastolic recoil

   (a) increases venous return
   (b) decreases venous return
   (c) has no effect on venous return

9. Using the answer code on the right, indicate what kind of compensatory changes occur in the factors in question to restore blood pressure to normal in response to hypovolemic hypotension resulting from severe hemorrhage:
   ___ 1. rate of afferent firing generated by the carotid sinus and aortic arch baroreceptors
   ___ 2. sympathetic output by the cardiovascular center
   ___ 3. parasympathetic output by the cardiovascular center
   ___ 4. heart rate
   ___ 5. stroke volume
   ___ 6. cardiac output
   ___ 7. arteriolar radius
   ___ 8. total peripheral resistance
   ___ 9. venous radius
   ___ 10. venous return
   ___ 11. urinary output
   ___ 12. fluid retention within the body
   ___ 13. fluid movement from interstitial fluid into plasma across the capillaries

   (a) increased
   (b) decreased
   (c) no effect

Essay Questions

1. Compare blood flow through reconditioning organs and through organs that do not recondition the blood.

2. Discuss the relationships among flow rate, pressure gradient, and vascular resistance. What is the major determinant of resistance to flow?

3. Describe the structure and major functions of each segment of the vascular tree.

4. How do the arteries serve as a pressure reservoir?

5. Describe the indirect technique of measuring arterial blood pressure by means of a sphygmomanometer.

6. Define *vasoconstriction* and *vasodilation.*

7. Discuss the local and extrinsic controls that regulate arteriolar resistance.

8. What is the primary means by which individual solutes are exchanged across capillary walls? What forces produce bulk flow across capillary walls? Of what importance is bulk flow?

9. How is lymph formed? What are the functions of the lymphatic system?

10. Define *edema,* and discuss its possible causes.

11. How do veins serve as a blood reservoir?

12. Compare the effect of vasoconstriction on the rate of blood flow in arterioles and veins.

13. Discuss the factors that determine mean arterial pressure.

14. Review the effects on the cardiovascular system of parasympathetic and sympathetic stimulation.

15. Differentiate between secondary hypertension and primary hypertension. What are the potential consequences of hypertension?

16. Define *circulatory shock.* What are its consequences and compensations? What is irreversible shock?

Quantitative Exercises (Solutions on p. A-46)

1. Recall that the flow rate of blood equals the pressure gradient divided by the total peripheral resistance of the vascular system. The conventional unit of resistance in physiological systems is expressed in PRU (peripheral resistance unit), which is defined as (1 liter/min)/(1 mm Hg). At rest, Tom's total peripheral resistance is about 20 PRU. Last week while playing racquetball, his cardiac output increased to 30 liters/min and his mean arterial pressure increased to 120 mm Hg. What was his total peripheral resistance during the game?

2. Systolic pressure rises as a person ages. By age 85, an average male (untreated for hypertension) has a systolic pressure of 180 mm Hg and a diastolic pressure of 90 mm Hg.
   a. What is the mean arterial pressure of this average 85-year-old male?
   b. From your knowledge of capillary dynamics, predict the result at the capillary level of this age-related change in mean arterial pressure if no homeostatic mechanisms were operating. (Recall that mean arterial pressure is about 93 mm Hg at age 20.)

3. Compare the flow rates in the systemic and the pulmonary circulations of an individual with the following measurements:
   systemic mean arterial pressure = 95 mm Hg
   systemic resistance = 19 PRU
   pulmonary mean arterial pressure = 20 mm Hg
   pulmonary resistance = 4 PRU

4. Which of the following changes would increase the resistance in an arteriole? Explain.
   a. a longer length
   b. a smaller caliber
   c. increased sympathetic stimulation
   d. increased blood viscosity
   e. all of the above

POINTS TO PONDER

(Explanations on p. A-46)

1. During coronary bypass surgery, a piece of vein is often removed from the patient's leg and surgically attached within the coronary circulatory system so that blood detours, through the vein, around an occluded coronary artery segment. Why must the patient wear, for an extended period of time after surgery, an elastic support stocking on the limb from which the vein was removed?

2. Assume a person has a blood pressure recording of 125/77:
   a. What is the systolic pressure?
   b. What is the diastolic pressure?
   c. What is the pulse pressure?
   d. What is the mean arterial pressure?

   e. Would any sound be heard when the pressure in an external cuff around the arm was 130 mm Hg? *(Yes or no?)*
   f. Would any sound be heard when cuff pressure was 118 mm Hg?
   g. Would any sound be heard when cuff pressure was 75 mm Hg?

3. A classmate who has been standing still for several hours working on a laboratory experiment suddenly faints. What is the probable explanation? What would you do if the person next to him tried to get him up?

4. A drug applied to a piece of excised arteriole causes the vessel to relax, but an isolated piece of arteriolar muscle stripped from the other layers of the vessel fails to respond to the same drug. What is the probable explanation?

5. Explain how each of the following antihypertensive drugs would lower arterial blood pressure:
   a. drugs that block α_1-adrenergic receptors (for example, *phentolamine*)
   b. drugs that block β_1-adrenergic receptors (for example, *metoprolol*) (*Hint:* Review *adrenergic receptors* on p. 243.)
   c. drugs that directly relax arteriolar smooth muscle (for example, *hydralazine*)
   d. diuretic drugs that increase urinary output (for example, *furosemide*)
   e. drugs that block release of norepinephrine from sympathetic endings (for example, *guanethidine*)
   f. drugs that act on the brain to reduce sympathetic output (for example, *clonidine*)
   g. drugs that block Ca^{2+} channels (for example, *verapamil*)
   h. drugs that interfere with the production of angiotensin II (for example, *captopril*)
   i. drugs that block angiotensin receptors (for example, *losartan*)

CLINICAL CONSIDERATION

(Explanation on p. A-47)

Li-Ying C. has just been diagnosed as having hypertension secondary to a *pheochromocytoma,* a tumor of the adrenal medulla that secretes excessive epinephrine. Explain how this condition leads to secondary hypertension, by describing the effect that excessive epinephrine would have on various factors that determine arterial blood pressure.

Blood

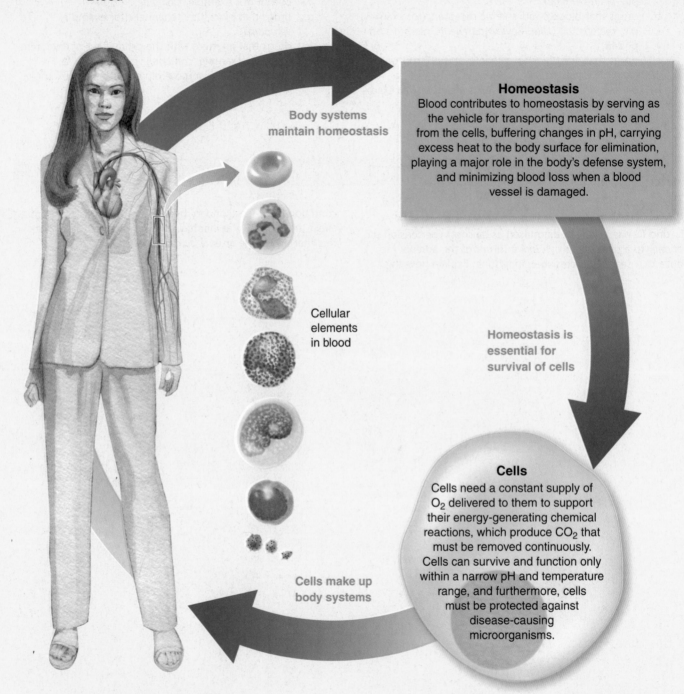

Body systems maintain homeostasis

Homeostasis
Blood contributes to homeostasis by serving as the vehicle for transporting materials to and from the cells, buffering changes in pH, carrying excess heat to the body surface for elimination, playing a major role in the body's defense system, and minimizing blood loss when a blood vessel is damaged.

Cellular elements in blood

Homeostasis is essential for survival of cells

Cells
Cells need a constant supply of O_2 delivered to them to support their energy-generating chemical reactions, which produce CO_2 that must be removed continuously. Cells can survive and function only within a narrow pH and temperature range, and furthermore, cells must be protected against disease-causing microorganisms.

Cells make up body systems

Blood is the vehicle for long-distance, mass transport of materials between the cells and external environment or between the cells themselves. Such transport is essential for maintaining homeostasis. Blood consists of a complex liquid **plasma** in which the cellular elements—*erythrocytes, leukocytes,* and *platelets*—are suspended. **Erythrocytes (red blood cells, or RBCs)** are essentially plasma membrane—enclosed bags of hemoglobin that transport O_2 in the blood. **Leukocytes (white blood cells,** or **WBCs),** the immune system's mobile defense units, are transported in the blood to sites of injury or of invasion by disease-causing microorganisms. **Platelets (thrombocytes)** are important in hemostasis, the stopping of bleeding from an injured vessel.

11

The Blood

CENGAGENOW™ Log on to CengageNOW at **http://www
.cengage.com/sso/** for an opportunity to explore animations
and interactive quizzes to help you learn, review, and master
physiology concepts.

Plasma

The hematocrit represents the packed cell volume of whole blood; plasma accounts for the rest of the volume.

Blood represents about 8% of total body weight and has an average volume of 5 liters in women and 5.5 liters in men. It consists of three types of specialized cellular elements, *erythrocytes (red blood cells), leukocytes (white blood cells),* and *platelets (thrombocytes),* suspended in the complex liquid *plasma* (● Figure 11-1 and ▲ Table 11-1). Erythrocytes and leukocytes are both whole cells, whereas platelets are cell fragments.

The constant movement of blood as it flows through the blood vessels keeps the cellular elements rather evenly dispersed within the plasma. However, if you put a sample of whole blood in a test tube and treat it to prevent clotting, the heavier cells slowly settle to the bottom and the lighter plasma rises to the top. This process can be speeded up by centrifuging, which quickly packs the cells in the bottom of the tube (● Figure 11-1). Because more than 99% of the cells are erythrocytes, the **hematocrit,** or **packed cell volume,** essentially represents the percentage of erythrocytes in the total blood volume. The hematocrit averages 42% for women and slightly higher, 45%, for men. Plasma accounts for the remaining volume. Accordingly, the average volume of plasma in the blood is 58% for women and 55% for men. White blood cells and platelets, which are colorless and less dense than red cells, are packed in a thin, cream-colored layer, the *buffy coat,* on top of the packed red cell column. They are less than 1% of the total blood volume.

Let us first consider the properties of the largest portion of the blood, the plasma, before turning attention to the cellular elements.

Plasma water is a transport medium for many inorganic and organic substances.

Plasma, being a liquid, consists of 90% water. Plasma water serves as a medium for materials being carried in the blood. Also, because water has a high capacity to hold heat, plasma can absorb and distribute much of the heat generated metabolically within tissues, with only minimal changes in the temperature of the blood itself. As blood travels close to the surface of the skin, heat energy not needed to maintain body temperature is eliminated to the environment.

● **FIGURE 11-1 Hematocrit and types of blood cells.** The values given are for men. The average hematocrit for women is 42%, with plasma occupying 58% of the blood volume. Note the biconcave shape of the erythrocytes.

A large number of inorganic and organic substances are dissolved in the plasma. Inorganic constituents account for about 1% of plasma weight. The most abundant electrolytes (ions) in the plasma are Na^+ and Cl^-, the components of common salt. Smaller amounts of HCO_3^-, K^+, Ca^{2+}, and others are present. The most notable functions of these ions are their roles in membrane excitability, osmotic distribution of fluid between the ECF and cells, and buffering of pH changes; these functions are discussed elsewhere.

The most plentiful organic constituents by weight are the plasma proteins, which make up 6% to 8% of plasma's total weight. We examine them more thoroughly in the next section. The remaining small percentage of plasma consists of other organic substances, including nutrients (such as glucose, amino acids, lipids, and vitamins), waste products (creatinine, bilirubin, and nitrogenous substances such as urea), dissolved gases (O_2 and CO_2), and hormones. Most of these substances are merely being transported in the plasma. For example, endocrine glands secrete hormones into the plasma, which transports these chemical messengers to their sites of action.

Many of the functions of plasma are carried out by plasma proteins.

Plasma proteins are the one group of plasma constituents that are not along just for the ride. These important components normally stay in the plasma, where they perform many valuable functions. Here are the most important of these functions, which are elaborated on elsewhere in the text:

1. Unlike other plasma constituents that are dissolved in the plasma water, plasma proteins are dispersed as a colloid (see p. A-10). Furthermore, because they are the largest of the plasma constituents, plasma proteins usually do not exit through the narrow pores in the capillary walls to enter the interstitial fluid. By their presence as a colloidal dispersion in the plasma and their absence in the interstitial fluid, plasma proteins establish an osmotic gradient between blood and interstitial fluid. This colloid osmotic pressure is the primary force preventing excessive loss of plasma from the capillaries into the interstitial fluid and thus helps maintain plasma volume (see p. 366).

2. Plasma proteins are partially responsible for plasma's capacity to buffer changes in pH (see p. 575).

3. The three groups of plasma proteins—*albumins, globulins, and fibrinogen*—are classified according to their various physical and chemical properties. In addition to the general functions just listed, each type of plasma protein performs specific tasks as follows:

   a. **Albumins,** the most abundant plasma proteins, contribute most extensively to the colloid osmotic pressure by virtue of their numbers. They also nonspecifically bind many substances that are poorly soluble in plasma (such as bilirubin, bile salts, and penicillin) for transport in the plasma.

   b. There are three subclasses of **globulins: alpha (α), beta (β),** and **gamma (γ).**
      (1) Like albumins, some alpha and beta globulins bind poorly water-soluble substances for transport in the plasma, but these globulins are highly specific as to which passenger they will bind and carry. Examples of substances carried by specific globulins include thyroid hormone (see p. 694), cholesterol (see p. 336), and iron (see p. 632).
      (2) Many of the factors involved in the blood-clotting process are alpha or beta globulins.
      (3) Inactive, circulating proteins, which are activated as needed by specific regulatory inputs, belong to the alpha globulin group (for example, the alpha globu-

Blood Constituents and Their Functions

Constituent	Functions
Plasma	
Water	Transport medium; carries heat
Electrolytes	Membrane excitability; osmotic distribution of fluid between ECF and ICF; buffer pH changes
Nutrients, wastes, gases, hormones	Transported in blood; the blood CO_2 plays a role in acid–base balance
Plasma proteins	In general, exert an osmotic effect important in the distribution of ECF between the vascular and interstitial compartments; buffer pH changes
Albumins	Transport many substances; contribute most to colloid osmotic pressure
Globulins	
Alpha and beta	Transport many water-insoluble substances; clotting factors; inactive precursor molecules
Gamma	Antibodies
Fibrinogen	Inactive precursor for the fibrin meshwork of a clot
Cellular Elements	
Erythrocytes	Transport O_2 and CO_2 (mainly O_2)
Leukocytes	
Neutrophils	Phagocytes that engulf bacteria and debris
Eosinophils	Attack parasitic worms; important in allergic reactions
Basophils	Release histamine, which is important in allergic reactions, and heparin, which helps clear fat from the blood
Monocytes	In transit to become tissue macrophages
Lymphocytes	
B lymphocytes	Produce antibodies
T lymphocytes	Cell-mediated immune responses
Platelets	Hemostasis

lin *angiotensinogen* is activated to *angiotensin*, which plays an important role in regulating salt balance in the body; see p. 527).

(4) The gamma globulins are the *immunoglobulins* (antibodies), which are crucial to the body's defense mechanism (see p. 429).

c. **Fibrinogen** is a key factor in blood clotting.

Plasma proteins are synthesized by the liver, with the exception of gamma globulins, which are produced by lymphocytes, one of the types of white blood cells.

Erythrocytes

Each milliliter of blood contains about 5 billion **erythrocytes (red blood cells, or RBCs),** on average, commonly reported clinically in a **red blood cell count** as 5 million cells per cubic millimeter (mm^3).

The structure of erythrocytes is well suited to their main function of O_2 transport in the blood.

The shape and content of erythrocytes are ideally suited to carry out their primary function, namely, transporting O_2 and to a lesser extent CO_2 and hydrogen ions (H^+) in the blood.

ERYTHROCYTE STRUCTURE Erythrocytes are flat, disc-shaped cells indented in the middle on both sides, like a doughnut with a flattened center instead of a hole (that is, they are biconcave discs 8 μm in diameter, 2 μm thick at the outer edges, and 1 μm thick in the center) (● Figure 11-1). This unique shape contributes in two ways to the efficiency with which red blood cells perform their main function of O_2 transport in the blood: (1) The biconcave shape provides a larger surface area for diffusion of O_2 across the membrane than a spherical cell of the same volume would. (2) The thinness of the cell enables O_2 to diffuse rapidly between the exterior and innermost regions of the cell.

Another structural feature that facilitates RBCs' transport function is the flexibility of their membrane. Red blood cells, whose diameter is normally 8 μm, can deform amazingly as they squeeze single file through capillaries as narrow as 3 μm in diameter. Because they are extremely pliant, RBCs can travel through the narrow, tortuous capillaries to deliver their O_2 cargo at the tissue level without rupturing in the process.

The most important anatomic feature that enables RBCs to transport O_2 is the hemoglobin they contain. Let us look at this unique molecule in more detail.

PRESENCE OF HEMOGLOBIN Hemoglobin is found only in red blood cells. A **hemoglobin** molecule has two parts: (1) the **globin** portion, a protein made up of four highly folded polypeptide chains (two α subunits and two β subunits); and (2) four iron-containing, nonprotein groups known as **heme groups,** each of which is bound to one of the polypeptides (● Figure 11-2). Each of the four iron atoms can combine reversibly with one molecule of O_2; thus, each hemoglobin molecule can pick up four O_2 passengers in the lungs. Because O_2 is poorly soluble

β polypeptide chain

β polypeptide chain

α polypeptide chain

Heme groups

α polypeptide chain

(a) Hemoglobin molecule

(b) Iron-containing heme group

● **FIGURE 11-2 Hemoglobin molecule.** A hemoglobin molecule consists of four highly folded polypeptide chains (the globin portion) and four iron-containing heme groups.

in the plasma, 98.5% of the O_2 carried in the blood is bound to hemoglobin (see p. 490).

Hemoglobin is a pigment (that is, it is naturally colored). Because of its iron content, it appears reddish when combined with O_2 and bluish when deoxygenated. Thus, fully oxygenated arterial blood is red in color, and venous blood, which has lost some of its O_2 load at the tissue level, has a bluish cast.

In addition to carrying O_2, hemoglobin can also combine with the following:

1. *Carbon dioxide.* Hemoglobin helps transport this gas from the tissue cells back to the lungs (see p. 494).

2. *The acidic hydrogen-ion portion (H^+) of ionized carbonic acid,* which is generated at the tissue level from CO_2. Hemoglobin buffers this acid so that it minimally alters the pH of the blood (see p. 575).

3. *Carbon monoxide (CO).* This gas is not normally in the blood, but if inhaled, it preferentially occupies the O_2-binding sites on hemoglobin, causing CO poisoning (see p. 495).

4. *Nitric oxide (NO).* In the lungs, the vasodilator nitric oxide binds to hemoglobin. This NO is released at the tissues, where it relaxes and dilates the local arterioles (see p. 356). Vasodilation helps ensure that the O_2-rich blood can make its vital rounds and also helps stabilize blood pressure.

Therefore, hemoglobin plays the key role in O_2 transport while contributing significantly to CO_2 transport and the pH-buffering capacity of blood. Furthermore, by toting along its own vasodilator, hemoglobin helps deliver the O_2 it is carrying.

LACK OF NUCLEUS AND ORGANELLES To maximize its hemoglobin content, a single erythrocyte is stuffed with more than 250 million hemoglobin molecules, excluding almost everything else. (That means each RBC can carry more than a billion O_2 molecules!) Red blood cells contain no nucleus or organelles. During the cell's development these structures are extruded to make room for more hemoglobin (● Figure 11-3). Thus, an RBC is mainly a plasma membrane–enclosed sac full of hemoglobin.

KEY ERYTHROCYTE ENZYMES Only a few crucial, nonrenewable enzymes remain within a mature erythrocyte: *glycolytic enzymes* and *carbonic anhydrase.* The **glycolytic enzymes** are necessary for generating the energy needed to fuel the active-transport mechanisms involved in maintaining proper ionic concentrations within the cell. Ironically, even though erythrocytes are the vehicles for transporting O_2 to all other tissues of the body, for energy production they themselves cannot use the O_2 they are carrying. Lacking the mitochondria that house the enzymes for oxidative phosphorylation, erythrocytes must rely entirely on glycolysis for ATP formation (see p. 33).

The other important enzyme within RBCs, **carbonic anhydrase,** is critical in CO_2 transport. This enzyme catalyzes a key reaction that leads to the conversion of metabolically produced CO_2 into **bicarbonate ion (HCO_3^-)**, which is the primary form in which CO_2 is transported in the blood. Thus, erythrocytes contribute to CO_2 transport in two ways—by means of its carriage on hemoglobin and its carbonic anhydrase–induced conversion to HCO_3^-.

Nucleus and organelles   Remnants of organelles   No nucleus or organelles

Pluripotent stem cell   Myeloid stem cell   Erythroblast   Reticulocyte   Erythrocyte

● **FIGURE 11-3 Major steps in erythrocyte production (erythropoiesis).** Erythrocytes are derived in the red bone marrow from pluripotent stem cells that give rise to all the types of blood cells. Myeloid stem cells are partially differentiated cells that give rise to erythrocytes and several other types of blood cells. Nucleated erythroblasts are committed to becoming mature erythrocytes. These cells extrude their nucleus and organelles, making more room for hemoglobin. Reticulocytes are immature red blood cells that contain organelle remnants. Mature erythrocytes are released into the abundant capillaries in the bone marrow.

The bone marrow continuously replaces worn-out erythrocytes.

Each of us has a total of 25 trillion to 30 trillion RBCs streaming through our blood vessels at any given time (100,000 times more in number than the entire U.S. population)! Yet these vital gas-transport vehicles are short-lived and must be replaced at the average rate of 2 million to 3 million cells per second.

ERYTHROCYTES' SHORT LIFE SPAN The price erythrocytes pay for their generous content of hemoglobin to the exclusion of the usual specialized intracellular machinery is a short life span. Without DNA, RNA, and ribosomes, red blood cells cannot synthesize proteins for cell repair, growth, and division or for renewing enzyme supplies. Equipped only with initial supplies synthesized before they extrude their nucleus and organelles, RBCs survive an average of only 120 days, in contrast to nerve and muscle cells, which last a person's entire life. During its short life span of four months, each erythrocyte travels about 700 miles as it circulates through the vasculature.

As a red blood cell ages, its nonreparable plasma membrane becomes fragile and prone to rupture as the cell squeezes through tight spots in the vascular system. Most old RBCs meet their final demise in the **spleen,** because this organ's narrow, winding capillary network is a tight fit for these fragile cells. The spleen lies in the upper left part of the abdomen. In addition to removing most of the old erythrocytes from circulation, the spleen has a limited ability to store healthy erythrocytes in its pulpy interior, serves as a reservoir for platelets, and contains an abundance of lymphocytes, a type of white blood cell.

ERYTHROPOIESIS Because erythrocytes cannot divide to replenish their own numbers, the old ruptured cells must be replaced by new cells produced in an erythrocyte factory—the **bone marrow**—which is the soft, highly cellular tissue that fills the internal cavities of bones. The bone marrow normally generates new red blood cells, a process known as **erythropoiesis,** at a rate to keep pace with the demolition of old cells.

During intrauterine development, erythrocytes are produced first by the yolk sac and then by the developing liver and spleen, until the bone marrow is formed and takes over eryth-

rocyte production exclusively. In children, most bones are filled with **red bone marrow** that is capable of blood cell production. As a person matures, however, fatty **yellow bone marrow** that is incapable of erythropoiesis gradually replaces red marrow, which remains only in a few isolated places, such as the sternum (breastbone), ribs, pelvis, and upper ends of the long limb bones. These are sites where bone marrow is extracted for examination or for use in bone marrow transplants.

Red marrow not only produces RBCs but also is the ultimate source for leukocytes and platelets. Undifferentiated **pluripotent stem cells,** the source of all blood cells, reside in the red marrow, where they continuously divide and differentiate to give rise to each of the types of blood cells (● Figure 11-3; also see ● Figure 11-9, p. 404). The different types of immature blood cells, along with the stem cells, are intermingled in the red marrow at various stages of development. Once mature, the blood cells are released into the rich supply of capillaries that permeate the red marrow. Bone marrow capillaries are of the less common discontinuous type that has large gaps between the endothelial cells (see p. 363). Mature red blood cells can pass through these very large pores to enter the blood, but, once circulating, these cells cannot exit the blood through the much narrower pores in ordinary capillaries. Regulatory factors act on the *hemopoietic* ("blood-producing") red marrow to govern the type and number of cells generated and discharged into the blood. Of the blood cells, the mechanism for regulating RBC production is the best understood. We consider it next.

Erythropoiesis is controlled by erythropoietin from the kidneys.

Because O_2 transport in the blood is the erythrocytes' primary function, you might logically suspect that the primary stimulus for increased erythrocyte production would be reduced O_2 delivery to the tissues. You would be correct, but low O_2 levels do not stimulate erythropoiesis by acting directly on the red bone marrow. Instead, reduced O_2 delivery to the kidneys stimulates them to secrete the hormone **erythropoietin** into the blood, and this hormone in turn stimulates erythropoiesis by the bone marrow (● Figure 11-4).

Erythropoietin acts on derivatives of undifferentiated stem cells that are already committed to becoming RBCs, stimulating their proliferation and maturation into mature erythrocytes. This increased erythropoietic activity elevates the number of circulating RBCs, thereby increasing O_2-carrying capacity of the blood and restoring O_2 delivery to the tissues to normal. Once normal O_2 delivery to the kidneys is achieved, erythropoietin secretion is turned down until needed again. In this way, erythrocyte production is normally balanced

Blood Doping: Is More of a Good Thing Better?

Exercising muscles require a continual supply of O_2 for generating energy to sustain endurance activities (see p. 278). **Blood doping** is a technique designed to temporarily increase the O_2-carrying capacity of blood in an attempt to gain a competitive advantage. Blood doping involves removing blood from an athlete, then promptly reinfusing the plasma but freezing the RBCs for reinfusion one to seven days before a competitive event. One to four units of blood (one unit equals 450 ml) are usually withdrawn at three- to eight-week intervals before the competition. In the periods between blood withdrawals, increased erythropoietic activity restores the RBC count to a normal level.

Reinfusion of the stored RBCs temporarily increases the red blood cell count and hemoglobin level above normal. Theoretically, blood doping would benefit endurance athletes by improving the blood's O_2-carrying capacity. If too many red cells were infused, however, performance could suffer because the increased blood viscosity would decrease blood flow.

Research indicates that in a standard laboratory exercise test, athletes who have used blood doping may realize a 5% to 13% increase in aerobic capacity; a reduction in heart rate during exercise compared to the rate during the same exercise in the absence of blood doping; improved performance; and reduced lactate levels in the blood. (Lactate is produced when muscles resort to less efficient anaerobic glycolysis for energy production; see page 278.)

Blood doping, although effective, is illegal in both collegiate athletics and Olympic competition for ethical and medical reasons. Of concern, as with the use of any banned performance-enhancing product, is the loss of fair competition. Furthermore, the practice has even been implicated in the deaths of some athletes. The prohibitive regulations are very difficult to enforce, however. Blood doping cannot be detected by currently available testing procedures. The only way of exposing the practice of blood doping is through witnesses or self-admission.

The recent development of synthetic erythropoietin exacerbates the problem of blood doping. Injection of this product stimulates RBC production and thus temporarily increases the O_2-carrying capacity of the blood. Rigorous studies have demonstrated that injected erythropoietin may improve an endurance athlete's performance by 7% to 10%. Although formally banned, a black market for erythropoietin developed among sports cheats when this product became available as a drug to treat anemia. Erythropoietin is now widely used among competitors in cycling, cross-country skiing, and long-distance running and swimming. This practice is ill advised, however, not only because of legal and ethical implications but because of the dangers of increasing blood viscosity. Synthetic erythropoietin is believed responsible for the deaths of 20 European cyclists since 1987. Unfortunately, too many athletes are willing to take the risks. The recent development of a test to detect erythropoietin abuse hopefully will curb its use.

Kidney

Erythropoietin

1 Reduced oxygen-carrying capacity

Relieves **5**

Developing erythrocytes in red bone marrow

4 Increased oxygen-carrying capacity

Erythrocytes

1 Kidneys detect reduced O_2-carrying capacity of blood.

2 When less O_2 is delivered to the kidneys, they secrete erythropoietin into blood.

3 Erythropoietin stimulates erythropoiesis by bone marrow.

4 Additional circulating erythrocytes increase O_2-carrying capacity of blood.

5 Increased O_2-carrying capacity relieves initial stimulus that triggered erythropoietin secretion.

● **FIGURE 11-4 Control of erythropoiesis.**

against destruction or loss of these cells so that O_2-carrying capacity in the blood stays fairly constant. In severe loss of RBCs, as in hemorrhage or abnormal destruction of young circulating erythrocytes, the rate of erythropoiesis can be increased to more than six times the normal level. (For a discussion of erythropoietin abuse by some athletes, see the accompanying boxed feature, ■ A Closer Look at Exercise Physiology.)

The preparation of an erythrocyte for its departure from the marrow involves several steps, such as synthesis of hemoglobin and extrusion of the nucleus and organelles. Cells closest to maturity need a few days to be "finished off" and released into the blood in response to erythropoietin, and less developed and newly proliferated cells may take up to several weeks to reach maturity. Therefore, the time required for complete replacement of lost RBCs depends on how many are needed to return the number to normal. (When you donate blood, your circulating erythrocyte supply is replenished in less than a week.)

KEY

☐ = Plasma   ■ = Erythrocytes

● **FIGURE 11-5 Hematocrit under various circumstances.** (a) Normal hematocrit. (b) The hematocrit is lower than normal in anemia, because of too few circulating erythrocytes; and (c) above normal in polycythemia, because of excess circulating erythrocytes. (d) The hematocrit can also be elevated in dehydration when the normal number of circulating erythrocytes is concentrated in a reduced plasma volume.

RETICULOCYTES When demands for RBC production are high (for example, following hemorrhage), the bone marrow may release large numbers of immature erythrocytes, known as **reticulocytes,** into the blood to quickly meet the need (see ● Figure 11-3). These immature cells can be recognized by staining techniques that make visible the residual organelle remnants that have not yet been extruded. Their presence above the normal level of 0.5% to 1.5% of the total number of circulating erythrocytes indicates a high rate of erythropoietic activity. At very rapid rates, more than 30% of the circulating red cells may be at the immature reticulocyte stage.

SYNTHETIC ERYTHROPOIETIN Researchers have identified the gene that directs erythropoietin synthesis, so this hormone can now be produced in a laboratory. Laboratory-produced erythropoietin (Epogen, Procrit) has currently become biotechnology's single-biggest money-maker, with sales exceeding $1 billion annually. This hormone is often used to boost RBC production in patients with suppressed erythropoietic activity, such as those with kidney failure or those undergoing chemotherapy for cancer. (Chemotherapy drugs interfere with the rapid cell division characteristic of both cancer cells and developing RBCs.) Furthermore, the ready availability of this hormone has diminished the need for blood transfusions. For example, transfusion of a surgical patient's own previously collected blood, coupled with erythropoietin to stimulate further RBC production, has reduced the use of donor blood by as much as 50% in some hospitals.

Anemia can be caused by a variety of disorders.

Despite control measures, O_2-carrying capacity cannot always be maintained to meet tissue needs. A below-normal O_2-carrying capacity of the blood is known as **anemia,** which is characterized by a low hematocrit (● Figure 11-5a and b). Anemia can be brought about by a decreased rate of erythropoiesis, excessive losses of erythrocytes, or a deficiency in the hemoglobin content of erythrocytes. The various causes of anemia can be grouped into six categories:

1. **Nutritional anemia** is caused by a dietary deficiency of a factor needed for erythropoiesis. The production of RBCs depends on an adequate supply of essential raw ingredients, some of which are not synthesized in the body but must be provided by dietary intake. For example, *iron deficiency anemia* occurs when not enough iron is available for synthesizing hemoglobin.

2. **Pernicious anemia** is caused by an inability to absorb enough ingested vitamin B_{12} from the digestive tract. Vitamin B_{12} is essential for normal RBC production and maturation. It is abundant in a variety of foods. The problem is a deficiency of *intrinsic factor,* a special substance secreted by the lining of the stomach (see p. 608). Vitamin B_{12} can be absorbed from the intestinal tract only when this nutrient is bound to intrinsic factor. When intrinsic factor is deficient, not enough ingested vitamin B_{12} is absorbed. The resulting impairment of RBC production and maturation leads to anemia.

3. **Aplastic anemia** is caused by failure of the bone marrow to produce enough RBCs, even though all ingredients necessary

for erythropoiesis are available. Reduced erythropoietic capability can be caused by destruction of red bone marrow by toxic chemicals (such as benzene), heavy exposure to radiation (fallout from a nuclear bomb explosion, for example, or excessive exposure to X rays), invasion of the marrow by cancer cells, or chemotherapy for cancer. The destructive process may selectively reduce the marrow's output of erythrocytes, or it may reduce the productive capability for leukocytes and platelets as well. The anemia's severity depends on the extent to which erythropoietic tissue is destroyed; severe losses are fatal.

4. **Renal anemia** may result from kidney disease. Because erythropoietin from the kidneys is the primary stimulus for promoting erythropoiesis, inadequate erythropoietin secretion by diseased kidneys leads to insufficient RBC production.

5. **Hemorrhagic anemia** is caused by losing a lot of blood. The loss can be either acute, such as a bleeding wound, or chronic, such as excessive menstrual flow.

6. **Hemolytic anemia** is caused by the rupture of too many circulating erythrocytes. **Hemolysis,** the rupture of RBCs, occurs either because otherwise normal cells are induced to rupture by external factors, as in the invasion of RBCs by malaria parasites, or because the cells are defective, as in sickle cell disease.

Malaria is caused by protozoan parasites introduced into a victim's blood by the bite of a carrier mosquito (hence the use of mosquito nets over beds at night in tropical regions to reduce the incidence of malaria). These parasites selectively invade RBCs, where they multiply to the point that the mass of malarial organisms ruptures the cells, releasing hundreds of new active parasites that quickly invade other RBCs. As this cycle continues and more erythrocytes are destroyed, the anemic condition progressively worsens.

Sickle cell disease is the best-known example among various hereditary abnormalities of erythrocytes that make these cells very fragile. It affects about 1 in 650 African Americans. Sickle cell disease is caused by a genetic mutation that changes a single amino acid in the 146-long amino acid chain that makes up the β chain of hemoglobin (valine replaces glutamate at position 6 in this amino acid chain). This defective type of hemoglobin joins together to form rigid chains that make the RBC stiff and unnaturally shaped, like a crescent or sickle (● Figure 11-6). Unlike normal erythrocytes, these deformed RBCs tend to clump together. The resultant logjam blocks blood flow through small vessels, leading to pain and tissue damage at the affected site. Furthermore, the defective erythrocytes are fragile and prone to rupture, even as young cells, as they travel through the narrow splenic capillaries. Despite an accelerated rate of erythropoiesis triggered by the constant excessive loss of RBCs, production may not be able to keep pace with the rate of destruction, and anemia may result. Interestingly, patients who have sickle cell anemia are more likely to survive malaria, because defective cells invaded by malaria parasites are more apt to be destroyed as they travel through the spleen, eliminating the infected cells before the parasites have a chance to multiply and spread. This advantage may explain the increased incidence of sickle cell disease among the descendants of people who lived in tropical regions where malaria is prevalent. In an evolution-

● **FIGURE 11-6 Sickle-shaped red blood cell.**

ary "trade-off," the benefit conferred by surviving malaria may have offset the disadvantages of sickle cell disease.

Polycythemia is an excess of circulating erythrocytes.

Polycythemia, in contrast to anemia, is characterized by too many circulating RBCs and an elevated hematocrit (see ● Figure 11-5c). There are two general types of polycythemia, depending on the circumstances that trigger the excess RBC production: *primary polycythemia* and *secondary polycythemia.*

Primary polycythemia is caused by a tumorlike condition of the bone marrow in which erythropoiesis proceeds at an excessive, uncontrolled rate instead of being subject to the normal erythropoietin regulatory mechanism. The RBC count may reach 11 million cells/mm^3 (normal is 5 million cells/mm^3), and the hematocrit may be as high as 70% to 80% (normal is 42% to 45%). No benefit is derived from the extra O_2-carrying capacity of the blood, because O_2 delivery is more than adequate with normal RBC numbers. Inappropriate polycythemia has harmful effects, however. The excessive number of red cells increases blood's viscosity up to five to seven times normal (that is, makes the blood "thicker"), causing the blood to flow very sluggishly, which may actually reduce O_2 delivery to the tissues (see p. 345). The increased viscosity also increases the total peripheral resistance, which may elevate the blood pressure, thus increasing the workload of the heart, unless blood-pressure control mechanisms can compensate (see ● Figure 10-14, p. 360).

Secondary polycythemia, in contrast, is an appropriate erythropoietin-induced adaptive mechanism to improve the blood's O_2-carrying capacity in response to a prolonged reduction in O_2 delivery to the tissues. It occurs normally in people living at high altitudes, where less O_2 is available in the air, or in people for whom O_2 delivery to the tissues is impaired by chronic lung disease or cardiac failure. The red cell count in secondary polycythemia is usually lower than that in primary polycythemia, typically averaging 6 million to 8 million cells/mm^3. The price paid for improved O_2 delivery is an increased viscosity of the blood.

An elevated hematocrit can occur when the body loses fluid but not erythrocytes, as in dehydration accompanying heavy sweating or profuse diarrhea (see ● Figure 11-5d). This is not a true polycythemia, however, because the number of circulating RBCs is not increased. A normal number of erythrocytes are simply concentrated in a smaller plasma volume. This condition is sometimes termed **relative polycythemia.**

We next shift attention to the different blood types, which depend on special carbohydrate markers on the surface membrane of erythrocytes.

Blood types depend on surface antigens on erythrocytes.

An **antigen** is a large, complex molecule that triggers a specific immune response against itself when it gains entry to the body. For example, antigens are found on the surface of foreign cells such as bacterial invaders. Certain types of white blood cells recognize antigens and produce specific antibodies against them. An **antibody** binds with the specific antigen against which it is produced and leads to the antigen's destruction by various means. Thus the body rejects cells bearing antigens that do not match its own. You will learn more about these immune responses in the next chapter on body defenses. For now, we focus on the special antigen/antibody reaction that forms the basis of different blood types.

ABO BLOOD TYPES The surface membranes of human erythrocytes contain inherited antigens that vary depending on blood type. Within the major blood group system, the **ABO system,** the erythrocytes of people with type A blood contain A antigens, those with type B blood contain B antigens, those with type AB blood have both A and B antigens, and those with type O blood do not have any A or B red blood cell surface antigens.

Antibodies against erythrocyte antigens not present on the body's own erythrocytes begin to appear in human plasma after a baby is about six months of age. Accordingly, the plasma of type A blood contains anti-B antibodies, type B blood contains anti-A antibodies, no antibodies related to the ABO system are present in type AB blood, and both anti-A and anti-B antibodies are present in type O blood. Typically, one would expect antibody production against A or B antigen to be induced only if blood containing the alien antigen were injected into the body. However, high levels of these antibodies are found in the plasma of people who have never been exposed to a different type of blood. Consequently, these were considered naturally occurring antibodies, that is, produced without any known exposure to the antigen. Scientists now know that people are unknowingly exposed at an early age to small amounts of A- and B-like antigens associated with common intestinal bacteria. Antibodies produced against these foreign antigens coincidentally also interact with a nearly identical antigen for a foreign blood group, even on first exposure to it.

TRANSFUSION REACTION If a person is given blood of an incompatible type, two different antigen–antibody interactions take place. By far the more serious consequences arise from the effect of the antibodies in the recipient's plasma on the incoming donor erythrocytes. The effect of the donor's antibodies on the recipient's erythrocyte-bound antigens is less important unless a large amount of blood is transfused, because the donor's antibodies are so diluted by the re-

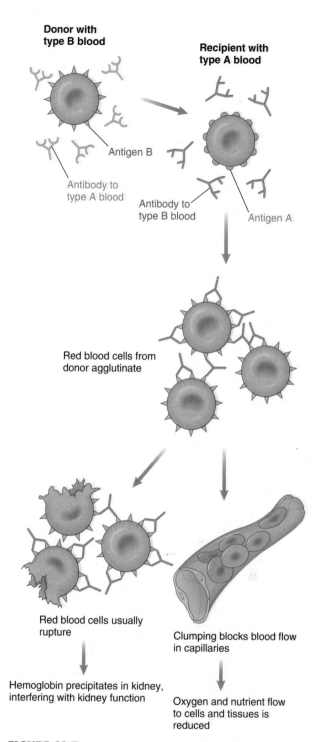

● FIGURE 11-7 Transfusion reaction. A transfusion reaction resulting from type B blood being transfused into a recipient with type A blood.

cipient's plasma that little red blood cell damage takes place in the recipient.

Antibody interaction with erythrocyte-bound antigen may result in agglutination (clumping) or hemolysis (rupture) of the attacked red blood cells. Agglutination and hemolysis of donor red blood cells by antibodies in the recipient's plasma can lead to a sometimes fatal **transfusion reaction** (● Figure 11-7). Agglutinated clumps of incoming donor cells can plug small blood vessels. In addition, one of the most lethal consequences of mismatched transfusions is acute kidney failure caused by the release of large amounts of hemoglobin from ruptured donor erythrocytes. If the free hemoglobin in the plasma rises above a critical level, it will precipitate in the kidneys and block the urine-forming structures, leading to acute kidney shutdown.

UNIVERSAL BLOOD DONORS AND RECIPIENTS Because type O individuals have no A or B antigens, their erythrocytes will not be attacked by either anti-A or anti-B antibodies, so they are considered **universal donors.** Their blood can be transfused into people of any blood type. However, type O individuals can receive only type O blood, because the anti-A and anti-B antibodies in their plasma will attack either A or B antigens in incoming blood. In contrast, type AB individuals are called **universal recipients.** Lacking both anti-A and anti-B antibodies, they can accept donor blood of any type, although they can donate blood only to other AB people. Because their erythrocytes have both A and B antigens, their cells would be attacked if transfused into individuals with antibodies against either of these antigens.

The terms *universal donor* and *universal recipient* are misleading, however. In addition to the ABO system, many other erythrocyte antigens and plasma antibodies can cause transfusion reactions, the most important of which is the Rh factor.

Rh BLOOD-GROUP SYSTEM People who have the **Rh factor** (an erythrocyte antigen first observed in rhesus monkeys, hence the designation Rh) are said to have *Rh-positive* blood, whereas those lacking the Rh factor are considered *Rh-negative*. In contrast to the ABO system, no naturally occurring antibodies develop against the Rh factor.

Clinical Note Anti-Rh antibodies are produced only by Rh-negative individuals when (and if) such people are first exposed to the foreign Rh antigen present in Rh-positive blood. A subsequent transfusion of Rh-positive blood could produce a transfusion reaction in such a sensitized Rh-negative person. Rh-positive individuals, in contrast, never produce antibodies against the Rh factor that they themselves possess. Therefore, Rh-negative people should be given only Rh-negative blood, whereas Rh-positive people can safely receive either Rh-negative or Rh-positive blood. The Rh factor is of particular medical importance when an Rh-negative mother develops antibodies against the erythrocytes of an Rh-positive fetus she is carrying, a condition known as **erythroblastosis fetalis,** or **hemolytic disease of the newborn.** Because the maternal antibodies destroy many fetal erythrocytes, the fetal bone marrow cannot keep pace with the rate of destruction and releases immature precursors of erythrocytes, such as reticulocytes and eventually erythroblasts (see ● Figure 11-3), hence the name of the condition. (See the Clinical Consideration on p. 414 for further discussion of this disorder.)

Except in extreme emergencies, it is safest to individually cross-match blood before a transfusion is undertaken even though the ABO and Rh typing is already known, because there are approximately 23 other minor human erythrocyte antigen systems, with hundreds of different subtypes. Compatibility is determined by mixing the red blood cells from the potential donor with plasma from the recipient. If no clumping occurs, the blood is considered an adequate match for transfusion. (See the boxed feature on pp. 402–403, ▇ Concepts, Challenges, and Controversies, for an update on alternatives to whole-blood transfusions under investigation.)

In addition to being an important consideration in transfusions, the various blood-group systems are also of legal importance in disputed paternity cases because the erythrocyte antigens are inherited. In recent years, however, DNA "fingerprinting" has become a more definitive test.

Leukocytes

Leukocytes (white blood cells, or **WBCs)** are the mobile units of the body's immune defense system. **Immunity** is the body's ability to resist or eliminate potentially harmful foreign materials or abnormal cells. Leukocytes and their derivatives, along with a variety of plasma proteins, make up the **immune system,** an internal defense system that recognizes and either destroys or neutralizes materials within the body that are foreign to the "normal self." Specifically, the immune system (1) defends against invading **pathogens** (disease-producing microorganisms such as bacteria and viruses); (2) identifies and destroys cancer cells that arise in the body; and (3) functions as a "cleanup crew" that removes worn-out cells (such as aged red blood cells) and tissue debris (for example, tissue damaged by trauma or disease). The latter is essential for wound healing and tissue repair.

Leukocytes primarily function as defense agents outside the blood.

To carry out their functions, leukocytes largely use a "seek out and attack" strategy; that is, they go to sites of invasion or tissue damage. The main reason WBCs are in the blood is to be rapidly transported from their site of production or storage to wherever they are needed. Unlike erythrocytes, leukocytes are able to exit the blood by assuming amoeba-like behavior to wriggle through narrow capillary pores and crawl to assaulted areas (see ● Figure 12-2, p. 421). For this reason, we introduce the specific circulating leukocytes here to round out our discussion of blood but leave a more detailed discussion of their phagocytic and immunologic functions, which take place primarily in the tissues, for the next chapter.

There are five types of leukocytes.

Leukocytes lack hemoglobin (in contrast to erythrocytes), so they are colorless (that is, "white") unless specifically stained for microscopic visibility. Unlike erythrocytes, which are of uniform structure, identical function, and constant number, leukocytes vary in structure, function, and number. There are five different types of circulating leukocytes—neutrophils, eosinophils, basophils, monocytes, and lymphocytes—each with a characteristic structure and function. They are all somewhat larger than erythrocytes.

The five types of leukocytes fall into two main categories, depending on the appearance of their nuclei and the presence

			Leukocytes				
Polymorphonuclear granulocytes			Mononuclear agranulocytes		Erythrocyte	Platelets	
Neutrophil	Eosinophil	Basophil	Monocyte	Lymphocyte			
60%–70%	1%–4%	0.25%–0.5%	2%–6%	25%–33%	Erythrocyte concentration = 5 billion/ ml blood	Platelet concentration = 250 million/ ml blood	
Differential WBC count (percentage distribution of types of leukocytes)							
Leukocyte concentration = 7 million/ml blood					RBC count = 5,000,000/mm³	Platelet count = 250,000/mm³	
WBC count = 7000/mm³							

● **FIGURE 11-8 Normal blood cellular elements and typical human blood cell count.**

or absence of granules in their cytoplasm when viewed microscopically (● Figure 11-8). Neutrophils, eosinophils, and basophils are categorized as **polymorphonuclear** ("many-shaped nucleus") **granulocytes** ("granule-containing cells"). Their nuclei are segmented into several lobes of varying shapes, and their cytoplasm contains an abundance of membrane-enclosed granules. The three types of granulocytes are distinguished on the basis of the varying affinity of their granules for dyes: *Eosinophils* have an affinity for the red dye eosin, *basophils* preferentially take up a basic blue dye, and *neutrophils* are neutral, showing no dye preference. Monocytes and lymphocytes are known as **mononuclear** ("single nucleus") **agranulocytes** ("cells lacking granules"). Both have a single, large, nonsegmented nucleus and few granules. Monocytes are the larger of the two and have an oval or kidney-shaped nucleus. *Lymphocytes,* the smallest of the leukocytes, characteristically have a large spherical nucleus that occupies most of the cell.

Leukocytes are produced at varying rates depending on the changing defense needs of the body.

All leukocytes ultimately originate from the same undifferentiated pluripotent stem cells in the red bone marrow that also give rise to erythrocytes and platelets (● Figure 11-9, p. 404). The cells destined to become leukocytes eventually differentiate into various committed cell lines and proliferate under the influence of appropriate stimulating factors. Granulocytes and monocytes are produced only in the bone marrow, which releases these mature leukocytes into the blood. Lymphocytes are originally derived from precursor cells in the bone marrow, but most new lymphocytes are actually produced by lymphocytes already in the **lymphoid** (lymphocyte-containing) **tissues,** such as the lymph nodes and tonsils.

The total number of leukocytes normally ranges from 5 million to 10 million cells per milliliter of blood, with an average of 7 million cells/ml, expressed as an average **white blood cell count** of 7000/mm³. Leukocytes are the least numerous of the blood cells (about 1 white blood cell for every 700 red blood cells), not because fewer are produced but because they are merely in transit while in the blood. Normally, about two thirds of the circulating leukocytes are granulocytes, mostly neutrophils, whereas one third are agranulocytes, predominantly lymphocytes (see ● Figure 11-8). However, the total number of white cells and the percentage of each type may vary considerably to meet changing defense needs. Depending on the type and extent of assault the body is combating, different types of leukocytes are selectively produced at varying rates. Chemical messengers arising from invaded or damaged tissues or from activated leukocytes themselves govern the rates of production of the various leukocytes. Specific messengers analogous to erythropoietin direct the differentiation and proliferation of each cell type. Some of these messengers have been identified and can be produced in the laboratory; an example is **granulocyte colony–stimulating factor,** which stimulates increased replication and release of granulocytes, especially neutrophils, from the bone marrow. Marketed under the name Neulasta, this synthetic agent is a powerful new therapeutic tool that can be used to bolster defense and thus decrease the incidence of infection in cancer patients being treated with chemotherapy drugs, which suppress all rapidly dividing cells, including hemopoietic cells in the bone marrow as well as the targeted cancer cells.

FUNCTIONS AND LIFE SPANS OF LEUKOCYTES Following are the functions and life spans of the granulocytes:

▪ **Neutrophils** are phagocytic specialists. Furthermore, scientists recently discovered that neutrophils can also act like "suicide bombers," undergoing an unusual type of programmed cell death in which they release into the ECF a web of prepared fibers dubbed *neutrophil extracellular traps (NETs)* as the plasma membrane ruptures. These fibers, which consist of granule pro-

In Search of a Blood Substitute

One of the hottest medical contests of the last three decades has been the race to develop a universal substitute for human blood that is safe, inexpensive, and disease free and that has a long shelf life.

Need for a Blood Transfusion

A blood transfusion is administered, on average, every two seconds in the United States alone. As a general rule of thumb, a transfusion is warranted when the hemoglobin concentration falls below 7 grams/100 ml whole blood (normal range is 13 to 18 grams/100 ml for men and 12 to 16 grams/100 ml for women), or when another specific component of the blood, such as platelets or clotting factors, is in short supply. The need can arise with victims of traumatic emergencies, such as automobile accidents or gunshot wounds; with surgery patients; with cancer patients on chemotherapy that suppresses blood cell production; with people who have blood disorders like leukemia or sickle cell anemia; with patients undergoing bone marrow transplants; and with other illnesses. Some patients need ongoing transfusions to survive. Sometimes whole blood is separated so that different components may be used for multiple recipients. Collection specialists have even developed a technique known as *apheresis* in which blood is collected into an instrument that separates the blood into separate portions, then the rest of the blood is returned to the donor after a selected portion, such as platelets, is recovered. With this technique, enough platelets can be collected from one donor for a platelet transfusion; otherwise, platelets from 5 to 10 whole-blood donors must be pooled to collect enough platelets for a platelet transfusion.

Need for a Blood Substitute

With only about 5% of the population now donating blood, regional shortages of particular blood types necessitate the shipping and sharing of blood among areas. (Almost everyone over 17 years of age who weighs over 110 pounds, is healthy, and has not donated blood in the last 8 weeks is eligible to be a blood donor, but most people do not take this opportunity to give the gift of life.) Medical personnel anticipate serious widespread shortages of blood in the near future because the number of blood donors continues to decline at the same time that the number of senior citizens, the group of people who most often need transfusions, continues to grow. The benefits for society of a safe blood substitute that could be administered without regard for the recipient's blood type are great, as will be the profits for the manufacturer of the first successful product. Experts estimate that the world market for a good blood substitute, more accurately called an *oxygen therapeutic,* may be as much as $10 billion per year.

Researchers began working on blood substitutes in the 1960s, but the search for an alternative to whole-blood transfusions was given new impetus in the 1980s by the rising incidence of AIDS and the concomitant concern over the safety of the nation's blood supply. Infectious diseases such as AIDS, viral hepatitis, and West Nile virus infection can be transmitted from infected blood donors to recipients of blood transfusions. Furthermore, restrictions have been placed on potential donors who lived or traveled in Europe during the time mad cow disease hit the beef industry. Although careful screening of our blood supply minimizes the possibility that infectious diseases will be transmitted through transfusion, the public remains wary and would welcome a safe substitute.

Eliminating the risk of disease transmission is only one advantage of finding an alternative to whole-blood transfusion. Whole blood must be kept refrigerated, and even then it has a shelf life of only 42 days. Also, transfusion of whole blood requires blood typing and cross matching, which cannot be done at the scene of an accident or on a battlefield.

Major Approaches

The goal is not to find a replacement for whole blood but to duplicate its O_2-carrying capacity. The biggest need for blood transfusions is to replace acute blood loss in accident victims, surgical patients, and wounded soldiers. These individuals require short-term replenishment of blood's O_2-carrying capacity until their own bodies synthesize replacement erythrocytes. The many other important elements in blood are not as immediately critical in sustaining life as the hemoglobin within the RBCs is. Problematically, RBCs are the whole-blood component that requires refrigeration, has a short shelf life, and bears the markers for the various blood types.

Therefore, the search for a blood substitute has focused on two major possibilities: (1) hemoglobin products that exist outside an RBC and can be stored at room temperature for up to six months to a year, and (2) chemically synthesized products that serve as artificial hemoglobin by dissolving large amounts of O_2 when O_2 levels are high (as in the lungs) and releasing it when O_2 levels are low (as in the tissues). A variety of potential blood substitutes are in various stages of development. Some have reached the stage of clinical trials, but no products have yet reached the market, although they are getting close. Let us examine each of these major approaches.

Hemoglobin Products

By far the greatest number of research efforts have focused on manipulating the structure of hemoglobin so that it can be effectively and safely administered as a substitute for whole-blood transfusions. If appropriately stabilized and suspended in saline solution,

---

teins from the neutrophil's cytoplasm and chromatin from its nucleus, bind with bacteria and contain bacteria-killing chemicals, enabling NETs to trap, then destroy bacteria extracellularly. Thus, neutrophils can destroy bacteria both intracellularly by phagocytosis and extracellularly via the NETs they release on their death. Neutrophils invariably are the first defenders on the scene of bacterial invasion and, accordingly, are very important in inflammatory responses. Furthermore, they scavenge to clean up debris.

 As might be expected in view of these functions, an increase in circulating neutrophils (**neutrophilia**) typically accompanies acute bacterial infections. In fact, a hemoglobin could be injected to bolster O_2-carrying capacity of the recipients' blood no matter what their blood type. The following strategies are among those being pursued to develop a hemoglobin product:

- One problem is that hemoglobin behaves differently when it is outside RBCs. "Naked" hemoglobin splits into halves that do not release O_2 for tissue use as readily as normal hemoglobin does. Also, these hemoglobin fragments can cause kidney damage. A cross-binding reagent has been developed that keeps hemoglobin molecules intact when they are outside the confines of RBCs, thus surmounting one major obstacle to administering free hemoglobin.
- Some products under investigation are derived from outdated, donated human blood. Instead of the blood being discarded, its hemoglobin is extracted, purified, sterilized, and chemically stabilized. However, this strategy relies on the continued practice of collecting human blood donations.
- Several products use cows' blood as a starting point. Bovine hemoglobin is readily available from slaughterhouses, is cheap, and can be treated for administration to humans. A big concern with these products is the potential of introducing into humans unknown disease-causing microbes that might be lurking in the bovine products.
- A potential candidate as a blood substitute is genetically engineered hemoglobin, which bypasses the ongoing need for human donors or the risk of spreading disease from cows to humans. Genetic engineers can insert the gene for human hemoglobin into bacteria, which act as a "factory" to produce the desired hemoglobin product. A drawback for genetically engineered hemoglobin is the high cost involved in operating the facilities.

- One promising strategy encapsulates hemoglobin within liposomes—membrane-wrapped containers—similar to real hemoglobin-stuffed, plasma membrane–enclosed RBCs. These so-called neo red cells await further investigation.

Synthetic O_2 Carriers

Other researchers are pursuing the development of chemical-based strategies that rely on *perfluorocarbons (PFCs)*, which are artificial O_2-carrying compounds composed of carbon and fluorine. PFCs are completely inert, chemically synthesized molecules that can dissolve large quantities of O_2 in direct proportion to the amount of O_2 breathed in. Because they are derived from a nonbiological source, PFCs cannot transmit disease. This, coupled with their low cost, makes them attractive as a blood substitute. Yet use of PFCs is not without risk. Their administration can cause flulike symptoms, and because of poor excretion, they may be retained and accumulate in the body. Ironically, PFC administration poses the danger of causing O_2 toxicity by delivering too much O_2 to the tissues in uncontrolled fashion (see p. 497).

A more recent strategy under development is a water soluble, synthetic plastic version of hemoglobin. The molecule has the same size and shape of hemoglobin, reversibly binds with O_2, and is made from a substance known to be safe in the body, but it has not yet undergone biological testing.

Tactics to Reduce Need for Donated Blood

Other tactics besides blood substitutes aimed toward reducing the need for donated blood include the following:

- By changing surgical practices, the medical community has reduced the need for transfusions. These blood-saving methods include recycling a patient's own blood during surgery (collecting lost blood, then reinfusing it); using less invasive and therefore less bloody surgical techniques; and treating the patient with blood-enhancing erythropoietin prior to surgery.
- The necessity of matching blood types for transfusions is a major reason for waste at blood banks. Transfusion of mismatched blood causes a serious, even fatal reaction (see p. 399). Therefore, a blood bank may be discarding stocks of one blood type that has gone unused, while facing a serious shortage of another type. The various blood types in the major ABO system are distinguished by differences in specific antigens on the red blood cell membrane surface (see p. 399). Mismatched RBC antigens are the target for attack in a transfusion reaction. After much searching, researchers recently zeroed in on two enzymes from different bacteria, one enzyme that can cleave A antigens and the other that can cleave B antigens away from RBCs, thus converting all donated blood into Type O, which could be safely transfused into anyone. If clinical trials using this enzyme-converted blood are successful, such a product would reduce the current wastage and would be highly beneficial to battlefield surgeons who often do not have time to type blood.
- Other investigators are seeking ways to prolong the life of RBCs, either in a blood bank or in patients, thus reducing the need for fresh, transfusable blood.

As this list of strategies attests, considerable progress has been made toward developing a safe, effective alternative to whole-blood transfusions. Yet after three decades of intense effort, considerable challenges remain, and no ideal solution has been found.

---

differential WBC count (a determination of the proportion of each type of leukocyte present) can be useful in making an immediate, reasonably accurate prediction of whether an infection, such as pneumonia or meningitis, is of bacterial or viral origin. Obtaining a definitive answer as to the causative agent by culturing a sample of the infected tissue's fluid takes several days. Because an elevated neutrophil count is highly indicative of bacterial infection, it is appropriate to initiate antibiotic therapy long before the true causative agent is actually known. (Bacteria generally succumb to antibiotics, whereas viruses do not.)

● FIGURE 11-9 Blood cell production (hemopoiesis). All the blood cell types ultimately originate from the same undifferentiated pluripotent stem cells in the red bone marrow.

■ **Eosinophils** are specialists of another type. An increase in circulating eosinophils (**eosinophilia**) is associated with allergic conditions (such as asthma and hay fever) and with internal parasite infestations (for example, worms). Eosinophils obviously cannot engulf a much larger parasitic worm, but they do attach to the worm and secrete substances that kill it.

■ **Basophils** are the least numerous and most poorly understood of the leukocytes. They are quite similar structurally and functionally to **mast cells,** which never circulate in the blood but instead are dispersed in connective tissue throughout the body. Scientists once believed that basophils became mast cells by migrating from the circulatory system, but they now know that basophils arise from the bone marrow, whereas mast cells are derived from precursor cells in connective tissue. The mast-cell precursor cells are originally derived from bone-marrow cells that migrated during development to connective tissue, where they settle down and give rise to mast cells. Both basophils and mast cells synthesize and store *histamine* and *heparin,* powerful chemical substances that can be released on appropriate stimulation. Histamine release is important in allergic reactions, whereas heparin speeds up removal of fat particles from the blood after a fatty meal. Heparin can also prevent clotting (coagulation) of blood samples drawn for clinical analysis and is used extensively as an anticoagulant drug, but whether it plays a physiologic role in clot prevention is still debated.

Once released into the blood from the bone marrow, a granulocyte usually stays in transit in the blood for less than a day before leaving the blood vessels to enter the tissues, where it survives another three to four days unless it dies sooner in the line of duty.

By comparison, the functions and life spans of the agranulocytes are as follows:

■ **Monocytes,** like neutrophils, become professional phagocytes. They emerge from the bone marrow while still immature and circulate for only a day or two before settling down in various tissues throughout the body. At their new residences, monocytes continue to mature and greatly enlarge, becoming the large tissue phagocytes known as **macrophages** (*macro* means "large"; *phage* means "eater") (● Figure 11-9). A macrophage's life span may range from months to years unless it is destroyed sooner while performing its phagocytic activity. A phagocytic cell can ingest only a limited amount of foreign material before it succumbs.

■ **Lymphocytes** provide immune defense against targets for which they are specifically programmed. There are two functional types of lymphocytes, B lymphocytes and T lymphocytes (B and T cells), which look alike. **B lymphocytes** produce antibodies, which circulate in the blood and are responsible for *antibody-mediated,* or *humoral, immunity.* An antibody binds with and marks for destruction (by phagocytosis or other means) the specific kinds of foreign matter, such as bacteria, that induced production of the antibody. **T lymphocytes** do not produce circulating antibodies; instead, they directly destroy their specific target cells by releasing chemicals that punch holes in the victim cell, a process called *cell-mediated immunity*. The target cells of T cells include body cells invaded by viruses and cancer cells. Lymphocytes live for about 100 to 300 days. During this period, most continually recycle among the lymphoid tissues, lymph, and blood, spending only a few hours at a time in the blood. Therefore, only a small part of the total lymphocytes are in transit in the blood at any given moment.

ABNORMALITIES IN LEUKOCYTE PRODUCTION Even though levels of circulating leukocytes may vary, changes in these levels are normally controlled and adjusted according to the body's needs. However, abnormalities in leukocyte production can occur that are not subject to control; that is, either too few or too many WBCs can be produced. The bone marrow can greatly slow down or even stop production of white blood cells when it is exposed to certain toxic chemical agents (such as benzene and anticancer drugs) or to excessive radiation. The most serious consequence is the reduction in professional phagocytes (neutrophils and macrophages), which greatly reduces the body's defense capabilities against invading microorganisms. The only defense still available when the bone marrow fails is the immune capabilities of the lymphocytes produced by the lymphoid organs.

In **infectious mononucleosis,** not only does the number of lymphocytes (but not other leukocytes) in the blood increase but also many of the lymphocytes are atypical in structure. This condition, which is caused by the *Epstein-Barr virus,* is characterized by pronounced fatigue, a mild sore throat, and low-grade fever. Full recovery usually requires a month or more.

Surprisingly, one of the major consequences of **leukemia,** a cancerous condition that involves uncontrolled proliferation of WBCs, is inadequate defense capabilities against foreign invasion. In leukemia, the WBC count may reach as high as 500,000/mm³, compared with the normal 7,000/mm³; but because most of these cells are abnormal or immature, they cannot perform their normal defense functions. Another devastating consequence of leukemia is displacement of the other blood cell lines in the bone marrow. This results in anemia because of a reduction in erythropoiesis and in internal bleeding because of a deficit of platelets. Platelets play a critical role in preventing bleeding from the myriad tiny breaks that normally occur in small blood vessel walls. Consequently, overwhelming infections and hemorrhage are the most common causes of death in leukemic patients. The next section examines platelets' role in greater detail to show how they normally minimize the threat of hemorrhage.

Platelets and Hemostasis

In addition to erythrocytes and leukocytes, **platelets (thrombocytes)** are a third type of cellular element present in the blood. An average of 250 million platelets are normally present in each milliliter of blood (range of 150,000 to 350,000/mm³).

Megakaryocyte   Clusters of platelets about to shed off

Developing leukocyte

Cluster of developing erythrocytes

Carolina Biological/Visuals Unlimited

● **FIGURE 11-10** A megakaryocyte forming platelets.

Platelets are cell fragments shed from megakaryocytes.

Platelets are not whole cells but small cell fragments (about 2 to 4 μm in diameter) that are shed off the outer edges of extraordinarily large (up to 60 μm in diameter) bone marrow–bound cells known as **megakaryocytes** (● Figure 11-10). A single megakaryocyte typically produces about 1000 platelets. Megakaryocytes are derived from the same undifferentiated stem cells that give rise to the erythrocytic and leukocytic cell lines (see ● Figure 11-9). Platelets are essentially detached vesicles containing pieces of megakaryocyte cytoplasm wrapped in plasma membrane.

Platelets remain functional for an average of 10 days, at which time they are removed from circulation by the tissue macrophages, especially those in the spleen and liver, and are replaced by newly released platelets from the bone marrow. The hormone **thrombopoietin,** produced by the liver, increases the number of megakaryocytes in the bone marrow and stimulates each megakaryocyte to produce more platelets. The factors that control thrombopoietin secretion and regulate the platelet level are currently under investigation.

Platelets do not leave the blood as WBCs do, but at any given time about one third of them are in storage in blood-filled spaces in the spleen. These stored platelets can be released from the spleen into the circulating blood as needed (for example, during hemorrhage) by sympathetically induced splenic contraction.

Because platelets are cell fragments, they lack nuclei. However, they have organelles and cytosolic enzyme systems for generating energy and synthesizing secretory products, which they store in numerous granules dispersed throughout the cytosol. Furthermore, platelets contain high concentrations of actin and myosin, which enable them to contract. Their secretory and contractile abilities are important in hemostasis, a topic to which we now turn.

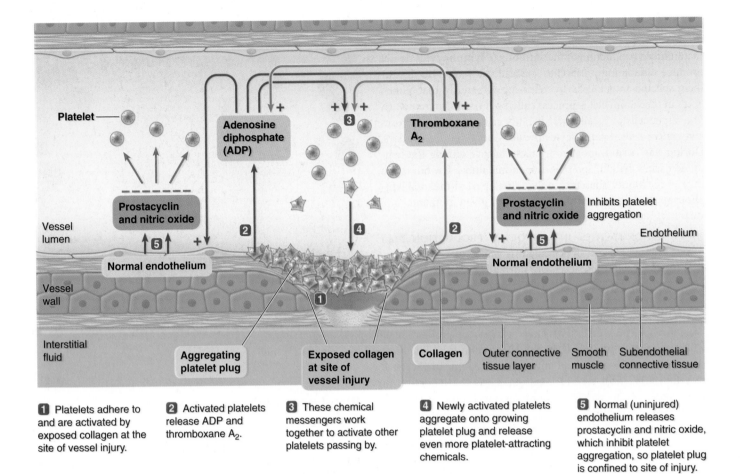

Platelet —

Adenosine diphosphate (ADP)

3

Thromboxane A₂

Prostacyclin and nitric oxide

Vessel lumen

5 **2** **4** **2**

Normal endothelium

1

Prostacyclin and nitric oxide

Inhibits platelet aggregation

Endothelium

5

Normal endothelium

Vessel wall

Interstitial fluid

Aggregating platelet plug

Exposed collagen at site of vessel injury

Collagen

Outer connective tissue layer

Smooth muscle

Subendothelial connective tissue

1 Platelets adhere to and are activated by exposed collagen at the site of vessel injury.

2 Activated platelets release ADP and thromboxane A₂.

3 These chemical messengers work together to activate other platelets passing by.

4 Newly activated platelets aggregate onto growing platelet plug and release even more platelet-attracting chemicals.

5 Normal (uninjured) endothelium releases prostacyclin and nitric oxide, which inhibit platelet aggregation, so platelet plug is confined to site of injury.

● **FIGURE 11-11 Formation of a platelet plug.** Platelets aggregate at a vessel defect through a positive-feedback mechanism involving the release of adenosine diphosphate (ADP) and thromboxane A₂ from platelets, which stick to exposed collagen at the site of the injury. Platelets are prevented from aggregating at the adjacent normal vessel lining by the release of prostacyclin and nitric oxide from the undamaged endothelial cells.

Hemostasis prevents blood loss from damaged small vessels.

Hemostasis is the arrest of bleeding from a broken blood vessel—that is, the stopping of hemorrhage (*hemo* means "blood"; *stasis* means "standing"). For bleeding to take place from a vessel, there must be a break in the vessel wall and the pressure inside the vessel must be greater than the pressure outside it to force blood out through the defect.

The small capillaries, arterioles, and venules are often ruptured by minor traumas of everyday life; such traumas are the most common source of bleeding, although we often are not even aware that any damage has taken place. The body's inherent hemostatic mechanisms normally are adequate to seal defects and stop blood loss through these small microcirculatory vessels.

Clinical Note The much rarer occurrence of bleeding from medium- to large-size vessels usually cannot be stopped by the body's hemostatic mechanisms alone. Bleeding from a severed artery is more profuse and therefore more dangerous than venous bleeding, because the outward driving pressure is greater in the arteries (that is, arterial blood pressure is much higher than venous pressure). First-aid measures for a severed

artery include applying external pressure to the wound that is greater than the arterial blood pressure to temporarily halt the bleeding until the torn vessel can be surgically closed. Hemorrhage from a traumatized vein can often be stopped simply by elevating the bleeding body part to reduce gravity's effects on pressure in the vein (see p. 374). If the accompanying drop in venous pressure is not enough to stop the bleeding, mild external compression is usually adequate.

Hemostasis involves three major steps: (1) *vascular spasm*, (2) *formation of a platelet plug*, and (3) *blood coagulation (clotting)*. Platelets play a pivotal role in hemostasis. They obviously play a major part in forming a platelet plug, but they also contribute significantly to the other two steps.

Vascular spasm reduces blood flow through an injured vessel.

A cut or torn blood vessel immediately constricts. The underlying mechanism is unclear but is thought to be an intrinsic response triggered by a paracrine released locally from the inner lining (endothelium) of the injured vessel (see p. 114). This constriction, or **vascular spasm,** slows blood flow through the defect and thus minimizes blood loss. Also, as the opposing endo-thelial surfaces of the vessel are pressed together by this initial vascular spasm, they become sticky and adhere to each other, further sealing off the damaged vessel. These physical measures alone cannot completely prevent further blood loss, but they minimize blood flow through the break in the vessel until the other hemostatic measures can actually plug up the hole.

Platelets aggregate to form a plug at a vessel tear or cut.

Platelets normally do not stick to the smooth endothelial surface of blood vessels, but when this lining is disrupted because of vessel injury, platelets adhere to and are activated by the exposed *collagen,* which is a fibrous protein in the underlying connective tissue (see pp. 8 and 348). (Connective tissue lies between the endothelium and circular smooth muscle layer in all types of vessels except for capillaries; this connective tissue contains variable amounts of collagen fibers and elastin fibers, depending on the vessel type.) When platelets passing by in the blood are exposed to collagen, they become tethered to the collagen by means of specific membrane proteins, such as integrin, a cell adhesion molecule (CAM; see p. 57) that binds with the collagen. This adhesion prevents these platelets from being swept forward in the circulation. This layer of stuck platelets forms the foundation of a hemostatic **platelet plug** at the site of the defect. Collagen activates the bound platelets. Activated platelets quickly reorganize their actin cytoskeletal elements to develop spiky processes, which help them adhere to the collagen and to other platelets. Activated platelets also release several important chemicals from their storage granules. Among these chemicals is *adenosine diphosphate (ADP),* which activates other nearby circulating platelets and causes their surfaces to become sticky so that they adhere to the first layer of aggregated platelets. These newly aggregated platelets release more ADP, which causes more platelets to pile on, and so on; thus, a plug of platelets is rapidly built up at the defect site, in a positive-feedback fashion (● Figure 11-11). This aggregating process is reinforced by the ADP-stimulated formation of a paracrine similar to prostaglandins, *thromboxane A₂,* from a component of the platelet plasma membrane (see p. 758). Thromboxane A₂ directly promotes platelet aggregation and further enhances it indirectly by triggering the release of even more ADP from the platelet granules. Thus formation of a platelet plug involves the three successive, closely integrated events of adhesion, activation, and aggregation.

Given the self-perpetuating nature of platelet aggregation, why does the platelet plug not continue to develop and expand over the surface of the adjacent normal vessel lining? A key reason is that ADP and other chemicals released by the activated platelets stimulate the release of *prostacyclin* and *nitric oxide* from the adjacent normal endothelium. Both these chemicals profoundly inhibit platelet aggregation. Thus, the platelet plug is limited to the defect and does not spread to the nearby undamaged vascular tissue (● Figure 11-11).

The aggregated platelet plug not only physically seals the break in the vessel but also performs three other important roles. (1) The actin-myosin complex within the aggregated platelets contracts to compact and strengthen what was originally a fairly

● **FIGURE 11-12 Erythrocytes trapped in the fibrin mesh-work of a clot.**

loose plug. (2) The chemicals released from the platelet plug include several powerful vasoconstrictors (serotonin, epinephrine, and thromboxane A₂), which induce profound constriction of the affected vessel to reinforce the initial vascular spasm. (3) The platelet plug releases other chemicals that enhance blood coagulation, the next step of hemostasis. Although the platelet-plugging mechanism alone is often enough to seal the myriad minute tears in capillaries and other small vessels that occur many times daily, larger holes in vessels require the formation of a blood clot to completely stop the bleeding.

Clot formation results from a triggered chain reaction involving plasma clotting factors.

Blood coagulation, or **clotting,** is the transformation of blood from a liquid into a solid gel. Formation of a clot on top of the platelet plug strengthens and supports the plug, reinforcing the seal over a break in a vessel. Furthermore, as blood in the vicinity of the vessel defect solidifies, it can no longer flow. Clotting is the body's most powerful hemostatic mechanism. It is required to stop bleeding from all but the most minute defects.

CLOT FORMATION The ultimate step in clot formation is the conversion of **fibrinogen,** a large, soluble plasma protein produced by the liver and normally always present in the plasma, into **fibrin,** an insoluble, threadlike molecule. This conversion into fibrin is catalyzed by the enzyme **thrombin** at the site of the injury. Fibrin molecules adhere to the damaged vessel surface, forming a loose, netlike meshwork that traps blood cells, including aggregating platelets. The resulting mass, or **clot,** typically appears red because of the abundance of trapped RBCs, but the foundation of the clot is formed of fibrin derived from the plasma (● Figure 11-12). Except for platelets, which play an important role in ultimately

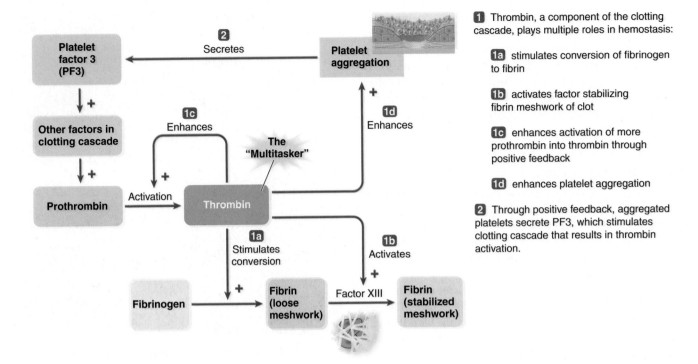

1 Thrombin, a component of the clotting cascade, plays multiple roles in hemostasis:

1a stimulates conversion of fibrinogen to fibrin

1b activates factor stabilizing fibrin meshwork of clot

1c enhances activation of more prothrombin into thrombin through positive feedback

1d enhances platelet aggregation

2 Through positive feedback, aggregated platelets secrete PF3, which stimulates clotting cascade that results in thrombin activation.

● **FIGURE 11-13 Roles of thrombin in hemostasis.**

bringing about the conversion of fibrinogen to fibrin, clotting can take place in the absence of all other blood cells.

The original fibrin web is rather weak, because the fibrin strands are only loosely interlaced. However, chemical linkages rapidly form between adjacent strands to strengthen and stabilize the clot meshwork. This cross-linkage process is catalyzed by a clotting factor known as **factor XIII (fibrin-stabilizing factor),** which normally is present in the plasma in inactive form.

Fibrin is the stretchiest natural protein that scientists have ever studied. On average fibrin fibers can be passively stretched to 2.8 times their original length and still snap back to their starting length and can be stretched to 4.3 times their original length before they break. This highly elastic property accounts for the extraordinary stretchiness of blood clots.

ROLES OF THROMBIN Thrombin, in addition to converting fibrinogen into fibrin (step 1a in Figure 11-13), also activates factor XIII to stabilize the resultant fibrin mesh (step 1b), acts in a positive-feedback fashion to facilitate its own formation (step 1c), and enhances platelet aggregation (step 1d), which in turn is essential to the clotting process (step 2).

Because thrombin converts the ever-present fibrinogen molecules in the plasma into a blood-stanching clot, thrombin must normally be absent from the plasma except in the vicinity of vessel damage. Otherwise, blood would always be coagulated—a situation incompatible with life. How can thrombin normally be absent from the plasma, yet be readily available to trigger fibrin formation when a vessel is injured? The solution lies in thrombin's existence in the plasma in the form of an inactive precursor called **prothrombin.** What converts prothrombin into thrombin

when blood clotting is desirable? This conversion involves the clotting cascade.

THE CLOTTING CASCADE Yet another activated plasma clotting factor, **factor X,** converts prothrombin to thrombin; factor X itself is normally present in the blood in inactive form and must be converted into its active form by still another activated factor, and so on. Altogether, 12 plasma clotting factors participate in essential steps that lead to the final conversion of fibrinogen into a stabilized fibrin mesh (● Figure 11-14). These factors are designated by roman numerals in the order in which the factors were discovered, not the order in which they participate in the clotting process.[1] Most of these clotting factors are plasma proteins synthesized by the liver. One consequence of liver disease is that clotting time is prolonged by reduced production of clotting factors. Normally, they are always present in the plasma in an inactive form, such as fibrinogen and prothrombin. In contrast to fibrinogen, which is converted into insoluble fibrin strands, prothrombin and the other precursors, when converted to their active form, act as proteolytic (protein-splitting) enzymes. These enzymes activate another specific factor in the clotting sequence. Once the first factor in the sequence is activated, it in turn activates the next factor, and so on, in a series of sequential reactions known as the **clotting cascade,** until thrombin catalyzes the final conversion of fibrinogen into fibrin. Several of these steps require the presence of plasma Ca^{2+}

---

[1]The term *factor VI* is no longer used. What once was considered a separate factor VI has now been determined to be an activated form of factor V.

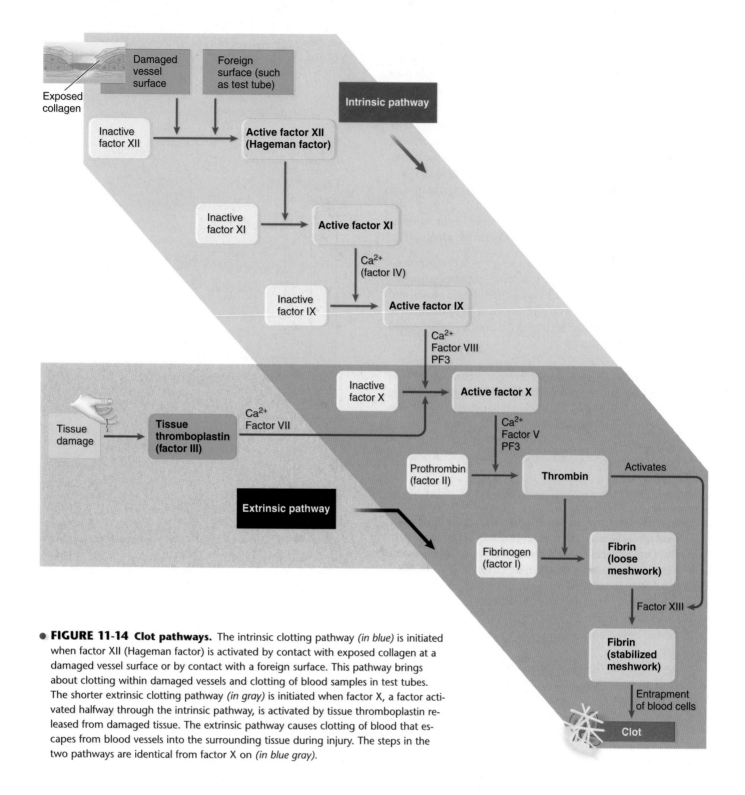

● **FIGURE 11-14 Clot pathways.** The intrinsic clotting pathway *(in blue)* is initiated when factor XII (Hageman factor) is activated by contact with exposed collagen at a damaged vessel surface or by contact with a foreign surface. This pathway brings about clotting within damaged vessels and clotting of blood samples in test tubes. The shorter extrinsic clotting pathway *(in gray)* is initiated when factor X, a factor activated halfway through the intrinsic pathway, is activated by tissue thromboplastin released from damaged tissue. The extrinsic pathway causes clotting of blood that escapes from blood vessels into the surrounding tissue during injury. The steps in the two pathways are identical from factor X on *(in blue gray)*.

and *platelet factor 3 (PF3),* a phospholipid secreted by the aggregated platelet plug. Thus, platelets also contribute to clot formation.

INTRINSIC AND EXTRINSIC PATHWAYS The clotting cascade may be triggered by the *intrinsic pathway* or the *extrinsic pathway:*

▪ The **intrinsic pathway** precipitates clotting within damaged vessels as well as clotting of blood samples in test tubes. All ele-

ments necessary to bring about clotting by means of the intrinsic pathway are present in the blood. This pathway, which involves seven separate steps (shown in blue in ● Figure 11-14), is set off when **factor XII (Hageman factor)** is activated by coming into contact with either exposed collagen in an injured vessel or a foreign surface such as a glass test tube. Remember that exposed collagen also initiates platelet aggregation. Thus, formation of a platelet plug and the chain reaction leading to clot formation are simultaneously set in motion when a vessel is damaged. Furthermore, these complementary hemostatic mechanisms reinforce each other. The aggregated platelets secrete PF3, which is essential for the clotting cascade that in turn enhances further platelet aggregation (● Figure 11-15; also see ● Figure 11-13).

■ The **extrinsic pathway** takes a shortcut and requires only four steps (shown in gray in ● Figure 11-14). This pathway, which requires contact with tissue factors external to the blood, initiates clotting of blood that has escaped into the tissues. When a tissue is traumatized, it releases a protein complex known as **tissue thromboplastin.** Tissue thromboplastin directly activates factor X, thereby bypassing all preceding steps of the intrinsic pathway. From this point on, the two pathways are identical.

The intrinsic and extrinsic mechanisms usually operate simultaneously. When tissue injury involves rupture of vessels, the intrinsic mechanism stops blood in the injured vessel, and the extrinsic mechanism clots blood that escaped into the tissue before the vessel was sealed off. Typically, clots are fully formed in three to six minutes.

CLOT RETRACTION Once a clot is formed, contraction of the platelets trapped within the clot shrinks the fibrin mesh, pulling the edges of the damaged vessel closer together. During **clot retraction,** fluid is squeezed from the clot. This fluid, which is essentially plasma minus fibrinogen and other clotting precursors that have been removed during the clotting process, is called **serum.**

AMPLIFICATION DURING CLOTTING PROCESS Although a clotting process that involves so many steps may seem inefficient, the advantage is the amplification accomplished during many of the steps. One molecule of an activated factor can activate perhaps a hundred molecules of the next factor in the sequence, each of which can activate many more molecules of the next factor, and so on. In this way, large numbers of the final factors involved in clotting are rapidly activated as a result of the initial activation of only a few molecules in the beginning step of the sequence.

How then is the clotting process, once initiated, confined to the site of vessel injury? If the activated clotting factors were allowed to circulate, they would induce inappropriate widespread clotting that would plug up vessels throughout the body. Fortunately, after participating in the local clotting process, the massive number of factors activated in the vicinity of vessel injury are rapidly inactivated by enzymes and other factors present in the plasma or tissue.

Fibrinolytic plasmin dissolves clots.

A clot is not meant to be a permanent solution to vessel injury. It is a transient device to stop bleeding until the vessel can be repaired.

VESSEL REPAIR The aggregated platelets secrete a chemical that helps promote the invasion of fibroblasts ("fiber formers") from the surrounding connective tissue into the wounded area of the vessel. Fibroblasts form a scar at the vessel defect.

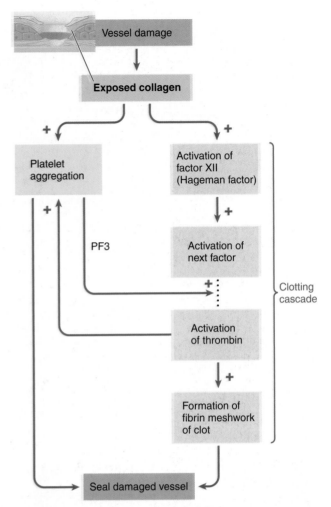

● **FIGURE 11-15 Concurrent platelet aggregation and clot formation.** Exposed collagen at the site of vessel damage simultaneously initiates platelet aggregation and the clotting cascade. These two hemostatic mechanisms positively reinforce each other as they seal the damaged vessel.

CLOT DISSOLUTION Simultaneous with the healing process, the clot, which is no longer needed to prevent hemorrhage, is slowly dissolved by a fibrinolytic (fibrin-splitting) enzyme called **plasmin.** If clots were not removed after they performed their hemostatic function, the vessels, especially the small ones that endure tiny ruptures on a regular basis, would eventually become obstructed by clots.

Plasmin, like the clotting factors, is a plasma protein produced by the liver and present in the blood in an inactive precursor form, **plasminogen.** Plasmin is activated in a fast cascade of reactions involving many factors, among them factor XII (Hageman factor), which also triggers the chain reaction leading to clot formation (● Figure 11-16). When a clot is rapidly being formed, activated plasmin becomes trapped in the clot and later dissolves it by slowly breaking down the fibrin meshwork.

Phagocytic white blood cells gradually remove the products of clot dissolution. You have observed the slow removal of blood that has clotted after escaping into the tissue layers of

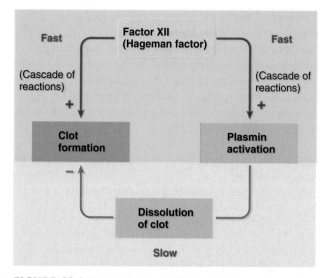

● **FIGURE 11-16 Role of factor XII in clot formation and dissolution.** Activation of factor XII (Hageman factor) simultaneously initiates a fast cascade of reactions that result in clot formation and a fast cascade of reactions that result in plasmin activation. Plasmin, which is trapped in the clot, subsequently slowly dissolves the clot. This action removes the clot when it is no longer needed after the vessel has been repaired.

your skin following an injury. The black-and-blue marks of such bruised skin result from deoxygenated clotted blood within the skin; this blood is eventually cleared by plasmin action, followed by the phagocytic cleanup crew.

PREVENTING INAPPROPRIATE CLOT FORMATION In addition to removing clots that are no longer needed, plasmin functions continually to prevent clots from forming inappropriately. Throughout the vasculature, small amounts of fibrinogen are constantly being converted into fibrin, triggered by unknown mechanisms. Clots do not develop, however, because the fibrin is quickly disposed of by plasmin activated by **tissue plasminogen activator (tPA)** from the tissues, especially the lungs. Normally, the low level of fibrin formation is counterbalanced by a low level of fibrinolytic activity, so inappropriate clotting does not occur. Only when a vessel is damaged do additional factors precipitate the explosive chain reaction that leads to more extensive fibrin formation and results in local clotting at the site of injury.

Genetically engineered tPA and other similar chemicals that trigger clot dissolution are frequently used to limit damage to cardiac muscle during heart attacks. Administering a clot-busting drug within the first hours after a clot has blocked a coronary (heart) vessel often dissolves the clot in time to restore blood flow to the cardiac muscle supplied by the blocked vessel before the muscle dies of O_2 deprivation. In recent years, tPA and related drugs have also been used successfully to promptly dissolve a stroke-causing clot within a cerebral (brain) blood vessel, minimizing loss of irreplaceable brain tissue after a stroke (see p. 142).

Inappropriate clotting produces thromboembolism.

Despite protective measures, clots occasionally form in intact vessels. Abnormal or excessive clot formation within blood vessels—what has been dubbed "hemostasis in the wrong place"—can compromise blood flow to vital organs. The body's clotting and anticlotting systems normally function in a check-and-balance manner. Acting in concert, they permit prompt formation of "good" blood clots, thus minimizing blood loss from damaged vessels, while preventing "bad" clots from forming and blocking blood flow in intact vessels. An abnormal intravascular clot attached to a vessel wall is known as a **thrombus,** and freely floating clots are called **emboli.** An enlarging thrombus narrows and can eventually completely occlude the vessel in which it forms. By entering and completely plugging a smaller vessel, a circulating embolus can suddenly block blood flow (see ● Figure 9-29, p. 335).

Several factors, acting independently or simultaneously, can cause *thromboembolism:* (1) Roughened vessel surfaces associated with atherosclerosis can lead to thrombus formation (see p. 335). (2) Imbalances in the clotting–anticlotting systems can trigger clot formation. (3) Slow-moving blood is more apt to clot, probably because small quantities of fibrin accumulate in the stagnant blood, for example, in blood pooled in varicosed leg veins (see p. 375). (4) Widespread clotting is occasionally triggered by the release of tissue thromboplastin into the blood from large amounts of traumatized tissue. Similar widespread clotting can occur in **septicemic shock,** in which bacteria or their toxins initiate the clotting cascade.

Anticoagulant drugs are sometimes given to prevent thromboembolism in people with conditions that make them more prone to the development of clots. For example, low-molecular weight *heparin*, which must be injected, accelerates the action of a normal blood-borne inhibitor of thrombin, halting this all-purpose clot promoter. *Warfarin* (Coumadin), which can be taken orally, is a vitamin K antagonist. Vitamin K, commonly known as the blood-clotting vitamin, is essential for normal clot formation. In a complex sequence of biochemical events, vitamin K combines with O_2, releasing free energy that is ultimately used in the activation processes of the clotting cascade. By interfering with vitamin K's action, Coumadin leads to the formation of structurally incomplete clotting factors II, VII, IX, and X.

Hemophilia is the primary condition that produces excessive bleeding.

In contrast to inappropriate clot formation in intact vessels, the opposite hemostatic disorder is failure of clots to form promptly in injured vessels, resulting in life-threatening hemorrhage from even relatively mild traumas. The most common cause of excessive bleeding is **hemophilia,** which is caused by a deficiency of one of the factors in the clotting cascade. Although a deficiency of any of the clotting factors could block the clotting process, 80% of all hemophiliacs lack the genetic ability to synthesize factor VIII.

People with a platelet deficiency, in contrast to the more profuse bleeding that accompanies defects in the clotting mechanism, continuously develop hundreds of small, confined hemorrhagic areas throughout the body tissues as blood leaks from tiny breaks in the small blood vessels before coagulation takes place. Platelets normally are the primary sealers of these ever-occurring minute ruptures. In the skin of a platelet-deficient person, the diffuse capillary hemorrhages are visible as small, purplish blotches, giving rise to the term **thrombocytopenia purpura** ("the purple of thrombocyte deficiency") for this condition. (Recall that *thrombocyte* is another name for platelet.)

Vitamin K deficiency can also cause a bleeding tendency because of incomplete activation of the vitamin-K dependent clotting factors.

Chapter in Perspective: Focus on Homeostasis

Blood contributes to homeostasis in a variety of ways. First, the composition of interstitial fluid, the true internal environment that surrounds and directly exchanges materials with the cells, depends on the composition of blood plasma. Because of the thorough exchange that occurs between the interstitial and vascular compartments, interstitial fluid has the same composition as plasma with the exception of plasma proteins, which cannot escape through the capillary walls. Thus, blood serves as the vehicle for rapid, long-distance, mass transport of materials to and from the cells, and interstitial fluid serves as the go-between.

Homeostasis depends on the blood carrying essential supplies such as O_2 and nutrients to the cells as rapidly as the cells consume them and carrying materials such as metabolic wastes away from the cells as rapidly as the cells produce these products. It also depends on the blood carrying hormonal messengers from their site of production to their distant site of action. Once a substance enters the blood, it can be transported throughout the body within seconds, whereas diffusion of the substance over long distances in a large multicellular organism such as a human would take months to years—a situation incompatible with life. Diffusion can, however, effectively accomplish short local exchanges of materials between blood and surrounding cells through the intervening interstitial fluid.

Blood has special transport capabilities that enable it to move its cargo efficiently throughout the body. For example, life-sustaining O_2 is poorly soluble in water, but blood is equipped with O_2-carrying specialists, the erythrocytes (red blood cells), which are stuffed full of hemoglobin, a complex molecule that transports O_2. Likewise, homeostatically important water-insoluble hormonal messengers are shuttled in the blood by plasma protein carriers.

Specific components of the blood perform the following additional homeostatic activities that are unrelated to blood's transport function:

- Blood helps maintain the proper pH in the internal environment by buffering changes in the body's acid–base load.
- Blood helps maintain body temperature by absorbing heat produced by heat-generating tissues such as contracting skeletal muscles and distributing it throughout the body. Excess heat is carried by the blood to the body surface for elimination to the external environment.
- Electrolytes in the plasma are important in membrane excitability, which is critical for nerve and muscle function.
- Electrolytes in the plasma are important in osmotic distribution of fluid between the extracellular and intracellular fluid. Plasma proteins play a critical role in distributing extracellular fluid between the plasma and interstitial fluid.
- Through their hemostatic functions, the platelets and clotting factors minimize the loss of life-sustaining blood after vessel injury.
- The leukocytes (white blood cells), their secretory products, and certain types of plasma proteins, such as antibodies, constitute the immune defense system. This system defends the body against invading disease-causing agents, destroys cancer cells, and paves the way for wound healing and tissue repair by clearing away debris from dead or injured cells. These actions indirectly contribute to homeostasis by helping the organs that directly maintain homeostasis stay healthy. We could not survive beyond early infancy were it not for the body's defense mechanisms.

Objective Questions (Answers on p. A-47)

1. Blood can absorb metabolic heat while undergoing only small changes in temperature. *(True or false?)*

2. Hemoglobin can carry only O_2. *(True or false?)*

3. Erythrocytes, leukocytes, and platelets all originate from the same undifferentiated stem cells. *(True or false?)*

4. Erythrocytes are unable to use the O_2 they contain for their own ATP formation. *(True or false?)*

5. White blood cells spend the majority of their time in the blood. *(True or false?)*

6. The type of leukocyte produced primarily in lymphoid tissue is _____.

7. Most clotting factors are synthesized by the _____.

8. Which of the following is *not* a function of plasma proteins?
   a. facilitating retention of fluid in the blood vessels
   b. playing an important role in blood clotting
   c. transporting water-insoluble substances in the blood
   d. transporting O_2 in the blood
   e. serving as antibodies
   f. contributing to the buffering capacity of the blood

9. Which of the following is *not* directly triggered by exposed collagen in an injured vessel?
   a. initial vascular spasm
   b. platelet aggregation
   c. activation of the clotting cascade
   d. activation of plasminogen

10. Match the following *(an answer may be used more than once)*:
    ___ 1. causes platelets to aggregate in positive-feedback fashion
    ___ 2. activates prothrombin
    ___ 3. fibrinolytic enzyme
    ___ 4. inhibits platelet aggregation
    ___ 5. first factor activated in intrinsic clotting pathway
    ___ 6. forms meshwork of the clot
    ___ 7. stabilizes the clot
    ___ 8. activates fibrinogen
    ___ 9. activated by tissue thromboplastin

    (a) prostacyclin
    (b) plasmin
    (c) ADP
    (d) fibrin
    (e) thrombin
    (f) factor X
    (g) factor XII
    (h) factor XIII

11. Match the following blood abnormalities with their causes:
    ___ 1. deficiency of intrinsic factor
    ___ 2. insufficient amount of iron to synthesize adequate hemoglobin
    ___ 3. destruction of bone marrow
    ___ 4. abnormal loss of blood
    ___ 5. tumorlike condition of bone marrow
    ___ 6. inadequate erythropoietin secretion
    ___ 7. excessive rupture of circulating erythrocytes
    ___ 8. associated with living at high altitudes

    (a) hemolytic anemia
    (b) aplastic anemia
    (c) nutritional anemia
    (d) hemorrhagic anemia
    (e) pernicious anemia
    (f) renal anemia
    (g) primary polycythemia
    (h) secondary polycythemia

Essay Questions

1. What is the average blood volume in women and in men?

2. What is the normal percentage of blood occupied by erythrocytes and by plasma? What is the hematocrit? What is the buffy coat?

3. What is the composition of plasma?

4. List the three major groups of plasma proteins, and state their functions.

5. Describe the structure and functions of erythrocytes.

6. Why can erythrocytes survive for only about 120 days?

7. Describe the process and control of erythropoiesis.

8. List the antigens and antibodies present in the blood of the four major ABO blood types. Describe what happens in a transfusion reaction. Does the recipient have to have had prior exposure to incompatible blood of the ABO type for a transfusion reaction to occur? Does the recipient have to have had prior exposure to incompatible blood of the Rh type for a transfusion reaction to occur? Explain.

9. Compare the structure, functions, and life spans of the five types of leukocytes.

10. Discuss the derivation of platelets.

11. Describe the three steps of hemostasis, including a comparison of the intrinsic and extrinsic pathways by which the clotting cascade is triggered.

12. Compare plasma and serum.

13. What normally prevents inappropriate clotting in vessels?

Quantitative Exercises (Solutions on p. A-47)

1. The normal concentration of hemoglobin in blood (as measured clinically) is 15 g/100 ml of blood.
   a. Given that 1 mole of hemoglobin weighs 66,000 grams, what is the concentration of hemoglobin in millimoles (mM)?
   b. Each hemoglobin molecule can bind four molecules of O_2. What is the concentration of O_2 bound to hemoglobin at maximal saturation (in mM)?
   c. Given that 1 mole of an ideal gas occupies 22.4 liters, what is the maximal carrying capacity of normal blood for O_2 (usually expressed in ml of O_2/liter of blood)?

2. Assume that the blood sample in ● Figure 11-5b, p. 397, is from a patient with hemorrhagic anemia. Given a normal blood volume of 5 liters, a normal red blood cell concentration of 5 billion/ml, and an RBC production rate of 3 million cells/second, how long will it take the body to return the hematocrit to normal?

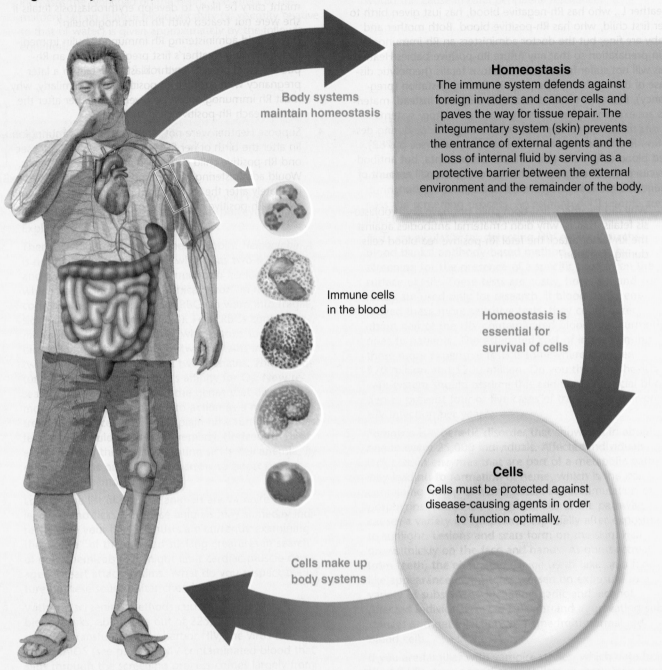

Immune System
Integumentary System (Skin)

Body systems maintain homeostasis

Homeostasis
The immune system defends against foreign invaders and cancer cells and paves the way for tissue repair. The integumentary system (skin) prevents the entrance of external agents and the loss of internal fluid by serving as a protective barrier between the external environment and the remainder of the body.

Immune cells in the blood

Homeostasis is essential for survival of cells

Cells
Cells must be protected against disease-causing agents in order to function optimally.

Cells make up body systems

Humans constantly come into contact with external agents that could be harmful if they entered the body. The most serious are disease-causing microorganisms. If bacteria or viruses do gain entry, the body is equipped with a complex, multifaceted internal defense system—the **immune system**—that provides continual protection against invasion by foreign agents. Furthermore, body surfaces exposed to the external

environment, such as the **integumentary system (skin)**, serve as a first line of defense to resist penetration by foreign microorganisms. The immune system also protects against cancer and paves the way for repair of damaged tissues.

The immune system indirectly contributes to homeostasis by helping maintain the health of organs that directly contribute to homeostasis.

12

Body Defenses

CENGAGENOW™ Log on to CengageNOW at **http://www
.cengage.com/sso/** for an opportunity to explore animations
and interactive quizzes to help you learn, review, and master
physiology concepts.

Immune System: Targets, Effectors, Components

Immunity is the body's ability to resist or eliminate potentially harmful foreign materials or abnormal cells. The first line of defense against foreign invaders is the epithelial barriers that surround the outer surface of the body (the skin) and line the body cavities (such as the digestive tract and lungs) that are in contact with the external environment. These epithelial barriers are not part of the immune system. We will discuss their roles in the body's overall defense mechanisms after examining the immune system in detail. As an overview, the following activities are attributable to the **immune system,** an internal defense system that plays a key role in recognizing and either destroying or neutralizing materials that are not "normal self," either foreign materials that have entered the body or abnormal cells that have arisen within the body:

1. Defending against invading **pathogens** (disease-producing microorganisms such as bacteria and viruses)

2. Removing "worn-out" cells and tissue damaged by trauma or disease, paving the way for wound healing and tissue repair

3. Identifying and destroying abnormal or mutant cells that have originated in the body. This function, termed *immune surveillance,* is the primary internal-defense mechanism against cancer

4. Mounting inappropriate immune responses that lead either to *allergies,* which occur when the body turns against a normally harmless environmental chemical entity, or to *autoimmune diseases,* which happen when the defense system erroneously produces antibodies against a particular type of the body's own cells

Pathogenic bacteria and viruses are the major targets of the immune system.

The primary foreign enemies against which the immune system defends are bacteria and viruses. Comparing their relative sizes, if an average bacterium were the size of a pitcher's mound, a virus would be the size of a baseball. **Bacteria** are nonnucleated, single-celled microorganisms self-equipped with all the machinery essential for their own survival and reproduction. Pathogenic bacteria that invade the body cause tissue damage and produce disease largely by releasing enzymes or toxins that physically injure or functionally disrupt affected cells and organs. The disease-producing power of a pathogen is known as its **virulence.**

In contrast to bacteria, **viruses** are not self-sustaining cellular entities. They consist only of nucleic acids (genetic material—DNA or RNA) enclosed by a protein coat. Because they lack cellular machinery for energy production and protein synthesis, viruses cannot carry out metabolism and reproduce unless they invade a **host cell** (a body cell of the infected individual) and take over the cell's biochemical facilities for their own uses. Not only do viruses sap the host cell's energy resources but the viral nucleic acids also direct the host cell to synthesize proteins needed for viral replication.

When a virus becomes incorporated into a host cell, the body's own defense mechanisms may destroy the cell because they no longer recognize it as a "normal self" cell. Other ways in which viruses can lead to cell damage or death are by depleting essential cell components, dictating that the cell produce substances toxic to the cell, or transforming the cell into a cancer cell.

Leukocytes are the effector cells of the immune system.

Leukocytes (white blood cells, or WBCs) and their derivatives, along with a variety of plasma proteins, are responsible for the different immune defense strategies.

LEUKOCYTE FUNCTIONS As a brief review, the functions of the five types of leukocytes are as follows (see pp. 401–405 and ● Figures 11-8 and 11-9):

1. **Neutrophils** are highly mobile phagocytic specialists that engulf and destroy unwanted materials.

2. **Eosinophils** secrete chemicals that destroy parasitic worms and are involved in allergic reactions.

3. **Basophils** release histamine and heparin and also are involved in allergic reactions.

4. **Monocytes** are transformed into **macrophages,** which are large, tissue-bound phagocytic specialists.

5. **Lymphocytes** are of two types.
   a. **B lymphocytes (B cells)** are transformed into plasma cells, which secrete antibodies that indirectly lead to the destruction of foreign material (antibody-mediated immunity).
   b. **T lymphocytes (T cells)** directly destroy virus-invaded cells and mutant cells by releasing chemicals that punch lethal holes in the victim cells (cell-mediated immunity).

A given leukocyte is present in the blood only transiently. Most leukocytes are out in the tissues, on defense missions. As a result, the immune system's effector cells are widely dispersed throughout the body and can defend in any location.

LYMPHOID TISSUES Almost all leukocytes originate from common precursor stem cells in the bone marrow and are subsequently released into the blood. The only exception is lympho-

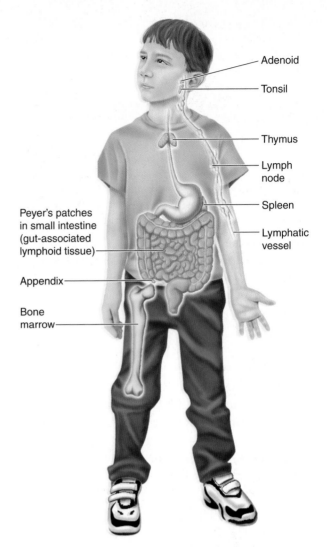

● **FIGURE 12-1 Lymphoid tissues.** The lymphoid tissues, which are dispersed throughout the body, produce, store, or process lymphocytes.

cytes, which arise in part from lymphocyte colonies in various lymphoid tissues originally populated by cells derived from bone marrow (see p. 401).

Lymphoid tissues, collectively, are the tissues that produce, store, or process lymphocytes. These include the bone marrow, lymph nodes, spleen, thymus, tonsils, adenoids, appendix, and aggregates of lymphoid tissue in the lining of the digestive tract called **Peyer's patches** or **gut-associated lymphoid tissue (GALT)** (● Figure 12-1). Lymphoid tissues are strategically located to intercept invading microorganisms before they have a chance to spread very far. For example, lymphocytes populating the *tonsils* and *adenoids* are situated advantageously to respond to inhaled microbes, whereas microorganisms invading through the digestive system immediately encounter lymphocytes in the *appendix* and *GALT*. Potential pathogens that gain access to lymph are filtered through *lymph nodes*, where they are exposed to lymphocytes as well as to macrophages that line lymphatic passageways. The *spleen*, the largest lymphoid tissue, performs immune functions on blood similar to those that lymph nodes perform on lymph. Through actions of its lymphocyte and macrophage population, the spleen clears blood that passes through it of microorganisms and other foreign matter and also removes worn-out red blood cells (see p. 395). The *thymus* and *bone marrow* play important roles in processing T and B lymphocytes, respectively, to prepare them to carry out their specific immune strategies. ▲ Table 12-1 summarizes the major functions of the various lymphoid tissues, some described in Chapter 11 and others of which are discussed later in this chapter.

We now turn attention to the two major components of the immune system's response to foreign invaders and other targets—innate and adaptive immune responses. In the process, we will further examine the roles of each type of leukocyte.

Immune responses may be either innate and nonspecific, or adaptive and specific.

Protective immunity is conferred by the complementary actions of two separate but interdependent components of the immune system: the *innate immune system* and the *adaptive, or acquired, immune system*. The responses of these two systems differ in timing and in the selectivity of the defense mechanisms.

The **innate immune system** encompasses the body's *nonspecific* immune responses that come into play immediately on exposure to a threatening agent. These nonspecific responses are inherent (innate or built-in) defense mechanisms that nonselectively defend against foreign or abnormal material of any type, even on initial exposure to it. Such responses provide a first line of internal defense against a wide range of threats, including infectious agents, chemical irritants, and tissue injury from mechanical trauma and burns. Everyone is born with essentially the same innate immune-response mechanisms, although there are some subtle genetic differences. The **adaptive, or acquired, immune system,** in contrast, relies on *specific* immune responses selectively targeted against a particular foreign material to which the body has already been exposed and has had an opportunity to prepare for an attack aimed discriminatingly at the enemy. The adaptive immune system thus takes considerably more time to mount and takes on specific foes. The innate and adaptive immune systems work in harmony to contain, then eliminate, harmful agents.

INNATE IMMUNE SYSTEM The components of the innate system are always on guard, ready to unleash a limited, rather crude, repertoire of defense mechanisms at any and every invader. Of the immune effector cells, the neutrophils and macrophages—both phagocytic specialists—are especially important in innate defense. Several groups of plasma proteins also play key roles, as you will see shortly. The various nonspecific immune responses are set in motion in response to generic molecular patterns associated with threatening agents, such as the carbohydrates typically found in bacterial cell walls but not found in human cells. The responding phagocytic cells are studded with plasma membrane proteins known as **toll-like receptors (TLRs),** which recognize these pathogen-associated molecular patterns. There are about a dozen different TLRs, each of which recognizes a different specific set of molecular patterns on pathogens. For example, some recognize gram-positive bacteria

Lymphoid Tissue	Functions
Bone Marrow	Origin of all blood cells
	Site of maturational processing for B lymphocytes
Lymph Nodes, Tonsils, Adenoids, Appendix, Gut-Associated Lymphoid Tissue	Exchange lymphocytes with the lymph (remove, store, produce, and add them)
	Resident lymphocytes produce antibodies and activated T cells, which are released into the lymph
	Resident macrophages remove microbes and other particulate debris from the lymph
Spleen	Exchanges lymphocytes with the blood (removes, stores, produces, and adds them)
	Resident lymphocytes produce antibodies and activated T cells, which are released into the blood
	Resident macrophages remove microbes and other particulate debris, most notably worn-out red blood cells, from the blood
	Stores a small percentage of red blood cells, which can be added to the blood by splenic contraction as needed
Thymus	Site of maturational processing for T lymphocytes
	Secretes the hormone thymosin

▲ TABLE 12-1 **Functions of Lymphoid Tissues**

(bacteria that can be stained with the dark blue Gram stain), others recognize gram-negative bacteria (bacteria whose cell wall does not pick up the Gram stain), still others recognize viral DNA or RNA, and so on. TLRs have been dubbed the "eyes of the innate immune system" because these immune sensors recognize and bind with the unique, telltale pathogen markers, allowing the effector cells of the innate system to "see" pathogens as distinct from "self" cells. TLRs function at the cell surface to recognize pathogens in the ECF, but most viruses are hidden inside host cells instead of being free in the ECF. Scientists recently discovered another type of innate pattern-recognition receptor, **retinoic acid inducible gene I (RIG-I),** which is located inside cells and recognizes viral RNA.

A TLR's recognition of a pathogen triggers the phagocyte to engulf and destroy the infectious microorganism. Moreover, activation of the TLR induces the phagocytic cell to secrete chemicals, some of which contribute to inflammation, an important innate response to microbial invasion. TLRs link the innate and adaptive branches of the immune system, because still other chemicals secreted by the phagocytes are important in recruiting cells of the adaptive immune system. Because of their pivotal role in the immune system, TLRs are targets for many new drugs and vaccines under development.

As another link between the innate and adaptive branches of the immune system, foreign particles are deliberately marked for phagocytic ingestion by being coated with antibodies produced by the B cells of the adaptive immune system. These are but a few examples of how various components of the immune system are highly interactive and interdependent. The most significant cooperative relationships among the immune effectors are pointed out throughout this chapter.

The innate mechanisms give us all a rapid but limited and nonselective response to unfriendly challenges of all kinds, much like medieval guardsmen lashing out with brute-force weapons at any enemy approaching the walls of the castle they are defending. Innate immunity largely contains and limits the spread of infection. These nonspecific responses are important for keeping the foe at bay until the adaptive immune system, with its highly selective weapons, can be prepared to take over and mount strategies to eliminate the villain.

ADAPTIVE IMMUNE SYSTEM The responses of the adaptive, or acquired, immune system are mediated by the B and T lymphocytes. Each B and T cell can recognize and defend against only one particular type of foreign material, such as one kind of bacterium. Among the millions of B and T cells in the body, only the ones specifically equipped to recognize the unique molecular features of a particular infectious agent are called into action to discriminatingly defend against this agent. This specialization is similar to modern, specially trained military personnel called into active duty to accomplish a very specific task. The chosen lymphocytes multiply, expanding the pool of specialists that can launch a highly targeted attack against the invader.

The adaptive immune system is the ultimate weapon against most pathogens. The repertoire of activated and expanded B and T cells is constantly changing in response to the various pathogens encountered. Thus, the adaptive immune system adapts to wage battle against the specific pathogens in each person's environment. The targets of the adaptive immune system vary among people, depending on the types of immune assaults each individual meets. Furthermore, this system acquires an ability to more efficiently eradicate a particular foe when rechallenged by the same pathogen in the future. It does so by establishing a pool of memory cells as a result of an encounter with a given pathogen so that when later exposed to the same agent, it can more swiftly defend against the invader.

We first examine in more detail the innate immune responses before looking more closely at adaptive immunity.

Innate Immunity

Innate defenses include the following:

1. *Inflammation,* a nonspecific response to tissue injury in which the phagocytic specialists—neutrophils and macrophages—play a major role, along with supportive input from other immune cell types

2. *Interferon,* a family of proteins that nonspecifically defend against viral infection

3. *Natural killer cells,* a special class of lymphocyte-like cells that spontaneously and nonspecifically lyse (rupture) and thereby destroy virus-infected host cells and cancer cells

4. The *complement system,* a group of inactive plasma proteins that, when sequentially activated, bring about destruction of foreign cells by attacking their plasma membranes

We will discuss each of these in turn, beginning with inflammation.

Inflammation is a nonspecific response to foreign invasion or tissue damage.

The term **inflammation** refers to an innate, nonspecific series of highly interrelated events set into motion in response to foreign invasion, tissue damage, or both. The ultimate goal of inflammation is to bring to the invaded or injured area phagocytes and plasma proteins that can (1) isolate, destroy, or inactivate the invaders; (2) remove debris; and (3) prepare for subsequent healing and repair. The overall inflammatory response is remarkably similar no matter what the triggering event (bacterial invasion, chemical injury, or mechanical trauma), although some subtle differences may be evident, depending on the injurious agent or the site of damage. The following sequence of events typically occurs during inflammation. As an example we will use bacterial entry into a break in the skin (● Figure 12-2).

DEFENSE BY RESIDENT TISSUE MACROPHAGES When bacteria invade through a break in the external barrier of skin, macrophages already in the area immediately begin phagocytizing the foreign microbes. Although usually not enough resident macrophages are present to meet the challenge alone, they defend against infection during the first hour or so, before other mechanisms can be mobilized. Macrophages are usually rather stationary, gobbling debris and contaminants that come their way, but when necessary, they become mobile and migrate to sites of battle against invaders.

LOCALIZED VASODILATION Almost immediately on microbial invasion, arterioles within the area dilate, increasing blood flow to the site of injury. This localized vasodilation is mainly induced by **histamine** released from mast cells in the area of tissue damage (the connective tissue–bound "cousins" of circulating basophils; see p. 404). Increased local delivery of blood brings to the site more phagocytic leukocytes and plasma proteins, both crucial to the defense response.

INCREASED CAPILLARY PERMEABILITY Released histamine also increases the capillaries' permeability by enlarging the capillary pores (the spaces between the endothelial cells), so that plasma proteins normally prevented from leaving the blood can escape into the inflamed tissue (see p. 364).

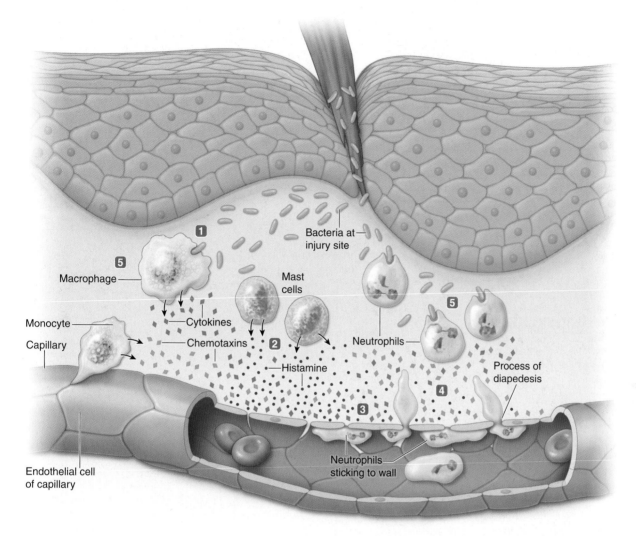

1 A break in the skin introduces bacteria, which reproduce at the wound site. Activated resident macrophages engulf the pathogens and secrete cytokines and chemotaxins.

2 Activated mast cells release histamine.

3 Histamine and cytokines dilate local blood vessels and widen the capillary pores. The cytokines also make the blood vessel wall sticky, causing neutrophils and monocytes to attach.

4 Chemotaxins attract neutrophils and monocytes, which squeeze out between cells of the blood vessel wall, a process called diapedesis, and migrate to the infection site.

5 Monocytes enlarge into macrophages. Newly arriving macrophages and neutrophils engulf the pathogens and destroy them.

● **FIGURE 12-2 Steps producing inflammation.** Chemotaxins released at the site of damage attract phagocytes to the scene. Note the leukocytes emigrating from the blood into the tissues by assuming amoeba-like behavior and squeezing through the capillary pores, a process known as *diapedesis.* Mast cells secrete vessel-dilating, pore-widening histamine. Macrophages secrete cytokines that exert multiple local and systemic effects.

LOCALIZED EDEMA Accumulation of leaked plasma proteins in the interstitial fluid raises the local interstitial fluid–colloid osmotic pressure. Furthermore, the increased local blood flow elevates capillary blood pressure. Because both these pressures tend to move fluid out of the capillaries, these changes favor enhanced ultrafiltration and reduced reabsorption of fluid across the involved capillaries. The end result of this shift in fluid balance is localized edema (see p. 367). Thus, the familiar swelling that accompanies inflammation is the result of histamine-induced vascular changes. Likewise, the other well-known gross manifestations of inflammation, such as redness and heat, are largely caused by the enhanced flow of warm arterial blood to the damaged tissue (*inflammare* means "to set on fire"). Pain is caused both by local distension within the swollen tissue and by the direct effect of locally produced substances on the receptor endings of afferent neurons that supply the area. These observable characteristics of the inflammatory process (swelling, redness, heat, and pain) are coincidental to the pri-

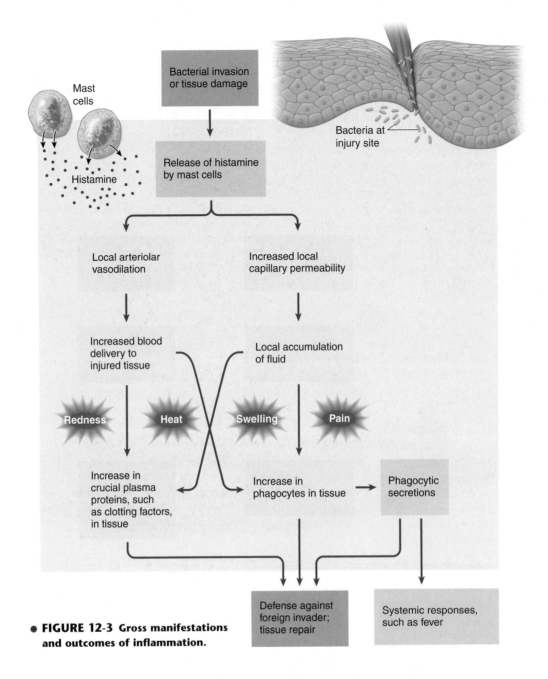

● **FIGURE 12-3 Gross manifestations and outcomes of inflammation.**

mary purpose of the vascular changes in the injured area—to increase the number of leukocytic phagocytes and crucial plasma proteins in the area (● Figure 12-3).

WALLING OFF THE INFLAMED AREA The leaked plasma proteins most critical to the immune response are those in the complement system as well as clotting and anticlotting factors. On exposure to tissue thromboplastin in the injured tissue and to specific chemicals secreted by phagocytes on the scene, fibrinogen—the final factor in the clotting system—is converted into fibrin (see p. 407). Fibrin forms interstitial fluid clots in the spaces around the bacterial invaders and damaged cells. This walling off of the injured region from the surrounding tissues prevents, or at least delays, the spread of bacterial invaders and their toxic products. Later, the more slowly activated anticlotting factors gradually dissolve the clots after they are no longer needed (see p. 410).

EMIGRATION OF LEUKOCYTES Within an hour after injury, the area is teeming with leukocytes that have left the vessels. Neutrophils arrive first, followed during the next 8 to 12 hours by the slower-moving monocytes. The latter swell and mature into macrophages during another 8- to 12-hour period. Once neutrophils or monocytes leave the bloodstream, they never recycle back to the blood.

Leukocytes can emigrate from the blood into the tissues by means of the following steps:

■ Blood-borne leukocytes, especially neutrophils and monocytes, stick to the inner endothelial lining of capillaries in the affected tissue, a process called **margination.** *Selectins,* a type of cell adhesion molecule (CAM; see p. 57) that protrudes from the inner endothelial lining, cause leukocytes flowing by in the blood to slow down and roll along the interior of the vessel, much as the nap of a carpet slows down a child's rolling toy car.

This slowing-down allows the leukocytes enough time to check for local activating factors—"SOS signals" from nearby injured or infected tissues. When present, these activating factors cause the leukocytes to adhere firmly to the endothelial lining by means of interaction with another type of CAM, the *integrins*.

■ Soon the adhered leukocytes start leaving by a mechanism known as **diapedesis.** Assuming *amoeba-like behavior* (see p. 48), an adhered leukocyte pushes a long, narrow projection through a capillary pore; then the remainder of the cell flows forward into the projection (see ● Figure 12-2). In this way, the leukocyte is able to wriggle its way through the capillary pore—even though it is much larger than the pore. Outside the vessel, the leukocyte crawls toward the assaulted area. Neutrophils arrive on the inflammatory scene earliest because they are more mobile than monocytes.

■ **Chemotaxis** guides phagocytic cells in the direction of migration; that is, the cells are attracted to certain chemical mediators, known as *chemotaxins,* released at the site of damage (*taxis* means "attraction"). Binding of chemotaxins with protein receptors on the plasma membrane of a phagocytic cell increases Ca^{2+} entry into the cell. Calcium, in turn, switches on the cellular contractile apparatus that leads to amoeba-like crawling. Because the concentration of chemotaxins progressively increases toward the site of injury, phagocytic cells move unerringly toward this site along a chemotaxin concentration gradient.

LEUKOCYTE PROLIFERATION Resident tissue macrophages as well as leukocytes that exited from the blood and migrated to the inflammatory site are soon joined by new phagocytic recruits from the bone marrow. Within a few hours after the onset of the inflammatory response, the number of neutrophils in the blood may increase up to four to five times that of normal. This increase is the result of both transfer into the blood of large numbers of preformed neutrophils stored in the bone marrow and increased production of new neutrophils by the bone marrow. A slower-commencing but longer-lasting increase in monocyte production by the bone marrow also occurs, making available larger numbers of these macrophage precursor cells. In addition, multiplication of resident macrophages adds to the pool of these important immune cells. Proliferation of new neutrophils, monocytes, and macrophages and mobilization of stored neutrophils are stimulated by various chemical mediators (*colony-stimulating factors;* see p. 401) released from the inflamed region.

MARKING OF BACTERIA FOR DESTRUCTION BY OPSONINS Obviously, phagocytes must be able to distinguish between normal cells and foreign or abnormal cells before accomplishing their destructive mission. Otherwise they could not selectively engulf and destroy only unwanted materials. First, as you have already learned, phagocytes, by means of their TLRs, recognize and subsequently engulf infiltrators that have standard bacterial cell wall components not found in human cells. Second, foreign particles are deliberately marked for phagocytic ingestion by being coated with chemical mediators generated by the immune system. Such body-produced chemicals that make bacte-

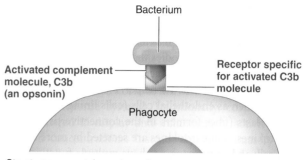

Structures are not drawn to scale.

● **FIGURE 12-4 Mechanism of opsonin action.** One of the activated complement molecules, C3b, links a foreign cell, such as a bacterium, and a phagocytic cell by nonspecifically binding with the foreign cell and specifically binding with a receptor on the phagocyte. This link ensures that the foreign victim does not escape before it can be engulfed by the phagocyte.

ria more susceptible to phagocytosis are known as **opsonins** (*opsonin* means "to prepare for eating"). The most important opsonins are antibodies and one of the activated proteins of the complement system.

An opsonin enhances phagocytosis by linking the foreign cell to a phagocytic cell (● Figure 12-4). One portion of an opsonin molecule binds nonspecifically to the surface of an invading bacterium, whereas another portion of the opsonin molecule binds to receptor sites specific for it on the phagocytic cell's plasma membrane. This link ensures that the bacterial victim does not have a chance to "get away" before the phagocyte can perform its lethal attack.

LEUKOCYTIC DESTRUCTION OF BACTERIA Neutrophils and macrophages clear the inflamed area of infectious and toxic agents as well as tissue debris by both phagocytic and nonphagocytic means; this clearing action is the main function of the inflammatory response.

Phagocytosis involves the engulfment and intracellular degradation (breakdown) of foreign particles and tissue debris. Macrophages can engulf a bacterium in less than 0.01 second. Recall that phagocytic cells contain an abundance of lysosomes, which are organelles filled with hydrolytic enzymes. After a phagocyte has internalized targeted material, a lysosome fuses with the membrane enclosing the engulfed matter and releases its hydrolytic enzymes within the confines of the vesicle, where the enzymes begin breaking down the entrapped material (see p. 31). Phagocytes eventually die from the accumulation of toxic by-products from foreign particle degradation or from inadvertent release of destructive lysosomal chemicals into the cytosol. Neutrophils usually succumb after phagocytizing from 5 to 25 bacteria, whereas macrophages survive much longer and can engulf up to 100 or more bacteria. Indeed, the longer-lived macrophages even clear the area of dead neutrophils in addition to other tissue debris. The **pus** that forms in an infected wound is a collection of these phagocytic cells, both living and dead; necrotic (dead) tissue liquefied by lysosomal enzymes released from the phagocytes; and bacteria.

Vaccination: A Victory over Many Dreaded Diseases

Modern society has come to hope and even expect that vaccines can be developed to protect us from almost any dreaded infectious disease. This expectation has been brought into sharp focus by our current frustration over the inability to date to develop a successful vaccine against HIV, the virus that causes AIDS.

Nearly 2500 years ago, our ancestors were aware of the existence of immune protection. Writing about a plague that struck Athens in 430 BC, Thucydides observed that the same person was never attacked twice by this disease. However, the ancients did not understand the basis of this protection, so they could not manipulate it to their advantage.

Early attempts to deliberately acquire lifelong protection against smallpox, a dreaded disease that was highly infectious and frequently fatal (up to 40% of the sick died), consisted of intentionally exposing oneself by coming into direct contact with a person suffering from a milder form of the disease. The hope was to protect against a future fatal bout of smallpox by deliberately inducing a mild case of the disease. By the beginning of the

17th century, this technique had evolved into using a needle to extract small amounts of pus from active smallpox pustules (the fluid-filled bumps on the skin, which leave a characteristic depressed scar or "pock" mark after healing) and introducing this infectious material into healthy individuals. This inoculation process was done by applying the pus directly to slight cuts in the skin or by inhaling dried pus.

Edward Jenner, an English physician, was the first to demonstrate that immunity against cowpox, a disease similar to but less serious than smallpox, could also protect humans against smallpox. Having observed that milkmaids who got cowpox seemed to be protected from smallpox, Jenner in 1796 inoculated a healthy boy with pus he had extracted from cowpox boils (in fact, *vacca* as in *vaccination* means "cow"). After the boy recovered, Jenner (not being restricted by modern ethical standards of research on human subjects) deliberately inoculated him with what was considered a normally fatal dose of smallpox infectious material. The boy survived.

Jenner's results were not taken seriously, however, until a century later when, in the 1880s, Louis Pasteur, the first great experimental immunologist, extended Jenner's technique. Pasteur demonstrated that the disease-inducing capability of organisms could be greatly reduced (attenuated) so they could no longer produce disease but would still induce antibody formation when introduced into the body—the basic principle of modern vaccines. His first vaccine was against anthrax, a deadly disease of sheep and cows. Pasteur isolated and heated anthrax bacteria, then injected these attenuated organisms into a group of healthy sheep. A few weeks later at a gathering of fellow scientists, Pasteur injected these vaccinated sheep as well as a group of unvaccinated sheep with fully potent anthrax bacteria. The result was dramatic—all the vaccinated sheep survived, but all the unvaccinated sheep died. Pasteur's notorious public demonstrations such as this, coupled with his charismatic personality, caught the attention of physicians and scientists of the time, sparking the development of modern immunology.

mune system does not "remember," regardless of the number of prior exposures.

Active immunity is self-generated; passive immunity is "borrowed."

The production of antibodies as a result of exposure to an antigen is referred to as **active immunity** against that antigen. A second way in which an individual can acquire antibodies is by the direct transfer of antibodies actively formed by another person (or animal). The immediate "borrowed" immunity conferred on receipt of preformed antibodies is known as **passive immunity.** Such transfer of antibodies of the IgG class normally occurs from the mother to the fetus across the placenta during intrauterine development. In addition, a mother's colostrum (first milk) contains IgA antibodies that provide further protection for breast-fed babies. Passively transferred antibodies are usually broken down in less than a month, but meanwhile the newborn is provided important immune protection (essentially the same as its moth-

er's) until it can begin actively mounting its own immune responses. Antibody-synthesizing ability does not develop for about a month after birth.

Clinical Note Passive immunity is sometimes employed clinically to provide immediate protection or to bolster resistance against an extremely virulent infectious agent or potentially lethal toxin to which a person has been exposed (for example, rabies virus, tetanus toxin in nonimmunized individuals, and poisonous snake venom). Typically, the administered preformed antibodies have been harvested from another source (often nonhuman) that has been exposed to an attenuated form of the antigen. Frequently, horses or sheep are used in the deliberate production of antibodies to be collected for passive immunizations. Although injection of serum containing these antibodies (**antiserum** or **antitoxin**) is beneficial in providing immediate protection against the specific disease or toxin, the recipient may develop an immune response against the injected antibodies themselves, because they are foreign proteins. The result may be a severe allergic reaction to the treatment, a condition known as **serum sickness.**

The huge repertoire of B cells is built by reshuffling a small set of gene fragments.

Considering the millions of different antigens against which each of us has the potential to actively produce antibodies, how is it possible for an individual to have such a tremendous diversity of B lymphocytes, each capable of producing a different antibody? Antibodies are proteins synthesized in accordance with a nuclear DNA blueprint. Because all cells of the body, including the antibody-producing cells, contain the same nuclear DNA, it is hard to imagine how enough DNA could be packaged within the nuclei of every cell to code for the 100 million different antibodies (a different portion of the genetic code being used by each B-cell clone), along with all the other genetic instructions used by other cells. Actually, only a relatively small number of gene fragments code for antibody synthesis, but during B-cell development these fragments are cut, reshuffled, and spliced in a vast number of different combinations. Each different combination gives rise to a unique B-cell clone. Antibody genes are later even further diversified by somatic mutation (see p. A-31). The antibody genes of already-formed B cells are highly prone to mutations in the region that codes for the variable antigen-binding sites on the antibodies. Each different mutant cell in turn gives rise to a new clone. Thus, the great diversity of antibodies is made possible by the reshuffling of a small set of gene fragments during B-cell development, as well as by further somatic mutation in already-formed B cells. In this way, a huge antibody repertoire is possible using only a modest share of the genetic blueprint.

We now turn attention to T cells.

T Lymphocytes: Cell-Mediated Immunity

As important as B lymphocytes and their antibody products are in specific defense against invading bacteria and other foreign material, they represent only half of the body's specific immune defenses. The T lymphocytes are equally important in defense against most viral infections and also play an important regulatory role in immune mechanisms.

T cells bind directly with their targets.

Whereas B cells and antibodies defend against conspicuous invaders in the ECF, T cells defend against covert invaders that hide inside cells where antibodies and the complement system cannot reach them. Unlike B cells, which secrete antibodies that can attack antigen at long distances, T cells do not secrete antibodies. Instead, they must directly contact their targets, a process known as *cell-mediated immunity*. T cells of the killer type release chemicals that destroy targeted cells that they contact, such as virus-infected cells and cancer cells.

Like B cells, T cells are clonal and exquisitely antigen specific. On its plasma membrane, each T cell bears unique receptor proteins called *T-cell receptors (TCRs)*, similar although not identical to the surface receptors on B cells (see ● Figure 12-8b, p. 430). Immature lymphocytes acquire their TCRs in the thymus during their differentiation into T cells. Unlike B cells,

T cells are activated by foreign antigen only when it is on the surface of a cell that also carries a marker of the individual's own identity; that is, both foreign antigens and **self-antigens** known as **major histocompatibility complex (MHC) molecules** must be on a cell's surface before a T cell can bind with it. During thymic education, T cells learn to recognize foreign antigens only in combination with the person's own tissue antigens—a lesson passed on to all T cells' future progeny. The importance of this dual antigen requirement and the nature of the MHC self-antigens will be described shortly.

A delay of a few days generally follows exposure to the appropriate antigen before **activated T cells** are prepared to launch a cell-mediated immune attack. When exposed to a specific antigen combination, cells of the complementary T-cell clone proliferate and differentiate for several days, yielding large numbers of activated effector T cells that carry out various cell-mediated responses.

Like B cells, T cells form a memory pool and display both primary and secondary responses. Primary responses tend to be initiated in the lymphoid tissues. During a few-week period after the infection is cleared, more than 90% of the huge number of effector T cells generated during the primary response die by means of *apoptosis* (cell suicide; see p. 124). To stay alive, activated T lymphocytes require the continued presence of their specific antigen and appropriate stimulatory signals. Once the foe succumbs, the vast majority of the now superfluous T lymphocytes commit suicide because their supportive antigen and stimulatory signals are withdrawn. Elimination of most of the effector T cells following a primary response is essential to prevent congestion in the lymphoid tissues. (Such paring down is not needed for B cells—those that become plasma cells and not memory B cells on antigen stimulation rapidly work themselves to death producing antibodies.) The remaining surviving effector T cells become long-lived memory T cells that migrate to all areas of the body, where they are poised for a swift secondary response to the same pathogen in the future.

The three types of T cells are cytotoxic T cells, helper T cells, and regulatory T cells.

There are three subpopulations of T cells, which can be designated in two ways: by their functions or by specific proteins associated with their outer membrane. By function, the three types of T cells are *cytotoxic T cells, helper T cells,* and *regulatory T cells.* By specific membrane protein type, these same cells are *CD8+ T cells, CD4+ T cells,* and *CD4+CD25+ T cells,* respectively. The various types of immune cells have a number of specific immune-related surface membrane proteins given official *cluster designation (CD) numbers* that help characterize them.

■ **Cytotoxic,** or **killer, T cells** destroy host cells harboring anything foreign and thus bearing foreign antigen, such as body cells invaded by viruses, cancer cells that have mutated proteins resulting from malignant transformations, and transplanted cells. The T-cell receptors for cytotoxic T cells are associated with *coreceptors* designated CD8, which are inserted into the plasma membrane as these cells pass through the thymus. Therefore, these cells are also known as **CD8+ T cells.**

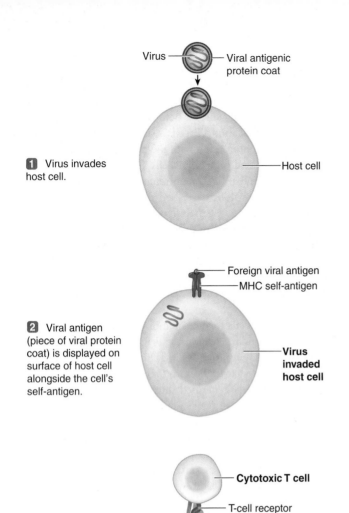

1 Virus invades host cell.

Virus — Viral antigenic protein coat

— Host cell

2 Viral antigen (piece of viral protein coat) is displayed on surface of host cell alongside the cell's self-antigen.

— Foreign viral antigen
— MHC self-antigen

Virus invaded host cell

3 Cytotoxic T cell recognizes and binds with a specific foreign antigen (viral antigen) in association with the self-antigen.

— **Cytotoxic T cell**

— T-cell receptor
— MHC self-antigen and foreign antigen complex

Virus invaded host cell

4 Cytotoxic T cell releases chemicals that destroy attacked cell before virus can enter its nucleus and start to replicate.

● **FIGURE 12-15 A cytotoxic T cell lysing a virus-invaded cell.**

Helper T cells do not directly participate in immune destruction of invading pathogens. Instead, they modulate activities of other immune cells. Because of the important role they play in "turning on" the full power of all the other activated lymphocytes and macrophages, helper T cells constitute the immune system's "master switch." Helper T cells are by far the most numerous T cells, making up 60% to 80% of circulating T cells. The T-cell receptors for helper T cells are associated with coreceptors designated CD4. Accordingly, helper T cells are also called **CD4+ cells.**

Regulatory T cells (T_{regs}), originally called **suppressor T cells,** are a recently identified small subset of CD4+ cells. They have the same CD4 coreceptors as the helper T cells, but in addition they also have CD25, a component of a receptor for IL-2, which promotes T_{reg} activities. Thus, these cells are also referred to as **CD4+CD25+ T cells.** Regulatory T cells, which represent 5% to 10% of CD4 cells, suppress immune responses. They keep the rest of the immune system under tight control. T_{regs} are specialized to inhibit both innate and adaptive immune responses in check-and-balance fashion to minimize harmful immune pathology. T_{regs} contain large amounts of the intracellular protein **Foxp3,** which somehow turns developing T cells into T_{regs} and gives these suppressor cells the ability to quiet other immune cells. Researchers hope that the ability of T_{regs} to put the brakes on helper T cells, B cells, NK cells, and dendritic (macrophage-like) cells can somehow be used therapeutically to curb autoimmune diseases and prevent rejection of transplanted organs.

We now examine the functions of the two most abundant and best known T cells, cytotoxic T cells and helper T cells, in further detail.

Cytotoxic T cells secrete chemicals that destroy target cells.

Cytotoxic T cells are microscopic "hit men." The targets of these destructive cells most frequently are host cells infected with viruses. When a virus invades a body cell, as it must to survive, the cell breaks down the envelope of proteins surrounding the virus and loads a fragment of this viral antigen piggyback onto a newly synthesized MHC self-antigen. This self-antigen and viral antigen complex is inserted into the host cell's surface membrane, where it serves as a red flag indicating the cell is harboring the invader (● Figure 12-15, steps **1** and **2**). To attack the intracellular virus, cytotoxic T cells must destroy the infected host cell in the process. Cytotoxic T cells of the clone specific for this particular virus recognize and bind to the viral-antigen and self-antigen on the surface of an infected cell (● Figure 12-15, step **3**). Thus activated by viral antigen, a cytotoxic T cell can kill the infected cell by either direct or indirect means, depending on the type of lethal chemicals the activated T cell releases. Let us elaborate.

An activated cytotoxic T cell may directly kill the victim cell by releasing chemicals that lyse the attacked cell before viral replication can begin (● Figure 12-15, step **4**). Specifically, cytotoxic T cells as well as NK cells destroy a targeted cell by releasing **perforin** molecules, which penetrate the target cell's surface membrane and join to form porelike channels (● Figure 12-16). This

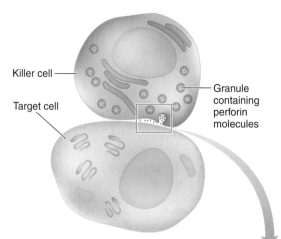

Killer cell

Target cell

Granule containing perforin molecules

1. Killer cell binds to its target.

2. As a result of binding, killer cell's perforin-containing granules fuse with plasma membrane.

3. Granules disgorge their perforin by exocytosis into a small pocket of intercellular space between killer cell and its target.

4. On exposure to Ca^{2+} in this space, individual perforin molecules change from spherical to cylindrical shape.

5. Remodeled perforin molecules bind to target cell membrane and insert into it.

6. Individual perforin molecules group together like staves of a barrel to form pores.

7. Pores admit salt and H_2O, causing target cell to swell and burst.

Perforin molecule

Granule

Salt, H_2O

Ca^{2+}

Killer cell

Target cell

(a) Details of the killing process for cytotoxic T cells and NK cells

Perforin molecule

(b) Enlargement of perforin-formed pores in a target cell

● **FIGURE 12-16 Mechanism of killing by killer cells.** Note the similarity of the perforin-formed pores in a target cell to the membrane attack complex formed by complement molecules (see ● Figure 12-6b, p. 427).

(*Source:* Adapted from the illustration by Dana Burns-Pizer in "How Killer Cells Kill," by John Ding-E Young and Zanvil A. Cohn in *Scientific America,* 1988.)

technique of killing a cell by punching holes in its membrane is similar to the method employed by the membrane attack complex of the complement cascade. This contact-dependent mechanism of killing has been nicknamed the "kiss of death."

■ A cytotoxic T cell can also indirectly bring about death of an infected host cell by releasing **granzymes,** which are enzymes similar to digestive enzymes. Granzymes enter the target cell through the perforin channels. Once inside, these chemicals trigger the virus-infected cell to self-destruct through apoptosis.

The virus released on destruction of the host cell by either of these methods is directly destroyed in the ECF by phagocytic

cells, neutralizing antibodies, and the complement system. Meanwhile the cytotoxic T cell, which has not been harmed in the process, can move on to kill other infected host cells.

The surrounding healthy cells replace the lost cells by means of cell division. Usually, to halt a viral infection, only some of the host cells must be destroyed. If the virus has had a chance to multiply, however, with replicated virus leaving the original cell and spreading to other host cells, the cytotoxic T-cell defense mechanism may sacrifice so many of the host cells that serious malfunction may ensue.

Recall that other nonspecific defense mechanisms also come into play to combat viral infections, most notably NK cells, interferon, macrophages, and the complement system. As usual, an intricate web of interplay exists among the immune defenses that are launched against viral invaders (▲ Table 12-2).

ANTIVIRAL DEFENSE IN THE NERVOUS SYSTEM The usual method of destroying virus-infected host cells is not appropriate for the nervous system. If cytotoxic T cells destroyed virus-infected neurons, the lost cells could not be replaced, because neurons cannot reproduce. Fortunately, virus-infected neurons are spared from extermination by the immune system, but how, then, are neurons protected from viruses? Immunologists long thought that the only antiviral defenses for neurons were those aimed at free viruses in the ECF. Surprising new research has revealed, however, that antibodies not only target viruses for destruction in the ECF but can also eliminate viruses inside neurons. It is unclear whether antibodies actually enter the neurons and interfere directly with viral replication (neurons have been shown to take up antibodies near their synaptic endings) or bind with the surface of nerve cells and trigger intracellular changes that stop viral replication.

Clinical Note The fact that some viruses, such as the herpes virus, persist for years in nerve cells, occasionally "flaring up" to produce symptoms, demonstrates that the antibodies' intraneuronal mechanism does not provide a foolproof antiviral defense for neurons.

Helper T cells secrete chemicals that amplify the activity of other immune cells.

In contrast to cytotoxic T cells, helper T cells are not killer cells. Instead, helper T cells secrete cytokines that "help," or augment, nearly all aspects of the immune response. Most cytokines are produced by helper T cells. B cells stimulated by T-dependent antigen cannot develop into antibody-secreting plasma cells without the help of helper T cells. In the usual web of interaction among immune effectors, antibodies from B cells may influence cytotoxic T cells' ability to destroy a victim cell. Furthermore, helper T cells secrete cytokines that enhance activity of the appropriate cytotoxic T cells and activate macrophages.

Clinical Note This is why **acquired immunodeficiency syndrome (AIDS),** caused by the **human immunodeficiency virus (HIV),** is so devastating to the immune defense system. The AIDS virus selectively invades helper T cells, destroying or incapacitating the cells that normally orchestrate much of the immune response (● Figure 12-17). The virus also invades macrophages, further crippling the immune system, and sometimes enters brain cells as well, leading to the dementia (severe impairment of intellectual capacity) noted in some AIDS victims.

The following are among the best known of helper T-cell cytokines, although the list is rapidly growing as researchers uncover more cytokines with other roles:

1. Helper T cells secrete several interleukins that serve collectively as a **B-cell growth factor (IL-4, IL-5,** and **IL-6),** which contributes to B-cell function in concert with the IL-1 secreted by macrophages. Antibody secretion is greatly reduced or absent without the assistance of helper T cells, especially in defense against T-dependent antigen.

When the virus is free in the ECF,

Macrophages

Destroy the free virus by phagocytosis

Process and present the viral antigen to helper T cells

Secrete interleukin 1, which activates B and T cell clones specific to the viral antigen

Plasma Cells Derived from B Cells Specific to the Viral Antigen Secrete Antibodies That

Neutralize the virus to prevent its entry into a host cell

Activate the complement cascade that directly destroys the free virus and enhances phagocytosis of the virus by acting as an opsonin

When the virus has entered a host cell (which it must do to survive and multiply, with the replicated viruses leaving the original host cell to enter the ECF in search of other host cells),

Interferon

Is secreted by virus-infected cells

Binds with and prevents viral replication in other host cells

Enhances the killing power of macrophages, natural killer cells, and cytotoxic T cells

Natural Killer Cells

Nonspecifically lyse virus-infected host cells

Cytotoxic T Cells

Are specifically activated by the viral antigen and lyse the infected host cells before the virus has a chance to replicate

Helper T Cells

Secrete cytokines, which enhance cytotoxic T-cell activity and B-cell antibody production

When a virus-infected cell is destroyed, the free virus is released into the ECF, where it is attacked directly by macrophages, antibodies, and the activated complement components.

---

2. Helper T cells similarly secrete **T-cell growth factor (IL-2),** which augments the activity of cytotoxic T cells and even of other helper T cells responsive to the invading antigen. In typical interplay fashion, IL-1 secreted by macrophages not only enhances the activity of both the appropriate B- and T-cell clones but also stimulates secretion of IL-2 by activated helper T cells.

3. Some chemicals secreted by T cells act as *chemotaxins* to lure more neutrophils and macrophages-to-be to the invaded

area. Cytokines that act as chemotaxins are specifically called **chemokines.**

4. Once macrophages are attracted to the area, **macrophage-migration inhibition factor,** another important cytokine released from helper T cells, keeps these large phagocytic cells in the region by inhibiting their outward migration. As a result, a great number of chemotactically attracted macrophages accumulate in the infected area. This factor also confers greater phagocytic power on the gathered macrophages. These so-called **angry macrophages** have more powerful destructive ability. They are especially important in defending against the bacteria that cause tuberculosis because such microbes can survive simple phagocytosis by nonactivated macrophages.

5. One cytokine secreted by helper T cells (**IL-5**) activates *eosinophils* and another (**IL-4**) promotes the development of *IgE antibodies* for defense against parasitic worms.

T HELPER 1 AND T HELPER 2 CELLS Not all helper cells secrete the same cytokines. Two subsets of helper T cells—**T helper 1 (T$_H$1)** and **T helper 2 (T$_H$2)** cells—augment different patterns of immune responses by secreting different types of cytokines. T$_H$1 cells rally a cell-mediated (cytotoxic T-cell) response, which is appropriate for infections with intracellular microbes, such as viruses, whereas T$_H$2 cells promote antibody-mediated immunity by B cells and rev up eosinophil activity for defense against parasitic worms.

Helper T cells produced in the thymus are in a naive state until they encounter the antigen they are primed to recognize. Whether a naive helper T cell becomes a T$_H$1 or T$_H$2 cell depends on which cytokines are secreted by macrophages and dendritic cells (macrophage-like cells) of the innate immune system that "present" the antigen to the uncommitted T cell. You will learn about antigen presentation in the next section. **IL-12** drives a naive T cell specific for the antigen to become a T$_H$1 cell, whereas **IL-4** favors the development of a naive cell into a T$_H$2 cell. Thus, the antigen-presenting cells of the nonspecific immune system can influence the whole tenor of the specific immune response by determining whether the T$_H$1 or T$_H$2 cellular subset dominates. In the usual case, the secreted cytokines promote the appropriate specific immune response against the particular threat at hand.

Scientists recently discovered a third, much smaller subset of helper T cells in some individuals: T$_H$17 cells, which produce IL-17. IL-6 from macrophages appears to guide the development of T$_H$17 cells, but the underlying factors that trigger this pathway are unclear. T$_H$17 cells promote inflammation and are effector molecules in the development of inflammatory autoimmune diseases.

We next examine how T cells are activated by antigen-presenting cells and the roles of MHC molecules.

T lymphocytes respond only to antigens presented to them by antigen-presenting cells.

T cells cannot perform their tasks without assistance from antigen-presenting cells. That is, relevant T cells cannot recognize "raw" foreign antigens entering the body; before reacting

● **FIGURE 12-17 AIDS virus.** Human immunodeficiency virus (HIV) (in gray), the AIDS-causing virus, on a helper T lymphocyte, HIV's primary target.

to it, a T-cell clone must be formally "introduced" to the antigen. **Antigen-presenting cells (APCs)** handle the formal introduction; they process and present antigen, complexed with MHC molecules, on their surface to the T cells. APCs include macrophages and closely related dendritic cells, both of which are effector cells of the innate defense system but also play a key role in activating cell-mediated adaptive immunity. The helper T cells of the cell-mediated branch of adaptive immunity in turn enhance the antibody-mediated branch of adaptive immunity. Thus APCs are important to both the innate and adaptive components of the immune system.

You are already familiar with macrophages. **Dendritic cells** are specialized APCs that act as sentinels in almost every tissue. They are so named because they have many surface projections, or branches, that resemble the dendrites of neurons (*dendros* means "tree") (● Figure 12-18). Dendritic cells are especially abundant in the skin and mucosal linings of the lungs and digestive tract—strategic locations where microbes are likely to enter the body. After exposure to the appropriate antigen, dendritic cells leave their tissue home and migrate through the lymphatic system to nearby lymph nodes, where they cluster and activate T cells.

We will use a dendritic cell engulfing a bacterium and presenting it to a helper T cell as an example of an APC (● Figure 12-19). Invading bacteria are phagocytized by dendritic cells (or macrophages) (step **1**). Within the dendritic cell, the endocytic vesicle containing the engulfed bacterium fuses with a lysosome, which enzymatically breaks down the bacterium's proteins into antigenic peptides (small protein fragments) (step **2**). Each antigenic peptide then binds to an MHC molecule, which has been newly synthesized within the endoplasmic reticulum/ Golgi complex (steps **3** and **4**). An MHC molecule has a deep groove into which a variety of antigenic peptides can bind, depending on what the macrophage has engulfed. Loading of the antigenic peptide onto an MHC molecule takes place in a newly discovered specialized organelle within APCs, the **compartment for peptide loading.** The MHC molecule then transports

Activation of helper T cells by antigen presentation

1. Bacterium is taken up by phagocytosis and degraded in a lysosome.

2. Bacterial antigenic peptides are displayed on APC cell surface bound to class II MHC proteins and presented to helper (CD4+) T cells with TCRs that recognize the antigen.

3. APC secretes interleukins, which activates T cell.

4. Activated T cell secretes cytokines, which stimulate T cell to proliferate to expand clone of selected cells.

5. Cloned helper T cells are ready to activate B cells and enhance other immune activities.

● **FIGURE 12-21 Interactions among large phagocytic cells (APCs), helper T cells, and B cells responsive to T-dependent antigen.**

class II MHC molecules. Thus, the specific binding requirements for cytotoxic T cells and helper T cells ensure the appropriate T-cell responses. A helper T cell can bind with foreign antigen only when it is found on the surfaces of immune cells with which the helper T cell directly interacts—*macrophages, dendritic cells,* and *B cells.* Class II MHC molecules are found on macrophages and dendritic cells, which present antigens to helper T cells, as well as on B cells, whose activities are enhanced by cytokines secreted by helper T cells. B cells do not phagocytize antigenic particles like macrophages and dendritic cells do, but they can internalize T-dependent antigen bound to their surface receptor by receptor-mediated endocytosis (see p. 31), then display the antigens complexed with class II MHC on their surface. Binding of the antigen-bearing B cell with the matching helper T cell causes the T cell to secrete cytokines that activate this specific B cell, leading to clonal expansion and conversion of this B cell clone into antibody-producing plasma cells and memory cells (● Figure 12-21). The capabilities of helper T cells would be squandered if these cells were able to bind with body cells other than these special APC and B immune cells. This is the primary pathway by which the adaptive immune system fights bacteria. ▲ Table 12-3 summarizes the innate and adaptive immune strategies that defend against bacterial invasion.

TRANSPLANT REJECTION T cells bind with MHC antigens present on the surface of *transplanted cells* in the absence of foreign viral antigen. The ensuing destruction of the transplanted cells triggers the rejection of transplanted or grafted tissues. Presumably, some of the recipient's

T cells "mistake" the MHC antigens of the donor cells for a closely resembling combination of a conventional viral foreign antigen complexed with the recipient's MHC self-antigens.

To minimize the rejection phenomenon, technicians match the tissues of donor and recipient according to MHC antigens as closely as possible. Therapeutic procedures to suppress the immune system then follow. In the past, the primary immunosuppressive tools included radiation therapy and drugs aimed at destroying the actively multiplying lymphocyte populations, plus anti-inflammatory drugs that suppressed growth of all lymphoid tissue. However, these measures not only suppressed the T cells that were primarily responsible for rejecting transplanted tissue but also depleted the antibody-secreting B cells. Unfortunately, the treated individual was left with little specific immune protection against bacterial and viral infections. In recent years, new therapeutic agents have become extremely useful in selectively depressing T-cell-mediated immune activity while leaving B-cell antibody-mediated immunity essentially intact. For example, *cyclosporin* blocks IL-2, the cytokine secreted by helper T cells that is required for expansion of the selected cytotoxic T-cell clone. Furthermore, a new technique under investigation may completely prevent rejection of transplanted tissues even from an unmatched donor. This technique involves the use of tailor-made antibodies that block specific facets of the rejection process. If proven safe and effective, the technique will have a tremendous impact on tissue transplantation.

What factors normally prevent the adaptive immune system from unleashing its powerful defense capabilities against the body's own self-antigens? We examine this issue next.

3. Exposure o
structurally alm
production of a
only interact w
with the closel
molecular mim
for "strep throa
very similar to
valves of some i
duced against
with this heart t
responsible for
matic fever.

4. New studies
diseases, one tha
orders are more
scientists have s
eases was some
findings sugges
self-destructive
nancy. Research
gain access to t
labor and delive
after the pregna
cells from the r
involving autoir
these condition
persistent fetal
but not identica
as being foreign
immune attack
closely related a
be the underlyi
autoimmune at
many organs, n

Let us now
fending against

Immune sur
involves an
and interfer

Besides destro
tant function o
ing newly aris
they have a ch
as **immune su**
immune syster
become cancer
a cancer cell if
cell division a
chance alone
genic (cancer
certain enviror
natively, a few
turn the cells t
tem recognizes

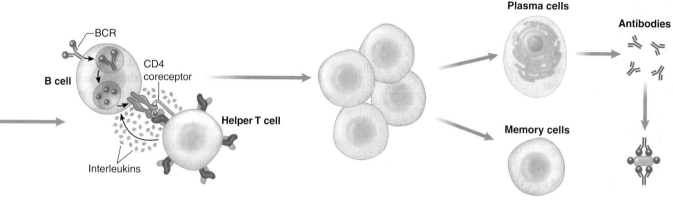

Activation of B cells responsive to T-dependent antigen

BCR

B cell

CD4 coreceptor

Helper T cell

Interleukins

Plasma cells

Antibodies

Memory cells

6 BCR binds to antigen. Antigen is internalized by receptor-mediated endocytosis and its macromolecules degraded. Antigenic peptides produced are displayed on cell surface bound to class II MHC proteins.

7 TCR of a helper T cell recognizes specific antigen on B cell, and CD4 coreceptor links the two cells together.

8 Helper T cell secretes interleukins, which stimulate B cell proliferation to produce clone of selected cells.

9 Some cloned B cells differentiate into plasma cells, which secrete antibodies specific for the antigen, while a few differentiate into memory B cells.

10 Antibodies bind with antigen, targeting antigenic invader for destruction by the innate immune system.

▲ TABLE 12-3 Innate and Adaptive Immune Responses to Bacterial Invasion

Innate Immune Mechanisms	Adaptive Immune Mechanisms
Inflammation	
Resident tissue macrophages engulf invading bacteria.	B cells specific to T-independent antigen are activated on binding with the antigen.
Histamine-induced vascular responses increase blood flow to the area, bringing in additional immune-effector cells and plasma proteins.	B cells specific to T-dependent antigen present antigen to helper T cells. On binding with the B cells, helper T cells activate the B cells.
A fibrin clot walls off the invaded area.	The activated B-cell clone proliferates and differentiates into plasma cells and memory cells.
Neutrophils and monocytes/macrophages migrate from the blood to the area to engulf and destroy foreign invaders and to remove cell debris.	Plasma cells secrete customized antibodies, which specifically bind to invading bacteria. Plasma cell activity is enhanced by
Phagocytic cells secrete cytokines, which enhance both innate and adaptive immune responses and induce local and systemic symptoms associated with an infection.	Interleukin 1 secreted by macrophages
	Helper T cells, which have been activated by the same bacterial antigen processed and presented to them by macrophages or dendritic cells
Nonspecific Activation of the Complement System	Antibodies bind to invading bacteria and enhance innate mechanisms that lead to the bacteria's destruction. Specifically, antibodies
Complement components form a hole-punching membrane attack complex that lyses bacterial cells.	Act as opsonins to enhance phagocytic activity
Complement components enhance many steps of inflammation.	Activate the lethal complement system
	Stimulate natural killer cells, which directly lyse bacteria
	Memory cells persist that are capable of responding more rapidly and more forcefully should the same bacteria be encountered again.

The term **toler**
preventing the
tissues. During
goes on durin;
are by chance
tissue antigen:
function, they
nately, the imn
ies or activated
instead directs

At least eig

1. *Clonal dele*
antigens early
cifically capab
are permanent
tion is accom|
that would re;
elimination is
developed.

2. *Clonal ane*
phocyte must
activated (turr
stimulatory cc
surface of an /
which are intrc
on by finding
cosignal, the c
other cells. Foi
any virus-inva
the infected c
dual signals—
antigens becai
bearing APCs
antigen turns
unresponsive
ring the cell tc
anergy (*anerg;*
inactivated (tł
antigens. Clon
T lymphocyte

3. *Receptor ε*
body of self-re
nism, once a E
antigens encoi
swiftly changi
way, an origii
tated" so that

4. *Active sup*
in tolerance
clones specifi

5. *Immunolc*
sequestering. ¹
immune syste
with the ECF

Specialized cells in the epidermis produce keratin and melanin and participate in immune defense.

The epidermis contains four distinct resident cell types—*melanocytes, keratinocytes, Langerhans cells,* and *Granstein cells*—plus transient T lymphocytes that are scattered throughout the epidermis and dermis. Each of these resident cell types performs specialized functions.

MELANOCYTES Melanocytes produce the pigment **melanin,** which they disperse to surrounding skin cells. The amount and type of melanin, which can vary among black, brown, yellow, and red pigments, are responsible for the different shades of skin color of the various races. Fair-skinned people have about the same number of melanocytes as dark-skinned people; the difference in skin color depends on the amount of melanin produced by each melanocyte. Melanin is produced through complex biochemical pathways in which the melanocyte enzyme *tyrosinase* plays a key role. Most people, regardless of skin color, have enough tyrosinase that, if fully functional, could result in enough melanin to make their skin very black. In those with lighter skin, however, two genetic factors prevent this melanocyte enzyme from functioning at full capacity: (1) much of the tyrosinase produced is in an inactive form, and (2) various inhibitors that block tyrosinase action are produced. As a result, less melanin is produced. One cosmetic company has applied for a patent for a technique that they have developed to switch off the tyrosinase gene, which means that genetic skin-lightening agents might soon be on the market. The social impacts of such a product could be far-reaching.

In addition to hereditary determination of melanin content, the amount of this pigment can be increased transiently in response to exposure to ultraviolet (UV) light rays from the sun. This additional melanin, the outward appearance of which constitutes a "tan," performs the protective function of absorbing harmful UV rays.

KERATINOCYTES The most abundant epidermal cells are the **keratinocytes,** which, as the name implies, are specialists in keratin production. As they die, they form the outer protective keratinized layer. They also generate hair and nails. A surprising, recently discovered function is that keratinocytes are also important immunologically. They secrete IL-1 (a product also secreted by macrophages), which influences the maturation of T cells that tend to localize in the skin. Interestingly, the epithelial cells of the thymus have been shown to bear anatomic, molecular, and functional similarities to those of the skin. Apparently, some postthymic steps in T-cell maturation take place in the skin under keratinocyte guidance.

OTHER IMMUNE CELLS OF THE SKIN The two other epidermal cell types also play a role in immunity. **Langerhans cells,** which migrate to the skin from the bone marrow, are dendritic cells that serve as antigen-presenting cells. Thus, the skin not only is a mechanical barrier but actually alerts lymphocytes if the barrier is breached by invading microorganisms. Langerhans cells present antigen to helper T cells, facilitating their responsiveness to skin-associated antigens. In contrast, **Granstein cells** seem to act as a "brake" on skin-activated immune responses. These cells are the most recently discovered and least understood of the skin's immune cells. Significantly, Langerhans cells are more susceptible to damage by UV radiation (as from the sun) than Granstein cells are. Losing Langerhans cells as a result of exposure to UV radiation can detrimentally lead to a predominant suppressor signal rather than the normally dominant helper signal, leaving the skin more vulnerable to microbial invasion and cancer cells.

The various epidermal components of the immune system are collectively termed **skin-associated lymphoid tissue,** or **SALT.** Recent research suggests that the skin probably plays an even more elaborate role in adaptive immune defense than described here. This is appropriate, because the skin serves as a major interface with the external environment.

VITAMIN D SYNTHESIS BY THE SKIN The epidermis also synthesizes vitamin D in the presence of sunlight. The cell type that produces vitamin D is undetermined. Vitamin D, which is derived from a precursor molecule closely related to cholesterol, promotes the absorption of Ca^{2+} from the digestive tract into the blood (Chapter 16). Dietary supplements of vitamin D are usually required because typically the skin is not exposed to sufficient sunlight to produce adequate amounts of this essential chemical.

Protective measures within body cavities discourage pathogen invasion into the body.

The human body's defense system must guard against entry of potential pathogens not only through the outer surface of the body but also through the internal cavities that communicate directly with the external environment—namely, the digestive system, the urogenital system, and the respiratory system. These systems use various tactics to destroy microorganisms entering through these routes.

DEFENSES OF THE DIGESTIVE SYSTEM Saliva secreted into the mouth at the entrance of the digestive system contains an enzyme that lyses certain ingested bacteria. "Friendly" bacteria that live on the back of the tongue convert food-derived nitrate into nitrite, which is swallowed. Acidification of nitrite on reaching the highly acidic stomach generates nitric oxide, which is toxic to a variety of microorganisms. Furthermore, many of the surviving bacteria that are swallowed are killed directly by the strongly acidic gastric juice in the stomach. Farther down the tract, the intestinal lining is endowed with gut-associated lymphoid tissue. These defensive mechanisms are not 100% effective, however. Some bacteria do manage to survive and reach the large intestine (the last portion of the digestive tract), where they continue to flourish. Surprisingly, this normal microbial population provides a natural barrier against infection within the lower intestine. These harmless resident microbes competitively suppress the growth of potential pathogens that have managed to escape the antimicrobial measures of earlier parts of the digestive tract.

Clinical Note On occasion, orally administered antibiotic therapy against an infection elsewhere within the body may actually induce an intestinal infection. By knocking out some of the normal intestinal bacteria, an antibiotic may permit an antibiotic-resistant pathogenic species to overgrow in the intestine.

DEFENSES OF THE UROGENITAL SYSTEM Within the urogenital system, would-be invaders encounter hostile conditions in the acidic urine and acidic vaginal secretions. The urogenital organs also produce a sticky mucus, which, like flypaper, entraps small invading particles. Subsequently, the particles are either engulfed by phagocytes or are swept out as the organ empties (for example, they are flushed out with urine flow).

DEFENSES OF THE RESPIRATORY SYSTEM The respiratory system is likewise equipped with several important defense mechanisms against inhaled particulate matter. The respiratory system is the largest surface of the body that comes into direct contact with the increasingly polluted external environment. The surface area of the respiratory system exposed to the air is 30 times that of the skin. Larger airborne particles are filtered out of the inhaled air by hairs at the entrance of the nasal passages. Lymphoid tissues, the *tonsils* and *adenoids,* provide immunological protection against inhaled pathogens near the beginning of the respiratory system. Farther down in the respiratory airways, millions of tiny hairlike projections known as *cilia* constantly beat in an outward direction (see p. 45). The respiratory airways are coated with a layer of thick, sticky mucus secreted by epithelial cells within the airway lining. This mucus sheet, laden with any inspired particulate debris (such as dust) that adheres to it, is constantly moved upward to the throat by ciliary action. This moving "staircase" of mucus is known as the **mucus escalator.** The dirty mucus is either expectorated (spit out) or in most cases swallowed without the person even being aware of it; any indigestible foreign particulate matter is later eliminated in the feces. Besides keeping the lungs clean, this mechanism is an important defense against bacterial infection, because many bacteria enter the body on dust particles. Also contributing to defense against respiratory infections are antibodies secreted in the mucus. In addition, an abundance of phagocytic specialists called the **alveolar macrophages** scavenge within the air sacs (alveoli) of the lungs. Further respiratory defenses include coughs and sneezes. These commonly experienced reflex mechanisms involve forceful outward expulsion of material in an attempt to remove irritants from the trachea (*coughs*) or nose (*sneezes*). A sneeze can propel air and particles from the respiratory airways up to 30 feet away at rates over 90 miles per hour.

Clinical Note Cigarette smoking suppresses these normal respiratory defenses. The smoke from a single cigarette can paralyze the cilia for several hours, with repeated exposure eventually leading to ciliary destruction. Failure of ciliary activity to sweep out a constant stream of particulate-laden mucus enables inhaled carcinogens to remain in contact with the respiratory airways for prolonged periods. Furthermore, cigarette smoke incapacitates alveolar macrophages.

Not only do particulates in cigarette smoke overwhelm the macrophages but certain components of cigarette smoke have a direct toxic effect on the macrophages, reducing their ability to engulf foreign material. In addition, noxious agents in tobacco smoke irritate the mucous linings of the respiratory tract, resulting in excess mucus production, which may partially obstruct the airways. "Smoker's cough" is an attempt to dislodge this excess stationary mucus. These and other direct toxic effects on lung tissue lead to the increased incidence of lung cancer and chronic respiratory diseases associated with cigarette smoking. Air pollutants include some of the same substances found in cigarette smoke and can similarly affect the respiratory system.

We will examine the respiratory system in greater detail in the next chapter.

Chapter in Perspective: Focus on Homeostasis

We could not survive beyond early infancy were it not for the body's defense mechanisms. These mechanisms resist and eliminate potentially harmful foreign agents with which we continuously come into contact in our hostile external environment and also destroy abnormal cells that often arise within the body. Homeostasis can be optimally maintained, and thus life sustained, only if the body cells are not physically injured or functionally disrupted by pathogenic microorganisms or are not replaced by abnormally functioning cells, such as traumatized cells or cancer cells. The immune defense system—a complex, multifaceted, interactive network of leukocytes, their secretory products, and plasma proteins—contributes indirectly to homeostasis by keeping other cells alive so that they can perform their specialized activities to maintain a stable internal environment. The immune system protects the other healthy cells from foreign agents that have gained entrance to the body, eliminates newly arisen cancer cells, and clears away dead and injured cells to pave the way for replacement with healthy new cells.

The skin contributes indirectly to homeostasis by serving as a protective barrier between the external environment and the rest of the body cells. It helps prevent harmful foreign agents such as pathogens and toxic chemicals from entering the body and helps prevent the loss of precious internal fluids from the body. The skin also contributes directly to homeostasis by helping maintain body temperature by means of the sweat glands and adjustments in skin blood flow. The amount of heat carried to the body surface for dissipation to the external environment is determined by the volume of warmed blood flowing through the skin.

Other systems that have internal cavities in contact with the external environment, such as the digestive, urogenital, and respiratory systems, also have defense capabilities to prevent harmful external agents from entering the body through these avenues.

Objective Questions (Answers on p. A-48)

1. The complement system can be activated only by antibodies. (*True or false?*)

2. Specific adaptive immune responses are accomplished by neutrophils. (*True or false?*)

3. Damaged tissue is always replaced by scar tissue. (*True or false?*)

4. Active immunity against a particular disease can be acquired only by actually having the disease. (*True or false?*)

5. A secondary response has a more rapid onset, is more potent, and has a longer duration than a primary response. (*True or false?*)

6. _____ are receptors on the plasma membrane of phagocytes that recognize and bind with telltale molecular patterns present on the surface of microorganisms but absent from human cells.

7. The complement system's _____ forms a doughnut-shaped complex that embeds in a microbial surface membrane, causing osmotic lysis of the victim cell.

8. _____ is a collection of phagocytic cells, necrotic tissue, and bacteria.

9. _____ is the localized response to microbial invasion or tissue injury that is accompanied by swelling, heat, redness, and pain.

10. A chemical that enhances phagocytosis by serving as a link between a microbe and the phagocytic cell is known as a(n) _____.

11. _____, collectively, are all the chemical messengers other than antibodies secreted by lymphocytes.

12. Which of the following statements concerning leukocytes is/are *incorrect*?
    a. Monocytes are transformed into macrophages.
    b. T lymphocytes are transformed into plasma cells that secrete antibodies.
    c. Neutrophils are highly mobile phagocytic specialists.
    d. Basophils release histamine.
    e. Lymphocytes arise in large part from lymphoid tissues.

13. Match the following:
    ___ 1. a family of proteins that nonspecifically defend against viral infection
    ___ 2. a response to tissue injury in which neutrophils and macrophages play a major role
    ___ 3. a group of plasma proteins that, when activated, bring about destruction of foreign cells by attacking their plasma membranes
    ___ 4. lymphocyte-like entities that spontaneously lyse tumor cells and virus-infected host cells

    (a) complement system
    (b) natural killer cells
    (c) interferon
    (d) inflammation

14. Using the answer code on the right, indicate whether the numbered characteristics of the adaptive immune system apply to antibody-mediated immunity or cell-mediated immunity (or both):
    ___ 1. involves secretion of antibodies
    ___ 2. mediated by B cells
    ___ 3. mediated by T cells
    ___ 4. accomplished by thymus-educated lymphocytes
    ___ 5. triggered by the binding of specific antigens to complementary lymphocyte receptors
    ___ 6. involves formation of memory cells in response to initial exposure to an antigen
    ___ 7. primarily aimed against virus-infected host cells
    ___ 8. protects primarily against bacterial invaders
    ___ 9. directly destroys targeted cells
    ___ 10. involved in rejection of transplanted tissue
    ___ 11. requires binding of a lymphocyte to a free extracellular antigen
    ___ 12. requires dual binding of a lymphocyte with both foreign antigen and self-antigens present on the surface of a host cell

    (a) antibody-mediated immunity
    (b) cell-mediated immunity
    (c) both antibody-mediated and cell-mediated immunity

15. Using the answer code on the right, indicate whether the numbered characteristics apply to the epidermis or dermis:
    ___ 1. is the inner layer of skin
    ___ 2. has layers of epithelial cells that are dead and flattened
    ___ 3. has no direct blood supply
    ___ 4. contains sensory nerve endings
    ___ 5. contains keratinocytes
    ___ 6. contains melanocytes
    ___ 7. contains rapidly dividing cells
    ___ 8. is mostly connective tissue

    (a) epidermis
    (b) dermis

Essay Questions

1. Distinguish between bacteria and viruses.
2. Summarize the functions of each of the lymphoid tissues.
3. Distinguish between innate and adaptive immune responses.
4. Compare the life histories of B cells and T cells.
5. What is an antigen?
6. Describe the structure of an antibody. List and describe the five subclasses of immunoglobulins.
7. In what ways do antibodies exert their effect?
8. Describe the clonal selection theory.
9. Compare the functions of B cells and T cells. What are the roles of the three types of T cells?

10. Summarize the functions of macrophages in immune defense.

11. What mechanisms are involved in tolerance?

12. What is the importance of class I and class II MHC glycoproteins?

13. Describe the factors that contribute to immune surveillance against cancer cells.

14. Distinguish among immunodeficiency disease, autoimmune disease, immune complex disease, immediate hypersensitivity, and delayed hypersensitivity.

15. What are the immune functions of the skin?

Quantitative Exercises (Solutions on p. A-49)

1. As a result of the innate immune response to an infection, for example from a cut on the skin, capillary walls near the site of infection become very permeable to plasma proteins that normally remain in the blood. These proteins diffuse into the interstitial fluid, raising the interstitial fluid–colloid osmotic pressure. This increased colloid osmotic pressure causes fluid to leave the circulation and accumulate in the tissue, forming a welt, or wheal. This process is referred to as the *wheal response.* The wheal response is mediated in part by histamine secreted from mast cells in the area of infection. The histamine binds to receptors, called *H−1 receptors,* on capillary endothelial cells. The histamine signal is transduced via the Ca^{2+} second-messenger pathway involving phospholipase C (see p. 122). In response to this signal, the capillary endothelial cells contract (via internal actin–myosin interaction), which causes a widening of the intercellular gaps (pores) between the capillary endothelial cells (see p. 364). In addition, substance P (see p. 192) also contributes to pore widening. Plasma proteins can pass through these widened pores and leave the capillaries. Looking at ● Figure 10-22, p. 367, compare the magnitude of the wheal response (that is, the extent of localized edema) if P_{IF} were raised (a) from 0 mm Hg to 5 mm Hg and (b) from 0 mm Hg to 10 mm Hg. In both cases, compare the net exchange pressure (NEP) at the arteriolar end of the capillary, the venular end of the capillary, and the average NEP. (Assume the other forces acting across the capillary wall remain unchanged.)

POINTS TO PONDER

(Explanations on p. A-49)

1. Compare the defense mechanisms that come into play in response to bacterial and viral pneumonia.

2. Nearly 30 years have passed since the first cases of AIDS were reported in the United States and millions of research dollars have been spent studying this disease. Much has been learned, and drugs have been developed that delay or manage the condition, but no AIDS vaccine has been approved despite many unsuccessful attempts.

Why does the frequent mutation of HIV (the AIDS virus) make it difficult to develop a vaccine against this virus?

3. What impact would failure of the thymus to develop embryonically have on the immune system after birth?

4. Medical researchers are currently working on ways to "teach" the immune system to view foreign tissue as "self." What useful clinical application will the technique have?

5. When someone looks at you, are the cells of your body that person is viewing dead or alive?

CLINICAL CONSIDERATION

(Explanation on p. A-49)

Linda P. is allergic to molds and dust mites. She takes allergy medication as needed to keep her symptoms under control, but in the hopes of ridding herself of the troublesome allergies, she has been taking allergy (desensitization) shots. Allergy shots consist of weekly injections of minute, gradually increasing doses of the offending allergens. How can deliberately injecting the offending agent lead to a reduction in allergic response to the allergen? The leading theory regarding the mechanism of action of allergy shots is that the immune response to the allergen is gradually shifted from production of IgE antibodies to production of IgG antibodies against this antigen. How would this switch lead to a reduction in allergic symptoms on environmental exposure to higher doses of the antigen?

Respiratory System

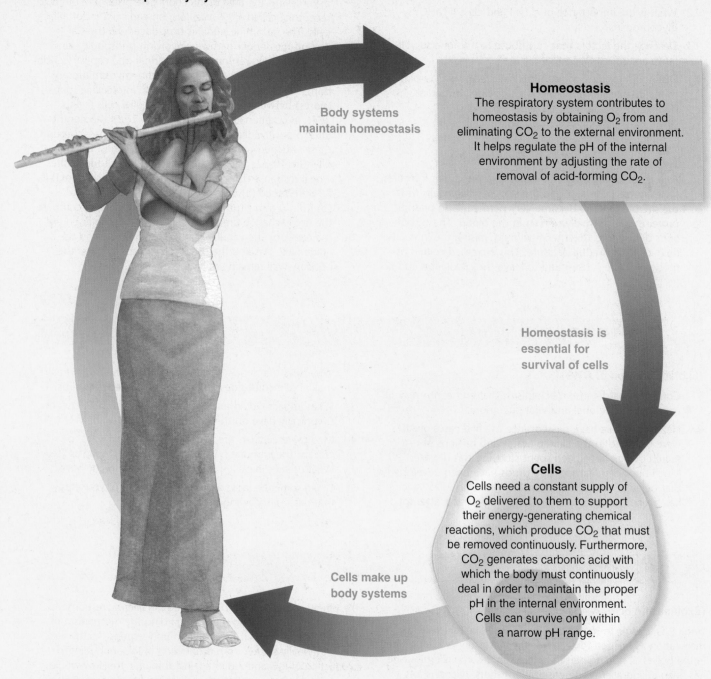

Body systems maintain homeostasis

Homeostasis
The respiratory system contributes to homeostasis by obtaining O_2 from and eliminating CO_2 to the external environment. It helps regulate the pH of the internal environment by adjusting the rate of removal of acid-forming CO_2.

Homeostasis is essential for survival of cells

Cells
Cells need a constant supply of O_2 delivered to them to support their energy-generating chemical reactions, which produce CO_2 that must be removed continuously. Furthermore, CO_2 generates carbonic acid with which the body must continuously deal in order to maintain the proper pH in the internal environment. Cells can survive only within a narrow pH range.

Cells make up body systems

Energy is essential for sustaining life-supporting cellular activities, such as protein synthesis and active transport across plasma membranes. The cells of the body need a continuous supply of O_2 to support their energy-generating chemical reactions. The CO_2 produced during these reactions must be eliminated from the body at the same rate as it is produced to prevent dangerous fluctuations in pH (that is, to maintain the acid–base balance), because CO_2 generates carbonic acid.

Respiration involves the sum of the processes that accomplish ongoing passive movement of O_2 from the atmosphere to the tissues to support cell metabolism, as well as the continual passive movement of metabolically produced CO_2 from the tissues to the atmosphere. The **respiratory system** contributes to homeostasis by exchanging O_2 and CO_2 between the atmosphere and blood. The blood transports O_2 and CO_2 between the respiratory system and tissues.

The Respiratory System

CENGAGENOW™ Log on to CengageNOW at **http://www .cengage.com/sso/** for an opportunity to explore a learning module that illustrates difficult concepts with self-study tutorials, animations, and interactive quizzes to help you learn, review, and master physiology concepts.

Respiratory Anatomy

The primary function of respiration is to obtain O_2 for use by the body cells and to eliminate the CO_2 the cells produce.

The respiratory system does not participate in all steps of respiration.

Most people think of respiration as the process of breathing in and breathing out. In physiology, however, respiration has a much broader meaning. Respiration encompasses two separate but related processes: cellular respiration and external respiration.

CELLULAR RESPIRATION The term **cellular respiration** refers to the intracellular metabolic processes carried out within the mitochondria, which use O_2 and produce CO_2 while deriving energy from nutrient molecules (see p. 33). The **respiratory quotient (RQ)**, the ratio of CO_2 produced to O_2 consumed, varies depending on the foodstuff consumed. When carbohydrate is being used, the RQ is 1; that is, for every molecule of O_2 consumed, one molecule of CO_2 is produced. For fat utilization, the RQ is 0.7; for protein, it is 0.8. On a typical American diet consisting of a mixture of these three nutrients, resting O_2 consumption averages about 250 ml/min, and CO_2 production averages about 200 ml/min, for an average RQ of 0.8:

$$RQ = \frac{CO_2 \text{ produced}}{O_2 \text{ consumed}} = \frac{(200 \text{ ml/min})}{(250 \text{ ml/min})} = 0.8$$

EXTERNAL RESPIRATION The term **external respiration** refers to the entire sequence of events in the exchange of O_2 and CO_2 between the external environment and the cells of the body. External respiration, the topic of this chapter, encompasses four steps (● Figure 13-1):

1. Air is alternately moved into and out of the lungs so that air can be exchanged between the atmosphere (external environment) and air sacs (*alveoli*) of the lungs. This exchange is accomplished by the mechanical act of **breathing**, or **ventilation.** The rate of ventilation is regulated to adjust the flow of air between the atmosphere and alveoli according to the body's metabolic needs for O_2 uptake and CO_2 removal.

2. Oxygen and CO_2 are exchanged between air in the alveoli and blood within the pulmonary (*pulmonary* means "lung") capillaries by the process of diffusion.

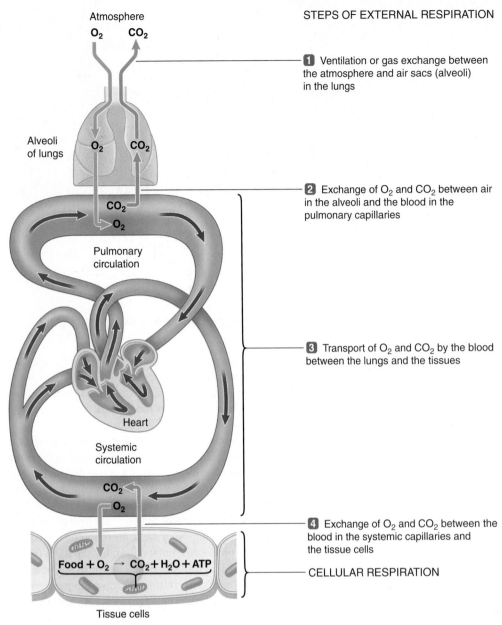

STEPS OF EXTERNAL RESPIRATION

1 Ventilation or gas exchange between the atmosphere and air sacs (alveoli) in the lungs

2 Exchange of O_2 and CO_2 between air in the alveoli and the blood in the pulmonary capillaries

3 Transport of O_2 and CO_2 by the blood between the lungs and the tissues

4 Exchange of O_2 and CO_2 between the blood in the systemic capillaries and the tissue cells

CELLULAR RESPIRATION

• **FIGURE 13-1 External and cellular respiration.** External respiration encompasses the steps involved in the exchange of O_2 and CO_2 between the external environment and tissue cells (steps 1 through 4). Cellular respiration encompasses the intracellular metabolic reactions involving the use of O_2 to derive energy (ATP) from food, producing CO_2 as a by-product.

3. The blood transports O_2 and CO_2 between the lungs and tissues.

4. Oxygen and CO_2 are exchanged between the tissue cells and blood by the process of diffusion across the systemic (tissue) capillaries.

The respiratory system does not accomplish all the steps of respiration; it is involved only with ventilation and the exchange of O_2 and CO_2 between the lungs and blood (steps **1** and **2**). The circulatory system carries out the remaining steps.

NONRESPIRATORY FUNCTIONS OF THE RESPIRATORY SYSTEM
The respiratory system also fills these nonrespiratory functions:

■ It is a route for water loss and heat elimination. Inspired (inhaled) atmospheric air is humidified and warmed by the respiratory airways before it is expired. Moistening of inspired air is essential to prevent the alveolar linings from drying out. Oxygen and CO_2 cannot diffuse through dry membranes.

■ It enhances venous return (see the "respiratory pump," p. 375).

■ It helps maintain normal acid–base balance by altering the amount of H^+-generating CO_2 exhaled (see p. 575).

■ It enables speech, singing, and other vocalization.

■ It defends against inhaled foreign matter (see p. 457).

■ It removes, modifies, activates, or inactivates various materials passing through the pulmonary circulation. All blood returning to the heart from the tissues must pass through the lungs before being returned to the systemic circulation. The lungs, therefore, are uniquely situated to act on specific materials that have been added to the blood at the tissue level before they have a chance to reach other parts of the body by means of the arterial system. For example, prostaglandins, a collection of chemical messengers released in many tissues to mediate particular local responses (see p. 758), may spill into the blood, but they are *inactivated* during passage through the lungs so that they cannot exert systemic effects. By contrast, the lungs *activate* angiotensin II, a hormone that plays an important role in regulating the concentration of Na^+ in the ECF (see p. 527).

■ The nose, a part of the respiratory system, serves as the organ of smell (see p. 230).

The respiratory airways conduct air between the atmosphere and alveoli.

The **respiratory system** includes the respiratory airways leading into the lungs, the lungs themselves, and the structures of the thorax (chest) involved in producing movement of air through the airways into and out of the lungs. The **respiratory**

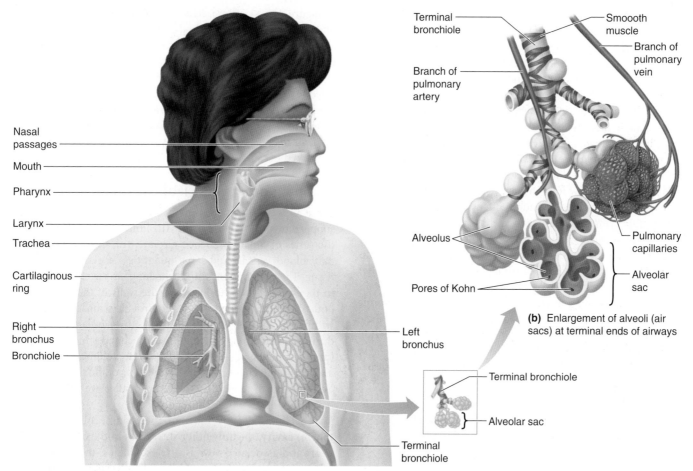

Nasal passages

Mouth

Pharynx

Larynx

Trachea

Cartilaginous ring

Right bronchus

Bronchiole

Terminal bronchiole

Branch of pulmonary artery

Terminal bronchiole

Smoooth muscle

Branch of pulmonary vein

Left bronchus

Alveolus

Pores of Kohn

Pulmonary capillaries

Alveolar sac

Terminal bronchiole

Alveolar sac

(b) Enlargement of alveoli (air sacs) at terminal ends of airways

(a) Respiratory airways

● **FIGURE 13-2 Anatomy of the respiratory system.** (a) The respiratory airways include the nasal passages, pharynx, larynx, trachea, bronchi, and bronchioles. (b) Most alveoli (air sacs) are clustered in grapelike arrangements at the end of the terminal bronchioles.

airways are tubes that carry air between the atmosphere and air sacs, the latter being the only site where gases can be exchanged between air and blood. The airways begin with the **nasal passages (nose)** (● Figure 13-2a). The nasal passages open into the **pharynx (throat),** which serves as a common passageway for both the respiratory and digestive systems. Two tubes lead from the pharynx—the **trachea (windpipe),** through which air is conducted to the lungs, and the **esophagus,** the tube through which food passes to the stomach. Air normally enters the pharynx through the nose, but it can enter by the mouth as well when the nasal passages are congested; that is, you can breathe through your mouth when you have a cold. Because the pharynx serves as a common passageway for food and air, reflex mechanisms close off the trachea during swallowing so that food enters the esophagus and not the airways. The esophagus stays closed except during swallowing to keep air from entering the stomach during breathing.

The **larynx,** or **voice box,** is located at the entrance of the trachea. The anterior protrusion of the larynx forms the "Adam's apple." The **vocal folds,** two bands of elastic tissue that lie across the opening of the larynx, can be stretched and positioned in different shapes by laryngeal muscles (● Figure 13-3a). Air

Vocal fold

Glottis

(a) Glottis open

(b) Glottis closed

Kay Pentax, a division of Pentax Medical

● **FIGURE 13-3 Vocal folds.** Photograph of the vocal folds as viewed from above at the laryngeal opening, showing the vocal folds (a) positioned apart when the glottis is open and (b) in tight apposition when the glottis is closed.

passes into the larynx through the space between the vocal folds. This laryngeal opening is known as the **glottis.** As air moves through the open glottis past the variably positioned, taut vocal folds, they vibrate to produce the many different sounds of speech. The lips, tongue, and soft palate modify the sounds into recognizable sound patterns. During swallowing, the vocal folds assume a function not related to speech: They close the glottis.

That is, laryngeal muscles bring the vocal folds into tight apposition to each other to close off the entrance to the trachea so that food does not get into the airways (● Figure 13-3b).

Beyond the larynx, the trachea divides into two main branches, the right and left **bronchi,** which enter the right and left lungs, respectively. Within each lung, the bronchus continues to branch into progressively narrower, shorter, and more numerous airways, much like the branching of a tree. The smaller branches are known as **bronchioles.** Clustered at the ends of the terminal bronchioles are the **alveoli,** the tiny air sacs where gases are exchanged between air and blood (see ● Figure 13-2b).

To permit airflow into and out of the gas-exchanging portions of the lungs, the continuum of conducting airways from the entrance through the terminal bronchioles to the alveoli must remain open. The trachea and larger bronchi are fairly rigid, nonmuscular tubes encircled by a series of cartilaginous rings that prevent these tubes from compressing. The smaller bronchioles have no cartilage to hold them open. Their walls contain smooth muscle that is innervated by the autonomic nervous system and is sensitive to certain hormones and local chemicals. These factors, by varying the degree of contraction of bronchiolar smooth muscle and hence the caliber of these small terminal airways, regulate the amount of air passing between the atmosphere and each cluster of alveoli.

The gas-exchanging alveoli are thin-walled, inflatable air sacs encircled by pulmonary capillaries.

The lungs are ideally structured for gas exchange. According to Fick's law of diffusion, the shorter the distance through which diffusion must take place, the greater the rate of diffusion is. Also, the greater the surface area across which diffusion can take place, the greater the rate of diffusion is (see p. 62).

The alveoli are clusters of thin-walled, inflatable, grapelike sacs at the terminal branches of the conducting airways. The alveolar walls consist of a single layer of flattened, fried-egg appearing **Type I alveolar cells** (● Figure 13-4a). Each alveolus is surrounded by a network of pulmonary capillaries, the walls of which are also only one cell thick (● Figure 13-4b). The interstitial space between an alveolus and the surrounding capillary network forms an extremely thin barrier, with only 0.5 μm separating air in the alveoli from blood in the pulmonary capillaries. (A sheet of tracing paper is about 50 times thicker than this air-to-blood barrier.) The thinness of this barrier facilitates gas exchange.

Furthermore, the alveolar air–blood interface presents a tremendous surface area for exchange. The lungs contain about 500 million alveoli, each about 300 μm in diameter. So dense are the pulmonary capillary networks that each alveolus is encircled by an almost continuous sheet of blood (● Figure 13-4c). The total surface area thus exposed between alveolar air and pulmonary capillary blood is about 75 m² (about the size of a tennis court). In contrast, if the lungs consisted of a single hollow chamber of the same dimensions instead of being divided into a myriad alveolar units, the total surface area would be only about 0.01 m².

In addition to the thin, wall-forming Type I cells, 5% of the alveolar surface epithelium is covered by **Type II alveolar cells**

(● Figure 13-4a). These cells secrete *pulmonary surfactant,* a phospholipoprotein complex that facilitates lung expansion (described later). Furthermore, defensive alveolar macrophages stand guard within the lumen of the air sacs (see p. 457).

Minute **pores of Kohn** exist in the walls between adjacent alveoli (see ● Figure 13-2b). Their presence permits airflow between adjoining alveoli, a process known as **collateral ventilation.** These passageways are especially important in allowing fresh air to enter an alveolus whose terminal conducting airway is blocked because of disease.

The lungs occupy much of the thoracic cavity.

There are two **lungs,** each divided into several lobes and each supplied by one of the bronchi. The lung tissue itself consists of the series of highly branched airways, the alveoli, the pulmonary blood vessels, and large quantities of elastic connective tissue. The only muscle within the lungs is the smooth muscle in the walls of the arterioles and the walls of the bronchioles, both of which are subject to control. No muscle is present within the alveolar walls to cause them to inflate and deflate during the breathing process. Instead, changes in lung volume (and accompanying changes in alveolar volume) are brought about through changes in the dimensions of the thoracic cavity. You will learn about this mechanism after we complete our discussion of respiratory anatomy.

The lungs occupy most of the volume of the **thoracic (chest) cavity,** the only other structures in the chest being the heart and associated vessels, the esophagus, the thymus, and some nerves. The outer chest wall (**thorax**) is formed by 12 pairs of curved **ribs,** which join the **sternum** (breastbone) anteriorly and the **thoracic vertebrae** (backbone) posteriorly. The rib cage provides bony protection for the lungs and heart. The **diaphragm,** which forms the floor of the thoracic cavity, is a large, dome-shaped sheet of skeletal muscle that completely separates the thoracic cavity from the abdominal cavity. It is penetrated only by the esophagus and blood vessels traversing the thoracic and abdominal cavities. At the neck, muscles and connective tissue enclose the thoracic cavity. The only communication between the thorax and the atmosphere is through the respiratory airways into the alveoli. Like the lungs, the chest wall contains considerable amounts of elastic connective tissue.

A pleural sac separates each lung from the thoracic wall.

A double-walled, closed sac called the **pleural sac** separates each lung from the thoracic wall and other surrounding structures (● Figure 13-5). The interior of the pleural sac is known as the **pleural cavity.** In the illustration, the dimensions of the pleural cavity are greatly exaggerated to aid visualization; in reality the layers of the pleural sac are in close contact with one another. The surfaces of the pleura secrete a thin **intrapleural fluid** (*intra* means "within"), which lubricates the pleural surfaces as they slide past each other during respiratory movements.

 Pleurisy, an inflammation of the pleural sac, is accompanied by painful breathing, because each inflation and each deflation of the lungs cause a "friction rub."

Alveolus surrounded by pulmonary capillaries

Pulmonary capillary networks

Alveolus with capillary cut away

Dr. Richard Kessel and Dr. Randy Kardon, Tissues and Organs, Visuals Unlimited, Inc.

(b) Scanning electron micrograph of alveoli and surrounding pulmonary capillaries

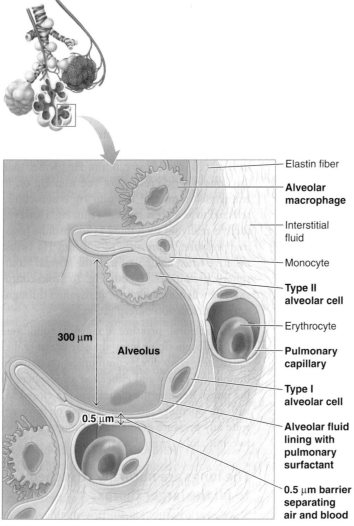

Elastin fiber

Alveolar macrophage

Interstitial fluid

Monocyte

Type II alveolar cell

Erythrocyte

Pulmonary capillary

Type I alveolar cell

Alveolar fluid lining with pulmonary surfactant

0.5 μm barrier separating air and blood

300 μm

Alveolus

0.5 μm

(a) Alveolus and surrounding pulmonary capillaries

Pulmonary capillary

Alveolus

Erythrocyte

© Don W. Fawcett

(c) Transmission electron micrograph of several alveoli and surrounding pulmonary capillaries

● **FIGURE 13-4 Alveolus and associated pulmonary capillaries.** (a) A single layer of flat-tened Type I alveolar cells forms the alveolar walls. Type II alveolar cells embedded within the alveolar wall secrete pulmonary surfactant. Wandering alveolar macrophages are found within the alveolar lumen. The size of the cells and respiratory membrane is exaggerated compared to the size of the alveolar and pulmonary capillary lumens. The diameter of an alveolus is actually about 600 times larger (300 μm) than the intervening space between air and blood (0.5 μm). (b) Each alveolus is encircled with a dense network of pulmonary capillaries. (c) Note that each alveolus is surrounded by an almost continuous sheet of blood.

Respiratory Mechanics

Air tends to move from a region of higher pressure to a region of lower pressure, that is, down a **pressure gradient.**

Interrelationships among pressures inside and outside the lungs are important in ventilation.

Air flows into and out of the lungs during the act of breathing by moving down alternately reversing pressure gradients established between the alveoli and atmosphere by cyclic respiratory muscle activity. Three different pressure considerations are important in ventilation (● Figure 13-6):

1. Atmospheric (barometric) pressure is the pressure exerted by the weight of the air in the atmosphere on objects on Earth's surface. At sea level it equals 760 mm Hg (● Figure 13-7). Atmospheric pressure diminishes with increasing altitude above sea level as the layer of air above Earth's surface correspondingly decreases in thickness. Minor fluctuations in atmospheric pressure occur at any height because of changing weather conditions (that is, when barometric pressure is rising or falling).

2. Intra-alveolar pressure, also known as **intrapulmonary pressure,** is the pressure within the alveoli. Because the alveoli communicate with the atmosphere through the conducting airways, air quickly flows down its pressure gradient any time intra-alveolar pressure differs from atmospheric pressure; air-

(a) Traumatic pneumothorax

Puncture wound in chest wall

(b) Collapsed lung

Hole in lung

(c) Spontaneous pneumothorax

Numbers are mm Hg pressure.

● **FIGURE 13-9 Pneumothorax.** (a) In *traumatic pneumothorax,* a puncture in the chest wall permits air from the atmosphere to flow down its pressure gradient and enter the pleural cavity, abolishing the transmural pressure gradient. (b) When the transmural pressure gradient is abolished, the lung collapses to its unstretched size, and the chest wall springs outward. (c) In *spontaneous pneumothorax,* a hole in the lung wall permits air to move down its pressure gradient and enter the pleural cavity from the lungs, abolishing the transmural pressure gradient. As with traumatic pneumothorax, the lung collapses to its unstretched size.

REASON WHY THE INTRAPLEURAL PRESSURE IS SUBATMOSPHERIC Because neither the thoracic wall nor the lungs are in their natural position when they are held in apposition to each other, they constantly try to assume their own inherent dimensions. The stretched lungs have a tendency to pull inward away from the thoracic wall, whereas the compressed thoracic wall tends to move outward away from the lungs. The transmural pressure gradient and intrapleural fluid's cohesiveness, however, prevent these structures from pulling away from each other except to the slightest degree. Even so, the resultant ever-so-slight expansion of the pleural cavity is sufficient to drop the pressure in this cavity by 4 mm Hg, bringing the intrapleural pressure to the subatmospheric level of 756 mm Hg. This pressure drop occurs because the pleural cavity is filled with fluid, which cannot expand to fill the slightly larger volume. Therefore, a vacuum exists in the infinitesimal space in the slightly expanded pleural cavity not occupied by intrapleural fluid, producing a small drop in intrapleural pressure below atmospheric pressure.

Note the interrelationship between the transmural pressure gradient and the subatmospheric intrapleural pressure. The lungs are stretched and the thorax is compressed by a transmural pressure gradient that exists across their walls because of the presence of a subatmospheric intrapleural pressure. The intrapleural pressure, in turn, is subatmospheric because the stretched lungs and compressed thorax tend to pull away from each other, slightly expanding the pleural cavity and dropping the intrapleural pressure below atmospheric pressure.

PNEUMOTHORAX Normally, air does not enter the pleural cavity, because there is no communication between the cavity and either the atmosphere or the alveoli. However, if the chest wall is punctured (for example, by a stab wound or a broken rib), air flows down its pressure gradient from the higher atmospheric pressure and rushes into the pleural space (● Figure 13-9a). The abnormal condition of air entering the pleural cavity is known as **pneumothorax** ("air in the chest"). Intrapleural and intra-alveolar pressure are now both equilibrated with atmospheric pressure, so a transmural pressure gradient no longer exists across either the lung wall or the chest wall. With no force present to stretch the lung, it collapses to its unstretched size (● Figure 13-9b). (The intrapleural fluid's cohesiveness cannot hold the lungs and thoracic wall in apposition in the absence of the transmural pressure gradient.) The thoracic wall likewise springs outward to its unrestricted dimensions, but this has much less serious consequences than collapse of the lung. Similarly, pneumothorax and lung collapse can occur if air enters the pleural cavity through a hole in the lung produced, for example, by a disease process (● Figure 13-9c).

Flow of air into and out of the lungs occurs because of cyclic changes in intra-alveolar pressure.

Because air flows down a pressure gradient, the intra-alveolar pressure must be less than atmospheric pressure for air to flow into the lungs during inspiration. Similarly, the intra-alveolar pressure must be greater than atmospheric pressure for air to flow out of the lungs during expiration. Intra-alveolar pressure can be changed by altering the volume of the lungs, in accordance with Boyle's law. **Boyle's law** states that at any constant temperature, the pressure exerted by a gas varies inversely with the volume of the gas (● Figure 13-10); that is, as the volume of a gas increases, the pressure exerted by the gas decreases proportionately. Conversely, the pressure increases proportionately as the volume decreases. Changes in lung volume, and accordingly intra-alveolar pressure, are brought about indirectly by respiratory muscle activity.

The respiratory muscles that accomplish breathing do not act directly on the lungs to change their volume. Instead, these muscles change the volume of the thoracic cavity, causing a corresponding change in lung volume because the thoracic wall and lungs are linked by the intrapleural fluid's cohesiveness and the transmural pressure gradient.

Let us follow the changes that occur during one respiratory cycle—that is, one breath in (**inspiration**) and out (**expiration**).

ONSET OF INSPIRATION: CONTRACTION OF INSPIRATORY MUSCLES The major **inspiratory muscles**—the muscles that contract to accomplish an inspiration during quiet breathing—include the *diaphragm* and *external intercostal muscles* (● Figure 13-11). Before the beginning of inspiration, the respiratory muscles are relaxed, no air is flowing, and intra-alveolar pressure is equal to atmospheric pressure (● Figure 13-12a). At the onset of inspiration, these muscles are stimulated to contract, enlarging the thoracic cavity. The major inspiratory muscle is the **diaphragm,** a sheet of skeletal muscle that forms the floor of the thoracic cavity and is innervated by the **phrenic nerve.** The relaxed diaphragm has a dome shape that protrudes upward into the thoracic cavity. When the diaphragm contracts (on stimulation by the phrenic nerve), it descends downward, enlarging the volume of the thoracic cavity by increasing its vertical (top-to-bottom) dimension (● Figure 13-12b). During quiet breathing, the diaphragm descends about 1 cm during inspiration, but during heavy breathing, it may descend as much as 10 cm. The abdominal wall, if relaxed, bulges outward during inspiration as the descending diaphragm pushes the abdominal contents downward and forward. Seventy-five percent of the enlargement of the thoracic cavity during quiet inspiration is accomplished by contraction of the diaphragm.

Two sets of **intercostal muscles** lie between the ribs (*inter* means "between"; *costa* means "rib"). The *external intercostal muscles* lie on top of the *internal intercostal muscles*. Contraction of the **external intercostal muscles,** whose fibers run downward and forward between adjacent ribs, enlarges the thoracic cavity in both the lateral (side-to-side) and anteroposterior (front-to-back) dimensions. When the external intercostals contract, they elevate the ribs and subsequently the sternum upward and outward (● Figure 13-12b). **Intercostal nerves** activate these intercostal muscles.

Before inspiration, at the end of the preceding expiration, intra-alveolar pressure is equal to atmospheric pressure, so no air is flowing into or out of the lungs (● Figure 13-13a). As the thoracic cavity enlarges, the lungs are also forced to expand to fill the

Container A   Container B   Container C

Closed container with given number of gas molecules

Piston

Pressure gauge

Volume = 1/2
Pressure = 2

Volume = 1
Pressure = 1

Volume = 2
Pressure = 1/2

● **FIGURE 13-10 Boyle's law.** Each container has the same number of gas molecules. Given the random motion of gas molecules, the likelihood of a gas molecule striking the interior wall of the container and exerting pressure varies inversely with the volume of the container at any constant temperature. The gas in container B exerts more pressure than the same gas in larger container C but less pressure than the same gas in smaller container A. This relationship is stated as Boyle's law: $P_1V_1 = P_2V_2$. As the volume of a gas increases, the pressure of the gas decreases proportionately; conversely, the pressure increases proportionately as the volume decreases.

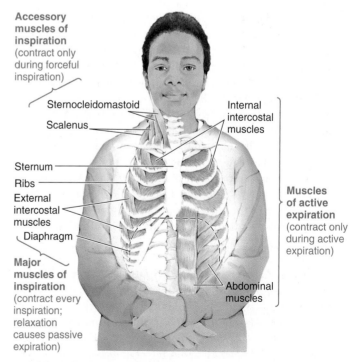

Accessory muscles of inspiration (contract only during forceful inspiration)

Sternocleidomastoid

Scalenus

Sternum

Ribs

External intercostal muscles

Diaphragm

Major muscles of inspiration (contract every inspiration; relaxation causes passive expiration)

Internal intercostal muscles

Muscles of active expiration (contract only during active expiration)

Abdominal muscles

● **FIGURE 13-11 Anatomy of the respiratory muscles.**

External
intercostal
muscles
(relaxed)

Elevated
rib cage

Contraction
of external
intercostal
muscles

Sternum

Diaphragm
(relaxed)

Elevation of ribs causes sternum
to move upward and outward, which
increases front-to-back dimension
of thoracic cavity

Contraction of
diaphragm

Contraction of external intercostal
muscles causes elevation of ribs,
which increases side-to-side
dimension of thoracic cavity

Lowering of diaphragm on
contraction increases vertical
dimension of thoracic cavity

(a) Before inspiration

(b) Inspiration

● **FIGURE 13-12 Respiratory muscle activity during inspiration and expiration.** (a) Before inspiration, all respiratory muscles are relaxed. (b) During *inspiration,* the diaphragm descends on contraction, increasing the vertical dimension of the thoracic cavity. Contraction of the external intercostal muscles elevates the ribs and subsequently the sternum to enlarge the thoracic cavity from front to back and from side to side.

larger thoracic cavity. As the lungs enlarge, the intra-alveolar pressure drops because the same number of air molecules now occupy a larger lung volume. In a typical inspiratory excursion, the intra-alveolar pressure drops 1 mm Hg to 759 mm Hg (● Figure 13-13b). Because the intra-alveolar pressure is now less than atmospheric pressure, air flows into the lungs down the pressure gradient from higher to lower pressure. Air continues to enter the lungs until no further gradient exists—that is, until intra-alveolar pressure equals atmospheric pressure. Thus, lung expansion is not caused by movement of air into the lungs; instead, air flows into the lungs because of the fall in intra-alveolar pressure brought about by lung expansion.

During inspiration, the intrapleural pressure falls to 754 mm Hg as a result of expansion of the thorax. The resultant increase in the transmural pressure gradient during inspiration ensures that the lungs are stretched to fill the expanded thoracic cavity.

ROLE OF ACCESSORY INSPIRATORY MUSCLES Deeper inspirations (more air breathed in) can be accomplished by contracting the diaphragm and external intercostal muscles more forcefully and by bringing the **accessory inspiratory muscles** into play to further enlarge the thoracic cavity. Contracting these accessory muscles, which are in the neck (see ● Figure 13-11), raises the sternum and elevates the first two ribs, enlarging the

upper portion of the thoracic cavity. As the thoracic cavity increases even further in volume than under resting conditions, the lungs likewise expand more, dropping the intra-alveolar pressure further. Consequently, a larger inward flow of air occurs before equilibration with atmospheric pressure is achieved; that is, a deeper breath occurs.

ONSET OF EXPIRATION: RELAXATION OF INSPIRATORY MUSCLES At the end of inspiration, the inspiratory muscles relax. The diaphragm assumes its original dome-shaped position when it relaxes. The elevated rib cage falls because of gravity when the external intercostals relax (see ● Figure 13-12c). With no forces causing expansion of the chest wall (and accordingly, expansion of the lungs), the chest wall and stretched lungs recoil to their preinspiratory size because of their elastic properties, much as a stretched balloon would on release. As the lungs recoil and become smaller in volume, the intra-alveolar pressure rises, because the greater number of air molecules contained within the larger lung volume at the end of inspiration are now compressed into a smaller volume. In a resting expiration, the intra-alveolar pressure increases about 1 mm Hg above atmospheric level to 761 mm Hg (● Figure 13-13c). Air now leaves the lungs down its pressure gradient from higher intra-alveolar pressure to lower atmospheric pressure. Outward flow of air ceases when intra-alveolar pressure

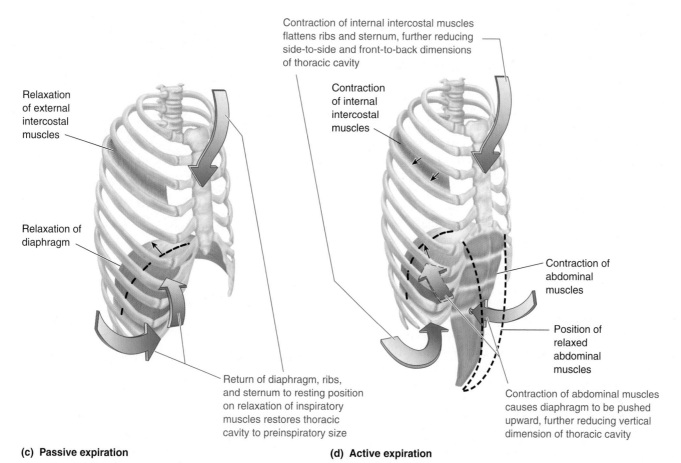

Relaxation of external intercostal muscles

Relaxation of diaphragm

Contraction of internal intercostal muscles flattens ribs and sternum, further reducing side-to-side and front-to-back dimensions of thoracic cavity

Contraction of internal intercostal muscles

Contraction of abdominal muscles

Position of relaxed abdominal muscles

Return of diaphragm, ribs, and sternum to resting position on relaxation of inspiratory muscles restores thoracic cavity to preinspiratory size

Contraction of abdominal muscles causes diaphragm to be pushed upward, further reducing vertical dimension of thoracic cavity

(c) Passive expiration

(d) Active expiration

● **FIGURE 13-12** *(continued)* (c) During *quiet passive expiration,* the diaphragm relaxes, reducing the volume of the thoracic cavity from its peak inspiratory size. As the external intercostal muscles relax, the elevated rib cage falls because of the force of gravity. This also reduces the volume of the thoracic cavity. (d) During *active expiration,* contraction of the abdominal muscles increases the intra-abdominal pressure, exerting an upward force on the diaphragm. This reduces the vertical dimension of the thoracic cavity further than it is reduced during quiet passive expiration. Contraction of the internal intercostal muscles decreases the front-to-back and side-to-side dimensions by flattening the ribs and sternum.

Equilibrated; no net movement of air

760

760

756

Preinspiratory size of thorax

Preinspiratory size of lungs

(a) Before inspiration

760

759

754

Size of thorax on contraction of inspiratory muscles

Size of lungs as they are stretched to fill the expanded thorax

(b) During inspiration

760

761

756

Size of thorax on relaxation of inspiratory muscles

Size of lungs as they recoil

(c) During expiration

Numbers are mm Hg pressure.

● **FIGURE 13-13 Changes in lung volume and intra-alveolar pressure during inspiration and expiration.** (a) Before inspiration, at the end of the preceding expiration, intra-alveolar pressure is equilibrated with atmospheric pressure, and no air is flowing. (b) As the lungs increase in volume during inspiration, the intra-alveolar pressure decreases, establishing a pressure gradient that favors the flow of air into the alveoli from the atmosphere; that is, an inspiration occurs. (c) As the lungs recoil to their preinspiratory size on relaxation of the inspiratory muscles, the intra-alveolar pressure increases, establishing a pressure gradient that favors the flow of air out of the alveoli into the atmosphere; that is, an expiration occurs.

becomes equal to atmospheric pressure and a pressure gradient no longer exists. ● Figure 13-14 summarizes the intra-alveolar and intrapleural pressure changes that take place during one respiratory cycle.

Clinical Note Because the diaphragm is the major inspiratory muscle and its relaxation also causes expiration, paralysis of the intercostal muscles alone does not seriously influence quiet breathing. Disruption of diaphragm activity caused by nerve or muscle disorders, however, leads to respiratory paralysis. Fortunately, the phrenic nerve arises from the spinal cord in the neck region (cervical segments 3, 4, and 5) and then descends to the diaphragm at the floor of the thorax, instead of arising from the thoracic region of the cord as might be expected. For this reason, individuals completely paralyzed below the neck by traumatic severance of the spinal cord are still able to breathe, even though they have lost use of all other skeletal muscles in their trunk and limbs.

FORCED EXPIRATION: CONTRACTION OF EXPIRATORY MUSCLES
During quiet breathing, expiration is normally a passive process because it is accomplished by elastic recoil of the lungs on relaxation of the inspiratory muscles, with no muscular exertion or energy expenditure required. In contrast, inspiration is *always* active, because it is brought about only by contraction of inspiratory muscles at the expense of energy use. Expiration does become active to empty the lungs more completely and more rapidly than is accomplished during quiet breathing, as during the deeper breaths accompanying exercise. To force more air out, the intra-alveolar pressure must be increased even further above atmospheric pressure than can be accomplished by simple relaxation of the inspiratory muscles and elastic recoil of the lungs. To produce such a **forced,** or **active, expiration,** expiratory muscles must contract to further reduce the volume of the thoracic cavity and lungs (see ● Figures 13-11 and 13-12d). The most important **expiratory muscles** are (unbelievable as it may seem at first) the *muscles of the abdominal wall.* As the abdominal muscles contract, the resultant increase in intra-abdominal pressure exerts an upward force on the diaphragm, pushing it farther up into the thoracic cavity than its relaxed position, thus decreasing the vertical dimension of the thoracic cavity even more. The other expiratory muscles are the **internal intercostal muscles,** whose contraction pulls the ribs downward and inward, flattening the chest wall and further decreasing the size of the thoracic cavity; this action is just the opposite of that of the external intercostal muscles.

As active contraction of the expiratory muscles further reduces the volume of the thoracic cavity, the lungs also become further reduced in volume because they do not have to be stretched as much to fill the smaller thoracic cavity; that is, they are permitted to recoil to an even smaller volume. The intra-alveolar pressure increases further as the air in the lungs is confined within this smaller volume. The differential between intra-alveolar and atmospheric pressure is even greater now than during passive expiration, so more air leaves down the pressure gradient before equilibration is achieved. In this way, the lungs are emptied more completely during forceful, active expiration than during quiet, passive expiration.

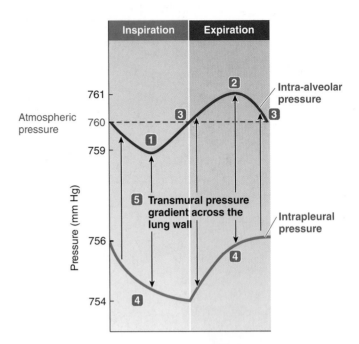

1. During inspiration, intra-alveolar pressure is less than atmospheric pressure.

2. During expiration, intra-alveolar pressure is greater than atmospheric pressure.

3. At the end of both inspiration and expiration, intra-alveolar pressure is equal to atmospheric pressure because the alveoli are in direct communication with the atmosphere, and air continues to flow down its pressure gradient until the two pressures equilibrate.

4. Throughout the respiratory cycle, intrapleural pressure is less than intra-alveolar pressure.

5. Thus, a transmural pressure gradient always exists, and the lung is always stretched to some degree, even during expiration.

● **FIGURE 13-14 Intra-alveolar and intrapleural pressure changes throughout the respiratory cycle.**

During forceful expiration, the intrapleural pressure exceeds atmospheric pressure, but the lungs do not collapse. Because the intra-alveolar pressure is also increased correspondingly, a transmural pressure gradient still exists across the walls of the lungs, keeping them stretched to fill the thoracic cavity. For example, if the pressure within the thorax increases 10 mm Hg, the intrapleural pressure becomes 766 mm Hg and the intra-alveolar pressure becomes 770 mm Hg—still a 4 mm Hg pressure difference.

Airway resistance influences airflow rates.

Thus far we have discussed airflow in and out of the lungs as a function of the magnitude of the pressure gradient between the alveoli and the atmosphere. However, just as flow of blood through the blood vessels depends not only on the pressure gradient but also on the resistance to the flow offered by the vessels, so it is with airflow:

$$F = \frac{\Delta P}{R}$$

TABLE 13-1 — Factors Affecting Airway Resistance

Status of Airways	Effect on Resistance	Factors Producing the Effect
Bronchoconstriction	↓ radius, ↑ resistance to airflow	*Pathological factors:*

		Allergy-induced spasm of the airways caused by
		    Slow-reactive substance of anaphylaxis (leukotrienes)
		    Histamine

		Physical blockage of the airways caused by
		    Excess mucus
		    Edema of the walls
		    Airway collapse

		Physiological control factors:
		Neural control: parasympathetic stimulation
		Local chemical control: ↓ CO_2 concentration
Bronchodilation	↑ radius, ↓ resistance to airflow	*Pathological factors:* none
		Physiological control factors:
		Neural control: sympathetic stimulation (minimal effect)
		Hormonal control: epinephrine
		Local chemical control: ↑ CO_2 concentration

where

F = airflow rate

ΔP = difference between atmospheric and intra-alveolar pressure (pressure gradient)

R = resistance of airways, determined by their radius

The primary determinant of resistance to airflow is the radius of the conducting airways. We ignored airway resistance in our preceding discussion of pressure gradient–induced airflow rates because in a healthy respiratory system, the radius of the conducting system is large enough that resistance remains extremely low. Therefore, the pressure gradient between the alveoli and the atmosphere is usually the primary factor determining the airflow rate. Indeed, the airways normally offer such low resistance that only very small pressure gradients of 1 to 2 mm Hg need be created to achieve adequate rates of airflow into and out of the lungs. (By comparison, it would take a pressure gradient 250 times greater to move air through a smoker's pipe than through the respiratory airways at the same flow rate.)

Normally, modest adjustments in airway size can be accomplished by autonomic nervous system regulation to suit the body's needs. Parasympathetic stimulation, which occurs in quiet, relaxed situations when the demand for airflow is low, promotes bronchiolar smooth muscle contraction, which increases airway resistance by producing **bronchoconstriction** (a decrease in the radius of bronchioles). In contrast, sympathetic stimulation and to a greater extent its associated hormone, epinephrine, bring about **bronchodilation** (an increase in bronchiolar radius) and decreased airway resistance by promoting bronchiolar smooth muscle relaxation (▲ Table 13-1). Thus, during periods of sympathetic domination, when increased demands for O_2 uptake are actually or potentially placed on the body, bronchodilation ensures that the pressure gradients established by respiratory muscle activity can achieve maximum airflow rates with minimum resistance. Because of this bronchodilator action, epinephrine or related drugs (such as albuterol, which is a more selective β_2 adrenergic agonist that relaxes bronchiolar smooth muscle without affecting β_1 adrenergic targets like the heart; see p. 244) are useful therapeutic tools to counteract airway constriction in patients with bronchial spasms.

Resistance becomes an extremely important impediment to airflow when airway lumens become abnormally narrowed by disease. We have all transiently experienced the effect that increased airway resistance has on breathing when we have a cold. We know how difficult it is to produce an adequate airflow rate through a "stuffy nose" when the nasal passages are narrowed by swelling and mucus accumulation. More serious is chronic obstructive pulmonary disease, to which we now turn attention.

Airway resistance is abnormally increased with chronic obstructive pulmonary disease.

Chronic obstructive pulmonary disease (COPD) is a group of lung diseases characterized by increased airway resistance resulting from narrowing of the lumen of the lower airways. When airway resistance increases, a larger pressure gradient must be established to maintain even a normal airflow rate. For example, if resistance is doubled by narrowing of airway lumens, ΔP must be doubled through increased respiratory muscle exertion to induce the same flow rate of air in and out of the lungs as a healthy person accomplishes during quiet breathing. Accordingly, patients with COPD must work harder to breathe.

Chronic obstructive pulmonary disease encompasses three chronic (long-term) diseases: *chronic bronchitis, asthma,* and *emphysema.*

CHRONIC BRONCHITIS Chronic bronchitis is a long-term inflammatory condition of the lower respiratory airways, generally triggered by frequent exposure to irritating cigarette smoke, polluted air, or allergens. In response to the chronic irritation, the airways become narrowed by prolonged edematous thickening of the airway linings, coupled with overproduction of thick mucus. Despite frequent coughing associated with the chronic irritation, the plugged mucus often cannot be satisfactorily removed, especially because the irritants immobilize the ciliary mucus escalator (see p. 457). Pulmonary bacterial infections frequently occur because the accumulated mucus serves as an excellent medium for bacterial growth.

ASTHMA In asthma, airway obstruction is the result of (1) thickening of airway walls, brought about by inflammation and histamine-induced edema (see p. 420); (2) plugging of the airways by excessive secretion of very thick mucus; and (3) airway hyperresponsiveness, characterized by profound constriction of the smaller airways caused by trigger-induced spasm of the smooth muscle in the walls of these airways (see p. 453). Triggers that lead to these inflammatory changes and/or the exaggerated bronchoconstrictor response include repeated exposure to allergens (such as dust mites or pollen), irritants (as in cigarette smoke), respiratory infections, and vigorous exercise. A growing number of studies suggest that long-term infections with *Chlamydia pneumoniae,* a common cause of lung infections, may underlie up to half of the adult cases of asthma. In severe asthmatic attacks, pronounced clogging and narrowing of the airways can cut off all airflow, leading to death. An estimated 15 million people in the United States have asthma, with the number steadily climbing. Asthma is the most common chronic childhood disease. Scientists are unsure why asthma incidence is rising.

EMPHYSEMA Emphysema is characterized by (1) collapse of the smaller airways and (2) breakdown of alveolar walls. This irreversible condition can arise in two different ways. Most commonly, emphysema results from excessive release of protein-digesting enzymes such as *trypsin* from alveolar macrophages as a defense mechanism in response to chronic exposure to inhaled cigarette smoke or other irritants. The lungs are normally protected from damage by these enzymes by α_1-*antitrypsin,* a protein that inhibits trypsin. Excessive secretion of these destructive enzymes in response to chronic irritation, however, can overwhelm the protective capability of α_1-antitrypsin so that these enzymes destroy not only foreign materials but lung tissue as well. Loss of lung tissue leads to the breakdown of alveolar walls and collapse of small airways, which characterize emphysema.

Less frequently, emphysema arises from a genetic inability to produce α_1-antitrypsin so that the lung tissue has no protection from trypsin. The unprotected lung tissue gradually disintegrates under the influence of even small amounts of macrophage-released enzymes, in the absence of chronic exposure to inhaled irritants.

DIFFICULTY IN EXPIRING When COPD of any type increases airway resistance, expiration is more difficult than inspiration. The smaller airways, lacking the cartilaginous rings that hold the larger airways open, are held open by the same transmural pressure gradient that distends the alveoli. Expansion of the thoracic cavity during inspiration indirectly dilates the airways even further than their expiratory dimensions, like alveolar expansion, so airway resistance is lower during inspiration than during expiration. In a healthy individual, the airway resistance is always so low that the slight variation between inspiration and expiration is not noticeable. When airway resistance has substantially increased, however, as during an asthmatic attack, the difference is quite noticeable. Thus, a person with asthma has more difficulty expiring than inspiring, giving rise to the characteristic "wheeze" as air is forced out through the narrowed airways.

Normally the smaller airways stay open during quiet breathing and even during active expiration when intrapleural pressure is elevated, as during exercise (● Figure 13-15a and b). In people without chronic obstructive pulmonary disease, the smaller airways collapse and further outflow of air is halted only at very low lung volumes during maximal forced expiration (● Figure 13-15c). Because of this airway collapse, the lungs can never be emptied completely. By contrast, in people who have COPD, especially emphysema, the smaller airways may routinely collapse during expiration, preventing further outflow of air through these passageways (● Figure 13-15d). Because of the extra air trapped behind these collapsed airways, people with emphysema have enlarged lungs and tend to appear barrel-chested. Compared to normal, those with pronounced emphysema suffer an ironic twist of having more air in the lungs but less gas exchange (because of the loss of surface area resulting from the breakdown of alveolar walls) and decreased ability to bring in fresh air (because the chest wall cannot expand as much from its larger-than-normal resting position to reduce the intra-alveolar pressure enough to accomplish adequate inspiration).

Elastic behavior of the lungs is a result of elastic connective tissue and alveolar surface tension.

During the respiratory cycle, the lungs alternately expand during inspiration and recoil during expiration. What properties of the lungs enable them to behave like balloons, being stretchable and then snapping back to their resting position when the stretching forces are removed? Two interrelated concepts are involved in pulmonary elasticity: *compliance* and *elastic recoil.*

The term **compliance** refers to how much effort is required to stretch or distend the lungs; it is analogous to how hard you have to work to blow up a balloon. (By comparison, 100 times

● **FIGURE 13-15 Airway collapse during forced expiration.** (a) During normal quiet breathing, airway resistance is low, so there is little frictional loss of pressure within the airways. Intrapleural pressure remains less than airway pressure throughout the length of the airways, so the airways remain open. (b) Even though intrapleural pressure is elevated during the active expiration accompanying routine vigorous activity, intra-alveolar pressure is also elevated and airway resistance is still low, so the friction-induced drop in airway pressure does not fall below the elevated intrapleural pressure until the level at which the airways are held open by cartilaginous rings. Therefore, airway collapse does not occur. (c) During maximal forced expiration, both intra-alveolar and intrapleural pressures are markedly increased. When frictional losses cause the airway pressure to fall below the surrounding elevated intrapleural pressure, the small nonrigid airways are compressed closed, blocking further expiration of air through the airway. In healthy individuals, this dynamic compression of airways occurs only at very low lung volumes. (d) In obstructive lung disease, premature airway collapse occurs for two reasons: (1) the pressure drop along the airways is magnified as a result of increased airway resistance, and (2) the intrapleural pressure is higher than normal because of the loss, as in emphysema, of lung tissue that is responsible for the lung's tendency to recoil and pull away from the thoracic wall. Excessive air trapped in the alveoli behind the compressed bronchiolar segments reduces the amount of gas exchanged between the alveoli and the atmosphere. Therefore, less alveolar air is "freshened" with each breath when airways collapse at higher lung volumes in patients with obstructive lung disease.

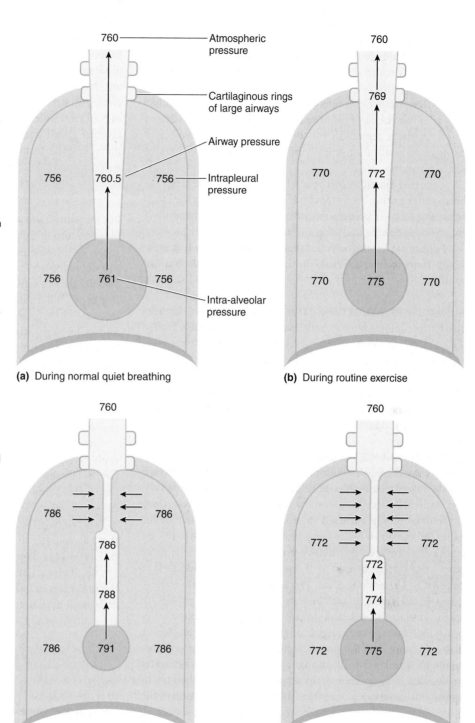

(a) During normal quiet breathing

(b) During routine exercise

(c) During maximal forced expiration

(d) In obstructive lung disease

Numbers are mm Hg pressure.

more distending pressure is required to inflate a child's toy balloon than to inflate the lungs.) Specifically, compliance is a measure of how much change in lung volume results from a given change in the transmural pressure gradient, the force that stretches the lungs. A highly compliant lung stretches farther for a given increase in the pressure difference than a less compliant lung does. Stated another way, the lower the compliance of the lungs, the larger is the transmural pressure gradient that must be created during inspiration to produce normal lung expansion. In turn, a greater-than-normal transmural pressure gradient during inspiration can be achieved only by making the intrapleural pressure more subatmospheric than usual. This is accomplished by greater expansion of the thorax through more vigorous contraction of the inspiratory muscles.

Therefore, the less compliant the lungs are, the more work is required to produce a given degree of inflation. A poorly compliant lung is referred to as a "stiff" lung because it lacks normal stretchability.

 Respiratory compliance can be decreased by a number of factors, as in **pulmonary fibrosis,** where normal lung tissue is replaced with scar-forming fibrous connective tissue as a result of chronically breathing in asbestos fibers or similar irritants.

The term **elastic recoil** refers to how readily the lungs rebound after having been stretched. It is responsible for the lungs returning to their preinspiratory volume when the inspiratory muscles relax at the end of inspiration.

Pulmonary elastic behavior depends mainly on two factors: *highly elastic connective tissue* in the lungs and *alveolar surface tension.*

PULMONARY ELASTIC CONNECTIVE TISSUE Pulmonary connective tissue contains large quantities of elastin fibers (see ● Figure 13-4a, p. 465; also see p. 58). Not only do these fibers have elastic properties themselves but they also are arranged into a meshwork that amplifies their elastic behavior, much like threads in stretch-knit fabric. The entire piece of fabric (or lung) is stretchier and tends to bounce back to its original shape more than the individual threads (elastin fibers) do.

ALVEOLAR SURFACE TENSION An even more important factor influencing elastic behavior of the lungs is the **alveolar surface tension** displayed by the thin liquid film that lines each alveolus. At an air–water interface, the water molecules at the surface are more strongly attracted to other surrounding water molecules (by means of hydrogen bonds; see p. A-7) than to the air above the surface. This unequal attraction produces a force known as *surface tension* at the surface of the liquid. Surface tension has a twofold effect. First, the liquid layer resists any force that increases its surface area; that is, it opposes expansion of the alveolus because the surface water molecules oppose being pulled apart. Accordingly, the greater the surface tension is, the less compliant the lungs. Second, the liquid surface area tends to shrink as small as possible because the surface water molecules, being preferentially attracted to one another, try to get as close together as possible. Thus, the surface tension of the liquid lining an alveolus tends to reduce alveolus size, squeezing in on the air inside. This property, along with the rebound of the stretched elastin fibers, produces the lungs' elastic recoil back to their preinspiratory size when inspiration is over.

 With emphysema, elastic recoil is decreased by loss of elastin fibers and the reduction in alveolar surface tension resulting from the breakdown of alveolar walls. The decrease in elastic recoil contributes, along with the increased airway resistance, to the patient's difficulty in expiration.

Pulmonary surfactant decreases surface tension and contributes to lung stability.

The cohesive forces between water molecules are so strong that if the alveoli were lined with water alone, surface tension would be so great that the lungs would collapse. The recoil force attributable to the elastin fibers and high surface tension would exceed the opposing stretching force of the transmural pressure gradient. Furthermore, the lungs would be very poorly compliant, so exhausting muscular efforts would be required to accomplish stretching and inflation of the alveoli. Two factors oppose the tendency for alveoli to collapse, thereby maintaining alveolar stability and reducing the work of breathing. These are *pulmonary surfactant* and *alveolar interdependence.* The tremendous surface tension of pure water is normally counteracted by pulmonary surfactant.

PULMONARY SURFACTANT Pulmonary surfactant is a complex mixture of lipids and proteins secreted by the Type II alveolar cells (see ● Figure 13-4a, p. 465). It intersperses between the water molecules in the fluid lining the alveoli and lowers alveolar surface tension (*surfactant* is derived from "*sur*face *active agent*"). Because the cohesive force between a water molecule and an adjacent pulmonary surfactant molecule is very low, pulmonary surfactant decreases the extent of hydrogen bonding between molecules at the alveolar air–water interface. By lowering alveolar surface tension, pulmonary surfactant provides two important benefits: (1) it increases pulmonary compliance, reducing the work of inflating the lungs; and (2) it reduces the lungs' tendency to recoil, so they do not collapse as readily.

Pulmonary surfactant's role in reducing the tendency of alveoli to recoil, thereby discouraging alveolar collapse, is important in helping maintain lung stability. The division of the lung into myriad tiny air sacs provides the advantage of a tremendously increased surface area for exchange of O_2 and CO_2, but it also presents the problem of maintaining the stability of all these alveoli. Recall that the pressure generated by alveolar surface tension is directed inward, squeezing in on the air in the alveoli. If you visualize the alveoli as spherical bubbles, according to the **law of LaPlace,** the magnitude of the inward-directed collapsing pressure is directly proportional to the surface tension and inversely proportional to the radius of the bubble:

$$P = \frac{2T}{r}$$

where

P = inward-directed collapsing pressure

T = surface tension

r = radius of bubble (alveolus)

Because the collapsing pressure is inversely proportional to the radius, the smaller the alveolus, the smaller its radius and the greater its tendency to collapse at a given surface tension. Accordingly, if two alveoli of unequal size but the same surface tension are connected by the same terminal airway, the smaller alveolus—because it generates a larger collapsing pressure—has a tendency to collapse and empty its air into the larger alveolus (● Figure 13-16a).

Small alveoli normally do not collapse and blow up larger alveoli, however, because pulmonary surfactant reduces the surface tension of small alveoli more than that of larger alveoli. Pulmonary surfactant decreases surface tension to a greater degree in small alveoli than in larger alveoli because the surfactant molecules are crowded closer together in the smaller alveoli. The larger an alveolus, the more spread out are its surfactant molecules and the less effect they have on reducing surface tension. The surfactant-induced lower surface tension of small alveoli offsets the effect of their smaller radius in determining the inward-directed pressure. Therefore, the presence of surfactant causes the collapsing pressure of small alveoli to become comparable to that of larger alveoli and minimizes the tendency for small alveoli to collapse and empty their contents into larger alveoli (● Figure 13-16b). Pulmonary surfactant therefore helps stabilize the sizes of the alveoli and helps keep them open and available to participate in gas exchange.

ALVEOLAR INTERDEPENDENCE A second factor that contributes to alveolar stability is the interdependence among neighboring alveoli. Each alveolus is surrounded by other alveoli and interconnected with them by connective tissue. If an alveolus starts to collapse, the surrounding alveoli are stretched as their walls are pulled in the direction of the caving-in alveolus (● Figure 13-17a). These neighboring alveoli, by recoiling in resistance to being stretched, exert expanding forces on the collapsing alveolus and thereby help keep it

open (● Figure 13-17b). This phenomenon, which can be likened to a stalemated "tug of war" between adjacent alveoli, is termed **alveolar interdependence.**

The opposing forces acting on the lung (that is, the forces keeping the alveoli open and the countering forces that promote alveolar collapse) are summarized in ▲ Table 13-2.

Clinical Note **NEWBORN RESPIRATORY DISTRESS SYNDROME** Developing fetal lungs normally cannot synthesize pulmonary surfactant until late in pregnancy. Especially in an infant born prematurely, not enough pulmonary surfactant may be produced to reduce the alveolar surface tension to manageable levels. The resulting collection of symptoms is

Law of LaPlace:
Magnitude of inward-directed pressure (P) in a bubble (alveolus) $= \dfrac{2 \times \text{Surface tension } (T)}{\text{Radius } (r) \text{ of bubble (alveolus)}}$

$P_1 = \dfrac{2 \times T}{r}$
$P_1 = \dfrac{2 \times T}{1}$
$P_1 = 2T$

$P_1 = 2T$

Radius = 1
Surface tension = T

Airways

Alveoli

$P_2 = 1T$

$P_2 = \dfrac{2 \times T}{r}$
$P_2 = \dfrac{2 \times T}{2}$
$P_2 = 1T$

Radius = 2
Surface tension = T

$P_1 > P_2$

KEY

T = A given surface tension

(a) Without pulmonary surfactant

● **FIGURE 13-16 Role of pulmonary surfactant in counteracting the tendency for small alveoli to collapse into larger alveoli.** (a) According to the law of LaPlace, if two alveoli of unequal size but the same surface tension are connected by the same terminal airway, the smaller alveolus—because it generates a larger inward-directed collapsing pressure—has a tendency (without pulmonary surfactant) to collapse and empty its air into the larger alveolus. (b) Pulmonary surfactant reduces the surface tension of a smaller alveolus more than that of a larger alveolus. This reduction in surface tension offsets the effect of the smaller radius in determining the inward-directed pressure. Consequently, the collapsing pressures of the small and large alveoli are comparable. Therefore, in the presence of pulmonary surfactant a small alveolus does not collapse and empty its air into the larger alveolus.

$P_1 = \dfrac{2 \times T}{r}$
$P_1 = \dfrac{2 \times \frac{1}{2}T}{1}$
$P_1 = 1T$

$P_1 = 1T$

Radius = 1
Surface tension = $\frac{1}{2}T$

Airways

Alveoli

$P_2 = 1T$

$P_2 = \dfrac{2 \times T}{r}$
$P_2 = \dfrac{2 \times T}{2}$
$P_2 = 1T$

Pulmonary surfactant molecule

Radius = 2
Surface tension = T

$P_1 = P_2$

(b) With pulmonary surfactant

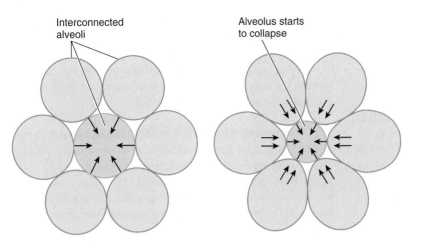

(a) Collapsing alveolus stretches surrounding alveoli

(b) Stretched surrounding alveoli recoil, pulling collapsing alveolus open

● **FIGURE 13-17 Alveolar interdependence.** (a) When an alveolus (in pink) in a group of interconnected alveoli starts to collapse, the surrounding alveoli are stretched by the collapsing alveolus. (b) As the neighboring alveoli recoil in resistance to being stretched, they pull outward on the collapsing alveolus. This expanding force pulls the collapsing alveolus open.

▲ TABLE 13-2	**Opposing Forces Acting on the Lung**
Forces Keeping the Alveoli Open	**Forces Promoting Alveolar Collapse**
Transmural pressure gradient	Elasticity of stretched pulmonary connective tissue fibers
Pulmonary surfactant (which opposes alveolar surface tension)	
Alveolar interdependence	Alveolar surface tension

termed **newborn respiratory distress syndrome.** The infant must make very strenuous inspiratory efforts to overcome the high surface tension in an attempt to inflate the poorly compliant lungs. Moreover, the work of breathing is further increased because the alveoli, in the absence of surfactant, tend to collapse almost completely during each expiration. It is more difficult (requires a greater transmural pressure differential) to expand a collapsed alveolus by a given volume than to increase an already partially expanded alveolus by the same volume. The situation is analogous to blowing up a new balloon. It takes more effort to blow in that first breath of air when starting to blow up a new balloon than to blow additional breaths into the already partially expanded balloon. With newborn respiratory distress syndrome, it is as though with every breath the infant must start blowing up a new balloon. Lung expansion may require transmural pressure gradients of 20 to 30 mm Hg (compared to the normal 4 to 6 mm Hg) to overcome the tendency of surfactant-deprived alveoli to collapse. Worse yet, the new-

born's muscles are still weak. The respiratory distress from surfactant deficiency may soon lead to death if breathing efforts become exhausting or inadequate to support sufficient gas exchange.

This life-threatening condition affects 30,000 to 50,000 newborns, primarily premature infants, each year in the United States. Until the surfactant-secreting cells mature sufficiently, the condition is treated by surfactant replacement. In addition, drugs can hasten the maturation process.

The work of breathing normally requires only about 3% of total energy expenditure.

During normal quiet breathing, the respiratory muscles must work during inspiration to expand the lungs against their elastic forces and to overcome airway resistance, whereas expiration is passive. Normally, the lungs are highly compliant and airway resistance is low, so only about 3% of the total energy expended by the body is used for quiet breathing.

 The work of breathing may be increased in four different situations:

1. *When pulmonary compliance is decreased,* such as with pulmonary fibrosis, more work is required to expand the lungs.

2. *When airway resistance is increased,* such as with COPD, more work is required to achieve the greater pressure gradients necessary to overcome the resistance so that adequate airflow can occur.

3. *When elastic recoil is decreased,* as with emphysema, passive expiration may be inadequate to expel the volume of air normally exhaled during quiet breathing. Thus, the abdominal muscles must work to aid in emptying the lungs, even when the person is at rest.

4. *When there is a need for increased ventilation*, such as during exercise, more work is required to accomplish both a greater depth of breathing (a larger volume of air moving in and out with each breath) and a faster rate of breathing (more breaths per minute).

During strenuous exercise, the amount of energy required to power pulmonary ventilation may increase up to 25-fold. However, because total energy expenditure by the body increases up to 15- to 20-fold during heavy exercise, the energy used for increased ventilation still represents only about 5% of total energy expended. In contrast, in patients with poorly compliant lungs or obstructive lung disease, the energy required for breathing even at rest may increase to 30% of total energy expenditure. In such cases, the individual's exercise ability is severely limited, as breathing itself becomes exhausting.

The lungs normally operate at about "half full."

On average, in healthy young adults, the maximum air that the lungs can hold is about 5.7 liters in males (4.2 liters in females). Anatomic build, age, the distensibility of the lungs, and the presence or absence of respiratory disease affect this total lung capacity. Normally, during quiet breathing the lungs are nowhere near maximally inflated nor are they deflated to their minimum volume. Thus, the lungs normally remain moderately inflated throughout the respiratory cycle. At the end of a normal quiet expiration, the lungs still contain about 2200 ml of air. During each typical breath under resting conditions, about 500 ml of air are inspired and the same quantity is expired, so during quiet breathing the lung volume varies between 2200 ml at the end of expiration to 2700 ml at the end of inspiration (● Figure 13-18a). During maximal expiration, lung volume can decrease to 1200 ml in males (1000 ml in females), but the lungs can never be completely deflated because the small airways collapse during forced expirations at low lung volumes, blocking further outflow (see ● Figure 13-15c, p. 475).

An important outcome of not being able to empty the lungs completely is that even during maximal expiratory efforts, gas exchange can still continue between blood flowing through the lungs and the remaining alveolar air. As a result, the gas content of the blood leaving the lungs for delivery to the tissues

(a) Normal range and extremes of lung volume in a healthy young adult male

KEY

TV	= Tidal volume (500 ml)
IRV	= Inspiratory reserve volume (3000 ml)
IC	= Inspiratory capacity (3500 ml)
ERV	= Expiratory reserve volume (1000 ml)
RV	= Residual volume (1200 ml)
FRC	= Functional residual capacity (2200 ml)
VC	= Vital capacity (4500 ml)
TLC	= Total lung capacity (5700 ml)

(b) Normal spirogram of a healthy young adult male

● **FIGURE 13-18 Variations in lung volume in a healthy young adult male.** Values for females are somewhat lower. (Note that residual volume cannot be measured with a spirometer but must be determined by another means.)

normally remains remarkably constant throughout the respiratory cycle. By contrast, if the lungs completely filled and emptied with each breath, the amount of O_2 taken up and CO_2 dumped off by the blood would fluctuate widely. Another advantage of the lungs not completely emptying with each breath is a reduction in the work of breathing. Recall that it takes less effort to inflate a partially inflated alveolus than a totally collapsed one.

The changes in lung volume that occur with different respiratory efforts can be measured using a spirometer, a device for determining the various lung volumes and capacities.

Fresh air
from inspiration

150 — Airway dead space volume (150 ml)

Alveolar air

(a) After inspiration, before expiration

500 ml "old" alveolar air expired
350 ml expired to atmosphere

150 ml remain in dead space

500 ml expired to atmosphere
150 ml fresh air from dead space (left from preceding inspiration)

350 ml "old" alveolar air

150
350
150

(b) During expiration

150
350
150

500 ml enter alveoli
150 ml "old" air from dead space (left from preceding expiration)
350 ml fresh air from atmosphere

500 ml fresh air enter from atmosphere
350 ml fresh air reach alveoli
150 ml fresh air remain in dead space

(c) During inspiration

KEY

"Old" alveolar air that has exchanged O_2 and CO_2 with the blood

Fresh atmospheric air that has not exchanged O_2 and CO_2 with the blood

● **FIGURE 13-22 Effect of dead space volume on exchange of tidal volume between the atmosphere and the alveoli.** Even though 500 ml of air move in and out between the atmosphere and the respiratory system and 500 ml move in and out of the alveoli with each breath, only 350 ml are actually exchanged between the atmosphere and the alveoli because of the anatomic dead space (the volume of air in the respiratory airways).

Alveolar ventilation is less than pulmonary ventilation because of dead space.

Various changes in lung volume represent only one factor in determining **pulmonary,** or **minute, ventilation,** which is the volume of air breathed in and out in one minute. The other important factor is **respiratory rate,** which averages 12 breaths per minute.

$$\text{Pulmonary ventilation} = \text{tidal volume} \times \text{respiratory rate}$$
$$\text{(ml/min)} \qquad \text{(ml/breath)} \qquad \text{(breaths/min)}$$

At an average tidal volume of 500 ml/breath and a respiratory rate of 12 breaths/minute, pulmonary ventilation is 6000 ml, or 6 liters, of air breathed in and out in one minute under resting conditions. For a brief period of time, a healthy young adult male can voluntarily increase his total pulmonary ventilation 25-fold, to 150 liters/min. To increase pulmonary ventilation, both tidal volume and respiratory rate increase, but depth of breathing increases more than frequency of breathing. It is usually more advantageous to have a greater increase in tidal volume than in respiratory rate because of anatomic dead space.

ANATOMIC DEAD SPACE Not all the inspired air gets down to the site of gas exchange in the alveoli. Part remains in the conducting airways, where it is not available for gas exchange. The volume of the conducting passages in an adult averages about 150 ml. This volume is considered **anatomic dead space** because air within these conducting airways is useless for exchange. Anatomic dead space greatly affects the efficiency of pulmonary ventilation. In effect, even though 500 ml of air are moved in and out with each breath, only 350 ml are actually exchanged between the atmosphere and alveoli because of the 150 ml occupying the anatomic dead space.

Looking at ● Figure 13-22, note that at the end of inspiration, the respiratory airways are filled with 150 ml of fresh atmospheric air from the inspiration. During the subsequent expiration, 500 ml of air are expired to the atmosphere. The first 150 ml expired are the fresh air that was retained in the airways and never used. The remaining 350 ml expired are "old" alveolar air that has participated in gas exchange with the blood. During the same expiration, 500 ml of gas also leave the alveoli. The first 350 ml are expired to the atmosphere; the other 150 ml of old alveolar air never reach the outside but remain in the conducting airways.

Effect of Different Breathing Patterns on Alveolar Ventilation

Breathing Pattern	Tidal Volume (ml/breath)	Respiratory Rate (breaths/min)	Dead Space Volume (ml)	Pulmonary Ventilation (ml/min)*	Alveolar Ventilation (ml/min)**
Quiet breathing at rest	500	12	150	6000	4200
Deep, slow breathing	1200	5	150	6000	5250
Shallow, rapid breathing	150	40	150	6000	0

*Equals tidal volume × respiratory rate.
**Equals (tidal volume − dead space volume) × respiratory rate.

On the next inspiration, 500 ml of gas enter the alveoli. The first 150 ml to enter the alveoli are the old alveolar air that remained in the dead space during the preceding expiration. The other 350 ml entering the alveoli are fresh air inspired from the atmosphere. Simultaneously, 500 ml of air enter from the atmosphere. The first 350 ml of atmospheric air reach the alveoli; the other 150 ml remain in the conducting airways to be expired without benefit of being exchanged with the blood, as the cycle repeats itself.

ALVEOLAR VENTILATION Because the amount of atmospheric air that reaches the alveoli and is actually available for exchange with blood is more important than the total amount breathed in and out, **alveolar ventilation**—the volume of air exchanged between the atmosphere and alveoli per minute—is more important than pulmonary ventilation. In determining alveolar ventilation, the amount of wasted air moved in and out through the anatomic dead space must be taken into account, as follows:

Alveolar ventilation = (tidal volume − dead space volume) × respiratory rate

With average resting values,

Alveolar ventilation = (500 ml/breath − 150 ml dead space volume) × 12 breaths/min

= 4200 ml/min

Thus, with quiet breathing, alveolar ventilation is 4200 ml/min, whereas pulmonary ventilation is 6000 ml/min.

EFFECT OF BREATHING PATTERNS ON ALVEOLAR VENTILATION
To understand how important dead space volume is in determining the magnitude of alveolar ventilation, examine the effect of various breathing patterns on alveolar ventilation, as shown in ▲ Table 13-3. If a person deliberately breathes deeply (for example, a tidal volume of 1200 ml) and slowly (for example, a respiratory rate of 5 breaths/min), pulmonary ventilation is 6000 ml/min, the same as during quiet breathing at rest, but alveolar ventilation increases to 5250 ml/min compared to the resting rate of 4200 ml/min. In contrast, if a person deliberately breathes shallowly (for example, a tidal volume of 150 ml) and rapidly (a frequency of 40 breaths/min), pulmonary ventilation

would still be 6000 ml/min; however, alveolar ventilation would be 0 ml/min. In effect, the person would only be drawing air in and out of the anatomic dead space without any atmospheric air being exchanged with the alveoli, where it could be useful. The individual could voluntarily maintain such a breathing pattern for only a few minutes before losing consciousness, at which time normal breathing would resume.

The value of reflexly bringing about a larger increase in depth of breathing than in rate of breathing when pulmonary ventilation increases during exercise should now be apparent. It is the most efficient means of elevating alveolar ventilation. When tidal volume is increased, the entire increase goes toward elevating alveolar ventilation, whereas an increase in respiratory rate does not go entirely toward increasing alveolar ventilation. When respiratory rate is increased, the frequency with which air is wasted in the dead space also increases, because a portion of *each breath* must move in and out of the dead space. As needs vary, ventilation is normally adjusted to a tidal volume and respiratory rate that meet those needs most efficiently in terms of energy cost.

ALVEOLAR DEAD SPACE We have assumed that all the atmospheric air entering the alveoli participates in exchanges of O_2 and CO_2 with pulmonary blood. However, the match between air and blood is not always perfect, because not all alveoli are equally ventilated with air and perfused with blood. Any ventilated alveoli that do not participate in gas exchange with blood because they are inadequately perfused are considered **alveolar dead space.** In healthy people, alveolar dead space is quite small and of little importance, but it can increase to even lethal levels in several types of pulmonary disease.

Next you will learn why alveolar dead space is minimal in healthy individuals.

Local controls act on the smooth muscle of the airways and arterioles to match airflow to blood flow.

When discussing the role of airway resistance in determining airflow rate into and out of the lungs, we were referring to the overall resistance of all the airways collectively. However, the resistance of individual airways supplying specific alveoli can be

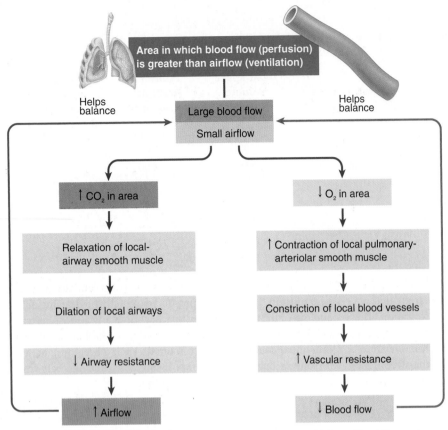

(a) Local controls to adjust ventilation and perfusion to lung area with large blood flow and small airflow

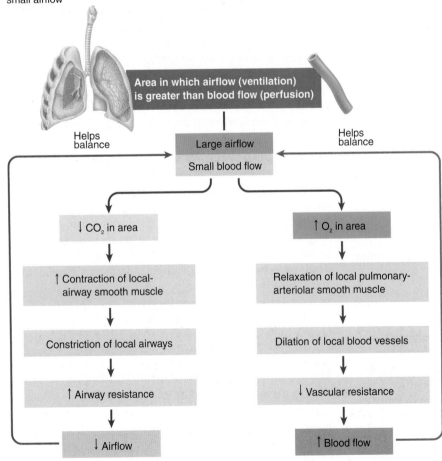

(b) Local controls to adjust ventilation and perfusion to a lung area with large airflow and small blood flow

adjusted independently in response to changes in the airway's local environment. This situation is comparable to the control of systemic arterioles. Recall that overall systemic arteriolar resistance (that is, total peripheral resistance) is an important determinant of the blood pressure gradient that drives blood flow throughout the systemic circulatory system (see p. 358). Yet the radius of individual arterioles supplying various tissues can be adjusted locally to match the tissues' differing metabolic needs (see p. 353).

EFFECT OF CO_2 ON BRONCHIOLAR SMOOTH MUSCLE Similar to arteriolar smooth muscle, bronchiolar smooth muscle is sensitive to local changes within its immediate environment, particularly to local CO_2 levels. If an alveolus is receiving too little airflow in comparison to its blood flow, CO_2 levels will increase in the alveolus and surrounding tissue as the blood drops off more CO_2 than is exhaled into the atmosphere. This local increase in CO_2 directly promotes relaxation of the bronchiolar smooth muscle, bringing about dilation of the airway supplying the underaerated alveolus. The resultant decrease in airway resistance leads to an increased airflow (for the same ΔP) to the involved alveolus, so its airflow now matches its blood supply (● Figure 13-23a). The converse is also true. A localized decrease in CO_2 associated with an alveolus that is receiving too much air for its blood supply directly increases contractile activity of the airway smooth muscle involved, constricting the airway supplying this overaerated alveolus. The result is reduced airflow to the overaerated alveolus (● Figure 13-23b).

EFFECT OF O_2 ON PULMONARY ARTERIOLAR SMOOTH MUSCLE Simultaneously, a similar locally induced effect on pulmonary vascular smooth muscle takes place to maximally match blood flow to airflow. The two mechanisms for matching airflow

● **FIGURE 13-23 Local controls to match airflow and blood flow to an area of the lung.** CO_2 acts locally on bronchiolar smooth muscle and O_2 acts locally on arteriolar smooth muscle to adjust ventilation and perfusion, respectively, to match airflow and blood flow to an area of the lung.

| TABLE 13-4 | Effects of Local Changes in O_2 on the Pulmonary and Systemic Arterioles |

	EFFECT OF A LOCAL CHANGE IN O_2	
Vessels	**Decreased O_2**	**Increased O_2**
Pulmonary Arterioles	Vasoconstriction	Vasodilation
Systemic Arterioles	Vasodilation	Vasoconstriction

and blood flow function concurrently, so normally very little air or blood is wasted in the lung. Just as in the systemic circulation, distribution of the cardiac output to different alveolar capillary networks can be controlled by adjusting the resistance to blood flow through specific pulmonary arterioles. If blood flow is greater than airflow to a given alveolus, the O_2 level in the alveolus and surrounding tissues falls below normal as the overabundance of blood extracts more O_2 than usual from the alveolus. The local decrease in O_2 concentration causes vasoconstriction of the pulmonary arteriole supplying this particular capillary bed, thus reducing blood flow to match the smaller airflow (● Figure 13-23a). Conversely, an increase in alveolar O_2 concentration caused by a mismatched large airflow and small blood flow brings about pulmonary vasodilation, which increases blood flow to match the larger airflow (● Figure 13-23b).

This local effect of O_2 on pulmonary arteriolar smooth muscle is, appropriately, just the opposite of its effect on systemic arteriolar smooth muscle (▲ Table 13-4). In the systemic circulation, a decrease in O_2 in a tissue causes localized vasodilation to increase blood flow to the deprived area, and vice versa, which is important in matching blood supply to local metabolic needs.

Because of gravitational effects, some regional differences in ventilation and perfusion exist from the top to the bottom of the lung. When a person is standing upright, ventilation and perfusion are both less at the top of the lung and greater at the bottom of the lung, but gravity exerts a more marked effect on blood flow than on airflow. Therefore, the ventilation–perfusion ratio (the rate of airflow compared to the rate of blood flow) decreases from the top to the bottom of the lung (● Figure 13-24). In other words, the top of the lung receives less air and blood than the bottom of the lung, but it receives relatively more air than blood; the bottom of the lung receives more air and blood than the top of the lung, but it receives relatively less air than blood. In healthy lungs, the effect of this mismatch between air and blood has a negligible effect on overall O_2 uptake and CO_2 elimination. Airflow and blood flow at a particular alveolar interface are usually matched as much as possible by local controls to accomplish efficient exchange of O_2 and CO_2. However, in pathologic conditions, ventilation-perfusion mismatches can exceed the capability of local controls to compensate, so the impact on O_2 uptake and CO_2 elimination can become significant, as with widespread plugging of airways with inflammatory mucus secretion or widespread damage to pulmonary vessels.

We have now completed our discussion of respiratory mechanics—all the factors involved in ventilation. We next examine gas exchange between alveolar air and the blood and then between the blood and systemic tissues.

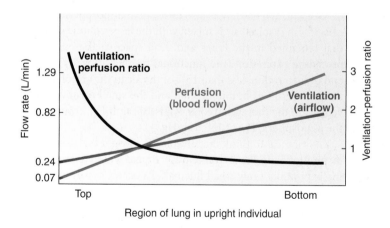

(a) Regional ventilation and perfusion rates and ventilation-perfusion ratios in the lungs

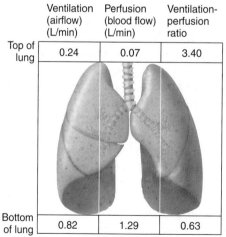

	Ventilation (airflow) (L/min)	Perfusion (blood flow) (L/min)	Ventilation-perfusion ratio
Top of lung	0.24	0.07	3.40
Bottom of lung	0.82	1.29	0.63

(b) Ventilation and perfusion rates and ventilation-perfusion ratios at top and bottom of lungs

● **FIGURE 13-24 Differences in ventilation, perfusion, and ventilation-perfusion ratios at the top and bottom of the lungs as a result of gravitational effects.** Note that the top of the lungs receives less air and blood than the bottom of the lungs, but the top of the lungs receives relatively more air than blood and the bottom of the lungs receives relatively less air than blood.

Gas Exchange

The ultimate purpose of breathing is to provide a continual supply of fresh O_2 for pickup by the blood and to constantly remove CO_2 unloaded from the blood. Blood acts as a transport system for O_2 and CO_2 between the lungs and tissues, with the tissue cells extracting O_2 from the blood and eliminating CO_2 into it.

Gases move down partial pressure gradients.

Gas exchange at both the pulmonary capillary and the tissue capillary levels involves simple passive diffusion of O_2 and CO_2 down *partial pressure gradients*. No active transport mechanisms exist for these gases. Let us see what partial pressure gradients are and how they are established.

PARTIAL PRESSURES Atmospheric air is a mixture of gases; typical dry air contains about 79% nitrogen (N_2) and 21% O_2, with almost negligible percentages of CO_2, H_2O vapor, other gases, and pollutants. Altogether, these gases exert a total atmospheric pressure of 760 mm Hg at sea level. This total pressure is equal to the sum of the pressures that each gas in the mixture partially contributes. The pressure exerted by a particular gas is directly proportional to the percentage of that gas in the total air mixture. Every gas molecule, no matter what its size, exerts the same amount of pressure; for example, a N_2 molecule exerts the same pressure as an O_2 molecule. Because 79% of the air consists of N_2 molecules, 79% of the 760 mm Hg atmospheric pressure, or 600 mm Hg, is exerted by the N_2 molecules. Similarly, because O_2 represents 21% of the atmosphere, 21% of the 760 mm Hg atmospheric pressure, or 160 mm Hg, is exerted by O_2 (● Figure 13-25). The individual pressure exerted independently by a particular gas within a mixture of gases is known as its **partial pressure**, designated by P_{gas}. Thus, the partial pressure of O_2 in atmospheric air, P_{O_2}, is normally 160 mm Hg. The atmospheric partial pressure of CO_2, P_{CO_2}, is negligible at 0.23 mm Hg.

Gases dissolved in a liquid such as blood or another body fluid also exert a partial pressure. The greater the partial pressure of a gas in a liquid, the more of that gas is dissolved.

PARTIAL PRESSURE GRADIENTS A difference in partial pressure between capillary blood and surrounding structures is known as a **partial pressure gradient.** Partial pressure gradients exist between the alveolar air and pulmonary capillary blood. Similarly, partial pressure gradients exist between systemic capillary blood and surrounding tissues. A gas always diffuses down its partial pressure gradient from the area of higher partial pressure to the area of lower partial pressure, similar to diffusion down a concentration gradient.

Oxygen enters and CO_2 leaves the blood in the lungs passively down partial pressure gradients.

We first consider the magnitude of alveolar P_{O_2} and P_{CO_2} and then look at the partial pressure gradients that move these two gases between the alveoli and incoming pulmonary capillary blood.

● **FIGURE 13-25 Concept of partial pressures.** The partial pressure exerted by each gas in a mixture equals the total pressure times the fractional composition of the gas in the mixture.

ALVEOLAR P_{O_2} AND P_{CO_2} Alveolar air is not of the same composition as inspired atmospheric air, for two reasons. First, as soon as atmospheric air enters the respiratory passages, exposure to the moist airways saturates it with H_2O. Like any other gas, water vapor exerts a partial pressure. At body temperature, the partial pressure of H_2O vapor is 47 mm Hg. Humidification of inspired air in effect "dilutes" the partial pressure of the inspired gases by 47 mm Hg because the sum of the partial pressures must total the atmospheric pressure of 760 mm Hg. In moist air, $P_{H_2O} = 47$ mm Hg, $P_{N_2} = 563$ mm Hg, and $P_{O_2} = 150$ mm Hg.

Second, alveolar P_{O_2} is also lower than atmospheric P_{O_2} because fresh inspired air is mixed with the large volume of old air that remained in the lungs and dead space at the end of the preceding expiration (the functional residual capacity). At the end of inspiration, less than 15% of the air in the alveoli is fresh air. As a result of humidification and the small turnover of alveolar air, the average alveolar P_{O_2} is 100 mm Hg, compared to the atmospheric P_{O_2} of 160 mm Hg.

It is logical to think that alveolar P_{O_2} would increase during inspiration with the arrival of fresh air and would decrease during expiration. Only small fluctuations occur, however, for two reasons. First, only a small proportion of the total alveolar air is exchanged with each breath. The relatively small volume of inspired, high-P_{O_2} air is quickly mixed with the much larger volume of retained alveolar air, which has a lower P_{O_2}. Thus, the O_2 in the inspired air can only slightly elevate the level of the total alveolar P_{O_2}. Even this potentially small elevation of P_{O_2} is diminished for another reason. Oxygen is continually moving by passive diffusion down its partial pressure gradient from the alveoli into the blood. The O_2 arriving in the alveoli in the newly inspired air simply replaces the O_2 diffusing out of the alveoli into the pulmonary capillaries. Therefore, alveolar P_{O_2} remains

Numbers are mm
Hg pressure.

● **FIGURE 13-26 Oxygen and CO_2 exchange across pulmonary and systemic capillaries caused by partial pressure gradients.**

The figure contains the following labels:

Atmospheric air
P_{O_2} 160
P_{CO_2} 0.23

Inspiration   Expiration

3
Gradients across pulmonary capillaries:

100 40
Alveolar sacs

Pulmonary capillaries

40 46
O_2 CO_2

6
Gradients across systemic capillaries:

40 46
Tissue cells

Systemic capillaries

100 40
O_2 CO_2

Tissue cell

Net diffusion gradients for O_2 and CO_2, between the lungs and tissues

High P_{O_2} Low P_{CO_2}

Low P_{O_2} High P_{CO_2}

O_2 CO_2 Alveoli
1 P_{O_2} 100 P_{CO_2} 40

2 P_{CO_2} 46 P_{O_2} 100 **4**
P_{O_2} 40 O_2 CO_2 P_{CO_2} 40

Pulmonary circulation

8

Heart

Systemic circulation

P_{O_2} 40 O_2 CO_2 P_{O_2} 100
7 P_{CO_2} 46 P_{CO_2} 40 **4**

5 P_{O_2} < 40; P_{CO_2} > 46

Food + O_2 → CO_2 + H_2O + ATP

1 Alveolar P_{O_2} remains relatively high and alveolar P_{CO_2} remains relatively low because a portion of the alveolar air is exchanged for fresh atmospheric air with each breath.

2 In contrast, the systemic venous blood entering the lungs is relatively low in O_2 and high in CO_2, having given up O_2 and picked up CO_2 at the systemic capillary level.

3 The partial pressure gradients established between the alveolar air and pulmonary capillary blood induce passive diffusion of O_2 into the blood and CO_2 out of the blood until the blood and alveolar partial pressures become equal.

4 The blood leaving the lungs is thus relatively high in O_2 and low in CO_2. It arrives at the tissues with the same blood-gas content as when it left the lungs.

5 The partial pressure of O_2 is relatively low and that of CO_2 is relatively high in the O_2-consuming, CO_2-producing tissue cells.

6 Consequently, partial pressure gradients for gas exchange at the tissue level favor the passive movement of O_2 out of the blood into cells to support their metabolic requirements and also favor the simultaneous transfer of CO_2 into the blood.

7 Having equilibrated with the tissue cells, the blood leaving the tissues is relatively low in O_2 and high in CO_2.

8 The blood then returns to the lungs to once again fill up on O_2 and dump off CO_2.

relatively constant at about 100 mm Hg throughout the respiratory cycle. Because pulmonary blood P_{O_2} equilibrates with alveolar P_{O_2}, the P_{O_2} of the blood leaving the lungs likewise remains fairly constant at this same value. Accordingly, the amount of O_2 in the blood available to the tissues varies only slightly during the respiratory cycle.

A similar situation exists in reverse for CO_2. Carbon dioxide, which is continuously produced by the body tissues as a metabolic waste product, is constantly added to the blood at the level of the systemic capillaries. In the pulmonary capillaries, CO_2 diffuses down its partial pressure gradient from the blood into the alveoli and is subsequently removed from the body during expiration. As with O_2, alveolar P_{CO_2} remains fairly constant throughout the respiratory cycle but at a lower value of 40 mm Hg.

P_{O_2} AND P_{CO_2} GRADIENTS ACROSS THE PULMONARY CAPILLARIES As blood passes through the lungs, it picks up O_2 and gives up CO_2 simply by diffusion down partial pressure gradients that exist between the blood and alveoli. Ventilation constantly replenishes alveolar O_2 and removes CO_2, thus maintaining the appropriate partial pressure gradients between the blood and alveoli. Blood entering the pulmonary capillaries is systemic venous blood pumped to the lungs through the pulmonary arteries. This blood, having just returned from the body tissues, is relatively low in O_2, with a P_{O_2} of 40 mm Hg, and is relatively high in CO_2, with a P_{CO_2} of 46 mm Hg. As this blood flows through the pulmonary capillaries, it is exposed to alveolar air (● Figure 13-26). Because the alveolar P_{O_2} at 100 mm Hg is higher than the P_{O_2} of 40 mm Hg in the blood entering the lungs, O_2 diffuses down its partial pressure gradient from the transported by the blood until the P_{O_2} falls below 60 mm Hg, because of the plateau region of the curve. If the arterial P_{O_2} falls 40%, from 100 to 60 mm Hg, the concentration of dissolved O_2 as reflected by the P_{O_2} is likewise reduced 40%. At a blood P_{O_2} of 60 mm Hg, however, the % Hb saturation is still remarkably high at 90%. Accordingly, the total O_2 content of the blood is only slightly decreased despite the 40% reduction in P_{O_2} because Hb is still carrying an almost full load of O_2, and, as mentioned before, the vast majority of O_2 is transported by Hb rather than dissolved. However, even if the blood P_{O_2} is greatly increased—say, to 600 mm Hg—by breathing pure O_2, very little additional O_2 is added to the blood. A small extra amount of O_2 dissolves, but the % Hb saturation can be maximally increased by only another 2.5%, to 100% saturation. Therefore, in the P_{O_2} range between 60 and 600 mm Hg or even higher, there is only a 10% difference in the amount of O_2 carried by Hb. Thus, the plateau portion of the O_2–Hb curve provides a good margin of safety in O_2-carrying capacity of the blood.

Arterial P_{O_2} may be reduced by pulmonary diseases accompanied by inadequate ventilation or defective gas exchange or by circulatory disorders that result in inadequate blood flow to the lungs. It may also fall in healthy people under two circumstances: (1) at high altitudes, where total atmospheric pressure and hence the P_{O_2} of the inspired air are reduced; or (2) in O_2-deprived environments at sea level, such as if someone were accidentally locked in a vault. Unless the arterial P_{O_2} becomes markedly reduced (falls below 60 mm Hg) in either pathological conditions or abnormal circumstances, near-normal amounts of O_2 can still be carried to the tissues.

SIGNIFICANCE OF THE STEEP PORTION OF THE O_2–Hb CURVE The steep portion of the curve between 0 and 60 mm Hg is in the blood P_{O_2} range that exists at the systemic capillaries, where O_2 is unloaded from Hb. In the systemic capillaries, the blood equilibrates with the surrounding tissue cells at an average P_{O_2} of 40 mm Hg. Note in ● Figure 13-28 that at a P_{O_2} of 40 mm Hg, the % Hb saturation is 75%. The blood arrives in the tissue capillaries at a P_{O_2} of 100 mm Hg with 97.5% Hb saturation. Because Hb can only be 75% saturated at the P_{O_2} of 40 mm Hg in the systemic capillaries, nearly 25% of the HbO_2 must dissociate, yielding reduced Hb and O_2. This released O_2 is free to diffuse down its partial pressure gradient from the red blood cells through the plasma and interstitial fluid into the tissue cells.

The Hb in the venous blood returning to the lungs is still normally 75% saturated. If the tissue cells are metabolizing more actively, the P_{O_2} of the systemic capillary blood falls (for example, from 40 to 20 mm Hg) because the cells are consuming O_2 more rapidly. Note on the curve that this drop of 20 mm Hg in P_{O_2} decreases the % Hb saturation from 75% to 30%; that is, about 45% more of the total HbO_2 than normal gives up its O_2 for tissue use. The normal 60 mm Hg drop in P_{O_2} from 100 to 40 mm Hg in the systemic capillaries causes about 25% of the total HbO_2 to unload its O_2. In comparison, a further drop in P_{O_2} of only 20 mm Hg results in an additional 45% of the total HbO_2 unloading its O_2 because the O_2 partial pressures in this range are operating in the steep portion of the curve. In this range, only a small drop in systemic capillary P_{O_2} can auto-

matically make large amounts of O_2 immediately available to meet the O_2 needs of more actively metabolizing tissues. As much as 85% of the Hb may give up its O_2 to actively metabolizing cells during strenuous exercise. In addition to this more thorough withdrawal of O_2 from the blood, even more O_2 is made available to actively metabolizing cells, such as exercising muscles, by circulatory and respiratory adjustments that increase the flow rate of oxygenated blood through the active tissues.

Hemoglobin promotes the net transfer of O_2 at both the alveolar and tissue levels.

We still have not really clarified the role of Hb in gas exchange. Because blood P_{O_2} depends entirely on the concentration of *dissolved* O_2, we could ignore the O_2 bound to Hb in our earlier discussion of O_2 being driven from the alveoli to the blood by a P_{O_2} gradient. However, Hb does play a crucial role in permitting the transfer of large quantities of O_2 before blood P_{O_2} equilibrates with the surrounding tissues (● Figure 13-29).

ROLE OF HEMOGLOBIN AT THE ALVEOLAR LEVEL Hemoglobin acts as a "storage depot" for O_2, removing O_2 from solution as soon as it enters the blood from the alveoli. Because only dissolved O_2 contributes to P_{O_2}, the O_2 stored in Hb cannot contribute to blood P_{O_2}. When systemic venous blood enters the pulmonary capillaries, its P_{O_2} is considerably lower than alveolar P_{O_2}, so O_2 immediately diffuses into the blood, raising blood P_{O_2}. As soon as the blood P_{O_2} increases, the percentage of Hb that can bind with O_2 likewise increases, as indicated by the O_2–Hb curve. Consequently, most of the O_2 that has diffused into the blood combines with Hb and no longer contributes to blood P_{O_2}. As O_2 is removed from solution by combining with Hb, blood P_{O_2} falls to about the same level it was when the blood entered the lungs, even though the total quantity of O_2 in the blood actually has increased. Because blood P_{O_2} is once again considerably below alveolar P_{O_2}, more O_2 diffuses from the alveoli into the blood, only to be soaked up by Hb again.

Even though we have considered this process stepwise for clarity, net diffusion of O_2 from alveoli to blood occurs continuously until Hb becomes as completely saturated with O_2 as it can be at that particular P_{O_2}. At a normal P_{O_2} of 100 mm Hg, Hb is 97.5% saturated. Thus, by soaking up O_2, Hb keeps blood P_{O_2} low and prolongs the existence of a partial pressure gradient so that a large net transfer of O_2 into the blood can take place. Not until Hb can store no more O_2 (that is, Hb is maximally saturated for that P_{O_2}) does all the O_2 transferred into the blood remain dissolved and directly contribute to the P_{O_2}. Only now does blood P_{O_2} rapidly equilibrate with alveolar P_{O_2} and bring further O_2 transfer to a halt, but this point is not reached until Hb is already loaded to the maximum extent possible. Once blood P_{O_2} equilibrates with alveolar P_{O_2}, no further O_2 transfer can take place, no matter how little or how much total O_2 has already been transferred.

ROLE OF HEMOGLOBIN AT THE TISSUE LEVEL The reverse situation occurs at the tissue level. Because the P_{O_2} of blood entering

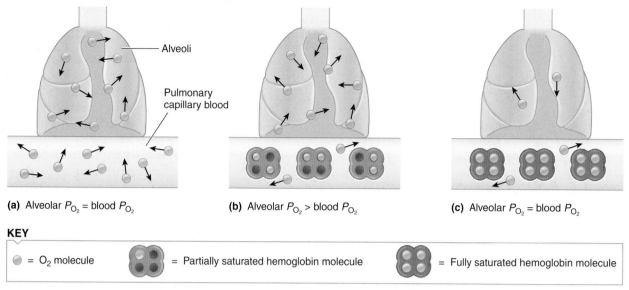

(a) Alveolar P_{O_2} = blood P_{O_2}

(b) Alveolar P_{O_2} > blood P_{O_2}

(c) Alveolar P_{O_2} = blood P_{O_2}

KEY

= O_2 molecule = Partially saturated hemoglobin molecule = Fully saturated hemoglobin molecule

● **FIGURE 13-29 Hemoglobin facilitating a large net transfer of O₂ by acting as a storage depot to keep P_{O_2} low.** (a) In the hypothetical situation in which no hemoglobin is present in the blood, the alveolar P_{O_2} and the pulmonary capillary blood P_{O_2} are at equilibrium. (b) Hemoglobin has been added to the pulmonary capillary blood. As the Hb starts to bind with O_2, it removes O_2 from solution. Because only dissolved O_2 contributes to blood P_{O_2}, the blood P_{O_2} falls below that of the alveoli, even though the same number of O_2 molecules are present in the blood as in part (a). By "soaking up" some of the dissolved O_2, Hb favors the net diffusion of more O_2 down its partial pressure gradient from the alveoli to the blood. (c) Hemoglobin is fully saturated with O_2, and the alveolar and blood P_{O_2} are at equilibrium again. The blood P_{O_2} resulting from dissolved O_2 is equal to the alveolar P_{O_2}, despite the fact that the total O_2 content in the blood is much greater than in part (a) when blood P_{O_2} was equal to alveolar P_{O_2} in the absence of Hb.

the systemic capillaries is considerably higher than the P_{O_2} of the surrounding tissue, O_2 immediately diffuses from the blood into the tissues, lowering blood P_{O_2}. When blood P_{O_2} falls, Hb must unload some stored O_2 because the % Hb saturation is reduced. As the O_2 released from Hb dissolves in the blood, blood P_{O_2} increases and once again exceeds the P_{O_2} of the surrounding tissues. This favors further movement of O_2 out of the blood, although the total quantity of O_2 in the blood has already fallen. Only when Hb can no longer release any more O_2 into solution (when Hb is unloaded to the greatest extent possible for the P_{O_2} existing at the systemic capillaries) can blood P_{O_2} fall as low as in surrounding tissue. At this time, further transfer of O_2 stops. Hemoglobin, because it stores a large quantity of O_2 that can be freed by a slight reduction in P_{O_2} at the systemic capillary level, permits the transfer of tremendously much more O_2 from the blood into the cells than would be possible in its absence.

Thus, Hb plays an important role in the *total quantity* of O_2 that the blood can pick up in the lungs and drop off in the tissues. If Hb levels fall to one half of normal, as in a severely anemic patient (see p. 397), the O_2-carrying capacity of the blood falls by 50% even though the arterial P_{O_2} is the normal 100 mm Hg with 97.5% Hb saturation. Only half as much Hb is available to be saturated, emphasizing once again how critical

Hb is in determining how much O_2 can be picked up at the lungs and made available to tissues.

Factors at the tissue level promote the unloading of O₂ from hemoglobin.

Even though the main factor determining the % Hb saturation is the P_{O_2} of the blood, other factors can affect the affinity, or bond strength, between Hb and O_2 and, accordingly, can shift the O_2–Hb curve (that is, change the % Hb saturation at a given P_{O_2}). These other factors are CO_2, acidity, temperature, and 2,3-bisphosphoglycerate, which we will examine separately. The O_2–Hb dissociation curve with which you are already familiar (see ● Figure 13-28) is a typical curve at normal arterial CO_2 and acidity levels, normal body temperature, and normal 2,3-bisphosphoglycerate concentration.

EFFECT OF CO₂ ON % Hb SATURATION An increase in P_{CO_2} shifts the O_2–Hb curve to the right (● Figure 13-30). The % Hb saturation still depends on the P_{O_2}, but for any given P_{O_2}, less O_2 and Hb can be combined. This effect is important because the P_{CO_2} of the blood increases in the systemic capillaries as CO_2 diffuses down its gradient from the cells into the blood. The presence of this additional CO_2 in the blood in effect decreases the affinity

Arterial P_{CO_2} and acidity, normal body temperature (as at pulmonary level)

↑ P_{CO_2}
↑ Acid (H^+)
↑ Temperature
or
↑ 2,3-Bisphosphoglycerate

(as at tissue level)

● **FIGURE 13-30 Effect of increased P_{CO_2}, H^+, temperature, and 2,3-bisphosphoglycerate on the O_2–Hb curve.** Increased P_{O_2}, acid, and temperature, as found at the tissue level, shift the O_2–Hb curve to the right. As a result, less O_2 and Hb can be combined at a given P_{O_2} so that more O_2 is unloaded from Hb for use by the tissues. Similarly, 2,3-bisphosphoglycerate, whose production is increased in red blood cells when arterial HbO_2 is chronically below normal, shifts the O_2–Hb curve to the right, making more of the limited O_2 available at the tissue level.

of Hb for O_2, so Hb unloads even more O_2 at the tissue level than it would if the reduction in P_{O_2} in the systemic capillaries were the only factor affecting % Hb saturation.

EFFECT OF ACID ON % Hb SATURATION An increase in acidity also shifts the curve to the right. Because CO_2 generates carbonic acid (H_2CO_3), the blood becomes more acidic at the systemic capillary level as it picks up CO_2 from the tissues. The resulting reduction in Hb affinity for O_2 in the presence of increased acidity aids in releasing even more O_2 at the tissue level for a given P_{O_2}. In actively metabolizing cells, such as exercising muscles, not only is more carbonic acid–generating CO_2 produced, but lactate (lactic acid) also may be produced if the cells resort to anaerobic metabolism (see pp. 37 and 278). The resultant local elevation of acid in the working muscles facilitates further unloading of O_2 in the very tissues that need the most O_2.

BOHR EFFECT The influence of CO_2 and acid on the release of O_2 is known as the **Bohr effect.** Both CO_2 and the hydrogen ion (H^+) component of acids can combine reversibly with Hb at sites other than the O_2-binding sites. The result is a change in the molecular structure of Hb that reduces its affinity for O_2. (Note that the % Hb saturation refers only to the extent to which Hb is combined with O_2, not the extent to which it is bound with CO_2, H^+, or other molecules. Indeed, the % Hb saturation decreases when CO_2 and H^+ bind with Hb, be-

cause their presence on Hb facilitates increased release of O_2 from Hb.)

EFFECT OF TEMPERATURE ON % Hb SATURATION In a similar manner, a rise in temperature shifts the O_2–Hb curve to the right, resulting in more unloading of O_2 at a given P_{O_2}. An exercising muscle or other actively metabolizing cell produces heat. The resulting local rise in temperature enhances O_2 release from Hb for use by more active tissues.

COMPARISON OF THESE FACTORS AT THE TISSUE AND PULMONARY LEVELS As you just learned, increases in CO_2, acidity, and temperature at the tissue level, all of which are associated with increased cellular metabolism and increased O_2 consumption, enhance the effect of a drop in P_{O_2} in facilitating the release of O_2 from Hb. These effects are largely reversed at the pulmonary level, where the extra acid-forming CO_2 is blown off and the local aerated environment is cooler. Appropriately, therefore, Hb has a higher affinity for O_2 in the pulmonary capillary environment, enhancing the effect of raised P_{O_2} in loading O_2 onto Hb.

EFFECT OF 2,3-BISPHOSPHOGLYCERATE ON % Hb SATURATION The preceding changes take place in the *environment* of the red blood cells, but a factor *inside* the red blood cells can also affect the degree of O_2–Hb binding: **2,3-bisphosphoglycerate (BPG).** This erythrocyte constituent, which is produced during red blood cell metabolism, can bind reversibly with Hb and reduce its affinity for O_2, just as CO_2 and H^+ do. Thus, an increased level of BPG, like the other factors, shifts the O_2–Hb curve to the right, enhancing O_2 unloading as the blood flows through the tissues.

BPG production by red blood cells gradually increases whenever Hb in the arterial blood is chronically undersaturated—that is, when arterial HbO_2 is below normal. This condition may occur in people living at high altitudes or in those suffering from certain types of circulatory or respiratory diseases or anemia. By helping unload O_2 from Hb at the tissue level, increased BPG helps maintain O_2 availability for tissue use even though arterial O_2 supply is chronically reduced.

However, unlike the other factors—which normally are present only at the tissue level and thus shift the O_2–Hb curve to the right only in the systemic capillaries, where the shift is advantageous in unloading O_2—BPG is present in the red blood cells throughout the circulatory system and, accordingly, shifts the curve to the right to the same degree in both the tissues and the lungs. As a result, BPG decreases the ability to load O_2 at the pulmonary level, which is the negative side of increased BPG production.

(a) Alveolar P_{O_2} = blood P_{O_2}　　**(b)** Alveolar P_{O_2} > blood P_{O_2}　　**(c)** Alveolar P_{O_2} = blood P_{O_2}

KEY

　= O_2 molecule　　　= Partially saturated hemoglobin molecule　　　= Fully saturated hemoglobin molecule

● **FIGURE 13-29 Hemoglobin facilitating a large net transfer of O_2 by acting as a storage depot to keep P_{O_2} low.** (a) In the hypothetical situation in which no hemoglobin is present in the blood, the alveolar P_{O_2} and the pulmonary capillary blood P_{O_2} are at equilibrium. (b) Hemoglobin has been added to the pulmonary capillary blood. As the Hb starts to bind with O_2, it removes O_2 from solution. Because only dissolved O_2 contributes to blood P_{O_2}, the blood P_{O_2} falls below that of the alveoli, even though the same number of O_2 molecules are present in the blood as in part (a). By "soaking up" some of the dissolved O_2, Hb favors the net diffusion of more O_2 down its partial pressure gradient from the alveoli to the blood. (c) Hemoglobin is fully saturated with O_2, and the alveolar and blood P_{O_2} are at equilibrium again. The blood P_{O_2} resulting from dissolved O_2 is equal to the alveolar P_{O_2}, despite the fact that the total O_2 content in the blood is much greater than in part (a) when blood P_{O_2} was equal to alveolar P_{O_2} in the absence of Hb.

the systemic capillaries is considerably higher than the P_{O_2} of the surrounding tissue, O_2 immediately diffuses from the blood into the tissues, lowering blood P_{O_2}. When blood P_{O_2} falls, Hb must unload some stored O_2 because the % Hb saturation is reduced. As the O_2 released from Hb dissolves in the blood, blood P_{O_2} increases and once again exceeds the P_{O_2} of the surrounding tissues. This favors further movement of O_2 out of the blood, although the total quantity of O_2 in the blood has already fallen. Only when Hb can no longer release any more O_2 into solution (when Hb is unloaded to the greatest extent possible for the P_{O_2} existing at the systemic capillaries) can blood P_{O_2} fall as low as in surrounding tissue. At this time, further transfer of O_2 stops. Hemoglobin, because it stores a large quantity of O_2 that can be freed by a slight reduction in P_{O_2} at the systemic capillary level, permits the transfer of tremendously much more O_2 from the blood into the cells than would be possible in its absence.

Thus, Hb plays an important role in the *total quantity* of O_2 that the blood can pick up in the lungs and drop off in the tissues. If Hb levels fall to one half of normal, as in a severely anemic patient (see p. 397), the O_2-carrying capacity of the blood falls by 50% even though the arterial P_{O_2} is the normal 100 mm Hg with 97.5% Hb saturation. Only half as much Hb is available to be saturated, emphasizing once again how critical Hb is in determining how much O_2 can be picked up at the lungs and made available to tissues.

Factors at the tissue level promote the unloading of O_2 from hemoglobin.

Even though the main factor determining the % Hb saturation is the P_{O_2} of the blood, other factors can affect the affinity, or bond strength, between Hb and O_2 and, accordingly, can shift the O_2–Hb curve (that is, change the % Hb saturation at a given P_{O_2}). These other factors are CO_2, acidity, temperature, and 2,3-bisphosphoglycerate, which we will examine separately. The O_2–Hb dissociation curve with which you are already familiar (see ● Figure 13-28) is a typical curve at normal arterial CO_2 and acidity levels, normal body temperature, and normal 2,3-bisphosphoglycerate concentration.

EFFECT OF CO_2 ON % Hb SATURATION An increase in P_{CO_2} shifts the O_2–Hb curve to the right (● Figure 13-30). The % Hb saturation still depends on the P_{O_2}, but for any given P_{O_2}, less O_2 and Hb can be combined. This effect is important because the P_{CO_2} of the blood increases in the systemic capillaries as CO_2 diffuses down its gradient from the cells into the blood. The presence of this additional CO_2 in the blood in effect decreases the affinity

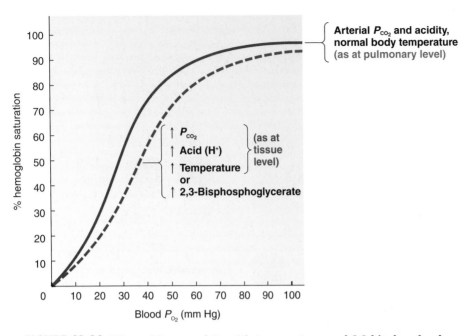

Arterial P_{CO_2} and acidity,
normal body temperature
(as at pulmonary level)

↑ P_{CO_2}
↑ Acid (H⁺)
↑ Temperature
or
↑ 2,3-Bisphosphoglycerate
(as at tissue level)

• **FIGURE 13-30 Effect of increased P_{CO_2}, H⁺, temperature, and 2,3-bisphosphoglycerate on the O_2–Hb curve.** Increased P_{O_2}, acid, and temperature, as found at the tissue level, shift the O_2–Hb curve to the right. As a result, less O_2 and Hb can be combined at a given P_{O_2} so that more O_2 is unloaded from Hb for use by the tissues. Similarly, 2,3-bisphosphoglycerate, whose production is increased in red blood cells when arterial HbO$_2$ is chronically below normal, shifts the O_2–Hb curve to the right, making more of the limited O_2 available at the tissue level.

cause their presence on Hb facilitates increased release of O_2 from Hb.)

EFFECT OF TEMPERATURE ON % Hb SATURATION In a similar manner, a rise in temperature shifts the O_2–Hb curve to the right, resulting in more unloading of O_2 at a given P_{O_2}. An exercising muscle or other actively metabolizing cell produces heat. The resulting local rise in temperature enhances O_2 release from Hb for use by more active tissues.

COMPARISON OF THESE FACTORS AT THE TISSUE AND PULMONARY LEVELS As you just learned, increases in CO_2, acidity, and temperature at the tissue level, all of which are associated with increased cellular metabolism and increased O_2 consumption, enhance the effect of a drop in P_{O_2} in facilitating the release of O_2 from Hb. These effects are largely reversed at the pulmonary level, where the extra acid-forming CO_2 is blown off and the local aerated environment is cooler. Appropriately, therefore, Hb has a higher affinity for O_2 in the pulmonary capillary environment, enhancing the effect of raised P_{O_2} in loading O_2 onto Hb.

of Hb for O_2, so Hb unloads even more O_2 at the tissue level than it would if the reduction in P_{O_2} in the systemic capillaries were the only factor affecting % Hb saturation.

EFFECT OF ACID ON % Hb SATURATION An increase in acidity also shifts the curve to the right. Because CO_2 generates carbonic acid (H_2CO_3), the blood becomes more acidic at the systemic capillary level as it picks up CO_2 from the tissues. The resulting reduction in Hb affinity for O_2 in the presence of increased acidity aids in releasing even more O_2 at the tissue level for a given P_{O_2}. In actively metabolizing cells, such as exercising muscles, not only is more carbonic acid–generating CO_2 produced, but lactate (lactic acid) also may be produced if the cells resort to anaerobic metabolism (see pp. 37 and 278). The resultant local elevation of acid in the working muscles facilitates further unloading of O_2 in the very tissues that need the most O_2.

BOHR EFFECT The influence of CO_2 and acid on the release of O_2 is known as the **Bohr effect.** Both CO_2 and the hydrogen ion (H⁺) component of acids can combine reversibly with Hb at sites other than the O_2-binding sites. The result is a change in the molecular structure of Hb that reduces its affinity for O_2. (Note that the % Hb saturation refers only to the extent to which Hb is combined with O_2, not the extent to which it is bound with CO_2, H⁺, or other molecules. Indeed, the % Hb saturation decreases when CO_2 and H⁺ bind with Hb, be-

EFFECT OF 2,3-BISPHOSPHOGLYCERATE ON % Hb SATURATION The preceding changes take place in the *environment* of the red blood cells, but a factor *inside* the red blood cells can also affect the degree of O_2–Hb binding: **2,3-bisphosphoglycerate (BPG).** This erythrocyte constituent, which is produced during red blood cell metabolism, can bind reversibly with Hb and reduce its affinity for O_2, just as CO_2 and H⁺ do. Thus, an increased level of BPG, like the other factors, shifts the O_2–Hb curve to the right, enhancing O_2 unloading as the blood flows through the tissues.

BPG production by red blood cells gradually increases whenever Hb in the arterial blood is chronically undersaturated—that is, when arterial HbO$_2$ is below normal. This condition may occur in people living at high altitudes or in those suffering from certain types of circulatory or respiratory diseases or anemia. By helping unload O_2 from Hb at the tissue level, increased BPG helps maintain O_2 availability for tissue use even though arterial O_2 supply is chronically reduced.

However, unlike the other factors—which normally are present only at the tissue level and thus shift the O_2–Hb curve to the right only in the systemic capillaries, where the shift is advantageous in unloading O_2—BPG is present in the red blood cells throughout the circulatory system and, accordingly, shifts the curve to the right to the same degree in both the tissues and the lungs. As a result, BPG decreases the ability to load O_2 at the pulmonary level, which is the negative side of increased BPG production.

ca = Carbonic anhydrase

● **FIGURE 13-31 Carbon dioxide transport in the blood.** Carbon dioxide (CO_2) picked up at the tissue level is transported in the blood to the lungs in three ways: (1) physically dissolved, (2) bound to hemoglobin (Hb), and (3) as bicarbonate ion (HCO_3^-). Hemoglobin is present only in the red blood cells, as is carbonic anhydrase, the enzyme that catalyzes the production of HCO_3^-. The H^+ generated during the production of HCO_3^- also binds to Hb. Bicarbonate moves by facilitated diffusion down its concentration gradient out of the red blood cell into the plasma, and chloride (Cl^-) moves by means of the same passive carrier into the red blood cell down the electrical gradient created by the outward diffusion of HCO_3^-. The reactions that occur at the tissue level are reversed at the pulmonary level, where CO_2 diffuses out of the blood to enter the alveoli.

Hemoglobin has a much higher affinity for carbon monoxide than for O_2.

 Carbon monoxide (CO) and O_2 compete for the same binding sites on Hb, but Hb's affinity for CO is 240 times that of its affinity for O_2. The combination of CO and Hb is known as **carboxyhemoglobin (HbCO).** Because Hb preferentially latches onto CO, even small amounts of CO can tie up a disproportionately large share of Hb, making Hb unavailable for O_2 transport. Even though the Hb concentration and P_{O_2} are normal, the O_2 content of the blood is seriously reduced.

Fortunately, CO is not a normal constituent of inspired air. It is a poisonous gas produced during the incomplete combustion (burning) of carbon products such as automobile gasoline, coal, wood, and tobacco. Carbon monoxide is especially dangerous because it is so insidious. If CO is being produced in a closed environment so that its concentration continues to increase (for example, in a parked car with the motor running and windows closed), it can reach lethal levels without the victim ever being aware of the danger. Because it is odorless, colorless,

tasteless, and nonirritating, CO is not detectable. Furthermore, for reasons described later the victim has no sensation of breathlessness and makes no attempt to increase ventilation, even though the cells are O_2 starved.

Most CO_2 is transported in the blood as bicarbonate.

When arterial blood flows through the tissue capillaries, CO_2 diffuses down its partial pressure gradient from the tissue cells into the blood. Carbon dioxide is transported in the blood in three ways (● Figure 13-31 and ▲ Table 13-6, p. 490):

1. *Physically dissolved.* As with dissolved O_2, the amount of CO_2 physically dissolved in the blood depends on the P_{CO_2}. Because CO_2 is more soluble than O_2 in plasma water, a greater proportion of the total CO_2 than of O_2 in the blood is physically dissolved. Even so, only 10% of the blood's total CO_2 content is carried this way at the normal systemic venous P_{CO_2} level.

2. *Bound to hemoglobin.* Another 30% of the CO_2 combines with Hb to form **carbamino hemoglobin (HbCO$_2$).** Carbon dioxide binds with the globin portion of Hb, in contrast to O_2, which combines with the heme portions. Reduced Hb has a greater affinity for CO_2 than HbO_2 does. The unloading of O_2 from Hb in the tissue capillaries therefore facilitates the picking up of CO_2 by Hb.

3. *As bicarbonate.* By far the most important means of CO_2 transport is as **bicarbonate (HCO_3^-)**, with 60% of the CO_2 being converted into HCO_3^- by the following chemical reaction:

$$CO_2 + H_2O \rightleftharpoons H_2CO_3 \rightleftharpoons H^+ + HCO_3^-$$

In the first step, CO_2 combines with H_2O to form **carbonic acid (H_2CO_3).** As is characteristic of acids, some of the carbonic acid molecules spontaneously dissociate into hydrogen ions (H^+) and bicarbonate ions (HCO_3^-). The one carbon and two oxygen atoms of the original CO_2 molecule are thus present in the blood as an integral part of HCO_3^-. This is beneficial because HCO_3^- is more soluble in the blood than CO_2 is.

This reaction takes place slowly in the plasma, but it proceeds swiftly within the red blood cells because of the presence of the erythrocyte enzyme **carbonic anhydrase,** which catalyzes (speeds up) the reaction. In fact, under the influence of carbonic anhydrase, the reaction proceeds directly from $CO_2 + H_2O$ to $H^+ + HCO_3^-$ without the intervening H_2CO_3 step:

$$CO_2 + H_2O \xrightarrow{\text{carbonic anhydrase}} H^+ + HCO_3^-$$

CHLORIDE SHIFT As this reaction proceeds, HCO_3^- and H^+ start to accumulate within the red blood cells in the systemic capillaries. The red blood cell membrane has a HCO_3^-–Cl^- carrier that passively facilitates the diffusion of these ions in opposite directions across the membrane. The membrane is relatively impermeable to H^+. Consequently, HCO_3^-, but not H^+, diffuses down its concentration gradient out of the erythrocytes into the plasma. Because HCO_3^- is a negatively charged ion, the efflux of HCO_3^- unaccompanied by a comparable outward diffusion of positively charged ions creates an electrical gradient (see p. 63). Chloride ions (Cl^-), the dominant plasma anions, diffuse into the red blood cells down this electrical gradient to restore electric neutrality. This inward shift of Cl^- in exchange for the efflux of CO_2-generated HCO_3^- is known as the **chloride (Cl^-) shift.**

HALDANE EFFECT Hemoglobin binds with most of the H^+ that is formed within the erythrocytes. As with CO_2, reduced Hb has a greater affinity for H^+ than HbO_2 does. Therefore, unloading O_2 facilitates Hb pickup of CO_2-generated H^+. Because only free, dissolved H^+ contributes to the acidity of a solution, the venous blood would be considerably more acidic than the arterial blood if Hb did not mop up most of the H^+ generated at the tissue level.

The fact that removing O_2 from Hb increases the ability of Hb to pick up CO_2 and CO_2-generated H^+ is known as the **Haldane effect.** The Haldane effect and Bohr effect work in synchrony to facilitate O_2 liberation and the uptake of CO_2 and CO_2-generated H^+ at the tissue level. Increased CO_2 and H^+

cause increased O_2 release from Hb by the Bohr effect; increased O_2 release from Hb, in turn, causes increased CO_2 and H^+ uptake by Hb through the Haldane effect. The entire process is very efficient. Reduced Hb must be carried back to the lungs to refill on O_2 anyway. Meanwhile, after O_2 is released, Hb picks up new passengers—CO_2 and H^+—that are going in the same direction to the lungs.

The reactions at the tissue level as CO_2 enters the blood from the tissues are reversed once the blood reaches the lungs and CO_2 leaves the blood to enter the alveoli (● Figure 13-31).

Various respiratory states are characterized by abnormal blood-gas levels.

▲ Table 13-7 is a glossary of terms used to describe various states associated with respiratory abnormalities.

ABNORMALITIES IN ARTERIAL P_{O_2} The term **hypoxia** refers to the condition of having insufficient O_2 at the cell level. There are four general categories of hypoxia:

1. *Hypoxic hypoxia* is characterized by a low arterial blood P_{O_2} accompanied by inadequate Hb saturation. It is caused by (a) a respiratory malfunction involving inadequate gas exchange, typified by a normal alveolar P_{O_2} but a reduced arterial P_{O_2}, or (b) exposure to high altitude or to a suffocating environment where atmospheric P_{O_2} is reduced so that alveolar and arterial P_{O_2} are likewise reduced.

2. *Anemic hypoxia* is a reduced O_2-carrying capacity of the blood. It can result from (a) a decrease in circulating red blood cells, (b) an inadequate amount of Hb within the red blood cells, or (c) CO poisoning. In all cases of anemic hypoxia, the arterial P_{O_2} is normal but the O_2 content of the arterial blood is lower than normal because of the reduction in available Hb.

3. *Circulatory hypoxia* arises when too little oxygenated blood is delivered to the tissues. Circulatory hypoxia can be restricted to a limited area by a local vascular spasm or blockage. Or the body may experience circulatory hypoxia in general, from congestive heart failure or circulatory shock. The arterial P_{O_2} and O_2 content are typically normal, but too little oxygenated blood reaches the cells.

4. In *histotoxic hypoxia,* O_2 delivery to the tissues is normal, but the cells cannot use the O_2 available to them. The classic example is *cyanide poisoning.* Cyanide blocks cellular enzymes essential for cellular respiration (enzymes in the electron transport system; see p. 35).

Hyperoxia, an above-normal arterial P_{O_2}, cannot occur when a person is breathing atmospheric air at sea level. However, breathing supplemental O_2 can increase alveolar and, consequently, arterial P_{O_2}. Because more of the inspired air is O_2, more of the total pressure of the inspired air is attributable to the O_2 partial pressure, so more O_2 dissolves in the blood before arterial P_{O_2} equilibrates with alveolar P_{O_2}. Even though arterial P_{O_2} increases, the total blood O_2 content does not significantly increase because Hb is nearly fully saturated at the normal arterial P_{O_2}. In certain pulmonary diseases associated with a reduced arterial P_{O_2}, however, breathing supplemental O_2 can help establish a larger alveoli-to-blood driving gradient, improving arterial P_{O_2}. Far from being advantageous, in contrast, a markedly elevated arterial P_{O_2} can be dangerous. If the arterial P_{O_2} is too high, **oxygen toxicity** can occur. Even though the total O_2 content of the blood is only slightly increased, exposure to a high P_{O_2} can damage some cells. In particular, brain damage and blindness-causing damage to the retina are associated with O_2 toxicity. Therefore, O_2 therapy must be administered cautiously.

ABNORMALITIES IN ARTERIAL P_{CO_2} The term **hypercapnia** refers to the condition of having excess CO_2 in arterial blood; it is caused by **hypoventilation** (ventilation inadequate to meet metabolic needs for O_2 delivery and CO_2 removal). With most lung diseases, CO_2 accumulates in arterial blood concurrently with an O_2 deficit because both O_2 and CO_2 exchange between lungs and atmosphere are equally affected

● **FIGURE 13-32 Effects of hyperventilation and hypoventilation on arterial P_{O_2} and P_{CO_2}.**

(● Figure 13-32). However, when a decrease in arterial P_{O_2} is the result of reduced pulmonary diffusing capacity, as in pulmonary edema or emphysema, O_2 transfer suffers more than CO_2 transfer because the diffusion constant for CO_2 is 20 times that of O_2. As a result, in these circumstances hypoxic hypoxia occurs much more readily than hypercapnia.

Hypocapnia, below-normal arterial P_{CO_2} levels, is brought about by hyperventilation. **Hyperventilation** occurs when a person "overbreathes," that is, when the rate of ventilation exceeds the body's metabolic needs for CO_2 removal. As a result, CO_2 is blown off to the atmosphere more rapidly than it is produced in the tissues, and arterial P_{CO_2} falls. Hyperventilation can be triggered by anxiety states, fever, and aspirin poisoning. Alveolar P_{O_2} increases during hyperventilation as more fresh O_2 is delivered to the alveoli from the atmosphere than the blood extracts from the alveoli for tissue consumption, and arterial P_{O_2} increases correspondingly (● Figure 13-32). However, because Hb is almost fully saturated at the normal arterial P_{O_2}, very little additional O_2 is added to the blood. Except for the small extra amount of dissolved O_2, blood O_2 content remains essentially unchanged during hyperventilation.

Increased ventilation is not synonymous with hyperventilation. Increased ventilation that matches an increased metabolic demand, such as the increased need for O_2 delivery and CO_2 elimination during exercise, is termed **hyperpnea.** During exercise, alveolar and arterial P_{O_2} and P_{CO_2} remain constant, with the increased atmospheric exchange just keeping pace with the increased O_2 consumption and CO_2 production.

Effects of Heights and Depths on the Body

Our bodies are optimally equipped for existence at normal atmospheric pressure. Ascent into mountains high above sea level or descent into the depths of the ocean can have adverse effects on the body.

Effects of High Altitude on the Body

Atmospheric pressure progressively declines as altitude increases. At 18,000 ft above sea level, atmospheric pressure is only 380 mm Hg—half of its normal sea-level value. Because the proportion of O_2 and N_2 in the air remains the same, the P_{O_2} of inspired air at this altitude is 21% of 380 mm Hg, or 80 mm Hg, with alveolar P_{O_2} being even lower at 45 mm Hg. At any altitude above 10,000 ft, the arterial P_{O_2} falls into the steep portion of the O_2–Hb curve, below the safety range of the plateau region. As a result, the % Hb saturation in the arterial blood declines precipitously with further increases in altitude.

People who rapidly ascend to altitudes of 10,000 ft or more experience symptoms of **acute mountain sickness** attributable to hypoxic hypoxia and the resultant hypocapnia-induced alkalosis. The increased ventilatory drive to obtain more O_2 causes respiratory alkalosis because acid-forming CO_2 is blown off more rapidly than

it is produced. Symptoms of mountain sickness include fatigue, nausea, loss of appetite, labored breathing, rapid heart rate (triggered by hypoxia as a compensatory measure to increase circulatory delivery of available O_2 to the tissues), and nerve dysfunction characterized by poor judgment, dizziness, and incoordination.

Despite these acute responses to high altitude, millions of people live at elevations above 10,000 ft, with some villagers even residing in the Andes at altitudes higher than 16,000 ft. How do they live and function normally? They do so through the process of **acclimatization.** When a person remains at high altitude, the acute compensatory responses of increased ventilation and increased cardiac output are gradually replaced over a period of days by more slowly developing compensatory measures that permit adequate oxygenation of the tissues and restoration of normal acid–base balance. Red blood cell (RBC) production is increased, stimulated by erythropoietin in response to reduced O_2 delivery to the kidneys (see p. 395). The rise in the number of RBCs increases the O_2-carrying capacity of the blood. Hypoxia also promotes the synthesis of BPG within the RBCs, so that O_2 is unloaded from Hb more easily at the tissues. The number of capillaries within the tissues is increased, reducing the distance that O_2 must diffuse from the blood to reach the cells. Also, the high inhabitants' endothelial cells release up to 10 times more nitric oxide (NO) than is released in those dwelling near sea level. This extra NO more than doubles blood flow in the acclimatized individuals.

© F. Westmorland/Photo Researchers, Inc.

Clinical Note **CONSEQUENCES OF ABNORMALITIES IN ARTERIAL BLOOD GASES** The consequences of reduced O_2 availability to the tissues during hypoxia are apparent. The cells need adequate O_2 supplies to sustain energy-generating metabolic activities. The consequences of abnormal blood CO_2 levels are less obvious. Changes in blood CO_2 concentration primarily affect acid–base balance. Hypercapnia elevates production of CO_2-generated H^+. The subsequent generation of excess H^+ produces an acidic condition termed *respiratory acidosis*. Conversely, less-than-normal amounts of H^+ are generated from CO_2 in conjunction with hypocapnia. The resultant alkalotic (less acidic than normal) condition is called *respiratory alkalosis* (Chapter 15). (To learn about the effects of mountain climbing and deep-sea diving on blood gases, see the

accompanying boxed feature, ■ Concepts, Challenges, and Controversies.)

Control of Respiration

Like the heartbeat, breathing must occur in a continuous, cyclic pattern to sustain life processes. Cardiac muscle must rhythmically contract and relax to alternately empty blood from the heart and fill it again. Similarly, inspiratory muscles must rhythmically contract and relax to alternately fill the lungs with air and empty them. Both these activities are accomplished automatically, without conscious effort. However, the underlying mechanisms and control of these two systems are remarkably different.

Furthermore, acclimatized cells are able to use O_2 more efficiently through an increase in the number of mitochondria, the energy organelles (see p. 32). The kidneys restore the arterial pH to nearly normal by conserving acid that normally would have been lost in the urine.

These compensatory measures are not without undesirable trade-offs. For example, the greater number of RBCs increases blood viscosity (makes the blood "thicker"), thereby increasing resistance to blood flow. As a result, the heart has to work harder to pump blood through the vessels (see p. 345).

Effects of Deep-Sea Diving on the Body

When a deep-sea diver descends underwater, the body is exposed to greater than atmospheric pressure. Pressure rapidly increases with sea depth as a result of the weight of the water. Pressure is already doubled at about 30 ft below sea level. The air provided by scuba equipment is delivered to the lungs at these high pressures. Recall that (1) the amount of a gas in solution is directly proportional to the partial pressure of the gas and (2) air is composed of 79% N_2. Nitrogen is poorly soluble in body tissues, but the high P_{N_2} that occurs during deep-sea diving causes more of this gas than normal to dissolve in the body tissues. The small amount of N_2 dissolved in the tissues at sea level has no known effect, but as more N_2 dissolves at greater depths, **nitrogen narcosis,** or **"rapture of the deep,"** ensues. Nitrogen narcosis is believed to result from a reduction in the excitability of neurons because of the highly lipid-soluble N_2 dissolving in their lipid membranes. At 150 ft underwater, divers experience a feeling of euphoria and become drowsy, similar to the effect of having a few cocktails. At lower depths, divers become weak and clumsy, and at 350 to 400 ft, they lose consciousness. **Oxygen toxicity** resulting from the high P_{O_2} is another possible detrimental effect of descending deep underwater.

Another problem associated with deep-sea diving occurs during ascent. If a diver who has been submerged long enough for a significant amount of N_2 to dissolve in the tissues suddenly ascends to the surface, the rapid reduction in P_{N_2} causes N_2 to quickly come out of solution and form bubbles of gaseous N_2 in the body, much as bubbles of gaseous CO_2 form in a bottle of champagne when the cork is popped. The consequences depend on the amount and location of N_2 bubble formation in the body. This condition is called **decompression sickness** or **"the bends"** because the victim often bends over in pain. Decompression sickness can be prevented by ascending slowly to the surface or by decompressing gradually in a decompression tank so that the excess N_2 can slowly escape through the lungs without bubble formation.

© Dave B. Fleetham/Visuals Unlimited

Respiratory centers in the brain stem establish a rhythmic breathing pattern.

Whereas the heart can generate its own rhythm by means of its intrinsic pacemaker activity, the respiratory muscles, being skeletal muscles, contract only when stimulated by their nerve supply. The rhythmic pattern of breathing is established by cyclic neural activity to the respiratory muscles. In other words, the pacemaker activity that establishes breathing rhythm resides in the respiratory control centers in the brain, not in the lungs or respiratory muscles themselves. The nerve supply to the heart, not being needed to initiate the heartbeat, only modifies the rate and strength of cardiac contraction. In contrast, the nerve supply to the respiratory system is absolutely essential in maintaining breathing and in reflexly adjusting the level of ventilation to match changing needs for O_2 uptake and CO_2 removal. Furthermore, unlike cardiac activity, which is not subject to voluntary control, respiratory activity can be voluntarily modified to accomplish speaking, singing, whistling, playing a wind instrument, or holding one's breath while swimming.

COMPONENTS OF NEURAL CONTROL OF RESPIRATION Neural control of respiration involves three distinct components: (1) factors that generate the alternating inspiration/expiration rhythm, (2) factors that regulate the magnitude of ventilation (that is, the rate and depth of breathing) to match body needs, and (3) factors that modify respiratory activity to serve other purposes. The latter modifications may be either voluntary, as in the breath control required for speech, or involuntary, as in the respiratory maneuvers involved in a cough or sneeze.

Respiratory control centers housed in the brain stem generate the rhythmic pattern of breathing. The primary respiratory control center, the *medullary respiratory center,* consists of several aggregations of neuronal cell bodies within the medulla that provide output to the respiratory muscles. In addition, two other respiratory centers lie higher in the brain stem in the pons—the *pneumotaxic center* and *apneustic center.* These pontine centers influence output from the medullary respiratory center (● Figure 13-33). Here is a description of how these various regions interact to establish respiratory rhythmicity.

INSPIRATORY AND EXPIRATORY NEURONS IN THE MEDULLARY CENTER We rhythmically breathe in and out during quiet breathing because of alternate contraction and relaxation of the inspiratory muscles, namely the diaphragm and external intercostal muscles, supplied by the phrenic nerve and intercostal nerves, respectively. The cell bodies for the neuronal fibers composing these nerves are located in the spinal cord. Impulses originating in the medullary center end on these motor-neuron cell bodies. When these motor neurons are activated, they in turn stimulate the inspiratory muscles, leading to inspiration; when these neurons are not firing, the inspiratory muscles relax, and expiration takes place.

The **medullary respiratory center** consists of two neuronal clusters known as the *dorsal respiratory group* and the *ventral respiratory group* (● Figure 13-33).

■ The **dorsal respiratory group (DRG)** consists mostly of *inspiratory neurons* whose descending fibers terminate on the motor neurons that supply the inspiratory muscles. When the DRG inspiratory neurons fire, inspiration takes place; when they cease firing, expiration occurs. Expiration is brought to an end as the inspiratory neurons once again reach threshold and fire. The DRG has important interconnections with the ventral respiratory group.

■ The **ventral respiratory group (VRG)** is composed of *inspiratory neurons* and *expiratory neurons,* both of which remain inactive during normal quiet breathing. This region is called into play by the DRG as an "overdrive" mechanism during periods when demands for ventilation are increased. It is especially important in active expiration. No impulses are generated in the descending pathways from the expiratory neurons during quiet breathing. Only during active expiration do the expiratory neurons stimulate the motor neurons supplying the expiratory muscles (the abdominal and internal intercostal muscles). Furthermore, the VRG inspiratory neurons, when stimulated by the DRG, rev up inspiratory activity when demands for ventilation are high.

GENERATION OF RESPIRATORY RHYTHM Until recently, the DRG was generally thought to generate the basic rhythm of ventilation. However, generation of respiratory rhythm is now widely believed to lie in the **pre-Bötzinger complex,** a region located near the upper (head) end of the medullary respiratory center (● Figure 13-33). A network of neurons in this region display pacemaker activity, undergoing self-induced action potentials similar to those of the SA node of the heart. Scientists believe the rate at which the DRG inspiratory neurons rhythmically fire is driven by synaptic input from this complex.

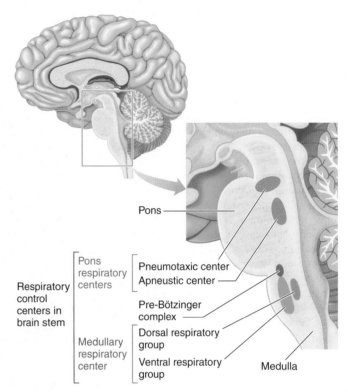

● **FIGURE 13-33 Respiratory control centers in the brain stem.**

INFLUENCES FROM THE PNEUMOTAXIC AND APNEUSTIC CENTERS The respiratory centers in the pons exert "fine-tuning" influences over the medullary center to help produce normal, smooth inspirations and expirations. The **pneumotaxic center** sends impulses to the DRG that help "switch off" the inspiratory neurons, limiting the duration of inspiration. In contrast, the **apneustic center** prevents the inspiratory neurons from being switched off, thus providing an extra boost to the inspiratory drive. In this check-and-balance system, the pneumotaxic center dominates over the apneustic center, helping halt inspiration and letting expiration occur normally. Without the pneumotaxic brakes, the breathing pattern consists of prolonged inspiratory gasps abruptly interrupted by very brief expirations. This abnormal breathing pattern is known as **apneusis;** hence, the center that promotes this type of breathing is the apneustic center. Apneusis occurs in certain types of severe brain damage.

HERING–BREUER REFLEX When the tidal volume is large (greater than 1 liter), as during exercise, the **Hering–Breuer reflex** is triggered to prevent overinflation of the lungs. **Pulmonary stretch receptors** within the smooth muscle layer of the airways are activated by stretching of the lungs at large tidal volumes. Action potentials from these stretch receptors travel through afferent nerve fibers to the medullary center and inhibit the inspiratory neurons. This negative feedback from the highly stretched lungs helps cut inspiration short before the lungs become overinflated.

Influence of Chemical Factors on Respiration

Chemical Factor	Effect on the Peripheral Chemoreceptors	Effect on the Central Chemoreceptors
↓ P_{O_2} in the Arterial Blood	Stimulates only when the arterial P_{O_2} has fallen to the point of being life threatening (< 60 mm Hg); an emergency mechanism	Directly depresses the central chemoreceptors and the respiratory center itself when < 60 mm Hg
↑ P_{CO_2} in the Arterial Blood (↑ H⁺ in the Brain ECF)	Weakly stimulates	Strongly stimulates; is the dominant control of ventilation (Levels > 70–80 mm Hg directly depress the respiratory center and central chemoreceptors)
↑ H⁺ in the Arterial Blood	Stimulates; important in acid–base balance	Does not affect; cannot penetrate the blood–brain barrier

The magnitude of ventilation is adjusted in response to three chemical factors: P_{O_2}, P_{CO_2}, and H⁺.

No matter how much O_2 is extracted from the blood or how much CO_2 is added to it at the tissue level, the P_{O_2} and P_{CO_2} of the systemic arterial blood leaving the lungs are held remarkably constant, indicating that arterial blood-gas content is precisely regulated. Arterial blood gases are maintained within the normal range by varying the magnitude of ventilation (rate and depth of breathing) to match the body's needs for O_2 uptake and CO_2 removal. If more O_2 is extracted from the alveoli and more CO_2 is dropped off by the blood because the tissues are metabolizing more actively, ventilation increases correspondingly to bring in more fresh O_2 and blow off more CO_2.

The medullary respiratory center receives inputs that provide information about the body's needs for gas exchange. It responds by sending appropriate signals to the motor neurons supplying the respiratory muscles, to adjust the rate and depth of ventilation to meet those needs. The two most obvious signals to increase ventilation are a decreased arterial P_{O_2} or an increased arterial P_{CO_2}. Intuitively, you would suspect that if O_2 levels in the arterial blood declined or if CO_2 accumulated, ventilation would be stimulated to obtain more O_2 or to eliminate excess CO_2. These two factors do indeed influence the magnitude of ventilation, but not to the same degree nor through the same pathway. Also, a third chemical factor, H⁺, notably influences the level of respiratory activity. We will examine the role of each of these important chemical factors in the control of ventilation (▲ Table 13-8).

Decreased arterial P_{O_2} increases ventilation only as an emergency mechanism.

Arterial P_{O_2} is monitored by **peripheral chemoreceptors** known as the **carotid bodies** and **aortic bodies,** which lie at the fork of the common carotid arteries on both the right and left sides and in the arch of the aorta, respectively (● Figure 13-34). These

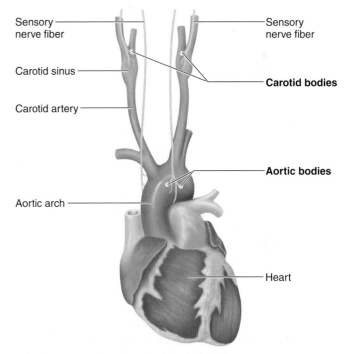

Sensory nerve fiber

Sensory nerve fiber

Carotid sinus

Carotid bodies

Carotid artery

Aortic bodies

Aortic arch

Heart

● **FIGURE 13-34 Location of the peripheral chemoreceptors.** The carotid bodies are located in the carotid sinus, and the aortic bodies are located in the aortic arch.

chemoreceptors respond to specific changes in the chemical content of the arterial blood that bathes them. They are distinctly different from the carotid sinus and aortic arch baroreceptors located in the same vicinity. The latter monitor pressure changes rather than chemical changes and are important in regulating systemic arterial blood pressure (see p. 378).

EFFECT OF A LARGE DECREASE IN P_{O_2} ON THE PERIPHERAL CHEMORECEPTORS The peripheral chemoreceptors are not sensitive to modest reductions in arterial P_{O_2}. The arterial P_{O_2} must fall below 60 mm Hg ($>$ 40% reduction) before the peripheral chemoreceptors respond by sending afferent impulses to the medullary inspiratory neurons, thereby reflexly increasing ventilation. Because arterial P_{O_2} falls below 60 mm Hg only in the unusual circumstances of severe pulmonary disease or reduced atmospheric P_{O_2}, it does not play a role in the normal ongoing regulation of respiration. This fact may seem surprising at first because a primary function of ventilation is to provide enough O_2 for uptake by the blood. However, there is no need to increase ventilation until the arterial P_{O_2} falls below 60 mm Hg, because of the safety margin in % Hb saturation afforded by the plateau portion of the O_2–Hb curve. Hemoglobin is still 90% saturated at an arterial P_{O_2} of 60 mm Hg, but the % Hb saturation drops precipitously when the P_{O_2} falls below this level. Therefore, reflex stimulation of respiration by the peripheral chemoreceptors serves as an important emergency mechanism in dangerously low arterial P_{O_2} states. Indeed, this reflex mechanism is a lifesaver because a low arterial P_{O_2} tends to directly depress the respiratory center, as it does all the rest of the brain.

Clinical Note Because the peripheral chemoreceptors respond to the P_{O_2} of the blood, *not* the total O_2 content of the blood, O_2 content in the arterial blood can fall to dangerously low or even fatal levels without the peripheral chemoreceptors ever responding to reflexly stimulate respiration. Remember that only physically dissolved O_2 contributes to blood P_{O_2}. The total O_2 content in the arterial blood can be reduced in anemic states, in which O_2-carrying Hb is reduced, or in CO poisoning, when Hb preferentially binds to this molecule rather than to O_2. In both cases, arterial P_{O_2} is normal, so respiration is not stimulated, even though O_2 delivery to the tissues may be so reduced that the person dies from cellular O_2 deprivation.

DIRECT EFFECT OF A LARGE DECREASE IN P_{O_2} ON THE RESPIRATORY CENTER Except for the peripheral chemoreceptors, the level of activity in all nervous tissue falls in O_2 deprivation. Were it not for stimulatory intervention of the peripheral chemoreceptors when the arterial P_{O_2} falls threateningly low, a vicious cycle ending in cessation of breathing would ensue. Direct depression of the respiratory center by the markedly low arterial P_{O_2} would further reduce ventilation, leading to an even greater fall in arterial P_{O_2}, which would even further depress the respiratory center until ventilation ceased and death occurred.

Carbon dioxide–generated H^+ in the brain is normally the main regulator of ventilation.

In contrast to arterial P_{O_2}, which does not contribute to the minute-to-minute regulation of respiration, arterial P_{CO_2} is the most important input regulating the magnitude of ventilation under resting conditions. This role is appropriate, because changes in alveolar ventilation have an immediate and pronounced effect on arterial P_{CO_2}. By contrast, changes in ventilation have little effect on % Hb saturation and O_2 availability to the tissues until the arterial P_{O_2} falls by more than 40%. Even slight alterations from normal in arterial P_{CO_2} induce a significant reflex effect on ventilation. An increase in arterial P_{CO_2} reflexly stimulates the respiratory center, with the resultant in-

crease in ventilation promoting elimination of the excess CO_2 to the atmosphere. Conversely, a fall in arterial P_{CO_2} reflexly reduces the respiratory drive. The subsequent decrease in ventilation lets metabolically produced CO_2 accumulate so that P_{CO_2} can be returned to normal.

EFFECT OF INCREASED P_{CO_2} ON THE CENTRAL CHEMORECEPTORS Surprisingly, given the key role of arterial P_{CO_2} in regulating respiration, no important receptors monitor arterial P_{CO_2} per se. The carotid and aortic bodies are only weakly responsive to changes in arterial P_{CO_2}, so they play only a minor role in reflexly stimulating ventilation in response to an elevation in arterial P_{CO_2}. More important in linking changes in arterial P_{CO_2} to compensatory adjustments in ventilation are the **central chemoreceptors,** located in the medulla near the respiratory center. These central chemoreceptors do not monitor CO_2 itself; however, they are sensitive to changes in CO_2-induced H^+ concentration in the brain extracellular fluid (ECF) that bathes them.

Movement of materials across the brain capillaries is restricted by the blood–brain barrier (see p. 141). Because this barrier is readily permeable to CO_2, any increase in arterial P_{CO_2} causes a similar rise in brain-ECF P_{CO_2} as CO_2 diffuses down its pressure gradient from the cerebral blood vessels into the brain ECF. Under the influence of carbonic anhydrase, the increased P_{CO_2} within the brain ECF correspondingly raises the concentration of H^+ according to the law of mass action as it applies to this reaction: $CO_2 + H_2O \rightleftharpoons H^+ + HCO_3^-$. An elevation in H^+ concentration in the brain ECF directly stimulates the central chemoreceptors, which in turn increase ventilation by stimulating the respiratory center through synaptic connections ($\bullet$ Figure 13-35). As the excess CO_2 is subsequently blown off, the arterial P_{CO_2} and the P_{CO_2} and H^+ concentration of the brain ECF return to normal. Conversely, a decline in arterial P_{CO_2} below normal is paralleled by a fall in P_{CO_2} and H^+ in the brain ECF, the result of which is a central chemoreceptor–mediated decrease in ventilation. As CO_2 produced by cell metabolism is consequently allowed to accumulate, arterial P_{CO_2} and P_{CO_2} and H^+ of the brain ECF are restored toward normal.

Unlike CO_2, H^+ cannot readily permeate the blood–brain barrier, so H^+ in the plasma cannot gain access to the central chemoreceptors. Accordingly, the central chemoreceptors respond only to H^+ generated within the brain ECF itself as a result of CO_2 entry. Thus, the major mechanism controlling ventilation under resting conditions is specifically aimed at regulating the brain-ECF H^+ concentration, which in turn directly reflects the arterial P_{CO_2}. Unless there are extenuating circumstances such as reduced availability of O_2 in the inspired air, arterial P_{O_2} is coincidentally also maintained at its normal value by the brain-ECF H^+ ventilatory driving mechanism.

The powerful influence of the central chemoreceptors on the respiratory center is responsible for your inability to deliberately hold your breath for more than about a minute. While you hold your breath, metabolically produced CO_2 continues to accumulate in your blood and then to build up the H^+ concentration in your brain ECF. Finally, the increased P_{CO_2}–H^+ stimulant to respiration becomes so powerful that central chemoreceptor excitatory input overrides voluntary inhibitory input to respiration, so breathing resumes despite deliberate at-

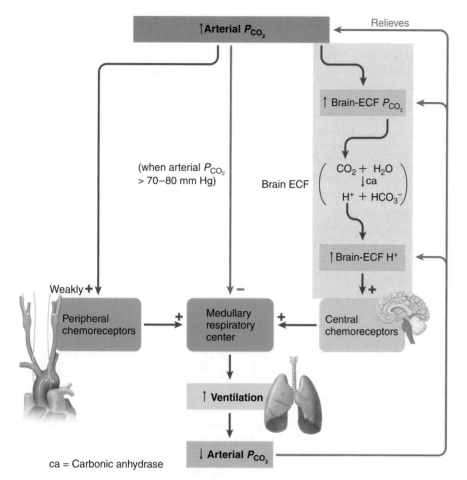

↑ Arterial P_{CO_2}

Relieves

↑ Brain-ECF P_{CO_2}

(when arterial P_{CO_2} > 70–80 mm Hg)

Brain ECF

$$\begin{pmatrix} CO_2 + H_2O \\ \downarrow ca \\ H^+ + HCO_3^- \end{pmatrix}$$

↑ Brain-ECF H^+

Weakly **+**

Peripheral chemoreceptors **+** → Medullary respiratory center **−**

+ Central chemoreceptors

↑ **Ventilation**

↓ **Arterial P_{CO_2}**

ca = Carbonic anhydrase

● **FIGURE 13-35 Effect of increased arterial P_{CO_2} on ventilation.**

tempts to prevent it. Breathing resumes long before arterial P_{O_2} falls to the threateningly low levels that trigger the peripheral chemoreceptors. Therefore, you cannot deliberately hold your breath long enough to create a dangerously high level of CO_2 or low level of O_2 in the arterial blood.

DIRECT EFFECT OF A LARGE INCREASE IN P_{CO_2} ON THE RESPIRATORY CENTER In contrast to the normal reflex stimulatory effect of the increased P_{CO_2}–H^+ mechanism on respiratory activity, very high levels of CO_2 directly depress the entire brain, including the respiratory center, just as very low levels of O_2 do. Up to a P_{CO_2} of 70 to 80 mm Hg, progressively higher P_{CO_2} levels induce correspondingly more vigorous respiratory efforts in an attempt to blow off the excess CO_2. A further increase in P_{CO_2} beyond 70 to 80 mm Hg, however, does not further increase ventilation but actually depresses the respiratory neurons. For this reason, CO_2 must be removed and O_2 supplied in closed environments such as closed-system anesthesia machines, submarines, or space capsules. Otherwise, CO_2 could reach lethal levels, not only because it depresses respiration but also because it produces severe respiratory acidosis.

 LOSS OF SENSITIVITY TO P_{CO_2} WITH LUNG DISEASE During prolonged hypoventilation caused by certain types of chronic lung disease, an elevated P_{CO_2} occurs simul-

taneously with a markedly reduced P_{O_2}. In most cases, the elevated P_{CO_2} (acting via the central chemoreceptors) and the reduced P_{O_2} (acting via the peripheral chemoreceptors) are *synergistic;* that is, the combined stimulatory effect on respiration exerted by these two inputs together is greater than the sum of their independent effects.

However, some patients with severe chronic lung disease lose their sensitivity to an elevated arterial P_{CO_2}. In a prolonged increase in H^+ generation in the brain ECF, from longstanding CO_2 retention, enough HCO_3^- may cross the blood–brain barrier to buffer, or "neutralize," the excess H^+. The additional HCO_3^- combines with the excess H^+, removing it from solution so that it no longer contributes to free H^+ concentration. When brain-ECF HCO_3^- concentration rises, brain-ECF H^+ concentration returns to normal although arterial P_{CO_2} and brain-ECF P_{CO_2} remain high. The central chemoreceptors are no longer aware of the elevated P_{CO_2}, because the brain-ECF H^+ is normal. Because the central chemoreceptors no longer reflexly stimulate the respiratory center in response to the elevated P_{CO_2}, the drive to eliminate CO_2 is blunted in such patients; that is, their level of ventilation is abnormally low considering their high arterial P_{CO_2}. In these patients, the hypoxic drive to ventilation becomes their primary respiratory stimulus, in contrast to normal individuals, in whom the arterial P_{CO_2} level is the dominant factor governing the magnitude of ventilation. Ironically, administering O_2 to such patients to relieve the hypoxic condition can markedly depress their drive to breathe by elevating the arterial P_{O_2} and removing the primary driving stimulus for respiration. Thus, O_2 therapy must be administered cautiously in patients with long-term pulmonary diseases.

Adjustments in ventilation in response to changes in arterial H^+ are important in acid–base balance.

Changes in arterial H^+ concentration cannot influence the central chemoreceptors because H^+ does not readily cross the blood–brain barrier. However, the aortic and carotid body peripheral chemoreceptors are highly responsive to fluctuations in arterial H^+ concentration, in contrast to their weak sensitivity to deviations in arterial P_{CO_2} and their unresponsiveness to arterial P_{O_2} until it falls 40% below normal.

Any change in arterial P_{CO_2} brings about a corresponding change in the H^+ concentration of the blood as well as of the brain ECF. These CO_2-induced H^+ changes in the arterial blood are detected by the peripheral chemoreceptors; the result is reflexly stimulated ventilation in response to increased arterial H^+ concentration and depressed ventilation in association with decreased arterial H^+ concentration. However, these changes in ventilation mediated by the peripheral chemoreceptors are far less important than the powerful central-chemoreceptor mechanism in adjusting ventilation in response to changes in CO_2-generated H^+ concentration.

The peripheral chemoreceptors do play a major role in adjusting ventilation in response to alterations in arterial H^+ concentration unrelated to fluctuations in P_{CO_2}. In many situations, even though P_{CO_2} is normal, arterial H^+ concentration is changed by the addition or loss of non-CO_2-generated acid from the body. For example, arterial H^+ concentration increases during diabetes mellitus because excess H^+-generating keto acids are abnormally produced and added to the blood. A rise in arterial H^+ concentration reflexly stimulates ventilation by means of the peripheral chemoreceptors. Conversely, the peripheral chemoreceptors reflexly suppress respiratory activity in response to a fall in arterial H^+ concentration resulting from nonrespiratory causes. Changes in ventilation by this mechanism are extremely important in regulating the body's acid–base balance. Changing the magnitude of ventilation can vary the amount of H^+-generating CO_2 eliminated. The resulting adjustment in the amount of H^+ added to the blood from CO_2 can compensate for the nonrespiratory-induced abnormality in arterial H^+ concentration that first elicited the respiratory response. (See Chapter 15 for further details.)

Exercise profoundly increases ventilation, but the mechanisms involved are unclear.

Alveolar ventilation may increase up to 20-fold during heavy exercise to keep pace with the increased demand for O_2 uptake and CO_2 output. (▲ Table 13-9 highlights changes in O_2- and CO_2-related variables during exercise.) The cause of increased ventilation during exercise is still largely speculative. It would seem logical that changes in the "big three" chemical factors—decreased P_{O_2}, increased P_{CO_2}, and increased H^+—could account for the increase in ventilation. This does not appear to be the case, however.

■ Despite the marked increase in O_2 use during exercise, arterial P_{O_2} does not decrease but remains normal or may actually increase slightly because the increase in alveolar ventilation keeps pace with or even slightly exceeds the stepped-up rate of O_2 consumption.

■ Likewise, despite the marked increase in CO_2 production during exercise, arterial P_{CO_2} does not increase but remains normal or decreases slightly because the extra CO_2 is removed as rapidly or even more rapidly than it is produced by the increase in ventilation.

■ During mild or moderate exercise, H^+ concentration does not increase, because H^+-generating CO_2 is held constant. During heavy exercise, H^+ concentration does increase somewhat from release of H^+-generating lactate (lactic acid) into the blood by anaerobic metabolism in the exercising muscles. Even

so, the elevation in H^+ concentration resulting from lactic acid formation is not enough to account for the large increase in ventilation accompanying exercise.

Some investigators argue that the constancy of the three chemical regulatory factors during exercise shows that ventilatory responses to exercise are actually being controlled by these factors—particularly by P_{CO_2}, because it is normally the dominant control during resting conditions. According to this reasoning, how else could alveolar ventilation be increased in exact proportion to CO_2 production, thereby keeping the P_{CO_2} constant? This proposal, however, cannot account for the observation that during heavy exercise, alveolar ventilation may increase relatively more than CO_2 production increases, thereby actually causing a slight decline in P_{CO_2}. Also, ventilation increases abruptly at the onset of exercise (within seconds), long before changes in arterial blood gases could become important influences on the respiratory center (which requires a matter of minutes).

A number of factors may increase ventilation during exercise.

Researchers have suggested that a number of other factors, including the following, play a role in the ventilatory response to exercise:

1. *Reflexes originating from body movements.* Joint and muscle receptors excited during muscle contraction reflexly stimulate the respiratory center, abruptly increasing ventilation. Even passive movement of the limbs (for example, someone else alternately flexing and extending a person's knee) may increase ventilation severalfold through activation of these receptors, although no actual exercise is occurring. Thus, the mechanical events of exercise are believed to play an important role in coordinating respiratory activity with the increased metabolic requirements of the active muscles.

2. *Increase in body temperature.* Much of the energy generated during muscle contraction is converted to heat rather than to actual mechanical work. Heat-loss mechanisms such as sweating frequently cannot keep pace with the increased heat production that accompanies increased physical activity, so body temperature often rises slightly during exercise (see p. 657). Because raised body temperature stimulates ventilation, this exercise-related heat production undoubtedly contributes to the respiratory response to exercise. For the same reason, increased ventilation often accompanies a fever.

3. *Epinephrine release.* The adrenal medullary hormone epinephrine also stimulates ventilation. The level of circulating epinephrine rises during exercise in response to the sympathetic nervous system discharge that accompanies increased physical activity.

4. *Impulses from the cerebral cortex.* Especially at the onset of exercise, the motor areas of the cerebral cortex are believed to simultaneously stimulate the medullary respiratory neurons and activate the motor neurons of the exercising muscles. This is similar to the cardiovascular adjustments initiated by the motor cortex at the onset of exercise. In this way, the motor region of the brain calls forth increased ventilatory and circulatory

Oxygen and Carbon Dioxide–Related Variables during Exercise

O_2- Or CO_2-Related Variable	Change	Comment
O_2 Use	Marked increase	Active muscles are oxidizing nutrient molecules more rapidly to meet their increased energy needs.
CO_2 Production	Marked increase	More actively metabolizing muscles produce more CO_2.
Alveolar Ventilation	Marked increase	By mechanisms not completely understood, alveolar ventilation keeps pace with or even slightly exceeds the increased metabolic demands during exercise.
Arterial P_{O_2}	Normal or slight ↑	Despite a marked increase in O_2 use and CO_2 production during exercise, alveolar ventilation keeps pace with or even slightly exceeds the stepped-up rate of O_2 consumption and CO_2 production.
Arterial P_{CO_2}	Normal or slight ↓	
O_2 Delivery to Muscles	Marked increase	Although arterial P_{O_2} remains normal, O_2 delivery to muscles is greatly increased by the increased blood flow to exercising muscles accomplished by increased cardiac output coupled with local vasodilation of active muscles.
O_2 Extraction by Muscles	Marked increase	Increased use of O_2 lowers the P_{O_2} at the tissue level, which results in more O_2 unloading from hemoglobin; this is enhanced by ↑ P_{CO_2}, ↑ H^+, and ↑ temperature.
CO_2 Removal from Muscles	Marked increase	The increased blood flow to exercising muscles removes the excess CO_2 produced by these more actively metabolizing tissues.
Arterial H^+ Concentration		
Mild to moderate exercise	Normal	Because H^+-generating CO_2 is held constant in arterial blood, arterial H^+ concentration does not change.
Heavy exercise	Modest increase	In heavy exercise, when muscles resort to anaerobic metabolism, lactic acid is added to the blood.

responses to support the increased physical activity it is about to orchestrate. These anticipatory adjustments are feedforward regulatory mechanisms; that is, they occur *before* any homeostatic factors actually change (see p. 16). This is in contrast to the more common case in which regulatory adjustments to restore homeostasis take place *after* a factor has altered.

None of these factors or combinations of factors are fully satisfactory in explaining the abrupt and profound effect exercise has on ventilation, nor can they completely account for the high degree of correlation between respiratory activity and the body's needs for gas exchange during exercise. (For a discussion of how O_2 consumption during exercise can be measured to determine a person's maximum work capacity, see the boxed feature on p. 506, ■ A Closer Look at Exercise Physiology.)

Ventilation can be influenced by factors unrelated to the need for gas exchange.

Respiratory rate and depth can be modified for reasons other than the need to supply O_2 or remove CO_2. Here are some examples of involuntary influences in this category:

■ Protective reflexes such as sneezing and coughing temporarily govern respiratory activity in an effort to expel irritant materials from respiratory passages.

■ Inhalation of particularly noxious agents frequently triggers immediate cessation of ventilation.

■ Pain originating anywhere in the body reflexly stimulates the respiratory center (for example, one "gasps" with pain).

■ Involuntary modification of breathing also occurs during the expression of various emotional states, such as laughing, crying, sighing, and groaning. The emotionally induced modifications are mediated through connections between the limbic system in the brain (which is responsible for emotions) and the respiratory center.

■ Hiccups occur when involuntary, spasmodic contractions of the diaphragm occur, each causing rapid intake of air, which is suddenly halted by abrupt closure of the glottis, resulting in the *"hic"* sound. The underlying trigger for hiccups is not known.

■ The respiratory center is reflexly inhibited during swallowing, when the airways are closed to prevent food from entering the lungs (see p. 599).

How to Find Out How Much Work You're Capable of Doing

The best single predictor of a person's work capacity is the determination of **maximal O_2 consumption,** or **max VO_2,** which is the maximum volume of O_2 the person is capable of using per minute to oxidize nutrient molecules for energy production. Max VO_2 is measured by having the person engage in exercise, usually on a treadmill or bicycle ergometer (a stationary bicycle with variable resistance). The workload is incrementally increased until the person becomes exhausted. Expired air samples collected during the last minutes of exercise, when O_2 consumption is at a maximum because the person is working as hard as possible, are analyzed for the percentage of O_2 and CO_2 they contain. Furthermore, the volume of air expired is measured. Equations are then employed to determine the amount of O_2 consumed, taking into account the percentages of O_2 and CO_2 in the inspired air, the total volume of air expired, and the percentages of O_2 and CO_2 in the exhaled air.

Maximal O_2 consumption depends on three systems. The respiratory system is essential for ventilation and exchange of O_2 and CO_2 between the air and blood in the lungs. The circulatory system is required to deliver O_2 to the working muscles. Finally, the muscles must have the oxidative enzymes available to use the O_2 once it has been delivered.

Regular aerobic exercise can improve max VO_2 by making the heart and respiratory system more efficient, thereby delivering more O_2 to the working muscles. Exercised muscles themselves become better equipped to use O_2 once it is delivered. The number of functional capillaries increases, as do the number and size of mitochondria, which contain the oxidative enzymes.

Maximal O_2 consumption is measured in liters per minute and then converted into milliliters per kilogram of body weight per minute so that large and small people can be compared. As would be expected, athletes have the highest values for maximal O_2 consumption. The max VO_2 for male cross-country skiers has been recorded to be as high as 94 ml O_2/kg/min. Distance runners maximally consume between 65 and 85 ml O_2/kg/min, and football players have max VO_2 values between 45 and 65 ml O_2/kg/min, depending on the position they play. Sedentary young men maximally consume between 25 and 45 ml O_2/kg/min. Female values for max VO_2 are 20% to 25% lower than for males when expressed as ml/kg/min of total body weight. The difference in max VO_2 between females and males is only 8% to 10% when expressed as ml/kg/min of lean body weight, however, because females generally have a higher percentage of body fat (the female sex hormone estrogen promotes fat deposition).

Available norms are used to classify people as being low, fair, average, good, or excellent in aerobic capacity for their age group. Exercise physiologists use max VO_2 measurements to prescribe or adjust training regimens to help people achieve their optimal level of aerobic conditioning.

Humans also have considerable voluntary control over ventilation. Voluntary control of breathing is accomplished by the cerebral cortex, which does not act on the respiratory center in the brain stem but instead sends impulses directly to the motor neurons in the spinal cord that supply the respiratory muscles. We can voluntarily hyperventilate ("overbreathe") or, at the other extreme, hold our breath, but only for a brief period of time. The resulting chemical changes in the arterial blood directly and reflexly influence the respiratory center, which in turn overrides the voluntary input to the motor neurons of the respiratory muscles. Other than these extreme forms of deliberately controlling ventilation, we also control our breathing to perform such voluntary acts as speaking, singing, and whistling.

During apnea, a person "forgets to breathe"; during dyspnea, a person feels "short of breath."

Apnea is the transient interruption of ventilation, with breathing resuming spontaneously. If breathing does not resume, the condition is called **respiratory arrest.** Because ventilation is normally decreased and the central chemoreceptors are less sensitive to the arterial P_{CO_2} drive during sleep, especially paradoxical sleep (see p. 170), apnea is most likely to occur during this time. Victims of **sleep apnea** may stop breathing for a few seconds or up to one or two minutes as many as 500 times a night. Mild sleep apnea is not dangerous unless the sufferer has pulmonary or circulatory disease, which can be exacerbated by recurrent bouts of apnea.

 SUDDEN INFANT DEATH SYNDROME In exaggerated cases of sleep apnea, the victim may be unable to recover from an apneic period, and death results. This is the case in **sudden infant death syndrome (SIDS),** or "crib death," the leading cause of death in the first year of life. With this tragic form of sleep apnea, a previously healthy two- to four-month-old infant is found dead in his or her crib for no apparent reason. The underlying cause of SIDS is the subject of intense investigation. Most evidence suggests that the baby "forgets to breathe" because the respiratory control mechanisms are immature, either in the brain stem or in the chemoreceptors that monitor the body's respiratory status. For example, on autopsy more than half the victims have poorly developed carotid bodies, the more important of the peripheral chemoreceptors. Abnormal lung development has been suggested as being responsible for at least some cases. Some infants who died of SIDS were found to have reduced levels of a growth factor critical for the lungs. Alternatively, some researchers believe the condition may be triggered by an initial cardiovascular failure rather than by an initial cessation of breathing. Still other investigators propose that some cases may be due to aspiration of gastric (stomach) juice containing the bacterium *Helicobacter pylori*. In one study, this microorganism was present in 88% of infants who died of SIDS. Scientists speculate that *H. pylori* may lead to the production of ammonia, which can be lethal if it gains access to the blood from the lungs. Perhaps a combination of factors might be involved, or maybe SIDS is a collection of early infancy deaths from a variety of causes.

Whatever the underlying cause, certain risk factors make babies more vulnerable to SIDS. Among them are sleeping position (an almost 40% higher incidence of SIDS is associated with sleeping on the abdomen rather than on the back or side) and exposure to nicotine during fetal life or after birth. Infants whose mothers smoked during pregnancy or who breathe cigarette smoke in the home are three times more likely to die of SIDS than those not exposed to smoke.

Clinical Note **DYSPNEA** People who have dyspnea have the subjective sensation that they are not getting enough air; that is, they feel "short of breath." **Dyspnea** is the mental anguish associated with the unsatiated desire for more adequate ventilation. It often accompanies the labored breathing characteristic of obstructive lung disease or the pulmonary edema associated with congestive heart failure. In contrast, during exercise a person can breathe very hard without experiencing dyspnea because such exertion is not accompanied by a sense of anxiety over the adequacy of ventilation. Surprisingly, dyspnea is not directly related to chronic elevation of arterial P_{CO_2} or reduction of P_{O_2}. The subjective feeling of air hunger may occur even when alveolar ventilation and the blood gases are normal. Some people experience dyspnea when they *perceive* that they are short of air even though this is not actually the case, such as in a crowded elevator.

Chapter in Perspective: Focus on Homeostasis

The respiratory system contributes to homeostasis by obtaining O_2 from and eliminating CO_2 to the external environment. All body cells ultimately need an adequate supply of O_2 to use in oxidizing nutrient molecules to generate ATP. Brain cells, which especially depend on a continual supply of O_2, die if deprived of O_2 for more than 4 minutes. Even cells that can resort to anaerobic ("without O_2") metabolism for energy production, such as strenuously exercising muscles, can do so only transiently by incurring an O_2 deficit that ultimately must be made up during the period of excess postexercise O_2 consumption (see p. 279).

As a result of these energy-yielding metabolic reactions, the body produces large quantities of CO_2 that must be eliminated. Because CO_2 and H_2O form carbonic acid, adjustments in the rate of CO_2 elimination by the respiratory system are important in regulating acid–base balance in the internal environment. Cells can survive only within a narrow pH range.

REVIEW EXERCISES

Objective Questions (Answers on p. A-49)

1. Breathing is accomplished by alternate contraction and relaxation of muscles within the lung tissue. *(True or false?)*

2. Normally the alveoli empty completely during maximal expiratory efforts. *(True or false?)*

3. Alveolar ventilation does not always increase when pulmonary ventilation increases. *(True or false?)*

4. Oxygen and CO_2 have equal diffusion constants. *(True or false?)*

5. Hemoglobin has a higher affinity for O_2 than for any other substance. *(True or false?)*

6. Rhythmicity of breathing is brought about by pacemaker activity displayed by the respiratory muscles. *(True or false?)*

7. The expiratory neurons send impulses to the motor neurons controlling the expiratory muscles during normal quiet breathing. *(True or false?)*

8. The three forces that tend to keep the alveoli open are ____, ____, and ____.

9. The two forces that promote alveolar collapse are ____ and ____.

10. ____ is a measure of the magnitude of change in lung volume accomplished by a given change in the transmural pressure gradient.

11. ____ is the phenomenon of the lungs snapping back to their resting size after having been stretched.

12. ____ is the erythrocytic enzyme that catalyzes the conversion of CO_2 into HCO_3^-.

13. Which of the following reactions take(s) place at the pulmonary capillaries?
    a. $Hb + O_2 \rightarrow HbO_2$
    b. $CO_2 + H_2O \rightarrow H^+ + HCO_3^-$
    c. $Hb + CO_2 \rightarrow HbCO_2$
    d. $Hb + H^+ \rightarrow HbH$

14. Indicate the O_2 and CO_2 partial pressure relationships important in gas exchange by circling > (greater than), < (less than), or = (equal to) as appropriate in each of the following statements:
    a. P_{O_2} in blood entering the pulmonary capillaries is (>, <, =) P_{O_2} in the alveoli.
    b. P_{CO_2} in blood entering the pulmonary capillaries is (>, <, =) P_{CO_2} in the alveoli.

c. P_{O_2} in the alveoli is (>, <, =) P_{O_2} in blood leaving the pulmonary capillaries.

d. P_{CO_2} in the alveoli is (>, <, =) P_{CO_2} in blood leaving the pulmonary capillaries.

e. P_{O_2} in blood leaving the pulmonary capillaries is (>, <, =) P_{O_2} in blood entering the systemic capillaries.

f. P_{CO_2} in blood leaving the pulmonary capillaries is (>, <, =) P_{CO_2} in blood entering the systemic capillaries.

g. P_{O_2} in blood entering the systemic capillaries is (>, <, =) P_{O_2} in the tissue cells.

h. P_{CO_2} in blood entering the systemic capillaries is (>, <, =) P_{CO_2} in the tissue cells.

i. P_{O_2} in the tissue cells is (>, <, approximately =) P_{O_2} in blood leaving the systemic capillaries.

j. P_{CO_2} in the tissue cells is (>, <, approximately =) P_{CO_2} in blood leaving the systemic capillaries.

k. P_{O_2} in blood leaving the systemic capillaries is (>, <, =) P_{O_2} in blood entering the pulmonary capillaries.

l. P_{CO_2} in blood leaving the systemic capillaries is (>, <, =) P_{CO_2} in blood entering the pulmonary capillaries.

15. Using the answer code on the right, indicate which chemoreceptors are being described:

___ 1. stimulated by an arterial P_{O_2} of 80 mm Hg

___ 2. stimulated by an arterial P_{O_2} of 55 mm Hg

___ 3. directly depressed by an arterial P_{O_2} of 55 mm Hg

___ 4. weakly stimulated by an elevated arterial P_{CO_2}

___ 5. strongly stimulated by an elevated brain-ECF H^+ concentration induced by an elevated arterial P_{CO_2}

___ 6. stimulated by an elevated arterial H^+ concentration

(a) peripheral chemoreceptors
(b) central chemoreceptors
(c) both peripheral and central chemoreceptors
(d) neither peripheral nor central chemoreceptors

Essay Questions

1. Distinguish between internal and external respiration. List the steps in external respiration.

2. Describe the components of the respiratory system. What is the site of gas exchange?

3. Compare atmospheric, intra-alveolar, and intrapleural pressures.

4. Why are the lungs normally stretched even during expiration?

5. Explain why air enters the lungs during inspiration and leaves during expiration.

6. Why is inspiration normally active and expiration normally passive?

7. Why does airway resistance become an important determinant of airflow rates in chronic obstructive pulmonary disease?

8. Explain pulmonary elasticity in terms of compliance and elastic recoil.

9. State the source and function of pulmonary surfactant.

10. Define the various lung volumes and capacities.

11. Compare pulmonary ventilation and alveolar ventilation. What is the consequence of anatomic and alveolar dead space?

12. Compare ventilation, perfusion, and the ventilation–perfusion ratio at the top and bottom of the lung. Explain what accounts for this difference.

13. What determines the partial pressures of a gas in air and in blood?

14. List the methods of O_2 and CO_2 transport in the blood.

15. What is the primary factor that determines the % hemoglobin saturation? What are the significances of the plateau and the steep portions of the O_2–Hb dissociation curve?

16. How does hemoglobin promote the net transfer of O_2 from the alveoli to the blood?

17. Explain the Bohr and Haldane effects.

18. Define the following: *hypoxic hypoxia, anemic hypoxia, circulatory hypoxia, histotoxic hypoxia, hypercapnia, hypocapnia, hyperventilation, hypoventilation, hyperpnea, apnea,* and *dyspnea.*

19. What are the locations and functions of the three respiratory control centers? Distinguish between the DRG and the VRG.

20. What brain region establishes the rhythmicity of breathing?

Quantitative Exercises (Solutions on p. A-49)

1. The two curves in ● Figure 13-32 (p. 497) show partial pressures for O_2 and CO_2 at various alveolar ventilation rates. These curves can be calculated from the following two equations:

$$P_{AO_2} = P_{IO_2} - (V_{O_2}/V_A)\ 863\ \text{mm Hg}$$

$$P_{ACO_2} = (V_{CO_2}/V_A)\ 863\ \text{mm Hg}$$

In these equations, P_{AO_2} = the partial pressure of O_2 in the alveoli, P_{ACO_2} = the partial pressure of CO_2 in the alveoli, P_{IO_2} = the partial pressure of O_2 in the inspired air, V_{O_2} = the rate of O_2 consumption by the body, V_{CO_2} = the rate of CO_2 production by the body, V_A = the rate of alveolar ventilation, and 863 mm Hg is a constant that accounts for atmospheric pressure and temperature.

John is in training for a marathon tomorrow and just ate a meal of pasta (assume this is pure carbohydrate, which is metabolized with an RQ of 1). His alveolar ventilation rate is 3.0 liters/min, and he is consuming O_2 at a rate of 300 ml/min. What is the value of John's P_{ACO_2}?

2. Assume you are flying in an airplane that is cruising at 18,000 ft, where the pressure outside the plane is 380 mm Hg.

a. Calculate the partial pressure of O_2 in the air outside the plane, ignoring water vapor pressure.

b. If the plane depressurized, what would be the value of your P_{AO_2}? Assume that the ratio of your O_2 consumption to ventilation was not changed (that is, equaled 0.06), and note that under these conditions the constant in the equation that accounts for atmospheric pressure and temperature decreases from 863 mm Hg to 431.5 mm Hg.

c. Calculate your P_{ACO_2}, assuming that your CO_2 production and ventilation rates remained unchanged at 200 ml/min and 4.2 liters/min, respectively.

3. A student has a tidal volume of 350 ml. While breathing at a rate of 12 breaths/min, her alveolar ventilation is 80% of her pulmonary ventilation. What is her anatomic dead space volume?

(Explanations on p. A-49)

POINTS TO PONDER

1. Why is it important that airplane interiors are pressurized (that is, the pressure is maintained at sea-level atmospheric pressure even though the atmospheric pressure surrounding the plane is substantially lower)? Explain the physiological value of using O_2 masks if the pressure in the airplane interior cannot be maintained.

2. Would hypercapnia accompany the hypoxia produced in each of the following situations? Explain why or why not.
   a. cyanide poisoning
   b. pulmonary edema
   c. restrictive lung disease
   d. high altitude
   e. severe anemia
   f. congestive heart failure
   g. obstructive lung disease

3. If a person lives one mile above sea level in Denver, Colorado, where the atmospheric pressure is 630 mm Hg, what would the P_{O_2} of the inspired air be once it is humidified in the respiratory airways before it reaches the alveoli?

4. Based on what you know about the control of respiration, explain why it is dangerous to voluntarily hyperventilate to lower the arterial P_{CO_2} before going underwater. The purpose of the hyperventilation is to stay under longer before P_{CO_2} rises above normal and drives the swimmer to surface for a breath of air.

5. If a person whose alveolar membranes are thickened by disease has an alveolar P_{O_2} of 100 mm Hg and an alveolar P_{CO_2} of 40 mm Hg, which of the following values of systemic arterial blood gases are most likely to exist?
   a. P_{O_2} = 105 mm Hg, P_{CO_2} = 35 mm Hg
   b. P_{O_2} = 100 mm Hg, P_{CO_2} = 40 mm Hg
   c. P_{O_2} = 90 mm Hg, P_{CO_2} = 45 mm Hg

   If the person is administered 100% O_2, will the arterial P_{O_2} increase, decrease, or remain the same? Will the arterial P_{CO_2} increase, decrease, or remain the same?

CLINICAL CONSIDERATION

(Explanation on p. A-50)

Keith M., a former heavy cigarette smoker, has severe emphysema. How does this condition affect his airway resistance? How does this change in airway resistance influence Keith's inspiratory and expiratory efforts? Describe how his respiratory muscle activity and intra-alveolar pressure changes compare to normal to accomplish a normal tidal volume. How would his spirogram compare to normal? What influence would Keith's condition have on gas exchange in his lungs? What blood-gas abnormalities are likely to be present? Would it be appropriate to administer O_2 to Keith to relieve his hypoxic condition?

Urinary System

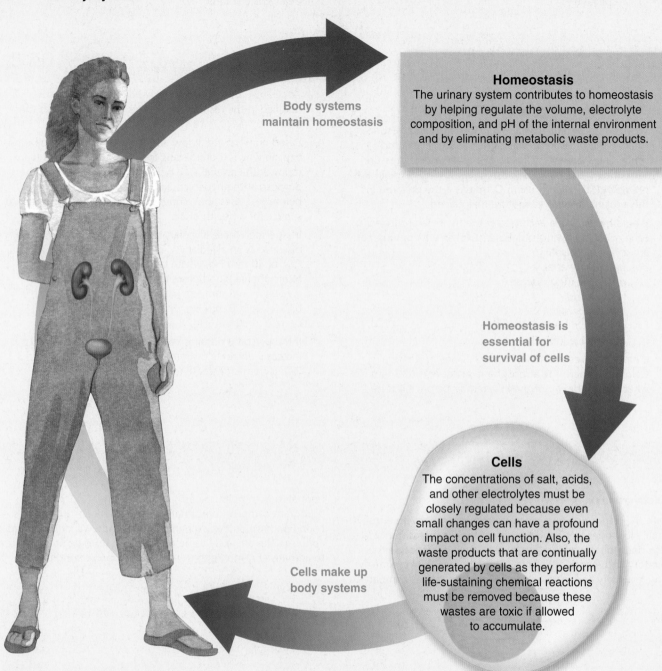

Body systems maintain homeostasis

Homeostasis
The urinary system contributes to homeostasis by helping regulate the volume, electrolyte composition, and pH of the internal environment and by eliminating metabolic waste products.

Homeostasis is essential for survival of cells

Cells
The concentrations of salt, acids, and other electrolytes must be closely regulated because even small changes can have a profound impact on cell function. Also, the waste products that are continually generated by cells as they perform life-sustaining chemical reactions must be removed because these wastes are toxic if allowed to accumulate.

Cells make up body systems

The survival and proper functioning of cells depend on maintaining stable concentrations of salt, acids, and other electrolytes in the internal fluid environment. Cell survival also depends on continuous removal of toxic metabolic wastes that cells produce as they perform life-sustaining chemical reactions. The kidneys play a major role in maintaining homeostasis by regulating the concentration of many plasma constituents, especially electrolytes and water, and by eliminating all metabolic wastes (except CO_2, which is removed by the lungs). As plasma repeatedly filters through the kidneys, they retain constituents of value for the body and eliminate undesirable or excess materials in the urine. Of special importance is the kidneys' ability to regulate the volume and osmolarity (solute concentration) of the internal fluid environment by controlling salt and water balance. Also crucial is their ability to help regulate pH by controlling elimination of acid and base in the urine.

The Urinary System

CENGAGENOW™ Log on to CengageNOW at **http://www .cengage.com/sso/** for an opportunity to explore a learning module that illustrates difficult concepts with self-study tutorials, animations, and interactive quizzes to help you learn, review, and master physiology concepts.

Kidneys: Functions, Anatomy, and Basic Processes

The composition of the fluid bathing all the cells could be notably altered by exchanges between the cells and this internal fluid environment if mechanisms did not exist to keep the ECF stable.

The kidneys perform a variety of functions aimed at maintaining homeostasis.

The kidneys, in concert with hormonal and neural inputs that control their function, are the organs primarily responsible for maintaining the stability of ECF volume, electrolyte composition, and osmolarity (solute concentration). By adjusting the quantity of water and various plasma constituents that are either conserved for the body or eliminated in the urine, the kidneys can maintain water and electrolyte balance within the very narrow range compatible with life, despite a wide range of intake and losses of these constituents through other avenues. The kidneys not only adjust for wide variations in ingestion of water (H_2O), salt, and other electrolytes but they also adjust urinary output of these ECF constituents to compensate for abnormal losses through heavy sweating, vomiting, diarrhea, or hemorrhage. Thus, as the kidneys do what they can to maintain homeostasis, urine composition varies widely.

When the ECF has a surplus of H_2O or a particular electrolyte such as salt (NaCl), the kidneys can eliminate the excess in the urine. If a deficit exists, the kidneys cannot provide additional quantities of the depleted constituent, but they can limit urinary losses of the material in short supply and thus conserve it until the person can take in more of the depleted substance. Accordingly, the kidneys can compensate more efficiently for excesses than for deficits. In fact, in some instances the kidneys cannot completely halt the loss of a valuable substance in the urine, even though the substance may be in short supply. A prime example is the case of a H_2O deficit. Even if a person is not consuming any H_2O, the kidneys must put out about half a liter of H_2O in the urine each day to fill another major role as the body's cleaners.

In addition to the kidneys' important regulatory role in maintaining fluid and electrolyte balance, they are the main route for eliminating potentially toxic metabolic wastes and foreign compounds from the body. These wastes cannot be

511

eliminated as solids; they must be excreted in solution, thus obligating the kidneys to produce a minimum volume of around 500 ml of waste-filled urine per day. Because H_2O eliminated in the urine is derived from the blood plasma, a person stranded without H_2O eventually urinates himself or herself to death: The plasma volume falls to a fatal level as H_2O is inexorably removed to accompany the wastes.

OVERVIEW OF KIDNEY FUNCTIONS The kidneys perform the following specific functions, most of which help preserve the constancy of the internal fluid environment:

1. *Maintaining H_2O balance in the body* (Chapter 15).

2. *Maintaining the proper osmolarity of body fluids, primarily through regulating H_2O balance.* This function is important to prevent osmotic fluxes into or out of the cells, which could lead to detrimental swelling or shrinking of the cells, respectively (Chapter 15).

3. *Regulating the quantity and concentration of most ECF ions,* including sodium (Na^+), chloride (Cl^-), potassium (K^+), calcium (Ca^{2+}), hydrogen ion (H^+), bicarbonate (HCO_3^-), phosphate (PO_4^{3-}), sulfate (SO_4^{2-}), and magnesium (Mg^{2+}). Even minor fluctuations in the ECF concentrations of some of these electrolytes can have profound influences. For example, changes in the ECF concentration of K^+ can potentially lead to fatal cardiac dysfunction.

4. *Maintaining proper plasma volume,* which is important in the long-term regulation of arterial blood pressure. This function is accomplished through the kidneys' regulatory role in salt (Na^+ and Cl^-) and H_2O balance (Chapter 15).

5. *Helping maintain the proper acid–base balance* of the body by adjusting urinary output of H^+ and HCO_3^- (Chapter 15).

6. *Excreting (eliminating) the end products (wastes) of bodily metabolism,* such as urea (from proteins), uric acid (from nucleic acids), creatinine (from muscle creatine), bilirubin (from hemoglobin), and hormone metabolites. If allowed to accumulate, many of these wastes are toxic, especially to the brain.

7. *Excreting many foreign compounds,* such as drugs, food additives, pesticides, and other exogenous nonnutritive materials that have entered the body.

8. *Producing erythropoietin,* a hormone that stimulates red blood cell production (Chapter 11).

9. *Producing renin,* an enzymatic hormone that triggers a chain reaction important in salt conservation by the kidneys.

10. *Converting vitamin D into its active form* (Chapter 19).

The kidneys form the urine; the rest of the urinary system carries the urine to the outside.

The **urinary system** consists of the urine-forming organs—the **kidneys**—and the structures that carry the urine from the kidneys to the outside for elimination from the body (● Figure 14-1a). The kidneys are a pair of bean-shaped organs about 4 in. to 5 in. long that lie in the back of the abdominal cavity, one on each side of the vertebral column, slightly above the waistline. Each kidney is supplied by a **renal artery** and a **renal vein,**

which, respectively, enter and leave the kidney at the medial indentation that gives this organ its beanlike form. The kidney acts on the plasma flowing through it to produce urine, conserving materials to be retained in the body and eliminating unwanted materials into the urine.

After urine is formed, it drains into a central collecting cavity, the **renal pelvis,** located at the medial inner core of each kidney (● Figure 14-1b). From there urine is channeled into the **ureter,** a smooth muscle–walled duct that exits at the medial border in close proximity to the renal artery and vein. There are two ureters, one carrying urine from each kidney to the single urinary bladder.

The **urinary bladder,** which temporarily stores urine, is a hollow, distensible, smooth muscle–walled sac. Periodically, urine is emptied from the bladder to the outside through another tube, the **urethra,** as a result of bladder contraction. The urethra in females is straight and short, passing directly from the neck of the bladder to the outside (● Figure 14-2; see also ● Figure 20-2, p. 744). In males the urethra is much longer and follows a curving course from the bladder to the outside, passing through both the prostate gland and the penis (● Figures 14-1a and 14-2b; see also ● Figure 20-1, p. 743). The male urethra serves the dual function of providing both a route for eliminating urine from the bladder and a passageway for semen from the reproductive organs. The prostate gland lies below the neck of the bladder and completely encircles the urethra.

 Prostatic enlargement, which often occurs during middle to older age, can partially or completely occlude the urethra, impeding the flow of urine.

The parts of the urinary system beyond the kidneys merely serve as ductwork to transport urine to the outside. Once formed by the kidneys, urine is not altered in composition or volume as it moves downstream through the rest of the tract.

The nephron is the functional unit of the kidney.

Each kidney consists of about 1 million microscopic functional units known as **nephrons,** which are bound together by connective tissue. Recall that a functional unit is the smallest unit within an organ capable of performing all of that organ's functions. Because the main function of the kidneys is to produce urine and, in so doing, maintain constancy in the ECF composition, a nephron is the smallest unit capable of forming urine.

The arrangement of nephrons within the kidneys gives rise to two distinct regions—an outer region called the **renal cortex,** which looks granular, and an inner region, the **renal medulla,** which is made up of striated triangles, the **renal pyramids** (see ● Figure 14-1b).

Knowledge of the structural arrangement of an individual nephron is essential for understanding the distinction between the cortical and medullary regions of the kidney and, more important, for understanding renal function. Each nephron consists of a *vascular component* and a *tubular component,* both of which are intimately related structurally and functionally (● Figure 14-3).

VASCULAR COMPONENT OF THE NEPHRON The dominant part of the nephron's vascular component is the **glomerulus,** a ball-

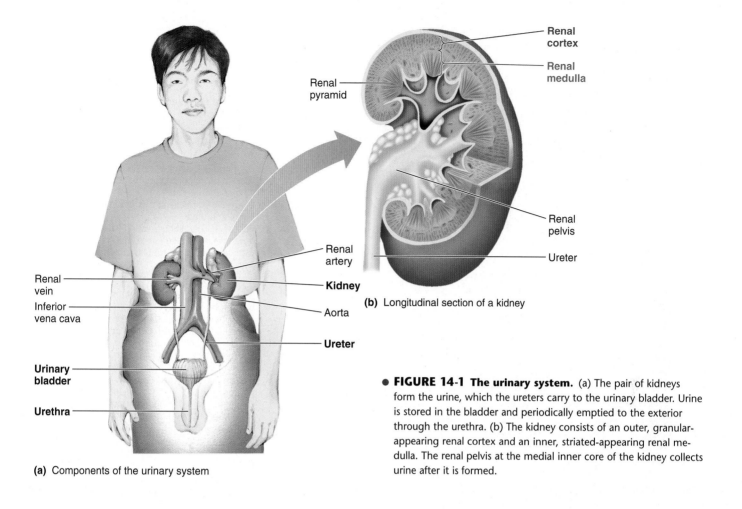

Renal pyramid

Renal cortex

Renal medulla

Renal artery

Kidney

Aorta

Ureter

Renal vein

Inferior vena cava

Urinary bladder

Urethra

Renal pelvis

Ureter

(b) Longitudinal section of a kidney

(a) Components of the urinary system

● **FIGURE 14-1 The urinary system.** (a) The pair of kidneys form the urine, which the ureters carry to the urinary bladder. Urine is stored in the bladder and periodically emptied to the exterior through the urethra. (b) The kidney consists of an outer, granular-appearing renal cortex and an inner, striated-appearing renal medulla. The renal pelvis at the medial inner core of the kidney collects urine after it is formed.

like tuft of capillaries through which part of the water and solutes is filtered from blood passing through. This filtered fluid, which is almost identical in composition to plasma, then passes through the nephron's tubular component, where various transport processes convert it into urine.

On entering the kidney, the renal artery subdivides to ultimately form many small vessels known as **afferent arterioles,** one of which supplies each nephron. The afferent arteriole delivers blood to the glomerulus. The glomerular capillaries rejoin to form another arteriole, the **efferent arteriole,** through which blood that was not filtered into the tubular component leaves the glomerulus (● Figures 14-3 and 14-4). The efferent arterioles are the only arterioles in the body that drain from capillaries. Typically, arterioles break up into capillaries that rejoin to form venules. At the glomerular capillaries, no O_2 or nutrients are extracted from the blood for use by the kidney tissues, nor are waste products picked up from the surrounding tissue. Thus, arterial blood enters the glomerular capillaries through the afferent arteriole, and arterial blood leaves the glomerulus through the efferent arteriole.

The efferent arteriole quickly subdivides into a second set of capillaries, the **peritubular capillaries,** which supply the renal tissue with blood and are important in

Ureter

Smooth muscle of bladder wall

Ureteral openings

Internal sphincter

Pelvic diaphragm

External sphincter

Urethra

External urethral orifice

(a) Female

Prostate gland (an accessory sex gland)

Bulbourethral glands (accessory sex glands)

Urethra

External urethral orifice

(b) Male

● **FIGURE 14-2 Comparison of the urethra in females and males.** (a) In females, the urethra is straight and short. (b) In males, the urethra, which is much longer, passes through the prostate gland and penis.

Proximal tubule

Juxtaglomerular apparatus

Efferent arteriole

Afferent arteriole

Artery

Vein

Peritubular capillaries

Distal tubule

Collecting duct

Bowman's capsule

Glomerulus

Cortex
- - - - - - - -
Medulla

Loop of Henle

To renal pelvis

Overview of Functions of Parts of a Nephron

Vascular component
- Afferent arteriole—carries blood to the glomerulus
- Glomerulus—a tuft of capillaries that filters a protein-free plasma into the tubular component
- Efferent arteriole—carries blood from the glomerulus
- Peritubular capillaries—supply the renal tissue; involved in exchanges with the fluid in the tubular lumen

Tubular component
- Bowman's capsule—collects the glomerular filtrate
- Proximal tubule—uncontrolled reabsorption and secretion of selected substances occur here
- Loop of Henle—establishes an osmotic gradient in the renal medulla that is important in the kidney's ability to produce urine of varying concentration
- Distal tubule and collecting duct— variable, controlled reabsorption of Na^+ and H_2O and secretion of K^+ and H^+ occur here; fluid leaving the collecting duct is urine, which enters the renal pelvis

Combined vascular/tubular component
- Juxtaglomerular apparatus—produces substances involved in the control of kidney function

● **FIGURE 14-3 A nephron.** Components of a cortical nephron, the most abundant type of nephron in humans.

exchanges between the tubular system and blood during conversion of the filtered fluid into urine. These peritubular capillaries, as their name implies, are intertwined around the tubular system (*peri* means "around"). The peritubular capillaries rejoin to form venules that ultimately drain into the renal vein, by which blood leaves the kidney.

TUBULAR COMPONENT OF THE NEPHRON The nephron's tubular component is a hollow, fluid-filled tube formed by a single layer of epithelial cells. Even though the tubule is continuous from its beginning near the glomerulus to its ending at the renal pelvis, it is arbitrarily divided into various segments based on differences in structure and function along its length (see ● Figure 14-3). The tubular component begins with **Bowman's**

capsule, an expanded, double-walled invagination that cups around the glomerulus to collect fluid filtered from the glomerular capillaries.

From Bowman's capsule, the filtered fluid passes into the **proximal tubule,** which lies entirely within the cortex and is highly coiled or convoluted throughout much of its course. The next segment, the **loop of Henle,** forms a sharp U-shaped or hairpin loop that dips into the renal medulla. The *descending limb* of the loop of Henle plunges from the cortex into the medulla; the *ascending limb* traverses back up into the cortex. The ascending limb returns to the glomerular region of its own nephron, where it passes through the fork formed by the afferent and efferent arterioles. Both the tubular and vascular cells at this point are specialized to form the **juxtaglomerular appara-**

Small branch of renal artery   Peritubular capillaries

Afferent arteriole   Efferent arteriole   Glomerulus

© Richard G. Kessel/Visuals Unlimited

● **FIGURE 14-4 Scanning electron micrograph of a glomerulus and associated arterioles.**

tus, a structure that lies next to the glomerulus (*juxta* means "next to"). This specialized region plays an important role in regulating kidney function. Beyond the juxtaglomerular apparatus, the tubule once again coils tightly to form the **distal tubule,** which also lies entirely within the cortex. The distal tubule empties into a **collecting duct** or **tubule,** with each collecting duct draining fluid from up to eight separate nephrons. Each collecting duct plunges down through the medulla to empty its fluid contents (now converted into urine) into the renal pelvis.

CORTICAL AND JUXTAMEDULLARY NEPHRONS Two types of nephrons—*cortical nephrons* and *juxtamedullary nephrons*—are distinguished by the location and length of some of their structures (● Figure 14-5). All nephrons originate in the cortex, but the glomeruli of **cortical nephrons** lie in the outer layer of the cortex, whereas the glomeruli of **juxtamedullary nephrons** lie in the inner layer of the cortex, next to the medulla. (Note the distinction between *juxtamedullary* nephrons and *juxtaglomerular* apparatus.) The presence of all glomeruli and associated Bowman's capsules in the cortex is responsible for this region's granular appearance. These two nephron types differ most markedly in their loops of Henle. The hairpin loop of cortical nephrons dips only slightly into the medulla. In contrast, the loop of juxtamedullary nephrons plunges through the entire depth of the medulla. Furthermore, the peritubular capillaries of juxtamedullary nephrons form hairpin vascular loops known as **vasa recta** ("straight vessels"), which run in close association with the long loops of Henle. In cortical nephrons, the peritubular capillaries do not form vasa recta but instead entwine around these nephrons' short loops of Henle. As they course through the medulla, the collecting ducts of both cortical and juxtamedullary nephrons run parallel to the ascending and descending limbs of the juxtamedullary nephrons' long loops of Henle and vasa recta. The parallel arrangement of tubules and vessels in the medulla creates this region's striated appearance. More important, as you will see, this arrangement—coupled with the perme-

ability and transport characteristics of the long loops of Henle and vasa recta—plays a key role in the kidneys' ability to produce urine of varying concentrations, depending on the needs of the body. About 80% of the nephrons in humans are of the cortical type. Species with greater urine-concentrating abilities than humans, such as the desert rat, have a greater proportion of juxtamedullary nephrons.

The three basic renal processes are glomerular filtration, tubular reabsorption, and tubular secretion.

Three basic processes are involved in forming urine: *glomerular filtration, tubular reabsorption,* and *tubular secretion.* To aid in visualizing the relationships among these renal processes, it is useful to unwind the nephron schematically, as in ● Figure 14-6.

GLOMERULAR FILTRATION As blood flows through the glomerulus, protein-free plasma filters through the glomerular capillaries into Bowman's capsule. Normally, about 20% of the plasma that enters the glomerulus is filtered. This process, known as **glomerular filtration,** is the first step in urine formation. On average, 125 ml of glomerular filtrate (filtered fluid) are formed collectively through all the glomeruli each minute. This amounts to 180 liters (about 47.5 gallons) each day. Considering that the average plasma volume in an adult is 2.75 liters, this means that the kidneys filter the entire plasma volume about 65 times per day. If everything filtered passed out in the urine, the total plasma volume would be urinated in less than half an hour! This does not happen, however, because the kidney tubules and peritubular capillaries are intimately related throughout their lengths, so that materials can be transferred between the fluid inside the tubules and the blood within the peritubular capillaries.

TUBULAR REABSORPTION As the filtrate flows through the tubules, substances of value to the body are returned to the peritubular capillary plasma. This selective movement of substances from inside the tubule (the tubular lumen) into the blood is called **tubular reabsorption.** Reabsorbed substances are not lost from the body in the urine but instead are carried by the peritubular capillaries to the venous system and then to the heart to be recirculated. Of the 180 liters of plasma filtered per day, 178.5 liters, on average, are reabsorbed. The remaining 1.5 liters left in the tubules pass into the renal pelvis to be eliminated as urine. In general, substances the body needs to conserve are selectively reabsorbed, whereas unwanted substances that must be eliminated stay in the urine.

TUBULAR SECRETION The third renal process, **tubular secretion,** is the selective transfer of substances from the peritubular capillary blood into the tubular lumen. It provides a second route for substances to enter the renal tubules from the blood, the first being by glomerular filtration. Only about 20% of the plasma flowing through the glomerular capillaries is filtered into Bowman's capsule; the remaining 80% flows on through the efferent arteriole into the peritubular capillaries. Tubular secretion provides a mechanism for more rapidly eliminating

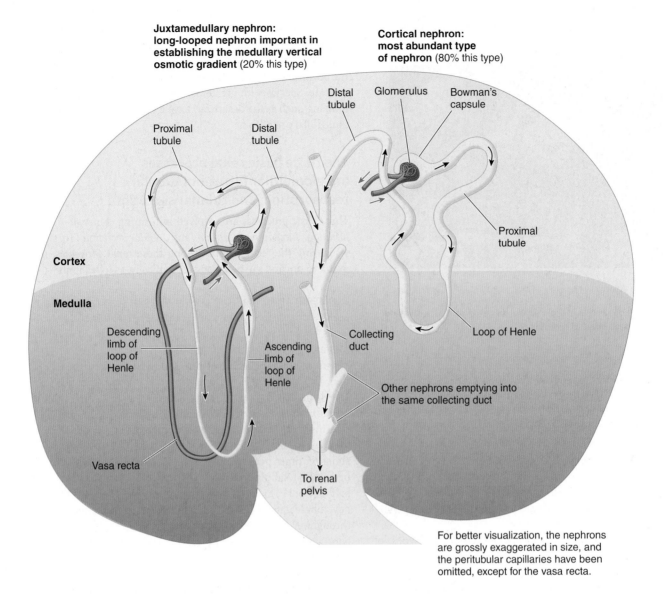

Juxtamedullary nephron:
long-looped nephron important in
establishing the medullary vertical
osmotic gradient (20% this type)

Cortical nephron:
most abundant type
of nephron (80% this type)

Proximal tubule

Distal tubule

Distal tubule

Glomerulus

Bowman's capsule

Cortex

Medulla

Descending limb of loop of Henle

Ascending limb of loop of Henle

Collecting duct

Proximal tubule

Loop of Henle

Other nephrons emptying into the same collecting duct

Vasa recta

To renal pelvis

For better visualization, the nephrons are grossly exaggerated in size, and the peritubular capillaries have been omitted, except for the vasa recta.

● **FIGURE 14-5 Comparison of juxtamedullary and cortical nephrons.** The glomeruli of cortical nephrons lie in the outer cortex, whereas the glomeruli of juxtamedullary nephrons lie in the inner part of the cortex next to the medulla. The loops of Henle of cortical nephrons dip only slightly into the medulla, but the juxtamedullary nephrons have long loops of Henle that plunge deep into the medulla. The juxtamedullary nephrons' peritubular capillaries form hairpin loops known as *vasa recta*.

selected substances from the plasma by extracting an additional quantity of a particular substance from the 80% of unfiltered plasma in the peritubular capillaries and adding it to the quantity of the substance already present in the tubule as a result of filtration.

URINE EXCRETION Urine excretion is the elimination of substances from the body in the urine. It is not really a separate process but the result of the first three processes. All plasma constituents filtered or secreted but not reabsorbed remain in the tubules and pass into the renal pelvis to be excreted as urine and eliminated from the body (● Figure 14-6). (Do not confuse excretion with secretion.) Note that anything filtered and subsequently reabsorbed, or not filtered at all, enters the venous

blood from the peritubular capillaries and thus is conserved for the body instead of being excreted in urine, despite passing through the kidneys.

THE BIG PICTURE OF THE BASIC RENAL PROCESSES Glomerular filtration is largely an indiscriminate process. With the exception of blood cells and plasma proteins, all constituents within the blood—H_2O, nutrients, electrolytes, wastes, and so on—nonselectively enter the tubular lumen as a bulk unit during filtration. That is, of the 20% of the plasma filtered at the glomerulus, everything in that part of the plasma enters Bowman's capsule except for the plasma proteins. The highly discriminating tubular processes then work on the filtrate to return to the blood a fluid of the composition and volume necessary to maintain constancy

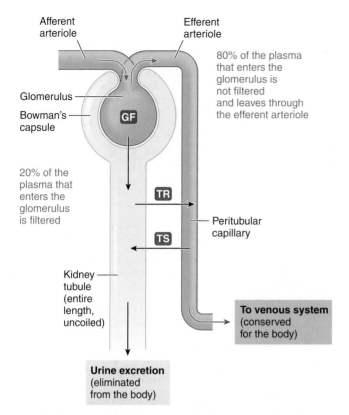

Afferent arteriole

Efferent arteriole

Glomerulus

Bowman's capsule

GF

80% of the plasma that enters the glomerulus is not filtered and leaves through the efferent arteriole

20% of the plasma that enters the glomerulus is filtered

TR

TS

Peritubular capillary

Kidney tubule (entire length, uncoiled)

To venous system (conserved for the body)

Urine excretion (eliminated from the body)

GF = Glomerular filtration—nondiscriminant filtration of a protein-free plasma from the glomerulus into Bowman's capsule

TR = Tubular reabsorption—selective movement of filtered substances from the tubular lumen into the peritubular capillaries

TS = Tubular secretion—selective movement of nonfiltered substances from the peritubular capillaries into the tubular lumen

● **FIGURE 14-6 Basic renal processes.** Anything filtered or secreted but not reabsorbed is excreted in the urine and lost from the body. Anything filtered and subsequently reabsorbed, or not filtered at all, enters the venous blood and is saved for the body.

of the internal fluid environment. The unwanted filtered material is left behind in the tubular fluid to be excreted as urine. Glomerular filtration can be thought of as pushing a part of the plasma, with all its essential components as well as those that need to be eliminated from the body, onto a tubular "conveyor belt" that terminates at the renal pelvis, which is the collecting point for urine within the kidney. All plasma constituents that enter this conveyor belt and are not subsequently returned to the plasma by the end of the line are spilled out of the kidney as urine. It is up to the tubular system to salvage by reabsorption the filtered materials that need to be preserved for the body while leaving behind substances that must be excreted. In addition, some substances are not only filtered but are also secreted onto the tubular conveyor belt, so the amounts of these substances excreted in the urine are greater than the amounts that were filtered. For many substances, these renal processes are subject to physiologic control. Thus, the kidneys handle each constituent in the plasma in a characteristic manner by a particular combination of filtration, reabsorption, and secretion.

The kidneys act only on the plasma, yet the ECF consists of both plasma and interstitial fluid. The interstitial fluid is actually the true internal fluid environment of the body, because it is the only component of the ECF that comes into direct contact with the cells. However, because of the free exchange between plasma and interstitial fluid across the capillary walls (with the exception of plasma proteins), interstitial fluid composition reflects the composition of plasma. Thus, by performing their regulatory and excretory roles on the plasma, the kidneys maintain the proper interstitial fluid environment for optimal cell function. Most of the rest of this chapter will be devoted to considering how the basic renal processes are accomplished and the mechanisms by which they are carefully regulated to help maintain homeostasis.

Glomerular Filtration

Fluid filtered from the glomerulus into Bowman's capsule must pass through the three layers that make up the **glomerular membrane** (● Figure 14-7): (1) the glomerular capillary wall, (2) the basement membrane, and (3) the inner layer of Bowman's capsule. Collectively, these layers function as a fine molecular sieve that retains the blood cells and plasma proteins but permits H_2O and solutes of small molecular dimension to filter through. Let us consider each layer in more detail.

The glomerular membrane is considerably more permeable than capillaries elsewhere.

The *glomerular capillary wall* consists of a single layer of flattened endothelial cells. It is perforated by many large pores (fenestrations; see p. 363) that make it more than 100 times more permeable to H_2O and solutes than capillaries elsewhere in the body.

The *basement membrane* is an acellular (lacking cells) gelatinous layer composed of collagen and glycoproteins that is sandwiched between the glomerulus and Bowman's capsule. The collagen provides structural strength, and the glycoproteins discourage the filtration of small plasma proteins. The larger plasma proteins cannot be filtered, because they cannot fit through the capillary pores, but the pores are just barely large enough to permit passage of albumin, the smallest of plasma proteins. However, because the glycoproteins are negatively charged, they repel albumin and other plasma proteins, which are also negatively charged. Therefore, plasma proteins are almost completely excluded from the filtrate, with less than 1% of the albumin molecules escaping into Bowman's capsule. The small proteins that do slip into the filtrate are picked up by the proximal tubule by endocytosis, then degraded into constituent amino acids that are returned to the blood. Thus, urine is normally protein free.

Clinical Note Some renal diseases characterized by excessive albumin in the urine (*albuminuria*) are the result of disruption of the negative charges within the basement membrane, which makes the glomerular membrane more permeable to albumin even though the size of the capillary pores remains constant.

The final layer of the glomerular membrane is the *inner layer of Bowman's capsule*. It consists of **podocytes,** octopus-like cells that encircle the glomerular tuft. Each podocyte bears

To be filtered, a substance must pass through

1 the pores between the endothelial cells of the glomerular capillary

2 an acellular basement membrane

3 the filtration slits between the foot processes of the podocytes in the inner layer of Bowman's capsule

● **FIGURE 14-7 Layers of the glomerular membrane.**

Dr. Donald Fawcett & D. Friend/Visuals Unlimited

many elongated foot processes (*podo* means "foot"; a *process* is a projection or appendage) that interdigitate with foot processes of adjacent podocytes, much as you interlace your fingers between each other when you cup your hands around a ball (● Figure 14-8). The narrow slits between adjacent foot processes, known as **filtration slits,** provide a pathway through which fluid leaving the glomerular capillaries can enter the lumen of Bowman's capsule.

Thus, the route that filtered substances take across the glomerular membrane is completely extracellular—first through capillary pores, then through the acellular basement membrane, and finally through capsular filtration slits (see ● Figure 14-7).

Glomerular capillary blood pressure is the major force that causes glomerular filtration.

To accomplish glomerular filtration, a force must drive a part of the plasma in the glomerulus through the openings in the glomerular membrane. No active transport mechanisms or local energy expenditures are involved in moving fluid from the plasma across the glomerular membrane into Bowman's capsule. Passive physical forces similar to those acting across capillaries elsewhere accomplish glomerular filtration.

Because the glomerulus is a tuft of capillaries, the same principles of fluid dynamics apply here that cause ultrafiltration across other capillaries (see p. 366), except for two important differences: (1) The glomerular capillaries are much more permeable than capillaries elsewhere, so more fluid is filtered for a given filtration pressure; and (2) the balance of forces across the glomerular membrane is such that filtration occurs throughout the entire length of the capillaries. In contrast, the balance of forces in other capillaries shifts so that filtration occurs in the beginning part of the vessel but reabsorption occurs toward the vessel's end (see ● Figure 10-23, p. 368).

FORCES INVOLVED IN GLOMERULAR FILTRATION Three physical forces are involved in glomerular filtration (▲ Table 14-1): glomerular capillary blood pressure, plasma-colloid osmotic pressure, and Bowman's capsule hydrostatic pressure. Let us examine the role of each.

Cell body of podocyte

Foot processes   Filtration slits

© F. Spinelli, Don W. Fawcett/Visuals Unlimited

● **FIGURE 14-8 Bowman's capsule podocytes with foot processes and filtration slits.** Note the filtration slits between adjacent foot processes on this scanning electron micrograph. The podocytes and their foot processes encircle the glomerular capillaries.

1. *Glomerular capillary blood pressure* is the fluid (hydrostatic) pressure exerted by the blood within the glomerular capillaries. It ultimately depends on contraction of the heart (the source of energy that produces glomerular filtration) and the resistance to blood flow offered by the afferent and efferent arterioles. Glomerular capillary blood pressure, at an estimated average value of 55 mm Hg, is higher than capillary blood pressure elsewhere. The reason for the higher pressure in the glomerular capillaries is the larger diameter of the afferent arteriole compared to that of the efferent arteriole. Because more blood can enter the glomerulus through the wide afferent arteriole than can leave through the narrower efferent arteriole, glomerular capillary blood pressure is maintained high as a result of blood damming up in the glomerular capillaries. Furthermore, because of the high resistance offered by the efferent arterioles, blood pressure does not have the same tendency to fall along the length of the glomerular capillaries as it does along other capillaries. This elevated, nondecremental glomerular blood pressure tends to push fluid out of the glomerulus into Bowman's capsule along the glomerular capillaries' entire length, and it is the major force producing glomerular filtration.

Whereas glomerular capillary blood pressure *favors* filtration, the two other forces acting across the glomerular membrane (plasma-colloid osmotic pressure and Bowman's capsule hydrostatic pressure) *oppose* filtration.

2. *Plasma-colloid osmotic pressure* is caused by the unequal distribution of plasma proteins across the glomerular membrane. Because plasma proteins cannot be filtered, they are in the glomerular capillaries but not in Bowman's capsule. Accordingly, the concentration of H_2O is higher in Bowman's capsule than in the glomerular capillaries. The resulting tendency for H_2O to move by osmosis down its own concentration gradient from Bowman's capsule into the glomerulus opposes glomerular fil-

▲ **TABLE 14-1** | **Forces Involved in Glomerular Filtration**

Force	Effect	Magnitude (mm Hg)
Glomerular Capillary Blood Pressure	Favors filtration	55
Plasma-Colloid Osmotic Pressure	Opposes filtration	30
Bowman's Capsule Hydrostatic Pressure	Opposes filtration	15
Net Filtration Pressure (Difference between Force Favoring Filtration and Forces Opposing Filtration)	Favors filtration	10

$$55 - (30 + 15) = 10$$

tration. This opposing osmotic force averages 30 mm Hg, which is slightly higher than across other capillaries. It is higher because much more H_2O is filtered out of the glomerular blood, so the concentration of plasma proteins is higher than elsewhere.

3. *Bowman's capsule hydrostatic pressure,* the pressure exerted by the fluid in this initial part of the tubule, is estimated to be about 15 mm Hg. This pressure, which tends to push fluid out of Bowman's capsule, opposes the filtration of fluid from the glomerulus into Bowman's capsule.

GLOMERULAR FILTRATION RATE As can be seen in ▲ Table 14-1, the forces acting across the glomerular membrane are not in balance. The total force favoring filtration is the glomerular capillary blood pressure at 55 mm Hg. The total of the two forces opposing filtration is 45 mm Hg. The net difference favoring filtration (10 mm Hg of pressure) is called the **net filtration pressure.** This modest pressure forces large volumes of fluid from the blood through the highly permeable glomerular membrane. The actual rate of filtration, the **glomerular filtration rate (GFR),** depends not only on the net filtration pressure but also on how much glomerular surface area is available for penetration and how permeable the glomerular membrane is (that is, how "holey" it is). These properties of the glomerular membrane are collectively referred to as the **filtration coefficient (K_f)**. Accordingly,

$$\text{GFR} = K_f \times \text{net filtration pressure}$$

Normally, about 20% of the plasma that enters the glomerulus is filtered at the net filtration pressure of 10 mm Hg, producing collectively through all glomeruli 180 liters of glomerular filtrate each day for an average GFR of 125 ml/min in males (160 liters of filtrate per day for an average GFR of 115 ml/min in females).

● **FIGURE 14-9 Direct effect of arterial blood pressure on the glomerular filtration rate (GFR).**

Changes in the GFR result primarily from changes in glomerular capillary blood pressure.

Because the net filtration pressure that accomplishes glomerular filtration is simply the result of an imbalance of opposing physical forces between the glomerular capillary plasma and Bowman's capsule fluid, alterations in any of these physical forces can affect the GFR. We will examine the effect that changes in each of these physical forces have on the GFR.

UNREGULATED INFLUENCES ON THE GFR Plasma-colloid osmotic pressure and Bowman's capsule hydrostatic pressure are not subject to regulation and, under normal conditions, do not vary much.

Clinical Note However, they can change pathologically and thus inadvertently affect the GFR. Because plasma-colloid osmotic pressure opposes filtration, a decrease in plasma protein concentration, by reducing this pressure, leads to an increase in the GFR. An uncontrollable reduction in plasma protein concentration might occur, for example, in severely burned patients who lose a large quantity of protein-rich, plasma-derived fluid through the exposed burned surface of their skin. Conversely, in situations in which the plasma-colloid osmotic pressure is elevated, such as in cases of dehydrating diarrhea, the GFR is reduced.

Bowman's capsule hydrostatic pressure can become uncontrollably elevated, and filtration subsequently can decrease, given a urinary tract obstruction, such as a kidney stone or enlarged prostate. The damming up of fluid behind the obstruction elevates capsular hydrostatic pressure.

CONTROLLED ADJUSTMENTS IN THE GFR Unlike plasma-colloid osmotic pressure and Bowman's capsule hydrostatic pressure—which may be uncontrollably altered in various disease states and thereby inappropriately alter the GFR—glomerular capillary blood pressure can be controlled to adjust the GFR to suit the body's needs. Assuming that all other factors stay constant, as the glomerular capillary blood pressure goes up, the net filtration pressure increases and the GFR increases correspondingly. The magnitude of the glomerular capillary blood pressure depends on the rate of blood flow within each of the glomeruli. The amount of blood flowing into a glomerulus per minute is determined largely by the magnitude of the mean systemic arterial blood pressure and

the resistance offered by the afferent arterioles. If resistance increases in the afferent arteriole, less blood flows into the glomerulus, decreasing the GFR. Conversely, if afferent arteriolar resistance is reduced, more blood flows into the glomerulus and the GFR increases. Two major control mechanisms regulate the GFR, both directed at adjusting glomerular blood flow by regulating the radius and thus the resistance of the afferent arteriole. These mechanisms are (1) autoregulation, which is aimed at preventing spontaneous changes in GFR; and (2) extrinsic sympathetic control, which is aimed at long-term regulation of arterial blood pressure.

MECHANISMS RESPONSIBLE FOR AUTOREGULATION OF THE GFR Because arterial blood pressure is the force that drives blood into the glomerulus, the glomerular capillary blood pressure and, accordingly, the GFR would increase in direct proportion to an increase in arterial pressure if everything else remained constant (● Figure 14-9). Similarly, a fall in arterial blood pressure would be accompanied by a decline in GFR. Such spontaneous, inadvertent changes in GFR are largely prevented by intrinsic regulatory mechanisms initiated by the kidneys themselves, a process known as **autoregulation** (*auto* means "self"). The kidneys can, within limits, maintain a constant blood flow into the glomerular capillaries (and thus a constant glomerular capillary blood pressure and a stable GFR) despite changes in the driving arterial pressure. They do so primarily by altering afferent arteriolar caliber, thereby adjusting resistance to flow through these vessels. For example, if the GFR increases as a direct result of a rise in arterial pressure, the net filtration pressure and GFR can be reduced to normal by constriction of the afferent arteriole, which decreases the flow of blood into the glomerulus (● Figure 14-10a). This local adjustment lowers the glomerular blood pressure and the GFR to normal.

Conversely, when GFR falls in the presence of a decline in arterial pressure, glomerular pressure can be increased to normal by vasodilation of the afferent arteriole, which allows more blood to enter despite the reduction in driving pressure (● Figure 14-10b). The resultant buildup of glomerular blood volume increases glomerular blood pressure, which in turn brings the GFR back up to normal.

Two intrarenal mechanisms contribute to autoregulation: (1) a *myogenic* mechanism, which responds to changes in pres-

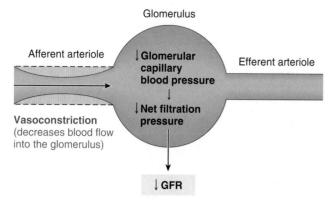

(a) Arteriolar vasoconstriction decreases the GFR

(b) Arteriolar vasodilation increases the GFR

● **FIGURE 14-10 Adjustments of afferent arteriole caliber to alter the GFR.**

sure within the nephron's vascular component; and (2) a *tubuloglomerular feedback* mechanism, which senses changes in the salt level in the fluid flowing through the nephron's tubular component.

■ The **myogenic** mechanism is a common property of vascular smooth muscle (*myogenic* means "muscle produced"). Arteriolar vascular smooth muscle contracts inherently in response to the stretch accompanying increased pressure within the vessel (see p. 357). Accordingly, the afferent arteriole automatically constricts on its own when it is stretched because of an increased arterial driving pressure. This response helps limit blood flow into the glomerulus to normal despite the elevated arterial pressure. Conversely, inherent relaxation of an unstretched afferent arteriole when pressure within the vessel is reduced increases blood flow into the glomerulus despite the fall in arterial pressure.

■ The **tubuloglomerular feedback (TGF)** mechanism involves the *juxtaglomerular apparatus,* which is the specialized combination of tubular and vascular cells where the tubule, after having bent back on itself, passes through the angle formed by the afferent and efferent arterioles as they join the glomerulus (● Figure 14-11; see also ● Figure 14-3, p. 514). The smooth muscle cells within the wall of the afferent arteriole in this region are specialized to form **granular cells,** so called because

they contain many secretory granules. Specialized tubular cells in this region are collectively known as the **macula densa.** The macula densa cells detect changes in the salt level of the fluid flowing past them through the tubule.

If the GFR is increased secondary to an elevation in arterial pressure, more fluid than normal is filtered and flows through the distal tubule. In response to the resultant rise in salt delivery to the distal tubule, the macula densa cells release *ATP* and *adenosine,* both of which act locally as a paracrine on the adjacent afferent arteriole, causing it to constrict, thus reducing glomerular blood flow and returning GFR to normal. In the opposite situation, when less salt is delivered to the distal tubule because of a spontaneous decline in GFR accompanying a fall in arterial pressure, less ATP and adenosine are released by the macula densa cells. The resultant afferent arteriolar vasodilation increases the glomerular flow rate, restoring the GFR to normal. To exert even more exquisite control over tubuloglomerular feedback, the macula densa cells also secrete the vasodilator *nitric oxide,* which puts the brakes on the action of ATP and adenosine at the afferent arteriole. By means of the TGF mechanism, the tubule of a nephron is able to monitor the salt level in the fluid flowing through it and adjust the rate of filtration through its own glomerulus accordingly to keep the early distal tubular fluid and salt delivery constant.

IMPORTANCE OF AUTOREGULATION OF THE GFR The myogenic and tubuloglomerular feedback mechanisms work in unison to autoregulate the GFR within the mean arterial blood pressure range of 80 to 180 mm Hg. Within this wide range, intrinsic autoregulatory adjustments of afferent arteriolar resistance can compensate for changes in arterial pressure, thus preventing inappropriate fluctuations in GFR, even though glomerular pressure tends to change in the same direction as arterial pressure. Normal mean arterial pressure is 93 mm Hg, so this range encompasses the transient changes in blood pressure that accompany daily activities unrelated to the need for the kidneys to regulate H_2O and salt excretion, such as the normal elevation in blood pressure accompanying exercise. Autoregulation is important because unintentional shifts in GFR could lead to dangerous imbalances of fluid, electrolytes, and wastes. Because at least a certain portion of the filtered fluid is always excreted, the amount of fluid excreted in the urine is automatically increased as the GFR increases. If autoregulation did not occur, the GFR would increase and H_2O and solutes would be lost needlessly as a result of the rise in arterial pressure accompanying heavy exercise. If, by contrast, the GFR were too low, the kidneys could not eliminate enough wastes, excess electrolytes, and other materials that should be excreted. Autoregulation thus greatly blunts the direct effect that changes in arterial pressure would otherwise have on GFR and subsequently on H_2O, solute, and waste excretion.

When changes in mean arterial pressure fall outside the autoregulatory range, these mechanisms cannot compensate. Therefore, dramatic changes in mean arterial pressure (<80 mm Hg or >180 mm Hg) directly cause the glomerular capillary pressure and, accordingly, the GFR to decrease or increase in proportion to the change in arterial pressure.

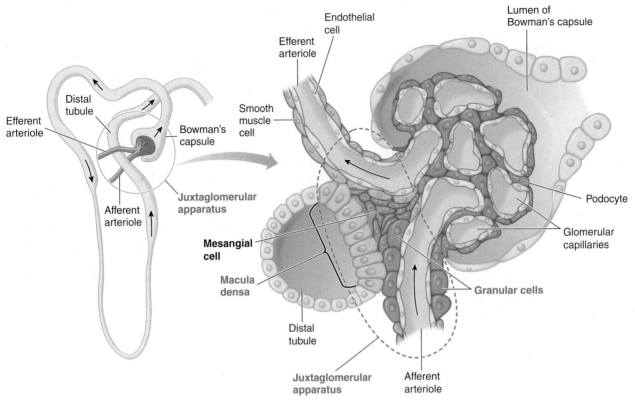

● **FIGURE 14-11 The juxtaglomerular apparatus.** The juxtaglomerular apparatus consists of specialized vascular cells (the granular cells) and specialized tubular cells (the macula densa) at a point where the distal tubule passes through the fork formed by the afferent and efferent arterioles of the same nephron.

IMPORTANCE OF EXTRINSIC SYMPATHETIC CONTROL OF THE GFR In addition to the intrinsic autoregulatory mechanisms designed to keep the GFR constant in the face of fluctuations in arterial blood pressure, the GFR can be *changed on purpose*—even when the mean arterial blood pressure is within the autoregulatory range—by extrinsic control mechanisms that override the autoregulatory responses. Extrinsic control of GFR, which is mediated by sympathetic nervous system input to the afferent arterioles, is aimed at regulating arterial blood pressure. The parasympathetic nervous system does not exert any influence on the kidneys.

If plasma volume is decreased—for example, by hemorrhage—the resulting fall in arterial blood pressure is detected by the arterial carotid sinus and aortic arch baroreceptors, which initiate neural reflexes to raise blood pressure toward normal (see p. 378). These reflex responses are coordinated by the cardiovascular control center in the brain stem and are mediated primarily through increased sympathetic activity to the heart and blood vessels. Although the resulting increase in both cardiac output and total peripheral resistance helps raise blood pressure toward normal, plasma volume is still reduced. In the long term, plasma volume must be restored to normal. One compensation for a depleted plasma volume is reduced urine output so that more fluid than normal is conserved for the body. Urine output is reduced in part by reducing the GFR; if less fluid is filtered, less is available to excrete.

ROLE OF THE BARORECEPTOR REFLEX IN EXTRINSIC CONTROL OF THE GFR No new mechanism is needed to decrease the GFR. It is reduced by the baroreceptor reflex response to a fall in blood pressure (● Figure 14-12). During this reflex, sympathetically induced vasoconstriction occurs in most arterioles throughout the body (including the afferent arterioles) as a compensatory mechanism to increase total peripheral resistance. The afferent arterioles have α_1 adrenergic receptors (see p. 243) and are innervated with sympathetic vasoconstrictor fibers to a far greater extent than the efferent arterioles are. When the afferent arterioles carrying blood to the glomeruli constrict from increased sympathetic activity, less blood flows into the glomeruli than normal, lowering glomerular capillary blood pressure (see ● Figure 14-10a). The resulting decrease in GFR, in turn, reduces urine volume. In this way, some of the H_2O and salt that would otherwise have been lost in the urine are saved for the body, helping in the long term to restore plasma volume to normal so that short-term cardiovascular adjustments that have been made are no longer necessary. Other mechanisms, such as increased tubular reabsorption of H_2O and salt as well as increased thirst (described more thoroughly elsewhere), also contribute to long-term maintenance of blood pressure, despite a loss of plasma volume, by helping to restore plasma volume.

Conversely, if blood pressure is elevated (for example, because of an expansion of plasma volume following ingestion of excessive fluid), the opposite responses occur. When the baroreceptors detect a rise in blood pressure, sympathetic vasoconstrictor activity to the arterioles, including the renal afferent arterioles, is reflexly reduced, allowing afferent arteriolar vasodilation to occur. As more blood enters the glomeruli through the dilated afferent arterioles, glomerular capillary blood pressure rises, increasing the GFR (see ● Figure 14-10b). As more fluid is filtered, more fluid is available to be eliminated in the urine. A hormonally adjusted reduction in the tubular reabsorption of H_2O and salt also contributes to the increase in urine volume. These two renal mechanisms—increased glomerular filtration and decreased tubular reabsorption of H_2O and salt—increase urine volume and eliminate the excess fluid from the body. Reduced thirst and fluid intake also help restore an elevated blood pressure to normal.

The GFR can be influenced by changes in the filtration coefficient.

Thus far we have discussed changes in the GFR as a result of changes in net filtration pressure. The rate of glomerular filtration, however, depends on the filtration coefficient (K_f) as well as on the net filtration pressure. For years K_f was considered a constant, except in disease situations in which the glomerular membrane becomes leakier than usual. Research to the contrary indicates that K_f is subject to change under physiologic control. Both factors on which K_f depends—the surface area and the permeability of the glomerular membrane—can be modified by contractile activity within the membrane.

The surface area available for filtration within the glomerulus is represented by the inner surface of the glomerular capillaries that comes into contact with blood. Each tuft of glomerular capillaries is held together by **mesangial cells** (see ● Figure 14-11). These cells contain contractile elements (that is, actinlike filaments). Contraction of these mesangial cells closes off a portion of the filtering capillaries, reducing the surface area available for filtration within the glomerular tuft. When the net filtration pressure remains unchanged, this reduction in K_f decreases GFR. Sympathetic stimulation causes the mesangial cells to contract, thus providing a second mechanism (besides promoting afferent arteriolar vasoconstriction) by which sympathetic activity can decrease the GFR.

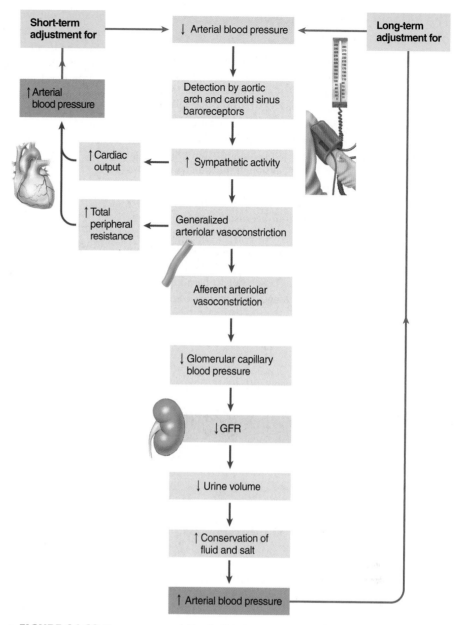

● **FIGURE 14-12 Baroreceptor reflex influence on the GFR in long-term regulation of arterial blood pressure.**

Podocytes also possess actinlike contractile filaments, whose contraction or relaxation can, respectively, decrease or increase the number of filtration slits open in the inner membrane of Bowman's capsule by changing the shapes and proximities of the foot processes (● Figure 14-13). The number of slits is a determinant of permeability; the more open slits, the greater the permeability. Contractile activity of the podocytes, which in turn affects permeability and the K_f, is under physiologic control by incompletely understood mechanisms.

Before turning our attention to the process of tubular reabsorption, we are first going to examine the percentage of cardiac output that goes to the kidneys. This will further reinforce the concept of how much blood flows through the kidneys and how much of that fluid is filtered and subsequently acted on by the tubules.

The kidneys normally receive 20% to 25% of the cardiac output.

At the average net filtration pressure and K_f, 20% of the plasma that enters the kidneys is converted into glomerular filtrate. That means at an average GFR of 125 ml/min, the total renal plasma flow must average about 625 ml/min. Because 55% of whole blood consists of plasma (that is, hematocrit = 45; see p. 391), the total flow of blood through the kidneys averages 1140 ml/min. This quantity is about 22% of the total cardiac output of 5 liters (5000 ml)/min, although the kidneys compose less than 1% of total body weight.

The kidneys need to receive such a seemingly disproportionate share of the cardiac output because they must continuously perform their regulatory and excretory functions on the huge volumes of plasma delivered to them to maintain stability in the internal fluid environment. Most of the blood goes to the kidneys not to supply the renal tissue but to be adjusted and purified by the kidneys. On average, 20% to 25% of the blood pumped out by the heart each minute "goes to the cleaners" instead of serving its normal purpose of exchanging materials with the tissues. Only by continuously processing such a large proportion of the blood can the kidneys precisely regulate the volume and electrolyte composition of the internal environment and adequately eliminate the large quantities of metabolic waste products that are constantly produced.

Tubular Reabsorption

All plasma constituents except the proteins are indiscriminately filtered together through the glomerular capillaries. In addition to waste products and excess materials that the body must eliminate, the filtered fluid contains nutrients, electrolytes, and other substances that the body cannot afford to lose in the urine. Indeed, through ongoing glomerular filtration, greater quantities of these materials are filtered per day than are even present in the entire body. The essential materials that are filtered are returned to the blood by *tubular reabsorption,* the discrete transfer of substances from the tubular lumen into the peritubular capillaries.

Tubular reabsorption is tremendous, highly selective, and variable.

Tubular reabsorption is a highly selective process. All constituents except plasma proteins are at the same concentration in the glomerular filtrate as in plasma. In most cases, the quantity reabsorbed of each substance is the amount required to maintain the proper composition and volume of the internal fluid environment. In general, the tubules have a high reabsorptive capacity for substances needed by the body and little or no reabsorptive capacity for substances of no value (▲ Table 14-2). Accordingly, only a small percentage, if any, of filtered plasma constituents that are useful to the body are present in the urine, most having been reabsorbed and returned to the blood. Only excess amounts of essential materials such as elec-

(a) Increased K_f on podocyte relaxation

(b) Decreased K_f on podocyte contraction

● **FIGURE 14-13 Change in the number of open filtration slits caused by podocyte relaxation and contraction.**
(a) Podocyte relaxation narrows the bases of the foot processes, increasing the number of fully open intervening filtration slits spanning a given area. (b) Podocyte contraction flattens the foot processes and thus decreases the number of intervening filtration slits.

(*Source:* Adapted from *Federation Proceedings,* vol. 42, pp. 3046–3052, 1983. Reprinted by permission.)

▲ TABLE 14-2	Fate of Various Substances Filtered by the Kidneys	
Substance	**Average Percentage of Filtered Substance Reabsorbed**	**Average Percentage of Filtered Substance Excreted**
Water	99	1
Sodium	99.5	0.5
Glucose	100	0
Urea (a waste product)	50	50
Phenol (a waste product)	0	100

trolytes are excreted in the urine. For the essential plasma constituents regulated by the kidneys, absorptive capacity may vary depending on the body's needs. In contrast, a large percentage of filtered waste products are present in the urine. These wastes, which are useless or even potentially harmful to the body if allowed to accumulate, are not reabsorbed to any extent. Instead, they stay in the tubules to be eliminated in the urine. As H_2O and other valuable constituents are reabsorbed, the waste products remaining in the tubular fluid become highly concentrated.

Of the 125 ml/min filtered, typically 124 ml/min are reabsorbed. Considering the magnitude of glomerular filtration, the extent of tubular reabsorption is tremendous: The tubules typically reabsorb 99% of the filtered H_2O (47 gal/day), 100% of the filtered sugar (2.5 lb/day), and 99.5% of the filtered salt (0.36 lb/day).

Tubular reabsorption involves transepithelial transport.

Throughout its entire length, the tubule wall is one cell thick and is in close proximity to a surrounding peritubular capillary (● Figure 14-14). Adjacent tubular cells do not come into contact with each other except where they are joined by tight junctions (see p. 59) at their lateral edges near their *luminal membranes*, which face the tubular lumen. Interstitial fluid lies in the gaps between adjacent cells—the **lateral spaces**—as well as between the tubules and capillaries. The *basolateral membrane* faces the interstitial fluid at the base and lateral edges of the cell. The tight junctions largely prevent substances from moving *between* the cells, so materials must pass *through* the cells to leave the tubular lumen and gain entry to the blood.

TRANSEPITHELIAL TRANSPORT To be reabsorbed, a substance must traverse five distinct barriers:

■ It must leave the tubular fluid by crossing the luminal membrane of the tubular cell (● Figure 14-14, step **1**).

■ It must pass through the cytosol from one side of the tubular cell to the other (step **2**).

■ It must cross the basolateral membrane of the tubular cell to enter the interstitial fluid (step **3**).

■ It must diffuse through the interstitial fluid (step **4**).

■ It must penetrate the capillary wall to enter the blood plasma (step **5**).

This entire sequence of steps is known as **transepithelial** ("across the epithelium") **transport.**

PASSIVE VERSUS ACTIVE REABSORPTION The two types of tubular reabsorption—*passive reabsorption* and *active reabsorption*—depend on whether local energy expenditure is needed for reabsorbing a particular substance. In **passive reabsorption,** all steps in the transepithelial transport of a substance from the tubular lumen to the plasma are passive; that is, no energy is spent for the substance's net movement, which occurs down electrochemical or osmotic gradients (see p. 63). In contrast, **active reabsorption** takes place if any one of the steps in the transepithelial transport of a substance requires energy, even if the four other steps are passive. With active reabsorption, net movement of the substance from the tubular lumen to the plasma occurs *against* an electrochemical gradient. Substances that are actively reabsorbed are of particular importance to the body, such as glucose, amino acids, and other organic nutrients, as well as Na^+ and other electrolytes such as PO_4^{3-}. Rather than specifically describe the reabsorptive

To be reabsorbed (to move from the filtrate to the plasma), a substance must cross five distinct barriers:

1 the luminal cell membrane   **3** the basolateral cell membrane   **5** the capillary wall

2 the cytosol   **4** the interstital fluid

● **FIGURE 14-14 Steps of transepithelial transport.**

process for each of the many filtered substances returned to the plasma, we will provide illustrative examples of the general mechanisms involved, after first highlighting the unique and important case of Na$^+$ reabsorption.

An active Na$^+$–K$^+$ ATPase pump in the basolateral membrane is essential for Na$^+$ reabsorption.

Sodium reabsorption is unique and complex. Of the total energy spent by the kidneys, 80% is used for Na$^+$ transport, indicating the importance of this process. Unlike most filtered solutes, Na$^+$ is reabsorbed throughout most of the tubule, but to varying extents in different regions. Of the Na$^+$ filtered, 99.5% is normally reabsorbed. Of the Na$^+$ reabsorbed, on average 67% is reabsorbed in the proximal tubule, 25% in the loop of Henle, and 8% in the distal and collecting tubules. Sodium reabsorption plays different important roles in each of these segments, as will become apparent as our discussion continues. Here is a preview of these roles.

■ Sodium reabsorption in the *proximal tubule* plays a pivotal role in reabsorbing glucose, amino acids, H$_2$O, Cl$^-$, and urea and is linked in part to K$^+$ secretion.

■ Sodium reabsorption in the ascending limb of the *loop of Henle,* along with Cl$^-$ reabsorption, plays a critical role in the kidneys' ability to produce urine of varying concentrations and volumes, depending on the body's need to conserve or eliminate H$_2$O.

■ Sodium reabsorption in the *distal and collecting tubules* is variable and subject to hormonal control. It plays a key role in regulating ECF volume, which is important in long-term control of arterial blood pressure.

Sodium is reabsorbed throughout the tubule with the exception of the descending limb of the loop of Henle. You will learn about the significance of this exception later. Throughout all Na$^+$-reabsorbing segments of the tubule, the active step in Na$^+$ reabsorption involves the energy-dependent Na$^+$–K$^+$ ATPase carrier located in the tubular cell's basolateral membrane (● Figure 14-15). This carrier is the same Na$^+$–K$^+$ pump present in all cells that actively extrudes Na$^+$ from the cell (see p. 70). As this basolateral pump transports Na$^+$ out of the tubular cell into the lateral space, it keeps the intracellular Na$^+$ concentration low while it simultaneously builds up the concentration of Na$^+$ in the lateral space; that is, it moves Na$^+$ against a concentration gradient. Because the intracellular Na$^+$ concentration is kept low by basolateral pump activity, a concentration gradient is established that favors the passive movement of Na$^+$ from its higher concentration in the tubular lumen across the luminal border into the tubular cell. The nature of the luminal Na$^+$ channels and/or transport carriers that permit movement of Na$^+$ from the lumen into the cell varies for differ-

KEY

→ = Active transport of ion against concentration gradient
→ = Passive movement of ion down concentration gradient

● **FIGURE 14-15 Sodium reabsorption.** The basolateral Na$^+$–K$^+$ pump actively transports Na$^+$ from the tubular cell into the interstitial fluid within the lateral space. This process establishes a concentration gradient for passive movement of Na$^+$ from the lumen into the tubular cell and from the lateral space into the peritubular capillary, accomplishing net transport of Na$^+$ from the tubular lumen into the blood at the expense of energy.

ent parts of the tubule, but in each case, movement of Na$^+$ across the luminal membrane is always a passive step. For example, in the proximal tubule, Na$^+$ crosses the luminal border by a symport carrier that simultaneously moves Na$^+$ and an organic nutrient such as glucose from the lumen into the cell. You will learn more about this cotransport process shortly. By contrast, in the collecting duct, Na$^+$ crosses the luminal border through a Na$^+$ leak channel. Once Na$^+$ enters the cell across the luminal border by whatever means, it is actively extruded to the lateral space by the basolateral Na$^+$–K$^+$ pump. This step is the same throughout the tubule. Sodium continues to diffuse down a concentration gradient from its high concentration in the lateral space into the surrounding interstitial fluid and finally into the peritubular capillary blood. Thus, net transport of Na$^+$ from the tubular lumen into the blood occurs at the expense of energy.

First let us consider the importance of regulating Na$^+$ reabsorption in the distal portion of the nephron and examine how this control is accomplished. Later we will explore in further detail the roles of Na$^+$ reabsorption in the proximal tubule and in the loop of Henle.

Aldosterone stimulates Na$^+$ reabsorption in the distal and collecting tubules.

In the proximal tubule and loop of Henle, a constant percentage of the filtered Na$^+$ is reabsorbed regardless of the **Na$^+$ load** (*total amount* of Na$^+$ in the body fluids, *not the concentration* of Na$^+$ in the body fluids). In the distal part of the tubule, the reabsorption of a small percentage of the filtered Na^+ is subject to hormonal control. The extent of this controlled, discretionary reabsorption is inversely related to the magnitude of the Na^+ load in the body. If there is too much Na^+, little of this controlled Na^+ is reabsorbed; instead, it is lost in the urine, thereby removing excess Na^+ from the body. If Na^+ is depleted, however, most or all of this controlled Na^+ is reabsorbed, conserving for the body Na^+ that otherwise would be lost in the urine.

The Na^+ load in the body is reflected by the ECF volume. Sodium and its accompanying anion Cl^- account for more than 90% of the ECF's osmotic activity. Whenever we speak of Na^+ load, we tacitly mean salt load too, because Cl^- goes along with Na^+. (NaCl is common table salt.) The Na^+ load is subject to regulation; Cl^- passively follows along. Recall that osmotic pressure can be thought of loosely as a "pulling" force that attracts and holds H_2O (see p. 64). When the Na^+ load is above normal and the ECF's osmotic activity is therefore increased, the extra Na^+ "holds" extra H_2O, expanding the ECF volume. Conversely, when the Na^+ load is below normal, thereby decreasing ECF osmotic activity, less H_2O than normal can be held in the ECF, so the ECF volume is reduced. Because plasma is part of the ECF, the most important result of a change in ECF volume is the matching change in blood pressure with expansion ($\uparrow$ blood pressure) or reduction ($\downarrow$ blood pressure) of the plasma volume. Thus, long-term control of arterial blood pressure ultimately depends on Na^+-regulating mechanisms. We now turn attention to these mechanisms.

ACTIVATION OF THE RENIN–ANGIOTENSIN–ALDOSTERONE SYSTEM The most important and best-known hormonal system involved in regulating Na^+ is the **renin–angiotensin–aldosterone system (RAAS)**. The granular cells of the juxtaglomerular apparatus (see ● Figure 14-11) secrete an enzymatic hormone, **renin,** into the blood in response to a fall in NaCl/ECF volume/blood pressure. This function is in addition to the role the macula densa cells of the juxtaglomerular apparatus play in autoregulation. Specifically, the following three inputs to the granular cells increase renin secretion:

1. The granular cells themselves function as *intrarenal baroreceptors.* They are sensitive to pressure changes within the afferent arteriole. When the granular cells detect a fall in blood pressure, they secrete more renin.

2. The macula densa cells in the tubular portion of the juxtaglomerular apparatus are sensitive to the NaCl moving past them through the tubular lumen. In response to a fall in NaCl, the macula densa cells trigger the granular cells to secrete more renin.

3. The granular cells are innervated by the sympathetic nervous system. When blood pressure falls below normal, the baroreceptor reflex increases sympathetic activity. As part of this reflex response, increased sympathetic activity stimulates the granular cells to secrete more renin.

These interrelated signals for increased renin secretion all indicate the need to expand the plasma volume to increase the arterial pressure to normal on a long-term basis. Through a complex series of events involving the RAAS, increased renin

secretion brings about increased Na^+ reabsorption by the distal and collecting tubules. Chloride always passively follows Na^+ down the electrical gradient established by sodium's active movement. The ultimate benefit of this salt retention is that it osmotically promotes H_2O retention, which helps restore the plasma volume, thus being important in the long-term control of blood pressure.

Let us examine in further detail the RAAS mechanism by which renin secretion ultimately leads to increased Na^+ reabsorption (● Figure 14-16). Once secreted into the blood, renin acts as an enzyme to activate **angiotensinogen** into **angiotensin I.** Angiotensinogen is a plasma protein synthesized by the liver and always present in the plasma in high concentration. On passing through the lungs via the pulmonary circulation, angiotensin I is converted into **angiotensin II** by **angiotensin-converting enzyme (ACE),** which is abundant in the pulmonary capillaries. ACE is located in small pits in the luminal surface of the pulmonary capillary endothelial cells. Angiotensin II is the main stimulus for secretion of the hormone *aldosterone* from the adrenal cortex. The *adrenal cortex* is an endocrine gland that produces several different hormones, each secreted in response to different stimuli.

FUNCTIONS OF THE RENIN–ANGIOTENSIN–ALDOSTERONE SYSTEM Two distinct types of tubular cells are located in the distal and collecting tubules: *principal cells* and *intercalated cells.* The more abundant **principal cells** are the site of action of aldosterone and vasopressin, and thus are involved in Na^+ reabsorption and K^+ secretion (both regulated by aldosterone) and in H_2O reabsorption (regulated by vasopressin). **Intercalated cells,** by contrast, are concerned with acid–base balance.

Among its actions, **aldosterone** increases Na^+ reabsorption by the principal cells of the distal and collecting tubules. It does so by promoting the insertion of additional Na^+ channels into the luminal membranes and additional Na^+–K^+ pumps into the basolateral membranes of these cells. The net result is a greater passive inward flux of Na^+ into these distal and collecting tubular cells from the lumen and increased active pumping of Na^+ out of the cells into the plasma—that is, an increase in Na^+ reabsorption, with Cl^- following passively. RAAS thus promotes salt retention and a resulting H_2O retention and rise in arterial blood pressure. Acting in negative-feedback fashion, this system alleviates the factors that triggered the initial release of renin—namely, salt depletion, plasma volume reduction, and decreased arterial blood pressure (● Figure 14-16).

In addition to stimulating aldosterone secretion, angiotensin II is a potent constrictor of the systemic arterioles, directly increasing blood pressure by increasing total peripheral resistance (see p. 359). Furthermore, it stimulates thirst (increasing fluid intake) and stimulates vasopressin (a hormone that increases H_2O retention by the kidneys), both of which contribute to plasma volume expansion and elevation of arterial pressure. (As you will learn later, other mechanisms related to the long-term regulation of blood pressure and ECF osmolarity are also important in controlling thirst and vasopressin secretion.)

The opposite situation exists when the Na^+ load, ECF and plasma volume, and arterial blood pressure are above normal. Under these circumstances, renin secretion is inhibited. There-

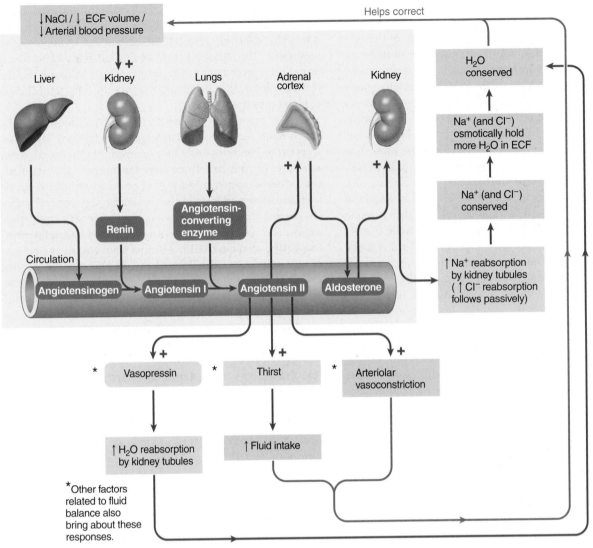

● **FIGURE 14-16 Renin–angiotensin–aldosterone system (RAAS).** The kidneys secrete the enzymatic hormone renin in response to reduced NaCl, ECF volume, and arterial blood pressure. Renin activates angiotensinogen, a plasma protein produced by the liver, into angiotensin I. Angiotensin I is converted into angiotensin II by angiotensin-converting enzyme (ACE) produced in the lungs. Angiotensin II stimulates the adrenal cortex to secrete the hormone aldosterone, which stimulates Na^+ reabsorption by the kidneys. The resulting retention of Na^+ exerts an osmotic effect that holds more H_2O in the ECF. Together, the conserved Na^+ and H_2O help correct the original stimuli that activated this hormonal system. Angiotensin II also exerts other effects that help rectify the original stimuli, such as by promoting arteriolar vasoconstriction.

fore, because angiotensinogen is not activated to angiotensin I and II, aldosterone secretion is not stimulated. Without aldosterone, the small aldosterone-dependent part of Na^+ reabsorption in the distal segments of the tubule does not occur. Instead, this nonreabsorbed Na^+ is lost in the urine. In the absence of aldosterone, the ongoing loss of this small percentage of filtered Na^+ can rapidly remove excess Na^+ from the body. Even though only about 8% of the filtered Na^+ depends on aldosterone for reabsorption, this small loss, multiplied many times as the entire plasma volume is filtered through the kidneys many times per day, can lead to a sizable loss of Na^+.

In the complete absence of aldosterone, 20 g of salt may be excreted per day. With maximum aldosterone secretion, all the filtered Na^+ (and, accordingly, all the filtered Cl^-) is reabsorbed, so salt excretion in the urine is zero. The amount of aldosterone secreted, and consequently the relative amount of salt conserved versus salt excreted, usually varies between these extremes, depending on the body's needs. For example, an average salt consumer typically excretes about 10 g of salt per day in the urine, a heavy salt consumer excretes more, and someone who has lost considerable salt during heavy sweating excretes less urinary salt. By varying the amount of renin and aldosterone secreted in accordance with the salt-determined fluid load in the body, the kidneys can finely adjust the amount of salt conserved or eliminated. In doing so, they maintain the salt load and ECF volume/arterial blood pressure at a relatively constant level despite wide variations in salt consumption and abnormal losses of salt-laden fluid.

ROLE OF THE RENIN–ANGIOTENSIN–ALDOSTERONE SYSTEM IN VARIOUS DISEASES Some cases of hypertension (high blood pressure) are the result of abnormal increases in RAAS activity. This system is also responsible in part for the fluid retention and edema accompanying congestive heart failure. Because of the failing heart, cardiac output is reduced and arterial blood pressure is low despite a normal or even expanded plasma volume. When a fall in blood pressure is the result of a failing heart rather than a reduced salt/fluid load in the body, the salt- and fluid-retaining reflexes triggered by the low blood pressure are inappropriate. Sodium excretion may fall to virtually zero despite continued salt ingestion and accumulation in the body. The resulting expansion of the ECF produces edema and exacerbates the congestive heart failure because the weakened heart cannot pump the additional plasma volume.

DRUGS THAT AFFECT Na⁺ REABSORPTION Because their salt-retaining mechanisms are being inappropriately triggered, patients with congestive heart failure are placed on low-salt diets. Often they are treated with **diuretics,** therapeutic agents that cause *diuresis* (increased urinary output) and thus promote loss of fluid from the body. Many of these drugs function by inhibiting tubular reabsorption of Na^+. As more Na^+ is excreted, more H_2O is also lost from the body, helping remove the excess ECF.

ACE inhibitor drugs, which block the action of angiotensin-converting enzyme (ACE), and **aldosterone receptor blockers,** are both also beneficial in treating hypertension and congestive heart failure. By respectively blocking the generation of angiotensin II or by blocking the binding of aldosterone with its renal receptors, these two classes of drugs halt the ultimate salt- and fluid-conserving actions and arteriolar constrictor effects of RAAS.

Atrial natriuretic peptide inhibits Na⁺ reabsorption.

Whereas RAAS exerts the most powerful influence on the renal handling of Na^+, this Na^+-retaining, blood pressure–raising system is opposed by a Na^+-losing, blood pressure–lowering system that in-

volves the hormones **atrial natriuretic peptide (ANP)** and **brain natriuretic peptide (BNP)** (*natriuretic* means "inducing excretion of large amounts of sodium in the urine"). The heart, in addition to its pump action, produces ANP and BNP. As its name implies, ANP is produced in the atrial cardiac muscle cells. BNP was first discovered in the brain (hence its name) but is produced primarily in the ventricular cardiac muscle cells. ANP and BNP are stored in granules and released when the heart muscle cells are mechanically stretched by an expansion of the circulating plasma volume when the ECF volume is increased. This expansion, which occurs as a result of Na^+ and H_2O retention, increases arterial blood pressure. In turn, the NPs promote natriuresis and accompanying diuresis, decreasing the plasma volume, and also influence the cardiovascular system to lower the blood pressure (● Figure 14-17).

The main action of ANP and BNP is to directly inhibit Na^+ reabsorption in the distal parts of the nephron, thus increasing Na^+ and accompanying osmotic H_2O excretion in the urine.

● **FIGURE 14-17 Atrial and brain natriuretic peptide.** The cardiac atria secrete the hormone atrial natriuretic peptide (ANP) and the cardiac ventricles secrete brain natriuretic peptide (BNP) in response to being stretched by Na^+ retention, expansion of the ECF volume, and increase in arterial blood pressure. ANP and BNP, in turn, promote natriuretic, diuretic, and hypotensive effects to help correct the original stimuli that resulted in their release.

They further increase Na^+ excretion in the urine by inhibiting two steps of the Na^+-conserving RAAS. The NPs inhibit renin secretion by the kidneys and act on the adrenal cortex to inhibit aldosterone secretion. In addition, they inhibit the secretion and actions of vasopressin, the H_2O-conserving hormone. ANP and BNP also promote natriuresis and accompanying diuresis by increasing the GFR. They dilate the afferent arterioles and constrict the efferent arterioles, thus raising glomerular capillary blood pressure and increasing the GFR. They further increase the GFR by relaxing the glomerular mesangial cells, leading to an increase in K_f. As more salt and water are filtered, more salt and water are excreted in the urine. ANP is more powerful in producing natriuresis and diuresis than BNP is. Besides indirectly lowering blood pressure by reducing the Na^+ load and hence the fluid load in the body, ANP and BNP directly lower blood pressure by decreasing the cardiac output and reducing peripheral vascular resistance by inhibiting sympathetic nervous activity to the heart and blood vessels.

The relative contributions of ANP and BNP in maintaining salt and H_2O balance and blood pressure regulation are presently being intensively investigated. Importantly, derangements of this system could logically contribute to hypertension. In fact, recent studies suggest that a deficiency of the counterbalancing natriuretic system may underlie some cases of long-term hypertension by leaving the powerful Na^+-conserving system unopposed. The resulting salt retention, especially in association with high salt intake, could expand ECF volume and elevate blood pressure.

We now shift attention to the reabsorption of other filtered solutes. However, we will continue to discuss Na^+ reabsorption because the reabsorption of many other solutes is linked in some way to Na^+ reabsorption.

Glucose and amino acids are reabsorbed by Na^+-dependent secondary active transport.

Large quantities of nutritionally important organic molecules such as glucose and amino acids are filtered each day. Because these substances normally are completely reabsorbed into the blood by energy- and Na^+-dependent mechanisms located in the proximal tubule, none of these materials are usually excreted in the urine. This rapid and thorough reabsorption early in the tubules protects against the loss of these important organic nutrients.

Even though glucose and amino acids are moved uphill against their concentration gradients from the tubular lumen into the blood until their concentration in the tubular fluid is virtually zero, no energy is directly used to operate the glucose or amino acid carriers. Glucose and amino acids are transferred by **secondary active transport.** With this process, specialized *symport carriers,* such as the *sodium and glucose cotransporter (SGLT),* located only in the proximal tubule simultaneously transfer both Na^+ and the specific organic molecule from the lumen into the cell (see p. 73). This luminal cotransport carrier is the means by which Na^+ passively crosses the luminal membrane in the proximal tubule. The lumen-to-cell Na^+ concentration gradient maintained by the energy-consuming basolateral Na^+–K^+ pump drives this co-

transport system and pulls the organic molecule against its concentration gradient without the direct expenditure of energy. Because the overall process of glucose and amino acid reabsorption depends on the use of energy, these organic molecules are considered to be actively reabsorbed, even though energy is not used directly to transport them across the membrane. Secondary active transport requires the presence of Na^+ in the lumen; without Na^+ the cotransport carrier is inoperable. Once transported into the tubular cells, glucose and amino acids passively diffuse down their concentration gradients across the basolateral membrane into the plasma, facilitated by a carrier, such as the *glucose transporter (GLUT),* which is not dependent on energy (see p. 69).

In general, actively reabsorbed substances exhibit a tubular maximum.

All actively reabsorbed substances bind with plasma membrane carriers that transfer them across the membrane against a concentration gradient. Each carrier is specific for the types of substances it can transport; for example, the glucose cotransport carrier cannot transport amino acids, or vice versa. Because a limited number of each carrier type is present in the cells lining the tubules, there is an upper limit on how much of a particular substance can be actively transported from the tubular fluid in a given period of time. The maximum reabsorption rate is reached when all of the carriers specific for a particular substance are fully occupied or saturated, so that they cannot handle any additional passengers at that time (see p. 67). This transport maximum is designated as the **tubular maximum,** or T_m. Any quantity of a substance filtered beyond its T_m is not reabsorbed and escapes instead into the urine. With the exception of Na^+, all actively reabsorbed substances have a tubular maximum. (Even though individual Na^+ transport carriers can become saturated, the tubules as a whole do not display a tubular maximum for Na^+, because aldosterone promotes the synthesis of more active Na^+–K^+ carriers in the distal and collecting tubular cells as needed.)

The plasma concentrations of some but not all substances that display carrier-limited reabsorption are regulated by the kidneys. How can the kidneys regulate some actively reabsorbed substances but not others, when the renal tubules limit the quantity of each of these substances that can be reabsorbed and returned to the plasma? We will compare glucose, a substance that has a T_m but *is not regulated* by the kidneys, with phosphate, a T_m-limited substance that *is regulated* by the kidneys.

Glucose is an example of an actively reabsorbed substance that is not regulated by the kidneys.

The normal plasma concentration of glucose is 100 mg of glucose/100 ml of plasma. Because glucose is freely filterable at the glomerulus, it passes into Bowman's capsule at the same concentration it has in the plasma. Accordingly, 100 mg of glucose are present in every 100 ml of plasma filtered. With 125 ml of plasma normally being filtered each minute (average GFR = 125 ml/min), 125 mg of glucose pass into Bowman's capsule with this filtrate every minute. The quantity of any substance filtered per minute, known as its **filtered load,** can be calculated as follows:

$$\text{Filtered load of a substance} = \text{plasma concentration of the substance} \times \text{GFR}$$

$$\text{Filtered load of glucose} = 100 \text{ mg} / 100 \text{ ml} \times 125 \text{ ml/min}$$

$$= 125 \text{ mg/min}$$

At a constant GFR, the filtered load of glucose is directly proportional to the plasma glucose concentration. Doubling the plasma glucose concentration to 200 mg/100 ml doubles the filtered load of glucose to 250 mg/min, and so on (● Figure 14-18).

TUBULAR MAXIMUM FOR GLUCOSE The T_m for glucose averages 375 mg/min; that is, the glucose carrier mechanism is capable of actively reabsorbing up to 375 mg of glucose per minute before it reaches its maximum transport capacity. At a normal plasma glucose concentration of 100 mg/100 ml, the 125 mg of glucose filtered per minute can readily be reabsorbed by the glucose carrier mechanism, because the filtered load is well below the T_m for glucose. Ordinarily, therefore, no glucose appears in the urine. Not until the filtered load of glucose exceeds 375 mg/min is the T_m reached. When more glucose is filtered per minute than can be reabsorbed because the T_m is exceeded, the maximum amount is reabsorbed, while the rest stays in the filtrate to be excreted. Accordingly, the plasma glucose concentration must be greater than 300 mg/100 ml—more than three times the normal value—before the amount filtered exceeds 375 mg/min and glucose starts spilling into the urine.

RENAL THRESHOLD FOR GLUCOSE The plasma concentration at which the T_m of a particular substance is reached and the substance first starts appearing in urine is called the **renal threshold.** At the average T_m of 375 mg/min and GFR of 125 ml/min, the renal threshold for glucose is 300 mg/100 ml.[1] Beyond the T_m, reabsorption stays constant at its maximum rate, and any further increase in the filtered load leads to a directly proportional increase in the amount of the substance excreted. For example, at a plasma glucose concentration of 400 mg/100 ml, the filtered load of glucose is 500 mg/min, 375 mg/min of which can be reabsorbed (a T_m worth) and 125 mg/min of which are excreted in the urine. At a plasma glucose concentration of 500 mg/100 ml, the filtered load is 625 mg/min, still only 375 mg/min can be reabsorbed, and 250 mg/min spill into the urine (● Figure 14-18).

 The plasma glucose concentration can become extremely high in *diabetes mellitus,* an endocrine disorder involving inadequate insulin action. Insulin is a pancre-

● **FIGURE 14-18 Renal handling of glucose as a function of plasma glucose concentration.** At a constant GFR, the quantity of glucose filtered per minute is directly proportional to the plasma concentration of glucose. All the filtered glucose can be reabsorbed up to the tubular maximum (T_m). If the amount of glucose filtered per minute exceeds the T_m, the maximum amount of glucose is reabsorbed (a T_m worth) and the rest stays in the filtrate to be excreted in urine. The renal threshold is the plasma concentration at which the T_m is reached and glucose first starts appearing in the urine.

atic hormone that facilitates transport of glucose into many body cells. When cellular glucose uptake is impaired, glucose that cannot gain entry into cells stays in the plasma, elevating the plasma glucose concentration. Consequently, although glucose does not normally appear in urine, it is found in the urine of people with untreated diabetes when the plasma glucose concentration exceeds the renal threshold, even though renal function has not changed.

What happens when plasma glucose concentration falls below normal? The renal tubules, of course, reabsorb all the filtered glucose because the glucose reabsorptive capacity is far from being exceeded. The kidneys cannot do anything to raise a low plasma glucose level to normal. They simply return all the filtered glucose to the plasma.

REASON WHY THE KIDNEYS DO NOT REGULATE GLUCOSE The kidneys do not influence plasma glucose concentration over a wide range of values from abnormally low levels up to three times the normal level. Because the T_m for glucose is well above the normal filtered load, the kidneys usually conserve all the glucose, thereby protecting against loss of this important nutrient in the urine. The kidneys do not regulate glucose, because they do not maintain glucose at some specific plasma concentration. Instead, this concentration is normally regulated by endocrine and liver mechanisms, with the kidneys merely maintaining whatever plasma glucose concentration is set by these other mechanisms (except when excessively high levels

[1]This is an idealized situation. In reality, glucose often starts spilling into the urine at glucose concentrations of 180 mg/100 ml and above. Glucose is often excreted before the average renal threshold of 300 mg/100 ml is reached for two reasons. First, not all nephrons have the same T_m, so some nephrons may have exceeded their T_m and be excreting glucose while others have not yet reached their T_m. Second, the efficiency of the glucose cotransport carrier may not be working at its maximum capacity at elevated values less than the true T_m, so some of the filtered glucose may fail to be reabsorbed and spill into the urine even though the average renal threshold has not been reached.

overwhelm the kidneys' reabsorptive capacity). The same general principle holds true for other organic plasma nutrients, such as amino acids and water-soluble vitamins.

Phosphate is an example of an actively reabsorbed substance that is regulated by the kidneys.

The kidneys do directly contribute to the regulation of many electrolytes, such as phosphate (PO_4^{3-}) and calcium (Ca^{2+}), because the renal thresholds of these inorganic ions equal their normal plasma concentrations. The transport carriers for these electrolytes are located in the proximal tubule. We will use PO_4^{3-} as an example. Our diets are generally rich in PO_4^{3-}, but because the tubules can reabsorb up to the normal plasma concentration's worth of PO_4^{3-} and no more, the excess ingested PO_4^{3-} is quickly spilled into the urine, restoring the plasma concentration to normal. The greater the amount of PO_4^{3-} ingested beyond the body's needs, the greater will be the amount excreted. In this way, the kidneys maintain the desired plasma PO_4^{3-} concentration while eliminating any excess PO_4^{3-} ingested.

Unlike the reabsorption of organic nutrients, the reabsorption of PO_4^{3-} and Ca^{2+} is also subject to hormonal control. Parathyroid hormone can alter the renal thresholds for PO_4^{3-} and Ca^{2+}, thus adjusting the quantity of these electrolytes conserved, depending on the body's momentary needs (Chapter 19).

Active Na⁺ reabsorption is responsible for the passive reabsorption of Cl⁻, H₂O, and urea.

Not only is secondary active reabsorption of glucose and amino acids linked to the basolateral Na^+-K^+ pump, but passive reabsorption of Cl^-, H_2O, and urea also depends on this active Na^+ reabsorption mechanism.

CHLORIDE REABSORPTION The negatively charged chloride ions are passively reabsorbed down the electrical gradient created by the active reabsorption of the positively charged sodium ions. For the most part, chloride ions pass between, not through, the tubular cells (through "leaky" tight junctions). The amount of Cl^- reabsorbed is determined by the rate of active Na^+ reabsorption instead of being directly controlled by the kidneys.

WATER REABSORPTION Water is passively reabsorbed throughout the length of the tubule as H_2O osmotically follows Na^+ that is actively reabsorbed. Of the H_2O filtered, 65%—117 liters per day—is passively reabsorbed by the end of the proximal tubule. Another 15% of the filtered H_2O is obligatorily reabsorbed from the loop of Henle. This 80% of the filtered H_2O is reabsorbed in the proximal tubule and Henle's loop regardless of the H_2O load in the body and is not subject to regulation. Variable amounts of the remaining 20% are reabsorbed in the distal portions of

KEY

⇒ = Passive movement of H₂O by osmosis or hydrostatic pressure

→ = Active transport of ion ⟶ = Passive diffusion of ion

● **FIGURE 14-19 Water reabsorption in the proximal tubule.** The force for H_2O reabsorption is the compartment of hypertonicity in the lateral spaces established by active extrusion of Na^+ by the basolateral pump. The aqua blue arrows show the direction of osmotic movement of H_2O.

the tubule; the extent of reabsorption in the distal and collecting tubules is under direct hormonal control, depending on the body's state of hydration. No part of the tubule directly requires energy for this tremendous reabsorption of H_2O.

During reabsorption, H_2O passes primarily through **aquaporins (AQPs)**, or **water channels**, formed by specific plasma membrane proteins in the tubular cells. Different types of water channels are present in various parts of the nephron. The water channels in the proximal tubule, *AQP-1*, are always open, accounting for the high H_2O permeability of this region. The *AQP-2* channels in the principal cells in the distal parts of the nephron, in contrast, are regulated by the hormone *vasopressin*, accounting for the variable H_2O reabsorption in this region.

The main driving force for H_2O reabsorption in the proximal tubule is a compartment of hypertonicity in the lateral spaces between the tubular cells established by the basolateral pump's active extrusion of Na^+ (● Figure 14-19). As a result of this pump activity, the concentration of Na^+ rapidly diminishes in the tubular fluid and tubular cells while it simultaneously increases in the localized region within the lateral spaces. This osmotic gradient induces the passive net flow of H_2O from the lumen into the lateral spaces, either through the cells or intercellularly through "leaky" tight junctions. The accumulation of fluid in the lateral spaces results in a buildup of hydrostatic (fluid) pressure, which flushes H_2O out of the lateral spaces into the interstitial fluid and finally into the peritubular capillaries. Water also osmotically follows other preferentially reabsorbed solutes such as glucose (which is also Na^+ dependent), but the direct influence of Na^+ reabsorption on passive H_2O reabsorption is quantitatively more important.

This return of filtered H_2O to the plasma is enhanced by the fact that the plasma-colloid osmotic pressure is greater in the peritubular capillaries than elsewhere. The concentration of plasma proteins, which is responsible for the plasma-colloid osmotic pressure, is elevated in the blood entering the peritubular capillaries because of the extensive filtration of H_2O through the glomerular capillaries upstream. The plasma proteins left behind in the glomerulus are concentrated into a smaller volume of plasma H_2O, increasing the plasma-colloid osmotic pressure of the unfiltered blood that leaves the glomerulus and enters the peritubular capillaries. This force tends to "pull" H_2O into the peritubular capillaries simultaneously with the "push" of the hydrostatic pressure in the lateral spaces that drives H_2O toward the capillaries. By these means, 65% of the filtered H_2O—117 liters per day—is passively reabsorbed by the end of the proximal tubule.

The mechanisms responsible for H_2O reabsorption beyond the proximal tubule will be described later.

UREA REABSORPTION Passive reabsorption of urea, in addition to Cl^- and H_2O, is also indirectly linked to active Na^+ reabsorption. **Urea** is a waste product from the breakdown of protein. The osmotically induced reabsorption of H_2O in the proximal tubule secondary to active Na^+ reabsorption produces a concentration gradient for urea that favors passive reabsorption of this waste, as follows (● Figure 14-20). Extensive reabsorption of H_2O in the proximal tubule gradually reduces the original 125 ml/min of filtrate until only 44 ml/min of fluid remain in the lumen by the end of the proximal tubule (65% of the H_2O in the original filtrate, or 81 ml/min, has been reabsorbed). Substances that have been filtered but not reabsorbed become progressively more concentrated in the tubular fluid as H_2O is reabsorbed while they are left behind. Urea is one such substance. Urea's concentration as it is filtered at the glomerulus is identical to its concentration in the plasma entering the peritubular capillaries. The quantity of urea present in the 125 ml of filtered fluid at the beginning of the proximal tubule, however, is concentrated almost threefold in the 44 ml left at the end of the proximal tubule. As a result, the urea concentration within the tubular fluid becomes much greater than the urea concentration in the adjacent capillaries. Therefore, a concentration gradient is created for urea to passively diffuse from the tubular lumen into the peritubular capillary plasma. Because the walls of the proximal tubules are only somewhat permeable to urea, only about 50% of the filtered urea is passively reabsorbed by this means.

Clinical Note Even though only half of the filtered urea is eliminated from the plasma with each pass through the nephrons, this removal rate is adequate. The urea concentration in the plasma becomes elevated only in impaired kidney function, when much less than half of the urea is removed. An elevated urea level was one of the first chemical characteristics to be identified in the plasma of patients with severe renal failure. Accordingly, clinical measurement of **blood urea nitrogen (BUN)** came into use as a crude assessment of kidney function. It is now known that the most serious consequences of renal failure are not attributable to the retention of urea, which itself is not especially toxic, but rather to the accumulation of other

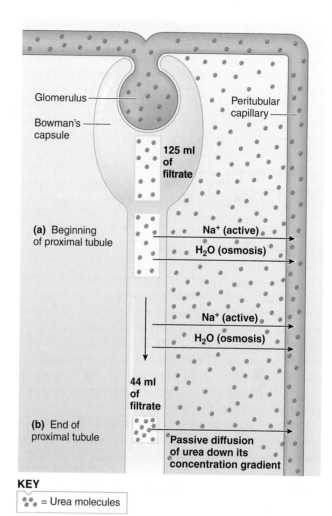

KEY

●●● = Urea molecules

● **FIGURE 14-20 Passive reabsorption of urea at the end of the proximal tubule.** (a) In Bowman's capsule and at the beginning of the proximal tubule, urea is at the same concentration as in the plasma and surrounding interstitial fluid. (b) By the end of the proximal tubule, 65% of the original filtrate has been reabsorbed, concentrating the filtered urea in the remaining filtrate. This establishes a concentration gradient favoring passive reabsorption of urea.

substances that are not adequately excreted because of their failure to be properly secreted—most notably H^+ and K^+. Health professionals still often refer to renal failure as **uremia** ("urea in the blood"), indicating excess urea in the blood, even though urea retention is not this condition's major threat.

In general, unwanted waste products are not reabsorbed.

The other filtered waste products besides urea, such as *uric acid, creatinine,* and *phenols* (derived from many foods) are likewise concentrated in the tubular fluid as H_2O leaves the filtrate to enter the plasma, but they are not passively reabsorbed, as urea is. Urea molecules, being the smallest of the waste products, are the only wastes passively reabsorbed by this concentrating effect. Even though the other wastes are also concentrated in the tubular fluid, they cannot leave the lumen down their concentration gradients to be passively reabsorbed, because they cannot permeate the tubular wall. Therefore, the waste products, not being reabsorbed, generally remain in the tubules and are excreted in the urine in highly concentrated form. This excretion of metabolic wastes is not subject to physiologic control. When renal function is normal, however, the excretory processes proceed at a satisfactory rate even though they are not controlled.

We have now completed discussion of tubular reabsorption and are going to shift attention to the other basic renal process carried out by the tubules—tubular secretion.

Tubular Secretion

Like tubular reabsorption, tubular secretion involves transepithelial transport, but now the steps are reversed. By providing a second route of entry into the tubules for selected substances, *tubular secretion*, the discrete transfer of substances from the peritubular capillaries into the tubular lumen, is a supplemental mechanism that hastens elimination of these compounds from the body. Anything that gains entry to the tubular fluid, whether by glomerular filtration or tubular secretion, and fails to be reabsorbed is eliminated in the urine.

The most important substances secreted by the tubules are *hydrogen ion (H⁺)*, *potassium ion (K⁺)*, and *organic anions and cations*, many of which are compounds foreign to the body.

Hydrogen ion secretion is important in acid–base balance.

Renal H^+ secretion is extremely important in regulating acid–base balance in the body. Hydrogen ion secreted into the tubular fluid is eliminated from the body in the urine. Hydrogen ion can be secreted by the proximal, distal, and collecting tubules, with the extent of H^+ secretion depending on the acidity of the body fluids. When the body fluids are too acidic, H^+ secretion increases. Conversely, H^+ secretion is reduced when the H^+ concentration in the body fluids is too low. (See Chapter 15 for further detail.)

Potassium ion secretion is controlled by aldosterone.

Potassium is one of the most abundant cations in the body, but about 98% of the K^+ is in the ICF, because the Na^+–K^+ pump actively transports K^+ into the cells. Because only a relatively small amount of K^+ is in the ECF, even slight changes in the ECF K^+ load can have a pronounced effect on the plasma K^+ concentration. Changes in the plasma K^+ concentration have a marked influence on membrane excitability. Therefore, plasma K^+ concentrations are tightly controlled, primarily by the kidneys.

Potassium ion is selectively moved in opposite directions in different parts of the tubule; it is actively reabsorbed in the proximal tubule and actively secreted by principal cells in the distal and collecting tubules. Furthermore, one type of intercalated cell actively secretes K^+ and another type actively reabsorbs K^+ in the distal and collecting tubules in conjunction with H^+ transport (Chapter 15). Early in the tubule K^+ is reabsorbed in a constant, unregulated fashion, whereas K^+ secretion later in the tubule by the principal cells is variable and subject to regulation. Because the filtered K^+ is almost completely reabsorbed in the proximal tubule, most K^+ in the urine is derived from controlled K^+ secretion in the distal parts of the nephron rather than from filtration.

During K^+ depletion, K^+ secretion in the distal parts of the nephron is reduced to a minimum, so only the small percentage of filtered K^+ that escapes reabsorption in the proximal tubule is excreted in the urine. In this way, K^+ that normally would have been lost in urine is conserved for the body. Conversely, when plasma K^+ levels are elevated, K^+ secretion is adjusted so that just enough K^+ is added to the filtrate for elimination to reduce the plasma K^+ concentration to normal. Thus, K^+ secretion, not the filtration or reabsorption of K^+, is varied in a controlled fashion to regulate the rate of K^+ excretion and maintain the desired plasma K^+ concentration.

MECHANISM OF K⁺ SECRETION Potassium ion secretion in the principal cells of the distal and collecting tubules is coupled to Na^+ reabsorption by the energy-dependent basolateral Na^+–K^+ pump (● Figure 14-21). This pump not only moves Na^+ out of the cell into the lateral space but also transports K^+ from the lateral space into the tubular cells. The resulting high intracellular K^+ concentration favors net movement of K^+ from the cells into the tubular lumen. Movement across the luminal membrane occurs passively through the large number of K^+ leak channels in this barrier in the distal and collecting tubules. By keeping the interstitial fluid concentration of K^+ low as it transports K^+ into the tubular cells from the surrounding interstitial fluid, the basolateral pump encourages passive movement of K^+ out of the peritubular capillary plasma into the interstitial fluid. A potassium ion leaving the plasma in this manner is later pumped into the cells, from which it passively moves into the lumen. In this way, the basolateral pump actively induces the net secretion of K^+ from the peritubular capillary plasma into the tubular lumen in the distal parts of the nephron.

Because K^+ secretion is linked with Na^+ reabsorption by the Na^+–K^+ pump, why isn't K^+ secreted throughout the Na^+-reabsorbing segments of the tubule instead of taking place only in the distal parts of the nephron? The answer lies in the location of the passive K^+ channels. In the distal and collecting tubules, the K^+ channels are concentrated in the luminal membrane, providing a route for K^+ pumped into the cell to exit into the lumen, thus being secreted. In the other tubular segments, the K^+ leak channels are located primarily in the basolateral membrane. As a result, K^+ pumped into the cell from the lateral space by the Na^+–K^+ pump simply moves back out into the lateral space through these channels. This K^+ recycling permits the ongoing operation of the Na^+–K^+ pump to accomplish Na^+ reabsorption with no local net effect on K^+.

CONTROL OF K⁺ SECRETION Several factors can alter the rate of K^+ secretion, the most important being aldosterone. This hormone stimulates K^+ secretion by the principal tubular cells late in the nephron while simultaneously enhancing these cells' reabsorption of Na^+. A rise in plasma K^+ concentration directly stimulates the adrenal cortex to increase its output of aldosterone, which in turn promotes the secretion and ultimate urinary excretion and elimination of excess K^+. Conversely, a decline in plasma K^+ concentration causes a reduction in aldosterone secretion and a corresponding decrease in aldosterone-stimulated renal K^+ secretion. The amount of filtered K^+ excreted in the urine varies from 80% to 1%, depending on the body's momentary needs.

Note that a rise in plasma K^+ concentration directly stimulates aldosterone secretion by the adrenal cortex, whereas a fall in plasma Na^+ concentration stimulates aldosterone secretion by means of the complex RAAS pathway. Thus, aldosterone secretion can be stimulated by two separate pathways (● Figure 14-22). No matter what the stimulus, however, increased aldosterone secretion always promotes simultaneous Na^+ reabsorption and K^+ secretion. For this reason, K^+ secretion can be inadvertently stimulated as a result of increased aldosterone activity brought about by Na^+ depletion, ECF volume reduction, or a fall in arterial blood pressure totally unrelated to K^+ balance. The resulting inappropriate loss of K^+ can lead to K^+ deficiency.

EFFECT OF H^+ SECRETION ON K^+ SECRETION Another factor that can inadvertently alter the magnitude of K^+ secretion is the acid–base status of the body. The intercalated cells in the distal portions of the nephron secrete either K^+ or H^+. An increased rate of secretion of either K^+ or H^+ is accompanied by a decreased rate of secretion of the other ion. Normally the kidneys secrete a preponderance of K^+, but when the body fluids are too acidic and H^+ secretion is increased as a compensatory measure, K^+ secretion is correspondingly reduced. This reduced secretion leads to inappropriate K^+ retention in the body fluids.

IMPORTANCE OF REGULATING PLASMA K^+ CONCENTRATION Except in the overriding circumstances of K^+ imbalances inadvertently induced during renal compensations for Na^+ or ECF volume deficits or acid–base imbalances, the kidneys usually exert a fine degree of control over plasma K^+ concentration. This is extremely important because even minor fluctuations in plasma K^+ concentration can have detrimental consequences. Potassium plays a key role in the membrane electrical activity of excitable tissues. Both increases and decreases in the plasma (ECF) K^+ concentration can alter the intracellular-to-extracellular K^+ concentration gradient, which in turn can change the resting membrane potential. The most serious consequences of both K^+ excess and K^+ deficiency are related to their impact on the heart. Both conditions result in decreased cardiac excitability, for different reasons. A rise in ECF K^+ concentration reduces resting potential (makes it less negative), which decreases the excitability of neurons, skeletal

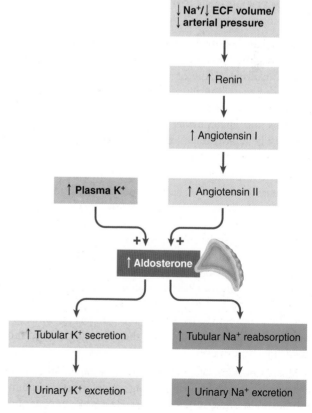

● **FIGURE 14-21 Potassium ion secretion.** The basolateral pump simultaneously transports Na^+ into the lateral space and K^+ into the tubular cell. In the parts of the tubule that secrete K^+, this ion leaves the cell through channels located in the luminal border, thus being secreted. (In the parts of the tubule that do not secrete K^+, the K^+ pumped into the cell during Na^+ reabsorption leaves the cell through channels located in the basolateral border, thus being retained in the body.)

● **FIGURE 14-22 Dual control of aldosterone secretion by K^+ and Na^+.**

muscle cells, and, most importantly, cardiac muscle cells, by keeping the voltage-gated Na^+ channels responsible for the rising phase of the cardiac action potential in their inactive (closed and not capable of opening) state (see p. 92). The cell membrane is unable to repolarize completely after depolarization to return the channel to its closed and capable of opening conformation. Some Na^+ channels are more sensitive than others to the depolarizing effect. As more and more Na^+ channels are inactivated by rising K^+ levels, cardiac excitability progressively decreases. In mild cases, the rate of rise and amplitude of cardiac action potentials decrease, and conduction velocity slows. In serious cases, cardiac arrhythmias possibly leading to fatal fibrillation ensue (see p. 320), or alternately death may result from cardiac arrest. A fall in ECF K^+ concentration results in hyperpolarization of nerve and muscle cell membranes, which also reduces their excitability. A greater depolarization than normal is needed to bring the membrane to threshold potential. The manifestations of ECF K^+ depletion are skeletal muscle weakness, diarrhea and abdominal distension caused by smooth muscle dysfunction, and abnormalities in cardiac rhythm and impulse conduction.

Organic anion and cation secretion helps efficiently eliminate foreign compounds from the body.

The proximal tubule contains two distinct types of secretory carriers, one for the secretion of organic anions and a separate system for secretion of organic cations.

FUNCTIONS OF ORGANIC ION SECRETORY SYSTEMS The organic ion secretory systems serve three important functions:

1. By adding more of a particular type of organic ion to the quantity that has already gained entry to the tubular fluid by glomerular filtration, the organic secretory pathways facilitate excretion of these substances. Included among these organic ions are certain blood-borne chemical messengers such as prostaglandins and epinephrine, which, having served their purpose, must be rapidly removed from the blood so that their biological activity is not unduly prolonged.

2. In some important instances, organic ions are poorly soluble in water. To be transported in blood, they are extensively but not irreversibly bound to plasma proteins. Because they are attached to plasma proteins, these substances cannot be filtered through the glomeruli. Tubular secretion facilitates elimination of these nonfilterable organic ions in urine. Even though a given organic ion is largely bound to plasma proteins, a small percentage of the ion always exists in free or unbound form in the plasma. Removal of this free organic ion by secretion permits "unloading" of some of the bound ion, which is then free to be secreted. This, in turn, encourages the unloading of even more organic ion, and so on.

3. Most important, the proximal tubule organic ion secretory systems play a key role in eliminating many foreign compounds from the body. These systems can secrete a large number of different organic ions, both those produced endogenously (within the body) and those foreign organic ions that have gained ac-

cess to the body fluids. This nonselectivity permits the organic ion secretory systems to hasten removal of many foreign organic chemicals, including food additives, environmental pollutants (for example, pesticides), drugs, and other nonnutritive organic substances that have entered the body. Even though this mechanism helps rid the body of potentially harmful foreign compounds, it is not subject to physiologic adjustments. The carriers cannot pick up their secretory pace when confronting an elevated load of these organic ions.

The liver plays an important role in helping rid the body of foreign compounds. Many foreign organic chemicals are not ionic in their original form, so they cannot be secreted by the organic ion systems. The liver converts these foreign substances into an anionic form that facilitates their secretion by the organic anion system and thus accelerates their elimination.

Clinical Note Many drugs, such as penicillin and nonsteroidal anti-inflammatory drugs (NSAIDs), are eliminated from the body by the organic ion secretory systems. To keep the plasma concentration of these drugs at effective levels, the dosage must be repeated on a regular, frequent basis that keeps pace with the rapid removal of these compounds in the urine.

Because the organic ion secretory carriers are not very selective, different drugs can compete for binding sites on the same carrier. For example, cimetidine (a drug used to treat stomach ulcers; see p. 612) and procainamide (a drug used to treat cardiac arrhythmias) both are secreted by the organic cation secretory carriers. If these drugs were given to the same patient, the urinary excretion rate of both substances would be decreased because they would compete for elimination by the secretory carriers. Thus coadministration of these drugs would lead to much higher blood concentrations of both substances than when each is given alone. Thus, to avoid potential drug toxicity, drugs eliminated by the same secretory pathway should not be taken together.

SUMMARY OF REABSORPTIVE AND SECRETORY PROCESSES This completes our discussion of the reabsorptive and secretory processes that occur across the proximal and distal portions of the nephron. These processes are summarized in ▲ Table 14-3. To generalize, the proximal tubule does most of the reabsorbing. This mass reabsorber transfers much of the filtered water and needed solutes back into the blood in unregulated fashion. Similarly, the proximal tubule is the major site of secretion, with the exception of K^+ secretion. The distal and collecting tubules then determine the final amounts of H_2O, Na^+, K^+, and H^+ excreted in the urine and thus eliminated from the body. They do so by fine-tuning the amount of Na^+ and H_2O reabsorbed and the amount of K^+ and H^+ secreted. These processes in the distal part of the nephron are all subject to control, depending on the body's momentary needs. The unwanted filtered waste products are left behind to be eliminated in the urine, along with excess amounts of filtered or secreted nonwaste products that fail to be reabsorbed.

We will next focus on the end result of the basic renal processes—what's left in the tubules to be excreted in urine, and, as a consequence, what has been cleared from plasma.

▲ TABLE 14-3	**Summary of Transport across Proximal and Distal Portions of the Nephron**

PROXIMAL TUBULE

Reabsorption	Secretion
67% of filtered Na^+ actively reabsorbed; not subject to control; Cl^- follows passively	Variable H^+ secretion depending on acid–base status of body
All filtered glucose and amino acids reabsorbed by secondary active transport; not subject to control	Organic ion secretion; not subject to control
Variable amounts of filtered PO_4^{3-} and other electrolytes reabsorbed; subject to control	
65% of filtered H_2O osmotically reabsorbed; not subject to control	
50% of filtered urea passively reabsorbed; not subject to control	
Almost all filtered K^+ reabsorbed; not subject to control	

DISTAL TUBULE AND COLLECTING DUCT

Reabsorption	Secretion
Variable Na^+ reabsorption, controlled by aldosterone; Cl^- follows passively	Variable H^+ secretion, depending on acid–base status of body
Variable H_2O reabsorption, controlled by vasopressin	Variable K^+ secretion, controlled by aldosterone

Urine Excretion and Plasma Clearance

Of the 125 ml of plasma filtered per minute, typically 124 ml/min are reabsorbed, so the final quantity of urine formed averages 1 ml/min. Thus, of the 180 liters filtered per day, 1.5 liters of urine are excreted.

Urine contains high concentrations of various waste products plus variable amounts of the substances regulated by the kidneys, with any excess quantities having spilled into the urine. Useful substances are conserved by reabsorption, so they do not appear in the urine.

A relatively small change in the quantity of filtrate reabsorbed can bring about a large change in the volume of urine formed. For example, a reduction of less than 1% in the total reabsorption rate, from 124 to 123 ml/min, increases the urinary excretion rate by 100%, from 1 to 2 ml/min.

Plasma clearance is the volume of plasma cleared of a particular substance per minute.

Compared to plasma entering the kidneys through the renal arteries, plasma leaving the kidneys through the renal veins lacks the materials that were left behind to be eliminated in the urine. By excreting substances in the urine, the kidneys clean or "clear" the plasma flowing through them of these substances. The **plasma clearance** of any substance is defined as the volume of plasma completely cleared of that substance by the kidneys per minute.[2] It refers not to the *amount of the substance* removed but to the *volume of plasma* from which that amount was removed. Plasma clearance is actually a more useful measure than urine excretion; it is more important to know what effect urine excretion has on removing materials from body fluids than to know the volume and composition of discarded urine. Plasma clearance expresses the kidneys' effectiveness in removing various substances from the internal fluid environment.

Plasma clearance can be calculated for any plasma constituent as follows:

$$\text{Clearance rate of a substance (ml/min)} = \frac{\begin{matrix}\text{urine concentration}\\\text{of the substance}\\\text{(quantity/ml urine)}\end{matrix} \times \begin{matrix}\text{urine flow rate}\\\text{(ml/min)}\end{matrix}}{\begin{matrix}\text{plasma concentration of the substance}\\\text{(quantity/ml plasma)}\end{matrix}}$$

The plasma clearance rate varies for different substances, depending on how the kidneys handle each substance.

If a substance is filtered but not reabsorbed or secreted, its plasma clearance rate equals the GFR.

Assume that a plasma constituent, substance X, is freely filterable at the glomerulus but is not reabsorbed or secreted. As 125 ml/min of plasma are filtered and subsequently reabsorbed, the quantity of substance X originally contained within the 125 ml is left behind in the tubules to be excreted. Thus, 125 ml of plasma are cleared of substance X each minute (● Figure 14-23a). (Of the 125 ml/min of plasma filtered, 124 ml/min of the filtered fluid are returned, through reabsorption, to the plasma minus substance X, thus clearing this 124 ml/min of substance X. In addition, the 1 ml/min of fluid lost in urine is eventually replaced by an equivalent volume of ingested H_2O that is already clear of substance X. Therefore, 125 ml of plasma cleared of substance X are, in effect, returned to the plasma for every 125 ml of plasma filtered per minute.)

---

[2]Actually, plasma clearance is an artificial concept, because when a particular substance is excreted in the urine, that substance's concentration in the plasma as a whole is uniformly decreased as a result of thorough mixing in the circulatory system. However, it is useful for comparative purposes to consider clearance in effect as the volume of plasma that would have contained the total quantity of the substance (at the substance's concentration prior to excretion) that the kidneys excreted in one minute; that is, the hypothetical volume of plasma completely cleared of that substance per minute.

Glomerulus

Tubule

Peritubular
capillary

In
urine

(a) For a substance filtered and not reabsorbed or
secreted, such as inulin, all of the filtered plasma
is cleared of the substance.

(b) For a substance filtered, not secreted, and
completely reabsorbed, such as glucose, none of
the filtered plasma is cleared of the substance.

(c) For a substance filtered, not secreted, and
partially reabsorbed, such as urea, only a portion
of the filtered plasma is cleared of the substance.

(d) For a substance filtered and secreted but not
reabsorbed, such as hydrogen ion, all of the filtered
plasma is cleared of the substance, and the
peritubular plasma from which the substance is
secreted is also cleared.

● **FIGURE 14-23 Plasma clearance for substances handled in different ways by the
kidneys.**

Clinical Note There is no endogenous chemical with the characteristics of substance X. All substances naturally present in the plasma, even wastes, are reabsorbed or secreted to some extent. However, **inulin** (do not confuse with insulin), a harmless foreign carbohydrate produced abundantly by Jerusalem artichokes and to a lesser extent by other root vegetables such as onions and garlic, is freely filtered and not reabsorbed or secreted—an ideal substance X. Inulin can be injected and its plasma clearance determined as a clinical means of ascertaining the GFR. Because all glomerular filtrate formed is cleared of inulin, the volume of plasma cleared of inulin per minute equals the volume of plasma filtered per minute—that is, the GFR.

$$\frac{\text{Clearance rate}}{\text{for inulin}} = \frac{30 \text{ mg/ml urine} \times 1.25 \text{ ml urine/min}}{0.30 \text{ mg/ml plasma}}$$

$$= 125 \text{ ml plasma/min}$$

Although determination of inulin plasma clearance is accurate and straightforward, it is not very convenient, because inulin must be infused continuously throughout the determination to maintain a constant plasma concentration. Therefore, the plasma clearance of an endogenous substance, **creatinine,** is often used instead to find a rough estimate of the GFR. Creatinine, an end product of muscle metabolism, is produced at a relatively constant rate. It is freely filtered and not reabsorbed but is slightly secreted. Accordingly, creatinine clearance is not a completely accurate reflection of the GFR, but it does provide a close approximation and can be more readily determined than inulin clearance.

If a substance is filtered and reabsorbed but not secreted, its plasma clearance rate is always less than the GFR.

Some or all of a reabsorbable substance that has been filtered is returned to the plasma. Because less than the filtered volume of plasma will have been cleared of the substance, the plasma clearance rate of a reabsorbable substance is always less than the GFR. For example, the plasma clearance for glucose is normally zero. All the filtered glucose is reabsorbed along with the rest of the returning filtrate, so none of the plasma is cleared of glucose (● Figure 14-23b).

For a substance that is partially reabsorbed, such as urea, only part of the filtered plasma is cleared of that substance. With about 50% of the filtered urea being passively reabsorbed, only half of the filtered plasma, or 62.5 ml, is cleared of urea each minute (● Figure 14-23c).

If a substance is filtered and secreted but not reabsorbed, its plasma clearance rate is always greater than the GFR.

Tubular secretion allows the kidneys to clear certain materials from the plasma more efficiently. Only 20% of the plasma entering the kidneys is filtered. The remaining 80% passes unfiltered into the peritubular capillaries. The only means by which this unfiltered plasma can be cleared of any substance during the trip through the kidneys before being returned to the general circulation is by secretion. An example is H^+. Not only is filtered plasma cleared of nonreabsorbable H^+, but the plasma from which H^+ is secreted is also cleared of H^+. For example, if the quantity of H^+ secreted is equivalent to the quantity of H^+ present in 25 ml of plasma, the clearance rate for H^+ will be 150 ml/min at the normal GFR of 125 ml/min. Every minute 125 ml of plasma will lose its H^+ through filtration and failure of reabsorption, and 25 more ml of plasma will lose its H^+ through secretion. The plasma clearance for a secreted substance is always greater than the GFR (● Figure 14-23d).

Clinical Note Just as inulin can be used clinically to determine the GFR, plasma clearance of another foreign compound, the organic anion **para-aminohippuric acid (PAH),** can be used to measure renal plasma flow. Like inulin, PAH is freely filterable and nonreabsorbable. It differs, however, in that all the PAH in the plasma that escapes filtration is secreted from the peritubular capillaries by the organic anion secretory pathway in the proximal tubule. Thus, PAH is removed from all the plasma that flows through the kidneys—both from plasma that is filtered and subsequently reabsorbed without its PAH, and from unfiltered plasma that continues on in the peritubular capillaries and loses its PAH by active secretion into the tubules. Because all the plasma that flows through the kidneys is cleared of PAH, the plasma clearance for PAH is a reasonable estimate of the rate of plasma flow through the kidneys. Typically, renal plasma flow averages 625 ml/min, for a renal blood flow (plasma plus blood cells) of 1140 ml/min—more than 20% of the cardiac output.

FILTRATION FRACTION If you know PAH clearance (renal plasma flow) and inulin clearance (GFR), you can easily determine the **filtration fraction,** the fraction of plasma flowing through the glomeruli that is filtered into the tubules.

$$\frac{\text{Filtration}}{\text{fraction}} = \frac{\text{GFR (plasma inulin clearance)}}{\text{renal plasma flow (plasma PAH clearance)}}$$

$$= \frac{125 \text{ ml/min}}{625 \text{ ml/min}} = 20\%$$

Thus, 20% of the plasma that enters the glomeruli is typically filtered.

The kidneys can excrete urine of varying concentrations depending on the body's state of hydration.

Having considered how the kidneys deal with a variety of solutes in the plasma, we will now concentrate on renal handling of plasma H_2O. The ECF osmolarity (solute concentration) depends on the relative amount of H_2O compared to solute. At normal fluid balance and solute concentration, the body fluids are **isotonic** at an osmolarity of 300 milliosmols/liter (mOsm) (see pp. 66 and A-9). If too much H_2O is present relative to the solute load, the body fluids are **hypotonic,** which means they are too dilute at an osmolarity less than 300 mOsm. However, if a H_2O deficit exists relative to the solute load, the body fluids are too concentrated or are **hypertonic,** having an osmolarity greater than 300 mOsm.

Medulla

Cortex

300 300 300 300 600 900 1200
300 300 300 600 900 1200
300 300 600 900 1200
300 600 900 1200
300 1200 900 600 300
1200 900 600 300
1200 900 600 300
300

● **FIGURE 14-24 Vertical osmotic gradient in the renal medulla.** Schematic representation of the kidney rotated 90° from its normal position in an upright person for better visualization of the vertical osmotic gradient in the renal medulla. All values are in mOsm. The osmolarity of the interstitial fluid throughout the renal cortex is isotonic at 300 mOsm, but the osmolarity of the interstitial fluid in the renal medulla increases progressively from 300 mOsm at the boundary with the cortex to a maximum of 1200 mOsm at the junction with the renal pelvis.

Knowing that the driving force for H_2O reabsorption throughout the entire length of the tubules is an osmotic gradient between the tubular lumen and surrounding interstitial fluid, you would expect, given osmotic considerations, that the kidneys could not excrete urine more or less concentrated than the body fluids. Indeed, this would be the case if the interstitial fluid surrounding the tubules in the kidneys were identical in osmolarity to the remaining body fluids. Water reabsorption would proceed only until the tubular fluid equilibrated osmotically with the interstitial fluid, and the body would have no way to eliminate excess H_2O when the body fluids were hypotonic or to conserve H_2O in the presence of hypertonicity.

Fortunately, a large **vertical osmotic gradient** is uniquely maintained in the interstitial fluid of the medulla of each kidney. The concentration of the interstitial fluid progressively increases from the cortical boundary down through the depth of the renal medulla until it reaches a maximum of 1200 mOsm in humans at the junction with the renal pelvis (● Figure 14-24).

By a mechanism described shortly, this gradient enables the kidneys to produce urine that ranges in concentration from 100 to 1200 mOsm, depending on the body's state of hydration. When the body is in ideal fluid balance, 1 ml/min of isotonic urine is formed. When the body is overhydrated (too much H_2O), the kidneys can produce a large volume of dilute urine (up to 25 ml/min and hypotonic at 100 mOsm), eliminating the excess H_2O in the urine. Conversely, the kidneys can put out a small volume of concentrated urine (down to 0.3 ml/min and hypertonic at 1200 mOsm) when the body is dehydrated (too little H_2O), conserving H_2O for the body.

Unique anatomic arrangements and complex functional interactions between the various nephron components in the renal medulla establish and use the vertical osmotic gradient. Recall that the hairpin loop of Henle dips only slightly into the medulla in cortical nephrons, but in juxtamedullary nephrons the loop plunges through the entire depth of the medulla so that the tip of the loop lies near the renal pelvis (see ● Figure 14-5, p. 516). Also, the vasa recta of juxtamedullary nephrons follow the same deep hairpin loop as the long loop of Henle. Flow in both the long loops of Henle and the vasa recta is considered countercurrent because the flow in the two closely adjacent limbs of the loop moves in opposite directions. Also running through the medulla in the descending direction only, on their way to the renal pelvis, are the collecting ducts that serve both types of nephrons. This arrangement, coupled with the permeability and transport characteristics of these tubular segments, plays a key role in the kidneys' ability to produce urine of varying concentrations, depending on the body's needs for water conservation or elimination. Briefly, the juxtamedullary nephrons' long loops of Henle *establish* the vertical osmotic gradient, their vasa recta *preserve* this gradient while providing blood to the renal medulla, and the collecting ducts of all nephrons *use* the gradient, in conjunction with the hormone vasopressin, to produce urine of varying concentrations. Collectively, this entire functional organization is known as the **medullary countercurrent system.** We will examine each of its facets in greater detail.

The medullary vertical osmotic gradient is established by countercurrent multiplication.

We will follow the filtrate through a long-looped nephron to see how this structure establishes a vertical osmotic gradient in the medulla. Immediately after the filtrate is formed, uncontrolled osmotic reabsorption of filtered H_2O occurs in the proximal tubule secondary to active Na^+ reabsorption. As a result, by the end of the proximal tubule about 65% of the filtrate has been reabsorbed, but the 35% remaining in the tubular lumen still has the same osmolarity as the body fluids. Therefore, the fluid entering the loop of Henle is still isotonic. An additional 15% of the filtered H_2O is obligatorily reabsorbed from the loop of Henle during the establishment and maintenance of the vertical osmotic gradient, with the osmolarity of the tubular fluid being altered in the process.

PROPERTIES OF THE DESCENDING AND ASCENDING LIMBS OF A LONG HENLE'S LOOP The following functional distinctions between the descending limb of a long Henle's loop (which carries fluid from the proximal tubule down into the depths of the medulla) and the ascending limb (which carries fluid up and out of the medulla into the distal tubule) are crucial for establishing the incremental osmotic gradient in the medullary interstitial fluid.

The *descending limb*

1. is highly permeable to H_2O (via abundant, always-open AQP-1 water channels).

2. does not actively extrude Na^+. (That is, it does not reabsorb Na^+. It is the only segment of the entire tubule that does not do so.)

The *ascending limb*

1. actively transports NaCl out of the tubular lumen into the surrounding interstitial fluid.

2. is always impermeable to H_2O, so salt leaves the tubular fluid without H_2O osmotically following along.

MECHANISM OF COUNTERCURRENT MULTIPLICATION The close proximity and countercurrent flow of the two limbs allow important interactions between them. Even though the flow of fluids is continuous through the loop of Henle, we will visualize what happens step by step, much like an animated film run so slowly that each frame can be viewed.

Initially, before the vertical osmotic gradient is established, the medullary interstitial fluid concentration is uniformly 300 mOsm, as are the rest of the body fluids (● Figure 14-25).

The active salt pump in the ascending limb can transport NaCl out of the lumen until the surrounding interstitial fluid is 200 mOsm more concentrated than the tubular fluid in this limb. When the ascending limb pump starts actively extruding salt, the medullary interstitial fluid becomes hypertonic. Water cannot follow osmotically from the ascending limb, because this limb is impermeable to H_2O. However, net diffusion of H_2O does occur from the descending limb into the interstitial fluid. The tubular fluid entering the descending limb from the proximal tubule is isotonic. Because the descending limb is highly permeable to H_2O, net diffusion of H_2O occurs by osmosis out of the descending limb into the more concentrated interstitial fluid. The passive movement of H_2O out of the descending limb continues until the osmolarities of the fluid in the descending limb and interstitial fluid become equilibrated. Thus, the tubular fluid entering the loop of Henle immediately starts to become more concentrated as it loses H_2O. At equilibrium, the osmolarity of the ascending limb fluid is 200 mOsm and the osmolarities of the interstitial fluid and descending limb fluid are equal at 400 mOsm (● Figure 14-25, step **1**).

If we now advance the entire column of fluid in the loop of Henle several frames (step **2**), a mass of 200 mOsm fluid exits from the top of the ascending limb into the distal tubule, and a new mass of isotonic fluid at 300 mOsm enters the top of the descending limb from the proximal tubule. At the bottom of the loop, a comparable mass of 400 mOsm fluid from the descending limb moves forward around the tip into the ascending limb, placing it opposite a 400 mOsm region in the descending limb. Note that the 200 mOsm concentration difference has been lost at both the top and the bottom of the loop.

The ascending limb pump again transports NaCl out while H_2O passively leaves the descending limb until a 200 mOsm difference is reestablished between the ascending limb and both the interstitial fluid and descending limb at each horizontal level (step **3**). Note, however, that the concentration of tubular fluid is progressively increasing in the descending limb and progressively decreasing in the ascending limb.

As the tubular fluid is advanced still further (step **4**), the 200 mOsm concentration gradient is disrupted once again at all horizontal levels. Again, active extrusion of NaCl from the ascending limb, coupled with the net diffusion of H_2O out of the descending limb, reestablishes the 200 mOsm gradient at each horizontal level (step **5**).

As the fluid flows slightly forward again and this stepwise process continues (step **6**), the fluid in the descending limb becomes progressively more hypertonic until it reaches a maximum concentration of 1200 mOsm at the bottom of the loop, four times the normal concentration of body fluids. Because the interstitial fluid always achieves equilibrium with the descending limb, an incremental vertical concentration gradient ranging from 300 to 1200 mOsm is likewise established in the medullary interstitial fluid. In contrast, the concentration of the tubular fluid progressively decreases in the ascending limb as salt is pumped out but H_2O is unable to follow. In fact, the tubular fluid even becomes hypotonic before leaving the ascending limb to enter the distal tubule at a concentration of 100 mOsm, one third the normal concentration of body fluids.

Note that although a gradient of only 200 mOsm exists between the ascending limb and the surrounding fluids at each medullary horizontal level, a much larger vertical gradient exists from the top to the bottom of the medulla. Even though the ascending limb pump can generate a gradient of only 200 mOsm, this effect is multiplied into a large vertical gradient because of the countercurrent flow within the loop. This concentrating mechanism accomplished by the loop of Henle is known as **countercurrent multiplication.**

We have artificially described countercurrent multiplication in a stop-and-flow, stepwise fashion to facilitate understanding. It is important to realize that once the incremental medullary gradient is established, it stays constant because of the continuous flow of fluid coupled with the ongoing ascending limb active transport and accompanying descending limb passive fluxes.

BENEFITS OF COUNTERCURRENT MULTIPLICATION If you consider only what happens to the tubular fluid as it flows through the loop of Henle, the whole process seems an exercise in futility. The isotonic fluid that enters the loop becomes progressively more concentrated as it flows down the descending limb, achieving a maximum concentration of 1200 mOsm, only to become progressively more dilute as it flows up the ascending limb, finally leaving the loop at a minimum concentration of 100 mOsm. What is the point of concentrating the fluid fourfold and then turning around and diluting it until it leaves at one third the concentration at which it entered? Such a mechanism offers two benefits. First, it establishes a vertical osmotic gradient in the medullary interstitial fluid. This gradient, in turn, is used by the collecting ducts to concentrate the tubular fluid so that a urine *more concentrated* than normal body fluids can be excreted. Second, the fact that the fluid is hypotonic as it enters the distal parts of the tubule enables the kidneys to excrete a urine *more dilute* than normal body fluids. Let us see how.

Vasopressin-controlled, variable H_2O reabsorption occurs in the final tubular segments.

After obligatory H_2O reabsorption from the proximal tubule (65% of the filtered H_2O) and loop of Henle (15% of the filtered H_2O), 20% of the filtered H_2O remains in the lumen to enter the distal

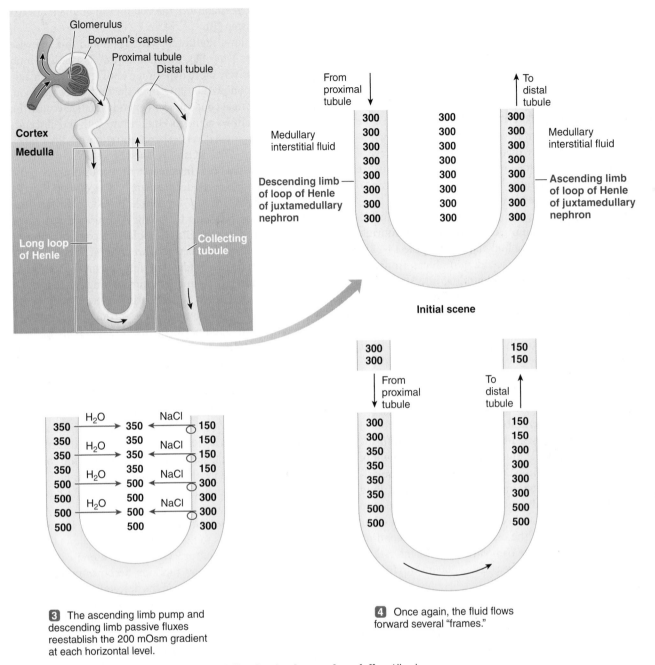

FIGURE 14-25 Countercurrent multiplication in the renal medulla. All values are in mOsm.

and collecting tubules for variable reabsorption that is under hormonal control. This is still a large volume of filtered H_2O subject to regulated reabsorption; 20% × GFR (180 liters/day) = 36 liters per day to be reabsorbed to varying extents, depending on the body's state of hydration. This is more than 13 times the amount of plasma H_2O in the entire circulatory system.

The fluid leaving the loop of Henle enters the distal tubule at 100 mOsm, so it is hypotonic to the surrounding isotonic (300 mOsm) interstitial fluid of the renal cortex through which the distal tubule passes. The distal tubule then empties into the collecting duct, which is bathed by progressively increasing concentrations (300 to 1200 mOsm) of surrounding interstitial fluid as it descends through the medulla.

ROLE OF VASOPRESSIN For H_2O absorption to occur across a segment of the tubule, two criteria must be met: (1) an osmotic gradient must exist across the tubule, and (2) the tubular segment must be permeable to H_2O. The distal and collecting tubules are *impermeable* to H_2O except in the presence of **vasopressin**, also known as **antidiuretic hormone** (*anti* means "against"; *diuretic* means "increased urine output"),[3] which increases their permeability to H_2O. Vasopressin is produced by

---

[3]Even though textbooks traditionally have tended to use the name *antidiuretic hormone* for this hormone, especially when discussing its actions on the kidney, investigators in the field now prefer *vasopressin*.

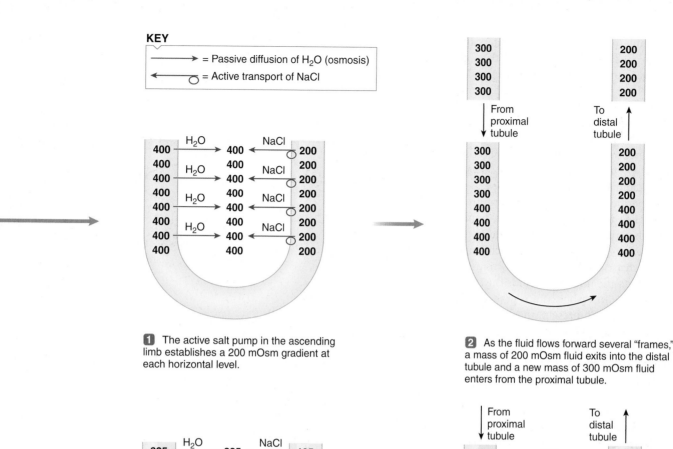

KEY

→ = Passive diffusion of H_2O (osmosis)
← ⊙ = Active transport of NaCl

1 The active salt pump in the ascending limb establishes a 200 mOsm gradient at each horizontal level.

2 As the fluid flows forward several "frames," a mass of 200 mOsm fluid exits into the distal tubule and a new mass of 300 mOsm fluid enters from the proximal tubule.

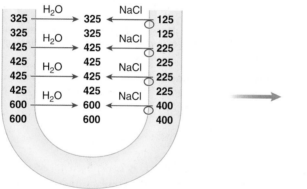

5 The 200 mOsm gradient at each horizontal level is established once again.

6 The final vertical osmotic gradient is established and maintained by the ongoing countercurrent multiplication of the long loops of Henle.

several specific neuronal cell bodies in the *hypothalamus,* then stored in the *posterior pituitary gland,* which is attached to the hypothalamus by a thin stalk. The hypothalamus controls release of vasopressin from the posterior pituitary into the blood. In negative-feedback fashion, vasopressin secretion is stimulated by a H_2O deficit when the ECF is too concentrated (that is, hypertonic) and H_2O must be conserved for the body, and inhibited by a H_2O excess when the ECF is too dilute (that is, hypotonic) and surplus H_2O must be eliminated in urine.

Vasopressin reaches the basolateral membrane of the principal cells lining the distal and collecting tubules through the circulatory system. Here it binds with V_2 *receptors* specific for it (● Figure 14-26). (Vasopressin binds with different V_1 *receptors* on vascular smooth muscle to exert its vasoconstrictor effects; see p. 359.) Binding of vasopressin with its V_2 receptors, which are G-protein-coupled receptors (see p. 117), activates the cyclic AMP (cAMP) second-messenger system within these tubular cells (see p. 121). This binding ultimately increases permeability of the opposite luminal membrane to H_2O by promoting insertion of aquaporins (specifically, AQP-2) in this membrane by means of exocytosis. Without these aquaporins, the luminal membrane is impermeable to H_2O. Once H_2O enters the tubular cells from the filtrate through these vasopressin-regulated luminal water channels, it passively leaves the cells down the osmotic gradient across the cells' basolateral membrane to enter the interstitial fluid. The aquaporins in the basolateral membrane of the distal and collecting tubule (*AQP-3* and *AQP-4*) are always present and open, so this membrane is always permeable to H_2O. By permitting more H_2O to permeate from the lumen into the tubular cells, the additional vasopressin-regulated luminal channels thus increase H_2O reabsorption from the filtrate into the interstitial fluid. The tubular response to vasopressin is graded: The more vasopressin present, the more luminal water channels inserted, and the greater the permeability of the distal and collecting tubules to H_2O. The increase in luminal membrane water channels is not permanent, however. The channels are retrieved by endocytosis when vasopressin secretion decreases and cAMP activity is similarly decreased. Accordingly, H_2O permeability is reduced when vasopressin secretion decreases. These H_2O channels are stored in the internalized vesicles ready for reinsertion the next time vasopressin secretion increases. This shuttling of AQP-2 into and out of the luminal membrane under vasopressin command provides a means of rapidly controlling H_2O permeability of the distal and collecting tubules, depending on the body's momentary needs.

Vasopressin influences H_2O permeability only in the distal part of the nephron, especially the collecting ducts. It has no influence over the 80% of the filtered H_2O that is obligatorily reabsorbed without control in the proximal tubule and loop of Henle. The ascending limb of Henle's loop is always impermeable to H_2O, even in the presence of vasopressin.

REGULATION OF H_2O REABSORPTION IN RESPONSE TO A H_2O DEFICIT When vasopressin secretion increases in response to a H_2O deficit and the permeability of the distal and collecting tubules to H_2O accordingly increases, the hypotonic tubular fluid entering the distal part of the nephron can lose progressively more H_2O by osmosis into the interstitial fluid as the tubular fluid first flows through the isotonic cortex and then is exposed to the ever-increasing osmolarity of the medullary interstitial fluid as it plunges toward the renal pelvis (● Figure 14-27a). As the 100 mOsm tubular fluid enters the distal tubule and is exposed to a surrounding interstitial fluid of 300 mOsm, H_2O leaves the tubular fluid by osmosis across the now permeable tubular cells until the tubular fluid reaches a maximum concentration of 300 mOsm by the end of the distal tubule. As this 300 mOsm tubular fluid progresses farther into the collect-

① Blood-borne vasopressin binds with its receptor sites on the basolateral membrane of a principal cell in the distal or collecting tubule.

② This binding activates the cyclic AMP (cAMP) second-messenger system within the cell.

③ Cyclic AMP increases the opposite luminal membrane's permeability to H_2O by promoting the insertion of AQP-2 water channels into the membrane. This membrane is impermeable to water in the absence of vasopressin.

④ Water enters the tubular cell from the tubular lumen through the inserted water channels.

⑤ Water exits the cell through a different water channel (either AQP-3 or AQP-4) permanently positioned at the basolateral border, and then enters the blood, in this way being reabsorbed.

● **FIGURE 14-26 Mechanism of action of vasopressin.**

ing duct, it is exposed to even higher osmolarity in the surrounding medullary interstitial fluid. Consequently, the tubular fluid loses more H_2O by osmosis and becomes further concentrated, only to move farther forward and be exposed to an even higher interstitial fluid osmolarity and lose even more H_2O, and so on.

Under the influence of maximum levels of vasopressin, it is possible to concentrate the tubular fluid up to 1200 mOsm by the end of the collecting ducts. No further modification of the tubular fluid occurs beyond the collecting duct, so what remains in the tubules at this point is urine. As a result of this extensive vasopressin-promoted reabsorption of H_2O in the late segments of the tubule, a small volume of urine concentrated up to 1200 mOsm can be excreted. As little as 0.3 ml of urine may be formed each minute, less than one third the normal urine flow rate of 1 ml/min. The reabsorbed H_2O entering the medullary interstitial fluid is picked up by the peritubular capillaries and returned to the general circulation, thus being conserved for the body.

Although vasopressin promotes H_2O conservation by the body, it cannot completely halt urine production, even when a person is not taking in any H_2O, because a minimum volume of H_2O must be excreted with the solute wastes. Collectively, the

Vasopressin present: distal and collecting tubules permeable to H₂O

H_2O

Small volume of concentrated urine excreted; reabsorbed H₂O picked up by peritubular capillaries and conserved for body

(a) In the face of a water deficit

From proximal tubule

Filtrate has concentration of 100 mOsm as it enters distal and collecting tubules

Distal tubule

Cortex

Medulla

Loop of Henle

Collecting duct

Concentration of urine may be up to 1200 mOsm as it leaves collecting tubule

No vasopressin present: distal and collecting tubules impermeable to H₂O

Large volume of dilute urine; no H₂O reabsorbed in distal portion of nephron; excess H₂O eliminated

(b) In the face of a water excess

From proximal tubule

Filtrate has concentration of 100 mOsm as it enters distal and collecting tubules

Distal tubule

Cortex

Medulla

Loop of Henle

Collecting duct

Concentration of urine may be as low as 100 mOsm as it leaves collecting tubule

● **FIGURE 14-27 Excretion of urine of varying concentration depending on the body's needs.** All values are in mOsm.

waste products and other constituents eliminated in the urine average 600 milliosmols each day. Because the maximum urine concentration is 1200 mOsm (milliosmols/liter), the minimum volume of urine required to excrete these wastes is 500 ml/day (600 milliosmols of wastes/day ÷ 1200 milliosmols/liter of urine = 0.5 liter, or 500 ml/day, or 0.3 ml/min). Thus, under

maximal vasopressin influence, 99.7% of the 180 liters of plasma H_2O filtered per day is returned to the blood, with an obligatory H_2O loss of half a liter.

The kidneys' ability to tremendously concentrate urine to minimize H_2O loss when necessary is possible only because of the presence of the vertical osmotic gradient in the medulla. If this gradient did not exist, the kidneys could not produce a urine more concentrated than the body fluids no matter how much vasopressin was secreted, because the only driving force for H_2O reabsorption is a concentration differential between the tubular fluid and the interstitial fluid.

REGULATION OF H_2O REABSORPTION IN RESPONSE TO A H_2O EXCESS Conversely, when a person consumes large quantities of H_2O, the excess H_2O must be removed from the body without simultaneously losing solutes that are critical for maintaining homeostasis. Under these circumstances, no vasopressin is secreted, so the distal and collecting tubules remain impermeable to H_2O. The tubular fluid entering the distal tubule is hypotonic (100 mOsm), having lost salt without an accompanying loss of H_2O in the ascending limb of Henle's loop. As this hypotonic fluid passes through the distal and collecting tubules (● Figure 14-27b), the medullary osmotic gradient cannot exert any influence because of the late tubular segments' impermeability to H_2O. In other words, none of the H_2O remaining in the tubules can leave the lumen to be reabsorbed, even though the tubular fluid is less concentrated than the surrounding interstitial fluid. Thus, in the absence of vasopressin, the 20% of the filtered fluid that reaches the distal tubule is not reabsorbed. Meanwhile, excretion of wastes and other urinary solutes remains constant. The net result is a large volume of dilute urine, which helps rid the body of excess H_2O. Urine osmolarity may be as low as 100 mOsm, the same as in the fluid entering the distal tubule. Urine flow may be increased up to 25 ml/min in the absence of vasopressin, compared to the normal urine production of 1 ml/min.

The ability to produce urine less concentrated than the body fluids depends on the fact that the tubular fluid is hypotonic as it enters the distal part of the nephron. This dilution is accomplished in the ascending limb as NaCl is actively extruded but H_2O cannot follow. Therefore, the loop of Henle, by simultaneously establishing the medullary osmotic gradient and diluting the tubular fluid before it enters the distal segments, plays a key role in allowing the kidneys to excrete urine that ranges in concentration from 100 to 1200 mOsm.

Countercurrent exchange within the vasa recta conserves the medullary vertical osmotic gradient.

Obviously, the renal medulla must be supplied with blood to nourish the tissues in this area as well as to transport water that is reabsorbed by the loops of Henle and collecting ducts back to the general circulation. In doing so, however, it is important that circulation of blood through the medulla does not disturb the vertical gradient of hypertonicity established by the loops of Henle. Consider the situation if blood were to flow straight through from the cortex to the inner medulla and then directly into the renal vein (● Figure 14-28a). Because capillaries are freely permeable to NaCl and H_2O, the blood would progressively pick up salt and lose H_2O through passive fluxes down concentration and osmotic gradients as it flowed through the depths of the medulla. Isotonic blood entering the medulla, on equilibrating with each medullary level,

would leave the medulla very hypertonic at 1200 mOsm. It would be impossible to establish and maintain the medullary hypertonic gradient because the NaCl pumped into the medullary interstitial fluid would continuously be carried away by the circulation.

This dilemma is avoided by the hairpin construction of the vasa recta, which, by looping back through the concentration gradient in reverse, allows the blood to leave the medulla and enter the renal vein essentially isotonic to incoming arterial blood (● Figure 14-28b). As blood passes down the descending limb of the vasa recta, equilibrating with the progressively increasing concentration of the surrounding interstitial fluid, it picks up salt and loses H_2O until it is very hypertonic by the bottom of the loop. Then, as blood flows up the ascending limb, salt diffuses back out into the interstitium, and H_2O reenters the vasa recta as progressively decreasing concentrations are encountered in the surrounding interstitial fluid. This passive exchange of solutes and H_2O between the two limbs of the vasa recta and the interstitial fluid is known as **countercurrent exchange.** Unlike countercurrent multiplication, it does not *establish* the concentration gradient. Rather, it *preserves* (*prevents the dissolution of*) the gradient. Because blood enters and leaves the medulla at the same osmolarity as a result of countercurrent exchange, the medullary tissue is nourished with blood, yet the incremental gradient of hypertonicity in the medulla is preserved.

Water reabsorption is only partially linked to solute reabsorption.

It is important to distinguish between H_2O reabsorption that mandatorily follows solute reabsorption and reabsorption of "free" H_2O not linked to solute reabsorption.

■ *In the tubular segments permeable to H_2O, solute reabsorption is* always *accompanied by comparable H_2O reabsorption because of osmotic considerations.* Therefore, the total volume of H_2O reabsorbed is determined in large part by the total mass of solute reabsorbed; this is especially true of NaCl, because it is the most abundant solute in the ECF.

■ *Solute excretion is* always *accompanied by comparable H_2O excretion because of osmotic considerations.* This fact is responsible for the obligatory excretion of at least a minimal volume of H_2O, even when a person is severely dehydrated.

For the same reason, when excess unreabsorbed solute is present in the tubular fluid, its presence exerts an osmotic effect to hold excessive H_2O in the lumen. This phenomenon is known as osmotic diuresis. Diuresis is increased urinary excretion, of which there are two types: osmotic diuresis and water diuresis.

 Osmotic diuresis involves increased excretion of both H_2O and solute caused by excess unreabsorbed solute in the tubular fluid, such as occurs in diabetes mellitus. The large quantity of unreabsorbed glucose that remains in the tubular fluid in people with diabetes osmotically drags H_2O with it into the urine. Some diuretic drugs act by blocking specific solute reabsorption so that extra H_2O spills into the urine along with the unreabsorbed solute.

(a) Hypothetical pattern of blood flow

(b) Actual pattern of blood flow

● **FIGURE 14-28 Countercurrent exchange in the renal medulla.** All values are in mOsm. (a) If the blood supply to the renal medulla flowed straight through from the cortex to the inner medulla, the blood would be isotonic on entering but very hypertonic on leaving, having picked up salt and lost H_2O as it equilibrated with the surrounding interstitial fluid at each incremental horizontal level. It would be impossible to maintain the vertical osmotic gradient, because the salt pumped out by the ascending limb of Henle's loop would be continuously flushed away by blood flowing through the medulla. (b) Blood equilibrates with the interstitial fluid at each incremental horizontal level in both the descending limb and the ascending limb of the vasa recta, so blood is isotonic as it enters and leaves the medulla. This countercurrent exchange prevents dissolution of the medullary osmotic gradient while providing blood to the renal medulla.

Water diuresis, in contrast, is increased urinary output of H_2O with little or no increase in excretion of solutes.

■ *A loss or gain of pure H_2O that is not accompanied by comparable solute deficit or excess in the body (that is, "free" H_2O) leads to changes in ECF osmolarity.* Such an imbalance between H_2O and solute is corrected by partially dissociating H_2O reabsorption from solute reabsorption in the distal portions of the nephron through the combined effects of vasopressin secretion and the medullary osmotic gradient. Through this mechanism, free H_2O can be reabsorbed without comparable solute reabsorption to correct for hypertonicity of the body fluids. Conversely, a large quantity of free H_2O can be excreted unaccompanied by comparable solute excretion (that is, water diuresis) to rid the body of excess pure H_2O, thus correcting for hypotonicity of the body fluids. Water diuresis is normally a compensation for ingesting too much H_2O.

Excessive water diuresis follows alcohol ingestion. Because alcohol inhibits vasopressin secretion, the kidneys inappropriately lose too much H_2O. Typically, more fluid is lost in the urine than is consumed in the alcoholic beverage, so the body becomes dehydrated despite substantial fluid ingestion.

▲ Table 14-4 summarizes how various tubular segments of the nephron handle Na^+ and H_2O and the significance of these processes.

Renal failure has wide-ranging consequences.

Urine excretion and the resulting clearance of wastes and excess electrolytes from the plasma are crucial for maintaining homeostasis. When the functions of both kidneys are so disrupted that they cannot perform their regulatory and excretory functions sufficiently to maintain homeostasis, **renal failure** has set in. Renal failure has a variety of causes, some of which begin elsewhere in the body and affect renal function secondarily. Among the causes are the following:

1. *Infectious organisms,* either blood-borne or gaining entrance to the urinary tract through the urethra

2. *Toxic agents,* such as lead, arsenic, pesticides, or even long-term exposure to high doses of aspirin

3. *Inappropriate immune responses,* such as **glomerulonephritis,** which occasionally follows streptococcal throat infections as antigen–antibody complexes leading to localized inflammatory damage are deposited in the glomeruli (see p. 433)

4. *Obstruction of urine flow* by kidney stones, tumors, or an enlarged prostate gland, with back pressure reducing glomerular filtration as well as damaging renal tissue

5. An *insufficient renal blood supply* that leads to inadequate filtration pressure, which can occur secondary to circulatory disorders such as heart failure, hemorrhage, shock, or narrowing and hardening of the renal arteries by atherosclerosis

▲ TABLE 14-4	**Handling of Sodium and Water by Various Tubular Segments of the Nephron**			
	Na⁺ REABSORPTION		**H₂O REABSORPTION**	
Tubular Segment	**Percentage of Reabsorption in This Segment**	**Distinguishing Features**	**Percentage of Reabsorption in This Segment**	**Distinguishing Features**
Proximal Tubule	67	Active; uncontrolled; plays a pivotal role in the reabsorption of glucose, amino acids, Cl⁻, H₂O, and urea	65	Passive; obligatory osmotic reabsorption following active Na⁺ reabsorption
Loop of Henle	25	Active, uncontrolled; Na⁺ along with Cl⁻ reabsorption from the ascending limb helps establish the medullary interstitial vertical osmotic gradient, which is important in the kidneys' ability to produce urine of varying concentrations and volumes, depending on the body's needs	15	Passive; obligatory osmotic reabsorption from the descending limb as the ascending limb extrudes NaCl into the interstitial fluid (that is, reabsorbs NaCl)
Distal and Collecting Tubules	8	Active; variable and subject to aldosterone control; important in the regulation of ECF volume and long-term control of blood pressure; linked to K⁺ secretion and H⁺ secretion	20	Passive; not linked to solute reabsorption; variable quantities of "free" H₂O reabsorption subject to vasopressin control; driving force is the vertical osmotic gradient in the medullary interstitial fluid established by the long loops of Henle; important in regulating ECF osmolarity

Although these conditions may have different origins, almost all can cause some degree of nephron damage. The glomeruli and tubules may be independently affected, or both may be dysfunctional. Regardless of cause, renal failure can manifest itself either as *acute renal failure*, characterized by a sudden onset with rapidly reduced urine formation until less than the essential minimum of around 500 ml of urine is being produced per day; or as *chronic renal failure*, characterized by slow, progressive, insidious loss of renal function. A person may die from acute renal failure, or the condition may be reversible and lead to full recovery. Chronic renal failure, in contrast, is not reversible. Gradual, permanent destruction of renal tissue eventually proves fatal. Chronic renal failure is insidious because up to 75% of the kidney tissue can be destroyed before the loss of kidney function is even noticeable. Because of the abundant reserve of kidney function, only 25% of kidney tissue is needed to adequately maintain all the essential renal excretory and regulatory functions. With less than 25% of functional kidney tissue remaining, however, renal insufficiency becomes apparent. *End-stage renal failure* ensues when 90% of kidney function has been lost. Over 20 million people in the United States have some extent of kidney disease, which leads to more than 80,000 deaths per year.

We will not sort out the different stages and symptoms associated with various renal disorders, but ▲ Table 14-5, which summarizes the potential consequences of renal failure, gives you an idea of the broad effects that kidney impairment can have. The extent of these effects should not be surprising, considering the central role the kidneys play in maintaining homeostasis. When the kidneys cannot maintain a normal internal environment, widespread disruption of cell activities can bring about abnormal function in other organ systems as well. By the time end-stage renal failure occurs, literally every body system has become impaired to some extent. (One symptom of renal disease occurs during strenuous exercise, but it is transient and harmless. See the boxed feature on p. 550, ● A Closer Look at Exercise Physiology.)

Because chronic renal failure is irreversible and eventually fatal, treatment is aimed at maintaining renal function by alternative methods, such as dialysis and kidney transplantation. (For further explanation of these procedures, see the boxed feature on p. 551, ● Concepts, Challenges, and Controversies.)

This finishes our discussion of kidney function. For the remainder of the chapter, we focus on the plumbing that stores and carries the urine formed by the kidneys to the outside.

Potential Ramifications of Renal Failure

Uremic toxicity caused by retention of waste products

Nausea, vomiting, diarrhea, and ulcers caused by a toxic effect on the digestive system

Bleeding tendency arising from a toxic effect on platelet function

Mental changes—such as reduced alertness, insomnia, and shortened attention span, progressing to convulsions and coma—caused by toxic effects on the central nervous system

Abnormal sensory and motor activity caused by a toxic effect on the peripheral nerves

Metabolic acidosis caused by the inability of the kidneys to adequately secrete H^+ that is continually being added to the body fluids as a result of metabolic activity *(among most life-threatening consequences of renal failure)*

Altered enzyme activity caused by the action of too much acid on enzymes

Depression of the central nervous system caused by the action of too much acid interfering with neuronal excitability

Potassium retention resulting from inadequate tubular secretion of K^+ *(among most life-threatening consequences of renal failure)*

Altered cardiac and neural excitability as a result of changing the resting membrane potential of excitable cells

Sodium imbalances caused by the inability of the kidneys to adjust Na^+ excretion to balance changes in Na^+ consumption

Elevated blood pressure, generalized edema, and congestive heart failure if too much Na^+ is consumed

Hypotension and, if severe enough, circulatory shock if too little Na^+ is consumed

Phosphate and calcium imbalances arising from impaired reabsorption of these electrolytes

Disturbances in skeletal structures caused by abnormalities in deposition of calcium phosphate crystals, which harden bone

Loss of plasma proteins as a result of increased "leakiness" of the glomerular membrane

Edema caused by a reduction in plasma-colloid osmotic pressure

Inability to vary urine concentration as a result of impairment of the countercurrent system

Hypotonicity of body fluids if too much H_2O is ingested

Hypertonicity of body fluids if too little H_2O is ingested

Hypertension arising from the combined effects of salt and fluid retention and vasoconstrictor action of excess angiotensin II

Anemia caused by inadequate erythropoietin production

Depression of the immune system, most likely caused by toxic levels of wastes and acids
Increased susceptibility to infections

Urine is temporarily stored in the bladder, from which it is emptied by micturition.

Once urine has been formed by the kidneys, it is transmitted through the ureters to the urinary bladder. Urine does not flow through the ureters by gravitational pull alone. Peristaltic (forward-pushing) contractions of the smooth muscle within the ureteral wall propel the urine forward from the kidneys to the bladder. The ureters penetrate the wall of the bladder obliquely, coursing through the wall several centimeters before they open into the bladder cavity. This anatomic arrangement prevents backflow of urine from the bladder to the kidneys when pressure builds up in the bladder. As the bladder fills, the ureteral ends within its wall are compressed closed. Urine can still enter, however, because ureteral contractions generate enough pressure to overcome the resistance and push urine through the occluded ends.

ROLE OF THE BLADDER The bladder can accommodate large fluctuations in urine volume. The bladder wall consists of smooth muscle lined by a special type of epithelium. It was once assumed that the bladder was an inert sac. However, both the epithelium and the smooth muscle actively participate in the bladder's ability to accommodate large changes in urine volume. The epithelial lining can increase and decrease in surface area by the orderly process of membrane recycling as the bladder alternately fills and empties. Membrane-enclosed cytoplasmic vesicles are inserted by exocytosis into the surface area during bladder filling; then the vesicles are withdrawn by endocytosis to shrink the surface area following emptying (see pp. 28 and 75). As is characteristic of smooth muscle, bladder muscle can stretch tremendously without building up bladder wall tension (see p. 298). In addition, the highly folded bladder wall flattens out during filling to increase bladder storage capacity. Because the kidneys continuously form urine, the bladder must have enough storage capacity to preclude the need to continuously get rid of the urine.

The bladder smooth muscle is richly supplied by parasympathetic fibers, stimulation of which causes bladder contraction. If the passageway through the urethra to the outside is open, bladder contraction empties urine from the bladder. The exit from the bladder, however, is guarded by two sphincters, the *internal urethral sphincter* and the *external urethral sphincter*.

When Protein in the Urine Does Not Mean Kidney Disease

Urinary loss of proteins usually signifies kidney disease (nephritis). However, a urinary protein loss similar to that of nephritis often occurs following exercise, but the condition is harmless, transient, and reversible. The term *athletic pseudonephritis* is used to describe this postexercise (after exercise) proteinuria (protein in the urine). Studies indicate that 70% to 80% of athletes have proteinuria after very strenuous exercise. This condition occurs in participants in both noncontact and contact sports, so it does not arise from physical trauma to the kidneys. In one study, subjects who engaged in maximal short-term running excreted more protein than when they were bicycling, rowing, or swimming at the same work intensity. The reason for this difference is unknown.

Usually, only a very small fraction of the plasma proteins that enter the glomerulus is filtered; the plasma proteins that are filtered are reabsorbed in the tubules, so normally no plasma proteins appear in the urine. Two basic mechanisms can cause proteinuria: (1) increased glomerular permeability with no change in tubular reabsorption; or (2) impairment of tubular reabsorption. Research has shown that the proteinuria occurring during mild to moderate exercise results from changes in glomerular permeability, whereas the proteinuria occurring during short-term exhaustive exercise seems to be caused by both increased glomerular permeability and tubular dysfunction.

This reversible kidney dysfunction is believed to result from circulatory and hormonal changes that occur with exercise. Several studies have shown that renal blood flow is reduced during exercise as the renal vessels are constricted and blood is diverted to the exercising muscles. This reduction is positively correlated with exercise intensity. With intense exercise, the renal blood flow may be reduced to 20% of normal. As a result, glomerular blood flow is also reduced, but not to the same extent as renal blood flow, presumably because of autoregulatory mechanisms (see p. 520). Some investigators propose that decreased glomerular blood flow enhances diffusion of proteins into the tubular lumen because as the more slowly flowing blood spends more time in the glomerulus, a greater proportion of the plasma proteins has time to escape through the glomerular membrane. Hormonal changes that occur with exercise may also affect glomerular permeability. For example, renin injection is a well-recognized way to experimentally induce proteinuria. Plasma renin activity increases during strenuous exercise and may contribute to postexercise proteinuria. Researchers also hypothesize that maximal tubular reabsorption is reached during severe exercise, which could impair protein reabsorption.

ROLE OF THE URETHRAL SPHINCTERS A **sphincter** is a ring of muscle that, when contracted, closes off passage through an opening. The **internal urethral sphincter**—which is smooth muscle and, accordingly, is under involuntary control—is not really a separate muscle but instead consists of the last part of the bladder. Although it is not a true sphincter, it performs the same function as a sphincter. When the bladder is relaxed, the anatomic arrangement of the internal urethral sphincter region closes the outlet of the bladder.

Farther down the passageway, the urethra is encircled by a layer of skeletal muscle, the **external urethral sphincter.** This sphincter is reinforced by the entire **pelvic diaphragm,** a skeletal muscle sheet that forms the floor of the pelvis and helps support the pelvic organs (see ● Figure 14-2, p. 513). The motor neurons that supply the external sphincter and pelvic diaphragm fire continuously at a moderate rate unless they are inhibited, keeping these muscles tonically contracted so that they prevent urine from escaping through the urethra. Normally, when the bladder is relaxed and filling, both the internal and external urethral sphincters are closed to keep urine from dribbling out. Furthermore, because the external sphincter and pelvic diaphragm are skeletal muscle and thus under voluntary control, the person can deliberately tighten them to prevent urination from occurring even when the bladder is contracting and the internal sphincter is open.

MICTURITION REFLEX Micturition, or **urination,** the process of bladder emptying, is governed by two mechanisms: the micturition reflex and voluntary control. The **micturition reflex** is initiated when stretch receptors within the bladder wall are stimulated (● Figure 14-29). The bladder in an adult can accommodate up to 250 to 400 ml of urine before the tension within its walls begins to rise sufficiently to activate the stretch receptors (● Figure 14-30). The greater the distension beyond this, the greater is the extent of receptor activation. Afferent fibers from the stretch receptors carry impulses into the spinal cord and eventually, via interneurons, stimulate the parasympathetic supply to the bladder and inhibit the motor-neuron supply to the external sphincter. Parasympathetic stimulation of the bladder causes it to contract. No special mechanism is required to open the internal sphincter; changes in the shape of the bladder during contraction mechanically pull the internal sphincter open. Simultaneously, the external sphincter relaxes as its motor neuron supply is inhibited. Now both sphincters are open, and urine is expelled through the urethra by the force of bladder contraction. This micturition reflex, which is entirely a spinal reflex, governs bladder emptying in infants. As soon as the bladder fills enough to trigger the reflex, the baby automatically wets.

Dialysis: Cellophane Tubing or Abdominal Lining as an Artificial Kidney

Because chronic renal failure is irreversible and eventually fatal, treatment is aimed at maintaining renal function by alternative methods, such as dialysis and kidney transplantation. Over 300,000 people in the United States are currently on dialysis, and this number is expected to climb as the population ages and the incidence of diabetes mellitus, one of the leading causes of kidney failure, continues to rise. End-stage renal failure (less than 10% kidney function) caused by diabetes mellitus is increasing at a rate of over 11% annually.

The process of dialysis bypasses the kidneys to maintain normal fluid and electrolyte balance and remove wastes artificially. In the original method of dialysis, **hemodialysis,** a patient's blood is pumped through cellophane tubing that is surrounded by a large volume of fluid similar in composition to normal plasma. After dialysis, the blood is returned to the patient's circulatory system. During hemodialysis, about 250 ml of blood is outside of the body at any given time.

Like capillaries, cellophane is highly permeable to most plasma constituents but is impermeable to plasma proteins. As blood flows through the tubing, solutes move across the cellophane down their individual concentration gradients; plasma proteins, however, stay in the blood. Urea and other wastes, which are absent in the dialysis fluid, diffuse out of the plasma into the surrounding fluid, cleaning the blood of these wastes. Plasma constituents that are not regulated by the kidneys and are at normal concentration, such as glucose, do not move across the cellophane into the dialysis fluid, because there is no driving force to produce their movement. (The dialysis fluid's glucose concentration is the same as normal plasma glucose concentration.) Electrolytes, such as K^+ and PO_4^{3-}, which are higher than their normal plasma concentrations because the diseased kidneys cannot eliminate excess quantities of these substances, move out of the plasma until equilibrium is achieved between the plasma and the dialysis fluid. Because the dialysis fluid's solute concentrations are maintained at normal plasma values, the solute concentration of the blood returned to the patient after dialysis is essentially normal.

Hemodialysis is repeated as often as necessary to maintain the plasma composition within an acceptable level. Conventionally, it is done three times per week for several hours at each session at a treatment center or at home, but newer, more user-friendly, at-home methods dialyze the blood up to six times per week during the day or at night while the person is sleeping. The more frequent methods maintain better stability in plasma constituents than the less frequent methods do.

Another method of dialysis, **continuous ambulatory peritoneal dialysis (CAPD),** uses the peritoneal membrane (the lining of the abdominal cavity) as the dialysis membrane. With this method, 2 liters of dialysis fluid are inserted into the patient's abdominal cavity through a permanently implanted catheter. Urea, K^+, and other wastes and excess electrolytes diffuse from the plasma across the peritoneal membrane into the dialysis fluid, which is drained off and replaced several times a day. The CAPD method offers several advantages: The patient can self-administer it, the patient's blood is continuously purified and adjusted, and the patient can engage in normal activities while dialysis is being accomplished. One drawback is the increased risk of peritoneal infections.

Although dialysis can remove metabolic wastes and foreign compounds and help maintain fluid and electrolyte balance within acceptable limits, this plasma-cleansing technique cannot make up for the failing kidneys' reduced ability to produce hormones (erythropoietin and renin) and to activate vitamin D. One new experimental technique incorporates living kidney cells derived from pigs within a dialysis-like machine. Standard ultrafiltration technology like that used in hemodialysis purifies and adjusts the plasma as usual. Importantly, the living cells not only help maintain even better control of plasma constituents, especially K^+, but also add the deficient renal hormones to the plasma passing through the machine and activate vitamin D. This promising new technology has not yet been tested in large-scale clinical trials.

For now, transplanting a healthy kidney from a donor is another option for treating chronic renal failure. A kidney is one of the few transplants that can be provided by a living donor. Because 25% of the total kidney tissue can maintain the body, both the donor and the recipient have ample renal function with only one kidney each. The biggest problem with transplants is the possibility that the patient's immune system will reject the organ. Risk of rejection can be minimized by matching the tissue types of the donor and the recipient as closely as possible (the best donor choice is usually a close relative), coupled with immunosuppressive drugs. Over 15,000 kidney transplants are performed in the United States each year, with 60,000 more on waiting lists for a donor kidney.

Another new technique on the horizon for treating end-stage renal failure is a continuously functioning artificial kidney that mimics natural renal function. Using nanotechnology (very small-scale devices), researchers are working on a device that contains two membranes, the first for filtering blood like the glomerulus does and the second for mimicking the renal tubules by selectively altering the filtrate. The device, which will directly process the blood on an ongoing basis without using dialysis fluid, will return important substances to the body while discharging unneeded substances to a discardable bag that will serve as an external bladder. Scientists have developed computer models for such a device and thus far have created the filtering membrane.

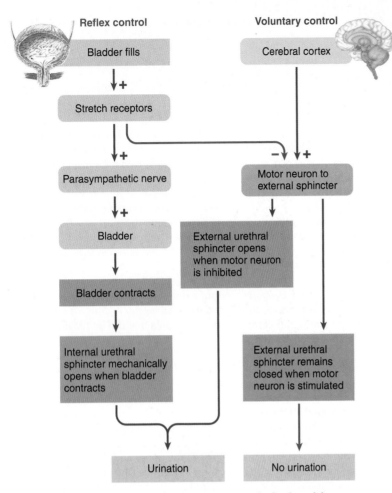

Reflex control

Bladder fills

↓ +

Stretch receptors

↓ +

Parasympathetic nerve

↓ +

Bladder

↓

Bladder contracts

↓

Internal urethral sphincter mechanically opens when bladder contracts

Voluntary control

Cerebral cortex

− +

Motor neuron to external sphincter

↓

External urethral sphincter opens when motor neuron is inhibited

External urethral sphincter remains closed when motor neuron is stimulated

↓

Urination

No urination

● **FIGURE 14-29 Reflex and voluntary control of micturition.**

● **FIGURE 14-30 Pressure changes within the urinary bladder as the bladder fills with urine.**

VOLUNTARY CONTROL OF MICTURITION In addition to triggering the micturition reflex, bladder filling also gives rise to the conscious urge to urinate. The perception of bladder fullness appears before the external sphincter reflexly relaxes, warning that micturition is imminent. As a result, voluntary control of micturition, learned during toilet training in early childhood, can override the micturition reflex so that bladder emptying can take place at the person's convenience rather than when bladder filling first activates the stretch receptors. If the time when the micturition reflex is initiated is inopportune for urination, the person can voluntarily prevent bladder emptying by deliberately tightening the external sphincter and pelvic diaphragm. Voluntary excitatory impulses from the cerebral cortex override the reflex inhibitory input from the stretch receptors to the involved motor neurons (the relative balance of EPSPs and IPSPs), keeping these muscles contracted so that no urine is expelled (see p. 106).

Urination cannot be delayed indefinitely. As the bladder continues to fill, reflex input from the stretch receptors increases with time. Finally, reflex inhibitory input to the external sphincter motor neuron becomes so powerful that it can no longer be overridden by voluntary excitatory input, so the sphincter relaxes and the bladder uncontrollably empties.

Micturition can also be deliberately initiated, even though the bladder is not distended, by voluntarily relaxing the external sphincter and pelvic diaphragm. Lowering of the pelvic floor allows the bladder to drop downward, which simultaneously pulls open the internal urethral sphincter and stretches the bladder wall. The subsequent activation of the stretch receptors brings about bladder contraction by the micturition reflex. Voluntary bladder emptying may be further assisted by contracting the abdominal wall and respiratory diaphragm. The resulting increase in intra-abdominal pressure squeezes down on the bladder to facilitate its emptying.

Clinical Note **URINARY INCONTINENCE Urinary incontinence,** or inability to prevent discharge of urine, occurs when descending pathways in the spinal cord that mediate voluntary control of the external sphincter and pelvic diaphragm are disrupted, as in spinal-cord injury. Because the components of the micturition reflex arc are still intact in the lower spinal cord, bladder emptying is governed by an uncontrollable spinal reflex, as in infants. A lesser degree of incontinence characterized by urine escaping when bladder pressure suddenly increases transiently, such as during coughing or sneezing, can result from impaired sphincter function. This is common in women who have borne children or in men whose sphincters have been injured during prostate surgery.

Chapter in Perspective: Focus on Homeostasis

The kidneys contribute to homeostasis more extensively than any other single organ. They regulate the electrolyte composition, volume, osmolarity, and pH of the internal environment

and eliminate all the waste products of bodily metabolism except for respiration-removed CO_2. They accomplish these regulatory functions by eliminating in the urine substances the body doesn't need, such as metabolic wastes and excess quantities of ingested salt or water, while conserving useful substances. The kidneys can maintain the plasma constituents they regulate within the narrow range compatible with life, despite wide variations in intake and losses of these substances through other avenues. Illustrating the magnitude of the kidneys' task, about a quarter of the blood pumped into the systemic circulation goes to the kidneys to be adjusted and purified, with only three quarters of the blood being used to supply all the other tissues.

The kidneys contribute to homeostasis in the following specific ways:

Regulatory Functions

■ The kidneys regulate the quantity and concentration of most ECF electrolytes, including those important in maintaining proper neuromuscular excitability.

■ They help maintain proper pH by eliminating excess H^+ (acid) or HCO_3^- (base) in the urine.

■ They help maintain proper plasma volume, which is important in long-term regulation of arterial blood pressure, by controlling salt balance in the body. The ECF volume, including plasma volume, reflects total salt load in the ECF, because Na^+ and its attendant anion, Cl^-, are responsible for more than 90% of the ECF's osmotic (water-holding) activity.

■ The kidneys maintain water balance in the body, which is important in maintaining proper ECF osmolarity (concentration of solutes). This role is important in maintaining stability of cell volume by keeping water from osmotically moving into or out of the cells, thus preventing them from swelling or shrinking, respectively.

Excretory Functions

■ The kidneys excrete the end products of metabolism in the urine. If allowed to accumulate, these wastes are toxic to cells.

■ The kidneys also excrete many foreign compounds that enter the body.

Hormonal Functions

■ The kidneys produce erythropoietin, the hormone that stimulates the bone marrow to produce red blood cells. This action contributes to homeostasis by helping maintain the optimal O_2 content of blood. More than 98% of O_2 in the blood is bound to hemoglobin within red blood cells.

■ They also produce renin, the hormone that initiates the renin–angiotensin–aldosterone pathway for controlling renal tubular Na^+ reabsorption, which is important in long-term maintenance of plasma volume and arterial blood pressure.

Metabolic Functions

■ The kidneys help convert vitamin D into its active form. Vitamin D is essential for Ca^{2+} absorption from the digestive tract. Calcium, in turn, exerts a wide variety of homeostatic functions.

REVIEW EXERCISES

Objective Questions (Answers on p. A-51)

1. Part of the kidneys' energy supply is used to accomplish glomerular filtration. (True or false?)

2. Sodium reabsorption is under hormonal control throughout the length of the tubule. (True or false?)

3. Glucose and amino acids are reabsorbed by secondary active transport. (True or false?)

4. Solute excretion is always accompanied by comparable H_2O excretion. (True or false?)

5. Water excretion can occur without comparable solute excretion. (True or false?)

6. The functional unit of the kidneys is the _____.

7. _____ is the only ion actively reabsorbed in the proximal tubule and actively secreted in the distal and collecting tubules.

8. The daily minimum volume of obligatory H_2O loss that must accompany excretion of wastes is _____ ml.

9. Indicate whether each of the following factors would (a) increase or (b) decrease the GFR, if everything else remained constant.

___ 1. a rise in Bowman's capsule pressure resulting from ureteral obstruction by a kidney stone
___ 2. a fall in plasma protein concentration resulting from loss of these proteins from a large burned surface of skin
___ 3. a dramatic fall in arterial blood pressure following severe hemorrhage (<80 mm Hg)
___ 4. afferent arteriolar vasoconstriction
___ 5. tubuloglomerular feedback response to decreased salt delivery to the distal tubule
___ 6. myogenic response of an afferent arteriole stretched as a result of an increased driving blood pressure
___ 7. ↑ sympathetic activity to the afferent arterioles
___ 8. contraction of mesangial cells
___ 9. contraction of podocytes

10. Which of the following filtered substances is normally not present in the urine at all?
   a. Na^+
   b. PO_4^{3-}
   c. urea
   d. H^+
   e. glucose

11. Reabsorption of which of the following substances is *not* linked in some way to active Na^+ reabsorption?
    a. glucose
    b. PO_4^{3-}
    c. H_2O
    d. urea
    e. Cl^-

In questions 12–14, indicate, by writing the identifying letters in the proper order in the blanks, the proper sequence through which fluid flows as it traverses the structures in question.

12. a. ureter   ___ ___ ___ ___ ___
    b. kidney
    c. urethra
    d. bladder
    e. renal pelvis

13. a. efferent arteriole   ___ ___ ___ ___ ___ ___
    b. peritubular capillaries
    c. renal artery
    d. glomerulus
    e. afferent arteriole
    f. renal vein

14. a. loop of Henle   ___ ___ ___ ___ ___ ___ ___
    b. collecting duct
    c. Bowman's capsule
    d. proximal tubule
    e. renal pelvis
    f. distal tubule
    g. glomerulus

15. Using the answer code on the right, indicate what the osmolarity of the tubular fluid is at each of the designated points in a nephron:
    ___ 1. Bowman's capsule
    ___ 2. end of proximal tubule
    ___ 3. tip of Henle's loop of juxtamedullary nephron (at the bottom of the U-turn)
    ___ 4. end of Henle's loop of juxtamedullary nephron (before entry into distal tubule)
    ___ 5. end of collecting duct

    (a) isotonic (300 mOsm)
    (b) hypotonic (100 mOsm)
    (c) hypertonic (1200 mOsm)
    (d) ranging from hypotonic to hypertonic (100 mOsm to 1200 mOsm)

Essay Questions

1. List the functions of the kidneys.
2. Describe the anatomy of the urinary system. Describe the components of a nephron.
3. Describe the three basic renal processes; indicate how they relate to urine excretion.
4. Distinguish between *secretion* and *excretion*.
5. Discuss the forces involved in glomerular filtration. What is the average GFR?
6. How is GFR regulated as part of the baroreceptor reflex?
7. Why do the kidneys receive a seemingly disproportionate share of the cardiac output? What percentage of renal blood flow is normally filtered?
8. List the steps in transepithelial transport.
9. Distinguish between active and passive reabsorption.
10. Describe all the tubular transport processes that are linked to the basolateral Na^+–K^+ ATPase carrier.
11. Describe the renin–angiotensin–aldosterone system. What are the functions of aldosterone and angiotensin II? Discuss the source and functions of ANP and BNP.
12. To what do the terms *tubular maximum* (T_m) and *renal threshold* refer? Compare two substances that display a T_m, one substance that *is* and one that *is not* regulated by the kidneys.
13. What is the importance of tubular secretion? What are the most important secretory processes?
14. What is the average rate of urine formation?
15. Define *plasma clearance.*
16. What establishes a vertical osmotic gradient in the medullary interstitial fluid? Of what importance is this gradient?
17. Discuss the function and mechanism of action of vasopressin.
18. Describe the transfer of urine to, the storage of urine in, and the emptying of urine from the bladder.

Quantitative Exercises (Solutions on p. A-51)

1. Two patients are voiding protein in their urine. To determine whether or not this proteinuria indicates a serious problem, a physician injects small amounts of inulin and PAH into each patient. Recall that inulin is freely filtered and neither secreted nor reabsorbed in the nephron and that PAH at this concentration is completely removed from the blood by tubular secretion. The data collected are given in the following table, where $[I]_u$ and $[PAH]_u$ are the concentrations of inulin or PAH in the urine (in mM), respectively; $[I]_p$ and $[PAH]_p$ are the concentrations of these substances in the plasma; and v_u is the flow rate of urine (in ml/min).

Patient	$[I]_u$	$[I]_p$	$[PAH]_u$	$[PAH]_p$	v_u
1	25	2	186	3	10
2	31	1.5	300	4.5	6

   a. Calculate each patient's GFR and renal plasma flow.
   b. Calculate the renal blood flow for each patient, assuming both have a hematocrit of 0.45.
   c. Calculate the filtration fraction for each patient.
   d. Which of the values calculated for each patient are within the normal range? Which values are abnormal? What could be causing these deviations from normal?

2. What is the filtered load of sodium if inulin clearance is 125 ml/min and the sodium concentration in plasma is 145 mM?

3. Calculate a patient's rate of urine production, given that his inulin clearance is 125 ml/min and his urine and plasma concentrations of inulin are 300 mg/liter and 3 mg/liter, respectively.

4. If the urine concentration of a substance is 7.5 mg/ml of urine, its plasma concentration is 0.2 mg/ml of plasma, and the urine flow rate is 2 ml/min, what is the clearance rate of the substance? Is the substance being reabsorbed or secreted by the kidneys?

POINTS TO PONDER

(Explanations on p. A-51)

1. The juxtamedullary nephrons of animals adapted to survive with minimal water consumption, such as desert rats, have relatively much longer loops of Henle than humans have. Of what benefit would these longer loops be?

2. If the plasma concentration of substance X is 200 mg/100 ml and the GFR is 125 ml/min, what is the filtered load of this substance? If the T_m for substance X is 200 mg/min, how much of the substance will be reabsorbed at a plasma concentration of 200 mg/100 ml and a GFR of 125 ml/min? How much of substance X will be excreted?

3. *Conn's syndrome* is an endocrine disorder brought about by a tumor of the adrenal cortex that secretes excessive aldosterone in uncontrolled fashion. Given what you know about the functions of aldosterone, describe what the most prominent features of this condition would be.

4. Because of a mutation, a child was born with an ascending limb of Henle that was water permeable. What would be the minimum/maximum urine osmolarities (in units of mOsm) the child could produce?
   a. 100/300
   b. 300/1200
   c. 100/100
   d. 1200/1200
   e. 300/300

5. An accident victim suffers permanent damage of the lower spinal cord and is paralyzed from the waist down. Describe what governs bladder emptying in this individual.

CLINICAL CONSIDERATION

(Explanation on p. A-51)

Marcus T. has noted a gradual decrease in his urine flow rate and is now experiencing difficulty in initiating micturition. He needs to urinate frequently, and often he feels as if his bladder is not empty even though he has just urinated. Analysis of Marcus's urine reveals no abnormalities. Are his urinary tract symptoms most likely caused by kidney disease, a bladder infection, or prostate enlargement?

Systems of Major Importance in Maintaining Fluid and Acid-Base Balance

Body systems maintain homeostasis

Homeostasis

The kidneys, in conjunction with hormones involved in salt and water balance, are responsible for maintaining the volume and osmolarity of the extracellular fluid (internal environment). The kidneys, along with the respiratory system and chemical buffer systems in the body fluids, also contribute to homeostasis by maintaining the proper pH in the internal environment.

$$CO_2 + H_2O \rightleftharpoons H_2CO_3 \rightleftharpoons H^+ + HCO_3^-$$
Chemical buffer systems in body fluids

Homeostasis is essential for survival of cells

Cells

The volume of circulating blood must be maintained to help ensure adequate pressure to drive life-sustaining blood to the cells. The osmolarity of fluid surrounding the cells must be closely regulated to prevent detrimental osmotic movement of water between the cells and ECF. The pH of the internal environment must remain stable because pH changes alter neuromuscular excitability and enzyme activity, among other serious consequences.

Cells make up body systems

Homeostasis depends on maintaining a balance between the input and the output of all constituents in the internal fluid environment. Regulation of **fluid balance** involves two separate components: control of ECF volume, of which circulating plasma volume is a part, and control of ECF osmolarity (solute concentration). The kidneys control ECF volume by maintaining **salt balance** and control ECF osmolarity by maintaining **water balance**. The kidneys maintain this balance by adjusting the output of salt and water in the urine as needed to compensate for variable input and abnormal losses of these constituents.

Similarly, the kidneys help maintain **acid–base balance** by adjusting the urinary output of hydrogen ion (acid) and bicarbonate ion (base) as needed. Also contributing to acid–base balance are the lungs, which can adjust the rate at which they excrete hydrogen ion–generating CO_2, and the chemical buffer systems in the body fluids.

Fluid and Acid–Base Balance

Balance Concept

The cells of complex multicellular organisms are able to survive and function only within a very narrow range of composition of the extracellular fluid (ECF), the internal fluid environment that bathes them.

The internal pool of a substance is the amount of that substance in the ECF.

The quantity of any particular substance in the ECF is considered a readily available internal **pool.** The amount of the substance in the pool may be increased either by transferring more in from the external environment (most commonly by ingestion) or by metabolically producing it within the body (● Figure 15-1). Substances may be removed from the body by being excreted to the outside or by being used up in a metabolic reaction. If the quantity of a substance is to remain stable within the body, its **input** through ingestion or metabolic production must be balanced by an equal **output** through excretion or metabolic consumption. This relationship, known as the **balance concept,** is extremely important in maintaining homeostasis. Not all input and output pathways are applicable for every body-fluid constituent. For example, salt is not synthesized or consumed by the body, so the stability of salt concentration in the body fluids depends entirely on a balance between salt ingestion and salt excretion.

The ECF pool can further be altered by transferring a particular ECF constituent into storage within the body. If the body as a whole has a surplus or deficit of a particular stored substance, the storage site can be expanded or partially depleted to maintain the ECF concentration of the substance within homeostatically prescribed limits. For example, after absorption of a meal, when more glucose is entering the plasma than is being consumed by the cells, the extra glucose can be temporarily stored, in the form of glycogen, in muscle and liver cells. This storage depot can then be tapped between meals as needed to maintain the plasma glucose level when no new nutrients are being added to the blood by eating. This internal storage capacity is limited, however. Although an internal exchange between the ECF and a storage depot can temporarily restore the plasma concentration of a particular substance to normal, in the long run any excess or deficit of that constituent must be compensated for by appropriate adjustments in total body input or output.

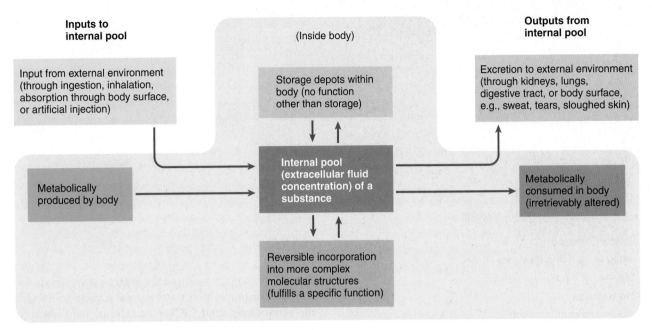

• FIGURE 15-1 Inputs to and outputs from the internal pool of a body constituent.

Another possible internal exchange between the pool and the remainder of the body is the reversible incorporation of certain plasma constituents into more complex molecular structures. For example, iron is incorporated into hemoglobin within the red blood cells during their synthesis but is released intact back into the body fluids when the red cells degenerate. This process differs from metabolic consumption of a substance, in which the substance is irretrievably converted into another form—for example, glucose converted into CO_2 plus H_2O plus energy. It also differs from storage in that the latter serves no purpose other than storage, whereas reversible incorporation into a more complex structure serves a specific purpose.

To maintain stable balance of an ECF constituent, its input must equal its output.

When total body input of a particular substance equals its total body output, a **stable balance** exists. When the gains via input for a substance exceed its losses via output, a **positive balance** exists. The result is an increase in the total amount of the substance in the body. In contrast, when losses for a substance exceed its gains, a **negative balance** exists and the total amount of the substance in the body decreases.

Changing the magnitude of any of the input or output pathways for a given substance can alter its plasma concentration. To maintain homeostasis, any change in input must be balanced by a corresponding change in output (for example, increased salt intake must be matched by a corresponding increase in salt output in the urine), and, conversely, increased losses must be compensated for by increased intake. Thus, maintaining a stable balance requires control. However, not all input and output pathways are regulated to maintain balance. Generally, input of various plasma constituents is poorly controlled or not controlled at all. We frequently ingest salt and H_2O, for example,

not because we need them but because we want them, so the intake of salt and H_2O is highly variable. Likewise, hydrogen ion (H^+) is uncontrollably generated internally and added to the body fluids. Salt, H_2O, and H^+ can also be lost to the external environment to varying degrees through the digestive tract (vomiting), skin (sweating), and elsewhere without regard for the salt, H_2O, or H^+ balance in the body. Compensatory adjustments in the urinary excretion of these substances maintain the body fluids' volume and salt and acid composition within the extremely narrow homeostatic range compatible with life despite the wide variations in input and unregulated losses of these plasma constituents.

The rest of this chapter is devoted to discussing the regulation of fluid balance (maintaining salt and H_2O balance) and acid–base balance (maintaining H^+ balance).

Fluid Balance

Water is by far the most abundant component of the human body, constituting 60% of body weight, on average, but ranging from 40% to 80%. The H_2O content of an individual remains fairly constant, largely because the kidneys efficiently regulate H_2O balance, but the percentage of body H_2O varies from person to person. The reason for the wide range in body H_2O among individuals is the variability in the amount of their adipose tissue (fat). Adipose tissue has a low H_2O percentage compared to other tissues. Plasma, as you might suspect, is more than 90% H_2O. Even the soft tissues such as skin, muscles, and internal organs consist of 70% to 80% H_2O. The relatively drier skeleton is only 22% H_2O. Fat, however, is the driest tissue of all, having only 10% H_2O content. Accordingly, a high body H_2O percentage is associated with leanness and a low body H_2O percentage with obesity because a larger proportion of the overweight body consists of relatively dry fat.

The percentage of body H_2O is also influenced by the sex and age of the individual. Women have a lower body H_2O percentage than men, primarily because the female sex hormone, estrogen, promotes fat deposition in the breasts, buttocks, and elsewhere. This not only gives rise to the typical female figure but also endows women with a higher proportion of adipose tissue and, therefore, a lower body H_2O proportion. The percentage of body H_2O also decreases progressively with age.

Body water is distributed between the ICF and ECF compartments.

Body H_2O is distributed between two major fluid compartments: fluid within the cells, or **intracellular fluid (ICF),** and fluid surrounding the cells, or **extracellular fluid (ECF)** (▲ Table 15-1). (The terms "H_2O" and "fluid" are commonly used interchangeably. Although this usage is not entirely accurate, because it ignores the solutes in body fluids, it is acceptable when discussing total volume of fluids because the major proportion of these fluids consists of H_2O.)

PROPORTION OF H_2O IN THE MAJOR FLUID COMPARTMENTS The ICF compartment composes about two thirds of the total body H_2O. Even though each cell contains its own unique mixture of constituents, these trillions of minute fluid compartments are similar enough to be considered collectively as one large fluid compartment.

The remaining third of the body H_2O found in the ECF compartment is further subdivided into plasma and interstitial fluid. The **plasma,** which makes up about a fifth of the ECF volume, is the fluid portion of blood. The **interstitial fluid,** which represents the other four fifths of the ECF compartment, is the fluid in the spaces between cells. It bathes and makes exchanges with tissue cells.

MINOR ECF COMPARTMENTS Two other minor categories are included in the ECF compartment: lymph and transcellular fluid. **Lymph** is the fluid being returned from the interstitial fluid to the plasma by means of the lymphatic system, where it is filtered through lymph nodes for immune defense purposes (see pp. 369 and 418). **Transcellular fluid** consists of a number of small, specialized fluid volumes, all of which are secreted by specific cells into a particular body cavity to perform some specialized function. Transcellular fluid includes *cerebrospinal fluid* (surrounding, cushioning, and nourishing the brain and spinal cord); *intraocular fluid* (maintaining the shape of and nourishing the eye); *synovial fluid* (lubricating and serving as a shock absorber for the joints); *pericardial, intrapleural,* and *peritoneal fluids* (lubricating movements of the heart, lungs, and intestines, respectively); and the *digestive juices* (digesting ingested foods).

Although these fluids are extremely important functionally, they represent an insignificant fraction of total body H_2O. Furthermore, the transcellular compartment as a whole usually does not reflect changes in the body's fluid balance. For example, cerebrospinal fluid does not decrease in volume when the body as a whole is experiencing a negative H_2O balance. This is not to say that these fluid volumes never change. Localized changes in a particular transcellular fluid compart-

▲ TABLE 15-1	Classification of Body Fluid		
Compartment	**Volume of Fluid (in Liters)**	**Percentage of Body Fluid**	**Percentage of Body Weight**
Total Body Fluid	42	100	60
Intracellular Fluid (ICF)	28	67	40
Extracellular Fluid (ECF)	14	33	20
Plasma	2.8	6.6 (20% of ECF)	4
Interstitial Fluid	11.2	26.4 (80% of ECF)	16
Lymph	Negligible	Negligible	Negligible
Transcellular Fluid	Negligible	Negligible	Negligible

ment can occur pathologically (such as too much intraocular fluid accumulating in the eyes of people with glaucoma; see p. 196), but such a localized fluid disturbance does not affect the fluid balance of the body. Therefore, the transcellular compartment can usually be ignored when dealing with problems of fluid balance. The main exception to this generalization occurs when digestive juices are abnormally lost from the body during heavy vomiting or diarrhea, which can bring about a fluid imbalance.

The plasma and interstitial fluid are similar in composition, but the ECF and ICF differ markedly.

Several barriers separate the body-fluid compartments, limiting the movement of H_2O and solutes between the various compartments to differing degrees.

THE BARRIER BETWEEN PLASMA AND INTERSTITIAL FLUID: BLOOD VESSEL WALLS The two components of the ECF—plasma and interstitial fluid—are separated by the walls of the blood vessels. However, H_2O and all plasma constituents except for plasma proteins are continuously and freely exchanged between plasma and interstitial fluid by passive means across the thin, pore-lined capillary walls. Accordingly, plasma and interstitial fluid are nearly identical in composition, except that interstitial fluid lacks plasma proteins. Any change in one of these ECF compartments is quickly reflected in the other compartment because they are constantly mixing.

THE BARRIER BETWEEN THE ECF AND ICF: CELLULAR PLASMA MEMBRANES In contrast to the very similar composition of vascular and interstitial fluid compartments, the composition of

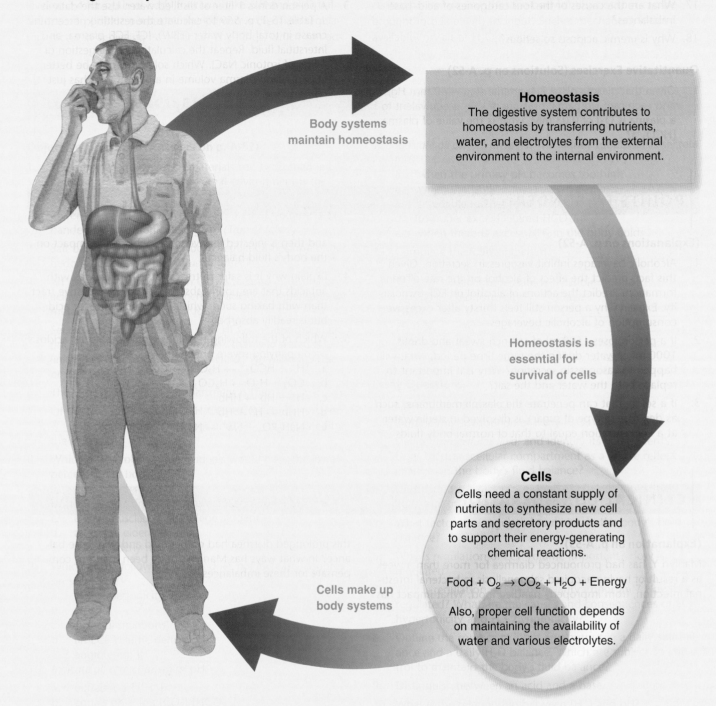

Digestive System

Body systems maintain homeostasis

Homeostasis

The digestive system contributes to homeostasis by transferring nutrients, water, and electrolytes from the external environment to the internal environment.

Homeostasis is essential for survival of cells

Cells

Cells need a constant supply of nutrients to synthesize new cell parts and secretory products and to support their energy-generating chemical reactions.

Food + O_2 → CO_2 + H_2O + Energy

Also, proper cell function depends on maintaining the availability of water and various electrolytes.

Cells make up body systems

To maintain homeostasis, nutrient molecules used for energy production must continually be replaced by new, energy-rich nutrients. Also, nutrient molecules, especially proteins, are needed for ongoing synthesis of new cells and cell parts in the course of tissue turnover and growth. Similarly, water and electrolytes constantly lost in urine and sweat and through other avenues must be replenished regularly. The **digestive system** contributes to homeostasis by transferring nutrients, water, and electrolytes from the external environment to the internal environment. The digestive system does not directly regulate the concentration of any of these constituents in the internal environment. It does not vary nutrient, water, or electrolyte uptake based on body needs (with few exceptions); rather, it optimizes conditions for digesting and absorbing what is ingested.

CHAPTER 6

The Digestive System

CENGAGENOW Log on to CengageNOW at **http://www
.cengage.com/sso/** for an opportunity to explore a learning
module that illustrates difficult concepts with self-study
tutorials, animations, and interactive quizzes to help you learn,
review, and master physiology concepts.

General Aspects of Digestion

The primary function of the **digestive system** is to transfer nutrients, water, and electrolytes from the food we eat into the
body's internal environment. Ingested food is essential as an
energy source, or fuel, from which the cells can generate ATP to
carry out their particular energy-dependent activities, such as
active transport, contraction, synthesis, and secretion. Food is
also a source of building supplies for the renewal and addition
of body tissues.

The act of eating does not automatically make the preformed organic molecules in food available to the body cells as
a source of fuel or as building blocks. The food first must be
digested, or biochemically broken down, into small, simple
molecules that can be absorbed from the digestive tract into the
circulatory system for distribution to the cells. Normally, about
95% of the ingested food is made available for the body's use.
Thus, the sequence in nutrient acquisition is ingestion, digestion, absorption, distribution, and usage.

We will first provide an overview of the digestive system,
examining the common features of the various components of
the system, before we begin a detailed tour of the tract from
beginning to end.

The digestive system performs four basic digestive processes.

There are four basic digestive processes: *motility, secretion, digestion,* and *absorption.*

MOTILITY The term **motility** refers to the muscular contractions that mix and move forward the contents of the digestive
tract. Although the smooth muscle in the walls of the digestive
tract is phasic smooth muscle that displays action-potential-
induced bursts of contraction (see p. 293), it also maintains a
constant low level of contraction known as **tone.** Tone is important in maintaining a steady pressure on the contents of the digestive tract as well as in preventing its walls from remaining
permanently stretched following distension.

Two basic types of phasic digestive motility are superimposed on this ongoing tonic activity: propulsive movements
and mixing movements. *Propulsive movements* propel or push
the contents forward through the digestive tract, with the rate
of propulsion varying depending on the functions accom-

plished by the different regions. That is, the contents are moved forward in a given segment at an appropriate velocity to allow that segment to do its job. For example, transit of food through the esophagus is rapid, which is appropriate because this structure merely serves as a passageway from the mouth to the stomach. In comparison, in the small intestine—the main site of digestion and absorption—the contents are moved forward slowly, allowing time for the breakdown and absorption of food.

Mixing movements serve a twofold function. First, by mixing food with the digestive juices, these movements promote digestion of the food. Second, they facilitate absorption by exposing all parts of the intestinal contents to the absorbing surfaces of the digestive tract.

Contraction of the smooth muscle within the walls of the digestive organs accomplishes movement of material through most of the digestive tract. The exceptions are at the ends of the tract—the mouth through the early part of the esophagus at the beginning and the external anal sphincter at the end—where motility involves skeletal muscle rather than smooth muscle activity. Accordingly, the acts of chewing, swallowing, and defecation have voluntary components, because skeletal muscle is under voluntary control. By contrast, motility accomplished by smooth muscle throughout the rest of the tract is controlled by complex involuntary mechanisms.

SECRETION A number of digestive juices are secreted into the digestive tract lumen by exocrine glands (see p. 4) along the route, each with its own specific secretory product. Each **digestive secretion** consists of water, electrolytes, and specific organic constituents important in the digestive process, such as enzymes, bile salts, or mucus. The secretory cells extract from the plasma large volumes of water and the raw materials necessary to produce their particular secretion. Secretion of all digestive juices requires energy, both for active transport of some of the raw materials into the cell (others diffuse in passively) and for synthesis of secretory products by the endoplasmic reticulum. On appropriate neural or hormonal stimulation, the secretions are released into the digestive tract lumen. Normally, the digestive secretions are reabsorbed in one form or another back into the blood after their participation in digestion. Failure to do so (because of vomiting or diarrhea, for example) results in loss of this fluid that has been "borrowed" from the plasma.

Furthermore, endocrine cells located in the digestive tract wall secrete gastrointestinal hormones into the blood that help control digestive motility and exocrine gland secretion.

DIGESTION Humans consume three different biochemical categories of energy-rich foodstuffs: *carbohydrates, proteins,* and *fats.* These large molecules cannot cross plasma membranes intact to be absorbed from the lumen of the digestive tract into the blood or lymph. The term **digestion** refers to the biochemical breakdown of the structurally complex foodstuffs of the diet into smaller, absorbable units by the enzymes produced within the digestive system as follows:

1. The simplest **carbohydrates** are the simple sugars or **monosaccharides** ("one-sugar" molecules), such as **glucose, fructose,** and **galactose,** very few of which are normally found in

the diet (see p. A-12). Most ingested carbohydrate is in the form of **polysaccharides** ("many-sugar" molecules), which consist of chains of interconnected glucose molecules. The most common polysaccharide consumed is **starch** derived from plant sources. In addition, meat contains **glycogen,** the polysaccharide storage form of glucose in muscle. **Cellulose,** another dietary polysaccharide, found in plant walls, cannot be digested into its constituent monosaccharides by the digestive juices humans secrete; thus, it represents the indigestible *fiber* or "bulk" of our diets. Besides polysaccharides, a lesser source of dietary carbohydrate is in the form of **disaccharides** ("two-sugar" molecules), including **sucrose** (table sugar, which consists of one glucose and one fructose molecule) and **lactose** (milk sugar made up of one glucose and one galactose molecule). Through the process of digestion, starch, glycogen, and disaccharides are converted into their constituent monosaccharides, principally glucose with small amounts of fructose and galactose. These monosaccharides are the absorbable units for carbohydrates.

2. Dietary **proteins** consist of various combinations of **amino acids** held together by peptide bonds (see p. A-14). Through the process of digestion, proteins are degraded primarily into their constituent amino acids as well as a few **small polypeptides** (several amino acids linked by peptide bonds), both of which are the absorbable units for protein.

3. Most dietary **fats** are in the form of **triglycerides,** which are neutral fats, each consisting of a glycerol with three **fatty acid** molecules attached (*tri* means "three") (see p. A-13). During digestion, two of the fatty acid molecules are split off, leaving a **monoglyceride,** a glycerol molecule with one fatty acid molecule attached (*mono* means "one"). Thus, the end products of fat digestion are monoglycerides and free fatty acids, which are the absorbable units of fat.

Digestion is accomplished by enzymatic **hydrolysis** ("breakdown by water"; see p. A-15). By adding H_2O at the bond site, enzymes in the digestive secretions break down the bonds that hold the small molecular subunits within the nutrient molecules together, thus setting the small molecules free (● Figure 16-1). The removal of H_2O at the bond sites originally joined these small subunits to form nutrient molecules. Hydrolysis replaces the H_2O and frees the small absorbable units. Digestive enzymes are specific in the bonds they can hydrolyze. As food moves through the digestive tract, it is subjected to various enzymes, each of which breaks down the food molecules even further. In this way, large food molecules are converted to simple absorbable units in a progressive, stepwise fashion, like an assembly line in reverse, as the digestive tract contents are propelled forward.

ABSORPTION In the small intestine, digestion is completed and most absorption occurs. Through the process of **absorption,** the small absorbable units that result from digestion, along with water, vitamins, and electrolytes, are transferred from the digestive tract lumen into the blood or lymph. As we examine the digestive tract from beginning to end, we will discuss the four processes of motility, secretion, digestion, and absorption as they take place within each digestive organ (▲ Table 16-1).

● **FIGURE 16-1 An example of hydrolysis.** In this example, the disaccharide maltose (the intermediate breakdown product of polysaccharides) is broken down into two glucose molecules by the addition of H_2O at the bond site.

The digestive tract and accessory digestive organs make up the digestive system.

The digestive system consists of the *digestive* (or *gastrointestinal*) *tract* plus the accessory digestive organs (*gastro* means "stomach"). The **accessory digestive organs** include the *salivary glands,* the *exocrine pancreas,* and the *biliary system,* which is composed of the *liver* and *gallbladder.* These exocrine organs lie outside the digestive tract and empty their secretions through ducts into the digestive tract lumen.

The **digestive tract** is essentially a tube about 4.5 m (15 feet) in length in its normal contractile state.[1] Running through the middle of the body, the digestive tract includes the following organs (▲ Table 16-1): *mouth; pharynx* (throat); *esophagus; stomach; small intestine* (consisting of the *duodenum, jejunum,* and *ileum*); *large intestine* (the *cecum, appendix, colon,* and *rectum*); and *anus.* Although these organs are continuous with one another, they are considered as separate entities because of their regional modifications, which allow them to specialize in particular digestive activities.

Because the digestive tract is continuous from the mouth to the anus, the lumen of this tube, like the lumen of a straw, is continuous with the external environment. As a result, the contents within the lumen of the digestive tract are technically outside the body, just as the soda you suck through a straw is not a part of the straw. Only after a substance has been absorbed from the lumen across the digestive tract wall is it considered part of the body. This is important, because conditions essential to the digestive process can be tolerated in the digestive tract lumen that could not be tolerated in the body proper. Consider the following examples:

■ The pH of the stomach contents falls as low as 2 as a result of the gastric secretion of hydrochloric acid (HCl), yet in the body fluids the range of pH compatible with life is 6.8 to 8.0.

■ The digestive enzymes that hydrolyze the protein in food could also destroy the body's own tissues that produce them.

(Protein is the main structural component of cells.) Therefore, once these enzymes are synthesized in inactive form, they are not activated until they reach the lumen, where they actually attack the food outside the body (that is, within the lumen), thereby protecting the body tissues against self-digestion.

■ In the lower part of the intestine exist quadrillions of living microorganisms that are normally harmless and even beneficial, yet if these same microorganisms enter the body proper (as may happen with a ruptured appendix), they may be extremely harmful or even lethal.

■ Foodstuffs are complex foreign particles that would be attacked by the immune system if they were in contact with the body proper. However, the foodstuffs are digested within the lumen into absorbable units such as glucose, amino acids, and fatty acids that are indistinguishable from these simple energy-rich molecules already present in the body.

The digestive tract wall has four layers.

The digestive tract wall has the same general structure throughout most of its length from the esophagus to the anus, with some local variations characteristic for each region. A cross section of the digestive tube reveals four major tissue layers (● Figure 16-2). From the innermost layer outward they are the *mucosa,* the *submucosa,* the *muscularis externa,* and the *serosa.*

MUCOSA The **mucosa** lines the luminal surface of the digestive tract. It is divided into three layers:

■ The primary component of the mucosa is a **mucous membrane,** an inner epithelial layer that serves as a protective surface. It is also modified in particular areas for secretion and absorption. The mucous membrane contains *exocrine gland cells* for secretion of digestive juices, *endocrine gland cells* for secretion of blood-borne gastrointestinal hormones, and *epithelial cells* specialized for absorbing digested nutrients.

■ The **lamina propria** is a thin middle layer of connective tissue on which the epithelium rests. It houses the **gut-associated lymphoid tissue (GALT),** which is important in the defense against disease-causing intestinal bacteria (see p. 418).

■ The **muscularis mucosa,** a sparse layer of smooth muscle, is the outermost mucosal layer that lies adjacent to the submucosa.

The mucosal surface is generally highly folded, with many ridges and valleys that greatly increase the surface area available for absorption. The degree of folding varies in different areas of the digestive tract, being most extensive in the small intestine, where maximum absorption occurs, and least extensive in the esophagus, which merely serves as a transit tube. The pattern of surface folding can be modified by contraction of the muscularis mucosa. This is important in exposing different areas of the absorptive surface to the luminal contents.

---

[1]Because the uncontracted digestive tract in a cadaver is about twice as long as the contracted tract in a living person, anatomy texts indicate that the digestive tract is 30 feet long compared to the length of 15 feet indicated in physiology texts.

Anatomy and Functions of Components of the Digestive System

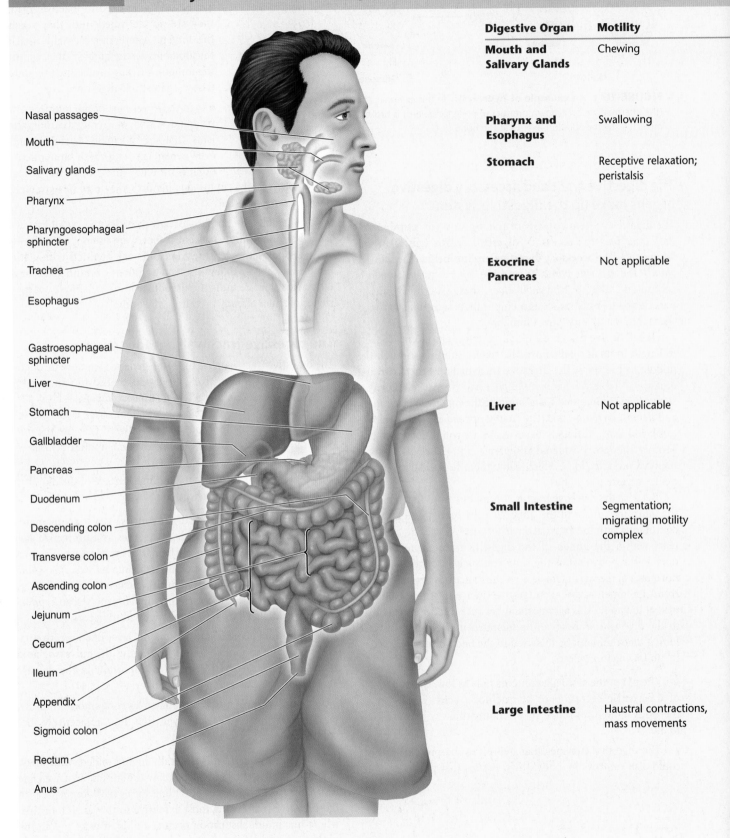

Digestive Organ	Motility
Mouth and Salivary Glands	Chewing
Pharynx and Esophagus	Swallowing
Stomach	Receptive relaxation; peristalsis
Exocrine Pancreas	Not applicable
Liver	Not applicable
Small Intestine	Segmentation; migrating motility complex
Large Intestine	Haustral contractions, mass movements

Nasal passages
Mouth
Salivary glands
Pharynx
Pharyngoesophageal sphincter
Trachea
Esophagus
Gastroesophageal sphincter
Liver
Stomach
Gallbladder
Pancreas
Duodenum
Descending colon
Transverse colon
Ascending colon
Jejunum
Cecum
Ileum
Appendix
Sigmoid colon
Rectum
Anus

Secretion	Digestion	Absorption
Saliva ■ Amylase ■ Mucus ■ Lysozyme	Carbohydrate digestion begins	No foodstuffs; a few medications—for example, nitroglycerin
Mucus	None	None
Gastric juice ■ HCl ■ Pepsin ■ Mucus ■ Intrinsic factor	Carbohydrate digestion continues in body of stomach; protein digestion begins in antrum of stomach	No foodstuffs; a few lipid-soluble substances, such as alcohol and aspirin
Pancreatic digestive enzymes ■ Trypsin, chymotrypsin, carboxypeptidase ■ Amylase ■ Lipase Pancreatic aqueous NaHCO₃ secretion	These pancreatic enzymes accomplish digestion in duodenal lumen	Not applicable
Bile ■ Bile salts ■ Alkaline secretion ■ Bilirubin	Bile does not digest anything, but bile salts facilitate fat digestion and absorption in duodenal lumen	Not applicable
Succus entericus ■ Mucus ■ Salt (Small intestine enzymes—disaccharidases and aminopeptidases—are not secreted but function within the brush-border membrane)	In lumen, under influence of pancreatic enzymes and bile, carbohydrate and protein digestion continues and fat digestion is completely accomplished; in brush border, carbohydrate and protein digestion completed	All nutrients, most electrolytes, and water
Mucus	None	Salt and water, converting contents to feces

SUBMUCOSA The **submucosa** ("under the mucosa") is a thick layer of connective tissue that provides the digestive tract with its distensibility and elasticity. It contains the larger blood and lymph vessels, both of which send branches inward to the mucosal layer and outward to the surrounding thick muscle layer. Also, a nerve network known as the *submucosal plexus* lies within the submucosa (*plexus* means "network").

MUSCULARIS EXTERNA The **muscularis externa,** the major smooth muscle coat of the digestive tube, surrounds the submucosa. In most parts of the tract, the muscularis externa consists of two layers: an *inner circular layer* and an *outer longitudinal layer.* The fibers of the inner smooth muscle layer (adjacent to the submucosa) encircle the tube. Contraction of these circular fibers decreases the diameter of the lumen, constricting the tube at the point of contraction. Contraction of the fibers in the outer layer, which run longitudinally along the length of the tube, shortens the tube. Together, contractile activity of these smooth muscle layers produces the propulsive and mixing movements. Another nerve network, the *myenteric plexus,* lies between the two muscle layers (*myo* means "muscle"; *enteric* means "intestine"). Together the submucosal and myenteric plexuses, along with hormones and local chemical mediators, help regulate local gut activity.

SEROSA The outer connective tissue covering of the digestive tract is the **serosa,** which secretes a watery, slippery fluid (**serous fluid**) that lubricates and prevents friction between the digestive organs and surrounding viscera. Throughout much of the tract, the serosa is continuous with the **mesentery,** which suspends the digestive organs from the inner wall of the abdominal cavity like a sling (● Figure 16-2). This attachment provides relative fixation, supporting the digestive organs in proper position, while still allowing them freedom for mixing and propulsive movements.

Regulation of digestive function is complex and synergistic.

Digestive motility and secretion are carefully regulated to maximize digestion and absorption of ingested food. Four factors are involved in regulating digestive system function: (1) autonomous smooth muscle function, (2) intrinsic nerve plexuses, (3) extrinsic nerves, and (4) gastrointestinal hormones.

AUTONOMOUS SMOOTH MUSCLE FUNCTION Like self-excitable cardiac muscle cells, some specialized smooth muscle cells are pacemaker cells that display rhythmic, spontaneous variations in membrane potential. The prominent type of self-induced electrical activity in digestive smooth muscle is **slow-wave potentials** (see p. 295), alternatively referred to as the digestive tract's **basic electrical rhythm (BER).** Musclelike but noncontractile cells known as the **interstitial cells of Cajal** are the pacemaker cells that instigate cyclic slow-wave activity. These pacemaker cells lie at the boundary between the longitudinal and circular smooth muscle layers. Slow waves are not action potentials and do not directly induce muscle contraction; they are rhythmic, wavelike fluctuations in membrane potential that cyclically bring the

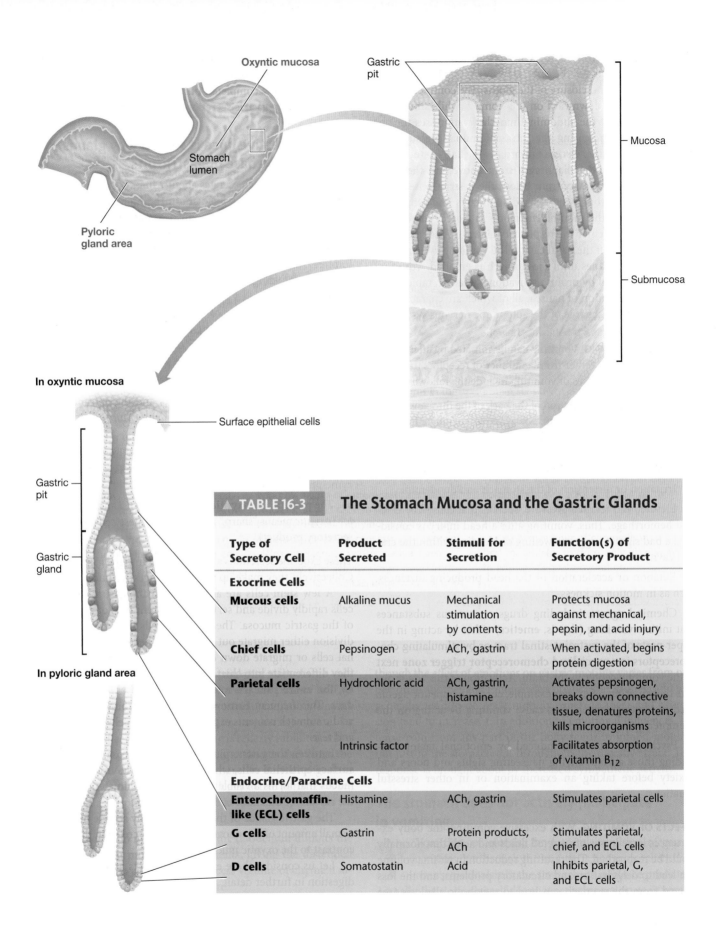

Oxyntic mucosa

Gastric pit

Stomach lumen

Pyloric gland area

Mucosa

Submucosa

In oxyntic mucosa

Surface epithelial cells

Gastric pit

Gastric gland

In pyloric gland area

▲ TABLE 16-3	The Stomach Mucosa and the Gastric Glands		
Type of Secretory Cell	**Product Secreted**	**Stimuli for Secretion**	**Function(s) of Secretory Product**
Exocrine Cells			
Mucous cells	Alkaline mucus	Mechanical stimulation by contents	Protects mucosa against mechanical, pepsin, and acid injury
Chief cells	Pepsinogen	ACh, gastrin	When activated, begins protein digestion
Parietal cells	Hydrochloric acid	ACh, gastrin, histamine	Activates pepsinogen, breaks down connective tissue, denatures proteins, kills microorganisms
	Intrinsic factor		Facilitates absorption of vitamin B_{12}
Endocrine/Paracrine Cells			
Enterochromaffin-like (ECL) cells	Histamine	ACh, gastrin	Stimulates parietal cells
G cells	Gastrin	Protein products, ACh	Stimulates parietal, chief, and ECL cells
D cells	Somatostatin	Acid	Inhibits parietal, G, and ECL cells

falls as low as 2. Hydrogen ion (H^+) and chloride ion (Cl^-) are actively transported by separate pumps in the parietal cells' plasma membrane. Hydrogen ion is actively transported against a tremendous concentration gradient, with the H^+ concentration being as much as 3 million times greater in the lumen than in the blood. Chloride is secreted by a secondary active-transport mechanism against a much smaller concentration gradient of only 1.5 times.

MECHANISM OF H^+ AND Cl^- SECRETION The secreted H^+ is not transported from the plasma but is derived instead from metabolic processes within the parietal cells (● Figure 16-9). Specifically, the H^+ to be secreted is derived from the breakdown of H_2O molecules into H^+ and OH^- (hydroxyl ions) within the parietal cells. This H^+ is secreted into the lumen by a H^+–K^+ ATPase pump in the parietal cell's luminal membrane. This primary active-transport carrier also pumps K^+ into the cell from the lumen (see p. 69). The transported K^+ then passively leaks back into the lumen through luminal K^+ channels, thus leaving K^+ levels unchanged by the process of H^+ secretion.

The parietal cells contain an abundance of the enzyme carbonic anhydrase (ca) (see pp. 496 and 572). In the presence of carbonic anhydrase, the OH^- generated by the breakdown of H_2O readily combines with CO_2 (which either has been produced within the parietal cell by metabolic processes or has diffused in from the blood) to form HCO_3^-.

The generated HCO_3^- is moved into the plasma by a Cl^-–HCO_3^- antiporter in the basolateral membrane of the parietal cells (see p. 71). Driven by the HCO_3^- gradient, this carrier moves HCO_3^- out of the cell into the plasma down its electrochemical gradient and simultaneously transports Cl^- from the plasma into the parietal cell against its electrochemical gradient. By building up the concentration of Cl^- inside the parietal cell, the Cl^-–HCO_3^- antiporter establishes a Cl^- concentration gradient between the parietal cell and gastric lumen. Because of this concentration gradient and because the cell interior is negative compared to the luminal contents, the negatively charged Cl^- pumped into the cell by the basolateral antiporter diffuses out of the cell down its electrochemical gradient through Cl^- channels in the luminal membrane into the gastric lumen, completing the Cl^- secretory process. In the meantime, the blood leaving the stomach is alkaline, because HCO_3^- has been added to it.

FUNCTIONS OF HCl Although HCl does not actually digest anything (that is, it does not break apart nutrient chemical bonds), it performs these specific functions that aid digestion:

1. HCl activates the enzyme precursor pepsinogen to an active enzyme, pepsin, and provides an acid medium that is optimal for pepsin activity.

2. It aids in the breakdown of connective tissue and muscle fibers, reducing large food particles into smaller particles.

KEY

(ATP) →	= Active transport	**ca**	= Carbonic anhydrase
		→	= Passive diffusion
⬭	= Secondary active transport	→	= Chemical reaction

● **FIGURE 16-9 Mechanism of HCl secretion.** The stomach's parietal cells actively secrete H^+ and Cl^- by the actions of two separate pumps. Hydrogen ion is secreted into the lumen by a primary H^+–K^+ ATPase active-transport pump at the parietal cell's luminal border. The K^+ transported into the cell by the pump promptly exits through a luminal K^+ channel, thus being recycled between the cell and lumen. The secreted H^+ is derived from the breakdown of H_2O into H^+ and OH^-. Catalyzed by carbonic anhydrase, the OH^- combines with CO_2 (that is either metabolically produced in the cell or diffuses in from the plasma) to form HCO_3^-. Chloride is secreted by secondary active transport. Driven by the HCO_3^- concentration gradient, a Cl^-–HCO_3^- antiporter in the basolateral membrane transports HCO_3^- down its concentration gradient into the plasma and simultaneously transports Cl^- into the parietal cell against its concentration gradient. Chloride secretion is completed as the Cl^- that entered from the plasma diffuses out of the cell down its electrochemical gradient through a luminal Cl^- channel into the lumen.

3. It denatures protein; that is, it uncoils proteins from their highly folded final form, thus exposing more of the peptide bonds for enzymatic attack.

4. Along with salivary lysozyme, HCl kills most of the microorganisms ingested with food, although some do escape and continue to grow and multiply in the large intestine.

Pepsinogen, once activated, begins protein digestion.

The major digestive constituent of gastric secretion is **pepsinogen,** an inactive enzymatic molecule produced by the chief cells. Pepsinogen is stored in the chief cells' cytoplasm within

GASTRIC PHASE The *gastric phase of gastric secretion* begins when food actually reaches the stomach. Stimuli acting in the stomach—namely *protein*, especially peptide fragments; *distension; caffeine;* and *alcohol*—increase gastric secretion by overlapping efferent pathways. For example, protein in the stomach, the most potent stimulus, stimulates chemoreceptors that activate the intrinsic nerve plexuses, which in turn stimulate the secretory cells. Furthermore, protein brings about activation of the extrinsic vagal fibers to the stomach. Vagal activity further enhances intrinsic nerve stimulation of the secretory cells and triggers the release of gastrin. Protein also directly stimulates the release of gastrin. Gastrin, in turn, is a powerful stimulus for further HCl and pepsinogen secretion and also calls forth release of histamine, which further increases HCl secretion. Through these synergistic and overlapping pathways, protein induces the secretion of a highly acidic, pepsin-rich gastric juice, which continues the digestion of the protein that first initiated the process (▲ Table 16-4).

When the stomach is distended with protein-rich food that needs to be digested, these secretory responses are appropriate. Caffeine and, to lesser extent, alcohol also stimulate the secretion of a highly acidic gastric juice, even when no food is present. This unnecessary acid can irritate the linings of the stomach and duodenum. For this reason, people with ulcers or gastric hyperacidity should avoid caffeinated and alcoholic beverages.

INTESTINAL PHASE The *intestinal phase of gastric secretion* encompasses the factors originating in the small intestine that influence gastric secretion. Whereas the other phases are excitatory, this phase is inhibitory. The intestinal phase is important in helping shut off the flow of gastric juices as chyme begins to be emptied into the small intestine, a topic to which we now turn.

Gastric secretion gradually decreases as food empties from the stomach into the intestine.

You now know what factors turn on gastric secretion before and during a meal, but how is the flow of gastric juices shut off when they are no longer needed? Gastric secretion is gradually reduced in three different ways as the stomach empties (▲ Table 16-5):

- As the meal is gradually emptied into the duodenum, the major stimulus for enhanced gastric secretion—the presence of protein in the stomach—is withdrawn.

- After foods leave the stomach, gastric juices accumulate to such an extent that gastric pH falls very low. This fall in pH within the stomach lumen comes about largely because food proteins that had been buffering HCl are no longer present in the lumen as the stomach empties. (Recall that proteins serve as excellent buffers; see p. 575.) Somatostatin is released in response to this high gastric acidity (pH less than 3). In negative-feedback fashion, gastric secretion declines as a result of somatostatin's inhibitory effects.

- The same stimuli that inhibit gastric motility (fat, acid, hypertonicity, or distension in the duodenum brought about by the emptying of stomach contents into the duodenum) inhibit gastric secretion as well. The enterogastric reflex and the enterogastrones suppress the gastric secretory cells while they simultaneously reduce the excitability of the gastric smooth muscle cells. This inhibitory response is the intestinal phase of gastric secretion.

The gastric mucosal barrier protects the stomach lining from gastric secretions.

How can the stomach contain strong acid contents and proteolytic enzymes without destroying itself? You already learned that mucus provides a protective physical coating. Furthermore,

▲ TABLE 16-5	Inhibition of Gastric Secretion

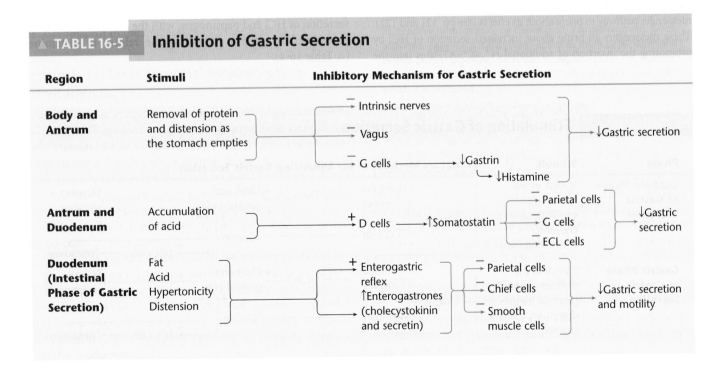

Region	Stimuli	Inhibitory Mechanism for Gastric Secretion
Body and Antrum	Removal of protein and distension as the stomach empties	Intrinsic nerves / Vagus / G cells → ↓Gastrin → ↓Histamine → ↓Gastric secretion
Antrum and Duodenum	Accumulation of acid	D cells → ↑Somatostatin → Parietal cells / G cells / ECL cells → ↓Gastric secretion
Duodenum (Intestinal Phase of Gastric Secretion)	Fat, Acid, Hypertonicity, Distension	Enterogastric reflex ↑Enterogastrones (cholecystokinin and secretin) → Parietal cells / Chief cells / Smooth muscle cells → ↓Gastric secretion and motility

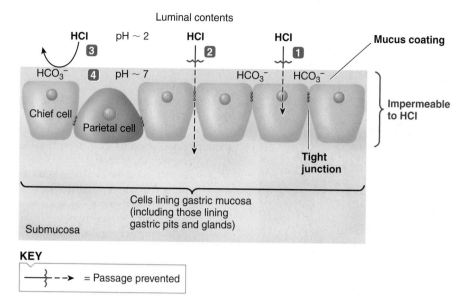

Luminal contents

HCl   pH ~ 2   HCl   HCl   Mucus coating

HCO₃⁻ [4] pH ~ 7   HCO₃⁻ HCO₃⁻

Chief cell
Parietal cell

Impermeable
to HCl

Tight
junction

Cells lining gastric mucosa
(including those lining
gastric pits and glands)

Submucosa

KEY

⊣⟩--→ = Passage prevented

The components of the gastric mucosal barrier enable the stomach to contain acid without injuring itself:

1 The luminal membranes of the gastric mucosal cells are impermeable to H⁺ so that HCl cannot penetrate into the cells.

2 The cells are joined by tight junctions that prevent HCl from penetrating between them.

3 A mucus coating over the gastric mucosa serves as a physical barrier to acid penetration.

4 The HCO₃⁻-rich mucus also serves as a chemical barrier that neutralizes acid in the vicinity of the mucosa. Even when luminal pH is 2, the mucus pH is 7.

● **FIGURE 16-11 Gastric mucosal barrier.**

the surface mucus-secreting cells secrete HCO_3^- that is trapped in the mucus and neutralizes acid in the vicinity. In addition, other barriers to mucosal acid damage are provided by the mucosal lining itself. First, the luminal membranes of the gastric mucosal cells are almost impermeable to H^+, so acid cannot penetrate into the cells and damage them. Furthermore, the lateral edges of these cells are joined near their luminal borders by tight junctions, so acid cannot diffuse between the cells from the lumen into the underlying submucosa (see p. 59). The properties of the gastric mucosa that enable the stomach to contain acid without injuring itself constitute the **gastric mucosal barrier** (● Figure 16-11). These protective mechanisms are further enhanced by the fact that the entire stomach lining is replaced every three days. Because of rapid mucosal turnover, cells are usually replaced before they are exposed to the wear and tear of harsh gastric conditions long enough to suffer damage.

Clinical Note Despite the protection provided by mucus, by the gastric mucosal barrier, and by the frequent turnover of cells, the barrier occasionally is broken and the gastric wall is injured by its acidic and enzymatic contents. When this occurs, an erosion, or **peptic ulcer,** of the stomach wall results. Excessive gastric reflux into the esophagus and dumping of excessive acidic gastric contents into the duodenum can lead to peptic ulcers in these locations as well. (For a further discussion of ulcers, see the boxed feature on p. 612, ■ Concepts, Challenges, and Controversies.)

We now turn to the remaining two digestive processes in the stomach, gastric digestion and absorption.

Carbohydrate digestion continues in the body of the stomach; protein digestion begins in the antrum.

Two separate digestive processes take place within the stomach. In the body of the stomach, food remains in a semisolid mass because peristaltic contractions in this region are too weak for

mixing to occur. Because food is not mixed with gastric secretions in the body of the stomach, very little protein digestion occurs here. In the interior of the mass, however, carbohydrate digestion continues under the influence of salivary amylase. Even though acid inactivates salivary amylase, the unmixed interior of the food mass is free of acid.

Digestion by the gastric juice itself is accomplished in the antrum of the stomach, where the food is thoroughly mixed with HCl and pepsin, beginning protein digestion.

The stomach absorbs alcohol and aspirin but no food.

No food or water is absorbed into the blood through the stomach mucosa. However, two noteworthy nonnutrient substances are absorbed directly by the stomach—*ethyl alcohol* and *aspirin*. Alcohol is somewhat lipid soluble, so it can diffuse through the lipid membranes of the epithelial cells that line the stomach and can enter the blood through the submucosal capillaries. Although alcohol can be absorbed by the gastric mucosa, it can be absorbed even more rapidly by the small-intestine mucosa because the surface area for absorption in the small intestine is much greater than in the stomach. Thus, alcohol absorption occurs more slowly if gastric emptying is delayed so that the alcohol remains in the more slowly absorbing stomach longer. Because fat is the most potent duodenal stimulus for inhibiting gastric motility, consuming fat-rich foods (for example, whole milk, pizza, or nuts) before or during alcohol ingestion delays gastric emptying and prevents the alcohol from producing its effects as rapidly.

Clinical Note Another category of substances absorbed by the gastric mucosa includes weak acids, most notably *acetylsalicylic acid* (aspirin). In the highly acidic environment of the stomach lumen, weak acids are almost totally un-ionized; that is, the H^+ and associated anion of the acid are bound together. In an un-ionized form, these weak acids are lipid soluble,

Ulcers: When Bugs Break the Barrier

Peptic ulcers are erosions that typically begin in the mucosal lining of the stomach and may penetrate into the deeper layers of the stomach wall. They occur when the gastric mucosal barrier is disrupted, and thus pepsin and HCl act on the stomach wall instead of food in the lumen. Frequent backflow of acidic gastric juices into the esophagus or excess unneutralized acid from the stomach in the duodenum can lead to peptic ulcers in these sites as well.

In a surprising discovery in the early 1990s, the bacterium *Helicobacter pylori* was pinpointed as the cause of more than 80% of all peptic ulcers. Thirty percent of the population in the United States harbors *H. pylori*. Those who have this slow bacterium have a 3 to 12 times greater risk of developing an ulcer within 10 to 20 years of acquiring the infection than those without the bacterium. They are also at increased risk of developing stomach cancer.

For years, scientists had overlooked the possibility that ulcers could be triggered by an infectious agent, because bacteria typically cannot survive in a strongly acidic environment such as the stomach lumen. An exception, *H. pylori* exploits several strategies to survive in this hostile environment. First, these organisms are motile, being equipped with four to six flagella (whiplike appendages; see the accompanying figure), which enable them to tunnel through and take up residence under the stomach's thick layer of alkaline mucus. Here they are protected from the highly acidic gastric contents. Furthermore, *H. pylori* preferentially settles in the antrum, which has no acid-producing parietal cells, although HCl from the upper parts of the stomach does reach the antrum. Also, these bacteria produce urease, an enzyme that breaks down urea, an end product of protein metabolism, into ammonia (NH_3) and CO_2. Ammonia serves as a buffer (see p. 580) that neutralizes stomach acid locally in the vicinity of the *H. pylori*.

H. pylori contributes to ulcer formation in part by secreting toxins that cause a persistent inflammation, or *chronic superficial*

Helicobacter pylori. *Helicobacter pylori,* the bacterium responsible for most cases of peptic ulcers, has flagella that enable it to tunnel beneath the protective layer of mucus that coats the stomach lining.

A. B. Dowsett/Science Photo Library/Photo Researchers, Inc.

gastritis, at the site it colonizes. *H. pylori* further weakens the gastric mucosal barrier by disrupting the tight junctions between the gastric epithelial cells, thereby making the gastric mucosa leakier than normal.

Alone or in conjunction with this infectious culprit, other factors are known to contribute to ulcer formation. Frequent exposure to some chemicals can break the gastric mucosal barrier; the most important of these are ethyl alcohol and nonsteroidal anti-inflammatory drugs (NSAIDs), such as aspirin, ibuprofen, or more potent medications for the treatment of arthritis or other chronic inflammatory processes. The barrier frequently breaks in patients with preexisting debilitating conditions, such as severe injuries or infections. Persistent stressful situations are frequently associated with ulcer formation, presumably because emotional response to stress can stimulate excessive gastric secretion.

When the gastric mucosal barrier is broken, acid and pepsin diffuse into the mucosa and underlying submucosa, with serious pathophysiologic consequences. The surface erosion, or ulcer, progressively enlarges as increasing levels of acid and pepsin continue to damage the stomach wall. Two of the most serious consequences of ulcers are (1) hemorrhage resulting from damage to submucosal capillaries and (2) perforation, or complete erosion through the stomach wall, resulting in the escape of potent gastric contents into the abdominal cavity.

Treatment of ulcers includes antibiotics, H-2 histamine receptor blockers, and proton pump inhibitors. With the discovery of the infectious component of most ulcers, antibiotics are now a treatment of choice. The other drugs are also used alone or in combination with antibiotics.

Two decades before the discovery of *H. pylori*, researchers discovered an antihistamine *(cimetidine)* that specifically blocks H-2 receptors, the type of receptors that bind histamine released from the stomach. These receptors differ from H-1 receptors, which bind the histamine involved in allergic respiratory disorders. Accordingly, traditional antihistamines used for respiratory allergies (such as hay fever and asthma) are not effective against ulcers, nor is cimetidine useful for respiratory problems. Because histamine potentiates the acid-promoting actions of ACh and gastrin, treatment with H-2 histamine blockers significantly suppresses acid secretion despite the fact that they do not directly interfere with the actions of these other two stimulatory messengers.

Another recent class of drugs used in treating ulcers inhibits acid secretion by directly blocking the pump that transports H^+ into the stomach lumen. These so-called proton-pump inhibitors (H^+ is a naked proton without its electron) help reduce the corrosive effect of HCl on the exposed tissue.

so they can be absorbed quickly by crossing the plasma membranes of the epithelial cells that line the stomach. Most other drugs are not absorbed until they reach the small intestine, so they do not begin to take effect as quickly.

Having completed our coverage of the stomach, we will move to the next part of the digestive tract, the small intestine and the accessory digestive organs that release their secretions into the small-intestine lumen.

Pancreatic and Biliary Secretions

When gastric contents are emptied into the small intestine, they are mixed not only with juice secreted by the small-intestine mucosa but also with the secretions of the exocrine pancreas and liver that are released into the duodenal lumen. We will discuss the roles of each of these accessory digestive organs before we examine the contributions of the small intestine itself.

The pancreas is a mixture of exocrine and endocrine tissue.

The **pancreas** is an elongated gland that lies behind and below the stomach, above the first loop of the duodenum (● Figure 16-12). This mixed gland contains both exocrine and endocrine tissue. The predominant exocrine part consists of grapelike clusters of secretory cells that form sacs known as **acini,** which connect to ducts that eventually empty into the duodenum. The smaller endocrine part consists of isolated islands of endocrine tissue, the **islets of Langerhans,** which are dispersed throughout the pancreas. The most important hormones secreted by the islet cells are insulin and glucagon (Chapter 19). The exocrine and endocrine tissues of the pancreas are derived from different tissues during embryonic development and share only their location in common. Although both are involved with the metabolism of nutrient molecules, they have different functions under the control of different regulatory mechanisms.

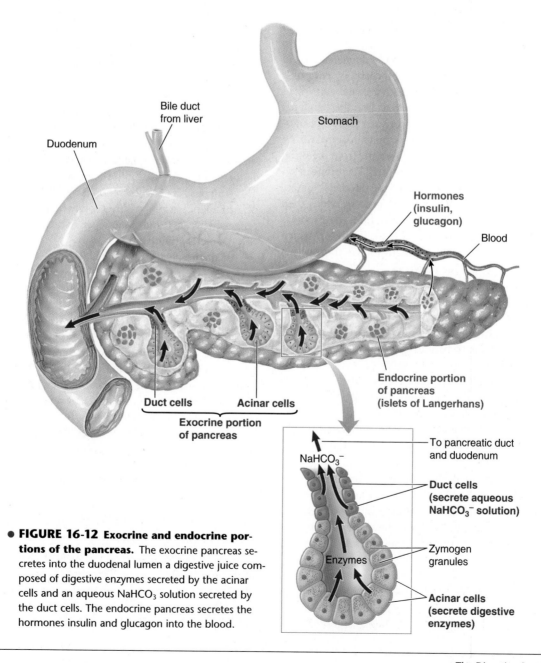

● **FIGURE 16-12 Exocrine and endocrine portions of the pancreas.** The exocrine pancreas secretes into the duodenal lumen a digestive juice composed of digestive enzymes secreted by the acinar cells and an aqueous NaHCO₃ solution secreted by the duct cells. The endocrine pancreas secretes the hormones insulin and glucagon into the blood.

The exocrine pancreas secretes digestive enzymes and an aqueous alkaline fluid.

The **exocrine pancreas** secretes a pancreatic juice consisting of two components: (1) *pancreatic enzymes* actively secreted by the *acinar cells* that form the acini and (2) an *aqueous alkaline solution* actively secreted by the *duct cells* that line the pancreatic ducts. The aqueous (watery) alkaline component is rich in sodium bicarbonate ($NaHCO_3$).

Like pepsinogen, pancreatic enzymes are stored within zymogen granules after being produced, then are released by exocytosis as needed. These pancreatic enzymes are important because they can almost completely digest food in the absence of all other digestive secretions. The acinar cells secrete three different types of pancreatic enzymes capable of digesting all three categories of foodstuffs: (1) **proteolytic enzymes** for protein digestion, (2) **pancreatic amylase** for carbohydrate digestion, and (3) **pancreatic lipase** for fat digestion.

PANCREATIC PROTEOLYTIC ENZYMES The three major pancreatic proteolytic enzymes are *trypsinogen, chymotrypsinogen,* and *procarboxypeptidase,* each of which is secreted in an inactive form. When **trypsinogen** is secreted into the duodenal lumen, it is activated to its active enzyme form, **trypsin,** by **enterokinase** (also known as **enteropeptidase**), an enzyme embedded in the luminal membrane of the cells that line the duodenal mucosa. Trypsin then autocatalytically activates more trypsinogen. Like pepsinogen, trypsinogen must remain inactive within the pancreas to prevent this proteolytic enzyme from digesting the proteins of the cells in which it is formed. Trypsinogen remains inactive, therefore, until it reaches the duodenal lumen, where enterokinase triggers the activation process, which then proceeds autocatalytically. As further protection, the pancreas also produces a chemical known as **trypsin inhibitor,** which blocks trypsin's actions if spontaneous activation of trypsinogen inadvertently occurs within the pancreas.

Chymotrypsinogen and **procarboxypeptidase,** the other pancreatic proteolytic enzymes, are converted by trypsin to their active forms, **chymotrypsin** and **carboxypeptidase,** respectively, within the duodenal lumen. Thus, once enterokinase has activated some of the trypsin, trypsin then carries out the rest of the activation process.

Each of these proteolytic enzymes attacks different peptide linkages. The end products that result from this action are a mixture of small peptide chains and amino acids. Mucus secreted by the intestinal cells protects against digestion of the small-intestine wall by the activated proteolytic enzymes.

PANCREATIC AMYLASE Like salivary amylase, pancreatic amylase contributes to carbohydrate digestion by converting polysaccharides into the disaccharide maltose. Amylase is secreted in the pancreatic juice in an active form because active amylase does not endanger the secretory cells. These cells do not contain any polysaccharides.

PANCREATIC LIPASE Pancreatic lipase is extremely important because it is the only enzyme secreted throughout the entire digestive system that can digest fat. (In humans, insignificant amounts of lipase are secreted in the saliva and gastric juice— *lingual lipase* and *gastric lipase.*) Pancreatic lipase hydrolyzes dietary triglycerides into monoglycerides and free fatty acids, which are the absorbable units of fat. Like amylase, lipase is secreted in its active form because there is no risk of pancreatic self-digestion by lipase. Triglycerides are not a structural component of pancreatic cells.

PANCREATIC INSUFFICIENCY When pancreatic enzymes are deficient, digestion of food is incomplete. Because the pancreas is the only significant source of lipase, pancreatic enzyme deficiency results in serious maldigestion of fats. The main clinical manifestation of pancreatic exocrine insufficiency is **steatorrhea,** or excessive undigested fat in the feces. Up to 60% to 70% of the ingested fat may be excreted in the feces. Digestion of protein and carbohydrates is impaired to a lesser degree because salivary, gastric, and small-intestinal enzymes contribute to the digestion of these two foodstuffs.

PANCREATIC AQUEOUS ALKALINE SECRETION Pancreatic enzymes function best in a neutral or slightly alkaline environment, yet the highly acidic gastric contents are emptied into the duodenal lumen in the vicinity of pancreatic enzyme entry into the duodenum. This acidic chyme must be neutralized quickly in the duodenal lumen, not only to allow optimal functioning of the pancreatic enzymes but also to prevent acid damage to the duodenal mucosa. The alkaline ($NaHCO_3$-rich) fluid secreted by the pancreatic duct cells into the duodenal lumen serves the important function of neutralizing the acidic chyme as the latter is emptied into the duodenum from the stomach. This aqueous $NaHCO_3$ secretion is by far the largest component of pancreatic secretion. The volume of pancreatic secretion ranges between 1 and 2 liters per day, depending on the type and degree of stimulation.

All of the details of pancreatic $NaHCO_3$ secretion are not known, but carbonic anhydrase is involved and the Na^+-K^+ pump provides the driving energy for secondary active transport. According to the current model, under the influence of carbonic anhydrase, CO_2 in the pancreatic duct cell combines with OH^- generated from H_2O to produce HCO_3^-, which exits across the luminal membrane to enter the duct lumen via a $Cl^--HCO_3^-$ antiporter. Sodium diffuses through the "leaky" tight junctions into the lumen. Together these actions accomplish $NaHCO_3$ secretion. The H^+ generated from H_2O within the duct cell enters the blood across the basolateral border either by active transport or a Na^+-H^+ antiporter. Thus the pancreatic duct cells secrete HCO_3^- and absorb H^+, whereas the gastric parietal cells secrete H^+ and absorb HCO_3^-, so the acid−base status of the body is not altered by digestive secretion.

Pancreatic exocrine secretion is regulated by secretin and CCK.

Pancreatic exocrine secretion is regulated primarily by hormonal mechanisms. A small amount of parasympathetically induced pancreatic secretion occurs during the cephalic phase of digestion, with a further token increase occurring during the gastric phase in response to gastrin. However, the predominant stimulation of pancreatic secretion occurs during the intestinal phase of digestion when chyme is in the small intestine. The release of the two major enterogastrones, secretin and cholecystokinin (CCK), in response to chyme in the duodenum plays the central role in controlling pancreatic secretion (● Figure 16-13).

ROLE OF SECRETIN IN PANCREATIC SECRETION Of the factors that stimulate enterogastrone release (fat, acid, hypertonicity, and distension), the primary stimulus specifically for secretin release is acid in the duodenum. Secretin, in turn, is carried by the blood to the pancreas, where it stimulates the duct cells to markedly increase their secretion of a $NaHCO_3$-rich aqueous fluid into the duodenum. Even though other stimuli may cause the release of secretin, it is appropriate that the most potent stimulus is acid, because secretin promotes the alkaline pancreatic secretion that neutralizes the acid. This mechanism provides a control system for maintaining neutrality of the chyme in the intestine. The amount of secretin released is proportional to the amount of acid that enters the duodenum, so the amount of $NaHCO_3$ secreted parallels the duodenal acidity.

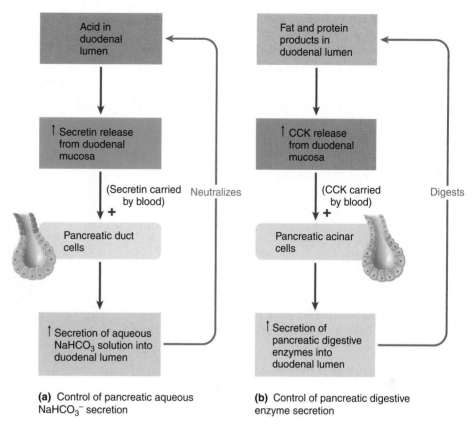

(a) Control of pancreatic aqueous $NaHCO_3^-$ secretion

(b) Control of pancreatic digestive enzyme secretion

● **FIGURE 16-13 Hormonal control of pancreatic exocrine secretion.**

ROLE OF CCK IN PANCREATIC SECRETION Cholecystokinin is important in regulating pancreatic digestive enzyme secretion. The main stimulus for release of CCK from the duodenal mucosa is the presence of fat and, to a lesser extent, protein products. The circulatory system transports CCK to the pancreas where it stimulates the pancreatic acinar cells to increase digestive enzyme secretion. Among these enzymes are lipase and the proteolytic enzymes, which appropriately further digest the fat and protein that initiated the response and also help digest carbohydrate. In contrast to fat and protein, carbohydrate does not have any direct influence on pancreatic digestive enzyme secretion.

All three types of pancreatic digestive enzymes are packaged together in the zymogen granules, so all the pancreatic enzymes are released together on exocytosis of the granules. Therefore, even though the *total amount* of enzymes released varies depending on the type of meal consumed (the most being secreted in response to fat), the *proportion* of enzymes released does not vary on a meal-to-meal basis. That is, a high-protein meal does not cause the release of a greater proportion of proteolytic enzymes. Evidence suggests, however, that long-term adjustments in the proportion of the types of enzymes produced may occur as an adaptive response to a prolonged change in diet. For example, with a long-term switch to a high-

protein diet, a greater proportion of proteolytic enzymes are produced. Cholecystokinin may play a role in pancreatic digestive enzyme adaptation to changes in diet.

Just as gastrin is trophic to the stomach and small intestine, CCK and secretin exert trophic effects on the exocrine pancreas to maintain its integrity.

We will now look at the contributions of the remaining accessory digestive unit, the liver and gallbladder.

The liver performs various important functions, including bile production.

Besides pancreatic juice, the other secretory product emptied into the duodenal lumen is **bile.** The **biliary system** includes the *liver,* the *gallbladder,* and associated ducts.

LIVER FUNCTIONS The **liver** is the largest and most important metabolic organ in the body; it can be viewed as the body's major biochemical factory. Its importance to the digestive system is its secretion of *bile salts,* which aid fat digestion and absorption. The liver also performs a wide variety of functions not related to digestion, including the following:

1. Metabolic processing of the major categories of nutrients (carbohydrates, proteins, and lipids) after their absorption from the digestive tract (see p. 632)

2. Detoxifying or degrading body wastes and hormones as well as drugs and other foreign compounds (see p. 27)

3. Synthesizing plasma proteins, including those needed for blood clotting (see p. 393), those that transport steroid and thyroid hormones and cholesterol in the blood (see pp. 120 and 336), and angiotensinogen important in the salt-conserving renin–angiotensin–aldosterone system (see p. 527)

4. Storing glycogen, fats, iron, copper, and many vitamins (see p. 42)

5. Activating vitamin D, which the liver does in conjunction with the kidneys (see p. 734)

6. Removing bacteria and worn-out red blood cells, thanks to its resident macrophages

7. Secreting the hormones thrombopoietin (stimulates platelet production; see p. 405), hepcidin (inhibits iron uptake from the intestine; see p. 632), and insulin-like growth factor-I (stimulates growth; see p. 678)

8. Producing acute phase proteins important in inflammation (see p. 424)

9. Excreting cholesterol and bilirubin, the latter being a breakdown product derived from the destruction of worn-out red blood cells (see p. 619).

Given this wide range of complex functions, there is amazingly little specialization among cells within the liver. Each liver cell, or **hepatocyte,** performs the same wide variety of metabolic and secretory tasks (*hepato* means "liver"; *cyte* means "cell"). The specialization comes from the highly developed organelles within each hepatocyte. The only liver function not accomplished by the hepatocytes is the phagocytic activity carried out by the resident macrophages, which are known as **Kupffer cells.**

LIVER BLOOD FLOW To carry out these wide-ranging tasks, the anatomic organization of the liver permits each hepatocyte to be in direct contact with blood from two sources: arterial blood coming from the aorta and venous blood coming directly from the digestive tract. Like other cells, the hepatocytes receive fresh arterial blood via the hepatic artery, which supplies their oxygen and delivers blood-borne metabolites for hepatic processing. Venous blood also enters the liver by the **hepatic portal system,** a unique and complex vascular connection between the digestive tract and liver (● Figure 16-14). The veins draining the digestive tract do not directly join the inferior vena cava, the large vein that returns blood to the heart. Instead, the veins from the stomach and intestine enter the hepatic portal vein, which carries the products absorbed from the digestive tract directly to the liver for processing, storage, or detoxification before they gain access to the general circulation. Within the liver, the portal vein once again breaks up into a capillary network (the liver *sinusoids*) to permit exchange between the blood and hepatocytes before draining into the hepatic vein, which joins the inferior vena cava.

The liver lobules are delineated by vascular and bile channels.

The liver is organized into functional units known as **lobules,** which are hexagonal arrangements of tissue surrounding a central vein (● Figure 16-15a). At each of the six outer corners of

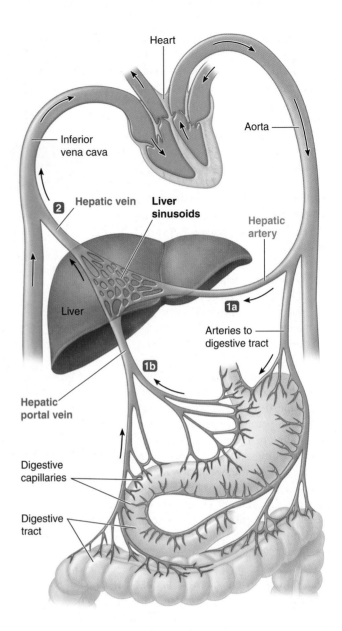

The liver receives blood from two sources:

1a Arterial blood, which provides the liver's O_2 supply and carries blood-borne metabolites for hepatic processing, is delivered by the **hepatic artery**.

1b Venous blood draining the digestive tract is carried by the **hepatic portal vein** to the liver for processing and storage of newly absorbed nutrients.

2 Blood leaves the liver via the **hepatic vein**.

● **FIGURE 16-14 Schematic representation of liver blood flow.**

the lobule are three vessels: a branch of the hepatic artery, a branch of the hepatic portal vein, and a bile duct. Blood from the branches of both the hepatic artery and the portal vein flows from the periphery of the lobule into large, expanded capillary spaces called **sinusoids** (see p. 364), which run between rows of liver cells to the central vein like spokes on a bicycle wheel (● Figure 16-15b). The Kupffer cells line the sinusoids and en-

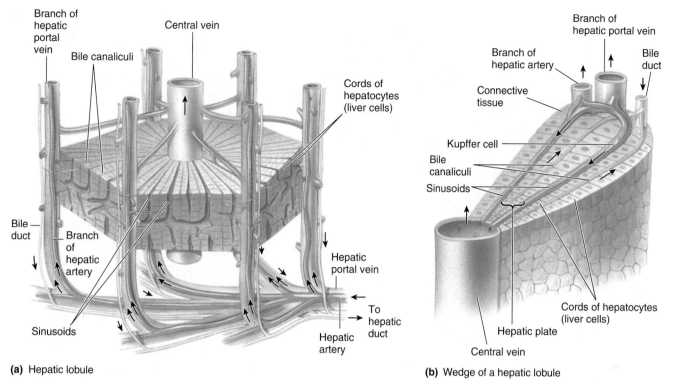

Branch of hepatic portal vein

Bile canaliculi

Central vein

Cords of hepatocytes (liver cells)

Bile duct

Branch of hepatic artery

Sinusoids

Hepatic portal vein

To hepatic duct

Hepatic artery

(a) Hepatic lobule

Branch of hepatic portal vein

Branch of hepatic artery

Bile duct

Connective tissue

Kupffer cell

Bile canaliculi

Sinusoids

Cords of hepatocytes (liver cells)

Hepatic plate

Central vein

(b) Wedge of a hepatic lobule

● **FIGURE 16-15 Anatomy of the liver.**

gulf and destroy old red blood cells and bacteria that pass through in the blood. The hepatocytes are arranged between the sinusoids in plates two cell layers thick, so that each lateral edge faces a sinusoidal pool of blood. The central veins of all the liver lobules converge to form the hepatic vein, which carries the blood away from the liver. The thin bile-carrying channel, a **bile canaliculus,** runs between the cells within each hepatic plate. Hepatocytes continuously secrete bile into these thin channels, which carry the bile to a bile duct at the periphery of the lobule. The bile ducts from the various lobules converge to eventually form the *common bile duct,* which transports the bile from the liver to the duodenum. Each hepatocyte is in contact with a sinusoid on one side and a bile canaliculus on the other side.

Bile is continuously secreted by the liver and is diverted to the gallbladder between meals.

The opening of the bile duct into the duodenum is guarded by the **sphincter of Oddi,** which prevents bile from entering the duodenum except during digestion of meals (● Figure 16-16). When this sphincter is closed, most of the bile secreted by the liver is diverted back up into the **gallbladder,** a small, saclike structure tucked beneath but not directly connected to the liver. Thus, bile is not transported directly from the liver to the gallbladder. The bile is subsequently stored and concentrated in the gallbladder between meals. After a meal, bile enters the duodenum as a result of the combined effects of gallbladder emptying and increased bile secretion by the liver. The amount of bile secreted per day ranges from 250 ml to 1 liter, depending on the degree of stimulation.

Bile salts are recycled through the enterohepatic circulation.

Bile contains several organic constituents, namely *bile salts, cholesterol, lecithin* (a phospholipid), and *bilirubin* (all derived from hepatocyte activity) in an *aqueous alkaline fluid* (added by the duct cells) similar to the pancreatic $NaHCO_3$ secretion. Even though bile does not contain any digestive enzymes, it is important for the digestion and absorption of fats, primarily through the activity of bile salts.

Bile salts are derivatives of cholesterol. They are actively secreted into the bile and eventually enter the duodenum along with the other biliary constituents. Following their participation in fat digestion and absorption, most bile salts are reabsorbed into the blood by special active-transport mechanisms located in the terminal ileum. From here bile salts are returned by the hepatic portal system to the liver, which resecretes them into the bile. This recycling of bile salts (and some of the other biliary constituents) between the small intestine and liver is called the **enterohepatic circulation** (*entero* means "intestine"; *hepatic* means "liver") (● Figure 16-16).

The total amount of bile salts in the body averages about 3 to 4 g, yet 3 to 15 g of bile salts may be emptied into the duodenum in a single meal. On average, bile salts cycle between the liver and small intestine twice during the digestion of a typical meal. Usually, only about 5% of the secreted bile escapes into the feces daily. These lost bile salts are replaced by new bile salts synthesized by the liver; thus, the size of the pool of bile salts is kept constant.

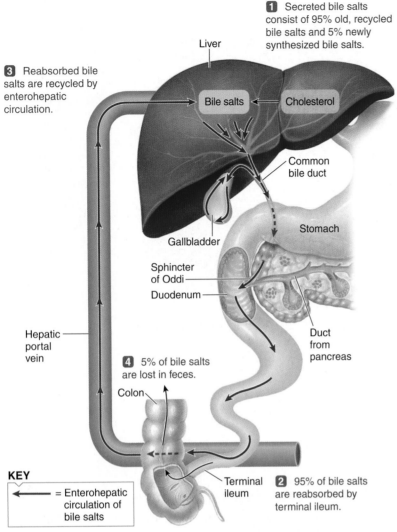

Liver

3 Reabsorbed bile salts are recycled by enterohepatic circulation.

Bile salts ← Cholesterol

Common bile duct

Stomach

Gallbladder

Sphincter of Oddi

Duodenum

Hepatic portal vein

Duct from pancreas

4 5% of bile salts are lost in feces.

Colon

KEY

⟵ = Enterohepatic circulation of bile salts

Terminal ileum

2 95% of bile salts are reabsorbed by terminal ileum.

● **FIGURE 16-16 Enterohepatic circulation of bile salts.** The majority of bile salts are recycled between the liver and small intestine through the enterohepatic circulation (blue arrows). After participating in fat digestion and absorption, most bile salts are reabsorbed by active transport in the terminal ileum and returned through the hepatic portal vein to the liver, which resecretes them in the bile.

Bile salts aid fat digestion and absorption.

Bile salts aid fat digestion through their detergent action (emulsification) and facilitate fat absorption by participating in the formation of micelles. Both functions are related to the structure of bile salts. Let us see how.

DETERGENT ACTION OF BILE SALTS The term **detergent action** refers to bile salts' ability to convert large fat globules into a **lipid emulsion** consisting of many small fat droplets suspended in the aqueous chyme, thus increasing the surface area available for attack by pancreatic lipase. Fat globules, no matter their size, are made up primarily of undigested triglyceride molecules. To digest fat, lipase must come into direct contact with the triglyceride molecule. Because triglycerides are not soluble in water,

they tend to aggregate into large droplets in the watery environment of the small-intestine lumen. If bile salts did not emulsify these large droplets, lipase could act on the triglyceride molecules only at the surface of the large droplets, and fat digestion would be greatly prolonged.

Bile salts exert a detergent action similar to that of the detergent you use to break up grease when you wash dishes. A bile salt molecule contains a lipid-soluble part (a steroid derived from cholesterol) plus a negatively charged, water-soluble part. Bile salts *adsorb* on the surface of a fat droplet; that is, the lipid-soluble part of the bile salt dissolves in the fat droplet, leaving the charged water-soluble part projecting from the surface of the droplet (● Figure 16-17a). Intestinal mixing movements break up large fat droplets into smaller ones. These small droplets would quickly recoalesce were it not for bile salts adsorbing on their surface and creating a shell of water-soluble negative charges on the surface of each little droplet. Because like charges repel, these negatively charged groups on the droplet surfaces cause the fat droplets to repel one another (● Figure 16-17b). This electrical repulsion prevents the small droplets from recoalescing into large fat droplets and thus produces a lipid emulsion that increases the surface area available for lipase action. The small emulsified fat droplets range in diameter from 200 to 5000 nm, with an average of about 1000 nm (1 μm).

Although bile salts increase the surface area available for attack by pancreatic lipase, lipase alone cannot penetrate the layer of bile salts adsorbed on the surface of the small emulsified fat droplets. To solve this dilemma, the pancreas secretes the polypeptide **colipase** along with lipase. Like bile salts, colipase has both a lipid-soluble part and a water-soluble part. Colipase displaces some bile salts and lodges at the surface of the fat droplets, where it binds to lipase, thus anchoring this enzyme to its site of action amid the bile-salt coating.

MICELLAR FORMATION Bile salts—along with cholesterol and lecithin, which are also constituents of bile—play an important role in facilitating fat absorption through micellar formation. Like bile salts, lecithin (a phospholipid similar to the ones in the lipid bilayer of the plasma membrane) has both a lipid-soluble and a water-soluble part, whereas cholesterol is almost totally insoluble in water. In a **micelle,** the bile salts and lecithin aggregate in small clusters with their fat-soluble parts huddled together in the middle to form a hydrophobic ("water-fearing") core, while their water-soluble parts form an outer hydrophilic ("water-loving") shell (● Figure 16-18). A micelle is 3 to 10 nm in diameter, compared to an average diameter of 1000 nm for an emulsified lipid droplet. Micelles are water soluble by virtue of their hydrophilic shells, but they can dissolve water-insoluble (and hence lipid-soluble) substances in their lipid-soluble cores.

Micelles thus provide a handy vehicle for carrying water-insoluble substances through the watery luminal contents. The most important lipid-soluble substances carried within micelles are the products of fat digestion (monoglycerides and free fatty acids) as well as fat-soluble vitamins, which are all transported to their sites of absorption by this means. If they did not hitch a ride in the water-soluble micelles, these nutrients would float on the surface of the aqueous chyme (just as oil floats on top of water), never reaching the absorptive surfaces of the small intestine.

In addition, cholesterol, a highly water-insoluble substance, dissolves in the micelle's hydrophobic core. This mechanism is important in cholesterol homeostasis. The amount of cholesterol that can be carried in micellar formation depends on the relative amount of bile salts and lecithin in comparison to cholesterol.

Clinical Note When cholesterol secretion by the liver is out of proportion to bile salt and lecithin secretion (either too much cholesterol or too little bile salts and lecithin), the excess cholesterol in the bile precipitates into microcrystals that can aggregate into **gallstones.** One treatment of cholesterol-containing gallstones involves ingestion of bile salts to increase the bile salt pool in an attempt to dissolve the cholesterol stones. Only about 75% of gallstones are derived from cholesterol, however. The other 25% are made up of abnormal precipitates of another bile constituent, bilirubin.

Bilirubin is a waste product excreted in the bile.

Bilirubin, the other major constituent of bile, does not play a role in digestion at all but instead is a waste product excreted in the bile. Bilirubin is the primary bile pigment derived from the breakdown of worn-out red blood cells. The typical life span of a red blood cell in the circulatory system is 120 days. Worn-out red blood cells are removed from the blood by the macrophages that line the liver sinusoids and reside in other areas in the body. Bilirubin is the end product from degradation of the heme (iron-containing) part of the hemoglobin contained within these old red blood cells (see p. 393). This bilirubin is extracted from the blood by the hepatocytes and is actively excreted into the bile. Bilirubin is not entirely a useless waste product. Scientists recently discovered that it is a fleeting but potent antioxidant. Because bilirubin is lipid soluble whereas other natural antioxidants in the body are water soluble, it may play a role in protecting lipid membranes from free radical damage (see p. 142).

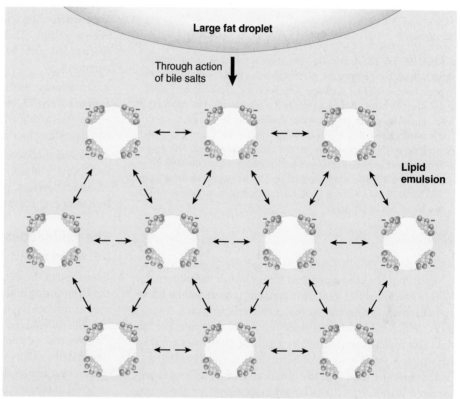

(a) Structure of bile salts and their adsorption on the surface of a small lipid droplet

(b) Formation of a lipid emulsion through the action of bile salts

● **FIGURE 16-17 Schematic structure and function of bile salts.** (a) A bile salt consists of a lipid-soluble part that dissolves in the fat droplet and a negatively charged, water-soluble part that projects from the surface of the droplet. (b) When a large fat droplet is broken up into smaller fat droplets by intestinal contractions, bile salts adsorb on the surface of the small droplets, creating shells of negatively charged, water-soluble bile salt components that cause the fat droplets to repel one another. This emulsifying action holds the fat droplets apart and prevents them from recoalescing, increasing the surface area of exposed fat available for digestion by pancreatic lipase.

KEY

All lipid-soluble	Water-soluble portion / Lipid-soluble portion	Water-soluble portion / Lipid-soluble portion
Cholesterol	**Bile salt**	**Lecithin**

● **FIGURE 16-18 A micelle.** Bile constituents (bile salts, lecithin, and cholesterol) aggregate to form micelles that consist of a hydrophilic (water-soluble) shell and a hydrophobic (lipid-soluble) core. Because the outer shell of a micelle is water soluble, the products of fat digestion, which are not water soluble, can be carried through the watery luminal contents to the absorptive surface of the small intestine by dissolving in the micelle's lipid-soluble core. This figure is not drawn to scale compared to the lipid emulsion droplets in ● Figure 16-17b. An emulsified fat droplet ranges in diameter from 200 to 5000 nm (average 1000 nm) compared to a micelle, which is 3 to 10 nm in diameter.

Bilirubin is a yellow pigment that gives bile its yellow color. Within the intestinal tract, this pigment is modified by bacterial enzymes, giving rise to the characteristic brown color of feces. When bile secretion does not occur, as when the bile duct is completely obstructed by a gallstone, the feces are grayish white. A small amount of bilirubin is normally reabsorbed by the intestine back into the blood, and when it is eventually excreted in the urine, it is largely responsible for the urine's yellow color. The kidneys cannot excrete bilirubin until after it has been modified during its passage through the liver and intestine.

Clinical Note If bilirubin is formed more rapidly than it can be excreted, it accumulates in the body and causes **jaundice.** Patients with this condition appear yellowish, with this color being seen most easily in the whites of their eyes. Jaundice can be brought about in three different ways:

1. *Prehepatic* (the problem occurs "before the liver"), or *hemolytic, jaundice* arises from excessive breakdown (hemolysis) of red blood cells, which results in the liver being presented with more bilirubin than it is capable of excreting.

2. *Hepatic* (the problem is the "liver") *jaundice* occurs when the liver is diseased and cannot deal with even the normal load of bilirubin.

3. *Posthepatic* (the problem occurs "after the liver"), or *obstructive, jaundice* occurs when the bile duct is obstructed, such

as by a gallstone, so that bilirubin cannot be eliminated in the feces.

Bile salts are the most potent stimulus for increased bile secretion.

Bile secretion may be increased by chemical, hormonal, and neural mechanisms:

■ *Chemical mechanism (bile salts).* Any substance that increases bile secretion by the liver is called a **choleretic.** The most potent choleretic is bile salts themselves. Between meals, bile is stored in the gallbladder, but during a meal bile is emptied into the duodenum as the gallbladder contracts. After bile salts participate in fat digestion and absorption, they are reabsorbed and returned by the enterohepatic circulation to the liver, where they act as potent choleretics to stimulate further bile secretion. Therefore, during a meal, when bile salts are needed and being used, bile secretion by the liver is enhanced.

■ *Hormonal mechanism (secretin).* Besides increasing the aqueous $NaHCO_3$ secretion by the pancreas, secretin stimulates an aqueous alkaline bile secretion by the liver ducts without any corresponding increase in bile salts.

■ *Neural mechanism (vagus nerve).* Vagal stimulation of the liver plays a minor role in bile secretion during the cephalic phase of digestion, promoting an increase in liver bile flow before food ever reaches the stomach or intestine.

The gallbladder stores and concentrates bile between meals and empties during meals.

Even though the factors just described increase bile secretion by the liver during and after a meal, bile secretion by the liver occurs continuously. Between meals the secreted bile is shunted into the gallbladder, where it is stored and concentrated. Active transport of salt out of the gallbladder, with water following osmotically, results in a 5 to 10 times concentration of the organic constituents. The gallbladder can hold up to a maximum of 50 ml of bile, but the liver may secrete more than that volume of bile during a long fast between meals (that is, between periods of gallbladder emptying). However, because the gallbladder concentrates the bile, it can accommodate the greater volume secreted by the liver.

Clinical Note Because the gallbladder stores this concentrated bile, it is the primary site for precipitation of concentrated bile constituents into gallstones. Fortunately, the gallbladder does not play an essential digestive role, so its removal as a treatment for gallstones or other gallbladder disease presents no particular problem. The bile secreted between meals is stored instead in the common bile duct, which becomes dilated.

During digestion of a meal, when chyme reaches the small intestine, the presence of food, especially fat products, in the duodenal lumen triggers the release of CCK. This hormone stimulates contraction of the gallbladder and relaxation of the sphincter of Oddi, so bile is discharged into the duodenum, where it appropriately aids in the digestion and absorption of the fat that initiated the release of CCK.

Hepatitis and cirrhosis are the most common liver disorders.

Hepatitis is an inflammatory disease of the liver that results from a variety of causes, including viral infection, obesity (*fatty liver disease),* or most commonly exposure to toxic agents, including alcohol, carbon tetrachloride, and certain tranquilizers. Hepatitis ranges in severity from mild, reversible symptoms to acute massive liver damage, with possible imminent death resulting from acute hepatic failure.

Repeated or prolonged hepatic inflammation, usually in association with chronic alcoholism, can lead to **cirrhosis,** a condition in which damaged hepatocytes are permanently replaced by connective tissue. Liver tissue has the ability to regenerate, normally undergoing a gradual turnover of cells. If part of the hepatic tissue is destroyed, the lost tissue can be replaced by an increase in the rate of cell division. There is a limit, however, to how rapidly hepatocytes can be replaced. In addition to hepatocytes, a small number of fibroblasts (connective tissue cells) are dispersed between the hepatic plates and form a supporting framework for the liver. If the liver is exposed to toxic substances such as alcohol so often that new hepatocytes cannot be generated rapidly enough to replace the damaged cells, the sturdier fibroblasts take advantage of the situation and overproduce. This extra connective tissue leaves little space for the hepatocytes' regrowth. Thus, as cirrhosis develops slowly over time, active liver tissue is gradually reduced, leading eventually to chronic liver failure. Fueled by exciting new discoveries revealing the underlying mechanisms of hepatic fibrosis, scientists are currently working on ways to hopefully prevent and even reverse cirrhosis.

Having looked at the accessory digestive organs that empty their exocrine products into the small-intestine lumen, we now examine the contributions of the small intestine itself.

Small Intestine

The **small intestine** is the site where most digestion and absorption take place. No further digestion is accomplished after the luminal contents pass beyond the small intestine, and no further absorption of ingested nutrients occurs, although the large intestine does absorb small amounts of salt and water. The small intestine lies coiled within the abdominal cavity, extending between the stomach and large intestine. It is arbitrarily divided into three segments—the **duodenum,** the **jejunum,** and the **ileum.**

As usual, we will examine motility, secretion, digestion, and absorption in the small intestine, in that order. Small-intestine motility includes *segmentation* and the *migrating motility complex.* Let us consider segmentation first.

Segmentation contractions mix and slowly propel the chyme.

Segmentation, the small intestine's primary method of motility during digestion of a meal, both mixes and slowly propels the chyme. Segmentation consists of oscillating, ringlike contrac-

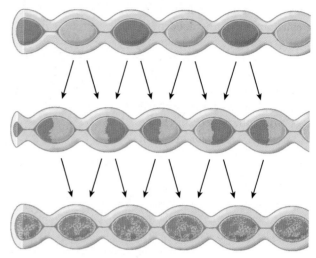

● **FIGURE 16-19 Segmentation.** Segmentation consists of ring-like contractions along the length of the small intestine. Within a matter of seconds, the contracted segments relax and the previously relaxed areas contract. These oscillating contractions thoroughly mix the chyme within the small-intestine lumen.

tions of the circular smooth muscle along the small intestine's length; between the contracted segments are relaxed areas containing a small bolus of chyme. The contractile rings occur every few centimeters, dividing the small intestine into segments like a chain of sausages. These contractile rings do not sweep along the length of the intestine as peristaltic waves do. Rather, after a brief period of time, the contracted segments relax, and ringlike contractions appear in the previously relaxed areas (● Figure 16-19). The new contraction forces the chyme in a previously relaxed segment to move in both directions into the now relaxed adjacent segments. A newly relaxed segment therefore receives chyme from both the contracting segment immediately ahead of it and the one immediately behind it. Shortly thereafter, the areas of contraction and relaxation alternate again. In this way, the chyme is chopped, churned, and thoroughly mixed. These contractions can be compared to squeezing a pastry tube with your hands to mix the contents.

INITIATION AND CONTROL OF SEGMENTATION Segmentation contractions are initiated by the small intestine's pacemaker cells, which produce a basic electrical rhythm (BER) similar to the gastric BER that governs peristalsis in the stomach. If the small-intestine BER brings the circular smooth muscle layer to threshold, segmentation contractions are induced, with the frequency of segmentation following the frequency of the BER.

The circular smooth muscle's degree of responsiveness and thus the intensity of segmentation contractions can be influenced by distension of the intestine, by the hormone gastrin, and by extrinsic nerve activity. All these factors influence the excitability of the small-intestine smooth muscle cells by moving the starting potential around which the BER oscillates closer to or farther from the threshold. Segmentation is slight or absent between meals but becomes very vigorous immediately after a meal. Both the duodenum and the ileum start to segment simultaneously when the meal first enters the small intestine. The duodenum starts to segment primarily in response to local distension caused by the presence of chyme. Segmentation of the empty ileum, in contrast, is brought about by gastrin secreted in response to the presence of chyme in the stomach, a mechanism known as the **gastroileal reflex.** Extrinsic nerves can modify the strength of these contractions. Parasympathetic stimulation enhances segmentation, whereas sympathetic stimulation depresses segmental activity.

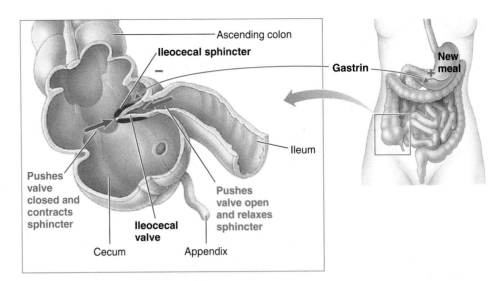

● **FIGURE 16-20 Control of the ileocecal valve/sphincter.** The juncture between the ileum and large intestine is the ileocecal valve, which is surrounded by thickened smooth muscle, the ileocecal sphincter. Pressure on the cecal side pushes the valve closed and contracts the sphincter, preventing the bacteria-laden colonic contents from contaminating the nutrient-rich small intestine. The valve/sphincter opens and allows ileal contents to enter the large intestine in response to pressure on the ileal side of the valve and to the hormone gastrin secreted as a new meal enters the stomach.

FUNCTIONS OF SEGMENTATION

The mixing accomplished by segmentation serves the dual functions of mixing the chyme with the digestive juices secreted into the small-intestine lumen and exposing all the chyme to the absorptive surfaces of the small-intestine mucosa.

Segmentation not only accomplishes mixing but also slowly moves chyme through the small intestine. How can this be, when each segmental contraction propels chyme both forward and backward? The chyme slowly progresses forward because the frequency of segmentation declines along the length of the small intestine. The pacemaker cells in the duodenum spontaneously depolarize faster than those farther down the tract, with segmentation contractions occurring in the duodenum at a rate of 12 per minute, compared to only 9 per minute in the terminal ileum. Because segmentation occurs with greater frequency in the upper part of the small intestine than in the lower part, more chyme, on average, is pushed forward than is pushed backward. As a result, chyme is moved very slowly from the upper to the lower part of the small intestine, being shuffled back and forth to accomplish thorough mixing and absorption in the process. This slow propulsive mechanism is advantageous because it allows ample time for the digestive and absorptive processes to take place. The contents usually take 3 to 5 hours to move through the small intestine.

The migrating motility complex sweeps the intestine clean between meals.

When most of the meal has been absorbed, segmentation contractions cease and are replaced between meals by the **migrating motility complex,** or **"intestinal housekeeper."** This between-meal motility consists of weak, repetitive peristaltic waves that move a short distance down the intestine before dying out. The waves start at the stomach and migrate down the intestine; that is, each new peristaltic wave is initiated at a site a little farther down the small intestine. These short peristaltic waves take about 100 to 150 minutes to gradually migrate from the stomach to the end of the small intestine, with each contraction sweeping any remnants of the preceding meal plus mucosal debris and bacteria forward toward the colon, just like a good "intestinal housekeeper." After the end of the small intestine is reached, the cycle begins again and continues to repeat itself until the next meal. The migrating motility complex is regulated between meals by the hormone **motilin,** which is secreted during the unfed state by endocrine cells of the small-intestine mucosa. When the next meal arrives, segmental activity is triggered again, and the migrating motility complex ceases. Motilin release is inhibited by feeding.

The ileocecal juncture prevents contamination of the small intestine by colonic bacteria.

At the juncture between the small and large intestines, the last part of the ileum empties into the cecum (● Figure 16-20). Two factors contribute to this region's ability to act as a barrier between the small and large intestines. First, the anatomic arrangement is such that valvelike folds of tissue protrude from the ileum into the lumen of the cecum. When the ileal contents are pushed forward, this **ileocecal valve** is easily pushed open, but the folds of tissue are forcibly closed when the cecal contents attempt to move backward. Second, the smooth muscle within the last several centimeters of the ileal wall is thickened, forming a sphincter that is under neural and hormonal control. Most of the time this **ileocecal sphincter** remains at least mildly constricted. Pressure on the cecal side of the sphincter causes it to contract more forcibly; distension of the ileal side causes the sphincter to relax, a reaction mediated by the intrinsic plexuses in the area. In this way, the ileocecal juncture prevents the bacteria-laden contents of the large intestine from contaminating the small intestine and at the same time lets the ileal contents pass into the colon. If the colonic bacteria gained access to the nutrient-rich small intestine, they would multiply rapidly. Relaxation of the sphincter is enhanced by the release of gastrin at the onset of a meal, when increased gastric activity is taking place. This relaxation allows the undigested fibers and unabsorbed solutes from the preceding meal to be moved forward as the new meal enters the tract.

Small-intestine secretions do not contain any digestive enzymes.

Each day the exocrine gland cells in the small-intestine mucosa secrete into the lumen about 1.5 liters of an aqueous salt and mucus solution called **succus entericus** ("juice of intestine"). Secretion increases after a meal in response to local stimulation of the small-intestine mucosa by the presence of chyme.

The mucus in the secretion provides protection and lubrication. Furthermore, this aqueous secretion provides plenty of H_2O to participate in the enzymatic digestion of food. Recall that digestion involves hydrolysis—bond breakage by reaction with H_2O—which proceeds most efficiently when all the reactants are in solution.

No digestive enzymes are secreted into this intestinal juice. The small intestine does synthesize digestive enzymes, but they act within the brush-border membrane of the epithelial cells that line the lumen instead of being secreted directly into the lumen.

The small-intestine enzymes complete digestion within the brush-border membrane.

Digestion within the small-intestine lumen is accomplished by the pancreatic enzymes, with fat digestion being enhanced by bile secretion. As a result of pancreatic enzymatic activity, fats are completely reduced to their absorbable units of monoglycerides and free fatty acids, proteins are broken down into small peptide fragments and some amino acids, and carbohydrates are reduced to disaccharides and some monosaccharides. Thus, fat digestion is completed within the small-intestine lumen, but carbohydrate and protein digestion have not been brought to completion.

Special hairlike projections on the luminal surface of the small-intestine epithelial cells, the **microvilli,** form the **brush border** (see p. 48). The brush-border plasma membrane contains three categories of integral proteins that function as enzymes:

1. **Enterokinase,** which activates the pancreatic proteolytic enzyme trypsinogen

2. The **disaccharidases (maltase, sucrase,** and **lactase),** which complete carbohydrate digestion by hydrolyzing the remaining disaccharides (maltose, sucrose, and lactose, respectively) into their constituent monosaccharides

3. The **aminopeptidases,** which hydrolyze most of the small peptide fragments into their amino acid components, thereby completing protein digestion

Thus, carbohydrate and protein digestion are completed within the confines of the brush border. (▲ Table 16-6 provides a summary of the digestive processes for the three major categories of nutrients.)

Clinical Note A fairly common disorder, **lactose intolerance,** involves a deficiency of lactase, the disaccharidase specific for the digestion of lactose, or milk sugar. Most children under 4 years of age have adequate lactase, but this may be gradually lost so that in many adults, lactase activity is diminished or absent. When lactose-rich milk or dairy products are consumed by a person with lactase deficiency, the undigested lactose remains in the lumen and has several related consequences. First, accumulation of undigested lactose creates an osmotic gradient that draws H_2O into the intestinal lumen. Second, bacteria living in the large intestine have lactose-splitting ability, so they eagerly attack the lactose as an energy source, producing large quantities of CO_2 and methane gas in the process. Distension of the intestine by both fluid and gas produces pain (cramping) and diarrhea. Infants with lactose intolerance may also suffer from malnutrition.

Finally we are ready to discuss absorption of nutrients. Up to this point, no food, water, or electrolytes have been absorbed.

The small intestine is remarkably well adapted for its primary role in absorption.

All products of carbohydrate, protein, and fat digestion, as well as most of the ingested electrolytes, vitamins, and water, are normally absorbed by the small intestine indiscriminately. Usually only the absorption of calcium and iron is adjusted to the body's needs. Thus, the more food consumed, the more that will be digested and absorbed, as people who are trying to control their weight are all too painfully aware.

Most absorption occurs in the duodenum and jejunum; very little occurs in the ileum, not because the ileum does not have absorptive capacity but because most absorption has already been accomplished before the intestinal contents reach the ileum. The small intestine has an abundant reserve absorptive capacity. About 50% of the small intestine can be removed with little interference to absorption—with one exception. If the terminal ileum is removed, vitamin B_{12} and bile salts are not properly absorbed, because the specialized transport mechanisms for these two substances are located only in this region. All other substances can be absorbed throughout the small intestine's length.

The mucous lining of the small intestine is remarkably well adapted for its special absorptive function for two reasons: (1) it has a very large surface area, and (2) the epithelial cells in this lining have a variety of specialized transport mechanisms.

ADAPTATIONS THAT INCREASE THE SMALL INTESTINE'S SURFACE AREA The following special modifications of the small-intestine mucosa greatly increase the surface area available for absorption (● Figure 16-21):

■ The inner surface of the small intestine is thrown into permanent circular folds that are visible to the naked eye and increase the surface area threefold.

Digestive Processes for the Three Major Categories of Nutrients

Nutrients	Enzymes for Digesting Nutrient	Source of Enzymes	Site of Action of Enzymes	Action of Enzymes	Absorbable Units of Nutrients
Carbohydrate	Amylase	Salivary glands	Mouth and (mostly) body of stomach	Hydrolyzes polysaccharides to disaccharides	
		Exocrine pancreas	Small-intestine lumen		
	Disaccharidases (maltase, sucrase, lactase)	Small-intestine epithelial cells	Small-intestine brush border	Hydrolyze disaccharides to monosaccharides	Monosaccharides, especially glucose
Protein	Pepsin	Stomach chief cells	Stomach antrum	Hydrolyzes protein to peptide fragments	
	Trypsin, chymotrypsin carboxypeptidase	Exocrine pancreas	Small-intestine lumen	Attack different peptide fragments	
	Aminopeptidases	Small-intestine epithelial cells	Small-intestine brush border	Hydrolyze peptide fragments to amino acids	Amino acids and a few small peptides
Fat	Lipase	Exocrine pancreas	Small-intestine lumen	Hydrolyzes triglycerides to fatty acids and monoglycerides	Fatty acids and monoglycerides
	Bile salts (not an enzyme)	Liver	Small-intestine lumen	Emulsify large fat globules for attack by pancreatic lipase	

■ Projecting from this folded surface are microscopic finger-like projections known as **villi,** which give the lining a velvety appearance and increase the surface area by another 10 times (● Figure 16-22). The surface of each villus is covered by epithelial cells interspersed occasionally with mucous cells.

■ Even smaller hairlike projections, the brush border or microvilli, arise from the luminal surface of these epithelial cells, increasing the surface area another 20-fold. Each epithelial cell has as many as 3000 to 6000 of these microvilli, which are visible only with an electron microscope. The small-intestine enzymes perform their functions within the membrane of this brush border.

Altogether, the folds, villi, and microvilli provide the small intestine with a luminal surface area 600 times greater than if it were a tube of the same length and diameter lined by a flat surface. In fact, if the surface area of the small intestine were spread out flat, it would cover an entire tennis court.

Clinical Note **Malabsorption** (impairment of absorption) may be caused by damage to or reduction of the surface area of the small intestine. One of the most common causes is **gluten enteropathy,** also known as **celiac disease.** In this condition, the person's small intestine is abnormally sensitive to *gluten,* a protein constituent of wheat, barley, and rye. These grain products are widely prevalent in processed foods. This condition is a complex immunologic disorder in which exposure to gluten erroneously activates a T-cell response that damages the intestinal villi: The normally luxuriant array of villi is reduced, the mucosa becomes flattened, and the brush border becomes short and stubby (● Figure 16-23). Because this loss of villi decreases the surface area available for absorption, absorption of all nutrients is impaired. The condition is treated by eliminating gluten from the diet.

STRUCTURE OF A VILLUS Absorption across the digestive tract wall involves transepithelial transport similar to movement of material across the kidney tubules (see p. 525). Each villus has the following major components (see ● Figure 16-21c):

■ *Epithelial cells that cover the surface of the villus.* The epithelial cells are joined at their lateral borders by tight junctions, which limit passage of luminal contents between the cells, although the tight junctions in the small intestine are leakier than those in the stomach. Within their luminal brush borders, these epithelial cells have carriers for absorption of specific nutrients and electrolytes from the lumen as well as the membrane-bound digestive enzymes that complete carbohydrate and protein digestion.

■ *A connective tissue core.* This core is formed by the lamina propria.

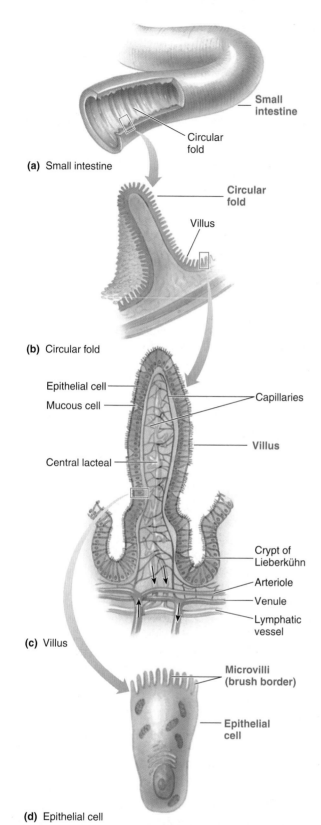

(a) Small intestine

Circular fold

(b) Circular fold

Circular fold

Villus

Epithelial cell

Mucous cell

Capillaries

Villus

Central lacteal

Crypt of Lieberkühn

Arteriole

Venule

Lymphatic vessel

(c) Villus

Microvilli (brush border)

Epithelial cell

(d) Epithelial cell

Small intestine

● **FIGURE 16-21 Small-intestine absorptive surface.** (a) Gross structure of the small intestine. (b) The circular folds of the small-intestine mucosa collectively increase the absorptive surface area threefold. (c) Microscopic fingerlike projections known as villi collectively increase the surface area another 10-fold. (d) Each epithelial cell on a villus has microvilli on its luminal border; the microvilli increase the surface area another 20-fold. Together, these surface modifications increase the small intestine's absorptive surface area 600-fold.

● **FIGURE 16-22 Scanning electron micrograph of villi projecting from the small-intestine mucosa.**

Brush border

(a) Normal

Brush border

(b) Gluten enteropathy

● **FIGURE 16-23 Reduction in the brush border with gluten enteropathy.** (a) Electron micrograph of the brush border of a small-intestine epithelial cell in a normal individual. (b) Electron micrograph of the short, stubby brush border of a small-intestine epithelial cell in a patient with gluten enteropathy.

(*Source:* Thomas W. Sheehy, M.D.; Robert L. Slaughter, M.D.: *The Malabsorption Syndrome* by Medcom, Inc. Reproduced by permission of Medcom, Inc.)

- *A capillary network.* Each villus is supplied by an arteriole that breaks up into a capillary network within the villus core. The capillaries rejoin to form a venule that drains away from the villus.

- *A terminal lymphatic vessel.* Each villus is supplied by a single blind-ended lymphatic vessel known as the **central lacteal,** which occupies the center of the villus core.

During the process of absorption, digested substances enter the capillary network or the central lacteal. To be absorbed, a substance must pass completely through the epithelial cell, diffuse through the interstitial fluid within the connective tissue core of the villus, and then cross the wall of a capillary or lymph vessel. Like renal transport, intestinal absorption may be active or passive, with active absorption involving energy expenditure during at least one of the transepithelial transport steps.

The mucosal lining experiences rapid turnover.

Dipping down into the mucosal surface between the villi are shallow invaginations known as the **crypts of Lieberkühn** (see ● Figure 16-21c). Unlike the gastric pits, these intestinal crypts do not secrete digestive enzymes, but they do secrete water and electrolytes, which, along with the mucus secreted by the cells on the villus surface, constitute the succus entericus.

Furthermore, the crypts function as nurseries. The epithelial cells lining the small intestine slough off and are replaced at a rapid rate as a result of high mitotic activity of *stem cells* in the crypts. New cells that are continually being produced in the crypts migrate up the villi and, in the process, push off the older cells at the tips of the villi into the lumen. In this manner, more than 100 million intestinal cells are shed per minute. The entire trip from crypt to tip averages about three days, so the epithelial lining of the small intestine is replaced approximately every three days. Because of this high rate of cell division, the crypt stem cells are very sensitive to damage by radiation and anticancer drugs, both of which may inhibit cell division.

The new cells undergo several changes as they migrate up the villus. The concentration of brush-border enzymes increases and the capacity for absorption improves, so the cells at the tip of the villus have the greatest digestive and absorptive capability. Just at their peak, these cells are pushed off by the newly migrating cells. Thus, the luminal contents are constantly exposed to cells that are optimally equipped to complete the digestive and absorptive functions efficiently. Furthermore, just as in the stomach, the rapid turnover of cells in the small intestine is essential because of the harsh luminal conditions. Cells exposed to the abrasive and corrosive luminal contents are easily damaged and cannot live for long, so they must be continually replaced by a fresh supply of newborn cells.

The old cells sloughed off into the lumen are not entirely lost to the body. These cells are digested, with the cell constituents being absorbed into the blood and reclaimed for synthesis of new cells, among other things.

In addition to the stem cells, Paneth cells are also found in the crypts. **Paneth cells** serve a defensive function, safeguarding the stem cells. They produce two chemicals that thwart bacteria: (1) *lysozyme,* the bacterial-lysing enzyme also found in saliva; and (2) *defensins,* small proteins with antimicrobial powers (see p. 454).

We now turn attention to the ways in which the epithelial lining of the small intestine is specialized to accomplish absorption of luminal contents and the mechanisms through which the specific dietary constituents are normally absorbed.

Energy-dependent Na$^+$ absorption drives passive H$_2$O absorption.

Sodium may be absorbed both passively and actively. When the electrochemical gradient favors movement of Na$^+$ from the lumen to the blood, passive diffusion of Na$^+$ can occur *between* the intestinal epithelial cells through the "leaky" tight junctions into the interstitial fluid within the villus. Movement of Na$^+$ *through* the cells is energy dependent and involves different carriers or channels at the luminal and basolateral membranes, similar to the process of Na$^+$ reabsorption across the kidney tubules (see pp. 526 and 530). Sodium enters the epithelial cells across the luminal border either by itself passively through Na$^+$ channels or in the company of another ion or a nutrient molecule by secondary active transport via three different carriers: Na$^+$–Cl- symporter, Na$^+$–H$^+$ antiporter, or Na$^+$–glucose (or amino acid) symporter. Sodium is actively pumped out of the cell by the Na$^+$–K$^+$ pump at the basolateral membrane into the interstitial fluid in the lateral spaces between the cells where they are not joined by tight junctions. From the interstitial fluid, Na$^+$ diffuses into the capillaries.

As with the renal tubules in the early part of the nephron, the absorption of Cl$^-$, H$_2$O, glucose, and amino acids from the small intestine is linked to this energy-dependent Na$^+$ absorption. Chloride passively follows down the electrical gradient created by Na$^+$ absorption and can be absorbed by secondary active transport as well if needed. Most H$_2$O absorption in the digestive tract depends on the active carrier that pumps Na$^+$ into the lateral spaces, resulting in a concentrated area of high osmotic pressure in that localized region between the cells, similar to the situation in the kidneys (see p. 532). This localized high osmotic pressure induces H$_2$O to move from the lumen through the cell (and possibly from the lumen through the leaky tight junction) into the lateral space. Water entering the space reduces the osmotic pressure but raises the hydrostatic (fluid) pressure. As a result, H$_2$O is flushed out of the lateral space into the interior of the villus, where it is picked up by the capillary network. Meanwhile, more Na$^+$ is pumped into the lateral space to encourage more H$_2$O absorption.

Carbohydrate and protein are both absorbed by secondary active transport and enter the blood.

Absorption of the digestion end products of both carbohydrates and proteins is accomplished by Na$^+$-dependent symport, and both categories of end products are absorbed into the blood.

CARBOHYDRATE ABSORPTION Dietary carbohydrate is presented to the small intestine for absorption mainly in the forms of the disaccharides maltose (the product of polysaccharide digestion), sucrose, and lactose (● Figure 16-24a). The disacchari-

(a) Carbohydrate digestion

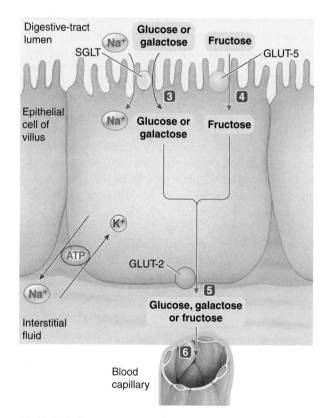

(b) Carbohydrate absorption

1 The dietary polysaccharides starch and glycogen are converted into the disaccharide maltose through the action of salivary and pancreatic amylase.

2 Maltose and the dietary disaccharides lactose and sucrose are converted to their respective monosaccharides by the disaccharidases (maltase, lactase, and sucrase) located in the brush borders of the small-intestine epithelial cells.

3 The monosaccharides glucose and galactose are absorbed into the epithelial cells by Na$^+$- and energy-dependent secondary active transport (via the symporter SGLT) located at the luminal membrane.

4 The monosaccharide fructose enters the cell by passive facilitated diffusion via GLUT-5.

5 Glucose, galactose, and fructose exit the cell at the basal membrane by passive facilitated diffusion via GLUT-2.

6 These monosaccharides enter the blood by simple diffusion.

KEY

(ATP)	= Active transport
	= Symport
	= Facilitated diffusion
→	= Simple diffusion

● **FIGURE 16-24 Carbohydrate digestion and absorption.**

dases located in the brush-border membrane of the small intestine cells further reduce these disaccharides into the absorbable monosaccharide units of glucose, galactose, and fructose.

Glucose and galactose are both absorbed by secondary active transport, in which symport carriers, such as the *sodium and glucose cotransporter (SGLT; see p. 73)* on the luminal membrane transport both the monosaccharide and Na^+ from the lumen into the interior of the intestinal cell (Figure 16-24b). The operation of these symporters, which do not directly use energy themselves, depends on the Na^+ concentration gradient established by the energy-consuming basolateral Na^+-K^+ pump (see p. 70). Glucose (or galactose), having been concentrated in the cell by these symporters, leaves the cell down its concentration gradient by facilitated diffusion (passive carrier-mediated transport; see p. 69) via the *glucose transporter GLUT-2* in the basal border to enter the blood within the villus. In addition to glucose being absorbed through the cells by means of the symporter, recent evidence suggests that a significant amount of glucose crosses the epithelial barrier through the leaky tight junctions between the epithelial cells.

Fructose is absorbed into the blood solely by facilitated diffusion. It enters the epithelial cells from the lumen via GLUT-5 and exits these cells to enter the blood via GLUT-2 (● Figure 16-24b).

PROTEIN ABSORPTION Both ingested proteins and endogenous (within the body) proteins that have entered the digestive tract lumen from the three following sources are digested and absorbed:

1. Digestive enzymes, all of which are proteins, that have been secreted into the lumen

2. Proteins within the cells that are pushed off from the villi into the lumen during the process of mucosal turnover

3. Small amounts of plasma proteins that normally leak from the capillaries into the digestive tract lumen

About 20 to 40 g of endogenous protein enter the lumen each day from these three sources. This quantity can amount to more than the quantity of protein in ingested food. All endogenous proteins must be digested and absorbed along with the dietary proteins to prevent depletion of the body's protein stores. The amino acids absorbed from both food and endogenous protein are used primarily to synthesize new protein in the body.

The protein presented to the small intestine for absorption is primarily in the form of amino acids and a few small peptide fragments (● Figure 16-25a). Amino acids are absorbed across the intestinal cells by symporters, similar to glucose and galactose absorption (● Figure 16-25b). The sugar symporters are distinct from the amino-acid symporters, and the amino-acid symporters are selective for different amino acids. Small peptides gain entry by means of yet another Na^+-dependent carrier in a process known as **tertiary active transport** (*tertiary* meaning "third," in reference to a third linked step ultimately being driven by energy used in the first step). In this case, the symporter simultaneously transports both H^+ and the peptide from the lumen into the cell, driven by H^+ moving down its concentration gradient and the peptide moving against its concentra-

tion gradient (● Figure 16-25b). The H^+ gradient is established by an antiporter in the luminal membrane that is driven by Na^+ moving into the cell down its concentration gradient and H^+ moving out of the cell against its concentration gradient. The Na^+ concentration gradient that drives the antiporter in turn is established by the energy-dependent Na^+-K^+ pump at the basolateral membrane. Thus, glucose, galactose, amino acids, and small peptides all get a "free ride" in on the energy expended for Na^+ transport. The small peptides are broken down into their constituent amino acids by the aminopeptidases in the brush-border membrane or by intracellular peptidases (● Figure 16-25a). Like monosaccharides, amino acids enter the capillary network within the villus.

Digested fat is absorbed passively and enters the lymph.

Fat absorption is quite different from carbohydrate and protein absorption because the insolubility of fat in water presents a special problem. Fat must be transferred from the watery chyme through the watery body fluids, even though fat is not water soluble. Therefore, fat must undergo a series of physical and chemical transformations to circumvent this problem during its digestion and absorption (● Figure 16-26).

A REVIEW OF FAT EMULSIFICATION AND DIGESTION When the stomach contents are emptied into the duodenum, the ingested fat is aggregated into large, oily triglyceride droplets that float in the chyme. Recall that through the bile salts' detergent action in the small-intestine lumen, the large droplets are dispersed into a lipid emulsification of small droplets, exposing a much greater surface area of fat for digestion by pancreatic lipase. The products of lipase digestion (monoglycerides and free fatty acids) are also not very water soluble, so very little of these end products of fat digestion can diffuse through the aqueous chyme to reach the absorptive lining. However, biliary components facilitate absorption of these fatty end products by forming micelles.

FAT ABSORPTION Remember that micelles are water-soluble particles that can carry the end products of fat digestion within their lipid-soluble interiors. Once these micelles reach the luminal membranes of the epithelial cells, the monoglycerides and free fatty acids passively diffuse from the micelles through the lipid component of the epithelial cell membranes to enter the interior of these cells. As these fat products leave the micelles and are absorbed across the epithelial cell membranes, the micelles can pick up more monoglycerides and free fatty acids, which have been produced from digestion of other triglyceride molecules in the fat emulsion.

Bile salts continuously repeat their fat-solubilizing function down the length of the small intestine until all fat is absorbed. Then the bile salts themselves are reabsorbed in the terminal ileum by special active transport. This is an efficient process because relatively small amounts of bile salts can facilitate digestion and absorption of large amounts of fat, with each bile salt performing its ferrying function repeatedly before it is reabsorbed.

(a) Protein digestion

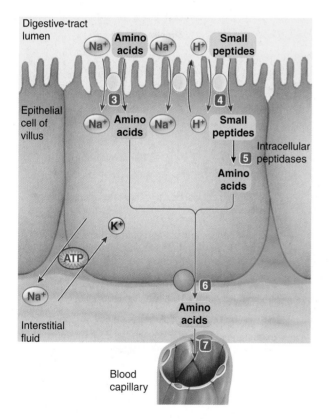

(b) Protein absorption

● **FIGURE 16-25 Protein digestion and absorption.**

1 Dietary and endogenous proteins are hydrolyzed into their constituent amino acids and a few small peptide fragments by gastric pepsin and the pancreatic proteolytic enzymes.

2 Many small peptides are converted into their respective amino acids by the aminopeptidases located in the brush borders of the small-intestine epithelial cells.

3 Amino acids are absorbed into the epithelial cells by means of Na$^+$- and energy-dependent secondary active transport via a symporter. Various amino acids are transported by carriers specific for them.

4 Some small peptides are absorbed by a different type of symporter driven by H$^+$, Na$^+$-, and energy-dependent tertiary active transport.

5 Most absorbed small peptides are broken down into their amino acids by intracellular peptidases.

6 Amino acids exit the cell at the basal membrane via various passive carriers.

7 Amino acids enter the blood by simple diffusion. (A small percentage of di- and tripeptides also enter the blood intact.)

KEY

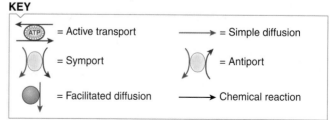

= Active transport   → = Simple diffusion

= Symport   = Antiport

= Facilitated diffusion   → = Chemical reaction

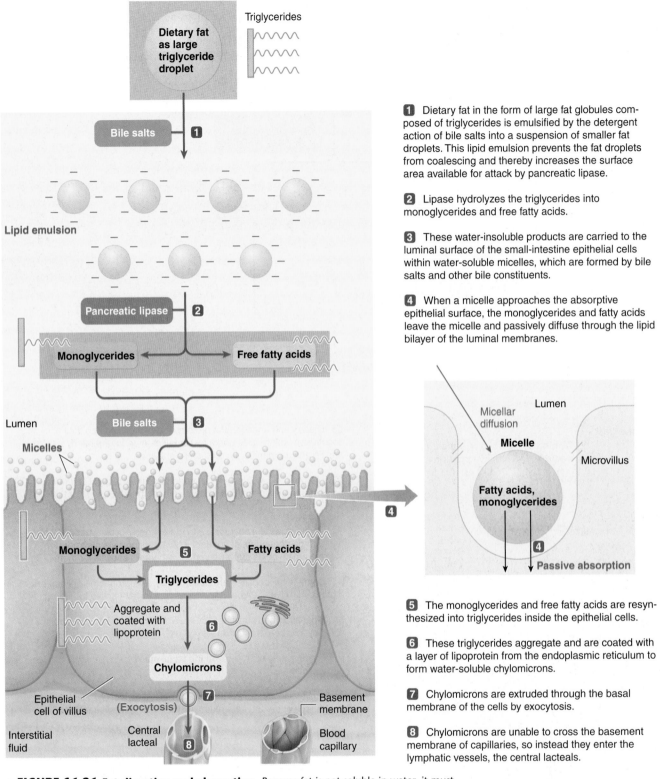

1 Dietary fat in the form of large fat globules composed of triglycerides is emulsified by the detergent action of bile salts into a suspension of smaller fat droplets. This lipid emulsion prevents the fat droplets from coalescing and thereby increases the surface area available for attack by pancreatic lipase.

2 Lipase hydrolyzes the triglycerides into monoglycerides and free fatty acids.

3 These water-insoluble products are carried to the luminal surface of the small-intestine epithelial cells within water-soluble micelles, which are formed by bile salts and other bile constituents.

4 When a micelle approaches the absorptive epithelial surface, the monoglycerides and fatty acids leave the micelle and passively diffuse through the lipid bilayer of the luminal membranes.

5 The monoglycerides and free fatty acids are resynthesized into triglycerides inside the epithelial cells.

6 These triglycerides aggregate and are coated with a layer of lipoprotein from the endoplasmic reticulum to form water-soluble chylomicrons.

7 Chylomicrons are extruded through the basal membrane of the cells by exocytosis.

8 Chylomicrons are unable to cross the basement membrane of capillaries, so instead they enter the lymphatic vessels, the central lacteals.

● **FIGURE 16-26 Fat digestion and absorption.** Because fat is not soluble in water, it must undergo a series of transformations in order to be digested and absorbed.

Once within the interior of the epithelial cells, the mono-glycerides and free fatty acids are resynthesized into triglycer-ides. These triglycerides conglomerate into droplets and are coated with a layer of lipoprotein (synthesized by the endoplas-mic reticulum of the epithelial cell), which makes the fat drop-lets water soluble. The large, coated fat droplets, known as **chylomicrons,** are extruded by exocytosis from the epithelial cells into the interstitial fluid within the villus. Chylomicrons are 75 to 500 nm in diameter, compared to micelles, which are 3 to 10 nm in diameter. The chylomicrons subsequently enter

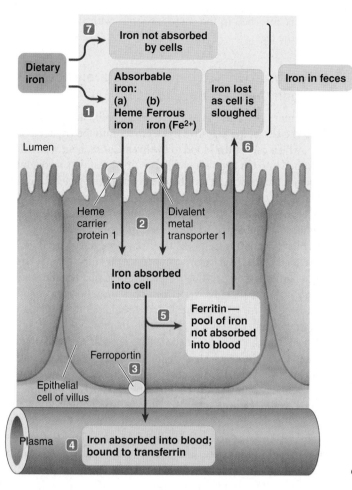

1 Only a portion of ingested iron is in a form that can be absorbed, either heme iron or ferrous iron (Fe^{2+}).

2 Iron is absorbed across the luminal membrane of small-intestine epithelial cells by different energy-dependent carriers for heme and Fe^{2+}.

3 Dietary iron that is absorbed into the small-intestine epithelial cells and is immediately needed for red blood cell production is transferred into the blood by the membrane iron transporter ferroportin.

4 In the blood, the absorbed iron is carried to the bone marrow bound to transferrin, a plasma protein carrier.

5 Absorbed dietary iron that is not immediately needed is stored in the epithelial cells as ferritin, which cannot be transferred into the blood.

6 This unused iron is lost in the feces as the ferritin-containing epithelial cells are sloughed.

7 Dietary iron that was not absorbed is also lost in the feces.

● **FIGURE 16-27 Iron absorption.**

the central lacteals rather than the capillaries because of the structural differences between these two vessels. Capillaries have a basement membrane (an outer layer of polysaccharides) (see p. 361) that prevents chylomicrons from entering, but the lymph vessels do not have this barrier. Thus, fat can be absorbed into the lymphatics but not directly into the blood.

The actual absorption or transfer of monoglycerides and free fatty acids from the chyme across the luminal membrane of the small-intestinal epithelial cells is traditionally considered a passive process because the lipid-soluble fatty end products merely dissolve in and pass through the lipid part of the membrane. However, the overall sequence of events needed for fat absorption does require energy. For example, bile salts are actively secreted by the liver, the resynthesis of triglycerides and formation of chylomicrons within the epithelial cells are active processes, and the exocytosis of chylomicrons requires energy.

Vitamin absorption is largely passive.

Water-soluble vitamins are primarily absorbed passively with water, whereas fat-soluble vitamins are carried in the micelles and absorbed passively with the end products of fat digestion. Some of the vitamins can also be absorbed by carriers, if necessary. Vitamin B_{12} is unique in that it must be in combination with gastric intrinsic factor for absorption by receptor-mediated endocytosis in the terminal ileum.

Iron and calcium absorption is regulated.

In contrast to the almost complete, unregulated absorption of other ingested electrolytes, dietary iron and calcium may not be absorbed completely because their absorption is subject to regulation, depending on the body's needs for these electrolytes.

IRON ABSORPTION Iron is essential for hemoglobin production. The normal iron intake is typically 15 to 20 mg/day, yet a man usually absorbs about 0.5 to 1 mg/day into the blood, and a woman takes up slightly more, at 1.0 to 1.5 mg/day (women need more iron because they periodically lose iron in menstrual blood flow).

Two main steps are involved in absorption of iron into blood: (1) absorption of iron from the lumen into small-intestine epithelial cells and (2) absorption of iron from the epithelial cells into the blood (● Figure 16-27).

Iron is actively transported from the lumen into the epithelial cells, with women having about four times more active-transport sites for iron than men. The extent to which ingested iron is taken up by the epithelial cells depends on the type of iron consumed. Dietary iron exists in two forms: *heme iron*, in which iron is bound as part of a heme group found in hemoglobin (see p. 393) and is present in meat, and *inorganic iron*, which is present in plants. Dietary heme is absorbed more efficiently than inorganic iron is. Dietary inorganic iron exists primarily in the oxidized ferric iron (Fe^{3+}) form, but the reduced ferrous iron (Fe^{2+}) form is absorbed more easily. Dietary Fe^{3+} is reduced to Fe^{2+} by a membrane-bound enzyme at the luminal membrane prior to absorption. The presence of other substances in the lumen can either promote or reduce iron absorption. For example, vitamin C increases iron absorption, primarily by reducing Fe^{3+} to Fe^{2+}. Phosphate and oxalate, in contrast, combine with ingested iron to form insoluble iron salts that cannot be absorbed.

Heme iron and Fe^{2+} are transported across the luminal membrane by separate energy-dependent carriers in the brush border: Heme iron enters the intestinal cell by *heme carrier protein 1* and Fe^{2+} is carried in via *divalent metal transporter 1*, which also transports other metals that have a valence of $+2$. An enzyme within the cell frees the iron from the heme complex.

After absorption into the small-intestine epithelial cells, iron has two possible fates:

1. Iron needed immediately for production of red blood cells is absorbed into the blood for delivery to the bone marrow, the site of red blood cell production. Iron exits the small-intestine epithelial cell via a membrane iron transporter known as **ferroportin.** Iron absorption is largely controlled by a recently discovered hormone, **hepcidin,** which is released from the liver when iron levels in the body become too high. Hepcidin prevents further iron export from the small-intestine epithelial cell into the blood by binding with ferroportin and promoting its internalization into the cell by endocytosis and its subsequent degradation by lysosomes. Thus hepcidin is the primary regulator of iron homeostasis. A deficiency of hepcidin leads to tissue iron overload, because ferroportin continues to transfer iron into the body without control.

Iron that exits the small-intestine epithelial cell is transported in the blood by a plasma protein carrier known as **transferrin.** The absorbed iron is then used in the synthesis of hemoglobin for the newly produced red blood cells.

2. Iron not immediately needed is irreversibly stored within the small-intestine epithelial cells in a granular form called **ferritin,** which cannot be absorbed into the blood. Iron stored as ferritin is lost in the feces within three days as the epithelial cells containing these granules are sloughed off during mucosal regeneration. Large amounts of iron in the feces give them a dark, almost black color.

CALCIUM ABSORPTION The amount of calcium (Ca^{2+}) absorbed is also regulated. Calcium enters the luminal membrane of the small-intestine epithelial cells down its electrochemical gradient through a specialized Ca^{2+} channel, is ferried within the cell by a Ca^{2+}-binding protein, **calbindin,** and exits the basolateral membrane by two energy-dependent mechanisms: a primary active transport Ca^{2+} ATPase pump and a secondary active transport Na^+-Ca^{2+} antiporter. Vitamin D greatly enhances all of these steps in Ca^{2+} absorption. Vitamin D can exert this effect only after it has been activated in the liver and kidneys, a process that is enhanced by parathyroid hormone. Appropriately, secretion of parathyroid hormone increases in response to a fall in Ca^{2+} concentration in the blood. Normally, of the average 1000 mg of Ca^{2+} taken in daily, only about two thirds is absorbed in the small intestine, with the rest passing out in the feces.

Most absorbed nutrients immediately pass through the liver for processing.

The venules that leave the small-intestine villi, along with those from the rest of the digestive tract, empty into the hepatic portal vein, which carries the blood to the liver. Consequently, anything absorbed into the digestive capillaries first must pass through the hepatic biochemical factory before entering the general circulation. Thus, the products of carbohydrate and protein digestion are channeled into the liver, where many of these energy-rich products are subjected to immediate metabolic processing. Furthermore, harmful substances that may have been absorbed are detoxified by the liver before gaining access to the general circulation. After passing through the portal circulation, the venous blood from the digestive system empties into the vena cava and returns to the heart to be distributed throughout the body, carrying glucose and amino acids for use by the tissues.

Fat, which cannot penetrate the intestinal capillaries, is picked up by the central lacteal and enters the lymphatic system instead, bypassing the hepatic portal system. Contractions of the villi, accomplished by the muscularis mucosa, periodically compress the central lacteal and "milk" the lymph out of this vessel. The lymph vessels eventually converge to form the *thoracic duct*, a large lymph vessel that empties into the venous system within the chest. In this way fat ultimately gains access to the blood. The absorbed fat is carried by the systemic circulation to the liver and to other tissues of the body. Therefore, the liver does have a chance to act on the digested fat, but not until the fat has been diluted by the blood in the general circulatory system. This dilution of fat protects the liver from being inundated with more fat than it can handle at one time.

Extensive absorption by the small intestine keeps pace with secretion.

The small intestine normally absorbs about 9 liters of fluid per day in the form of H_2O and solutes, including the absorbable units of nutrients, vitamins, and electrolytes. How can that be, when humans normally ingest only about 1250 ml of fluid and consume 1250 g of solid food (80% of which is H_2O) per day (see p. 566)? ▲ Table 16-7 illustrates the tremendous daily absorption performed by the small intestine. Each day about 9500 ml of H_2O and solutes enter the small intestine. Note that of this 9500 ml, only 2500 ml are ingested from the external environment. The remaining 7000 ml (7 liters) of fluid are digestive juices derived from the plasma. Recall that plasma is the ultimate source of digestive secretions because the secretory cells extract from the plasma the necessary raw materials for their secretory product. Considering that the entire plasma volume is only about 2.75 liters, absorption must closely parallel secretion to keep the plasma volume from falling sharply.

Of the 9500 ml of fluid entering the small-intestine lumen per day, about 95%, or 9000 ml of fluid, is normally absorbed by the small intestine back into the plasma, with only 500 ml of the small-intestine contents passing on into the colon. Thus, the

▲ TABLE 16-7 **Volumes Absorbed by the Small and Large Intestine per Day**

Volume entering the small intestine per day			
Sources	Ingested	Food eaten	1250 g*
		Fluid drunk	1250 ml
	Secreted from the plasma	Saliva	1500 ml
		Gastric juice	2000 ml
		Pancreatic juice	1500 ml
		Bile	500 ml
		Intestinal juice	1500 ml
			9500 ml
Volume absorbed by the small intestine per day			9000 ml
Volume entering the colon from the small intestine per day			500 ml
Volume absorbed by the colon per day			350 ml
Volume of feces eliminated from the colon per day			150 g*

*One milliliter of H₂O weighs 1 g. Therefore, because a high percentage of food and feces is H₂O, we can roughly equate grams of food or feces with milliliters of fluid.

body does not lose the digestive juices. After the constituents of the juices are secreted into the digestive tract lumen and perform their function, they are returned to the plasma. The only secretory product that escapes from the body is bilirubin, a waste product that must be eliminated.

Biochemical balance among the stomach, pancreas, and small intestine is normally maintained.

Because the secreted juices are normally absorbed back into the plasma, the acid–base balance of the body is not altered by digestive processes. Within the small-intestine lumen, the HCl secreted by the parietal cells of the stomach is neutralized by the NaHCO₃ secreted by the pancreatic duct cells:

$$HCl + NaHCO_3 \rightarrow NaCl + H_2CO_3$$

The resultant H_2CO_3 decomposes into $CO_2 + H_2O$:

$$H_2CO_3 \rightarrow CO_2 + H_2O$$

The end products of these reactions—Na^+, Cl^-, CO_2, and H_2O—are all absorbed by the intestinal epithelium into the blood. Thus, through these interactions, the body normally does not experience a net gain or loss of acid or base during digestion.

Diarrhea results in loss of fluid and electrolytes.

When secretion and absorption do not parallel each other, however, acid–base abnormalities can result because these normal neutralization processes cannot

take place. We have already described vomiting and the subsequent loss of acidic gastric contents leading to metabolic alkalosis in the section on gastric motility. The other common digestive tract disturbance that can lead to a loss of fluid and an acid–base imbalance is **diarrhea.** This condition is characterized by passage of a highly fluid fecal matter, often with increased frequency of defecation. Just as with vomiting, the effects of diarrhea can be either beneficial or harmful. Diarrhea is beneficial when rapid emptying of the intestine hastens elimination of harmful material from the body. However, not only are some of the ingested materials lost but some of the secreted materials that normally would have been reabsorbed are lost as well. Excessive loss of intestinal contents causes dehydration, loss of nutrient material, and metabolic acidosis resulting from loss of HCO_3^- (see p. 584). The abnormal fluidity of the feces usually occurs because the small intestine is unable to absorb fluid as extensively as normal. This extra unabsorbed fluid passes out in the feces.

The causes of diarrhea are as follows:

1. The most common cause of diarrhea is excessive small-intestinal motility, which arises either from local irritation of the gut wall by bacterial or viral infection of the small intestine or from emotional stress. Rapid transit of the small-intestine contents does not allow enough time for adequate absorption of fluid to occur.

2. Diarrhea also occurs when excess osmotically active particles, such as those found in lactase deficiency, are present in the digestive tract lumen. These particles cause excessive fluid to enter and be retained in the lumen, thus increasing the fluidity of the feces.

3. Toxins of the bacterium *Vibrio cholera* (the causative agent of cholera) and certain other microorganisms promote the secretion of excessive amounts of fluid by the small-intestine mucosa, resulting in profuse diarrhea. Diarrhea produced in response to toxins from infectious agents is the leading cause of death of small children in developing nations. Fortunately, a low-cost, effective *oral rehydration therapy* that takes advantage of the intestine's glucose symport carrier is saving the lives of millions of children. (For details about oral rehydration therapy, see the boxed feature on p. 634, ■ Concepts, Challenges, and Controversies.)

Large Intestine

The **large intestine** consists of the colon, cecum, appendix, and rectum (● Figure 16-28). The **cecum** forms a blind-ended pouch below the junction of the small and large intestines at the ileocecal valve. The small, fingerlike projection at the bottom of the cecum is the **appendix,** a lymphoid tissue that houses lymphocytes (see p. 418). The **colon,** which makes up most of the large intestine, is not coiled like the small intestine but consists of three relatively straight parts—the *ascending colon,* the *transverse colon,* and the *descending colon.* The end part of the descending colon becomes S shaped, forming the *sigmoid colon* (*sigmoid* means "S shaped"), then straightens out to form the **rectum** (*rectum* means "straight").

Oral Rehydration Therapy: Sipping a Simple Solution Saves Lives

Diarrhea-inducing microorganisms such as *Vibrio cholera*, which causes cholera, are the leading cause of death in children under five worldwide. The problem is especially pronounced in developing countries, refugee camps, and elsewhere where poor sanitary conditions encourage the spread of the microorganisms, and medical supplies and health-care personnel are scarce. Fortunately, a low-cost, easily obtainable, uncomplicated remedy—oral rehydration therapy—has been developed to combat potentially fatal diarrhea. This treatment exploits the symporters located at the luminal border of the villus epithelial cells.

Let us examine the pathophysiology of life-threatening diarrhea and then see how simple oral rehydration therapy can save lives. During digestion of a meal, the crypt cells of the small intestine normally secrete succus entericus, a salt and mucus solution, into the lumen. These cells actively transport Cl^- into the lumen, promoting the parallel passive transport of Na^+ and H_2O from the blood into the lumen. The fluid provides the watery environment needed for enzymatic breakdown of ingested nutrients into absorbable units. Glucose and amino acids, the absorbable units of dietary carbohydrates and proteins, respectively, are absorbed by secondary active transport. This absorption mechanism uses the Na^+–glucose (or amino acid) cotransport carriers (SGLT) located at the luminal membrane of the villus epithelial cells (see p. 73). In addition, separate active Na^+ carriers not linked with nutrient absorption transfer Na^+, passively accompanied by Cl^- and H_2O, from the lumen into the blood.

The net result of these various carrier activities is absorption of the secreted salt and H_2O along with the digested nutrients. Normally, absorption of salt and H_2O exceeds their secretion, so not only are the secreted fluids salvaged but additional ingested salt and H_2O are absorbed as well.

Cholera and most diarrhea-inducing microbes cause diarrhea by stimulating the secretion of Cl^- and/or impairing the absorption of Na^+ (see p. 85). As a result, more fluid is secreted from the blood into the lumen than is subsequently transferred back into the blood. The excess fluid is lost in the feces, producing the watery stool characteristic of diarrhea. More important, the loss of fluids and electrolytes that came from the blood leads to dehydration. The subsequent reduction in effective circulating plasma volume can cause death in a matter of days or even hours, depending on the severity of the fluid loss.

About 50 years ago, physicians learned that replacing the lost fluids and electrolytes intravenously saves the lives of most diarrhea victims. In many parts of the world, however, adequate facilities, equipment, and personnel are not available to administer intravenous rehydration therapy. Consequently, millions of children still succumbed to diarrhea annually.

In 1966, researchers learned that SGLT is not affected by diarrhea-causing microbes. This discovery led to the development of **oral rehydration therapy (ORT).** When both Na^+ and glucose are present in the lumen, this symporter transports them both from the lumen into the villus epithelial cells, from which they enter the blood. Because H_2O osmotically follows the absorbed Na^+, ingestion of a glucose and salt solution promotes the uptake of fluid into the blood from the intestinal tract without the need for intravenous replacement of fluids.

The first proof of ORT's life-saving ability in the field came in 1971 when several million refugees poured into India from war-ravaged Bangladesh. Of the thousands of refugees who fell victim to cholera and other diarrheal diseases, more than 30% died because of the scarcity of sterile fluids and needles for intravenous therapy. In one refugee camp, however, under the supervision of a group of scientists who had been experimenting with oral rehydration therapy, families were taught to administer ORT to diarrhea victims, most of whom were small children. The scarce intravenous solutions were reserved for those unable to drink. Death from diarrhea was reduced to 3% in this camp, compared with a 10-fold higher mortality among refugees elsewhere.

Based on this evidence, the World Health Organization (WHO) started aggressively promoting ORT. Packets of dry ingredients for ORT are now manufactured locally in more than 60 countries. The WHO estimates that about 30% of the world's children who contract diarrhea are treated with the prepackaged mixture or home-prepared versions. In the United States, commercially prepared oral solutions are widely available at pharmacies and supermarkets. An estimated 1 million children worldwide are saved annually as a result of ORT.

The large intestine is primarily a drying and storage organ.

The colon normally receives about 500 ml of chyme from the small intestine each day. Because most digestion and absorption have been accomplished in the small intestine, the contents delivered to the colon consist of indigestible food residues (such as cellulose), unabsorbed biliary components, and the remaining fluid. The colon extracts more H_2O and salt from the contents. What remains to be eliminated is known as **feces.** The primary function of the large intestine is to store feces before defecation. Cellulose and other indigestible substances in the diet provide bulk and help maintain regular bowel movements by contributing to the volume of the colonic contents.

Haustral contractions slowly shuffle the colonic contents back and forth.

The outer longitudinal smooth muscle layer does not completely surround the large intestine. Instead, it consists only of three separate, conspicuous, longitudinal bands of muscle, the **taeniae coli,** which run the length of the large intestine. These taeniae coli are shorter than the underlying circular smooth muscle and mucosal layers would be if these layers were stretched out flat. Be-

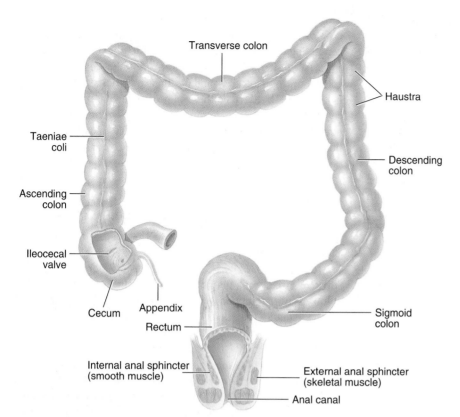

• FIGURE 16-28 Anatomy of the large intestine.

cause of this, the underlying layers are gathered into pouches or sacs called **haustra,** much as the material of a full skirt is gathered at the narrower waistband. The haustra are not merely passive permanent gathers, however; they actively change location as a result of contraction of the circular smooth muscle layer.

Most of the time, movements of the large intestine are slow and nonpropulsive, as is appropriate for its absorptive and storage functions. The colon's main motility is **haustral contractions** initiated by the autonomous rhythmicity of colonic smooth muscle cells. These contractions, which throw the large intestine into haustra, are similar to small-intestine segmentations but occur much less frequently. Thirty minutes may elapse between haustral contractions, whereas segmentation contractions in the small intestine occur at rates of between 9 and 12 per minute. The location of the haustral sacs gradually changes as a relaxed segment that has formed a sac slowly contracts while a previously contracted area simultaneously relaxes to form a new sac. These movements slowly shuffle the contents in a back-and-forth mixing movement that exposes the colonic contents to the absorptive mucosa. Haustral contractions are largely controlled by locally mediated reflexes involving the intrinsic plexuses.

Mass movements propel feces long distances.

Three to four times a day, generally after meals, a marked increase in motility takes place during which large segments of the ascending and transverse colon contract simultaneously, driving the feces one third to three fourths of the length of the

colon in a few seconds. These massive contractions, appropriately called **mass movements,** drive the colonic contents into the distal part of the large intestine, where material is stored until defecation occurs.

When food enters the stomach, mass movements are triggered in the colon primarily by the **gastrocolic reflex,** which is mediated from the stomach to the colon by gastrin and by the extrinsic autonomic nerves. In many people, this reflex is most evident after the first meal of the day and is often followed by the urge to defecate. Thus, when a new meal enters the digestive tract, reflexes are initiated to move the existing contents farther along down the tract to make way for the incoming food. The gastroileal reflex moves the remaining small-intestine contents into the large intestine, and the gastrocolic reflex pushes the colonic contents into the rectum, triggering the defecation reflex.

Feces are eliminated by the defecation reflex.

When mass movements of the colon move feces into the rectum, the resultant distension of the rectum stimulates stretch receptors in the rectal wall, initiating the **defecation reflex.** This reflex causes the **internal anal sphincter** (which is smooth muscle) to relax and the rectum and sigmoid colon to contract more vigorously. If the **external anal sphincter** (which is skeletal muscle) is also relaxed, defecation occurs. Being skeletal muscle, the external anal sphincter is under voluntary control. The initial distension of the rectal wall is accompanied by the conscious urge to defecate. If circumstances are unfavorable for defecation, voluntary tightening of the external anal sphincter can prevent defecation despite the defecation reflex. If defecation is delayed, the distended rectal wall gradually relaxes, and the urge to defecate subsides until the next mass movement propels more feces into the rectum, once again distending the rectum and triggering the defecation reflex. During periods of inactivity, both anal sphincters remain contracted to ensure fecal continence.

When defecation does occur, it is usually assisted by voluntary straining movements that involve simultaneous contraction of the abdominal muscles and a forcible expiration against a closed glottis. This maneuver greatly increases intra-abdominal pressure, which helps expel the feces.

Constipation occurs when the feces become too dry.

If defecation is delayed too long, **constipation** may result. When colonic contents are retained for longer periods of time than normal, more than the usual amount of H_2O is absorbed from the feces, so they become hard and dry. Normal variations in frequency of defecation among individuals range from after every meal to up to once a week. When the frequency is delayed beyond what is normal for a particular person, constipation and its attendant symptoms may occur. These symptoms include abdominal discomfort, dull headache, loss of appetite sometimes accompanied by nausea, and mental depression. Contrary to popular belief, these symptoms are not caused by toxins absorbed from the retained fecal material. Although bacterial metabolism produces some potentially toxic substances in the colon, these substances normally pass through the portal system and are removed by the liver before they can reach the systemic circulation. Instead, the symptoms associated with constipation are caused by prolonged distension of the large intestine, particularly the rectum; the symptoms promptly disappear after relief from distension.

Possible causes for delayed defecation that might lead to constipation include (1) ignoring the urge to defecate; (2) decreased colon motility accompanying aging, emotion, or a low-bulk diet; (3) obstruction of fecal movement in the large bowel caused by a local tumor or colonic spasm; and (4) impairment of the defecation reflex, such as through injury of the nerve pathways involved.

Clinical Note If hardened fecal material becomes lodged in the appendix, it may obstruct normal circulation and mucus secretion in this narrow, blind-ended appendage. This blockage leads to **appendicitis**. The inflamed appendix often becomes swollen and filled with pus, and the tissue may die as a result of local circulatory interference. If not surgically removed, the diseased appendix may rupture, spewing its infectious contents into the abdominal cavity.

Large-intestine secretion is entirely protective.

The large intestine does not secrete any digestive enzymes. None are needed, because digestion is completed before chyme ever reaches the colon. Colonic secretion consists of an alkaline ($NaHCO_3$) mucus solution, whose function is to protect the large-intestine mucosa from mechanical and chemical injury. The mucus provides lubrication to facilitate passage of the feces, whereas the $NaHCO_3$ neutralizes irritating acids produced by local bacterial fermentation. Secretion increases in response to mechanical and chemical stimulation of the colonic mucosa mediated by short reflexes and parasympathetic innervation.

No digestion takes place within the large intestine because there are no digestive enzymes. However, the colonic bacteria do digest some of the cellulose for their use.

The colon contains myriad beneficial bacteria.

Because of slow colonic movement, bacteria have time to grow and accumulate in the large intestine. In contrast, in the small intestine the contents are moved through too rapidly for bacterial growth to occur. Furthermore, the mouth, stomach, and small intestine secrete antibacterial agents, but the colon does not. Not all ingested bacteria are destroyed by lysozyme and HCl, however. The surviving bacteria continue to thrive in the large intestine. About 10 times more bacteria live in the human colon than the human body has cells. An estimated 500 to 1000 different species of bacteria typically live in the colon. These colonic microorganisms not only are typically harmless but, in fact are beneficial. For example, indigenous bacteria (1) enhance intestinal immunity by competing with potentially pathogenic microbes for nutrients and space (see p. 456); (2) promote colonic motility; (3) help maintain colonic mucosal integrity; and (4) make nutritional contributions. For example, bacteria synthesize absorbable vitamin K and raise colonic acidity, thereby promoting the absorption of calcium, magnesium, and zinc. Furthermore, contrary to earlier assumptions, some of the glucose released during bacterial processing of dietary fiber is absorbed by the colonic mucosa.

The large intestine absorbs salt and water, converting the luminal contents into feces.

Some absorption takes place within the colon, but not to the same extent as in the small intestine. Because the luminal surface of the colon is fairly smooth, it has considerably less absorptive surface area than the small intestine. Furthermore, the colon is not equipped with extensive specialized transport mechanisms as the small intestine is. When excessive small-intestine motility delivers the contents to the colon before absorption of nutrients has been completed, the colon cannot absorb most of these materials and they are lost in diarrhea.

The colon normally absorbs salt and H_2O. Sodium is actively absorbed, Cl^- follows passively down the electrical gradient, and H_2O follows osmotically. The colon absorbs token amounts of other electrolytes as well as vitamin K synthesized by colonic bacteria.

Through absorption of salt and H_2O, a firm fecal mass is formed. Of the 500 ml of material entering the colon per day from the small intestine, the colon normally absorbs about 350 ml, leaving 150 g of feces to be eliminated from the body each day (see ▲ Table 16-7). This fecal material normally consists of 100 g of H_2O and 50 g of solid, including undigested cellulose, bilirubin, bacteria, and small amounts of salt. Thus, contrary to popular thinking, the digestive tract is not a major excretory passageway for eliminating wastes from the body. The main waste product excreted in the feces is bilirubin. The other fecal constituents are unabsorbed food residues and bacteria, which were never actually a part of the body. Bacteria account for nearly one third the dry weight of feces.

Intestinal gases are absorbed or expelled.

Occasionally, instead of feces passing from the anus, intestinal gas, or **flatus**, passes out. This gas is derived primarily from two sources: (1) swallowed air (as much as 500 ml of air may be swallowed during a meal) and (2) gas produced by bacterial fermentation in the colon. The presence of gas percolating through the luminal contents gives rise to gurgling sounds known as **borborygmi**. Eructation (burping) removes most of the swallowed air from the stomach, but some passes on into the intestine. Usually, very little gas is present in the small intestine because the gas is either quickly absorbed or passes on into the colon. Most gas in the colon is the result of bacterial activity, with the quantity and nature of the gas depending on the type of food eaten and the characteristics of the colonic bacteria. Some foods, such as beans, contain types of carbohydrates that humans cannot digest but that can be attacked by gas-producing bacteria. Much of the gas is absorbed through the intestinal mucosa. The rest is expelled through the anus.

To selectively expel gas when feces are also present in the rectum, the person voluntarily contracts the abdominal muscles and external anal sphincter at the same time. When abdominal contraction raises the pressure against the contracted anal sphincter sufficiently, the pressure gradient forces air out at a high velocity through a slitlike anal opening that is too narrow for solid feces to escape through. This passage of air at high velocity causes the edges of the anal opening to vibrate, giving rise to the characteristic low-pitched sound accompanying passage of gas.

Overview of the Gastrointestinal Hormones

We have repeatedly mentioned different functions of the three major gastrointestinal hormones: gastrin, secretin, and CCK. Let us now fit all of these functions together so you can appreciate the overall adaptive importance of these interactions. Furthermore, we will introduce a more recently identified gastrointestinal hormone, GIP. All of these hormones are small peptides and perform their functions by binding to G protein-coupled receptors on the plasma membrane of their target cells, thereby activating second-messenger pathways that bring about the desired responses (see p. 117).

GASTRIN Protein in the stomach stimulates the release of gastrin, which performs the following functions:

1. It acts in multiple ways to increase secretion of HCl and pepsinogen, two substances of primary importance in initiating digestion of the protein that promoted their secretion.

2. It enhances gastric motility, stimulates ileal motility, relaxes the ileocecal sphincter, and induces mass movements in the colon—functions that are all aimed at keeping the contents moving through the tract on the arrival of a new meal.

3. It also is trophic not only to the stomach mucosa but also to the small-intestine mucosa, helping maintain a well-developed, functionally viable digestive tract lining.

Predictably, gastrin secretion is inhibited by an accumulation of acid in the stomach and by the presence in the duodenal lumen of acid and other constituents that necessitate a delay in gastric secretion.

SECRETIN As the stomach empties into the duodenum, the presence of acid in the duodenum stimulates the release of secretin, which performs the following interrelated functions:

1. It inhibits gastric emptying to prevent further acid from entering the duodenum until the acid already present is neutralized.

2. It inhibits gastric secretion to reduce the amount of acid being produced.

3. It stimulates the pancreatic duct cells to produce a large volume of aqueous $NaHCO_3$ secretion, which is emptied into the duodenum to neutralize the acid.

4. It stimulates secretion by the liver of a $NaHCO_3$-rich bile, which likewise is emptied into the duodenum to assist in the neutralization process. Neutralization of the acidic chyme in the duodenum helps prevent damage to the duodenal walls and provides a suitable environment for the optimal functioning of the pancreatic digestive enzymes, which are inhibited by acid.

5. Secretin and CCK are both trophic to the exocrine pancreas.

CCK As chyme empties from the stomach, fat and other nutrients enter the duodenum. These nutrients, especially fat and, to a lesser extent, protein products, cause the release of CCK, which performs the following interrelated functions:

1. It inhibits gastric motility and secretion, thereby allowing adequate time for the nutrients already in the duodenum to be digested and absorbed.

2. It stimulates the pancreatic acinar cells to increase secretion of pancreatic enzymes, which continue the digestion of these nutrients in the duodenum (this action is especially important for fat digestion, because pancreatic lipase is the only enzyme that digests fat).

3. It causes contraction of the gallbladder and relaxation of the sphincter of Oddi so that bile is emptied into the duodenum to aid fat digestion and absorption. Bile salts' detergent action is particularly important in enabling pancreatic lipase to perform its digestive task. Once again, the multiple effects of CCK are remarkably well adapted to dealing with the fat whose presence in the duodenum triggered this hormone's release.

4. CCK has also been implicated in long-term adaptive changes in the proportion of pancreatic enzymes produced in response to prolonged changes in diet.

5. Besides facilitating the digestion of ingested nutrients, CCK is an important regulator of food intake. It plays a key role in satiety, the sensation of having had enough to eat (see p. 647).

GIP A more recently recognized hormone released by the duodenum, **GIP,** helps promote metabolic processing of the nutrients once they are absorbed. This hormone was originally named *gastric inhibitory peptide* (*GIP*) for its presumed role as an enterogastrone. It was believed to inhibit gastric motility and secretion, similar to secretin and CCK. GIP's contribution in this regard is now considered minimal. Instead, this hormone stimulates insulin release by the pancreas, so it is now called **glucose-dependent insulinotrophic peptide** (once again, **GIP**). Again, this is remarkably adaptive. As soon as the meal is absorbed, the body has to shift its metabolic gears to use and store the newly arriving nutrients. The metabolic activities of this absorptive phase are largely under the control of insulin (see pp. 713 and 717). Stimulated by the presence of a meal, especially glucose, in the digestive tract, GIP initiates the release of insulin in anticipation of absorption of the meal, in a feedforward fashion. Insulin is especially important in promoting the uptake and storage of glucose.

This overview of the multiple, integrated, adaptive functions of the gastrointestinal hormones provides an excellent example of the remarkable efficiency of the human body.

Chapter in Perspective: Focus on Homeostasis

To maintain constancy in the internal environment, materials that are used up in the body (such as energy-rich nutrients and O_2) or uncontrollably lost from the body (such as evaporative H_2O loss from the airways or salt loss in sweat) must constantly be replaced by new supplies of these materials from the external environment. All these replacement supplies except O_2 are acquired through the digestive system. Fresh supplies of O_2 are transferred to the internal environment by the respiratory system, but all the nutrients, H_2O, and various electrolytes needed to maintain homeostasis are acquired through the digestive system. The large, complex food that is ingested is broken down by the digestive system into small absorbable units. These small energy-rich nutrient molecules are transferred across the small-intestine epithelium into the blood for delivery to the cells to replace the nutrients constantly used for ATP production and for repair and growth of body tissues. Likewise, ingested H_2O, salt, and other electrolytes are absorbed by the intestine into the blood.

Unlike regulation in most body systems, regulation of digestive system activities is not aimed at maintaining homeostasis. The quantity of nutrients and H_2O ingested is subject to control, but the quantity of ingested materials absorbed by the digestive tract is not subject to control, with few exceptions. The hunger mechanism governs food intake to help maintain energy balance (Chapter 17), and the thirst mechanism controls H_2O intake to help maintain H_2O balance (Chapter 15). However, we often do not heed these control mechanisms, and often eat and drink even when we are not hungry or thirsty. Once these materials are in the digestive tract, the digestive system does not vary its rate of nutrient, H_2O, or electrolyte uptake according to body needs (with the exception of iron and calcium); rather, it optimizes conditions for digesting and absorbing what is ingested. Truly, what you eat is what you get. The digestive system is subject to many regulatory processes, but these are not influenced by the nutritional or hydration state of the body. Instead, these control mechanisms are governed by the composition and volume of digestive tract contents so that the rate of motility and secretion of digestive juices are optimal for digestion and absorption of the ingested food.

If excess energy-rich nutrients are ingested and absorbed, the extra is placed in storage, such as in adipose tissue (fat), so that the blood level of nutrient molecules is kept at a constant level. Excess ingested H_2O and electrolytes are eliminated in the urine to homeostatically maintain the blood levels of these constituents.

REVIEW EXERCISES

Objective Questions (Answers on p. A-53)

1. The extent of nutrient uptake from the digestive tract depends on the body's needs. *(True or false?)*

2. The stomach is relaxed during vomiting. *(True or false?)*

3. Acid cannot normally penetrate into or between the cells lining the stomach, which enables the stomach to contain acid without injuring itself. *(True or false?)*

4. Protein is continually lost from the body through digestive secretions and sloughed epithelial cells, which pass out in the feces. *(True or false?)*

5. Foodstuffs not absorbed by the small intestine are absorbed by the large intestine. *(True or false?)*

6. The endocrine pancreas secretes secretin and CCK. *(True or false?)*

7. A digestive reflex involving the autonomic nerves is known as a _____ reflex, whereas a reflex in which all elements of the reflex arc are located within the gut wall is known as a _____ reflex.

8. When food is mechanically broken down and mixed with gastric secretions, the resultant thick, liquid mixture is known as _____.

9. The entire lining of the small intestine is replaced approximately every _____ days.

10. The two substances absorbed by specialized transport mechanisms located only in the terminal ileum are _____ and _____.

11. The most potent choleretic is _____.

12. Which of the following is *not* a function of saliva?
    a. begins digestion of carbohydrate
    b. facilitates absorption of glucose across the oral mucosa
    c. facilitates speech
    d. exerts an antibacterial effect
    e. plays an important role in oral hygiene

13. Match the following:
    ___ 1. prevents reentry of food into the mouth during swallowing
    ___ 2. triggers the swallowing reflex
    ___ 3. seals off the nasal passages during swallowing
    ___ 4. prevents air from entering the esophagus during breathing
    ___ 5. closes off the respiratory airways during swallowing
    ___ 6. prevents gastric contents from backing up into the esophagus

    (a) closure of the pharyngoesophageal sphincter
    (b) elevation of the uvula
    (c) position of the tongue against the hard palate
    (d) closure of the gastroesophageal sphincter
    (e) bolus pushed to the rear of the mouth by the tongue
    (f) tight apposition of the vocal folds

14. Use the answer code on the right to identify the characteristics of the listed substances:
    ___ 1. activates pepsinogen
    ___ 2. inhibits amylase
    ___ 3. is essential for vitamin B$_{12}$ absorption
    ___ 4. can act autocatalytically
    ___ 5. is a potent stimulant for acid secretion
    ___ 6. breaks down connective tissue and muscle fibers
    ___ 7. begins protein digestion
    ___ 8. serves as a lubricant
    ___ 9. kills ingested bacteria
    ___ 10. is alkaline
    ___ 11. is deficient in pernicious anemia
    ___ 12. coats the gastric mucosa

    (a) pepsin
    (b) mucus
    (c) HCl
    (d) intrinsic factor
    (e) histamine

Essay Questions

1. Describe the four basic digestive processes.
2. List the three categories of energy-rich foodstuffs and the absorbable units of each.
3. List the components of the digestive system. Describe the cross-sectional anatomy of the digestive tract.
4. What four general factors are involved in regulating digestive system function? What is the role of each?
5. Describe the types of motility in each component of the digestive tract. What factors control each type of motility?
6. State the composition of the digestive juice secreted by each component of the digestive system. Describe the factors that control each digestive secretion.
7. List the enzymes involved in digesting each category of foodstuff. Indicate the source and control of secretion of each of the enzymes.
8. Why are some digestive enzymes secreted in inactive form? How are they activated?
9. What absorption processes take place within each component of the digestive tract? What special adaptations of the small intestine enhance its absorptive capacity?
10. Describe the absorptive mechanisms for salt, water, carbohydrate, protein, and fat.
11. What are the contributions of the accessory digestive organs? What are the nondigestive functions of the liver?
12. Summarize the functions of each of the three major gastrointestinal hormones. What is the role of GIP?
13. What waste product is excreted in the feces?
14. How is vomiting accomplished? What are the causes and consequences of vomiting, diarrhea, and constipation?
15. Describe the process of mucosal turnover in the stomach and small intestine.

Quantitative Exercises (Solutions on p. A-53)

1. Suppose a lipid droplet in the gut is essentially a sphere with a diameter of 1 cm.
   a. What is the surface area–to–volume ratio of the droplet? (*Hint:* The area of a sphere is $4\pi r^2$, and the volume is $4/3\pi r^3$.)
   b. Now, suppose that this sphere were emulsified into 100 essentially equal-sized droplets. What is the average surface area–to–volume ratio of each droplet?
   c. How much greater is the total surface area of these 100 droplets compared to the original single droplet?
   d. How much did the total volume change as a result of emulsification?

POINTS TO PONDER

(Explanations on p. A-53)

1. Why do patients who have had a large part of their stomachs removed for treatment of stomach cancer or severe peptic ulcer disease have to eat small quantities of food frequently instead of consuming three meals a day?

2. The number of immune cells in the *gut-associated lymphoid tissue (GALT)* housed in the mucosa is estimated to be equal to the total number of these defense cells in the rest of the body. Speculate on the adaptive significance of this extensive defense capability of the digestive system.

3. How would defecation be accomplished in a patient paralyzed from the waist down by a lower spinal-cord injury?

4. After bilirubin is extracted from the blood by the liver, it is conjugated (combined) with glucuronic acid by the enzyme glucuronyl transferase within the liver. Only when conjugated can bilirubin be actively excreted into the bile. For the first few days of life, the liver does not make adequate quantities of glucuronyl transferase. Explain how this transient enzyme deficiency leads to the common condition of jaundice in newborns.

5. Explain why removal of either the stomach or the terminal ileum leads to pernicious anemia.

CLINICAL CONSIDERATION

(Explanation on p. A-53)

Thomas W. experiences a sharp pain in his upper right abdomen after eating a high-fat meal. Also, he has noted that his feces are grayish white instead of brown. What is the most likely cause of his symptoms? Explain why each of these symptoms occurs with this condition.

Components Important in Energy Balance and Temperature Regulation

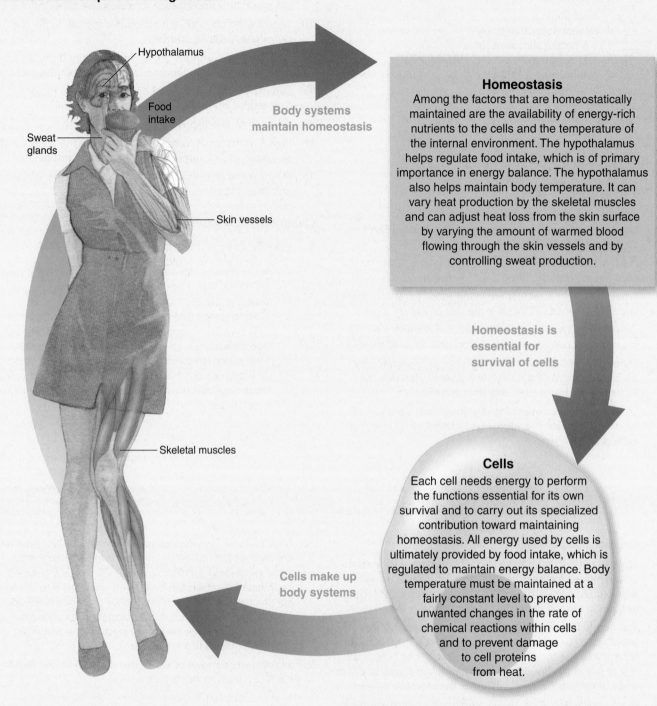

Hypothalamus

Food intake

Sweat glands

Skin vessels

Skeletal muscles

Body systems maintain homeostasis

Homeostasis
Among the factors that are homeostatically maintained are the availability of energy-rich nutrients to the cells and the temperature of the internal environment. The hypothalamus helps regulate food intake, which is of primary importance in energy balance. The hypothalamus also helps maintain body temperature. It can vary heat production by the skeletal muscles and can adjust heat loss from the skin surface by varying the amount of warmed blood flowing through the skin vessels and by controlling sweat production.

Homeostasis is essential for survival of cells

Cells
Each cell needs energy to perform the functions essential for its own survival and to carry out its specialized contribution toward maintaining homeostasis. All energy used by cells is ultimately provided by food intake, which is regulated to maintain energy balance. Body temperature must be maintained at a fairly constant level to prevent unwanted changes in the rate of chemical reactions within cells and to prevent damage to cell proteins from heat.

Cells make up body systems

Food intake is essential to power cell activities. For body weight to remain constant, the caloric value of food must equal total energy needs. **Energy balance** and thus body weight are maintained primarily by controlling food intake.

Energy expenditure generates heat, which is important in **temperature regulation.** Humans, usually in environments cooler than their bodies, must constantly generate heat to maintain their body temperatures. Also, they must have mech-

anisms to cool the body if it gains too much heat from heat-generating skeletal muscle activity or from a hot external environment. Body temperature must be regulated because the rate of cellular chemical reactions depends on temperature and because overheating damages cell proteins.

The hypothalamus is the major integrating center for maintaining both energy balance and body temperature.

Energy Balance and Temperature Regulation

CENGAGENOW™ Log on to CengageNOW at **http://www .cengage.com/sso/** for an opportunity to explore animations and interactive quizzes to help you learn, review, and master physiology concepts.

Energy Balance

Each cell in the body needs energy to perform the functions essential for the cell's own survival (such as active transport and cellular repair) and to carry out its specialized contributions toward maintaining homeostasis (such as gland secretion or muscle contraction). All energy used by cells is ultimately provided by food intake.

Most food energy is ultimately converted into heat in the body.

According to the **first law of thermodynamics,** energy can be neither created nor destroyed. Therefore, energy is subject to the same kind of input–output balance as are the chemical components of the body, such as H_2O and salt (see p. 557).

ENERGY INPUT AND OUTPUT The energy in ingested food constitutes *energy input* to the body. Chemical energy locked in the bonds that hold the atoms together in nutrient molecules is released when these molecules are broken down in the body. Cells capture a portion of this nutrient energy in the high-energy phosphate bonds of ATP (see pp. 32 and A-17). Energy harvested from biochemical processing of ingested nutrients is either used immediately to perform biological work or stored in the body for later use as needed during periods when food is not being digested and absorbed.

Energy output or expenditure by the body falls into two categories (● Figure 17-1): external work and internal work. **External work** is the energy expended when skeletal muscles are contracted to move external objects or to move the body in relation to the environment. **Internal work** constitutes all other forms of biological energy expenditure that do not accomplish mechanical work outside the body. Internal work encompasses two types of energy-dependent activities: (1) skeletal muscle activity used for purposes other than external work, such as the contractions associated with postural maintenance and shivering; and (2) all the energy-expending activities that must go on all the time just to sustain life. The latter include the work of pumping blood and breathing; the energy required for active transport of critical materials across plasma membranes; and the energy used during synthetic reactions essential for the maintenance, repair, and growth of cellular structures—in short, the "metabolic cost of living."

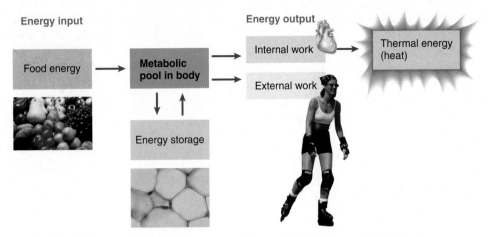

FIGURE 17-1 Energy input and output. (Photos: *left,* Ralph Pleasant/FPG/Getty Images; *center,* Ed Reschke; *right,* Gary Head.)

CONVERSION OF NUTRIENT ENERGY TO HEAT Not all energy in nutrient molecules can be harnessed to perform biological work. Energy cannot be created or destroyed, but it can be converted from one form to another. The energy in nutrient molecules not used to energize work is transformed into **thermal energy,** or **heat.** During biochemical processing, only about 50% of the energy in nutrient molecules is transferred to ATP; the other 50% of nutrient energy is immediately lost as heat. During ATP expenditure by the cells, another 25% of the energy derived from ingested food becomes heat. Because the body is not a heat engine, it cannot convert heat into work. Therefore, not more than 25% of nutrient energy is available for work, either external or internal. The remaining 75% is lost as heat during the sequential transfer of energy from nutrient molecules to ATP to cellular systems.

Furthermore, of the energy actually captured for use by the body, almost all expended energy eventually becomes heat. To exemplify, energy expended by the heart to pump blood is gradually changed into heat by friction as blood flows through the vessels. Likewise, energy used in synthesizing structural protein eventually appears as heat when that protein is degraded during the normal course of turnover of bodily constituents. Even in performing external work, skeletal muscles convert chemical energy into mechanical energy inefficiently; as much as 75% of the expended energy is lost as heat. Thus, all energy liberated from ingested food that is not directly used for moving external objects or stored in fat (adipose tissue) deposits (or, in the case of growth, as protein) eventually becomes body heat. This heat is not entirely wasted energy, however, because much of it is used to maintain body temperature.

The metabolic rate is the rate of energy use.

The rate at which energy is expended by the body during both external and internal work is known as the **metabolic rate:**

Metabolic rate = energy expenditure/unit of time

Because most of the body's energy expenditure eventually appears as heat, the metabolic rate is normally expressed in terms of the rate of heat production in kilocalories per hour. The basic unit of heat energy is the **calorie,** which is the amount of heat required to raise the temperature of 1 g of H_2O by 1°C. This unit is too small to be convenient when discussing the human body because of the magnitude of heat involved, so the **kilocalorie** or **Calorie,** which is equivalent to 1000 calories, is used. When nutritionists speak of "calories" in quantifying the energy content of various foods, they are actually referring to kilocalories or Calories. Four kilocalories of heat energy are released when 1 g of glucose is oxidized or "burned," whether the oxidation takes place inside or outside the body.

CONDITIONS FOR MEASURING THE BASAL METABOLIC RATE The metabolic rate and, consequently, the amount of heat produced vary depending on a variety of factors, such as exercise, anxiety, shivering, and food intake. Increased skeletal muscle activity is the factor that can increase metabolic rate to the greatest extent. Even slight increases in muscle tone notably elevate the metabolic rate, and various levels of physical activity markedly alter energy expenditure and heat production (▲ Table 17-1). For this reason, a person's metabolic rate is determined under standardized basal conditions established to control as many as possible of the variables that can alter metabolic rate. In this way, the metabolic activity necessary to maintain the basic body functions at rest can be determined. The so-called **basal metabolic rate (BMR)** is a reflection of the body's "idling speed," or the minimal waking rate of internal energy expenditure. The BMR is measured under the following specified conditions:

1. The person should be at physical rest, having refrained from exercise for at least 30 minutes to eliminate any contribution of muscular exertion to heat production.

2. The person should be at mental rest to minimize skeletal muscle tone (people "tense up" when they are nervous) and to prevent a rise in epinephrine, a hormone secreted in response to stress that increases metabolic rate.

3. The measurement should be performed at a comfortable room temperature so the person does not shiver. Shivering can markedly increase heat production.

4. The subject should not have eaten any food within 12 hours before the BMR determination to avoid **diet-induced thermogenesis** (*thermo* means "heat"; *genesis* means "production"), or the obligatory increase in metabolic rate that occurs as a consequence of food intake. This short-lived (less than 12-hour) rise in metabolic rate is the result not of digestive activities but of the increased metabolic activity associated with the processing and storage of ingested nutrients, especially by the major biochemical factory, the liver.

TABLE 17-1 Rate of Energy Expenditure for a 70 kg Person during Different Types of Activity

Form of Activity	Energy Expenditure (kcal/hr)
Sleeping	65
Awake, lying still	77
Sitting at rest	100
Standing relaxed	105
Getting dressed	118
Typewriting	140
Walking slowly on level (2.6 mi/hr)	200
Carpentry, painting a house	240
Sexual intercourse	280
Bicycling on level (5.5 mi/hr)	304
Shoveling snow, sawing wood	480
Swimming	500
Jogging (5.3 mi/hr)	570
Rowing (20 strokes/min)	828
Walking up stairs	1100

METHODS OF MEASURING THE BASAL METABOLIC RATE The rate of heat production in BMR determinations can be measured directly or indirectly. With **direct calorimetry,** the person sits in an insulated chamber with water circulating through the walls. The difference in the temperature of the water entering and leaving the chamber reflects the amount of heat liberated by the person and picked up by the water as it passes through the chamber. Even though this method provides a direct measurement of heat production, it is not practical, because a calorimeter chamber is costly and takes up a lot of space. Therefore, a more practical method of indirectly determining the rate of heat production was developed for widespread use. With **indirect calorimetry,** only the person's O_2 uptake per unit of time is measured, which is a simple task using minimal equipment. Recall that

$$\text{Food} + O_2 \rightarrow CO_2 + H_2O + \text{energy (mostly transformed into heat)}$$

Accordingly, a direct relationship exists between the volume of O_2 used and the quantity of heat produced. This relationship also depends on the type of food being oxidized. Although carbohydrates, proteins, and fats require different amounts of O_2 for their oxidation and yield different amounts of kilocalories when oxidized, an average estimate can be made of the quantity of heat produced per liter of O_2 consumed on a typical mixed American diet. This approximate value, known as the **energy equivalent of O_2**, is 4.8 kilocalories of energy liberated per liter of O_2 consumed. Using this method, the metabolic rate of a person consuming 15 liters/hr of O_2 can be estimated as follows:

15	liters/hr	= O_2 consumption
× 4.8	kilocalories/liter	= energy equivalent of O_2
72	kilocalories/hr	= estimated metabolic rate

In this way, a simple measurement of O_2 consumption can be used to reasonably approximate heat production in determining metabolic rate.

Once the rate of heat production is determined under the prescribed basal conditions, it must be compared with normal values for people of the same sex, age, height, and weight, because these factors all affect the basal rate of energy expenditure. For example, a large man actually has a higher rate of heat production than a smaller man; but expressed in terms of total surface area (which is a reflection of height and weight), the output in kilocalories per hour per square meter of surface area is normally about the same.

FACTORS INFLUENCING THE BASAL METABOLIC RATE Thyroid hormone is the primary but not sole determinant of the rate of basal metabolism. As thyroid hormone increases, the BMR increases correspondingly. As mentioned, epinephrine also increases the BMR.

Surprisingly, the BMR is not the body's lowest metabolic rate. The rate of energy expenditure during sleep is 10% to 15% lower than the BMR, presumably because of the more complete muscle relaxation that occurs during the paradoxical stage of sleep (see p. 170).

Energy input must equal energy output to maintain a neutral energy balance.

Because energy cannot be created or destroyed, energy input must equal energy output, as follows:

Energy input = energy output

Energy in food consumed	=	external work	+	internal heat production	±	stored energy

There are three possible states of energy balance:

■ *Neutral energy balance.* If the amount of energy in food intake exactly equals the amount of energy expended by the muscles in performing external work plus the basal internal energy expenditure that eventually appears as body heat, then energy input and output are exactly in balance, and body weight remains constant.

■ *Positive energy balance.* If the amount of energy in food intake is greater than the amount of energy expended by means of external work and internal functioning, the extra energy taken in but not used is stored in the body, primarily as adipose tissue, so body weight increases.

■ *Negative energy balance.* Conversely, if the energy derived from food intake is less than the body's immediate energy requirements, the body must use stored energy to supply energy needs, and body weight decreases accordingly.

For a person to maintain a constant body weight (with the exception of minor fluctuations caused by changes in H_2O content), energy acquired through food intake must equal energy expenditure by the body. Because the average adult maintains a fairly constant weight over long periods of time, this implies that precise homeostatic mechanisms exist to maintain a long-term balance between energy intake and energy expenditure. Theoretically, total body energy content could be maintained at a constant level by regulating the magnitude of food intake, physical activity, or internal work and heat production. Control of food intake to match changing metabolic expenditures is the major means of maintaining a neutral energy balance. The level of physical activity is principally under voluntary control, and mechanisms that alter the degree of internal work and heat production are aimed primarily at regulating body temperature rather than total energy balance.

However, after several weeks of eating less or more than desired, small counteracting changes in metabolism may occur. For example, a compensatory increase in the body's efficiency of energy use in response to underfeeding partially explains why some dieters become stuck at a plateau after having lost the first 10 or so pounds of weight fairly easily. Similarly, a compensatory reduction in the efficiency of energy use in response to overfeeding accounts in part for the difficulty experienced by very thin people who are deliberately trying to gain weight. Despite these modest compensatory changes in metabolism, regulation of food intake is the most important factor in the long-term maintenance of energy balance and body weight.

Food intake is controlled primarily by the hypothalamus.

Even though food intake is adjusted to balance changing energy expenditures over a period of time, there are no calorie receptors per se to monitor energy input, energy output, or total body energy content. Instead, various blood-borne chemical factors that signal the body's nutritional state, such as how much fat is stored or the feeding status, are important in regulating food intake. Control of food intake does not depend on changes in a single signal but is determined by the integration of many inputs that provide information about the body's energy status. Multiple molecular signals together ensure that feeding behavior is synchronized with the body's immediate and long-term energy needs. Some information is used for short-term regulation of food intake, helping to control meal size and frequency. Even so, over a 24-hour period the energy in ingested food rarely matches energy expenditure for that day. The correlation between total caloric intake and total energy output is excellent, however, over long periods of time. As a result, the total energy content of the body—and, consequently, body weight—remains relatively constant on a long-term basis. Thus, energy homeostasis, that is, energy balance, is carefully regulated.

ROLE OF THE ARCUATE NUCLEUS: NPY AND MELANOCORTINS Control of energy balance and food intake is primarily a function of the hypothalamus. The **arcuate nucleus** of the hypothalamus plays a central role in both the long-term control of

energy balance and body weight and the short-term control of food intake on a meal-to-meal basis. The arcuate nucleus is an arc-shaped collection of neurons located adjacent to the floor of the third ventricle. Multiple, highly integrated, redundant pathways crisscross into and out of the arcuate nucleus, indicative of the complex systems involved in feeding and satiety. **Feeding,** or **appetite, signals** give rise to the sensation of **hunger,** driving us to eat. By contrast, **satiety** is the feeling of being full. **Satiety signals** tell us when we have had enough and suppress the desire to eat.

The arcuate nucleus has two subsets of neurons that function in an opposing manner. One subset releases *neuropeptide Y,* and the other releases *melanocortins* derived from pro-opiomelanocortin (POMC), a precursor molecule that can be cleaved in different ways to produce several different hormone products (see p. 672).[1] **Neuropeptide Y (NPY),** one of the most potent appetite stimulators ever found, leads to increased food intake, thus promoting weight gain. **Melanocortins,** a group of hormones traditionally known to be important in varying the skin color for the purpose of camouflage in some species, have been shown to exert an unexpected role in energy homeostasis in humans. Melanocortins, most notably *α melanocyte-stimulating hormone (α-MSH)* from the hypothalamus, suppress appetite, thus leading to reduced food intake and weight loss. Melanocortins do not play a role in determining inherited skin coloration in humans, but α-MSH produced in the skin in response to ultraviolet light from the sun acts locally on melanin (pigment)-producing cells to cause tanning. Melanocortin's major importance in our species, however, is the role of hypothalamic α-MSH in toning down appetite.

But NPY and melanocortins are not the final effectors in appetite control. These arcuate-nucleus chemical messengers, in turn, influence the release of neuropeptides in other parts of the brain that exert more direct control over food intake. Scientists are currently trying to unravel the other factors that act upstream and downstream from NPY and melanocortins to regulate appetite.

Based on current evidence, the following regulatory inputs to the arcuate nucleus and beyond are important in the long-term maintenance of energy balance and the short-term control of food intake at meals (● Figure 17-2).

LONG-TERM MAINTENANCE OF ENERGY BALANCE: LEPTIN AND INSULIN Scientists' notion of fat cells (**adipocytes**) in adipose tissue as merely storage space for triglyceride fat underwent a dramatic change late in the last century with the discovery of their active role in energy homeostasis. Adipocytes secrete sev-

---

[1]The two subsets of neurons in the arcuate nucleus are the **NPY/AgRP** population and the **POMC/CART** population. *AgRP* stands for *agouti-related protein.* Both NPY and AgRP stimulate appetite. *CART* stands for *cocaine- and amphetamine-related transcript.* Melanocortins and CART peptide both suppress appetite. For simplicity's sake, we will discuss only the role of NPY and melanocortins but recognize that other chemical signals released from the arcuate nucleus exert similar functions.

● **FIGURE 17-2 Factors that influence food intake.**

KEY

CCK = Cholecystokinin
LHA = Lateral hypothalamus area
NPY = Neuropeptide Y
NTS = Nucleus tractus solitarius
POMC = Pro-opiomelanocortin
PYY = Peptide YY
PVN = Paraventricular nucleus

Signals important in long-term matching of food intake to energy expenditure to control body weight

Signals important in short-term control of the timing and size of meals

Psychosocial and environmental factors that influence food intake

*Other chemicals also released from this area that exert similar functions.

eral hormones, collectively termed **adipokines,** that play important roles in energy balance and metabolism. Thus, adipose tissue is now considered an endocrine gland. ▲ Table 17-2 summarizes the functions of the best-known adipokines. Some adipokines are released only from stored fat; *leptin,* for example, plays an important role in energy balance. Some, like inflammatory-related *tumor necrosis factor* and *interleukin 6,* are released from adipose tissue and immune cells and contribute to chronic, low-grade inflammation in excessive fat stores. Some, like *visfatin,* are released only from **visceral fat,** the deep, "bad" fat that

surrounds the abdominal organs. Visceral fat is more likely to be chronically inflamed and is associated with increased heart disease, in contrast to the more superficial and less harmful **subcutaneous fat** that is deposited under the skin. (Subcutaneous fat is the fat you can pinch.) Some, like *adiponectin,* are "good" adipokines. Adiponectin increases sensitivity to insulin (which helps protect against Type 2 diabetes mellitus), decreases body weight, and exerts anti-inflammatory actions. Unfortunately, obesity suppresses adiponectin secretion. By contrast, some, like *resistin,* are "bad" adipokines. Resistin,

TABLE 17-2 Major Adipokines

Adipokine	Function
Leptin	Released from stored fat; suppresses appetite; dominant long-term regulator of energy balance and body weight
Adiponectin	Secretion from adipocytes suppressed in obesity; promotes fatty acid oxidation by muscle; increases sensitivity to insulin; decreases body weight by increasing energy expenditure; has anti-inflammatory actions
Resistin	Released primarily in obesity; leads to insulin resistance
Visfatin	Released primarily from visceral fat; stimulates glucose uptake; binds with insulin receptor at a site distinct from the insulin-binding site
Tumor necrosis factor α (TNF-α) and interleukin 6 (IL-6)	Promote low-level inflammation in fat and throughout body

which is released primarily in obesity, leads to insulin resistance (thus increasing the risk of developing Type 2 diabetes).

One of the most important adipokines is **leptin,** a hormone essential for normal body-weight regulation (*leptin* means "thin"). The amount of leptin in the blood is an excellent indicator of the total amount of triglyceride fat stored in adipose tissue: The larger the fat stores, the more leptin released into the blood. This blood-borne signal, discovered in the mid-1990s, was the first molecular satiety signal identified. This finding was the breakthrough that touched off a flurry of research responsible for greatly expanding our knowledge in recent years of the complex interplay of chemical signals that regulate food intake and body size.

The arcuate nucleus is the major site for leptin action. Acting in negative-feedback fashion, increased leptin from burgeoning fat stores serves as a "trim-down" signal. Leptin suppresses appetite, thus decreasing food consumption and promoting weight loss, by inhibiting hypothalamic output of appetite-stimulating NPY and stimulating output of appetite-suppressing melanocortins. Conversely, a decrease in fat stores and the resultant decline in leptin secretion bring about an increase in appetite, leading to weight gain. The leptin signal is generally considered the dominant factor responsible for the long-term matching of food intake to energy expenditure so that total body energy content remains balanced and body weight remains constant.

Interestingly, leptin has recently been shown to also be important in reproduction. It is one of the triggers for the onset of puberty, signaling that the female has enough long-term energy (adipose) stores to sustain a pregnancy.

Another blood-borne signal besides leptin that plays an important role in long-term control of body weight is **insulin.** Insulin, a hormone secreted by the pancreas in response to a rise in the concentration of glucose and other nutrients in the blood following a meal, stimulates cellular uptake, use, and storage of these nutrients. Thus, the increase in insulin secretion that accompanies nutrient abundance, use, and storage appropriately inhibits the NPY-secreting cells of the arcuate nucleus, thus suppressing further food intake.

In addition to the importance of leptin, insulin, and perhaps other so-called **adiposity** (fat-related) **signals** in the long-term control of body weight, other factors play a role in controlling the timing and size of meals. Several blood-borne messengers from the digestive tract and pancreas are important in regulating how often and how much we eat in a given day.

SHORT-TERM EATING BEHAVIOR: GHRELIN AND PYY$_{3-36}$ SECRETION Two peptides important in the short-term control of food intake have recently been identified: *ghrelin* and *peptide YY$_{3-36}$* (*PYY$_{3-36}$*), which signify hunger and fullness, respectively. Both are secreted by the digestive tract. **Ghrelin** (GRELL-in), the so-called hunger hormone, is a potent appetite stimulator produced by the stomach and regulated by the feeding status (*ghrelin* is the Hindu word for "growth"). Secretion of this mealtime stimulator peaks before meals and makes people feel like eating, then falls once food is eaten. Ghrelin stimulates appetite by activating the hypothalamic NPY-secreting neurons.

PYY$_{3-36}$ is a counterpart of ghrelin. The secretion of PYY$_{3-36}$, which is produced by the small and large intestines, is at its lowest level before a meal but rises during meals and signals satiety. This peptide acts by inhibiting the appetite-stimulating NPY-secreting neurons in the arcuate nucleus. By thwarting appetite, PYY$_{3-36}$ is believed to be an important mealtime terminator.

The following other factors are also involved in signaling where the body is on the hunger–satiety scale.

BEYOND THE ARCUATE NUCLEUS: OREXINS AND OTHERS Two hypothalamic areas are richly supplied by axons from the NPY- and melanocortin-secreting neurons of the arcuate nucleus. These second-order neuronal areas involved in energy balance and food intake are the **lateral hypothalamic area (LHA)** and **paraventricular hypothalamic nucleus (PVN).** In a recently proposed model, the LHA and PVN release chemical messengers in response to input from the arcuate nucleus neurons. These messengers act downstream from the NPY and melanocortin signals to regulate appetite. The LHA produces two closely related neuropeptides known as **orexins** (ore-EKS-ins), which are potent stimulators of food intake (*orexis* means "appetite"). NPY stimulates and melanocortins inhibit the release of orexins, leading to an increase in appetite and greater food intake. By contrast, the PVN releases chemical messengers, for example, **corticotropin-releasing hormone,** that decrease appetite and food intake. (As its name implies, corticotropin-releasing hormone is better known for its role as a hormone.

You will learn more about this chemical's endocrine function in the next chapter.) Melanocortins stimulate and NPY inhibits the release of these appetite-suppressing neuropeptides.

In contrast to the key role of the hypothalamus in maintaining energy balance and long-term control of body weight, a region in the brain stem known as the **nucleus tractus solitarius (NTS)** processes signals important in the feeling of being full. It is considered the *satiety center*. Not only does the NTS receive input from the higher hypothalamic neurons involved in energy homeostasis but it also receives afferent inputs from the digestive tract and elsewhere that signal satiety (for example, afferent neural input indicating the extent of stomach distension). We now turn attention to cholecystokinin, the most important of these satiety signals.

CHOLECYSTOKININ AS A SATIETY SIGNAL Cholecystokinin **(CCK),** one of the gastrointestinal hormones released from the duodenal mucosa during digestion of a meal, is an important satiety signal for regulating the size of meals. CCK is secreted in response to the presence of nutrients in the small intestine. Through multiple effects on the digestive system, CCK facilitates digestion and absorption of these nutrients (see p. 637). It is appropriate that this blood-borne signal, whose rate of secretion is correlated with the amount of nutrients ingested, also contributes to the sense of being filled after a meal has been consumed but before it has actually been digested and absorbed. We feel satisfied when adequate food to replenish the stores is in the digestive tract, even though the body's energy stores are still low. This explains why we stop eating before the ingested food is made available to meet the body's energy needs.

▲ Table 17-3 summarizes the effect of involuntary regulatory signals on appetite.

PSYCHOSOCIAL AND ENVIRONMENTAL INFLUENCES Thus far we have described involuntary signals that automatically occur to control food intake. However, as with water intake, people's eating habits are also shaped by psychological, social, and environmental factors. Often our decision to eat or stop eating is not determined merely by whether we are hungry or full, respectively. Frequently, we eat out of habit (eating three meals a day on schedule no matter what our status on the hunger–satiety continuum) or because of social custom (food often plays a prime role in entertainment, leisure, and business activities). Even well-intentioned family pressure—"Clean your plate before you leave the table"—can have an impact on the amount consumed.

Furthermore, the amount of pleasure derived from eating can reinforce feeding behavior. Eating foods with an enjoyable taste, smell, and texture can increase appetite and food intake. This has been demonstrated in an experiment in which rats were offered their choice of highly palatable human foods. They overate by as much as 70% to 80% and became obese. When the rats returned to eating their regular monotonous but nutritionally balanced rat chow, their obesity was rapidly reversed, as their food intake was controlled once again by physiologic drives rather than by hedonistic urges for the tastier offerings.

Stress, anxiety, depression, and boredom have also been shown to alter feeding behavior in ways that are unrelated to energy needs in both experimental animals and humans. People often eat to satisfy psychological needs rather than to satisfy hunger. Furthermore, environmental influences, such as the amount of food available, play an important role in determining extent of food intake. Thus, any comprehensive explanation of how food intake is controlled must take into account these voluntary eating acts that can reinforce or override the internal signals governing feeding behavior.

Obesity occurs when more kilocalories are consumed than are burned up.

Clinical Note **Obesity** is defined as excessive fat content in the adipose tissue stores; the arbitrary boundary for obesity is generally considered to be greater than 20% overweight compared to normal standards. More than two thirds of the adults in the United States are now clinically overweight, with one third being clinically obese. To make matters worse, obesity is on the rise. The number of obese adults in the United States is 75% higher now than it was 15 years ago. And much of the world is following the same trend, recently leading the World Health Organization to coin the new word *globesity* to describe the worldwide situation.

Obesity occurs when, over a period of time, more kilocalories are ingested in food than are used to support the body's energy needs, with the excessive energy being stored as triglycerides in adipose tissue. Early in the development of obesity, existing fat cells get larger. An average adult has between 40 billion and 50 billion adipocytes. Each fat cell can store the maxi-

▲ TABLE 17-3	Effect of Involuntary Regulatory Signals on Appetite	
Regulatory Signal	**Source of Signal**	**Effect of Signal on Appetite**
Neuropeptide Y	Arcuate nucleus of hypothalamus	Increases
Melanocortins	Arcuate nucleus of hypothalamus	Decreases
Leptin	Adipose tissue	Decreases
Insulin	Endocrine pancreas	Decreases
Ghrelin	Stomach	Increases
PPY$_{3-36}$	Small and large intestines	Decreases
Orexins	Lateral hypothalamus	Increases
Corticotropin-releasing hormone	Paraventricular nucleus of hypothalamus	Decreases
Cholecystokinin	Small intestine	Decreases
Stomach distension	Stomach	Decreases

mum of about 1.2 μg of triglycerides. Once the existing fat cells are full, if people continue to consume more calories than they expend, they make more adipocytes, contrary to earlier beliefs.

The causes of obesity are many, and some remain obscure. Some factors that may be involved include the following:

■ *Disturbances in the leptin-signaling pathway.* Some cases of obesity have been linked to leptin resistance. For many overweight people, excess energy input occurs only during the time that obesity is actually developing. Some investigators suggest that the hypothalamic centers involved in maintaining energy homeostasis are "set at a higher level" in obese people. Once obesity is developed, all that is required to maintain the condition is that energy input equals energy output. For example, the problem may lie with faulty leptin receptors in the brain that do not respond appropriately to the high levels of circulating leptin from abundant adipose stores. Thus, the brain does not detect leptin as a signal to turn down appetite until a higher set point (and accordingly greater fat storage) is achieved. This could explain why overweight people do tend to maintain their weight but at a heavier-than-normal level. Instead of faulty leptin receptors, other disturbances in the leptin pathway may be at fault, such as defective transport of leptin across the blood–brain barrier or a deficiency of one of the chemical messengers in the leptin pathway.

■ *Lack of exercise.* Numerous studies have shown that, on average, fat people do not eat any more than thin people. One possible explanation is that overweight persons do not overeat but "underexercise"—the "couch potato" syndrome. Very low levels of physical activity typically are not accompanied by comparable reductions in food intake.

For this reason, modern technology is partly to blame for the current obesity epidemic. Our ancestors had to exert physical effort to eke out a subsistence. By comparison, we now have machines to replace much manual labor, remote controls to operate our machines with minimal effort, and computers that encourage long hours of sitting. We have to make a conscious effort to exercise.

■ *Differences in the "fidget factor."* **Nonexercise activity thermogenesis (NEAT),** or the "fidget factor," might explain some variation in fat storage among people. NEAT refers to energy expended by physical activities other than planned exercise. Those who engage in toe tapping or other types of repetitive, spontaneous physical activity expend a substantial number of kilocalories throughout the day without a conscious effort.

■ *Differences in extracting energy from food.* Another reason why lean people and obese people may have dramatically different body weights despite consuming the same number of kilocalories may lie in the efficiency with which each extracts energy from food. Studies suggest that leaner individuals tend to derive less energy from the food they consume because they convert more of the food's energy into heat than into energy for immediate use or for storage. For example, slimmer individuals have more **uncoupling proteins,** which allow their cells to convert more of the nutrient calories into heat instead of fat. These are the people who can eat a lot without gaining weight. By contrast, obese people may have more efficient metabolic systems for extracting energy from food—a useful trait in times of food shortage but a hardship when trying to maintain a desirable weight when food is plentiful.

■ *Hereditary tendencies.* Often differences in the regulatory pathways for energy balance—either those governing food intake or those influencing energy expenditure—arise from genetic variations. For example, individuals who have one faulty copy of a gene called *FTO* are 30% more likely to be obese than those who have two normal copies of the FTO gene. For people who have two faulty copies of FTO, the risk of becoming obese jumps to 70%. Scientists do not know yet what the FTO gene governs.

■ *Development of an excessive number of fat cells as a result of overfeeding.* One of the problems in fighting obesity is that once fat cells are created, they do not disappear with dieting and weight loss. Even if a dieter loses a large portion of the triglyceride fat stored in these cells, the depleted cells remain, ready to refill. Therefore, rebound weight gain after losing weight is difficult to avoid and discouraging for the dieter.

■ *The existence of certain endocrine disorders such as hypothyroidism* (see p. 695). Hypothyroidism involves a deficiency of thyroid hormone, the main factor that bumps up the BMR so that the body burns more calories in its idling state.

■ *An abundance of convenient, highly palatable, energy-dense, relatively inexpensive foods.* The current availability and convenience of food compared to most of human history make overeating easier than ever.

■ *Emotional disturbances in which overeating replaces other gratifications.*

■ *Stress.* Evidence suggests that chronic stress leads to increased release of NPY from sympathetic nerves, in turn causing increased deposition of visceral fat.

■ *Too little sleep.* Some studies suggest that decreased time sleeping may be a contributing factor in the recent rise of obesity. On average, people in the United States are sleeping one to two hours less per night now than they did 40 years ago. Researchers found that those who typically sleep six hours a night are 23% more likely to be obese, those who average five hours of sleep are 50% more likely to be obese, and those who sleep four hours nightly are 75% more likely to be obese than "traditional" sleepers who sleep for seven to nine hours. Studies have shown that leptin (a signal to stop eating) levels are lower and ghrelin (a signal to start eating) levels are higher in people who sleep less compared with those who sleep the traditional eight hours.

■ *A possible virus link.* One intriguing new proposal links a relatively common cold virus to a propensity to become overweight and may account for a portion of the current obesity epidemic. One study showed that the cold virus *adenovirus-36* might lead to obesity by transforming adult tissue-specific stem cells into fat-storing adipocytes.

■ *Composition of colonic bacterial communities.* Studies demonstrate that obese people have a greater proportion of a type of bacteria in their colon that breaks down indigestible fiber more efficiently for absorption from the digestive tract compared to the bacterial communities in lean people. By making more absorbable units available for uptake from the digestive tract, the fat-promoting bacteria help their human host obtain more en-

What the Scales Don't Tell You

Body composition is the percentage of body weight that is composed of lean tissue and adipose tissue. Assessing body composition is an important step in evaluating a person's health status. One crude means of assessing body composition is by calculating the **body mass index (BMI)** using the following formula:

$$BMI = \frac{(\text{weight in pounds}) \times 700}{(\text{height in inches})^2}$$

A BMI of 25 or less is considered healthy, whereas a BMI of 30 or higher is considered to place the person at increased risk for various diseases and premature death. BMIs between this range are considered borderline.

BMI determinations and the age–height–weight tables used by insurance companies can be misleading for determining healthy body weight. By these charts, many athletes, for example, would be considered overweight. A football player may be 6 feet 5 inches tall and weigh 300 pounds but have only 12% body fat. This player's extra weight is muscle, not fat, and therefore is not a detriment to his health. A sedentary person, in contrast, may be normal on the height–weight charts but have 30% body fat. This person should maintain body weight while increasing muscle mass and decreasing fat. Ideally, men should have 15% fat or less and women should have 20% fat or less.

The most accurate method for assessing body composition is underwater weighing. This technique is based on the fact that lean tissue is denser than water and fat tissue is less dense than water. (You can readily demonstrate this for yourself by dropping a piece of lean meat and a piece of fat into a glass of water; the lean meat will sink and the fat will float.) In underwater weighing, the person breathes out as much air as possible and then completely submerges in a tank of water while sitting in a swing attached to a scale. The results are used to determine body density using equations that take into consideration the density of water, the difference between the person's weight in air and underwater, and the residual volume of air remaining in the lungs. Because of the difference in density between lean and fat tissue, people who have more fat have a lower density and weigh relatively less underwater than in air compared to their lean counterparts. Body composition is then determined by means of an equation that correlates percentage fat with body density.

Another common way to assess body composition is skinfold thickness. Because approximately half of the body's total fat content is located just beneath the skin, total body fat can be estimated from measurements of skinfold thickness taken at various sites on the body. Skinfold thickness is determined by pinching up a fold of skin at one of the designated sites and measuring its thickness by means of a caliper, a hinged instrument that fits over the fold and is calibrated to measure thickness. Mathematical equations specific for the person's age and sex can be used to predict the percentage of fat from the skinfold thickness scores. A major criticism of skinfold assessments is that accuracy depends on the investigator's skill.

There are different ways to be fat, and one way is more dangerous than the other. Obese patients can be classified into two categories—*android,* a male-type of adipose tissue distribution, and *gynoid,* a female-type distribution—based on the anatomic distribution of adipose tissue measured as the ratio of waist circumference to hip circumference. **Android obesity** is characterized by abdominal fat distribution (people shaped like "apples"), whereas **gynoid obesity** is characterized by fat distribution in the hips and thighs (people shaped like "pears"). Both sexes can display either android or gynoid obesity.

Android obesity is associated with a number of disorders, including insulin resistance, Type 2 (adult-onset) diabetes mellitus, excess blood lipid levels, high blood pressure, coronary heart disease, and stroke. Gynoid obesity is not associated with high risk for these diseases. Because android obesity is associated with increased risk for disease, it is most important for apple-shaped overweight individuals to reduce their fat stores. Among the apples, a greater proportion of visceral fat is more worrisome than accumulation of subcutaneous fat. **Visceral fat** is the adipose tissue that surrounds the abdominal viscera, or organs. **Subcutaneous fat** is the adipose tissue that lies below the skin (see p. 645). Visceral fat releases more of the "bad" chemical signals that promote insulin resistance and boost the low-level inflammation that underlies the development of atherosclerosis (see p. 333).

Research on the success of weight-reduction programs indicates that it is very difficult for people to lose weight, but when weight loss occurs, it is from the areas of increased stores. Because very-low-calorie diets are difficult to maintain, an alternative to severely cutting caloric intake to lose weight is to increase energy expenditure through physical exercise. Exercise physiologists often assess body composition as an aid in prescribing and evaluating exercise programs. Exercise generally reduces the percentage of body fat and, by increasing muscle mass, increases the percentage of lean tissue. An aerobic exercise program further helps reduce the risk of the disorders associated with android obesity.

---

ergy from the same number of kilocalories consumed than leaner people who have a preponderance of less energy-efficient colonic bacteria. Although the colon does not have much absorptive capacity, even a little bit more energy absorbed from "indigestible" food for which humans do not have digestive enzymes over the course of time can lead to added pounds.

Despite this rather lengthy list, our knowledge about the causes and control of obesity is still rather limited, as evidenced by the number of people who are constantly trying to stabilize their weight at a more desirable level. This is important from more than an aesthetic viewpoint. It is known that obesity, especially of the android type, can predispose an individual to illness and premature death from a multitude of diseases. (To learn about the differences between android and gynoid obesity, see the accompanying boxed feature, ■ A Closer Look at Exercise Physiology.)

People suffering from anorexia nervosa have a pathological fear of gaining weight.

The converse of obesity is generalized nutritional deficiency. The obvious causes for reduction of food intake below energy needs are lack of availability of food, interference with swallowing or digestion, and impairment of appetite.

One poorly understood disorder in which lack of appetite is a prominent feature is **anorexia nervosa.** Patients with this disorder, most commonly adolescent girls and young women, have a morbid fear of becoming fat. They have a distorted body image, tending to visualize themselves as being much heavier than they actually are. Because they have an aversion to food, they eat very little and consequently lose considerable weight, perhaps even starving themselves to death. Other characteristics of the condition include altered secretion of many hormones, absence of menstrual periods, and low body temperature. It is unclear whether these symptoms occur secondarily as a result of general malnutrition or arise independently of the eating disturbance as a part of a primary hypothalamic malfunction. Many investigators think the underlying problem may be psychological rather than biological. Some experts suspect that anorexics may suffer from addiction to endogenous opiates, self-produced morphinelike substances (see p. 192) thought to be released during prolonged starvation.

Temperature Regulation

Humans are usually in environments cooler than their bodies, but they constantly generate heat internally, which helps maintain body temperature. Heat production ultimately depends on the oxidation of metabolic fuel derived from food.

Changes in body temperature in either direction alter cell activity—an increase in temperature speeds up cellular chemical reactions, whereas a fall in temperature slows down these reactions. Because cell function is sensitive to fluctuations in internal temperature, humans homeostatically maintain body temperature at a level optimal for cellular metabolism to proceed in a stable fashion. Overheating is more serious than cooling. Even moderate elevations of body temperature begin to cause nerve malfunction and irreversible protein denaturation. Most people suffer convulsions when the internal body temperature reaches about 106°F (41°C); 110°F (43.3°C) is considered the upper limit compatible with life.

By contrast, most of the body's tissues can transiently withstand substantial cooling. This characteristic is useful during cardiac surgery when the heart must be stopped. The patient's body temperature is deliberately lowered. The cooled tissues need less nourishment than they do at normal body temperature because of their reduced metabolic activity. However, a pronounced, prolonged fall in body temperature slows metabolism to a fatal level.

Internal core temperature is homeostatically maintained at 100°F (37.8°C).

Normal body temperature taken orally (by mouth) has traditionally been considered 98.6°F (37°C). However, a recent study indicates that normal body temperature varies among individuals and varies throughout the day, ranging from 96.0°F (35.5°C) in the morning to 99.9°F (37.7°C) in the evening, with an overall average of 98.2°F (36.7°C).

Furthermore, there is no one body temperature because the temperature varies from organ to organ. From a thermoregulatory viewpoint, the body may conveniently be viewed as a *central core* surrounded by an *outer shell.* The temperature within the inner core, which consists of the abdominal and thoracic organs, the central nervous system, and the skeletal muscles, generally remains fairly constant. This internal **core temperature** is subject to precise regulation to maintain its homeostatic constancy. The core tissues function best at a relatively constant temperature of around 100°F (37.8°C).

The skin and subcutaneous fat constitute the outer shell. In contrast to the constant high temperature in the core, the temperature within the shell is generally cooler and may vary substantially. For example, skin temperature may fluctuate between 68° and 104°F (20° and 40°C) without damage. In fact, as you will see, the temperature of the skin is deliberately varied as a control measure to help maintain the core's thermal constancy.

SITES FOR MONITORING BODY TEMPERATURE Several easily accessible sites are used for monitoring body temperature. The oral and axillary (under the armpit) temperatures are comparable, whereas rectal temperature averages about 1°F (0.56°C) higher. Also available is a temperature-monitoring instrument that scans the heat generated by the eardrum and converts this temperature into an oral equivalent. A more recent device is the temporal scanner that measures the temperature of the blood in the temporal artery. The temporal scanner senses heat naturally radiated from the body. The computerized instrument is gently stroked from one side to the other across the forehead over the temporal artery, which lies less than 2 mm below the skin surface in this region, then is momentarily touched in the neck area behind the ear lobe for comparison. Temporal temperature is closer to rectal temperature than to oral temperature and is the best determinant of core temperature because it is nearly identical to the temperature of the blood exiting the heart. However, none of these measurements is an absolute indication of the internal core temperature, which is a bit higher at 100°F than the monitored sites.

NORMAL VARIATIONS IN CORE TEMPERATURE Even though the core temperature is held relatively constant, several factors cause it to vary slightly:

1. Most people's core temperature normally varies about 1.8°F (1°C) during the day, with the lowest level early in the morning before rising (6 to 7 A.M.) and the highest point in late afternoon (5 to 7 P.M.). This variation is the result of an innate biological rhythm, or "biological clock" (see p. 685).

2. Women also experience a monthly rhythm in core temperature in connection with their menstrual cycle. The core temperature averages 0.9°F (0.5°C) higher during the last half of the cycle from the time of ovulation to menstruation. The mild sustained elevation in temperature during this period was once thought to be caused by the increased secretion of progesterone,

one of the ovarian hormones, but this is no longer believed to be the case. The actual cause is still undetermined.

3. The core temperature increases during exercise because of the tremendous increase in heat production by the contracting muscles. During hard exercise, the core temperature may increase to as much as 104°F (40°C). In a resting person, this temperature would be considered a fever, but it is normal during strenuous exercise.

3. Older is colder. The elderly naturally have lower temperatures, with a midday average of 97.7°F (36.4°C).

4. Because the temperature-regulating mechanisms are not 100% effective, the core temperature may vary slightly with exposure to extremes of temperature. For example, the core temperature may fall several degrees in cold weather or rise a degree or so in hot weather.

Thus, the core temperature can vary at the extremes between about 96° and 104°F but usually deviates less than a few degrees. This relative constancy is made possible by multiple thermoregulatory mechanisms coordinated by the hypothalamus.

Heat input must balance heat output to maintain a stable core temperature.

The core temperature is a reflection of the body's total heat content. Heat input to the body must balance heat output to maintain a constant total heat content and thus a stable core temperature (● Figure 17-3). *Heat input* occurs by way of heat gain from the external environment and internal heat production, the latter being the most important source of heat for the body. Recall that most of the body's energy expenditure ultimately appears as heat. This heat is important in maintaining core temperature. In fact, usually more heat is generated than required to maintain normal body temperature, so the excess heat must be eliminated. *Heat output* occurs by way of heat loss from exposed body surfaces to the external environment.

Balance between heat input and output is frequently disturbed by (1) changes in internal heat production for purposes unrelated to regulation of body temperature, most notably by exercise, which markedly increases heat production; and (2) changes in the external environmental temperature that influence the degree of heat gain or heat loss that occurs between the body and its surroundings. Compensatory adjustments must take place in heat-loss and heat-gain mechanisms to maintain body temperature within narrow limits despite changes in metabolic heat production and changes in environmental temperature. If the core temperature starts to fall, heat production is increased and heat loss is minimized so that normal temperature can be restored. Conversely, if the temperature starts to rise above normal, it can be corrected by increasing heat loss while simultaneously reducing heat production.

We will now elaborate on the means by which heat gains and losses can be adjusted to maintain body temperature.

Heat exchange takes place by radiation, conduction, convection, and evaporation.

All heat loss or heat gain between the body and external environment must take place between the body surface and its surroundings. The same physical laws of nature that govern heat transfer between inanimate objects also control the transfer of heat between the body surface and environment. The temperature of an object is a measure of the concentration of heat within the object. Accordingly, heat always moves down its concentration gradient—that is, down a **thermal gradient** from a warmer to a cooler region (*thermo* means "heat"). The body uses four mechanisms of heat transfer: *radiation, conduction, convection,* and *evaporation.*

RADIATION **Radiation** is the emission of heat energy from the surface of a warm body in the form of **electromagnetic waves,** or **heat waves,** which travel through space (● Figure 17-4a). When radiant energy strikes an object and is absorbed, the energy of the wave motion is transformed into heat within the object. The human body both emits (source of heat loss) and absorbs (source of heat gain) radiant energy. Whether the body loses or gains heat by radiation depends on the difference in temperature between the skin surface and the surfaces of other objects in the body's environment. Because net transfer of heat by radiation is always from warmer objects to cooler ones, the body gains heat by radiation from objects warmer than the skin surface, such as the sun, a radiator, or burning logs. By contrast, the body loses heat by radiation to objects in its environment whose surfaces are cooler than the surface of the skin, such as building walls, furniture, or trees. On average, humans lose close to half of their heat energy through radiation.

CONDUCTION **Conduction** is the transfer of heat between objects of differing temperatures that are in direct contact with each other, with heat moving down its thermal gradient from the warmer to the cooler object by being transferred from molecule to molecule. All molecules are constantly in vibratory motion, with warmer molecules moving faster than cooler ones. When molecules of differing heat content touch each other, the faster-moving, warmer molecule agitates the cooler molecule into more rapid motion, thereby warming up the cooler molecule. During this process, the original warmer molecule loses some of its thermal energy as it slows down and

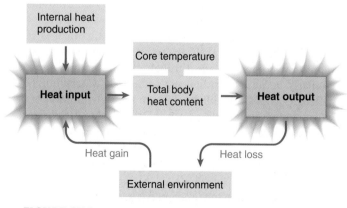

● **FIGURE 17-3 Heat input and output.**

(a) Radiation—the transfer of heat energy from a warmer object to a cooler object in the form of electromagnetic waves ("heat waves"), which travel through space.

Snowball   Heating pad

(b) Conduction—the transfer of heat from a warmer to a cooler object that is in direct contact with the warmer one. The heat is transferred through the movement of thermal energy from molecule to adjacent molecule.

Convection current

(c) Convection—the transfer of heat energy by air currents. Cool air warmed by the body through conduction rises and is replaced by more cool air. This process is enhanced by the forced movement of air across the body surface.

Liquid converted to gaseous vapor

(d) Evaporation—conversion of a liquid such as sweat into a gaseous vapor, a process that requires heat (the heat of vaporization), which is absorbed from the skin.

● **FIGURE 17-4 Mechanisms of heat transfer.** The direction of the arrows depicts the direction of heat transfer.

CONVECTION The term **convection** refers to the transfer of heat energy by *air* (or *water*) *currents*. As the body loses heat by conduction to the surrounding cooler air, the air in immediate contact with the skin is warmed. Because warm air is lighter (less dense) than cool air, the warmed air rises while cooler air moves in next to the skin to replace the vacating warm air. The process is then repeated (● Figure 17-4c). These air movements, known as *convection currents,* help carry heat away from the body. Without convection currents, no further heat could be dissipated from the skin by conduction once the temperature of the layer of air immediately around the body equilibrated with skin temperature.

The combined conduction–convection process of dissipating heat from the body is enhanced by forced movement of air across the body surface, either by external air movements, such as those caused by the wind or a fan, or by movement of the body through the air, as during bicycle riding. Because forced air movement sweeps away the air warmed by conduction and replaces it with cooler air more rapidly, a greater total amount of heat can be carried away from the body over a given time period. Thus, wind makes us feel cooler on hot days, and windy days in the winter are more chilling than calm days at the same cold temperature. For this reason, weather forecasters have developed the concept of *wind chill factor* (how cold it feels).

EVAPORATION During **evaporation** from the skin surface, the heat required to transform water from a liquid to a gaseous state is absorbed from the skin, thereby cooling the body (● Figure 17-4d). Evaporative heat loss makes you feel cooler when your bathing suit is wet than when it is dry. Evaporative heat loss occurs continually from the linings of the respiratory airways and from the surface of the skin. Heat is continuously lost in expired air as a result of the air gaining water vapor during its passage through the respiratory system. Similarly, because the skin is not completely waterproof, H_2O molecules constantly diffuse through the skin and evaporate. This ongoing evaporation from the skin is completely unrelated to the sweat glands. These passive evaporative heat-loss processes are not subject to physiologic control and go on even in very cold weather, when the problem is one of conserving body heat.

Sweating is an active evaporative heat-loss process under sympathetic nervous control. The rate of evaporative heat loss

cools off a bit. Given enough time, therefore, the temperature of the two touching objects eventually equalizes.

The rate of heat transfer by conduction depends on the *temperature difference* between the touching objects and the *thermal conductivity* of the substances involved (that is, how easily heat is conducted by the molecules of the substances). Heat can be lost or gained by conduction when the skin is in contact with a good conductor (● Figure 17-4b). When you hold a snowball, for example, your hand becomes cold because heat moves by conduction from your hand to the snowball. Conversely, when you apply a heating pad to a body part, the part is warmed up as heat is transferred directly from the pad to the body.

Similarly, you either lose or gain heat by conduction to the layer of air in direct contact with your body. The direction of heat transfer depends on whether the air is cooler or warmer, respectively, than your skin. Only a small percentage of total heat exchange between the skin and environment takes place by conduction alone, however, because air is not a very good conductor of heat. (For this reason, swimming pool water at 80°F (26.7°C) feels cooler than air at the same temperature; heat is conducted more rapidly from the body surface into the water, which is a good conductor, than into the air, which is a poor conductor.)

can be deliberately adjusted by varying the extent of sweating, which is an important homeostatic mechanism to eliminate excess heat as needed. In fact, when the environmental temperature exceeds the skin temperature, sweating is the only avenue for heat loss because the body is gaining heat by radiation and conduction under these circumstances. At normal temperature, an average of 100 ml of sweat is produced per day; this value increases to 1.5 liters during hot weather and climbs to 4 liters during heavy exercise.

Sweat is a dilute salt solution actively extruded to the surface of the skin by **eccrine sweat glands** dispersed all over the body. This clear, salty sweat is the kind important in cooling the body. **Apocrine sweat glands,** which are located primarily in the armpits and genital area, produce a thick, milky sweat that is rich in organic constituents, like proteins and lipids. Body odor is generated when bacteria in the vicinity break down these organic compounds. There is no known physiological importance of the apocrine sweat glands. They are remnants of sexual scent glands found in other species.

Eccrine sweat must be evaporated from the skin for heat loss to occur. If sweat merely drips from the surface of the skin or is wiped away, no heat loss is accomplished. The most important factor determining the extent of evaporation of sweat is the *relative humidity* of the surrounding air (the percentage of H_2O vapor actually present in the air compared to the greatest amount that the air can possibly hold at that temperature; for example, a relative humidity of 70% means that the air contains 70% of the H_2O vapor it is capable of holding). When the relative humidity is high, the air is already almost fully saturated with H_2O, so it has limited ability to take up additional moisture from the skin. Thus, little evaporative heat loss can occur on hot, humid days. The sweat glands continue to secrete, but the sweat simply remains on the skin or drips off instead of evaporating and producing a cooling effect. As a measure of the discomfort associated with combined heat and high humidity, meteorologists have devised the *temperature–humidity index,* or *heat index* (how hot it feels).

The hypothalamus integrates a multitude of thermosensory inputs.

The hypothalamus is the body's thermostat. The home thermostat keeps track of the temperature in a room and triggers a heating mechanism (the furnace) or a cooling mechanism (the air conditioner) as necessary to maintain room temperature at the indicated setting. Similarly, the hypothalamus, as the body's thermoregulatory integrating center, receives afferent information about the temperature in various regions of the body and initiates extremely complex, coordinated adjustments in heat-gain and heat-loss mechanisms as necessary to correct any deviations in core temperature from normal. The hypothalamus is far more sensitive than your home thermostat. It can respond to changes in blood temperature as small as 0.01°C.

To appropriately adjust the delicate balance between the heat-loss mechanisms and the opposing heat-producing and heat-conserving mechanisms, the hypothalamus must be apprised continuously of both the core and the skin temperature by specialized temperature-sensitive receptors called **thermoreceptors.** The core temperature is monitored by *central ther-*

moreceptors, which are located in the hypothalamus itself as well as in the abdominal organs and elsewhere. *Peripheral thermoreceptors* monitor skin temperature throughout the body.

Two centers for temperature regulation are in the hypothalamus. The *posterior region,* activated by cold, triggers reflexes that mediate heat production and heat conservation. The *anterior region,* activated by warmth, initiates reflexes that mediate heat loss. Let us examine the means by which the hypothalamus fulfills its thermoregulatory functions (● Figure 17-5).

Shivering is the primary involuntary means of increasing heat production.

The body can gain heat as a result of internal heat production generated by metabolic activity or from the external environment if the latter is warmer than body temperature. Because body temperature usually is higher than environmental temperature, metabolic heat production is the primary source of body heat. In a resting person, most body heat is produced by the thoracic and abdominal organs as a result of ongoing, cost-of-living metabolic activities. Above and beyond this basal level, the rate of metabolic heat production can be variably increased primarily by changes in skeletal muscle activity or, to a lesser extent, by certain hormonal actions. Thus, changes in skeletal muscle activity constitute the major way heat gain is controlled for temperature regulation.

ADJUSTMENTS IN HEAT PRODUCTION BY SKELETAL MUSCLES In response to a fall in core temperature caused by exposure to cold, the hypothalamus takes advantage of the fact that increased skeletal muscle activity generates more heat. Acting through descending pathways that terminate on the motor neurons controlling the skeletal muscles, the hypothalamus first gradually increases skeletal muscle tone. (Muscle tone is the constant level of tension within the muscles.) Soon shivering begins. **Shivering** consists of rhythmic, oscillating skeletal muscle contractions that occur at a rapid rate of 10 to 20 per second. This mechanism is very effective in increasing heat production; all the energy liberated during these muscle tremors is converted to heat because no external work is accomplished. Within a matter of seconds to minutes, internal heat production may increase two- to fivefold as a result of shivering.

Frequently, these reflex changes in skeletal muscle activity are augmented by increased voluntary, heat-producing actions such as bouncing up and down or hand clapping. The hypothalamus influences these behavioral responses as well as the involuntary physiologic responses. As part of the limbic system, the hypothalamus is extensively involved with controlling motivated behavior (see p. 156).

In the opposite situation—a rise in core temperature caused by heat exposure—two mechanisms reduce heat-producing skeletal muscle activity: Muscle tone is reflexly decreased, and voluntary movement is curtailed. When the air becomes very warm, people often complain it is "too hot even to move." These responses are not as effective at reducing heat production during heat exposure as are the muscular responses that increase heat production during cold exposure, for two reasons. First, because muscle tone is normally quite low, the capacity to re-

Change in skin temperature

Change in core temperature

Peripheral thermoreceptors in skin

Central thermoreceptors in hypothalamus, abdominal organs, and elsewhere

Hypothalamic centers for thermoregulation (body's thermostat)

Motor neurons

Sympathetic nerves

Sympathetic nerves

Skeletal muscles

Smooth muscle in arterioles in skin

Sweat glands

Voluntary changes in behavior

Muscle tone, shivering

Vasoconstriction, vasodilation

Sweating

Adjustments in heat production or heat loss

Adjustments in muscle activity (in metabolic heat production)

Adjustment in loss or conservation of heat

Adjustment in heat loss

● **FIGURE 17-5 Major thermoregulatory pathways.**

Having examined the mechanisms for adjusting heat production, we now turn to the other side of the equation, adjustments in heat loss.

The magnitude of heat loss can be adjusted by varying the flow of blood through the skin.

Heat-loss mechanisms are also subject to control, again largely by the hypothalamus. When we are hot, we need to increase heat loss to the environment; when we are cold, we need to decrease heat loss. The amount of heat lost to the environment by radiation and conduction-convection is largely determined by the temperature gradient between the skin and external environment. The body's central core is a heat-generating chamber in which the temperature must be maintained at approximately 100°F. Surrounding the core is an insulating shell through which heat exchanges between the body and external environment take place. To maintain a constant core temperature, the insulative capacity and temperature of the shell can be adjusted to vary the temperature gradient between the skin and external environment, thereby influencing the extent of heat loss.

The insulative capacity of the shell can be varied by controlling the amount of blood flowing through the skin. Skin blood flow serves two functions. First, it provides a nutritive blood supply to the skin. Second, most skin blood flow is for the function of temperature regulation; at normal room temperature, 20 to 30 times more blood flows through the skin than is needed for skin nutrition. In the process of thermoregulation, skin blood flow can vary tremendously, from 400 ml/min up to 2500 ml/min. The more blood that reaches the skin from the warm core, the closer the skin's temperature is to the core temperature. The skin's blood vessels diminish the effectiveness of the skin as an insulator by carrying heat to the surface, where it can be lost from the body by radiation and conduction-convection. Accordingly, skin arteriolar vasodilation, which permits increased flow of heated blood through the skin, increases heat loss. Conversely, skin vasoconstriction, which reduces skin blood flow, decreases heat loss by keeping the warm blood in the central core, where it is insulated from the external environment. This response conserves heat that otherwise would have been lost. Cold, relatively bloodless skin provides excellent insulation between the core and environment. However, the skin is not a perfect insulator, even with maximum vasoconstriction. Despite minimal blood flow to the skin, some heat can still be transferred by conduction from the deeper organs to the skin surface and then can be lost from the skin to the environment.

These skin vasomotor responses are coordinated by the hypothalamus by means of sympathetic nervous system output.

duce it further is limited. Second, the elevated body temperature tends to increase the rate of metabolic heat production.

NONSHIVERING THERMOGENESIS Although reflex and voluntary changes in muscle activity are the major means of increasing the rate of heat production, **nonshivering (chemical) thermogenesis** also plays a role in thermoregulation. In most experimental animals, chronic cold exposure brings about an increase in metabolic heat production that is independent of muscle contraction, instead being brought about by changes in heat-generating chemical activity. In humans, nonshivering thermogenesis is most important in newborns because they lack the ability to shiver. Nonshivering thermogenesis is mediated by the hormones epinephrine and thyroid hormone, both of which increase heat production by stimulating fat metabolism. Newborns have deposits of a special type of adipose tissue known as **brown fat,** which is especially capable of converting chemical energy into heat. The role of nonshivering thermogenesis in adults remains controversial.

Coordinated Adjustments in Response to Cold or Heat Exposure

IN RESPONSE TO COLD EXPOSURE (COORDINATED BY THE POSTERIOR HYPOTHALAMUS)		IN RESPONSE TO HEAT EXPOSURE (COORDINATED BY THE ANTERIOR HYPOTHALAMUS)	
Increased Heat Production	**Decreased Heat Loss (Heat Conservation)**	**Decreased Heat Production**	**Increased Heat Loss**
Increased muscle tone	Skin vasoconstriction	Decreased muscle tone	Skin vasodilation
Shivering	Postural changes to reduce exposed surface area (hunching shoulders, etc.)*	Decreased voluntary exercise*	Sweating
Increased voluntary exercise*	Warm clothing*		Cool clothing*
Nonshivering thermogenesis			

*Behavioral adaptations

Increased sympathetic activity to the skin arterioles produces heat-conserving vasoconstriction in response to cold exposure, whereas decreased sympathetic activity produces heat-losing vasodilation of these vessels in response to heat exposure.

Recall that the cardiovascular control center in the medulla also exerts control over the skin arterioles (as well as arterioles throughout the body) by means of adjusting sympathetic activity to these vessels for the purpose of blood pressure regulation (see p. 378). Hypothalamic control over the skin arterioles for the purpose of temperature regulation takes precedence over the cardiovascular control center's control of these same vessels (see p. 380). Thus, changes in blood pressure can result from pronounced thermoregulatory skin vasomotor responses. For example, blood pressure can fall on exposure to a very hot environment because the skin vasodilator response set in motion by the hypothalamic thermoregulatory center overrides the skin vasoconstrictor response called forth by the medullary cardiovascular control center.

The hypothalamus simultaneously coordinates heat-production and heat-loss mechanisms.

Let us now pull together the coordinated adjustments in heat production and heat loss in response to exposure to either a cold or a hot environment (▲ Table 17-4).

COORDINATED RESPONSES TO COLD EXPOSURE In response to cold exposure, the posterior region of the hypothalamus directs increased heat production, such as by shivering, while simultaneously decreasing heat loss (that is, conserving heat) by skin vasoconstriction and other measures.

Because there is a limit to the body's ability to reduce skin temperature through vasoconstriction, even maximum vasoconstriction is not sufficient to prevent excessive heat loss when the external temperature falls too low. Accordingly, other measures must be instituted to further reduce heat loss. In animals with dense fur or feathers, the hypothalamus, acting through the sympathetic nervous system, brings about contraction of the tiny muscles at the base of the hair or feather shafts to lift the hair or feathers off the skin surface. This puffing up traps a layer of poorly conductive air between the skin surface and environment, thus increasing the insulating barrier between the core and the cold air and reducing heat loss. Even though the hair-shaft muscles contract in humans in response to cold exposure, this heat-retention mechanism is ineffective because of the low density and fine texture of most human body hair. The result instead is useless *goose bumps*.

After maximum skin vasoconstriction has been achieved as a result of exposure to cold, further heat dissipation in humans can be prevented only by behavioral adaptations, such as postural changes that reduce as much as possible the exposed surface area from which heat can escape. These postural changes include maneuvers such as hunching over, clasping the arms in front of the chest, or curling up in a ball.

Putting on warmer clothing further insulates the body from too much heat loss. Clothing entraps layers of poorly conductive air between the skin surface and the environment, thereby diminishing loss of heat by conduction from the skin to the cold external air and curtailing the flow of convection currents.

COORDINATED RESPONSES TO HEAT EXPOSURE Under the opposite circumstance—heat exposure—the anterior part of the hypothalamus reduces heat production by decreasing skeletal muscle activity and promotes increased heat loss by inducing skin vasodilation. When even maximal skin vasodilation is inadequate to rid the body of excess heat, sweating is brought into play to accomplish further heat loss through evaporation. In fact, if the air temperature rises above the temperature of maximally vasodilated skin, the temperature gradient reverses itself so that heat is gained from the environment. Sweating is the only means of heat loss under these conditions.

Humans also employ voluntary measures, such as using fans, wetting the body, drinking cold beverages, and wearing cool clothing, to further enhance heat loss. Contrary to popular belief, wearing light-colored, loose clothing is cooler than being nude. Naked skin absorbs almost all the radiant energy that strikes it, whereas light-colored clothing reflects almost all the radiant energy that falls on it. Thus, if light-colored clothing is loose and thin enough to permit convection currents and evaporative heat loss to occur, wearing it is actually cooler than going without any clothes at all.

THERMONEUTRAL ZONE Skin vasomotor activity is highly effective in controlling heat loss in environmental temperatures between the upper 60s and mid-80s. This range, within which core temperature can be kept constant by vasomotor responses without calling supplementary heat-production or heat-loss mechanisms into play, is called the **thermoneutral zone.** When external air temperature falls below the lower limits of the ability of skin vasoconstriction to reduce heat loss further, increased heat production, especially by shivering, becomes necessary to maintain core temperature. At the other extreme, when external air temperature exceeds the upper limits of the ability of skin vasodilation to increase heat loss further, sweating becomes essential to maintain core temperature. (For a discussion of the effects of extreme heat or cold exposure, see the accompanying boxed feature, ■ Concepts, Challenges, and Controversies.)

During a fever, the hypothalamic thermostat is "reset" at an elevated temperature.

Clinical Note The term **fever** refers to an elevation in body temperature as a result of infection or inflammation. In response to microbial invasion, certain phagocytic cells (macrophages) release chemicals that act as an **endogenous pyrogen,** which, among its many infection-fighting effects (see p. 424), acts on the hypothalamic thermoregulatory center to raise the thermostat setting (● Figure 17-6). The hypothalamus now maintains the temperature at the new set level instead of maintaining normal body temperature. If, for example, endogenous pyrogen raises the set point to 102°F (38.9°C), the hypothalamus senses that the normal prefever temperature is too cold, so it initiates the cold-response mechanisms to raise the temperature to 102°F. Specifically, it initiates shivering to rapidly increase heat production, and promotes skin vasoconstriction to rapidly reduce heat loss, both of which drive the temperature upward. These events account for the sudden cold chills often experienced at the onset of a fever. Feeling cold, the person may voluntarily put on more blankets to help raise body temperature by conserving body heat. Once the new temperature is achieved, body temperature is regulated as normal in response to cold and heat but at a higher setting. Thus, fever production in response to an infection is a deliberate outcome and is not caused by a breakdown of thermoregulation. Although the physiologic significance of a fever is still unclear, many medical experts believe that a rise in body temperature has a beneficial role in fighting infection. A fever augments the inflammatory response and may interfere with bacterial multiplication.

During fever production, endogenous pyrogen raises the set point of the hypothalamic thermostat by triggering the local release of *prostaglandins,* which are local chemical mediators that act directly on the hypothalamus. Aspirin reduces a fever by inhibiting the synthesis of prostaglandins. Aspirin does not lower the temperature in a nonfebrile person, because in the absence of endogenous pyrogen, prostaglandins are not present in the hypothalamus in appreciable quantities.

● **FIGURE 17-6 Fever production.** (Photo: L. J. Le Beau/ Biological Photo Service.)

The exact molecular cause of a fever "breaking" naturally is unknown, although it presumably results from reduced pyrogen release or decreased prostaglandin synthesis. When the hypothalamic set point is restored to normal, the temperature at 102°F (in this example) is too high. The heat-response mechanisms are instituted to cool down the body. Skin vasodilation occurs, and sweating commences. The person feels hot and throws off extra covers. The gearing up of these heat-loss mechanisms by the hypothalamus reduces the temperature to normal.

Hyperthermia can occur unrelated to infection.

Hyperthermia denotes any elevation in body temperature above the normally accepted range. The term *fever* is usually reserved for an elevation in temperature caused by endogenous pyrogen resetting the hypothalamic set point during infection or inflammation; *hyperthermia* refers to all other imbalances between heat gain and heat loss that increase body temperature. Hyperthermia has a variety of causes, some of which are normal and harmless, others pathologic and fatal.

EXERCISE-INDUCED HYPERTHERMIA The most common cause of hyperthermia is sustained exercise. As a physical consequence of the tremendous heat load generated by exercising muscles, body temperature rises during the initial stage of exercise be-

The Extremes of Heat and Cold Can Be Fatal

Prolonged exposure to temperature extremes in either direction can overtax the body's thermoregulatory mechanisms, leading to disorders and even death.

Heat-Related Disorders

Heat exhaustion is a state of collapse, usually manifested by fainting, that is caused by reduced blood pressure brought about as a result of overtaxing the heat-loss mechanisms. Extensive sweating reduces cardiac output by depleting the plasma volume, and pronounced skin vasodilation causes a drop in total peripheral resistance. Because blood pressure is determined by cardiac output times total peripheral resistance, blood pressure falls, an insufficient amount of blood is delivered to the brain, and fainting takes place. Thus, heat exhaustion is a consequence of overactivity of the heat-loss mechanisms rather than a breakdown of these mechanisms. Because the heat-loss mechanisms have been very active, body temperature is only mildly elevated in heat exhaustion. By forcing the cessation of activity when the heat-loss mechanisms are no longer able to cope with heat gain through exercise or a hot environment, heat exhaustion serves as a safety valve to help prevent the more serious consequences of heatstroke.

Heatstroke is an extremely dangerous situation that arises from the complete breakdown of the hypothalamic thermoregulatory systems. Heat exhaustion may progress into heatstroke if the heat-loss mechanisms continue to be overtaxed. Heatstroke is more likely to occur on overexertion during a prolonged exposure to a hot, humid environment. The elderly, in whom thermoregulatory responses are generally slower and less efficient, are particularly vulnerable to heatstroke during prolonged, stifling heat waves. So too are individuals who are taking certain common tranquilizers, such as Valium, because these drugs interfere with the hypothalamic thermoregulatory centers' neurotransmitter activity.

The most striking feature of heatstroke is a lack of compensatory heat-loss measures, such as sweating, in the face of a rapidly rising body temperature **(hyperthermia).** No sweating occurs, despite a markedly elevated body temperature, because the hypothalamic thermoregulatory control centers are not functioning properly and cannot initiate heat-loss mechanisms. During the development of heatstroke, body temperature starts to climb as the heat-loss mechanisms are eventually overwhelmed by prolonged, excessive heat gain. Once the core temperature reaches the point at which the hypothalamic temperature-control centers are damaged by the heat, the body temperature rapidly rises even higher because of the complete shutdown of heat-loss mechanisms. Furthermore, as the body temperature increases, the rate of metabolism increases correspondingly because higher temperatures speed up the rate of all chemical reactions; the result is even greater heat production. This positive-feedback state sends the temperature spiraling upward. Heatstroke is a very dangerous situation that is rapidly fatal if untreated. Even with treatment to halt and reverse the rampant rise in body temperature, the mortality rate is still high. The rate of permanent disability in survivors is also high because of irreversible protein denaturation caused by the high internal heat.

Cold-Related Disorders

At the other extreme, the body can be harmed by cold exposure in two ways: frostbite and generalized hypothermia. **Frostbite** involves excessive cooling of a particular part of the body to the point where tissue in that area is damaged. If exposed tissues actually freeze, tissue damage results from disruption of the cells by formation of ice crystals or by lack of liquid water.

Hypothermia, a fall in body temperature, occurs when generalized cooling of the body exceeds the ability of the normal heat-producing and heat-conserving regulatory mechanisms to match the excessive heat loss. As hypothermia sets in, the rate of all metabolic processes slows down because of the declining temperature. Higher cerebral functions are the first affected by body cooling, leading to loss of judgment and to apathy, disorientation, and tiredness, all of which diminish the cold victim's ability to initiate voluntary mechanisms to reverse the falling body temperature. As body temperature continues to plummet, depression of the respiratory center occurs, reducing the ventilatory drive so that breathing becomes slow and weak. Activity of the cardiovascular system also is gradually reduced. The heart is slowed and cardiac output decreased. Cardiac rhythm is disturbed, eventually leading to ventricular fibrillation and death.

cause heat gain exceeds heat loss (● Figure 17-7). The elevation in core temperature reflexly triggers heat-loss mechanisms (skin vasodilation and sweating), which eliminate the discrepancy between heat production and heat loss. As soon as the heat-loss mechanisms are stepped up sufficiently to equalize heat production, the core temperature stabilizes at a level slightly above the set point despite continued heat-producing exercise. Thus, during sustained exercise, body temperature initially rises, then is maintained at the higher level as long as the exercise continues.

 PATHOLOGICAL HYPERTHERMIA Hyperthermia can also be brought about in a completely different way: excessive heat production in connection with abnormally high circulating levels of thyroid hormone or epinephrine that result from dysfunctions of the thyroid gland or adrenal medulla, respectively. Both these hormones elevate the core temperature by increasing the overall rate of metabolic activity and heat production.

Hyperthermia can also result from malfunction of the hypothalamic control centers. Certain brain lesions, for example, destroy the normal regulatory capacity of the hypothalamic thermostat. When the thermoregulatory mechanisms are not functional, lethal hyperthermia may occur very rapidly. Normal metabolism produces enough heat to kill a person in less than five hours if the heat-loss mechanisms are completely shut down. In addition to causing brain lesions, exposure to severe, prolonged heat stress can also break down the function of hy-

Core temperature

Several degrees above resting set point

Resting set-point temperature

1 Hyperthermia

2

← Exercise →

1 At the onset of exercise, the rate of heat production initially exceeds the rate of heat loss so the core temperature rises.

2 When heat loss mechanisms are reflexly increased sufficiently to equalize the elevated heat production, the core temperature stabilizes slightly above the resting point for the duration of the exercise.

● **FIGURE 17-7 Hyperthermia in sustained exercise.**

pothalamic thermoregulation. Similarly, the body can be harmed by extreme cold exposure. (For a discussion of the effects of extreme heat or cold exposure, see the boxed feature on p. 657, ■ Concepts, Challenges, and Controversies.)

Chapter in Perspective: Focus on Homeostasis

Because energy can be neither created nor destroyed, for body weight and body temperature to remain constant, input must equal output in the case of, respectively, the body's total energy balance and its heat energy balance. If total energy input exceeds total energy output, the extra energy is stored in the body and body weight increases. Similarly, if the input of heat energy exceeds its output, body temperature increases. Conversely, if output exceeds input, body weight decreases or body temperature falls. The hypothalamus is the major integrating center for maintaining both a constant total energy balance (and thus a constant body weight) and a constant heat energy balance (and thus a constant body temperature).

Body temperature, which is one of the homeostatically regulated factors of the internal environment, must be maintained within narrow limits, because the structure and reactivity of the chemicals that compose the body are temperature sensitive. Deviations in body temperature outside a limited range result in protein denaturation and death of the individual if the temperature rises too high, or metabolic slowing and death if the temperature falls too low.

Body weight, in contrast, varies widely among individuals. Only the extremes of imbalances between total energy input and output become incompatible with life. For example, in the face of insufficient energy input in the form of ingested food during prolonged starvation, the body resorts to breaking down muscle protein to meet its needs for energy expenditure once the adipose stores are depleted. Body weight dwindles because of this self-cannibalistic mechanism until death finally occurs as a result of loss of heart muscle, among other things. At the other extreme, when the food energy consumed greatly exceeds the energy expended, the extra energy input is stored as adipose tissue and body weight increases. The resultant gross obesity can also lead to heart failure. Not only must the heart work harder to pump blood to the excess adipose tissue but obesity also predisposes the person to atherosclerosis and heart attacks (see p. 333).

REVIEW EXERCISES

Objective Questions (Answers on p. A-54)

1. If more food energy is consumed than is expended, the excess energy is lost as heat. (True or false?)

2. All the energy within nutrient molecules can be harnessed to perform biological work. (True or false?)

3. Each liter of O_2 contains 4.8 kilocalories of heat energy. (True or false?)

4. A body temperature greater than 98.2°F is always indicative of a fever. (True or false?)

5. Core temperature is relatively constant, but skin temperature can vary markedly. (True or false?)

6. Sweat that drips off the body has no cooling effect. (True or false?)

7. Production of "goose bumps" has no value in regulating body temperature. (True or false?)

8. The posterior region of the hypothalamus triggers shivering and skin vasoconstriction. (True or false?)

9. The _____ of the hypothalamus contains two populations of neurons, one that secretes appetite-enhancing NPY and another that secretes appetite-suppressing melanocortins.

10. The primary means of involuntarily increasing heat production is _____.

11. Increased heat production independent of muscle contraction is known as _____.

12. The only means of heat loss when the environmental temperature exceeds the core temperature is _____.

13. Which of the following statements concerning heat exchange between the body and the external environment is *incorrect*?
    a. Heat gain is primarily by means of internal heat production.
    b. Radiation serves as a means of heat gain but not of heat loss.
    c. Heat energy always moves down its concentration gradient from warmer to cooler objects.
    d. The temperature gradient between the skin and the external air is subject to control.
    e. Very little body heat is lost by conduction alone.

14. Using the answer code on the right, indicate which mechanism of heat transfer is being described:

    __ 1. sitting on a cold metal chair
    __ 2. sunbathing on the beach
    __ 3. being in a gentle breeze
    __ 4. sitting in front of a fireplace
    __ 5. sweating
    __ 6. riding in a car with the windows open
    __ 7. lying under an electric blanket
    __ 8. sitting in a wet bathing suit
    __ 9. fanning yourself
   __ 10. immersing yourself in cold water

(a) radiation
(b) conduction
(c) convection
(d) evaporation

Essay Questions

1. Differentiate between external and internal work.
2. Define *metabolic rate* and *basal metabolic rate.* Explain the process of indirect calorimetry.
3. Describe the three states of energy balance.
4. By what means is energy balance primarily maintained?
5. List the sources of heat input and output for the body.
6. Describe the source and role of the following in the long-term regulation of energy balance and the short-term control of the timing and size of meals: neuropeptide Y, melanocortins, leptin, insulin, ghrelin, PYY_{3-36}, orexins, corticotropin-releasing hormone, cholecystokinin, and stomach distension.

7. Discuss the compensatory measures that occur in response to a fall in core temperature as a result of cold exposure and in response to a rise in core temperature as a result of heat exposure.
8. Describe the sequence of events in fever production.

Quantitative Exercises (Solutions on p. A-54)

1. The basal metabolic rate (BMR) is a measure of how much energy the body consumes to maintain its "idling speed." Normal BMR = about 72 kcal/hr. The vast majority of this energy is converted to heat. Our thermoregulatory systems function to eliminate this heat to keep body temperature constant. If our bodies were not able to lose this heat, our temperature would rise until we boiled (of course, a person would die before reaching that temperature). It is relatively easy to calculate how long it would take to reach the hypothetical boiling point. If an amount of energy ΔU is put into a liquid of mass m, the temperature change ΔT (in °C) is given by the following formula:

$$\Delta T = \Delta U/m \times C$$

In this equation, C is the specific heat of the liquid. For water, $C = 1.0$ kcal/kg-°C. Calculate how long it would take for the heat from the BMR to boil your body fluids (assume 42 liters of water in your body and a starting point of normal body temperature at 37°C). When exercising maximally, a person consumes about 1000 kcal/hr. How long would it take to boil in this case?

POINTS TO PONDER

(Explanations on p. A-54)

1. Explain how drugs that selectively inhibit CCK increase feeding behavior in experimental animals.
2. What advice would you give an overweight friend who asks for your help in designing a safe, sensible, inexpensive program for losing weight?
3. Why is it dangerous to engage in heavy exercise on a hot, humid day?
4. Describe the avenues for heat loss in a person soaking in a hot bath.
5. Consider the difference between you and a fish in a local pond with regard to control of body temperature. Humans are thermoregulators; they can maintain a remarkably constant, rather high internal body temperature

despite the body's exposure to a wide range of environmental temperatures. To maintain thermal homeostasis, humans physiologically manipulate mechanisms within their bodies to adjust heat production, heat conservation, and heat loss. In contrast, fish are thermoconformers; their body temperatures conform to the temperature of their surroundings. Thus, their body temperatures vary capriciously with changes in the environmental temperature. Even though fish produce heat, they cannot physiologically regulate internal heat production, nor can they control heat exchange with their environment to maintain a constant body temperature when the temperature in their surroundings rises or falls. Knowing this, do you think fish run a fever when they have a systemic infection? Why or why not?

CLINICAL CONSIDERATION

(Explanation on p. A-54)

Michael F., a near-drowning victim, was pulled from the icy water by rescuers 15 minutes after he fell through thin ice when skating. Michael is now alert and recuperating in the hospital. How can you explain his "miraculous" survival even

though he was submerged for 15 minutes and irreversible brain damage, soon followed by death, normally occurs if the brain is deprived of O_2 for more than 4 or 5 minutes?

Endocrine System

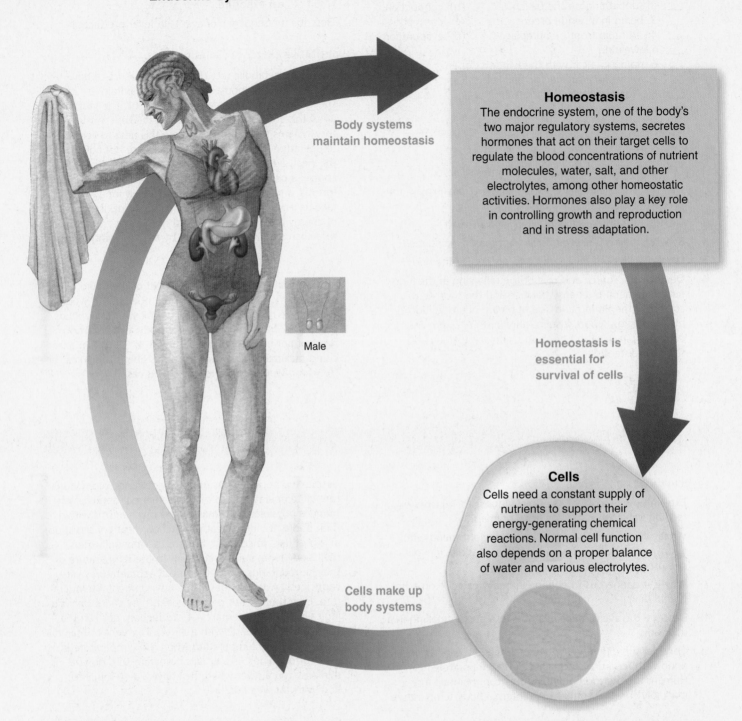

Male

Body systems maintain homeostasis

Homeostasis
The endocrine system, one of the body's two major regulatory systems, secretes hormones that act on their target cells to regulate the blood concentrations of nutrient molecules, water, salt, and other electrolytes, among other homeostatic activities. Hormones also play a key role in controlling growth and reproduction and in stress adaptation.

Homeostasis is essential for survival of cells

Cells
Cells need a constant supply of nutrients to support their energy-generating chemical reactions. Normal cell function also depends on a proper balance of water and various electrolytes.

Cells make up body systems

The **endocrine system** regulates activities that require duration rather than speed. Endocrine glands release **hormones,** blood-borne chemical messengers that act on target cells located a long distance from the endocrine gland. Most target-cell activities under hormonal control are directed toward maintaining homeostasis. The central endocrine glands, which are in or closely associated with the brain, include the hypothalamus, the pituitary gland, and the pineal gland. The **hypothalamus** (a part of the brain) and **posterior pituitary gland** act as a unit to release hormones essential for maintain-ing water balance, for giving birth, and for breast-feeding. The hypothalamus also secretes regulatory hormones that control the hormonal output of the **anterior pituitary gland,** which secretes six hormones that, in turn, largely control the hormonal output of several peripheral endocrine glands. One anterior pituitary hormone, growth hormone, promotes growth and influences nutrient homeostasis. The **pineal gland** is a part of the brain that secretes a hormone important in establishing the body's biological rhythms.

18

Principles of Endocrinology; The Central Endocrine Glands

CENGAGENOW™ Log on to CengageNOW at **http://www .cengage.com/sso/** for an opportunity to explore animations and interactive quizzes to help you learn, review, and master physiology concepts.

General Principles of Endocrinology

The **endocrine system** consists of the ductless endocrine glands (see p. 4) that are scattered throughout the body (● Figure 18-1). Even though the endocrine glands for the most part are not connected anatomically, they constitute a system in a functional sense. They all accomplish their functions by secreting hormones into the blood, and many functional interactions take place among the various endocrine glands. Once secreted, a hormone travels in the blood to its distant target cells, where it regulates or directs a particular function. **Endocrinology** is the study of the homeostatic chemical adjustments and other activities that hormones accomplish.

Even though the blood distributes hormones throughout the body, only specific **target cells** can respond to each hormone because only the target cells have receptors for binding with the particular hormone (see p. 113).

The binding of a hormone with its specific target-cell receptors initiates a chain of events within the target cells to bring about the hormone's final effect. Recall that the means by which a hormone brings about its ultimate physiologic effect depends on whether the hormone is hydrophilic (peptide hormones, catecholamines, and indoleamines) or lipophilic (steroid hormones and thyroid hormone). *Peptide hormones,* the most abundant chemical category of hormone, are chains of amino acids of varying length. *Catecholamines,* produced by the adrenal medulla, are derived from the amino acid tyrosine. *Indoleamines* are produced by the pineal gland and are derived from the amino acid tryptophan. *Steroid hormones,* produced by the adrenal cortex and reproductive endocrine glands, are neutral lipids derived from cholesterol. *Thyroid hormone,* produced by the thyroid gland, is an iodinated tyrosine derivative.

Let us review the characteristics of the two major groups of hormones, peptide hormones and steroid hormones. Hydrophilic peptide hormones are produced and processed by the endoplasmic reticulum and Golgi complex of the endocrine cell and are stored in secretory vesicles until signaled for release by exocytosis. On binding with surface membrane receptors of their target cells, hydrophilic hormones act primarily through

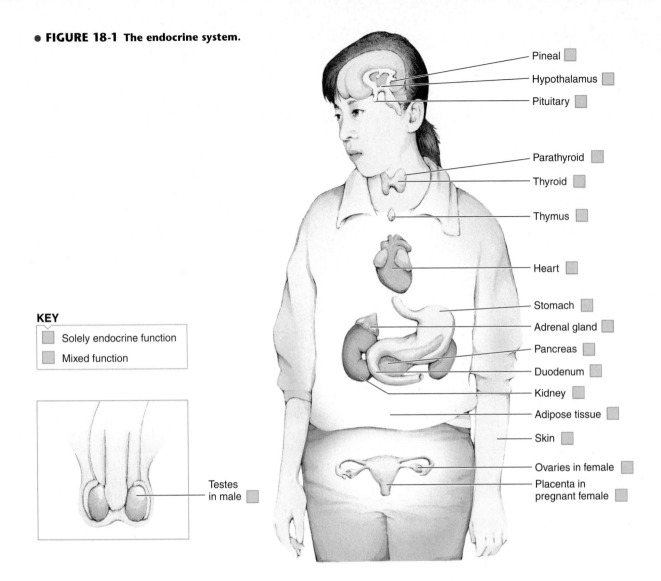

Pineal ☐
Hypothalamus ☐
Pituitary ☐

Parathyroid ☐
Thyroid ☐

Thymus ☐

Heart ☐

Stomach ☐
Adrenal gland ☐
Pancreas ☐
Duodenum ☐
Kidney ☐
Adipose tissue ☐
Skin ☐

Ovaries in female ☐
Placenta in pregnant female ☐

Testes in male ☐

KEY

☐ Solely endocrine function
☐ Mixed function

second-messenger pathways to alter the activity of preexisting proteins, such as enzymes, to produce their physiologic response. Catecholamines and indoleamines are produced by different mechanisms (to be described later) than peptide hormones, but these hydrophilic hormones act at their target cells by the same mechanism as hydrophilic peptides. Lipophilic steroid hormones, by contrast, are produced within the endocrine cell by stepwise modifications of a basic cholesterol precursor molecule. Steroidogenic ("steroid-producing") organs specialize in the type of hormones they produce because each of these organs has the enzymes necessary to produce only one or several, not all, of the steroid hormones. Steroid hormones act primarily by activating genes on binding with receptors inside the cell, thus bringing about formation of new proteins in the target cell that carry out the desired response. Lipophilic thyroid hormone is synthesized by a unique pathway within the thyroid gland (which will be covered in detail later) but functions at its target cells by means similar to those used by lipophilic steroids. Hydrophilic hormones circulate in the blood largely dissolved in the plasma, whereas lipophilic hormones are largely bound to plasma proteins. (See pp. 119–126 for further detail.)

Hormones exert a variety of regulatory effects throughout the body.

The endocrine system is one of the body's two major regulatory systems, the other being the nervous system, with which you are already familiar (Chapters 4 through 7). Recall that the endocrine and nervous systems are specialized for controlling different types of activities. In general, the nervous system coordinates rapid, precise responses and is especially important in mediating the body's interactions with the external environment. The endocrine system, by contrast, primarily controls activities that require duration rather than speed. It regulates, coordinates, and integrates cellular and organ function at a distance.

OVERALL FUNCTIONS OF THE ENDOCRINE SYSTEM In its regulatory role, the endocrine system exerts wide-ranging effects throughout the body, including:

1. Regulating organic metabolism and H_2O and electrolyte balance, which are important collectively in maintaining a constant internal environment

2. Inducing adaptive changes to help the body cope with stressful situations

3. Promoting smooth, sequential growth and development

4. Controlling reproduction

5. Regulating red blood cell production

6. Along with the autonomic nervous system, controlling and integrating activities of both the circulatory and digestive systems

TROPIC HORMONES Some hormones regulate the production and secretion of another hormone. A hormone that has as its primary function the regulation of hormone secretion by another endocrine gland is classified functionally as a **tropic hormone** (trō′-pick) (*tropic* means "nourishing"). Tropic hormones stimulate and maintain their endocrine target tissues. For example, the tropic hormone thyroid-stimulating hormone (TSH), from the anterior pituitary, stimulates thyroid hormone secretion by the thyroid gland and also maintains the structural integrity of this gland. In the absence of TSH, the thyroid gland atrophies (shrinks) and produces very low levels of its hormones.

COMPLEXITY OF ENDOCRINE FUNCTION The following factors add to the complexity of the system:

■ A single endocrine gland may produce multiple hormones. The anterior pituitary, for example, secretes six different hormones, each under a different control mechanism and having distinct functions.

■ A single hormone may be secreted by more than one endocrine gland. For example, both the hypothalamus and pancreas secrete the hormone somatostatin, and somatostatin acts as a paracrine in the stomach.

■ Frequently, a single hormone has more than one type of target cell and therefore can induce more than one type of effect, typically by binding with different subtypes of receptors. For example, vasopressin promotes H_2O reabsorption by the kidney tubules by binding with V_2 (vasopressin 2) receptors on the distal and collecting tubular cells and causes vasoconstriction of arterioles throughout the body by binding with V_1 receptors on arteriolar smooth muscle. Sometimes hormones that have multiple target-cell types can coordinate and integrate the activities of various tissues toward a common end. For example, the effects of insulin on muscle, liver, and fat all act in concert to store nutrients after absorption of a meal.

■ The rate of secretion of some hormones varies considerably over the course of time in a cyclic pattern. Therefore, endocrine systems also provide temporal (time) coordination of function. This is particularly apparent in endocrine control of reproductive cycles, such as the menstrual cycle, in which normal function requires highly specific patterns of change in the secretion of various hormones.

■ A single target cell may be influenced by more than one hormone. Some cells contain an array of receptors for responding in different ways to different hormones. To illustrate, insulin promotes the conversion of glucose into glycogen within liver cells by stimulating one particular hepatic enzyme, whereas

another hormone, glucagon, by activating yet another hepatic enzyme, enhances the degradation of glycogen into glucose within liver cells.

■ The same chemical messenger may be either a hormone or a neurotransmitter, depending on its source and mode of delivery to the target cell. Norepinephrine, which is secreted as a hormone by the adrenal medulla and released as a neurotransmitter from sympathetic postganglionic nerve fibers, is a prime example.

■ Some organs are exclusively endocrine in function (they specialize in hormone secretion alone, the anterior pituitary being an example), whereas other organs of the endocrine system perform nonendocrine functions in addition to secreting hormones. For example, the testes produce sperm and also secrete the male sex hormone testosterone.

The effective plasma concentration of a hormone is influenced by the hormone's secretion, peripheral conversion, transport, inactivation, and excretion.

The primary function of most hormones is the regulation of various homeostatic activities. Because hormones' effects are proportional to their concentrations in the plasma, these concentrations are subject to control according to homeostatic need. Furthermore, the magnitude of the hormonal response depends on the availability and sensitivity of the target cells' receptors for the hormone. We will first examine the factors that influence the plasma concentration of the hormone before turning attention to the target cells' responsiveness to the hormone.

The effective plasma concentration of free, biologically active hormone—and thus the hormone's availability to its receptors—depends on several factors:

■ *The hormone's rate of secretion into the blood by the endocrine gland.* The rate of secretion, a factor that increases the plasma concentration of the hormone, is subject to control to maintain the hormone concentration at a desired set point.

■ *For a few hormones, its rate of metabolic activation or conversion.* After being secreted into the blood by the endocrine gland, lipophilic hormones in particular are often modified in other organs. Sometimes this peripheral (away from the endocrine gland) modification results in a more active form of the hormone. For example, the most abundant form of thyroid hormone secreted by the thyroid gland is thyroxine (which contains four iodines), but the most powerful form of thyroid hormone in the blood is tri-iodothyronine (which contains three iodines). Once secreted, thyroxine is converted to the more active form as a result of one of its iodines being stripped peripherally, primarily by the liver and kidneys. Usually the rate of such hormone activation is itself under hormonal control. Sometimes peripheral action actually converts one hormone into a functionally different hormone. For example, a small proportion of testosterone, a potent male sex hormone, is converted peripherally by action of the enzyme aromatase in adipose tissue and elsewhere into estrogen, a potent female sex hormone.

■ *For lipophilic hormones, its extent of binding to plasma proteins.* Because lipophilic hormones are poorly soluble in water, they circulate in the plasma bound to specific plasma proteins. Only the small, unbound portion of the hormone is free to interact with its target cells. The magnitude of this free pool rather than the total pool of hormone is monitored and adjusted to maintain normal endocrine function. Clinical assays used to determine the plasma concentration of a given hormone measure the total concentration of the hormone and do not take into account the extent of hormone binding. These results can sometimes be misleading. For example, during pregnancy more of the specific plasma protein that binds thyroid hormone is produced. Because more thyroid hormone is bound to this plasma protein, the total concentration of thyroid hormone in the plasma is increased (as much as doubled), yet the concentration of free, active hormone remains unchanged, so the pregnant woman has normal thyroid function despite elevated plasma levels of thyroid hormone.

■ *Its rate of removal from the blood by metabolic inactivation and excretion in the urine.* All hormones are eventually inactivated by enzymes in the liver, kidneys, blood, or target cells. The amount of time after a hormone is secreted before it is inactivated, and the means by which this takes place, differ for different classes of hormones. Hydrophilic peptides most commonly are inactivated by hydrolysis of peptide bonds (see p. 29). In the case of some peptide hormones, such as insulin, the target cell actually engulfs the bound hormone by receptor-mediated endocytosis and degrades it intracellularly (see p. 31). Catecholamines are enzymatically converted to related biologically inactive molecules. Lipophilic steroid hormones and thyroid hormone are inactivated by alteration of the active portion of the molecule by various biochemical means. After lipophilic hormones are inactivated, the liver typically adds charged groups to make them more water soluble so they can be freed from their plasma protein carriers and be eliminated in the urine.

In general, the hydrophilic peptides and catecholamines are easy targets for blood and tissue enzymes, so they remain in the blood only briefly (a few minutes to a few hours) before being enzymatically inactivated. In contrast, binding of lipophilic hormones to plasma proteins makes them less vulnerable to metabolic inactivation and keeps them from escaping into the urine. Therefore, lipophilic hormones are removed from plasma much more slowly. They may persist in the blood for hours (steroids) or up to a week (thyroid hormone).

Hormones and their metabolites are typically eliminated from the plasma by urinary excretion. In contrast to the tight controls on hormone secretion, hormone inactivation and excretion are not regulated.

Clinical Note Because the liver and kidneys are important in removing hormones from the plasma, patients with liver or kidney disease may suffer from excess activity of certain hormones solely because hormone inactivation and/or elimination are reduced.

When liver and kidney function are normal, measuring urinary concentrations of hormones and their metabolites provides a useful, noninvasive way to assess endocrine function because the rate of excretion of these products in the urine directly reflects their rate of secretion by the endocrine glands.

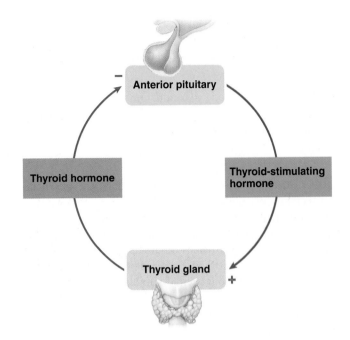

● **FIGURE 18-2 Negative-feedback control.**

The effective plasma concentration of a hormone is normally regulated by changes in the rate of its secretion.

Normally, the effective plasma concentration of a hormone is regulated by appropriate adjustments in the rate of its secretion. Endocrine glands do not secrete their hormones at a constant rate; the secretion rates of all hormones vary, subject to control often by a combination of several complex mechanisms. The regulatory system for each hormone is considered in detail in later sections. For now, we will address these general mechanisms of controlling secretion that are common to many different hormones: negative-feedback control, neuroendocrine reflexes, and diurnal (circadian) rhythms.

NEGATIVE-FEEDBACK CONTROL Negative feedback is a prominent feature of hormonal control systems. Stated simply, *negative feedback exists when the output of a system counteracts a change in input*, maintaining a controlled variable within a narrow range around a set level (see p. 15). Negative feedback maintains the plasma concentration of a hormone at a given level, similar to the way in which a home heating system maintains the room temperature at a given set point. Control of hormonal secretion provides some classic physiologic examples of negative feedback. For example, when the plasma concentration of free circulating thyroid hormone falls below a given "set point," the anterior pituitary secretes thyroid-stimulating hormone (TSH), which stimulates the thyroid to increase its secretion of thyroid hormone (● Figure 18-2). Thyroid hormone, in turn, inhibits further secretion of TSH by the anterior pituitary. Negative feedback ensures that once thyroid gland secretion has been "turned on" by TSH, it will not continue unabated but will

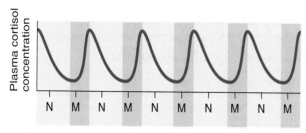

KEY

N = noon	
M = midnight	Light ☐ Dark ▨

● **FIGURE 18-3 Diurnal rhythm of cortisol secretion.**
(*Source:* Adapted from George A. Hedge, Howard D. Colby, and Robert L. Goodman, *Clinical Endocrine Physiology,* Figure 1-13, p. 28. © 1987, with permission from Elsevier.)

be "turned off" when the appropriate level of free circulating thyroid hormone has been achieved. Thus, the effect of a particular hormone's actions can inhibit its own secretion. The feedback loops often become quite complex.

NEUROENDOCRINE REFLEXES Many endocrine control systems involve **neuroendocrine reflexes,** which include neural as well as hormonal components. The purpose of such reflexes is to produce a sudden increase in hormone secretion (that is, "turn up the thermostat setting") in response to a specific stimulus, frequently a stimulus external to the body. In some instances, neural input to the endocrine gland is the only factor regulating secretion of the hormone. For example, secretion of epinephrine by the adrenal medulla is solely controlled by the sympathetic nervous system. Some endocrine control systems, in contrast, include both feedback control (which maintains a constant basal level of the hormone) and neuroendocrine reflexes (which cause sudden bursts in secretion in response to a sudden increased need for the hormone). An example is the increased secretion of cortisol, the "stress hormone," by the adrenal cortex during a stress response (see ● Figure 19-9, p. 701).

DIURNAL (CIRCADIAN) RHYTHMS The secretion rates of many hormones rhythmically fluctuate up and down as a function of time. The most common endocrine rhythm is the **diurnal** ("day–night"), or **circadian** ("around a day") **rhythm,** which is characterized by repetitive oscillations in hormone levels that are very regular and cycle once every 24 hours. This rhythmicity is caused by endogenous oscillators similar to the self-paced respiratory neurons in the brain stem that control the rhythmic motions of breathing, except the timekeeping oscillators cycle on a much longer time scale. Furthermore, unlike the rhythmicity of breathing, endocrine rhythms are locked on, or **entrained,** to external cues such as the light–dark cycle. That is, the inherent 24-hour cycles of peak and ebb of hormone secretion are set to "march in step" with cycles of light and dark. For example, cortisol secretion rises during the night, reaches its peak secretion in the morning before a person gets up, then falls throughout the

day to its lowest level at bedtime (● Figure 18-3). Inherent hormonal rhythmicity and entrainment are not accomplished by the endocrine glands themselves but result from the central nervous system changing the set point of these glands. We discuss the master biological clock further in a later section. Negative-feedback control mechanisms operate to maintain whatever set point is established for that time of day. Some endocrine cycles operate on time scales other than a circadian rhythm, a well-known example being the monthly menstrual cycle.

Endocrine disorders result from hormone excess or deficiency or decreased target-cell responsiveness.

Clinical Note Abnormalities in a hormone's effective plasma concentration can arise from a variety of factors (▲ Table 18-1). Endocrine disorders most commonly result from abnormal plasma concentrations of a hormone caused by inappropriate rates of secretion—that is, too little hormone secreted (**hyposecretion**) or too much hormone secreted (**hypersecretion**). Occasionally, endocrine dysfunction arises because target-cell responsiveness to the hormone is abnormally low, even though plasma concentration of the hormone is normal.

HYPOSECRETION *Primary hyposecretion* occurs when an endocrine gland is secreting too little of its hormone because of an abnormality within that gland. *Secondary hyposecretion* takes place when an endocrine gland is normal but is secreting too little hormone because of a deficiency of its tropic hormone.

The following are among the many different factors (each listed with an example) that may cause primary hyposecretion: (1) genetic (inborn absence of an enzyme that catalyzes synthesis of the hormone, such as the inability to synthesize cortisol because of the lack of a specific enzyme in the adrenal cortex); (2) dietary (lack of iodine, which is needed for synthesis of thyroid hormone); (3) chemical or toxic (certain insecticide residues may destroy the adrenal cortex); (4) immunologic (autoimmune antibodies may destroy the body's own thyroid tissue); (5) other disease processes (cancer or tuberculosis may coincidentally destroy endocrine glands); (6) *iatrogenic* (physician induced, such as surgical removal of a cancerous thyroid gland); and (7) *idiopathic* (meaning the cause is not known).

The most common method of treating hormone hyposecretion is to administer a hormone that is the same as (or similar to, such as from another species) the deficient or missing one. Such replacement therapy seems straightforward, but the hormone's source and means of administration involve some practical problems. The sources of hormone preparation for clinical use include (1) endocrine tissues from domestic livestock, (2) placental tissue and urine of pregnant women, (3) laboratory synthesis of hormones, and (4) "hormone factories," or bacteria into which genes coding for the production of human hormones have been introduced. The method of choice for a given hormone is determined largely by its structural complexity and degree of species specificity.

HYPERSECRETION Like *hypo*secretion, *hyper*secretion by a particular endocrine gland is designated as primary or secondary depending on whether the defect lies in that gland or results from excessive stimulation from the outside, respectively. Hypersecretion may be caused by (1) tumors that ignore the normal regulatory input and continuously secrete excess hormone and (2) immunologic factors, such as excessive stimulation of the thyroid gland by an abnormal antibody that mimics the action of TSH, the thyroid tropic hormone. Excessive levels of a particular hormone may also arise from substance abuse, such as the outlawed practice among athletes of using certain steroids that increase muscle mass by promoting protein synthesis in muscle cells (see p. 282).

There are several ways of treating hormonal hypersecretion. If a tumor is the culprit, it may be surgically removed or destroyed with radiation treatment. In some instances, hypersecretion can be limited by drugs that block hormone synthesis or inhibit hormone secretion. Sometimes the condition may be treated by giving drugs that inhibit the action of the hormone without actually reducing the excess hormone secretion.

ABNORMAL TARGET-CELL RESPONSIVENESS Endocrine dysfunction can also occur because target cells do not respond adequately to the hormone, even though the effective plasma concentration of a hormone is normal. This unresponsiveness may be caused, for example, by an inborn lack of receptors for the hormone, as in **testicular feminization syndrome.** In this condition, receptors for testosterone, a masculinizing hormone produced by the male testes, are not produced because of a specific genetic defect. Although adequate testosterone is available, masculinization does not take place, just as if no testosterone were present. Abnormal responsiveness may also occur if the target cells for a particular hormone lack an enzyme essential to carrying out the response.

The responsiveness of a target cell can be varied by regulating the number of hormone-specific receptors.

In contrast to endocrine dysfunction caused by *unintentional* receptor abnormalities, the target-cell receptors for a particular hormone can be *deliberately altered* as a result of physiologic control mechanisms. A target cell's response to a hormone is correlated with the number of the cell's receptors occupied by molecules of that hormone, which in turn depends not only on the plasma concentration of the hormone but also on the number of receptors in the target cell for that hormone. Thus, the response of a target cell to a given plasma concentration can be fine-tuned up or down by varying the number of receptors available for hormone binding.

DOWN REGULATION As an illustration of this fine-tuning, when the plasma concentration of insulin is chronically elevated, the total number of target-cell receptors for insulin is gradually reduced as a direct result of the effect a sustained elevation of insulin has on the insulin receptors. This phenomenon, known as **down regulation,** constitutes an important locally acting negative-feedback mechanism that prevents the target cells from overreacting to a prolonged high concentration of insulin; that is, the target cells are *desensitized* to insulin, helping blunt the effect of insulin hypersecretion.

Down regulation of insulin is accomplished by the following mechanism. The binding of insulin to its surface receptors first triggers the dictated cellular response, then induces receptor-mediated endocytosis of the hormone receptor complex, which is subsequently attacked by intracellular lysosomal enzymes (see p. 28). This internalization serves a twofold purpose: It provides a pathway for degrading the hormone after it has exerted its effect and helps regulate the number of receptors available for binding on the target cell's surface. At high plasma insulin concentrations, the number of surface receptors for insulin is gradually reduced by the accelerated rate of receptor internalization and degradation brought about by increased hormonal binding. The rate of synthesis of new receptors within the endoplasmic reticulum and their insertion in the plasma membrane do not keep pace with their rate of destruction. Over time, this self-induced loss of target-cell receptors for insulin reduces the target cell's sensitivity to the elevated hormone concentration.

PERMISSIVENESS, SYNERGISM, AND ANTAGONISM A given hormone's effects are influenced not only by the concentration of the hormone itself but also by the concentrations of other hormones that interact with it. Because hormones are widely distributed through the blood, target cells may be exposed simultaneously to many different hormones, giving rise to numerous complex hormonal interactions on target cells. Hormones frequently alter the receptors for other kinds of hormones as part of their normal physiologic activity. A hormone can influence the activity of another hormone at a given target cell in one of three ways: permissiveness, synergism, and antagonism.

With **permissiveness,** one hormone must be present in adequate amounts for the full exertion of another hormone's effect.

In essence, the first hormone, by enhancing a target cell's responsiveness to another hormone, "permits" this other hormone to exert its full effect. For example, thyroid hormone increases the number of receptors for epinephrine in epinephrine's target cells, increasing the effectiveness of epinephrine. In the absence of thyroid hormone, epinephrine is only marginally effective.

Synergism occurs when the actions of several hormones are complementary and their combined effect is greater than the sum of their separate effects. An example is the synergistic action of follicle-stimulating hormone and testosterone, both of which are required for maintaining the normal rate of sperm production. Synergism results from each hormone's influence on the number or affinity of receptors for the other hormone.

Antagonism occurs when one hormone causes the loss of another hormone's receptors, reducing the effectiveness of the second hormone. To illustrate, progesterone (a hormone secreted during pregnancy that decreases contractions of the uterus) inhibits uterine responsiveness to estrogen (another hormone secreted during pregnancy that increases uterine contractions). By causing loss of estrogen receptors on uterine smooth muscle, progesterone prevents estrogen from exerting its excitatory effects during pregnancy and thus keeps the uterus a quiet (noncontracting) environment suitable for the developing fetus.

This has been a brief overview of the general functions of the endocrine system. ▲ Table 18-2 summarizes the most important specific functions of the major hormones. As extensive as the table appears, it leaves out a variety of "candidate" or potential hormones that have not fully qualified as hormones, either because they do not quite fit the classic definition of a hormone or because they have been discovered so recently that

▲ TABLE 18-2	Summary of the Major Hormones		
Endocrine Gland	**Hormones**	**Target Cells**	**Major Functions of Hormones**
Hypothalamus	Releasing and inhibiting hormones (TRH, CRH, GnRH, GHRH, somatostatin, PRH, dopamine)	Anterior pituitary	Controls release of anterior pituitary hormones
Posterior Pituitary (hormones stored in)	Vasopressin (antidiuretic hormone, ADH)	Kidney tubules	Increases H_2O reabsorption
		Arterioles	Produces vasoconstriction
	Oxytocin	Uterus	Increases contractility
		Mammary glands (breasts)	Causes milk ejection
Anterior Pituitary	Thyroid-stimulating hormone (TSH)	Thyroid follicular cells	Stimulates T_3 and T_4 secretion
	Adrenocorticotropic hormone (ACTH)	Zona fasciculata and zona reticularis of adrenal cortex	Stimulates cortisol secretion
	Growth hormone (GH)	Bone; soft tissues	Essential but not solely responsible for growth; by means of IGF-I, indirectly stimulates protein anabolism and growth of bones and soft tissues; direct metabolic effects include fat mobilization and glucose conservation
		Liver	Stimulates IGF-I secretion
	Follicle-stimulating hormone (FSH)	Females: ovarian follicles	Promotes follicular growth and development; stimulates estrogen secretion
		Males: seminiferous tubules in testes	Stimulates sperm production
	Luteinizing hormone (LH)	Females: ovarian follicle and corpus luteum	Stimulates ovulation, corpus luteum development, and estrogen and progesterone secretion

(continued)

Endocrine Gland	Hormones	Target Cells	Major Functions of Hormones
Anterior Pituitary (continued)	Luteinizing hormone (LH) (continued)	Males: interstitial cells of Leydig in testes	Stimulates testosterone secretion
	Prolactin (PRL)	Females: mammary glands	Promotes breast development; stimulates milk secretion
		Males	Uncertain
Pineal Gland	Melatonin	Brain; anterior pituitary; reproductive organs; immune system; possibly others	Entrains body's biological rhythm with external cues; inhibits gonadotropins; its reduction likely initiates puberty; acts as an antioxidant; enhances immunity
Thyroid Gland Follicular Cells	Tetraiodothyronine (T_4 or thyroxine); tri-iodothyronine (T_3)	Most cells	Increases metabolic rate; essential for normal growth and nerve development
Thyroid Gland C Cells	Calcitonin	Bone	Decreases plasma Ca^{2+} concentration
Adrenal Cortex			
Zona glomerulosa	Aldosterone (mineralocorticoid)	Kidney tubules	Increases Na^+ reabsorption and K^+ secretion
Zona fasciculata and zona reticularis	Cortisol (glucocorticoid)	Most cells	Increases blood glucose at the expense of protein and fat stores; contributes to stress adaptation
	Androgens (dehydroepiandrosterone)	Females: bone and brain	Responsible for pubertal growth spurt and sex drive in females
Adrenal Medulla	Epinephrine and norepinephrine	Sympathetic receptor sites throughout body	Reinforces sympathetic nervous system; contributes to stress adaptation and blood pressure regulation
Endocrine Pancreas (Islets of Langerhans)	Insulin (β cells)	Most cells	Promotes cellular uptake, use, and storage of absorbed nutrients
	Glucagon (α cells)	Most cells	Important for maintaining nutrient levels in blood during postabsorptive state
	Somatostatin (D cells)	Digestive system	Inhibits digestion and absorption of nutrients
		Pancreatic islet cells	Inhibits secretion of all pancreatic hormones
Parathyroid Gland	Parathyroid hormone (PTH)	Bone, kidneys, intestine	Increases plasma Ca^{2+} concentration; decreases plasma PO_4^{3-} concentration; stimulates vitamin D activation
Gonads			
Female: ovaries	Estrogen (estradiol)	Female sex organs; body as a whole	Promotes follicular development; governs development of secondary sexual characteristics; stimulates uterine and breast growth

Endocrine Gland	Hormones	Target Cells	Major Functions of Hormones
Female: ovaries (*continued*)	Estrogen (estradiol) (*continued*)	Bone	Promotes closure of the epiphyseal plate
	Progesterone	Uterus	Prepares for pregnancy
Male: testes	Testosterone	Male sex organs; body as a whole	Stimulates sperm production; governs development of secondary sexual characteristics; promotes sex drive
		Bone	Enhances pubertal growth spurt; promotes closure of the epiphyseal plate
Testes and ovaries	Inhibin	Anterior pituitary	Inhibits secretion of follicle-stimulating hormone
Placenta	Estrogen (estriol); progesterone	Female sex organs	Help maintain pregnancy; prepare breasts for lactation
	Human chorionic gonadotropin (hCG)	Ovarian corpus luteum	Maintains corpus luteum of pregnancy
Kidneys	Renin (by activating angiotensin)	Zona glomerulosa of adrenal cortex (acted on by angiotensin, which is activated by renin)	Stimulates aldosterone secretion; angiotensin II is also a potent vasoconstrictor and stimulates thirst
	Erythropoietin	Bone marrow	Stimulates erythrocyte production
Stomach	Ghrelin	Hypothalamus	Signals hunger; stimulates appetite
	Gastrin	Digestive tract exocrine glands and smooth muscles; pancreas; liver; gallbladder	Control motility and secretion to facilitate digestive and absorptive processes
Small intestine	Secretin; cholecystokinin (CCK)		
	Glucose-dependent insulinotropic peptide (GIP)	Endocrine pancreas	Stimulates insulin secretion
	PYY_{3-36}	Hypothalamus	Signals satiety; suppresses appetite
Liver	Insulin-like growth factor I (IGF-I)	Bone; soft tissues	Promotes growth
	Thrombopoietin	Bone marrow	Stimulates platelet production
	Hepcidin	Intestine	Inhibits absorption of iron into blood
Skin	Vitamin D	Intestine	Increases absorption of ingested Ca^{2+} and PO_4^{3-}
Thymus	Thymosin	T lymphocytes	Enhances T lymphocyte proliferation and function
Heart	Atrial and brain natriuretic peptides (ANP; BNP)	Kidney tubules	Inhibit Na^+ reabsorption
Adipose Tissue	Leptin	Hypothalamus	Suppresses appetite; important in long-term control of body weight
	Other adipokines	Multiple sites	Play role in metabolism and inflammation

their hormonal status has not yet been conclusively documented. The table also excludes the cytokines secreted by the effector cells of the defense system (white blood cells and macrophages; see p. 424) and a variety of growth factors that promote growth of specific tissues, such as *epidermal growth factor* and *nerve growth factor*. Furthermore, new hormones are likely to be discovered, and additional functions may be found for known hormones. As an example, vasopressin's role in conserving H_2O during urine formation was determined first, followed later by the discovery of its constrictor effect on arterioles. More recently, vasopressin has also been found to play roles in fever, learning, memory, and behavior.

Some of the hormones listed in the table have been introduced elsewhere and are not discussed further; these are the renal hormones (erythropoietin in Chapter 11 and renin in Chapter 14), the hepatic hormones (thrombopoietin in Chapter 11 and hepcidin in Chapter 16), the gastrointestinal hormones (Chapter 16), thymosin from the thymus (Chapter 12), atrial and brain natriuretic peptides from the heart (Chapter 15), and leptin and other adipokines from adipose tissue (Chapter 17). The remainder of the hormones are described in greater detail in this and the next two chapters. We start in this chapter with the central endocrine glands—those in the brain itself or in close association with the brain—namely, the hypothalamus, the pituitary gland, and the pineal gland. The peripheral endocrine glands are discussed in the following chapters.

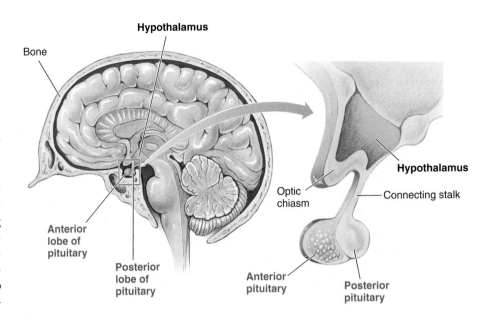

(a) Relation of pituitary gland to hypothalamus and rest of brain

(b) Enlargement of pituitary gland and its connection to hypothalamus

● **FIGURE 18-4 Anatomy of the pituitary gland.**

Hypothalamus and Pituitary

The **pituitary gland,** or **hypophysis,** is a small endocrine gland located in a bony cavity at the base of the brain just below the hypothalamus (● Figure 18-4). The pituitary is connected to the hypothalamus by a thin connecting stalk. If you point one finger between your eyes and another finger toward one of your ears, the imaginary point where these lines would intersect is about where your pituitary is located.

The pituitary gland consists of anterior and posterior lobes.

The pituitary has two anatomically and functionally distinct lobes, the **posterior pituitary** and the **anterior pituitary.** The posterior pituitary is composed of nervous tissue and thus is also termed the **neurohypophysis.** The anterior pituitary consists of glandular epithelial tissue and accordingly is also called the **adenohypophysis** (*adeno* means "glandular"). The posterior and anterior pituitary lobes have only their location in common. They arise from different tissues embryonically, serve different functions, and are subject to different control mechanisms.

The release of hormones from both the posterior and the anterior pituitary is directly controlled by the hypothalamus, but the nature of the relationship is entirely different. The posterior pituitary connects to the hypothalamus by a neural pathway, whereas the anterior pituitary connects to the hypothalamus by a unique vascular link. We will look first at the posterior pituitary.

The hypothalamus and posterior pituitary act as a unit to secrete vasopressin and oxytocin.

The hypothalamus and posterior pituitary form a neuroendocrine system that consists of a population of neurosecretory neurons whose cell bodies lie in two well-defined clusters in the hypothalamus, the **supraoptic** and **paraventricular nuclei.** The axons of these neurons pass down through the thin connecting stalk to terminate on capillaries in the posterior pituitary (● Figure 18-5). The posterior pituitary consists of these neuronal terminals plus glial-like supporting cells known as **pituicytes.** Functionally as well as anatomically, the posterior pituitary is simply an extension of the hypothalamus.

The posterior pituitary does not actually produce any hormones. It simply stores and, on appropriate stimulation, releases into the blood two small peptide hormones, *vasopressin* and *oxytocin,* which are synthesized by the neuronal cell bodies in the hypothalamus. Both these hydrophilic peptides are made in both the supraoptic and the paraventricular nuclei, but a single neuron can produce only one of these hormones. The synthesized hormones are packaged in secretory granules that are transported by molecular motors down the cytoplasm of the axon (see p. 43) and stored in the neuronal terminals within the posterior pituitary. Each terminal stores either vasopressin or oxytocin, but not both. Thus, these hormones can be released independently as needed. On stimulatory input to the hypo-

KEY
- = Vasopressin
- = Oxytocin

Vasopressin		Oxytocin	
Nephrons in kidneys	Arterioles throughout body	Uterus	Mammary glands
Increases permeability of distal and collecting tubules to H_2O	Causes vasoconstriction	Stimulates uterine contractions	Stimulates milk ejection during breast-feeding

1 The paraventricular and supraoptic nuclei both contain neurons that produce vasopressin and oxytocin. The hormone, either vasopressin or oxytocin depending on the neuron, is synthesized in the neuronal cell body in the hypothalamus.

2 The hormone travels down the axon to be stored in the neuronal terminals within the posterior pituitary.

3 When the neuron is excited, the stored hormone is released from the terminals into the systemic blood for distribution throughout the body.

● **FIGURE 18-5 Relationship of the hypothalamus and posterior pituitary.**

thalamus, either vasopressin or oxytocin is released into the systemic blood from the posterior pituitary by exocytosis of the appropriate secretory granules. This hormonal release is triggered in response to action potentials that originate in the hypothalamic cell body and sweep down the axon to the neuronal

terminal in the posterior pituitary. As in any other neuron, action potentials are generated in these neurosecretory neurons in response to synaptic input to their cell bodies.

The actions of vasopressin and oxytocin are briefly summarized here to make our endocrine story complete. They are described more thoroughly elsewhere—vasopressin in Chapters 14 and 15 and oxytocin in Chapter 20.

VASOPRESSIN Vasopressin (antidiuretic hormone, ADH) has two major effects that correspond to its two names: (1) it enhances retention of H_2O by the kidney nephrons during urine formation (an antidiuretic effect), and (2) it causes contraction of arteriolar smooth muscle (a vessel pressor effect). The first effect has more physiologic importance. Under normal conditions, vasopressin is the primary endocrine factor that regulates urinary H_2O loss and overall H_2O balance. In contrast, typical levels of vasopressin play only a minor role in regulating blood pressure by means of the hormone's pressor effect.

The major control for hypothalamic-induced release of vasopressin from the posterior pituitary is input from hypothalamic osmoreceptors, which increase vasopressin secretion in response to a rise in plasma osmolarity. A less powerful input from the left atrial volume receptors increases vasopressin secretion in response to a fall in ECF volume and arterial blood pressure (see p. 567). (For further information on the importance of vasopressin secretion when exercising in the heat, see the boxed feature on p. 672, ■ A Closer Look at Exercise Physiology.)

OXYTOCIN Oxytocin stimulates contraction of the uterine smooth muscle to help expel the infant during childbirth, and it promotes ejection of the milk from the mammary glands (breasts) during breast-feeding. Appropriately, oxytocin secretion is increased by reflexes that originate within the birth canal during childbirth and by reflexes that are triggered when the infant suckles the breast.

In addition to these two major physiologic effects, oxytocin has recently been shown to influence a variety of behaviors, especially maternal behaviors. For example, this hormone fittingly facilitates bonding, or attachment, between a mother and her infant.

Most anterior pituitary hormones are tropic.

Unlike the posterior pituitary, which releases hormones synthesized by the hypothalamus, the anterior pituitary itself synthesizes the hormones it releases into the blood. Five different cell populations within the anterior pituitary secrete six major peptide hormones. The actions of each of these hormones are described in detail in later sections. For now, here is a brief statement of their source and primary effects to provide a rationale for their names (● Figure 18-6):

1. The type of anterior pituitary cells known as **somatotropes** secrete **growth hormone (GH, somatotropin),** the primary hormone responsible for regulating overall body growth (*somato* means "body"). GH also exerts important metabolic actions.

2. **Thyrotropes** secrete **thyroid-stimulating hormone (TSH, thyrotropin),** which stimulates secretion of thyroid hormone and growth of the thyroid gland.

Signal Transduction Pathways Used by Major Hormones

Signal Transduction Pathway	Receptor Type Involved	Outcome of Hormone Binding to Receptor	Hormones That Use This Pathway
Cyclic AMP/PKA	G-protein coupled receptor in surface membrane	Alters preexisting designated protein	LH, FSH, TSH, ACTH, vasopressin, epinephrine, norepinephrine, glucagon, PTH, CRH, GHRH, somatostatin, calcitonin
IP$_3$/Ca^{2+} and DAG/PKC	G-protein coupled receptor in surface membrane	Alters preexisting designated protein	TRH, GnRH, oxytocin
Tyrosine kinase	Surface-membrane receptor that exerts tyrosine kinase activity	Alters preexisting designated protein	Insulin, insulin-like growth factors (IGF-I and IGF-II)
JAK/STAT	Surface-membrane receptor that is bound to and activates JAK enzymes, which phosphorylate STAT	Turns on gene transcription, leading to synthesis of designated protein	Growth hormone, prolactin
Hormone-response elements on DNA	Intracellular cytoplasmic or nuclear receptors	Turns on gene transcription, leading to synthesis of designated protein	All lipophilic hormones: thyroid hormone, cortisol, aldosterone, testosterone, estrogen, progesterone, vitamin D

way with which you are not familiar—the **JAK/STAT** pathway. Binding of these hormones to their target-cell surface membrane receptors on the ECF side activates the enzyme *Janus kinase (JAK)* attached to the cytosolic side of the receptor. JAK phosphorylates *signal transducers and activators of transcription (STAT)* within the cytosol. Phosphorylated STAT moves to the nucleus and turns on gene transcription, resulting in the synthesis of new proteins that carry out the cellular response. ▲ Table 18-3 summarizes the signal transduction pathways by which the major hormones bring about their effects. Some of these pathways will be discussed in this and upcoming endocrine-related chapters. Others were discussed when these hormones were covered in preceding chapters. This table is a means of collecting these diverse discussions in a single place.

Hypothalamic releasing and inhibiting hormones help regulate anterior pituitary hormone secretion.

None of the anterior pituitary hormones are secreted at a constant rate. Even though each of these hormones has a unique control system, there are some common regulatory patterns. The two most important factors that regulate anterior pituitary hormone secretion are (1) hypothalamic hormones and (2) feedback by target-gland hormones.

Because the anterior pituitary secretes hormones that control the secretion of various other hormones, it long held the undeserved title of "master gland." Scientists now know that the release of each anterior pituitary hormone is largely controlled by still other hormones produced by the hypothalamus. The secretion of these regulatory neurohormones, in turn, is controlled by a variety of neural and hormonal inputs to the hypothalamic neurosecretory cells.

Major Hypophysiotropic Hormones

Hormone	Effect on the Anterior Pituitary
Thyrotropin-Releasing Hormone (TRH)	Stimulates release of TSH (thyrotropin) and prolactin
Corticotropin-Releasing Hormone (CRH)	Stimulates release of ACTH (corticotropin)
Gonadotropin-Releasing Hormone (GnRH)	Stimulates release of FSH and LH (gonadotropins)
Growth Hormone–Releasing Hormone (GHRH)	Stimulates release of growth hormone
Somatostatin (Growth Hormone–Inhibiting Hormone; GHIH)	Inhibits release of growth hormone and TSH
Prolactin-Releasing Hormone (PRH)	Stimulates release of prolactin
Dopamine (Prolactin-Inhibiting Hormone; PIH)	Inhibits release of prolactin

ROLE OF THE HYPOTHALAMIC RELEASING AND INHIBITING HORMONES The secretion of each anterior pituitary hormone is stimulated or inhibited by one or more of the seven hypothalamic **hypophysiotropic hormones** (*hypophysis* means "pituitary"; *tropic* means "nourishing"). These small peptide hormones are listed in ▲ Table 18-4. Depending on their actions, these hormones are called **releasing hormones** or **inhibiting hormones.** In each case, the primary action of the hormone is apparent from its name. For example, **thyrotropin-releasing hormone (TRH)** stimulates the release of TSH (alias thyrotropin) from the anterior pituitary, whereas **prolactin-inhibiting hormone (PIH),** which is **dopamine** (the same as the neurotransmitter in the "pleasure" pathways in the brain; see p. 157), inhibits the release of prolactin from the anterior pituitary. Note that hypophysiotropic hormones in most cases are involved in a three-hormone hierarchic chain of command (● Figure 18-7): The hypothalamic hypophysiotropic hormone (*hormone 1*) controls the output of an anterior-pituitary tropic hormone (*hormone 2*). This tropic hormone, in turn, regulates secretion of the target endocrine gland's hormone (*hormone 3*), which exerts the final physiologic effect. This three-hormone sequence is called an **endocrine axis,** as in the hypothalamus–pituitary–thyroid axis.

Although endocrinologists originally speculated that there was one hypophysiotropic hormone for each anterior pituitary hormone, many hypothalamic hormones have more than one effect, so their names indicate only the function first identified. Moreover, a single anterior pituitary hormone may be regulated by two or more hypophysiotropic hormones, which may even exert opposing effects. For example, **growth hormone–releasing hormone (GHRH)** stimulates growth hormone secretion, whereas **growth hormone–inhibiting hormone (GHIH),** also known as **somatostatin,** inhibits it. The output of the anterior-pituitary somatotropes (that is, the rate of growth hormone secretion) in response to two such opposing inputs depends on the relative concentrations of these hypothalamic hormones as well as on the intensity of other regulatory inputs.

Chemical messengers that are identical in structure to the hypothalamic releasing and inhibiting hormones and to vasopressin are produced in many areas of the brain outside the hypothalamus. Instead of being released into the blood, these messengers act locally as neurotransmitters and as neuromodu-

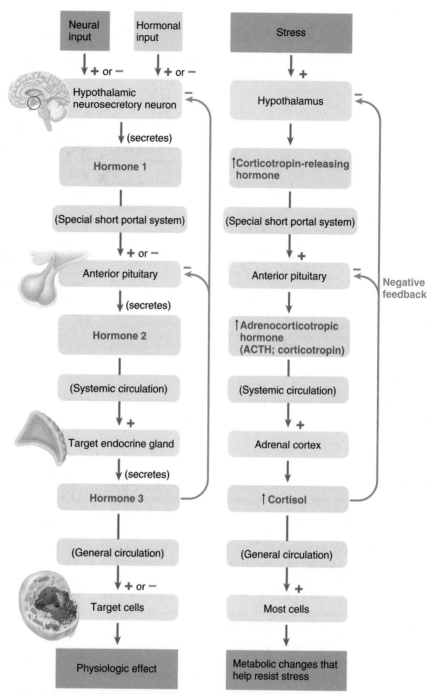

● **FIGURE 18-7 Hierarchic chain of command and negative feedback in endocrine control.** The general pathway involved in the hierarchic chain of command in the hypothalamus–anterior pituitary–peripheral target endocrine-gland axis is depicted on the left. The pathway on the right leading to cortisol secretion provides a specific example of this endocrine chain of command. The hormone ultimately secreted by the target endocrine gland, such as cortisol, acts in negative-feedback fashion to reduce secretion of the regulatory hormones higher in the chain of command.

lators in these other sites. For example, PIH is identical to dopamine, a major neurotransmitter in the basal nuclei and elsewhere. Others are thought to modulate a variety of functions that range from motor activity (TRH) to libido (GnRH) to learning (vasopressin). These examples further illustrate the multiplicity of ways chemical messengers function.

Neurosecretory neurons in hypothalamus (secrete releasing and inhibiting hormones into portal system)

Hypothalamus

Capillaries in hypothalamus

Systemic arterial blood in

Endocrine cells of anterior pituitary (secrete anterior pituitary hormones into systemic blood)

Capillaries in anterior pituitary

Systemic venous blood out

Releasing and inhibiting hormones

Hypothalamic-hypophyseal portal system

Posterior pituitary

Anterior pituitary

KEY

•‒• = Hypophysiotropic hormones   • = Anterior pituitary hormone

1 Hypophysiotropic hormones (releasing hormones and inhibiting hormones) produced by neurosecretory neurons in the hypothalamus enter the hypothalamic capillaries.

2 These hypothalamic capillaries rejoin to form the hypothalamic-hypophyseal portal system, a vascular link to the anterior pituitary.

3 The portal system branches into the capillaries of the anterior pituitary.

4 The hypophysiotropic hormones, which leave the blood across the anterior pituitary capillaries, control the release of anterior pituitary hormones.

5 When stimulated by the appropriate hypothalamic releasing hormone, the anterior pituitary secretes a given hormone into these capillaries.

6 The anterior pituitary capillaries rejoin to form a vein, through which the anterior pituitary hormones leave for ultimate distribution throughout the body by the systemic circulation.

● **FIGURE 18-8 Vascular link between the hypothalamus and anterior pituitary.**

ROLE OF THE HYPOTHALAMIC–HYPOPHYSEAL PORTAL SYSTEM The hypothalamic regulatory hormones reach the anterior pituitary by means of a unique vascular link. In contrast to the direct neural connection between the hypothalamus and posterior pituitary, the anatomic and functional link between the hypothalamus and anterior pituitary is an unusual capillary-to-capillary connection, the **hypothalamic–hypophyseal portal system.** A portal system is a vascular arrangement in which venous blood flows directly from one capillary bed through a connecting vessel to another capillary bed. The largest and best-known portal system is the hepatic portal system, which drains intestinal venous blood directly into the liver for immediate processing of absorbed nutrients (see p. 616). Although much smaller, the hypothalamic–hypophyseal portal system is no less important, because it provides a critical link between the brain and much of the endocrine system. It begins in the base of the hypothalamus with a group of capillaries that recombine into small portal vessels, which pass down through the connecting stalk into the anterior pituitary. Here the portal vessels branch to form most of the anterior pituitary capillaries, which in turn drain into the systemic venous system (● Figure 18-8).

As a result, almost all the blood supplied to the anterior pituitary must first pass through the hypothalamus. Because materials can be exchanged between the blood and surrounding tissue only at the capillary level, the hypothalamic–hypophyseal portal system provides a "private" route where releasing and inhibiting hormones can be picked up at the hypothalamus and delivered immediately and directly to the anterior pituitary at relatively high concentrations, completely bypassing the general

circulation. If the portal system did not exist, once the hypophysiotropic hormones were picked up in the hypothalamus, they would be returned to the heart by the systemic venous system. From here, they would travel to the lungs and back to the heart through the pulmonary circulation, and finally enter the systemic arterial system for delivery throughout the body, including the anterior pituitary. This process would not only take longer but the hypophysiotropic hormones would be considerably diluted by the much larger volume of blood flowing through this usual circulatory route.

The axons of the neurosecretory neurons that produce the hypothalamic regulatory hormones terminate on the capillaries at the origin of the portal system. These hypothalamic neurons secrete their hormones in the same way as the hypothalamic neurons that produce vasopressin and oxytocin. The hormone is synthesized in the cell body and then transported in vesicles by molecular motors to the axon terminal. It is stored there until its release by exocytosis into an adjacent capillary on appropriate stimulation. The major difference is that the hypophysiotropic hormones are released into the portal vessels, which deliver them to the anterior pituitary where they control the release of anterior pituitary hormones into the general circulation. In contrast, the hypothalamic hormones stored in the posterior pituitary are themselves released into the general circulation.

CONTROL OF HYPOTHALAMIC RELEASING AND INHIBITING HORMONES What regulates secretion of these hypophysiotropic hormones? Like other neurons, the neurons secreting these regulatory hormones receive abundant input of information (both neural and hormonal and both excitatory and inhibitory) that they must integrate. Studies are still in progress to unravel the complex neural input from many diverse areas of the brain to the hypophysiotropic secretory neurons. Some of these inputs carry information about a variety of environmental conditions. One example is the marked increase in secretion of corticotropin-releasing hormone (CRH) in response to stress (see ● Figure 18-7b). Numerous neural connections also exist between the hypothalamus and the portions of the brain concerned with emotions (the limbic system; see p. 155). Thus, emotions greatly influence secretion of hypophysiotropic hormones. The menstrual irregularities sometimes experienced by women who are emotionally upset are a common manifestation of this relationship.

In addition to being regulated by different regions of the brain, the hypophysiotropic neurons are controlled by various chemical inputs that reach the hypothalamus through the blood. Unlike other regions of the brain, portions of the hypothalamus are not guarded by the blood–brain barrier, so the hypothalamus can easily monitor chemical changes in the blood. The most common blood-borne factors that influence hypothalamic neurosecretion are the negative-feedback effects of target-gland hormones, to which we now turn our attention.

Target-gland hormones inhibit hypothalamic and anterior pituitary hormone secretion via negative feedback.

In most cases, hypophysiotropic hormones initiate a three-hormone sequence: (1) hypophysiotropic hormone, (2) anterior-pituitary tropic hormone, and (3) hormone from the peripheral target endocrine gland. Typically, in addition to producing its physiologic effects, the target-gland hormone suppresses secretion of the tropic hormone that is driving it. This negative feedback is accomplished by the target-gland hormone acting either directly on the pituitary itself or on the release of hypothalamic hormones, which in turn regulate anterior pituitary function (see ● Figure 18-7). As an example, consider the CRH–ACTH–cortisol system. Hypothalamic CRH (corticotropin-releasing hormone) stimulates the anterior pituitary to secrete ACTH (adrenocorticotropic hormone, alias corticotropin), which in turn stimulates the adrenal cortex to secrete cortisol. The final hormone in the system, cortisol, inhibits the hypothalamus to reduce CRH secretion and also acts directly on the corticotropes in the anterior pituitary to reduce synthesis of pro-opiomelanocortin and ACTH secretion. Through this double-barreled approach, cortisol exerts negative-feedback control to stabilize its own plasma concentration. If plasma cortisol levels start to rise above a prescribed set level, cortisol suppresses its own further secretion by its inhibitory actions at the hypothalamus and anterior pituitary. This mechanism ensures that once a hormonal system is activated, its secretion does not continue unabated. If plasma cortisol levels fall below the desired set point, cortisol's inhibitory actions at the hypothalamus and anterior pituitary are reduced, so the driving forces for cortisol secretion (CRH–ACTH) increase accordingly. The other target-gland hormones act by similar negative-feedback loops to maintain their plasma levels relatively constant at the set point.

Diurnal rhythms are superimposed on this type of stabilizing negative-feedback regulation; that is, the set point changes as a function of the time of day. Furthermore, other controlling inputs may break through the negative-feedback control to alter hormone secretion (that is, change the set level) at times of special need. For example, stress raises the set point for cortisol secretion.

The detailed functions and control of all the anterior pituitary hormones except growth hormone are discussed elsewhere in conjunction with the target tissues that they influence; for example, thyroid-stimulating hormone is covered in the next chapter with the discussion of the thyroid gland. Accordingly, growth hormone is the only anterior pituitary hormone we elaborate on at this time.

Endocrine Control of Growth

In growing children, continuous net protein synthesis occurs under the influence of growth hormone as the body steadily gets larger. Weight gain alone is not synonymous with growth, because weight gain may occur from retaining excess H_2O or storing fat without true structural growth of tissues. Growth requires net synthesis of proteins and includes lengthening of the long bones (the bones of the extremities) as well as increases in the size and number of cells in the soft tissues.

Growth depends on growth hormone but is influenced by other factors as well.

Although, as the name implies, growth hormone (GH) is absolutely essential for growth, it is not wholly responsible for determining the rate and final magnitude of growth in a given individual. The following factors affect growth:

■ *Genetic determination* of an individual's maximum growth capacity. Attaining this full growth potential further depends on the other factors listed here.

■ *An adequate diet,* including enough total protein and ample essential amino acids to accomplish the protein synthesis necessary for growth. Malnourished children never achieve their full growth potential. The growth-stunting effects of inadequate nutrition are most profound when they occur in infancy. In severe cases, the child may be locked in to irreversibly stunted body growth and brain development. About 70% of total brain growth occurs in the first two years of life. By contrast, a person cannot exceed his or her genetically determined maximum by eating a more than adequate diet. The excess food intake produces obesity instead of growth.

■ *Freedom from chronic disease and stressful environmental conditions.* Stunted growth under adverse circumstances is a result in large part of the prolonged stress-induced secretion of cortisol from the adrenal cortex. Cortisol exerts several potent antigrowth effects, such as promoting protein breakdown, inhibiting growth in the long bones, and blocking the secretion of GH. Even though sickly or stressed children do not grow well, if the underlying condition is corrected before adult size is achieved, they can rapidly catch up to their normal growth curve through a remarkable spurt in growth.

■ *Normal levels of growth-influencing hormones.* In addition to the absolutely essential GH, other hormones, including thyroid hormone, insulin, and the sex hormones, play secondary roles in promoting growth.

The rate of growth is not continuous, nor are the factors responsible for promoting growth the same throughout the growth period. *Fetal growth* is promoted largely by certain hormones from the placenta (the hormone-secreting organ of exchange between the fetal and maternal circulatory systems; see p. 782), with the size at birth being determined principally by genetic and environmental factors. GH plays no role in fetal development. After birth, GH and other nonplacental hormonal factors begin to play an important role in regulating growth. Genetic and nutritional factors also strongly affect growth during this period.

Children display two periods of rapid growth—a *postnatal growth spurt* during their first two years of life and a *pubertal growth spurt* during adolescence (● Figure 18-9). From age 2 until puberty, the *rate* of linear growth progressively declines, even though the child is still growing. Before puberty there is little sexual difference in height or weight. During puberty, a marked acceleration in linear growth takes place because the long bones lengthen. Puberty begins at about age 11 in girls and 13 in boys and lasts for several years in both sexes. The mechanisms responsible for the pubertal growth spurt are not clearly understood. Apparently both genetic and hormonal factors are involved. Some evidence indicates that GH secretion is elevated during puberty and thus may contribute to growth acceleration during this time. Furthermore, **androgens** ("male" sex hormones), whose secretion increases dramatically at puberty, also contribute to the pubertal growth spurt by promoting protein synthesis and bone growth. The potent androgen from the male testes, testosterone, is of greatest importance in promoting a sharp increase in height in adolescent boys; whereas the less potent adrenal androgens from the adrenal gland, which also show a sizable increase in secretion during adolescence, are most likely important in the female pubertal growth spurt. Although estrogen secretion by the ovaries also begins during puberty, it is unclear what role this "female" sex hormone may play in the pubertal growth spurt in girls. Testosterone and estrogen both ultimately act on bone to halt its further growth so that full adult height is attained by the end of adolescence.

Growth hormone is essential for growth, but it also directly exerts metabolic effects not related to growth.

GH is the most abundant hormone produced by the anterior pituitary, even in adults in whom growth has already ceased, although GH secretion typically starts to decline after middle age. The continued high secretion of GH beyond the growing period implies that this hormone has important influences other than on growth. In addition to promoting growth, GH has important metabolic effects and enhances the immune system. We will briefly describe GH's metabolic actions before turning our attention to its growth-promoting actions.

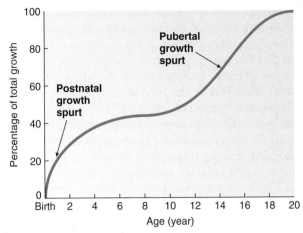

● **FIGURE 18-9 Normal growth curve.**

To exert its metabolic effects, GH binds directly with its target organs, namely adipose tissue, skeletal muscles, and liver. GH increases fatty acid levels in the blood by enhancing the breakdown of triglyceride fat stored in adipose tissue, and it increases blood glucose levels by decreasing glucose uptake by muscles and increasing glucose output by the liver. Muscles use the mobilized fatty acids instead of glucose as a metabolic fuel. Thus, the overall metabolic effect of GH is to mobilize fat stores as a major energy source while conserving glucose for glucose-dependent tissues such as the brain. The brain can use only glucose as its metabolic fuel, yet nervous tissue cannot store glycogen (stored glucose) to any extent. This metabolic pattern is suitable for maintaining the body during prolonged fasting or other situations when the body's energy needs exceed available glucose stores. GH also stimulates amino acid uptake and protein synthesis, but it does not act directly to accomplish these growth-promoting metabolic actions or any other growth-related actions. Before examining the means by which GH promotes growth, let us summarize its metabolic effects: It increases blood fatty acids, increases blood glucose and spares glucose for the brain, and stimulates protein synthesis (decreasing blood amino acids in the process).

Growth hormone exerts its growth-promoting effects indirectly by stimulating insulin-like growth factors.

GH does not act directly to bring about its growth-producing actions (increased cell division, enhanced protein synthesis, and bone growth). Instead, GH's growth-promoting actions are directly mediated by **insulin-like growth factors (IGFs),** which act on the target cells to cause growth of both soft tissues and bones. IGFs are produced in many tissues and have endocrine, paracrine, and autocrine actions. Originally called **somatomedins,** these peptide mediators are now preferentially called insulin-like growth factors because they are structurally and functionally similar to insulin. Like insulin, IGFs exert their effects largely by binding with receptor-enzymes that activate designated effector proteins within the target cell by phosphorylating the tyrosines (a type of amino acid) in the protein (the tyrosine kinase pathway; see p. 115). There are two IGFs—**IGF-I** and **IGF-II**.

IGF-I IGF-I synthesis is stimulated by GH and mediates this hormone's growth-promoting actions. The major source of circulating IGF-I is the liver, which releases this peptide product into the blood in response to GH stimulation. IGF-I is also produced by most other tissues, although they do not release it into the blood to any extent. Researchers propose that IGF-I produced locally in target tissues may act through paracrine means (see p. 114). Such a mechanism could explain why blood levels of GH are no higher, and indeed circulating IGF-I levels are lower, during the first several years of life compared to adult values, even though growth is quite rapid during the postnatal period. Local production of IGF-I in target tissues may possibly be more important than delivery of blood-borne IGF-I or GH during this time.

Production of IGF is controlled by a number of factors other than GH, including nutritional status, age, and tissue-specific factors as follows:

- IGF-I production depends on adequate nutrition. Inadequate food intake reduces IGF-I production. As a result, changes in circulating IGF-I levels do not always coincide with changes in GH secretion. For example, fasting decreases IGF-I levels even though it increases GH secretion.

- Age-related factors influence IGF-I production. A dramatic increase in circulating IGF-I levels accompanies the moderate increase in GH at puberty, which may, of course, be an impetus to the pubertal growth spurt.

- Finally, various tissue-specific stimulatory factors can increase IGF-I production in particular tissues. To illustrate, the gonadotropins and sex hormones stimulate IGF-I production within reproductive organs such as the testes in males and the ovaries and uterus in females.

Thus, control of IGF-I production is complex and subject to a variety of systemic and local factors.

IGF-II In contrast to IGF-I, IGF-II production is not influenced by GH. IGF-II is primarily important during fetal development. Although IGF-II continues to be produced during adulthood, its role in adults remains unclear.

We now describe GH's growth-promoting effects, mediated by IGF-I.

Growth hormone/IGF-I promote growth of soft tissues by stimulating hyperplasia and hypertrophy.

When tissues are responsive to its growth-promoting effects, GH (acting through IGF-I) stimulates growth of both soft tissues and the skeleton. GH promotes growth of soft tissues by (1) increasing the number of cells (**hyperplasia**) and (2) increasing the size of cells (**hypertrophy**). GH increases the number of cells by stimulating cell division and by preventing apoptosis (programmed cell death; see p. 124). GH increases the size of cells by favoring synthesis of proteins, the main structural

component of cells. GH stimulates almost all aspects of protein synthesis while it simultaneously inhibits protein degradation. It promotes the uptake of amino acids (the raw materials for protein synthesis) by cells, decreasing blood amino acid levels in the process. Furthermore, it stimulates the cellular machinery responsible for accomplishing protein synthesis according to the cell's genetic code. It increases DNA and RNA synthesis and increases incorporation of amino acids into the new proteins at the ribosomal level (see p. 39).

Growth of the long bones resulting in increased height is the most dramatic effect of GH. Before you can understand the means by which GH stimulates bone growth, you must first become familiar with the structure of bone and how growth of bone is accomplished.

Bone grows in thickness and in length by different mechanisms, both stimulated by growth hormone.

Bone is a living tissue. Being a form of connective tissue, it consists of cells and an extracellular organic matrix known as **osteoid** produced by the cells. The bone cells that produce the organic matrix are known as **osteoblasts** ("bone formers"). Osteoid is composed of collagen fibers (see p. 58) in a semisolid gel. This organic matrix has a rubbery consistency and is responsible for bone's tensile strength (the resilience of bone to breakage when tension is applied). Bone is made hard by precipitation of *calcium phosphate crystals* within the osteoid. These inorganic crystals provide the bone with compressional strength (the ability of bone to hold its shape when squeezed or compressed). If bones consisted entirely of inorganic crystals, they would be brittle, like pieces of chalk. Bones have structural strength approaching that of reinforced concrete, yet they are not brittle and are much lighter in weight because they have the structural blending of an organic scaffolding hardened by inorganic crystals. **Cartilage** is similar to bone, except that living cartilage is not calcified.

A long bone basically consists of a fairly uniform cylindrical shaft, the **diaphysis,** with a flared articulating knob at either end, an **epiphysis.** In a growing bone, the diaphysis is separated at each end from the epiphysis by a layer of cartilage known as the **epiphyseal plate** (● Figure 18-10a). The central cavity of the bone is filled with bone marrow, the site of blood cell production (see p. 395).

BONE GROWTH Growth in *thickness* of bone is achieved by adding new bone on top of the outer surface of already existing bone. This growth is produced by osteoblasts within the **periosteum,** a connective tissue sheath that covers the outer bone. As osteoblast activity deposits new bone on the external surface, other cells within the bone, the **osteoclasts** ("bone breakers"), dissolve the bony tissue on the inner surface next to the marrow cavity. In this way, the marrow cavity enlarges to keep pace with the increased circumference of the bone shaft.

Growth in *length* of long bones is accomplished by a different mechanism. Bones grow in length as a result of activity of the cartilage cells, or **chondrocytes,** in the epiphyseal plates (● Figure 18-10b). During growth, cartilage cells on the outer edge of

(a) Anatomy of a long bone

KEY

	Cartilage
	Calcified cartilage
	Bone

1. Chondrocytes undergo cell division.

2. The older chondrocytes grow larger.

Causes thickening of epiphyseal plate

3. As the extracellular matrix calcifies, the entrapped chondrocytes die.

4. The dead chondrocytes are cleared away by osteoclasts.

5. Osteoblasts swarm up from diaphysis and deposit bone over persisting remnants of disintegrating cartilage.

(b) Two sections of the same epiphyseal plate at different times, depicting the lengthening of long bones

● **FIGURE 18-10 Anatomy and growth of long bones.**

the plate next to the epiphysis divide and multiply, temporarily widening the epiphyseal plate. As new chondrocytes are formed on the epiphyseal border, the older cartilage cells toward the diaphyseal border are enlarging. This combination of proliferation of new cartilage cells and hypertrophy of maturing chondrocytes temporarily widens the epiphyseal plate. This thickening of the intervening cartilaginous plate pushes the bony epiphysis farther away from the diaphysis. Soon the matrix surrounding the oldest hypertrophied cartilage becomes calcified. Because cartilage lacks its own capillary network, the survival of cartilage cells depends on diffusion of nutrients and O_2 through the matrix, a process prevented by the deposition of calcium salts. As a result, the old nutrient-deprived cartilage cells on the diaphyseal border die. As osteoclasts clear away dead chondro-

cytes and the calcified matrix that imprisoned them, the area is invaded by osteoblasts, which swarm upward from the diaphysis, trailing their capillary supply with them. These new tenants lay down bone around the persisting remnants of disintegrating cartilage until bone entirely replaces the inner region of cartilage on the diaphyseal side of the plate. When this **ossification** ("bone formation") is complete, the bone on the diaphyseal side has lengthened and the epiphyseal plate has returned to its original thickness. The cartilage that bone has replaced on the diaphyseal end of the plate is as thick as the new cartilage on the epiphyseal end of the plate. Thus, bone growth is made possible by the growth and death of cartilage, which acts like a "spacer" to push the epiphysis farther out while it provides a framework for future bone formation on the end of the diaphysis.

MATURE, NONGROWING BONE As the extracellular matrix produced by an osteoblast calcifies, the osteoblast—like its chondrocyte predecessor—becomes entombed by the matrix it has deposited around itself. Unlike chondrocytes, however, osteoblasts trapped within a calcified matrix do not die, because they are supplied by nutrients transported to them through small canals that the osteoblasts themselves form by sending out cytoplasmic extensions around which the bony matrix is deposited. Thus, within the final bony product a network of permeating tunnels radiates from each entrapped osteoblast, serving as a lifeline system for nutrient delivery and waste removal. The entrapped osteoblasts, now called **osteocytes,** retire from active bone-forming duty because their imprisonment prevents them from laying down new bone. However, they are involved in the hormonally regulated exchange of calcium between bone and the blood. This exchange is under the control of parathyroid hormone (discussed in the next chapter), not GH.

Growth hormone/IGF-I promote growth of bone in both length and thickness.

GH causes bones to grow both in length and thickness. GH, via IGF-I, stimulates the proliferation of epiphyseal cartilage, thereby making space for more bone formation, and stimulates osteoblast activity. GH can promote lengthening of long bones as long as the epiphyseal plate remains cartilaginous, or is "open." At the end of adolescence, under the influence of the sex hormones, these plates completely ossify, or "close," so that the bones cannot lengthen any further despite the presence of GH. Thus, after the plates are closed, the individual does not grow any taller.

*These factors all increase growth hormone secretion, but it is unclear whether they do so by stimulating GHRH or inhibiting GHIH somatostatin, or both.

● **FIGURE 18-11 Control of growth hormone secretion.**

Growth hormone secretion is regulated by two hypophysiotropic hormones.

The control of GH secretion is complex, with two hypothalamic hypophysiotropic hormones playing a key role.

GROWTH HORMONE–RELEASING HORMONE AND GROWTH HORMONE–INHIBITING HORMONE Two antagonistic regulatory hormones from the hypothalamus are involved in controlling growth hormone secretion: growth hormone–releasing hormone (GHRH), which is stimulatory and the dominant influence, and growth hormone–inhibiting hormone (GHIH, or somatostatin), which is inhibitory (● Figure 18-11). (Note the distinctions among *somatotropin,* alias growth hormone; *somatomedin,* a liver hormone (alias IGF-I) that directly mediates the effects of GH; and *somatostatin,* which inhibits GH secretion.) Both GHRH and somatostatin act on the anterior pituitary somatotropes by binding with G-protein-coupled receptors linked to the cAMP second-messenger pathway, with GHRH increasing cAMP and somatostatin decreasing cAMP.

As with control of other anterior pituitary hormones, negative-feedback loops participate in controlling GH secretion. Complicating the feedback loops for the hypothalamus–pituitary–liver axis is the fact that GH secretion is directly regulated by both stimulatory and inhibitory factors. Therefore negative-feedback loops involve both inhibition of stimulatory factors and stimulation of inhibitory factors. GH stimulates IGF-I secretion by the liver, and IGF-I in turn is the primary inhibitor of GH secretion by the anterior pituitary. IGF-I inhibits the somatotropes in the pituitary directly and further decreases GH secretion by inhibiting GHRH-secreting cells and stimulating the somatostatin-secreting cells in the hypothalamus, thus decreasing hypothalamic stimulation of the somatotropes. Furthermore, GH itself inhibits hypothalamic GHRH secretion and stimulates somatostatin release.

FACTORS THAT INFLUENCE GH SECRETION A number of factors influence GH secretion by acting on the hypothalamus. GH secretion displays a well-characterized diurnal rhythm. Through most of the day, GH levels tend to be low and fairly constant. About an hour after the onset of deep sleep, however, GH secretion markedly increases up to five times the low daytime value.

Superimposed on this diurnal fluctuation in GH secretion are further bursts in secretion that occur in response to exercise, stress, and low blood glucose, the major stimuli for increased secretion. The benefits of increased GH secretion during these situations when energy demands outstrip the body's glucose reserves are presumably that glucose is conserved for the brain and fatty acids are provided as an alternative energy source for muscle.

Because GH uses up fat stores and promotes synthesis of body proteins, it encourages a change in body composition away from adipose deposition toward an increase in muscle protein. Accordingly, the increase in GH secretion that accompanies exercise may at least in part mediate the effects of exercise in reducing the percentage of body fat while increasing the lean body mass.

A rise in blood amino acids after a high-protein meal also enhances GH secretion. In turn, GH promotes the use of these amino acids for protein synthesis. GH is also stimulated by a decline in blood fatty acids. Because GH mobilizes fat, such regulation helps maintain fairly constant blood fatty acid levels.

Finally, ghrelin, a potent appetite stimulator released from the stomach, also stimulates GH secretion (see p. 646). This "hunger hormone" may play a role in coordinating growth with nutrient acquisition.

Note that the known regulatory inputs for GH secretion are aimed at adjusting the levels of glucose, amino acids, and fatty acids in the blood. No known growth-related signals influence GH secretion. The whole issue of what really controls growth is complicated by the fact that GH levels during early childhood, a period of quite rapid linear growth, are similar to those in normal adults. As mentioned earlier, the poorly understood control of IGF-I activity may be important in this regard. Another related question is, Why aren't adult tissues still responsive to GH's growth-promoting effects? We know we do not

● **FIGURE 18-12 Examples of the effect of abnormalities in growth hormone secretion on growth.** The man at the right displays pituitary dwarfism resulting from underproduction of growth hormone in childhood. The man at the center of the photograph has gigantism caused by excessive growth hormone secretion in childhood. The man at the left is of average height.

grow any taller after adolescence, because the epiphyseal plates have closed, but why do soft tissues not continue to grow through hypertrophy and hyperplasia under the influence of GH? One speculation is that levels of GH may only be high enough to produce its growth-promoting effects during the secretion bursts that occur in deep sleep. It is interesting to note that time spent in deep sleep is greatest in infancy and gradually declines with age. Still, even as we age, we spend some time in deep sleep, yet we do not gradually get larger. Further research is needed to unravel these mysteries.

Abnormal growth hormone secretion results in aberrant growth patterns.

 Diseases related to both deficiencies and excesses of growth hormone can occur. The effects on the pattern of growth are much more pronounced than the metabolic consequences.

GROWTH HORMONE DEFICIENCY GH deficiency may be caused by a pituitary defect (lack of GH) or may occur secondary to hypothalamic dysfunctions (lack of GHRH). Hyposecretion of GH in a child is one cause of **dwarfism.** The predominant feature is short stature caused by retarded skeletal growth (● Figure 18-12). Less obvious characteristics include poorly developed muscles (reduced muscle protein synthesis) and excess subcutaneous fat (less fat mobilization).

In addition, growth may be thwarted because the tissues fail to respond normally to GH. An example is **Laron dwarfism,** which is characterized by abnormal GH receptors that are unresponsive to the hormone. The symptoms of this condition resemble those of severe GH deficiency even though blood levels of GH are actually high.

In some instances, GH levels are adequate and target-cell responsiveness is normal, but IGFs are lacking. African pygmies are an interesting example. Their usual short stature is attributable to a lack of IGF-I.

The onset of GH deficiency in adulthood after growth is already complete produces relatively few symptoms. GH-deficient adults tend to have reduced skeletal muscle mass and strength (less muscle protein) as well as decreased bone density (less osteoblast activity during ongoing bone remodeling). Furthermore, because GH is essential for maintaining cardiac muscle mass and performance in adulthood, GH-deficient adults may be at increased risk of developing heart failure. (For a discussion of GH therapy, see the boxed feature on p. 684, ■ Concepts, Challenges, and Controversies.)

GROWTH HORMONE EXCESS Hypersecretion of GH is most often caused by a tumor of the GH-producing cells of the anterior pituitary. The symptoms depend on the age of the individual when the abnormal secretion begins. If overproduction of GH begins in childhood before the epiphyseal plates close, the principal manifestation is a rapid growth in height without distortion of body proportions. Appropriately, this condition is known as **gigantism** (● Figure 18-12). If not treated by removal of the tumor or by drugs that block the effect of GH, the person may reach a height of eight feet or more. All the soft tissues grow correspondingly, so the body is still well proportioned.

If GH hypersecretion occurs after adolescence when the epiphyseal plates have already closed, further growth in height is prevented. Under the influence of excess GH, however, the bones become thicker and the soft tissues, especially connective tissue and skin, proliferate. This disproportionate growth pattern produces a disfiguring condition known as **acromegaly** (*acro* means "extremity"; *megaly* means "large"). Bone thickening is most obvious in the extremities and face. A marked coarsening of the features to an almost apelike appearance gradually develops as the jaws and cheekbones become more prominent because of the thickening of the facial bones and the skin (● Figure 18-13). The hands and feet enlarge, and the fingers and toes become greatly thickened. Peripheral nerve disorders often occur as nerves are entrapped by overgrown connective tissue or bone.

Age 13   Age 21   Age 35

Dean Warchak@www.acromegalysupport.com.

● **FIGURE 18-13 Progressive development of acromegaly.** In this series of photos from childhood to the present, note how the patient's brow bones, cheekbones, and jawbones are becoming progressively more prominent as a result of ongoing thickening of the bones and skin caused by excessive GH secretion.

Other hormones besides growth hormone are essential for normal growth.

Several other hormones in addition to GH contribute in special ways to overall growth:

■ *Thyroid hormone* is essential for growth but is not itself directly responsible for promoting growth. It plays a permissive role in skeletal growth; the actions of GH fully manifest only when enough thyroid hormone is present. As a result, growth is severely stunted in hypothyroid children, but hypersecretion of thyroid hormone does not cause excessive growth.

■ *Insulin* is an important growth promoter. Insulin deficiency often blocks growth, and hyperinsulinism frequently spurs excessive growth. Because insulin promotes protein synthesis, its growth-promoting effects should not be surprising. However, these effects may also arise from a mechanism other than insulin's direct effect on protein synthesis. Insulin structurally resembles the IGFs and may interact with the IGF-I receptor, which is very similar to the insulin receptor.

■ *Androgens,* which are believed to play an important role in the pubertal growth spurt, powerfully stimulate protein synthesis in many organs. Androgens stimulate linear growth, promote weight gain, and increase muscle mass. The most potent androgen, testicular testosterone, is responsible for men developing heavier musculature than women do. These androgenic growth-promoting effects depend on the presence of GH. Androgens have virtually no effect on body growth in the absence of GH, but in its presence they synergistically enhance linear growth. Although androgens stimulate growth, they ultimately stop further growth by promoting closure of the epiphyseal plates.

■ *Estrogens,* like androgens, ultimately terminate linear growth by stimulating complete conversion of the epiphyseal plates to

Growth and Youth in a Bottle?

Unlike most hormones, growth hormone's structure and activity differ among species. As a result, growth hormone (GH) from animal sources is ineffective for treating GH deficiency in humans. In the past, the only source of human GH was pituitary glands from human cadavers. This supply was never adequate, and it was finally removed from the market by the Food and Drug Administration (FDA) because of fear of viral contamination. Now, however, the technique of genetic engineering has made available an unlimited supply of human GH. The gene that directs synthesis of GH in humans has been introduced into bacteria, converting the bacteria into "factories" that synthesize human GH.

Even though adequate human GH is now available through genetic engineering, a new problem for the medical community is to determine under which circumstances synthetic GH treatment is appropriate. Until recently, the FDA had approved treatment only for the following uses: (1) for GH-deficient children, (2) for adults with a pituitary tumor or other disease that causes severe GH deficiency, and (3) for AIDS patients suffering from severe muscle wasting. Although not approved by the FDA for this use, GH therapy is also widely used to promote faster healing of skin in severely burned patients. In 2003, amid emotionally charged debates, the FDA approved GH shots for another group, the shortest 1.2% of children who are unusually short for no apparent reason. This therapy involves multiple GH injections per week for a number of years under careful supervision of pediatric endocrinologists for an average gain in height of one to three inches. GH-deficient children experience more dramatic gains of about six to eight inches on GH therapy.

Another group whom replacement GH therapy may benefit is the elderly. GH secretion typically peaks during a person's 20s, then in many people may start to dwindle after age 40. This decline may contribute to some of the characteristic signs of aging:

- Decreased muscle mass (GH promotes synthesis of proteins, including muscle protein)
- Increased fat deposition (GH promotes leanness by mobilizing fat stores for use as an energy source)

- Reduced bone density (GH promotes the activity of bone-forming cells)
- Thinner, sagging skin (GH promotes proliferation of skin cells)

(However, inactivity is also believed to play a major role in age-associated reductions in muscle mass, bone density, and strength.)

Several studies in the early 1990s suggested that some of these consequences of aging may be counteracted through the use of synthetic GH in people over age 60. Elderly men treated with supplemental GH showed increased muscle mass, reduced fatty tissue, and thickened skin. In similar studies in elderly women, supplemental GH therapy did not increase muscle mass significantly but did reduce fat mass and protect against bone loss.

Even though these early results were exciting, further studies have been more discouraging. For example, despite increased lean body mass, many treated people surprisingly do not have increased muscle strength or exercise capabilities. Also, when GH is supplemented for an extended time or in large doses, harmful side effects include an increased likelihood of diabetes, kidney stones, high blood pressure, headaches, joint pain, and carpal tunnel syndrome. (Carpal tunnel syndrome involves thickening and narrowing of the tunnel in the wrist through which the nerve supply to the hand muscles passes; *carpal* means "wrist.") Furthermore, synthetic GH is costly ($15,000 to $20,000 annually) and must be injected regularly. Also, some scientists worry that sustained administration of synthetic GH may raise the risk of developing cancer by promoting uncontrolled cell proliferation. For these reasons, many investigators no longer view synthetic GH as a potential "fountain of youth." Instead, they hope it can be used in a more limited way to strengthen muscle and bone sufficiently in the many elderly who have GH deficits, to help reduce the incidence of bone-breaking falls that often lead to disability and loss of independence. The National Institute of Aging is currently sponsoring a nationwide series of studies involving GH therapy in the elderly to help sort out potentially legitimate roles of this supplemental hormone. Unwilling to wait,

an estimated 30,000 elderly are currently taking GH.

An ethical dilemma is whether the drug should be used by others who have normal GH levels but want the product's growth-promoting actions for cosmetic or athletic reasons, such as normally growing teenagers who wish to attain even greater height. The drug is already being used illegally by some athletes and bodybuilders. Furthermore, a recent study found that only 4 out of 10 children who are legitimately receiving GH therapy under medical supervision are actually GH deficient or in the bottom 1.2% of height. The others are receiving the treatment because parents, physicians, and the children are being swayed by perceived cultural pressures that favor height rather than by medical factors.

Using the drug in children with normal GH levels may be problematic because synthetic GH is a double-edged product. Although it promotes growth and muscle mass, it also has negative effects, such as potentially troubling side effects. Furthermore, one study revealed that supplemental GH therapy in children who do not lack the hormone redistributes the body's fat and protein. The investigators compared two groups of otherwise healthy six- to eight-year-olds who were among the shortest for their age. One group consisted of children who were receiving GH, the other of children who were not. At the end of six months, the children taking the synthetic hormone had outpaced the untreated group in growth by more than 1.5 inches per year. However, the untreated children added both muscle and fat as they grew, whereas the treated children became unusually muscular and lost up to 76% of their body fat. The loss of fat became especially obvious in their faces and limbs, giving them a rawboned, gangly appearance. It is unclear what long-term effects—either deleterious or desirable—these dramatic changes in body composition might have. Scientists also express concern that these readily observable physical changes may be accompanied by more subtle abnormalities in organs and cells. Thus, the debate about whether to use GH in normal but short children is likely to continue.

bone. However, the effects of estrogen on growth prior to bone maturation are not well understood. Some studies suggest that large doses of estrogen may even hinder further body growth by inhibiting chondrocyte proliferation while the epiphyseal plates are still open.

Several factors contribute to the average height differences between men and women. First, because puberty occurs about two years earlier in girls than in boys, on the average boys have two more years of prepubertal growth than girls do. As a result, boys are usually several inches taller than girls at the start of their respective growth spurts. Second, as already mentioned, boys experience a greater androgen-induced growth spurt than girls before their respective gonadal steroids seal their long bones from further growth; this results in greater heights in men than in women on average. Third, the pubertal rise in estrogen may reduce the pubertal growth spurt in females. Fourth, recent evidence suggests that androgens "imprint" the brains of males during development, giving rise to a "masculine" secretory pattern of GH characterized by higher cyclic peaks, which are speculated to contribute to the greater height of males.

In addition to these hormones that exert overall effects on body growth, a number of poorly understood peptide *growth factors* have been identified that stimulate mitotic activity of specific tissues (for example, epidermal growth factor).

We now shift attention to the other central endocrine gland—the pineal gland.

Pineal Gland and Circadian Rhythms

The **pineal gland,** a tiny, pinecone-shaped structure located in the center of the brain (see ● Figure 5-7b, p. 146, and ● Figure 18-1, p. 662), secretes the hormone **melatonin,** an indoleamine hormone derived from the amino acid tryptophan. (Do not confuse melatonin with the skin-darkening pigment, *melanin*.) Although melatonin was discovered in 1959, investigators have only recently begun to unravel its many functions. One of melatonin's most widely accepted roles is helping to keep the body's inherent circadian rhythms in synchrony with the light–dark cycle. We will first examine circadian rhythms in general before looking at the role of melatonin in this regard and considering other functions of this hormone.

The suprachiasmatic nucleus is the master biological clock.

Hormone secretion rates are not the only factor in the body that fluctuates cyclically over a 24-hour period. Humans have similar biological clocks for many other bodily functions, ranging from gene expression to physiologic processes such as temperature regulation (see p. 650) to behavior. The master biological clock that serves as the pacemaker for the body's circadian rhythms is the **suprachiasmatic nucleus (SCN).** It consists of a cluster of nerve cell bodies in the hypothalamus above the optic chiasm, the point at which part of the nerve fibers from each

eye cross to the opposite half of the brain (*supra* means "above"; *chiasm* means "cross") (see p. 210; ● Figure 5-7b, p. 146; and ● Figure 18-4, p. 670). The self-induced rhythmic firing of the SCN neurons plays a major role in establishing many of the body's inherent daily rhythms.

ROLE OF CLOCK PROTEINS Scientists have now unraveled the underlying molecular mechanisms responsible for the SCN's circadian oscillations. Specific self-starting genes within the nuclei of SCN neurons set in motion a series of events that brings about the synthesis of **clock proteins** in the cytosol surrounding the nucleus. As the day wears on, these clock proteins continue to accumulate, finally reaching a critical mass, at which time they are transported into the nucleus. Here they block the genetic process responsible for their own production. The level of clock proteins gradually dwindles as they degrade within the nucleus, thus removing their inhibitory influence from the clock-protein genetic machinery. No longer being blocked, these genes once again rev up the production of more clock proteins as the cycle repeats itself. Each cycle takes about a day. The fluctuating levels of clock proteins bring about cyclic changes in neural output from the SCN that, in turn, lead to cyclic changes in effector organs throughout the day. An example is the diurnal variation in cortisol secretion (see ● Figure 18-3, p. 665). Circadian rhythms are thus linked to fluctuations in clock proteins, which use a feedback loop to control their own production. In this way, internal timekeeping is a self-sustaining mechanism built into the genetic makeup of the SCN neurons.

SYNCHRONIZATION OF THE BIOLOGICAL CLOCK WITH ENVIRONMENTAL CUES On its own, this biological clock generally cycles a bit slower than the 24-hour environmental cycle. Without any external cues, the SCN sets up cycles that average about 25 hours. The cycles are consistent for a given individual but vary somewhat among different people. If this master clock were not continually adjusted to keep pace with the world outside, the body's circadian rhythms would become progressively out of sync with the cycles of light (periods of activity) and dark (periods of rest). Thus, the SCN must be reset daily by external cues so that the body's biological rhythms are synchronized with the activity levels driven by the surrounding environment. The effect of not maintaining the internal clock's relevance to the environment is well known by people who experience **jet lag** when their inherent rhythm is out of step with external cues. The SCN works in conjunction with the pineal gland and its hormonal product melatonin to synchronize the various circadian rhythms with the 24-hour day–night cycle. (For a discussion of problems associated with being out of sync with environmental cues, see the boxed feature on p. 686, ■ Concepts, Challenges, and Controversies.)

Melatonin helps keep the body's circadian rhythms in time with the light–dark cycle.

Daily changes in light intensity are the major environmental cue used to adjust the SCN master clock. Special photoreceptors in the retina pick up light signals and transmit them di-

Tinkering with Our Biological Clocks

Research shows that the hectic pace of modern life, stress, noise, pollution, and the irregular schedules many workers follow can upset internal rhythms, illustrating how a healthy external environment affects our own internal environment—and our health.

Dr. Richard Restak, a neurologist and author, notes that the "usual rhythms of wakefulness and sleep . . . seem to exert a stabilizing effect on our physical and psychological health." The greatest disrupter of our natural circadian rhythms is the variable work schedule, surprisingly common in industrialized countries. Today, one out of every four working men and one out of every six working women has a variable work schedule—shifting frequently between day and night work. In many industries, to make optimal use of equipment and buildings, workers are on the job day and night. As a spin-off, more restaurants and stores stay open 24 hours a day and more health-care workers must be on duty at night to care for accident victims.

To spread the burden, many companies that maintain shifts round the clock alter their workers' schedules. One week, employees work the day shift. The next week, they move to the "graveyard shift" from midnight to 8 A.M. The next week, they work the night shift from 4 P.M. to midnight. Many shift workers feel tired most of the time and have trouble staying awake at the job. Work performance suffers because of the workers' fatigue. When workers arrive home for bed, they're exhausted but cannot sleep, because they are trying to doze off at a time when the body is trying to wake them up. Unfortunately, the weekly changes in schedule never permit workers' internal alarm clocks to fully adjust. Most people require 4 to 14 days to adjust to a new schedule.

Workers on alternating shifts suffer more ulcers, insomnia, irritability, depression, and tension than workers on unchanging shifts. Their lives are never the same. To make matters worse, tired, irritable workers whose judgment is impaired by fatigue pose a threat to society as a whole. Consider an example.

At 4 A.M. in the control room of the Three Mile Island nuclear reactor in Pennsylvania, three operators made the first mistake in a series of errors that led to the worst nuclear accident in U.S. history. The operators did not notice warning lights and failed to observe that a crucial valve had remained open. When the morning-shift operators entered the control room the next day, they quickly discovered the problems, but it was too late. Pipes in the system had burst, sending radioactive steam and water into the air and into two buildings. Ironically, this 1979 accident occurred 12 days after release of the popular movie, *The China Syndrome,* about a nuclear reactor accident, pumping up public awareness and alarm. Fortunately, no surrounding residents were immediately injured and no long-term studies have found a conclusive link between an increased incidence of cancer and the low level of radiation to which those living near the troubled reactor were exposed. Cleanup of the accident lasted from 1970 until 1993 and cost 975 million dollars.

Late in April 1986, another nuclear power plant went amok. This accident, in Chernobyl in the former Soviet Union, was far more severe. In the early hours of the morning, two engineers were testing the reactor. Violating standard operational protocol, they deactivated key safety systems. This single error in judgment (possibly caused by fatigue) led to the largest and

most costly nuclear accident in world history. Steam built up inside the reactor and blew the roof off the containment building. A thick cloud of radiation rose skyward and then spread throughout Europe and the world. For 10 days the reactor burned, disgorging 400 times more radiation than released from the atomic bomb dropped on Hiroshima in World War II. While workers battled to cover the molten radioactive core that spewed radiation into the sky, the whole world watched in horror.

The Chernobyl disaster, like the accident at Three Mile Island, may have been the result of workers operating at a time unsuitable for clear thinking. One has to wonder how many plane crashes, auto accidents, and acts of medical malpractice can be traced to judgment errors resulting from our insistence on working against inherent body rhythms.

Thanks to studies of biological rhythms, researchers are finding ways to reset biological clocks, which could help lessen the misery and suffering of shift workers and could improve the performance of the graveyard-shift workers. For instance, one simple measure is to put shift workers on three-week cycles to give their clocks time to adjust. And instead of shifting workers from daytime to graveyard shifts, transfer them forward, rather than backward (for example, from a daytime to a nighttime shift). It is a much easier adjustment. Bright lights can also be used to reset the biological clock. It is a small price to pay for a healthy workforce and a safer society. Furthermore, use of supplemental melatonin, the hormone that sets the internal clock to march in step with environmental cycles, may prove useful in resetting the body's clock when that clock is out of synchrony with external cues.

rectly to the SCN. These photoreceptors are distinct from the rods and cones used to perceive, or see, light (see p. 204). **Melanopsin,** a protein found in a special retinal ganglion cell (see p. 201), is the receptor for light that keeps the body in tune with external time. The vast majority of retinal ganglion cells receive input from the rod and cone photoreceptors. The axons of these ganglion cells form the *optic nerve* that carries information to the visual cortex in the occipital lobe (see p. 144). Intermingled among the visually oriented retinal ganglion cells, about 1% to 2% of the retinal ganglion cells instead form an entirely independent light-detection system that responds to levels of illumination, like a light meter on a camera, rather than the contrasts, colors, and contours detected by the image-forming visual system. The melanopsin-containing, illumination-detecting retinal ganglion cells cue the pineal gland about the absence or presence of light by sending their signals along the **retinohypothalamic tract** to the SCN. This pathway is distinct from the neural systems that result in vision perception. The SCN relays the message regarding light status to the pineal gland. This is the major way the internal clock is coordinated to a 24-hour day. Melatonin is the hormone of darkness. Melatonin secretion increases up to 10-fold during the darkness of night and then falls to low levels during the light of day. Fluctuations in melatonin secretion, in turn, help entrain the body's biological rhythms with the external light–dark cues.

Proposed roles of melatonin besides regulating the body's biological clock include the following:

■ Taken exogenously (in a pill), melatonin induces a natural sleep without the side effects that accompany hypnotic sedatives, so it may play a normal role in promoting sleep.

■ Melatonin is believed to inhibit the hormones that stimulate reproductive activity. Puberty may be initiated by a reduction in melatonin secretion.

■ In a related role, in some species, seasonal fluctuations in melatonin secretion associated with changes in the number of daylight hours are important triggers for seasonal breeding, migration, and hibernation.

■ In another related role, melatonin is being used in clinical trials as a method of birth control, because at high levels it shuts down ovulation (egg release). A male contraceptive using melatonin to stop sperm production is also under development.

■ Melatonin appears to be a very effective **antioxidant,** a defense tool against biologically damaging free radicals. *Free radicals* are very unstable electron-deficient particles that are highly reactive and destructive. Free radicals have been implicated in several chronic diseases such as coronary artery disease (see p. 333) and cancer and are believed to contribute to the aging process.

■ Evidence suggests that melatonin may slow the aging process, perhaps by removing free radicals or by other means.

■ Melatonin appears to enhance immunity and has been shown to reverse some of the age-related shrinkage of the thymus, the source of T lymphocytes (see p. 428), in old experimental animals.

Because of melatonin's many proposed roles, use of supplemental melatonin for a variety of conditions is very promising. However, most researchers are cautious about recommending supplemental melatonin until its effectiveness as a drug is further substantiated. Meanwhile, many people are turning to melatonin as a health food supplement; as such, it is not regulated by the Food and Drug Administration for safety and effectiveness. The two biggest self-prescribed uses of melatonin are as a prevention for jet lag and as a sleep aid.

Chapter in Perspective: Focus on Homeostasis

The endocrine system is one of the body's two major regulatory systems; the other is the nervous system. Through its relatively slowly acting hormone messengers, the endocrine system generally regulates activities that require duration rather than speed. Most of these activities are directed toward maintaining homeostasis. The specific contributions of the central endocrine organs to homeostasis are as follows:

■ The hypothalamus–posterior pituitary unit secretes vasopressin, which acts on the kidneys during urine formation to help maintain H_2O balance. Control of H_2O balance, in turn, is essential for maintaining ECF osmolarity and proper cell volume.

■ For the most part, the hormones secreted by the anterior pituitary do not directly contribute to homeostasis. Instead, most are tropic; that is, they stimulate the secretion of other hormones.

■ However, growth hormone from the anterior pituitary, in addition to its growth-promoting actions, also exerts metabolic effects that help maintain the plasma concentration of amino acids, glucose, and fatty acids.

■ The pineal gland secretes melatonin, which helps entrain the body's circadian rhythm to the environmental cycle of light (period of activity) and dark (period of inactivity).

The peripheral endocrine glands further help maintain homeostasis in the following ways:

■ Hormones help maintain the proper concentration of nutrients in the internal environment by directing chemical reactions involved in the cellular uptake, storage, and release of these molecules. Furthermore, the rate at which these nutrients are metabolized is controlled in large part by the endocrine system.

■ Salt balance, which is important in maintaining the proper ECF volume and arterial blood pressure, is achieved by hormonally controlled adjustments in salt reabsorption by the kidneys during urine formation.

■ Likewise, hormones act on various target cells to maintain the plasma concentration of calcium and other electrolytes. These electrolytes, in turn, play key roles in homeostatic activities. For example, maintenance of calcium levels within narrow limits is critical for neuromuscular excitability and blood clotting, among other life-supporting actions.

■ The endocrine system orchestrates a wide range of adjustments that help the body maintain homeostasis in response to stressful situations.

■ The endocrine and nervous systems work in concert to control the circulatory and digestive systems, which in turn carry out important homeostatic activities.

Unrelated to homeostasis, hormones direct the growing process and control most aspects of the reproductive system.

Objective Questions (Answers on p. A-54)

1. One endocrine gland may secrete more than one hormone. *(True or false?)*

2. One hormone may influence more than one type of target cell. *(True or false?)*

3. All endocrine glands are exclusively endocrine in function. *(True or false?)*

4. A single target cell may be influenced by more than one hormone. *(True or false?)*

5. Hyposecretion or hypersecretion of a specific hormone can occur even though its endocrine gland is perfectly normal. *(True or false?)*

6. Growth hormone levels in the blood are no higher during the early childhood growing years than during adulthood. *(True or false?)*

7. A hormone that has as its primary function the regulation of another endocrine gland is classified functionally as a _____ hormone.

8. Self-induced reduction in the number of receptors for a specific hormone is known as _____.

9. Activity within the cartilaginous layer of bone known as the _____ is responsible for lengthening of long bones.

10. The _____ in the hypothalamus is the body's master biological clock.

11. Indicate the relationships among the hormones in the hypothalamic/anterior pituitary/adrenal cortex system by using the following answer code to identify which hormone belongs in each blank:
    (a) cortisol
    (b) ACTH
    (c) CRH

(1) _____ from the hypothalamus stimulates the secretion of (2) _____ from the anterior pituitary. (3) _____, in turn, stimulates the secretion of (4) _____ from the adrenal cortex. In negative-feedback fashion, (5) _____ inhibits secretion of the releasing hormone (6) _____ and furthermore inhibits secretion of the tropic hormone (7) _____.

Essay Questions

1. List the overall functions of the endocrine system.

2. How is the plasma concentration of a hormone normally regulated?

3. List and briefly state the source and functions of the posterior pituitary hormones.

4. List and briefly state the source and functions of the anterior pituitary hormones.

5. Compare the relationship between the hypothalamus and posterior pituitary with the relationship between the hypothalamus and anterior pituitary. Describe the role of the hypothalamic–hypophyseal portal system and the hypothalamic releasing and inhibiting hormones.

6. Describe the actions of growth hormone that are unrelated to growth. What are growth hormone's growth-promoting actions? What is the role of IGFs?

7. Discuss the control of growth hormone secretion.

8. Describe the role of clock proteins.

9. What are the source, functions, and stimulus for secretion of melatonin?

POINTS TO PONDER

(Explanations on p. A-54)

1. Would you expect the concentration of hypothalamic releasing and inhibiting hormones in a systemic venous blood sample to be higher, lower, or the same as the concentration of these hormones in a sample of hypothalamic–hypophyseal portal blood?

2. Thinking about the feedback control loop among TRH, TSH, and thyroid hormone, would you expect the concentration of TSH to be normal, above normal, or below normal in a person whose diet is deficient in iodine (an element essential for synthesizing thyroid hormone)?

3. A patient displays symptoms of excess cortisol secretion. What factors could be measured in a blood sample to determine whether the condition is caused by a defect at the hypothalamic/anterior pituitary level or the adrenal cortex level?

4. Why would males with testicular feminization syndrome be unusually tall?

5. A black market for growth hormone abuse already exists among weight lifters and other athletes. What actions of growth hormone would induce a full-grown athlete to take supplemental doses of this hormone? What are the potential detrimental side effects?

(Explanation on p. A-55)

At 18 years of age and eight feet tall, Anthony O. was diagnosed with gigantism caused by a pituitary tumor. The condition was treated by surgically removing his pituitary gland. What hormonal replacement therapy would Anthony need?

Endocrine System

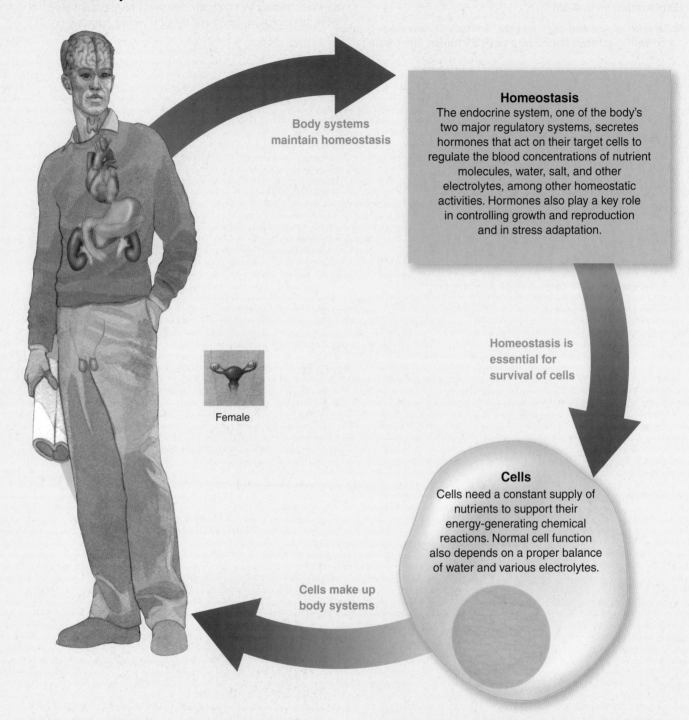

Body systems maintain homeostasis

Homeostasis

The endocrine system, one of the body's two major regulatory systems, secretes hormones that act on their target cells to regulate the blood concentrations of nutrient molecules, water, salt, and other electrolytes, among other homeostatic activities. Hormones also play a key role in controlling growth and reproduction and in stress adaptation.

Homeostasis is essential for survival of cells

Female

Cells

Cells need a constant supply of nutrients to support their energy-generating chemical reactions. Normal cell function also depends on a proper balance of water and various electrolytes.

Cells make up body systems

The **endocrine system,** by means of the blood-borne **hormones** it secretes, generally regulates activities that require duration rather than speed. The peripheral endocrine glands include the **thyroid gland,** which controls the body's basal metabolic rate; the **adrenal glands,** which secrete hormones important in metabolizing nutrient molecules, in adapting to stress, and in maintaining salt balance; the **endocrine pancreas,** which secretes hormones important in metabolizing nutrient molecules; and the **parathyroid glands,** which secrete a hormone important in Ca^{2+} metabolism.

The Peripheral Endocrine Glands

CENGAGENOW™ Log on to CengageNOW at **http://www .cengage.com/sso/** for an opportunity to explore animations and interactive quizzes to help you learn, review, and master physiology concepts.

Thyroid Gland

The **thyroid gland** consists of two lobes of endocrine tissue joined in the middle by a narrow portion of the gland, the *isthmus,* giving it a bow-tie shape (● Figure 19-1a). The gland is even located in the appropriate place for a bow tie, lying over the trachea just below the larynx.

The major cells that secrete thyroid hormone are organized into colloid-filled follicles.

The major thyroid secretory cells, known as **follicular cells,** are arranged into hollow spheres, each of which forms a functional unit called a **follicle.** On a microscopic section (● Figure 19-1b), the follicles appear as rings consisting of a single layer of follicular cells enclosing an inner lumen filled with **colloid,** a substance that serves as an extracellular storage site for thyroid hormone. Note that the colloid within the follicular lumen is extracellular (that is, outside the thyroid cells), even though it is located within the interior of the follicle. Colloid is not in direct contact with the extracellular fluid that surrounds the follicle, similar to an inland lake that is not in direct contact with the oceans that surround a continent.

The chief constituent of colloid is a large glycoprotein molecule known as **thyroglobulin (Tg),** within which are incorporated the thyroid hormones in their various stages of synthesis. The follicular cells produce two iodine-containing hormones derived from the amino acid tyrosine: **tetraiodothyronine (T_4, or thyroxine)** and **tri-iodothyronine (T_3). The prefixes *tetra* and *tri* and the subscripts *4* and *3* denote the number of iodine atoms incorporated into each of these hormones. These two hormones, collectively referred to as **thyroid hormone,** are important regulators of overall basal metabolic rate.

Interspersed in the interstitial spaces between the follicles is another secretory cell type, the **C cells,** which secrete the peptide hormone **calcitonin.** Calcitonin plays a role in calcium metabolism and is not related in any way to the two other major thyroid hormones. We will here discuss T_4 and T_3 and talk about calcitonin later, in a section dealing with endocrine control of calcium balance.

Thyroid gland

Right lobe   Trachea   Isthmus   Left lobe

(a) Gross anatomy of thyroid gland

Follicular cell   Colloid   C cell

Elizabeth R. Walker and Dennis O. Overman

(b) Light-microscopic appearance of thyroid gland

● **FIGURE 19-1 Anatomy of the thyroid gland.** (a) Gross anatomy of the thyroid gland, anterior view. The thyroid gland lies over the trachea just below the larynx and consists of two lobes connected by a thin strip called the isthmus. (b) Light-microscope appearance of the thyroid gland. The thyroid gland is composed primarily of colloid-filled spheres enclosed by a single layer of follicular cells.

Thyroid hormone is synthesized and stored on the thyroglobulin molecule.

The basic ingredients for thyroid hormone synthesis are tyrosine and iodine, both of which must be taken up from the blood by the follicular cells. Tyrosine, an amino acid, is synthesized in sufficient amounts by the body, so it is not essential to the diet.

By contrast, the iodine needed for thyroid hormone synthesis must be obtained from dietary intake. Dietary iodine (I) is reduced to iodide (I^-) prior to absorption by the small intestine. We now examine the steps involved in the synthesis, storage, secretion, and transport of thyroid hormone.

Most steps of thyroid hormone synthesis take place on the thyroglobulin molecules within the colloid. Thyroglobulin itself is produced by the endoplasmic reticulum/Golgi complex of the thyroid follicular cells. The amino acid tyrosine becomes incorporated in the much larger thyroglobulin molecules as the latter are being produced. Once produced, tyrosine-containing thyroglobulin is exported in vesicles from the follicular cells into the colloid by exocytosis (step **1** in ● Figure 19-2). The thyroid captures I^- from the blood and transfers it into the colloid by an *iodide pump*—the powerful, energy-requiring carrier proteins in the outer membranes of the follicular cells (step **2**). The iodide pump is a symporter driven by the Na^+ concentration gradient established by the $Na^+–K^+$ pump at the basolateral membrane (the outer membrane of the follicular cell in contact with the interstitial fluid). The iodide pump transports Na^+ into the follicular cell down its concentration gradient and I^- into the cell against its concentration gradient. Almost all the I^- in the body is moved against its concentration gradient to become trapped in the thyroid for thyroid hormone synthesis. Iodide is usually about 30 times more concentrated in the thyroid follicular cells than in the blood. Iodide serves no other function in the body.

Inside the follicular cell, iodide is oxidized to "active" iodide by a membrane-bound enzyme, **thyroperoxidase (TPO),** located at the luminal membrane, the membrane of the follicular cell in contact with the colloid (step **3**). This active iodide exits through a channel in the luminal membrane to enter the colloid (step **4**).

Within the colloid, TPO, still membrane-bound, quickly attaches iodide to a tyrosine within the thyroglobulin molecule. Attachment of one iodine to tyrosine yields **monoiodotyrosine (MIT)** (step **5a**). Attachment of two iodines to tyrosine yields **di-iodotyrosine (DIT)** (step **5b**). After MIT and DIT are formed, a coupling process occurs within the thyroglobulin molecule between the iodinated tyrosine molecules to form the thyroid hormones. Coupling of one MIT (with one iodide) and one DIT (with two iodides) yields **tri-iodothyronine,** or T_3 (with three iodides) (step **6a**). Coupling of two DITs (each bearing two iodide atoms) yields **tetraiodothyronine (T_4, or thyroxine),** the four-iodide form of thyroid hormone (step **6b**). Coupling does not occur between two MIT molecules. All these products remain attached to thyroglobulin by peptide bonds. Thyroid hormones remain stored in this form in the colloid until they are split off and secreted. Sufficient thyroid hormone is normally stored to supply the body's needs for several months.

To secrete thyroid hormone, the follicular cells phagocytize thyroglobulin-laden colloid.

The release of thyroid hormone into the systemic circulation is a rather complex process, for two reasons. First, before their release, T_3 and T_4 are still bound within the thyroglobulin molecule. Second, these hormones are stored at an inland ex-

Comparison of Type 1 and Type 2 Diabetes Mellitus

Characteristic	Type 1 Diabetes	Type 2 Diabetes
Level of Insulin Secretion	None or almost none	May be normal or exceed normal
Typical Age of Onset	Childhood	Adulthood
Percentage of Diabetics	10%–20%	80%–90%
Basic Defect	Destruction of β cells	Reduced sensitivity of insulin's target cells
Treatment	Insulin injections; dietary management; exercise	Dietary control and weight reduction; exercise; sometimes oral hypoglycemic drugs

crine pancreas to reduce the anticipated rise in blood glucose. The first drug on the market of this type, *Byetta*, mimics the gut-released hormone *glucagon-like peptide 1* (*GLP-1*; see p. 717). GLP-1 is released from the small intestine L cells in response to food intake and has multiple glucose-lowering effects. GLP-1 itself is too short-lived to be suitable as a drug. Byetta, which is a version of a peptide found in the venom of a poisonous Gila monster, must be injected. Like GLP-1, this drug stimulates insulin secretion when blood glucose is high but not when glucose is in the normal range. It also suppresses production of glucose-raising glucagon and slows gastric emptying. By promoting satiety, Byetta decreases food intake and in the long term causes weight loss (see p. 644). Evidence suggests that Byetta even stimulates regeneration of pancreatic β cells.

6. By increasing endogenous GLP-1 levels (*dipeptidyl peptidase-4 or DPP-4 inhibitors;* for example, Januvia). The newest approved class of drugs, the DPP-4 inhibitors increase endogenous GLP-1 levels by blocking DPP-4, an enzyme that breaks down GLP-1, thus prolonging action of this incretin. Prolonged activity of GLP-1 boosts insulin secretion until glucose levels return to normal. Januvia also suppresses the release of glucose by the liver and slows digestion.

Because none of these drugs deliver new insulin to the body, they cannot re-place insulin therapy for people with Type 1 diabetes. Furthermore, sometimes the weakened β cells of Type 2 diabetics eventually burn out and can no longer produce insulin. In such a case, the previously non-insulin-dependent patient must be placed on insulin therapy for life.

New Approaches to Managing Diabetes

Several newer approaches are currently available for insulin-dependent diabetics that preclude the need for the one or two insulin injections daily.

- Implanted insulin pumps can deliver a prescribed amount of insulin on a regular basis, but the recipient must time meals with care to match the automatic insulin delivery.
- Pancreas transplants are also being performed more widely now, with increasing success rates. On the downside, recipients of pancreas transplants must take immunosuppressive drugs for life to prevent rejection of their donated organs. Also, donor organs are in short supply.

Current research on several fronts may dramatically change the approach to diabetic therapy in the near future. The following new treatments are on the horizon:

- Some methods under development circumvent the need for insulin injections by using alternative routes of administration that bypass the destructive digestive tract enzymes, such as by using ultrasound to force insulin into the skin from an insulin-impregnated patch. Some researchers are seeking methods to protect swallowed insulin from destruction by the digestive tract, for example, by attaching oral insulin to vitamin B_{12}, which protects the insulin from digestive enzymes until the vitamin–insulin complex is absorbed by intrinsic factor-induced endocytosis in the terminal ileum (see p. 608).
- Others have identified a potential oral substitute for insulin—namely, a nonpeptide chemical that binds with the insulin receptors and brings about the same intracellular responses as insulin does. Because this insulin mimetic is not a protein, it would not be destroyed by the proteolytic digestive enzymes if taken as a pill.
- Another hope is pancreatic islet transplants. Scientists have developed several types of devices that isolate donor islet cells from the recipient's immune system. Such immunoisolation of islet cells permits use of grafts from other animals, circumventing the shortage of human donor cells. Pig islet cells are an especially good source, because pig insulin is nearly identical to human insulin.
- Some researchers have coaxed stem cells to develop into insulin-secreting cells that hopefully can be implanted.
- In a related approach, others are turning to genetic engineering in the hope of developing surrogates for pancreatic β cells. An example is the potential reprogramming of the small-intestine endocrine cells that produce GIP. The goal is to cause these non-β cells to cosecrete both insulin and GIP on feeding.
- Another approach under development is an implanted, glucose-detecting, insulin-releasing "artificial pancreas" that would continuously monitor the patient's blood glucose level and deliver insulin in response to need.
- On another front, scientists are hopeful of one day developing immunotherapies that specifically block the attack of the immune system against the β cells, thus curbing or preventing Type 1 diabetes.
- Nearly 400 new drugs for diabetes, mostly for the Type 2 form of the disease, are under development.

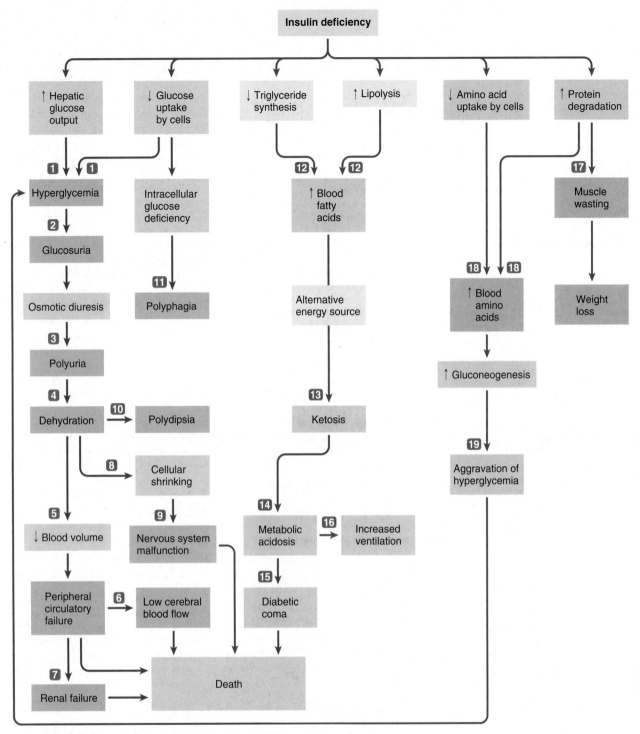

● **FIGURE 19-19 Acute effects of diabetes mellitus.** The acute consequences of diabetes mellitus can be grouped according to the effects of inadequate insulin action on carbohydrate, fat, and protein metabolism. These effects ultimately cause death through a variety of pathways. See pp. 718–719 for an explanation of the numbers.

greater incidence than in nondiabetics. Because vascular lesions often develop in the kidneys and retinas of the eyes, diabetes is the leading cause of both kidney failure and blindness in the United States. Impaired delivery of blood to the extremities may cause these tissues to become gangrenous, and toes or even whole limbs may have to be amputated. In addition to circula-

tory problems, degenerative lesions in nerves lead to multiple neuropathies that result in dysfunction of the brain, spinal cord, and peripheral nerves. The latter is most often characterized by pain, numbness, and tingling, especially in the extremities.

Regular exposure of tissues to excess blood glucose over a prolonged time leads to tissue alterations responsible for the

15. Which of the following are characteristic of the postabsorptive state? *(Indicate all that apply.)*
    a. glycogenolysis
    b. gluconeogenesis
    c. lipolysis
    d. glycogenesis
    e. protein synthesis
    f. triglyceride synthesis
    g. protein degradation
    h. increased insulin secretion
    i. increased glucagon secretion
    j. glucose sparing

Essay Questions

1. Describe the steps of thyroid hormone synthesis.

2. What are the effects of T_3 and T_4? Which is the more potent? What is the source of most circulating T_3?

3. Describe the regulation of thyroid hormone.

4. Discuss the causes and symptoms of both hypothyroidism and hyperthyroidism. For each cause, indicate whether or not a goiter occurs, and explain why.

5. What hormones are secreted by the adrenal cortex? What are the functions and control of each of these hormones?

6. Discuss the causes and symptoms of each type of adrenocortical dysfunction.

7. What is the relationship of the adrenal medulla to the sympathetic nervous system? What are the functions of epinephrine? How is epinephrine release controlled?

8. Define stress. Describe the neural and hormonal responses to a stressor.

9. Define *fuel metabolism, anabolism,* and *catabolism.*

10. Indicate the primary circulating form and storage form of each of the three classes of organic nutrients.

11. Distinguish between the absorptive and postabsorptive states with regard to the handling of nutrient molecules.

12. Name the two major cell types of the islets of Langerhans, and indicate the hormonal product of each.

13. Compare the functions and control of insulin secretion with those of glucagon secretion.

14. What are the consequences of diabetes mellitus? Distinguish between Type 1 and Type 2 diabetes mellitus.

15. Why must plasma Ca^{2+} be closely regulated?

16. Explain how osteoblasts influence osteoclast function.

17. Discuss the contributions of parathyroid hormone, calcitonin, and vitamin D to Ca^{2+} metabolism. Describe the source and control of each of these hormones.

18. Discuss the major disorders in Ca^{2+} metabolism.

POINTS TO PONDER

(Explanations on p. A-55)

1. Iodine is naturally present in salt water and is abundant in soil along coastal regions. Fish and shellfish living in the ocean and plants grown in coastal soil take up iodine from their environment. Fresh water does not contain iodine, and the soil becomes more iron poor the farther inland it is. Knowing this, explain why the midwestern United States was once known as an endemic goiter belt. Why is this region no longer an endemic goiter belt even though the soil is still iodine poor?

2. Why do doctors recommend that people who are allergic to bee stings and thus are at risk for anaphylactic shock (see p. 453) carry a vial of epinephrine for immediate injection in case of a sting?

3. Why would an infection tend to raise the blood glucose level of a diabetic individual?

4. Tapping the facial nerve at the angle of the jaw in a patient with moderate hyposecretion of a particular hormone elicits a characteristic grimace on that side of the face. What endocrine abnormality could give rise to this so-called Chvostek's sign?

5. Soon after a technique to measure plasma Ca^{2+} levels was developed in the 1920s, physicians observed that hypercalcemia accompanied a broad range of cancers. Early researchers proposed that malignancy-associated hypercalcemia arose from metastatic (see p. 447) tumor cells that invaded and destroyed bone, releasing Ca^{2+} into the blood. This conceptual framework was overturned when physicians noted that hypercalcemia often appeared in the absence of bone lesions. Furthermore, cancer patients often manifested hypophosphatemia in addition to hypercalcemia. This finding led investigators to suspect that the tumors might be producing a PTH-like substance. Explain how they reached this conclusion. In 1987, this substance was identified and named parathyroid hormone–related peptide (PTHrP), which binds to and activates PTH receptors.

CLINICAL CONSIDERATION

(Explanation on p. A-56)

Najma G. sought medical attention after her menstrual periods ceased and she started growing excessive facial hair. Also, she had been thirstier than usual and urinated more frequently. A clinical evaluation revealed that Najma was hyperglycemic. Her physician told her that she had an endocrine disorder dubbed "diabetes of bearded ladies." What underlying defect do you think is responsible for Najma's condition?

Reproductive System

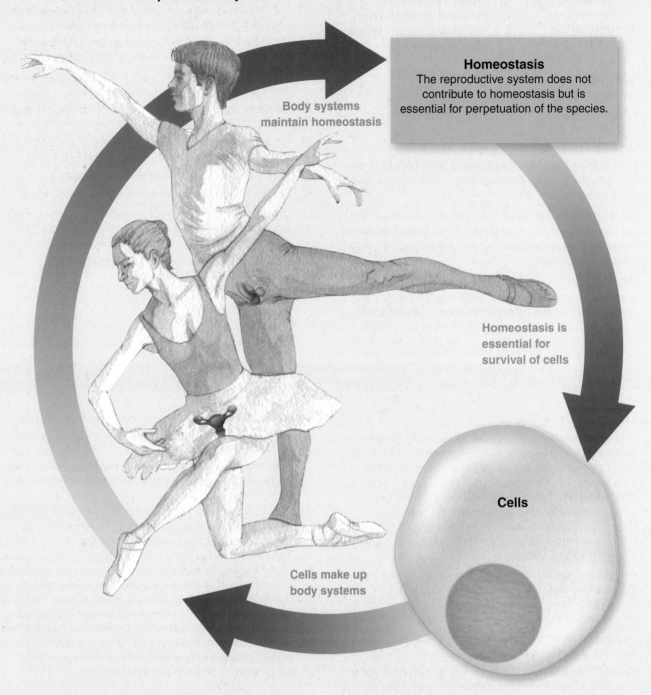

Body systems maintain homeostasis

Homeostasis
The reproductive system does not contribute to homeostasis but is essential for perpetuation of the species.

Homeostasis is essential for survival of cells

Cells

Cells make up body systems

Normal functioning of the **reproductive system** is not aimed toward homeostasis and is not necessary for survival of an individual, but it is essential for survival of the species. Only through reproduction can the complex genetic blueprint of each species survive beyond the lives of individual members of the species.

The Reproductive System

CENGAGENOW Log on to CengageNOW at **http://www.cengage.com/sso/** for an opportunity to explore animations and interactive quizzes to help you learn, review, and master physiology concepts.

Uniqueness of the Reproductive System

The central theme of this book has been the physiologic processes aimed at maintaining homeostasis to ensure survival of the individual. We are now going to disengage from this theme to discuss the reproductive system, which serves primarily the purpose of perpetuating the species.

Unique among body systems, the reproductive system does not contribute to homeostasis but exerts other important effects.

Even though the reproductive system does not contribute to homeostasis and is not essential for survival of an individual, it still plays an important role in a person's life. For example, the manner in which people relate as sexual beings contributes in significant ways to psychosocial behavior and has important influences on how people view themselves and how they interact with others. Reproductive function also has a profound effect on society. The universal organization of societies into family units provides a stable environment that is conducive for perpetuating our species. On the other hand, the population explosion and its resultant drain on dwindling resources have led to worldwide concern with the means by which reproduction can be limited.

Reproductive capability depends on intricate relationships among the hypothalamus, anterior pituitary, reproductive organs, and target cells of the sex hormones. These relationships employ many of the regulatory mechanisms used by other body systems for maintaining homeostasis, such as negative-feedback control. In addition to these basic biological processes, sexual behavior and attitudes are deeply influenced by emotional factors and the socio-cultural mores of the society in which the individual lives. We will concentrate on the basic sexual and reproductive functions that are under nervous and hormonal control and will not examine the psychological and social ramifications of sexual behavior.

The reproductive system includes the gonads, reproductive tract, and accessory sex glands, all of which are different in males and females.

Reproduction depends on the union of male and female **gametes (reproductive,** or **germ, cells)**, each with a half set of chromosomes, to form a new individual with a full, unique set of chromosomes. Unlike the other body systems, which are essentially identical in the two sexes, the reproductive systems of males and females are markedly different, befitting their different roles in the reproductive process. The **male** and **female reproductive systems** are designed to enable union of genetic material from the two sexual partners, and the female system is equipped to house and nourish the offspring to the developmental point at which it can survive independently in the external environment.

The **primary reproductive organs,** or **gonads,** consist of a pair of **testes** in the male and a pair of **ovaries** in the female. In both sexes, the mature gonads perform the dual function of (1) producing gametes (**gametogenesis**), that is, **spermatozoa (sperm)** in the male and **ova (eggs)** in the female; and (2) secreting sex hormones, specifically, **testosterone** in males and **estrogen** and **progesterone** in females.

In addition to the gonads, the reproductive system in each sex includes a **reproductive tract** encompassing a system of ducts that are specialized to transport or house the gametes after they are produced, plus **accessory sex glands** that empty their supportive secretions into these passageways. In females, the *breasts* are also considered accessory reproductive organs. The externally visible portions of the reproductive system are known as **external genitalia.**

SECONDARY SEXUAL CHARACTERISTICS The **secondary sexual characteristics** are the many external characteristics not directly involved in reproduction that distinguish males and females, such as body configuration and hair distribution. In humans, for example, males have broader shoulders, whereas females have curvier hips; and males have beards, whereas females do not. Testosterone in the male and estrogen in the female govern the development and maintenance of these characteristics. Progesterone has no influence on secondary sexual characteristics. Even though growth of axillary and pubic hair at puberty is promoted in both sexes by androgens—testosterone in males and adrenocortical dehydroepiandrosterone in females (see p. 702)—this hair growth is not a secondary sexual characteristic, because both sexes display this feature. Thus, testosterone and estrogen alone govern the nonreproductive distinguishing features.

In some species, the secondary sexual characteristics are of great importance in courting and mating behavior; for example, the rooster's headdress or comb attracts the female's attention, and the stag's antlers are useful to ward off other males. In humans, the differentiating marks between males and females do serve to attract the opposite sex, but attraction is also strongly influenced by the complexities of human society and cultural behavior.

OVERVIEW OF MALE REPRODUCTIVE FUNCTIONS AND ORGANS The essential reproductive functions of the male are as follows:

1. Production of sperm (*spermatogenesis*)

2. Delivery of sperm to the female

The sperm-producing organs, the testes, are suspended outside the abdominal cavity in a skin-covered sac, the **scrotum,** which lies within the angle between the legs. The male reproductive system is designed to deliver sperm to the female reproductive tract in a liquid vehicle, *semen,* which is conducive to sperm viability. The major **male accessory sex glands,** whose secretions provide the bulk of the semen, are the *seminal vesicles, prostate gland,* and *bulbourethral glands* (● Figure 20-1). The **penis** is the organ used to deposit semen in the female. Sperm exit each testis through the **male reproductive tract,** consisting on each side of an *epididymis, ductus (vas) deferens,* and *ejaculatory duct.* These pairs of reproductive tubes empty into a single *urethra,* the canal that runs the length of the penis and empties to the exterior. These parts of the male reproductive system are described more thoroughly later when their functions are discussed.

OVERVIEW OF FEMALE REPRODUCTIVE FUNCTIONS AND ORGANS The female's role in reproduction is more complicated than the male's. The essential female reproductive functions include the following:

1. Production of ova (*oogenesis*)

2. Reception of sperm

3. Transport of the sperm and ovum to a common site for union (*fertilization,* or *conception*)

4. Maintenance of the developing fetus until it can survive in the outside world (*gestation,* or *pregnancy*), including formation of the *placenta,* the organ of exchange between mother and fetus

5. Giving birth to the baby (*parturition*)

6. Nourishing the infant after birth by milk production (*lactation*)

The product of fertilization is known as an **embryo** during the first two months of intrauterine development when tissue differentiation is taking place. Beyond this time, the developing living being is recognizable as human and is known as a **fetus** during the remainder of gestation. Although no further tissue differentiation takes place during fetal life, it is a time of tremendous tissue growth and maturation.

The ovaries and female reproductive tract lie within the pelvic cavity (● Figure 20-2a and b). The **female reproductive tract** consists of the following components. Two **oviducts (uterine,** or **Fallopian tubes),** which are in close association with the two ovaries, pick up ova on ovulation (ovum release from an ovary) and serve as the site for fertilization. The thick-walled, hollow **uterus** is primarily responsible for maintaining the fetus during its development and expelling it at the end of pregnancy. The **vagina** is a muscular, expandable tube that connects the uterus to the external environment. The lowest portion of the uterus, the cervix, projects into the vagina and contains a single, small opening, the **cervical canal.** Sperm are deposited in the vagina by the penis during sexual intercourse. The cervical canal serves as a pathway for sperm through the uterus to the site of fertilization in the oviduct and, when greatly dilated during parturition, serves as the passageway for delivery of the baby from the uterus.

The **vaginal opening** is located in the **perineal region** between the urethral opening anteriorly and the anal opening posteriorly (● Figure 20-2c). It is partially covered by a thin

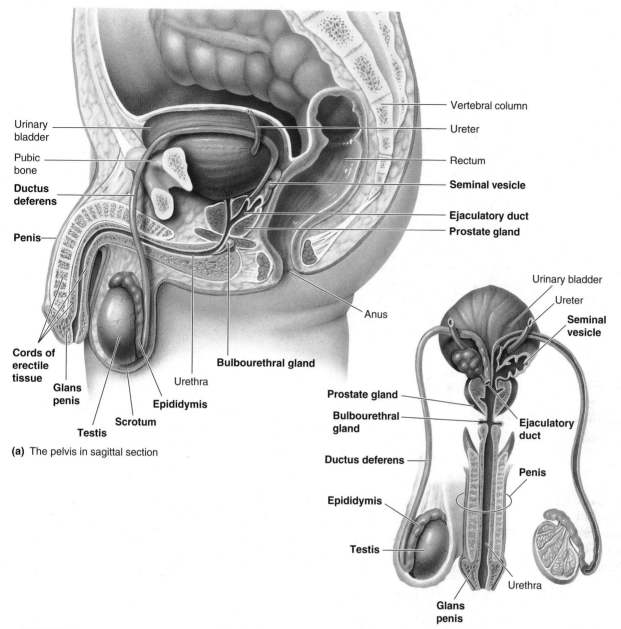

Urinary bladder

Pubic bone

Ductus deferens

Penis

Cords of erectile tissue

Glans penis

Testis

Scrotum

Epididymis

Urethra

Bulbourethral gland

Vertebral column

Ureter

Rectum

Seminal vesicle

Ejaculatory duct

Prostate gland

Anus

(a) The pelvis in sagittal section

Urinary bladder

Ureter

Seminal vesicle

Prostate gland

Bulbourethral gland

Ductus deferens

Epididymis

Testis

Ejaculatory duct

Penis

Urethra

Glans penis

(b) Posterior view of the reproductive organs

● **FIGURE 20-1 The male reproductive system.**

mucous membrane, the **hymen,** which typically is physically disrupted by the first sexual intercourse. The vaginal and urethral openings are surrounded laterally by two pairs of skin folds, the **labia minora** and **labia majora.** The smaller labia minora are located medially to the more prominent labia majora. The externally visible part of the **clitoris,** an erotic structure composed of tissue similar to that of the penis, lies at the anterior end of the folds of the labia minora. The female external genitalia are collectively referred to as the **vulva.**

Reproductive cells each contain a half set of chromosomes.

The DNA molecules that carry the cell's genetic code are not randomly crammed into the nucleus but are precisely organized into **chromosomes** (see p. A-19). Each chromosome consists of a dif-

ferent DNA molecule that contains a unique set of genes. Somatic (body) cells contain 46 chromosomes (the **diploid number**), which can be sorted into 23 pairs on the basis of various distinguishing features. Chromosomes composing a matched pair are termed **homologous chromosomes,** one member of each pair having been derived from the individual's maternal parent and the other member from the paternal parent. Gametes (that is, sperm and eggs) contain only one member of each homologous pair for a total of 23 chromosomes (the **haploid number**).

Gametogenesis is accomplished by meiosis, resulting in genetically unique sperm and ova.

Most cells in the human body have the ability to reproduce themselves, a process important in growth, replacement, and repair of tissues. Cell division involves two components: division

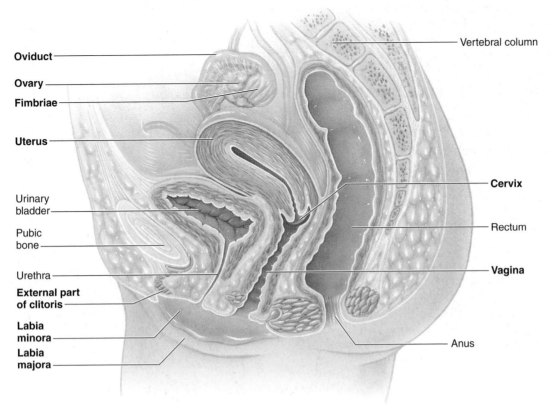

Oviduct

Ovary

Fimbriae

Uterus

Urinary
bladder

Pubic
bone

Urethra

External part
of clitoris

Labia
minora

Labia
majora

Vertebral column

Cervix

Rectum

Vagina

Anus

(a) The pelvis in sagittal section

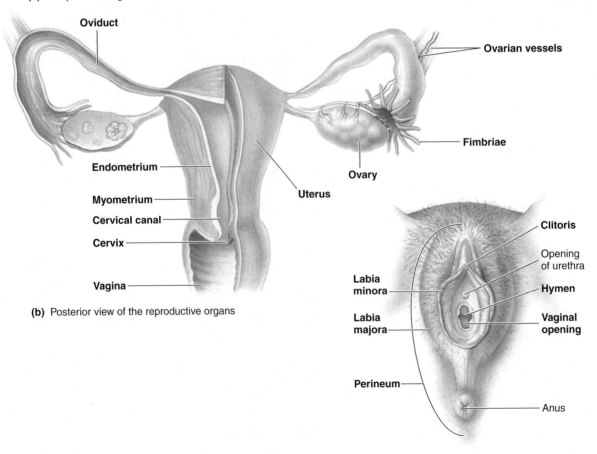

Oviduct

Ovarian vessels

Endometrium

Myometrium

Cervical canal

Cervix

Uterus

Ovary

Fimbriae

Vagina

(b) Posterior view of the reproductive organs

Clitoris

Opening
of urethra

Hymen

Vaginal
opening

Labia
minora

Labia
majora

Perineum

Anus

(c) Perineal view of the external genitalia

● **FIGURE 20-2 The female reproductive system.**

of the nucleus and division of the cytoplasm. Nuclear division in somatic cells is accomplished by **mitosis.** In mitosis, the chromosomes replicate (make duplicate copies of themselves); then the identical chromosomes are separated so that a complete set of genetic information (that is, a diploid number of chromosomes) is distributed to each of the two new daughter cells. Nuclear division in the specialized case of gametes is accomplished by **meiosis,** in which only a half set of genetic information (that is, a haploid number of chromosomes) is distributed to each of four new daughter cells (see p. A-30).

During meiosis, a specialized diploid germ cell undergoes one chromosome replication followed by two nuclear divisions. In the first meiotic division, the replicated chromosomes do not separate into two individual, identical chromosomes but remain joined. The doubled chromosomes sort themselves into homologous pairs, and the pairs separate so that each of two daughter cells receives a half set of doubled chromosomes. During the second meiotic division, the doubled chromosomes within each of the two daughter cells separate and are distributed into two cells, yielding four daughter cells, each containing a half set of chromosomes, a single member of each pair. During this process, the maternally and paternally derived chromosomes of each homologous pair are distributed to the daughter cells in random assortments containing one member of each chromosome pair without regard for its original derivation. That is, not all of the mother-derived chromosomes go to one daughter cell and the father-derived chromosomes to the other cell. More than 8 million (2^{23}) different mixtures of the 23 paternal and maternal chromosomes are possible. This genetic mixing provides novel combinations of chromosomes. Crossing-over contributes even further to genetic diversity. *Crossing-over* refers to the physical exchange of chromosome material between the homologous pairs prior to their separation during the first meiotic division (see p. A-31).

Thus, sperm and ova each have a unique haploid number of chromosomes. When fertilization takes place, a sperm and ovum fuse to form the start of a new individual with 46 chromosomes, one member of each chromosomal pair having been inherited from the mother and the other member from the father (● Figure 20-3).

The sex of an individual is determined by the combination of sex chromosomes.

Whether individuals are destined to be males or females is a genetic phenomenon determined by the sex chromosomes they possess. As the 23 chromosome pairs are separated during meiosis, each sperm or ovum receives only one member of each chromosome pair. Of the chromosome pairs, 22 are **autosomal chromosomes** that code for general human characteristics as well as for specific traits such as eye color. The remaining pair of chromosomes consists of the **sex chromosomes,** of which there are two genetically different types—a larger **X chromosome** and a smaller **Y chromosome.**

Sex determination depends on the combination of sex chromosomes: **Genetic males** have both an X and a Y sex chromosome; **genetic females** have two X sex chromosomes. Thus, the genetic difference responsible for all the anatomic and func-

● **FIGURE 20-3 Chromosomal distribution in sexual reproduction.**

tional distinctions between males and females is the single Y chromosome. Males have it; females do not.

As a result of meiosis during gametogenesis, all chromosome pairs are separated so that each daughter cell contains only one member of each pair, including the sex chromosome pair. When the XY sex chromosome pair separates during sperm formation, half the sperm receive an X chromosome and the other half a Y chromosome. In contrast, during oogenesis, every ovum receives an X chromosome because separation of the XX sex chromosome pair yields only X chromosomes. During fertilization, combination of an X-bearing sperm with an X-bearing ovum produces a genetic female, XX, whereas union of a Y-bearing sperm with an X-bearing ovum results in a genetic male, XY. Thus, genetic sex is determined at the time of conception and depends on which type of sex chromosome is contained within the fertilizing sperm.

Sexual differentiation along male or female lines depends on the presence or absence of masculinizing determinants.

Differences between males and females exist at three levels: genetic, gonadal, and phenotypic (anatomic) sex (● Figure 20-4).

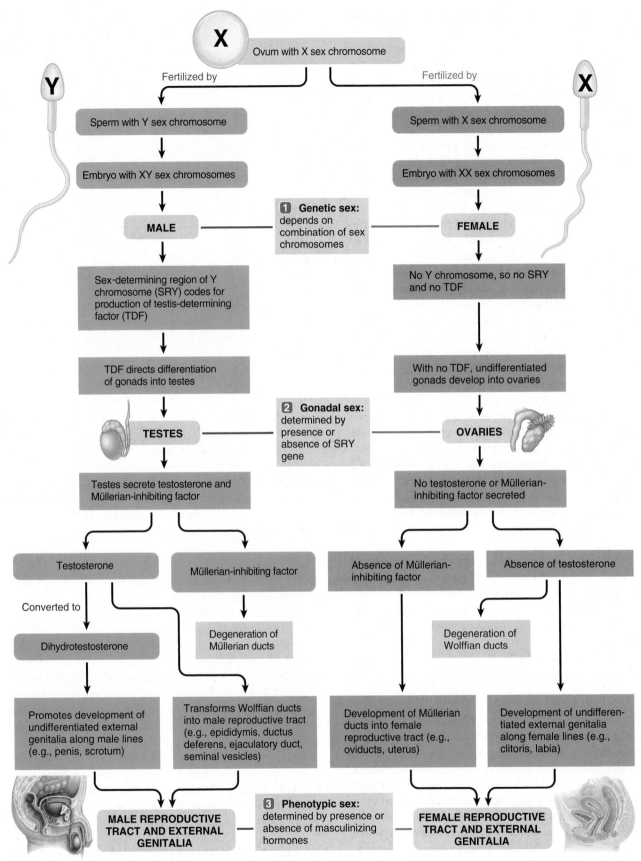

(a) Male sex determination and sexual differentiation

(b) Female sex determination and sexual differentiation

● **FIGURE 20-4 Sex determination and sexual differentiation.**

GENETIC AND GONADAL SEX Genetic sex, which depends on the combination of sex chromosomes at the time of conception, in turn determines **gonadal sex,** that is, whether testes or ovaries develop. The presence or absence of a Y chromosome determines gonadal differentiation. For the first month and a half of gestation, all embryos have the potential to differentiate along either male or female lines because the developing reproductive tissues of both sexes are identical and indifferent. Gonadal specificity appears during the seventh week of intrauterine life when the indifferent gonadal tissue of a genetic male begins to differentiate into testes under the influence of the **sex-determining region** of the Y chromosome (**SRY**), the single gene responsible for sex determination. This gene triggers a chain of reactions that leads to physical development of a male. SRY "masculinizes" the gonads by coding for production of **testis-determining factor (TDF)** (also known as **SRY protein**) within primitive gonadal cells. TDF directs a series of events that leads to differentiation of the gonads into testes.

Because genetic females lack the SRY gene and consequently do not produce TDF, their gonadal cells never receive a signal for testes formation, so by default during the ninth week the undifferentiated gonadal tissue starts developing into ovaries instead.

PHENOTYPIC SEX Phenotypic sex, the apparent anatomic sex of an individual, is hormonally mediated and depends on the genetically determined gonadal sex. The term **sexual differentiation** refers to the embryonic development of the external genitalia and reproductive tract along either male or female lines. As with the undifferentiated gonads, embryos of both sexes have the potential to develop either male or female external genitalia and reproductive tracts. Differentiation into a male-type reproductive system is induced by **androgens,** which are masculinizing hormones secreted by the developing testes. Testosterone is the most potent androgen. The absence of these testicular hormones in female fetuses results in the development of a female-type reproductive system. By 10 to 12 weeks of gestation, the sexes can easily be distinguished by the anatomic appearance of the external genitalia.

SEXUAL DIFFERENTIATION OF THE EXTERNAL GENITALIA Male and female external genitalia develop from the same embryonic tissue. In both sexes, the undifferentiated external genitals consist of a *genital tubercle,* paired *urethral folds* surrounding a urethral groove, and, more laterally, *genital (labioscrotal) swellings* (● Figure 20-5). The **genital tubercle** gives rise to exquisitely sensitive erotic tissue—in males the **glans penis** (the cap at the distal end of the penis) and in females the glans clitoris. The major distinctions between the glans penis and glans clitoris are the smaller size of the glans clitoris and the penetration of the glans penis by the urethral opening. The urethra is the tube through which urine is transported from the bladder to the outside and also serves in males as a passageway for exit of semen through the penis to the outside. In males, the **urethral folds** fuse around the urethral groove to form the penis, which encircles the urethra. The **genital swellings** similarly fuse to form the scrotum and **prepuce,** a fold of skin that extends over the end of the penis

and more or less completely covers the glans penis. In females, the urethral folds and genital swellings do not fuse at midline but develop instead into the labia minora and labia majora, respectively. The urethral groove remains open, providing access to the interior through both the urethral opening and the vaginal opening.

SEXUAL DIFFERENTIATION OF THE REPRODUCTIVE TRACT Although the male and female external genitalia develop from the same undifferentiated embryonic tissue, this is not the case with the reproductive tracts. Two primitive duct systems—the Wolffian ducts and the Müllerian ducts—develop in all embryos. In males, the reproductive tract develops from the **Wolffian ducts** and the Müllerian ducts degenerate, whereas in females the **Müllerian ducts** differentiate into the reproductive tract and the Wolffian ducts regress (● Figure 20-6). Because both duct systems are present before sexual differentiation occurs, the early embryo has the potential to develop either a male or a female reproductive tract.

Development of the reproductive tract along male or female lines is determined by the presence or absence of two hormones secreted by two different cell types in the fetal testes—*testosterone* produced by the newly developed Leydig cells and *Müllerian-inhibiting factor* (also known as *anti-Müllerian hormone*), produced by the early Sertoli cells (see ● Figure 20-4). A hormone released by the placenta, *human chorionic gonadotropin,* is the stimulus for this early testicular secretion. Testosterone induces development of the Wolffian ducts into the male reproductive tract (epididymis, ductus deferens, and seminal vesicles). This hormone, after being converted into **dihydrotestosterone (DHT),** is also responsible for differentiating the external genitalia into the penis and scrotum. Meanwhile, Müllerian-inhibiting factor causes regression of the Müllerian ducts.

In the absence of testosterone and Müllerian-inhibiting factor in females, the Wolffian ducts regress, the Müllerian ducts develop into the female reproductive tract (oviducts, uterus, and upper part of vagina), and the external genitalia differentiate into the clitoris and labia.

Note that the undifferentiated embryonic reproductive tissue passively develops into a female structure unless actively acted on by masculinizing factors. In the absence of male testicular hormones, a female reproductive tract and external genitalia develop regardless of the genetic sex of the individual. For feminization of the fetal genital tissue, ovaries do not even need to be present. Such a control pattern for determining sexual differentiation is appropriate, considering that fetuses of both sexes are exposed to high concentrations of female sex hormones throughout gestation. If female sex hormones influenced the development of the reproductive tract and external genitalia, all fetuses would be feminized.

ERRORS IN SEXUAL DIFFERENTIATION Genetic sex and phenotypic sex are usually compatible; that is, a genetic male anatomically appears to be a male and functions as a male, and the same compatibility holds true for females. Occasionally, however, discrepancies occur between genetic and anatomic sexes because of errors in sexual differentiation, as the following examples illustrate:

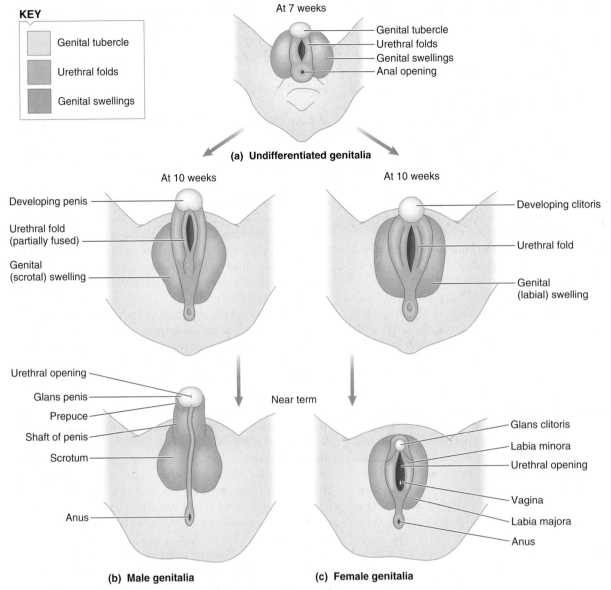

KEY
- Genital tubercle
- Urethral folds
- Genital swellings

At 7 weeks
- Genital tubercle
- Urethral folds
- Genital swellings
- Anal opening

(a) Undifferentiated genitalia

At 10 weeks
- Developing penis
- Urethral fold (partially fused)
- Genital (scrotal) swelling

At 10 weeks
- Developing clitoris
- Urethral fold
- Genital (labial) swelling

Near term

- Urethral opening
- Glans penis
- Prepuce
- Shaft of penis
- Scrotum
- Anus

- Glans clitoris
- Labia minora
- Urethral opening
- Vagina
- Labia majora
- Anus

(b) Male genitalia   **(c) Female genitalia**

● **FIGURE 20-5 Sexual differentiation of the external genitalia.**

■ If testes in a genetic male fail to properly differentiate and secrete hormones, the result is the development of an apparent anatomic female in a genetic male, who, of course, will be sterile. Similarly, genetic males whose target cells lack receptors for testosterone are feminized, even though their testes secrete adequate testosterone (see p. 666, *testicular feminization syndrome*).

■ Because testosterone acts on the Wolffian ducts to convert them into a male reproductive tract but the testosterone derivative DHT is responsible for masculinization of the external genitalia, a genetic deficiency of the enzyme that converts testosterone into DHT results in a genetic male with testes and a male reproductive tract but with female external genitalia.

■ The adrenal gland normally secretes a weak androgen, *dehydroepiandrosterone,* in insufficient quantities to masculinize females. However, pathologically excessive secretion of this hormone in a genetically female fetus during critical developmental

stages imposes differentiation of the reproductive tract and genitalia along male lines (see *adrenogenital syndrome,* p. 704).

Sometimes these discrepancies between genetic sex and apparent sex are not recognized until puberty, when the discovery produces a psychologically traumatic gender identity crisis. For example, a masculinized genetic female with ovaries but with male-type external genitalia may be reared as a boy until puberty, when breast enlargement (caused by estrogen secretion by the awakening ovaries) and lack of beard growth (caused by lack of testosterone secretion in the absence of testes) signal an apparent problem. Therefore, it is important to diagnose any problems in sexual differentiation in infancy. Once a sex has been assigned, it can be reinforced, if necessary, with surgical and hormonal treatment so that psychosexual development can proceed as normally as possible. Less dramatic cases of inappropriate sexual differentiation often appear as sterility problems.

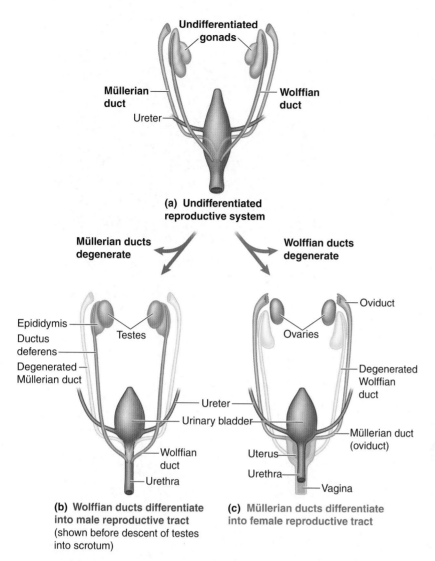

(a) Undifferentiated reproductive system

Undifferentiated gonads

Müllerian duct

Wolffian duct

Ureter

Müllerian ducts degenerate

Wolffian ducts degenerate

Epididymis

Ductus deferens

Degenerated Müllerian duct

Testes

Ureter

Urinary bladder

Wolffian duct

Urethra

Oviduct

Ovaries

Degenerated Wolffian duct

Müllerian duct (oviduct)

Uterus

Urethra

Vagina

(b) Wolffian ducts differentiate into male reproductive tract (shown before descent of testes into scrotum)

(c) Müllerian ducts differentiate into female reproductive tract

● **FIGURE 20-6 Sexual differentiation of the reproductive tract.**

Male Reproductive Physiology

In the embryo, the testes develop from the gonadal ridge located at the rear of the abdominal cavity. In the last months of fetal life, they begin a slow descent, passing out of the abdominal cavity through the **inguinal canal** into the scrotum, one testis dropping into each pocket of the scrotal sac. Testosterone from the fetal testes induces descent of the testes into the scrotum.

After the testes descend into the scrotum, the opening in the abdominal wall through which the inguinal canal passes closes snugly around the sperm-carrying duct and blood vessels that traverse between each testis and the abdominal cavity. Incomplete closure or rupture of this opening permits abdominal viscera to slip through, resulting in an **inguinal hernia.**

Although the time varies somewhat, descent is usually complete by the seventh month of gestation. As a result, descent is complete in 98% of full-term baby boys.

 However, in a substantial percentage of premature male infants the testes are still within the inguinal canal at birth. In most instances of retained testes, descent occurs

naturally before puberty or can be encouraged with administration of testosterone. Rarely, a testis remains undescended into adulthood, a condition known as **cryptorchidism** ("hidden testis").

The scrotal location of the testes provides a cooler environment essential for spermatogenesis.

The temperature within the scrotum averages several degrees Celsius less than normal body (core) temperature. Descent of the testes into this cooler environment is essential because spermatogenesis is temperature sensitive and cannot occur at normal body temperature. Therefore, a cryptorchid is unable to produce viable sperm.

The position of the scrotum in relation to the abdominal cavity can be varied by a spinal reflex mechanism that plays an important role in regulating testicular temperature. Reflex contraction of scrotal muscles on exposure to a cold environment raises the scrotal sac to bring the testes closer to the warmer abdomen. Conversely, relaxation of the muscles on exposure to heat permits the scrotal sac to become more pendulous, moving the testes farther from the warm core of the body.

The testicular Leydig cells secrete masculinizing testosterone.

The testes perform the dual function of producing sperm and secreting testosterone. About 80% of the testicular mass consists of highly coiled **seminiferous tubules,** within which spermatogenesis takes place. The endocrine cells that produce testosterone—the **Leydig,** or **interstitial, cells**—lie in the connective tissue (interstitial tissue) between the seminiferous tubules (● Figure 20-7b). Thus, the portions of the testes that produce sperm and secrete testosterone are structurally and functionally distinct.

Testosterone is a steroid hormone derived from a cholesterol precursor molecule, as are the female sex hormones, estrogen and progesterone (see ● Figure 19-8, p. 699). Once produced, some of the testosterone is secreted into the blood, where it is transported, primarily bound to plasma proteins, to its target sites of action. A substantial portion of the newly synthesized testosterone goes into the lumen of the seminiferous tubules, where it plays an important role in sperm production.

To exert its effects, testosterone (and other androgens) bind with androgen receptors in the cytoplasm of target cells. The androgen-receptor complex moves to the nucleus, where it binds with the androgen-response element on DNA, leading to transcription of genes that direct synthesis of new proteins that carry out the desired cellular response.

Most but not all of testosterone's actions ultimately function to ensure delivery of sperm to the female. The effects of testos-

(a) Longitudinal section of testis showing location and arrangement of seminiferous tubules

(b) Light micrograph of cross section of seminiferous tubules

(c) Scanning electron micrograph of cross section of seminiferous tubules

(d) Relationship of Sertoli cells to developing sperm cells

● **FIGURE 20-7 Anatomy of testis depicting the site of spermatogenesis.** (a) The seminiferous tubules are the sperm-producing portion of the testis. (b) The undifferentiated germ cells (the spermatogonia) lie in the periphery of the tubule, and the differentiated spermatozoa are in the lumen, with the various stages of sperm development in between. (c) Note the presence of the highly differentiated spermatozoa (recognizable by their tails) in the lumen of the seminiferous tubules. (d) Relationship of the Sertoli cells to the developing sperm cells.

terone can be grouped into five categories: (1) effects on the reproductive system before birth; (2) effects on sex-specific tissues after birth; (3) other reproduction-related effects; (4) effects on secondary sexual characteristics; and (5) nonreproductive actions (▲ Table 20-1).

EFFECTS ON THE REPRODUCTIVE SYSTEM BEFORE BIRTH Before birth, testosterone secretion by the Leydig cells of the fetal testes masculinizes the reproductive tract and external genitalia and promotes descent of the testes into the scrotum, as already described. After birth, testosterone secretion ceases, and the testes

▲ TABLE 20-1 Effects of Testosterone

Effects before Birth

Masculinizes the reproductive tract and external genitalia

Promotes descent of the testes into the scrotum

Effects on Sex-Specific Tissues after Birth

Promotes growth and maturation of the reproductive system at puberty

Is essential for spermatogenesis

Maintains the reproductive tract throughout adulthood

Other Reproduction-Related Effects

Develops the sex drive at puberty

Controls gonadotropin hormone secretion

Effects on Secondary Sexual Characteristics

Induces the male pattern of hair growth (e.g., beard)

Causes the voice to deepen because of thickening of the vocal folds

Promotes muscle growth responsible for the male body configuration

Nonreproductive Actions

Exerts a protein anabolic effect

Promotes bone growth at puberty

Closes the epiphyseal plates after being converted to estrogen by aromatase

May induce aggressive behavior

and remainder of the reproductive system remain small and nonfunctional until puberty.

EFFECTS ON SEX-SPECIFIC TISSUES AFTER BIRTH Puberty is the period of arousal and maturation of the previously nonfunctional reproductive system, culminating in sexual maturity and the ability to reproduce. It usually begins sometime between the ages of 10 and 14; on average, it begins about two years earlier in females than in males. Usually lasting three to five years, puberty encompasses a complex sequence of endocrine, physical, and behavioral events. **Adolescence** is a broader concept that refers to the entire transition period between childhood and adulthood, not just to sexual maturation. In both sexes, the reproductive changes that take place during puberty are (1) enlargement and maturation of gonads, (2) development of secondary sexual characteristics, (3) achievement of fertility (gamete production), (4) growth and maturation of the reproductive tract, and (5) attainment of libido (sex drive). The pubertal growth spurt also occurs.

At puberty in males, the Leydig cells start secreting testosterone once again. Testosterone is responsible for growth and maturation of the entire male reproductive system. Under the influence of the pubertal surge in testosterone secretion, the testes enlarge and start producing sperm for the first time, the accessory sex glands enlarge and become secretory, and the penis and scrotum enlarge.

Ongoing testosterone secretion is essential for spermatogenesis and for maintaining a mature male reproductive tract throughout adulthood. Once initiated at puberty, testosterone secretion and spermatogenesis occur continuously throughout the male's life. Testicular efficiency gradually declines after 45 to 50 years of age, however, even though men in their 70s and beyond may continue to enjoy an active sex life, and some even father a child at this late age. The gradual reduction in circulating testosterone levels and in sperm production is not caused by a decrease in stimulation of the testes but probably arises instead from degenerative changes associated with aging that occur in the small testicular blood vessels. This age-related gradual decline is sometimes mistakenly termed "male menopause" or "andropause," but it is not comparable to female menopause, which is preprogrammed and results in complete, abrupt cessation of reproductive capability. Recently the androgen decline in males has been more aptly termed **androgen deficiency in aging males (ADAM).**

 Following **castration** (surgical removal of the testes) or testicular failure caused by disease, the other sex organs regress in size and function.

OTHER REPRODUCTION-RELATED EFFECTS Testosterone governs the development of sexual libido at puberty and helps maintain the sex drive in the adult male. Stimulation of this behavior by testosterone is important for facilitating delivery of sperm to females. In humans, libido is also influenced by many interacting social and emotional factors. Once libido has developed, testosterone is no longer absolutely required for its maintenance. Castrated males often remain sexually active but at a reduced level.

In another reproduction-related function, testosterone participates in the normal negative-feedback control of gonadotropin hormone secretion by the anterior pituitary, a topic covered more thoroughly later.

EFFECTS ON SECONDARY SEXUAL CHARACTERISTICS All male secondary sexual characteristics depend on testosterone for their development and maintenance. These nonreproductive male characteristics induced by testosterone include (1) the male pattern of hair growth (for example, beard and chest hair and, in genetically predisposed men, baldness); (2) a deep voice caused by enlargement of the larynx and thickening of the vocal folds; (3) thick skin; and (4) the male body configuration (for example, broad shoulders and heavy arm and leg musculature) as a result of protein deposition. A male castrated before puberty (a **eunuch**) does not mature sexually, nor does he develop secondary sexual characteristics.

NONREPRODUCTIVE ACTIONS Testosterone exerts several important effects not related to reproduction. It has a general protein anabolic (synthesis) effect and promotes bone growth, thus contributing to the more muscular physique of males and to the pubertal growth spurt. Ironically, testosterone not only stimulates bone growth but eventually prevents further growth by sealing the growing ends of the long bones (that is, ossifying, or "closing," the epiphyseal plates; see p. 680). Testosterone also stimulates oil secretion by the sebaceous glands. This effect is most striking during the adolescent surge of testosterone secretion, predisposing the young man to develop acne.

In animals, testosterone induces aggressive behavior, but whether it influences human behavior other than in the area of sexual behavior is an unresolved issue. Even though some athletes and bodybuilders who take testosterone-like anabolic androgenic steroids to increase muscle mass have been observed to display more aggressive behavior (see p. 282), it is unclear to what extent general behavioral differences between the sexes are hormonally induced or result from social conditioning.

CONVERSION OF TESTOSTERONE TO ESTROGEN IN MALES Although testosterone is classically considered the male sex hormone and estrogen a female sex hormone, the distinctions are not as clear-cut as once thought. In addition to the small amount of estrogen produced by the adrenal cortex (see p. 702), a portion of the testosterone secreted by the testes is converted to estrogen outside the testes by the enzyme **aromatase,** which is widely distributed but most abundant in adipose tissue. Because of this conversion, it is sometimes difficult to distinguish effects of testosterone itself and testosterone-turned-estrogen inside cells. For example, scientists recently learned that closure of the epiphyseal plates in males is induced not by testosterone per se but by testosterone turned into estrogen by aromatization. Estrogen receptors have been identified in the testes, prostate, bone, and elsewhere in males. Recent findings suggest that estrogen plays an essential role in male reproductive health; for example, it is important in spermatogenesis and surprisingly contributes to male heterosexuality. Also, it likely contributes to bone homeostasis (see p. 729). The depth, breadth, and mechanisms of action of estrogen in males are only beginning to be explored. (Likewise, in addition to the weak androgenic hormone DHEA produced by the adrenal cortex in both sexes, the ovaries in females secrete a small amount of testosterone, the functions of which remain unclear.)

We now shift attention from testosterone secretion to the other function of the testes—sperm production.

Spermatogenesis yields an abundance of highly specialized, mobile sperm.

About 250 m (800 ft) of sperm-producing seminiferous tubules are packed within the testes (● Figure 20-7a). Two functionally important cell types are present in these tubules: *germ cells,* most of which are in various stages of sperm development, and *Sertoli cells,* which provide crucial support for spermatogenesis (● Figure 20-7b, c, and d). **Spermatogenesis** is a complex process by which relatively undifferentiated primordial germ cells, the **spermatogonia** (each of which contains a diploid complement of 46 chromosomes), proliferate and are converted into extremely specialized, motile spermatozoa (sperm), each bearing a randomly distributed haploid set of 23 chromosomes.

Microscopic examination of a seminiferous tubule reveals layers of germ cells in an anatomic progression of sperm develop-

ment, starting with the least differentiated in the outer layer and moving inward through various stages of division to the lumen, where the highly differentiated sperm are ready for exit from the testis (● Figure 20-7b, c, and d). Spermatogenesis takes 64 days for development from a spermatogonium to a mature sperm. The steps in spermatogenesis do not occur synchronously throughout the seminiferous tubules. At any given time, different seminiferous tubules are in different stages of spermatogenesis, so that up to several hundred million sperm may reach maturity daily. Spermatogenesis encompasses three major stages: *mitotic proliferation, meiosis,* and *packaging* (● Figure 20-8).

MITOTIC PROLIFERATION Spermatogonia located in the outermost layer of the tubule continuously divide mitotically, with all new cells bearing the full complement of 46 chromosomes identical to those of the parent cell. Such proliferation provides a continual supply of new germ cells. Following mitotic division of a spermatogonium, one of the daughter cells remains at the outer edge of the tubule as an undifferentiated spermatogonium, thus maintaining the germ-cell line. The other daughter cell starts moving toward the lumen while undergoing the various steps required to form sperm, which will be released into the lumen. In humans, the sperm-forming daughter cell divides mitotically twice more to form four identical **primary spermatocytes.** After the last mitotic division, the primary spermatocytes enter a resting phase during which the chromosomes are duplicated and the doubled strands remain together in preparation for the first meiotic division.

MEIOSIS During meiosis, each primary spermatocyte (with a diploid number of 46 doubled chromosomes) forms two **secondary spermatocytes** (each with a haploid number of 23 doubled chromosomes) during the first meiotic division, finally yielding four **spermatids** (each with 23 single chromosomes) as a result of the second meiotic division.

No further division takes place beyond this stage of spermatogenesis. Each spermatid is remodeled into a single spermatozoon. Because each sperm-producing spermatogonium mitotically produces four primary spermatocytes and each primary spermatocyte meiotically yields four spermatids (spermatozoa-to-be), the spermatogenic sequence in humans can theoretically produce 16 spermatozoa each time a spermatogonium initiates this process. Usually, however, some cells are lost at various stages, so the efficiency of productivity is rarely this high.

PACKAGING Even after meiosis, spermatids still resemble undifferentiated spermatogonia structurally, except for their half complement of chromosomes. Production of extremely specialized, mobile spermatozoa from spermatids requires extensive remodeling, or **packaging,** of cell elements, a process alternatively known as **spermiogenesis.** Sperm are essentially "stripped-down" cells in which most of the cytosol and any organelles not needed for delivering the sperm's genetic information to an ovum have been extruded. Thus, sperm travel lightly, taking with them only the bare essentials to accomplish fertilization.

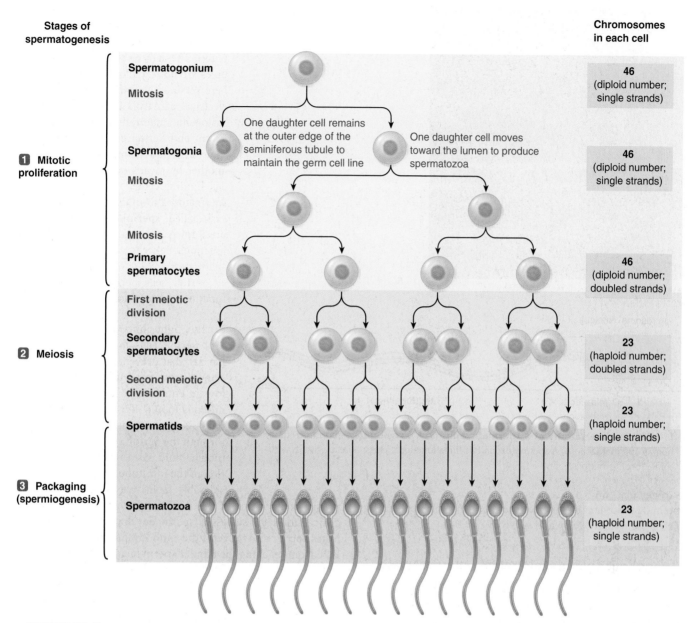

● **FIGURE 20-8 Spermatogenesis.**

A **spermatozoon** has three parts (● Figure 20-9): a head capped with an acrosome, a midpiece, and a tail. The **head** consists primarily of the nucleus, which contains the sperm's complement of genetic information. The **acrosome,** an enzyme-filled vesicle that caps the tip of the head, is used as an "enzymatic drill" for penetrating the ovum. The acrosome, a modified lysosome (see p. 28), is formed by aggregation of vesicles produced by the endoplasmic reticulum/Golgi complex before these organelles are discarded. The acrosomal enzymes remain inactive until the sperm contacts an egg, at which time the enzymes are released. Mobility for the spermatozoon is provided by a long, whiplike **tail** (a flagellum; see p. 45), movement of which is powered by energy generated by the mitochondria concentrated within the **midpiece** of the sperm.

Until sperm maturation is complete, the developing germ cells arising from a single primary spermatocyte remain joined by cytoplasmic bridges. These connections, which result from incomplete cytoplasmic division, permit the four developing sperm to exchange cytoplasm. This linkage is important, because the X chromosome, but not the Y chromosome, contains genes that code for cell products essential for sperm development. (Whereas the large X chromosome contains several thousand genes, the small Y chromosome has only a few dozen, the most important of which are the SRY gene and others that play critical roles in male fertility.) During meiosis, half the sperm receive an X and the other half a Y chromosome. Were it not for the sharing of cytoplasm so that all the haploid cells are provided with the products coded for by X chromosomes until

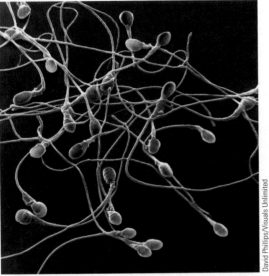

(a) Photomicrograph of human spermatozoa

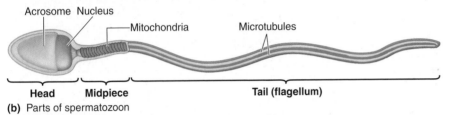

Acrosome Nucleus
Mitochondria
Microtubules

Head Midpiece Tail (flagellum)
(b) Parts of spermatozoon

● **FIGURE 20-9 Anatomy of a spermatozoon.** (a) A phase-contrast photomicrograph of human spermatozoa. (b) A spermatozoon has three functional parts: a head with its acrosome "cap," a midpiece, and a tail.

spread of action potentials from one cell to the next as charge-carrying ions pass through these connecting tunnels (see p. 59). The cells in the seminiferous tubules are not excitable, so gap junctions here serve a role other than transfer of electrical activity. At all stages of spermatogenic maturation, the developing sperm and Sertoli cells exchange small molecules and communicate with one another by means of this direct cell-to-cell binding. Final release of a mature spermatozoa from the Sertoli cell, a process called **spermiation**, requires the breakdown of the tight junctions and gap junctions between the Sertoli cell and spermatozoa.

Sertoli cells perform the following functions essential for spermatogenesis:

1. The tight junctions between adjacent Sertoli cells form a **blood–testes barrier** that prevents blood-borne substances from passing between the cells to gain entry to the lumen of the seminiferous tubule. Because of this barrier, only selected molecules that can pass through the Sertoli cells reach the intratubular fluid. As a result, the composition of the intratubular fluid varies considerably from that of the blood. The unique composition of this fluid that bathes the germ cells is critical for later stages of sperm development. The blood–testes barrier also prevents the antibody-producing cells in the ECF from reaching the tubular sperm factory, thus preventing the formation of antibodies against the highly differentiated spermatozoa.

2. Because the secluded developing sperm cells do not have direct access to blood-borne nutrients, the "nurse" Sertoli cells provide nourishment for them. Developing sperm cells cannot efficiently use glucose. The Sertoli cells take up glucose via a GLUT-1 symporter, metabolize the glucose to lactate, then transfer the lactate to the sperm cells, which can use lactate as an energy source.

3. The Sertoli cells have an important phagocytic function. They engulf the cytoplasm extruded from the spermatids during their remodeling, and they destroy defective germ cells that fail to successfully complete all stages of spermatogenesis.

4. The Sertoli cells secrete into the lumen **seminiferous tubule fluid,** which "flushes" the released sperm from the tubule into the epididymis for storage and further processing.

5. An important component of this Sertoli secretion is **androgen-binding protein.** As the name implies, this protein binds androgens (that is, testosterone), thus maintaining a very high level of this hormone within the seminiferous tubule lumen. Testosterone is 100 times more concentrated in the seminiferous tubule fluid than in the blood. This high local concen-

sperm development is complete, the Y-bearing, male-producing sperm could not develop and survive.

Throughout their development, sperm remain intimately associated with Sertoli cells.

The seminiferous tubules house the **Sertoli cells** in addition to the spermatogonia and developing sperm cells. The Sertoli cells, which are epithelial cells, lie side by side and form a ring that extends from the outer surface of the tubule to the lumen. Each Sertoli cell spans the entire distance from the outer surface membrane of the seminiferous tubule to the fluid-filled lumen (see ● Figure 20-7b and d). Adjacent Sertoli cells are joined by tight junctions at a point slightly beneath the outer membrane (see p. 59). Developing sperm cells are tucked between adjacent Sertoli cells, with spermatogonia lying at the outer perimeter of the tubule, outside the tight junction (● Figure 20-7b and d). During spermatogenesis, developing sperm cells arising from spermatogonial mitotic activity pass through the tight junctions, which transiently separate to make a path for them, then migrate toward the lumen in close association with the adjacent Sertoli cells, undergoing their further divisions during this migration. The cytoplasm of the Sertoli cells envelops the migrating sperm cells, which remain buried within these cytoplasmic recesses throughout their development. Sertoli cells form tight junctions and gap junctions with the developing sperm cells. Recall that gap junctions between excitable cells permit the tration of testosterone is essential for sustaining sperm production. Androgen-binding protein is necessary to retain testosterone within the lumen because this steroid hormone is lipid soluble and could easily diffuse across the plasma membranes and leave the lumen. Testosterone itself stimulates production of androgen-binding protein.

6. The Sertoli cells are the site of action for control of spermatogenesis by both testosterone and follicle-stimulating hormone (FSH). Sertoli cells have distinct receptors for each of these hormones. The Sertoli cells themselves release another hormone, *inhibin,* which acts in negative-feedback fashion to regulate FSH secretion.

7. During fetal development, Sertoli cells also secrete Müllerian-inhibiting factor.

LH and FSH from the anterior pituitary control testosterone secretion and spermatogenesis.

The testes are controlled by the two gonadotropic hormones secreted by the anterior pituitary, **luteinizing hormone (LH)** and **follicle-stimulating hormone (FSH),** both of which are produced by the same cell type, the gonadotrope. Both hormones in both sexes act on the gonads by activating cAMP.

FEEDBACK CONTROL OF TESTICULAR FUNCTION LH and FSH act on separate components of the testes (● Figure 20-10). LH acts on the Leydig cells to regulate testosterone secretion. FSH acts on the Sertoli cells to enhance spermatogenesis. Secretion of both LH and FSH from the anterior pituitary is stimulated in turn by a single hypothalamic hormone, **gonadotropin-releasing hormone (GnRH)** (see p. 674). However, FSH and LH are segregated to a large extent into separate secretory vesicles in the gonadotrope and are not secreted in equal amounts because other regulatory factors influence how much of each gonadotropin is secreted.

Two factors—*testosterone* and *inhibin*—differentially influence the secretory rate of LH and FSH. Testosterone, the product of LH stimulation of the Leydig cells, acts in negative-feedback fashion to inhibit LH secretion in two ways. The predominant negative-feedback effect of testosterone is to decrease GnRH release by acting on the hypothalamus, thus indirectly decreasing both LH and FSH release by the anterior pituitary. In addition, testosterone acts directly on the anterior pituitary to selectively reduce LH secretion. The latter action explains why testosterone exerts a greater inhibitory effect on LH secretion than on FSH secretion.

The testicular inhibitory signal specifically directed at controlling FSH secretion is the peptide hormone **inhibin,** which is secreted by the Sertoli cells. Inhibin acts directly on the anterior pituitary to selectively inhibit FSH secretion. This feedback inhibition of FSH by a Sertoli cell product is appropriate, because FSH stimulates spermatogenesis by acting on the Sertoli cells.

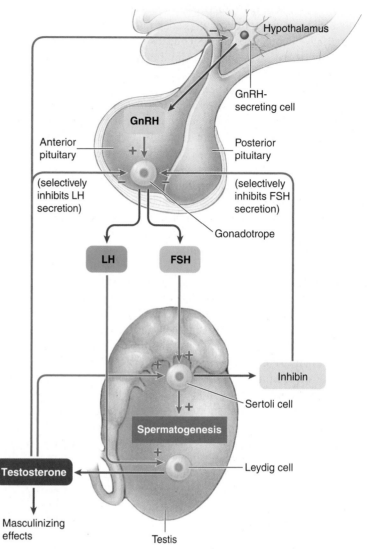

● **FIGURE 20-10 Control of testicular function.**

ROLES OF TESTOSTERONE AND FSH IN SPERMATOGENESIS Both testosterone and FSH play critical roles in controlling spermatogenesis, each exerting its effect by acting on the Sertoli cells. Testosterone is essential for both mitosis and meiosis of the germ cells, whereas FSH is needed for spermatid remodeling. Testosterone concentration is much higher in the testes than in the blood because a substantial portion of this hormone produced locally by the Leydig cells is retained in the intratubular fluid complexed with androgen-binding protein secreted by the Sertoli cells. Only this high concentration of testicular testosterone is adequate to sustain sperm production.

Gonadotropin-releasing hormone activity increases at puberty.

Even though the fetal testes secrete testosterone, which directs masculine development of the reproductive system, after birth the testes become dormant until puberty. During the prepubertal period, LH and FSH are not secreted at adequate levels to stimulate any significant testicular activity. The prepubertal delay in the onset of reproductive capability allows time for the individual to mature physically (although not necessarily psychologically) enough to handle child rearing. (This physical maturation is especially important in the female, whose body must support the developing fetus.)

During the prepubertal period, GnRH activity is inhibited. The pubertal process is initiated by an increase in GnRH activity sometime between 8 and 12 years of age. Early in puberty, GnRH secretion occurs only at night, causing a brief nocturnal increase in LH secretion and accordingly, testosterone secretion. The extent of GnRH secretion gradually increases as puberty progresses until the adult pattern of GnRH, FSH, LH, and testosterone secretion is established. Under the influence of the rising levels of testosterone during puberty, the physical changes that encompass the secondary sexual characteristics and reproductive maturation become evident.

The factors responsible for initiating puberty in humans remain unclear. The leading proposal focuses on a potential role for the hormone *melatonin,* which is secreted by the *pineal gland* within the brain (see p. 685). Melatonin, whose secretion decreases during exposure to the light and increases during exposure to the dark, has an antigonadotropic effect in many species. Light striking the eyes inhibits the nerve pathways that stimulate melatonin secretion. In many seasonally breeding species, the overall decrease in melatonin secretion in connection with longer days and shorter nights initiates the mating season. Some researchers suggest that an observed reduction in the overall rate of melatonin secretion at puberty in humans— particularly during the night, when the peaks in GnRH secretion first occur—is the trigger for the onset of puberty.

Having completed our discussion of testicular function, we now shift attention to the roles of the other components of the male reproductive system.

The reproductive tract stores and concentrates sperm and increases their fertility.

The remainder of the male reproductive system (besides the testes) is designed to deliver sperm to the female reproductive tract. Essentially, it consists of (1) a tortuous pathway of tubes (the reproductive tract), which transports sperm from the testes to outside the body; (2) several accessory sex glands, which contribute secretions important to the viability and motility of the sperm; and (3) the penis, which is designed to penetrate and deposit the sperm within the vagina of the female. We will examine each of these parts in greater detail, beginning with the reproductive tract.

COMPONENTS OF THE MALE REPRODUCTIVE TRACT A comma-shaped **epididymis** is loosely attached to the rear surface of each testis (see Figures 20-1, p. 743, and 20-7a, p. 750). After sperm are produced in the seminiferous tubules, they are swept into the epididymis as a result of the pressure created by the continual secretion of tubular fluid by the Sertoli cells. The epididymal ducts from each testis converge to form a large, thick-walled, muscular duct called the **ductus (vas) deferens.** The ductus deferens from each testis passes up out of the scrotal sac and runs back through the inguinal canal into the abdominal cavity, where it eventually empties into the urethra at the neck of the bladder (see ● Figure 20-1). The urethra carries sperm out of the penis during ejaculation, the forceful expulsion of semen from the body.

FUNCTIONS OF THE EPIDIDYMIS AND DUCTUS DEFERENS These ducts perform several important functions. The epididymis and ductus deferens serve as the sperm's exit route from the testis. As they leave the testis, the sperm are capable of neither movement nor fertilization. They gain both capabilities during their passage through the epididymis. This maturational process is stimulated by the testosterone retained within the tubular fluid bound to androgen-binding protein. Sperm's capacity to fertilize is enhanced even further by exposure to secretions of the female reproductive tract. This enhancement of sperm's capacity in the male and female reproductive tracts is known as **capacitation.** Scientists believe that *defensin,* a protein secreted by the epididymis that defends sperm from microorganisms, may serve a second role by boosting sperm's motility. The epididymis also concentrates the sperm a hundredfold by absorbing most of the fluid that enters from the seminiferous tubules. The maturing sperm are slowly moved through the epididymis into the ductus deferens by rhythmic contractions of the smooth muscle in the walls of these tubes.

The ductus deferens serves as an important site for sperm storage. Because the tightly packed sperm are relatively inactive and their metabolic needs are accordingly low, they can be stored in the ductus deferens for up to two months, even though they have no nutrient blood supply and are nourished only by simple sugars present in the tubular secretions.

Clinical Note **VASECTOMY** In a **vasectomy,** a common sterilization procedure in males, a small segment of each ductus deferens (alias vas deferens, hence the term *vasectomy*) is surgically removed after it passes from the testis but before it enters the inguinal canal, thus blocking the exit of sperm from the testes. The sperm that build up behind the tied-off testicular end of the severed ductus are removed by phagocytosis. Although this procedure blocks sperm exit, it does not interfere with testosterone activity, because the Leydig cells secrete testosterone into the blood, not through the ductus deferens. Thus, testosterone-dependent masculinity or libido should not diminish after a vasectomy.

The accessory sex glands contribute the bulk of the semen.

Several accessory sex glands—the seminal vesicles and prostate—empty their secretions into the duct system before it joins the urethra (see ● Figure 20-1). A pair of saclike *seminal vesicles* empty into the last portion of the two ductus deferens, one on each side. The short segment of duct that passes beyond the entry point of the seminal vesicle to join the urethra is called the *ejaculatory duct.* The *prostate* is a large, single gland that completely surrounds the ejaculatory ducts and urethra. In a significant number of men, the prostate enlarges in middle to older age (a condition called **benign prostatic hypertrophy**). Difficulty in urination is often encountered as the enlarging

Location and Functions of the Components of the Male Reproductive System

Component	Number and Location	Functions
Testis	Pair; located in the scrotum, a skin-covered sac suspended within the angle between the legs	Produce sperm Secrete testosterone
Epididymis and Ductus Deferens	Pair; one epididymis attached to the rear of each testis; one ductus deferens travels from each epididymis up out of the scrotal sac through the inguinal canal and empties into the urethra at the neck of the bladder	Serve as the sperm's exit route from the testis Serve as the site for maturation of the sperm for motility and fertility Concentrate and store the sperm
Seminal Vesicle	Pair; both empty into the last portion of the ductus deferens, one on each side	Supply fructose to nourish the ejaculated sperm Secrete prostaglandins that stimulate motility to help transport the sperm within the male and female Provide the bulk of the semen Provide precursors for the clotting of semen
Prostate Gland	Single; completely surrounds the urethra at the neck of the bladder	Secretes an alkaline fluid that neutralizes the acidic vaginal secretions Triggers clotting of the semen to keep the sperm in the vagina during penis withdrawal
Bulbourethral Gland	Pair; both empty into the urethra, one on each side, just before the urethra enters the penis	Secrete mucus for lubrication

prostate impinges on the portion of the urethra that passes through the prostate. Another pair of accessory sex glands, the *bulbourethral glands,* drain into the urethra after it has passed through the prostate and just before it enters the penis. Numerous mucus-secreting glands also lie along the length of the urethra.

SEMEN During ejaculation, the accessory sex glands contribute secretions that provide support for the continuing viability of the sperm inside the female reproductive tract. These secretions constitute the bulk of the **semen,** which is a mixture of accessory sex gland secretions, sperm, and mucus. Sperm make up only a small percentage of the total ejaculated fluid.

FUNCTIONS OF THE MALE ACCESSORY SEX GLANDS Although the accessory sex gland secretions are not absolutely essential for fertilization, they do greatly facilitate the process:

■ The **seminal vesicles** (1) supply fructose, which serves as the primary energy source for ejaculated sperm; (2) secrete *prostaglandins,* which stimulate contractions of the smooth muscle in both the male and female reproductive tracts, thereby helping to transport sperm from their storage site in the male to the site of fertilization in the female oviduct; (3) provide about 60% of the semen volume, which helps wash the sperm into the urethra and also dilutes the thick mass of sperm, enabling them to become mobile; and (4) secrete fibrinogen, a precursor of fibrin, which forms the meshwork of a clot (see p. 407).

■ The **prostate gland** (1) secretes an alkaline fluid that neutralizes the acidic vaginal secretions, an important function because sperm are more viable in a slightly alkaline environment; (2) provides clotting enzymes; and (3) releases prostate-specific antigen. The prostatic clotting enzymes act on fibrinogen from the seminal vesicles to produce fibrin, which "clots" the semen, thus helping keep the ejaculated sperm in the female reproductive tract during withdrawal of the penis. Shortly thereafter, the seminal clot is broken down by **prostate-specific antigen (PSA),** a fibrin-degrading enzyme from the prostate, thus releasing mobile sperm within the female tract.

Clinical Note Because PSA is produced only in the prostate gland, measurement of PSA levels in the blood is used as one type of screening test for possible prostate cancer. Elevated levels of PSA in the blood are associated with prostate cancer, benign prostatic hypertrophy, or prostate infections.

■ During sexual arousal, the **bulbourethral glands** secrete a mucuslike substance that provides lubrication for sexual intercourse.

▲ Table 20-2 summarizes the locations and functions of the components of the male reproductive system.

Before turning to the act of delivering sperm to the female (sexual intercourse), we are going to briefly digress to discuss the diverse roles of prostaglandins, which were first discovered in semen but are abundant throughout the body.

Prostaglandins are ubiquitous, locally acting chemical messengers.

Although **prostaglandins** were first identified in the semen and were believed to be of prostate gland origin (hence their name, even though they are actually secreted into the semen by the seminal vesicles), their production and actions are by no means limited to the reproductive system. These 20-carbon fatty acid derivatives are among the most ubiquitous chemical messengers in the body. They are produced in virtually all tissues from arachidonic acid, a fatty acid constituent of the phospholipids within the plasma membrane. On appropriate stimulation, arachidonic acid is split from the plasma membrane by a membrane-bound enzyme and then is converted into the appropriate prostaglandin, which acts as a paracrine locally within or near its site of production. After prostaglandins act, they are rapidly inactivated by local enzymes before they gain access to the blood; or if they do reach the circulatory system, they are swiftly degraded on their first pass through the lungs so that they are not dispersed through the systemic arterial system.

Prostaglandins are designated as belonging to one of three groups—PGA, PGE, or PGF—according to structural variations in the five-carbon ring that they contain at one end (● Figure 20-11). Within each group, prostaglandins are further identified by the number of double bonds present in the two side chains that project from the ring structure (for example, PGE_1 has one double bond and PGE_2 has two double bonds).

Prostaglandins as well as *prostacyclins, thromboxanes,* and *leukotrienes* (other closely related arachidonic-acid derivatives) are collectively known as **eicosanoids** and are among the most biologically active compounds known. Prostaglandins exert a bewildering variety of effects. Not only are slight variations in prostaglandin structure accompanied by profound differences in biological action, but the same prostaglandin molecule may even exert opposite effects in different tissues. Besides enhancing sperm transport in semen, these abundant chemical messengers are known or suspected to exert other actions in the female reproductive system and in the respiratory, urinary, digestive, nervous, and endocrine systems, in addition to affecting platelet aggregation, fat metabolism, and inflammation (▲ Table 20-3).

As prostaglandins' various actions are better understood, new ways of manipulating them therapeutically are becoming available. A classic example is the use of aspirin, which blocks the conversion of arachidonic acid into prostaglandins, for fever reduction and pain relief. Prostaglandin action is also therapeutically inhibited in the treatment of premenstrual symptoms and menstrual cramping. Furthermore, specific prostaglandins have been medically administered in such diverse situations as inducing labor, treating asthma, and treating gastric ulcers.

Next, before considering the female in greater detail, we will examine the means by which males and females come together to accomplish reproduction.

● **FIGURE 20-11 Structure and nomenclature of prostaglandins.**

Letter designation (PGA, PGE, PGF) denotes structural variations in the five-carbon ring

Number designation (e.g., PGE_1, PGE_2) denotes number of double bonds present in the two side chains

▲ TABLE 20-3	Actions of Prostaglandins
Body System Affected	**Actions of Prostaglandins**
Reproductive System	Promote sperm transport by action on smooth muscle in the male and female reproductive tracts
	Play a role in ovulation
	Play important role in menstruation
	Contribute to preparation of the maternal portion of the placenta
	Contribute to parturition
Respiratory System	Some promote bronchodilation, others bronchoconstriction
Urinary System	Increase the renal blood flow
	Increase excretion of water and salt
Digestive System	Inhibit HCl secretion by the stomach
	Stimulate intestinal motility
Nervous System	Influence neurotransmitter release and action
	Act at the hypothalamic "thermostat" to increase body temperature
	Exacerbate sensation of pain
Endocrine System	Enhance cortisol secretion
	Influence tissue responsiveness to hormones in many instances
Circulatory System	Influence platelet aggregation
Fat Metabolism	Inhibit fat breakdown
Defense System	Promote many aspects of inflammation, including development of fever

Sexual Intercourse Between Males and Females

Ultimately, union of male and female gametes to accomplish reproduction in humans requires delivery of sperm-laden semen into the female vagina through the **sex act,** also known as **sexual intercourse, coitus,** or **copulation.**

The male sex act is characterized by erection and ejaculation.

The *male sex act* involves two components: (1) **erection,** or hardening of the normally flaccid penis to permit its entry into the vagina, and (2) **ejaculation,** or forceful expulsion of semen into the urethra and out of the penis (▲ Table 20-4). In addition to these strictly reproduction-related components, the **sexual response cycle** encompasses broader physiologic responses that can be divided into four phases:

1. The *excitement phase* includes erection and heightened sexual awareness.

2. The *plateau phase* is characterized by intensification of these responses, plus more generalized body responses, such as steadily increasing heart rate, blood pressure, respiratory rate, and muscle tension.

3. The *orgasmic phase* includes ejaculation as well as other responses that culminate the mounting sexual excitement and are collectively experienced as an intense physical pleasure.

4. The *resolution phase* returns the genitalia and body systems to their prearousal state.

The human sexual response is a multicomponent experience that, in addition to these physiologic phenomena, encompasses emotional, psychological, and sociological factors. We will examine only the physiologic aspects of sex.

Erection is accomplished by penis vasocongestion.

Erection is accomplished by engorgement of the penis with blood. The penis consists almost entirely of **erectile tissue** made up of three columns or cords of spongelike vascular spaces extending the length of the organ (● Figure 20-12a). In the absence of sexual excitation, the erectile tissues contain little blood, because the arterioles that supply these vascular chambers are constricted. As a result, the penis remains small and flaccid. During sexual arousal, these arterioles reflexly dilate and the erectile tissue fills with blood, causing the penis to enlarge both in length and width and to become more rigid. The veins that drain the erectile tissue are mechanically compressed by this engorgement and expansion of the vascular spaces, reducing venous outflow and thereby contributing even further to the buildup of blood, or *vasocongestion.* These local vascular responses transform the penis into a hardened, elongated organ capable of penetrating the vagina.

ERECTION REFLEX The erection reflex is a spinal reflex triggered by stimulation of highly sensitive mechanoreceptors located in the *glans penis,* which caps the tip of the penis. A recently identified **erection-generating center** lies in the lower spinal cord. Tactile stimulation of the glans reflexly triggers, by means of this center, increased parasympathetic vasodilator activity and decreased sympathetic vasoconstrictor activity to the penile arterioles. The result is rapid, pronounced vasodilation of these arterioles and an ensuing erection (● Figure 20-13). As long as this spinal reflex arc remains intact, erection is possible even in men paralyzed by a higher spinal-cord injury.

This parasympathetically induced vasodilation is the major instance of direct parasympathetic control over blood vessel diameter in the body. Parasympathetic stimulation brings about relaxation of penile arteriolar smooth muscle by nitric oxide, which causes arteriolar vasodilation in response to local tissue changes elsewhere in the body (see p. 356). Arterioles are typi-

▲ TABLE 20-4	Components of the Male Sex Act	
Components of the Male Sex Act	**Definition**	**How Accomplished**
Erection	Hardening of the normally flaccid penis to permit its entry into the vagina	Engorgement of the penis erectile tissue with blood as a result of marked parasympathetically induced vasodilation of the penile arterioles and mechanical compression of the veins
Ejaculation		
Emission phase	Emptying of sperm and accessory sex gland secretions (semen) into the urethra	Sympathetically induced contraction of the smooth muscle in the walls of the ducts and accessory sex glands
Expulsion phase	Forceful expulsion of semen from the penis	Motor-neuron-induced contraction of the skeletal muscles at the base of the penis

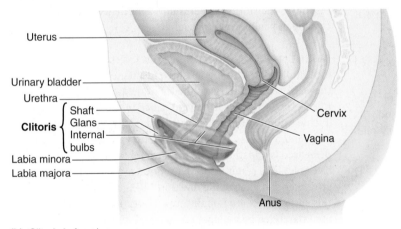

(a) Penis in males

(b) Clitoris in females

● **FIGURE 20-12 Erectile tissue in males and females.**

cally supplied only by sympathetic nerves, with increased sympathetic activity producing vasoconstriction and decreased sympathetic activity resulting in vasodilation (see p. 358). Concurrent parasympathetic stimulation and sympathetic inhibition of penile arterioles accomplish vasodilation more rapidly and in greater magnitude than is possible in other arterioles supplied only by sympathetic nerves. Through this efficient means of rapidly increasing blood flow into the penis, the penis can become completely erect in as little as 5 to 10 seconds. At the same time, parasympathetic impulses promote secretion of lubricating mucus from the bulbourethral glands and the urethral glands in preparation for coitus.

A flurry of recent research has led to the discovery of numerous regions throughout the brain that can influence the male sexual response. The erection-influencing brain sites appear extensively interconnected and function as a unified network to either facilitate or inhibit the basic spinal erection reflex, depending on the momentary circumstances. As an example of facilitation, psychic stimuli, such as viewing something sexually exciting, can induce an erection in the complete absence of tactile stimulation of the penis. In contrast, failure to achieve an erection despite appropriate stimulation may result from inhibition of the

erection reflex by higher brain centers. Let us examine erectile dysfunction in more detail.

ERECTILE DYSFUNCTION A pattern of failing to achieve or maintain an erection suitable for sexual intercourse—**erectile dysfunction** or **impotence**—may be attributable to psychological or physical factors. An occasional episode of a failed erection does not constitute impotence, but a man who becomes overly anxious about his ability to perform the sex act may well be on his way to chronic failure. Anxiety can lead to erectile dysfunction, which fuels the man's anxiety level and thus perpetuates the problem. Impotence may also arise from physical limitations, including nerve damage, certain medications that interfere with autonomic function, and problems with blood flow through the penis.

Erectile dysfunction is widespread. More than 50% of men between ages 40 and 70 experience some impotence, climbing to nearly 70% by age 70. No wonder, then, that more prescriptions were written for the much-publicized drug *sildenafil* (Viagra) during its first year on the market after its approval in 1998 for treating erectile dysfunction than for any other new drug in history. Sildenafil does not produce an erection, but it amplifies and prolongs an erectile response triggered by usual means of stimulation. Here is how the drug works. Nitric oxide released in response to parasympathetic stimulation activates a membrane-bound enzyme, *guanylate cyclase*, within nearby arteriolar smooth muscle cells. This enzyme activates *cyclic guanosine monophosphate (cGMP)*, an intracellular second messenger similar to cAMP. Cyclic GMP, in turn, leads to relaxation of the penile arteriolar smooth muscle, bringing about pronounced local vasodilation. Under normal circumstances, once cGMP is activated and brings about an erection, this second messenger is broken down by the intracellular enzyme *phosphodiesterase 5 (PDE5)*. Sildenafil inhibits PDE5. As a result, cGMP remains active longer so that penile arteriolar vasodilation continues and the erection is sustained long enough for a formerly impotent man to accomplish the sex act. Just as pushing a pedal on a piano will not cause a note to be played but will prolong a played note, sildenafil cannot cause the release of nitric oxide and subsequent activation of erection-producing cGMP, but it can prolong the triggered response. The drug has no benefit for those who do not have erectile dysfunction, but its success rate has been high among sufferers of the condition. Side effects have been limited because the drug concentrates in the penis, thus having more impact on this organ than elsewhere in the body.

Ejaculation includes emission and expulsion.

The second component of the male sex act is *ejaculation*. Like erection, ejaculation is a spinal reflex. The same types of tactile and psychic stimuli that induce erection cause ejaculation when

Thoughts about sex

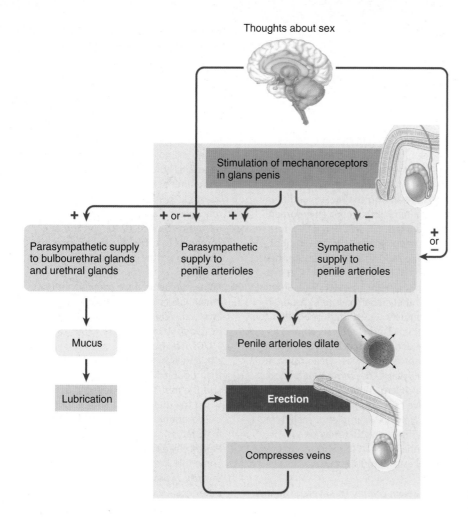

● **FIGURE 20-13 Erection reflex.**

ORGASM The rhythmic contractions that occur during semen expulsion are accompanied by involuntary rhythmic throbbing of pelvic muscles and peak intensity of the overall body responses that were climbing during the earlier phases. Heavy breathing, a heart rate of up to 180 beats per minute, marked generalized skeletal muscle contraction, and heightened emotions are characteristic. These pelvic and overall systemic responses that culminate the sex act are associated with an intense pleasure characterized by a feeling of release and complete gratification, an experience known as **orgasm.**

RESOLUTION During the resolution phase following orgasm, sympathetic vasoconstrictor impulses slow the inflow of blood into the penis, causing the erection to subside. A deep relaxation ensues, often accompanied by a feeling of fatigue. Muscle tone returns to normal, while the cardiovascular and respiratory systems return to their prearousal level of activity. Once ejaculation has occurred, a temporary refractory period of variable duration ensues before sexual stimulation can trigger another erection. Males therefore cannot experience multiple orgasms within a matter of minutes, as females sometimes do.

VOLUME AND SPERM CONTENT OF THE EJACULATE The volume and sperm content of the ejaculate depend on the length of time between ejaculations. The average volume of semen is 2.75 ml, ranging from 2 to 6 ml, the higher volumes following periods of abstinence. An average human ejaculate contains about 180 million sperm (66 million/ml), but some ejaculates contain as many as 400 million sperm.

Both quantity and quality of sperm are important determinants of fertility. A man is considered clinically infertile if his sperm concentration falls below 20 million/ml of semen. Even though only one spermatozoon actually fertilizes the ovum, large numbers of accompanying sperm are needed to provide sufficient acrosomal enzymes to break down the barriers surrounding the ovum until the victorious sperm penetrates into the ovum's cytoplasm. The quality of sperm also must be taken into account when assessing the fertility potential of a semen sample. The presence of substantial numbers of sperm with abnormal motility or structure, such as sperm with distorted tails, reduces the chances of fertilization. (For a discussion of how environmental estrogens may be decreasing sperm counts as well as negatively impacting the male and female reproductive systems in other ways, see the boxed feature on pp. 762–763, ■ Concepts, Challenges, and Controversies.)

the level of excitation intensifies to a critical peak. The overall ejaculatory response occurs in two phases: *emission* and *expulsion* (see ▲ Table 20-4).

EMISSION First, sympathetic impulses cause sequential contraction of smooth muscles in the prostate, reproductive ducts, and seminal vesicles. This contractile activity delivers prostatic fluid, then sperm, and finally seminal vesicle fluid (collectively, semen) into the urethra. This phase of the ejaculatory reflex is called **emission.** During this time, the sphincter at the neck of the bladder is tightly closed to prevent semen from entering the bladder and urine from being expelled along with the ejaculate through the urethra.

EXPULSION Second, filling of the urethra with semen triggers nerve impulses that activate a series of skeletal muscles at the base of the penis. Rhythmic contractions of these muscles occur at 0.8-second intervals and increase the pressure within the penis, forcibly expelling the semen through the urethra to the exterior. This is the **expulsion** phase of ejaculation.

Environmental "Estrogens": Bad News for the Reproductive System

Unknowingly, during the last 60 years we humans have been polluting our environment with synthetic endocrine-disrupting chemicals as an unintended side effect of industrialization. Known as **endocrine disrupters,** these hormonelike pollutants bind with the receptor sites normally reserved for the naturally occurring hormones. Depending on how they interact with the receptors, these disrupters can either mimic or block normal hormonal activity. Most endocrine disrupters exert feminizing effects. Many of these environmental contaminants mimic or alter the action of estrogen, the feminizing steroid hormone produced by the female ovaries. Although not yet conclusive, laboratory and field studies suggest that these estrogen disrupters might be responsible for some disturbing trends in reproductive health problems, such as falling sperm counts in males and an increased incidence of breast cancer in females.

Estrogenic pollutants are everywhere. They contaminate our food, drinking water, and air. Proved feminizing synthetic compounds include (1) certain weed killers and insecticides, (2) some detergent breakdown products, (3) petroleum by-products found in car exhaust, (4) a common food preservative used to retard rancidity, and (5) softeners that make plastics flexible. These plastic softeners are commonly found in food packaging and can readily leach into food with which they come in contact, especially during heating. They have also

been found to leach from some babies' plastic teething toys into the saliva. They are in numerous medical products, such as the bags in which blood is stored. Plastic softeners are among the most plentiful industrial contaminants in our environment.

Investigators are only beginning to identify and understand the implications for reproductive health of myriad synthetic chemicals that have become such an integral part of modern societies. An estimated 87,000 synthetic chemicals are already in our environment. Scientists suspect that the estrogen-mimicking chemicals among these may underlie a spectrum of reproductive disorders that have been rising in the past 60 years—the same time period during which large amounts of these pollutants have been introduced into our environment. Here are examples of male reproductive dysfunctions that may be circumstantially linked to exposure to environmental estrogen disrupters:

- *Falling sperm counts.* The average sperm count fell from 113 million sperm/ml of semen in 1940 to 66 million/ml in 1990. Making matters worse, the volume of a single ejaculate has declined from 3.40 to 2.75 ml. This means that men, on average, are now ejaculating less than half the number of sperm as men did 50 years ago—a drop from more than 380 million sperm to about 180 million sperm per ejaculate. Furthermore, the number of motile sperm

has also dipped. Importantly, the sperm count has not declined in the less polluted areas of the world during the same time period.

- *Increased incidence of testicular and prostate cancer.* Cases of testicular cancer have tripled since 1940, and the rate continues to climb. Prostate cancer has also been on the rise over the same time period.
- *Rising number of male reproductive tract abnormalities at birth.* The incidence of cryptorchidism (undescended testis) nearly doubled from the 1950s to the 1970s. The number of cases of *hypospadia,* a malformation of the penis, more than doubled between the mid-1960s and the mid-1990s. Hypospadia results when the urethral fold fails to fuse closed during the development of a male fetus.
- *Evidence of gender bending in animals.* Some fish and wild animal populations severely exposed to environmental estrogens—such as those living in or near water heavily polluted with hormone-mimicking chemical wastes—display a high rate of grossly impaired reproductive systems. Examples include male fish that are hermaphrodites (having both male and female reproductive parts) and male alligators with abnormally small penises. Similar reproductive abnormalities have been identified in land mammals. Presumably, excessive estrogen exposure is emasculating these populations.

The female sexual cycle is very similar to the male cycle.

Both sexes experience the same four phases of the sexual cycle—excitement, plateau, orgasm, and resolution. Furthermore, the physiologic mechanisms responsible for orgasm are fundamentally the same in males and females.

The excitement phase in females can be initiated by either physical or psychological stimuli. Tactile stimulation of the externally exposed portion of the glans clitoris and surrounding perineal area is an especially powerful sexual stimulus. These stimuli trigger spinal reflexes that bring about parasympathetically induced vasodilation of arterioles throughout the vagina and exter-

nal genitalia, especially the clitoris. The resultant inflow of blood becomes evident as swelling of the labia and erection of the clitoris. The latter—like its male homolog, the penis—is composed largely of erectile tissue. Contrary to a common misconception, the clitoris is much larger than its externally visible portion. Most of the clitoris is located internally and consists of large, highly vascular internal bulbs that surround the urethra and vagina (see ● Figure 20-12b). These bulbs engorge with blood during erection. The functional significance of this erection is uncertain. Scientists speculate that it may (1) squeeze the urethra closed to prevent contamination of the urinary tract during sexual activity, (2) support the vaginal wall during penile penetration, and/or (3) increase pleasure signaling.

- *Decline in male births.* Many countries are reporting a slight decline in the ratio of baby boys to baby girls being born. In the United States, 17 fewer males were born per 10,000 births in 2007 compared to 1970, and Japan has seen an overall drop of 37 males per 10,000 births during the same time period. Although several plausible explanations have been put forth, many researchers attribute this troubling trend to disruption of normal male fetal development by environmental estrogens. In one compelling piece of circumstantial evidence, people inadvertently exposed to the highest level of an endocrine-disrupting agent during an industrial accident have subsequently had all daughters and no sons, whereas those least exposed have had the normal ratio of girls and boys. Similarly, a 2004 study in the Russian Arctic found a remarkable ratio of 2.5:1 female to male births among women who had blood concentrations of 4μg/L or greater of a known estrogen-mimicking pollutant.

Environmental estrogens are also implicated in the rising incidence of breast cancer in females. Breast cancer is 25% to 30% more prevalent now than in the 1940s. Many of the established risk factors for breast cancer, such as starting to menstruate earlier than usual and undergoing menopause later than usual, are associated with an elevation in the total lifetime exposure to estrogen. Because increased exposure to natural estrogen bumps up the risk for breast cancer, prolonged exposure to environmental estrogens may be contributing to the rising prevalence of this malignancy among women (and men).

In addition to the estrogen disrupters, scientists have recently identified a new class of chemical offenders—androgen disrupters that either mimic or suppress the action of male hormones. For example, studies suggest that bacteria in wastewater from pulp mills can convert the sterols in pine pulp into androgens. By contrast, anti-androgen compounds have been found in the fungicides commonly sprayed on vegetable and fruit crops. Yet another cause for concern are the androgens used by the livestock industry to enhance the production of muscle (that is, meat) in feedlot cattle. (Androgens have a protein anabolic effect.) These drugs do not end up in the meat, but they can get into drinking water and other food as hormone-laden feces contaminate rivers and streams.

In response to the growing evidence that has emerged circumstantially linking a wide variety of environmental pollutants to disturbing reproductive abnormalities, the U.S. Congress legally mandated the Environmental Protection Agency (EPA) in 1996 to determine which synthetic chemicals might be endocrine disrupters. In response, the EPA formed an advisory committee, which in 1998 proposed an ambitious plan to begin comprehensive testing of manufactured compounds for their potential to disrupt hormones in humans and wildlife. Although eventually all the 87,000 existing synthetic compounds will be tested, the initial screening was narrowed to evaluating the endocrine-disrupting potential of 15,000 widely used chemicals. Declaring this a national health priority, the government has allocated millions of dollars for this research. Yet in this time-consuming process, only a few thousand chemicals were tested in the first 10 years of investigation as EPA's *Toxic Release Inventory* slowly grows and other chemicals have been deemed safe. Complicating the situation even further, scientists recently learned that individual chemicals present at harmless levels in the body may have synergistic effects and exert detrimental effects when they interact. Therefore, it might not be sufficient to evaluate the risk of synthetic chemicals one by one. Scientists and regulatory agencies might have to take into account the cumulative risk of mixtures of chemicals. Increasingly, environmental watchdogs are calling for increased measures to limit exposure to synthetic chemicals (such as switching from plastic to glass baby bottles) and better labeling from manufacturers so that consumers can make more informed decisions about products they use.

Vasocongestion of the vaginal capillaries forces fluid out of the vessels into the vaginal lumen. This fluid, which is the first positive indication of sexual arousal, serves as the primary lubricant for intercourse. Additional lubrication is provided by the mucus secretions from the male and by mucus released during sexual arousal from glands located at the outer opening of the vagina. Also during the excitement phase in the female, the nipples become erect and the breasts enlarge as a result of vasocongestion. In addition, the majority of women show a *sex flush* during this time, which is caused by increased blood flow through the skin.

During the plateau phase, the changes initiated during the excitement phase intensify, while systemic responses similar to those in the male (such as increased heart rate, blood pressure, respiratory rate, and muscle tension) occur. Further vasocongestion of the lower third of the vagina during this time reduces its inner capacity so that it tightens around the thrusting penis, heightening tactile sensation for both the female and the male. Simultaneously, the uterus raises upward, lifting the cervix and enlarging the upper two thirds of the vagina. This ballooning, or **tenting effect,** creates a space for ejaculate deposition.

If erotic stimulation continues, the sexual response culminates in orgasm as sympathetic impulses trigger rhythmic contractions of the pelvic musculature at 0.8-second intervals, the same rate as in males. The contractions occur most intensely in the engorged lower third of the vaginal canal. Systemic responses identical to those of the male orgasm also occur. In fact, the orgasmic experience in females parallels that of males with two exceptions. First, there is no female counterpart to ejaculation. Second, females do not become refractory following an orgasm, so they can respond immediately to continued erotic stimulation and achieve multiple orgasms. If stimulation continues, the sexual intensity only diminishes to the plateau level following orgasm and can quickly be brought to a peak again. Women have been known to achieve as many as 12 successive orgasms in this manner.

During resolution, pelvic vasocongestion and the systemic manifestations gradually subside. As with males, this is a time of great physical relaxation for females.

We will now examine how females fulfill their part of the reproductive process.

Female Reproductive Physiology

Female reproductive physiology is much more complex than male reproductive physiology.

Complex cycling characterizes female reproductive physiology.

Unlike the continuous sperm production and essentially constant testosterone secretion characteristic of the male, release of ova is intermittent, and secretion of female sex hormones displays wide cyclic swings. The tissues influenced by these sex hormones also undergo cyclic changes, the most obvious of which is the monthly menstrual cycle (menstruus means "monthly"). During each cycle, the female reproductive tract is prepared for the fertilization and implantation of an ovum released from the ovary at ovulation. If fertilization does not occur, the cycle repeats. If fertilization does occur, the cycles are interrupted while the female system adapts to nurture and protect the newly conceived human being until it has developed into an individual capable of living outside the maternal environment. Furthermore, the female continues her reproductive functions after birth by producing milk (lactation) for the baby's nourishment. Thus, the female reproductive system is characterized by complex cycles that are interrupted by even more complex changes should pregnancy occur.

The ovaries, as the primary female reproductive organs, perform the dual function of producing ova (oogenesis) and secreting the female sex hormones, estrogen and progesterone. These hormones act together to promote fertilization of the ovum and to prepare the female reproductive system for pregnancy. Estrogen in the female governs many functions similar to those carried out by testosterone in the male, such as maturation and maintenance of the entire female reproductive system and establishment of female secondary sexual characteristics. In general, the actions of estrogen are important to preconception events. Estrogen is essential for ova maturation and release, development of physical characteristics that are sexually attractive to males, and transport of sperm from the vagina to the site of fertilization in the oviduct. Furthermore, estrogen contributes to breast development in anticipation of lactation. The other ovarian steroid, progesterone, is important

in preparing a suitable environment for nourishing a developing embryo/fetus and for contributing to the breasts' ability to produce milk.

Being steroids, estrogen and progesterone exert their multiple effects by binding with their respective receptors in the cytoplasm of their target cells, with the hormone-receptor complex moving to the nucleus where it binds with a hormone-specific DNA hormone-response element. This binding leads to gene transcription and synthesis of designated proteins that exert the hormone's dictated response in the target cells. Estrogen has two different cytoplasmic receptors, which have a differential distribution in various tissues and permit selective actions in specific target tissues. **Selective estrogen receptor modulators (SERMs),** like raloxifene, are drugs that selectively bind with a specific estrogen receptor. Raloxifene is approved to treat osteoporosis because it selectively binds with estrogen receptors in bone, where it mimics estrogen's beneficial effect on maintaining bone density, while not exerting any effect on reproductive organs, where extra estrogen-like influence could increase the risk of cancer (see p. 731). Estrogen also binds with surface membrane receptors, where it acts via cAMP to elicit rapid, nongenomic effects (see p. 126).

As in males, reproductive capability begins at puberty in females, but unlike males, who have reproductive potential throughout life, female reproductive potential ceases during middle age at menopause.

The steps of gametogenesis are the same in both sexes, but the timing and outcome differ sharply.

Oogenesis contrasts sharply with spermatogenesis in several important aspects, even though the identical steps of chromosome replication and division take place during gamete production in both sexes. The undifferentiated primordial germ cells in the fetal ovaries, the **oogonia** (comparable to the spermatogonia), divide mitotically to give rise to 6 million to 7 million oogonia by the fifth month of gestation, when mitotic proliferation ceases.

FORMATION OF PRIMARY OOCYTES AND PRIMARY FOLLICLES During the last part of fetal life, the oogonia begin the early steps of the first meiotic division but do not complete it. Known now as **primary oocytes,** they contain the diploid number of 46 replicated chromosomes, which are gathered into homologous pairs but do not separate. The primary oocytes remain in this state of **meiotic arrest** for years until they are prepared for ovulation.

Before birth, each primary oocyte is surrounded by a single layer of connective tissue-derived **granulosa cells.** Together, an oocyte and surrounding granulosa cells make up a **primary follicle.** Oocytes that are not incorporated into follicles self-destruct by apoptosis (cell suicide). At birth only about 2 million primary follicles remain, each containing a single primary oocyte capable of producing a single ovum. No new oocytes or follicles appear after birth, with the follicles already present in the ovaries at birth serving as a reservoir from which all ova throughout the reproductive life of a female must arise. The follicular pool gradually dwindles away as a result of processes that "use up" the oocyte-containing follicles.

The pool of primary follicles gives rise to an ongoing trickle of developing follicles, stimulated by poorly understood paracrine factors produced by both oocytes and granulosa cells. Once it starts to develop, a follicle is destined for one of two fates: It will reach maturity and ovulate, or it will degenerate to form scar tissue, a process known as **atresia**. Until puberty, all the follicles that start to develop undergo atresia in the early stages without ever ovulating. Even for the first few years after puberty, many of the cycles are **anovulatory** (that is, no ovum is released). Of the original total pool of follicles at birth, about 300,000 remain at puberty, and only about 400 will mature and release ova; 99.97% never ovulate but instead undergo atresia at some stage in development. By menopause, which occurs on average in a woman's early 50s, few primary follicles remain, having either already ovulated or become atretic. From this point on, the woman's reproductive capacity ceases.

This limited gamete potential in females is in sharp contrast to the continual process of spermatogenesis in males, who have the potential to produce several hundred million sperm in a single day. Furthermore, considerable chromosome wastage occurs in oogenesis compared with spermatogenesis. Let us see how.

FORMATION OF SECONDARY OOCYTES AND SECONDARY FOLLICLES The primary oocyte within a primary follicle is still a diploid cell that contains 46 doubled chromosomes. From puberty until menopause, a portion of the resting pool of follicles develops into more advanced follicles on a cyclic basis. The mechanisms determining which follicles in the pool will develop during a given cycle are unknown. Development of a follicle is characterized by growth of the primary oocyte and by expansion and differentiation of the surrounding cell layers. The oocyte enlarges about a thousandfold. This oocyte enlargement is caused by a buildup of cytoplasmic materials that will be needed by the early embryo.

Just before ovulation, the primary oocyte, whose nucleus has been in meiotic arrest for years, completes its first meiotic division. This division yields two daughter cells, each receiving a haploid set of 23 doubled chromosomes, analogous to the formation of secondary spermatocytes (● Figure 20-14). However, almost all the cytoplasm remains with one of the daughter cells, now called the **secondary oocyte,** which is destined to become the ovum. The chromosomes of the other daughter cell together with a small share of cytoplasm form the **first polar body.** In this way, the ovum-to-be loses half of its chromosomes to form a haploid gamete but retains all of its nutrient-rich cytoplasm. The nutrient-poor polar body soon degenerates.

FORMATION OF A MATURE OVUM Actually, the secondary oocyte, and not the mature ovum, is ovulated and fertilized, but common usage refers to the developing female gamete as an *ovum* even in its primary and secondary oocyte stages. Sperm entry into the secondary oocyte is needed to trigger the second meiotic division. Secondary oocytes that are not fertilized never complete this final division. During this division, a half set of chromosomes along with a thin layer of cytoplasm is extruded

as the **second polar body.** The other half set of 23 unpaired chromosomes remains behind in what is now the **mature ovum** (sometimes called the *ootid,* comparable to the spermatid, until the polar bodies disintegrate and the mature ovum alone remains). These 23 maternal chromosomes unite with the 23 paternal chromosomes of the penetrating sperm to complete fertilization. If the first polar body has not already degenerated, it too undergoes the second meiotic division at the same time the fertilized secondary oocyte is dividing its chromosomes.

COMPARISON OF STEPS IN OOGENESIS AND SPERMATOGENESIS
The steps involved in chromosome distribution during oogenesis parallel those of spermatogenesis, except that the cytoplasmic distribution and time span for completion sharply differ (● Figure 20-15). Just as four haploid spermatids are produced by each primary spermatocyte, four haploid daughter cells are produced by each primary oocyte (if the first polar body does not degenerate before it completes the second meiotic division). In spermatogenesis, each daughter cell develops into a highly specialized, motile spermatozoon unencumbered by unessential cytoplasm and organelles, its only destiny being to supply half of the genes for a new individual. In oogenesis, however, of the four daughter cells only the one destined to become the mature ovum receives cytoplasm. This uneven distribution of cytoplasm is important because the ovum, in addition to providing half the genes, provides all of the cytoplasmic components needed to support early development of the fertilized ovum. The large, relatively undifferentiated ovum contains numerous nutrients, organelles, and structural and enzymatic proteins. The three other cytoplasm-scarce daughter cells, the polar bodies, rapidly degenerate, their chromosomes being deliberately wasted.

Note also the considerable difference in time to complete spermatogenesis and oogenesis. It takes about two months for a spermatogonium to develop into fully remodeled spermatozoa. In contrast, development of an oogonium (present before birth) to a mature ovum requires anywhere from 11 years (beginning of ovulation at onset of puberty) to 50 years (end of ovulation at onset of menopause). The actual length of the active steps in meiosis is the same in both males and females, but in females the developing eggs remain in meiotic arrest for a variable number of years.

 The older age of ova released by women in their late 30s and 40s is believed to account for the higher incidence of genetic abnormalities, such as Down syndrome, in children born to women in this age range.

The ovarian cycle consists of alternating follicular and luteal phases.

After the onset of puberty, the ovary constantly alternates between two phases: the **follicular phase,** which is dominated by the presence of *maturing follicles;* and the **luteal phase,** which is characterized by the presence of the *corpus luteum* (to be described shortly). Normally this cycle is interrupted only if pregnancy occurs and is finally terminated by menopause. The average ovarian cycle lasts 28 days, but this varies among women and among cycles in any particular woman. The folli-

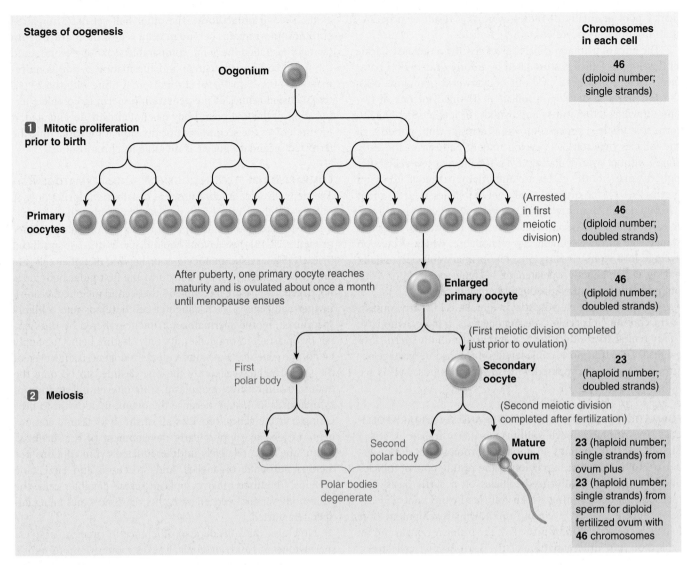

Stages of oogenesis

Chromosomes in each cell

Oogonium

46 (diploid number; single strands)

1 Mitotic proliferation prior to birth

Primary oocytes

(Arrested in first meiotic division)

46 (diploid number; doubled strands)

After puberty, one primary oocyte reaches maturity and is ovulated about once a month until menopause ensues

Enlarged primary oocyte

46 (diploid number; doubled strands)

(First meiotic division completed just prior to ovulation)

First polar body

Secondary oocyte

23 (haploid number; doubled strands)

2 Meiosis

(Second meiotic division completed after fertilization)

Second polar body

Mature ovum

23 (haploid number; single strands) from ovum plus **23** (haploid number; single strands) from sperm for diploid fertilized ovum with **46** chromosomes

Polar bodies degenerate

● **FIGURE 20-14 Oogenesis.** Compare with ● Figure 20-8, p. 753, spermatogenesis.

cle operates in the first half of the cycle to produce a mature egg ready for ovulation at midcycle. The corpus luteum takes over during the last half of the cycle to prepare the female reproductive tract for pregnancy in case fertilization of the released egg occurs.

The follicular phase is characterized by the development of maturing follicles.

At any given time throughout the cycle, a portion of the primary follicles is starting to develop (● Figure 20-16b, step **1**). However, only those that do so during the follicular phase, when the hormonal environment is right to promote their maturation, continue beyond the early stages of development. The others, lacking hormonal support, undergo atresia. During follicular development, as the primary oocyte is synthesizing and storing materials for future use if fertilized, important changes take place in the cells surrounding the reactivated oocyte in preparation for the egg's release from the ovary.

PROLIFERATION OF GRANULOSA CELLS AND FORMATION OF THE ZONA PELLUCIDA First, the single layer of granulosa cells in a primary follicle proliferates to form several layers that surround the oocyte. These granulosa cells secrete a thick, gel-like "rind" that covers the oocyte and separates it from the surrounding granulosa cells. This intervening membrane is known as the **zona pellucida** (step **2**).

Gap junctions penetrate the zona pellucida and extend between the oocyte and surrounding granulosa cells in a developing follicle. Ions and small molecules can travel through these connecting tunnels. Glucose, amino acids, and other important molecules are delivered to the oocyte from the granulosa cells through these tunnels, enabling the egg to stockpile these critical nutrients. Also, signaling molecules pass through the gap junctions in both directions, helping coordinate the changes that take place in the oocyte and surrounding cells as both mature and prepare for ovulation. The nurturing relationship between granulosa cells and a developing egg is similar in many ways to the relationship between Sertoli cells and developing sperm.

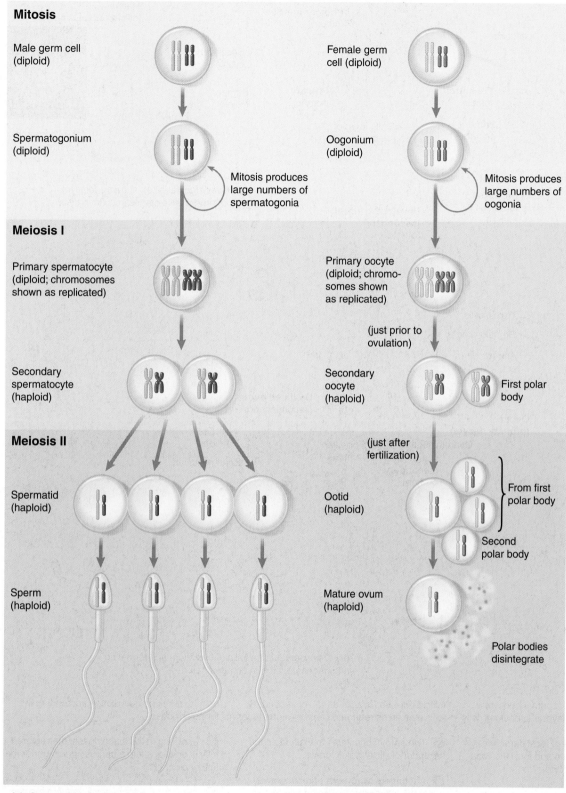

Mitosis

Male germ cell (diploid)

Spermatogonium (diploid)

Mitosis produces large numbers of spermatogonia

Female germ cell (diploid)

Oogonium (diploid)

Mitosis produces large numbers of oogonia

Meiosis I

Primary spermatocyte (diploid; chromosomes shown as replicated)

Secondary spermatocyte (haploid)

Primary oocyte (diploid; chromosomes shown as replicated)

(just prior to ovulation)

Secondary oocyte (haploid)

First polar body

Meiosis II

Spermatid (haploid)

Sperm (haploid)

(just after fertilization)

Ootid (haploid)

From first polar body

Second polar body

Mature ovum (haploid)

Polar bodies disintegrate

(a) **Spermatogenesis**

(b) **Oogenesis**

● **FIGURE 20-15 Comparison of mitotic and meiotic divisions producing spermatozoa and eggs from germ cells.**

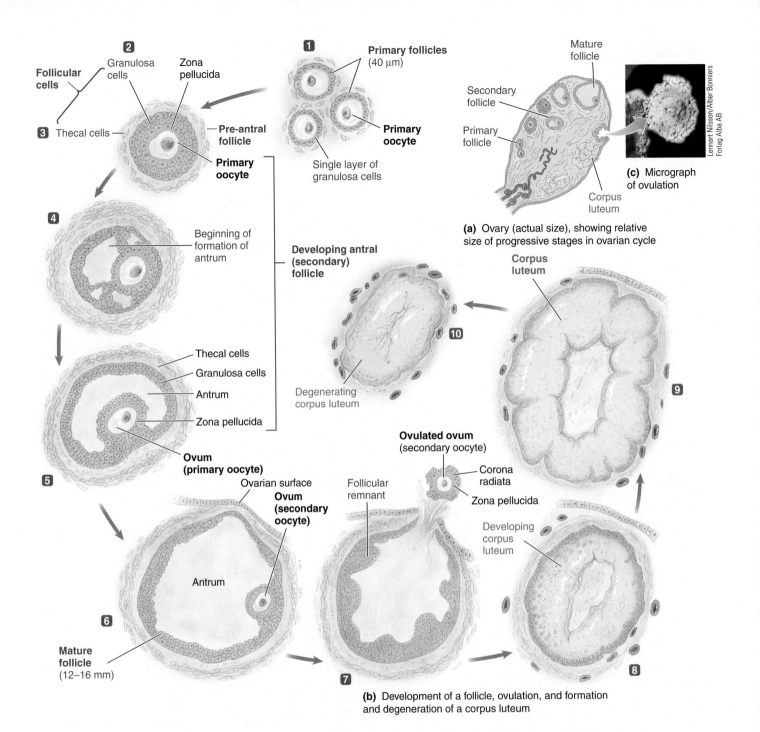

Follicular cells { **2** Granulosa cells, Zona pellucida, **3** Thecal cells }

1 **Primary follicles** (40 µm)

Single layer of granulosa cells

Primary oocyte

Pre-antral follicle

Primary oocyte

Mature follicle

Secondary follicle

Primary follicle

Corpus luteum

(c) Micrograph of ovulation

Lennart Nilsson/Alber Bonniers
Förlag Alba AB

(a) Ovary (actual size), showing relative size of progressive stages in ovarian cycle

4 Beginning of formation of antrum

Developing antral (secondary) follicle

Thecal cells
Granulosa cells
Antrum
Zona pellucida

Ovum (primary oocyte)

5

Ovarian surface

Ovum (secondary oocyte)

Antrum

6

Mature follicle (12–16 mm)

Follicular remnant

7

Ovulated ovum (secondary oocyte)

Corona radiata
Zona pellucida

Developing corpus luteum

8

Degenerating corpus luteum

10

Corpus luteum

9

(b) Development of a follicle, ovulation, and formation and degeneration of a corpus luteum

1 In a primary follicle, a primary oocyte is surrounded by a single layer of granulosa cells.

2 Under the influence of local paracrines, granulosa cells proliferate and form the zona pellucida around the oocyte.

3 Surrounding ovarian connective tissue differentiates into thecal cells, converting a primary follicle into a pre-antral secondary follicle.

4 Follicles reaching the pre-antral stage are recruited for further development under the influence of FSH at the beginning of the follicular phase of the ovarian cycle. A recruited follicle develops into an antral, or secondary, follicle as an estrogen-rich antrum starts to form.

5 The antrum continues to expand as the secondary follicle rapidly grows.

6 After about two weeks of rapid growth under the influence of FSH, the follicle has developed into a mature follicle, which has a greatly expanded antrum; the oocyte, which by now has developed into a secondary oocyte, is displaced to one side.

7 At midcycle, in response to a burst in LH secretion, the mature follicle, bulging on the ovarian surface, ruptures and releases the oocyte, resulting in ovulation and ending the follicular phase.

8 Ushering in the luteal phase, the ruptured follicle develops into a corpus luteum under the influence of LH.

9 The corpus luteum continues to grow and secrete progesterone and estrogen that prepare the uterus for implantation of a fertilized ovum.

10 After 14 days, if a fertilized ovum does not implant in the uterus, the corpus luteum degenerates, the luteal phase ends, and a new follicular phase begins under the influence of a changing hormonal milieu.

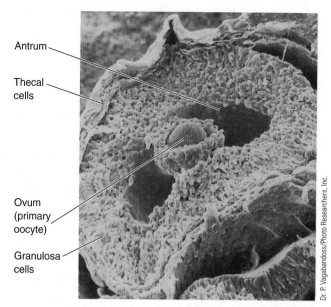

Antrum

Thecal cells

Ovum (primary oocyte)

Granulosa cells

Dr. P. Vagabandoss/Photo Researchers, Inc.

● **FIGURE 20-17 Scanning electron micrograph of a developing secondary follicle.**

● **FIGURE 20-16 Ovarian cycle.** (a) Ovary showing progressive stages in one ovarian cycle. All of these stages occur sequentially at one site, but the stages are represented in a loop in the periphery of the ovary so that all of the stages can be seen in progression simultaneously. (b) Enlarged view of the stages in one ovarian cycle. (c) Micrograph of a secondary oocyte being released (ovulation).

PROLIFERATION OF THECAL CELLS; ESTROGEN SECRETION At the same time as the oocyte is enlarging and the granulosa cells are proliferating, specialized ovarian connective tissue cells in contact with the expanding granulosa cells proliferate and differentiate to form an outer layer of **thecal cells** in response to paracrines secreted by the granulosa cells (step ❸). The thecal and granulosa cells, collectively known as **follicular cells,** function as a unit to secrete estrogen. Of the three physiologically important estrogens in order of potency—*estradiol, estrone,* and *estriol*—**estradiol** is the principal ovarian estrogen.

FORMATION OF THE ANTRUM The early stages of follicular development that occur without gonadotropin influence take about two months and are not part of the follicular phase of the ovarian cycle. Only preantral follicles that have developed sufficiently to respond to FSH stimulation are "recruited" at the beginning of the follicular phase when FSH levels rise. Typically during each cycle about 15 to 20 follicles are recruited. The hormonal environment of the follicular phase promotes rapid enlargement and development of the recruited follicular cells' secretory capacity, converting the primary follicle into an **antral,** or **secondary, follicle** capable of estrogen secretion (step ❹). During this stage of follicular development, a fluid-filled cavity, or **antrum,** forms in the middle of the granulosa cells (● Figure 20-17). The follicular fluid originates partially from transudation (passage through capillary pores) of plasma and partially from follicular cell secretions. As the follicular cells start producing estrogen, some of this hormone is secreted into the blood for distribution throughout the body. However, a portion of the estrogen collects in the hormone-rich antral fluid.

The oocyte has reached full size by the time the antrum begins to form. The shift to an antral follicle initiates a period of rapid follicular growth (● Figure 20-16, step ❺). During this time, the follicle increases in size from a diameter of less than 1 mm to 12 to 16 mm shortly before ovulation. Part of the follicular growth is the result of continued proliferation of the granulosa and thecal cells, but most results from a dramatic expansion of the antrum. As the follicle grows, estrogen is produced in increasing quantities.

FORMATION OF A MATURE FOLLICLE One of the follicles, the "dominant" follicle, usually grows more rapidly than the others, developing into a **mature (preovulatory, tertiary,** or **Graafian) follicle** within about 14 days after the onset of follicular development (step ❻). The dominant follicle that develops into a mature follicle generally has the most FSH receptors and therefore is most responsive to hormonal stimulation. The antrum occupies most of the space in a mature follicle. The oocyte, surrounded by the zona pellucida and a single layer of granulosa cells, is displaced asymmetrically at one side of the growing follicle, in a little mound that protrudes into the antrum.

OVULATION The greatly expanded mature follicle bulges on the ovarian surface, creating a thin area that ruptures to release the oocyte at **ovulation** (step ❼). Rupture of the follicle is facilitated by the release from the follicular cells of enzymes (triggered by a burst in LH secretion, which will be described later) that digest the connective tissue in the follicular wall. The bulging wall is thus weakened so that it balloons out even farther, to the point that it can no longer contain the rapidly expanding follicular contents.

Just before ovulation, the oocyte completes its first meiotic division. The ovum (secondary oocyte), still surrounded by its tightly adhering zona pellucida and granulosa cells (now called the **corona radiata,** meaning "radiating crown"), is swept out of the ruptured follicle into the abdominal cavity by the leaking antral fluid. The released ovum is quickly drawn into the oviduct, where fertilization may or may not take place.

The other developing follicles that failed to reach maturation and ovulate undergo degeneration, never to be reactivated. Occasionally, two (or perhaps more) follicles reach maturation and ovulate at about the same time. If both are fertilized, **fraternal twins** result. Because fraternal twins arise from separate ova fertilized by separate sperm, they share no more in common than any other two siblings except for the same birth date. **Identical twins,** in contrast, develop from a single fertilized ovum that completely divides into two separate, genetically identical embryos at a very early stage in development.

Rupture of the follicle at ovulation signals the end of the follicular phase and ushers in the luteal phase.

The luteal phase is characterized by the presence of a corpus luteum.

The ruptured follicle left behind in the ovary after release of the ovum changes rapidly. The granulosa and thecal cells remaining in the remnant follicle first collapse into the emptied antral space that has been partially filled by clotted blood.

FORMATION OF THE CORPUS LUTEUM; ESTROGEN AND PROGESTERONE SECRETION These old follicular cells soon undergo a dramatic structural transformation to form the **corpus luteum,** in a process called **luteinization** (step **8**). The follicular-turned-luteal cells enlarge and are converted into very active steroid hormone–producing tissue. Abundant storage of cholesterol, the steroid precursor molecule, in lipid droplets within the corpus luteum gives this tissue a yellowish appearance, hence its name (*corpus* means "body"; *luteum* means "yellow").

The corpus luteum becomes highly vascularized as blood vessels from the thecal region invade the luteinizing granulosa. These changes are appropriate for the corpus luteum's function: to secrete into the blood abundant quantities of progesterone along with smaller amounts of estrogen. Estrogen secretion in the follicular phase followed by progesterone secretion in the luteal phase is essential for preparing the uterus for implantation of a fertilized ovum. The corpus luteum becomes fully functional within four days after ovulation, but it continues to increase in size for another four or five days (step **9**).

DEGENERATION OF THE CORPUS LUTEUM If the released ovum is not fertilized and does not implant, the corpus luteum degenerates within about 14 days after its formation (step **10**). The luteal cells degenerate and are phagocytized, the vascular supply is withdrawn, and connective tissue rapidly fills in to form a fibrous tissue mass known as the **corpus albicans** ("white body"). The luteal phase is now over, and one ovarian cycle is complete. A new wave of follicular development, which begins when degeneration of the old corpus luteum is completed, signals the onset of a new follicular phase.

CORPUS LUTEUM OF PREGNANCY If fertilization and implantation do occur, the corpus luteum continues to grow and produce increasing quantities of progesterone and estrogen instead of degenerating. Now called the *corpus luteum of pregnancy,* this ovarian structure persists until pregnancy ends. It provides the hormones essential for maintaining pregnancy until the developing placenta can take over this crucial function. You will learn more about the role of these structures later.

The ovarian cycle is regulated by complex hormonal interactions.

The ovary has two related endocrine units: the estrogen-secreting follicle during the first half of the cycle and the corpus luteum, which secretes both progesterone and estrogen, during the last half of the cycle. These units are sequentially triggered by complex cyclic hormonal relationships among the hypothalamus, anterior pituitary, and these two ovarian endocrine units.

As in the male, gonadal function in the female is directly controlled by the anterior pituitary gonadotropic hormones, namely, follicle-stimulating hormone (FSH) and luteinizing hormone (LH). These hormones, in turn, are regulated by hypothalamic gonadotropin-releasing hormone (GnRH) and feedback actions of gonadal hormones. Unlike in the male, however, control of the female gonads is complicated by the cyclic nature of ovarian function. For example, the effects of FSH and LH on the ovaries depend on the stage of the ovarian

cycle. Furthermore, estrogen exerts negative-feedback effects during part of the cycle and positive-feedback effects during another part of the cycle, depending on the concentration of estrogen. Also in contrast to the male, FSH is not strictly responsible for gametogenesis, nor is LH solely responsible for gonadal hormone secretion. We will consider control of follicular function, ovulation, and the corpus luteum separately, using ● Figure 20-18 as a means of integrating the various concurrent and sequential activities that take place throughout the cycle. To facilitate correlation between this rather intimidating figure and the accompanying text description of this complex cycle, the numbers in the figure and its legend correspond to the numbers in the text description.

CONTROL OF FOLLICULAR FUNCTION We begin with the follicular phase of the ovarian cycle (● Figure 20-18, step **1**). The early stages of preantral follicular growth and oocyte maturation do not require gonadotropic stimulation. Hormonal support is required, however, for antrum formation, further follicular development (step **2**), and estrogen secretion (step **3**). Estrogen, FSH (step **4**), and LH (step **5**) are all needed. Antrum formation is induced by FSH. Both FSH and estrogen stimulate proliferation of the granulosa cells. Both LH and FSH are required for synthesis and secretion of estrogen by the follicle, but these hormones act on different cells and at different steps in the estrogen production pathway (● Figure 20-19). Both granulosa and thecal cells participate in estrogen production. The conversion of cholesterol into estrogen requires a number of sequential steps, the last of which is conversion of androgens into estrogens (see ● Figure 19-8, p. 699). Thecal cells readily produce androgens but have limited capacity to convert them into estrogens. Granulosa cells, in contrast, contain the enzyme *aromatase,* so they can readily convert androgens into estrogens, but they cannot produce androgens in the first place. LH acts on the thecal cells to stimulate androgen production, whereas FSH acts on the granulosa cells to promote conversion of thecal androgens (which diffuse into the granulosa cells from the thecal cells) into estrogens. Because low basal levels of FSH (● Figure 20-18, step **6**) are sufficient to promote this final conversion to estrogen, the rate of estrogen secretion by the follicle primarily depends on the circulating level of LH, which continues to rise during the follicular phase (step **7**). Furthermore, as the follicle continues to grow, more estrogen is produced simply because more estrogen-producing follicular cells are present.

Part of the estrogen produced by the growing follicle is secreted into the blood and is responsible for the steadily increasing plasma estrogen levels during the follicular phase (step **8**). The remainder of the estrogen remains within the follicle, contributing to the antral fluid and stimulating further granulosa cell proliferation (see ● Figure 20-19).

The secreted estrogen, in addition to acting on sex-specific tissues such as the uterus, inhibits the hypothalamus and anterior pituitary in typical negative-feedback fashion (● Figure 20-20). The rising, moderate levels of estrogen characterizing the follicular phase act directly on the hypothalamus to inhibit GnRH secretion, thus suppressing GnRH-prompted release of FSH and LH from the anterior pituitary. However, estrogen's

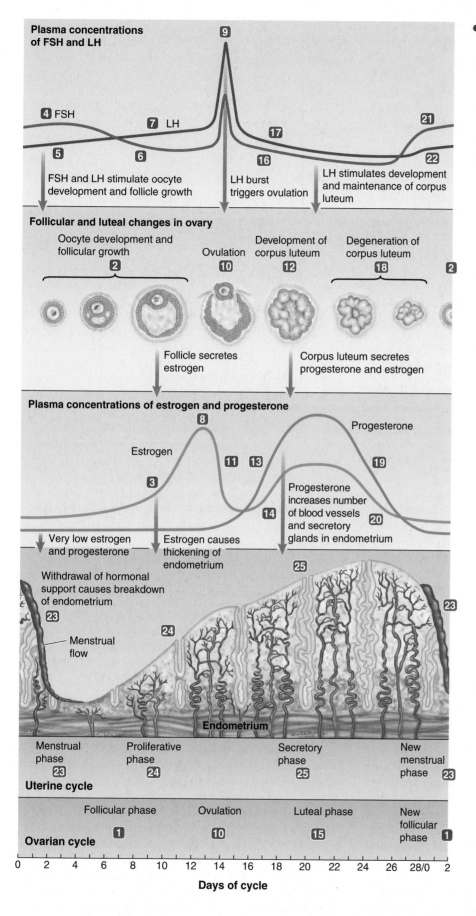

Plasma concentrations of FSH and LH

4 FSH

7 LH

9

5   6   17   21   22   16

FSH and LH stimulate oocyte development and follicle growth

LH burst triggers ovulation

LH stimulates development and maintenance of corpus luteum

Follicular and luteal changes in ovary

Oocyte development and follicular growth
2

Ovulation
10

Development of corpus luteum
12

Degeneration of corpus luteum
18

2

Follicle secretes estrogen

Corpus luteum secretes progesterone and estrogen

Plasma concentrations of estrogen and progesterone

8

Estrogen

Progesterone

11 13   19

3

14   20

Progesterone increases number of blood vessels and secretory glands in endometrium

Very low estrogen and progesterone

Estrogen causes thickening of endometrium

25

Withdrawal of hormonal support causes breakdown of endometrium
23

23

Menstrual flow

24

Endometrium

| Menstrual phase 23 | Proliferative phase 24 | Secretory phase 25 | New menstrual phase 23 |

Uterine cycle

| Follicular phase 1 | Ovulation 10 | Luteal phase 15 | New follicular phase 1 |

Ovarian cycle

0 2 4 6 8 10 12 14 16 18 20 22 24 26 28/0 2

Days of cycle

• **FIGURE 20-18 Correlation between hormonal levels and cyclic ovarian and uterine changes.** During the first half of the ovarian cycle, the follicular phase (step 1), the ovarian follicle (step 2) secretes estrogen (step 3) under the influence of FSH (step 4), LH (step 5), and estrogen itself. The rising, moderate levels of estrogen inhibit FSH secretion, which declines (step 6) during the last part of the follicular phase, and incompletely suppress tonic LH secretion, which continues to rise (step 7) throughout the follicular phase. When the follicular output of estrogen reaches its peak (step 8), the high levels of estrogen trigger a surge in LH secretion at midcycle (step 9). This LH surge brings about ovulation of the mature follicle (step 10). Estrogen secretion plummets (step 11) when the follicle meets its demise at ovulation.

The old follicular cells are transformed into the corpus luteum (step 12), which secretes progesterone (step 13) as well as estrogen (step 14) during the last half of the ovarian cycle, the luteal phase (step 15). Progesterone strongly inhibits both FSH (step 16) and LH (step 17), which continue to decrease throughout the luteal phase. The corpus luteum degenerates (step 18) in about two weeks if the released ovum has not been fertilized and implanted in the uterus. Progesterone (step 19) and estrogen (step 20) levels sharply decrease when the corpus luteum degenerates, removing the inhibitory influences on FSH and LH. As these anterior pituitary hormone levels start to rise again (steps 21 and 22) on the withdrawal of inhibition, they stimulate the development of a new batch of follicles as a new follicular phase is ushered in (steps 1 and 2).

Concurrent uterine phases reflect the influences of the ovarian hormones on the uterus. Early in the follicular phase, the highly vascularized, nutrient-rich endometrial lining is sloughed off (the uterine menstrual phase) (step 23). This sloughing results from the withdrawal of estrogen and progesterone (steps 19 and 20) when the old corpus luteum degenerated at the end of the preceding luteal phase (step 18). Late in the follicular phase, the rising levels of estrogen (step 3) cause the endometrium to thicken (the uterine proliferative phase) (step 24). After ovulation (step 10), progesterone from the corpus luteum (step 13) brings about vascular and secretory changes in the estrogen-primed endometrium to produce a suitable environment for implantation (the uterine secretory, or progestational, phase) (step 25). When the corpus luteum degenerates (step 18), a new ovarian follicular phase (steps 1 and 2) and uterine menstrual phase (step 23) begin.

1. LH stimulates the thecal cells in the ovarian follicle.

2. On stimulation, the thecal cells convert cholesterol into androgen.

3. Androgen diffuses from the thecal cells into the adjacent granulosa cells.

4. FSH stimulates the granulosa cells in the ovarian follicle.

5. On stimulation, the granulosa cells convert androgen into estrogen.

6a. Part of the estrogen is secreted into the blood, where it exerts systemic effects.

6b. Part of the estrogen remains within the follicle and contributes to antral formation.

7. Local estrogen, along with FSH, stimulates proliferation of the granulosa cells.

● **FIGURE 20-19 Production of estrogen by an ovarian follicle.**

primary effect is directly on the pituitary itself. Estrogen selectively inhibits FSH secretion by the gonadotropes.

This differential secretion of FSH and LH by the gonadotropes induced by estrogen is in part responsible for the declining plasma FSH level, unlike the rising plasma LH concentration, during the follicular phase as the estrogen level rises (● Figure 20-18, step **6**). Another contributing factor to the fall in FSH during the follicular phase is secretion of *inhibin* by the follicular cells. Inhibin preferentially inhibits FSH secretion by acting at the anterior pituitary, just as it does in the male (● Figure 20-20). The decline in FSH secretion brings about atresia of all but the single most mature of the developing follicles.

In contrast to FSH, LH secretion continues to rise slowly during the follicular phase (● Figure 20-18, step **7**) despite inhibition of GnRH (and thus, indirectly, LH) secretion. This seeming paradox is a result of the fact that estrogen alone cannot completely suppress **tonic** (low-level, ongoing) **LH secretion**; to completely inhibit tonic LH secretion, both estrogen and progesterone are required. Because progesterone does not appear until the luteal phase of the cycle, the basal level of circulating LH slowly increases during the follicular phase under incomplete inhibition by estrogen alone.

CONTROL OF OVULATION Ovulation and subsequent luteinization of the ruptured follicle are triggered by an abrupt, massive increase in LH secretion (step **9**). This **LH surge** brings about four major changes in the follicle:

1. It halts estrogen synthesis by the follicular cells (step **11**).

2. It reinitiates meiosis in the oocyte of the developing follicle by blocking release of an *oocyte maturation–inhibiting substance* produced by the granulosa cells. This substance is responsible for arresting meiosis in the primary oocytes once they are wrapped within granulosa cells in the fetal ovary.

3. It triggers production of local prostaglandins, which induce ovulation by promoting vascular changes that cause rapid follicular swelling while inducing enzymatic digestion of the follicular wall. Together these actions lead to rupture of the weakened wall that covers the bulging follicle (step **10**).

4. It causes differentiation of follicular cells into luteal cells. Because the LH surge triggers both ovulation and luteinization, formation of the corpus luteum automatically follows ovulation (step **12**). Thus, the midcycle burst in LH secretion is a dramatic point in the cycle; it terminates the follicular phase and initiates the luteal phase (step **15**).

The two different modes of LH secretion—the tonic secretion of LH (step **7**) responsible for promoting ovarian hormone secretion and the LH surge (step **9**) that causes ovulation—not only occur at different times and produce different effects but also are controlled by different mechanisms. Tonic LH secretion is partially suppressed (step **7**) by the inhibitory action of the rising, moderate levels of estrogen (step **3**) during the follicular phase and is completely suppressed (step **17**) by the increasing levels of progesterone during the luteal phase (step **13**). Because

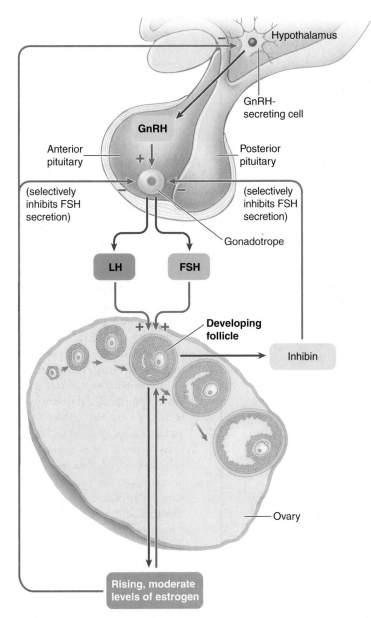

● FIGURE 20-20 Feedback control of FSH and tonic LH secretion during the follicular phase.

tonic LH secretion stimulates both estrogen and progesterone secretion, this is a typical negative-feedback effect.

In contrast, the LH surge is triggered by a *positive-feedback effect*. Whereas the rising, moderate levels of estrogen early in the follicular phase *inhibit* LH secretion, the high level of estrogen that occurs during peak estrogen secretion late in the follicular phase (step **8**) *stimulates* LH secretion and initiates the LH surge (● Figure 20-21). Thus, LH enhances estrogen production by the follicle, and the resultant peak estrogen concentration stimulates LH secretion. The high plasma concentration of estrogen acts directly on the hypothalamus to increase GnRH, thereby increasing both LH and FSH secretion. It also acts directly on the anterior pituitary to specifically increase LH secretion by the gonadotropes. The latter effect accounts in large part for the much greater surge in LH secretion compared to FSH secretion at midcycle (● Figure 20-18, step **9**). Also,

continued inhibin secretion by the follicular cells preferentially inhibits FSH secretion, keeping the FSH levels from rising as high as the LH levels. There is no known role for the modest midcycle surge in FSH that accompanies the pronounced and pivotal LH surge. Because only a mature, preovulatory follicle, not follicles in earlier stages of development, can secrete high-enough levels of estrogen to trigger the LH surge, ovulation is not induced until a follicle has reached the proper size and degree of maturation. In a way, then, the follicle lets the hypothalamus know when it is ready to be stimulated to ovulate. The LH surge lasts for about a day at midcycle, just before ovulation.

CONTROL OF THE CORPUS LUTEUM LH "maintains" the corpus luteum; that is, after triggering development of the corpus luteum, LH stimulates ongoing steroid hormone secretion by this ovarian structure. Under the influence of LH, the corpus luteum secretes both progesterone (step **13**) and estrogen (step **14**), with progesterone being its most abundant hormonal product. The plasma progesterone level increases for the first time during the luteal phase. No progesterone is secreted during the follicular phase. Therefore, the follicular phase is dominated by estrogen and the luteal phase by progesterone.

A transitory drop in the level of circulating estrogen occurs at midcycle (step **11**) as the estrogen-secreting follicle meets its demise at ovulation. The estrogen level climbs again during the luteal phase because of the corpus luteum's activity, although it does not reach the same peak as during the follicular phase. What keeps the modestly high estrogen level during the luteal phase from triggering another LH surge? Progesterone. Even though a high level of estrogen stimulates LH secretion, progesterone, which dominates the luteal phase, powerfully inhibits LH secretion (step **17**) as well as FSH secretion (step **16**) by acting at both the hypothalamus and anterior pituitary (● Figure 20-22). Inhibition of FSH and LH by progesterone prevents new follicular maturation and ovulation during the luteal phase. Under progesterone's influence, the reproductive system is gearing up to support the just-released ovum, should it be fertilized, instead of preparing other ova for release. No inhibin is secreted by the luteal cells.

The corpus luteum functions for an average of two weeks, then degenerates if fertilization does not occur (● Figure 20-18, step **18**). The mechanisms that govern degeneration of the corpus luteum are not fully understood. The declining level of circulating LH (step **17**), driven down by inhibitory actions of progesterone, undoubtedly contributes to the corpus luteum's downfall. Prostaglandins and estrogen released by the luteal cells themselves may play a role. Demise of the corpus luteum terminates the luteal phase and sets the stage for a new follicular phase. As the corpus luteum degenerates, plasma progesterone (step **19**) and estrogen (step **20**) levels fall rapidly because these hormones are no longer being produced. Withdrawal of the inhibitory effects of these hormones on the hypothalamus allows FSH (step **21**) and tonic LH (step **22**) secretion to mod-

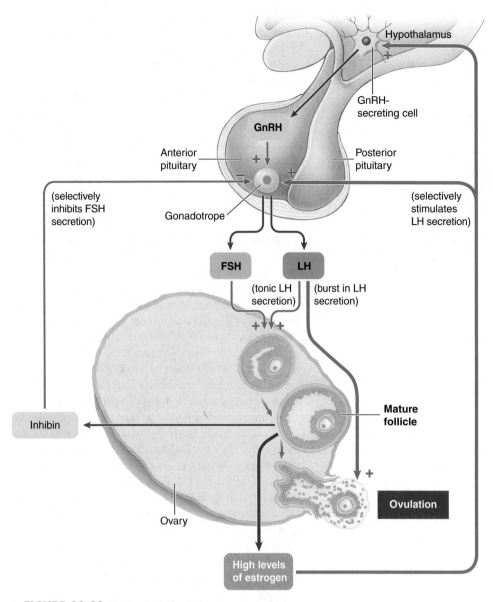

Hypothalamus

GnRH-
secreting cell

GnRH

Anterior
pituitary

Posterior
pituitary

(selectively
inhibits FSH
secretion)

Gonadotrope

(selectively
stimulates
LH secretion)

FSH   **LH**

(tonic LH
secretion)

(burst in LH
secretion)

Inhibin

**Mature
follicle**

Ovary

Ovulation

**High levels
of estrogen**

● **FIGURE 20-21 Control of the LH surge at ovulation.**

estly increase once again. Under the influence of these gonadotropic hormones, another batch of primary follicles (step **2**) is induced to mature as a new follicular phase begins (step **1**).

Cyclic uterine changes are caused by hormonal changes during the ovarian cycle.

The fluctuations in circulating levels of estrogen and progesterone during the ovarian cycle induce profound changes in the uterus, giving rise to the **menstrual,** or **uterine, cycle.** Because it reflects hormonal changes during the ovarian cycle, the menstrual cycle averages 28 days, as does the ovarian cycle, although even normal adults vary considerably from this mean. The outward manifestation of the cyclic changes in the uterus is the menstrual bleeding once during each menstrual cycle (that is, once a month). Less obvious changes take place throughout the cycle, however, as the uterus is prepared for implantation should a released ovum be fertilized, then is stripped clean of its

prepared lining (menstruation) if implantation does not occur, only to repair itself and start preparing for the ovum that will be released during the next cycle.

We will briefly examine the influences of estrogen and progesterone on the uterus and then consider the effects of cyclic fluctuations of these hormones on uterine structure and function.

INFLUENCES OF ESTROGEN AND PROGESTERONE ON THE UTERUS
The uterus consists of two main layers: the **myometrium,** the outer smooth muscle layer; and the **endometrium,** the inner lining that contains numerous blood vessels and glands. Estrogen stimulates growth of both the myometrium and the endometrium. It also induces the synthesis of progesterone receptors in the endometrium. Thus, progesterone can exert an effect on the endometrium only after it has been "primed" by estrogen. Progesterone acts on the estrogen-primed endometrium to convert it into a hospitable and nutritious lining suitable for implantation of a fertilized ovum. Under the influence of progesterone, the endometrial connective tissue becomes loose and edematous as a result of an accumulation of electrolytes and water, which facilitates implantation of the fertilized ovum. Progesterone further

prepares the endometrium to sustain an early-developing embryo by inducing the endometrial glands to secrete and store large quantities of glycogen and by causing tremendous growth of the endometrial blood vessels. Progesterone also reduces the contractility of the uterus to provide a quiet environment for implantation and embryonic growth.

The menstrual cycle consists of three phases: the *menstrual phase,* the *proliferative phase,* and the *secretory,* or *progestational, phase.*

MENSTRUAL PHASE The **menstrual phase** is the most overt phase, characterized by discharge of blood and endometrial debris from the vagina (● Figure 20-18, step **23**). By convention, the first day of menstruation is considered the start of a new cycle. It coincides with the end of the ovarian luteal phase and onset of a new follicular phase. As the corpus luteum degenerates because fertilization and implantation of the ovum released during the preceding cycle did not take place (step

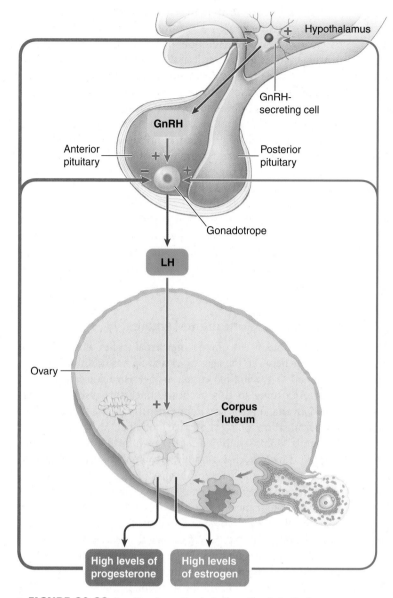

● **FIGURE 20-22 Feedback control during the luteal phase.**

trial debris from the uterine cavity out through the vagina as **menstrual flow.** Excessive uterine contractions caused by prostaglandin overproduction produce the menstrual cramps (**dysmenorrhea**) some women experience.

The average blood loss during a single menstrual period is 50 to 150 ml. Blood that seeps slowly through the degenerating endometrium clots within the uterine cavity, then is acted on by fibrinolysin, a fibrin dissolver that breaks down the fibrin forming the meshwork of the clot. Therefore, blood in the menstrual flow usually does not clot, because it has already clotted and the clot has been dissolved before it passes out of the vagina. When blood flows rapidly through the leaking vessels, however, it may not be exposed to sufficient fibrinolysin, so when the menstrual flow is most profuse, blood clots may appear. In addition to the blood and endometrial debris, large numbers of leukocytes are found in the menstrual flow. These white blood cells play an important defense role in helping the raw endometrium resist infection.

Menstruation typically lasts for about five to seven days after degeneration of the corpus luteum, coinciding in time with the early portion of the ovarian follicular phase (see steps 23 and 1). Withdrawal of progesterone and estrogen (steps 19 and 20) on degeneration of the corpus luteum leads simultaneously to sloughing of the endometrium (menstruation) (step 23) and development of new follicles in the ovary (steps 1 and 2) under the influence of rising gonadotropic hormone levels (steps 21 and 22). The drop in gonadal hormone secretion removes inhibitory influences from the hypothalamus and anterior pituitary, so FSH and LH secretion increases and a new follicular phase begins. After five to seven days under the influence of FSH and LH, the newly growing follicles are secreting enough estrogen (step 3) to induce repair and growth of the endometrium.

18), circulating levels of progesterone and estrogen drop precipitously (steps 19 and 20). Because the net effect of progesterone and estrogen is to prepare the endometrium for implantation of a fertilized ovum, withdrawal of these steroids deprives the highly vascular, nutrient-rich uterine lining of its hormonal support.

The fall in ovarian hormone levels also stimulates release of a uterine prostaglandin that causes vasoconstriction of the endometrial vessels, disrupting the blood supply to the endometrium. The subsequent reduction in O_2 delivery causes death of the endometrium, including its blood vessels. The resulting bleeding through the disintegrating vessels flushes the dying endometrial tissue into the uterine lumen. Most of the uterine lining sloughs during each menstrual period except for a deep, thin layer of epithelial cells and glands, from which the endometrium will regenerate. The same local uterine prostaglandin also stimulates mild rhythmic contractions of the uterine myometrium. These contractions help expel the blood and endome-

PROLIFERATIVE PHASE Thus, menstrual flow ceases, and the **proliferative phase** of the uterine cycle begins concurrent with the last portion of the ovarian follicular phase as the endometrium starts to repair itself and proliferate (step 24) under the influence of estrogen from the newly growing follicles. When the menstrual flow ceases, a thin endometrial layer less than 1 mm thick remains. Estrogen stimulates proliferation of epithelial cells, glands, and blood vessels in the endometrium, increasing this lining to a thickness of 3 to 5 mm. The estrogen-dominant proliferative phase lasts from the end of menstruation to ovulation. Peak estrogen levels (step 8) trigger the LH surge (step 9) responsible for ovulation (step 10).

SECRETORY, OR PROGESTATIONAL, PHASE After ovulation, when a new corpus luteum is formed (step 12), the uterus enters the **secretory, or progestational, phase,** which coincides in time with the ovarian luteal phase (steps 25 and 15). The corpus luteum secretes large amounts of progesterone (step 13) and estrogen (step **14**). Progesterone converts the thickened, estrogen-primed endometrium to a richly vascularized, glycogen-filled tissue. This period is called either the *secretory phase,* because the endometrial glands are actively secreting glycogen into the uterine lumen for early nourishment of a developing embryo before it implants, or the *progestational* ("before pregnancy") *phase,* referring to the development of a lush endometrial lining capable of supporting an early embryo after implantation. If fertilization and implantation do not occur, the corpus luteum degenerates and a new follicular phase and menstrual phase begin once again.

Various factors can interfere with the delicate balance of the hypothalamic-pituitary-ovarian-peripheral target organ axis, leading to menstrual irregularities and fertility problems. Among these factors are starvation (a problem for example with anorexia nervosa; see p. 650), stress, and heavy exercise. (For the effects of exercise on this cycle, see the accompanying boxed feature, ▦ A Closer Look at Exercise Physiology).

Fluctuating estrogen and progesterone levels produce cyclical changes in cervical mucus.

Hormonally induced changes also take place in the cervix during the ovarian cycle. Under the influence of estrogen during the follicular phase, the mucus secreted by the cervix becomes abundant, clear, and thin. This change, which is most pronounced when estrogen is at its peak and ovulation is approaching, facilitates passage of sperm through the cervical canal. After ovulation, under the influence of progesterone from the corpus luteum, the mucus becomes thick and sticky, essentially plugging up the cervical opening. This plug is an important defense mechanism, preventing bacteria (that might threaten a possible pregnancy) from entering the uterus from the vagina. Sperm also cannot penetrate this thick mucus barrier.

Pubertal changes in females are similar to those in males.

Regular menstrual cycles are absent in both young and aging females, but for different reasons. The female reproductive system does not become active until puberty. Unlike the fetal testes, the fetal ovaries need not be functional, because in the absence of fetal testosterone secretion in a female, the reproductive system is automatically feminized, without requiring the presence of female sex hormones. The female reproductive system remains quiescent from birth until puberty, which occurs at about 12 years of age when hypothalamic GnRH activity increases for the first time. As in the male, the mechanisms that govern the onset of puberty are not clearly understood but are believed to involve the pineal gland and melatonin secretion. Furthermore, recent scientific studies suggest that leptin, a hormone secreted by adipose tissue (see p. 646), may play a permissive role in hypothalamic maturation at puberty. This action might account at least in part for the fact that, on average, lean girls tend to enter puberty later than their heavier counterparts do. In evolutionary history, this mechanism could have been a way to ensure that females had sufficient energy stored to sustain a pregnancy when food supplies were unpredictable.

GnRH begins stimulating release of anterior pituitary gonadotropic hormones, which in turn stimulate ovarian activity. The resulting secretion of estrogen by the activated ovaries induces growth and maturation of the female reproductive tract as well as development of the female secondary sexual characteristics. Estrogen's prominent action in the latter regard is to promote fat deposition in strategic locations, such as the breasts, buttocks, and thighs, giving rise to the typical curvaceous female figure. Enlargement of the breasts at puberty is the result primarily of fat deposition in the breast tissue, not to functional development of the mammary glands. The pubertal rise in estrogen also closes the epiphyseal plates, halting further growth in height, similar to the effect of testosterone-turned-estrogen in males. Three other pubertal changes in females—growth of axillary and pubic hair, the pubertal growth spurt, and development of libido—are attributable to a spurt in adrenal androgen secretion at puberty, not to estrogen.

Menopause is unique to females.

The cessation of a woman's menstrual cycles at **menopause** sometime between the ages of 45 and 55 has traditionally been attributed to the limited supply of ovarian follicles present at birth. According to this proposal, once this reservoir is depleted, ovarian cycles, and hence menstrual cycles, cease. Thus, the termination of reproductive potential in a middle-aged woman is "preprogrammed" at her own birth. Recent evidence suggests, however, that a midlife hypothalamic change instead of aging ovaries may trigger the onset of menopause. Evolutionarily, menopause may have developed as a mechanism that prevented pregnancy in women beyond the time that they could likely rear a child before their own death.

Males do not experience complete gonadal failure as females do, for two reasons. First, a male's germ cell supply is unlimited because mitotic activity of the spermatogonia continues. Second, gonadal hormone secretion in males is not inextricably dependent on gametogenesis, as in females. If female sex hormones were produced by separate tissues unrelated to those governing gametogenesis, as are male sex hormones, estrogen and progesterone secretion would not automatically stop when oogenesis stopped.

Menopause is preceded by a period of progressive ovarian failure characterized by increasingly irregular cycles and dwindling estrogen levels. This entire period of transition from sexual maturity to cessation of reproductive capability is known as the **climacteric,** or **perimenopause.** Ovarian estrogen production declines from as much as 300 mg per day to essentially nothing. Postmenopausal women are not completely devoid of estrogen, however, because adipose tissue, the liver, and the adrenal cortex continue to produce up to 20 mg of estrogen per day. In addition to the ending of ovarian and menstrual cycles, the loss of ovarian estrogen following menopause brings about many physical and emotional changes. These changes include vaginal dryness, which can cause discomfort during sex, and gradual atrophy of the genital organs. However, postmenopausal women still have a sex drive because of their adrenal androgens.

Menstrual Irregularities: When Cyclists and Other Female Athletes Do Not Cycle

Since the 1970s, as women in growing numbers have participated in a variety of sports requiring vigorous training regimens, researchers have become increasingly aware that many women experience changes in their menstrual cycles as a result of athletic participation. These changes are referred to as *athletic menstrual cycle irregularity* (AMI). The menstrual cycle dysfunction can vary in severity from amenorrhea (cessation of menstrual periods) to oligomenorrhea (cycles at irregular or infrequent intervals) to cycles that are normal in length but are anovulatory (no ovulation) or that have a short or inadequate luteal phase.

In early research studies using surveys and questionnaires to determine the prevalence of the problem, the frequency of these sport-related disorders varied from 2% to 51%. In contrast, the rate of occurrence of menstrual cycle dysfunction in females of reproductive age in the general population is 2% to 5%. A major problem of using surveys to determine the frequency of menstrual cycle irregularity is the questionable accuracy of recall of menstrual periods. Furthermore, without blood tests to determine hormone levels throughout the cycle, a woman would not know whether she was anovulatory or had had a shortened luteal phase. Studies in which hormone levels have been determined through the menstrual cycle have demonstrated that seemingly normal cycles in athletes frequently have a short luteal phase (less than two days long with low progesterone levels).

In a study conducted to determine whether strenuous exercise spanning two menstrual cycles would induce menstrual disorders, 28 initially untrained college women with documented ovulation and luteal adequacy served as subjects. The women participated in an eight-week exercise program in which they initially ran 4 miles per day and progressed to 10 miles per day by the fifth week. They were expected to participate daily in 3.5 hours of moderate-intensity sports. Only four women had normal menstrual cycles during the training. Abnormalities resulting from training included abnormal bleeding, delayed menstrual periods, abnormal luteal function, and loss of LH surge. All the women returned to normal cycles within six months after training. The results of this study suggest that the frequency of AMI with strenuous exercise may be much greater than indicated by questionnaire alone. In other studies using low-intensity exercise regimens, AMI was much less frequent.

The mechanisms of AMI are unknown at present, although studies have implicated rapid weight loss, decreased percentage of body fat, dietary insufficiencies, prior menstrual dysfunction, stress, age at onset of training, and the intensity of training as factors that play a role. Epidemiologists have shown that if girls participate in vigorous sports before **menarche** (the first menstrual period), menarche is delayed. On average, athletes have their first period when they are about three years older than their nonathletic counterparts. Furthermore, females who participate in sports before menarche seem to have a higher frequency of AMI throughout their athletic careers than those who begin to train after menarche. Hormonal changes found in female athletes include (1) severely depressed FSH levels, (2) elevated LH levels, (3) low progesterone during the luteal phase, (4) low estrogen levels in the follicular phase, and (5) an FSH–LH environment totally unbalanced as compared to those of age-matched nonathletic women. The preponderance of evidence indicates that cy-cles return to normal once vigorous training is stopped.

The major problem associated with athletic amenorrhea is a reduction in bone mineral density. Studies have shown that the mineral density in the vertebrae of the lower spine of those with athletic amenorrhea is lower than in athletes with normal menstrual cycles and lower than in age-matched nonathletes. However, amenorrheic runners have higher bone mineral density than amenorrheic nonathletes, presumably because the mechanical stimulus of exercise helps retard bone loss. Studies have shown that amenorrheic athletes are at higher risk for stress-related fractures than athletes with normal menstrual cycles. One study, for example, found stress fractures in 6 of 11 amenorrheic runners but in only 1 of 6 runners with normal menstrual cycles. The mechanism for bone loss is probably the same as is found in postmenopausal osteoporosis—lack of estrogen (see p. 730). The problem is serious enough that an amenorrheic athlete should discuss the possibility of estrogen replacement therapy with her physician.

There may be some positive benefits of athletes' menstrual dysfunction. A recent epidemiological study to determine if the long-term reproductive and general health of women who had been college athletes differed from that of college nonathletes showed that former athletes had less than half the lifetime occurrence rate of cancers of the reproductive system and half the breast cancer occurrence compared to nonathletes. Because these are hormone-sensitive cancers, the delayed menarche and lower estrogen levels found in women athletes may play a key role in decreasing the risk of cancer of the reproductive system and breast.

Clinical Note Because estrogen has widespread physiologic actions beyond the reproductive system, the dramatic loss of ovarian estrogen in menopause affects other body systems, most notably the skeleton and the cardiovascular system. Estrogen helps build strong bones, shielding premenopausal women from the bone-thinning condition of osteoporosis (see p. 730). The postmenopausal reduction in estrogen increases activity of the bone-dissolving osteoclasts and diminishes activity of the bone-building osteoblasts. The result is decreased bone density and a greater incidence of bone fractures.

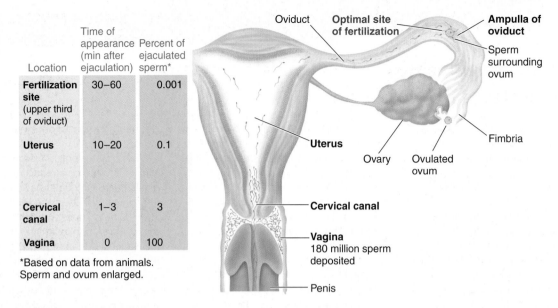

Location	Time of appearance (min after ejaculation)	Percent of ejaculated sperm*
Fertilization site (upper third of oviduct)	30–60	0.001
Uterus	10–20	0.1
Cervical canal	1–3	3
Vagina	0	100

*Based on data from animals.
Sperm and ovum enlarged.

● **FIGURE 20-23 Ovum and sperm transport to the site of fertilization.**

Estrogen also helps modulate the actions of epinephrine and norepinephrine on the arteriolar walls by promoting local release of the vasodilator nitric oxide (see p. 356). The menopausal diminution of estrogen leads to unstable control of blood flow, especially in the skin vessels. Transient increases in the flow of warm blood through these superficial vessels are responsible for the "**hot flashes**" that frequently accompany menopause. Vasomotor stability is gradually restored in postmenopausal women so that hot flashes eventually cease.

You have now learned about the events that take place if fertilization does not occur. Because the primary function of the reproductive system is, of course, reproduction, we will next turn attention to the sequence of events that take place when fertilization does occur.

The oviduct is the site of fertilization.

Fertilization, the union of male and female gametes, normally occurs in the **ampulla,** the upper third of the oviduct (● Figure 20-23). Thus, both the ovum and the sperm must be transported from their gonadal site of production to the ampulla.

OVUM TRANSPORT TO THE OVIDUCT Unlike the male reproductive tract, which has a continuous lumen from the site of sperm production in the seminiferous tubules to exit of the sperm from the urethra at ejaculation, the ovaries are not in direct contact with the reproductive tract. The ovum is released into the peritoneal cavity at ovulation. Normally, however, the egg is quickly picked up by the oviduct. The dilated end of the oviduct cups around the ovary and contains **fimbriae,** finger-like projections that contract in a sweeping motion to guide the released ovum into the oviduct (● Figures 20-2b, p. 744, and 20-23). Furthermore, the fimbriae are lined by cilia—fine, hair-like projections that beat in waves toward the interior of the

oviduct—further assuring the ovum's passage into the oviduct (see p. 45). Within the oviduct, the ovum is rapidly propelled by peristaltic contractions and ciliary action to the ampulla.

Conception can take place during a very limited time span in each cycle (the **fertile period**). If not fertilized, the ovum begins to disintegrate within 12 to 24 hours and is subsequently phagocytized by cells that line the reproductive tract. Fertilization must therefore occur within 24 hours after ovulation, when the ovum is still viable. Sperm typically survive about 48 hours but can survive up to five days in the female reproductive tract, so sperm deposited from five days before ovulation to 24 hours after ovulation may be able to fertilize the released ovum, although these times vary considerably.

 Occasionally, an ovum fails to be transported into the oviduct and remains instead in the peritoneal cavity. Rarely, such an ovum gets fertilized, resulting in an **ectopic abdominal pregnancy,** in which the fertilized egg implants in the rich vascular supply to the digestive organs rather than in its usual site in the uterus (*ectopic* means "out of place"). An abdominal pregnancy often leads to life-threatening hemorrhage because the digestive organ blood supply is not primed to respond appropriately to implantation as the endometrium is. If this unusual pregnancy proceeds to term, the baby must be delivered surgically because the normal vaginal exit is not available. The probability of maternal complications at birth is greatly increased because the digestive vasculature is not designed to "seal itself off" after birth as the endometrium does.

SPERM TRANSPORT TO THE OVIDUCT After sperm are deposited in the vagina at ejaculation, they must travel through the cervical canal, through the uterus, and then up to the egg in the upper third of the oviduct (● Figure 20-23). The first sperm arrive in the oviduct within half an hour after ejaculation. Even though sperm are mobile by means of whiplike contractions of their tails, 30 minutes is much too soon for a sperm's own mobility to transport it to the site of fertilization. To make this formidable journey, sperm need the help of the female reproductive tract.

The first hurdle is passage through the cervical canal. Throughout most of the cycle, because of high progesterone or low estrogen levels, the cervical mucus is too thick to permit sperm penetration. The cervical mucus becomes thin and watery enough to permit sperm to penetrate only when estrogen levels are high, as in the presence of a mature follicle about to ovulate. Sperm migrate up the cervical canal under their own power. The canal remains penetrable for only two or three days during each cycle, around the time of ovulation.

Once sperm have entered the uterus, contractions of the myometrium churn them around in "washing-machine" fashion. This action quickly disperses sperm throughout the uterine cavity. When sperm reach the oviduct, they are propelled to the fertilization site in the upper end of the oviduct by upward contractions of the oviduct smooth muscle. These myometrial and oviduct contractions that facilitate sperm transport are induced by the high estrogen level just before ovulation, aided by seminal prostaglandins.

New research indicates that when sperm reach the ampulla, ova are not passive partners in conception. Scientists have already identified a chemoattractant (see p. 423) released by mature eggs in frogs that attracts sperm and causes them to propel themselves toward the waiting female gamete. They have not yet identified a specific egg-released chemoattractant in humans, but they suspect that one might exist. They have found in human sperm a receptor comparable to the one that detects and responds to the ovum-released chemoattractant in frogs. Interestingly, this receptor, called **hOR17-4,** is an olfactory receptor (OR) similar to those found in the nose for smell perception (see p. 230). Therefore, sperm are believed to "smell" the egg. According to current thinking, activation of the hOR17-4 receptor on binding with a chemoattractant from the egg triggers a second-messenger pathway in sperm that brings about intracellular Ca^{2+} release. This Ca^{2+} turns on the microtubule sliding that brings about tail movement and sperm swimming in the direction of the chemical signal (see p. 46).

Even around ovulation time, when sperm can penetrate the cervical canal, of the nearly 200 million sperm typically deposited in a single ejaculate, only a few thousand make it to the oviduct (● Figure 20-23). That only a very small percentage of the deposited sperm ever reach their destination is one reason why sperm concentration must be so high (20 million/ml of semen) for a man to be fertile. The other reason is that the acrosomal enzymes of many sperm are needed to break down the barriers surrounding the ovum (● Figure 20-24).

FERTILIZATION The tail of the sperm is used to maneuver for final penetration of the ovum. To fertilize an ovum, a sperm must first pass through the corona radiata and zona pellucida surrounding it. The sperm penetrates the corona radiata by means of membrane-bound enzymes in the surface membrane that surrounds the head (● Figure 20-25a, step **1**). Sperm can penetrate the zona pellucida only after binding with specific binding sites on the surface of this layer. The binding partners between the sperm and ovum were recently identified. **Fertilin,**

● **FIGURE 20-24 Scanning electron micrograph of sperm amassed at the surface of an ovum.**

a protein found on the plasma membrane of the sperm, binds with glycoproteins known as **ZP3** in the outer layer of the zona pellucida. Only sperm of the same species can bind to these zona pellucida sites and pass through. Binding of sperm triggers the **acrosome reaction,** in which the acrosomal membrane disrupts and the acrosomal enzymes are released (step **2**). The acrosomal enzymes digest the zona pellucida, enabling the sperm, with its tail still beating, to tunnel a path through this protective barrier (step **3**). The first sperm to reach the ovum itself fuses with the plasma membrane of the ovum (actually a secondary oocyte), and its head enters the ovum's cytoplasm (step **4**). The sperm's tail is frequently lost in this process, but the head carries the crucial genetic information. Sperm–egg fusion triggers a chemical change in the ovum's surrounding membrane that makes this outer layer impenetrable to the entry of any more sperm. This phenomenon is known as **block to polyspermy** ("many sperm"). Specifically, fertilization-induced release of intracellular Ca^{2+} into the cytosol triggers the exocytosis of enzyme-filled **cortical granules** that are located in the outermost, or cortical, region of the egg into the space between the egg membrane and the zona pellucida (step **5**). These enzymes diffuse into the zona pellucida, where they inactivate the ZP3 receptors so that other sperm reaching the zona pellucida cannot bind with it. The enzymes also crosslink molecules in the zona pellucida, hardening it and sealing off tunnels in progress to keep other penetrating sperm from advancing. Furthermore, the released Ca^{2+} triggers the second meiotic division of the egg, which is now ready to unite with the sperm to complete the fertilization process.

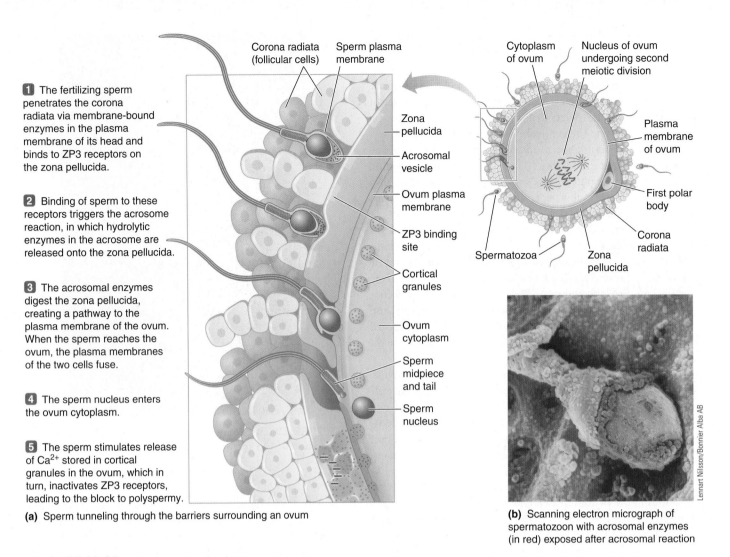

1 The fertilizing sperm penetrates the corona radiata via membrane-bound enzymes in the plasma membrane of its head and binds to ZP3 receptors on the zona pellucida.

2 Binding of sperm to these receptors triggers the acrosome reaction, in which hydrolytic enzymes in the acrosome are released onto the zona pellucida.

3 The acrosomal enzymes digest the zona pellucida, creating a pathway to the plasma membrane of the ovum. When the sperm reaches the ovum, the plasma membranes of the two cells fuse.

4 The sperm nucleus enters the ovum cytoplasm.

5 The sperm stimulates release of Ca²⁺ stored in cortical granules in the ovum, which in turn, inactivates ZP3 receptors, leading to the block to polyspermy.

Corona radiata (follicular cells)
Sperm plasma membrane
Zona pellucida
Acrosomal vesicle
Ovum plasma membrane
ZP3 binding site
Cortical granules
Ovum cytoplasm
Sperm midpiece and tail
Sperm nucleus

Cytoplasm of ovum
Nucleus of ovum undergoing second meiotic division
Plasma membrane of ovum
First polar body
Corona radiata
Spermatozoa
Zona pellucida

(a) Sperm tunneling through the barriers surrounding an ovum

Lennart Nilsson/Bonnier Alba AB

(b) Scanning electron micrograph of spermatozoon with acrosomal enzymes (in red) exposed after acrosomal reaction

● **FIGURE 20-25 Process of fertilization.**

Within an hour, the sperm and egg nuclei fuse, thanks to a centrosome (microtubule organizing center; see p. 41) provided by the sperm that forms microtubules to bring the male and female chromosome sets together for uniting. In addition to contributing its half of the chromosomes to the fertilized ovum, now called a **zygote**, the victorious sperm also activates ovum enzymes essential for the early embryonic developmental program. Thus fertilization accomplishes the dual events of combining genes from the two parents to form a genetically unique organism and setting in motion the development of that organism.

The blastocyst implants in the endometrium through the action of its trophoblastic enzymes.

During the first three to four days following fertilization, the zygote remains within the ampulla because a constriction between the ampulla and the remainder of the oviduct canal prevents further movement of the zygote toward the uterus.

THE BEGINNING STEPS IN THE AMPULLA The zygote is not idle during this time, however. It rapidly undergoes a number of mitotic cell divisions to form a solid ball of cells called the

morula (● Figure 20-26). Meanwhile, the rising levels of progesterone from the newly developing corpus luteum that formed after ovulation stimulate release of glycogen from the endometrium into the reproductive tract lumen for use as energy by the early embryo. The nutrients stored in the cytoplasm of the ovum can sustain the product of conception for less than a day. The concentration of secreted nutrients increases more rapidly in the small confines of the ampulla than in the uterine lumen.

DESCENT OF THE MORULA TO THE UTERUS About three to four days after ovulation, progesterone is being produced in sufficient quantities to relax the oviduct constriction, thus permitting the morula to be rapidly propelled into the uterus by oviductal peristaltic contractions and ciliary activity. The temporary delay before the developing embryo passes into the uterus lets enough nutrients accumulate in the uterine lumen to support the embryo until implantation can take place. If the morula arrives prematurely, it dies.

When the morula descends to the uterus, it floats freely within the uterine cavity for another three to four days, living on endometrial secretions and continuing to divide. During the first six to seven days after ovulation, while the developing em-

Blastocyst (cross section)

Inner cell mass

Destined to become fetus

Trophoblast

Accomplishes implantation and develops into fetal portions of placenta

Morula

Cleavage

Spermatozoa

Ovum (cross section)

Fertilization

Polar bodies

Secondary oocyte (ovum)

Ovulation

Ovary

Implantation

Endometrium of uterus

Structures not drawn to scale.

Actual size of blastocyst

● **FIGURE 20-26 Early stages of development from fertilization to implantation.** Note that the fertilized ovum progressively divides and differentiates into a blastocyst as it moves from the site of fertilization in the upper oviduct to the site of implantation in the uterus.

bryo is in transit in the oviduct and floating in the uterine lumen, the uterine lining is simultaneously being prepared for implantation under the influence of luteal-phase progesterone. During this time, the uterus is in its secretory, or progestational phase, storing up glycogen and becoming richly vascularized.

Clinical Note Occasionally, the morula fails to descend into the uterus and continues to develop and implant in the lining of the oviduct. This leads to an **ectopic tubal pregnancy,** which must be terminated. Ninety-five percent of ectopic pregnancies are tubal pregnancies. Such a pregnancy can never succeed, because the oviduct cannot expand as the uterus does to accommodate the growing embryo. The first warning of a tubal pregnancy is pain caused by the growing embryo stretching the oviduct. If not removed, the enlarging embryo will rupture the oviduct, causing possibly lethal hemorrhage.

IMPLANTATION OF THE BLASTOCYST IN THE PREPARED ENDO-METRIUM By the time the endometrium is suitable for implantation (about a week after ovulation), the morula has descended to the uterus and continued to proliferate and differentiate into a *blastocyst* capable of implantation. The week's delay after fertilization and before implantation allows time for both the endometrium and the developing embryo to prepare for implantation.

A **blastocyst** is a single-layer hollow ball of about 50 cells encircling a fluid-filled cavity, with a dense mass of cells grouped together at one side (● Figure 20-26). This dense mass, known as the **inner cell mass,** becomes the embryo/fetus itself. The rest of the blastocyst is never incorporated into the fetus, instead serving a supportive role during intrauterine life. The

thin outermost layer, the **trophoblast,** accomplishes implantation, after which it develops into the fetal portion of the placenta.

When the blastocyst is ready to implant, its surface becomes sticky. By this time the endometrium is ready to accept the early embryo and it too has become more adhesive through increased formation of cell adhesion molecules (CAMs) that help "Velcro" the blastocyst when it first contacts the uterine lining. The blastocyst adheres to the uterine lining on the side of its inner cell mass (● Figure 20-27, step **1**). **Implantation** begins when, on contact with the endometrium, the trophoblastic cells overlying the inner cell mass release protein-digesting enzymes. These enzymes digest pathways between the endometrial cells, permitting fingerlike cords of trophoblastic cells to penetrate into the depths of the endometrium, where they continue to digest uterine cells (step **2**). Through its cannibalistic actions, the trophoblast performs the dual functions of (1) accomplishing implantation as it carves out a hole in the endometrium for the blastocyst and (2) making metabolic fuel and raw materials available for the developing embryo as the advancing trophoblastic projections break down the nutrient-rich endometrial tissue. The plasma membranes of the advancing trophoblastic cells degenerate, forming a multinucleated syncytium that will eventually become the fetal portion of the placenta.

Stimulated by the invading trophoblast, the endometrial tissue at the contact site undergoes dramatic changes that enhance its ability to support the implanting embryo. In response to a chemical messenger released by the blastocyst, the underlying endometrial cells secrete prostaglandins, which locally increase vascularization, produce edema, and enhance nutrient

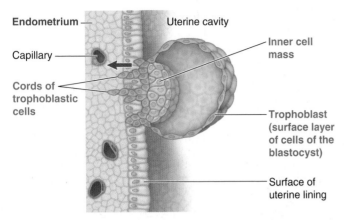

1 When the free-floating blastocyst adheres to the endometrial lining, cords of trophoblastic cells begin to penetrate the endometrium.

Endometrium —
Capillary —
Cords of trophoblastic cells —

Uterine cavity
Inner cell mass
Trophoblast (surface layer of cells of the blastocyst)
Surface of uterine lining

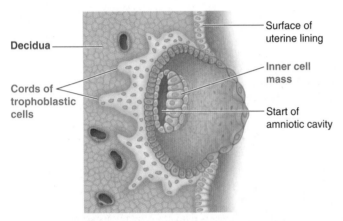

2 Advancing cords of trophoblastic cells tunnel deeper into the endometrium, carving out a hole for the blastocyst. The boundaries between the cells in the advancing trophoblastic tissue disintegrate.

Decidua —
Cords of trophoblastic cells —

Surface of uterine lining
Inner cell mass
Start of amniotic cavity

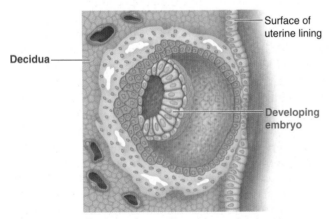

3 When implantation is finished, the blastocyst is completely buried in the endometrium.

Decidua —

Surface of uterine lining
Developing embryo

● **FIGURE 20-27 Implantation of the blastocyst.**

storage. The endometrial tissue so modified at the implantation site is called the **decidua.** It is into this superrich decidual tissue that the blastocyst becomes embedded. After the blastocyst burrows into the decidua by means of trophoblastic activity, a layer of endometrial cells covers over the surface of the hole, completely burying the blastocyst within the uterine lining (step **3**). The trophoblastic layer continues to digest the surrounding decidual cells, providing energy for the embryo until the placenta develops.

PREVENTING REJECTION OF THE EMBRYO/FETUS What prevents the mother from immunologically rejecting the embryo/fetus, which is actually a "foreigner" to the mother's immune system, being half derived from genetically different paternal chromosomes? Following are several proposals under investigation. New evidence indicates the trophoblasts produce **Fas ligand,** which binds with **Fas,** a specialized receptor on the surface of approaching activated maternal cytotoxic T cells. Cytotoxic T cells are the immune cells that carry out the job of destroying foreign cells (see p. 438). This binding triggers the immune cells that are targeted to destroy the developing foreigner to undergo apoptosis, sparing the embryo/fetus from immune rejection. Other researchers have found that the fetal portion of the placenta, which is derived from trophoblasts, produces an enzyme, **indoleamine 2,3-dioxygenase (IDO),** which destroys tryptophan. Tryptophan, an amino acid, is a critical factor in the activation of maternal cytotoxic T cells. Thus, the embryo/fetus, through its trophoblast connection, is believed to defend itself against rejection by shutting down the activity of the mother's cytotoxic T cells within the placenta that would otherwise attack the developing foreign tissues. Furthermore, recent studies demonstrate that production of **regulatory T cells** is doubled or tripled in pregnant experimental animals. Regulatory T cells suppress maternal cytotoxic T cells that might target the fetus (see p. 438).

CONTRACEPTION Couples wishing to engage in sexual intercourse but avoid pregnancy have available a number of methods of **contraception** ("against conception"). These methods act by blocking one of three major steps in the reproductive process: sperm transport to the ovum, ovulation, or implantation. (See the boxed feature on pp. 784–785, ■ Concepts, Challenges, and Controversies, for further details on the ways and means of contraception.)

Next let us examine the placenta in further detail.

The placenta is the organ of exchange between maternal and fetal blood.

The glycogen stores in the endometrium are sufficient to nourish the embryo only during its first few weeks. To sustain the growing embryo/fetus for the duration of its intrauterine life, the **placenta,** a specialized organ of exchange between the maternal and fetal blood, rapidly develops (● Figure 20-28). The placenta is derived from both trophoblastic and decidual tissue. It is an unusual organ because it is composed of tissues of two organisms: the embryo/fetus and the mother.

(a) Relationship between developing fetus and uterus as pregnancy progresses

4 weeks

8 weeks

12 weeks

Full term

(b) Representation of interlocking maternal and fetal structures that form the placenta

Umbilical cord
Amniotic fluid
Pool of maternal blood
Placental villus
Intervillus space
Uterine decidual tissue
Maternal arteriole
Maternal venule
Fetal vessels
Chorionic tissue
Placenta
Umbilical vein
Umbilical arteries
Chorion/amnion

● **FIGURE 20-28 Placentation.** Fingerlike projections of chorionic (fetal) tissue form the placental villi, which protrude into a pool of maternal blood. Decidual (maternal) capillary walls are broken down by the expanding chorion so that maternal blood oozes through the spaces between the placental villi. Fetal placental capillaries branch off the umbilical arteries and project into the placental villi. Fetal blood flowing through these vessels is separated from the maternal blood by only the capillary wall and thin chorionic layer that forms the placental villi. Maternal blood enters through the maternal arterioles, then percolates through the pool of blood in the intervillus spaces. Here, exchanges are made between the fetal and maternal blood before the fetal blood leaves through the umbilical vein and maternal blood exits through the maternal venules.

FORMATION OF THE PLACENTA AND AMNIOTIC SAC By day 12, the embryo is completely embedded in the decidua. By this time the trophoblastic layer is two cell layers thick and is called the **chorion.** As the chorion continues to release enzymes and expand, it forms an extensive network of cavities within the decidua. As the expanding chorion erodes decidual capillary walls, maternal blood leaks from the capillaries and fills these cavities. The blood is kept from clotting by an anticoagulant produced by the chorion. Fingerlike projections of chorionic tissue extend into the pools of maternal blood. Soon the developing embryo sends

The Ways and Means of Contraception

The term **contraception** refers to the process of avoiding pregnancy while engaging in sexual intercourse. A number of methods of contraception are available that range in ease of use and effectiveness (see the accompanying table). These methods can be grouped into three categories based on the means by which they prevent pregnancy: blockage of sperm transport to the ovum, prevention of ovulation, or blockage of implantation. After examining the most common ways in which contraception can be accomplished by each of these means, we will take a glimpse at future contraceptive possibilities on the horizon before concluding with a discussion of termination of unwanted pregnancies.

Blockage of Sperm Transport to the Ovum

■ *Natural contraception* or the *rhythm method* of birth control relies on abstinence from intercourse during the woman's fertile period. The woman can predict when ovulation is to occur based on keeping careful records of her menstrual cycles. Because of variability in cycles, this technique is only partially effective. The time of ovulation can be determined more precisely by recording body temperature each morning before getting up. Body temperature rises slightly about a day after ovulation has taken place. The temperature rhythm method is not useful in determining when it is safe to engage in intercourse before ovulation, but it can be helpful in determining when it is safe to resume sex after ovulation.

■ *Coitus interruptus* involves withdrawal of the penis from the vagina before ejaculation occurs. This method is only moderately effective, however, because timing is difficult, and some sperm may pass out of the urethra prior to ejaculation.

■ *Chemical contraceptives,* such as spermicidal ("sperm-killing") jellies, foams, creams, and suppositories, when inserted into the vagina are toxic to sperm for about an hour after application.

■ *Barrier methods* mechanically prevent sperm transport to the oviduct. For males, the *condom* is a thin, strong rubber or latex sheath placed over the erect penis before ejaculation to prevent sperm from entering the vagina. For females, the *diaphragm,* which must be fitted by a trained professional, is a flexible rubber dome that is inserted through the vagina and positioned over the cervix to block sperm entry into the cervical canal. It is held in position by lodging snugly against the vaginal wall and must be left in place for at least 6 hours but no longer than 24 hours after intercourse. Barrier methods are often used in conjunction with spermicidal agents for increased effectiveness. The *cervical cap* is an alternative to the diaphragm. Smaller than a diaphragm, the cervical cap, which is coated with a film of spermicide, cups over the cervix and is held in place by suction.

■ The *female condom* (or *vaginal pouch*) is the latest barrier method developed. It is a seven-inch-long, polyurethane, cylindrical pouch that is closed on one end and open on the other end, with a flexible ring at both ends. The ring at the closed end of the device is inserted into the vagina and fits over the cervix, similar to a diaphragm. The ring at the open end of the pouch is positioned outside the vagina over the external genitalia.

■ *Sterilization,* which involves surgical disruption of either the ductus deferens (*vasectomy*) in men or the oviduct (*tubal ligation*) in women, is considered a permanent method of preventing sperm and ovum from uniting.

Prevention of Ovulation

■ *Oral contraceptives,* or *birth control pills,* available only by prescription, prevent ovulation primarily by suppressing gonadotropin secretion. These pills, which contain synthetic estrogen-like and progesterone-like steroids, are taken for three weeks, either in combination or in sequence, and then are withdrawn for one week. These steroids, like the natural steroids produced during the ovarian cycle, inhibit GnRH and thus FSH and LH secretion. As a result, follicle maturation and ovulation do not take place, so conception is impossible. The endometrium responds to the exogenous steroids by thickening and developing secretory capacity, just as it would respond to the natural hormones. When these synthetic steroids are withdrawn after three weeks, the endometrial lining sloughs and menstruation occurs, as it normally would on degeneration of the corpus luteum. In addition to blocking ovulation, oral contraceptives prevent pregnancy by increasing the viscosity of cervical mucus, which makes sperm penetration more difficult, and by decreasing muscular contractions in the female reproductive tract, which reduces sperm transport to the oviduct. Oral contraceptives have been shown to increase the risk of intravascular clotting, especially in women who also smoke tobacco.

Average Failure Rate of Various Contraceptive Techniques

Contraceptive Method	Average Failure Rate (Annual Pregnancies/ 100 Women)
None	90
Natural (rhythm) methods	20–30
Coitus interruptus	23
Chemical contraceptives	20
Barrier methods	10–15
Intrauterine device	4
Oral contraceptives	2–2.5
Implanted contraceptives	1

- Several other contraceptive methods contain synthetic female sex hormones and act similarly to birth control pills to prevent ovulation. These include long-acting *subcutaneous* ("under the skin") *implantation* of hormone-containing capsules (not widely available) that gradually release hormones at a nearly steady rate for five years and *birth control patches* impregnated with hormones that are absorbed through the skin.

Blockage of Implantation

Medically, pregnancy is not considered to begin until implantation. According to this view, any mechanism that interferes with implantation is said to prevent pregnancy. Not all hold this view, however. Some consider pregnancy to begin at time of fertilization. To them, any interference with implantation is a form of abortion. Therefore, methods of contraception that rely on blockage of implantation are more controversial than methods that prevent fertilization from taking place.

- Blockage of implantation is most commonly accomplished by a physician inserting a small *intrauterine device (IUD)* into the uterus. The IUD's mechanism of action is not completely understood, although most evidence suggests that the presence of this foreign object in the uterus induces a local inflammatory response that prevents implantation of a fertilized ovum.
- Implantation can also be blocked by so-called *morning-after pills,* also known as *emergency contraception.* The first term is actually a misnomer, because these pills can prevent pregnancy if taken within 72 hours after, not just the morning after, unprotected sexual intercourse. The most common form of emergency contraception is a kit consisting of high doses of birth control pills. These pills, available only by prescription, work in different ways to prevent pregnancy depending on where the woman is in her cycle when she takes the pills. They can either suppress ovulation or cause premature degeneration of the corpus luteum, thus preventing implantation of a fertilized ovum by withdrawing the developing endometrium's hormonal support. These kits are for emergency use only—for instance, if a condom breaks or in the case of rape—and should not be used as a substitute for ongoing contraceptive methods.

Future Possibilities

- On the horizon are improved varieties of currently available contraceptive techniques, such as a new birth control pill that suppresses ovulation and menstrual periods for months at a time.
- A future birth control technique is *immunocontraception*—the use of vaccines that prod the immune system to produce antibodies targeted against a particular protein critical to the reproductive process. The contraceptive effects of the vaccines are expected to last about a year. For example, in the testing stage is a vaccine that induces the formation of antibodies against human chorionic gonadotropin so that this essential corpus luteum–supporting hormone is not effective if pregnancy occurs. Another promising immunocontraception approach is aimed at blocking the acrosomal enzymes so that sperm could not enter the ovum. Still other researchers have developed an experimental vaccine that targets a protein added by the epididymis to the surface of sperm during their maturation.
- Some researchers are exploring ways to block the union of sperm and egg by interfering with a specific interaction that normally occurs between the male and female gametes. For example, under study are chemicals introduced into the vagina that trigger premature release of the acrosomal enzymes, depriving the sperm of a means to fertilize an ovulated egg.
- Some scientists are seeking ways to manipulate hormones to block sperm production in males without depriving the man of testosterone. One example of a male contraception under development is a combination of testosterone and progestin that inhibits GnRH and the gonadotropic hormones, thereby turning off the signals that stimulate spermatogenesis. Still other researchers are trying to interfere with the bond between Sertoli cells and developing sperm so that spermatogenesis cannot proceed to completion.
- Another outlook for male contraception is chemical sterilization designed to be reversed, unlike surgical sterilization, which is considered irreversible. In this experimental technique, a nontoxic polymer is injected into the ductus deferens, where the chemical interferes with sperm's fertilizing capabilities. Flushing of the polymer from the ductus deferens by a solvent reverses the contraceptive effect.
- One interesting avenue being explored holds hope for a unisex contraceptive that would stop sperm in their tracks and could be used by either males or females. The idea is to use Ca^{2+}-blocking drugs to prevent the entry of Ca^{2+} into sperm tails. As in muscle cells, Ca^{2+} switches on the contractile apparatus responsible for the sperm's motility. With no Ca^{2+}, sperm would not be able to maneuver to accomplish fertilization.

Termination of Unwanted Pregnancies

- When contraceptive practices fail or are not used and an unwanted pregnancy results, women often turn to *abortion* to terminate the pregnancy. More than half of the approximately 6.4 million pregnancies in the United States each year are unintended, and about 1.6 million of them end with an abortion. Although surgical removal of an embryo/fetus is legal in the United States, the practice of abortion is fraught with emotional, ethical, and political controversy.
- In late 2000, the "abortion pill," *RU 486,* or *mifepristone,* was approved for use in the United States amid considerable controversy, even though it has been available in other countries since 1988. This drug terminates an early pregnancy by chemical interference rather than by surgery. RU 486, a progesterone antagonist, binds tightly with the progesterone receptors on the target cells but does not evoke progesterone's usual effects and prevents progesterone from binding and acting. Deprived of progesterone activity, the highly developed endometrial tissue sloughs off, carrying the implanted embryo with it. RU 486 administration is followed in 48 hours by a prostaglandin that induces uterine contractions to help expel the endometrium and embryo.

out capillaries into these chorionic projections to form **placental villi.** Some villi extend completely across the blood-filled spaces to anchor the fetal portion of the placenta to the endometrial tissue, but most simply project into the pool of maternal blood.

Each placental villus contains embryonic (later fetal) capillaries surrounded by a thin layer of chorionic tissue, which separates the embryonic/fetal blood from the maternal blood in the intervillus spaces. Maternal and fetal blood do not actually mingle, but the barrier between them is extremely thin. To visualize this relationship, think of your hands (the fetal capillary blood vessels) in rubber gloves (the chorionic tissue) immersed in water (the pool of maternal blood). Only the rubber gloves separate your hands from the water. In the same way, only the thin chorionic tissue (plus the capillary wall of the fetal vessels) separates the fetal and maternal blood. All exchanges between these two bloodstreams take place across this extremely thin barrier. This entire system of interlocking maternal (decidual) and fetal (chorionic) structures makes up the placenta.

Even though not fully developed, the placenta is well established and operational by five weeks after implantation. By this time, the heart of the developing embryo is pumping blood into the placental villi as well as to the embryonic tissues. Throughout gestation, fetal blood continuously traverses between the placental villi and the circulatory system of the fetus by means of two **umbilical arteries** and one **umbilical vein,** which are wrapped within the **umbilical cord,** a lifeline between the fetus and the placenta (● Figure 20-28). The maternal blood within the placenta is continuously replaced as fresh blood enters through the uterine arterioles; percolates through the intervillus spaces, where it exchanges substances with fetal blood in the surrounding villi; then exits through the uterine vein.

Meanwhile, during the time of implantation and early placental development, the inner cell mass forms a fluid-filled **amniotic cavity** between the chorion and the portion of the inner cell mass destined to become the fetus (see ● Figure 20-27, step ❷). The epithelial layer that encloses the amniotic cavity is called the **amniotic sac,** or **amnion.** As it continues to develop, the amniotic sac eventually fuses with the chorion, forming a single combined membrane that surrounds the embryo/fetus. The fluid in the amniotic cavity, the **amniotic fluid,** which is similar in composition to normal ECF, surrounds and cushions the fetus throughout gestation (● Figures 20-28 and 20-29).

FUNCTIONS OF THE PLACENTA During intrauterine life, the placenta performs the functions of the digestive system, the respiratory system, and the kidneys for the "parasitic" fetus. The fetus has these organ systems, but within the uterine environment they cannot (and do not need to) function. Nutrients and O_2 move from the maternal blood across the thin placental barrier into the fetal blood, whereas CO_2 and other metabolic wastes simultaneously move from the fetal blood into the maternal blood. The nutrients and O_2 brought to the fetus in the maternal blood are acquired by the mother's digestive and respiratory systems, and the CO_2 and wastes transferred into the maternal blood are eliminated by the mother's lungs and kidneys, respectively. Thus, the mother's digestive tract, respiratory system, and kidneys serve the fetus's needs as well as her own.

● **FIGURE 20-29 A human fetus surrounded by the amniotic sac.** The fetus is near the end of the first trimester of development.

Lennart Nilsson/Bonnier Alba AB

The means by which materials move across the placenta depends on the substance. Some substances that can permeate the placental membrane, such as O_2, CO_2, water, and electrolytes, cross by simple diffusion. Some traverse the placental barrier by special mediated-transport systems in the placental membranes, such as glucose by facilitated diffusion and amino acids by secondary active transport. Other substances such as cholesterol in the form of LDL (see p. 336) move across by receptor-mediated endocytosis.

Clinical Note Unfortunately, many drugs, environmental pollutants, other chemical agents, and microorganisms in the mother's bloodstream also can cross the placental barrier, and some of them may harm the developing fetus. Individuals born limbless as a result of exposure to *thalidomide,* a tranquilizer prescribed for pregnant women before this drug's devastating effects on the growing fetus were known, serve as a grim reminder of this fact. Similarly, newborns who have become "addicted" during gestation by their mother's abuse of a drug such as heroin suffer withdrawal symptoms after birth. Even more common chemicals such as aspirin, alcohol, and agents in cigarette smoke can reach the fetus and have adverse effects. Likewise, fetuses can acquire AIDS before birth if their mothers are infected with the virus. A pregnant woman should therefore be very cautious about potentially harmful exposure from any source.

Placental Hormones

Hormone	Function
Human Chorionic Gonadotropin (hCG)	Maintains the corpus luteum of pregnancy Stimulates secretion of testosterone by the developing testes in XY embryos
Estrogen (also secreted by the corpus luteum of pregnancy)	Stimulates growth of the myometrium, increasing uterine strength for parturition Helps prepare the mammary glands for lactation
Progesterone (also secreted by the corpus luteum of pregnancy)	Suppresses uterine contractions to provide a quiet environment for the fetus Promotes formation of a cervical mucus plug to prevent uterine contamination Helps prepare the mammary glands for lactation
Human Chorionic Somatomammotropin (has a structure similar to that of both growth hormone and prolactin)	Reduces maternal use of glucose and promotes the breakdown of stored fat (similar to growth hormone) so that greater quantities of glucose and free fatty acids may be shunted to the fetus Helps prepare the mammary glands for lactation (similar to prolactin)
Relaxin (also secreted by the corpus luteum of pregnancy)	Softens the cervix in preparation for cervical dilation at parturition Loosens the connective tissue between the pelvic bones in preparation for parturition
Placental PTHrp (parathyroid hormone–related peptide)	Increases maternal plasma Ca^{2+} level for use in calcifying fetal bones; if necessary, promotes localized dissolution of maternal bones, mobilizing their Ca^{2+} stores for use by the developing fetus

The placenta has yet another important responsibility—it becomes a temporary endocrine organ during pregnancy, a topic to which we now turn. During pregnancy, three endocrine systems interact to support and enhance the growth and development of the fetus, to coordinate the timing of parturition, and to prepare the mammary glands for nourishing the baby after birth: placental hormones, maternal hormones, and fetal hormones.

Hormones secreted by the placenta play a critical role in maintaining pregnancy.

The fetally derived portion of the placenta has the remarkable capacity to secrete a number of peptide and steroid hormones essential for maintaining pregnancy. The most important are *human chorionic gonadotropin, estrogen,* and *progesterone* (▲ Table 20-5). Serving as the major endocrine organ of pregnancy, the placenta is unique among endocrine tissues in two regards. First, it is a transient tissue. Second, secretion of its hormones is not subject to extrinsic control, in contrast to the stringent, often complex mechanisms that regulate the secretion of other hormones. Instead, the type and rate of placental hormone secretion depend primarily on the stage of pregnancy.

SECRETION OF HUMAN CHORIONIC GONADOTROPIN One of the first endocrine events is secretion by the developing chorion of **human chorionic gonadotropin (hCG),** a peptide hormone that prolongs the life span of the corpus luteum. Recall that during the ovarian cycle, the corpus luteum degenerates and the highly prepared, luteal-dependent uterine lining sloughs off if fertilization and implantation do not occur. When fertilization does occur, the implanted blastocyst saves itself from being flushed out in menstrual flow by producing hCG. This hormone, which is similar to LH and binds to the same receptor as LH, stimulates and maintains the corpus luteum so it does not degenerate. Now called the **corpus luteum of pregnancy,** this ovarian endocrine unit grows even larger and produces increasingly greater amounts of estrogen and progesterone for an additional 10 weeks until the placenta takes over secretion of these steroid hormones. Because of the persistence of estrogen and progesterone, the thick, pulpy endometrial tissue is maintained instead of sloughing. Accordingly, menstruation ceases during pregnancy.

Stimulation by hCG is necessary to maintain the corpus luteum of pregnancy because LH, which maintains the corpus luteum during the normal luteal phase of the uterine cycle, is suppressed through feedback inhibition by the high levels of progesterone.

Maintenance of a normal pregnancy depends on high concentrations of progesterone and estrogen. Thus, hCG production is critical during the first trimester to maintain ovarian output of these hormones. In a male fetus, hCG also stimulates the precursor Leydig cells in the fetal testes to secrete testosterone, which masculinizes the developing reproductive tract.

● **FIGURE 20-30** Secretion rates of placental hormones.

The secretion rate of hCG increases rapidly during early pregnancy to save the corpus luteum from demise. Peak secretion of hCG occurs about 60 days after the end of the last menstrual period (● Figure 20-30). By the 10th week of pregnancy, hCG output declines to a low rate of secretion that is maintained for the duration of gestation. The fall in hCG occurs at a time when the corpus luteum is no longer needed for its steroid hormone output because the placenta has begun to secrete substantial quantities of estrogen and progesterone. The corpus luteum of pregnancy partially regresses as hCG secretion dwindles, but it is not converted into scar tissue until after delivery of the baby.

Clinical Note Human chorionic gonadotropin is eliminated from the body in the urine. Pregnancy diagnosis tests can detect hCG in urine as early as the first month of pregnancy, about two weeks after the first missed menstrual period. Because this is before the growing embryo can be detected by physical examination, the test permits early confirmation of pregnancy.

A frequent early clinical sign of pregnancy is morning sickness, a daily bout of nausea and vomiting that often occurs in the morning but can take place at any time of day. Because this condition usually appears shortly after implantation and coincides with the time of peak hCG production, scientists speculate that this early placental hormone may trigger the symptoms, perhaps by acting on the chemoreceptor trigger zone in the vomiting center (see p. 605).

SECRETION OF ESTROGEN AND PROGESTERONE Why does the developing placenta not start producing estrogen and progesterone in the first place instead of secreting hCG, which in turn stimulates the corpus luteum to secrete these two critical hormones? The answer is that, for different reasons, the placenta cannot produce enough estrogen or progesterone in the first trimester of pregnancy. In the case of estrogen, the placenta does not have all the enzymes needed for complete synthesis of this hormone. Estrogen synthesis requires a complex interaction between the placenta and the fetus (● Figure 20-31). The placenta converts the androgen hormone produced by the fetal adrenal cortex, dehydroepiandrosterone (DHEA), into estrogen. The placenta cannot produce estrogen until the fetus has developed to the point that its adrenal cortex is secreting DHEA into the blood. The placenta extracts DHEA from the fetal blood and converts it into estrogen, which it then secretes into the maternal blood. The primary estrogen synthesized by this means is **estriol,** in contrast to the main estrogen product of the ovaries, estradiol. Consequently, measurement of estriol levels in the maternal urine can be used clinically to assess the viability of the fetus.

In the case of progesterone, the placenta can synthesize this hormone soon after implantation. Even though the early placenta has the enzymes necessary to convert cholesterol extracted from the maternal blood into progesterone, it does not produce much of this hormone, because the amount of progesterone produced is proportional to placental weight. The placenta is simply too small in the first 10 weeks of pregnancy to produce enough progesterone to maintain the endometrial tissue. The notable increase in circulating progesterone in the last seven months of gestation reflects placental growth during this period.

ROLES OF ESTROGEN AND PROGESTERONE DURING PREGNANCY
As noted earlier, high concentrations of estrogen and progesterone are essential to maintain a normal pregnancy. Estrogen stimulates growth of the myometrium, which increases in size throughout pregnancy. The stronger uterine musculature is needed to expel the fetus during labor. Estriol also promotes development of the ducts within the mammary glands, through which milk will be ejected during lactation.

Progesterone performs various roles throughout pregnancy. Its main function is to prevent miscarriage by suppressing contractions of the uterine myometrium. Progesterone also promotes formation of a mucus plug in the cervical canal to prevent vaginal contaminants from reaching the uterus. Finally, placental progesterone stimulates development of milk glands in the breasts in preparation for lactation.

Maternal body systems respond to the increased demands of gestation.

The period of **gestation (pregnancy)** is about 38 weeks from conception (40 weeks from the end of the last menstrual period). During gestation, the embryo/fetus develops and grows to the point of being able to leave its maternal life-support system. Meanwhile, a number of physical changes within the mother accommodate the demands of pregnancy. The most obvious change is uterine enlargement. The uterus expands and increases in weight more than 20 times, exclusive of its contents. The breasts enlarge and develop the ability to produce milk. Body systems other than the reproductive system also make needed adjustments. The volume of blood increases by

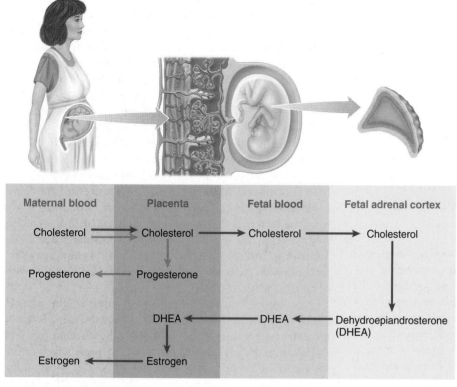

The hCS-induced metabolic changes in the mother make available greater quantities of glucose and fatty acids for shunting to the fetus. Also, if the mother does not consume enough Ca^{2+}, yet another placental hormone similar to parathyroid hormone, **parathyroid hormone-related peptide (PTHrp),** mobilizes Ca^{2+} from the maternal bones to ensure adequate calcification of the fetal bones (▲ Table 20-5).

Changes during late gestation prepare for parturition.

Parturition (labor, delivery, or **birth)** requires (1) dilation of the cervical canal to accommodate passage of the fetus from the uterus through the vagina and to the outside and (2) contractions of the uterine myometrium that are sufficiently strong to expel the fetus.

Several changes take place during late gestation in preparation for the onset of parturition. During the first two trimesters of gestation, the uterus remains relatively quiet because of the inhibitory effect of the high levels of progesterone on the uterine muscle. During the last trimester, however, the uterus becomes progressively more excitable, so that mild contractions (**Braxton–Hicks contractions**) are experienced with increasing strength and frequency. Sometimes these contractions become regular enough to be mistaken for the onset of labor, a phenomenon called "false labor."

KEY

→ Pathway for placental synthesis of progesterone

→ Pathway for placental synthesis of estrogen

● **FIGURE 20-31 Secretion of estrogen and progesterone by the placenta.** The placenta secretes increasing quantities of progesterone and estrogen into the maternal blood after the first trimester. The placenta itself can convert cholesterol into progesterone *(orange pathway)* but lacks some of the enzymes necessary to convert cholesterol into estrogen. However, the placenta can convert DHEA derived from cholesterol in the fetal adrenal cortex into estrogen when DHEA reaches the placenta by means of the fetal blood *(blue pathway)*.

30%, and the cardiovascular system responds to the increasing demands of the growing placental mass. Weight gain during pregnancy is the result only in part of the weight of the fetus. The remainder is mostly from increased weight of the uterus, including the placenta, and increased blood volume. Respiratory activity increases by about 20% to handle the additional fetal requirements for O_2 utilization and CO_2 removal. Urinary output increases, and the kidneys excrete the additional wastes from the fetus.

The increased metabolic demands of the growing fetus increase nutritional requirements for the mother. In general, the fetus takes what it needs from the mother, even if this leaves the mother with a nutritional deficit. For example, the placental hormone **human chorionic somatomammotropin (hCS)** is responsible for the decreased use of glucose by the mother and the mobilization of free fatty acids from maternal adipose stores, similar to the actions of growth hormone (see p. 678). (In fact, hCS has a structure similar to that of both growth hormone and prolactin and exerts similar actions.)

Throughout gestation, the exit of the uterus remains sealed by the rigid, tightly closed cervix. As parturition approaches, the cervix begins to soften (or "ripen") as a result of the dissociation of its tough connective tissue (collagen) fibers. Because of this softening, the cervix becomes malleable so that it can gradually yield, dilating the exit, as the fetus is forcefully pushed against it during labor. This cervical softening is caused largely by **relaxin,** a peptide hormone produced by the corpus luteum of pregnancy and by the placenta. Other factors to be described shortly contribute to cervical softening. Relaxin also "relaxes" the birth canal by loosening the connective tissue between pelvic bones.

Meanwhile, the fetus shifts downward (the baby "drops") and is normally oriented so that the head is in contact with the cervix in preparation for exiting through the birth canal. In a **breech birth,** any part of the body other than the head approaches the birth canal first.

Scientists are closing in on the factors that trigger the onset of parturition.

Rhythmic, coordinated contractions, usually painless at first, begin at the onset of true labor. As labor progresses, the contractions increase in frequency, intensity, and discomfort. These strong, rhythmic contractions force the fetus against the cervix, dilating the cervix. Then, after having dilated the cervix enough for the fetus to pass through, these contractions force the fetus out through the birth canal.

The exact factors triggering the increase in uterine contractility and thus initiating parturition are not fully established, although much progress has been made in recent years in unraveling the sequence of events. Let us look at what is known about this process.

ROLE OF HIGH ESTROGEN LEVELS During early gestation, maternal estrogen levels are relatively low, but as gestation proceeds, placental estrogen secretion continues to rise. In the immediate days before the onset of parturition, soaring levels of estrogen bring about changes in the uterus and cervix to prepare them for labor and delivery (● Figures 20-30 and 20-32). First, high levels of estrogen promote the synthesis of connexons within the uterine smooth muscle cells. These myometrial cells are not functionally linked to any extent throughout most of gestation. The newly manufactured connexons are inserted in the myometrial plasma membranes to form gap junctions that electrically link together the uterine smooth muscle cells so they become able to contract as a coordinated unit (see p. 59).

Simultaneously, high levels of estrogen dramatically and progressively increase the concentration of myometrial receptors for oxytocin. Together, these myometrial changes collectively bring about the increased uterine responsiveness to oxytocin that ultimately initiates labor.

In addition to preparing the uterus for labor, the increasing levels of estrogen promote production of local prostaglandins that contribute to cervical ripening by stimulating cervical enzymes that locally degrade collagen fibers. Furthermore, these prostaglandins themselves increase uterine responsiveness to oxytocin.

ROLE OF OXYTOCIN Oxytocin is a peptide hormone produced by the hypothalamus, stored in the posterior pituitary, and released into the blood from the posterior pituitary on nervous stimulation by the hypothalamus (see p. 670). Oxytocin exerts its effects via the $IP_3/Ca^{2+}/DAG$ pathway. A powerful uterine muscle stimulant, oxytocin plays the key role in the progression of labor. However, this hormone was once discounted as the trigger for parturition because the circulating levels of oxytocin remain constant prior to the onset of labor. The discovery that uterine responsiveness to oxytocin is 100 times greater at term than in nonpregnant women (because of the connexons and increased concentration of myometrial oxytocin receptors) led to the now widely accepted conclusion that labor is initiated when the myometrial responsiveness to oxytocin reaches a critical threshold that permits the onset of strong, coordinated contractions in response to ordinary levels of circulating oxytocin.

ROLE OF CORTICOTROPIN-RELEASING HORMONE Until recently, scientists were baffled by the factors that raise levels of placental estrogen secretion. Recent research has shed new light on the probable mechanism. Evidence suggests that *corticotropin-releasing hormone (CRH)* secreted by the fetal portion of the placenta into both the maternal and fetal circulations not only drives the manufacture of placental estrogen, thus ultimately dictating the timing of the onset of labor, but also promotes changes in the fetal lungs needed for breathing air (● Figure 20-32). Recall that CRH is normally secreted by the hypothalamus and regulates the output of ACTH by the anterior pituitary (see pp. 674 and 702). In turn, ACTH stimulates production of both cortisol and DHEA by the adrenal cortex. In the fetus, much of the CRH comes from the placenta rather than solely from the fetal hypothalamus. The additional cortisol secretion summoned by the extra CRH promotes fetal lung maturation. Specifically, cortisol stimulates the synthesis of pulmonary surfactant, which facilitates lung expansion and reduces the work of breathing (see p. 476).

The bumped-up rate of DHEA secretion by the adrenal cortex in response to placental CRH leads to the rising levels of placental estrogen secretion. Recall that the placenta converts DHEA from the fetal adrenal gland into estrogen, which enters the maternal bloodstream (see ● Figure 20-31). When sufficiently high, this estrogen sets in motion the events that initiate labor. Thus, pregnancy duration and delivery timing are determined largely by the placenta's rate of CRH production. That is, a **"placental clock"** ticks out the length of time until parturition. The timing of parturition is established early in pregnancy, with delivery at the end point of a maturational process that extends throughout the entire gestation. The ticking of the placental clock is measured by the rate of placental secretion. As the pregnancy progresses, CRH levels in maternal plasma rise. Researchers can accurately predict the timing of parturition by measuring the maternal plasma levels of CRH as early as the end of the first trimester. Higher-than-normal levels are associated with premature deliveries, whereas lower-than-normal levels indicate late deliveries. These and other data suggest that when a critical level of placental CRH is reached, parturition is triggered. This critical CRH level ensures that when labor begins, the infant is ready for life outside the womb. It does so by concurrently increasing the fetal cortisol needed for lung maturation and the estrogen needed for the uterine changes that bring on labor. The remaining unanswered puzzle regarding the placental clock is, What controls placental secretion of CRH?

ROLE OF INFLAMMATION Surprisingly, recent research suggests that inflammation plays a central role in the labor process, in the onset of both full-term labor and premature labor. Key to this inflammatory response is activation of **nuclear factor κB (NF-κB)** in the uterus. NF-κB boosts production of inflammatory cytokines such as *interleukin 8 (IL-8)* (see p. 424) and prostaglandins that increase the sensitivity of the uterus to contraction-inducing chemical messengers and help soften the cervix. What activates NF-κB, setting off an inflammatory cascade that helps prompt labor? Various factors associated with the onset of full-term labor and premature labor can cause an upsurge in NF-κB. These include stretching of the uterine

● **FIGURE 20-32 Initiation and progression of parturition.**

Essay Questions

1. What are the primary reproductive organs, gametes, sex hormones, reproductive tract, accessory sex glands, external genitalia, and secondary sexual characteristics in males and in females?

2. List the essential reproductive functions of the male and the female.

3. Discuss the differences between males and females with regard to genetic, gonadal, and phenotypic sex.

4. What parts of the male and female reproductive systems develop from each of the following: genital tubercle, urethral folds, genital swellings, Wolffian ducts, and Müllerian ducts?

5. Of what functional significance is the scrotal location of the testes?

6. Discuss the source and functions of testosterone.

7. Describe the three major stages of spermatogenesis. Discuss the functions of each part of a spermatozoon. What are the roles of Sertoli cells?

8. Discuss the control of testicular function.

9. Compare the sex act in males and females.

10. Compare oogenesis with spermatogenesis.

11. Describe the events of the follicular and luteal phases of the ovarian cycle. Correlate the phases of the uterine cycle with those of the ovarian cycle.

12. How are the ovum and spermatozoa transported to the site of fertilization? Describe the process of fertilization.

13. Describe the process of implantation and placenta formation.

14. What are the functions of the placenta? What hormones does the placenta secrete?

15. What is the role of human chorionic gonadotropin?

16. What factors contribute to the initiation of parturition? What are the stages of labor? What is the role of oxytocin?

17. Describe the hormonal factors that play a role in lactation.

18. Summarize the actions of estrogen and progesterone.

POINTS TO PONDER

(Explanations on p. A-56)

1. The hypothalamus releases GnRH in pulsatile bursts once every two to three hours, with no secretion occurring in between. The blood concentration of GnRH depends on the frequency of these bursts of secretion. A promising line of research for a new method of contraception involves administration of GnRH-like drugs. In what way could such drugs act as contraceptives when GnRH is the hypothalamic hormone that triggers the chain of events leading to ovulation? (*Hint:* The anterior pituitary is "programmed" to respond only to the normal pulsatile pattern of GnRH.)

2. Occasionally, testicular tumors composed of interstitial cells of Leydig may secrete up to 100 times the normal amount of testosterone. When such a tumor develops in young children, they grow up much shorter than their genetic potential. Explain why. What other symptoms would be present?

3. What type of sexual dysfunction might arise in men taking drugs that inhibit sympathetic nervous system activity as part of the treatment for high blood pressure?

4. Explain the physiologic basis for administering a posterior pituitary extract to induce or facilitate labor.

5. The symptoms of menopause are sometimes treated with supplemental estrogen and progesterone. Why wouldn't treatment with GnRH or FSH and LH also be effective?

CLINICAL CONSIDERATION

(Explanation on p. A-56)

Maria A., who is in her second month of gestation, has been experiencing severe abdominal cramping. Her physician has diagnosed her condition as a tubal pregnancy: The developing embryo is implanted in the oviduct instead of in the uterine endometrium. Why must this pregnancy be surgically terminated?

The Metric System

| ▲ TABLE A-1 | Metric Measures and English Equivalents | | |

Unit	Measure	Symbol	English Equivalent
Linear Measure			
1 kilometer	= 1000 meters	10^3 m km	0.62137 mile
1 meter		10^0 m m	39.37 inches
1 decimeter	= 1/10 meter	10^{-1} m dm	3.937 inches
1 centimeter	= 1/100 meter	10^{-2} m cm	0.3937 inch
1 millimeter	= 1/1000 meter	10^{-3} m mm	Not used
1 micrometer (or micron)	= 1/1,000,000 meter	10^{-6} m μm (or μ)	Not used
1 nanometer	= 1/1,000,000,000 meter	10^{-9} m nm	Not used
Measures of Capacity (for fluids and gases)			
1 liter		L	1.0567 U.S. liquid quarts
1 milliliter	= 1/1000 liter = volume of 1 g of water at stp*	ml	
Measures of Volume			
1 cubic meter		m^3	
1 cubic decimeter	= 1/1000 cubic meter = 1 liter (L)	dm^3	
1 cubic centimeter	= 1/1,000,000 cubic meter = 1 milliliter (ml)	cm^3 = ml	
1 cubic millimeter	= 1/100,000,000 cubic meter	mm^3	
Measures of Mass			
1 kilogram	= 1000 grams	kg	2.2046 pounds
1 gram		g	15.432 grains
1 milligram	= 1/1000 gram	mg	0.01 grain (about)
1 microgram	= 1/1,000,000 gram	μg (or mcg)	

*stp = standard temperature and pressure

Comparison of human height in meters with the sizes of some biological and molecular structures.

Metric–English Conversions

Length

English		Metric	
Inch	=	2.54	centimeters
Foot	=	0.30	meters
Yard	=	0.91	meter
Mile (5280 feet)	=	1.61	kilometers

To convert	multiply by	to obtain
Inches	2.54	centimeters
Feet	30.00	centimeters
Centimeters	0.39	inches
Millimeters	0.039	inches

Mass/Weight

English		Metric	
Grain	=	64.80	milligrams
Ounce	=	28.35	grams
Pound	=	453.60	grams
Ton (short) (2000 pounds)	=	0.91	metric ton

To convert	multiply by	to obtain
Ounces	28.3	grams
Pounds	453.6	grams
Pounds	0.45	kilograms
Grams	0.035	ounces
Kilograms	2.2	pounds

Volume and Capacity

English		Metric	
Cubic inch	=	16.39	cubic centimeters
Cubic foot	=	0.03	cubic meter
Cubic yard	=	0.765	cubic meter
Ounce	=	0.03	liter
Pint	=	0.47	liter
Quart	=	0.95	liter
Gallon	=	3.79	liters

To convert	multiply	to obtain
Fluid ounces	30.00	milliliters
Quart	0.95	liters
Milliliters	0.03	fluid ounces
Liters	1.06	quarts

Linear Measurement Comparison

Fahrenheit–Celsius Temperature Comparison

Temperature Conversions

To convert Fahrenheit to Celsius:

$Tc = (Tf - 32)/1.8$

To convert Celsius to Fahrenheit:

$Tf = (Tc \times 1.8) + 32$

Key

Tc = Temperature in Celsius
Tf = Temperature in Fahrenheit

A Review of Chemical Principles

By Spencer Seager, Weber State University, and Lauralee Sherwood

Chemical Level of Organization in the Body

Matter is anything that occupies space and has mass, including all living and nonliving things in the universe. **Mass** is the amount of matter in an object. **Weight,** in contrast, is the effect of gravity on that mass. The more gravity exerted on a mass, the greater the weight of the mass. An astronaut has the same mass whether on Earth or in space but is weightless in the zero gravity of space.

Atoms

All matter is made up of tiny particles called **atoms.** These particles are too small to be seen individually, even with the most powerful electron microscopes available today.

Even though extremely small, atoms consist of three types of even smaller subatomic particles. Different types of atoms vary in the number of these various subatomic particles they contain. **Protons** and **neutrons** are particles of nearly identical mass, with protons carrying a positive charge and neutrons having no charge. **Electrons** have a much smaller mass than protons and neutrons and are negatively charged. An atom consists of two regions—a dense, central *nucleus* made of protons and neutrons surrounded by a three-dimensional *electron cloud,* where electrons move rapidly around the nucleus in orbitals (● Figure B-1). The magnitude of the charge of a proton exactly matches that of an electron, but it is opposite in sign, being positive. In all atoms, the number of protons in the nucleus is equal to the number of electrons moving around the nucleus, so their charges balance and the atoms are neutral.

Elements and atomic symbols

A pure substance composed of only one type of atom is called an **element.** A pure sample of the element carbon contains only carbon atoms, even though the atoms might be arranged in the form of diamond or in the form of graphite (pencil "lead"). Each element is designated by an **atomic symbol,** a one- or two-letter chemical shorthand for the element's name. Usually

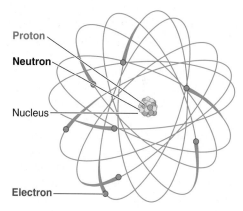

● **FIGURE B-1 The atom.** The atom consists of two regions. The central nucleus contains protons and neutrons and makes up 99.9% of the mass. Surrounding the nucleus is the electron cloud, where the electrons move rapidly around the nucleus. (Figure not drawn to scale.)

these symbols are easy to follow, because they are derived from the English name for the element. Thus H stands for *hydrogen,* C for *carbon,* and O for *oxygen.* In a few cases, the atomic symbol is based on the element's Latin name—for example, Na for *sodium* (*natrium* in Latin) and K for *potassium* (*kalium*). Of the 109 known elements, 26 are normally found in the body. Four elements—oxygen, carbon, hydrogen, and nitrogen—compose 96% of the body's mass.

Compounds and molecules

Pure substances composed of more than one type of atom are known as **compounds.** Pure water, for example, is a compound that contains atoms of hydrogen and atoms of oxygen in a 2-to-1 ratio, regardless of whether the water is in the form of liquid, solid (ice), or vapor (steam). A **molecule** is the smallest unit of a pure substance that has the properties of that substance and is capable of a stable, independent existence. For example, a molecule of water consists of two atoms of hydrogen and one atom of oxygen, held together by chemical bonds.

Atomic number

Exactly what are we talking about when we refer to a "type" of atom? That is, what makes carbon, hydrogen, and oxygen atoms different? The answer is the number of protons in the nucleus. Regardless of where they are found, all hydrogen atoms have 1 proton in the nucleus, all carbon atoms have 6, and all oxygen atoms have 8. Of course, these numbers also represent the number of electrons moving around each nucleus, because the number of electrons and number of protons in an atom are equal. The number of protons in the nucleus of an atom of an element is called the **atomic number** of the element.

Atomic weight

As expected, tiny atoms have tiny masses. For example, the actual mass of a hydrogen atom is 1.67×10^{-24} g, that of a carbon atom is 1.99×10^{-23} g, and that of an oxygen atom is 2.66×10^{-23} g. These very small numbers are inconvenient to work with in calculations, so a system of relative masses has been developed. These relative masses simply compare the actual masses of the atoms with each other. Suppose the actual masses of two people were determined to be 45.50 kg and 113.75 kg. Their relative masses are determined by dividing each mass by the smaller mass of the two: $45.50/45.50 = 1.00$, and $113.75/45.50 = 2.50$. Thus the relative masses of the two people are 1.00 and 2.50; these numbers simply express the fact that the mass of the heavier person is 2.50 times that of the other person. The relative masses of atoms are called **atomic masses,** or **atomic weights,** and are given in *atomic mass units (amu)*. In this system, hydrogen atoms, the least massive of all atoms, have an atomic weight of 1.01 amu. The atomic weight of carbon atoms is 12.01 amu, and that of oxygen atoms is 16.00 amu. Thus, oxygen atoms have a mass about 16 times that of hydrogen atoms. ▲ Table B-1 gives the atomic weights and some other characteristics of the elements that are most important physiologically.

▲ TABLE B-1	Characteristics of Selected Elements		
Name and Symbol	**Number of Protons**	**Atomic Number**	**Atomic Weight (amu)**
Hydrogen (H)	1	1	1.01
Carbon (C)	6	6	12.01
Nitrogen (N)	7	7	14.01
Oxygen (O)	8	8	16.00
Sodium (Na)	11	11	22.99
Magnesium (Mg)	12	12	24.31
Phosphorus (P)	15	15	30.97
Sulfur (S)	16	16	32.06
Chlorine (Cl)	17	17	35.45
Potassium (K)	19	19	39.10
Calcium (Ca)	20	20	40.08

Chemical Bonds

Because all matter is made up of atoms, atoms must somehow be held together to form matter. The forces holding atoms together are called **chemical bonds.** Not all chemical bonds are formed in the same way, but all involve the electrons of atoms. Whether one atom will bond with another depends on the number and arrangement of its electrons. An atom's electrons are arranged in electron shells, to which we now turn our attention.

Electron shells

Electrons tend to move around the nucleus in a specific pattern. The orbitals, or pathways traveled by electrons around the nucleus, are arranged in an orderly series of concentric layers known as **electron shells,** which consecutively surround the nucleus. Each electron shell can hold a specific number of electrons. The first (innermost) shell closest to the nucleus can contain a maximum of only 2 electrons, no matter what the element is. The second shell can hold a total of 8 more electrons. The third shell also can hold a maximum of 8 electrons. As the number of electrons increases with increasing atomic number, still more electrons occupy successive shells, each at a greater distance from the nucleus. Each successive shell from the nucleus has a higher **energy level.** Because the negatively charged electrons are attracted to the positively charged nucleus, it takes more energy for an electron to overcome the nuclear attraction and orbit farther from the nucleus. Thus the first electron shell has the lowest energy level and the outermost shell of an atom has the highest energy level.

In general, electrons belong to the lowest energy shell possible, up to the maximum capacity of each shell. For example, hydrogen atoms have only 1 electron, so it is in the first shell. Helium atoms have 2 electrons, which are both in the first shell and fill it. Carbon atoms have 6 electrons, 2 in the first shell and 4 in the second shell, whereas the 8 electrons of oxygen are arranged with 2 in the first shell and 6 in the second shell.

Bonding characteristics of an atom; valence

Atoms tend to undergo processes that result in a filled outermost electron shell. Thus the electrons of the outer or higher-energy shell determine the bonding characteristics of an atom and its ability to interact with other atoms. Atoms that have a vacancy in their outermost shell tend to either give up, accept, or share electrons with other atoms (whichever is most favorable energetically) so that all participating atoms have filled outer shells. For example, an atom that has only 1 electron in its outermost shell may empty this shell so its remaining shells are completely full. By contrast, another atom that lacks only 1 electron in its outer shell may acquire the deficient electron from the first atom to fill all its shells to the maximum. The number of electrons an atom loses, gains, or shares to achieve a filled outer shell is known as the atom's **valence.** A **chemical bond** is the force of attraction that holds participating atoms together as a result of an interaction between their outermost electrons.

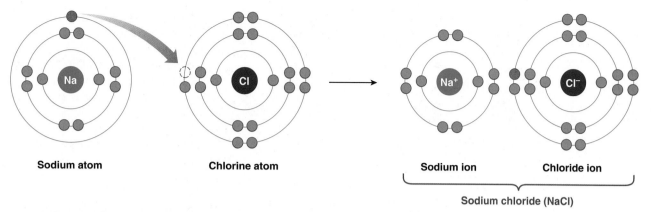

Sodium atom   Chlorine atom   Sodium ion   Chloride ion

Sodium chloride (NaCl)

● **FIGURE B-2 Ions and ionic bonds.** Sodium (Na) and chlorine (Cl) atoms both have partially filled outermost shells. Therefore, sodium tends to give up its lone electron in the outer shell to chlorine, filling chlorine's outer shell. As a result, sodium becomes a positively charged ion, and chlorine becomes a negatively charged ion known as chloride. The oppositely charged ions attract each other, forming an ionic bond.

Consider sodium atoms (Na) and chlorine atoms (Cl) (● Figure B-2). Sodium atoms have 11 electrons: 2 in the first shell, 8 in the second shell, and 1 in the third shell. Chlorine atoms have 17 electrons: 2 in the first shell, 8 in the second shell, and 7 in the third shell. Because 8 electrons are required to fill the second and third shells, sodium atoms have 1 electron more than is needed to provide a filled second shell, whereas chlorine atoms have 1 less electron than is needed to fill the third shell. Each sodium atom can lose an electron to a chlorine atom, leaving each sodium with 10 electrons, 8 of which are in the second shell, which is full and is now the outer shell occupied by electrons. By accepting 1 electron, each chlorine atom now has a total of 18 electrons, with 8 of them in the third, or outer, shell, which is now full.

Ions; ionic bonds

Recall that atoms are electrically neutral because they have an identical number of positively charged protons and negatively charged electrons. By giving up and accepting electrons, the sodium atoms and chlorine atoms have achieved filled outer shells, but now each atom is unbalanced electrically. Although each sodium now has 10 electrons, it still has 11 protons in the nucleus and a net electrical charge, or valence, of +1. Similarly, each chlorine now has 18 electrons, but only 17 protons. Thus each chlorine has a −1 charge. Such charged atoms are called **ions.** Positively charged ions are called **cations;** negatively charged ions are called **anions.** As a helpful hint to keep these terms straight, imagine the "t" in *cation* as standing for a "+" sign and the first "n" in *anion* as standing for "negative."

Note that both a cation and anion are formed whenever an electron is transferred from one atom to another. Because opposite charges attract, sodium ions (Na⁺) and charged chlorine atoms, now called *chloride* ions (Cl⁻), are attracted toward each other. This electrical attraction that holds cations and anions together is known as an **ionic bond.** Ionic bonds hold Na⁺ and Cl⁻ together in the compound **sodium chloride, NaCl,** which is common table salt. A sample of sodium chloride actually contains sodium and chloride ions in a three-dimensional geo-

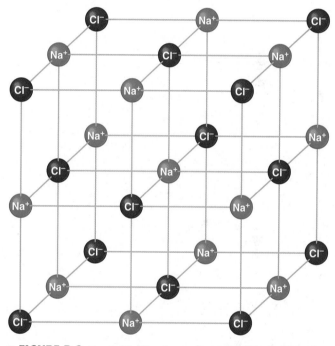

● **FIGURE B-3 Crystal lattice for sodium chloride (table salt).**

metric arrangement called a *crystal lattice.* The ions of opposite charge occupy alternate sites within the lattice (● Figure B-3).

Covalent bonds

It is not favorable, energywise, for an atom to give up or accept more than three electrons. Nevertheless, carbon atoms, which have four electrons in their outer shell, form compounds. They do so by another bonding mechanism, *covalent bonding.* Atoms that would have to lose or gain four or more electrons to achieve outer-shell stability usually bond by *sharing* electrons. Shared electrons actually orbit around *both* atoms. Thus a carbon atom can share its 4 outer electrons with the 4 electrons of 4 hydrogen

Molecular formula	Atomic structure	Structural formula with covalent bond

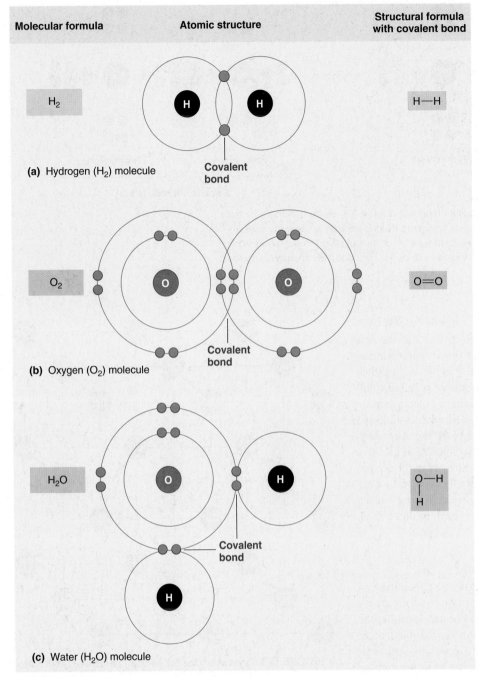

Each electron that is shared by two atoms is counted toward the number of electrons needed to fill the outer shell of each atom. Thus each carbon atom shares four pairs, or 8 electrons, and so has 8 in its outer shell. Each hydrogen shares one pair, or 2 electrons, and so has a filled outer shell. (Remember, hydrogen atoms need only two electrons to complete their outer shell, which is the first shell.) The sharing of a pair of electrons by atoms binds them together by means of a **covalent bond** (● Figure B-4). Covalent bonds are the strongest of chemical bonds; that is, they are the hardest to break.

Covalent bonds also form between some identical atoms. For example, two hydrogen atoms can complete their outer shells by sharing one electron pair made from the single electrons of each atom, as shown in Equation B-2:

$$H\cdot \ + \ \cdot H \rightarrow H:H \qquad \text{Eq. B-2}$$

Thus hydrogen gas consists of individual H_2 molecules (● Figure B-4a). (A subscript following a chemical symbol indicates the number of that type of atom present in the molecule.) Several other nonmetallic elements also exist as molecules because covalent bonds form between identical atoms; oxygen (O_2) is an example (● Figure B-4b).

Often, an atom can form covalent bonds with more than one atom. One of the most familiar examples is water (H_2O), consisting of two hydrogen atoms each forming a single covalent bond with one oxygen atom (● Figure B-4c). Equation B-3 represents the formation of water's covalent bonds:

$$\begin{array}{c} H\cdot \\ \ \ \ \ \ \ + \ \cdot \ddot{O}: \ \rightarrow \ H:\ddot{O}: \qquad \text{Eq. B-3}\\ H\cdot \ \ \ \ \ \ \ \ \ \ \ \ H \end{array}$$

The water molecule is sometimes represented as

$$\begin{array}{c} H—O \\ | \\ H \end{array}$$

where the nonshared electron pairs are not shown and the covalent bonds, or shared pairs, are represented by dashes.

● **FIGURE B-4 A covalent bond.** A covalent bond is formed when atoms that share a pair of electrons are both attracted toward the shared pair.

atoms, as shown in Equation B-1, where the outer-shell electrons are shown as dots around the symbol of each atom. (The resulting compound is methane, CH_4, a gas made up of individual CH_4 molecules.)

Shared electron pairs

$$\cdot \overset{\cdot}{\underset{\cdot}{C}} \ + \ 4\cdot H \rightarrow H\overset{H}{\underset{H}{:}}\overset{\cdot\cdot}{C}\overset{\cdot\cdot}{:}H \qquad \text{Eq. B-1}$$

Shared electron pairs

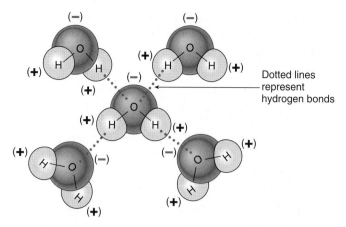

Dotted lines represent hydrogen bonds

● **FIGURE B-5 A polar molecule.** A water molecule is an example of a polar molecule, in which the distribution of shared electrons is not uniform. Because the oxygen atom pulls the shared electrons more strongly than the hydrogen atoms do, the oxygen side of the molecule is slightly negatively charged, and the hydrogen sides are slightly positively charged.

● **FIGURE B-6 A hydrogen bond.** A hydrogen bond is formed by the attraction of a positively charged hydrogen end of a polar molecule to the negatively charged end of another polar molecule.

Nonpolar and polar molecules

The electrons between two atoms in a covalent bond are not always shared equally. When the atoms sharing an electron pair are identical, such as two oxygen atoms, the electrons are attracted equally by both atoms and so are shared equally. The result is a **nonpolar molecule.** The term nonpolar implies no difference at the two ends (two "poles") of the bond. Because both atoms within the molecule exert the same pull on the shared electrons, each shared electron spends the same amount of time orbiting each atom. Thus both atoms remain electrically neutral in a nonpolar molecule such as O_2.

When the sharing atoms are not identical, unequal sharing of electrons occurs because atoms of different elements do not exert the same pull on shared electrons. For example, an oxygen atom strongly attracts electrons when it is bonded to other atoms. A **polar molecule** results from the unequal sharing of electrons between different types of atoms covalently bonded together. The water molecule is a good example of a polar molecule. The oxygen atom pulls the shared electrons more strongly than do the hydrogen atoms within each of the two covalent bonds. Consequently, the electron of each hydrogen atom tends to spend more time away orbiting around the oxygen atom than at home around the hydrogen atom. Because of this nonuniform distribution of electrons, the oxygen side of the water molecule where the shared electrons spend more time is slightly negative, and the two hydrogens that are visited less frequently by the electrons are slightly more positive (● Figure B-5). Note that the entire water molecule has the same number of electrons as it has protons, and so as a whole has no net charge. This is unlike ions, which have an electron excess or deficit. Polar molecules have a balanced number of protons and electrons but an unequal distribution of the shared electrons among the atoms making up the molecule.

Hydrogen bonds

Polar molecules are attracted to other polar molecules. In water, for example, an attraction exists between the positive hydrogen ends of some molecules and the negative oxygen ends of others. Hydrogen is not a part of all polar molecules, but when it is covalently bonded to an atom that strongly attracts electrons to form a covalent molecule, the attraction of the positive (hydro-

gen) end of the polar molecule to the negative end of another polar molecule is called a **hydrogen bond** (● Figure B-6). Thus, the polar attractions of water molecules to each other are an example of hydrogen bonding.

Chemical Reactions

Processes in which chemical bonds are broken and/or formed are called **chemical reactions.** Reactions are represented by equations in which the reacting substances (**reactants**) are typically written on the left, the newly produced substances (**products**) are written on the right, and an arrow meaning "yields" points from the reactants to the products. These conventions are illustrated in Equation B-4:

$$A + B \rightarrow C + D \qquad \text{Eq. B-4}$$
$$\text{Reactants} \qquad \text{Products}$$

Balanced equations

A chemical equation is a "chemical bookkeeping" ledger that describes what happens in a reaction. By the **law of conservation of mass,** the total mass of all materials entering a reaction equals the total mass of all the products. Thus, the total number of atoms of each element must always be the same on the left and right sides of the equation, because no atoms are lost. Such equations in which the same number of atoms of each type appears on both sides are called **balanced equations.** When writing a balanced equation, the number *preceding* a chemical symbol designates the number of independent (unjoined) atoms, ions, or molecules of that type, whereas a number written as a subscript *following* a chemical symbol denotes the number of a particular atom within a molecule. The absence of a number indicates "one" of that particular chemical. Let us look at a specific example, the oxidation of glucose (the sugar that cells use as fuel), as shown in Equation B-5:

$$C_6H_{12}O_6 + 6\,O_2 \rightarrow 6\,CO_2 + 6\,H_2O \qquad \text{Eq. B-5}$$
$$\text{glucose} \quad \text{oxygen} \quad \underset{\text{dioxide}}{\text{carbon}} \quad \text{water}$$

According to this equation, 1 molecule of glucose reacts with 6 molecules of oxygen to produce 6 molecules of carbon dioxide and 6 molecules of water. Note the following balance in this reaction:

- 6 carbon atoms on the left (in 1 glucose molecule) and 6 carbon atoms on the right (in 6 carbon dioxide molecules)

- 12 hydrogen atoms on the left (in 1 glucose molecule) and 12 on the right (in 6 water molecules, each containing 2 hydrogen atoms)

- 18 oxygen atoms on the left (6 in 1 glucose molecule plus 12 more in the 6 oxygen molecules) and 18 on the right (12 in 6 carbon dioxide molecules, each containing 2 oxygen atoms, and 6 more in the 6 water molecules, each containing 1 oxygen atom)

Reversible and irreversible reactions

Under appropriate conditions, the products of a reaction can be changed back to the reactants. For example, carbon dioxide gas dissolves in and reacts with water to form carbonic acid, H_2CO_3:

$$CO_2 + H_2O \rightarrow H_2CO_3 \qquad \text{Eq. B-6}$$

Carbonic acid is not very stable, however, and as soon as some is formed, part of it decomposes to give carbon dioxide and water:

$$H_2CO_3 \rightarrow CO_2 + H_2O \qquad \text{Eq. B-7}$$

Reactions that go in both directions are called **reversible reactions.** They are usually represented by double arrows pointing in both directions:

$$CO_2 + H_2O \rightleftharpoons H_2CO_3 \qquad \text{Eq. B-8}$$

Theoretically, every reaction is reversible. Often, however, conditions are such that a reaction, for all practical purposes, goes in only one direction; such a reaction is called **irreversible.** For example, an irreversible reaction takes place when an explosion occurs, because the products do not remain in the vicinity of the reaction site to get together to react.

Catalysts; enzymes

The rates (speeds) of chemical reactions are influenced by a number of factors, of which catalysts are one of the most important. A **catalyst** is a "helper" molecule that speeds up a reaction without being used up in the reaction. Living organisms use catalysts known as **enzymes.** These enzymes exert amazing influence on the rates of chemical reactions that take place in the organisms. Reactions that take weeks or even months to occur under normal laboratory conditions take place in seconds under the influence of enzymes in the body. One of the fastest-acting enzymes is *carbonic anhydrase,* which catalyzes the reaction between carbon dioxide and water to form carbonic acid. This reaction is important in the transport of carbon dioxide from tissue cells, where it is produced metabolically, to the lungs, where it is excreted. The equation for the reaction was shown in Equation B-6. (Carbonic anhydrase indirectly catalyzes this reaction by converting $CO_2 + H_2O$ di-

rectly to HCO_3^-, which can form H_2CO_3, but it is common practice to represent the combined result of these reactions in the summary form shown in Equation B-6; see p. 571 for specific details.) Each molecule of carbonic anhydrase catalyzes the conversion of 36 million CO_2 molecules per minute! Enzymes are important in essentially every chemical reaction that takes place in living organisms.

Molecular and Formula Weight and the Mole

Because molecules are made up of atoms, the relative mass of a molecule is simply the sum of the relative masses (atomic weights) of the atoms found in the molecule. The relative masses of molecules are called **molecular masses** or **molecular weights.** The molecular weight of water, H_2O, is thus the sum of the atomic weights of two hydrogen atoms and one oxygen atom, or 1.01 amu + 1.01 amu + 16.00 amu = 18.02 amu.

Not all compounds exist in the form of molecules. Ionically bonded substances such as sodium chloride consist of three-dimensional arrangements of sodium ions (Na^+) and chloride ions (Cl^-) in a 1-to-1 ratio. The formulas for ionic compounds reflect only the ratio of the ions in the compound and should not be interpreted in terms of molecules. Thus the formula for sodium chloride, NaCl, indicates that the ions combine in a 1-to-1 ratio. It is convenient to apply the concept of relative masses to ionic compounds even though they do not exist as molecules. The **formula weight** for such compounds is defined as the sum of the atomic weights of the atoms found in the formula. Thus the formula weight of NaCl is equal to the sum of the atomic weights of one sodium atom and one chlorine atom, or 22.99 amu + 35.45 amu = 58.44 amu.

As you have seen, chemical reactions can be represented by equations and discussed in terms of numbers of molecules, atoms, and ions reacting with each other. To carry out reactions in the laboratory, however, a scientist cannot count out numbers of reactant particles but instead must be able to weigh out the correct amount of each reactant. Using the mole concept makes this task possible. A **mole** (abbreviated *mol*) of a pure element or compound is the amount of material contained in a sample of the pure substance that has a mass in grams equal to the substance's atomic weight (for elements) or the molecular weight or formula weight (for compounds). Thus 1 mole of potassium, K, would be a sample of the element with a mass of 39.10 g. Similarly, a mole of H_2O would have a mass of 18.02 g, and a mole of NaCl would be a sample with a mass of 58.44 g.

The fact that atomic weights, molecular weights, and formula weights are relative masses leads to a fundamental characteristic of moles. One mole of oxygen atoms has a mass of 16.00 g, and 1 mole of hydrogen atoms has a mass of 1.01 g. Thus the ratio of the masses of 1 mole of each element is 16.00/1.01, the same as the ratio of the atomic weights for the two elements. Recall that these atomic weights compare the relative masses of oxygen and hydrogen. Accordingly, the number of oxygen atoms present in 16 grams of oxygen (1 mole of oxygen) is the same as the number of hydrogen atoms present in 1.01 grams of hydrogen. Therefore, 1 mole of oxygen contains exactly the same number of oxygen atoms as the number of hydrogen atoms in 1 mole of hydrogen. Thus, it is possible and sometimes useful to think of a mole as a specific number of particles. This number, called **Avogadro's number,** is equal to 6.02×10^{23}.

Solutions, Colloids, and Suspensions

In contrast to a compound, a **mixture** consists of two or more types of elements or molecules physically blended together (intermixed) instead of being linked by chemical bonds. A compound has very different properties from the individual elements of which it is composed. For example, the solid, white NaCl (table salt) crystals you use to flavor your food are very different from either sodium (a silvery white metal) or chlorine (a poisonous yellow-green gas found in bleach). By comparison, each component of a mixture retains its own chemical properties. If you mix salt and sugar together, each retains its own distinct taste and other individual properties. The constituents of a compound can only be separated by chemical means—bond breakage. By contrast, the components of a mixture can be separated by physical means, such as filtration or evaporation. The most common mixtures in the body are mixtures of water and various other substances. These mixtures are categorized as *solutions, colloids,* or *suspensions,* depending on the size and nature of the substance mixed with water.

Solutions

Most chemical reactions in the body take place between reactants that have dissolved to form solutions. **Solutions** are homogenous mixtures containing a relatively large amount of one substance called the **solvent** (the dissolving medium) and smaller amounts of one or more substances called **solutes** (the dissolved particles). Salt water, for example, contains mostly water, which is thus the solvent, and a smaller amount of salt, which is the solute. Water is the solvent in most solutions found in the human body.

Electrolytes; nonelectrolytes

When ionic solutes are dissolved in water to form solutions, the resulting solution will conduct electricity. This is not true for most covalently bonded solutes. For example, a salt–water solution conducts electricity, but a sugar–water solution does not. When salt dissolves in water, the solid lattice of Na^+ and Cl^- is broken down, and the individual ions are separated and distributed uniformly throughout the solution. These mobile, charged ions conduct electricity through the solution. Solutes that form ions in solution and conduct electricity are called **electrolytes.** Some very polar covalent molecules also behave this way. When sugar dissolves, however, individual covalently bonded sugar molecules leave the solid and become uniformly distributed throughout the solution. These uncharged molecules cannot conduct a current. Solutes that do not form conductive solutions are called **nonelectrolytes.**

Measures of concentration

The amount of solute dissolved in a specific amount of solution can vary. For example, a salt–water solution might contain 1 g of salt in 100 ml of solution, or it could contain 10 g of salt in 100 ml of solution. Both solutions are salt–water solutions, but they have different concentrations of solute. The **concentration** of a solution indicates the relationship between the amount of solute and the amount of solution. Concentrations can be given in a number of different units.

MOLARITY Concentrations given in terms of **molarity (M)** give the number of moles of solute in exactly 1 liter of solution. Thus a half molar (0.5 M) solution of NaCl would contain one-half mole, or 29.22 g, of NaCl in each liter of solution.

NORMALITY When the solute is an electrolyte, it is sometimes useful to express the concentration of the solution in a unit that gives information about the amount of ionic charge in the solution. This is done by expressing concentration in terms of **normality (N).** The normality of a solution gives the number of equivalents of solute in exactly 1 liter of solution. An **equivalent** of an electrolyte is the amount that produces 1 mole of positive (or negative) charges when it dissolves. The number of equivalents of an electrolyte can be calculated by multiplying the number of moles of electrolyte by the total number of positive charges produced when one formula unit of the electrolyte dissolves. Consider NaCl and calcium chloride ($CaCl_2$) as examples. The ionization reactions for one formula unit of each solute are:

$$NaCl \rightarrow Na^+ + Cl^- \qquad \text{Eq. B-9}$$

$$CaCl_2 \rightarrow Ca^{2+} + 2\ Cl^- \qquad \text{Eq. B-10}$$

Thus 1 mole of NaCl produces 1 mole of positive charges (Na^+) and so contains 1 equivalent:

$$(1 \text{ mole NaCl})(1) = 1 \text{ equivalent}$$

where the number 1 used to multiply the 1 mole of NaCl came from the +1 charge on Na^+.

One mole of $CaCl_2$ produces 1 mole of Ca^{2+}, which is 2 moles of positive charge. Thus 1 mole of $CaCl_2$ contains 2 equivalents:

$$(1 \text{ mole CaCl}_2)(2) = 2 \text{ equivalents}$$

where the number 2 used in the multiplication came from the +2 charge on Ca^{2+}.

If two solutions were made such that one contained 1 mole of NaCl per liter and the other contained 1 mole of $CaCl_2$ per liter, the NaCl solution would contain 1 equivalent of solute per liter and would be 1 normal (1 N). The $CaCl_2$ solution would contain 2 equivalents of solute per liter and would be 2 normal (2 N).

OSMOLARITY Another expression of concentration frequently used in physiology is **osmolarity (osm),** which indicates the total number of solute particles in a liter of solution instead of the relative weights of the specific solutes. The osmolarity of a solution is the product of M and *n,* where *n* is the number of moles of solute particles obtained when 1 mole of solute dissolves. Because nonelectrolytes such as glucose do not dissociate in solution, $n = 1$ and the osmolarity (n times M) is equal to the molarity of the solution. For electrolyte solutions, the osmolarity exceeds the molarity by a factor equal to the number of ions produced on dissociation of each molecule in solution. For example, because a NaCl molecule dissociates into two ions, Na^+ and Cl^-, the osmolarity of a 1 M solution of NaCl is $2 \times 1\ M = 2$ osm.

Colloids and suspensions

In solutions, solute particles are ions or small molecules. By contrast, the particles in colloids and suspensions are much larger than ions or small molecules. In colloids and suspensions, these particles are known as **dispersed-phase particles** instead of solutes. When the dispersed-phase particles are no more than about 100 times the size of the largest solute particles found in a solution, the mixture is called a **colloid.** The dispersed-phase particles of colloids generally do not settle out. All dispersed-phase particles of colloids carry electrical charges of the same sign. Thus they repel each other. The constant buffeting from these collisions keeps the particles from settling. The most abundant colloids in the body are small functional proteins that are dispersed in the body fluids. An example is the colloidal dispersion of the plasma proteins in the blood (see p. 392).

When dispersed-phase particles are larger than those in colloids, if the mixture is left undisturbed the particles will settle out because of the force of gravity. Such mixtures are usually called **suspensions.** The major example of a suspension in the body is the mixture of blood cells suspended in the plasma. The constant movement of blood as it circulates through the blood vessels keeps the blood cells rather evenly dispersed within the plasma. However, if a blood sample is placed in a test tube and treated to prevent clotting, the heavier blood cells gradually settle to the bottom of the tube.

Inorganic and Organic Chemicals

Chemicals are commonly classified into two categories: inorganic and organic.

Distinction between inorganic and organic chemicals

The original criterion used for this classification was the origin of the chemicals. Those that came from living or once-living sources were *organic,* and those that came from other sources were *inorganic.* Today the basis for classification is the element carbon. **Organic** chemicals are generally those that contain carbon. All others are classified as **inorganic.** A few carbon-containing chemicals are also classified as inorganic; the most common are pure carbon in the form of diamond and graphite, carbon dioxide (CO_2), carbon monoxide (CO), carbonates such as limestone ($CaCO_3$), and bicarbonates such as baking soda ($NaHCO_3$).

The unique ability of carbon atoms to bond to each other and form networks of carbon atoms results in an interesting fact. Even though organic chemicals all contain carbon, millions of these compounds have been identified. Some were isolated from natural plant or animal sources, and many have been synthesized in laboratories. Inorganic chemicals include all the other 108 elements and their compounds. The number of known inorganic chemicals made up of all these other elements is estimated to be about 250,000, compared to millions of organic compounds made up predominantly of carbon.

Monomers and polymers

Another result of carbon's ability to bond to itself is the large size of some organic molecules. Organic molecules range in size from methane, CH_4, a small, simple molecule with one carbon atom, to molecules such as DNA that contain as many as a million carbon atoms. Organic molecules that are essential for life are called **biological molecules,** or **biomolecules** for short. Some biomolecules are rather small organic compounds, including *simple sugars, fatty acids, amino acids,* and *nucleotides.* These small, single units, known as **monomers** (meaning "single unit"), are building blocks for the synthesis of larger biomolecules, including *complex carbohydrates, lipids, proteins, and nucleic acids,* respectively. These larger organic molecules are called **polymers** (meaning "many units"), reflecting the fact that they are made by the bonding together of a number of smaller monomers. For example, starch is formed by linking many glucose molecules together. Very large organic polymers are often referred to as **macromolecules,** reflecting their large size (macro means "large"). Macromolecules include many naturally occurring molecules, such as DNA and structural proteins, as well as many molecules that are synthetically produced, such as synthetic textiles (for example, nylon) and plastics.

Acids, Bases, and Salts

Acids, bases, and salts may be inorganic or organic compounds.

Acids and bases

Acids and bases are chemical opposites, and salts are produced when acids and bases react with each other. In 1887, Swedish chemist Svante Arrhenius proposed a theory defining acids and bases. He said that an *acid* is any substance that will dissociate, or break apart, when dissolved in water and in the process release a hydrogen ion, H^+. Similarly, bases are substances that dissociate when dissolved in water and in the process release a hydroxyl ion, OH^-. Hydrogen chloride (HCl) and sodium hydroxide (NaOH) are examples of Arrhenius acids and bases; their dissociations in water are represented in Equations B-11 and B-12, respectively:

$$HCl \rightarrow H^+ + Cl^- \qquad \text{Eq. B-11}$$

$$NaOH \rightarrow Na^+ + OH^- \qquad \text{Eq. B-12}$$

Note that the hydrogen ion is a bare proton, the nucleus of a hydrogen atom. Also note that both HCl and NaOH would behave as electrolytes.

Arrhenius did not know that free hydrogen ions cannot exist in water. They covalently bond to water molecules to form hydronium ions, as shown in Equation B-13:

$$H^+ + :\overset{\displaystyle ..}{\underset{\displaystyle |}{O}}-H \rightarrow \left[H-\overset{\displaystyle ..}{\underset{\displaystyle |}{O}}-H \right]^+ \qquad \text{Eq. B-13}$$

In 1923, Johannes Brønsted in Denmark and Thomas Lowry in England proposed an acid–base theory that took this behavior into account. They defined an **acid** as any hydrogen-containing substance that donates a proton (hydrogen ion) to another substance (an acid is a *proton donor*) and a **base** as any substance that accepts a proton (a base is a *proton acceptor*). According to these definitions, the acidic behavior of HCl given in Equation B-11 is rewritten as shown in Equation B-14:

$$HCl + H_2O \rightleftharpoons H_3O^+ + Cl^- \qquad \text{Eq. B-14}$$

Note that this reaction is reversible, and the hydronium ion is represented as H_3O^+. In Equation B-14, the HCl acts as an acid in the forward (left-to-right) reaction, whereas water acts as a base. In the reverse reaction (right-to-left), the hydronium ion gives up a proton and thus is an acid, whereas the chloride ion, Cl^-, accepts the proton and so is a base. It is still a common practice to use equations such as B-11 to simplify the representation of the dissociation of an acid, even though scientists recognize that equations like B-14 are more correct.

Salts; neutralization reactions

At room temperature, **inorganic salts** are crystalline solids that contain the positive ion (cation) of an Arrhenius base such as NaOH and the negative ion (anion) of an acid such as HCl. Salts can be produced by mixing solutions of appropriate acids and bases, allowing a neutralization reaction to occur. In **neutralization reactions,** the acid and base react to form a salt and water. Most salts that form are water soluble and can be recovered by evaporating the water. Equation B-15 is a neutralization reaction:

$$HCl + NaOH \rightarrow NaCl + H_2O \qquad \text{Eq. B-15}$$

When acids or bases are used as solutes in solutions, the concentrations can be expressed as normalities just as they were earlier for salts. An equivalent of acid is the amount that gives up 1 mole of H^+ in solution. Thus, 1 mole of HCl is also 1 equivalent, but 1 mole of H_2SO_4 is 2 equivalents. Bases are described in a similar way, but an equivalent is the amount of base that gives 1 mole of OH^-.

See Chapter 15 for a discussion of acid–base balance in the body.

Functional Groups of Organic Molecules

Organic molecules consist of carbon and one or more additional elements covalently bonded to one another in "Tinker Toy" fashion. The simplest organic molecules, *hydrocarbons,* such as methane and petroleum products, have only hydrogen atoms attached to carbon backbones of varying lengths. All biomolecules always have additional elements besides hydrogen added to the carbon backbone. The carbon backbone forms the stable portion of most biomolecules. Other atoms covalently bonded to the carbon backbone, either alone or in clusters, form what are termed *functional groups.* All organic compounds can be classified according to the functional group or groups they contain. **Functional groups** are specific combinations of atoms that generally react in the same way, regardless of the number of carbon atoms in the molecule to which they are attached. For example, all aldehydes contain a functional group that includes one carbon atom, one oxygen atom, and one hydrogen atom covalently bonded in a specific way:

$$\overset{\displaystyle O}{\underset{\displaystyle }{\overset{\displaystyle \|}{(-C-H)}}}$$

The carbon atom in an aldehyde group forms a single covalent bond with the hydrogen atom and a **double bond** (a bond in which two covalent bonds are formed between the same atoms, designated by a double line between the atoms) with the oxygen atom. The aldehyde group is attached to the rest of the molecule by a single covalent bond extending to the left of the carbon atom. Most aldehyde reactions are the same regardless of the size and nature of the rest of the molecule to which the aldehyde group is attached. Reactions of physiological importance often occur between two functional groups or between one functional group and a small molecule such as water.

Carbohydrates

Carbohydrates are organic compounds of tremendous biological and commercial importance. They are widely distributed in nature and include such familiar substances as starch, table sugar, and cellulose. Carbohydrates have five important functions in living organisms: They provide energy, serve as a stored form of chemical energy, provide dietary fiber, supply carbon atoms for the synthesis of cell components, and form part of the structural elements of cells.

Chemical composition of carbohydrates

Carbohydrates contain carbon, hydrogen, and oxygen. They acquired their name because most of them contain these three elements in an atomic ratio of one carbon to two hydrogens to one oxygen. This ratio suggests that the general formula is CH_2O and that the compounds are simply carbon hydrates ("watered" carbons), or carbohydrates. It is now known that they are not hydrates of carbon, but the name persists. All carbohydrates have a large number of functional groups per molecule. The most common functional groups in carbohydrates are *alcohol, ketone,* and *aldehyde*—

$$(-OH),\ \overset{\displaystyle O}{\overset{\displaystyle \|}{(-C-)}},\ \overset{\displaystyle O}{\overset{\displaystyle \|}{(-C-H)}}$$
Alcohol   Ketone   Aldehyde

—or functional groups formed by reactions between pairs of these three.

(a) Chain form of glucose **(b)** Ring form of glucose

● **FIGURE B-7 Forms of glucose.**

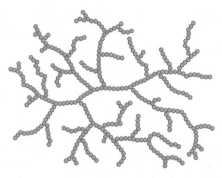

● **FIGURE B-8 A simplified representation of glycogen.** Each circle represents a glucose molecule.

Types of carbohydrates

The simplest carbohydrates are simple sugars, also called **mono-saccharides.** As their name indicates, they consist of single, simple-sugar units called saccharides (*mono* means "one"). The molecular structure of glucose, an important monosaccharide, is shown in ● Figure B-7a. In solution, most glucose molecules assume the ring form shown in ● Figure B-7b. Other common monosaccharides are *fructose, galactose,* and *ribose.*

Disaccharides are sugars formed by linking two monosaccharide molecules together through a covalent bond (*di* means "two"). Some common examples of disaccharides are *sucrose* (common table sugar) and *lactose* (milk sugar). Sucrose molecules are formed from one glucose and one fructose molecule. Lactose molecules each contain one glucose and one galactose unit.

Because of the many functional groups on carbohydrate molecules, large numbers of simple carbohydrate molecules are able to bond together and form long chains and branched networks. The resultant substances, **polysaccharides,** contain many saccharide units (*poly* means "many"). Three common polysaccharides made up entirely of glucose units are glycogen, starch, and cellulose.

■ *Glycogen* is a storage carbohydrate found in animals. It is a highly branched polysaccharide that averages a branch every 8 to 12 glucose units. The structure of glycogen is represented in ● Figure B-8, where each circle represents one glucose unit.

■ *Starch,* a storage carbohydrate of plants, consists of two fractions, amylose and amylopectin. Amylose consists of long, essentially unbranched chains of glucose units. Amylopectin is a highly branched network of glucose units averaging 24 to 30 glucose units per branch. Thus it is less highly branched than glycogen.

■ *Cellulose,* a structural carbohydrate of plants, exists in the form of long, unbranched chains of glucose units. The bonding between the glucose units of cellulose is slightly different from the bonding between the glucose units of glycogen and starch. Humans have digestive enzymes that catalyze the breaking (hydrolysis) of the glucose-to-glucose bonds in starch but lack the necessary enzymes to hydrolyze cellulose glucose-to-glucose bonds. Thus starch is a food for humans, but cellulose is not. Cellulose is the indigestible fiber in our diets.

Lipids

Lipids are a diverse group of organic molecules made up of substances with widely different compositions and molecular structures. Unlike carbohydrates, which are classified on the basis of their *molecular structure,* substances are classified as lipids on the basis of their *solubility.* **Lipids** are insoluble in water but soluble in nonpolar solvents such as alcohol.

Lipids are the waxy, greasy, or oily compounds found in plants and animals. Lipids repel water, a useful characteristic of the protective wax coatings found on some plants. Fats and oils are energy-rich and have relatively low densities. These properties account for the use of fats and oils as stored energy in plants and animals. Still other lipids occur as structural components, especially in cellular membranes. The oily plasma membrane that surrounds each cell serves as a barrier that separates the intracellular contents from the surrounding extracellular fluid (see p. 2, p. 23, p. 53, and p. 55).

Simple lipids

Simple lipids contain just two types of components, fatty acids and alcohols. **Fatty acid molecules** consist of a hydrocarbon chain with a *carboxyl* functional group (—COOH) on the end. The hydrocarbon chain can be of variable length, but natural fatty acids always contain an even number of carbon atoms. The hydrocarbon chain can also contain one or more double bonds between carbon atoms. Fatty acids with no double bonds are called **saturated fatty acids,** whereas those with double bonds are called unsaturated fatty acids. The more double bonds present, the higher the degree of unsaturation. Saturated fatty acids predominate in dietary animal products (for example, meat, eggs, and dairy products), whereas **unsaturated fatty acids** are more prevalent in plant products (for example, grains, vegetables, and fruits). Consumption of a greater proportion of saturated than unsaturated fatty acids is linked with a higher incidence of cardiovascular disease (see p. 337).

The most common alcohol found in simple lipids is **glycerol** (glycerin), a three-carbon alcohol that has three alcohol functional groups (—OH).

Simple lipids called fats and oils are formed by a reaction between the carboxyl group of three fatty acids and the three alcohol groups of glycerol. The resulting lipid is an E-shaped molecule called a **triglyceride.** Such lipids are classified as fats or oils on the basis of their melting points. *Fats* are solids at room temperature, whereas *oils* are liquids. Their melting points depend on the degree of unsaturation of the fatty acids of the triglyceride. The melting point goes down with increasing degree of unsaturation. Thus oils contain more unsaturated fatty acids than fats do. Examples of the components of fats and oils and a typical triglyceride molecule are shown in ● Figure B-9.

When triglycerides form, a molecule of water is released as each fatty acid reacts with glycerol. Adipose tissue in the body contains triglycerides. When the body uses adipose tissue as an energy source, the triglycerides react with water to release free fatty acids into the blood. The fatty acids can be used as an immediate energy source by many organs. In the liver, free fatty acids are converted into compounds called **ketone bodies.** Two of the ketone bodies are acids, and one is the ketone called acetone. Excess ketone bodies are produced during diabetes mellitus, a condition in which most cells resort to using fatty acids as an energy source because the cells are unable to take up adequate amounts of glucose in the face of inadequate insulin action (see p. 719).

● **FIGURE B-9 Triglyceride components and structure.**

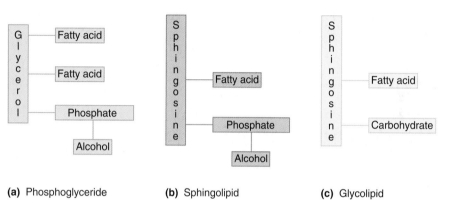

(a) Phosphoglyceride   (b) Sphingolipid   (c) Glycolipid

● **FIGURE B-10 Examples of complex lipids.** In parts (b) and (c), sphingosine is an alcohol similar to glycerol.

Complex lipids

Complex lipids have more than two types of components. The different complex lipids usually contain three or more of the following components: glycerol, fatty acids, a phosphate group, an alcohol other than glycerol, and a carbohydrate. Those that contain phosphate are called **phospholipids.** ● Figure B-10 contains representations of a few complex lipids; it emphasizes the components but does not give details of the molecular structures.

Steroids are lipids that have a unique structural feature consisting of a fused carbon ring system containing three six-membered rings and a single five-membered ring (● Figure B-11). Different steroids possess this characteristic ring structure but have different functional groups and carbon chains attached.

Cholesterol, a steroidal alcohol, is the most abundant steroid in the human body. It is a component of cell membranes and is used by the body to produce other important steroids that include bile salts, male and female sex hormones, and adrenocortical hormones. The structures of cholesterol and corti-

(a) Detailed steroid ring system

(b) Simplified steroid ring system

● **FIGURE B-11 The steroid ring system.**

● FIGURE B-12 Examples of steroidal compounds.

sol, an important adrenocortical hormone, are given in ● Figure B-12.

Proteins

The name *protein* is derived from the Greek word *proteios,* which means "of first importance." It is certainly an appropriate term for these very important biological compounds. Proteins are indispensable components of all living things, where they play crucial roles in all biological processes. Proteins are the main structural component of cells, and all chemical reactions in the body are catalyzed by enzymes, all of which are proteins.

Chemical composition of proteins

Proteins are macromolecules made up of monomers called amino acids. Hundreds of different **amino acids,** both natural and synthetic, are known, but only 20 are commonly found in natural proteins. From this limited pool of 20 amino acids, cells build thousands of different types of proteins, each with a distinctly different function, much the same way that composers create a diversity of unique music from a relatively small number of notes. Different proteins are constructed by varying the types and numbers of amino acids used and by varying the order in which they are linked together. Proteins are not built haphazardly, though, by randomly linking together amino acids. Every protein in the body is deliberately and precisely synthesized under the direction of the blueprint laid down in the person's genes. Thus amino acids are assembled in a specific pattern to produce a given protein to accomplish a particular structural or functional task in the body. (More information about protein synthesis can be found in Appendix C.)

● FIGURE B-13 The general structure of amino acids.

Peptide bonds

Each amino acid molecule has three important parts: an amino functional group ($-NH_2$), a carboxyl functional group ($-COOH$), and a characteristic side chain or R group. These components are shown in expanded form in ● Figure B-13. Amino acids form long chains as a result of reactions between the amino group of one amino acid and the carboxyl group of another amino acid. This reaction is illustrated in Equation B-16:

$$\text{Eq. B-16}$$

Notice that after the two molecules react, the ends of the product still have an amino group and a carboxyl group that can react to extend the chain length. The covalent bond formed in the reaction is called a **peptide bond** (● Figure B-14).

On a molecular scale, proteins are immense molecules. Their size can be illustrated by comparing a glucose molecule to a molecule of hemoglobin, a protein. Glucose has a molecular weight of 180 amu and a molecular formula of $C_6H_{12}O_6$. Hemoglobin, a relatively small protein, has a molecular weight of 65,000 amu and a molecular formula of $C_{2952}H_{4664}O_{832}N_{812}S_8Fe_4$.

Levels of protein structure

The many atoms in a protein are not arranged in a random way. In fact, proteins have a high degree of structural organization that plays an important role in their behavior in the body.

PRIMARY STRUCTURE The first level of protein structure is called the **primary structure.** It is simply the order in which amino acids are bonded together to form the protein chain. Amino acids are frequently represented by three-letter abbreviations, such as Gly for glycine and Arg for arginine. When this practice is followed, the primary structure of a protein can be represented as in ● Figure B-15, which shows part of the primary structure of human insulin, or as in ● Figure B-16a, which depicts a portion of the primary structure of hemoglobin.

SECONDARY STRUCTURE The second level of protein structure, called the **secondary structure,** results when hydrogen bond-

● **FIGURE B-14 A peptide bond.** In forming a peptide bond, the carboxyl group of one amino acid reacts with the amino group of another amino acid.

Thr—Lys—Pro—Thr—Tyr—Phe—Phe—Gly—Arg— · · · · ·

● **FIGURE B-15 A portion of the primary protein structure of human insulin.**

ing occurs between the amino hydrogen of one amino acid and the carboxyl oxygen

$$(-\overset{\overset{\textstyle O}{\|}}{C}-)$$

of another amino acid in the same chain. As a result of this hydrogen bonding, the involved portion of the chain typically assumes a coiled, helical shape called an alpha (α) helix, which is by far the most common secondary structure found in the body (● Figure B-16b). Other secondary structures such as beta (β) pleated sheets and random coils can also form, depending on the pattern of hydrogen bonding between amino acids located in different parts of the same chain.

TERTIARY AND QUATERNARY STRUCTURE The third level of structure in proteins is the **tertiary structure.** It results when functional groups of the side chains of amino acids in the protein chain react with each other. Several different types of interactions are possible, as shown in ● Figure B-17. Tertiary structures can be visualized by letting a length of wire represent the chain of amino acids in the primary structure of a protein. Next imagine that the wire is wound around a pencil to form a helix, which represents the secondary structure. The pencil is removed, and the helical structure is now folded back on itself or carefully wadded into a ball. Such folded or spherical structures represent the tertiary structure of a protein (see ● Figure B-16c).

All functional proteins exist in at least a tertiary structure. Sometimes, several polypeptides interact with each other to form a fourth level of protein structure, the **quaternary structure.** For example, hemoglobin contains four highly folded polypeptide chains (the **globin** portion) (see ● Figure B-16d). Four iron-containing *heme* groups, one tucked within the interior of each of the folded polypeptide subunits, completes the quaternary structure of hemoglobin (see ● Figure 11-2, p. 394).

Hydrolysis and denaturation

In addition to serving as enzymes that catalyze the many essential chemical reactions of the body, proteins can also undergo reactions themselves. Two of the most important are hydrolysis and denaturation.

HYDROLYSIS Notice that according to Equation B-16, the formation of peptide bonds releases water molecules. Under appropriate conditions, it is possible to reverse such reactions by adding water to the peptide bonds and breaking them. **Hydrolysis** ("breakdown by H_2O") reactions of this type convert large proteins into smaller fragments or even into individual amino acids. Hydrolysis is the means by which digestive enzymes break down ingested food into small units that can be absorbed from the digestive tract lumen into the blood.

DENATURATION **Denaturation** of proteins occurs when the bonds holding a protein in its characteristic conformation are broken so that the protein chain takes on a random, disorganized conformation. Denaturation can result when proteins are heated (including when body temperature rises too high; see p. 650), subjected to extremes of pH (see p. 571), or treated with specific chemicals such as alcohol. In some instances, denaturation is accompanied by coagulation or precipitation, as illustrated by the changes that occur in the white of an egg as it is fried.

Nucleic Acids

Nucleic acids are high-molecular-weight macromolecules responsible for storing and using genetic information in living cells and passing it on to future generations. These important biomolecules are classified into two categories: **deoxyribonucleic acids (DNA)** and **ribonucleic acids (RNA).** DNA is found primarily in the cell's nucleus, and RNA is found primarily in the cytoplasm that surrounds the nucleus.

Both types of nucleic acid are made up of units called **nucleotides,** which in turn are composed of three simpler components. Each nucleotide contains an organic nitrogenous base, a sugar, and a phosphate group. The three components are chemically bonded together with the sugar molecule lying between the base and the phosphate. In RNA, the sugar is ribose, whereas in DNA it is deoxyribose. When nucleotides bond together to form nucleic acid chains, the bonding is between the phosphate of one nucleotide and the sugar of another. The resulting nucleic acids consist of chains of alternating phosphates and sugar molecules, with a base molecule extending out of the chain from each sugar molecule (see ● Figure C-1, p. A-20).

(a) Primary structure

Peptide bonds

Amino acids

(b) Secondary structure

Hydrogen bonds

Beta pleated sheet

Alpha helix

Random coil

(c) Tertiary structure

(d) Quaternary structure

Hemoglobin molecule composed of four highly folded polypeptides

● **FIGURE B-16 Levels of protein structure.** Proteins can have four levels of structure. (a) The primary structure is a particular sequence of amino acids bonded in a chain. (b) At the secondary level, hydrogen bonding occurs between various amino acids within the chain, causing the chain to assume a particular shape. The most common secondary protein structure in the body is the alpha helix. (c) The tertiary structure is formed by the folding of the secondary structure into a functional three-dimensional configuration. (d) Many proteins have a fourth level of structure composed of several polypeptides, as exemplified by hemoglobin.

The chains of nucleic acids have structural features somewhat like those found in proteins. DNA takes the form of two chains that mutually coil around one another to form the well-known double helix. Some RNA occurs in essentially straight chains, whereas in other types the chain forms specific loops or helices. See Appendix C for further details.

High-Energy Biomolecules

Not all nucleotides are used to construct nucleic acids. One very important nucleotide—**adenosine triphosphate (ATP)**—is used as the body's primary energy carrier. Certain bonds in ATP temporarily store energy that is harnessed during the metabolism of foods and make it available to the parts of the cells where it is needed to do specific cellular work (see pp. 32–39). Let us see how ATP functions in this role. Structurally, ATP is a modified RNA (ribose-containing) nucleotide that has adenine as its base and two additional phosphates bonded in sequence to the original nucleotide phosphate. Thus adenosine triphosphate, as the name implies, has a total of three phosphates attached in a string to *adenosine*, the composite of ribose and adenine (● Figure B-18). Attaching these additional phosphates requires considerable energy input. The high-energy input used to create these **high-energy phosphate bonds** is "stored" in the bonds for later use. Most energy transfers in the body involve ATP's terminal phosphate bond. When energy is needed, the third phosphate is cleaved off by hydrolysis, yielding *adenosine diphosphate (ADP)* and an inorganic phosphate (P_i) and releasing energy in the process (Equation B-17):

$$ATP \rightarrow ADP + P_i + \text{energy for use by cell} \qquad \text{Eq. B-17}$$

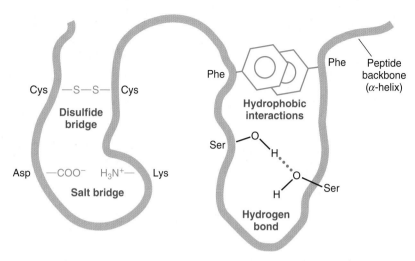

● **FIGURE B-17 Side chain interactions leading to the tertiary protein structure.**

● **FIGURE B-18 The structure of ATP.**

Why use ATP as an energy currency that cells can cash in by splitting the high-energy phosphate bonds as needed? Why not just directly use the energy released during the oxidation of nutrient molecules such as glucose? If all the chemical energy stored in glucose were to be released at once, most of the energy would be squandered because the cell could not capture much of the energy for immediate use. Instead, the energy trapped within the glucose bonds is gradually released and harnessed as cellular "bite-size pieces" in the form of the high-energy phosphate bonds of ATP.

Under the influence of an enzyme, ATP can be converted to a cyclic form of adenosine monophosphate, which contains only one phosphate group, the other two having been cleaved off. The resultant molecule, called **cyclic AMP** or **cAMP**, serves as an intracellular messenger, affecting the activities of a number of enzymes involved in important reactions in the body (see p. 121).

Storage, Replication, and Expression of Genetic Information

Deoxyribonucleic Acid (DNA) and Chromosomes

The nucleus of the cell houses **deoxyribonucleic acid (DNA)**, the genetic blueprint that is unique for each individual.

Functions of DNA

As genetic material, DNA serves two essential functions. First, it contains "instructions" for assembling the structural and enzymatic proteins of the cell. Cellular enzymes in turn control the formation of other cellular structures and also determine the functional activity of the cell by regulating the rate at which metabolic reactions proceed. The nucleus serves as the cell's control center by directly or indirectly controlling almost all cell activities through the role its DNA plays in governing protein synthesis. Because cells make up the body, the DNA code determines the structure and the function of the body as a whole. The DNA an organism has not only dictates whether the organism is a human, a toad, or a pea, but also determines the unique physical and functional characteristics of that individual, all of which ultimately depend on the proteins produced under DNA control.

Second, by replicating (making copies of itself), DNA perpetuates the genetic blueprint within all new cells formed within the body and is responsible for passing on genetic information from parents to children. We will first examine the structure of DNA and the coding mechanism it uses, then turn our attention to the means by which DNA replicates itself and controls protein synthesis.

Structure of DNA

Deoxyribonucleic acid is a huge molecule, composed in humans of millions of nucleotides arranged into two long, paired strands that spiral around each other to form a double helix. Each **nucleotide** has three components: (1) a *nitrogenous base,* a ring-shaped organic molecule containing nitrogen; (2) a 5-carbon ring-shaped sugar molecule, which in DNA is *deoxyribose;* and (3) a phosphate group. Nucleotides are joined end to end by linkages between the sugar of one nucleotide and the phosphate group of the adjacent nucleotide to form a long polynucleotide ("many nucleotide") strand with a sugar–phosphate backbone and bases projecting out one side (● Figure C-1). The four different bases in DNA are the double-ringed bases **adenine (A)** and **guanine (G)** and the single-ringed bases **cytosine (C)** and **thymine (T).** The two polynucleotide strands within a DNA molecule are wrapped around each other so that their bases all project to the interior of the helix. The strands are held together by weak hydrogen bonds formed between the bases of adjoining strands (see ● Figure B-6, p. A-7). Base pairing is highly specific: Adenine pairs only with thymine and guanine pairs only with cytosine (● Figure C-2).

GENES The composition of the repetitive sugar–phosphate backbones that form the "sides" of the DNA "ladder" is identical for every molecule of DNA, but the sequence of the linked bases that form the "rungs" varies among different DNA molecules. The particular sequence of bases in a DNA molecule serves as "instructions," or a "code," dictating the assembly of amino acids into a given order for the synthesis of specific **polypeptides** (chains of amino acids linked by peptide bonds; see p. A-14). A **gene** is a stretch of DNA that codes for the synthesis of a particular polypeptide. Polypeptides, in turn, are folded into a three-dimensional configuration to form a functional protein. Not all portions of a DNA molecule code for structural or enzymatic proteins. Some stretches of DNA code for proteins that regulate genes. Other segments appear important in organizing and packaging DNA within the nucleus. Still other regions are "nonsense" base sequences that have no apparent significance.

Packaging of DNA into chromosomes

The DNA molecules within each human cell, if lined up end to end, would extend more than 2 m (2,000,000 μm), yet these molecules are packed into a nucleus that is only 5 μm in diameter. These molecules are not randomly crammed into the nucleus but are precisely organized into **chromosomes.** Each chromosome consists of a different DNA molecule and contains a unique set of genes.

● FIGURE C-1 Polynucleotide strand. Sugar–phosphate bonds link adjacent nucleotides together to form a polynucleotide strand with bases projecting to one side. The sugar–phosphate backbone is identical in all polynucleotides, but the sequence of the bases varies.

KEY

= Sugar–phosphate backbone of polynucleotide strand

Somatic (body) **cells** contain 46 chromosomes (the **diploid number**), which can be sorted into 23 pairs on the basis of various distinguishing features. Chromosomes composing a matched pair are termed **homologous chromosomes,** one member of each pair having been derived from the individual's maternal parent and the other member from the paternal parent. **Germ** (reproductive) **cells** (that is, sperm and eggs) contain only one member of each homologous pair for a total of 23 chromosomes (the **haploid number**). Union of a sperm and an egg results in a new diploid cell with 46 chromosomes, consisting of a set of 23 chromosomes from the mother and another set of 23 from the father (see p. 745).

DNA molecules are packaged and compressed into discrete chromosomal units in part by nuclear proteins associated with DNA. Two classes of proteins—histone and nonhistone proteins—bind with DNA. **Histones** form bead-shaped bodies that play a key role in packaging DNA into its chromosomal structure. The **nonhistones** are important in gene regulation. The complex formed between the DNA and its associated proteins is known as **chromatin.** The long threads of DNA within a chromosome are wound around histones at regular intervals, thus compressing a given DNA molecule to about one-sixth of its fully extended length. This "beads-on-a-string" structure is further folded and supercoiled into higher and higher levels of organization to further condense DNA into rodlike chromosomes that are readily visible through a light microscope during cell division (● Figure C-3). When the cell is not dividing, the chromosomes partially "unravel" or decondense to a less compact form of chromatin that is indistinct under a light microscope but appears as thin strands and clumps with an electron microscope. The decondensed form of DNA is its working form; that is, it is the form used as a template for protein assembly. Let us turn our attention to this working form of DNA in operation.

Complementary Base Pairing, Replication, and Transcription

Complementary base pairing serves as the foundation for both DNA replication and the initial step of protein synthesis. We will examine the mechanism and significance of complementary base pairing in each of these circumstances, starting with DNA replication.

DNA replication

During DNA replication, the two decondensed DNA strands "unzip" as the weak bonds between the paired bases are enzymatically broken. Then **complementary base pairing** takes

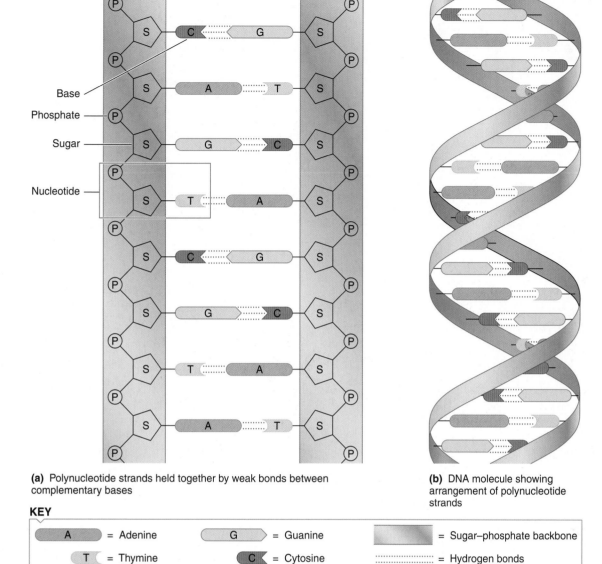

(a) Polynucleotide strands held together by weak bonds between complementary bases

(b) DNA molecule showing arrangement of polynucleotide strands

KEY

A = Adenine	G = Guanine		= Sugar–phosphate backbone	
T = Thymine	C = Cytosine		= Hydrogen bonds	

● **FIGURE C-2 Complementary base pairing in DNA.** (a) Two polynucleotide strands held together by weak hydrogen bonds formed between the bases of adjoining strands—adenine always paired with thymine and guanine always paired with cytosine. (b) Arrangement of the two bonded polynucleotide strands of a DNA molecule into a double helix.

place: New nucleotides present within the nucleus pair with the exposed bases from each unzipped strand (● Figure C-4). New adenine-bearing nucleotides pair with exposed thymine-bearing nucleotides in an old strand, and new guanine-bearing nucleotides pair with exposed cytosine-bearing nucleotides in an old strand. This complementary base pairing is initiated at one end of the two old strands and proceeds in an orderly fashion to the other end. The new nucleotides attracted to and thus aligned in a prescribed order by the old nucleotides are sequentially joined by sugar–phosphate linkages to form two new strands that are complementary to each of the old strands. This replication process results in two complete double-stranded DNA molecules, one strand within each molecule having come from the original DNA molecule and one strand having been newly formed by complementary base pairing.

These two DNA molecules are both identical to the original DNA molecule, with the "missing" strand in each of the original separated strands having been produced as a result of the imposed pattern of base pairing. This replication process, which occurs only during cell division, is essential for perpetuating the genetic code in both the new daughter cells. The duplicate copies of DNA are separated and evenly distributed to the two halves of the cell before it divides. We will cover the topic of cell division in more detail later.

DNA transcription and messenger RNA

At other times, when DNA is not replicating in preparation for cell division, it serves as a blueprint for dictating cellular protein synthesis. How is this accomplished when DNA is sequestered

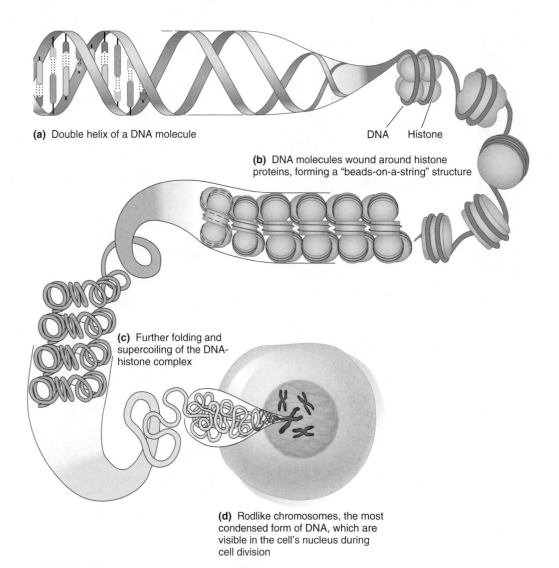

(a) Double helix of a DNA molecule

DNA   Histone

(b) DNA molecules wound around histone proteins, forming a "beads-on-a-string" structure

(c) Further folding and supercoiling of the DNA-histone complex

(d) Rodlike chromosomes, the most condensed form of DNA, which are visible in the cell's nucleus during cell division

● **FIGURE C-3 Levels of organization of DNA.**

within the nucleus and protein synthesis is carried out by ribosomes within the cytoplasm? Several types of another nucleic acid, **ribonucleic acid (RNA),** serve as the "go-between."

STRUCTURE OF RIBONUCLEIC ACID Ribonucleic acid differs structurally from DNA in three regards: (1) The five-carbon sugar in RNA is *ribose* instead of deoxyribose, the only difference between them being the presence in ribose of a single oxygen atom that is absent in deoxyribose; (2) RNA contains the closely related base **uracil** instead of thymine, with the three other bases being the same as in DNA; and (3) RNA is single-stranded and not self-replicating.

All RNA molecules are produced in the nucleus using DNA as a template, then exit the nucleus through openings in the nuclear membrane, called *nuclear pores* (see p. 23), which are large enough for passage of RNA molecules but block the much larger DNA molecules.

The DNA instructions for assembling a particular protein coded in the base sequence of a given gene are "transcribed"

into a molecule of **messenger RNA (mRNA).** The segment of the DNA molecule to be copied uncoils, and the base pairs separate to expose the particular sequence of bases in the gene. In any given gene, only one of the DNA strands is used as a template for transcribing RNA, with the copied strand varying for different genes along the same DNA molecule. The beginning and end of a gene within a DNA strand are designated by particular base sequences that serve as "start" and "stop" signals.

TRANSCRIPTION Transcription is accomplished by complementary base pairing of free RNA nucleotides with their DNA counterparts in the exposed gene (● Figure C-5). The same pairing rules apply except that uracil, the RNA nucleotide substitute for thymine, pairs with adenine in the exposed DNA nucleotides. As soon as the RNA nucleotides pair with their DNA counterparts, sugar–phosphate bonds are formed to join the nucleotides together into a single-stranded RNA molecule that is released from DNA once transcription is complete. The

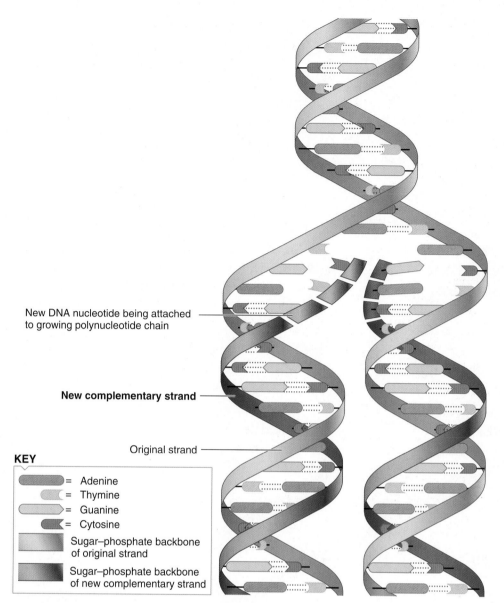

New DNA nucleotide being attached
to growing polynucleotide chain

New complementary strand

Original strand

KEY

=	Adenine
=	Thymine
=	Guanine
=	Cytosine
	Sugar–phosphate backbone of original strand
	Sugar–phosphate backbone of new complementary strand

● **FIGURE C-4 Complementary base pairing during DNA replication.** During DNA replication, the DNA molecule is unzipped, and each old strand directs the formation of a new strand; the result is two identical double-helix DNA molecules.

original conformation of DNA is then restored. The RNA strand is much shorter than a DNA strand because only a one-gene segment of DNA is transcribed into a single RNA molecule. The length of the finished RNA transcript varies, depending on the size of the gene. Within its nucleotide base sequence, this RNA transcript contains instructions for assembling a particular protein. Note that the message is coded in a base sequence that is *complementary to, not identical to,* the original DNA code.

Messenger RNA delivers the final coded message to the ribosomes for **translation** into a particular amino acid sequence to form a given protein. Thus genetic information flows from DNA (which can replicate itself) through RNA to protein. This

is accomplished first by *transcription* of the DNA code into a complementary RNA code, followed by *translation* of the RNA code into a specific protein (● Figure C-6). In the next section, you will learn more about the steps in translation. The structural and functional characteristics of the cell as determined by its protein composition can be varied, subject to control, depending on which genes are "switched on" to produce mRNA.

Free nucleotides present in the nucleus cannot be randomly joined together to form either DNA or RNA strands because the enzymes required to link together the sugar and phosphate components of nucleotides are active only when bound to DNA. This ensures that DNA, mRNA, and protein assembly occur only according to genetic plan.

Translation and Protein Synthesis

Three forms of RNA participate in protein synthesis. Besides messenger RNA, two other forms of RNA are required for translation of the genetic message into cellular protein: ribosomal RNA and transfer RNA (see p. 24).

■ **Messenger RNA (mRNA)** carries the coded message from nuclear DNA to a cytoplasmic ribosome, where it directs the synthesis of a particular protein.

■ **Ribosomal RNA (rRNA)** is an essential component of *ribosomes,* the "workbenches" for protein synthesis (see p. 40). Ribosomes "read" the base sequence code of mRNA and translate it into the appropriate amino-acid sequence during protein synthesis.

■ **Transfer RNA (tRNA)** transfers the appropriate amino acids in the cytosol to their designated site in the amino acid sequence of the protein under construction.

Triplet code; codon

Twenty different amino acids are used to construct proteins, yet only four different nucleotide bases are used to code for these 20 amino acids. In the "genetic dictionary," each different amino acid is specified by a **triplet code** that consists of a specific sequence of three bases in the DNA nucleotide chain. For example, the DNA sequence ACA (adenine, cytosine, adenine) specifies the amino acid cysteine, whereas the sequence ATA specifies the amino acid tyrosine. Each DNA triplet code is transcribed into mRNA as a complementary code word, or **codon,** consisting of a sequenced order of the three bases that pair with the DNA triplet. For example, the DNA triplet code ATA is transcribed as UAU (uracil, adenine, uracil) in mRNA.

Sixty-four different DNA triplet combinations (and, accordingly, 64 different mRNA codon combinations) are possible using the four different nucleotide bases (4^3). Of these possible combinations, 61 code for specific amino acids and the remain-

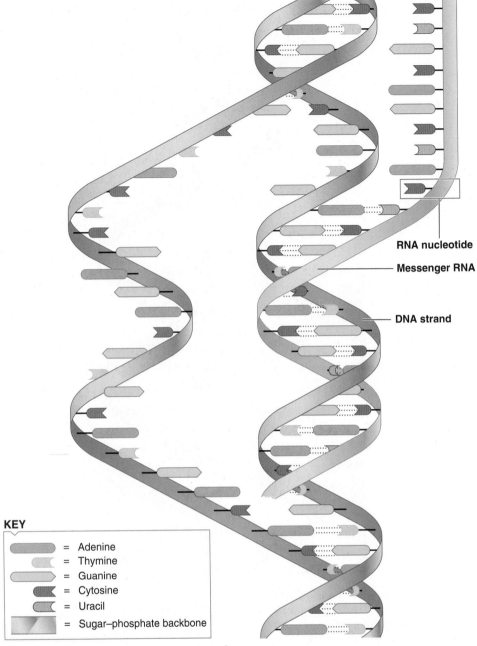

RNA nucleotide

Messenger RNA

DNA strand

KEY

	= Adenine
	= Thymine
	= Guanine
	= Cytosine
	= Uracil
	= Sugar–phosphate backbone

● **FIGURE C-5 Complementary base pairing during DNA transcription.** During DNA transcription, a messenger RNA molecule is formed as RNA nucleotides are assembled by complementary base pairing at a given segment of one strand of an unzipped DNA molecule (that is, a gene).

● **FIGURE C-6 Flow of genetic information from DNA through RNA to protein by transcription and translation.**

ing 3 serve as "stop signals." A stop signal acts as a "period" at the end of a "sentence." The sentence consists of a series of triplet codes that specify the amino acid sequence in a particular protein. When the stop codon is reached, ribosomal RNA releases the finished polypeptide product. Because 61 triplet codes each specify a particular amino acid and there are 20 different amino acids, a given amino acid may be specified by more than one base-triplet combination. For example, tyrosine is specified by the DNA sequence ATG as well as by ATA. In addition, one DNA triplet code, TAC (mRNA codon sequence AUG) functions as a "start signal" in addition to specifying the amino acid methionine. This code marks the place on mRNA where translation is to begin so that the message is started at the correct end and thus reads in the right direction. Interestingly, the same genetic dictionary is used universally; a given three-base code stands for the same amino acid in all living things, including microorganisms, plants, and animals.

Ribosomes

A **ribosome** brings together all components that participate in protein synthesis—mRNA, tRNA, and amino acids—and provides the enzymes and energy required for linking the amino acids together. The nature of the protein synthesized by a given ribosome is determined by the mRNA message being translated. Each mRNA serves as a code for only one particular polypeptide.

A ribosome is an rRNA-protein structure organized into two subunits of unequal size. These subunits are brought together only when a protein is being synthesized (● Figure C-7, step ❶; see also ● Figure 2-17, p. 40). During assembly of a ribosome, an mRNA molecule attaches to the smaller of the ribosomal subunits by means of a *leader sequence,* a section of mRNA that precedes the start codon. The small subunit with mRNA attached then binds to a large subunit to form a complete, functional ribosome. When the two subunits unite, a groove is formed that accommodates the mRNA molecule as it is being translated.

Transfer RNA and anticodons

Free amino acids in the cytosol cannot "recognize" and bind directly with their specific codons in mRNA. Transfer RNA must bring the appropriate amino acid to its proper codon. Even though tRNA is single-stranded, as are all RNA molecules, it is folded back onto itself into a T shape with looped ends (● Figure C-8). The open-ended stem portion recognizes and binds to a specific amino acid. There are at least 20 different varieties of tRNA, each able to bind with only one of the 20 different kinds of amino acids. A tRNA is said to be "charged" when it is carrying its passenger amino acid. The loop end of a tRNA opposite the amino-acid binding site contains a sequence of three exposed bases, known as the **anticodon,** which is complementary to the mRNA codon that specifies the amino acid being carried. Through complementary base pairing, a tRNA can bind with mRNA and insert its amino acid into the protein under construction only at the site designated by the codon for the amino acid. For example, the tRNA molecule that binds with tyrosine bears the anticodon AUA, which can pair only with the mRNA codon UAU, which specifies tyrosine. This dual binding function of tRNA molecules ensures that the correct amino acids are delivered to mRNA for assembly in the order specified by the genetic code. Transfer RNA can only bind with mRNA at a ribosome, so protein assembly does not occur except in the confines of a ribosome.

Steps of protein synthesis

The three steps of protein synthesis are initiation, elongation, and termination.

1. *Initiation.* Protein synthesis is initiated when a charged tRNA molecule bearing the anticodon specific for the start codon binds at this site on mRNA (see ● Figure C-7, step ❷; also see ● Figure 2-17, p. 40).

2. *Elongation.* A second charged tRNA bearing the anticodon specific for the next codon in the mRNA sequence then occupies the site next to the first tRNA (step ❸). At any given time, a ribosome can accommodate only two tRNA molecules bound to adjacent codons. Through enzymatic action, a peptide bond is formed between the two amino acids that are linked to the stems of the adjacent tRNA molecules (step ❹). The linkage is subsequently broken between the first tRNA and its amino acid passenger, leaving the second tRNA with a chain of two amino acids. The uncharged tRNA molecule (the one minus its amino acid passenger) is released from mRNA (step ❺). The ribosome then moves along the mRNA molecule by precisely three bases, a distance of one codon (step ❻), so that the tRNA bearing the chain of two amino acids is moved into the number one ribosomal site for tRNA. Then, an incoming charged tRNA with a complementary anticodon for the third codon in the mRNA sequence occupies the number two ribosomal site that was vacated by the second tRNA (step ❼). The chain of two amino acids subsequently binds with and is transferred to the third tRNA to form a chain of three amino acids (step ❽). Through repetition of this process, amino acids are subsequently added one at a time to a growing polypeptide chain in the order designated by the codon sequence as the ribosomal translation machinery moves stepwise along the mRNA molecule one codon at a time (step ❾). This process is rapid. As many as 10 to 15 amino acids can be added per second.

3. *Termination.* Elongation of the polypeptide chain continues until the ribosome reaches a stop codon in the mRNA molecule, at which time the polypeptide is released. The polypeptide is then folded and modified into a full-fledged protein. The ribosomal subunits dissociate and are free to reassemble into another ribosome for translation of other mRNA molecules.

Energy cost of protein synthesis

Protein synthesis is expensive in terms of energy. Attachment of each new amino acid to the growing polypeptide chain requires a total investment of splitting four high-energy phosphate bonds—two to charge tRNA with its amino acid, one to bind tRNA to the ribosomal-mRNA complex, and one to move the ribosome forward one codon.

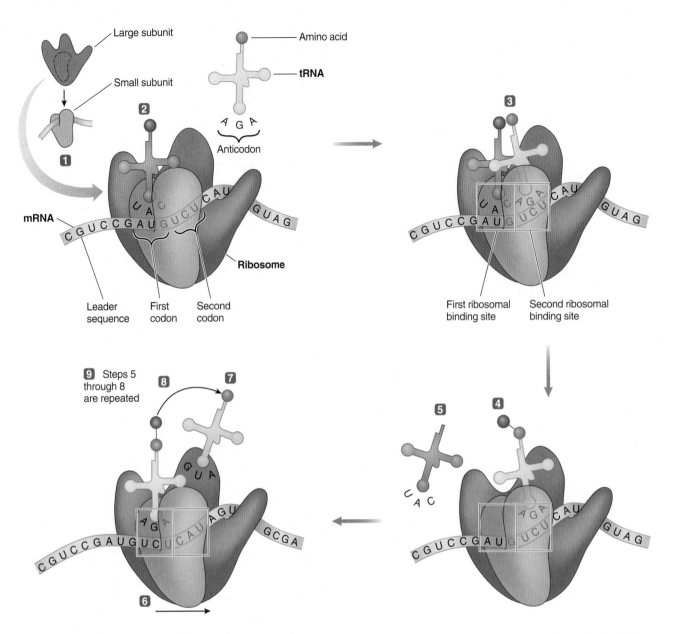

① On binding with a messenger RNA (mRNA) molecule, the small ribosomal subunit joins with the large subunit to form a functional ribosome.

② A transfer RNA (tRNA), charged with its specific amino acid passenger, binds to mNRA by means of complementary base pairing between the tRNA anticodon and the first mRNA codon positioned in the first ribosomal binding site.

③ Another tRNA molecule attaches to the next codon on mRNA positioned in the second ribosomal binding site.

④ The amino acid from the first tRNA is linked to the amino acid on the second tRNA.

⑤ The first tRNA detaches.

⑥ The ribosome shifts forward one codon (a distance of a three-base sequence) along the mRNA molecule.

⑦ Another charged tRNA moves in to attach with the next codon on mRNA, which has now moved into the second ribosomal binding site.

⑧ The amino acids from the tRNA in the first ribosomal site are linked with the amino acid in the second site.

⑨ This process continues (that is, steps 5 through 8 are repeated), with the polypeptide chain continuing to grow, until a stop codon is reached and the polypeptide chain is released.

● **FIGURE C-7 Ribosomal assembly and protein translation.**

● FIGURE C-9 A polyribosome. A polyribosome is formed by numerous ribosomes simultaneously translating mRNA.

● FIGURE C-8 Structure of a tRNA molecule. The open end of a tRNA molecule attaches to free amino acids. The anticodon loop of the tRNA molecule attaches to a complementary mRNA codon.

Polyribosomes

A number of copies of a given protein can be produced from a single mRNA molecule before the latter is chemically degraded. As one ribosome moves forward along the mRNA molecule, a new ribosome attaches at the starting point on mRNA and also starts translating the message. Attachment of many ribosomes to a single mRNA molecule results in a *polyribosome*. Multiple copies of the identical protein are produced as each ribosome moves along and translates the same message (● Figure C-9). The released proteins are used within the cytosol, except for the few that move into the nucleus through the nuclear pores.

Recall that, in contrast to the cytosolic polyribosomes, ribosomes directed to bind with the rough endoplasmic reticulum (ER) feed their growing polypeptide chains into the ER lumen (see p. 25). The resultant proteins are subsequently packaged for export out of the cell or for replacement of membrane components within the cell.

Control of gene activity and protein transcription

Because each somatic cell in the body has the identical DNA blueprint, you might assume that they would all produce the same proteins. This is not the case, however, because different cell types are able to transcribe different sets of genes and thus synthesize different sets of structural and enzymatic proteins. For example, only red blood cells can synthesize hemoglobin, even though all body cells carry the DNA instructions for hemoglobin synthesis. Only about 7% of the DNA sequences in a typical cell are ever transcribed into mRNA for ultimate expression as specific proteins.

Control of gene expression involves gene regulatory proteins that activate ("switch on") or repress ("switch off") the genes that code for specific proteins within a given cell. Various DNA segments that do not code for structural and enzymatic proteins code for synthesis of these regulatory proteins. The molecular mechanisms by which these regulatory genes in turn are controlled in human cells are only beginning to be understood. In some instances, regulatory proteins are controlled by **gene-signaling factors** that bring about differential gene activity among various cells to accomplish specialized tasks. The largest group of known gene-signaling factors in humans is the hormones. Some hormones exert their homeostatic effect by selectively altering the transcription rate of the genes that code for enzymes which are in turn responsible for catalyzing the reaction(s) regulated by the hormone. For example, the hormone cortisol promotes the breakdown of fat stores by stimulating synthesis of the enzyme that catalyzes the conversion of stored fat into its component fatty acids. In other cases, gene action appears to be time specific; that is, certain genes are expressed only at a certain developmental stage in the individual. This is especially important during embryonic development.

Cell Division

Most cells in the human body can reproduce themselves, a process important in growth, replacement, and repair of tissues. The rate at which cells divide is highly variable. Cells within the deeper layers of the intestinal lining divide every few days to replace cells that are continually sloughed off the surface of the lining into the digestive tract lumen. In this way, the entire intestinal lining is replaced about every three days (see p. 626). At the other extreme are nerve cells, which permanently lose the ability to divide beyond a certain period of fetal growth and development. Consequently, when nerve cells are lost through trauma or disease, they cannot be replaced (see p. 2). In between these two extremes are cells that divide infrequently ex-

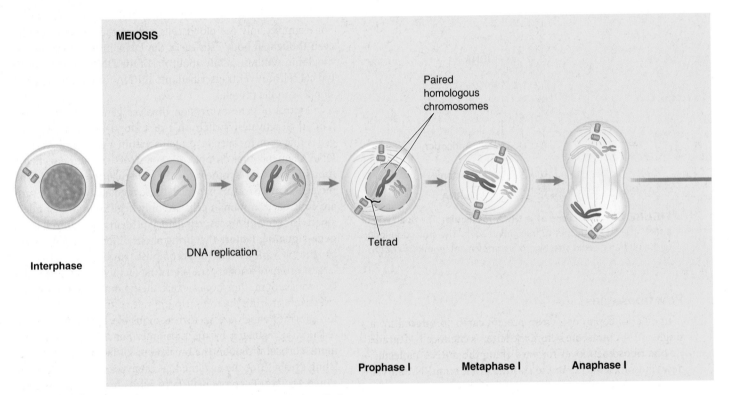

● **FIGURE C-10** A comparison of events in mitosis and meiosis.

cept when needed to replace damaged or destroyed tissue. The factors that control the rate of cell division remain obscure.

Mitosis

Recall that cell division involves two components: nuclear division and cytoplasmic division *(cytokinesis)* (see p. 48). Nuclear division in somatic cells is accomplished by **mitosis,** in which a complete set of genetic information (that is, a diploid number of chromosomes) is distributed to each of two new daughter cells.

A cell capable of dividing alternates between periods of mitosis and nondivision. The interval of time between cell divisions is known as **interphase.** Because mitosis takes less than an

hour to complete, the vast majority of cells in the body at any given time are in interphase.

Replication of DNA and growth of the cell take place during interphase in preparation for mitosis. Although mitosis is a continuous process, it displays four distinct phases: *prophase, metaphase, anaphase,* and *telophase* (● Figure C-10, top).

PROPHASE

1. Chromatin condenses and becomes microscopically visible as chromosomes. The condensed duplicate strands of DNA, known as *sister chromatids,* remain joined together within the chromosome at a point called the *centromere* (● Figure C-11).

2. Cells contain a pair of centrioles, short cylindrical structures that form the mitotic spindle during cell division (see ● Figure 2-19, p. 41). The centriole pair divides, and the daughter centrioles move to opposite ends of the cell, where they assemble between them a mitotic spindle made up of microtubules (see p. 47).

3. The membrane surrounding the nucleus starts to break down.

METAPHASE

1. The nuclear membrane completely disappears.

2. The 46 chromosomes, each consisting of a pair of sister chromatids, align themselves at the midline, or equator, of the cell. Each chromosome becomes attached to the spindle by means of several spindle fibers that extend from the centriole to the centromere of the chromosome.

ANAPHASE

1. The centromeres split, converting each pair of sister chromatids into two identical chromosomes, which separate and move toward opposite poles of the spindle. Molecular motors pull the chromosomes along the spindle fibers toward the poles (see p. 43).

FIGURE C-11 A scanning electron micrograph of human chromosomes from a dividing cell. The replicated chromosomes appear as double structures, with identical sister chromatids joined in the middle at a common centromere.

Dr. Christine J. Harrison and S. Karger AG, Basel.

2. At the end of anaphase, an identical set of 46 chromosomes is present at each of the poles, for a transient total of 92 chromosomes in the soon-to-be-divided cell.

TELOPHASE

1. The cytoplasm divides through formation and gradual tightening of an actin contractile ring at the midline of the cell, thus forming two separate daughter cells, each with a full diploid set of chromosomes (see ● Figure 2-21a, p. 43).

2. The spindle fibers disassemble.

3. The chromosomes uncoil to their decondensed chromatin form.

4. A nuclear membrane reforms in each new cell.

Cell division is complete with the end of telophase. Each of the new cells now enters interphase.

Meiosis

Nuclear division in the specialized case of germ cells is accomplished by **meiosis,** in which only half a set of genetic information (that is, a haploid number of chromosomes) is distributed to each daughter cell. Meiosis differs from mitosis in several important regards (see ● Figure C-10). Specialized diploid germ cells undergo one chromosome replication followed by two nuclear divisions to produce four haploid germ cells.

MEIOSIS I

1. During prophase of the first meiotic division (prophase I), the members of each homologous pair of chromosomes line up

side by side to form a **tetrad,** which is a group of four sister chromatids with two identical chromatids within each member of the pair.

2. The process of crossing over occurs during this period, when the maternal copy and the paternal copy of each chromosome are paired. **Crossing over** involves a physical exchange of chromosome material between nonsister chromatids within a tetrad (● Figure C-12). This process yields new chromosome combinations, thus contributing to genetic diversity.

3. During metaphase I, the 23 tetrads line up at the equator.

4. At anaphase I, homologous chromosomes, each consisting of a pair of sister chromatids joined at the centromere, separate and move toward opposite poles. Maternally and paternally derived chromosomes migrate to opposite poles in random assortments of one member of each chromosome pair without regard for its original derivation. This genetic mixing provides novel new combinations of chromosomes.

5. During telophase I, the cell divides into two cells. Each cell contains 23 chromosomes consisting of two sister chromatids.

MEIOSIS II

1. Following a brief interphase in which no further replication occurs, the 23 unpaired chromosomes line up at the equator, the centromeres split, and the sister chromatids separate for the first time into independent chromosomes that move to opposite poles (prophase II–telophase II in ● Figure C-10).

2. During cytokinesis, each of the daughter cells derived from the first meiotic division forms two new daughter cells. The end result is four daughter cells, each containing a haploid set of chromosomes.

Union of a haploid sperm and haploid egg results in a zygote (fertilized egg) that contains the diploid number of chromosomes. Development of a new multicellular individual from the zygote is accomplished by mitosis and cell differentiation. Because DNA is normally faithfully replicated in its entirety during each mitotic division, all cells in the body possess an identical aggregate of DNA molecules. Structural and functional variations between different cell types result from differential gene expression.

Mutations

An estimated 10^{16} cell divisions take place in the body during the course of a person's lifetime to accomplish growth, repair, and normal cell turnover. Because more than 3 billion nucleotides must be replicated during each cell division, no wonder "copying errors" occasionally occur. Any change in the DNA sequence is known as a **point (gene) mutation.** A point mutation arises when a base is inadvertently substituted, added, or deleted during the replication process.

When a base is inserted in the wrong position during DNA replication, the mistake can often be corrected by a built-in "proofreading" system. Repair enzymes remove the newly replicated strand back to the defective segment, at which time normal base pairing resumes to resynthesize a corrected strand. Not all mistakes can be corrected, however.

Centromere

(a) Formation of tetrad　　　　　**(b)** Crossing over　　　　　**(c)** New combinations of genetic material

● **FIGURE C-12 Crossing over.** (a) During prophase I of meiosis, each homologous pair of chromosomes lines up side by side to form a tetrad. (b) Physical exchange of chromosome material occurs between nonsister chromatids. (c) As a result of this crossing over, new combinations of genetic material are formed within the chromosomes.

Mutations can arise spontaneously by chance alone or they can be induced by **mutagens,** which are factors that increase the rate at which mutations take place. Mutagens include various chemical agents as well as ionizing radiation such as X rays and atomic radiation. Mutagens promote mutations either by chemically altering the DNA base code through a variety of mechanisms or by interfering with the repair enzymes so that abnormal base segments cannot be cut out.

Depending on the location and nature of a change in the genetic code, a given mutation may (1) have no noticeable effect if it does not alter a critical region of a cellular protein; (2) adversely alter cell function if it impairs the function of a crucial protein; (3) be incompatible with the life of the cell, in which case the cell dies and the mutation is lost with it; or (4) in rare cases, prove beneficial if a more efficient structural or enzymatic protein results. If a mutation occurs in a body cell (a **somatic mutation**), the outcome will be reflected as an alteration in all future copies of the cell in the affected individual, but it will not be perpetuated beyond the life of the individual. If, by contrast, a mutation occurs in a sperm- or egg-producing cell **(germ cell mutation),** the genetic alteration may be passed on to succeeding generations.

In most instances, **cancer** results from multiple somatic mutations that occur over a course of time within DNA segments known as **proto-oncogenes.** Proto-oncogenes are normal genes whose coded products are important in the regulation of cell growth and division. These genes have the potential of becoming overzealous **oncogenes** ("cancer genes"), which induce the uncontrolled cell proliferation characteristic of cancer. Proto-oncogenes can become cancer producing as a result of several sequential mutations in the gene itself or by changes in adjacent regions that regulate the proto-oncogenes. Less frequently, tumor viruses become incorporated in the DNA blueprint and act as oncogenes. Alternatively, cancer may arise from mutations that disable **tumor suppressor genes,** which normally restrain cell proliferation in check-and-balance fashion.

Principles of Quantitative Reasoning

*By Kim E. Cooper, Midwestern University, and John D. Nagy,
Scottsdale Community College*

Introduction

Historically, as a branch of science matures, it typically becomes more precise and usually more quantitative. This trend is becoming increasingly true of biology and especially of physiology. Most students, however, are uncomfortable with quantitative reasoning. Students are usually quite capable of doing the mechanical manipulations of mathematics but have trouble translating back and forth between words, concepts, and equations. This appendix is meant to help you become more comfortable working with equations.

Why Are Equations Useful?

A great deal of what we do in science involves establishing functional relations between variables of interest (for example, blood pressure and heart rate, transport rate and concentration gradient). Equations are simply a compact and exact way of expressing such relationships. The tools of mathematics then allow us to draw conclusions systematically from these relationships. Mathematics is a very powerful set of tools or, more generally, a very powerful way of thinking. Mathematics allows you to think extremely precisely, and therefore clearly, about complex relationships. Equations and quantitative notions are the keys to that precision. For example, a quantitative comparison of the predictions of a theory against the results of measurement forms the basis of statistics and of much of the hypothesis testing on which science is based. A scientific conclusion without adequate quantification and statistical backing may be little more than an impression or opinion.

It may seem odd to say that mathematics allows you to think more clearly about complex ideas. People unfamiliar with mathematical thinking often complain that even simple relationships produce complicated equations and that complex relationships are mathematically intractable. Certainly, many basic concepts require considerable mathematical expertise to be handled properly, but such concepts are in fact not simple. More commonly, however, many simple equations are seen as complex because many students are poorly trained in how to think about equations.

How to Think about an Equation

In this section we will take the first, and often overlooked, step in thinking quantitatively. How do we begin to think about some new equation presented to us? We start by becoming comfortable with the "meaning" of an equation. This step is absolutely necessary if you are to use an equation properly. As a specific example, consider the Nernst equation (see p. 79) for potassium. Here are several forms you will find in various books; they all say essentially the same thing:

$$E_{K^+} = (RT/zF) \log \{[K^+]_{out}/[K^+]_{in}\}$$

$$E_{K^+} = (RT/zF)\, 2.303 \log \{[K^+]_{out}/[K^+]_{in}\}$$

$$E_{K^+} = (61\ mV/z) \log \{[K^+]_{out}/[K^+]_{in}\}$$

For many students these equations may seem like meaningless strings of symbols. What are these equations trying to tell us? What do they represent? The following four steps may help you become comfortable with any new equation. Try them with the Nernst equation.

1. Be sure you can define the symbols and give dimensions and units. Check the equation for dimensional consistency.

One of the first steps is to figure out which symbols represent the variables of interest and which are simple constants. In this case, all the symbols are constants except two.

E_{K^+} is the Nernst (equilibrium) potential for potassium. It represents the concentration gradient (force of diffusion) on a mole of potassium ions. E_{K^+} has the dimensions of a voltage and is usually given in units of mV. This dimension is used so that the concentration gradient is expressed in the same dimensions as the other force acting on the ions, that is, the electrical gradient. Using the same dimensions makes it possible to compare the two forces.

$[K^+]$ represents the concentration of potassium. With the subscript "out," this symbol refers to the concentration of potassium outside the cell. With the subscript "in," this symbol refers to the concentration of potassium inside the cell. $[K^+]$ has dimensions of concentration and is usually expressed in units of mM (millimolars; millimols/liter).

2. Identify the dependent and independent variables. Try to find normal values and ranges for the variables. Before continuing, we should define "dependent" and "independent" variables.

Remember that equations represent relationships between variables. Whenever you hear the word *relationship,* think of a graph, as in ● Figure D-1, for example.

Graphs are often a good way to represent relationships and therefore equations. This graph says the value of variable 2 depends on the value of variable 1. Thus, for any value of variable 1 the corresponding value for variable 2 can be determined from the graph. In other words, variable 1 determines the value of variable 2. Because variable 2 depends on variable 1, we call variable 2 the *dependent variable.* Variable 1, in contrast, is independent of variable 2, so we call variable 1 the *independent variable.* There can be any number of dependent and independent variables.

How do you determine which variables are dependent and which are independent? The answer usually depends on cause and effect: "Effects" are dependent variables and "causes" are independent. For example, we know (see Chapter 10, p. 358) that mean arterial pressure *(MAP)* is the product of cardiac output *(CO)* and total peripheral resistance *(TPR);* that is,

$$MAP = CO \times TPR$$

MAP is on the left-hand side of this equation because we think of mean arterial pressure as a result of cardiac output and total peripheral resistance. Or, to put this another way, mean arterial pressure is a function of cardiac output and total peripheral resistance. As a cause–effect relationship, it seems backward to think of mean arterial pressure somehow "causing" cardiac output to be a certain value. Therefore, *MAP* is the effect, the dependent variable, and we place it on the left-hand side of the equality symbol. Conversely, *CO* and *TPR* are the causes, the independent variables, and we put them on the right.

In our Nernst equation example, the independent variables are the concentrations. The dependent variable is the Nernst potential because we think of the potential as being a result of the ion concentrations. We also know that E_{K^+} is about -90 mV, and $[K^+]_{out}$ and $[K^+]_{in}$ are about 5 mM and 150 mM, respectively.

3. Identify the constants and know their numeric values:

- *R* is the gas constant. It has dimensions of energy per mole per degree of temperature and the value of 8.31 joules/kelvin · mole. It is also convenient to note that a joule = volt × coulomb.
- *T* is temperature, with the dimension of temperature being in units of kelvins. Normal body temperature is around 37°C [= 308 kelvins (K)].
- *z* is the valence of the ion. Valence is the charge on an ion, including the sign. For potassium, $z = +1$.
- *F* is Faraday's constant, which has dimensions of charge per mole, units of coulombs per mole, and a value of 96,500 C/mol.

Refer back to the Nernst equations given on the preceding page. Note that the constants just defined appear in the first two equations, but not the third. In the third equation, the quantity *RT/F* has already been evaluated for you, as follows:

$$RT/F = [(8.31 \text{ V} \cdot \text{C/K} \cdot \text{mol})(308K)]/(96,500 \text{ C/mol})$$
$$= 26.5 \text{ mV}$$

We multiply this value by 2.303 to convert the natural logarithm to the base 10 logarithm. Note that 26.5 mV × 2.303 = 61 mV.

4. State the equation in words. Summarize it in a few sentences so that someone can understand what it is about. Don't just say the names of the symbols.

Just saying the names of the symbols would be equivalent to saying the following: "The Nernst potential is given by a constant times the logarithm of the ratio of the ion concentrations." This is certainly true but does nothing to aid our intuition. A preferable statement would be "The Nernst equation allows us to calculate the force pushing ions into or out of a cell via diffusion." This is valuable, because we can compare this force to the force moving ions in and out via the membrane voltage and see which is larger and hence in what direction the ions will actually move. The force is expressed in electrical units so we can compare it directly with the membrane voltage. The constants convert from concentration to electrical units.

Only when you understand what an equation means will you be able to use it to answer questions. The next section gives you some guidance in taking this next step.

How to Think with an Equation

Before you can use an equation to help you think, you need to develop a few basic skills. Luckily, these skills are not difficult to learn.

1. Be sure you know the algebraic rules for manipulating variables within any function (such as $\sqrt{}$, exp, or log) involved.

In the case of the Nernst equation, the tricky function is the logarithm. You should consult a college-level algebra book if you are hazy on the rules of working with logs or any other function. For instance, it is useful to know that

$$\log \{A\} = -\log \{1/A\}$$

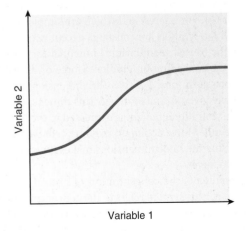

Variable 2 / Variable 1

● **FIGURE D-1**

and that the log operation is undone by taking it to the power of 10; that is,

$$10^{\log\{A\}} = A$$

2. Be able to solve for any variable in terms of the others.

Given just three variables (E_{K^+}, $[K^+]_{in}$, $[K^+]_{out}$), only a few types of questions can be asked. Two of the three variables must be given, and you must solve for the third. If the two concentrations are given, then the formula is already set to give you the Nernst potential. If the Nernst potential and one concentration are given, however, you must be able to solve for the other concentration. See if you can do this and obtain the two following equations:

$$[K^+]_{out} = [K^+]_{in}10^{(E_{K^+}/61 \text{ mV})}$$

$$[K^+]_{in} = [K^+]_{out}10^{(-E_{K^+}/61 \text{ mV})}$$

3. Be able to sketch, at least approximately, the dependence of any variable on any other variable.

The ability to do this is exceedingly valuable. Sketching helps you generate insight about equations; it helps you understand what an equation means. Therefore, sketching helps you understand the solution, as well as solve the problem. If you apply this technique consistently, you may find equations far simpler to handle than you previously suspected. In addition, be sure you can relate your sketch to experimental measurements and physiological situations.

For example, we can draw the relationships between the Nernst potential for potassium and the external potassium ion concentration predicted by the equations as in ● Figure D-2. This sketch makes clear that the Nernst potential, which can be measured physiologically, should decrease linearly as the log $[K^+]_{out}$ increases, which can be controlled experimentally. Therefore, this sketch suggests an experiment: vary $[K^+]_{out}$. If E_{K^+} does not decrease linearly with increasing log $[K^+]_{out}$, then we would have a flaw in our understanding. The Nernst equation would not describe the real situation, as we think it should. Scientific advances are almost always heralded by such contradictions.

4. Be able to combine several equations to find new relationships.

Combining separate pieces of information is always useful. In fact, some scientists have argued that this activity is all scientists ever do. To integrate knowledge for yourself, you must be able to combine the various relations you learn about into new combinations. This allows you to solve increasingly complex problems. As an example, consider the following relation:

$$I_{K^+} = G_{K^+}(V_m - E_{K^+})$$

This equation describes the number of potassium ions flowing across a membrane if both a concentration gradient (E_{K^+}) and an electrical gradient (V_m) are present. This equation can be combined with the Nernst equation for potassium to answer the following question. Suppose V_m, G_{K^+}, and $[K^+]_{in}$ are fixed. What would the external potassium ion concentration have to be such that no net flux of potassium ions occurs across the membrane? No net flux implies that $I_{K^+} = 0$. But if $G_{K^+} > 0$, $I_{K^+} = 0$ only when the membrane voltage equals the Nernst potential ($V_m = E_{K^+}$). To answer the question, then, we set $E_{K^+} = V_m$ in the Nernst equation and solve it for $[K^+]_{out}$.

5. Know the equation's underlying assumptions and limits of validity.

Every equation comes from some underlying theory or set of observations and, therefore, has some limited range of validity and rests on certain assumptions. Failure to understand this simple point often leads students to apply equations outside their realm of applicability. In that case, even though the math is done correctly, the results will be incorrect.

In the case of the Nernst equation, things are fairly simple. This equation is derived from a very powerful theory known as *equilibrium thermodynamics*, and hence it has very wide applicability. As another example, consider enzyme kinetics. The rate at which an enzyme catalyzes a reaction (v) is related to the concentration of substrate on which the enzyme works ($[S]$) by an equation called the Michaelis-Menton relationship. The graph of this relationship can be seen in ● Figure D-3.

● **FIGURE D-2**

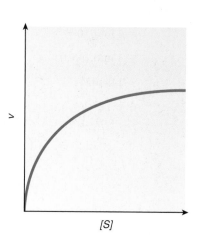

● **FIGURE D-3**

However, this relationship between reaction velocity and substrate concentration does not hold true for some real enzymes, such as lactate dehydrogenase. For this enzyme, the relationship between reaction velocity and substrate concentration is depicted in ● Figure D-4.

At high substrate concentrations, the enzyme actually is inhibited by too much substrate. For such enzymes, the Michaelis-Menton theory, which works well at low [S], is invalid at higher [S].

An Approach to Problem Solving

The final step is to apply these skills to solve a problem. As an example, calculate the concentration of potassium that must exist inside a cell if $E_{K^+} = -95$ mV and the interstitial fluid has a potassium concentration of 4 mM. Try using the following procedure to solve this problem:

1. Get a clear picture of what is being asked. State it out loud or write it down.

The question asks for the concentration of potassium in the cell, that is, $[K^+]_{in}$.

2. Determine what you need to know to answer the question.

To answer this, you need to know E_{K^+} and $[K^+]_{out}$.

3. Determine what information is given. Is it sufficient? Are other relevant facts or relationships not stated in the problem? Specifically, do you need any other equations?

E_{K^+} is given explicitly in the problem, but you have to translate the words to realize that $[K^+]_{out} = 4$ mM.

4. Manipulate the equation algebraically so that the unknown is on the left-hand side and everything else is on the right-hand side.

We now solve the Nernst equation for $[K^+]_{in}$. This was done on page A-35:

$$[K^+]_{in} = [K^+]_{out}10^{(-E_{K^+}/61 \text{ mV})}$$

Substituting for the values given, we obtain:

$$[K^+]_{in} = 4 \text{ mM } 10^{(95 \text{ mV}/61 \text{ mV})}$$

$$= 4 \text{ mM } 10^{(1.56)}$$

$$= 4 \text{ mM } (36.3)$$

$$= 145 \text{ mM}$$

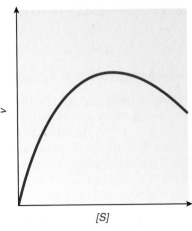

● **FIGURE D-4**

5. Is the answer dimensionally correct? Do not skip this step. It will tell you immediately if something went wrong.

Yes, the answer is in mM—the proper dimension and unit.

6. Does the answer make sense?

Yes, the value is not alarmingly low or high. Also, because the potassium ion is positively charged, if the potassium concentration outside the cell is lower than that inside the cell, then the interior of the cell would have to be negative at passive equilibrium, which it is.

Apply the approach to problem solving we have just outlined to the quantitative questions in the chapters. When you start out, apply the approach formally and carefully. For example, go through each step, and write everything out as we have done for the Nernst equation. After a time, you may not need to be so formal. Also, as you progress, you will develop your own style and approach to problem solving. Be prepared to spend some time and patience on some of the problems. Not every answer will be immediately apparent. This situation is normal. If you run into difficulties, relax, return to this appendix for guidance, and work through the problem again carefully. Do not go immediately to the answers if you are having difficulty with a problem. It may ease the frustration, but you will be cheating yourself of a valuable learning experience. Besides, being able to solve challenging problems has its own rewards.

Text References to Exercise Physiology

A Closer Look at Exercise Physiology Boxed Features by Chapter

Exercise References by Topic

Answers to End-of-Chapter Objective Questions, Quantitative Exercises, Points to Ponder, and Clinical Considerations

Chapter 1 Introduction to Physiology and Homeostasis

Objective Questions

(Questions on p. 17.)

1. e **2.** b **3.** c **4.** T **5.** F **6.** T **7.** muscle tissue, nervous tissue, epithelial tissue, connective tissue **8.** secretion **9.** exocrine, endocrine, hormones **10.** intrinsic, extrinsic **11.** 1.d, 2.g, 3.a, 4.e, 5.b, 6.j, 7.h, 8.i, 9.c, 10.f

Points to Ponder

(Questions on p. 18.)

1. The respiratory system eliminates internally produced CO_2 to the external environment. A decrease in CO_2 in the internal environment brings about a reduction in respiratory activity (that is, slower, shallower breathing) so that CO_2 produced within the body is allowed to accumulate instead of being blown off as rapidly as normal to the external environment. The extra CO_2 retained in the body increases the CO_2 levels in the internal environment to normal.

2. (b) (c) (b)

3. b

4. immune defense system

5. When a person is engaged in strenuous exercise, the temperature-regulating center in the brain will bring about widening of the blood vessels of the skin. The resultant increased blood flow through the skin will carry the extra heat generated by the contracting muscles to the body surface, where it can be lost to the surrounding environment.

Clinical Consideration

(Question on p. 19.)

Loss of fluids threatens the maintenance of proper plasma volume and blood pressure. Loss of acidic digestive juices threatens the maintenance of the proper pH in the internal fluid environment. The urinary system will help restore the proper plasma volume and pH by reducing the amount of water and acid eliminated in the urine. The respiratory system will help restore the pH by adjusting the rate of removal of acid-forming CO_2. Adjustments will be made in the circulatory system to help maintain blood pressure despite fluid loss. Increased thirst will encourage increased fluid intake to help restore plasma volume. These compensatory changes in the urinary, respiratory, and circulatory systems, as well as the sensation of thirst, will all be regulated by the two regulatory systems, the nervous and endocrine systems. Furthermore, the endocrine system will make internal adjustments to help maintain the concentration of nutrients in the internal environment even though no new nutrients are being absorbed from the digestive system.

Chapter 2 Cell Physiology

Objective Questions

(Questions on p. 50.)

1. plasma membrane **2.** deoxyribonucleic acid (DNA), nucleus **3.** organelles, cytosol, cytoskeleton **4.** endoplasmic reticulum, Golgi complex **5.** oxidative **6.** adenosine triphosphate (ATP) **7.** F **8.** F **9.** 1.b, 2.a, 3.b, 4.b **10.** 1.b, 2.c, 3.c, 4.a, 5.b, 6.c, 7.a, 8.c, 9.c

Quantitative Exercises

(Questions on p. 50.)

1. b

2. 24 moles O_2/day $\times$ 6 moles ATP/mole O_2
   = 144 moles ATP/day
   144 moles ATP/day $\times$ 507 g ATP/mole = 73,000 g ATP/day
   1000 g/2.2 lb = 73,000 g/x lb
   1000 x = 160,600
   x = approximately 160 lb

3. 144 mol/day (7300 cal/mol) =1,051,200 cal/day
   (1051 kilocal/day)

4. About 2/3 of the water in the body is intracellular. Because a person's mass is about 60% water, for a 150-pound (68 kg) person,

   68 kg(0.6)(2/3) = 27.2 kg

is the mass of water. Assume that 1 ml of body water weighs 1 g. Then the total volume in the person's cells is about 27.2 liters. The volume of an average cell is

$$\frac{4}{3}\pi(1 \times 10^{-3} \text{ cm})^3 \approx 4.2 \times 10^{-9} \text{ cm}^3$$

$$= 4.2 \times 10^{-9} \text{ ml}$$

So, the number of cells in a 68 kg person is about

$$27.2 \text{ liters} \left(\frac{1000 \text{ ml}}{1 \text{ liter}}\right)\left(\frac{1 \text{ cell}}{4.2 \times 10^{-9} \text{ ml}}\right)$$

$$= 6.476 \times 10^{12} \text{ cells}$$

5. $150 \text{ mg} \left(\dfrac{1 \text{ ml}}{0.015 \text{ mg}}\right) = 10,000 \text{ ml (10 liters)}$

Points to Ponder

(Questions on p. 51.)

1. The chief cells have an extensive rough endoplasmic reticulum, with this organelle being responsible for synthesizing these cells' protein secretory product, namely, pepsinogen. Because the parietal cells do not secrete a protein product to the cells' exterior, they do not need an extensive rough endoplasmic reticulum.

2. With cyanide poisoning, the cellular activities that depend on ATP expenditure, such as synthesis of new chemical compounds, membrane transport, and mechanical work, could not continue. The resultant inability of the heart to pump blood and failure of the respiratory muscles to accomplish breathing would lead to imminent death.

3. catalase

4. ATP is required for muscle contraction. Muscles are able to store limited supplies of nutrient fuel for use in the generation of ATP. During anaerobic exercise, muscles generate ATP from these nutrient stores by means of glycolysis, which yields 2 molecules of ATP per glucose molecule processed. During aerobic exercise, muscles can generate ATP by means of oxidative phosphorylation, which yields 32 molecules of ATP per glucose

molecule processed. Because glycolysis inefficiently generates ATP from nutrient fuels, it rapidly depletes the muscle's limited stores of fuel, and ATP can no longer be produced to sustain the muscle's contractile activity. Aerobic exercise, in contrast, can be sustained for prolonged periods. Not only does oxidative phosphorylation use far less nutrient fuel to generate ATP but it can be supported by nutrients delivered to the muscle by means of the blood instead of relying on stored fuel in the muscle. Intense anaerobic exercise outpaces the ability to deliver supplies to the muscle by the blood, so the muscle must rely on stored fuel and inefficient glycolysis, thus limiting anaerobic exercise to brief periods of time before energy sources are depleted.

5. skin. The mutant keratin weakens the skin cells of patients with epidermolysis bullosa so that the skin blisters in response to even a light touch.

Clinical Consideration

(Question on p. 51.)

Some hereditary forms of male sterility involving nonmotile sperm have been traced to defects in the cytoskeletal components of the sperm's flagella. These same individuals usually also have long histories of recurrent respiratory tract disease because the same types of defects are present in their respiratory cilia, which are unable to clear mucus and inhaled particles from the respiratory system.

Chapter 3   The Plasma Membrane and Membrane Potential

Objective Questions

(Questions on p. 83.)

1. T 2. T 3. T 4. T 5. 1.b, 2.a, 3.b, 4.a, 5.c, 6.b, 7.a, 8.b 6. 1.a, 2.a, 3.b, 4.a, 5.b, 6.a, 7.b 7. 1.c, 2.b, 3.a, 4.a, 5.c, 6.b, 7.c, 8.a, 9.b

Quantitative Exercises

(Questions on p. 84.)

1. $E = \dfrac{61 \text{ mV}}{z} \log \dfrac{C_o}{C_i}$

   a. $\dfrac{61 \text{ mV}}{2} \log \dfrac{1 \times 10^{-3}}{100 \times 10^{-9}} = +122 \text{ mV}$

   b. $\dfrac{61 \text{ mV}}{-1} \log \dfrac{110 \times 10^{-3}}{10 \times 10^{-3}} = -63.5 \text{ mV}$

2. $I_x = G_x(V_m - E_x)$

   $E_{Na^+} = 61 \text{ mV} \log \dfrac{145 \text{ mM}}{15 \text{ mM}} = 60.1 \text{ mV}$

   a. $= 1 \text{ ns } (-70 \text{ mV} - 60.1 \text{ mV})$
      $= 1 \text{ ns } (-130 \text{ mV})$
      $= -130 \text{ pA (A = amperes)}$

b. Entering

c. With concentration gradient; with electrical gradient

3. $V_m = 61 \log \dfrac{P_{K^+}[K^+]_o + P_{Na^+}[Na^+]_o}{P_{K^+}[K^+]_i + P_{Na^+}[Na^+]_i}$

$\quad = 61 \log \dfrac{(1)(10) + (0.04)(150)}{(1)(150) + (0.04)(15)}$

$\quad = 61 \log \dfrac{10 + 6}{150 + 0.6}$

$\quad = 61 \log 0.1062$

Because the log of 0.106 is -0.974

$\quad V_m = 61 (-0.974) = -59 \text{ mV}$

Therefore, the resting membrane potential is less than normal (that is, slightly depolarized compared to normal).

Points to Ponder

(Questions on p. 84.)

1. c. As Na^+ moves from side 1 to side 2 down its concentration gradient, Cl^- remains on side 1, unable to permeate the membrane. The resultant separation of charges produces a membrane potential, negative on side 1 because of unbalanced chloride ions and positive on side 2 because of unbalanced sodium ions. Sodium does not continue to move to side 2 until its concentration gradient is dissipated because of the development of an opposing electrical gradient.

2. Osmolarity refers to the concentration of all particles in a solution, both penetrating and nonpenetrating, yet only nonpenetrating solutes contribute to the tonicity of a solution. Therefore, a solution with a mixture of penetrating and nonpenetrating solutes may have an osmolarity of 300 mOsm, the same as in the ICF, but be hypotonic to the cells because the solution's concentration of nonpenetrating solutes is less than the concentration of nonpenetrating solutes inside the cells. The entire osmolarity of the ICF at 300 mOsm is attributable to nonpenetrating solutes.

3. more positive. Because the electrochemical gradient for Na^+ is inward, the membrane potential would become more positive as a result of an increased influx of Na^+ into the cell if the membrane were more permeable to Na^+ than to K^+. (Indeed, this is what happens during the rising phase of an action potential once threshold potential is reached; see Chapter 4.)

4. d. active transport. Leveling off of the curve designates saturation of a carrier molecule, so carrier-mediated transport is involved. The graph indicates that active transport is being used instead of facilitated diffusion, because the concentration of the substance in the intracellular fluid is greater than the concentration in the extracellular fluid at all points until after the transport maximum is reached. Thus, the substance is being moved *against* a concentration gradient, so active transport must be the method of transport being used.

5. vesicular transport. The maternal antibodies in the infant's digestive tract lumen are taken up by the intestinal cells by endocytosis and are extruded on the opposite side of the cell into the interstitial fluid by exocytosis. The antibodies are picked up from the intestinal interstitial fluid by the blood supply to the region.

6. accelerate. During an action potential, Na^+ enters and K^+ leaves the cell. Repeated action potentials would eventually "run down" the Na^+ and K^+ concentration gradients were it not for the Na^+–K^+ pump returning the Na^+ that entered back to the outside and the K^+ that left back to the inside. Indeed, the rate of pump activity is accelerated by the increase in both ICF Na^+ and ECF K^+ concentrations that occurs as a result of action potential activity, thus hastening the restoration of the concentration gradients.

Clinical Consideration

(Question on p. 85.)

As Cl^- is secreted by the intestinal cells into the intestinal tract lumen, Na^+ follows passively along the established electrical gradient. Water passively accompanies this salt (Na^+ and Cl^-) secretion by osmosis. The toxin produced by the cholera pathogen prevents the normal inactivation of the mechanism (cAMP pathway; see p. 121) that opens the Cl^- channels in the luminal membranes of the intestinal cells. Increased, ongoing secretion of Cl^- and the subsequent passively induced secretion of Na^+ and water are responsible for the severe diarrhea that characterizes cholera.

Chapter 4   Principles of Neural and Hormonal Communication

Objective Questions

(Questions on p. 129.)

1. T 2. F 3. F 4. F 5. T 6. F 7. refractory period
8. axon hillock 9. synapse 10. temporal summation
11. spatial summation 12. convergence, divergence
13. G protein 14. receptor-channel, receptor-enzyme, G-protein-coupled receptor 15. 1.b, 2.a, 3.a, 4.b, 5.b, 6.a
16. 1.a, 2.b, 3.a, 4.b, 5.d, 6.b, 7.b, 8.b, 9.a, 10.b, 11.a, 12.c

Quantitative Exercises

(Questions on p. 130.)

1. a. 0.6 m (1 sec/0.7 m) = 0.8571 sec
   b. 0.6 m (1 sec/120 m) = 0.005 sec
   c. unmyelinated: 0.8591 sec; myelinated: 0.007 sec
   d. unmyelinated: 0.8621 sec; myelinated: 0.01 sec

2. Total conduction time for the single axon is 1/60 sec. Let v m/sec be the unknown conduction velocity for the three neurons. Our equation for the total conduction time then is

$$\frac{1}{60} \sec = \left(\frac{1}{v} \times 1 \text{ m} \right) + 0.002 \sec$$

Solving for v, we obtain

$\quad v$ m/sec = 1 m/(1/60 sec − 0.002 sec) = 68.18 m/sec

3. $25 \times 10^{-3} \text{ V} \left[\dfrac{3.3 \ \mu\text{S/cm}^2 (240 \ \mu\text{S/cm}^2)}{(3.3 + 240) \ \mu\text{S/cm}^2} \right] \log \dfrac{240(145)}{3.3(4)}$

$= 25 \times 10^{-3} (11.1361) \text{V} \times \mu\text{S/cm}^2 = 0.2784 \ \mu\text{A/cm}^2$

Points to Ponder

(Questions on p. 131.)

1. c. The action potentials would stop as they met in the middle. As the two action potentials moving toward each other both reached the middle of the axon, the two adjacent patches of membrane in the middle would be in a refractory period, so further propagation of either action potential would be impossible.

2. A subthreshold stimulus would transiently depolarize the membrane but not sufficiently to bring the membrane to threshold, so no action potential would occur. Because a threshold stimulus would bring the membrane to threshold, an action potential would occur. An action potential of the same magnitude and duration would occur in response to a suprathreshold stimulus as to a threshold stimulus. Because of the all-or-none law, a stimulus larger than that necessary to bring the membrane to threshold would not produce a larger action potential. (The magnitude of the stimulus is coded in the *frequency* of action potentials generated in the neuron, not the *size* of the action potentials.)

3. The hand could be pulled away from the hot stove by flexion of the elbow accomplished by summation of EPSPs at the cell bodies of the neurons controlling the biceps muscle, thus bringing these neurons to threshold. The subsequent action potentials generated in these neurons would stimulate contraction of the biceps. Simultaneous contraction of the triceps muscle, which would oppose the desired flexion of the elbow, could be prevented by generation of IPSPs at the cell bodies of the neurons controlling this muscle. These IPSPs would keep the triceps neurons from reaching threshold and firing so that the triceps would not be stimulated to contract.

The arm could deliberately be extended despite a painful finger prick by voluntarily generating EPSPs to override the reflex IPSPs at the neuronal cell bodies controlling the triceps while simultaneously generating IPSPs to override the reflex EPSPs at the neuronal cell bodies controlling the biceps.

4. An EPSP, being a graded potential, spreads decrementally from its site of initiation in the postsynaptic neuron. If presynaptic neuron A (near the axon hillock of the postsynaptic cell) and presynaptic neuron B (on the opposite side of the postsynaptic cell body) both initiate EPSPs of the same magnitude and frequency, the EPSPs from A will be of greater strength when they reach the axon hillock than the EPSPs from B. An EPSP from B will decrease more in magnitude as it travels farther before reaching the axon hillock, the region of lowest threshold and thus the site of action potential initiation. Temporal summation of the larger EPSPs from A may bring the axon hillock to threshold and initiate an action potential in the postsynaptic neuron, whereas temporal summation of the weaker EPSPs from B at the axon hillock may not be sufficient to bring this region to threshold. Thus, the proximity of a presynaptic neuron to the axon hillock can bias its influence on the postsynaptic cell.

5. (1) Angiotensin receptor blockers (ARBs) decrease blood pressure by reducing the load of osmotically active (water-holding) salt in the body, thereby decreasing the volume of circulating plasma. The greater the plasma volume, and accordingly the blood volume, the higher the blood pressure is, all other factors being equal. (2) β_1-adrenergic receptor blockers suppress the action of epinephrine on the heart, thereby reducing the rate and strength of contraction of the heart. The more rapidly and more forcefully the heart beats, the more blood is pumped into the blood vessels per minute and the greater the pressure exerted by the blood on the vessel walls, all other factors being equal.

6. Estrogen acting on estrogen-dependent breast cancer cells promotes survival of these cells. By interfering with the ability of estrogen to bind with its receptors in the breast cancer cells, selective estrogen receptor modulators (SERMs) prevent estrogen from promoting survival of these cells. SERMs are taken for a number of years following removal of cancerous breast tissue with the goal of thwarting any cancer cells that may remain in the body.

Because SERMs interfere with estrogen (a lipophilic steroid hormone) binding with its receptors, which are located inside the target cell, one can infer that SERMs must also enter the target cell. Therefore, SERMs must be lipophilic, so they could be taken orally without risk of being destroyed by protein-digesting enzymes in the digestive tract.

Clinical Consideration

(Question on p. 131.)

Initiation and propagation of action potentials would not occur in nerve fibers acted on by local anesthetic because blockage of Na^+ channels by the local anesthetic would prevent the massive opening of voltage-gated Na^+ channels at threshold potential. As a result, pain impulses (action potentials in nerve fibers that carry pain signals) would not be initiated and propagated to the brain and would not reach the level of conscious awareness.

Chapter 5 The Central Nervous System

Objective Questions

(Questions on p.180.)

1. F **2.** F **3.** F **4.** T **5.** F **6.** F **7.** habituation **8.** consolidation **9.** dorsal, ventral **10.** 1.a, 2.c, 3.a and b, 4.b, 5.a, 6.c, 7.c **11.** 1.d, 2.c, 3.f, 4.e, 5.a, 6.b

Points to Ponder

(Questions on p. 181.)

1. Only the left hemisphere has language ability. When sharing of information between the two hemispheres is prevented as a result of severance of the corpus callosum, visual information presented only to the right hemisphere cannot be verbally identified by the left hemisphere, because the left hemisphere is unaware of the information. However, the information can be recognized by nonverbal means, of which the right hemisphere is capable.

2. Insulin excess drives too much glucose into insulin-dependent cells so that the blood glucose falls below normal and insufficient glucose is delivered to the non-insulin-dependent brain. Therefore, the brain, which depends on glucose as its energy source, does not receive adequate nourishment.

3. c. A severe blow to the back of the head is most likely to traumatize the visual cortex in the occipital lobe.

4. Salivation when seeing or smelling food, striking the appropriate letter on the keyboard when typing, and many of the actions involved in driving a car are conditioned reflexes. You undoubtedly will have many other examples.

5. Strokes occur when a portion of the brain is deprived of its vital O_2 and glucose supply because the cerebral blood vessel supplying the area either is blocked by a clot or has ruptured. Although a clot-dissolving drug could be helpful in restoring blood flow through a cerebral vessel blocked by a clot, such a drug would be detrimental in the case of a ruptured cerebral vessel sealed by a clot. Dissolution of a clot sealing a ruptured vessel would lead to renewed hemorrhage through the vessel and exacerbation of the problem.

Clinical Consideration

(Question on p. 181.)

The deficits following the stroke—numbness and partial paralysis on the upper right side of the body and inability to speak—are indicative of damage to the left somatosensory cortex and left primary motor cortex in the regions devoted to the upper part of the body plus Broca's area.

Chapter 6 The Peripheral Nervous System: Afferent Division; Special Senses

Objective Questions

(Questions on p. 233.)

1. transduction **2.** adequate stimulus **3.** F **4.** T **5.** T
6. T **7.** F **8.** F **9.** T **10.** F **11.** F **12.** 1.f, 2.h, 3.l,
4.d, 5.i, 6.e, 7.b, 8.j, 9.a, 10.g, 11.c, 12.k **13.** 1.a, 2.b, 3.c, 4.c,
5.c, 6.a, 7.b, 8.b

Quantitative Exercises

(Questions on p. 234.)

1. The slow pain pathway takes about (1.3 m) (1 sec/12 m) = 0.1083 sec. The fast pathway takes (1.3 m) (1 sec/30 m) = 0.0433 sec. The difference is 0.1083 sec − 0.0433 sec = 0.065 sec = 65 msec.

2. a. The amount of light entering the eye is proportional, approximately, to the area of the open pupil. Recall that the area

of a circle is πr^2. Let r be the pupil radius and A_1 be the original pupil area. Halving the diameter also halves the radius, so the new pupil area is

$$\pi\left(\frac{1}{2}r\right)^2 = \frac{1}{4}\pi r^2 = \frac{1}{4}A_1$$

Therefore, the amount of light allowed into the eye is a quarter of what it was originally.

b. The area of a rectangle is hw, where h is the height and w the width. Halving either dimension halves the area and hence the amount of light allowed into the eye.

c. The cat's pupil can be considered more precise. Think about the coarse and fine adjustments on a microscope. Fine adjustment translates rotations of the knob into much smaller movement of the stage than does coarse adjustment.

3. a. Solve the following for I:

$$\beta = (10 \text{ dB}) \log_{10}(I/I_0)$$
$$I = I_0 10^{\beta/10} W/m^2$$

Therefore,

$$I_1 = 10^{-12}(10^{20/10}) = 10^{-12}(10^2) = 10^{-10} \ W/m^2$$
$$I_2 = 10^{-12}(10^{70/10}) = 10^{-12}(10^7) = 10^{-5} \ W/m^2$$
$$I_3 = 10^{-12}(10^{120/10}) = 10^{-12}(10^{12}) = 1 \ W/m^2$$
$$I_4 = 10^{-12}(10^{170/10}) = 10^{-12}(10^{17}) = 10^5 \ W/m^2$$

b. Because of the logarithm in the definition of decibel, the sound intensity increases exponentially with respect to sound level. This fact should be clear from the definition of dB solved for I. This result implies that the human ear performs well throughout an enormous range of sound intensities.

Points to Ponder

(Questions on p. 234.)

1. Pain is a conscious warning that tissue damage is occurring or about to occur. A patient unable to feel pain because of a nerve disorder does not consciously take measures to withdraw from painful stimuli and thus prevent more serious tissue damage.

2. Pupillary dilation (mydriasis) can be deliberately induced by ophthalmic instillation of either an adrenergic drug (such as epinephrine or related compound) or a cholinergic blocking drug (such as atropine or related compounds). Adrenergic drugs produce mydriasis by causing contraction of the sympathetically supplied radial (dilator) muscle of the iris. Cholinergic blocking drugs cause pupillary dilation by blocking parasympathetic activity to the circular (constrictor) muscle of the iris so that action of the adrenergically controlled radial muscle of the iris is unopposed.

3. The defect would be in the left optic tract or optic radiation.

4. Fluid accumulation in the middle ear in accompaniment with middle ear infections impedes the normal movement of the tympanic membrane, ossicles, and oval window in response to sound. All these structures vibrate less vigorously in the presence of fluid, causing temporary hearing impairment. Chronic fluid accumulation in the middle ear is sometimes relieved by surgical implantation of drainage tubes in the eardrum. Hearing is restored to normal as the fluid drains to the exterior. Usually, the tube "falls out" as the eardrum heals and pushes out the foreign object.

5. The sense of smell is reduced when you have a cold, even though the cold virus does not directly adversely affect the olfactory receptor cells, because odorants do not reach the receptor cells as readily when the mucous membranes lining the nasal passageways are swollen and excess mucus is present.

Clinical Consideration

(Question on p. 235.)

Syncope most frequently occurs as a result of inadequate delivery of blood carrying sufficient oxygen and glucose supplies to the brain. Possible causes include circulatory disorders such as impaired pumping of the heart or low blood pressure; respiratory disorders resulting in poorly oxygenated blood; anemia, in which the oxygen-carrying capacity of the blood is reduced; or low blood glucose resulting from improper endocrine management of blood glucose levels. Vertigo, in contrast, typically results from a dysfunction of the vestibular apparatus, arising, for example, from viral infection or trauma, or abnormal neural processing of vestibular information, as, for example, with a brain tumor.

Chapter 7 The Peripheral Nervous System: Efferent Division

Objective Questions

(Questions on p. 254.)

1. T   **2.** F   **3.** c   **4.** c   **5.** sympathetic, parasympathetic
6. adrenal medulla   **7.** 1.a, 2.b, 3.a, 4.b, 5.a, 6.a, 7.b   **8.** 1.b,
2.b, 3.a, 4.a, 5.b, 6.b, 7.a   **9.** 1.c,f, 2.a, 3.d,f, 4.e, 5.e, 6.b,f

Quantitative Exercises

(Questions on p. 255.)

1.
$$t = \frac{x^2}{2D} = \frac{(200 \text{ nm})^2}{2 \times 10^{-5} \text{ cm}^2/\text{sec}}$$
$$= \frac{4 \times 10^{-14} \text{ m}^2 \cdot \text{sec}}{2 \times 10^{-5} \text{ cm}^2} \left(\frac{10^4 \text{ cm}^2}{\text{m}^2} \right) = 20 \text{ } \mu\text{sec}$$

Points to Ponder

(Questions on p. 255.)

1. By promoting arteriolar constriction, epinephrine administered in conjunction with local anesthetics reduces blood flow to the region and thus helps the anesthetic stay in the region instead of being carried away by the blood.

2. No. Atropine blocks the effect of acetylcholine at muscarinic receptors but does not affect nicotinic receptors. Nicotinic receptors are present on the motor end plates of skeletal muscle fibers.

3. The voluntarily controlled external urethral sphincter is composed of skeletal muscle and supplied by the somatic nervous system.

4. By interfering with normal acetylcholine activity at the neuromuscular junction, α bungarotoxin leads to skeletal muscle paralysis, with death ultimately occurring as a result of an inability to contract the diaphragm and breathe.

5. If the motor neurons that control the respiratory muscles, especially the diaphragm, are destroyed by poliovirus or amyotrophic lateral sclerosis, the person is unable to breathe and dies (unless breathing is assisted by artificial means).

Clinical Consideration

(Question on p. 255.)

Drugs that block β_1 receptors are useful for prolonged treatment of angina pectoris because they interfere with sympathetic stimulation of the heart during exercise or emotionally stressful situations. By preventing increased cardiac metabolism and thus an increased need for oxygen delivery to the cardiac muscle during these situations, beta blockers can reduce the frequency and severity of angina attacks.

Chapter 8 Muscle Physiology

Objective Questions

(Questions on p. 299.)

1. F   **2.** F   **3.** F   **4.** T   **5.** F   **6.** T   **7.** concentric, eccentric   **8.** alpha, gamma   **9.** denervation atrophy, disuse atrophy   **10.** a, b, e   **11.** b   **12.** 1.f, 2.d, 3.c, 4.e, 5.b, 6.g, 7.a   **13.** 1.a, 2.a, 3.a, 4.b, 5.b, 6.b

Quantitative Exercises

(Questions on p. 300.)

1. a. For the weekend athlete, the lever ratio is 70 cm/9 cm. So the velocity at the end of the arm is 2.6 m/sec (70/9) = 20.2 m/sec (about 45 mph).

b. For the professional ballplayer, the lever ratio is 90 cm/ 9 cm. So
$$10x = 85 \text{ mph}$$
$$x = 8.5 \text{ mi/hr}(1609 \text{ m/mi})(1 \text{ hr}/3600 \text{ sec})$$
$$= 3.8 \text{ m/sec}$$

2. The force–velocity curve is as follows:

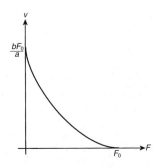

a. The shape of the curve indicates that it takes time to develop force and that the greater the force developed, the more time is needed.

b. The maximum velocity will not change when F_0 is increased, but the muscle is able to lift heavier loads or to generate more force. The maximum load will not increase when the cross-bridge cycling rate increases, but the muscle will be able to lift lighter loads faster. If the muscle increases in size, b increases, and the entire curve shifts up with respect to the v axis.

Points to Ponder

(Questions on p. 301.)

1. By placing increased demands on the heart to sustain increased delivery of O_2 and nutrients to working skeletal muscles, regular aerobic exercise induces changes in cardiac muscle that enable it to use O_2 more efficiently, such as increasing the number of capillaries supplying blood to the heart muscle. Intense exercise of short duration, such as weight training, in contrast, does not induce cardiac efficiency. Because this type of exercise relies on anaerobic glycolysis for ATP formation, no demands are placed on the heart for increased delivery of blood to the working muscles.

2. The power arm of the lever is 4 cm, and the load arm is 28 cm for a lever ratio of 1:7 (4 cm:28 cm). Thus, to lift an 8 kg stack of books with one hand, the child must generate an upward applied force in the biceps muscle of 56 kg. (With a lever ratio of 1:7, the muscle must exert seven times the force of the load; 7×8 kg = 56 kg.)

3. The length of the thin filaments is represented by the distance between a Z line and the edge of the adjacent H zone. This distance remains the same in a relaxed and contracted myofibril, leading to the conclusion that the thin filaments do not change in length during muscle contraction.

4. Regular bouts of anaerobic, short-duration, high-intensity resistance training would be recommended for competitive downhill skiing. By promoting hypertrophy of the fast glycolytic fibers, such exercise better adapts the muscles to activities that require intense strength for brief periods, such as a swift, powerful descent downhill. In contrast, regular aerobic exercise would be more beneficial for competitive cross-country skiers. Aerobic exercise induces metabolic changes within the oxidative fibers that enable the muscles to use O_2 more efficiently. These changes, which include an increase in mitochondria and capillaries within the oxidative fibers, adapt the muscles to better endure the prolonged activity of cross-country skiing without fatiguing.

5. Because the site of voluntary control to overcome the micturition reflex is at the external urethral sphincter and not the bladder, the external urethral sphincter must be skeletal muscle, which is innervated by the voluntarily controlled somatic nervous system, and the bladder must be smooth muscle, which is innervated by the involuntarily controlled autonomic nervous system. The only other type of involuntarily controlled muscle besides smooth muscle is cardiac muscle, which is found only in the heart. Therefore, the bladder must be smooth, not cardiac, muscle.

Clinical Consideration

(Question on p. 301.)

The muscles in the immobilized leg have undergone disuse atrophy. The physician or physical therapist can prescribe regular resistance-type exercises that specifically use the atrophied muscles to help restore them to their normal size.

Chapter 9   Cardiac Physiology

Objective Questions

(Questions on p. 338.)

1. intercalated discs, desmosomes, gap junctions
2. bradycardia, tachycardia   3. adenosine   4. F   5. F
6. F   7. F   8. T   9. d   10. d   11. e   12. 1.e, 2.a, 3.d, 4.b, 5.f, 6.c   13. AV, systole, semilunar, diastole   14. less than, greater than, less than, greater than, less than   15. 1.b, 2.c, 3.a, 4.b, 5.a, 6.b, 7.b, 8.c, 9.c, 10.a, 11.b, 12.c

Quantitative Exercises

(Questions on p. 340.)

1. CO = HR $\times$ SV
35 liters/min = HR $\times$ 0.07 liter
HR = (35 liters/min)/(0.07 liter) = 500 beats/min
This rate is not physiologically possible.

2. ESV = EDV − SV
    = 125 ml − 85 ml
    = 40 ml

3. Ejection fraction = SV/EDV
   a. Ejection fraction = 70/135 = 52%
   b. Ejection fraction = 100/135 = 74%
   c. Ejection fraction = 140/175 = 80%

Points to Ponder

(Questions on p. 340.)

1. Because, at a given heart rate, the interval between a premature beat and the next normal beat is longer than the interval between two normal beats, the heart fills for a longer period of time following a premature beat before the next period of contraction and emptying begins. Because of the longer filling time, the end-diastolic volume is larger, and, according to the Frank–Starling law of the heart, the subsequent stroke volume will also be correspondingly larger.

2. Trained athletes' hearts are stronger and can pump blood more efficiently so that the resting stroke volume is larger than in an untrained person. For example, if the resting stroke volume of a strong-hearted athlete is 100 ml, a resting heart rate of only 50 beats/minute produces a normal resting cardiac output of 5000 ml/minute. An untrained individual with a resting stroke volume of 70 ml, in contrast, must have a heart rate of about 70 beats/minute to produce a comparable resting cardiac output.

3. The direction of flow through a patent ductus arteriosus is the reverse of the flow that occurs through this vascular connection during fetal life. With a patent ductus arteriosus, some of the blood present in the aorta is shunted into the pulmonary artery because, after birth, the aortic pressure is greater than the pulmonary artery pressure. This abnormal blood flow produces a "machinery murmur," which lasts throughout the cardiac cycle but is more intense during systole and less intense during diastole. Thus, the murmur waxes and wanes with each beat of the heart, sounding somewhat like a washing machine as the agitator rotates back and forth. The murmur is present throughout the cardiac cycle because a pressure differential between the aorta and pulmonary artery is present during both systole and diastole. The murmur is more intense during systole because more blood is diverted through the patent ductus arteriosus as a result of the greater pressure differential between the aorta and pulmonary artery during ventricular systole than during ventricular diastole. Typically, the systolic aortic pressure is 120 mm Hg, and the systolic pulmonary arterial pressure is 24 mm Hg, for a pressure differential of 96 mm Hg. By contrast, the diastolic aortic pressure is normally 80 mm Hg, and the diastolic pulmonary arterial pressure is 8 mm Hg, for a pressure differential of 72 mm Hg.

4. A transplanted heart that does not have any innervation adjusts the cardiac output to meet the body's changing needs by means of both intrinsic control (the Frank–Starling mechanism) and extrinsic hormonal influences, such as the effect of epinephrine on the rate and strength of cardiac contraction.

5. In left bundle-branch block, the right ventricle becomes completely depolarized more rapidly than the left ventricle. As a result, the right ventricle contracts before the left ventricle, and the right AV valve is forced closed prior to closure of the left AV valve. Because the two AV valves do not close in unison, the first heart sound is "split"; that is, two distinct sounds in close succession can be detected as closure of the left valve lags behind closure of the right valve.

Clinical Consideration

(Question on p. 341.)

The most likely diagnosis is atrial fibrillation. This condition is characterized by rapid, irregular, uncoordinated depolarizations of the atria. Many of these depolarizations reach the AV node at a time when it is not in its refractory period, thus bringing about frequent ventricular depolarizations and a rapid heartbeat. However, because impulses reach the AV node erratically, the ventricular rhythm and thus the heartbeat are also very irregular as well as being rapid.

Ventricular filling is only slightly reduced despite the fact that the fibrillating atria are unable to pump blood because most ventricular filling occurs during diastole prior to atrial contraction. Because of the erratic heartbeat, variable lengths of time are available between ventricular beats for ventricular filling. However, the majority of ventricular filling occurs early in ventricular diastole after the AV valves first open, so even though the filling period may be shortened, the extent of filling may be near normal. Only when the ventricular filling period is very short is ventricular filling substantially reduced.

Cardiac output, which depends on stroke volume and heart rate, usually is not seriously impaired with atrial fibrillation. Because ventricular filling is only slightly reduced during most cardiac cycles, stroke volume, as determined by the Frank–Starling mechanism, is likewise only slightly reduced. Only when the ventricular filling period is very short and the cardiac muscle fibers are operating on the lower end of their length–tension curve is the resultant ventricular contraction weak. When the ventricular contraction becomes too weak, the ventricles eject a small or no stroke volume. During most cardiac cycles, however, the slight reduction in stroke volume is often offset by the increased heart rate so that cardiac output is usually near normal. Furthermore, if the mean arterial blood pressure falls because the cardiac output does decrease, increased sympathetic stimulation of the heart brought about by the baroreceptor reflex helps restore cardiac output to normal by shifting the Frank–Starling curve to the left.

On those cycles when ventricular contractions are too weak to eject enough blood to produce a palpable wrist pulse, if the heart rate is determined directly, either by the apex beat or via the ECG, and the pulse rate is taken concurrently at the wrist, the heart rate will exceed the pulse rate, producing a pulse deficit.

Chapter 10   The Blood Vessels and Blood Pressure

Objective Questions

(Questions on p. 387.)

1. T   **2.** F   **3.** T   **4.** T   **5.** F   **6.** T   **7.** a, c, d, e, f
8. 1.a, 2.a, 3.b, 4.a, 5.b, 6.a   **9.** 1.b, 2.a, 3.b, 4.a, 5.a, 6.a, 7.b, 8.a, 9.b, 10.a, 11.b, 12.a, 13.a

Quantitative Exercises

(Questions on p. 388.)

1. (120 mm Hg)/(30 liters/min) = 4 PRU

2. a. 90 mm Hg + (180 mm Hg − 90 mm Hg)/3
= 120 mm Hg

b. Because the other forces acting across the capillary wall, such as plasma colloid osmotic pressure, typically do not change with age, one would suspect fluid loss from the capillaries into the tissues as a result of the increase in capillary blood pressure.

3. systemic: (95 mm Hg)/(19 PRU)
= 95 mm Hg/(19 mm Hg/ liters/min) = 5 liters/min
pulmonary: (20 mm Hg)/(4 PRU) = 5 liters/min

4. e

Points to Ponder

(Questions on p. 388.)

1. An elastic support stocking increases external pressure on the remaining veins in the limb to produce a favorable pressure gradient that promotes venous return to the heart and minimizes swelling that would result from fluid retention in the extremity.

2. a. 125 mm Hg

b. 77 mm Hg

c. 48 mm Hg (125 mm Hg − 77 mm = 48 mm Hg)

d. 93 mm Hg $[77 + \frac{1}{3}(48) = 77 + 16 = 93$ mm Hg$]$

e. No; no blood would be able to get through the brachial artery, so no sound would be heard.

f. Yes; blood would flow through the brachial artery when the arterial pressure was between 118 and 125 mm Hg and would not flow through when the arterial pressure fell below 118 mm Hg. The turbulence created by this intermittent blood flow would produce sounds.

g. No; blood would flow continuously through the brachial artery in smooth, laminar fashion, so no sound would be heard.

3. The classmate has apparently fainted because of insufficient blood flow to the brain as a result of pooling of blood in the lower extremities brought about by standing still for a prolonged time. When the person faints and assumes a horizontal position, the pooled blood will quickly be returned to the heart, improving cardiac output and blood flow to his brain. Trying to get the person up would be counterproductive, so the classmate trying to get him up should be advised to let him remain lying down until he recovers on his own.

4. The drug is apparently causing the arteriolar smooth muscle to relax by causing the release of a local vasoactive chemical mediator from the endothelial cells that induces relaxation of the underlying smooth muscle.

5. a. Because activation of α_1-adrenergic receptors in vascular smooth muscle brings about vasoconstriction, blockage of α_1-adrenergic receptors reduces vasoconstrictor activity, thereby lowering the total peripheral resistance and arterial blood pressure.

b. Because activation of β_1-adrenergic receptors, which are found primarily in the heart, increases the rate and strength of cardiac contraction, drugs that block β_1-adrenergic receptors reduce cardiac output and thus arterial blood pressure by decreasing the rate and strength of the heartbeat.

c. Drugs that directly relax arteriolar smooth muscle lower arterial blood pressure by promoting arteriolar vasodilation and reducing total peripheral resistance.

d. Diuretic drugs reduce the plasma volume, thereby lowering arterial blood pressure, by increasing urinary output. Salt and water that normally would have been retained in the plasma are excreted in the urine.

e. Because sympathetic activity promotes generalized arteriolar vasoconstriction, thereby increasing total peripheral resistance and arterial blood pressure, drugs that block the release of norepinephrine from sympathetic endings lower blood pressure by preventing this vasoconstrictor effect.

f. Similarly, drugs that act on the brain to reduce sympathetic output lower blood pressure by preventing the effect of sympathetic activity on promoting arteriolar vasoconstriction and the resultant increase in total peripheral resistance and arterial blood pressure.

g. Drugs that block Ca^{2+} channels reduce the entry of Ca^{2+} into the vascular smooth muscle cells from the ECF in response to excitatory input. Because the level of contractile activity in vascular smooth muscle cells depends on their cytosolic Ca^{2+} concentration, drugs that block Ca^{2+} channels reduce the contractile activity of these cells by reducing Ca^{2+} entry and lowering their cytosolic Ca^{2+} concentration. Total peripheral resistance and, accordingly, arterial blood pressure are decreased as a result of reduced arteriolar contractile activity.

h. Drugs that interfere with the production of angiotensin II block activation of the hormonal pathway that promotes salt and water conservation (the renin–angiotensin–aldosterone system). As a result, more salt and water are lost in the urine, and less fluid is retained in the plasma. The resultant reduction in plasma volume lowers the arterial blood pressure.

i. Drugs that block angiotensin receptors prevent angiotensin II from causing arteriolar vasoconstriction, thereby decreasing total peripheral resistance, and also reduce the action of the salt- and water-conserving renin–angiotensin–aldosterone system, thereby decreasing plasma volume. Together these effects lower arterial blood pressure.

Clinical Consideration

(Question on p. 389.)

The abnormally elevated levels of epinephrine found with a pheochromocytoma bring about secondary hypertension by (1) increasing the heart rate; (2) increasing cardiac contractility, which increases stroke volume; (3) causing venous vasoconstriction, which increases venous return and subsequently stroke volume by means of the Frank–Starling mechanism; and (4) causing arteriolar vasoconstriction, which increases total peripheral resistance. Increased heart rate and stroke volume both lead to increased cardiac output. Increased cardiac output and increased total peripheral resistance both lead to increased arterial blood pressure.

Chapter 11   The Blood

Objective Questions

(Questions on p. 413.)

1. T   **2.** F   **3.** T   **4.** T   **5.** F   **6.** lymphocytes
7. liver   **8.** d   **9.** a   **10.** 1.c, 2.f, 3.b, 4.a, 5.g, 6.d, 7.h, 8.e,
9.f   **11.** 1.e, 2.c, 3.b, 4.d, 5.g, 6.f, 7.a, 8.h

Quantitative Exercises

(Questions on p. 413.)

1. a. (15 g)/(100 ml) = (150 g/liter)
   (150 g/liter) $\times$ (1 mole/66 $\times$ 10^3 g) = 2.27 mM

b. (2.27 mM) $\times$ (4 O$_2$/Hb) = 9.09 mM

c. (9.09 $\times$ 10^{-3} moles O$_2$/liter blood)
   $\times$ (22.4 liters O$_2$/l mole O$_2$) = 204 ml O$_2$/liter blood

2. Normal blood contains 5 $\times$ 10^9 RBCs/ml.
   Normal blood volume is 5 liters.

Thus, a normal person has $(5 \times 10^9$ RBCs/ml$) \times$ (5000 ml) $= 25 \times 10^{12}$ RBCs.

The normal hematocrit (Ht) is 45%, whereas the anemic has a Ht of 30%. This represents a loss of 1/3 of the RBCs, that is, 8.3×10^{12} RBCs. If RBCs are produced at a rate of 3×10^6 RBCs/sec, then the time to reestablish the Ht is 8.3×10^{12} RBCs/$(3 \times 10^6$ RBCs/sec$) = 2.77 \times 10^6$ sec $= 32$ days.

Thus, it takes about a month to replace a hemorrhagic loss of RBCs of this magnitude.

3. $v = 1.5 \times \exp(2h)$; calculate v for $h = 0.4$ and $h = 0.7$.
When $h = 0.4$, $v = 1.5 \times \exp(0.8) = 3.3$.
When $h = 0.7$, $v = 1.5 \times \exp(1.4) = 6.1$.
6.1/3.3 = 1.85, that is, an 85% increase in viscosity. Because resistance is directly proportional to viscosity, the resistance will also increase by 85%.

Points to Ponder

(Questions on p. 414.)

1. No, you cannot conclude that a person with a hematocrit of 62 definitely has polycythemia. With 62% of the whole-blood sample consisting of erythrocytes (normal being 45%), the number of erythrocytes compared to the plasma volume is definitely elevated. However, the person *may* have polycythemia, in which the number of erythrocytes is abnormally high, or may be dehydrated, in which case a normal number of erythrocytes is concentrated in a smaller-than-normal plasma volume.

2. If the genes that direct fetal hemoglobin-F synthesis could be reactivated in a patient with sickle cell anemia, a portion of the abnormal hemoglobin S that causes the erythrocytes to warp into defective sickle-shaped cells would be replaced by "healthy" hemoglobin F, thus sparing a portion of the RBCs from premature rupture. Hemoglobin F would not completely replace hemoglobin S because the gene for synthesis of hemoglobin S would still be active.

3. Most heart-attack deaths are attributable to the formation of abnormal clots that prevent normal blood flow. The sought-after chemicals in the "saliva" of bloodsucking creatures are agents that break up or prevent the formation of these abnormal clots.

Although genetically engineered tissue–plasminogen activator (tPA) is already being used as a clot-busting drug, this agent brings about degradation of fibrinogen as well as fibrin. Thus, even though the life-threatening clot in the coronary circulation is dissolved, the fibrinogen supplies in the blood are depleted for up to 24 hours until new fibrinogen is synthesized by the liver. If the patient sustains a ruptured vessel in the interim, insufficient fibrinogen might be available to form a blood-staunching clot. For example, many patients treated with tPA suffer hemorrhagic strokes within 24 hours of treatment due to incomplete sealing of a ruptured cerebral vessel. Therefore, scientists are searching for better alternatives to combat abnormal clot formation by examining the naturally occurring chemicals produced by bloodsucking creatures that permit them to suck a victim's blood without the blood clotting.

4. This question asks for your opinion, so there is no "right" answer. So far, the decision makers in our health-care system have felt that the additional financial burden of the more expensive testing of our nation's blood supply for HIV is not warranted.

5. When considering the symptoms of porphyria, one could imagine how tales of vampires—blood-craving, hairy, fanged, monstrous-looking creatures who roamed in the dark and were warded off by garlic—might easily have evolved from people's encounters with victims of this condition. This possibility is especially likely when considering how stories are embellished and distorted as they get passed along by word of mouth.

Clinical Consideration

(Question on p. 415.)

Heather's firstborn Rh-positive child did not have hemolytic disease of the newborn, because the fetal and maternal blood did not mix during gestation. Consequently, Heather did not produce any maternal antibodies against the fetus's Rh factor during gestation.

Because a small amount of the infant's blood likely entered the maternal circulation during the birthing process, Heather would produce antibodies against the Rh factor as she was first exposed to it at that time. During any subsequent pregnancies with Rh-positive fetuses, Heather's maternal antibodies against the Rh factor could cross the placental barrier and bring about destruction of fetal erythrocytes.

If, however, any Rh factor that accidentally mixed with the maternal blood during the birthing process were immediately tied up by Rh immunoglobulin administered to the mother, the Rh factor would not be available to induce maternal antibody production. Thus, no anti-Rh antibodies would be present in the maternal blood to threaten the RBCs of an Rh-positive fetus in a subsequent pregnancy. (The exogenously administered Rh immunoglobulin, being a passive form of immunity, is short-lived. In contrast, the active immunity that would result if Heather were exposed to Rh factor would be long-lived because of the formation of memory cells.)

Rh immunoglobulin must be administered following the birth of every Rh-positive child Heather bears to sop up any Rh factor before it can induce antibody production. Once an immune attack against Rh factor is launched, subsequent treatment with Rh immunoglobulin will not reverse the situation. Thus, if Heather were not treated with Rh immunoglobulin following the birth of a first Rh-positive child, and a second Rh-positive child developed hemolytic disease of the newborn, administration of Rh immunoglobulin following the second birth would not prevent the condition in a third Rh-positive child. Nothing could be done to eliminate the maternal antibodies already present.

Chapter 12   Body Defenses

Objective Questions

(Questions on p. 458.)

1. F 2. F 3. F 4. F 5. T 6. toll-like receptors
7. membrane-attack complex 8. pus 9. inflammation
10. opsonin 11. cytokines 12. b 13. 1.c, 2.d, 3.a, 4.b
14. 1.a, 2.a, 3.b, 4.b, 5.c, 6.c, 7.b, 8.a, 9.b, 10.b, 11.a, 12.b
15. 1.b, 2.a, 3.a, 4.b, 5.a, 6.a, 7.a, 8.b

Quantitative Exercises

(Question on p. 459.)

1. *NEP* = net outward pressure − net inward pressure

$$NEP = (P_C + \pi_{IF}) - (P_{IF} + \pi_P)$$

Note for this problem, $(P_{IF} + \pi_P) = (25 \text{ mm Hg} + 1 \text{ mm Hg}) = 26$ mm Hg, is constant for all cases.

Normal (π_{IF} = 0 mm Hg)

Arteriolar end NEP	$(37 + 0) - 26 = +11$ mm Hg
Venular end NEP	$(17 + 0) - 26 = -9$ mm Hg
Average NEP	$(+11 - 9)/2 = +1$ mm Hg (outward)

a. (π_{IF} = 5 mm Hg)

Arteriolar end NEP	$(37 + 5) - 26 = +16$ mm Hg
Venular end NEP	$(17 + 5) - 26 = -4$ mm Hg
Average NEP	$(+16 - 4)/2 = +6$ mm Hg (outward)
Condition	mild edema

b. (π_{IF} = 10 mm Hg)

Arteriolar end NEP	$(37 + 10) - 26 = +21$ mm Hg
Venular end NEP	$(17 + 10) - 26 = +1$ mm Hg
Average NEP	$(+21 + 1)/2 = +11$ mm Hg (outward)
Condition	Extreme edema

Points to Ponder

(Questions on p. 459.)

1. See p. 445 for a summary of immune responses to bacterial invasion and p. 440 for a summary of defenses against viral invasion.

2. A vaccine against a particular microbe can be effective only if it induces formation of antibodies and/or activated T cells against a stable antigen that is present on all microbes of this type. It has not been possible to produce a vaccine against HIV because it frequently mutates. Specific immune responses induced by vaccination against one form of HIV may prove to be ineffective against a slightly modified version of the virus.

3. Failure of the thymus to develop would lead to an absence of T lymphocytes and no cell-mediated immunity after birth. This outcome would seriously compromise the individual's ability to defend against viral invasion and cancer.

4. Researchers are working on ways to "teach" the immune system to view foreign tissue as "self" as a means of preventing the immune systems of organ-transplant patients from rejecting the foreign tissue while leaving the patients' immune defense capabilities fully intact. The immunosuppressive drugs now used to prevent transplant rejection cripple the recipients' immune defense systems and leave the patients more vulnerable to microbial invasion.

5. The skin cells visible on the body's surface are all dead.

Clinical Consideration

(Question on p. 459.)

If offending allergens bind with IgG antibodies instead of IgE antibodies, no allergy symptoms result because IgG antibodies do not attach to mast cells and basophils like IgE antibodies do.

Chapter 13 The Respiratory System

Objective Questions

(Questions on p. 507.)

1. F 2. F 3. T 4. F 5. F 6. F 7. F
8. transmural pressure gradient, pulmonary surfactant action, alveolar interdependence 9. pulmonary elasticity, alveolar surface tension 10. compliance 11. elastic recoil
12. carbonic anhydrase 13. a 14. a.<, b.>, c.=, d.=, e.=, f.=, g.>, h.<, i. approximately =, j. approximately =, k.=, l.= 15. 1.d, 2.a, 3.b, 4.a, 5.b, 6.a

Quantitative Exercises

(Questions on p. 508.)

For general reference for questions 1 and 2:
$$P_{AO_2} = P_{IO_2} - (V_{O_2}/V_A) \times 863 \text{ mm Hg}$$
$$P_{ACO_2} = (V_{CO_2}/V_A) \times 863 \text{ mm Hg}$$

1. V_A = 3 liters/min
   V_{O_2} = 0.3 liter/min, RQ = 1, therefore V_{CO_2} = 0.3 liter/min
   P_{ACO_2} = (0.3 liter/min/3 liters/min) × 863 mm Hg
   $\quad\quad$ = 86.3 mm Hg

2. a. 380 mm Hg × 0.21 = 79.8 mm Hg
   b. P_{AO_2} = 79.8 mm Hg − (0.06) × 431.5 mm Hg
   $\quad\quad$ = 79.8 mm Hg − 25.8 mm Hg
   $\quad\quad$ = 54 mm Hg
   c. P_{ACO_2} = (0.2 liter/min/4.2 liters/min) × 431.5 mm Hg
   $\quad\quad$ = 20.5 mm Hg

3. TV = 350 ml, BR = 12/min, V_A = 0.8 × V_E, DS = ?
   $V_A = BR \times (TV - DS)$
   $V_E = BR \times TV$
   $0.8 = V_A/V_E = [BR \times (TV - DS)]/(BR \times TV)$
   $\quad\quad = 1 - (DS/TV)$
   $0.8 = 1 - (DS/350 \text{ ml})$
   $DS/350 \text{ ml} = 0.2$
   $DS = 0.2(350 \text{ ml}) = 70 \text{ ml}$

Points to Ponder

(Questions on p. 509.)

1. Total atmospheric pressure decreases with increasing altitude, yet the percentage of O_2 in the air remains the same. At an altitude of 30,000 feet, the atmospheric pressure is only 226 mm Hg. Because 21% of atmospheric air consists of O_2, the P_{O_2} of inspired air at 30,000 feet is only 47.5 mm Hg, and alveolar P_{O_2} is even lower at about 20 mm Hg. At this low P_{O_2},

hemoglobin is only about 30% saturated with O_2—much too low to sustain tissue needs for O_2.

The P_{O_2} of inspired air can be increased by two means when flying at high altitude. First, by pressurizing the plane's interior to a pressure comparable to that of atmospheric pressure at sea level, the P_{O_2} of inspired air within the plane is 21% of 760 mm Hg, or the normal 160 mm Hg. Accordingly, alveolar and arterial P_{O_2} and percent hemoglobin saturation are likewise normal. In the emergency situation of failure to maintain internal cabin pressure, breathing pure O_2 can raise the P_{O_2} considerably above that accomplished by breathing normal air. When a person is breathing pure O_2, the entire pressure of inspired air is attributable to O_2. For example, with a total atmospheric pressure of 226 mm Hg at an altitude of 30,000 feet, the P_{O_2} of inspired pure O_2 is 226 mm Hg, which is more than adequate to maintain normal arterial hemoglobin saturation.

2. a. Hypercapnia would not accompany the hypoxia associated with cyanide poisoning. In fact, CO_2 levels decline, because oxidative metabolism is blocked by the tissue poisons so that CO_2 is not being produced.

b. Hypercapnia could but may not accompany the hypoxia associated with pulmonary edema. Pulmonary diffusing capacity is reduced in pulmonary edema, but O_2 transfer suffers more than CO_2 transfer because the diffusion constant for CO_2 is 20 times that for O_2. As a result, hypoxia occurs much more readily than hypercapnia in these circumstances. Hypercapnia does occur, however, when pulmonary diffusing capacity is severely impaired.

c. Hypercapnia would accompany the hypoxia associated with restrictive lung disease because ventilation is inadequate to meet the metabolic needs for both O_2 delivery and CO_2 removal. Both O_2 and CO_2 exchange between the lungs and atmosphere are equally affected.

d. Hypercapnia would not accompany the hypoxia associated with high altitude. In fact, arterial P_{CO_2} levels actually decrease. One of the compensatory responses in acclimatization to high altitudes is reflex stimulation of ventilation as a result of the reduction in arterial P_{O_2}. This compensatory hyperventilation to obtain more O_2 blows off too much CO_2 in the process, so arterial P_{CO_2} levels decline below normal.

e. Hypercapnia would not accompany the hypoxia associated with severe anemia. Reduced O_2-carrying capacity of the blood has no influence on blood CO_2 content, so arterial P_{CO_2} levels are normal.

f. Hypercapnia would accompany the circulatory hypoxia associated with congestive heart failure. Just as the diminished blood flow fails to deliver adequate O_2 to the tissues, it also fails to remove sufficient CO_2.

g. Hypercapnia would accompany the hypoxic hypoxia associated with obstructive lung disease because ventilation would be inadequate to meet the metabolic needs for both O_2 delivery and CO_2 removal. Both O_2 and CO_2 exchange between the lungs and atmosphere would be equally affected.

3. P_{O_2} = 122 mm Hg
0.21 (atmospheric pressure − partial pressure of H_2O)
= 0.21(630 mm Hg − 47 mm Hg)
= 0.21(583 mm Hg) = 122 mm Hg

4. Voluntarily hyperventilating before going underwater lowers the arterial P_{CO_2} but does not increase the O_2 content in the blood. Because the P_{CO_2} is below normal, the person can hold his or her breath longer than usual before the arterial P_{CO_2} increases to the point that he or she is driven to surface for a breath. Therefore, the person can stay underwater longer. The risk, however, is that the O_2 content of the blood, which was normal, not increased, before going underwater, continues to fall. Therefore, the O_2 level in the blood can fall dangerously low before the CO_2 level builds to the point of driving the person to take a breath. Low arterial P_{O_2} does not stimulate respiratory activity until it has plummeted to 60 mm Hg. Meanwhile, the person may lose consciousness and drown due to inadequate O_2 delivery to the brain. If the person does not hyperventilate so that both the arterial P_{CO_2} and O_2 content are normal before going underwater, the buildup of CO_2 will drive the person to the surface for a breath before the O_2 levels fall to a dangerous point.

5. c. The arterial P_{O_2} will be less than the alveolar P_{O_2}, and the arterial P_{CO_2} will be greater than the alveolar P_{CO_2}. Because pulmonary diffusing capacity is reduced, arterial P_{O_2} and P_{CO_2} do not equilibrate with alveolar P_{O_2} and P_{CO_2}.

If the person is administered 100% O_2, the alveolar P_{O_2} will increase and the arterial P_{O_2} will increase accordingly. Even though arterial P_{O_2} will not equilibrate with alveolar P_{O_2}, it will be higher than when the person is breathing atmospheric air.

The arterial P_{CO_2} will remain the same whether the person is administered 100% O_2 or is breathing atmospheric air. The alveolar P_{CO_2} and thus the blood-to-alveolar P_{CO_2} gradient are not changed by breathing 100% O_2 because the P_{CO_2} in atmospheric air and 100% O_2 are both essentially zero (P_{CO_2} in atmospheric air = 0.23 mm Hg).

Clinical Consideration

(Question on p. 509.)

Emphysema is characterized by a collapse of smaller respiratory airways and a breakdown of alveolar walls. Because of the collapse of smaller airways, airway resistance is increased with emphysema. As with other chronic obstructive pulmonary diseases, expiration is impaired to a greater extent than inspiration because airways are naturally dilated slightly more during inspiration than expiration as a result of the greater transmural pressure gradient during inspiration. Because airway resistance is increased, a patient with emphysema must produce larger-than-normal intra-alveolar pressure changes to accomplish a normal tidal volume. Unlike quiet breathing in a normal person, the accessory inspiratory muscles (neck muscles) and the muscles of active expiration (abdominal muscles and internal intercostal muscles) must be brought into play to inspire and expire a normal tidal volume of air.

The spirogram would be characteristic of chronic obstructive pulmonary disease. Because the patient experiences more difficulty emptying the lungs than filling them, the total lung capacity would be essentially normal, but the functional residual capacity and the residual volume would be elevated as a result of the additional air trapped in the lungs following expiration. Because the residual volume is increased, the inspiratory capacity and vital capacity will be reduced. Also, the FEV_1 will be markedly reduced because the airflow rate is decreased by the airway obstruction. The FEV_1–to–vital capacity ratio will be much lower than the normal 80%.

Because of the reduced surface area for exchange as a result of a breakdown of alveolar walls, gas exchange would be impaired. Therefore, arterial P_{CO_2} would be elevated and arterial P_{O_2} reduced compared to normal.

Ironically, administering O_2 to this patient to relieve his hypoxic condition would markedly depress his drive to breathe by elevating the arterial P_{O_2} and removing the primary driving stimulus for respiration. Because of this danger, O_2 therapy should either not be administered or administered extremely cautiously.

Chapter 14   The Urinary System

Objective Questions

(Questions on p. 553.)

1. F   **2.** F   **3.** T   **4.** T   **5.** T   **6.** nephron   **7.** potassium
8. 500   **9.** 1.b, 2.a, 3.b, 4.b, 5.a, 6.b, 7.b, 8.b, 9.b   **10.** e
11. b   **12.** b, e, a, d, c   **13.** c, e, d, a, b, f   **14.** g, c, d, a, f, b, e   **15.** 1.a, 2.a, 3.c, 4.b, 5.d

Quantitative Exercises

(Questions on p. 554.)

1.

	Patient 1	Patient 2
GFR	125 ml/min	124 ml/min
RPF	620 ml/min	400 ml/min
RBF	1127 ml/min	727 ml/min
FF	0.20	0.31

All of patient 1's values are within the normal range. Patient 2's GFR is normal, but he has a low renal plasma flow and a high filtration fraction. Therefore, his GFR is too high for that RPF. This might imply enlarged filtration slits or a "leaky" glomerulus in general. The low RPF may imply low renal blood pressure, perhaps from a partially blocked renal artery.

2. filtered load = GFR × plasma concentration
        = (0.125 liter/min) × (145 mmol/liter)
        = 18.125 mmol/min

3. $GFR = (U \times [I]_U)/[I]_B$
    $U = (GFR \times [I]_B)/[I]_U$
      = (125 ml/min)(3 mg/liter)/(300 mg/liter)
      = 1.25 ml/min

4. $\text{Clearance rate of a substance} = \dfrac{\begin{array}{c}\text{urine concentration} \\ \text{of a substance}\end{array} \times \begin{array}{c}\text{urine flow} \\ \text{rate}\end{array}}{\text{plasma concentration of the substance}}$

      $= \dfrac{7.5 \text{ mg/ml} \times 2 \text{ ml/min}}{0.2 \text{ mg/ml}}$

      = 75 ml/min

Because a clearance rate of 75 ml/min is less than the average GFR of 125 ml/min, the substance is being reabsorbed.

Points to Ponder

(Questions on p. 555.)

1. The longer loops of Henle in desert rats (known as *kangaroo rats*) permit a greater magnitude of countercurrent multiplication and thus a larger medullary vertical osmotic gradient. As a result, these rodents can produce urine that is concentrated up to an osmolarity of almost 6000 mOsm/liter, which is five times more concentrated than maximally concentrated human urine at 1200 mOsm/liter. Because of this tremendous concentrating ability, kangaroo rats never have to drink; the H_2O produced metabolically within their cells during oxidation of foodstuff (food + $O_2 \rightarrow CO_2 + H_2O$ + energy) is sufficient for their needs.

2. a. 250 mg/min filtered
      filtered load of substance = plasma concentration of sub-
                               stance × GFR
      filtered load of substance = 200 mg/100 ml × 125 ml/min
                               = 250 mg/min
   b. 200 mg/min reabsorbed
      A T_m worth of the substance will be reabsorbed.
   c. 50 mg/min excreted
      amount of substance excreted = amount of substance
      filtered − amount of substance reabsorbed =
      250 mg/min − 200 mg/min = 50 mg/min

3. Aldosterone stimulates Na^+ reabsorption and K^+ secretion by the renal tubules. Therefore, the most prominent features of Conn's syndrome (hypersecretion of aldosterone) are hypernatremia (elevated Na^+ levels in the blood) caused by excessive Na^+ reabsorption, hypophosphatemia (below-normal K^+ levels in the blood) caused by excessive K^+ secretion, and hypertension (elevated blood pressure) caused by excessive salt and water retention.

4. e. 300/300. If the ascending limb were permeable to water, it would not be possible to establish a vertical osmotic gradient in the interstitial fluid of the renal medulla, nor would the ascending-limb fluid become hypotonic before entering the distal tubule. As the ascending limb pumped NaCl into the interstitial fluid, water would osmotically follow, so both the interstitial fluid and the ascending limb would remain isotonic at 300 mOsm (mosm/liter). With the tubular fluid entering the distal tubule being 300 mosm/liter instead of the normal 100 mosm/liter, it would not be possible to produce urine with an osmolarity less than 300 mosm/liter. Likewise, in the absence of the medullary vertical osmotic gradient, it would not be possible to produce urine more concentrated than 300 mosm/liter, no matter how much vasopressin was present.

5. Because the descending pathways between the brain and the motor neurons supplying the external urethral sphincter and pelvic diaphragm are no longer intact, the accident victim can no longer voluntarily control micturition. Therefore, bladder emptying in this individual will be governed entirely by the micturition reflex.

Clinical Consideration

(Question on p. 555.)

prostate enlargement

Chapter 15 Fluid and Acid–Base Balance

Objective Questions

(Questions on p. 586.)

1. T 2. F 3. F 4. T 5. T 6. intracellular fluid
7. $[H_2CO_3]$, $[HCO_3^-]$ 8. d 9. b 10. a, d, e 11. b, e
12. c 13. 1. metabolic acidosis, 2. diabetes mellitus,
3. pH = 7.1, 4. respiratory alkalosis, 5. anxiety, 6. pH = 7.7,
7. respiratory acidosis, 8. pneumonia, 9. pH = 7.1,
10. metabolic alkalosis, 11. vomiting, 12. pH = 7.7

Quantitative Exercises

(Questions on p. 587.)

1. pH = 6.1 + log $[HCO_3^-]$/(0.03 mM/mm Hg × 40 mm Hg)
   7.4 = 6.1 + log $[HCO_3^-]$/1.2 mM
   log $[HCO_3^-]$/1.2 mM = 7.4 − 6.1 = 1.3
   $[HCO_3^-]$ = 1.2 mM × $(10^{1.3})$ = 24 mM

2. pH = − log $[H^+]$, $[H^+]$ = 10^{-pH}
   $[H^+]$ = $10^{-6.8}$ = 158 nM for pH = 6.8
   $[H^+]$ = $10^{-8.0}$ = 10 nM for pH = 8.0

3. Note that distilled water is permeable across all barriers, so it will distribute equally among all compartments. Because saline does not enter cells, it will stay in the ECF. The resultant distributions are summarized in the chart at the bottom of the page. Clearly, saline is better at expanding the plasma volume.

Points to Ponder

(Questions on p. 587.)

1. The rate of urine formation increases when alcohol inhibits vasopressin secretion and the kidneys are unable to reabsorb water from the distal and collecting tubules. Because extra free water that normally would have been reabsorbed from the distal parts of the tubule is lost from the body in the urine, the body becomes dehydrated and the ECF osmolarity increases following alcohol consumption. That is, more fluid is lost in the urine than is consumed in the alcoholic beverage as a result of alcohol's action on vasopressin. Thus, the imbibing person experiences a water deficit and still feels thirsty, despite the recent fluid consumption.

2. If a person loses 1500 ml of salt-rich sweat and drinks 1000 ml of water without replacing the salt during the same time period, there will still be a volume deficit of 500 ml, and the body fluids will have become hypotonic (the remaining salt in the body will be diluted by the ingestion of 1000 ml of free H_2O). As a result, the hypothalamic osmoreceptors (the dominant input) will signal the vasopressin-secreting cells to *decrease* vasopressin secretion and thus increase urinary excretion of the extra free water that is making the body fluids too dilute. Simultaneously, the left atrial volume receptors will signal the vasopressin-secreting cells to *increase* vasopressin secretion to conserve water during urine formation and thus help relieve the volume deficit. These two conflicting inputs to the vasopressin-secreting cells are counterproductive. For this reason, it is important to replace both water and salt following heavy sweating or abnormal loss of other salt-rich fluids. If salt is replaced along with water intake, the ECF osmolarity remains close to normal and the vasopressin-secreting cells receive signals only to increase vasopressin secretion to help restore the ECF volume to normal.

3. When a dextrose solution equal in concentration to that of normal body fluids is injected intravenously, the ECF volume is expanded but the ECF and ICF are still osmotically equal. Therefore, no net movement of water occurs between the ECF and ICF. When the dextrose enters the cell and is metabolized, however, the ECF becomes hypotonic as this solute leaves the plasma. If the excess free water is not excreted in the urine rapidly enough, water will move into the cells by osmosis.

4. Because baking soda ($NaHCO_3$) is readily absorbed from the digestive tract, treatment of gastric hyperacidity with baking soda can lead to metabolic alkalosis as too much HCO_3^- is absorbed. Treatment with antacids that are poorly absorbed is

Ingested Fluid	Compartment	Size of Compartment before Ingestion (liters)	Size of Compartment after Ingestion (liters)	% Increase in Size of Compartment after Ingestion
Distilled water	TBW	42	43	2%
	ICF (2/3 TBW)	28	28.667	2%
	ECF (1/3 TBW)	14	14.333	2%
	plasma (20% ECF)	2.8	2.866	2%
	ISF (80% ECF)	11.2	11.466	2%
Saline	TBW	42	43	2%
	ICF	28	28	0%
	ECF	14	15	7%
	plasma	2.8	3	7%
	ISF	11.2	12	7%

safer because these products remain in the digestive tract and do not produce an acid–base imbalance.

5. c. The hemoglobin buffer system buffers carbon dioxide–generated hydrogen ion. In the case of respiratory acidosis accompanying severe pneumonia, the $H^+ + Hb \rightarrow HHb$ reaction will be shifted toward the HHb side, thus removing some of the extra free H^+ from the blood.

Clinical Consideration

(Question on p. 587.)

The resultant prolonged diarrhea will lead to dehydration and metabolic acidosis as a result of, respectively, excessive loss in the feces of fluid and $NaHCO_3$ that normally would have been absorbed into the blood.

Compensatory measures for dehydration have included increased vasopressin secretion, resulting in increased water reabsorption by the distal and collecting tubules and a subsequent reduction in urine output. Simultaneously, fluid intake has been encouraged by increased thirst. The metabolic acidosis has been combated by removal of excess H^+ from the ECF by the HCO_3^- member of the H_2CO_3:HCO_3^- buffer system, by increased ventilation to reduce the amount of acid-forming CO_2 in the body fluids, and by the kidneys excreting extra H^+ and conserving HCO_3^-.

Chapter 16   The Digestive System

Objective Questions

(Questions on p. 638.)

1. F   **2.** T   **3.** T   **4.** F   **5.** F   **6.** F   **7.** long, short **8.** chyme   **9.** three   **10.** vitamin B_{12}, bile salts   **11.** bile salts   **12.** b   **13.** 1.c, 2.e, 3.b, 4.a, 5.f, 6.d   **14.** 1.c, 2.c, 3.d, 4.a, 5.e, 6.c, 7.a, 8.b, 9.c, 10.b, 11.d, 12.b

Quantitative Exercises

(Questions on p. 639.)

1. a. $r = 0.5$ cm, area $= 4\pi (0.25) = \pi$ cm^2
volume $= \left(\frac{4}{3}\pi\right)(0.5)^3 = 0.5236$ cm^3
area/volume $= 6$

b. Each new droplet's volume is 5.236×10^{-3} cm^3, so the average radius is therefore 0.1077 cm. The area of a sphere with that radius is
$4\pi(0.1077$ cm$)^2 = 0.1458$ cm^2
area/volume $= 27.8$

c. Area emulsified/area droplet $= (100)(0.1458$ cm$^2)/\pi$ cm^2
$= 4.64$
Thus, the total surface area of all 100 emulsified droplets is 4.64 times the area of the original larger lipid droplet.

d. Volume emulsified/volume droplet $= (100)(5.236 \times 10^{-3}$ cm$^3)/0.5236$ cm$^3 = 1.0$.

Thus, the total volume did not change as a result of emulsification, as would be expected because the total volume of the lipid is conserved during emulsification. The volume originally present in the large droplet is divided up among the 100 emulsified droplets.

Points to Ponder

(Questions on p. 639.)

1. Patients who have had their stomachs removed must eat small quantities of food frequently instead of consuming the typical three meals a day because they have lost the ability to store food in the stomach and meter it into the small intestine at an optimal rate. If a person without a stomach consumed a large meal that entered the small intestine all at once, the luminal contents would quickly become too hypertonic as digestion of the large nutrient molecules into a multitude of small, osmotically active, absorbable units outpaced the more slowly acting process of absorption of these units. As a consequence of this increased luminal osmolarity, water would enter the small-intestine lumen from the plasma by osmosis, resulting in circulatory disturbances as well as intestinal distension. To prevent this "dumping syndrome" from occurring, the patient must "feed" the small intestine only small amounts of food at a time so that absorption of the digestive end products can keep pace with their rate of production. The person has to consciously take over metering the delivery of food into the small intestine because the stomach is no longer present to assume this responsibility.

2. The gut-associated lymphoid tissue launches an immune attack against any pathogenic (disease-causing) microorganisms that enter the readily accessible digestive tract and escape destruction by salivary lysozyme or gastric HCl. This action defends against entry of these potential pathogens into the body proper. The large number of immune cells in the gut-associated lymphoid tissue is adaptive as a first line of defense against foreign invasion when considering that the surface area of the digestive tract lining represents the largest interface between the body proper and the external environment.

3. Defecation would be accomplished entirely by the defecation reflex in a patient paralyzed from the waist down because of lower spinal-cord injury. Voluntary control of the external anal sphincter would be impossible because of interruption in the descending pathway between the primary motor cortex and the motor neuron supplying this sphincter.

4. When insufficient glucuronyl transferase is available in the neonate to conjugate all of the bilirubin produced during erythrocyte degradation with glycuronic acid, the extra unconjugated bilirubin cannot be excreted into the bile. Therefore, this extra bilirubin remains in the body, giving rise to mild jaundice in the newborn.

5. Removal of the stomach leads to pernicious anemia because of the resultant lack of intrinsic factor, which is necessary for absorption of vitamin B_{12}. Removal of the terminal ileum leads to pernicious anemia because this is the only site where vitamin B_{12} can be absorbed.

Clinical Consideration

(Question on p. 639.)

A person whose bile duct is blocked by a gallstone experiences a painful "gallbladder attack" after eating a high-fat meal because the ingested fat triggers the release of cholecystokinin, which stimulates gallbladder contraction. As the gallbladder contracts and bile is squeezed into the blocked bile duct, the duct becomes distended prior to the blockage. This distension is painful.

The feces are grayish white because no bilirubin-containing bile enters the digestive tract when the bile duct is blocked. Bilirubin, when acted on by bacterial enzymes, is responsible for the brown color of feces, which are grayish white in its absence.

Chapter 17   Energy Balance and Temperature Regulation

Objective Questions

(Questions on p. 658.)

1. F **2.** F **3.** F **4.** F **5.** T **6.** T **7.** T **8.** T
9. arcuate nucleus **10.** shivering **11.** nonshivering thermogenesis **12.** sweating **13.** b **14.** 1.b, 2.a, 3.c, 4.a, 5.d, 6.c, 7.b, 8.d, 9.c, 10.b

Quantitative Exercises

(Questions on p. 659.)

1. From physics we know that $\Delta T(°C) = \Delta U/(C \times m)$. Also note that $\Delta U/t = BMR$; i.e., the rate of using energy is the basal metabolic rate. m represents the mass of body fluid; for a typical person, this is 42 liters.

(42 liters) $\times$ (1 kg/liter) = 42 kg
C = 1.0 kcal/(kg-°C)

Given that water boils at 100°C and normal body temperature is 37°C, we need to change the temperature by 63°C. Thus,

t = ($\Delta T \times C \times m$)/BMR = (63°C)[1.0 kcal/(kg-°C)]
(42 kg)/(75 kcal/hr) = 35 hr

At the higher metabolic rate during exercise,

t = (63°C)[1.0 kcal/(kg-°C)](42 kg)/(1000 kcal/hr) = 2.6 hr

Points to Ponder

(Questions on p. 659.)

1. Evidence suggests that CCK serves as a satiety signal. It is believed to serve as a signal to stop eating when enough food has been consumed to meet the body's energy needs, even though the food is still in the digestive tract. Therefore, when drugs that inhibit CCK release are administered to experimental animals, the animals overeat because this satiety signal is not released.

2. Don't go on a "crash diet." Be sure to eat a nutritionally balanced diet that provides all essential nutrients, but reduce total caloric intake, especially by cutting down on high-fat foods. Spread out consumption of the food throughout the day instead of just eating several large meals. Avoid bedtime snacks. Burn more calories through a regular exercise program.

3. Engaging in heavy exercise on a hot day is dangerous because of problems arising from trying to eliminate the extra heat generated by the exercising muscles. First, there will be conflicting demands for distribution of the cardiac output—temperature-regulating mechanisms will trigger skin vasodilation to promote heat loss from the skin surface, whereas metabolic changes within the exercising muscles will induce local vasodilation in the muscles to match the increased metabolic needs with increased blood flow. Further exacerbating the problem of conflicting demands for blood flow is the loss of effective circulating plasma volume resulting from the loss of a large volume of fluid through another important cooling mechanism, sweating. Therefore, it is difficult to maintain an effective plasma volume and blood pressure and simultaneously keep the body from overheating when engaging in heavy exercise in the heat, so heat exhaustion is likely to ensue.

4. When a person is soaking in a hot bath, loss of heat by radiation, conduction, convection, and evaporation is limited to the small surface area of the body exposed to the cooler air. Heat is being gained by conduction at the larger skin surface area exposed to the hotter water.

5. The thermoconforming fish would not run a fever when it has a systemic infection because it has no mechanisms for regulating internal heat production or for controlling heat exchange with its environment. The fish's body temperature varies capriciously with the external environment no matter whether it has a systemic infection or not. It is not able to maintain body temperature at a "normal" set point or an elevated set point (i.e., a fever).

Clinical Consideration

(Question on p. 659.)

Cooled tissues need less nourishment than they do at normal body temperature because of their pronounced reduction in metabolic activity. The lower O_2 need of cooled tissues accounts for the occasional survival of drowning victims who have been submerged in icy water considerably longer than one could normally survive without O_2.

Chapter 18   Principles of Endocrinology: The Central Endocrine Glands

Objective Questions

(Questions on p. 688.)

1. T **2.** T **3.** F **4.** T **5.** T **6.** T **7.** tropic
8. down regulation **9.** epiphyseal plate **10.** suprachiasmatic nucleus **11.** 1.c, 2.b, 3.b, 4.a, 5.a, 6.c, 7.b

Points to Ponder

(Questions on p. 688.)

1. The concentration of hypothalamic releasing and inhibiting hormones would be considerably lower (in fact, almost nonexistent) in a systemic venous blood sample compared to the concentration of these hormones in a sample of hypothalamic-hypophyseal portal blood. These hormones are secreted into the portal blood for local delivery between the hypothalamus and anterior pituitary. Any portion of these hormones picked up by the systemic blood at the anterior pituitary capillary level is greatly diluted by the much larger total volume of systemic blood compared to the extremely small volume of blood within the portal vessel.

2. Above normal. Without sufficient iodine, the thyroid gland is unable to synthesize enough thyroid hormone. The resultant reduction in negative-feedback activity by the reduced level of thyroid hormone would lead to increased TSH secretion. Despite the elevated TSH, however, the thyroid gland still could not secrete adequate thyroid hormone because of the iodine deficiency.

3. If CRH and/or ACTH is elevated in accompaniment with the excess cortisol secretion, the condition is secondary to a defect at the hypothalamic/anterior pituitary level. If CRH and ACTH levels are below normal in accompaniment with the excess cortisol secretion, the condition is due to a primary defect at the adrenal cortex level, with the excess cortisol inhibiting the hypothalamus and anterior pituitary in negative-feedback fashion.

4. Males with testicular feminization syndrome would be unusually tall because of the inability of testosterone to promote closure of the epiphyseal plates of the long bones in the absence of testosterone receptors.

5. Full-grown athletes sometimes illegally take supplemental doses of growth hormone because it promotes increased skeletal muscle mass through its IGF-I-induced protein anabolic effect. However, excessive growth hormone can have detrimental side effects, such as possibly causing diabetes or high blood pressure.

Clinical Consideration

(Question on p. 689.)

Hormonal replacement therapy following pituitary gland removal should include thyroid hormone (the thyroid gland will not produce sufficient thyroid hormone in the absence of TSH) and glucocorticoid (because of the absence of ACTH), especially in stress situations. If indicated, male or female sex hormones can be replaced, even though these hormones are not essential for survival. For example, testosterone in males plays an important role in libido. Growth hormone and prolactin need not be replaced because their absence will produce no serious consequences in this individual. Vasopressin may have to be replaced if insufficient quantities of this hormone are picked up by the blood at the hypothalamus in the absence of the posterior pituitary.

Chapter 19   The Peripheral Endocrine Glands

Objective Questions

(Questions on p. 738.)

1. F **2.** T **3.** T **4.** T **5.** T **6.** F **7.** F **8.** F
9. colloid, thyroglobulin **10.** glycogenesis, glycogenolysis,

gluconeogenesis **11.** brain, working muscles, liver
12. bone, kidneys, digestive tract **13.** osteoblasts, osteoclasts, osteocytes **14.** c **15.** a, b, c, g, i, j

Points to Ponder

(Questions on p. 739.)

1. The midwestern United States is no longer an endemic goiter belt even though the soil is still iodine poor, because individuals living in this region obtain iodine from iodine-supplemented nutrients, such as iodinated salt, and from seafood and other naturally iodine-rich foods shipped from coastal regions.

2. Anaphylactic shock is an extremely serious allergic reaction brought about by massive release of chemical mediators in response to exposure to a specific allergen—such as one associated with a bee sting—to which the individual has been highly sensitized. These chemical mediators bring about circulatory shock (severe hypotension) through a twofold effect: (1) by relaxing arteriolar smooth muscle, thus causing widespread arteriolar vasodilation and a resultant fall in total peripheral resistance and arterial blood pressure; and (2) by causing a generalized increase in capillary permeability, resulting in a shift of fluid from the plasma into the interstitial fluid. This shift decreases the effective circulating volume, further reducing arterial blood pressure. Additionally, these chemical mediators bring about pronounced bronchoconstriction, making it impossible for the victim to move sufficient air through the narrowed airways. Because these responses take place rapidly and can be fatal, people allergic to bee stings are advised to keep injectable epinephrine in their possession. By promoting arteriolar vasoconstriction through its action on α_1 receptors in arteriolar smooth muscle and promoting bronchodilation through its action on β_2 receptors in bronchiolar smooth muscle (see ▲ Table 7-3, p. 242), epinephrine counteracts the life-threatening effects of the anaphylactic reaction to the bee sting.

3. An infection elicits the stress response, which brings about increased secretion of cortisol and epinephrine, both of which increase the blood glucose level. This becomes a problem for diabetic patients who have to bring down the elevated blood glucose by injecting additional insulin or, preferably, by reducing carbohydrate intake and/or exercising to use up some of the extra blood glucose. In a normal individual, the check-and-balance system between insulin and the other hormones that oppose insulin's actions helps maintain the blood glucose within reasonable limits during the stress response.

4. The presence of Chvostek's sign results from increased neuromuscular excitability caused by moderate hyposecretion of parathyroid hormone.

5. If malignancy-associated hypercalcemia arose from metastatic tumor cells that invaded and destroyed bone, both hypercalcemia and hyperphosphatemia would result as calcium phosphate salts were released from the destructed bone. The fact that hypophosphatemia, not hyperphosphatemia, often accompanies malignancy-associated hypercalcemia led investigators to rule out bone destruction as the cause of the hypercalcemia. Instead, they suspected that the tumors produced a substance that mimics the actions of PTH in promoting concurrent hypercalcemia and hypophosphatemia.

Clinical Consideration

(Question on p. 739.)

"Diabetes of bearded ladies" is descriptive of both excess cortisol and excess adrenal androgen secretion. Excess cortisol secretion causes hyperglycemia and glucosuria. Glucosuria promotes osmotic diuresis, which leads to dehydration and a compensatory increased sensation of thirst. All these symptoms—hyperglycemia, glucosuria, polyuria, and polydipsia—mimic diabetes mellitus. Excess adrenal androgen secretion in females promotes masculinizing characteristics, such as beard growth. Simultaneous hypersecretion of both cortisol and adrenal androgen most likely occurs secondary to excess CRH/ACTH secretion, because ACTH stimulates both cortisol and androgen production by the adrenal cortex.

Chapter 20   The Reproductive System

Objective Questions

(Questions on p. 797.)
1. T **2.** T **3.** T **4.** F **5.** T **6.** F **7.** T
8. seminiferous tubules, FSH, testosterone **9.** thecal, LH, granulosa, FSH **10.** corpus luteum of pregnancy
11. human chorionic gonadotropin **12.** c **13.** e
14. 1.c, 2.a, 3.b, 4.c, 5.a, 6.a, 7.c, 8.b, 9.e **15.** 1.a, 2.c, 3.b, 4.a, 5.a, 6.b

Points to Ponder

(Questions on p. 798.)

1. The anterior pituitary responds only to the normal pulsatile pattern of GnRH and does not secrete gonadotropins in response to continuous exposure to GnRH. In the absence of FSH and LH secretion, ovulation and other events of the ovarian cycle do not ensue, so continuous GnRH administration may find use as a contraceptive technique.

2. Testosterone hypersecretion in a young boy causes premature closure of the epiphyseal plates so that he stops growing before he reaches his genetic potential for height. The child would also display signs of precocious pseudopuberty, characterized by premature development of secondary sexual characteristics, such as deep voice, beard, enlarged penis, and sex drive.

3. A potentially troublesome side effect of drugs that inhibit sympathetic nervous system activity as part of the treatment for high blood pressure is males' inability to carry out the sex act. Both divisions of the autonomic nervous system are required for the male sex act. Parasympathetic activity is essential for accomplishing erection, and sympathetic activity is important for ejaculation.

4. Posterior pituitary extract contains an abundance of stored oxytocin, which can be administered to induce or facilitate labor by increasing uterine contractility. Exogenous oxytocin is most successful in inducing labor if the woman is near term, presumably because of the increasing concentration of myometrial oxytocin receptors at that time.

5. GnRH or FSH and LH are not effective in treating the symptoms of menopause because the ovaries are no longer responsive to the gonadotropins. Thus, treatment with these hormones would not cause estrogen and progesterone secretion. In fact, GnRH, FSH, and LH levels are already elevated in postmenopausal women because of lack of negative feedback by the ovarian hormones.

Clinical Consideration

(Question on p. 798.)

The first warning of a tubal pregnancy is pain caused by stretching of the oviduct by the growing embryo. A tubal pregnancy must be surgically terminated because the oviduct cannot expand as the uterus does to accommodate the growing embryo. If not removed, the enlarging embryo will rupture the oviduct, causing possibly lethal hemorrhage.

Glossary

A band One of the dark bands that alternate with light (I) bands to create a striated appearance in a skeletal or cardiac muscle fiber when these fibers are viewed with a light microscope

absorption The transfer of digested nutrients and ingested liquids from the digestive tract lumen into the blood or lymph

absorptive state The metabolic state following a meal when nutrients are being absorbed and stored; fed state

accessory digestive organs Exocrine organs outside the wall of the digestive tract that empty their secretions through ducts into the digestive tract lumen

accessory sex glands Glands that empty their secretions into the reproductive tract

accommodation The ability to adjust the strength of the lens in the eye so that both near and far sources can be focused on the retina

acetylcholine (ACh) (as′-uh-teal-KŌ-lēn) The neurotransmitter released from all autonomic preganglionic fibers, parasympathetic postganglionic fibers, and motor neurons

acetylcholinesterase (AChE) (as′-uh-teal-kō-luh-NES-tuh-rās) An enzyme present in the motor end-plate membrane of a skeletal muscle fiber that inactivates acetylcholine

ACh See *acetylcholine*

AChE See *acetylcholinesterase*

acid A hydrogen-containing substance that yields a free hydrogen ion and anion on dissociation

acidosis (as-i-DŌ-sus) Blood pH of less than 7.35

acini (ĂS-i-nī) The secretory component of saclike exocrine glands, such as digestive enzyme–producing pancreatic glands or milk-producing mammary glands

acquired immune responses Responses that are selectively targeted against particular foreign material to which the body has previously been exposed; see also *antibody-mediated immunity* and *cell-mediated immunity*

ACTH See *adrenocorticotropic hormone*

actin The contractile protein forming the backbone of the thin filaments in muscle fibers

action potential A brief, rapid, large change in membrane potential that serves as a long-distance electrical signal in an excitable cell

active expiration Emptying of the lungs more completely than when at rest by contracting the expiratory muscles; also called *forced expiration*

active force A force that requires expenditure of cellular energy (ATP) in the transport of a substance across the plasma membrane

active reabsorption When any one of the five steps in the transepithelial transport of a substance reabsorbed across the kidney tubules requires energy expenditure

active transport Active carrier-mediated transport involving transport of a substance against its concentration gradient across the plasma membrane

acuity Discriminative ability; the ability to discern between two different points of stimulation

adaptation A reduction in receptor potential despite sustained stimulation of the same magnitude

adenosine diphosphate (ADP) (uh-DEN-uh-sēn) The two-phosphate product formed from the splitting of ATP to yield energy for the cell's use

adenosine triphosphate (ATP) The body's common energy "currency," which consists of an adenosine with three phosphate groups attached; splitting of the high-energy, terminal phosphate bond provides energy to power cellular activities

adenylyl cyclase (ah-DEN-il-il sī-klās) The membrane-bound enzyme activated by a G protein intermediary in response to binding of an extracellular messenger with a surface membrane receptor that, in turn, activates cyclic AMP, an intracellular second messenger

adequate stimulus The type of stimulus to which a specific receptor type responds, such as a photoreceptor responding to light

ADH See *vasopressin*

adipocytes Fat cells in adipose tissue; store triglyceride fat and secrete hormones termed *adipokines*

adipokines Hormones secreted by adipose tissue that play important roles in energy balance and metabolism

adipose tissue The tissue specialized for storage of triglyceride fat; found under the skin in the hypodermis

ADP See *adenosine diphosphate*

adrenal cortex (uh-DRĒ-nul) The outer portion of the adrenal gland; secretes three classes of steroid hormones: glucocorticoids, mineralocorticoids, and sex hormones

adrenal medulla (muh-DUL-uh) The inner portion of the adrenal gland; secretes the hormones epinephrine and norepinephrine into the blood in response to sympathetic stimulation

adrenergic fibers (ad′-ruh-NUR-jik) Nerve fibers that release norepinephrine as their neurotransmitter

adrenocorticotropic hormone (ACTH) (ad-rē′-nō-kor′-tuh-kō-TRŌP-ik) An anterior pituitary hormone that stimulates cortisol secretion by the adrenal cortex and promotes growth of the adrenal cortex

aerobic Referring to a condition in which oxygen is available

aerobic exercise Exercise that can be supported by ATP formation accomplished by oxidative phosphorylation because adequate O_2 is available to support the muscle's modest energy demands; also called *endurance-type exercise*

afferent arteriole (AF-er-ent ar-TIR-ē-ōl) The vessel that carries blood into the glomerulus of the kidney's nephron

afferent division The portion of the peripheral nervous system that carries information from the periphery to the central nervous system

afferent neuron Neuron that possesses a sensory receptor at its peripheral ending and carries information to the central nervous system

after hyperpolarization (hī′-pur-pō-luh-ruh-ZA-shun) A slight, transient hyperpolarization that sometimes occurs at the end of an action potential

agonist A substance that binds to a neurotransmitter's receptors and mimics the neurotransmitter's response

agranulocytes (ā-GRAN-yuh-lō-sīts′) Leukocytes that do not contain granules, including lymphocytes and monocytes

albumin (al-BEW-min) The smallest and most abundant plasma proteins; binds and transports many water-insoluble substances in the blood; contributes extensively to plasma-colloid osmotic pressure

aldosterone (al-dō-steer-OWN) or (al-DOS-tuh-rōn) The adrenocortical hormone that stimulates Na^+ reabsorption by the distal and collecting tubules of the kidney's nephron during urine formation

alkalosis (al′-kuh-LŌ-sus) Blood pH of greater than 7.45

allergy Acquisition of an inappropriate specific immune reactivity to a normally harmless environmental substance

all-or-none law An excitable membrane either responds to a stimulus with a maximal action potential that spreads nondecrementally throughout the membrane or does not respond with an action potential at all

alpha (α) cells The endocrine pancreatic cells that secrete the hormone glucagon

alpha motor neuron A motor neuron that innervates ordinary skeletal muscle fibers

alveolar surface tension (al-VĒ-ō-lur) The surface tension of the fluid lining the alveoli in the lungs; see *surface tension*

alveolar ventilation The volume of air exchanged between the atmosphere and alveoli per minute; equals (tidal volume minus dead space volume) times respiratory rate

alveoli (al-VĒ-ō-lī) The air sacs across which O_2 and CO_2 are exchanged between the blood and air in the lungs

amines (AH-mēnz) Hormones derived from the amino acid tyrosine, including thyroid hormone and catecholamines

amoeboid movement (uh-MĒ-boid) "Crawling" movement of white blood cells, similar to the means by which amoebas move

AMPA receptor One of two types of receptor-channels on a postsynaptic membrane to which the neurotransmitter glutamate binds, this one leading to the formation of EPSPs

anabolism (ah-NAB-ō-li-zum) The buildup, or synthesis, of larger organic molecules from small organic molecular subunits

anaerobic (an´-uh-RŌ-bik) Referring to a condition in which oxygen is not present

anaerobic exercise High-intensity exercise that can be supported by ATP formation accomplished by anaerobic glycolysis for brief periods of time when O_2 delivery to a muscle is inadequate to support oxidative phosphorylation

analgesic (an-al-JEE-zic) Pain relieving

anatomy The study of body structure

androgen A masculinizing "male" sex hormone; includes testosterone from the testes and dehydro-epiandrosterone from the adrenal cortex

anemia A reduction below normal in O_2-carrying capacity of the blood

anion (AN-ī-on) Negatively charged ion that has gained one or more electrons in its outer shell

ANP See *atrial natriuretic peptide*

antagonism Actions opposing each other; in the case of hormones, when one hormone causes the loss of another hormone's receptors, reducing the effectiveness of the second hormone

antagonist A substance that blocks a neurotransmitter's receptor, thus preventing the neurotransmitter from binding and producing a response

anterior pituitary The glandular portion of the pituitary that synthesizes, stores, and secretes growth hormone, TSH, ACTH, FSH, LH, and prolactin

antibody An immunoglobulin produced by a specific activated B lymphocyte (plasma cell) against a particular antigen; binds with the specific antigen against which it is produced and promotes the antigenic invader's destruction by augmenting nonspecific immune responses already initiated against the antigen

antibody-mediated immunity A specific immune response accomplished by antibody production by B cells

antidiuretic hormone (ADH) (an´-ti-dī´-yū-RET-ik) See *vasopressin*

antigen A large, complex molecule that triggers a specific immune response against itself when it gains entry into the body

antioxidant A substance that helps inactivate biologically damaging free radicals

antiport The form of secondary active transport in which the driving ion and transported solute move in opposite directions across the plasma membrane; also called *countertransport* or *exchange*

antrum (of ovary) The fluid-filled cavity formed within a developing ovarian follicle

antrum (of stomach) The lower portion of the stomach

aorta (a-OR-tah) The large vessel that carries blood from the left ventricle

aortic valve A one-way valve that permits the flow of blood from the left ventricle into the aorta during ventricular emptying but prevents the backflow of blood from the aorta to the left ventricle during ventricular relaxation

apoptosis (ā-pop-TŌ-sis) Programmed cell death; deliberate self-destruction of a cell.

aquaporin Water channel

aqueous humor (Ā-kwē-us) The clear, watery fluid in the anterior chamber of the eye; provides nourishment for the cornea and lens

arcuate nucleus (ARE-kyou-it´) The subcortical brain region housing neurons that secrete appetite-enhancing neuropeptide Y and those that secrete appetite-suppressing melanocortins

aromatase An enzyme that converts testosterone to estrogen outside of the testes

arterioles (ar-TIR-ē-ōlz) The highly muscular, high-resistance vessels that branch from arteries, the caliber of which can be changed to regulate the amount of blood distributed to the various tissues

artery A vessel that carries blood away from the heart

ascending tract A bundle of nerve fibers of similar function that travels up the spinal cord to transmit signals derived from afferent input to the brain

astrocyte A type of glial cell in the brain; major functions include holding the neurons together in proper spatial relationship, inducing the brain capillaries to form tight junctions important in the blood–brain barrier, and enhancing synaptic activity

atmospheric pressure The pressure exerted by the weight of the air in the atmosphere on objects on Earth's surface; equals 760 mm Hg at sea level

ATP See *adenosine triphosphate*

ATPase An enzyme that can split ATP

ATP synthase The mitochondrial enzyme that catalyzes the synthesis of ATP from ADP and inorganic phosphate

atrial natriuretic peptide (ANP) (Ā-trē-al NĀ-tree-ur-eh´tik) A peptide hormone released from the cardiac atria that promotes urinary loss of Na^+

atrioventricular (AV) node (ā´-trē-ō-ven-TRIK-yuh-lur) A small bundle of specialized cardiac cells at the junction of the atria and ventricles that is the only site of electrical contact between the atria and ventricles

atrioventricular (AV) valve A one-way valve that permits the flow of blood from the atrium to the ventricle during filling of the heart but prevents the backflow of blood from the ventricle to the atrium during emptying of the heart

atrium (atria, plural) (Ā-tree-um) An upper chamber of the heart that receives blood from the veins and transfers it to the ventricle

atrophy (AH-truh-fē) Decrease in mass of an organ

autonomic nervous system The portion of the efferent division of the peripheral nervous system that innervates smooth and cardiac muscle and exocrine glands; subdivided into the sympathetic nervous system and the parasympathetic nervous system

autophagy Selective self-digestion of cell parts such as worn-out organelles by lysosomes

autoregulation The ability of an organ to adjust its own rate of blood flow despite changes in the driving mean arterial blood pressure

autorhythmicity The ability of an excitable cell to rhythmically initiate its own action potentials

AV nodal delay The delay in impulse transmission between the atria and ventricles at the AV node, which allows enough time for the atria to become completely depolarized and contract, emptying their contents into the ventricles, before ventricular depolarization and contraction occur

AV valve See *atrioventricular valve*

axon A single, elongated tubular extension of a neuron that conducts action potentials away from the cell body; also known as a *nerve fiber*

axon hillock The first portion of a neuronal axon plus the region of the cell body from which the axon leaves; the site of action-potential initiation in most neurons

axon terminals The branched endings of a neuronal axon, which release a neurotransmitter that influences target cells in close association with the axon terminals

balance concept The balance between input of a substance through ingestion and/or metabolic production and its output through excretion and/or metabolic consumption

baroreceptor reflex An autonomically mediated reflex response that influences the heart and blood vessels to oppose a change in mean arterial blood pressure

baroreceptors Receptors located within the circulatory system that monitor blood pressure

basal ganglia See *basal nuclei*

basal metabolic rate (BĀ-sul) The minimal waking rate of internal energy expenditure; the body's "idling speed"

basal nuclei Several masses of gray matter located deep within the white matter of the cerebrum of the brain; play an important inhibitory role in motor control

base A substance that can combine with a free hydrogen ion and remove it from solution

basic electrical rhythm (BER) Self-induced electrical activity of the digestive-tract smooth muscle

basilar membrane (BAS-ih-lar) The membrane that forms the floor of the middle compartment of the cochlea and bears the organ of Corti, the sense organ for hearing

basophils (BAY-so-fills) White blood cells that synthesize, store, and release histamine, which is important in allergic responses, and heparin, which hastens the removal of fat particles from the blood

BER See *basic electrical rhythm*

beta (β) cells The endocrine pancreatic cells that secrete the hormone insulin

bicarbonate (HCO_3^-) The anion resulting from dissociation of carbonic acid, H_2CO_3

bile An alkaline solution containing bile salts and bilirubin secreted by the liver, stored in the gallbladder, and emptied into the small intestine lumen

bile salts Cholesterol derivatives secreted in the bile that facilitate fat digestion through their detergent action and facilitate fat absorption through their micellar formation

biliary system (BIL-ē-air´-ē) The bile-producing system, consisting of the liver, gallbladder, and associated ducts

bilirubin (bill-eh-RŪ-bin) A bile pigment, which is a waste product derived from the degradation of hemoglobin during the breakdown of old red blood cells

bipolar cells Neurons in the retina on which photoreceptors terminate and which in turn terminate on retinal ganglion cells

blastocyst The developmental stage of the fertilized ovum by the time it is ready to implant; consists of a single-layered sphere of cells encircling a fluid-filled cavity

blood–brain barrier (BBB) Special structural and functional features of the brain capillaries that limit access of materials from the blood into the brain tissue

B lymphocytes (B cells) White blood cells that produce antibodies against specific targets to which they have been exposed

body of the stomach The main, or middle, part of the stomach

body system A collection of organs that perform related functions and interact to accomplish a common activity that is essential for survival of the whole body; for example, the digestive system

bone marrow The soft, highly cellular tissue that fills the internal cavities of bones and is the source of most blood cells

Bowman's capsule The beginning of the tubular component of the kidney's nephron that cups around the glomerulus and collects the glomerular filtrate as it is formed

Boyle's law (boils) At any constant temperature, the pressure exerted by a gas varies inversely with the volume of the gas

brain stem The portion of the brain that is continuous with the spinal cord, serves as an integrating link between the spinal cord and higher brain levels, and controls many life-sustaining processes, such as breathing, circulation, and digestion

bronchioles (BRONG-kē-ōlz) The small, branching airways within the lungs

bronchoconstriction Narrowing of the respiratory airways

bronchodilation Widening of the respiratory airways

brush border The collection of microvilli projecting from the luminal border of epithelial cells lining the digestive tract and kidney tubules

buffer See *chemical buffer system*

bulbourethral glands (bul-bo-you-RĒTH-ral) Male accessory sex glands that secrete mucus for lubrication

bulk flow Movement in bulk of a protein-free plasma across the capillary walls between the blood and surrounding interstitial fluid; encompasses ultrafiltration and reabsorption

bundle of His (hiss) A tract of specialized cardiac cells that rapidly transmits an action potential down the interventricular septum of the heart

calcitonin (kal'-suh-TŌ-nun) A hormone secreted by the thyroid C cells that lowers plasma Ca^{2+} levels

calcium balance Maintenance of a constant total amount of Ca^{2+} in the body; accomplished by slowly responding adjustments in intestinal Ca^{2+} absorption and in urinary Ca^{2+} excretion

calcium homeostasis Maintenance of a constant free plasma Ca^{2+} concentration, accomplished by rapid exchanges of Ca^{2+} between the bone and ECF and, to a lesser extent, by modifications in urinary Ca^{2+} excretion

calcium-induced calcium release When in cardiac and smooth muscle cells the excitation-induced entry of a small amount of Ca^{2+} from the ECF through voltage-gated surface membrane receptors triggers the opening of Ca^{2+}-release channels in the sarcoplasmic reticulum, causing a much larger release of Ca^{2+} into the cytosol from this intracellular store

calmodulin (kal'-MA-jew-lin) An intracellular Ca^{2+} binding protein that, on activation by Ca^{2+}, induces a change in structure and function of another intracellular protein; especially important in smooth-muscle excitation–contraction coupling

cAMP See *cyclic adenosine monophosphate*

CAMs See *cell adhesion molecules*

capillaries The thin-walled, pore-lined smallest of blood vessels, across which exchange between the blood and surrounding tissues takes place

carbonic anhydrase (an-HĪ-drās) The enzyme found in erythrocytes, kidney tubular cells, gastric parietal cells, and pancreatic duct cells that catalyzes the conversion of CO_2 and H_2O into H^+ and HCO_3^-

cardiac cycle One period of systole and diastole

cardiac muscle The specialized muscle found only in the heart

cardiac output (CO) The volume of blood pumped by each ventricle each minute; equals stroke volume times heart rate

cardiovascular control center The integrating center located in the medulla of the brain stem that controls mean arterial blood pressure

carrier-mediated transport Transport of a substance across the plasma membrane facilitated by a carrier molecule

carrier molecules Membrane proteins that, by undergoing reversible changes in shape so that specific binding sites are alternately exposed at either side of the membrane, can bind with and transfer particular substances unable to cross the plasma membrane on their own

cascade A series of sequential reactions that culminates in a final product, such as a clot

catabolism (kuh-TAB-ō-li-zum) The breakdown, or degradation, of large, energy-rich molecules within cells

catalase (KAT-ah-lās) An antioxidant enzyme found in peroxisomes that decomposes potent hydrogen peroxide into harmless H_2O and O_2

catecholamines (kat'-uh-KŌ-luh-means) Amine hormones derived from tyrosine and secreted largely by the adrenal medulla

cations (KAT-ī-onz) Positively charged ions that have lost one or more electrons from their outer shell

C cells The thyroid cells that secrete calcitonin

CCK See *cholecystokinin*

cell The smallest unit capable of carrying out the processes associated with life; the basic unit of both structure and function in living organisms

cell adhesion molecules (CAMs) Proteins that protrude from the surface of the plasma membrane and form loops or other appendages that the cells use to grip one another and the surrounding connective tissue fibers

cell body The portion of a neuron that houses the nucleus and organelles

cell center See *centrosome*

cell-mediated immunity A specific immune response accomplished by activated T lymphocytes, which directly attack unwanted cells

cellular respiration The entire series of chemical reactions involving the intracellular breakdown of nutrient-rich molecules to yield energy, using O_2 and producing CO_2 in the process

center A functional collection of cell bodies within the central nervous system

central chemoreceptors (kē-mō-rē-SEP-turz) Receptors located in the medulla near the respiratory center that respond to changes in ECF H^+ concentration resulting from changes in arterial P_{CO_2} and adjust respiration accordingly

central lacteal (LAK-tē-ul) The initial lymphatic vessel that supplies each of the small-intestinal villi

central nervous system (CNS) The brain and spinal cord

central sulcus (SUL-kus) A deep infolding of the brain surface that runs roughly down the middle of the lateral surface of each cerebral hemisphere and separates the parietal and frontal lobes

centrioles (SEN-tree-ōls) A pair of short, cylindrical structures within a cell that form the mitotic spindle during cell division

centrosome The cell's microtubule organizing center located near the nucleus and consisting of the pair of centrioles surrounded by an amorphous mass of proteins; also called the *cell center*

cerebellum (ser'-uh-BEL-um) The part of the brain attached at the rear of the brain stem and concerned with maintaining proper position of the body in space and subconscious coordination of motor activity

cerebral cortex The outer shell of gray matter in the cerebrum; site of initiation of all voluntary motor output and final perceptual processing of all sensory input as well as integration of most higher neural activity

cerebral hemispheres The cerebrum's two halves, which are connected by a thick band of neuronal axons

cerebrospinal fluid (CSF) (ser'-uh-brō-SPĪ-nul) or (sah-REE-brō-SPĪ-nul) A special cushioning fluid that is produced by, surrounds, and flows through the central nervous system

cerebrum (SER-uh-brum) or (sah-REE-brum) The division of the brain that consists of the basal nuclei and cerebral cortex

channels Small, water-filled passageways through the plasma membrane; formed by membrane proteins that span the membrane and provide highly selective passage for small water-soluble substances such as ions

chemical bonds The forces holding atoms together

chemical buffer system A mixture in a solution of two or more chemical compounds that minimize pH changes when either an acid or a base is added to or removed from the solution

chemically gated channels Channels in the plasma membrane that open or close in response to the binding of a specific chemical messenger with a membrane receptor site that is in close association with the channel

chemical mediator A chemical that is secreted by a cell and that influences an activity outside the cell

chemical synapse The most abundant type of junction between two neurons in which an action potential in a presynaptic neuron alters the postsynaptic neuron's potential by means of a chemical messenger, a neurotransmitter

chemiosmosis (kē-ma-OS-mō-sis) ATP production in mitochondria catalyzed by ATP synthase, which is activated by flow of H^+ down a concentration gradient established by the electron transport system

chemoreceptor (kē-mo-rē-sep´-tur) A sensory receptor sensitive to specific chemicals

chemotaxin (kē-mō-TAK-sin) A chemical released at an inflammatory site that attracts phagocytes to the area

chief cells The cells in the gastric pits that secrete pepsinogen

cholecystokinin (CCK) (kō´-luh-sis-tuh-kī-nun) A hormone released from the duodenal mucosa primarily in response to the presence of fat; inhibits gastric motility and secretion, stimulates pancreatic enzyme secretion, stimulates gallbladder contraction, and acts as a satiety signal

cholesterol A type of fat molecule that serves as a precursor for steroid hormones and bile salts and is a stabilizing component of the plasma membrane

cholinergic fibers (kō´-lin-ER-jik) Nerve fibers that release acetylcholine as their neurotransmitter

chromaffin granules The granules that store catecholamines in adrenomedullary cells

chyme (kīm) A thick liquid mixture of food and digestive juices

cilia (SILL-ee-ah) Motile, hairlike protrusions from the surface of cells lining the respiratory airways and the oviducts

ciliary body The portion of the eye that produces aqueous humor and contains the ciliary muscle

ciliary muscle A circular ring of smooth muscle within the eye whose contraction increases the strength of the lens to accommodate for near vision

circadian rhythm (sir-KĀ-dē-un) Repetitive oscillations in the set point of various body activities, such as hormone levels and body temperature, that are very regular and have a frequency of one cycle every 24 hours, usually linked to light–dark cycles; diurnal rhythm; biological rhythm

circulatory shock When mean arterial blood pressure falls so low that adequate blood flow to the tissues can no longer be maintained

citric acid cycle A cyclic series of biochemical reactions that processes the intermediate breakdown products of nutrient molecules, generating carbon dioxide and preparing hydrogen carrier molecules for entry into the high-energy-yielding electron transport chain

CNS See *central nervous system*

cochlea (KOK-lē-uh) The snail-shaped portion of the inner ear that houses the receptors for sound

collecting tubule The last portion of tubule in the kidney's nephron that empties into the renal pelvis

colloid (KOL-oid) The thyroglobulin-containing substance enclosed within the thyroid follicles

colloid osmotic pressure The osmotic force across the capillary wall resulting from the uneven colloidal dispersion of plasma proteins between the blood and interstitial fluid

competition When several closely related substances compete for the same carrier binding sites

complement system A collection of plasma proteins that are activated in cascade fashion on exposure to invading microorganisms, ultimately producing a membrane attack complex that destroys the invaders

compliance The distensibility of a hollow, elastic structure, such as a blood vessel or the lungs; a measure of how easily the structure can be stretched

concave surface Curved in, as a surface of a lens that diverges light rays

concentration gradient A difference in concentration of a particular substance between two adjacent areas

conduction The transfer of heat between objects of differing temperature that are in direct contact

cones The eye's photoreceptors used for color vision in the light

connective tissue Tissue that serves to connect, support, and anchor various body parts; distinguished by relatively few cells dispersed within an abundance of extracellular material

contiguous conduction The means by which an action potential is propagated throughout a non-myelinated nerve fiber; local current flow between an active and adjacent inactive area brings the inactive area to threshold, triggering an action potential in a previously inactive area

contractile proteins Myosin and actin, whose interaction brings about shortening (contraction) of a muscle fiber

contractility (of heart) The strength of contraction of the heart at any given end-diastolic volume

control center See *integrator*

controlled variable Some factor that can vary but is held within a narrow range by a control system

convection Transfer of heat energy by air or water currents

convergence The converging of many presynaptic terminals from thousands of other neurons on a single neuronal cell body and its dendrites so that activity in the single neuron is influenced by the activity of many other neurons

convex surface Curved out, as a surface in a lens that converges light rays

core temperature The temperature within the inner core of the body (abdominal and thoracic organs, central nervous system, and skeletal muscles) that is homeostatically maintained at about 100°F

cornea (KOR-nee-ah) The clear, anteriormost outer layer of the eye through which light rays pass to the interior of the eye

coronary circulation The blood vessels that supply the heart muscle

corpus callosum (ka-LOW-sum) The thick band of nerve fibers that connects the two cerebral hemispheres structurally and functionally

corpus luteum (LOO-tē-um) The ovarian structure that develops from a ruptured follicle after ovulation

cortical nephrons The most abundant type of nephrons, whose glomeruli lie in the outer layer of the renal cortex and whose short loops of Henle dip only slightly into the renal medulla

corticotropes Anterior pituitary cells that secrete adrenocorticotropic hormone

cortisol (KORT-uh-sol) The adrenocortical hormone that plays an important role in carbohydrate, protein, and fat metabolism and helps the body resist stress

cotransport See *symport*

countercurrent system The means by which the long loops of Henle and vasa recta of juxtamedullary nephrons establish and maintain the vertical osmotic gradient in the renal medulla, making it possible to put out urine of variable concentration depending on the body's needs

countertransport See *antiport*

cranial nerves The 12 pairs of peripheral nerves, the majority of which arise from the brain stem

cross bridges The myosin molecules' globular heads that protrude from a thick filament within a muscle fiber and interact with the actin molecules in the thin filaments to bring about shortening of the muscle fiber during contraction

CSF See *cerebrospinal fluid*

current The flow of electrical charge, such as by movement of positive charges toward a more negatively charged area

cyclic adenosine monophosphate (cyclic AMP or cAMP) An intracellular second messenger derived from ATP

cyclic AMP See *cyclic adenosine monophosphate*

cyclic guanosine monophosphate (cyclic GMP or cGMP) An intracellular second messenger similar to cAMP

cytokines All chemicals other than antibodies that are secreted by lymphocytes

cytoplasm (SĪ-tō-plaz´-um) The portion of the cell interior not occupied by the nucleus

cytoskeleton The complex intracellular protein network that acts as the "bone and muscle" of the cell

cytosol (SĪ-tuh-sol´) The semiliquid portion of the cytoplasm not occupied by organelles

cytotoxic T cells (sī-tō-TOK-sik) The population of T cells that destroys host cells bearing foreign antigen, such as body cells invaded by viruses or cancer cells

dead-space volume The volume of air that occupies the respiratory airways as air is moved in and out and that is not available to participate in exchange of O_2 and CO_2 between the alveoli and atmosphere

dehydration A water deficit in the body

dehydroepiandrosterone (DHEA) (dē-HĪ-drō-ep-i-and-row-steer-own) The androgen secreted by the adrenal cortex in both sexes

dendrites Projections from the surface of a neuron's cell body that carry signals toward the cell body

deoxyribonucleic acid (DNA) (dē-OK-sē-rī-bō-new-klā-ik) The cell's genetic material, which is found within the nucleus and which provides codes for protein synthesis and serves as a blueprint for cell replication

depolarization (de´-pō-luh-ruh-ZĀ-shun) A reduction in membrane potential from resting potential; movement of the potential from resting toward 0 mV

dermis The connective tissue layer that lies under the epidermis in the skin; contains the skin's blood vessels and nerves

descending tract A bundle of nerve fibers of similar function that travels down the spinal cord to relay messages from the brain to efferent neurons

desmosome (dez´-muh-sōm) An adhering junction between two adjacent but nontouching cells formed by the extension of filaments between the cells' plasma membranes; most abundant in tissues that are subject to considerable stretching

DHEA See *dehydroepiandrosterone*

diacylglycerol (DAG) (die-ACE-sul-gli-sir-all) A component cleaved from phosphatidylinositol bisphosphate (PIP_2) in the plasma membrane that serves as a second messenger in response to binding of an extracellular (first) messenger with a G-protein coupled receptor

diaphragm (DIE-uh-fram) A dome-shaped sheet of skeletal muscle that forms the floor of the thoracic cavity; the major inspiratory muscle

diastole (dī-AS-tō-lē) The period of cardiac relaxation and filling

diencephalon (dī´-un-SEF-uh-lan) The division of the brain that consists of the thalamus and hypothalamus

differentiation The process of each type of cell becoming specialized during development of a multicellular organism to carry out a particular function

diffusion Random collisions and intermingling of molecules as a result of their continuous thermally induced random motion

digestion The process by which the structurally complex foodstuffs of the diet are broken down into smaller absorbable units by the enzymes produced within the digestive system

dihydropyridine receptors (die-HIGH-dro-PEER-ih-deen) Voltage-gated receptors on the T tubules that trigger the opening of adjoining ryanodine receptors on the sarcoplasmic reticulum during excitation-contraction coupling

diploid number (DIP-loid) A complete set of 46 chromosomes (23 pairs), as found in all human somatic cells

discriminative ability See *acuity*

distal tubule A highly convoluted tubule that extends between the loop of Henle and the collecting duct in the kidney's nephron

diurnal rhythm (dī-URN´-ul) Repetitive oscillations in hormone levels that are very regular and have a frequency of one cycle every 24 hours, usually linked to the light–dark cycle; circadian rhythm; biological rhythm

divergence The diverging, or branching, of a neuron's axon terminals so that activity in this single neuron influences the many other cells with which its terminals synapse

DNA See *deoxyribonucleic acid*

dorsal root ganglion A cluster of afferent neuronal cell bodies located adjacent to the spinal cord

down regulation A reduction in the number of receptors for (and thereby the target cell's sensitivity to) a particular hormone as a direct result of the effect that an elevated level of the hormone has on its own receptors

dynamic equilibrium When two opposing movements exactly counterbalance each other so that no net movement takes place, with no energy needed to maintain this constancy

dynein (DIE-neen) The molecular motor that "walks" along microtubular "highways" toward the cell center, such as in transporting debris from the axon terminal to the cell body for destruction by lysosomes

ECG See *electrocardiogram*

ECM See *extracellular matrix*

edema (i-DĒ-muh) Swelling of tissues as a result of excess interstitial fluid

EDV See *end-diastolic volume*

EEG See *electroencephalogram*

effector The component of a control system that accomplishes the output commanded by the integrator

effector organs The muscles or glands innervated by the nervous system that carry out the nervous system's orders to bring about a desired effect, such as a particular movement or secretion

efferent division (EF-er-ent) The portion of the peripheral nervous system that carries instructions from the central nervous system to effector organs

efferent neuron Neuron that carries information from the central nervous system to an effector organ

efflux (Ē-flux) Movement out of the cell

eicosanoids (ī-KŌ-sa-noydz) Derivatives of arachidonic acid in the phospholipid tails of the plasma membrane including the prostaglandins, prostacyclins, thromboxanes, and leukotrienes, that act as paracrines near their site of production throughout the body

elastic recoil Rebound of the lungs after having been stretched

electrical gradient A difference in charge between two adjacent areas

electrical synapse The least common type of junction between two neurons in which an action potential in the presynaptic neuron spreads directly to the postsynaptic neuron via gap junctions

electrocardiogram (ECG) The graphic record of the electrical activity that reaches the surface of the body as a result of cardiac depolarization and repolarization

electrochemical gradient The simultaneous existence of an electrical gradient and concentration (chemical) gradient for a particular ion

electroencephalogram (EEG) (i-lek´-trō-in-SEF-uh-luh-gram´) A graphic record of the collective postsynaptic potential activity in the cell bodies and dendrites located in the cortical layers under a recording electrode

electrolytes Solutes that form ions in solution and conduct electricity

electron transport system The series of electron carriers in the mitochondrial inner membrane that transfer electrons from higher to lower energy levels, with the released energy being used to establish the H^+ concentration gradient in the mitochondria that powers ATP synthesis

embolus (EM-bō-lus) A freely floating clot

embryonic stem cells (ESCs) Undifferentiated cells resulting from the early divisions of a fertilized egg that ultimately give rise to all the mature, specialized cells of the body while at the same time renewing themselves

end-diastolic volume (EDV) The volume of blood in the ventricle at the end of diastole, when filling is complete

endocrine axis A three-hormone sequence consisting of (1) a hypothalamic hypophysiotropic hormone that controls the output of (2) an anterior pituitary tropic hormone that regulates secretion of (3) a target endocrine gland hormone, which exerts the final physiologic effect

endocrine glands Ductless glands that secrete hormones into the blood

endocytic vesicle A small, intracellular, membrane-enclosed vesicle in which extracellular material is trapped

endocytosis (en´-dō-sī-TŌ-sis) Internalization of extracellular material within a cell as a result of the plasma membrane forming a pouch that contains the extracellular material, then sealing at the surface of the pouch to form an endocytic vesicle

endogenous opiates (en-DAJ´-eh-nus ō´-pē-āts) Endorphins and enkephalins, which bind with opiate receptors and are important in the body's natural analgesic system

endogenous pyrogen (pī´-ruh-jun) A chemical released from macrophages during inflammation that acts by means of local prostaglandins to raise the set point of the hypothalamic thermostat to produce a fever

endolymph The fluid within the cochlear duct and vestibular organs in the inner ear

endometrium (en´-dō-MĒ-trē-um) The lining of the uterus

endoplasmic reticulum (ER) (en´-dō-PLAZ-mik ri-TIK-yuh-lum) An organelle consisting of a continuous membranous network of fluid-filled tubules (smooth ER) and flattened sacs, partially studded with ribosomes (rough ER); synthesizes proteins and lipids for formation of new cell membrane and other cell components and manufactures products for secretion

endothelium (en´-dō-THĒ-lē-um) The thin, single-celled layer of epithelial cells lining the entire circulatory system

end-plate potential (EPP) The graded receptor potential that occurs at the motor end plate of a skeletal muscle fiber in response to binding with acetylcholine

end-systolic volume (ESV) The volume of blood in the ventricle at the end of systole, when emptying is complete

endurance-type exercise See *aerobic exercise*

energy balance The balance between energy input by means of food intake and energy output by means of external work and internal work

enteric nervous system The extensive network of nerve fibers consisting of the myenteric plexus and submucous plexus within the digestive tract wall that endows the tract with considerable self-regulation

enterogastrones (ent´-uh-rō-GAS-trōnz) Hormones secreted by the duodenal mucosa that inhibit gastric motility and secretion; include secretin and cholecystokinin

enterohepatic circulation (en´-tur-ō-hi-PAT-ik) The recycling of bile salts and other bile constituents between the small intestine and liver by means of the hepatic portal vein

enzyme A special protein molecule that speeds up a particular chemical reaction in the body

eosinophils (ē´-uh-SIN-uh-fils) White blood cells that are important in allergic responses and in combating internal parasite infestations

ependymal cells (eh-PEN-dim-ul) The glial cells lining the ventricles of the brain, which serve as neural stem cells

epidermis (ep´-uh-DER-mus) The outer layer of the skin, consisting of numerous layers of epithelial cells, with the outermost layers being dead and flattened

epididymis The male accessory reproductive organ that stores sperm and increases their motility and fertility prior to ejaculation

epinephrine (ep´-uh-NEF-rin) The primary hormone secreted by the adrenal medulla; important in preparing the body for "fight-or-flight" responses and in regulating arterial blood pressure; adrenaline

epiphyseal plate (eh-pif-i-SEE-al) A layer of cartilage that separates the diaphysis (shaft) of a long bone from the epiphysis (flared end); the site at which bones grow longer before the cartilage ossifies (turns into bone)

epithelial tissue (ep´-uh-THĒ-le-ul) A functional grouping of cells specialized in the exchange of materials between the cell and its environment; lines and covers various body surfaces and cavities and forms secretory glands

EPSP See *excitatory postsynaptic potential*

equilibrium potential (E_x) The potential that exists when the concentration gradient and opposing electrical gradient for a given ion exactly counterbalance each other so there is no net movement of the ion

erythrocytes (i-RITH-ruh-sīts) Red blood cells, which are plasma membrane–enclosed bags of hemoglobin that transport O_2 and, to a lesser extent, CO_2 and H^+ in the blood

erythropoiesis (i-rith´-rō-poi-Ē-sus) Erythrocyte production by the bone marrow

erythropoietin The hormone released from the kidneys in response to a reduction in O_2 delivery to the kidneys; stimulates the bone marrow to increase erythrocyte production

ESCs See *embryonic stem cells*

esophagus (i-SOF-uh-gus) A straight muscular tube that extends between the pharynx and stomach

estrogen Feminizing "female" sex hormone

ESV See *end-systolic volume*

evaporation The transfer of heat from the body surface by the transformation of water from a liquid to a gaseous state

exchange See *antiport*

excitable tissue Tissue capable of producing electrical signals when excited; includes nervous and muscle tissue

excitation–contraction coupling The series of events linking muscle excitation (the presence of an action potential) to muscle contraction (filament sliding and sarcomere shortening)

excitatory postsynaptic potential (EPSP) (pōst´-si-NAP-tik) A small depolarization of the postsynaptic membrane in response to neurotransmitter binding, bringing the membrane closer to threshold

excitatory synapse (SIN-aps´) Synapse in which the postsynaptic neuron's response to neurotransmitter release is a small depolarization of the postsynaptic membrane, bringing the membrane closer to threshold

exercise physiology The study of both the functional changes that occur in response to a single session of exercise and the adaptations that result from regular, repeated exercise sessions

exocrine glands Glands that secrete through ducts to the outside of the body or into a cavity that communicates with the outside

exocytosis (eks´-ō-sī-TŌ-sis) Fusion of a membrane-enclosed intracellular vesicle with the plasma membrane, followed by the opening of the vesicle and the emptying of its contents to the outside

expiration A breath out

expiratory muscles The skeletal muscles whose contraction reduces the size of the thoracic cavity and lets the lungs recoil to a smaller size, bringing about movement of air from the lungs to the atmosphere

external environment The environment surrounding the body

external genitalia The externally visible portions of the reproductive system

external intercostal muscles Inspiratory muscles whose contraction elevates the ribs, thereby enlarging the thoracic cavity

external work Energy expended by contracting skeletal muscles to move external objects or to move the body in relation to the environment

extracellular fluid (ECF) All the body's fluid outside the cells; consists of interstitial fluid and plasma

extracellular matrix (ECM) An intricate meshwork of fibrous proteins embedded in the interstitial fluid secreted by local cells

extrinsic controls Regulatory mechanisms initiated outside an organ that alter the activity of the organ; accomplished by the nervous and endocrine systems

extrinsic nerves The nerves originating outside the digestive tract that innervate the various digestive organs

facilitated diffusion Passive carrier-mediated transport involving transport of a substance down its concentration gradient across the plasma membrane

fatigue Inability to maintain muscle tension at a given level despite sustained stimulation

feedback A response that occurs after a change has been detected; may be *negative feedback* or *positive feedback*

feedforward mechanism A response designed to prevent an anticipated change in a controlled variable

feeding signals Appetite signals that give rise to the sensation of hunger and promote the desire to eat

fibrinogen (fī-BRIN-uh-jun) A large, soluble plasma protein that when converted into an insoluble, threadlike molecule forms the meshwork of a clot during blood coagulation

Fick's law of diffusion The rate of net diffusion of a substance across a membrane is directly proportional to the substance's concentration gradient, the membrane's permeability to the substance, and the surface area of the membrane and inversely proportional to the substance's molecular weight and the diffusion distance

fight-or-flight response The changes in activity of the various organs innervated by the autonomic nervous system in response to sympathetic stimulation, which collectively prepare the body for strenuous physical activity in the face of an emergency or stressful situation, such as a physical threat from the outside environment

firing When an excitable cell undergoes an action potential

first messenger An extracellular messenger, such as a hormone, that binds with a surface membrane receptor and activates an intracellular second messenger to carry out the desired cellular response

flagellum (fluh-JEL-um) The single, long, whip-like appendage that serves as the tail of a spermatozoon

flow rate (of blood or air) The volume of blood or air passing through a blood vessel or airway, respectively, per unit of time

fluid balance Maintenance of ECF volume (for long-term control of blood pressure) and ECF osmolarity (for maintaining normal cell volume)

follicle (of ovary) A developing ovum and the surrounding specialized cells

follicle-stimulating hormone (FSH) An anterior pituitary hormone that stimulates ovarian follicular development and estrogen secretion in females and stimulates sperm production in males

follicular cells (of ovary) (fah-LIK-you-lar) Collectively, the granulosa and thecal cells

follicular cells (of thyroid gland) The cells that form the walls of the colloid-filled follicles in the thyroid gland and secrete thyroid hormone

follicular phase The phase of the ovarian cycle dominated by the presence of maturing follicles prior to ovulation

Frank–Starling law of the heart Intrinsic control of the heart such that increased venous return resulting in increased end-diastolic volume leads to an increased strength of contraction and increased stroke volume; that is, the heart normally pumps out all the blood returned to it

free radicals Very unstable electron-deficient particles that are highly reactive and destructive

frontal lobes The lobes of the cerebral cortex at the top of the brain in front of the central sulcus, which are responsible for voluntary motor output, speaking ability, and elaboration of thought

FSH See *follicle-stimulating hormone*

fuel metabolism See *intermediary metabolism*

functional syncytium (sin-sish´-ē-um) A group of smooth or cardiac muscle cells that are interconnected by gap junctions and function electrically and mechanically as a single unit

functional unit The smallest component of an organ that can perform all the functions of the organ

gametes (GAM-ētz) Reproductive, or germ, cells, each containing a haploid set of chromosomes; sperm and ova

gamma motor neuron A motor neuron that innervates the fibers of a muscle–spindle receptor

ganglion (GAN-gle-un) A collection of neuronal cell bodies located outside the central nervous system

ganglion cells The nerve cells in the outermost layer of the retina whose axons form the optic nerve

gap junction A communicating junction formed between adjacent cells by small connecting tunnels that permit passage of charge-carrying ions between the cells so that electrical activity in one cell is spread to the adjacent cell

gastrin A hormone secreted by the pyloric gland area of the stomach that stimulates the parietal and chief cells to secrete a highly acidic gastric juice

gastrointestinal hormone Hormones secreted into the blood by the endocrine cells in the digestive tract mucosa that control motility and/or secretion in other parts of the digestive system

gestation Pregnancy

ghrelin (GRELL-in) The "hunger" hormone, a potent appetite stimulator secreted by the empty stomach

glands Epithelial tissue derivatives that are specialized for secretion

glial cells (glē-ul) Connective tissue cells of the CNS, which support the neurons both physically and metabolically, including astrocytes, oligodendrocytes, ependymal cells, and microglia

glomerular filtration (glow-MER-yū-lur) Filtration of a protein-free plasma from the glomerular capillaries into the tubular component of the kidney's nephron as the first step in urine formation

glomerular filtration rate (GFR) The rate at which glomerular filtrate is formed

glomerulus (in kidney) (glow-MER-yū-lus) A ball-like tuft of capillaries in the kidney's nephron that filters water and solute from the blood as the first step in urine formation

glomerulus (in olfactory bulb) A ball-like neural junction within the olfactory bulb that serves as a "smell file" sorting different scent components

glucagon (GLOO-kuh-gon) The pancreatic hormone that raises blood glucose and blood fatty-acid levels

glucocorticoids (gloo´-kō-KOR-ti-koidz) Adrenocortical hormones that are important in intermediary metabolism and in helping the body resist stress; primarily cortisol

gluconeogenesis (gloo´-kō-nē-ō-JEN-uh-sus) The conversion of amino acids into glucose

glycogen (GLĪ-kō-jen) The storage form of glucose in the liver and muscle

glycogenesis (glī´-kō-JEN-i-sus) The conversion of glucose into glycogen

glycogenolysis (glī´-kō-juh-NOL-i-sus) The conversion of glycogen to glucose

glycolysis (glī-KOL-uh-sus) A biochemical process taking place in the cell's cytosol that breaks down glucose into pyruvate molecules

GnRH See *gonadotropin-releasing hormone*

Golgi complex (GOL-jē) An organelle consisting of sets of stacked, flattened membranous sacs; processes raw materials transported to it from the endoplasmic reticulum into finished products and sorts and directs the finished products to their final destination

gonadotropes Anterior pituitary cells that secrete gonadotropin-releasing hormone

gonadotropin-releasing hormone (GnRH) (gō-nad´-uh-TRŌ-pin) The hypothalamic hormone that stimulates the release of FSH and LH from the anterior pituitary

gonadotropins FSH and LH; hormones that control secretion of the sex hormones by the gonads

gonads (GŌ-nadz) The primary reproductive organs, which produce the gametes and secrete the sex hormones; testes and ovaries

G protein A membrane-bound intermediary, which, when activated on binding of an extracellular first messenger to a surface receptor, activates effector proteins on the intracellular side of the membrane in the cAMP and IP₃/DAG second-messenger systems

G-protein coupled receptor A type of receptor that activates the associated G protein on binding with an extracellular chemical messenger

gradation of contraction Variable magnitudes of tension produced in a single whole muscle

graded potential A local change in membrane potential that occurs in varying grades of magnitude; serves as a short-distance signal in excitable tissues

grand postsynaptic potential (GPSP) The total composite potential in a postsynaptic neuron resulting from the sum of all EPSPs and IPSPs occurring at the same time

granulocytes (gran´-yuh-lō-sīts) Leukocytes that contain granules, including neutrophils, eosinophils, and basophils

granulosa cells (gran´-yuh-LO-suh) The layer of cells immediately surrounding a developing oocyte within an ovarian follicle

gray matter The portion of the central nervous system composed primarily of densely packaged neuronal cell bodies and dendrites

growth hormone (GH) An anterior pituitary hormone that is primarily responsible for regulating overall body growth and is also important in intermediary metabolism; somatotropin

H⁺ See *hydrogen ion*

haploid number (HAP-loid) The number of chromosomes found in gametes; a half set of chromosomes, one member of each pair, for a total of 23 chromosomes in humans

Hb See *hemoglobin*

hCG See *human chorionic gonadotropin*

helper T cells The population of T cells that enhances the activity of other immune-response effector cells

hematocrit (hi-MAT´-uh-krit) The percentage of blood volume occupied by erythrocytes as they are packed down in a centrifuged blood sample

hemoglobin (HĒ-muh-glō´-bun) **(Hb)** A large iron-bearing protein molecule in erythrocytes that binds with and transports most O_2 in the blood; also carries some of the CO_2 and H^+ in the blood

hemolysis (hē-MOL-uh-sus) Rupture of red blood cells

hemostasis (hē´-mō-STĀ-sus) The stopping of bleeding from an injured vessel

hepatic portal system (hi-PAT-ik) A complex vascular connection between the digestive tract and liver such that venous blood from the digestive system drains into the liver for processing of absorbed nutrients before being returned to the heart

hippocampus (hip-oh-CAM-pus) The elongated, medial portion of the temporal lobe that is a part of the limbic system and is especially crucial for forming long-term memories

histamine A chemical released from mast cells or basophils that brings about vasodilation and increased capillary permeability; important in allergic responses and inflammation

homeostasis (hō´-mē-ō-STĀ-sus) Maintenance by the highly coordinated, regulated actions of the body systems of relatively stable chemical and physical conditions in the internal fluid environment that bathes the body's cells

hormone A long-distance chemical mediator that is secreted by an endocrine gland into the blood, which transports it to its target cells

hormone response element (HRE) The specific attachment site on DNA for a given steroid hormone and its nuclear receptor

host cell A body cell infected by a virus

HRE See *hormone response element*

human chorionic gonadotropin (hCG) (kō-rē-ON-ik gō-nad´-uh-TRŌ-pin) A hormone secreted by the developing placenta that stimulates and maintains the corpus luteum of pregnancy

hydrogen ion (H⁺) The cationic portion of a dissociated acid

hydrolysis (hī-DROL-uh-sis) The digestion of a nutrient molecule by the addition of water at a bond site

hydrostatic (fluid) pressure (hī-dro-STAT-ik) The pressure exerted by fluid on the walls that contain it

hyperglycemia (hī´-pur-glī-SĒ-mē-uh) Elevated blood glucose concentration

hyperplasia (hī-pur-PLĀ-zē-uh) An increase in the number of cells

hyperpolarization An increase in membrane potential from resting potential; potential becomes even more negative than at resting potential

hypersecretion Too much of a particular hormone secreted

hypertension (hī´-pur-TEN-chun) Sustained, above-normal mean arterial blood pressure

hypertonic solution (hī´-pur-TON-ik) A solution with osmolarity greater than that of normal body fluids; more concentrated than normal

hypertrophy (hī-PUR-truh-fē) Increase in the size of an organ as a result of an increase in the size of its cells

hyperventilation Overbreathing; when the rate of ventilation is in excess of the body's metabolic needs for CO_2 removal

hypophysiotropic hormones (hi-PŌ-fiz-ē-oh-TRO-pik) Hormones secreted by the hypothalamus that regulate the secretion of anterior pituitary hormones; see also *releasing hormone* and *inhibiting hormone*

hyposecretion Too little of a particular hormone secreted

hypotension (hī-pō-TEN-chun) Sustained, below-normal mean arterial blood pressure

hypothalamic-hypophyseal portal system (hī-pō-thuh-LAM-ik hī-pō-FIZ-ē-ul) The vascular connection between the hypothalamus and anterior pituitary gland used for the pickup and delivery of hypophysiotropic hormones

hypothalamus (hī´-pō-THAL-uh-mus) The brain region beneath the thalamus that regulates many aspects of the internal fluid environment, such as water and salt balance and food intake; serves as an important link between the autonomic nervous system and endocrine system

hypotonic solution (hī´-pō-TON-ik) A solution with osmolarity less than that of normal body fluids; more dilute than normal

hypoventilation Underbreathing; ventilation inadequate to meet the metabolic needs for O_2 delivery and CO_2 removal

hypoxia (hī-POK-sē-uh) Insufficient O_2 at the cellular level

I band One of the light bands that alternate with dark (A) bands to create a striated appearance in a skeletal or cardiac muscle fiber when these fibers are viewed with a light microscope

ICF See *intracellular fluid*

IGF See *insulin-like growth factor*

immune surveillance Recognition and destruction of newly arisen cancer cells by the immune system

immunity The body's ability to resist or eliminate potentially harmful foreign materials or abnormal cells

immunoglobulins (im´-ū-nō-GLOB-yū-lunz) Antibodies; gamma globulins

impermeable Prohibiting passage of a particular substance through the plasma membrane

implantation The burrowing of a blastocyst into the endometrial lining

inclusion A nonpermanent mass of stored material, such as glygogen or triglycerides (fat), in a cell

incretin A hormone released by the digestive tract that stimulates insulin secretion by the pancreas

indoleamine A type of amine hormone derived from the amino acid tryptophan and secreted by the pineal gland

inflammation An innate, nonspecific series of highly interrelated events, especially involving neutrophils, macrophages, and local vascular changes, that are set into motion in response to foreign invasion or tissue damage

influx Movement into the cell

inhibin (in-HIB-un) A hormone secreted by the Sertoli cells of the testes or by the ovarian follicles that inhibits FSH secretion

inhibiting hormone A hypothalamic hormone that inhibits the secretion of a particular anterior pituitary hormone

inhibitory postsynaptic potential (IPSP) (pōst´-si-NAP-tik) A small hyperpolarization of the postsynaptic membrane in response to neurotransmitter binding, thereby moving the membrane farther from threshold

inhibitory synapse (SIN-aps´) Synapse in which the postsynaptic neuron's response to neurotransmitter release is a small hyperpolarization of the postsynaptic membrane, moving the membrane farther from threshold

innate immune responses Inherent defense responses that nonselectively defend against foreign or abnormal material, even on initial exposure to it; see also *inflammation, interferon, natural killer cells,* and *complement system*

inorganic Referring to substances that do not contain carbon; from nonliving sources

inositol trisphosphate (IP$_3$) A component cleaved from phosphatidylinositol bisphosphate (PIP$_2$) in the plasma membrane that mobilizes the Ca^{2+} second messenger system in response to binding of an extracellular (first) messenger to a G-protein coupled receptor

insensible loss Loss of water of which the person is not aware from the lungs or nonsweating skin

inspiration A breath in

inspiratory muscles The skeletal muscles whose contraction enlarges the thoracic cavity, bringing about lung expansion and movement of air into the lungs from the atmosphere

insulin (IN-suh-lin) The pancreatic hormone that lowers blood levels of glucose, fatty acids, and amino acids and promotes their storage

insulin-like growth factor (IGF) Synonymous with *somatomedins;* hormone secreted by the liver or other tissues that acts directly on target cells to promote growth; IGF-I is secreted in response to growth hormone and IGF-II acts independently of growth hormone

integrator A region that determines efferent output based on processing of afferent input; also called a *control center*

integument (in-TEG-yuh-munt) The skin and underlying connective tissue

intercalated cells Cells in the distal and collecting tubules of the kidney important in renal control of acid-base balance

intercostal muscles (int-ur-KOS-tul) The muscles that lie between the ribs; see also *external intercostal muscles* and *internal intercostal muscles*

interferon (in´-tur-FER-on) A chemical released from virus-invaded cells that provides nonspecific resistance to viral infections by transiently interfering with replication of the same or unrelated viruses in other host cells

intermediary metabolism The collective set of intracellular chemical reactions that involve the degradation, synthesis, and transformation of small nutrient molecules; also known as *fuel metabolism*

intermediate filaments Threadlike cytoskeletal elements that play a structural role in parts of the cells subject to mechanical stress

internal environment The body's aqueous extracellular environment, which consists of the plasma and interstitial fluid and which must be homeostatically maintained for the cells to make life-sustaining exchanges with it

internal intercostal muscles Expiratory muscles whose contraction pulls the ribs downward and inward, thereby reducing the size of the thoracic cavity

internal respiration The intracellular metabolic processes carried out within the mitochondria that use O$_2$ and produce CO$_2$ during the derivation of energy from nutrient molecules

internal work All forms of biological energy expenditure that do not accomplish mechanical work outside the body

interneuron Neuron that lies entirely within the central nervous system and is important for integrating peripheral responses to peripheral information as well as for the abstract phenomena associated with the "mind"

interstitial fluid (in´-tur-STISH-ul) The portion of the extracellular fluid that surrounds and bathes all the body's cells

intra-alveolar pressure (in´-truh-al-VĒ-uh-lur) The pressure within the alveoli

intracellular fluid (ICF) The fluid collectively contained within all the body's cells

intrapleural pressure (in´-truh-PLOOR-ul) The pressure within the pleural sac

intrinsic controls Local control mechanisms inherent to an organ

intrinsic factor A special substance secreted by the parietal cells of the stomach that must be combined with vitamin B$_{12}$ for this vitamin to be absorbed by the intestine; deficiency produces pernicious anemia

intrinsic nerve plexuses Interconnecting networks of nerve fibers within the digestive tract wall

involuntary muscle Muscle innervated by the autonomic nervous system and not subject to voluntary control; cardiac and smooth muscle

ion An atom that has gained or lost one or more of its electrons, so it is not electrically balanced

IP$_3$ See *inositol trisphosphate*

IPSP See *inhibitory postsynaptic potential*

iris A pigmented smooth muscle that forms the colored portion of the eye and controls pupillary size

islets of Langerhans (LAHNG-er-honz) The endocrine portion of the pancreas that secretes the hormones insulin and glucagon into the blood

isometric contraction (ī´-sō-MET-rik) A muscle contraction in which muscle tension develops at constant muscle length

isotonic contraction A muscle contraction in which muscle tension remains constant as the muscle fiber changes length

isotonic solution (ī´-sō-TON-ik) A solution with osmolarity equal to that of normal body fluids

juxtaglomerular apparatus (juks´-tuh-glō-MER-you-lur) A cluster of specialized vascular and tubular cells at a point where the ascending limb of the loop of Henle passes through the fork formed by the afferent and efferent arterioles of the same nephron in the kidney

juxtamedullary nephrons (juks´-tuh-MED-you-lair-ee) Nephrons whose glomeruli lie in the renal cortex next to the medulla and whose long loops of Henle dip deeply into the medulla; establish the medullary vertical osmotic gradient

keratin (CARE-uh-tin) The protein found in the intermediate filaments in skin cells that give the skin strength and help form a waterproof outer layer

killer (K) cells Cells that destroy a target cell that has been coated with antibodies by lysing its membrane

kinesin (kī-NĒ´-sin) The molecular motor that transports secretory vesicles along the microtubular highway within neuronal axons by "walking" along the microtubule

lactation Milk production by the mammary glands

larynx (LARE-inks) The "voice box" at the entrance of the trachea; contains the vocal cords

lateral inhibition The phenomenon in which the most strongly activated signal pathway originating from the center of a stimulus area inhibits the less excited pathways from the fringe areas by means of lateral inhibitory connections within sensory pathways

lateral sacs The expanded saclike regions of a muscle fiber's sarcoplasmic reticulum; store and release calcium, which plays a key role in triggering muscle contraction

law of mass action If the concentration of one of the substances involved in a reversible reaction is increased, the reaction is driven toward the opposite side, and if the concentration of one of the substances is decreased, the reaction is driven toward that side

leak channels Unregulated, ungated channels that are open all the time

left ventricle The heart chamber that pumps blood into the systemic circulation

length–tension relationship The relationship between the length of a muscle fiber at the onset of contraction and the tension the fiber can achieve on a subsequent tetanic contraction

lens A transparent, biconvex structure of the eye that refracts (bends) light rays and whose strength can be adjusted to accommodate for vision at different distances

leptin A hormone released from adipose tissue that plays a key role in long-term regulation of body weight by acting on the hypothalamus to suppress appetite

leukocytes (LOO-kuh-sīts) White blood cells, which are the immune system's mobile defense units

leukotrienes (loo-ko-TRY-eenz) Local chemical mediators derived from the plasma membrane that are especially important in development of asthma

Leydig cells (LĪ-dig) The interstitial cells of the testes that secrete testosterone

LH See *luteinizing hormone*

LH surge The burst in LH secretion that occurs at midcycle of the ovarian cycle and triggers ovulation

limbic system (LIM-bik) A functionally interconnected ring of forebrain structures that surrounds the brain stem and is concerned with emotions, basic survival and sociosexual behavioral patterns, motivation, and learning

lipase (LIE-payz) An enzyme secreted primarily by pancreatic acinar cells that digests dietary fat

lipid emulsion A suspension of small fat droplets held apart as a result of adsorption of bile salts on their surface

loop of Henle (HEN-lē) A hairpin loop that extends between the proximal and distal tubule of the kidney's nephron

lumen (LOO-men) The interior space of a hollow organ or tube

luteal phase (LOO-tē-ul) The phase of the ovarian cycle dominated by the presence of a corpus luteum

luteinization (loot´-ē-un-uh-ZĀ-shun) Formation of a postovulatory corpus luteum in the ovary

luteinizing hormone (LH) An anterior pituitary hormone that stimulates ovulation, luteinization, and secretion of estrogen and progesterone in females and testosterone secretion in males

lymph Interstitial fluid that is picked up by the lymphatic vessels and returned to the venous system, meanwhile passing through the lymph nodes for defense purposes

lymphocytes White blood cells that provide immune defense against targets for which they are specifically programmed

lymphoid tissues Tissues that produce and store lymphocytes, such as lymph nodes and tonsils

lysosomes (LĪ-sō-sōmz) Organelles consisting of membrane-enclosed sacs containing powerful hydrolytic enzymes that destroy unwanted material within the cell, such as internalized foreign material or cellular debris

macrophages (MAK-ruh-fājs) Large, tissue-bound phagocytes

mast cells Cells located within connective tissue that synthesize, store, and release histamine, as during allergic responses

mean arterial blood pressure The average pressure responsible for driving blood forward through the arteries into the tissues throughout the cardiac cycle; equals cardiac output times total peripheral resistance

mechanically gated channels Channels that open or close in response to stretching or other mechanical deformation

mechanoreceptor (meh-CAN-oh-rē-SEP-tur) or (mek´-uh-nō-rē-SEP-tur) A sensory receptor sensitive to mechanical energy, such as stretching or bending

medullary respiratory center (MED-you-LAIR-ē) Several aggregations of neuronal cell bodies within the medulla that provide output to the respiratory

muscles and receive input important for regulating the magnitude of ventilation

megakaryocyte A large bone-marrow bound cell that sheds off blood-borne platelets from its outer edges

meiosis (mī-ō-sis) Cell division in which the chromosomes replicate followed by two nuclear divisions so that only a half set of chromosomes is distributed to each of four new daughter cells

melanocyte-stimulating hormone (MSH) (mel-AH-nō-sīt) A hormone produced by the intermediate lobe of the pituitary in lower vertebrates that regulates skin coloration for camouflage in these species; in humans, is secreted as a paracrine by the hypothalamus for control of food intake and by keratinocytes in the skin to control dispersion of melanin granules from melanocytes during tanning

melatonin (mel-uh-TŌ-nin) A hormone secreted by the pineal gland during darkness that helps entrain the body's biological rhythms with the external light/dark cues

membrane attack complex A collection of the five final activated components of the complement system that aggregate to form a porelike channel in the plasma membrane of an invading microorganism, with the resultant leakage leading to destruction of the invader

membrane potential A separation of charges across the membrane; a slight excess of negative charges lined up along the inside of the plasma membrane and separated from a slight excess of positive charges on the outside

memory cells B or T cells that are newly produced in response to a microbial invader but that do not participate in the current immune response against the invader; instead, they remain dormant, ready to launch a swift, powerful attack should the same microorganism invade again in the future

meninges (men-IN-geez) Three membranes that wrap the brain and spinal cord; from outermost to innermost the dura mater, arachnoid mater, and pia mater

menstrual cycle (men´-stroo-ul) The cyclic changes in the uterus that accompany the hormonal changes in the ovarian cycle

menstrual phase The phase of the menstrual cycle characterized by sloughing of endometrial debris and blood out through the vagina

messenger RNA (mRNA) Carries the transcribed genetic blueprint for synthesis of a particular protein from nuclear DNA to the cytoplasmic ribosomes where the protein is synthesized

metabolic acidosis (met-uh-bol´-ik) Acidosis resulting from any cause other than excess accumulation of carbonic acid in the body

metabolic alkalosis (al´-kuh-LŌ-sus) Alkalosis caused by a relative deficiency of noncarbonic acid

metabolic rate Energy expenditure per unit of time

metabolism All chemical reactions that occur within the body cells

micelle (mī-SEL) A water-soluble aggregation of bile salts, lecithin, and cholesterol that has a hydrophilic shell and a hydrophobic core; carries the water-insoluble products of fat digestion to their site of absorption

microfilaments Cytoskeletal elements made of actin molecules (and myosin molecules in muscle cells); play a major role in various cellular contractile systems and serve as a mechanical stiffener for microvilli

microglia The type of glial cells that serve as the immune defense cells of the CNS

microtubules Cytoskeletal elements made of tubulin molecules arranged into long, slender, unbranched tubes that help maintain asymmetric cell shapes and coordinate complex cell movements

microvilli (mī´-krō-VIL-ī) Actin-stiffened, nonmotile, hairlike projections from the luminal surface of epithelial cells lining the digestive tract and kidney tubules; tremendously increase the surface area of the cell exposed to the lumen

micturition (mik-too-RISH-un) or (mik-chuh-RISH-un) The process of bladder emptying; urination

milk ejection The squeezing out of milk produced and stored in the alveoli of the breasts by means of contraction of the myoepithelial cells that surround each alveolus

mineralocorticoids (min-uh-rul-ō-KOR-ti-koidz) The adrenocortical hormones that are important in Na^+ and K^+ balance; primarily aldosterone

mitochondria (mī-tō-KON-drē-uh) The energy organelles, which contain the enzymes for oxidative phosphorylation

mitosis (mī-TŌ-sis) Cell division in which the chromosomes replicate before nuclear division so that each of the two daughter cells receives a full set of chromosomes

mitotic spindle The system of microtubules assembled during mitosis along which the replicated chromosomes are moved away from each other toward opposite sides of the cell prior to cell division

modality The energy form to which sensory receptors respond, such as heat, light, pressure, and chemical changes

molecular motor Specialized protein molecule with "feet" that can be alternately swung forward enabling the molecule to "walk" along a microtubular highway, carrying cargo from one part of the cell to another

molecule A chemical substance formed by the linking of atoms; the smallest unit of a given chemical substance

monocytes (MAH-nō-sīts) White blood cells that emigrate from the blood, enlarge, and become macrophages

monosaccharides (mah´-nō-SAK-uh-rīdz) Simple sugars, such as glucose; the absorbable unit of digested carbohydrates

motility Muscular contractions of the digestive tract wall that mix and propel forward the luminal contents

motor activity Movement of the body accomplished by contraction of skeletal muscles

motor end plate The specialized portion of a skeletal muscle fiber that lies immediately underneath the terminal button of the motor neuron and possesses receptor sites for binding acetylcholine released from the terminal button

motor neurons The neurons that innervate skeletal muscle and whose axons constitute the somatic nervous system

motor unit One motor neuron plus all the muscle fibers it innervates

motor unit recruitment The progressive activation of a muscle fiber's motor units to accomplish increasing gradations of contractile strength

mucosa (mew-KŌ-sah) The innermost layer of the digestive tract that lines the lumen

multiunit smooth muscle A smooth muscle mass consisting of multiple discrete units that function independently of one another and that must be separately stimulated by autonomic nerves to contract

muscarinic receptor (MUS-ka-rin´-ik) Type of cholinergic receptor found at the effector organs of all parasympathetic postganglionic fibers

muscle fiber A single muscle cell, which is relatively long and cylindrical in shape

muscle tension See *tension*

muscle tissue A functional grouping of cells specialized for contraction and force generation

myelin (MĪ-uh-lun) An insulative lipid covering that surrounds myelinated nerve fibers at regular intervals along the axon's length; each patch of myelin is formed by a separate myelin-forming cell that wraps itself jelly-roll fashion around the neuronal axon

myelinated fibers Neuronal axons covered at regular intervals with insulative myelin

myocardial ischemia (mī-ō-KAR-dē-ul is-KĒ-mē-uh) Inadequate blood supply to the heart tissue

myocardium (mī´-ō-KAR-dē-um) The cardiac muscle within the heart wall

myofibril (mī´-ō-FĪB-rul) A specialized intracellular structure of muscle cells that contains the contractile apparatus

myogenic activity Muscle-produced, nerve-independent contractile activity

myometrium (mī´-ō-mē-TRĒ-um) The smooth muscle layer of the uterus

myosin (MĪ-uh-sun) The contractile protein forming the thick filaments in muscle fibers

Na⁺–K⁺ pump A carrier that actively transports Na^+ out of the cell and K^+ into the cell

Na⁺ load See *sodium (Na⁺) load*

natural killer cells Naturally occurring, lymphocyte-like cells that nonspecifically destroy virus-infected cells and cancer cells by directly lysing their membranes on first exposure to them

negative balance Situation in which the losses for a substance exceed its gains so that the total amount of the substance in the body decreases

negative feedback A regulatory mechanism in which a change in a controlled variable triggers a response that opposes the change, thus maintaining a relatively steady set point for the regulated factor

nephron (NEF-ron´) The functional unit of the kidney; consisting of an interrelated vascular and tubular component, it is the smallest unit that can form urine

nerve A bundle of peripheral neuronal axons, some afferent and some efferent, enclosed by a connective tissue covering and following the same pathway

nervous system One of the two major regulatory systems of the body; in general, coordinates rapid activities of the body, especially those involving interactions with the external environment

nervous tissue A functional grouping of cells specialized for initiation and transmission of electrical signals

net diffusion The difference between the opposing movements of two types of molecules in a solution

net filtration pressure The net difference in the hydrostatic and osmotic forces acting across the glomerular membrane that favors the filtration of a protein-free plasma into Bowman's capsule

neurogenic activity Contractile activity in muscle cells initiated by nerves

neuroglia See *glial cells*

neurohormones Hormones released into the blood by neurosecretory neurons

neuromodulators (ner´ō-MA-jew-lā´-torz) Chemical messengers that bind to neuronal receptors at nonsynaptic sites and bring about long-term changes that subtly depress or enhance synaptic effectiveness

neuromuscular junction The juncture between a motor neuron and a skeletal muscle fiber

neuron (NER-on) A nerve cell specialized to initiate, propagate, and transmit electrical signals, typically consisting of a cell body, dendrites, and an axon

neuropeptides Large, slowly acting peptide molecules released from axon terminals along with classical neurotransmitters; most neuropeptides function as neuromodulators

neuropeptide Y (NPY) A potent appetite stimulator secreted by the hypothalamic arcuate nucleus

neurotransmitter The chemical messenger released from the axon terminal of a neuron in response to an action potential that influences another neuron or an effector with which the neuron is anatomically linked

neutrophils (new´-truh-filz) White blood cells that are phagocytic specialists and important in inflammatory responses and defense against bacterial invasion

nicotinic receptor (nick´-o-TIN-ik) Type of cholinergic receptor found at all autonomic ganglia and the motor end plates of skeletal muscle fibers

nitric oxide A local chemical mediator released from endothelial cells and other tissues; its effects range from causing local arteriolar vasodilation to acting as a toxic agent against foreign invaders to serving as a unique type of neurotransmitter

NMDA receptor One of two types of receptor-channels on a postsynaptic membrane to which the neurotransmitter glutamate binds, this one being both chemically mediated and voltage dependent and permitting Ca^{2+} entry when open

nociceptor (nō-sē-SEP-tur) A pain receptor, sensitive to tissue damage

nodes of Ranvier (RAN-vē-ā) The portions of a myelinated neuronal axon between the segments of insulative myelin; the axonal regions where the axonal membrane is exposed to the ECF and membrane potential exists

norepinephrine (nor´-ep-uh-NEF-run) The neurotransmitter released from sympathetic postganglionic fibers; noradrenaline

NPY See *neuropeptide Y*

nucleus (of brain) (NŪ-klē-us) A functional aggregation of neuronal cell bodies within the brain

nucleus (of cells) A distinct spherical or oval structure, usually located near the center of a cell, that contains the cell's genetic material, deoxyribonucleic acid (DNA)

occipital lobes (ok-SIP´-ut-ul) The posterior lobes of the cerebral cortex, which initially process visual input

O₂–Hb dissociation curve A graphic depiction of the relationship between arterial P_{O_2} and percent hemoglobin saturation

oligodendrocytes (ol-i-gō´-DEN-drō-sitz) The myelin-forming cells of the central nervous system

oogenesis (ō´-ō-JEN-uh-sus) Egg production

opsonin (OP´-suh-nun) Body-produced chemical that links bacteria to macrophages, thereby making the bacteria more susceptible to phagocytosis

optic nerve The bundle of nerve fibers leaving the retina that relay information about visual input

optimal length The length before the onset of contraction of a muscle fiber at which maximal force can be developed on a subsequent tetanic contraction

organ A distinct structural unit composed of two or more types of primary tissue organized to perform one or more particular functions; for example, the stomach

organelles (or´-gan-ELZ) Distinct, highly organized, membrane-bound intracellular compartments, each containing a specific set of chemicals for carrying out a particular cellular function

organic Referring to substances that contain carbon; originally from living or once-living sources

organism A living entity, whether unicellular or multicellular

organ of Corti (KOR-tē) The sense organ of hearing within the inner ear that contains hair cells whose hairs are bent in response to sound waves, setting up action potentials in the auditory nerve

osmolarity (oz´-mō-LAR-ut-ē) A measure of the concentration of a solution given in terms of milliosmoles/liter (mOsm), the number of millimoles of solute particles in a liter of solution

osmosis (os-MŌ-sis) Movement of water across a membrane down its own concentration gradient toward the area of higher solute concentration

osmotic pressure (os-MAH-tic) A measure of the tendency for osmotic flow of water into a solution resulting from its relative concentration of nonpenetrating solutes and water

osteoblasts (OS-tē-ō-blasts´) Bone cells that produce the organic matrix of bone

osteoclasts Bone cells that dissolve bone in their vicinity

osteocytes Retired osteoblasts entombed within the bone that they have laid down around themselves that continue to participate in calcium and phosphate exchange between the bone fluid and plasma

otolith organs (ŌT´-ul-ith) Sense organs in the inner ear that provide information about rotational changes in head movement; include the utricle and saccule

oval window The membrane-covered opening that separates the air-filled middle ear from the upper compartment of the fluid-filled cochlea in the inner ear

overhydration Water excess in the body

ovulation (ov´-yuh-LĀ-shun) Release of an ovum from a mature ovarian follicle

oxidative phosphorylation (fos´-for-i-LĀ-shun) The entire sequence of mitochondrial biochemical reactions that uses oxygen to extract energy from the nutrients in food and transforms it into ATP, producing CO_2 and H_2O in the process; includes the electron transport system and chemiosmosis

oxyhemoglobin (ok-si-HĒ-muh-glō-bun) Hemoglobin combined with O_2

oxyntic mucosa (ok-SIN-tic) The mucosa lining the body and fundus of the stomach, which contains gastric pits that lead to the gastric glands lined by mucous neck cells, parietal cells, and chief cells

oxytocin (ok´-sē-TŌ-sun) A hypothalamic hormone stored in the posterior pituitary that stimulates uterine contraction and milk ejection

pacemaker activity Self-excitable activity of an excitable cell in which its membrane potential gradually depolarizes to threshold on its own

pacemaker potential A self-induced slow depolarization to threshold occurring in a pacemaker cell as a result of shifts in passive ionic fluxes across the membrane accompanying automatic changes in channel permeability

pancreas (PAN-krē-us) A mixed gland composed of an exocrine portion that secretes digestive enzymes and an aqueous alkaline secretion into the duodenal lumen and an endocrine portion that secretes the hormones insulin and glucagon into the blood

paracrine (PEAR-uh-krin) A local chemical messenger whose effect is exerted only on neighboring cells in the immediate vicinity of its site of secretion

parasympathetic nervous system (pear´-uh-sim-puh-THET-ik) The subdivision of the autonomic nervous system that dominates in quiet, relaxed situations and promotes body maintenance activities such as digestion and emptying of the urinary bladder

parathyroid glands (pear´-uh-THĪ-roid) Four small glands located on the posterior surface of the thyroid gland that secrete parathyroid hormone

parathyroid hormone (PTH) A hormone that raises plasma Ca^{2+} levels

parietal cells (puh-RĪ-ut-ul) The stomach cells that secrete hydrochloric acid and intrinsic factor

parietal lobes The lobes of the cerebral cortex that lie at the top of the brain behind the central sulcus, which contain the somatosensory cortex

partial pressure The individual pressure exerted independently by a particular gas within a mixture of gases

partial pressure gradient A difference in the partial pressure of a gas between two regions that promotes the movement of the gas from the region of higher partial pressure to the region of lower partial pressure

parturition (par´-too-RISH-un) Delivery of a baby

passive expiration Expiration accomplished during quiet breathing as a result of elastic recoil of the lungs on relaxation of the inspiratory muscles, with no energy expenditure required

passive force A force that does not require expenditure of cellular energy to accomplish transport of a substance across the plasma membrane

passive reabsorption Reabsorption when none of the steps in the transepithelial transport of a substance across the kidney tubules requires energy expenditure

pathogens (PATH-uh-junz) Disease-causing microorganisms, such as bacteria or viruses

pathophysiology (path´-ō-fiz-ē-OL-ō-gē) Abnormal functioning of the body associated with disease

pepsin; pepsinogen (pep-SIN-uh-jun) An enzyme secreted in inactive form by the stomach that, once activated, begins protein digestion

peptide hormones Hormones that consist of a chain of specific amino acids of varying length

percent hemoglobin saturation A measure of the extent to which the hemoglobin present is combined with O_2

perception The conscious interpretation of the external world as created by the brain from a pattern of nerve impulses delivered to it from sensory receptors

peripheral chemoreceptors (kē´-mō-rē-SEP-turz) The carotid and aortic bodies, which respond to changes in arterial P_{O_2}, P_{CO_2}, and H^+ and adjust respiration accordingly

peripheral nervous system (PNS) Nerve fibers that carry information between the central nervous system and other parts of the body

peristalsis (per´-uh-STOL-sus) Ringlike contractions of the circular smooth muscle of a tubular organ that move progressively forward with a stripping motion, pushing the contents of the organ ahead of the contraction

peritubular capillaries (per´-i-TŪ-bū-lur) Capillaries that intertwine around the tubules of the kidney's nephron; they supply the renal tissue and participate in exchanges between the tubular fluid and blood during the formation of urine

permeable Permitting passage of a particular substance

permissiveness When one hormone must be present in adequate amounts for the full exertion of another hormone's effect

peroxisomes (puh-ROK´-suh-sōmz) Sac-like organelles containing powerful oxidative enzymes that detoxify various wastes produced within the cell or foreign compounds that have entered the cell

pH The logarithm to the base 10 of the reciprocal of the hydrogen ion concentration; $pH = \log 1/[H^+]$ or $pH = -\log[H^+]$

phagocytosis (fag´-oh-sī-TŌ-sus) A type of endocytosis in which large, multimolecular, solid particles are engulfed by a cell

pharynx (FARE-inks) The back of the throat, which serves as a common passageway for the digestive and respiratory systems

phasic smooth muscle Smooth muscle that contracts in bursts, triggered by the generation of action potentials

phosphorylation (fos´-fōr-i-LĀ-shun) Addition of a phosphate group to a molecule

photoreceptor A sensory receptor responsive to light

phototransduction The mechanism of converting light stimuli into electrical activity by the rods and cones of the eye

physiology (fiz-ē-OL-ō-gē) The study of body functions

pineal gland (PIN-ē-ul) A small endocrine gland located in the center of the brain that secretes the hormone melatonin

pinocytosis (pin-oh-cī-TŌ-sus) A type of endocytosis in which the cell internalizes fluid

pitch The tone of a sound, determined by the frequency of vibrations (that is, whether a sound is a C or G note)

pituitary gland (pih-TWO-ih-tair-ee) A small endocrine gland connected by a stalk to the hypothalamus; consists of the anterior pituitary and posterior pituitary

placenta (plah-SEN-tah) The organ of exchange between the maternal and fetal blood; also secretes hormones that support the pregnancy

plaque A deposit of cholesterol and other lipids, perhaps calcified, in thickened, abnormal smooth-muscle cells within blood vessels as a result of atherosclerosis

plasma The liquid portion of the blood

plasma cell An antibody-producing derivative of an activated B lymphocyte

plasma clearance The volume of plasma that is completely cleared of a given substance by the kidneys per minute

plasma-colloid osmotic pressure (KOL-oid os-MOT-ik) The force caused by the unequal distribution of plasma proteins between the blood and surrounding fluid that encourages fluid movement into the capillaries

plasma membrane A protein-studded lipid bilayer that encloses each cell, separating it from the extracellular fluid

plasma proteins The proteins in the plasma, which perform a number of important functions; include albumins, globulins, and fibrinogen

plasticity (plas-TIS-uh-tē) The ability of portions of the brain to assume new responsibilities in response to the demands placed on it

platelets (PLĀT-lets) Specialized cell fragments in the blood that participate in hemostasis by forming a plug at a vessel defect

pleural sac (PLOOR-ul) A double-walled, closed sac that separates each lung from the thoracic wall

pluripotent stem cells Precursor cells, for example those that reside in the bone marrow and continuously divide and differentiate to give rise to each of the types of blood cells

polarization The state of having membrane potential

polycythemia (pol-i-sī-THĒ-mē-uh) Excess circulating erythrocytes, accompanied by an elevated hematocrit

polysaccharides (pol´-ī-SAK-uh-rīdz) Complex carbohydrates, consisting of chains of interconnected glucose molecules

pool (of a substance) Total quantity of any particular substance in the ECF

positive balance Situation in which the gains via input for a substance exceed its losses via output, so that the total amount of the substance in the body increases

positive feedback A regulatory mechanism in which the input and the output in a control system continue to enhance each other so that the controlled variable is progressively moved farther from a steady state

postabsorptive state The metabolic state after a meal is absorbed during which endogenous energy stores must be mobilized and glucose must be spared for the glucose-dependent brain; fasting state

posterior pituitary The neural portion of the pituitary that stores and releases into the blood on hypothalamic stimulation two hormones produced by the hypothalamus, vasopressin and oxytocin

postganglionic fiber (pōst´-gan-glē-ON-ik) The second neuron in the two-neuron autonomic nerve pathway; originates in an autonomic ganglion and terminates on an effector organ

postsynaptic neuron (pōst´-si-NAP-tik) The neuron that conducts its action potentials away from a synapse

power stroke The ATP-powered cross-bridge binding and bending that pulls the thin filaments in closer together between the thick filaments during contraction of a muscle fiber

preganglionic fiber The first neuron in the two-neuron autonomic nerve pathway; originates in the central nervous system and terminates on an autonomic ganglion

pressure gradient A difference in pressure between two regions that drives the movement of blood or air from the region of higher pressure to the region of lower pressure

presynaptic facilitation Enhanced release of neurotransmitter from a presynaptic axon terminal as a result of excitation of another neuron that terminates on the axon terminal

presynaptic inhibition A reduction in the release of a neurotransmitter from a presynaptic axon terminal as a result of excitation of another neuron that terminates on the axon terminal

presynaptic neuron (prē-si-NAP-tik) The neuron that conducts its action potentials toward a synapse

primary active transport A carrier-mediated transport system in which energy is directly required to operate the carrier and move the transported substance against its concentration gradient

primary follicle A primary oocyte surrounded by a single layer of granulosa cells in the ovary

primary motor cortex The portion of the cerebral cortex that lies anterior to the central sulcus and is responsible for voluntary motor output

principal cells Cells in the distal and collecting tubules of the kidney that are the site of action of aldosterone and vasopressin

progestational phase See *secretory phase*

prolactin (PRL) (prō-LAK-tun) An anterior pituitary hormone that stimulates breast development and milk production in females

proliferative phase The phase of the menstrual cycle during which the endometrium repairs itself and thickens following menstruation; lasts from the end of the menstrual phase until ovulation

pro-opiomelanocortin (prō-op´Ē-ō-ma-LAN-oh-kor´-tin) A large precursor molecule that can be variably cleaved into adrenocorticotropic hormone, melanocyte-stimulating hormone, and endorphin

proprioception (prō´-prē-ō-SEP-shun) Awareness of position of body parts in relation to one another and to surroundings

prostaglandins (pros´-tuh-GLAN-dins) Local chemical mediators that are derived from a component of the plasma membrane, arachidonic acid

prostate gland A male accessory sex gland that secretes an alkaline fluid, which neutralizes acidic vaginal secretions

protein kinase (KĪ-nase) An enzyme that phosphorylates and thereby induces a change in the shape and function of a particular intracellular protein

proteolytic enzymes (prōt´-ē-uh-LIT-ik) Enzymes that digest protein

proximal tubule (PROKS-uh-mul) A highly convoluted tubule that extends between Bowman's capsule and the loop of Henle in the kidney's nephron

PTH See *parathyroid hormone*

pulmonary artery (PULL-mah-nair-ē) The large vessel that carries blood from the right ventricle to the lungs

pulmonary circulation The closed loop of blood vessels carrying blood between the heart and lungs

pulmonary surfactant (sur-FAK-tunt) A phospholipoprotein complex secreted by the Type II alveolar cells that intersperses between the water molecules that line the alveoli, thereby lowering the surface tension within the lungs

pulmonary valve A one-way valve that permits the flow of blood from the right ventricle into the pulmonary artery during ventricular emptying but prevents the backflow of blood from the pulmonary artery into the right ventricle during ventricular relaxation

pulmonary veins The large vessels that carry blood from the lungs to the heart

pulmonary ventilation The volume of air breathed in and out in one minute; equals tidal volume times respiratory rate

pupil An adjustable round opening in the center of the iris through which light passes to the interior portions of the eye

Purkinje fibers (pur-KIN-jē) Small terminal fibers that extend from the bundle of His and rapidly transmit an action potential throughout the ventricular myocardium

pyloric gland area (PGA) (pī-LŌR-ik) The specialized region of the mucosa in the antrum of the stomach that secretes gastrin

pyloric sphincter (pī-lōr´-ik SFINGK-tur) The juncture between the stomach and duodenum

PYY$_{3-36}$ A satiety signal secreted by the small and large intestines that inhibits appetite and serves as a mealtime terminator

RAAS See *renin-angiotensin-aldosterone system*

radiation Emission of heat energy from the surface of a warm body in the form of electromagnetic waves

reabsorption The net movement of interstitial fluid into the capillary

receptive field The circumscribed region surrounding a sensory neuron within which the neuron responds to stimulus information

receptor See *sensory receptor* or *receptor (in membrane)*

receptor-channel A type of receptor that is an integral part of a channel that opens (or closes) on binding with an extracellular messenger

receptor-enzyme A type of receptor that functions as an enzyme on binding with an extracellular chemical messenger

receptor (in membrane) Membrane protein that binds with a specific extracellular chemical messenger, bringing about membrane and intracellular events that alter the activity of the particular cell

receptor-mediated endocytosis Import of a particular large molecule from the ECF into a cell by formation and pinching off of an endocytic pouch in response to binding of the molecule to a surface membrane receptor specific for it

receptor potential The graded potential change that occurs in a sensory receptor in response to a stimulus; generates action potentials in the afferent neuron fiber

reduced hemoglobin Hemoglobin that is not combined with O_2

reflex Any response that occurs automatically without conscious effort; the components of a reflex arc include a receptor, afferent pathway, integrating center, efferent pathway, and effector

refraction Bending of a light ray

refractory period (rē-FRAK-tuh-rē) The time period when a recently activated patch of membrane is refractory (unresponsive) to further stimulation, which prevents the action potential from spreading backward into the area through which it has just passed and ensures the unidirectional propagation of the action potential away from the initial site of activation

regulatory proteins Troponin and tropomyosin, which play a role in regulating muscle contraction by either covering or exposing the sites of interaction between the contractile proteins

regulatory T cells A class of T lymphocytes that suppresses the activity of other lymphocytes

releasing hormone A hypothalamic hormone that stimulates the secretion of a particular anterior pituitary hormone

renal cortex An outer granular-appearing region of the kidney

renal medulla (RĒ-nul muh-DUL-uh) An inner striated-appearing region of the kidney

renal threshold The plasma concentration at which the T_m of a particular substance is reached and the substance first starts appearing in the urine

renin (RĒ-nin) An enzymatic hormone released from the kidneys in response to a decrease in NaCl/ECF volume/arterial blood pressure; activates angiotensinogen

renin–angiotensin–aldosterone system (RAAS) (an´jē-ō-TEN-sun al-dō-steer-OWN) The salt-conserving system triggered by the release of renin from the kidneys, which activates angiotensin, stimulating aldosterone secretion and Na^+ reabsorption by the kidney tubules during the formation of urine

repolarization (rē´-pō-luh-ruh-ZĀ-shun) Return of membrane potential to resting potential following a depolarization

reproductive tract The system of ducts specialized to transport or house the gametes after they are produced

residual volume The minimum volume of air remaining in the lungs even after a maximal expiration

resistance Hindrance of blood or air flow through a blood vessel or respiratory airway, respectively

respiration The sum of processes that accomplish ongoing passive movement of O_2 from the atmosphere to the tissues, as well as the continual passive movement of metabolically produced CO_2 from the tissues to the atmosphere

respiratory acidosis (as-i-DŌ-sus) Acidosis resulting from abnormal retention of CO_2 arising from hypoventilation

respiratory airways The system of tubes that conducts air between the atmosphere and the alveoli of the lungs

respiratory alkalosis (al´-kuh-LŌ-sus) Alkalosis caused by excessive loss of CO_2 from the body as a result of hyperventilation

respiratory rate Breaths per minute

resting membrane potential The membrane potential that exists when an excitable cell is not displaying an electrical signal

reticular activating system (RAS) (ri-TIK-ū-lur) Ascending fibers that originate in the reticular formation and carry signals upward to arouse and activate the cerebral cortex

reticular formation A network of interconnected neurons that runs throughout the brain stem and initially receives and integrates all synaptic input to the brain

retina The innermost layer in the posterior region of the eye that contains the eye's photoreceptors, the rods and cones

ribonucleic acid (RNA) (rī-bō-new-KLĀ-ik) A nucleic acid that exists in three forms (messenger RNA, ribosomal RNA, and transfer RNA), which participate in gene transcription and protein synthesis

ribosomes (RĪ-bō-sōms) Special ribosomal RNA–protein complexes that synthesize proteins under the direction of nuclear DNA

right atrium (Ā-trē´-um) The heart chamber that receives venous blood from the systemic circulation

right ventricle The heart chamber that pumps blood into the pulmonary circulation

RNA See *ribonucleic acid*

rods The eye's photoreceptors used for night vision

rough ER The flattened, ribosome-studded sacs of the endoplasmic reticulum that synthesize proteins for export or for use in membrane construction

round window The membrane-covered opening that separates the lower chamber of the cochlea in the inner ear from the middle ear

ryanodine receptors (rye-ah-NO-deen) Receptors on the sarcoplasmic reticulum that bind with dihydropyridine receptors on the adjoining T tubule and serve as Ca^{2+}-release channels during excitation–contraction coupling

salivary amylase (AM-uh-lās´) An enzyme produced by the salivary glands that begins carbohydrate digestion in the mouth and continues it in the stomach after the food and saliva have been swallowed

saltatory conduction (SAL-tuh-tōr´-ē) The means by which an action potential is propagated throughout a myelinated fiber, with the impulse jumping over the myelinated regions from one node of Ranvier to the next

salt balance Balance between salt intake and salt output; important in controlling ECF volume

SA node See *sinoatrial node*

sarcomere (SAR-kō-mir) The functional unit of skeletal muscle; the area between two Z lines within a myofibril

sarcoplasmic reticulum (ri-TIK-yuh-lum) A fine meshwork of interconnected tubules that surrounds a muscle fiber's myofibrils; contains expanded lateral sacs, which store calcium that is released into the cytosol in response to a local action potential

satiety signals (suh-TĪ-ut-ē) Signals that lead to the sensation of fullness and suppress the desire to eat

saturation The condition in which all binding sites on a carrier molecule are occupied

Schwann cells (shwahn) The myelin-forming cells of the peripheral nervous system

sclera The visible, white, outer layer of the eye

secondary active transport A transport mechanism in which a carrier molecule for glucose or an amino acid is driven by a Na^+ concentration gradient established by the energy-dependent Na^+ pump to transfer the glucose or amino acid uphill without directly expending energy to operate the carrier

secondary follicle A developing ovarian follicle that is secreting estrogen and forming an antrum

secondary sexual characteristics The many external characteristics that are not directly involved in reproduction but that distinguish males and females

second messenger An intracellular chemical that is activated by binding of an extracellular first messenger to a surface receptor site, triggering a preprogrammed series of biochemical events that alter activity of intracellular proteins controlling a particular cellular activity

secretin (si-KRĒT-´n) A hormone released from the duodenal mucosa primarily in response to the presence of acid; inhibits gastric motility and secretion and stimulates secretion of $NaHCO_3$ solution from the pancreas and liver

secretion Release to a cell's exterior, on appropriate stimulation, of substances produced by the cell

secretory phase The phase of the menstrual cycle characterized by the development of a lush endometrial lining capable of supporting a fertilized ovum; also known as the *progestational phase*

secretory vesicles (VES-i-kuls) Membrane-enclosed sacs containing proteins that have been synthesized and processed by the endoplasmic reticulum/Golgi complex of the cell and which will be released to the cell's exterior by exocytosis on appropriate stimulation

segmentation The small intestine's primary method of motility; consists of oscillating, ringlike contractions of the circular smooth muscle along the small intestine's length

selectively permeable membrane A membrane that permits some particles to pass through while excluding others

self-antigens Antigens that are characteristic of a person's own cells

semen (SĒ-men) A mixture of accessory sex gland secretions and sperm

semicircular canal Sense organ in the inner ear that detects rotational or angular acceleration or deceleration of the head

semilunar valves (sem´-ī-LEW-nur) The aortic and pulmonary valves

seminal vesicles (VES-i-kuls) Male accessory sex glands that supply fructose to ejaculated sperm and secrete prostaglandins

seminiferous tubules (sem´-uh-NIF-uh-rus) The highly coiled tubules within the testes that produce spermatozoa

sensor The component of a control system that monitors the magnitude of the controlled variable

sensory afferent Pathway into the central nervous system carrying information that reaches the level of consciousness

sensory input Input from somatic sensation and special senses

sensory receptor An afferent neuron's peripheral ending, which is specialized to respond to a particular stimulus in its environment

sensory transduction The conversion of stimulus energy into a receptor potential

Sertoli cells (sur-TO-lē) Cells located in the seminiferous tubules that support spermatozoa during their development

serum Plasma minus fibrinogen and other clotting precursors

set point The desired level at which homeostatic control mechanisms maintain a controlled variable

sex hormones The steroid hormones responsible for the development of masculine and feminine characteristics; testosterone in males and estrogens in females

signal molecule An extracellular chemical messenger that initiates signal transduction in a cell

signal transduction The sequence of events in which incoming signals from extracellular chemical messengers are conveyed into a target cell where they are transformed into the dictated cellular response

single-unit smooth muscle The most abundant type of smooth muscle; made up of muscle fibers that are interconnected by gap junctions so that they become excited and contract as a unit; also known as *visceral smooth muscle*

sinoatrial (SA) node (sī-nō-Ā-trē-ul) A small specialized autorhythmic region in the right atrial wall of the heart that has the fastest rate of spontaneous depolarizations and serves as the normal pacemaker of the heart

skeletal muscle Striated muscle, which is attached to the skeleton and is responsible for movement of the bones in purposeful relation to one another; innervated by the somatic nervous system and under voluntary control

slow-wave potentials Self-excitable activity of an excitable cell in which its membrane potential undergoes gradually alternating depolarizing and hyperpolarizing swings

smooth ER The tubules of the endoplasmic reticulum that package newly synthesized proteins in transport vesicles

smooth muscle Involuntary muscle in the walls of hollow organs and tubes innervated by the autonomic nervous system

sodium (Na^+) load The total amount of Na^+ in the body, which determines the ECF volume through its osmotic effect

somatic cells (sō-MAT-ik) Body cells, as contrasted with reproductive cells

somatic nervous system The portion of the efferent division of the peripheral nervous system that innervates skeletal muscles; consists of the axonal fibers of the alpha motor neurons

somatic sensation Sensory information arising from the body surface, including somesthetic sensation and proprioception

somatomedin (sō´-mat-uh-MĒ-din) See *insulin-like growth factor*

somatosensory cortex The region of the parietal lobe immediately behind the central sulcus; the site of initial processing of somesthetic and proprioceptive input

somatotropes Anterior pituitary cells that secrete growth hormone

somesthetic sensations (SŌ-mes-THEH-tik) Awareness of sensory input such as touch, pressure, temperature, and pain from the body's surface

sound waves Traveling vibrations of air in which regions of high pressure caused by compression of air molecules alternate with regions of low pressure caused by rarefaction of the molecules

spatial summation The summing of several post-synaptic potentials arising from the simultaneous activation of several excitatory (or several inhibitory) synapses

special senses Vision, hearing, taste, and smell

specificity Ability of carrier molecules to transport only specific substances across the plasma membrane

spermatogenesis (spur´-mat-uh-JEN-uh-sus) Sperm production

sphincter (sfink-tur) A voluntarily controlled ring of skeletal muscle that controls passage of contents through an opening into or out of a hollow organ or tube

spinal reflex A reflex that is integrated by the spinal cord

spleen A lymphoid tissue in the upper left part of the abdomen that stores lymphocytes and platelets and destroys old red blood cells

state of equilibrium State of a system in which no net change is occurring

steady state State of a system in which no net movement occurs because passive forces and active forces exactly counterbalance each other, with energy being used to maintain the constancy

stem cells Relatively undifferentiated cells that can give rise to highly differentiated, specialized cells while at the same time making new stem cells

stereocilia The auditory and vestibular hair cells that transduce mechanical movements into electrical signals

steroids (STEER-oidz) Hormones derived from cholesterol

stimulus A detectable physical or chemical change in the environment of a sensory receptor

stress The generalized, nonspecific response of the body to any factor that overwhelms, or threatens to overwhelm, the body's compensatory abilities to maintain homeostasis

stretch reflex A monosynaptic reflex in which an afferent neuron originating at a stretch-detecting receptor in a skeletal muscle terminates directly on the efferent neuron supplying the same muscle to cause it to contract and counteract the stretch

stroke volume (SV) The volume of blood pumped out of each ventricle with each contraction, or beat, of the heart

subcortical regions The brain regions that lie under the cerebral cortex, including the basal nuclei, thalamus, and hypothalamus

submucosa The connective tissue layer of the digestive tract that lies under the mucosa and contains the larger blood and lymph vessels and a nerve network

substance P The neurotransmitter released from pain fibers

subsynaptic membrane (sub-sih-NAP-tik) The portion of the postsynaptic cell membrane that lies immediately underneath a synapse and contains receptor sites for the synapse's neurotransmitter

suprachiasmatic nucleus (soup´-ra-kī-as-MAT-ik) A cluster of nerve cell bodies in the hypothalamus that serves as the master biological clock, acting as the pacemaker establishing many of the body's circadian rhythms

surface tension The force at the liquid surface of an air–water interface resulting from the greater attraction of water molecules to the surrounding water molecules than to the air above the surface; a force that tends to decrease the area of a liquid surface and resists stretching of the surface

sympathetic nervous system The subdivision of the autonomic nervous system that dominates in emergency ("fight-or-flight") or stressful situations and prepares the body for strenuous physical activity

symport The form of secondary active transport in which the driving ion and transported solute move in the same direction across the plasma membrane; also called *cotransport*

synapse (SIN-aps´) The specialized junction between two neurons where an action potential in the presynaptic neuron influences the membrane potential of the postsynaptic neuron, typically by releasing a chemical messenger that diffuses across the small cleft between the neurons

synergism (SIN-er-jiz´-um) The result of several complementary actions in which the combined effect is greater than the sum of the separate effects

systemic circulation (sis-TEM-ik) The closed loop of blood vessels carrying blood between the heart and body systems

systole (SIS-tō-lē) The period of cardiac contraction and emptying

T₃ See *tri-iodothyronine*

T₄ See *thyroxine*

T_m See *transport maximum* and *tubular maximum*

tactile (TACK-til) Referring to touch

target-cell receptors Receptors located on a target cell that are specific for a particular chemical mediator

target cells The cells that a particular extracellular chemical messenger, such as a hormone or a neurotransmitter, influences

temporal lobes The lateral lobes of the cerebral cortex, which initially process auditory input

temporal summation The summing of several postsynaptic potentials that occur very close together in time because of successive firing of a single presynaptic neuron

tension The force produced during muscle contraction by shortening of the sarcomeres, resulting in stretching and tightening of the muscle's elastic connective tissue and tendon, which transmit the tension to the bone to which the muscle is attached

terminal button A motor neuron's enlarged knob-like ending that terminates near a skeletal muscle fiber and releases acetylcholine in response to an action potential in the neuron

testosterone (tes-TOS-tuh-rōn) The male sex hormone, secreted by the Leydig cells of the testes

tetanus (TET´-n-us) A smooth, maximal muscle contraction that occurs when the fiber is stimulated so rapidly that it does not have a chance to relax at all between stimuli

tetraiodothyronine See *thyroxine*

thalamus (THAL-uh-mus) The brain region that serves as a synaptic integrating center for preliminary processing of all sensory input on its way to the cerebral cortex

thecal cells (THAY-kel) The outer layer of specialized ovarian connective tissue cells in a maturing follicle

thermoreceptor (thur´-mō-rē-SEP-tur) A sensory receptor sensitive to heat and cold

thick filaments Specialized cytoskeletal structures within skeletal muscle that are made up of myosin molecules and interact with the thin filaments to shorten the fiber during muscle contraction

thin filaments Specialized cytoskeletal structures within skeletal muscle that are made up of actin, tropomyosin, and troponin molecules and interact with the thick filaments to shorten the fiber during muscle contraction

thoracic cavity (thō-RAS-ik) Chest cavity

threshold potential The critical potential that must be reached before an action potential is initiated in an excitable cell

thrombus An abnormal clot attached to the inner lining of a blood vessel

thymus (THĪ-mus) A lymphoid gland located midline in the chest cavity that processes T lymphocytes and produces the hormone thymosin, which maintains the T-cell lineage

thyroglobulin (thī´-rō-GLOB-yuh-lun) A large, complex molecule on which all steps of thyroid hormone synthesis and storage take place

thyroid gland A bilobed endocrine gland that lies over the trachea and secretes the hormones thyroxine and tri-iodothyronine, which regulate overall basal metabolic rate, and calcitonin, which contributes to control of calcium balance

thyroid hormone Collectively, the hormones secreted by the thyroid follicular cells, namely, thyroxine and tri-iodothyronine

thyroid-stimulating hormone (TSH) An anterior pituitary hormone that stimulates secretion of thyroid hormone and promotes growth of the thyroid gland; thyrotropin

thyrotropes Anterior pituitary cells that secrete thyroid-stimulating hormone

thyroxine (thī-ROCKS-in) The most abundant hormone secreted by the thyroid gland; important in the regulation of overall metabolic rate; also known as *tetraiodothyronine* or *T₄*

tidal volume The volume of air entering or leaving the lungs during a single breath

tight junction An impermeable junction between two adjacent epithelial cells formed by the sealing together of the cells' lateral edges near their luminal borders; prevents passage of substances between the cells

tip links Cell adhesion molecules that link the tips of stereocilia of hair cells in adjacent rows in the cochlea and vestibular organs

tissue (1) A functional aggregation of cells of a single specialized type, such as nerve cells forming nervous tissue; (2) the aggregate of various cellular and extracellular components that make up a particular organ, such as lung tissue

tissue-specific stem cells Partially differentiated cells that can generate the highly differentiated, specialized cell types composing a particular tissue

T lymphocytes (T cells) White blood cells that accomplish cell-mediated immune responses against targets to which they have been previously exposed; see also *cytotoxic T cells, helper T cells,* and *regulatory T cells*

tone The ongoing baseline of activity in a given system or structure, as in muscle tone, sympathetic tone, or vascular tone

tonicity A measure of the effect a solution has on cell volume when the solution surrounds the cell

tonic smooth muscle Smooth muscle that is partially contracted at all times in the absence of action potentials

total peripheral resistance The resistance offered by all the peripheral blood vessels, with arteriolar resistance contributing most extensively

trachea (TRĀ-kē-uh) The "windpipe"; the conducting airway that extends from the pharynx and branches into two bronchi, each entering a lung

tract A bundle of nerve fibers (axons of long interneurons) with a similar function within the spinal cord

transduction Conversion of stimuli into action potentials by sensory receptors

transepithelial transport (tranz-ep-i-THĒ-lē-al) The entire sequence of steps involved in the transfer of a substance across the epithelium between either the renal tubular lumen or digestive tract lumen and the blood

transmural pressure gradient The pressure difference across the lung wall (intra-alveolar pressure is greater than intrapleural pressure) that stretches the lungs to fill the thoracic cavity, which is larger than the unstretched lungs

transporter recruitment The phenomenon of inserting additional transporters (carriers) for a particular substance into the plasma membrane, thereby increasing membrane permeability to the substance, in response to an appropriate stimulus

transport maximum (T_m) The maximum rate of a substance's carrier-mediated transport across the membrane when the carrier is saturated; known as *tubular maximum* in the kidney tubules

transport vesicle Membranous sac enclosing newly synthesized proteins that buds off the smooth ER and moves the proteins to the Golgi complex for further processing and packaging for their final destination

transverse tubule (T tubule) A perpendicular infolding of the surface membrane of a muscle fiber; rapidly spreads surface electric activity into the central portions of the muscle fiber

triglycerides (trī-GLIS-uh-rīdz) Neutral fats composed of one glycerol molecule with three fatty acid molecules attached

tri-iodothyronine (T_3) (trī-ī-ō-dō-THĪ-rō-nēn) The most potent hormone secreted by the thyroid follicular cells; important in the regulation of overall metabolic rate

trophoblast (TRŌF-uh-blast′) The outer layer of cells in a blastocyst that is responsible for accomplishing implantation and developing the fetal portion of the placenta

tropic hormone (TRŌ-pik) A hormone that regulates the secretion of another hormone

tropomyosin (trōp′-uh-MĪ-uh-sun) One of the regulatory proteins in the thin filaments of muscle fibers

troponin (tro-PŌ-nun) One of the regulatory proteins in the thin filaments of muscle fibers

TSH See *thyroid-stimulating hormone*

T tubule See *transverse tubule*

tubular maximum (T_m) The maximum amount of a substance that the renal tubular cells can actively transport within a given time period; the kidney cells' equivalent of transport maximum

tubular reabsorption The selective transfer of substances from tubular fluid into peritubular capillaries during urine formation

tubular secretion The selective transfer of substances from peritubular capillaries into the tubular lumen during urine formation

twitch A brief, weak contraction that occurs in response to a single action potential in a muscle fiber

twitch summation The addition of two or more muscle twitches as a result of rapidly repetitive stimulation, resulting in greater tension in the fiber than that produced by a single action potential

tympanic membrane (tim-PAN-ik) The eardrum, which is stretched across the entrance to the middle ear and which vibrates when struck by sound waves funneled down the external ear canal

Type I alveolar cells (al-VĒ-ō-lur) The single layer of flattened epithelial cells that forms the wall of the alveoli within the lungs

Type II alveolar cells The cells within the alveolar walls that secrete pulmonary surfactant

tyrosine kinase A type of receptor-enzyme that brings about the dictated cellular response on binding with an extracellular chemical messenger by phosphorylating the amino acid tyrosine in designated intracellular proteins

ultrafiltration The net movement of a protein-free plasma out of the capillary into the surrounding interstitial fluid

umami A meaty or savory taste

ureter (yū-RĒ-tur) A duct that transmits urine from the kidney to the bladder

urethra (yū-RĒ-thruh) A tube that carries urine from the bladder out of the body

urine excretion The elimination of substances from the body in the urine; anything filtered or secreted and not reabsorbed is excreted

vagus nerve (VĀ-gus) The tenth cranial nerve, which serves as the major parasympathetic nerve

varicosities Swellings in autonomic postganglionic fibers that simultaneously release neurotransmitter over a large area of an innervated organ

vascular tone The state of partial constriction of arteriolar smooth muscle that establishes a baseline of arteriolar resistance

vasoconstriction (vā′-zō-kun-STRIK-shun) The narrowing of a blood vessel lumen as a result of contraction of the vascular circular smooth muscle

vasodilation The enlargement of a blood vessel lumen as a result of relaxation of the vascular circular smooth muscle

vasopressin (vā-zō-PRES-sin) A hormone secreted by the hypothalamus, then stored and released from the posterior pituitary; increases the permeability of the distal and collecting tubules of the kidneys to water and promotes arteriolar vasoconstriction; also known as *antidiuretic hormone (ADH)*

vaults Nonmembranous organelles shaped like octagonal barrels; believed to serve as transporters for messenger RNA and/or the ribosomal subunits from the nucleus to sites of protein synthesis

vein A vessel that carries blood toward the heart

vena cava (venae cavae, plural) (VĒ-nah CĀV-ah; VĒ-nē cāv-ē) A large vein that empties blood into the right atrium

venous return (VĒ-nus) The volume of blood returned to each atrium per minute from the veins

ventilation The mechanical act of moving air in and out of the lungs; breathing

ventricle (VEN-tri-kul) **(of brain)** One of four interconnected chambers within the brain through which cerebrospinal fluid flows

ventricle (VEN-tri-kul) **(of heart)** A lower chamber of the heart that pumps blood into the arteries

vertical osmotic gradient A progressive increase in the concentration of the interstitial fluid in the renal medulla from the cortical boundary down to the renal pelvis; important in the ability of the kidneys to put out urine of variable concentration, depending on the body's needs

vesicle (VES-i-kul) A small, intracellular, fluid-filled, membrane-enclosed sac

vesicular transport Movement of large molecules or multimolecular materials into or out of the cell within a vesicle, as in endocytosis or exocytosis

vestibular apparatus (veh-STIB-yuh-lur) The component of the inner ear that provides information essential for the sense of equilibrium and for coordinating head movements with eye and postural movements; consists of the semicircular canals, utricle, and saccule

villus (villi, plural) (VIL-us) Microscopic fingerlike projections from the inner surface of the small intestine

virulence (VIR-you-lentz) The disease-producing power of a pathogen

visceral afferent A pathway into the central nervous system that carries subconscious information derived from the internal viscera

visceral smooth muscle (VIS-uh-rul) See *single-unit smooth muscle*

viscosity (vis-KOS-i-tē) The friction developed between molecules of a fluid as they slide over each other during flow of the fluid; the greater the viscosity, the greater the resistance to flow

visible light The portion of the electromagnetic spectrum to which the eyes' photoreceptors are responsive (wavelengths between 400 and 700 nanometers)

vital capacity The maximum volume of air that can be moved out during a single breath following a maximal inspiration

vitreous humor The jelly-like substance in the posterior cavity of the eye between the lens and retina

voltage-gated channels Channels in the plasma membrane that open or close in response to changes in membrane potential

voluntary muscle Muscle innervated by the somatic nervous system and subject to voluntary control; skeletal muscle

water balance The balance between water intake and water output; important in controlling ECF osmolarity

white matter The portion of the central nervous system composed of myelinated nerve fibers

Z line A flattened disclike cytoskeletal protein that connects the thin filaments of two adjoining sarcomeres

zona fasciculata (zō-nah fa-SIK-ū-lah-ta) The middle and largest layer of the adrenal cortex; major source of cortisol

zona glomerulosa (glō-MER-yū-lō-sah) The outermost layer of the adrenal cortex; sole source of aldosterone

zona reticularis (ri-TIK-yuh-lair-us) The innermost layer of the adrenal cortex; produces cortisol, along with the zona fasciculata

Index

A bands, 258, 262
A-delta fibers, 191
Abdominal wall, 472
ABO blood group system, 399
Abortion pill. *See* RU 486
Absolute refractory periods, 98–99
Absorption
  Ca^{2+}, 632
  carbohydrate, 626, 628
  fat, 628
  Fe^{2+}, 630
  protein, 628
  small intestine, 623–626
Accessory inspiratory muscles, 470
Acclimatization, 498, 563
Accommodation, 200–201
Acute myocardial infarction, 320–321
ACE (angiotensin-converting enzyme), 527, 529
Acetyl-CoA (acetyl coenzyme A)
  characterization, 33–34
  citric acid cycle, 33
  fuel metabolism and, 33
Acetylcholine (ACh)
  action potential initiation, 249–250
  black widow spider venom and, 251–252
  botulinum toxin and, 252
  characterization, 127–128
  curare and, 252–253
  digestion role, 595
  end-plate potential activity, 247, 250
  gastric function, 609
  muscle contraction role, 264
  at neurotransmitter junction, 247–250
  organophosphates and, 253
  receptor channel inactivation, 253
  release, 232
  release sites, 240
Acetylcholinesterase (AChE)
  inhibition, 253
  muscle contraction role, 264
  neuromuscular junction activity, 250–251
Acetylsalicylic acid
  fever reduction, 656
  gastric absorption, 611, 613
Acid–base balance
  ammonia role, 580–581
  in body, 570
  buffer systems, 573–575
  in chemistry, 570
  compensated disorders, 585
  concept, 557–558
  defined, 569
  H_2O_2, 534
  H^+ fluctuations, 569–584
  homeostasis, 556, 585–586
  intercalated cells, 577–578
  K^+ activity, 570–571
  pH, 569–570
  phosphate role, 580
  renal role, 579–581
  respiratory factors, 581–584
Acidic solutions, 570

Acidification, 456
Acidosis
  carbonic acid levels during, 579–580
  characterization, 570
  H^+ regulation during, 578–579
  metabolic, 549, 581, 583–584
  respiratory, 581–583
Acids
  arteriolar relaxation, 354
  carbonic formation, 571
  characterization, 569, A-10–A-11
  function, 569–570
  gastric emptying and, 603
  inorganic, 572
  organic, 572
  strong, 570
  weak, 570
Acinar cells, 614
Acini, 613
Acquired immune system, 419
Acquired reflexes, 176
Acromegaly, 683
Acrosome, 753
Acrosome reaction, 780
ACTH (adrenocorticotropic hormone)
  adrenal cortex and, 702
  adrenocortical insufficiency and, 705
  adrenogenital syndrome and, 704
  CRH link, 677
  function, 672
  parturition and, 790
  stress response and, 707–708
  target organs, 672, 674
Actin
  based assemblies, 48
  characterization, 47, 260
  in platelets, 405
  thin filaments and, 260–261
Action potentials
  axon hillock and, 109–110
  cardiac cells, 314
  changes during, 91–93
  characterization, 91
  conduction, 96
  contiguous conduction, 96
  EPP and, 249–250
  fiber diameter and, 102–104
  firing by, 91
  frequency limits, 96–99
  graded potentials *vs.*, 98
  initiation site, 109–110, 185
  myelination and, 100–102
  neurotransmitter function, 105–107
  olfactory receptors, 230
  one-way propagation of, 96–99
  pacemakers, 295, 309–310
  propagation, 94–96
  skeletal muscles, 267–268
  smooth muscles, 296
  spikes, 91
  stimulus strength, 99–100
  T tubules and, 264

threshold, 99
visual pathway, 206
Active area, 89
Active forces, 60
Active hyperemia, 354
Active immunity, 436
Active pumping, 81–82
Active reabsorption, 525–526
Active transport
  dephosphorylation, 70
  description, 69
  ion concentration gradient, 69
  model for, 71
  primary, 69–71
  secondary, 69, 71–73
Acuity, 189–190, 208
Acupuncture, 194
Acute mountain sickness, 498–499
Acute phase proteins, 424
Acute renal failure, 548
ADAM (androgen deficiency in aging males), 751
Adaptive immunity. *See* Antibody-mediated immunity; Cell-mediated immunity
ADCC (antibody-dependent cellular cytotoxicity), 433
Addiction, 112, 282
Addison's disease, 704–705
Adenine, A-19
Adenohypophysis, 670
Adenoids, 418, 457
Adenosine
  arousal center inhibition, 171
  cardiac release, 354
  GFR, 521
Adenylyl cyclase, 121
Adipocytes, 644, 648
Adiponectin, 720
Adipose tissues
  anticodons and, A-25
  characterization, 455, A-24
  fat storage by, 42
  fuel metabolism role, 714
  function, 24
  ribosomes, 40, A-24
Adolescence, 751
ADP (adenosine diphosphate)
  ATP production of, 32
  cross-bridge cycling and, 165
  energy production role, 39
  muscle fatigue in, 278–279
  platelet activation, 407
Adrenal cortex
  aldosterone secretion by, 700
  cortisol secretion by, 701–702
  dysfunctions, 703–705
  function, 698–700
  sex hormone secretion, 702–703
Adrenal glands
  anatomy, 698
  catecholamine-secreting medulla, 698–700, 706–707
  characterization, 690

Diastole
coronary circulation during, 332
events, 321–323
failure, 332
murmur, 324
Diastolic blood pressure, 350
Dicrotic notch, 323
Diencephalon, 143, 154
Diet-induced thermogenesis, 642
Dietary fats. *See also* Fat absorption
bile digestion of, 618–619
dietary, 590
digestion, 624, 630
emulsion, 618
gastric emptying and, 603
glucagon actions, 723
insulin effects, 717
metabolism, 719
process, 630–631
Dietary proteins
absorption, 628–629
characterization, 590
digestion, 600–601, 607–608
Differentiation. *See* Cell differentiation
Diffusion
CO_2, 62–63
concentration gradient, 60–61
facilitated model, 68–69
factors influencing, 62–63
Fick's law of, 62–63, 488–489
net, 62, 490
passive, 60–63
types, 61
Digestion
carbohydrate, 611
defined, 590
fat, 630
protein, 600–601, 628–629
saliva role, 597–598
small intestine role, 623
Digestive enzymes, 590
Digestive juices, 605
Digestive system
absorption, 590–591
autonomous smooth muscle function, 593–594
basic function, 589
basic process, 589–591
biliary section, 613–621
biochemical balance, 633
components, 6
defenses, 456–457
enzymes, A-12
esophagus, 598–600
function, 588
hepatic function, 615–617
homeostasis, 11, 588, 638
hormonal pathways, 595–596
hormones, 637
intrinsic nerve plexuses, 594–595
large intestine, 633–637
motility, 589–590
mouth, 595–598
neural reflexes, 595–596
pancreas, 613–615
pharynx, 598–600
regulation, 593–595
secretion, 590
small intestine, 621–633
stomach, 600–613
Digestive tract
anatomy, 592
blood flow to, 343
components, 591
function, 592–593
layers, 591, 593
mucosa, 591
Digitalis-like substances, 382

Dihydropyridine receptors, 264–265
Dilator muscle. *See* Radial muscle
Diploid number, 743, A-20
Diplopia, 211
Direct calorimetry, 643
Disaccharidases, 623
Disaccharide maltose, 591
Disaccharides, 590, A-12
Discriminative ability. *See* Acuity
Dispersed-phase particles, A-10
Distal tubule, 515, 577
Distension, 604
DIT (di-iodotyrosine), 692–694
Diuresis, 546–547
Diuretics, 529
Diurnal rhythms, 665, 702
Divergence, 112
DNA (deoxyribonucleic acid)
B cell reshuffling, 437
characterization, A-15, A-17
functions, 24, A-19
genes and, A-19
hormone receptor, 126
nuclear transfer, 9
organizational levels, A-22
packaging, A-19–A-20
replication, A-20–A21
structure, A-19
template, A-22
thyroid-response element, 694
transcription, A-21–A23
translation, A-24–A-27
DNA replication
anaphase, A-29–A-30
interphase, A-28
meiosis, A-30
metaphase, A-29
mitosis, 46–47, A-28–A-30
prophase, A-28–A-29
telophase, A-30
Docking markers, 27–28
Docking-marker acceptor, 28, 578
Dopamine
behavior and, 157
emotions and, 157
function, 675
lactation and, 795
Parkinson's disease treatment, 112
Dorsal horn, 174
Dorsal respiratory group (DRG), 500
Dorsal root, 175
Dorsal root ganglion, 175
Double bonds, A-11
Double vision. *See* Diplopia
DPP-4 (dipeptidyl peptidase-4) inhibitors, 721
Drugs
adrenergic, 453
anti-inflammatory (*See* NSAIDS)
chemotherapy, 41
epidermis absorption, 455
immunosuppressive, 8
inhibitory synapses affecting, 112
social abused, 112
statin, 731
tolerance, 112
Drunken sailor gait, 167
Dual innervation, 240, 243
Duchenne muscular dystrophy, 284–285
Ductus deferens, 756
Dura mater, 139
Dural sinuses, 139
Dwarfism, 682–683
Dynamic equilibrium, 61, 82
Dynein, 43, 46
Dynorphins, 192
Dyslexia, 150
Dysmenorrhea, 775

Dyspnea, 496, 506
Dystonias, 253
Dystrophin, 284

Ear. *See also* Hearing; Inner ear
anatomy, 215
bones, 217
canal, 216
components, 213, 228
external, 216
middle, 217–218
Ear stones. *See* Otoliths
Earwax, 216
Eccentric isotonic contractions, 274
Eccrine sweat glands, 653
ECF (extracellular fluid). *See also* Interstitial fluid; Plasma
adrenocortical insufficiency and, 705
antiviral defense in, 440
balance stability, 558
Ca^{2+} and, 315–316, 727
carbonic acid buffer, 574
cells mingling with, 23
components, 7, 559
distribution, capillary role, 366–368
hypertonicity, 564–565
ICF barrier between, 559–560
internal pool, 557–558
K^+ levels, 535–536
osmolarity, 563–569
plasma membrane contact, 54
RAAS and, 527–529
regulation, 560
release of, 28
surplus in H_2O, 511
ventilation and, 502–503
volume control, 560–563
ECG (electrocardiogram)
cardiac events and, 317–319
connections, 316–317
description, 316
diagnostic applications, 319–321
representations, 316–317
ECM (extracellular matrix), 58
Ectopic abdominal pregnancy, 778
Ectopic focus, 312
Ectopic tubal pregnancy, 781
Edema, 370–371, 421–422
EDV (end diastolic volume)
characterization, 314
Frank-Starling law, 328
volume increases, 328–329
EEG (electroencephalogram), 152–153
Effector, 15
Effector organs, 134
Effector proteins, 117
Effectors. *See* Efferent pathway
Efferent arterioles, 513
Efferent division, 134
Efferent fibers, 175–176
Efferent neurons, 135, 248
Efferent pathway, 177
Egg-sperm fusion, 779–780
Eicosanoids, 758
Ejaculation
characterization, 760–761
emission, 761
expulsion, 761
orgasm, 761
resolution, 761
sperm content, 761
Ejaculatory duct, 756
Ejection fraction, 329–330
Elastic connective tissue, 474–476
Elastic recoil, 476, 478
Elastin, 58
Electrical defibrillation, 313

Implantation, 781
Impotence, 760
Inclusions, 42
Incontinence, 552
Incretins, 718
Incus, 217
Indirect calorimetry, 643
Indoleamines, 118–119
Induced pluripotent stem cells (iPSCs), 9
Infants. *See also* Neonates
    breast-fed, 795–796
    respiratory distress, 477–478
    sudden death, 506–507
Infectious disease, 436
Infectious organisms, 547
Inflammation
    augmenting, 427–428
    capillary defense, 420
    chronic illnesses and, 425
    cortisol effects, 701
    defined, 420
    edema and, 421–422
    glucocorticoid suppression, 425
    IL-10 release, 446
    leukocyte defense, 422–423
    macrophage defense, 420
    manifestations, 422
    NSAIDS suppression, 425
    outcomes, 422
    parturition and, 790, 792
    phagocyte response, 424–425
    steps producing, 422
    tissue repair, 425
    vasodilation defense, 420
    walling off, 422
Inflammatory response, 333
Inguinal canal, 749
Inguinal hernia, 749
Inhibin, 755
Inhibiting hormones, 675
Inhibitory synapses, 106–107, 112
Initial lymphatics, 368–369
Innate immune response, 431–433
Innate immunity
    bacterial invasion response, 445
    characterization, 419
    complement system, 427–428
    compounds, 419–420
    inflammation defenses, 420–425
    interferon and, 425–426
    natural killer cells, 426
Inner cell mass, 781
Inner ear
    balance portion, 224–227
    characterization, 213
    hearing portion, 217–218, 220
    sensitivity, 222
Inner hair cells, 217, 220
Innervation
    defined, 104
    dual, 240, 243
    skeletal muscles, 246
Inorganic acids, 572
Inorganic chemicals
    acids, A-10–A-11
    bases, A-10–A-11
    organic *vs.*, A-10
Inorganic phosphate, 265, 278–279
Inorganic salts, A-11
Inositol trisphosphate (IP$_3$), 122
Insensible loss, 566
Insertion, defined, 269
Inspiration
    capacity, 480–481
    neurons associated with, 500
    onset, 469–470
    reserve volume, 480–481
    respiratory muscles during, 470–472

Insufficient immune responses, 449–450
Insufficient valve, 324
Insulators, 90
Insulin
    antagonists, 725
    deficient action, 68
    energy balance and, 646
    excess, 723
    glucagon *vs.*, 723–724
    glucose uptake and, 604
    growth and, 683
    metabolic regulation by, 714–715
    nutrient metabolism role, 715–717
    pumps, 721
    secretion factors, 719
    secretion stimulus, 717–718
    shock, 723
    stress response and, 708
    summary of actions, 717
Insulin-like growth factor (IGF-I), 282
Integral proteins, 54
Integrating center, 177
Integrator, 15
Integumentary system. *See also* Skin
    components, 7
    homeostasis, 416
    homeostasis role, 14
Intensity, 197
Intensity discrimination, 222
Intention tremor, 167
Interatrial pathway, 313
Intercalated cells, 577–578
Intercalated discs, 308
Intercellular communication, 113–117
Intercostal muscles, 469
Intercostal nerves, 469
Interferon
    anticancer effects, 426
    antiviral effects, 425
    characterization, 425
    defense mechanism, 440
    immune surveillance, 448
Interleukin 1ß, 792
Interleukin-1, 424
Interleukin-10, 446
Interleukin-12, 440
Interleukin-4, 440, 451
Interleukin-5, 440
Interleukin-6
    energy balance and, 645
    release, 424
    secretion, 440
Interleukin-8, 790
Intermediary metabolism, 42
Intermediate filaments, 42, 49
Internal anal sphincter, 635
Internal environment, 7, 10
Internal pool, 557
Internal respiration, 33, 461
Internal work, 461
Interneurons, 135
Internodal pathway, 313
Interphase, A-28
Interstitial cells. *See* Leydig cells
Interstitial fluid, 7
Intestinal gases, 636–637
Intestinal housekeeper. *See* Migrating motility
    complex
Intestinal phase, 610
Intestine. *See* Large intestine; Small intestine
Intoxication, water, 565
Intra-alveolar pressure
    characterization, 465
    during expiration, 471
    during inspiration, 471
Intracellular fluid (ICF), 23
Intracellular movement, 3

Intracellular proteins, 120
Intrafusal fibers, 285
Intrapleural fluids, 464, 467
Intrapleural pressure
    changes in, 468–473
    characterization, 466
    subatmospheric aspects, 468
Intrapulmonary pressure. *See* Intra-alveolar pressure
Intrathoracic pressure. *See* Intrapleural pressure
Intrinsic control, 328, 329
Intrinsic controls, 14
Intrinsic factor, 397, 608
Intrinsic nerve plexuses, 594–595
Inulin, 539
Involution, 792
Iodide pump, 692
Ion concentration gradient, 69
Ionic bonds, A-5
Ions. *See also* Cations
    concentration, 78
    membrane transport role, 63
    movement, 88, 90–93
    permeability, 78
IPSP (inhibitory postsynaptic potential)
    arousal sleep and, 171
    characterization, 106
    concurrent cancellation, 108
    glycine and, 107
    measuring, 152
    presynaptic inputs and, 107–109
    temporal summation, 108
Iris, 196–197
Iron (Fe), 394, 630
Irreversible reactions, A-8
IRV (inspiratory reserve volume), 480–481
Islets of Langerhans
    cells, 715
    characterization, 613
    metabolic regulation by, 714
Isometric contractions, 273–274
Isotonic contractions, 273–274
Isotonic fluids, 539
Isotonic solutions, 66
Isovolumetric ventricular contraction, 321
Isovolumetric ventricular relaxation, 323
IUDs (Intrauterine devices), 785

Janus kinase, 674
Jaundice, 620
Jet lag, 685
Joints, 275–276
Juxtaglomerular apparatus, 514–515, 521
Juxtamedullary nephrons, 515

Kallikrein, 424
Keratin, 49
Keratinized, 454
Keratinocytes, 456
Ketogenesis, 723
Ketone bodies
    characterization, 714
    defined, A-13
    types, 719
Kidneys
    acid–base balance and, 579–581
    anemia-associated, 398
    basic processes, 515–517
    blood flow to, 343–344
    blood supply, 547
    cardiac output, 524
    cells, 3
    diabetes and, 721
    dialysis, 551
    erythropoiesis control by, 395, 397
    excretory functions, 553
    failure, 547–549
    functional unit, 512–515

Myosin
    characterization, 47
    filament formation and, 260
    phosphorylation, 292
    in platelets, 405
    structure, 261
Myostatin, 282
Myxedema, 696

Na$^+$-K$^+$ pump
    blood pressure management and, 382
    Ca^{2+} absorption and, 632
    end-plate potential formation, 247, 249
    gradient levels restoration, 93–94
    iodide pump and, 692
    membrane potential and, 78
    mineralocorticoids and, 700
    in primary active transport, 70–71
    in secondary active transport, 71
    tubular reabsorption, 526
NAD$^+$ (nicotinamide adenine dinucleotide), 35
Nails, 455
Naive lymphocytes, 433
Narcolepsy, 172
Nasal passages, 463
Natural killer (NK) cells
    characterization, 426
    defense mechanism, 440
    immune surveillance, 448
    stimulation, 433
Nearsightedness. See Myopia
NEAT (nonexercise activity thermogenesis), 648
Necrosis, 124–125, 320
Negative energy balance, 643
Negative feedback
    anterior pituitary hormone secretion, 677
    characterization, 15
    defined, 15
    endocrine system, 675
    hormones, 664–665
Nemenda, 165
Neonates
    adrenogenital syndrome, 704
    breech birth, 789–790
    hemolytic disease, 399
    lactation and, 794
Nephrons
    anatomy, 514
    characterization, 512
    cortical, 515–516
    juxtamedullary, 515–516
    Na$^+$ reabsorption, 548
    regions, 512
    tubular component, 514–515
    vascular component, 512–514
    water reabsorption, 548
Nernst equation, 79–80
Nerve fiber. See Axons
Nervous system. See also Central nervous system;
        Neurons; Peripheral nervous system
    antiviral defense in, 440
    characterization, 127–128
    components, 7
    endocrine system vs., 126–129
    enteric, 595
    functions, 127, 132
    homeostasis, 86, 132, 180, 232–233, 236
    homeostasis role, 14
    immune system link, 448–449
    organization, 133–134
    thyroid hormone effects, 694–695
    tissues, 4
Net diffusion, 61
NETs (neutrophil extracellular traps), 401–402
Neural communication, 86
Neural networks, 153
Neural presbycusis, 223

Neural signals, 217–218
Neuregulin, 163
Neuritic plaque, 164
Neuroendocrine reflexes, 665
Neuroendocrinology, 129
Neurofibrillary tangles, 164
Neurofilaments, 49
Neurogenic properties, 294–295
Neurogenic shock, 384–385
Neuroglia. See Glial cells
Neuroglobin, 141
Neurohormones, 115, 129
Neurohypophysis, 670
Neurological disorders, 157
Neuromodulators, 110–111
Neuromuscular junction
    AChE function at, 250–251
    characterization, 246
    events at, 250
    neurotransmitter, 247–250
    synapse vs., 252
    vulnerability, 251–253
Neurons. See also specific nerves
    anatomy, 95
    axonal transport in, 45
    basic parts, 95
    brain, 138
    cerebellum, 166
    cerebral cortex, 143
    cranial, 168
    digestive, 594–596
    functional classes, 135
    GPSP integration, 108–109
    linkage, 112–113
    locations, 135
    membrane potential, 87
    postsynaptic, 104–105
    presynaptic, 104–105
    respiratory, 500
    rhythmic synchrony, 153
    short-term memory and, 159
    structure, 176
    termination, 104
    vestibular, 224
Neuropathic pain, 192
Neuropeptides, 110
Neurotransmitters. See also specific
        neurotransmitter
    -receptor response, 107
    action potential function, 105–107
    behavior and, 157
    common, 107
    emotion and, 157
    intracellular communication and, 114–115
    neuropeptides vs., 110
    storage, 104
    synaptic cleft removal, 107
Neutral energy balance, 643
Neutralization, 431
Neutralization reactions, A-11
Neutrons, A-3
Neutrophilia, 402–403
Neutrophils
    bacteria destruction, 423
    characterization, 401
    circulating, 402–403
    define, 418
    inflammatory response, 421
    release of, 401–402
    synthesis, 424
NF-κB (nuclear factor κB), 790
Nicotinic cholinergic receptors, 239
Nicotinic receptors, 243
Night blindness, 209
Nitric acid, 595
Nitric oxide (NO)
    abnormalities, 382

characterization, 161, 334
    functions, 356
    hemoglobin, 394
    high altitudes and, 498
    macrophage secretion, 424
    in platelet plug formation, 407
Nitrite, 456
Nitrogen, 533
Nitrogen narcosis, 499
Nitroglycerin, 335
NMDA receptors
    Alzheimer's disease and, 165
    LTP, 161
    pain and, 192
Nociceptors, 184
Nodes of Ranvier, 101
Noise. See Sound
Noncarbonic acids, 574
Nonhistones, A-20
Nonmembranous organelles, 24
Nonmuscle cells, 48
Nonpolar molecules, A-7
Nonrespiratory acidosis. See Metabolic acidosis
Nonshivering thermogenesis, 654
Norepinephrine
    adrenal medulla secretion, 705
    adrenergic receptor affinity, 706
    arterioles and, 359
    behavior and, 157
    characterization, 240
    emotions and, 157
    messenger function, 129
    release, 232, 240
Normality, A-9
Nose. See Nasal passages
NPY (neuropeptide Y), 644
NSAIDS (nonsteroidal anti-inflammatory drugs).
        See also Acetylsalicylic acid
    eliminations, 536
    gastric irritation from, 612
    inflammation suppression, 425
NTS (nucleus tractus solitarius), 647
Nuclear envelope, 23
Nuclear pores, 23
Nuclear transfer, 9
Nucleic acids, A-15, A-17
Nucleotides, A-15, A-19
Nucleus, 20, 23–24
Nutrients. See also Food; Fuel metabolism
    absorbed, hepatic processing of, 632
    absorption, digestive, 590–591, 598
    absorption, microvilli role, 48
    absorptive states, 713
    blood pressure lowering, 382
    body storage, 711
    brain, 139–141
    breakdown, 572
    cell uptake, 10
    circulatory distribution, 7, 10
    in colon, 536
    in cytoplasm, 780
    digestive process categories, 624
    essential, 712
    growth role, 677
    majority pathways, 711
    postabsorptive states, 713–714
    storage, 712–713
Nutritional anemia, 397

Obesity
    android, 649
    causes, 647–649
    colon bacteria and, 648–649
    defined, 647
    gynoid, 649
Occipital lobes, 144
Occlusion. See Dental caries

Oddi, sphincter of, 617
Odor eating enzymes, 232
Odorants, 230
Odors
    clearance, 232
    detection, 230–231
    discrimination, 232
    pheromones, 232
Off response, 186–187
Off-center bipolar cells, 205–207
Off-center ganglion cells, 205, 207
Olfactory bulb
    characterization, 230
    odor discrimination, 232
    scent processing in, 231
Olfactory mucosa, 230
Olfactory nerve, 230
Olfactory receptors
    action potentials, 230
    cells, 230, 232
    location, 231
    number, 230
    specialization, 230
Olfactory system, 232
Oligodendrocytes, 101, 138
On-center bipolar cells, 205, 207
On-center ganglion cells, 205, 207
Oncogenes, A-31
Onset of ventricular diastole, 323
Oocytes, 764–765, 772
Oogenesis
    characterization, 764
    meiotic divisions, 767
    mitotic divisions, 767
    process, 765–766
    spermatogenesis vs., 765
Oogonia, 764
OPG (osteoprotegerin), 728–729
Opiates, 192–193
Opsin, 205
Opsonins, 423, 428
Optic chiasm, 210
Optic disc, 201
Optic nerve, 201, 686
Optic radiation, 210
Optic tracts, 210
Optimal length, 272
Oral cavity, 596
Oral contraceptives, 784
Oral rehydration therapy (ORT), 633–634
Orexins, 646–647
Organ of Corti
    characterization, 217
    hair cells in, 217–218, 220
Organ printing, 8
Organelles. See also specific organelles
    defined, 24
    removal, 31
    types, 24
Organic acids, 572
Organic chemicals
    acids, A-10–A-11
    bases, A-10–A-11
    inorganic vs., A-10
    monomers, A-10
    polymers, A-10
    salts, A-11
Organic ion secretory system, 536
Organisms
    characterization, 2
    female, 763–764
    male, 761
    single-celled, 6
Organophosphates, 253
Origin, defined, 269
Oropharyngeal stage, 598–599
Orthostatic hypotension, 383–384

Osmolarity
    arteriolar, 354
    characterization, 563
    defined, A-9–A-10
    digestive, 603–604
    ECF, 563–569
    ICF, 564
    measurement, 64
Osmoreceptors, 184
Osmosis
    defined, 63
    diuresis, 546
    plasma-colloid, 366–367
    pressure, 64
    process of, 63–64
    tonicity and, 66–67
    water movement, 63–66
Ossicles, 217
Ossification, 680
Osteoblasts, 679
Osteoclasts, 679, 728–729
Osteocytes, 681
Osteocytic-osteoblastic bone membrane, 730
Osteomalacia, 737–738
Osteons, 729, 732
Osteoporosis
    characterization, 729, 730
    drug therapy for, 730
    exercise benefits, 731
Otolith organs, 224
Otoliths, 224–225
Outer hair cells, 217, 220
Oval window, 217–218
Ovarian cycle
    cyclic uterine changes, 774–776
    follicular phase, 765–766, 769
    hormonal regulation, 770–774
    luteal phase, 765, 769–774
    menstrual phase, 774–776
    progesterone in, 772
    proliferative phase, 775
    secretory phase, 775–776
Ovaries, 664, 771
Oviducts
    characterization, 742
    ovum transport to, 778
    sperm transport to, 778–779
Ovulation
    characterization, 769
    control of, 772–773
    development, 768
    LH surge, 772–774
Ovum, 745, 778
Oxidative enzymes, 31
Oxidative fibers, 280
Oxidative phosphorylation
    characterization, 35
    exercise and, 278
    mitochondrial inner membrane, 36
    muscle fibers, 280
    process, 276
    steps leading to, 37
Oxygen (O_2)
    arteriolar relaxation, 354
    blood carrying capacity, 398, 402–403
    brain needs, 141
    cardiac needs, 332–333
    cells' need for, 302
    circulatory distribution, 7, 10
    diffusion, 62–63
    dissolved, 490
    energy equivalent of, 643
    in glycolysis, 37
    in hemoglobin, 393–394
    homeostasis and, 11
    maximal, 506
    placental, 786

    post-exercise needs, 279
    pulmonary entrance, 486–488
    respiration and, 462
    smooth muscles and, 484–485
    synthetic carriers, 403
    toxicity, 497, 499
    transport in blood, 490–491
    ventilation magnitude and, 501–503
Oxygen (O_2)-hemoglobin (Hb) curve
    2,3-bisphosphoglycerate, 494
    acid and, 494
    alveolar level, 492
    arterial abnormalities, 496–498
    Bohr effect, 494
    dissociation, 491
    exercise-related variables, 505
    plateau portion, 491–492
    steep portion, 492
    storage, 493
    temperature and, 494
    tissue level, 492–495
    transport methods, 490–491
Oxyhemoglobin, 491
Oxyntic cells. See Parietal cells
Oxyntic mucosa, 605
Oxytocin
    function, 671
    homeostatic role, 16
    lactation and, 794–795
    parturition and, 790
    storage, 670–671

P wave, 317, 319
Pacemaker cells, 601
Pacemakers
    abnormal activity, 311–313
    activity analogy, 312
    artificial, 312
    normal activity, 310–311
    potentials, 295, 309–310
Pacinian corpuscle, 186–187
Packaging, 752–754
Packed cell volume. See Hematocrit
PAF (platelet-activating factor), 333
PAH (para-aminohippuric acid), 539
Pain
    brain analgesics, 192–194
    causes, 421
    characteristics, 191
    defined, 191
    fibers, 191–192
    higher-level processing, 192
    impulse origins, 191–192
    neuropathic, 192
    perception of, 191
    phantom, 189
    receptors, 184, 191
    referred, 176
Palate, 596
Pancreas. See also Endocrine pancreas; Exocrine pancreas
    bile production, 615–620
    biochemical balance, 622
    characterization, 690
    cystic fibrosis effects on, 56
    enzymes, 614
    metabolic regulation by, 714–715
    tissues, 613–615
Pancreatic amylase, 614
Pancreatic insufficiency, 614
Pancreatic islet transplants, 721
Pancreatic lipase, 614, 618
Paneth cells, 626
Papillary muscles, 307
Paracrines, 114. See also specific paracrines
Paradoxical sleep, 169–170
Paralysis, 283

Proximal tubule, 514, 577
Prozac, 107, 157
PSA (prostate-specific antigen), 757
Pseudomonas aeruginosa, 56
Pseudopods, 31
Psychiatric disorders, 157
PTH (parathyroid hormone)
    based osteoporosis therapy, 731
    bone metabolism and, 729–733
    Ca^{2+} metabolism and, 729–733
    characterization, 728
    feedback loop, 736
    function, 727–728
    hypersecretion, 736
    hyposecretion, 736–737
    renal functions, 733–734
    vitamin D and, 735
PTHrp (parathyroid hormone-related peptide), 789
Puberty
    characterization, 751
    female, changes induced by, 776
    female, initiation, 664–665
    follicle formation, 65
    GnRH activity at, 755–756
    male, initiation, 755–756
Pulse deficit, 320
Pulse pressure, 350
Punishments, 157
Pupils, 196–197
Purkinje fibers, 310–311
Pus, 423
PVC (premature ventricular contraction), 312
PVN (paraventricular hypothalamic nucleus), 646
Pyloric sphincter, 600
Pyramidal cells, 143, 283
Pyramidal motor system. *See* Corticospinal motor
    system
PYY secretion, 646

QRS complex, 317
Quadriplegia paralysis, 283
Quantitative reasoning, A-33–A-36
Quaternary protein structure, A-15

RAAS (renin-angiotensin-aldosterone system)
    diseases associated with, 529
    functions of, 527–528
    mechanism, 527
    Na^+ reabsorption, 562–563
    stress response and, 708–709
Radial muscle, 197
Radiation, 651
Raloxifene, 731
RANK (receptor activator of NF-κB) ligand,
    728–729
RAS (reticular activating system), 167
Reabsorption. *See* Tubular reabsorption
Reactants, A-7
Reactive hyperemia, 357
Reactive hypoglycemia, 723
Readiness potential, 146
Receptive field, 189–190
Receptor editing, 446
Receptor-mediated endocytosis, 31
Receptors. *See also specific receptors*
    -neurotransmitter response, 107
    adaptation speed, 186–187
    automatic nervous system, 243–245
    information detected by, 184–185
    lateral inhibition, 190
    opiate, 192
    pain, 191
    perception and, 190–191
    phasic, 186
    potential conversion, 184
    potential initiation site, 185–186

protein, 57
    sensory, 177
    stimuli effect, 185
    stimuli sensitivities, 183–185
    sustained stimulation, 186
    tactile, 187
    taste, 229
    tonic, 186
    types of, 184
Reciprocal innervation, 177
Reclast. *See* Zoledronic acid
Recognition markers, 28
Rectum, 633
Red blood cells. *See* Erythrocytes
Red bone marrow, 395
Red cones, 208
Red fibers, 280
Referred pain, 176
Reflexes
    arc, 177
    autonomic system and, 245
    crossed extensor, 179
    defecation, 635
    defined, 176
    digestive, 595–596
    enterogastric, 602
    exercise and, 504
    gastroileal, 622
    Hering-Breuer, 500
    knee-jerk, 286
    local responses, 179
    neuroendocrine, 665
    salivary, 597–598
    stretch, 285–287
    suckling, 794–795
    types, 176
    withdrawal, 177–179
Reflux, 600
Refraction
    concave lens, 199
    convex lens, 199
    defined, 198
    process, 198
    structures, 198–200
Refractory periods, 97–99
Regeneration, nerve, 103
Regeneration tube, 103
Regulatory proteins, 260
Regulatory T cells
    activation, 446
    characterization, 438
    function, 782
Regurgitation, 324
Rehearsal, 158
Relative humidity, 653
Relative refractory periods, 98–99
Relaxation
    gastric, 601
    isovolumetric ventricular, 323
    process, 267–268
    response to stress, 298
    skeletal muscles, 266
    smooth muscles, 297–298
    time of, 268
Relaxin, 789
Releasing hormones, 675
REM (rapid eye movement), 170–172
Remembering, 159
Renal anemia, 398
Renal artery, 512–513
Renal cortex, 512
Renal medulla, 512
Renal pelvis, 512
Renal pyramids, 512
Renal vein, 512
Renin. *See* RAAS (renin-angiotensin-aldosterone
    system)

Repolarization, 88
Reproductive cells. *See* Gametes
Reproductive system. *See also* Female reproductive
        system; Male reproductive system
    abnormalities at birth, 762
    anabolic steroid effects, 282
    chromosomes, 743, 745
    components, 7, 741–743
    DNA role, 24
    female, 742–743, 744
    gametogenesis, 743, 745
    homeostasis, 14, 740, 796–797
    hormones, 742
    primary organs, 742
    requirements, 741
    secondary sexual characteristics, 742
    sex determining region, 747
    sexual differentiation, 745–748
    sexual intercourse, 759–764
Reproductive tract
    female, 742
    male, 756
    sexual differentiation, 747
Residual volume (RV), 480–481
Resistin, 645, 720
Respiration. *See also* Expiration; Inspiration
    chemical factors and, 501
    control of, 598–501
    defined, 460
    external, 461–462
    internal, 461
    neural control components, 499–500
Respiratory arrest, 496
Respiratory distress syndrome, 477–478
Respiratory muscles
    anatomy, 469
    contractions, 469–470
    expiration, 470–472
    inspiration, 470–472
Respiratory pump, 375
Respiratory quotient (RQ), 461
Respiratory rate, 482
Respiratory states, 496–498
Respiratory system. *See also* Lungs
    acid–base balance and, 581–584
    acidosis, 498, 581–583
    alkalosis, 583
    anatomy, 461–465
    characterization, 460
    components, 6, 462–464
    cystic fibrosis effects on, 56
    defense mechanism, 457
    dysfunction, 481
    gas exchange, 486–490
    gas transport, 490–498
    H^+ regulation by, 575–576
    homeostasis, 11, 460, 507
    nonrespiratory functions, 462
    role of, 461
    venous return and, 375
Rest
    blood flow at, 344
    capillaries at, 364–365
    cardiac output, 355
    K^+ potentials, 91–92
    membrane potential at, 78, 81–82
    Na^+ potential, 91–92
Resting membrane potential, 78
Reticular formation
    activating system, 169
    defined, 167
    electrical stimulation, 192
Reticulocytes, 397
Retina
    cells, 201, 205, 210
    characterization, 195
    cones, 195, 201, 207–208

layers, 201
light input processing, 205
rods, 195, 207–208
Retinal, 205
Retinohypothalamic tract, 687
Reverse axonal transport, 43
Reverse vesicular traffic, 43
Reversible reactions, A-8
Rewards, 157
Rh blood group system, 400
Rheumatic fever, 324
Rheumatoid arthritis, 433, 701
Rhodopsin, 205
Rhythm
cardiac, 319, 704
electrical rhythm, 593
respiratory, 499–500
Rhythm method, 784
Ribosomal protein synthesis, 42
Ribosomal RNA, A-24
Ribosomes. *See also* Free ribosomes
assembly, A-26
function, 39–40, A-24
in rough ER, 40
Ribs, 464
Rickets, 737–738
RIG-I (retinoic acid inducible gene I), 419
Right cerebral hemisphere, 152
Rigor mortis, 267
RNA (ribonucleic acid)
characterization, A-15, A-17
defense mechanism, 425
function, 24
ribosomes, A-25
sugar in, A-22
transcription role, A-22–A-23
translation, A-24–A-27
types of, A-24
Rods
acuity, 208
characterization, 195
gray vision and, 207
location, 201
properties, 207
sensitivity, 207–208
Rough endoplasmic reticulum, 24–26, 40
Round window, 217
RU 486, 785
Ruffini endings, 186–187
Ryanodine receptors, 264

SA node, 325–328
Saccule, 224, 227
Salbutamol, 245–246
Saliva
characterization, 597
function, 597
secretion, 597–598
Salivary amylase, 597
Salivary center, 597
Salivary glands, 243
Salivary reflexes, 597–598
Salt. *See* Bile salts; Sodium (Na⁺)
SALT (skin-associated lymphoid tissues), 456
Salt balance
angiotensin and, 393
ECF volume and, 561–563
water balance and, 556
Saltatory conduction, 100–101
Salty taste, 229
Sarcomere, 258
Sarcoplasmic reticulum, 264–265
Satiety center, 647
Satiety signals, 644, 647
Saturated fatty acids, A-12
Scala media, 217
Scala tympani, 217

Scala vestibuli, 217
Scent processing, 231
Schwann cells, 101
Sclerosis, 102, 334
SCN (suprachiasmatic nucleus), 685–687
Scrotum, 72
Scurvy, 58
Sebaceous glands, 455
Sebum, 455
Second messengers
Ca^{2+} pathway, 122–123
characterization, 117
cyclic AMP, 120–122
LTP and, 161–162
pathway amplification, 123–124
Second-order sensory neuron, 188
Second polar body, 765
Secondary active transport, 69, 530
Secondary adrenocortical insufficiency, 704–705
Secondary endings, 285
Secondary follicles, 769
Secondary hyperaldosteronism, 703
Secondary oocytes, 765
Secondary peristaltic wave, 600
Secondary polycythemia, 398–399
Secondary protein structure, A-14–A-15
Secondary response, 434–436
Secondary spermatocytes, 752
Secretin, 614–615, 637
Secretion
ANF influence on, 238
bile, 617
digestive, 590
esophageal, 600
ghrelin, 646
HCL, 605, 607
large intestine, 636
proteins, 27
saliva, 597–598
small intestine, 623
vasopressin, 567–569
Secretory glands, 4
Secretory phase, 775–776
Secretory vesicles, 28, 42
Seed cells, 8
Segmentation
control of, 621–622
defined, 621
functions, 622
initiation, 621–622
Selectins, 422
Selectively permeable, 60
Self-antigens
autoimmunity and, 446–447
coding, 443
tolerance, 446
Self-recognition process, 55
Semantic memory, 163
Semen, 756–758
Semicircular canals, 224
Seminal vesicles, 756, 757
Seminiferous tubule fluid, 754
Senile dementia, 164
Sensitization, 159, 160–161
Sensor, 15
Sensorineural deafness, 223
Sensory homunculus, 145
Sensory information
categories, 188
coding, 189
vision *vs.* hearing, 214
Sensory nerves, 188–189
Septicemic shock, 411
Septum, 304
Series-elastic component, 269, 272
SERMs (Selective estrogen receptor modulators), 664–665

Serosa, 593
Serotonin, 157
Serous fluids, 593
Sertoli cells
characterization, 754
function, 754–755
secretion, 754–755
Serum sickness, 436
Set point, 15
Severe combined immunodeficiency, 449–450
Sex act. *See* Sexual intercourse
Sex chromosomes, 745
Sex determination
chromosomes, 745
masculinizing determinants, 745–749
sexual differentiation and, 746
Sex glands, 742, 756–758
Sex hormones
adrenal cortex secretion, 702–703
characterization, 699
female, 702–703
secretion, 698–700
Sexual arousal, 757–758
Sexual differentiation
defined, 474
errors in, 747–748
external genitalia, 747
organs, 747
reproductive tract, 747
sex determination and, 746
Sexual intercourse
defined, 759
ejaculation, 759, 760–762
erection, 759–760
female cycle, 762–764
male cycle, 759–763
Sexual nose. *See* Pheromones
Sexual response cycle, 759
SGLT (sodium and glucose cotransporter), 73, 530
Shear stress, 357
Shivering
controlling mechanism, 653–654
defined, 653
function, 653
Short reflexes, 595
Short-term behavior, 646
Short-term memory
characterization, 158
long-term memory *vs.*, 158–159
molecular mechanism, 159
synaptic activity and, 159–162
Sickle cell disease, 398
SIDS (sudden infant death syndrome), 506–507
Sigmoid colon, 633
Signal molecules. *See* Chemical messengers
Signal transduction, 124–125, 674
Sildenafil, 760
Simple cells, 211
Simple diffusion
carrier-mediated transport *vs.*, 69
characterization, 61
methods, 76
Simple reflexes, 176
Single-unit smooth muscles
cells function, 295
contractions, 296–297
myogenic properties, 295–296
pacemaker potentials, 295
slow-wave potentials, 295–296
Singulair, 453
Sinoatrial node, 310–313
Sinuses, 139
Sinusoids, 364, 616
Skeletal muscles. *See also* Muscle fibers
action potential, 264
bones attached to, 268–269
Ca^{2+} effects, 264–268

Introduction to Physiology (pp. 1–2)

■ Physiology is the study of body functions.

■ Physiologists explain body function in terms of the mechanisms of action involving cause-and-effect sequences of physical and chemical processes.

■ Physiology and anatomy are closely interrelated because body functions are highly dependent on the structure of the body parts that carry them out.

Levels of Organization in the Body (pp. 2–6)

■ The human body is a complex combination of specific atoms and molecules.

■ These nonliving chemicals are organized in a precise way to form cells, the smallest entities capable of carrying out life processes. Cells are the body's structural and functional living building blocks. (Review Figure 1-1.)

■ The basic functions performed by each cell for its own survival include (1) obtaining O_2 and nutrients, (2) performing energy-generating chemical reactions, (3) eliminating wastes, (4) synthesizing proteins and other cell components, (5) controlling movement of materials between the cell and its environment, (6) moving materials throughout the cell, (7) responding to the environment, and (8) reproducing.

■ In addition to its basic functions, each cell in a multicellular organism performs a specialized function.

■ Cells of similar structure and specialized function combine to form the four primary tissues of the body: muscle, nervous, epithelial, and connective. (Review Figure 1-2.)

■ Glands are derived from epithelial tissue and specialized for secretion. Exocrine glands secrete through ducts to the body surface or into a cavity that communicates with the outside; endocrine glands secrete hormones into the blood. (Review Figure 1-3.)

■ Organs are combinations of two or more types of tissues that act together to perform one or more functions. An example is the stomach. (Review Figure 1-2.)

■ Body systems are collections of organs that perform related functions and interact to accomplish a common activity essential for survival of the whole body. An example is the digestive system. (Review Figure 1-4.)

■ Body systems combine to form the organism, or whole body.

Concept of Homeostasis (pp. 6–14)

■ The fluid inside the cells of the body is intracellular fluid (ICF); the fluid outside the cells is extracellular fluid (ECF).

■ Because most body cells are not in direct contact with the external environment, cell survival depends on maintaining a relatively stable internal fluid environment with which the cells directly make life-sustaining exchanges.

■ The ECF serves as the body's internal environment. It consists of the plasma and interstitial fluid. (Review Figure 1-5.)

■ Homeostasis is the maintenance of a dynamic steady state in the internal environment.

■ The factors of the internal environment that must be homeostatically maintained are its (1) concentration of nutrient molecules; (2) concentration of O_2 and CO_2; (3) concentration of waste products; (4) pH; (5) concentration of water, salt, and other electrolytes; (6) volume and pressure; and (7) temperature. (Review Figure 1-7.)

■ The functions performed by the 11 body systems are directed toward maintaining homeostasis. These functions ultimately depend on the specialized activities of the cells that make up the system. Thus, homeostasis is essential for each cell's survival, and each cell contributes to homeostasis. (Review Figures 1-6 and 1-7.)

Homeostatic Control Systems (pp. 14–17)

■ A homeostatic control system is a network of body components working together to maintain a controlled variable in the internal environment relatively constant near an optimal set point despite changes in the variable.

■ Homeostatic control systems can be classified as (1) intrinsic (local) controls, which are inherent compensatory responses of an organ to a change; and (2) extrinsic (systemic) controls, which are responses of an organ triggered by factors external to the organ, namely, by the nervous and endocrine systems.

■ Both intrinsic and extrinsic control systems generally operate on the principle of negative feedback: A change in a controlled variable triggers a response that drives the variable in the opposite direction of the initial change, thus opposing the change. (Review Figure 1-8.)

Deviation in controlled variable
* relieves
(detected by)
Sensor
(informs)
Integrator
(sends instructions to)
(negative feedback shuts off system responsible for response)
Effector(s)
(brings about)
Compensatory response
(results in)
Controlled variable restored to normal *

(a) Components of a negative-feedback control system

■ In positive feedback, a change in a controlled variable triggers a response that drives the variable in the same direction as the initial change, thus amplifying the change. Positive feedback is uncommon in the body but is important in several instances, such as during childbirth.

■ Feedforward mechanisms are compensatory responses that occur in anticipation of a change.

BODY SYSTEMS
Made up of cells organized according to specialization to maintain homeostasis
See Chapter 1.

NERVOUS SYSTEM
Acts through electrical signals to control rapid responses of the body; also responsible for higher functions—e.g., consciousness, memory, and creativity
See Chapters 4, 5, 6, and 7.

ENDOCRINE SYSTEM
Acts by means of hormones secreted into the blood to regulate processes that require duration rather than speed—e.g., metabolic activities and water and electrolyte balance
See Chapters 4, 18, and 19.

RESPIRATORY SYSTEM
Obtains O_2 from and eliminates CO_2 to the external environment; helps regulate pH by adjusting the rate of removal of acid-forming CO_2
See Chapters 13 and 15.

INTEGUMENTARY SYSTEM
Serves as a protective barrier between the external environment and the remainder of the body; the sweat glands and adjustments in skin blood flow are important in temperature regulation
See Chapters 12 and 17.

URINARY SYSTEM
Important in regulating the volume, electrolyte composition, and pH of the internal environment; removes wastes and excess water, salt, acid, and other electrolytes from the plasma and eliminates them in the urine
See Chapters 14 and 15.

IMMUNE SYSTEM
Defends against foreign invaders and cancer cells; paves way for tissue repair
See Chapter 12.

DIGESTIVE SYSTEM
Obtains nutrients, water, and electrolytes from the external environment and transfers them into the plasma; eliminates undigested food residues to the external environment
See Chapter 16.

MUSCULAR AND SKELETAL SYSTEMS
Support and protect body parts and allow body movement; heat-generating muscle contractions are important in temperature regulation; calcium is stored in the bone
See Chapters 8, 17, and 19.

REPRODUCTIVE SYSTEM
Not essential for homeostasis, but essential for perpetuation of the species
See Chapter 20.

CIRCULATORY SYSTEM
Transports nutrients, O_2, CO_2, wastes, electrolytes, and hormones throughout the body
See Chapters 9, 10, and 11.

Regulate

Keeps internal fluids in

Keeps foreign material out

Protects against foreign invaders

Enables the body to interact with the external environment

Exchanges with all other systems

HOMEOSTASIS
A dynamic steady state of the constituents in the internal fluid environment that surrounds and exchanges materials with the cells
See Chapter 1.
Factors homeostatically maintained:
• Concentration of nutrient molecules
  See Chapters 16, 17, 18, and 19.
• Concentration of O_2 and CO_2
  See Chapter 13.
• Concentration of waste products
  See Chapter 14.
• pH   *See Chapter 15.*
• Concentration of water, salts, and other electrolytes
  See Chapters 14, 15, 18, and 19.
• Volume and pressure   *See Chapter 17.*
• Temperature   *See Chapter 17.*
  See Chapters 10, 14, and 15.

Body systems maintain homeostasis

Homeostasis is essential for survival of cells

CELLS
Need homeostasis for their own survival and for performing specialized functions essential for survival of the whole body
See Chapters 1, 2, and 3.
Need a continual supply of nutrients and O_2 and ongoing elimination of acid-forming CO_2 to generate the energy needed to power life-sustaining cellular activities as follows:
Food + $O_2 \longrightarrow CO_2 + H_2O$ + energy
See Chapter 17.

Cells make up body systems

EXTERNAL ENVIRONMENT

Information from the external environment relayed through the nervous system

O_2 CO_2

Urine containing wastes and excess water and electrolytes

Nutrients, water, electrolytes

Feces containing undigested food residue

Sperm leave male

Sperm enter female

Exchanges with all other systems

Cell Theory (p. 21)

■ The complex organization and interaction of the chemicals within a cell confer the unique characteristics of life. The cell is the smallest unit capable of carrying out life processes.

■ Cells are the living building blocks of the body. The structure and function of a multicellular organism ultimately depend on the structural and functional capabilities of its cells. (Review Table 2-1.)

Observations of Cells (p. 21–22)

■ Cells are too small for the unaided eye to see.

■ Using early microscopes, investigators learned that all plant and animal tissues consist of individual cells.

■ Scientists now know that a cell is a complex, highly organized, compartmentalized structure.

An Overview of Cell Structure (pp. 22–24)

■ Cells have three major subdivisions: the plasma membrane, the nucleus, and the cytoplasm. (Review Figure 2-1.)

■ The plasma membrane encloses the cell and separates the intracellular and extracellular fluid.

■ The nucleus contains deoxyribonucleic acid (DNA), the cell's genetic material.

■ Three types of RNA play a role in the protein synthesis coded by DNA: messenger RNA (mRNA), ribosomal RNA (rRNA), and transfer RNA (tRNA).

■ The cytoplasm consists of cytosol, a complex gel-like mass, which is laced with a cytoskeleton and organelles. Organelles are highly organized structures that serve a specific function.

■ There are two categories of organelles. Membranous organelles are bound by a membrane that separates the organelle's contents from the surrounding cytosol. They include the endoplasmic reticulum, Golgi complex, lysosomes, peroxisomes, and mitochondria. The nonmembranous organelles are not surrounded by membrane and include ribosomes, vaults, and centrioles. (Review Figure 2-1 and Table 2-2, p. 44.)

Endoplasmic Reticulum and Segregated Synthesis (pp. 24–27)

■ The endoplasmic reticulum (ER) is a single, complex membranous network that encloses a fluid-filled lumen.

■ The primary function of the ER is to synthesize proteins and lipids that are either (1) secreted to the exterior of the cell, such as enzymes and hormones, or (2) used to produce new cell components, particularly cellular membranes.

■ The two types of ER are rough ER (flattened interconnected sacs studded with ribosomes) and smooth ER (interconnected tubules with no ribosomes). (Review Figure 2-2.)

■ The rough-ER ribosomes synthesize proteins, which are released into the ER lumen so that they are separated from the cytosol. Also entering the lumen are lipids produced within the membranous walls of the ER.

■ Synthesized products move from the rough ER to the smooth ER, where they are packaged and discharged as transport vesicles. Transport vesicles are formed as a portion of the smooth ER "buds off." (Review Figure 2-3.)

Golgi Complex and Exocytosis (pp. 27–28)

■ Transport vesicles move to and fuse with the Golgi complex, which consists of a stack of separate, flattened, membrane-enclosed sacs. (Review Figures 2-3 and 2-4.)

■ The Golgi complex serves a twofold function: (1) to modify into finished products the newly synthesized molecules delivered to it in crude form from the ER; and (2) to sort, package, and direct molecular traffic to appropriate intracellular and extracellular destinations.

■ The Golgi complex of secretory cells packages proteins to be exported from the cell in secretory vesicles that are released by exocytosis on appropriate stimulation. (Review Figures 2-3, 2-5a, and 2-6.)

Lysosomes and Endocytosis (pp. 28–31)

■ Lysosomes are membrane-enclosed sacs that contain powerful hydrolytic (digestive) enzymes. (Review Figure 2-7.)

■ Serving as the intracellular digestive system, lysosomes destroy foreign materials such as bacteria that have been internalized by the cell and demolish worn-out cell parts to make way for new replacement parts.

■ Extracellular material is brought into the cell by endocytosis for attack by lysosomal enzymes. (Review Figure 2-5b.) The three forms of endocytosis are pinocytosis, receptor-mediated endocytosis, and phagocytosis. (Review Figure 2-8.)

Peroxisomes and Detoxification (p. 31)

■ Peroxisomes are small membrane-enclosed sacs containing powerful oxidative enzymes. (Review Figure 2-7.)

■ They carry out particular oxidative reactions that detoxify various wastes and toxic foreign compounds that have entered the cell. During these detoxification reactions, peroxisomes generate potent hydrogen peroxide, which they decompose into harmless water and oxygen by means of the catalase they contain.

Mitochondria and ATP Production (pp. 31–39)

■ The rod-shaped mitochondria are enclosed by two membranes, a smooth outer membrane and an inner membrane that forms a series of shelves, the cristae, which project into an interior gel-filled cavity, the matrix. *(Review Figure 2-9.)*

■ Mitochondria are the energy organelles of the cell. They efficiently convert the energy in food molecules to the usable energy stored in ATP molecules. Cells use ATP as an energy source for synthesis of new chemical compounds, for membrane transport, and for mechanical work.

■ *Cellular respiration* refers collectively to the intracellular reactions in which energy-rich molecules are broken down to form ATP, using O_2 and producing CO_2 in the process. Cellular respiration includes the sequential dismantling of nutrient molecules and subsequent ATP production in three stages: (1) glycolysis in the cytosol, (2) the citric acid cycle in the mitochondrial matrix, and (3) oxidative phosphorylation at the mitochondrial inner membrane. *(Review Figure 2-10.)*

■ *Oxidative phosphorylation* includes the electron transport system and chemiosmosis by ATP synthase. The electron transport system extracts high-energy electrons from hydrogens released during nutrient breakdown during glycolysis and the citric acid cycle and transfers them to successively lower energy levels. The free energy released during this process is used to create a H^+ gradient across the mitochondrial inner membrane. The flow of H^+ down this gradient activates ATP synthase, an enzyme that synthesizes ATP in a process called chemiosmosis. *(Review Figures 2-11 through 2-14.)*

■ A cell is more efficient at converting food energy into ATP when O_2 is available. Without O_2 (the anaerobic condition), a cell can produce only 2 molecules of ATP for every glucose molecule processed by glycolysis. With O_2 (the aerobic condition), the mitochondrial processes can yield another 30 molecules of ATP for every glucose molecule processed (2 from the citric acid cycle and 28 from oxidative phosphorylation). *(Review Figures 2-14 and 2-16.)*

Ribosomes and Protein Synthesis (pp. 39–40)

■ During protein synthesis, a large and a small ribosomal subunit unite to form a ribosome. *(Review Figure 2-17a.)*

■ Ribosomes translate mRNA into chains of amino acids assembled according to the DNA code carried by mRNA. Ribosomes have binding sites where tRNAs carrying specified amino acids link with mRNA during protein assembly. *(Review Figure 2-17b.)*

Vaults as Cellular Trucks (pp. 40–41)

■ Vaults are hollow, octagonal structures that are the same shape and size as the nuclear pores. *(Review Figure 2-18.)* They are believed to be cellular trucks that dock at the nuclear pores and pick up cargo for transport from the nucleus.

■ The leading proposals are that vaults may transport mRNA or the ribosomal subunits from the nucleus to the cytoplasmic sites of protein synthesis.

Centrosome, Centrioles, and Microtubule Organization (p. 41)

■ The centrosome (cell center) consists of a pair of centrioles surrounded by an amorphous mass. *(Review Figure 2-19.)*

■ The centrosome is the main microtubule organizing center of a cell. It forms and organizes the microtubule cytoskeleton, forms cilia and flagella, and forms the mitotic spindle.

Cytosol: Cell Gel (p. 42)

■ The cytosol contains the enzymes involved in intermediary metabolism and the ribosomal machinery essential for synthesis of these enzymes as well as other cytosolic proteins.

■ Many cells store unused nutrients within the cytosol in the form of glycogen granules or fat droplets. *(Review Figure 2-20.)*

■ Also present in the cytosol are various secretory, transport, and endocytic vesicles.

Cytoskeleton: Cell "Bone and Muscle" (pp. 42–49)

■ Extending throughout the cytosol is the cytoskeleton, which serves as the "bone and muscle" of the cell. *(Review Table 2-2, p. 44.)*

■ The three types of cytoskeletal elements—microtubules, microfilaments, and intermediate filaments—each consist of different proteins and perform various roles. *(Review Figure 2-21.)*

■ Microtubules, made of tubulin, maintain asymmetric cell shapes, serve as highways for intracellular transport by molecular motors, are the main component of cilia and flagella, and make up the mitotic spindle. *(Review Figures 2-22, 2-23, 2-24, and 2-25.)*

■ Microfilaments, made of actin in most cells, are important in various cellular contractile systems, including amoeboid movement and muscle contraction. They also serve as a mechanical stiffener for microvilli. *(Review Figures 2-26, 2-27, and 2-28.)*

■ Intermediate filaments are irregular threadlike proteins that help cells resist mechanical stress. They are most abundant in skin cells and nerve cells.

■ Collectively, the cytoskeletal elements give the cell shape and support, enable it to organize and move its internal structures as needed, and, in some cells, allow movement between the cell and its environment.

Membrane Structure and Functions (pp. 53–57)

■ All cells are bounded by a plasma membrane, a thin lipid bilayer interspersed with proteins and having carbohydrates attached on the outer surface.

■ The appearance of the plasma membrane in an electron microscope as a trilaminar structure (two dark lines separated by a light interspace) is produced by the arrangement of its molecules. The phospholipids orient themselves to form a bilayer with a hydrophobic interior (light interspace) sandwiched between hydrophilic outer and inner surfaces (dark lines). *(Review Figures 3-1, 3-2, and 3-3.)*

■ The lipid bilayer forms the structural boundary of the cell, serving as a barrier for water-soluble substances and being responsible for the fluid nature of the membrane. Cholesterol molecules tucked between the phospholipids contribute to the fluidity and stability of the membrane.

■ According to the fluid mosaic model of membrane structure, the lipid bilayer is embedded with proteins. *(Review Figure 3-3.)* Membrane proteins, which vary in type and distribution among cells, serve as (1) channels for passage of small ions across the membrane; (2) carriers for transport of specific substances in or out of the cell; (3) docking-marker acceptors for fusion with and subsequent exocytosis of secretory vesicles; (4) membrane-bound enzymes that govern specific chemical reactions; (5) receptors for detecting and responding to chemical messengers that alter cell function; and (6) cell adhesion molecules that help hold cells together and serve as a structural link between the extracellular surroundings and intracellular cytoskeleton.

■ Membrane carbohydrates on the outer surface of the cell serve as self-identity markers. *(Review Figure 3-3.)* They are important in recognition of "self" in cell-to-cell interactions such as tissue formation and tissue growth.

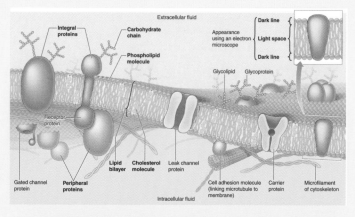

Cell-to-Cell Adhesions (pp. 57–60)

■ The extracellular matrix (ECM) serves as a biological "glue" between the cells of a tissue. The ECM consists of a watery, gel-like substance interspersed with three major types of protein fibers: collagen, elastin, and fibronectin.

■ Many cells are further joined by specialized cell junctions, of which there are three types: desmosomes, tight junctions, and gap junctions.

■ Desmosomes serve as adhering junctions to hold cells together mechanically and are especially important in tissues subject to a great deal of stretching. *(Review Figure 3-4.)*

■ Tight junctions actually fuse cells together, preventing the passage of materials between cells and thereby permitting only regulated passage of materials through the cells. These impermeable junctions are found in the epithelial sheets that separate compartments with very different chemical compositions. *(Review Figure 3-5.)*

■ Gap junctions are communicating junctions between two adjacent, but not touching, cells. They form small tunnels that permit exchange of ions and small molecules between the cells. Such movement of ions plays a key role in the spread of electrical activity to synchronize contraction in heart and smooth muscle. *(Review Figure 3-6.)*

Overview of Membrane Transport (p. 60)

■ Materials can pass between the ECF and ICF by unassisted and assisted means.

■ Transport mechanisms may also be passive (the particle moves across the membrane without the cell expending energy) or active (the cell expends energy to move the particle across the membrane). *(Review Table 3-2, p. 76.)*

Unassisted Membrane Transport (pp. 60–67)

■ Nonpolar (lipid-soluble) molecules of any size cross the membrane unassisted by dissolving in and passively moving through the lipid bilayer down concentration gradients. *(Review Figures 3-7 and 3-8.)* Small ions can traverse the membrane unassisted by passively moving down electrochemical gradients through open protein channels specific for the ion. *(Review Figure 3-3.)*

■ In osmosis, water moves passively down its own concentration gradient across a selectively permeable membrane to an area of higher concentration of nonpenetrating solutes. Penetrating solutes do not have an osmotic effect. *(Review Figures 3-9 through 3-12.)*

■ The *osmolarity* of a solution is a measure of its total number of solute particles, both penetrating and nonpenetrating, both molecules and ions, per liter. The *osmotic pressure* of a solution is the pressure that must be applied to the solution to completely stop osmosis. The *tonicity* of a solution refers to the effect the solution has on cell volume and depends on the solution's relative concentration of nonpenetrating solutes compared to the concentration of nonpenetrating solutes in the cell it surrounds. *(Review Figure 3-13.)*

Assisted Membrane Transport (pp. 67–75)

■ In carrier-mediated transport, small polar molecules and selected ions are transported across the membrane by specific membrane carrier proteins. Carriers open to one side of the membrane where a passenger binds to a binding site specific for it, then change shape so that the binding site is exposed to the opposite side of the membrane where the passenger is released. Carrier-mediated transport may be passive and move the particle down its concentration gradient (*facilitated diffusion*) (*review Figure 3-14*), or active and move the particle against its concentration gradient (*active transport*). Carriers exhibit a transport maximum (T_m) when saturated. (*Review Figure 3-15.*)

■ There are two forms of active transport: primary active transport and secondary active transport. *Primary active transport* requires the direct use of ATP to drive the pump. (*Review Figure 3-16.*) One of the most important examples of primary active transport is the Na^+–K^+ pump, which concentrates Na^+ in the ECF and K^+ in the ICF. (*Review Figure 3-17.*) *Secondary active transport* is driven by an ion concentration gradient established by a primary active-transport system. There are two types of secondary active transport: symport (or cotransport) and antiport (or countertransport or exchange). In *symport*, the cotransported solute is moved uphill in the same direction as the driving ion moves downhill. In *antiport*, the coupled solute is moved uphill in the opposite direction as the driving ion moves downhill. (*Review Figures 3-18 and 3-19.*)

■ Large polar molecules and multimolecular particles can leave or enter the cell by being wrapped in a piece of membrane to form vesicles that can be internalized (endocytosis) or externalized (exocytosis). (*Review Figures 2-5, 2-6, and 2-8.*)

■ Cells are differentially selective in what enters or leaves because they possess varying numbers and kinds of channels, carriers, and mechanisms for vesicular transport.

■ Large polar molecules (too large for channels and not lipid soluble) for which there are no special transport mechanisms are unable to cross the membrane.

Membrane Potential (pp. 75–82)

■ All cells have a membrane potential, which is a separation of opposite charges across the plasma membrane. (*Review Figure 3-20.*)

Resting membrane potential = −70 mV

■ The Na^+–K^+ pump makes a small direct contribution to membrane potential because it transports more Na^+ ions out than K^+ ions in. (*Review Figure 3-17.*) However, the primary role of the Na^+–K^+ pump is to actively maintain a greater concentration of Na^+ outside the cell and a greater concentration of K^+ inside the cell. These concentration gradients tend to passively move K^+ out of the cell and Na^+ into the cell. (*Review Table 3-3 and Figures 3-21 and 3-22.*)

■ Because the resting membrane is much more permeable to K^+ than to Na^+, substantially more K^+ leaves the cell than Na^+ enters, resulting in an excess of positive charges outside the cell. This leaves an unbalanced excess of negative charges, in the form of large protein anions (A^-), trapped inside the cell. (*Review Table 3-3 and Figure 3-23.*)

■ When the resting membrane potential of −70 mV is achieved, no further net movement of K^+ and Na^+ takes place, because any further leaking of these ions down their concentration gradients is quickly reversed by the Na^+–K^+ pump.

■ The distribution of Cl^- across the membrane is passively driven by the established membrane potential so that Cl^- is concentrated in the ECF.

Introduction to Neural Communication (pp. 87–88)

■ Nerve and muscle cells are excitable tissues because they can rapidly alter their membrane permeabilities and undergo transient membrane potential changes when excited. These rapid changes in potential serve as electrical signals.

■ Compared to resting potential, a membrane becomes depolarized when the magnitude of its negative potential is reduced (becomes less negative) and hyperpolarized when the magnitude of its negative potential is increased (becomes more negative). (*Review Figure 4-1.*)

■ Changes in potential are brought about by triggering events that alter membrane permeability, in turn leading to changes in ion movement across the membrane.

■ The two kinds of potential change are: (1) graded potentials, the short-distance signals, and (2) action potentials, the long-distance signals. (*Review Table 4-1, p. 98.*)

Graded Potentials (pp. 88–91)

■ A graded potential, usually a depolarization, occurs in a small, specialized region of an excitable cell membrane. The site undergoing a potential change is designated an active area. (*Review Figure 4-2.*)

■ The magnitude of a graded potential varies directly with the magnitude of the triggering event.

■ Graded potentials spread decrementally by local current flow between the active area and adjacent inactive areas and die out over a short distance. (*Review Figures 4-2 and 4-3.*)

Action Potentials (pp. 91–104)

■ During an action potential, depolarization of the membrane to threshold potential triggers sequential changes in permeability caused by conformational changes in voltage-gated Na^+ and K^+ channels. (*Review Figures 4-4 through 4-7.*)

■ These permeability changes bring about a brief reversal of membrane potential, with Na^+ influx causing the rising phase (from -70 to $+30$ mV), followed by K^+ efflux causing the falling phase (from peak back to resting). (*Review Figure 4-7.*)

■ Before an action potential returns to resting, it regenerates an identical new action potential in the area next to it by means of current flow that brings the previously inactive area to threshold. This self-perpetuating cycle continues until the action potential has spread throughout the cell membrane in undiminished fashion.

■ There are two types of action potential propagation: (1) contiguous conduction in unmyelinated fibers, in which the action potential spreads along every portion of the membrane; and (2) the more rapid, saltatory conduction in myelinated fibers, in which the impulse jumps from one node of Ranvier to the next over sections of the fiber covered with insulating myelin. (*Review Figures 4-9, 4-12, and 4-13.*)

■ The Na^+–K^+ pump gradually restores the ions that moved during propagation of the action potential to their original location, to maintain the concentration gradients.

■ It is impossible to restimulate the portion of the membrane where the impulse has just passed until it has recovered from its refractory period, ensuring the one-way propagation of action potentials. (*Review Figures 4-10 and 4-11.*)

■ Action potentials occur either maximally in response to stimulation or not at all (all or none law).

■ Variable strengths of stimuli are coded by varying the frequency of action potentials, not their magnitude, in an activated nerve fiber.

Synapses and Neuronal Integration (pp. 104–113)

■ One neuron directly interacts with another neuron primarily through a chemical synapse. (*Review Figures 4-14 and 4-15.*)

■ Most neurons have four different functional parts: (*Review Figure 4-8.*)
1. The dendrite/cell body region (the input zone) serves as the postsynaptic component that binds with and responds to neurotransmitters released from other neurons.
2. The axon hillock (the trigger zone) is where action potentials are initiated because it has the lowest threshold and thus reaches threshold first in response to an excitatory graded potential change.
3. The axon, or nerve fiber (the conducting zone), conducts action potentials in undiminished fashion from the axon hillock to the axon terminals.
4. The axon terminal (the output zone) serves as the presynaptic component, releasing a neurotransmitter that influences other postsynaptic cells in response to action potential propagation down the axon.

■ Released neurotransmitter combines with receptor-channels on the postsynaptic neuron. (*Review Figure 4-15.*) (1) If nonspecific cation channels that permit passage of both Na^+ and K^+ are opened, the resultant ionic fluxes cause an EPSP, a small depolarization that brings the postsynaptic cell closer to threshold. (2) If either K^+ or Cl^- channels are opened, the likelihood that the postsynaptic neuron will reach threshold is diminished when an IPSP, a small hyperpolarization, is produced. (*Review Figure 4-16.*)

■ If the dominant activity is in its excitatory inputs, the postsynaptic cell is likely to be brought to threshold and have an action potential. This can be accomplished by either (1) temporal summation (EPSPs from a single, repetitively firing, presynaptic input occurring so close together in time that they add together) or (2) spatial summation (adding of EPSPs occurring simultaneously from several different presynaptic inputs). (*Review Figure 4-17.*) If inhibitory inputs dominate, the postsynaptic potential is brought farther than usual from threshold. If excitatory and inhibitory activity to the postsynaptic neuron is balanced, the membrane remains close to resting.

■ Even though there are a number of different neurotransmitters, each synapse always releases the same neurotransmitter to produce a given response when combined with a particular receptor. (Review Table 4-2.)

■ Synaptic pathways between neurons are incredibly complex, due to convergence of neuronal input and divergence of its output. Usually, many presynaptic inputs converge on a single neuron and jointly control its level of excitability. This same neuron, in turn, diverges to synapse with and influence the excitability of many other cells. (Review Figure 4-19.)

■ Numerous factors may alter synaptic effectiveness: Some are built-in mechanisms to fine-tune neural responsiveness, for example presynaptic inhibition (Review Figure 4-18); some are pharmacologic manipulations to achieve a desired result; and some are caused by poisons or disease processes.

Intercellular Communication and Signal Transduction (pp. 113–117)

■ Intercellular communication is accomplished directly via (1) gap junctions or (2) transient direct linkup of cells'complementary surface markers. (Review Figure 4-20.)

■ More commonly cells communicate indirectly with one another to carry out various coordinated activities by dispatching extracellular chemical messengers, which act on particular target cells to bring about the desired response. There are four types of extracellular chemical messengers, which differ in their source and in the distance and means by which they get to their site of action: (1) paracrines (local chemical messengers); (2) neurotransmitters (very short-range chemical messengers released by neurons); (3) hormones (long-range chemical messengers secreted into the blood by endocrine glands); and (4) neurohormones (long-range chemical messengers secreted into the blood by neurosecretory neurons). (Review Figure 4-20.)

■ Transfer of the signal carried by the extracellular messenger into the cell for execution is known as signal transduction.

■ Attachment of an extracellular chemical messenger that cannot gain entry to the cell, such as a protein hormone (the first messenger), to a membrane triggers cellular responses by: (1) opening receptor-channels; (2) activating receptor-enzymes, such as tyrosine kinase; or (3) activating an intracellular second messenger via G-protein-coupled receptors. (Review Figures 4-21 and 4-22.)

Introduction to Hormonal Communication (pp. 117–129)

■ Hormones are long-distance chemical messengers secreted by the endocrine glands into the blood, which transports them to specific target sites where they control a particular function by altering protein activity within the target cells.

■ Hormones are grouped into two categories based on their solubility differences: (1) hydrophilic (water-soluble) hormones, which include peptides (the majority of hormones) and catecholamines (secreted by the adrenal medulla); and (2) lipophilic (lipid-soluble) hormones, which include steroid hormones (the sex hormones and those secreted by the adrenal cortex) and thyroid hormone. (Review Table 4-4.)

■ Hydrophilic peptide hormones are synthesized and packaged for export by the endoplasmic reticulum/Golgi complex, stored in secretory vesicles, and released by exocytosis on appropriate stimulation. They dissolve freely in the blood for transport to their target cells.

■ At their target cells, hydrophilic hormones bind with surface membrane receptors, triggering a chain of intracellular events by means of a second-messenger pathway that ultimately alters preexisting cell proteins, usually enzymes, leading to the target cell's response to the hormone. (Review Figures 4-24 and 4-25.) Through this cascade of reactions, the initial signal is greatly amplified. (Review Figure 4-26).

■ Steroids are synthesized by modifications of stored cholesterol through enzymes specific for each steroidogenic tissue. Steroids are not stored in the endocrine cells. Being lipophilic, they diffuse out through the lipid membrane barrier as soon as they are synthesized. Control of steroids is directed at their synthesis.

■ Lipophilic steroids and thyroid hormone are both transported in the blood largely bound to carrier plasma proteins, with only free, unbound hormone being biologically active.

■ Lipophilic hormones readily cross the lipid membrane barriers of their target cells and bind with receptors inside the cell. Once the hormone binds with the receptor, the hormone receptor complex binds with DNA and activates a gene, which leads to the synthesis of new enzymatic or structural intracellular proteins that carry out the hormone's effect on the target cell. (Review Figure 4-27.)

Comparison of the Nervous and Endocrine Systems (pp. 126–129)

■ The nervous and endocrine systems are the two main regulatory systems of the body. (Review Table 4-5.) The nervous system is anatomically "wired" to its target organs, whereas the "wireless" endocrine system secretes blood-borne hormones that reach distant target organs.

■ Specificity of neural action depends on the anatomic proximity of the neurotransmitter-releasing neuronal terminal to its target organ. Specificity of endocrine action depends on specialization of target cell receptors for a specific circulating hormone.

■ In general, the nervous system coordinates rapid responses, whereas the endocrine system regulates activities that require duration rather than speed.

Organization and Cells of the Nervous System (pp. 133–139)

■ The nervous system consists of the central nervous system (CNS), which includes the brain and spinal cord, and the peripheral nervous system, which includes the nerve fibers carrying information to (afferent division) and from (efferent division) the CNS. (Review Figure 5-1.)

■ Three functional classes of neurons—afferent neurons, efferent neurons, and interneurons—compose the excitable cells of the nervous system. (Review Figure 5-2.) (1) Afferent neurons inform the CNS about conditions in both the external and internal environment. (2) Efferent neurons carry instructions from the CNS to effector organs, namely, muscles and glands. (3) Interneurons are responsible for integrating afferent information and formulating an efferent response, as well as for all higher mental functions associated with the "mind."

■ Glial cells form the connective tissue within the CNS and physically, metabolically, and functionally support the neurons. The four types of glial cells are astrocytes, oligodendrocytes, microglia, and ependymal cells. (Review Figures 5-3 and 5-4 and Table 5-1.)

Protection and Nourishment of the Brain (pp. 139–141)

■ The brain is provided with several protective devices, which is important because neurons cannot divide to replace damaged cells. (1) The brain is wrapped in three layers of protective membranes—the meninges—and is further surrounded by a hard, bony covering. (2) Cerebrospinal fluid flows within and around the brain to cushion it against physical jarring. (Review Figure 5-6.) (3) Protection against chemical injury is conferred by a blood–brain barrier that limits access of blood-borne substances to the brain.

■ The brain depends on a constant blood supply for delivery of O_2 and glucose because it cannot generate ATP in the absence of either of these substances.

Overview of the Central Nervous System (pp. 141–143)

■ The parts of the brain from the lowest, most primitive level to the highest, most sophisticated level are the brain stem, cerebellum, hypothalamus, thalamus, basal nuclei, and cerebral cortex. (Review Table 5-2 and Figure 5-7.)

Cerebral Cortex (pp. 143–153)

■ The cerebral cortex is the outer shell of gray matter that caps an underlying core of white matter. The cortex itself consists primarily of neuronal cell bodies, dendrites, and glial cells. The white matter consists of bundles of nerve fibers that interconnect various areas. (Review Figure 5-14.)

■ Ultimate responsibility for many discrete functions is localized in particular regions of the cortex as follows: (1) the occipital lobes house the visual cortex; (2) the auditory cortex is in the temporal lobes; (3) the parietal lobes are responsible for reception and perceptual processing of somatosensory (somesthetic and proprioceptive) input; and (4) voluntary motor movement is set into motion by the motor areas in the frontal lobes. (Review Figures 5-8 through 5-10.)

■ Language ability depends on the integrated activity of two primary language areas—Broca's area and Wernicke's area—typically located only in the left cerebral hemisphere. (Review Figures 5-9 and 5-11.)

■ The association areas are regions of the cortex not specifically assigned to processing sensory input or commanding motor output or language ability. These areas provide an integrative link between diverse sensory information and purposeful action; they also play a key role in higher brain functions such as memory and decision making. The association areas include the prefrontal association cortex, the parietal-temporal-occipital association cortex, and the limbic association cortex. (Review Figures 5-9 and 5-12.)

Basal Nuclei, Thalamus, and Hypothalamus (pp. 153–155)

■ The subcortical brain structures include the basal nuclei, thalamus, and hypothalamus. (Review Figures 5-14 and 5-15 and Table 5-2.)

■ The basal nuclei inhibit muscle tone; coordinate slow, sustained postural contractions; and suppress useless patterns of movement.

■ The thalamus serves as a relay station for preliminary processing of sensory input. It also accomplishes a crude awareness of sensation and some degree of consciousness.

■ The hypothalamus regulates body temperature, thirst, urine output, and food intake; extensively controls the autonomic nervous system and endocrine system; and is part of the limbic system.

Emotion, Behavior, and Motivation (pp. 155–157)

■ The limbic system, which includes portions of the hypothalamus and other structures that encircle the brain stem, plays an important role in emotion, basic behavioral patterns, motivation, and learning. (Review Figure 5-16.)

■ Emotion refers to subjective feelings and moods and the physical responses associated with these feelings.

■ Basic behavioral patterns triggered by the limbic system are aimed at survival (such as attack) and perpetuation of the species (such as mating behavior). Higher cortical centers can reinforce, modify, or suppress these basic behaviors.

■ *Motivation* is the ability to direct behavior toward specific goals.

■ Norepinephrine, dopamine, and serotonin are the key neurotransmitters in pathways for emotions and behavior.

Learning and Memory (pp. 157–165)

■ *Learning* refers to acquiring knowledge or skills as a result of experience, instruction, or both. *Memory* is storage of acquired knowledge for later recall and use.

■ There are two types of memory: (1) a short-term memory with limited capacity and brief retention, coded by modification of activity at preexisting synapses; and (2) a long-term memory with large storage capacity and enduring retention, involving relatively permanent structural or functional changes, such as the formation of new synapses between existing neurons. Enhanced protein synthesis underlies these long-term changes. *(Review Table 5-3 and Figure 5-17.)*

■ *Consolidation* is the transfer of short-term memory to long-term memory. Long-term potentiation, a prolonged increase in the strength of existing synaptic connections in activated pathways, might be the link between short-term memory and consolidation of long-term memory. *(Review Figure 5-18.)*

■ The hippocampus and associated structures are especially important in declarative, or "what," memories of specific objects, facts, and events. The cerebellum and associated structures are especially important in procedural, or "how to," memories of motor skills gained through repetitive training.

■ The prefrontal association cortex is the site of working memory, which temporarily holds currently relevant data—both new information and knowledge retrieved from memory stores—and manipulates and relates them to accomplish the higher-reasoning processes of the brain.

Cerebellum (pp. 166–167)

■ The cerebellum, attached at the back of the brain stem beneath the cortex, consists of three functionally distinct parts. *(Review Figure 5-19.)*

■ The *vestibulocerebellum* helps maintain balance and controls eye movements. The *spinocerebellum* enhances muscle tone and helps coordinate voluntary movement, especially fast, phasic motor activities. The *cerebrocerebellum* plays a role in initiating voluntary movement and in storing procedural memories.

Brain Stem (pp. 167–172)

■ The brain stem is an important link between the spinal cord and higher brain levels.

■ The brain stem is the origin of the cranial nerves. *(Review Figure 5-20.)* It also contains centers that control cardiovascular, respiratory, and digestive function; regulates postural muscle reflexes; controls the overall degree of cortical alertness; and plays a key role in the sleep–wake cycle.

■ *Consciousness* is the subjective awareness of the external world and self. The states of consciousness in decreasing order of arousal level are (1) maximum alertness, (2) wakefulness, (3) several types of sleep, and (4) coma.

■ The prevailing state of consciousness depends on the cyclical interplay among an arousal system (the reticular activating system) originating in the brain stem and commanded by hypocretin-secreting neurons in the hypothalamus, (2) a slow-wave sleep center consisting of sleep-on neurons in the hypothalamus, and (3) an REM sleep center consisting of REM sleep-on neurons in the brain stem. *(Review Figure 5-21.)*

■ Sleep is an active process, not just the absence of wakefulness. While sleeping, a person cyclically alternates between slow-wave sleep and paradoxical (REM) sleep. *(Review Figure 5-23 and Table 5-4.)* *Slow-wave sleep* is characterized by slow waves on the EEG and little change in behavior pattern from the waking state except for not being consciously aware of the external world. *Paradoxical,* or *REM, sleep* is characterized by an EEG pattern similar to that of an alert, awake individual; rapid eye movements, dreaming, and abrupt changes in behavior pattern occur. *(Review Figure 5-22.)*

■ The leading theories of why we need sleep fall into the categories of (1) restoration and recovery and (2) memory consolidation.

Spinal Cord (pp. 172–179)

■ Extending from the brain stem, the spinal cord descends through a canal formed by surrounding protective vertebrae. *(Review Figures 5-24 and 5-25.)*

■ The spinal cord has two functions. (1) It serves as the neuronal link between the brain and the peripheral nervous system. All communication up and down the spinal cord is located in ascending and descending tracts in the cord's outer white matter. *(Review Figures 5-27 and 5-28.)* (2) It is the integrating center for spinal reflexes, including some basic protective and postural reflexes and those involved with the emptying of the pelvic organs. *(Review Figures 5-31 and 5-32.)*

■ The basic reflex arc includes a receptor, an afferent pathway, an integrating center, an efferent pathway, and an effector. *(Review Figure 5-31.)*

■ The centrally located gray matter of the spinal cord contains the interneurons interposed between the afferent input and efferent output as well as the cell bodies of efferent neurons. *(Review Figures 5-26 and 5-29.)*

■ A *nerve* is a bundle of peripheral neuronal axons, both afferent and efferent, wrapped in connective tissue and following the same pathway. *(Review Figure 5-30.)* Spinal nerves supply specific body regions and are attached to the spinal cord in paired fashion throughout its length. *(Review Figures 5-24, 5-25, and 5-26.)*

■ The 31 pairs of spinal nerves along with the 12 pairs of cranial nerves that arise from the brain stem constitute the peripheral nervous system. *(Review Figures 5-21 and 5-25.)*

Receptor Physiology (pp. 183–191)

■ The afferent division of the PNS carries information about the internal and external environment to the CNS.

■ Sensory receptors are specialized peripheral endings of afferent neurons. *(Review Figure 6-1.)* Each type of receptor (photoreceptor, mechanoreceptor, thermoreceptor, osmoreceptor, chemoreceptor, or nociceptor) responds to its adequate stimulus (a change in the energy form, or modality, to which it is responsive), translating the energy form of the stimulus into electrical signals.

■ A stimulus typically brings about a graded, depolarizing receptor potential by opening nonspecific cation channels, which leads to net Na^+ entry. Receptor potentials, if of sufficient magnitude, ultimately generate action potentials in the afferent fiber next to the receptor. These action potentials self-propagate along the afferent fiber to the CNS. *(Review Figures 6-1 and 6-2.)* The strength of the stimulus determines the magnitude of the receptor potential, which in turn determines the frequency of action potentials generated. *(Review Figure 6-3 and Table 6-1.)*

■ The magnitude of the receptor potential is also influenced by the extent of receptor adaptation, which is a reduction in receptor potential despite sustained stimulation. (1) Tonic receptors adapt slowly or not at all and thus provide continuous information about the stimuli they monitor. (2) Phasic receptors adapt rapidly and frequently exhibit off responses, thereby providing information about changes in the energy form they monitor. *(Review Figure 6-4.)*

■ Visceral afferent information remains mostly subconscious. Sensory afferent information reaches the level of conscious awareness, including (1) somatic sensation (somesthetic sensation and proprioception) and (2) special senses.

■ Discrete labeled-line pathways lead from the receptors to the CNS so that information about the type and location of stimuli can be deciphered by the CNS. *(Review Table 6-1.)*

■ The term *receptive field* refers to the area surrounding a receptor within which the receptor can detect stimuli. The acuity, or discriminative ability, of a body region varies inversely with the size of its receptive fields and also depends on the extent of lateral inhibition in the afferent pathways arising from receptors in the region. *(Review Figures 6-6 and 6-7.)*

■ Perception is the conscious interpretation of the external world that the brain creates from sensory input. What the brain perceives from its input is an abstraction and not reality. *(Review Figures 6-8 and 6-9.)* The only stimuli that can be detected are those for which receptors are present. Furthermore, as sensory signals ascend through progressively more complex processing, some of the information may be suppressed, whereas other parts of it may be enhanced.

Pain (pp. 191–194)

■ Painful experiences are elicited by noxious mechanical, thermal, or chemical stimuli and consist of two components: the perception of pain coupled with emotional and behavioral responses to it.

■ The three categories of pain receptors are mechanical, thermal, and polymodal nociceptors. The latter respond to all kinds of damaging stimuli, including chemicals released from injured tissues.

■ Pain signals are transmitted over two afferent pathways: a fast pathway that carries sharp, prickling pain signals; and a slow pathway that carries dull, aching, persistent pain signals. *(Review Table 6-2.)*

■ Afferent pain fibers terminate in the spinal cord on ascending pathways that transmit the signal to the brain for processing. Descending pathways from the brain use endogenous opiates to suppress the release of substance P, a pain-signaling neurotransmitter from the afferent pain-fiber terminal. Thus, these descending pathways block further transmission of the pain signal and serve as a built-in analgesic system. *(Review Figure 6-10.)*

Eye: Vision (pp. 195–213)

■ Light is a form of electromagnetic radiation, with visible light being only a small band in the total electromagnetic spectrum. *(Review Figures 6-14 and 6-15.)*

■ The eye houses the light-sensitive photoreceptors essential for vision—the rods and cones found in its retinal layer. *(Review Table 6-4, and Figures 6-11, 6-23, and 6-26.)*

■ The iris controls the size of the pupil to adjust the amount of light permitted to enter the eye. *(Review Figure 6-13.)*

■ The cornea and lens are the primary refractive structures that bend incoming light rays to focus the image on the retina. The cornea contributes most to the total refractive ability of the eye. The strength of the lens can be adjusted through action of the ciliary muscle to accommodate for differences in near and far vision. *(Review Figures 6-16 through 6-22.)*

■ Rods and cones have three parts: a photopigment-containing outer segment, a metabolically specialized inner segment, and a neurotransmitter-secreting synaptic terminal. *(Review Figures 6-23, 6-26, and 6-27.)*

■ Rods and cones secrete neurotransmitter in the dark. They are activated when their photopigments differentially absorb various wavelengths of light. Photopigments consist of opsin, a membrane protein, and retinal, a vitamin A derivative. During phototransduction, light absorption by retinal causes a biochemical change in the photopigment that, through a series of steps, hyperpolarizes the photoreceptor, leading to decreased neurotransmitter release. Further retinal processing by on-center and off-center bipolar and ganglion cells eventually converts this light-induced signal into a change in the rate of action potential propagation in the visual pathway leaving the retina. (*Review Figures 6-26, 6-27, and 6-28.*)

■ Cones display high acuity but can be used only for day vision because of their low sensitivity to light. Different ratios of stimulation of three cone types by varying wavelengths of light lead to color vision. (*Review Figure 6-29 and Table 6-3.*)

■ Rods provide only indistinct vision in shades of gray, but because they are very sensitive to light, they can be used for night vision. (*Review Table 6-3.*)

■ The visual message is transmitted via a complex crossed and uncrossed pathway to the visual cortex in the occipital lobe of the brain for perceptual processing. (*Review Figure 6-31.*)

Ear: Hearing and Equilibrium (pp. 213–228)

■ The ear performs two unrelated functions: (1) hearing, which involves the external ear, middle ear, and cochlea of the inner ear; and (2) sense of equilibrium, which involves the vestibular apparatus of the inner ear. The ear receptor cells located in the inner ear—the hair cells in the cochlea and vestibular apparatus—are mechanoreceptors. (*Review Table 6-6 and Figure 6-32.*)

■ Hearing depends on the ear's ability to convert airborne sound waves into mechanical deformations of auditory hair cells, thereby initiating neural signals. Sound waves consist of high-pressure regions of compression alternating with low-pressure regions of rarefaction of air molecules. The pitch (tone) of a sound is determined by the frequency of its waves, the loudness (intensity) by the amplitude of the waves, and the timbre (quality) by its characteristic overtones. (*Review Figures 6-33 and 6-34 and Table 6-5.*)

■ Sound waves are funneled through the ear canal to the tympanic membrane, which vibrates in synchrony with the waves. Middle ear bones bridging the gap between the tympanic membrane and the inner ear amplify the tympanic movements and transmit them to the oval window, whose movement sets up traveling waves in the cochlear fluid. (*Review Figures 6-35 and 6-36.*)

■ These waves, which are at the same frequency as the original sound waves, set the basilar membrane in motion. Various regions of this membrane selectively vibrate more vigorously in response to different frequencies of sound. The narrow, stiff end of the basilar membrane near the oval window vibrates best with high-frequency pitches, and the wide, flexible end near the helicotrema vibrates best with low-frequency pitches. (*Review Figure 6-36.*)

■ On top of the basilar membrane are the receptive inner hair cells of the organ of Corti, whose stereocilia ("hairs") are bent as the basilar membrane is deflected up and down in relation to the overhanging stationary tectorial membrane, which the hairs contact. (*Review Figures 6-35, 6-37, and 6-38.*)

■ Pitch discrimination depends on which region of the basilar membrane naturally vibrates maximally with a given frequency. Loudness discrimination depends on the amplitude of the vibrations. Hair bending in the region of maximal basilar membrane vibration is transduced into neural signals that are transmitted to the auditory cortex in the temporal lobe of the brain for sound perception. (*Review Figure 6-39.*)

■ The vestibular apparatus in the inner ear consists of (1) the semicircular canals, which detect rotational acceleration or deceleration in any direction; and (2) the utricle and the saccule, which collectively detect changes in the rate of linear movement in any direction and provide information important for determining head position in relation to gravity. Neural signals are generated in response to the mechanical deformation of vestibular hair cells by specific movement of fluid and related structures within these vestibular sense organs. (*Review Figures 6-41 and 6-42.*)

■ Vestibular input goes to the vestibular nuclei in the brain stem and to the cerebellum for use in maintaining balance and posture, controlling eye movement, and perceiving motion and orientation. (*Review Figure 6-43.*)

Chemical Senses: Taste and Smell (pp. 227–232)

■ Taste and smell are chemical senses. In both cases, attachment of specific dissolved molecules to binding sites on the receptor membrane causes receptor potentials that, in turn, set up neural impulses signaling the presence of the chemical.

■ Taste receptors are housed in taste buds on the tongue; olfactory receptors are located in the olfactory mucosa in the upper part of the nasal cavity. (*Review Figures 6-44 and 6-45.*)

■ Both sensory pathways include two routes: one to the limbic system for emotional and behavioral processing and one to the cortex for conscious perception and fine discrimination.

■ Taste and olfactory receptors are continuously renewed, unlike visual and hearing receptors, which are irreplaceable.

■ The five primary tastes are salty, sour, sweet, bitter, and umami (a meaty, "amino-acid" taste). Taste discrimination beyond the primary tastes depends on the patterns of stimulation of the taste buds, each of which responds in varying degrees to the different primary tastes. Salty and sour tastants bring about receptor potentials in taste buds by directly affecting membrane channels, whereas the other three categories of tastants act through second-messenger pathways to bring about receptor potentials.

■ There are 1000 different types of olfactory receptors, each of which responds to only one discrete component of an odor, an odorant. Odorants act through second-messenger pathways to trigger receptor potentials. The afferent signals arising from the olfactory receptors are sorted according to scent component by the glomeruli within the olfactory bulb. Odor discrimination depends on the patterns of activation of the glomeruli. (*Review Figure 6-46.*)

Autonomic Nervous System (pp. 237–245)

■ The CNS controls muscles and glands by transmitting signals to these effector organs through the efferent division of the PNS. *(Review Table 7-1.)*

■ There are two types of efferent output: the autonomic nervous system, which is under involuntary control and supplies cardiac and smooth muscle as well as most exocrine and some endocrine glands; and the somatic nervous system, which is subject to voluntary control and supplies skeletal muscle. *(Review Table 7-6, p. 247, and Table 7-7, p. 248.)*

■ The autonomic nervous system consists of two subdivisions—the sympathetic and parasympathetic nervous systems. *(Review Figures 7-2 and 7-3 and Tables 7-3 and 7-5.)*

■ An autonomic nerve pathway consists of a two-neuron chain. The preganglionic fiber originates in the CNS and synapses with the cell body of the postganglionic fiber in a ganglion outside the CNS. The postganglionic fiber ends on the effector organ. *(Review Figures 7-1, 7-2, and 7-3 and Table 7-5.)*

KEY
- Parasympathetic preganglionic fiber
- Parasympathetic postganglionic fiber
- Sympathetic preganglionic fiber
- Sympathetic postganglionic fiber
- ACh ● Acetylcholine
- NE ■ Norepinephrine
- E ▲ Epinephrine

■ All preganglionic fibers and parasympathetic postganglionic fibers release acetylcholine (ACh). Sympathetic postganglionic fibers release norepinephrine (NE). *(Review Figure 7-2 and Tables 7-2 and 7-4.)*

■ Postganglionic fibers have numerous swellings, or varicosities, that simultaneously release neurotransmitter over a large area of the innervated organ. *(Review Figures 7-1 and 8-32, p. 297.)*

■ The adrenal medulla, an endocrine gland, is a modified sympathetic ganglion that secretes the hormones epinephrine and to a lesser extent norepinephrine into the blood in response to stimulation by the sympathetic preganglionic fiber that innervates it. *(Review Figure 7-2.)*

■ The same neurotransmitter elicits different responses in different tissues. Thus, the response depends on specialization of the tissue cells, not on the properties of the messenger. *(Review Table 7-4.)*

■ Tissues innervated by the autonomic nervous system possess one or more of several different receptor types for the postganglionic chemical messengers (and for the related adrenomedullary hormone epinephrine). Cholinergic receptors include nicotinic and muscarinic receptors; adrenergic receptors include α_1, α_2, β_1, and β_2 receptors. *(Review Figure 7-2 and Tables 7-3, 7-4, and 7-5.)*

■ A given autonomic fiber either excites or inhibits activity in the organ it innervates. *(Review Tables 7-3 and 7-4.)*

■ Most visceral organs are innervated by both sympathetic and parasympathetic fibers, which in general produce opposite effects in a particular organ. Dual innervation of organs by both branches of the autonomic nervous system permits precise control over an organ's activity. *(Review Figure 7-3 and Table 7-3.)*

■ The sympathetic system dominates in emergency or stressful ("fight-or-flight") situations and promotes responses that prepare the body for strenuous physical activity. The parasympathetic system dominates in quiet, relaxed ("rest-and-digest") situations and promotes body-maintenance activities such as digestion. *(Review Tables 7-3 and 7-5.)*

■ Visceral afferent input is used by the CNS to direct appropriate autonomic output to maintain homeostasis. Autonomic activities are controlled by multiple areas of the CNS, including the spinal cord, medulla, hypothalamus, and prefrontal association cortex.

Somatic Nervous System (p. 246)

■ The somatic nervous system consists of the axons of motor neurons, which originate in the spinal cord or brain stem and end on skeletal muscle. *(Review Figure 7-4 and Table 7-6.)*

■ ACh, the neurotransmitter released from a motor neuron, stimulates muscle contraction.

■ Motor neurons are the final common pathway by which various regions of the CNS exert control over skeletal muscle activity. The areas of the CNS that influence skeletal muscle activity by acting through the motor neurons are the spinal cord, motor regions of the cortex, basal nuclei, cerebellum, and brain stem.

Neuromuscular Junction (pp. 246–253)

■ When a motor neuron reaches a muscle, it branches into axon terminals. Each axon terminal forms a neuromuscular junction with a single muscle cell (fiber). The axon terminal splits into multiple fine branches, each of which ends in an enlarged terminal button. *(Review Figure 7-4 and Table 7-8.)*

■ The specialized region of the muscle cell membrane underlying the axon-terminal complex is called the motor end plate. Because these structures do not make direct contact, signals are passed between a terminal button and muscle fiber by chemical means. *(Review Figure 7-5.)*

■ An action potential in the axon terminal causes the release of ACh from its storage vesicles in the terminal button. The released ACh diffuses across the space separating the nerve and muscle cells and binds to special receptor-channels on the underlying motor end plate. This binding triggers the opening of these nonspecific cation channels. The subsequent ion movements depolarize the motor end plate, producing the end-plate potential (EPP). *(Review Figure 7-5.)*

■ Local current flow between the depolarized end plate and adjacent muscle cell membrane brings these adjacent areas to threshold, initiating an action potential that is propagated throughout the muscle fiber. This muscle action potential triggers muscle contraction. *(Review Figure 7-5.)*

■ Membrane-bound acetylcholinesterase in the motor end plate inactivates ACh, ending the EPP and, subsequently, the action potential and resultant contraction. *(Review Figure 7-5.)*

1 An action potential in a motor neuron is propagated to the axon terminal (terminal button).

2 This local action potential triggers the opening of voltage-gated Ca^{2+} channels and the subsequent entry of Ca^{2+} into the terminal button.

3 Ca^{2+} triggers the release of acetylcholine (ACh) by exocytosis from a portion of the vesicles.

4 ACh diffuses across the space separating the nerve and muscle cells and binds with receptor-channels specific for it on the motor end plate of the muscle cell membrane.

5 This binding brings about the opening of these nonspecific cation channels, leading to a relatively large movement of Na^+ into the muscle cell compared to a smaller movement of K^+ outward.

6 The result is an end-plate potential. Local current flow occurs between the depolarized end plate and the adjacent membrane.

7 This local current flow opens voltage-gated Na^+ channels in the adjacent membrane.

8 The resultant Na^+ entry reduces the potential to threshold, initiating an action potential, which is propagated throughout the muscle fiber.

9 ACh is subsequently destroyed by acetylcholinesterase, an enzyme located on the motor end-plate membrane, terminating the muscle cell's response.

Structure of Skeletal Muscle (pp. 257–261)

■ Muscles, contraction specialists, can develop tension, shorten, produce movement, and accomplish work.

■ The three types of muscle are categorized in two different ways according to common characteristics. (1) Skeletal muscle and cardiac muscle are striated, whereas smooth muscle is unstriated. (2) Skeletal muscle is voluntary, whereas cardiac muscle and smooth muscle are involuntary. *(Review Figure 8-1 and Table 8-3, pp. 290–291.)*

■ Skeletal muscles are made up of bundles of long, cylindrical muscle cells known as muscle fibers, wrapped in connective tissue. Muscle fibers are packed with myofibrils, each myofibril consisting of alternating, slightly overlapping stacked sets of thick and thin filaments. This arrangement leads to a skeletal muscle fiber's striated microscopic appearance, which consists of alternating dark A bands and light I bands. A sarcomere, the area between two Z lines, is the functional unit of skeletal muscle. *(Review Figures 8-2 and 8-3.)*

■ Thick filaments consist of the protein myosin. Cross bridges made up of the myosin molecules' globular heads project from each thick filament toward the surrounding thin filaments. *(Review Figures 8-2 and 8-4.)*

■ Thin filaments consist primarily of the protein actin, which can bind and interact with the myosin cross bridges to bring about contraction. In the resting state two other proteins, tropomyosin and troponin, lie across the surface of the thin filament to prevent this cross-bridge interaction. *(Review Figures 8-2 and 8-5.)*

Molecular Basis of Skeletal Muscle Contraction (pp. 261–268)

■ Excitation of a skeletal muscle fiber by its motor neuron brings about contraction through a series of events that results in the thin filaments sliding closer together between the thick filaments. *(Review Figure 8-7.)*

■ This sliding filament mechanism of muscle contraction is switched on by Ca^{2+} release from the lateral sacs of the sarcoplasmic reticulum in response to the spread of a muscle fiber action potential into the central portions of the fiber via the T tubules. *(Review Figures 8-9, 8-10, and 8-11.)*

■ Released Ca^{2+} binds to troponin, slightly repositioning tropomyosin to uncover actin's cross-bridge binding sites. *(Review Figures 8-6 and 8-11.)*

■ Binding of actin to a myosin cross bridge triggers cross-bridge stroking, powered by energy stored in the myosin head from prior splitting of ATP by myosin ATPase. During a power stroke the cross bridge bends toward the thick filament's center, "rowing" in the thin filament to which it is attached. *(Review Figures 8-8, 8-11, and 8-12.)*

■ When a fresh ATP attaches to the cross bridge, myosin and actin detach, the cross bridge returns to its original shape, and the cycle is repeated. Repeated cycles of cross-bridge activity slide the thin filaments inward step by step. *(Review Figures 8-8 and 8-12.)*

■ When the action potential ends, the lateral sacs actively take up the Ca^{2+}, troponin and tropomyosin slip back into their blocking position, and relaxation occurs. *(Review Figure 8-11.)*

■ The entire contractile response lasts about 100 times longer than the action potential. *(Review Figure 8-13.)*

Skeletal Muscle Mechanics (pp. 268–276)

■ Tension is generated within a muscle by the contractile component (sarcomere shortening brought about by cross-bridge cycling). To move the bone to which the muscle is inserted, this internal tension is transmitted to the bone as the contractile component stretches and tightens the muscle's series-elastic component (intracellular titin, connective tissue, tendon). *(Review Figure 8-14.)*

■ Gradation of whole-muscle contraction can be accomplished by (1) varying the number of muscle fibers contracting within the muscle and (2) varying the tension developed by each contracting fiber. *(Review Table 8-2, p. 283.)*

■ The number of fibers contracting depends on (1) size of the muscle (number of muscle fibers present), (2) extent of motor unit recruitment (how many motor neurons supplying the muscle are active), and (3) size of each motor unit (how many muscle fibers are activated simultaneously by a single motor neuron). *(Review Figures 8-16 and 8-17 and Table 8-2.)*

■ Two variable factors that affect fiber tension are (1) frequency of stimulation, which determines the extent of twitch summation; and (2) length of the fiber before the onset of contraction (length-tension relationship). *(Review Table 8-2.)*

■ Twitch summation is the increase in tension accompanying repetitive stimulation of a muscle fiber. After undergoing an action potential, the muscle cell membrane recovers from its refractory period and can be restimulated while some contractile activity triggered by the first action potential still remains so that the twitches induced by the two rapidly successive action potentials sum. If the muscle fiber is stimulated so rapidly that it does not have a chance to start relaxing between stimuli, a smooth, sustained maximal contraction known as tetanus occurs. *(Review Figure 8-18.)*

■ The tension also depends on the length of the fiber at the onset of contraction. At optimal length (l_o) (the resting muscle length), there is maximal opportunity for cross-bridge interaction, because of optimal overlap of thick and thin filaments, so the greatest tension can develop. Less tension can develop at shorter or longer lengths. *(Review Figure 8-19.)*

■ The two primary types of muscle contraction—isometric (constant length) and isotonic (constant tension)—depend on the relationship between muscle tension and the load (the weight of an object being lifted). (1) If tension is less than the load, the muscle cannot shorten and lift the object but remains at constant length (an isometric contraction). (2) If the tension exceeds the load, the muscle can shorten and lift the object, maintaining constant tension while shortening (isotonic contraction).

■ The velocity, or speed, of shortening is inversely proportional to the load. (Review Figure 8-20.)

■ The amount of work accomplished by a contracting muscle equals the magnitude of the load times the distance the load is moved. The amount of energy consumed by a contracting muscle that is realized as external work varies from 0% to 25%; the remaining energy is converted to heat. (Review Figure 8-20.)

Skeletal Muscle Metabolism and Fiber Types (pp. 276–281)

■ Three pathways furnish the ATP needed for muscle contraction and relaxation: (1) the transfer of high-energy phosphates from stored creatine phosphate to ADP, providing the first source of ATP at the onset of exercise; (2) oxidative phosphorylation, which efficiently extracts large amounts of ATP from nutrients if enough O_2 is available to support this system; and (3) glycolysis, which can synthesize ATP in the absence of O_2 but uses large amounts of stored glycogen and produces lactate in the process. (Review Figure 8-22.)

■ The three types of skeletal muscle fibers are classified by the pathways they use for ATP synthesis (oxidative or glycolytic) and the rapidity with which they split ATP and subsequently contract (slow twitch or fast twitch): (1) slow-oxidative fibers, (2) fast-oxidative fibers, and (3) fast-glycolytic fibers. (Review Table 8-1.)

Control of Motor Movement (pp. 281–289)

■ Control of motor movement depends on activity in the three types of presynaptic inputs that converge on the motor neurons supplying various muscles: (1) spinal reflex pathways, which originate with afferent neurons; (2) the corticospinal (pyramidal) motor system, which originates in the primary motor cortex and is concerned with discrete, intricate movements of the hands; and (3) the multineuronal (extrapyramidal) motor system, which originates in the brain stem and is involved with postural adjustments and involuntary movements of the trunk and limbs. The final motor output from the brain stem is influenced by the cerebellum, basal nuclei, and cerebral cortex. (Review Figure 8-23.)

■ Establishment and adjustment of motor commands depend on continuous afferent input, especially feedback about changes in muscle length (monitored by muscle spindles) and muscle tension (monitored by Golgi tendon organs). (Review Figure 8-24.)

■ When a whole muscle is stretched, the stretch of its muscle spindles triggers the stretch reflex, which results in reflex contraction of that muscle. This reflex resists any passive changes in muscle length. (Review Figures 8-25 and 8-26.)

Smooth and Cardiac Muscle (pp. 289–299)

■ Smooth muscle cells are spindle shaped and much smaller than skeletal muscle fibers. The thick and thin filaments of smooth muscle are oriented diagonally in a diamond-shaped lattice instead of running longitudinally, so the fibers are not striated. (Review Figures 8-27 and 8-28.)

■ In smooth muscle, cytosolic Ca^{2+}, which enters from the ECF and is also released from sparse intracellular stores, activates cross-bridge cycling by initiating a series of biochemical reactions that result in phosphorylation of the light chains of the myosin cross bridges to enable them to bind with actin. (Review Figures 8-29 and 8-30.)

■ Smooth muscle in different organs is highly diversified and can be classified in various ways: phasic or tonic, multiunit or single-unit, and neurogenic or myogenic.

■ Phasic smooth muscle displays bursts of pronounced contraction in response to action potentials. Tonic smooth muscle is partially contracted at all times in the absence of action potentials because of ongoing Ca^{2+} entry through open surface-membrane Ca^{2+} channels.

■ Multiunit smooth muscle is neurogenic, requiring stimulation of individual muscle fibers by its autonomic nerve supply to trigger contraction. Single-unit smooth muscle is myogenic; it can initiate its own contraction. Phasic, single-unit smooth muscle spontaneously depolarizes to threshold as a result of pacemaker potentials or slow-wave potentials. Once an action potential is initiated, this electrical activity spreads by means of gap junctions to the surrounding cells within the functional syncytium, so the entire sheet becomes excited and contracts as a unit. (Review Figure 8-31 and Table 8-4.)

■ The level of tension in smooth muscle depends on the level of cytosolic Ca^{2+}. The autonomic nervous system (Review Figure 8-32), as well as hormones and local metabolites, can modify the rate and strength of contractions by altering cytosolic Ca^{2+} concentration.

■ Smooth muscle contractions are slow and energy efficient, enabling this type of muscle to economically sustain long-term contractions without fatigue. This economy, coupled with the fact that single-unit smooth muscle can exist at a variety of lengths with little change in tension, makes single-unit smooth muscle ideally suited for its task of forming the walls of hollow organs that can distend.

■ Cardiac muscle is found only in the heart. It has highly organized striated fibers, like skeletal muscle. Like single-unit smooth muscle, some cardiac muscle fibers can generate action potentials, which are spread throughout the heart with the aid of gap junctions. (Review Table 8-3.)

Anatomy of the Heart (pp. 303–309)

■ The circulatory system is the transport system of the body.

■ The three basic components of the circulatory system are the heart (the pump), the blood vessels (the passageways), and the blood (the transport medium).

■ The heart is positioned midline in the thoracic cavity at an angle with its wide base lying toward the right and its pointed apex toward the left.

■ The heart is basically a dual pump that provides the driving pressure for blood flow through the pulmonary circulation (between the heart and lungs) and systemic circulation (between the heart and other body systems). *(Review Figures 9-1 and 9-2.)*

■ The heart has four chambers: Each half of the heart consists of an atrium, or venous input chamber, and a ventricle, or arterial output chamber. The right atrium receives O_2-poor blood from the systemic circulation and the right ventricle pumps it into the pulmonary circulation. The left atrium receives O_2-rich blood from the pulmonary circulation and pumps it into the systemic circulation. *(Review Figures 9-1, 9-2, and 9-4.)*

■ Four heart valves direct blood in the right direction and keep it from flowing in the other direction. The right and left atrioventricular (AV) valves direct blood from the atria to the ventricles during diastole and prevent backflow of blood from the ventricles to the atria during systole. The aortic and pulmonary semilunar valves direct blood from the ventricles to the aorta and pulmonary artery, respectively, during systole and prevent backflow of blood from these major vessels to the ventricles during diastole. *(Review Figures 9-3, 9-4, and 9-5.)*

■ Contraction of the spirally arranged cardiac muscle fibers produces a wringing effect important for efficient pumping. Also important for efficient pumping is that the muscle fibers in each chamber act as a functional syncytium, contracting as a coordinated unit. *(Review Figure 9-6.)*

■ The branching cardiac muscle fibers are interconnected by intercalated discs, which contain (1) desmosomes that hold the cells together mechanically and (2) gap junctions that permit spread of electrical current between cells joined together as a functional syncytium. *(Review Figure 9-6.)*

Electrical Activity of the Heart (pp. 309–321)

■ The heart is self-excitable, initiating its own rhythmic contractions.

■ Autorhythmic cells are 1% of the cardiac muscle cells; they do not contract but are specialized to initiate and conduct action potentials. The other 99% of cardiac cells are contractile cells that contract in response to the spread of an action potential initiated by autorhythmic cells.

■ Autorhythmic cells display a pacemaker potential, a slow drift to threshold potential, as a result of a complex interplay of inherent changes in ion movement across the membrane. The first half of the pacemaker potential results from opening of unique funny channels that permit entry of Na^+ at the same time K^+ channels slowly close so that exit of K^+ slowly declines. Both of these actions gradually depolarize the membrane toward threshold. The final boost to threshold results from Ca^{2+} entry on opening of T-type Ca^{2+} channels. The rising phase of the action potential is the result of further Ca^{2+} entry on opening of L-type Ca^{2+} channels at threshold. The falling phase results from K^+ efflux on opening of K^+ channels at the peak of the action potential. Slow closure of these K^+ channels at the end of repolarization contributes to the next pacemaker potential. *(Review Figure 9-7.)*

■ The cardiac impulse originates at the SA node, the pacemaker of the heart, which has the fastest rate of spontaneous depolarization to threshold. *(Review Table 9-1 and Figures 9-8 and 9-9.)*

■ Once initiated, the action potential spreads throughout the right and left atria, partially facilitated by specialized conduction pathways but mostly by cell-to-cell spread of the impulse through gap junctions. *(Review Figure 9-8.)*

■ The impulse passes from the atria into the ventricles through the AV node, the only point of electrical contact between these chambers. The action potential is delayed briefly at the AV node, ensuring that atrial contraction precedes ventricular contraction to allow complete ventricular filling. *(Review Figure 9-8.)*

■ The impulse then travels rapidly down the interventricular septum via the bundle of His and rapidly disperses throughout the myocardium by means of the Purkinje fibers. The rest of the ventricular cells are activated by cell-to-cell spread of the impulse through gap junctions. *(Review Figure 9-8.)*

■ Thus, the atria contract as a single unit, followed after a brief delay by a synchronized ventricular contraction.

■ The action potentials of cardiac contractile cells exhibit a prolonged positive phase, or plateau, accompanied by a prolonged period of contraction, which ensures adequate ejection time. This plateau is primarily the result of activation of slow L-type Ca^{2+} channels. *(Review Figure 9-10.)*

■ Ca^{2+} entry though the L-type channels in the T tubules triggers a much larger release of Ca^{2+} from the sarcoplasmic reticulum. This Ca^{2+}-induced Ca^{2+} release leads to cross-bridge cycling and contraction. *(Review Figure 9-11.)*

■ Because a long refractory period occurs in conjunction with this prolonged plateau phase, summation and tetanus of cardiac muscle are impossible, ensuring the alternate periods of contraction and relaxation essential for pumping of blood. *(Review Figure 9-12.)*

■ The spread of electrical activity throughout the heart can be recorded from the body surface. In this electrocardiogram (ECG), the P wave represents atrial depolarization; the QRS complex, ventricular depolarization; and the T wave, ventricular repolarization. *(Review Figures 9-13, 9-14, and 9-15.)*

Mechanical Events of the Cardiac Cycle (pp. 321–324)

■ The cardiac cycle consists of three important events (*Review Figure 9-16*):
1. The generation of electrical activity as the heart autorhythmically depolarizes and repolarizes (*Review Figure 9-14.*)
2. Mechanical activity consisting of alternate periods of systole (contraction and emptying) and diastole (relaxation and filling), which are initiated by the rhythmic electrical cycle.
3. Directional flow of blood through the heart chambers, guided by valve opening and closing induced by pressure changes that are generated by mechanical activity.

■ The atrial pressure curve remains low throughout the entire cardiac cycle, with only minor fluctuations (normally varying between 0 and 8 mm Hg). The aortic pressure curve remains high the entire time, with moderate fluctuations (normally varying between a systolic pressure of 120 mm Hg and a diastolic pressure of 80 mm Hg). The ventricular pressure curve fluctuates dramatically, because ventricular pressure must be below the low atrial pressure during diastole to allow the AV valve to open for filling; and to force the aortic valve open to allow emptying, it must be above the high aortic pressure during systole. Therefore, ventricular pressure normally varies from 0 mm Hg during diastole to slightly more than 120 mm Hg during systole. During the periods of isovolumetric ventricular contraction and relaxation, ventricular pressure is above the low atrial pressure and below the high aortic pressure, so all valves are closed and no blood enters or leaves the ventricles. (*Review Figure 9-16.*)

■ The end-diastolic volume is the volume of blood in the ventricle when filling is complete at the end of diastole. The end-systolic volume is the volume of blood remaining in the ventricle when ejection is complete at the end of systole. The stroke volume is the volume of blood pumped out by each ventricle each beat. (*Review Figure 9-16.*)

■ Valve closing gives rise to two normal heart sounds. The first heart sound is caused by closing of the AV valves and signals the onset of ventricular systole. The second heart sound is the result of closing of the aortic and pulmonary valves at the onset of diastole. (*Review Figure 9-16.*)

■ Defective valve function produces turbulent blood flow, which is audible as a heart murmur. Abnormal valves may be either stenotic and not open completely or insufficient and not close completely. (*Review Figure 9-18 and Table 9-2.*)

Cardiac Output and Its Control (pp. 325–332)

■ Cardiac output, the volume of blood ejected by each ventricle each minute, is determined by heart rate times stroke volume. (*Review Figure 9-24.*)

■ Heart rate is varied by altering the balance of parasympathetic and sympathetic influence on the SA node. Parasympathetic stimulation slows the heart rate, and sympathetic stimulation speeds it up. (*Review Figure 9-19 and Table 9-3.*)

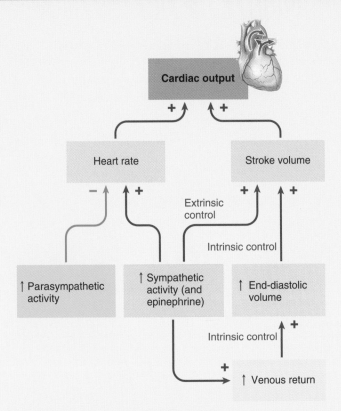

■ Stroke volume depends on (1) the extent of ventricular filling, with an increased end-diastolic volume resulting in a larger stroke volume by means of the length–tension relationship (Frank–Starling law of the heart, a form of intrinsic control); and (2) the extent of sympathetic stimulation, with increased sympathetic stimulation resulting in increased contractility of the heart, that is, increased strength of contraction and increased stroke volume at a given end-diastolic volume (extrinsic control). (*Review Figures 9-20 through 9-23.*)

■ The preload of the heart (the workload imposed on the heart before contraction begins) is the extent of filling. The afterload of the heart (the workload imposed on the heart after contraction has begun) is the arterial blood pressure.

Nourishing the Heart Muscle (pp. 332–338)

■ Cardiac muscle is supplied with oxygen and nutrients by blood delivered to it by the coronary circulation, not by blood within the heart chambers.

■ Most coronary blood flow occurs during diastole, because during systole the contracting heart muscle compresses the coronary vessels. (*Review Figure 9-26.*)

■ Coronary blood flow is normally varied to keep pace with cardiac oxygen needs. (*Review Figure 9-27.*)

■ Coronary blood flow may be compromised by development of atherosclerotic plaques, which can lead to ischemic heart disease ranging in severity from mild chest pain on exertion to fatal heart attacks. (*Review Figures 9-28 through 9-30 and Table 9-4.*)

Patterns and Physics of Blood Flow (pp. 343–347)

■ Materials can be exchanged between various parts of the body and with the external environment by means of the blood vessel network that transports blood to and from all organs. (Review Figure 10-1.)

■ Organs that replenish nutrient supplies and remove metabolic wastes from the blood receive a greater percentage of the cardiac output than is warranted by their metabolic needs. These "reconditioning" organs can better tolerate reductions in blood supply than can organs that receive blood solely for meeting their own metabolic needs. The reconditioning organs are the digestive organs, kidneys, and skin.

■ The brain is especially vulnerable to reductions in its blood supply. Therefore, maintaining adequate flow to this vulnerable organ is a high priority in circulatory function.

■ The flow rate of blood through a vessel (in volume per unit of time) is directly proportional to the pressure gradient and inversely proportional to the resistance. The higher pressure at the beginning of a vessel is established by the pressure imparted to the blood by cardiac contraction. The lower pressure at the end is the result of frictional losses as flowing blood rubs against the vessel wall. (Review Figure 10-2.)

■ Resistance, the hindrance to blood flow through a vessel, is influenced most by the vessel's radius. Resistance is inversely proportional to the fourth power of the radius, so small changes in radius profoundly influence flow. As the radius increases, resistance decreases and flow increases, and vice versa. (Review Figure 10-3.)

■ Blood flows in a closed loop between the heart and the organs. The arteries transport blood from the heart throughout the body. The arterioles regulate the amount of blood that flows through each organ. The capillaries are the actual site where materials are exchanged between blood and surrounding tissue cells. The veins return blood from the tissue level back to the heart. (Review Figure 10-4 and Table 10-1.)

Arteries (pp. 347–350)

■ Arteries are large-radius, low-resistance passageways from the heart to the organs. They also serve as a pressure reservoir. Because of their elasticity, owing to their abundant elastin fibers, arteries expand to accommodate the extra volume of blood pumped into them by cardiac contraction and then recoil to continue driving the blood forward when the heart is relaxing. (Review Table 10-1 and Figures 10-5 and 10-6.)

■ Systolic pressure (average 120 mm Hg) is the peak pressure exerted by the ejected blood against the vessel walls during cardiac systole. Diastolic pressure (average 80 mm Hg) is the minimum pressure in the arteries when blood is draining off into the vessels downstream during cardiac diastole. When blood pressure is 120/80, pulse pressure (the difference between systolic and diastolic pressures) is 40 mm Hg. (Review Figures 10-7 and 10-8.)

■ The average driving pressure throughout the cardiac cycle is the mean arterial pressure, which can be estimated using the following formula: mean arterial pressure = diastolic pressure + 1/3 pulse pressure. (Review Figure 10-9.)

Arterioles (pp. 350–360)

■ Arterioles are the major resistance vessels. Their high resistance produces a large drop in mean pressure between the arteries and capillaries. This decline enhances blood flow by contributing to the pressure differential between the heart and organs. (Review Figure 10-9.)

■ Arterioles have a thick layer of circular smooth muscle, variable contraction of which alters arteriolar caliber and resistance. (Review Table 10-1.) Tone, a baseline of contractile activity, is maintained in arterioles at all times. Arteriolar vasodilation (expansion of arteriolar caliber above tonic level) decreases resistance and increases blood flow through the vessel, whereas vasoconstriction (narrowing of the vessel) increases resistance and decreases flow. (Review Figure 10-10.)

■ Arteriolar caliber is subject to two types of control mechanisms: local (intrinsic) controls and extrinsic controls.

■ Local controls primarily involve local chemical changes associated with changes in the level of metabolic activity in an organ, such as local changes in O_2, which cause the release of vasoactive mediators from the endothelial cells in the vicinity. Examples include vasodilating nitric oxide and vasoconstricting endothelin. These vasoactive mediators act on the underlying arteriolar smooth muscle to bring about an appropriate change in the caliber of the arterioles supplying the organ. By adjusting the resistance to blood flow, the local control mechanism adjusts blood flow to the organ to match the momentary metabolic needs of the organ. (Review Figures 10-10, 10-11, and 10-14 and Tables 10-2 and 10-3.)

■ Arteriolar caliber can be adjusted independently in different organs by local control factors. Such adjustments are important in variably distributing cardiac output. (Review Figure 10-12.)

■ Other local influences include (1) histamine release (important in inflammatory and allergic reactions); (2) local application of heat or cold (important therapeutically); (3) chemical response to shear stress (which resists changes in the force exerted parallel to the vessel surface by flowing blood); and (4) myogenic response to stretch (which resists changes in the distending force exerted across the vessel wall by blood-pressure driven changes in blood flow).

■ Extrinsic control is accomplished primarily by sympathetic and to a lesser extent by hormonal influence over arteriolar smooth muscle. Extrinsic controls are important in maintaining mean arterial pressure. Arterioles are richly supplied with sympathetic nerve fibers, whose increased activity produces generalized vasoconstriction and a subsequent increase in total peripheral resistance, thus increasing mean arterial pressure. Decreased sympathetic activity produces generalized arteriolar vasodilation, which lowers mean arterial pressure. These extrinsically controlled adjustments of arteriolar caliber help maintain the appropriate pressure head for driving blood forward to the tissues. Most arterioles are not supplied by parasympathetic nerves. (*Review Figure 10-14.*)

■ Hormones that extrinsically influence arteriolar radius are norepinephrine, epinephrine, vasopressin, and angiotensin II, all of which cause generalized arteriolar vasoconstriction.

Capillaries (pp. 361–371)

■ The thin-walled, small-radius, extensively branched capillaries are ideally suited to serve as sites of exchange between the blood and surrounding tissue cells. Anatomically, the surface area for exchange is maximized and diffusion distance is minimized in the capillaries. Furthermore, because of their large total cross-sectional area, the velocity of blood flow through capillaries (in distance per unit of time) is relatively slow, providing adequate time for exchanges to take place. (*Review Figures 10-15 through 10-17 and Table 10-1, p. 348.*)

■ Two types of passive exchanges—diffusion and bulk flow—take place across capillary walls.

■ Individual solutes are exchanged primarily by diffusion down concentration gradients. Lipid-soluble substances pass directly through the single layer of endothelial cells lining a capillary, whereas water-soluble substances pass through water-filled pores between the endothelial cells. Plasma proteins generally do not escape. (*Review Figures 10-18 and 10-21.*)

■ Imbalances in physical pressures acting across capillary walls are responsible for bulk flow of fluid through the pores. (1) Fluid is forced out of the first portion of the capillary (ultrafiltration), where outward pressures (mainly capillary blood pressure) exceed inward pressures (mainly plasma-colloid osmotic pressure). (2) Fluid is returned to the capillary along its last half, when outward pressures fall below inward pressures. The reason for the shift in balance down the capillary's length is the continuous decline in capillary blood pressure while the plasma-colloid osmotic pressure remains constant. Bulk flow is responsible for the distribution of ECF between plasma and interstitial fluid. (*Review Figures 10-9, 10-22, and 10-23.*)

■ Normally, slightly more fluid is filtered than is reabsorbed. The extra fluid, any leaked proteins, and bacteria in the tissue are picked up by the lymphatic system. Bacteria are destroyed as lymph passes through lymph nodes on the way to being returned to the venous system. (*Review Figures 10-22, 10-24 and 10-25.*)

Veins (pp. 371–376)

■ Veins are large-radius, low-resistance passageways through which blood returns from the organs to the heart. In addition, the thin-walled, highly distensible veins, as capacitance vessels, can passively stretch to store a larger volume of blood and therefore act as a blood reservoir. The capacity of veins to hold blood can change markedly with little change in venous pressure. At rest, the veins contain more than 60% of the total blood volume. (*Review Table 10-10 and Figure 10-27.*)

■ The primary force that produces venous flow is the pressure gradient between the veins and atrium (that is, what remains of the driving pressure imparted to the blood by cardiac contraction). (*Review Figures 10-9 and 10-28.*)

■ Venous return is enhanced by sympathetically induced venous vasoconstriction and by external compression of the veins from contraction of surrounding skeletal muscles, both of which drive blood out of the veins. These actions help counter the effects of gravity on the venous system. (*Review Figures 10-28 through 10-31.*)

■ One-way venous valves ensure that blood is driven toward the heart and kept from flowing back toward the tissues. (*Review Figure 10-32.*)

■ Venous return is also enhanced by the respiratory pump and the cardiac suction effect. Respiratory activity produces a less-than-atmospheric pressure in the chest cavity, thus establishing an external pressure gradient that encourages flow from the lower veins that are exposed to atmospheric pressure to the chest veins that empty into the heart. In addition, slightly negative pressures created within the atria during ventricular systole and within the ventricles during ventricular diastole exert a suctioning effect that further enhances venous return and facilitates cardiac filling. (*Review Figures 10-28 and 10-33.*)

Blood Pressure (pp. 376–386)

■ Regulation of mean arterial pressure (MAP) depends on control of its two main determinants, cardiac output (CO) and total peripheral resistance (TPR). Control of CO, in turn, depends on regulation of heart rate and stroke volume, whereas TPR is determined primarily by the degree of arteriolar vasoconstriction. (*Review Figure 10-34.*)

■ Short-term regulation of blood pressure is accomplished mainly by the baroreceptor reflex. Carotid sinus and aortic arch baroreceptors continuously monitor MAP. When they detect a deviation from normal, they signal the medullary cardiovascular center, which responds by adjusting autonomic output to the heart and blood vessels to restore the blood pressure to normal. (*Review Figures 10-35 through 10-38.*)

■ Long-term control of blood pressure involves maintaining proper plasma volume through the kidneys' control of salt and water balance. (*Review Figure 10-34.*)

Plasma (p. 391–393)

■ Blood consists of three types of cellular elements—erythrocytes (red blood cells), leukocytes (white blood cells), and platelets (thrombocytes)—suspended in the liquid plasma. *(Review Figure 11-1 and Table 11-1.)*

■ The 5- to 5.5-liter volume of blood in an adult consists of 42% to 45% erythrocytes, less than 1% leukocytes and platelets, and 55% to 58% plasma. The percentage of whole-blood volume occupied by erythrocytes is the hematocrit. *(Review Figure 11-1.)*

Plasma = 55% of whole blood

Buffy coat: platelets and leukocytes = <1% of whole blood

Packed cell volume, or hematocrit

Platelets

Leukocytes (white blood cells)

Erythrocytes = 45% of whole blood

Erythrocytes (red blood cells)

■ Plasma is a complex liquid consisting of 90% water that serves as a transport medium for substances being carried in the blood. The most abundant inorganic constituents in plasma are Na^+ and Cl^-. The most plentiful organic constituents in plasma are plasma proteins.

■ All plasma constituents are freely diffusible across the capillary walls except the plasma proteins, which remain in the plasma, where they perform a variety of important functions. Plasma proteins include the albumins, globulins (α, β, and γ), and fibrinogen. *(Review Table 11-1.)*

Erythrocytes (pp. 393–400)

■ Erythrocytes are specialized for their primary function of O_2 transport in the blood. Their biconcave shape maximizes the surface area available for diffusion of O_2 into cells of this volume. *(Review Figure 11-1.)* Erythrocytes do not contain a nucleus or organelles (these are extruded during development) but instead are packed full of hemoglobin, an iron-containing molecule that can loosely and reversibly bind with O_2. Because O_2 is poorly soluble in blood, hemoglobin is indispensable for O_2 transport. Each hemoglobin molecule can carry four O_2 molecules. *(Review Figures 11-2 and 11-3.)*

■ Hemoglobin also contributes to CO_2 transport and buffering of blood by reversibly binding with CO_2 and H^+.

■ Unable to replace cell components, erythrocytes are destined to a short life span of about 120 days.

■ Undifferentiated pluripotent stem cells in the red bone marrow give rise to all cellular elements of the blood. *(Review Figures 11-3 and 11-9.)* Erythrocyte production (erythropoiesis) by the marrow normally keeps pace with the rate of erythrocyte loss, keeping the red cell count constant. Erythropoiesis is stimulated by erythropoietin, a hormone secreted by the kidneys in response to reduced O_2 delivery. *(Review Figure 11-4.)*

Kidney

Erythropoietin

Reduced oxygen-carrying capacity

Relieves 5

Developing erythrocytes in red bone marrow

Increased oxygen-carrying capacity

Erythrocytes

1 Kidneys detect reduced O_2-carrying capacity of blood.

2 When less O_2 is delivered to the kidneys, they secrete erythropoietin into blood.

3 Erythropoietin stimulates erythropoiesis by bone marrow.

4 Additional circulating erythrocytes increase O_2-carrying capacity of blood.

5 Increased O_2-carrying capacity relieves initial stimulus that triggered erythropoietin secretion.

■ The major ABO blood types depend on the presence of specific antigens on the surface of erythrocytes. The red blood cells of type A blood have A antigen, of type B blood have B antigen, of type AB blood have both A and B antigen, and of type O blood have no A or B antigen. Type A blood has anti-B antibodies, type B blood has anti-A antibodies, type AB blood has no anti-A or anti-B antibodies, and type O blood has both anti-A and anti-B antibodies. These antibodies cause the RBCs with the corresponding antigens to agglutinate (clump) and/or rupture, causing a transfusion reaction if incoming donor cells are exposed to corresponding antibodies in recipient blood. *(Review Figure 11-7.)*

Leukocytes (pp. 400–405)

■ Leukocytes are the defense corps of the body. They attack foreign invaders (the most common of which are bacteria and viruses), destroy cancer cells that arise in the body, and clean up cellular debris. Leukocytes as well as certain plasma proteins make up the immune system.

■ Each of the five types of leukocytes has a different task: (1) Neutrophils, the phagocytic specialists, are important in engulfing bacteria and debris. (2) Eosinophils specialize in attacking parasitic worms and play a role in allergic responses. (3) Basophils release two chemicals: histamine, which is also important in allergic responses; and heparin, which helps clear fat particles from the blood. (4) Monocytes, on leaving the blood, set up residence in the tissues and greatly enlarge to become the large tissue phagocytes known as macrophages. (5) Lymphocytes provide immune defense against bacteria, viruses, and other targets for which they are specifically programmed. Their defense tools include the production of antibodies that mark the victim for destruction by phagocytosis or other means (for B lymphocytes) and the release of chemicals that punch holes in the victim (for T lymphocytes). *(Review Figure 11-8 and Table 11-1.)*

■ Leukocytes are present in the blood only while in transit from their site of production and storage in the bone marrow (and also in the lymphoid tissues in the case of the lymphocytes) to their site of action in the tissues. *(Review Figure 11-9.)* At any given time, most leukocytes are out in the tissues on surveillance or performing actual combat missions.

■ All leukocytes have a limited life span and must be replenished by ongoing differentiation and proliferation of precursor cells. The total number and percentage of each of the different types of leukocytes are produced at variable rates depending on the momentary defense needs of the body. Factors that regulate production of the different types of leukocyte are released from invaded or injured tissues and/or from activated leukocytes.

Platelets and Hemostasis (pp. 405–412)

■ Platelets are cell fragments derived from large megakaryocytes in the bone marrow. *(Review Figures 11-8, 11-9, and 11-10.)*

■ Platelets play a role in hemostasis, the arrest of bleeding from an injured vessel. The three main steps in hemostasis are (1) vascular spasm, (2) platelet plugging, and (3) clot formation.

■ Vascular spasm reduces blood flow through an injured vessel.

■ Aggregation of platelets at the site of vessel injury quickly plugs the defect. Platelets start to aggregate on contact with exposed collagen in the damaged vessel wall. These aggregated platelets secrete ADP and thromboxane A_2, which to-

gether cause other passing-by platelets to pile on, setting up a positive-feedback cycle as the platelet plug grows to fill in the defect. Normal adjacent endothelium secretes inhibitory chemicals that prevent platelets from adhering to the surrounding undamaged part of the vessel. *(Review Figures 11-11 and 11-15.)*

■ Clot formation reinforces the platelet plug and converts blood in the vicinity of a vessel injury into a nonflowing gel.

■ Most factors necessary for clotting are always present in the plasma in inactive precursor form. When a vessel is damaged, exposed collagen initiates a cascade of reactions involving successive activation of these clotting factors, ultimately converting fibrinogen into fibrin via the intrinsic clotting pathway. *(Review Figures 11-13, 11-14, and 11-15.)*

■ Fibrin, an insoluble threadlike molecule, is laid down as the meshwork of the clot; the meshwork in turn entangles blood cellular elements to complete clot formation. *(Review Figure 11-12.)*

■ Platelet aggregation and clot formation mutually reinforce each other to seal the damaged vessel. Both processes are initiated simultaneously by exposure to collagen at a vessel defect. Thrombin converts fibrinogen to fibrin to form the clot and participates in hemostasis in multiple ways, including enhancing platelet aggregation. Platelets secrete PF3, which enhances the clotting cascade. *(Review Figures 11-13 and 11-15.)*

■ Blood that has escaped into the tissues clots on exposure to tissue thromboplastin, which sets the extrinsic clotting pathway into motion. *(Review Figure 11-14.)*

■ Clots form quickly. When no longer needed, they are slowly dissolved by plasmin, a fibrinolytic factor also activated by exposed collagen. *(Review Figure 11-16.)*

Immune System: Targets, Effectors, Components (pp. 417–420)

■ Foreign invaders and newly arisen mutant cells are immediately confronted with multiple interrelated defense mechanisms aimed at destroying and eliminating anything that is not part of the normal self. These mechanisms, collectively referred to as immunity, include both innate and adaptive immune responses. Innate immune responses are nonspecific responses that nonselectively defend against foreign material even on initial exposure to it. Adaptive immune responses are specific responses that selectively target particular invaders for which the body has been specially prepared after a prior exposure. (Review Table 12-3, p. 445.)

■ The most common invaders are bacteria and viruses. Bacteria are self-sustaining, single-celled organisms, which produce disease by virtue of the destructive chemicals they release. Viruses are protein-coated nucleic acid particles, which invade host cells and take over the cellular metabolic machinery for their own survival to the detriment of the host cell.

■ Leukocytes and their derivatives are the major effector cells of the immune system and are reinforced by a number of different plasma proteins. Leukocytes include neutrophils, eosinophils, basophils, monocytes, and lymphocytes.

■ Immune cells also clean up cellular debris, preparing the way for tissue repair.

Innate Immunity (pp. 420–428)

■ Innate immune responses include inflammation, interferon, natural killer cells, and the complement system.

■ Inflammation is a nonspecific response to foreign invasion or tissue damage mediated largely by the professional phagocytes (neutrophils and monocytes-turned-macrophages). The phagocytic cells destroy foreign and damaged cells both by phagocytosis and by release of lethal chemicals. (Review Figures 12-2 and 12-3.) Phagocytic secretions also augment inflammation, induce systemic manifestations such as fever, and enhance adaptive immune responses.

■ Histamine-induced vasodilation and increased permeability of local capillaries at the site of invasion or injury permit enhanced delivery of more phagocytic leukocytes and inactive plasma protein precursors crucial to defense, such as complement components. These vascular changes also largely produce the observable local manifestations of inflammation— swelling, redness, heat, and pain. (Review Figure 12-3.)

■ Interferon is nonspecifically released by virus-infected cells and transiently inhibits viral multiplication in other cells to which it binds. (Review Figure 12-5.)

■ Natural killer (NK) cells nonspecifically lyse and destroy virus-infected cells and cancer cells on first exposure to them. (Review Figure 12-11.)

■ On being activated by microbes themselves at the site of invasion or by antibodies produced against the microbes, the complement system directly destroys the foreign invaders by lysing their membranes and also augments other aspects of the inflammatory process, such as by acting as opsonins that enhance phagocytosis. The complement system lyses the targeted cells by forming a hole-punching membrane attack complex that inserts into the victim cell's membrane, leading to osmotic rupture of the cell. (Review Figures 12-4 and 12-6.)

Adaptive Immunity: General Concepts (pp. 428–429)

■ Not only is the adaptive immune system able to recognize foreign molecules as different from self-molecules—so that destructive immune reactions are not unleashed against the body itself—but it can also distinguish between millions of different foreign molecules. Lymphocytes, the effector cells of adaptive immunity, are each uniquely equipped with surface membrane receptors that can bind with only one specific complex foreign molecule, known as an antigen.

■ The two broad classes of adaptive immune responses are antibody-mediated immunity accomplished by plasma cells derived from B lymphocytes (B cells), and cell-mediated immunity accomplished by T lymphocytes (T cells). (Review Figure 12-7 and Table 12-4, p. 450.)

■ B cells develop from a lineage of lymphocytes that originally matured within the bone marrow. The T-cell lineage comes from lymphocytes that migrated from the bone marrow to the thymus to complete their maturation. New B and T cells arise from lymphocyte colonies in lymphoid tissues. (Review Figures 12-1 and 12-7 and Table 12-1.)

B Lymphocytes: Antibody-Mediated Immunity (pp. 429–437)

■ Each B cell recognizes specific free extracellular antigen, such as that found on the surface of bacteria.

■ After being activated by binding of its receptor (a B-cell receptor or BCR) with its specific antigen, a B cell rapidly proliferates, producing a clone of its own kind that can specifically wage battle against the invader. Most lymphocytes in the expanded B-cell clone become antibody-secreting plasma cells that participate in the primary response against the invader. Some of the new lymphocytes do not participate in the attack but become memory cells that lie in wait, ready to launch a swifter and more forceful secondary response should the same foreigner ever invade the body again. (Review Figures 12-8, 12-9, 12-12, 12-13, and 12-14.)

■ Antibodies are Y-shaped molecules. The antigen-binding sites on the tips of each arm of the antibody determine with what specific antigen the antibody can bind. Properties of the antibody's tail portion determine what the antibody does once it binds with antigen. There are five subclasses of antibodies, depending on differences in the biological activity of their tail portion: IgM, IgG, IgE, IgA, and IgD immunoglobulins. (Review Figure 12-10.)

■ Antibodies do not directly destroy antigenic material. Instead, they exert their protective effect by physically hindering antigens through neutralization or agglutination or by intensifying lethal innate immune responses already called into play by the foreign invasion. Antibodies activate the complement system, enhance phagocytosis, and stimulate killer cells. (Review Figure 12-11 and Table 12-3.)

T Lymphocytes: Cell-Mediated Immunity (pp. 437–449)

■ T cells accomplish cell-mediated immunity by being in direct contact with their targets and by releasing cytokines. Chemicals other than antibodies released by leukocytes are known as cytokines.

■ There are three types of T cells: cytotoxic, helper, and regulatory T cells.

■ The targets of cytotoxic (CD8+) T cells are virally invaded cells and cancer cells, which they destroy by releasing perforin molecules that form a lethal hole-punching complex that inserts into the membrane of the victim cell or by releasing granzymes that trigger the victim cell to undergo apoptosis. (Review Figures 12-15 and 12-16 and Table 12-2.)

■ Helper (CD4+) T cells bind with other immune cells and release cytokines that augment the activity of these other cells. B cells cannot convert to plasma cells and produce antibodies in response to T-dependent antigen without the help of helper cells. (Review Figure 12-21.)

■ Regulatory (CD4+CD25+) cells secrete cytokines that suppress other immune cells, putting the brake on immune responses in check-and-balance fashion.

■ Like B cells, T cells bear receptors (T-cell receptors or TCRs) that are antigen specific (Review Figure 12-8), undergo clonal selection, exert primary and secondary responses, and form memory pools for long-lasting immunity against targets to which they have already been exposed.

■ Helper T cells can recognize and bind with antigen only when it has been processed and presented to them by antigen-presenting cells (APCs), such as macrophages and dendritic cells. (Review Figures 12-18 and 12-19.)

■ Lymphocytes produced by chance that can attack the body's own cells are eliminated or suppressed so that they are prevented from functioning. In this way, the body is able to "tolerate" (not attack) its own antigens.

■ B and T cells have different targets because their requirements for antigen recognition differ. B cells recognize freely circulating antigen, such as bacteria, that can lead to antigen destruction at long distances via antibodies. T cells, in contrast, have a dual binding requirement of foreign antigen in association with self-antigens on the surface of one of the body's own cells. (Review Figures 12-20 and 12-21.)

■ The self-antigens on cell surfaces are class I or class II MHC molecules, which are unique for each individual. Cytotoxic T cells can bind only with virus-infected host cells or cancer cells, which always bear class I MHC self-antigen in association with foreign or abnormal antigen. Helper T cells can bind only with APCs and B cells that bear the class II MHC self-marker in association with foreign antigen. The APCs activate helper T cells, and helper T cells activate B cells. Thus, such differential binding ensures that the appropriate specific immune response ensues. (Review Figures 12-20 and 12-21.)

■ In the process of immune surveillance, natural killer cells, cytotoxic T cells, macrophages, and the interferon they collectively secrete normally eradicate newly arisen cancer cells before they have a chance to spread. (Review Figure 12-23.)

Immune Diseases (pp. 449–453)

■ Immune diseases are of two types: immunodeficiency diseases (insufficient immune responses) or inappropriate immune attacks (excessive or mistargeted immune responses).

■ Inappropriate attacks include autoimmune diseases, immune complex diseases, and allergies (hypersensitivities), of which there are two types: (1) Immediate hypersensitivities involving the production of IgE antibodies by B cells that trigger release of histamine from mast cells and basophils to bring about a swift response to the allergen, and (2) delayed hypersensitivities involving a more slowly responding cell-mediated, symptom-producing response by T cells against the allergen. (Review Figure 12-24 and Table 12-5.)

External Defenses (pp. 453–457)

■ The body surfaces exposed to the outside environment—both the outer covering of skin and the linings of internal cavities that communicate with the external environment—serve not only as mechanical barriers to deter would-be pathogenic invaders but also play an active role in thwarting entry of bacteria and other unwanted materials.

■ The skin consists of two layers: an outer vascular, keratinized epidermis and an inner, connective tissue dermis. The epidermis contains four cell types: pigment-producing melanocytes, keratin-producing keratinocytes, antigen-presenting Langerhans cells, and immune-suppressive Granstein cells. (Review Figure 12-25.)

■ The other main routes by which potential pathogens enter the body are the digestive system, the urogenital system, and the respiratory system, which are all defended by various antimicrobial strategies.

Respiratory Anatomy (pp. 461–464)

■ Cellular respiration refers to the intracellular metabolic reactions that use O_2 and produce CO_2 during energy-yielding oxidation of nutrient molecules. External respiration refers to the transfer of O_2 and CO_2 between the external environment and tissue cells. The respiratory and circulatory systems function together to accomplish external respiration. (Review Figure 13-1.)

■ The respiratory system exchanges air between the atmosphere and lungs. The airways conduct air from the atmosphere to the alveoli, across which O_2 and CO_2 are exchanged between air in these air sacs and blood in the surrounding pulmonary capillaries. The extremely thin alveolar walls are formed by Type I alveolar cells. Type II alveolar cells secrete pulmonary surfactant. (Review Figures 13-2 and 13-4.)

■ The lungs are housed within the closed compartment of the thorax, the volume of which can be changed by contractile activity of surrounding respiratory muscles.

■ Each lung is surrounded by a double-walled, closed sac, the pleural sac. (Review Figure 13-5.)

Respiratory Mechanics (pp. 465–485)

■ Ventilation, or breathing, is the process of cyclically moving air in and out of the lungs so that old alveolar air that has given up O_2 and picked up CO_2 can be exchanged for fresh atmospheric air.

■ Ventilation is mechanically accomplished by alternately shifting the direction of the pressure gradient for airflow between the atmosphere and alveoli through the cyclic expansion and recoil of the lungs. When intra-alveolar pressure decreases as a result of lung expansion during inspiration, air flows into the lungs from the higher atmospheric pressure. When intra-alveolar pressure increases as a result of lung recoil during expiration, air flows out of the lungs toward the lower atmospheric pressure. (Review Figures 13-6, 13-7, 13-10, 13-13, and 13-14.)

■ Alternate contraction and relaxation of the inspiratory muscles (primarily the diaphragm) indirectly produce periodic inflation and deflation of the lungs by cyclically expanding and compressing the thoracic cavity, with the lungs passively following its movements. (Review Figures 13-11 and 13-12.)

■ The lungs follow the movements of the thoracic cavity by virtue of the intrapleural fluid's cohesiveness and the transmural pressure gradient across the lung wall. The transmural pressure gradient exists because the intrapleural pressure is subatmospheric and thus less than the intra-alveolar pressure. (Review Figures 13-8 and 13-14.)

■ Because energy is required for contracting the inspiratory muscles, inspiration is an active process, but expiration is passive during quiet breathing because it is accomplished by elastic recoil of the lungs on relaxing inspiratory muscles, at no energy expense. (Review Figure 13-12a, b, and c.)

■ For more forceful active expiration, contraction of the expiratory muscles (namely, the abdominal muscles) further decreases the size of the thoracic cavity and lungs, which further increases the intra-alveolar–to–atmospheric-pressure gradient. (Review Figures 13-11 and 13-12d.)

■ The larger the gradient between the alveoli and atmosphere in either direction, the larger the airflow rate, because air flows until intra-alveolar pressure equilibrates with atmospheric pressure. (Review Figures 13-13 and 13-14.)

■ Besides being directly proportional to the pressure gradient, airflow rate is also inversely proportional to airway resistance. (Review Table 13-1.) Because airway resistance, which depends on the caliber of the conducting airways, is normally very low, airflow rate usually depends primarily on the pressure gradient between the alveoli and atmosphere.

■ The lungs can be stretched to varying degrees during inspiration and then recoil to their preinspiratory size during expiration because of their elastic behavior. Pulmonary compliance refers to the distensibility of the lungs—how much they stretch in response to a given change in the transmural pressure gradient. Elastic recoil refers to the snapping back of the lungs to their resting position during expiration.

■ Pulmonary elastic behavior depends on the elastic connective tissue within the lungs and on alveolar surface tension/pulmonary surfactant interaction. Alveolar surface tension, which is the result of attractive forces between the surface water molecules lining each alveolus, tends to resist the alveolus being stretched on inflation (decreases compliance) and tends to return it back to a smaller surface area during deflation (increases lung rebound). (Review Table 13-2.)

■ If the alveoli were lined by water alone, the surface tension would be so great that the lungs would be poorly compliant and would tend to collapse. Pulmonary surfactant intersperses between the water molecules and lowers alveolar surface tension, thereby increasing compliance and counteracting the tendency for alveoli to collapse. Alveolar interdependence also counteracts the tendency for alveoli to collapse. (Review Figures 13-16 and 13-17 and Table 13-2.)

■ The lungs can be filled to about 5.5 liters on maximal inspiration or emptied to about 1 liter on maximal expiration. Normally the lungs operate at "half full." Lung volume typically varies from about 2 to 2.5 liters as an average tidal volume of 500 ml of air is moved in and out with each breath. (Review Figures 13-18, 13-19, and 13-20.)

■ The amount of air moved in and out of the lungs in one minute, the pulmonary ventilation, is equal to tidal volume times respiratory rate.

■ Not all the air moved in and out is available for gas exchange with the blood, because part occupies the conducting airways (anatomic dead space). Alveolar ventilation, the volume of air exchanged between the atmosphere and alveoli in one minute, is a measure of the air actually available for gas exchange with the blood. Alveolar ventilation equals (tidal volume minus dead space volume) times respiratory rate. (Review Figure 13-22 and Table 13-3.)

Gas Exchange (pp. 486–490)

■ Oxygen and CO_2 move across body membranes by passive diffusion down partial pressure gradients. The partial pressure of a gas in air is that portion of the total atmospheric pressure contributed by this individual gas, which in turn is directly proportional to the percentage of this gas in the air. The partial pressure of a gas in blood depends on the amount of this gas dissolved in the blood. (Review Figure 13-25.)

■ Net diffusion of O_2 occurs first between the alveoli and blood and then between the blood and tissues as a result of the O_2 partial pressure gradients created by continuous use of O_2 in the cells and continuous replenishment of fresh alveolar O_2 provided by ventilation. Net diffusion of CO_2 occurs in the reverse direction, first between the tissues and blood and then between the blood and alveoli, as a result of the CO_2 partial pressure gradients created by continuous production of CO_2 in the cells and continuous removal of alveolar CO_2 through ventilation. (Review Figure 13-26.)

■ Other factors that influence the rate of gas exchange are surface area and thickness of the membrane across which the gas is diffusing and the diffusion constant of the gas in the membrane (Fick's law of diffusion). (Review Table 13-5.)

Gas Transport (pp. 490–498)

■ Because O_2 and CO_2 are not very soluble in blood, they must be transported primarily by mechanisms other than simply being physically dissolved. (Review Table 13-6.)

■ Only 1.5% of the O_2 is physically dissolved in the blood, with 98.5% chemically bound to hemoglobin (Hb).

■ The primary factor that determines the extent to which Hb and O_2 are combined (the % Hb saturation) is the P_{O_2} of the blood, depicted by an S-shaped curve known as the O_2–Hb dissociation curve. In the P_{O_2} range of the pulmonary capillaries (the plateau portion of the curve), Hb is still almost fully saturated even if the blood P_{O_2} falls as much as 40%. This provides a margin of safety by ensuring near-normal O_2 delivery to the tissues despite a substantial reduction in arterial P_{O_2}. In the P_{O_2} range in the systemic capillaries (the steep portion of the curve), Hb unloading increases greatly in response to a small local decline in blood P_{O_2} associated with increased cellular metabolism. In this way, more O_2 is provided to match the increased tissue needs. (Review Figure 13-28.)

■ Increased P_{CO_2}, increased acid, and increased temperature at the tissue level shift the O_2–Hb curve to the right, facilitating the unloading of O_2 from Hb for tissue use. (Review Figure 13-30.)

■ Hemoglobin facilitates a large net transfer of O_2 between the alveoli and blood and between the blood and tissue cells by acting as a storage depot to keep P_{O_2} (that is, dissolved O_2 concentration) low, despite a considerable increase in the total O_2 content of the blood. (Review Figure 13-29.)

■ Carbon dioxide picked up at the systemic capillaries is transported in the blood by three methods: (1) 10% is physically dissolved, (2) 30% is bound to Hb, and (3) 60% takes the form of bicarbonate (HCO_3^-). The erythrocyte enzyme carbonic anhydrase catalyzes conversion of CO_2 to HCO_3^- according to the reaction $CO_2 + H_2O \rightarrow H^+ + HCO_3^-$. These reactions are all reversed in the lungs as CO_2 is eliminated to the alveoli. (Review Table 13-6 and Figure 13-31.)

Control of Respiration (pp. 498–507)

■ Ventilation involves two aspects, both subject to neural control: (1) rhythmic cycling between inspiration and expiration and (2) regulation of ventilation magnitude, which depends on control of respiratory rate and depth of tidal volume.

■ Respiratory rhythm is established by the pre-Bötzinger complex, which displays pacemaker activity and drives the inspiratory neurons located in the dorsal respiratory group (DRG) of the medullary respiratory control center. When these neurons fire, impulses ultimately reach the inspiratory muscles to bring about inspiration. (Review Figure 13-33.)

■ When the inspiratory neurons stop firing, the inspiratory muscles relax and passive expiration takes place. For active expiration, the expiratory muscles are activated at this time by expiratory neurons in the ventral respiratory group (VRG) of the medullary respiratory control center.

■ This basic rhythm is smoothed out by the apneustic and pneumotaxic centers located in the pons. The apneustic center prolongs inspiration; the more powerful pneumotaxic center limits inspiration. (Review Figure 13-33.)

■ Three chemical factors play a role in determining the magnitude of ventilation: P_{CO_2}, P_{O_2}, and H^+ concentration of the arterial blood. (Review Table 13-8.)

■ The dominant factor in the ongoing regulation of ventilation is arterial P_{CO_2}, an increase of which is the most potent chemical stimulus for increasing ventilation. Changes in arterial P_{CO_2} alter ventilation by bringing about corresponding changes in the brain-ECF H^+ concentration, to which the central chemoreceptors are very sensitive. (Review Figure 13-35.)

■ The peripheral chemoreceptors are responsive to an increase in arterial H^+ concentration, which likewise reflexly brings about increased ventilation. The resulting adjustment in arterial H^+-generating CO_2 is important in maintaining the acid–base balance of the body. (Review Figure 13-34.)

■ The peripheral chemoreceptors also reflexly increase ventilation in response to a marked reduction in arterial P_{O_2} (<60 mm Hg), serving as an emergency mechanism to increase respiration when arterial P_{O_2} levels fall below the safety range provided by the plateau portion of the O_2–Hb curve.

■ Respiration activity can also be voluntarily modified.

Kidneys: Anatomy, Functions, and Basic Processes (pp. 511–517)

■ Each of the pair of kidneys consists of an outer renal cortex and inner renal medulla. The kidneys form urine. They eliminate unwanted plasma constituents in the urine while conserving materials of value to the body. Urine from each kidney is collected in the renal pelvis, then transmitted from both kidneys through the pair of ureters to the single urinary bladder, where urine is stored until emptied through the urethra to the outside. (Review Figures 14-1 and 14-2.)

■ The urine-forming functional unit of the kidneys, the nephron, is composed of interrelated vascular and tubular components. The vascular component consists of two capillary networks in series, the first being the glomerulus, a tuft of capillaries that filters large volumes of protein-free plasma into the tubular component. The second capillary network consists of the peritubular capillaries, which nourish the renal tissue and participate in exchanges between the tubular fluid and plasma. (Review Figures 14-3 and 14-4.)

■ The tubular component begins with Bowman's capsule, which cups around the glomerulus to catch the filtrate, then continues a specific tortuous course to ultimately empty into the renal pelvis. (Review Figure 14-3.) As the filtrate passes through various regions of the tubule, cells lining the tubules modify it, returning to the plasma only those materials necessary for maintaining proper ECF composition and volume. What is left behind in the tubules is excreted as urine.

■ The kidneys perform three basic processes: (1) glomerular filtration, the nondiscriminating movement of protein-free plasma from the blood into the tubules; (2) tubular reabsorption, the selective transfer of specific constituents in the filtrate back into the blood of the peritubular capillaries; and (3) tubular secretion, the highly specific movement of selected substances from peritubular capillary blood into the tubular fluid. Everything filtered or secreted but not reabsorbed is excreted as urine. (Review Figure 14-6.)

Glomerular Filtration (pp. 517–524)

■ Glomerular filtrate is produced when part of the plasma flowing through each glomerulus is passively forced under pressure through the glomerular membrane into the underlying Bowman's capsule. The net filtration pressure causing filtration results from a high glomerular capillary blood pressure that favors filtration outweighing the combined opposing forces of plasma-colloid osmotic pressure and Bowman's capsule hydrostatic pressure. (Review Figure 14-7 and Table 14-1.)

■ 20% to 25% of the cardiac output is delivered to the kidneys to be acted on by renal regulatory and excretory processes. Of the plasma flowing through the kidneys, normally 20% is filtered through the glomeruli, for an average glomerular filtration rate (GFR) of 125 ml/min.

■ Myogenic mechanisms and tubuloglomerular feedback, triggered by the juxtaglomerular apparatus, autoregulate glomerular blood flow and the GFR despite transient changes in the driving mean arterial blood pressure in the range of 80 to 180 mm Hg. (Review Figures 14-9, 14-10, and 14-11.)

■ The GFR can be deliberately altered by changing the glomerular capillary blood pressure via sympathetic influence on the afferent arterioles as part of the baroreceptor reflex response that compensates for changed arterial blood pressure. When blood pressure falls too low, sympathetically induced afferent arteriolar vasoconstriction lowers glomerular blood pressure and GFR. When blood pressure rises too high, reduced sympathetic activity causes afferent arteriolar vasodilation, leading to a rise in GFR. As the GFR is altered, the amount of fluid lost in urine changes correspondingly, adjusting plasma volume as needed to help restore blood pressure to normal on a long-term basis. (Review Figures 14-10 and 14-12.)

Tubular Reabsorption (pp. 524–534)

■ After the filtrate is formed, the tubules handle each filtered substance discretely, so that even though the initial glomerular filtrate is identical to plasma (with the exception of plasma proteins), the concentrations of different constituents are variously altered as the filtered fluid flows through the tubular system. (Review Tables 14-2 and 14-3, p. 537.)

■ The reabsorptive capacity of the tubular system is tremendous. More than 99% of the filtered plasma is returned to the blood through reabsorption. On average, 124 ml out of the 125 ml filtered per minute are reabsorbed. (Review Table 14-2.)

■ Tubular reabsorption involves transepithelial transport from the tubular lumen into the peritubular capillary plasma. This process may be active (requiring energy) or passive (using no energy). (Review Figure 14-14.)

■ The pivotal event to which most reabsorptive processes are linked is the active reabsorption of Na^+, driven by the energy-dependent Na^+–K^+ pump in the basolateral membrane of the tubular cells. The transport of Na^+ out of the cells into the lateral spaces between adjacent cells by this carrier induces the net reabsorption of Na^+ from the tubular lumen to the peritubular capillary plasma. (Review Figure 14-15.)

■ Most Na^+ reabsorption takes place early in the nephron in constant unregulated fashion, but in the distal and collecting tubules, the reabsorption of a small percentage of the filtered Na^+ is variable and controlled, primarily by the renin–angiotensin–aldosterone system. (Review Table 14-4, p. 548.)

■ Because Na^+ and its attendant anion, Cl^-, are the major osmotically active ions in the ECF, the ECF volume is determined by the Na^+ load in the body. In turn, the plasma volume, which reflects the total ECF volume, is important in the long-term determination of arterial blood pressure. Whenever the Na^+ load, ECF volume, plasma volume, and arterial blood pressure are below normal, the juxtaglomerular apparatus secretes renin, an enzymatic hormone that triggers a series of events ultimately leading to increased aldosterone secretion by the adrenal cortex. Aldosterone increases Na^+ reabsorption from the distal portions of the tubule, thus correcting for the original reduction in Na^+, ECF volume, and blood pressure. (*Review Figures 14-11 and 14-16.*)

■ By contrast, Na^+ reabsorption is inhibited by the natriuretic peptides, ANP and BNP, hormones released from the cardiac atria and ventricles, respectively, in response to expansion of the ECF volume and a subsequent increase in blood pressure. (*Review Figure 14-17.*)

■ In addition to driving the reabsorption of Na^+, the energy used by the Na^+–K^+ pump is ultimately responsible for the reabsorption of organic nutrients (glucose or amino acids) from the proximal tubule by secondary active transport. (*Review Figure 3-19, p. 74.*)

■ Other electrolytes actively reabsorbed by the tubules, such as PO_4^{3-} and Ca^{2+}, have their own independently functioning carrier systems within the proximal tubule.

■ Because the electrolyte and nutrient carriers can become saturated, each exhibits a maximal carrier-limited transport capacity (T_m). Once the filtered load of an actively reabsorbed substance exceeds the T_m, reabsorption proceeds at a constant maximal rate, and any additional filtered quantity of the substance is excreted in the urine. (*Review Figure 14-18.*)

■ Active Na^+ reabsorption also drives the passive reabsorption of Cl^- (via an electrical gradient), H_2O (by osmosis), and urea (down a urea concentration gradient created as a result of extensive osmotic-driven H_2O reabsorption). Sixty-five percent of the filtered H_2O is reabsorbed from the proximal tubule in unregulated fashion, driven by active Na^+ reabsorption. (*Review Figure 14-19 and Table 14-4.*) Reabsorption of H_2O increases the concentration of other substances remaining in the tubular fluid, most of which are filtered waste products. The small urea molecules are the only waste products that can passively permeate the tubular membranes, so urea is the only waste product partially (50%) reabsorbed as a result of being concentrated. (*Review Figure 14-20.*)

■ The other waste products, which are not reabsorbed, remain in the urine in highly concentrated form.

Tubular Secretion (pp. 534–537)

■ Tubular secretion involves transepithelial transport from the peritubular capillary plasma into the tubular lumen. By tubular secretion, the kidney tubules can selectively add some substances to the quantity already filtered. Secretion of substances hastens their excretion in the urine.

■ The most important secretory systems are for (1) H^+ (helps regulate acid–base balance); (2) K^+ (keeps the plasma K^+ concentration at the level needed to maintain normal membrane excitability in the heart, other muscles, and nerves); and (3) organic ions (accomplishes more efficient elimination

of foreign organic compounds from the body). H^+ is secreted in the proximal, distal, and collecting tubules. K^+ is secreted only in the distal and collecting tubules under control of aldosterone. Organic ions are secreted only in the proximal tubule. (*Review Table 14-3 and Figures 14-21 and 14-22.*)

Urine Excretion and Plasma Clearance (pp. 537–552)

■ Of the 125 ml/min of glomerular filtrate formed, normally only 1 ml/min remains in the tubules to be excreted as urine. Only wastes and excess electrolytes not wanted by the body are left behind, dissolved in a given volume of H_2O to be eliminated in the urine.

■ Because the excreted material is removed or "cleared" from the plasma, the term *plasma clearance* refers to the volume of plasma cleared of a particular substance each minute by renal activity. (*Review Figure 14-23.*)

■ The kidneys can excrete urine of varying volumes and concentrations to either conserve or eliminate H_2O, depending on whether the body has a H_2O deficit or excess, respectively. The kidneys can produce urine ranging from 0.3 ml/min at 1200 mOsm to 25 ml/min at 100 mOsm by reabsorbing variable amounts of H_2O from the distal portions of the nephron.

■ This variable reabsorption is made possible by a vertical osmotic gradient in the medullary interstitial fluid, established by the long loops of Henle of the juxtamedullary nephrons via countercurrent multiplication and preserved by the vasa recta of these nephrons via countercurrent exchange. (*Review Figures 14-5, 14-24, 14-25, and 14-28.*) This vertical osmotic gradient, to which the hypotonic (100 mOsm) tubular fluid is exposed as it passes through the distal portions of the nephron, establishes a passive driving force for progressive reabsorption of H_2O from the tubular fluid, but the actual extent of H_2O reabsorption depends on the amount of vasopressin (antidiuretic hormone) secreted. (*Review Figure 14-27.*)

■ Vasopressin increases the permeability of the distal and collecting tubules to H_2O; they are impermeable to H_2O in its absence. (*Review Figure 14-26.*) Vasopressin secretion increases in response to a H_2O deficit, increasing H_2O reabsorption. Its secretion is inhibited in response to a H_2O excess, reducing H_2O reabsorption. Thus vasopressin-controlled H_2O reabsorption helps correct any fluid imbalances.

■ Once formed, urine is propelled by peristaltic contractions through the ureters from the kidneys to the urinary bladder for temporary storage.

■ The bladder can accommodate up to 250 to 400 ml of urine before stretch receptors within its wall initiate the micturition reflex. (*Review Figure 14-30.*) This reflex causes involuntary emptying of the bladder by simultaneous bladder contraction and opening of both the internal and external urethral sphincters. Micturition can transiently be voluntarily prevented by deliberately tightening the external sphincter and pelvic diaphragm. (*Review Figure 14-29.*)

Balance Concept (pp. 557–558)

■ The internal pool of a substance is the quantity of that substance in the ECF. The inputs to the pool are by way of ingestion or metabolic production of the substance. The outputs from the pool are by way of excretion or metabolic consumption of the substance. *(Review Figure 15-1.)*

■ Input must equal output to maintain a stable balance of the substance.

Fluid Balance (pp. 558–569)

■ On average, the body fluids compose 60% of total body weight. This figure varies, depending on how much fat (a tissue with a low H_2O content) a person has. Two thirds of the body H_2O is in the ICF. The remaining third, in the ECF, is distributed between plasma (20% of ECF) and interstitial fluid (80% of ECF). *(Review Table 15-1.)*

■ Because all plasma constituents are freely exchanged across the capillary walls, the plasma and interstitial fluid are nearly identical in composition, except for the lack of plasma proteins in the interstitial fluid. In contrast, the ECF and ICF have markedly different compositions, because the plasma membrane barriers are highly selective as to what materials are transported into or out of the cells. *(Review Figure 15-2.)*

■ The essential components of fluid balance are control of ECF volume by maintaining salt balance and control of ECF osmolarity by maintaining water balance. *(Review Tables 15-2, 15-3, and 15-5.)*

■ Because of the osmotic holding power of Na^+, the major ECF cation, a change in the body's total Na^+ content, or load, causes a corresponding change in ECF volume, including plasma volume, which alters arterial blood pressure in the same direction. Appropriately, in the long run Na^+-regulating mechanisms compensate for changes in ECF volume and arterial blood pressure. *(Review Table 15-5.)*

■ Salt intake is not controlled in humans, but control of salt output in the urine is closely regulated to maintain salt balance. Blood pressure–regulating mechanisms can vary the GFR, and thus the amount of Na^+ filtered, by adjusting the radius of the afferent arterioles supplying the glomeruli. Blood pressure–regulating mechanisms can also vary aldosterone secretion to adjust Na^+ reabsorption by the renal tubules. Varying Na^+ filtration and Na^+ reabsorption can adjust how much Na^+ is excreted in the urine to regulate plasma volume and thus arterial blood pressure in the long term. *(Review Figure 15-3.)*

■ ECF osmolarity must be closely regulated to prevent osmotic shifts of H_2O between the ECF and ICF, because cell swelling or shrinking is harmful, especially to brain neurons. Excess free H_2O in the ECF dilutes ECF solutes; the resulting ECF hypotonicity drives H_2O into the cells. An ECF free H_2O deficit, by contrast, concentrates ECF solutes, so H_2O leaves the cells to enter the hypertonic ECF. *(Review Table 15-5.)*

■ To prevent these harmful fluxes, changes in ECF osmolarity are primarily detected and corrected by the systems that maintain free H_2O balance (H_2O without accompanying solute).

■ Free H_2O balance is regulated largely by vasopressin and, to a lesser degree, by thirst. Both of these factors are governed primarily by hypothalamic osmoreceptors, which monitor ECF osmolarity, and to a lesser extent by left atrial volume receptors, which monitor vascular "fullness." The amount of vasopressin secreted determines the extent of free H_2O reabsorption by distal portions of the nephrons, thereby determining the volume of urinary output. *(Review Figure 15-4 and Table 15-4.)*

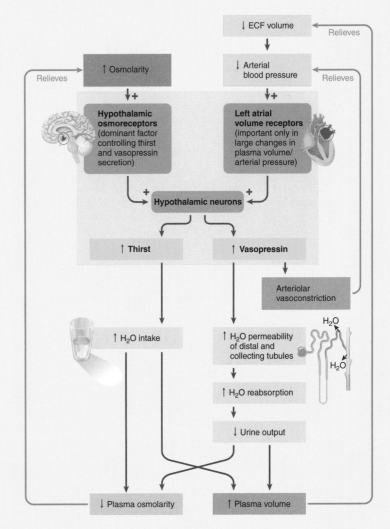

■ Simultaneously, intensity of thirst controls the volume of fluid intake. However, because the volume of fluid drunk is often not directly correlated with the intensity of thirst, control of urinary output by vasopressin is the most important regulatory mechanism for maintaining H_2O balance.

Acid–Base Balance (pp. 569–585)

■ Acids liberate free hydrogen ions (H^+) into solution; bases bind with free hydrogen ions and remove them from solution. (Review Figure 15-5.)

■ Acid–base balance refers to regulation of $[H^+]$ in the body fluids. To precisely maintain $[H^+]$, input of H^+ by metabolic production of acids within the body must continually be matched with H^+ output by urinary excretion of H^+ and respiratory removal of H^+-generating CO_2. Furthermore, between the time of this generation and its elimination, H^+ must be buffered within the body to prevent marked fluctuations in $[H^+]$.

■ Hydrogen ion concentration is often expressed in terms of pH, which is the logarithm of $1/[H^+]$.

■ The normal pH of the plasma is 7.4, slightly alkaline compared to neutral H_2O, which has a pH of 7.0. A pH lower than normal (higher $[H^+]$ than normal) indicates a state of acidosis. A pH higher than normal (lower $[H^+]$ than normal) characterizes a state of alkalosis. (Review Figure 15-6.)

■ Fluctuations in $[H^+]$ have profound effects, most notably (1) changes in neuromuscular excitability, with acidosis depressing excitability, especially in the CNS, and alkalosis producing overexcitability of both the PNS and CNS; (2) disruption of normal metabolic reactions by altering the structure and function of all enzymes; and (3) alterations in plasma $[K^+]$ (which affect cardiac function) brought about by H^+-induced changes in the rate of K^+ elimination by the kidneys.

■ The primary challenge in controlling acid–base balance is maintaining normal plasma alkalinity despite continual addition of H^+ to the plasma from ongoing metabolic activity. The major source of H^+ is from CO_2-generated H^+.

■ The three lines of defense for resisting changes in $[H^+]$ are *first* the chemical buffer systems, *second* respiratory control of pH, and *third* renal control of pH.

■ Chemical buffer systems each consist of a pair of chemicals involved in a reversible reaction, one that can liberate H^+ and the other that can bind H^+. By acting according to the law of mass action, a buffer pair acts immediately to minimize any changes in pH. The four chemical buffers are (1) H_2CO_3:HCO_3^-, (2) proteins, (3) hemoglobin, and (4) phosphate. (Review Figure 15-8 and Table 15-6.)

■ The relationship between pH and the members of the H_2CO_3:HCO_3^- buffer pair is represented in the Henderson-Hasselbalch equation: $pH = pK + \log [HCO_3^-]/[CO_2]$, with $[CO_2]$ reflecting $[H_2CO_3]$. $[HCO_3^-]$ is controlled by the kidneys; $[CO_2]$ is controlled by the lungs. pK is a constant at 6.1, and the normal ratio of $[HCO_3^-]/[CO_2]$ is 20/1 (the log of which is 1.3), for a normal pH of 7.4.

■ The respiratory system normally eliminates metabolically produced CO_2 so that CO_2-generated H^+ does not accumulate in the body fluids.

■ When chemical buffers alone have been unable to immediately minimize a pH change, the respiratory system responds within a few minutes by altering its rate of CO_2 removal. An increase in $[H^+]$ from sources other than CO_2 stimulates respiration so that more H^+-forming CO_2 is blown off, compensating for acidosis by reducing generation of CO_2-associated H^+. Conversely, a fall in $[H^+]$ depresses respiratory activity so that CO_2 and thus H^+ generated from this source can accumulate in the body fluids to compensate for alkalosis. (Review Table 15-7.)

■ The kidneys are the most powerful line of defense. They require hours to days to compensate for a deviation in body-fluid pH. However, not only can they eliminate the normal amount of H^+ produced from non-CO_2 sources but they can also alter their rate of H^+ removal in response to changes in both non-CO_2 and CO_2-generated acids. In contrast, the lungs can adjust only H^+ generated from CO_2. Furthermore, the kidneys can regulate $[HCO_3^-]$ in body fluids as well.

■ The kidneys compensate for acidosis by secreting the excess H^+ in the urine while adding new HCO_3^- to the plasma to expand the HCO_3^- buffer pool. During alkalosis, the kidneys conserve H^+ by reducing its secretion in urine. They also eliminate HCO_3^-, which is in excess because less HCO_3^- than usual is tied up buffering H^+ when H^+ is in short supply. (Review Figures 15-9 through 15-12 and Table 15-8.)

■ Secreted H^+ must be buffered in the tubular fluid to prevent the H^+ concentration gradient from becoming so great that it blocks further H^+ secretion. Normally, H^+ is buffered by the urinary phosphate buffer pair, which is abundant in the tubular fluid because excess dietary phosphate spills into the urine to be excreted from the body.

■ In acidosis, when all the phosphate buffer is already used up in buffering the extra secreted H^+, the kidneys secrete NH_3 into the tubular fluid to serve as a buffer so that H^+ secretion can continue.

■ The four types of acid–base imbalances are respiratory acidosis, respiratory alkalosis, metabolic acidosis, and metabolic alkalosis. Respiratory acid–base disorders stem from deviations from normal $[CO_2]$, whereas metabolic acid–base imbalances include all deviations in pH other than those caused by abnormal $[CO_2]$. (Review Figure 15-13 and Table 15-9.)

▲ TABLE 15-6	Chemical Buffers and Their Primary Roles
Buffer System	**Major Functions**
Carbonic Acid: Bicarbonate Buffer System	Primary ECF buffer against non-carbonic acid changes
Protein Buffer System	Primary ICF buffer; also buffers ECF
Hemoglobin Buffer System	Primary buffer against carbonic acid changes
Phosphate Buffer System	Important urinary buffer; also buffers ICF

General Aspects of Digestion (pp. 589–596)

■ The four basic digestive processes are motility, secretion, digestion, and absorption.

■ The three classes of energy-rich nutrients are digested into absorbable units as follows: (1) Dietary carbohydrates in the form of the polysaccharides starch and glycogen are digested into monosaccharides, mostly glucose. *(Review Figure 16-1.)* (2) Dietary proteins are digested into amino acids and a few small polypeptides. (3) Dietary fats (triglycerides) are digested into monoglycerides and free fatty acids.

■ The digestive system consists of the digestive tract and accessory digestive organs (salivary glands, exocrine pancreas, and biliary system). *(Review Table 16-1.)*

■ The lumen of the digestive tract (a tube that runs from the mouth to the anus) is continuous with the external environment, so its contents are technically outside the body; this arrangement permits digestion of food without self-digestion occurring in the process.

■ The digestive tract wall has four layers. From innermost outward, they are the mucosa, submucosa, muscularis externa, and serosa. *(Review Figure 16-2.)*

■ Digestive activities are carefully regulated by synergistic autonomous, neural (both intrinsic and extrinsic), and hormonal mechanisms to ensure that ingested food is maximally made available to the body. *(Review Figure 16-3.)*

Mouth (pp. 596–598)

■ *Motility:* Food enters the digestive system through the mouth, where it is chewed and mixed with saliva.

■ *Secretion and digestion:* The salivary enzyme, amylase, begins to digest polysaccharides into the disaccharide maltose, a process that continues in the stomach after swallowing. Salivary secretion is controlled by a salivary center in the medulla, mediated by autonomic nerves to the salivary glands. *(Review Figures 16-1 and 16-4.)*

■ *Absorption:* No food is absorbed from the mouth.

Pharynx and Esophagus (pp. 598–600)

■ *Motility:* The tongue propels the bolus of food to the rear of the throat, which initiates the swallowing reflex. The swallowing center in the medulla coordinates a complex group of activities that result in closure of the respiratory passages and propulsion of food through the pharynx and esophagus into the stomach. *(Review Figures 16-5 and 16-6.)*

■ *Secretion, digestion, and absorption:* The esophageal secretion, mucus, is protective. No nutrient digestion or absorption occurs here.

Stomach (pp. 600–613)

■ *Motility:* Gastric motility includes filling, storage, mixing, and emptying. Gastric filling is facilitated by vagally mediated receptive relaxation of the stomach. Gastric storage takes place in the body of the stomach, where peristaltic contractions of the thin muscle walls are too weak to mix the contents. Gastric mixing in the thick-muscled antrum results from vigorous peristaltic contractions. *(Review Figures 16-7 and 16-8.)*

■ Gastric emptying is influenced by factors in both the stomach and duodenum. (1) Increased volume and fluidity of chyme in the stomach promote emptying. (2) Fat, acid, hypertonicity, and distension in the duodenum (the dominant factors controlling gastric emptying) delay gastric emptying until the duodenum is ready to process more chyme. They do so by inhibiting stomach peristaltic activity via the enterogastric reflex and the enterogastrones, secretin and cholecystokinin (CCK), which are secreted by the duodenal mucosa. *(Review Figure 16-8 and Table 16-2.)*

■ *Secretion:* Gastric secretions into the stomach lumen include (1) HCl (from the parietal cells), which activates pepsinogen; (2) pepsinogen (from the chief cells), which, once activated, initiates protein digestion; (3) mucus (from the mucous cells), which provides a protective coating; and (4) intrinsic factor (from the parietal cells), which is needed for vitamin B_{12} absorption. *(Review Table 16-3 and Figures 16-9, 16-10, and 16-11.)*

■ The stomach also secretes the hormone gastrin, which plays a dominant role in stimulating gastric secretion, and the paracrines histamine and somatostatin, which stimulate and inhibit gastric secretion, respectively. *(Review Table 16-3.)*

■ Gastric secretion is increased before and during a meal via excitatory vagal and intrinsic nerve responses along with the stimulatory actions of gastrin and histamine. After the meal empties, gastric secretion is reduced by withdrawal of stimulatory factors, release of inhibitory somatostatin, and inhibitory actions of the enterogastric reflex and enterogastrones. *(Review Tables 16-4 and 16-5.)*

■ *Digestion and absorption:* Carbohydrate digestion continues by swallowed salivary amylase in the body of the stomach. Protein digestion is initiated by pepsin in the antrum of the stomach, where vigorous peristaltic contractions mix the food with gastric secretions, converting it to a thick liquid mixture known as chyme. *(Review Table 16-6, p. 624.)* No nutrients are absorbed from the stomach.

Pancreatic and Biliary Secretions (pp. 613–621)

■ Pancreatic exocrine secretions and bile from the liver both enter the duodenal lumen.

■ Pancreatic secretions include (1) potent digestive enzymes from the acinar cells, which digest all three categories of foodstuff; and (2) an aqueous NaHCO₃ solution from the duct cells, which neutralizes the acidic contents emptied into the duodenum from the stomach. Secretin stimulates the pancreatic duct cells, and CCK stimulates the acinar cells. (Review Figures 16-12 and 16-13.)

■ The pancreatic digestive enzymes include (1) the proteolytic enzymes trypsinogen, chymotrypsinogen, and procarboxypeptidase, which are secreted in inactive form and are activated in the duodenal lumen on exposure to enterokinase and activated trypsin; (2) pancreatic amylase, which continues carbohydrate digestion; and (3) lipase, which accomplishes fat digestion. (Review Table 16-6.)

■ The liver, the body's largest and most important metabolic organ, performs many varied functions. Its contribution to digestion is the secretion of bile, which contains bile salts. Bile salts aid fat digestion through their detergent action (forming a lipid emulsion) and facilitate fat absorption by forming water-soluble micelles that carry the water-insoluble products of fat digestion to their absorption site. (Review Figures 16-15 through 16-18 and 16-26)

■ Between meals, bile is stored and concentrated in the gallbladder, which is stimulated by CCK to contract and empty into the duodenum during meal digestion. After participating in fat digestion and absorption, bile salts are reabsorbed and returned via the hepatic portal system to the liver, where they are resecreted and also act as a potent choleretic to stimulate secretion of more bile. (Review Figures 16-14 and 16-16.)

■ Bile also contains bilirubin, a derivative of degraded hemoglobin, which is the major excretory product in the feces.

Small Intestine (pp. 621–633)

■ Motility: Segmentation, the small intestine's primary motility during digestion of a meal, thoroughly mixes the chyme with digestive juices to facilitate digestion; it also exposes the products of digestion to the absorptive surfaces. (Review Figure 16-19.) Between meals, the migrating motility complex sweeps the lumen clean.

■ Secretion: The juice secreted by the small intestine does not contain any digestive enzymes. The enzymes synthesized by the small intestine act within the brush-border membrane of the epithelial cells. (Review Figures 16-24a and 16-25a.)

■ Digestion: The small intestine is the main site for digestion and absorption. Carbohydrate and protein digestion continues in the small-intestine lumen by the pancreatic enzymes and is completed by the small-intestine brush-border enzymes (disaccharidases and aminopeptidases, respectively). Fat is digested entirely in the small-intestine lumen, by pancreatic lipase. (Review Table 16-6.)

■ Absorption: The small-intestine lining is remarkably adapted to its digestive and absorptive function. Its folds bear a rich array of fingerlike projections, the villi, which have a multi-tude of even smaller hairlike protrusions, the microvilli (brush border). Together, these surface modifications tremendously increase the area available to house the membrane-bound enzymes and to accomplish absorption. (Review Figures 16-21, 16-22, and 16-23.) This lining is replaced about every three days to ensure it is optimally healthy despite harsh lumen conditions.

■ The energy-dependent process of Na⁺ absorption provides the driving force for Cl⁻, water, glucose, and amino acid absorption. All these absorbed products enter the blood. (Review Figures 16-24b and 16-25b.)

■ Because they are not soluble in water, the products of fat digestion must undergo a series of transformations that enable them to be passively absorbed, eventually entering the lymph. (Review Figure 16-26.)

■ The small intestine absorbs almost everything presented to it, from ingested food to digestive secretions to sloughed epithelial cells. In contrast to the almost complete, unregulated absorption of ingested nutrients, water, and most electrolytes, the amount of iron and calcium absorbed is variable and subject to control. (Review Figure 16-27.) Only a small amount of fluid and indigestible food residue passes on to the large intestine. (Review Table 16-7.)

Large Intestine (pp. 633–637)

■ Motility: The colon (review Figure 16-28) concentrates and stores undigested food residues (fiber; that is, plant cellulose) and bilirubin until they can be eliminated in the feces. (Review Figure 16-28.) Haustral contractions slowly shuffle the colonic contents back and forth to mix and facilitate absorption of most of the remaining fluid and electrolytes. Mass movements several times a day, usually after meals, propel the feces long distances. Movement of feces into the rectum triggers the defecation reflex.

■ Secretion, digestion, and absorption: The alkaline mucus secretion is protective. No secretion of digestive enzymes or absorption of nutrients takes place in the colon. Absorption of some of the remaining salt and water converts the colonic contents into feces.

Overview of the Gastrointestinal Hormones (pp. 637–638)

■ The three major gastrointestinal hormones are gastrin from the stomach mucosa and secretin and cholecystokinin from the duodenal mucosa. Gastrin is released primarily in response to protein in the stomach, and its effects promote digestion of protein. Secretin is released primarily in response to acid in the duodenum, and its effects neutralize the acid. Cholecystokinin is released primarily in response to fat in the duodenum, and its effects optimize conditions for digesting fat.

Energy Balance (pp. 641–650)

■ Energy input to the body in the form of food energy must equal energy output, because energy cannot be created or destroyed. Energy output or expenditure includes (1) external work, performed by skeletal muscles to move an external object or move the body through the external environment; and (2) internal work, which consists of all other energy-dependent activities that do not accomplish external work, including active transport, smooth and cardiac muscle contraction, glandular secretion, and protein synthesis. (Review Figure 17-1.)

■ Only about 25% of the chemical energy in food is harnessed to do biological work. The rest is immediately converted to heat. Furthermore, all the energy expended to accomplish internal work is eventually converted into heat, and 75% of the energy expended by working skeletal muscles is lost as heat. Therefore, most of the energy in food ultimately appears as body heat.

■ The metabolic rate (energy expenditure per unit of time) is measured in kilocalories of heat produced per hour.

■ The basal metabolic rate (BMR) is a measure of the body's minimal waking rate of internal energy expenditure.

■ For a neutral energy balance, the energy in ingested food must equal energy expended in performing external work and transformed into heat. If more energy is consumed than is expended, the extra energy is stored in the body, primarily as adipose tissue, so body weight increases. By contrast, if more energy is expended than is available in the food, body energy stores are used to support energy expenditure, so body weight decreases.

■ Usually, body weight remains fairly constant over a prolonged period of time (except during growth) because food intake is adjusted to match energy expenditure on a long-term basis. Food intake is controlled primarily by the hypothalamus by means of complex regulatory mechanisms in which hunger and satiety are important components. Feeding or appetite signals give rise to the sensation of hunger and promote eating, whereas satiety signals lead to the sensation of fullness and suppress eating. (Review Table 17-3.)

■ The arcuate nucleus of the hypothalamus plays a key role in energy homeostasis by virtue of the two clusters of appetite-regulating neurons it contains: neurons that secrete neuropeptide Y (NPY), which increases appetite and food intake; and neurons that secrete melanocortins, which suppress appetite and food intake. (Review Figure 17-2.)

■ Adipocytes in fat stores secrete the hormone leptin, which reduces appetite and decreases food consumption by inhibiting the NPY-secreting neurons and stimulating the melanocortins-secreting neurons of the arcuate nucleus. This mechanism is important in the long-term matching of energy intake with energy output, thus maintaining body weight over the long term. (Review Table 17-2 and Figure 17-2.)

■ Insulin released by the endocrine pancreas in response to increased glucose and other nutrients in the blood also inhibits the NPY-secreting neurons and contributes to long-term control of energy balance and body weight.

■ NPY and melanocortins bring about their effects by acting on the lateral hypothalamus area (LHA) and paraventricular nucleus (PVN) to alter the release of chemical messengers from these areas. The LHA secretes orexins, which are potent stimulators of food intake, whereas the PVN releases neuropeptides such as corticotropin-releasing hormone, which decrease food intake. (Review Figure 17-2.)

■ Short-term control of the timing and size of meals is mediated primarily by the actions of two peptides secreted by the digestive tract. (1) Ghrelin, a mealtime initiator, is secreted by the stomach before a meal and signals hunger. Its secretion drops when food is consumed. Ghrelin stimulates appetite and promotes feeding behavior by stimulating the NPY-secreting neurons. (2) PYY_{3-36}, a mealtime terminator, is secreted by the small and large intestines during a meal and signals satiety. Its secretion is lowest before a meal. PYY_{3-36} inhibits the NPY-secreting neurons. (Review Figure 17-2.)

■ The nucleus tractus solitarius (NTS) in the brain stem serves as the satiety center and in this capacity also plays a key role in short-term control of meals. The NTS receives input from the higher hypothalamic areas concerned with control of energy balance and food intake as well as input from the digestive tract and pancreas. Satiety signals acting through the NTS to inhibit further food intake include stomach distension and increased CCK, a hormone released from the duodenum in response to the presence of nutrients, especially fat, in the digestive tract lumen. (Review Figure 17-2.)

■ Psychosocial and environmental factors can also influence food intake above and beyond the internal signals that govern feeding behavior. (Review Figure 17-2.)

■ The body can be thought of as a heat-generating core (internal organs, CNS, and skeletal muscles) surrounded by a shell of variable insulating capacity (the skin).

■ The skin exchanges heat energy with the external environment, with the direction and amount of heat transfer depending on the environmental temperature and the momentary insulating capacity of the shell. The four physical means by which heat is exchanged are (1) radiation (net movement of heat energy via electromagnetic waves); (2) conduction (exchange of heat energy by direct contact); (3) convection (transfer of heat energy by means of air currents); and (4) evaporation (extraction of heat energy from the body by the heat-requiring conversion of liquid H_2O to H_2O vapor). Because heat energy moves from warmer to cooler objects, radiation, conduction, and convection can be channels for either heat loss or heat gain, depending on whether surrounding objects are cooler or warmer, respectively, than the body surface. Normally, they are avenues for heat loss, along with evaporation resulting from sweating. *(Review Figure 17-4.)*

■ To prevent serious cell malfunction, the core temperature must be held constant at about 100°F (equivalent to an average oral temperature of 98.2°F) by continuously balancing heat gain and heat loss despite changes in environmental temperature and variation in internal heat production. *(Review Figure 17-3.)*

■ This thermoregulatory balance is controlled by the hypothalamus. The hypothalamus is apprised of the skin temperature by peripheral thermoreceptors and of the core temperature by central thermoreceptors, the most important of which are located in the hypothalamus itself. *(Review Figure 17-5.)*

■ The primary means of heat gain is heat production by metabolic activity, the biggest contributor being skeletal muscle contraction. *(Review Figure 17-5.)*

■ Heat loss is adjusted by sweating and by controlling to the greatest extent possible the temperature gradient between the skin and surrounding environment. The latter is accomplished by regulating the diameter of the skin's arterioles. (1) Skin vasoconstriction reduces the flow of warmed blood through the skin so that skin temperature falls. The layer of cool skin between the core and environment increases the insulating barrier between the warm core and the external air. (2) Skin vasodilation brings more warmed blood through the skin so that skin temperature approaches the core temperature, thus reducing the insulative capacity of the skin. *(Review Figure 17-5.)*

■ On exposure to cool surroundings, the core temperature starts to fall as heat loss increases, because of the larger-than-normal skin-to-air temperature gradient. The posterior hypothalamus responds to reduce the heat loss by inducing skin vasoconstriction while simultaneously increasing heat production through heat-generating shivering. *(Review Table 17-4.)*

■ Conversely, in response to a rise in core temperature (resulting either from excessive internal heat production accompanying exercise or from excessive heat gain on exposure to a hot environment), the anterior hypothalamus triggers heat-loss mechanisms, such as skin vasodilation and sweating, while simultaneously decreasing heat production, such as by reducing muscle tone. *(Review Table 17-4.)*

■ In both cold and heat responses, voluntary behavioral actions also help maintain thermal homeostasis.

■ A fever occurs when endogenous pyrogen released from macrophages in response to infection raises the hypothalamic set point. An elevated core temperature develops as the hypothalamus initiates cold-response mechanisms to raise the core temperature to the new set point. *(Review Figure 17-6.)*

General Principles of Endocrinology (pp. 661–670)

■ Hormones are long-distance chemical messengers secreted by the ductless endocrine glands into the blood, which transports the hormones to specific target cells where they control a particular function by altering protein activity.

■ Hormones are grouped into two categories based on differences in their solubility and are further grouped according to their chemical structure—hydrophilic hormones (peptide hormones, catecholamines, and indoleamines) and lipophilic hormones (steroid hormones and thyroid hormone).

■ The endocrine system is especially important in regulating organic metabolism, H_2O and electrolyte balance, growth, and reproduction and in helping the body cope with stress. (Review Figure 18-1 and Table 18-2, pp. 667–669.)

■ Some hormones are tropic, meaning their function is to stimulate and maintain other endocrine glands.

■ The effective plasma concentration of each hormone is normally controlled by regulated changes in the rate of hormone secretion. Secretory output of endocrine cells is primarily influenced by two types of direct regulatory inputs: (1) neural input, which increases hormone secretion in response to a specific need and governs diurnal variations in secretion; and (2) input from another hormone, which involves either stimulatory input from a tropic hormone or inhibitory input from a target-cell hormone in negative-feedback fashion. (Review Figures 18-2, 18-3, and 18-7, p. 675.)

■ The effective plasma concentration of a hormone can also be influenced by its rate of removal from the blood by metabolic inactivation and excretion and, for some hormones, by its rate of peripheral activation or its extent of binding to plasma proteins.

■ Endocrine dysfunction arises when too much or too little of any particular hormone is secreted or when there is decreased target-cell responsiveness to a hormone. (Review Table 18-1.)

■ Target-cell sensitivity to a given plasma concentration of a hormone to which the target cell is responsive can be modified by (1) down regulation (number of target-cell receptors decreases in the face of a prolonged increase in the hormone), (2) permissiveness (one hormone increases the effectiveness of another hormone), (3) synergism (combined effect of two hormones is greater than sum of their separate effects), and (4) antagonism (one hormone decreases the effectiveness of another hormone).

Hypothalamus and Pituitary (pp. 670–677)

■ The pituitary gland consists of two distinct lobes, the posterior pituitary and the anterior pituitary. (Review Figure 18-4.)

■ The hypothalamus, a portion of the brain, secretes nine peptide hormones. Two are stored in the posterior pituitary, and seven are carried through a special vascular link—the hypothalamic–hypophyseal portal system—to the anterior pituitary, where they regulate the release of particular anterior pituitary hormones. (Review Figures 18-5 and 18-8.)

■ The posterior pituitary is a neural extension of the hypothalamus. Cell bodies of neurosecretory neurons in the hypothalamus synthesize two small peptide hormones, vasopressin and oxytocin, which pass down the axon to be stored in nerve terminals within the posterior pituitary. These hormones are independently released from the posterior pituitary into the blood in response to action potentials originating in the hypothalamus. (Review Figure 18-5.)

■ The anterior pituitary secretes six different peptide hormones that it produces itself. Five anterior pituitary hormones are tropic. (1) Thyroid-stimulating hormone (TSH) stimulates secretion of thyroid hormone. (2) Adrenocorticotropic hormone (ACTH) stimulates secretion of cortisol by the adrenal cortex. (3 and 4) The gonadotropic hormones—follicle-stimulating hormone (FSH) and luteinizing hormone (LH)—stimulate production of gametes (eggs and sperm) as well as secretion of sex hormones. (5) Growth hormone (GH) stimulates growth indirectly by stimulating liver secretion of IGF-I, which in turn promotes growth. GH exerts metabolic effects as well. (6) Prolactin stimulates milk secretion and is not tropic to another endocrine gland. (Review Figure 18-6.)

■ The anterior pituitary releases its hormones into the blood at the bidding of releasing and inhibiting hormones from the hypothalamus. The hypothalamus, in turn, is influenced by a variety of neural and hormonal controlling inputs. (Review Table 18-4 and Figures 18-7 and 18-8.)

1 Hypophysiotropic hormones (releasing hormones and inhibiting hormones) produced by neurosecretory neurons in the hypothalamus enter the hypothalamic capillaries.

2 These hypothalamic capillaries rejoin to form the hypothalamic-hypophyseal portal system, a vascular link to the anterior pituitary.

3 The portal system branches into the capillaries of the anterior pituitary.

4 The hypophysiotropic hormones, which leave the blood across the anterior pituitary capillaries, control the release of anterior pituitary hormones.

5 When stimulated by the appropriate hypothalamic releasing hormone, the anterior pituitary secretes a given hormone into these capillaries.

6 The anterior pituitary capillaries rejoin to form a vein, through which the anterior pituitary hormones leave for ultimate distribution throughout the body by the systemic circulation.

Labels in figure: Neurosecretory neurons in hypothalamus (secrete releasing and inhibiting hormones into portal system); Hypothalamus; Capillaries in hypothalamus; Systemic arterial blood in; Endocrine cells of anterior pituitary (secrete anterior pituitary hormones into systemic blood); Capillaries in anterior pituitary; Systemic venous blood out; Releasing and inhibiting hormones; Hypothalamic-hypophyseal portal system; Posterior pituitary; Anterior pituitary; KEY — ● ○ = Hypophysiotropic hormones ○ = Anterior pituitary hormone

■ Both the hypothalamus and anterior pituitary are inhibited in negative-feedback fashion by the product of the target endocrine gland in the hypothalamus–anterior pituitary–target-gland axis. (Review Figure 18-7.)

Endocrine Control of Growth (pp. 677–685)

■ Growth depends not only on growth hormone and other growth-influencing hormones such as thyroid hormone, insulin, and the sex hormones but also on genetic determination, an adequate diet, and freedom from chronic disease or stress. Major growth spurts occur the first few years after birth and during puberty. *(Review Figure 18-9.)*

■ Growth hormone (GH) promotes growth indirectly by stimulating liver production of an insulin-like growth factor, IGF-I, which acts directly on soft tissues and bone to bring about growth-promoting actions. The GH/IGF-I pathway causes growth by stimulating protein synthesis, cell division, and the lengthening and thickening of bones. *(Review Figures 18-10 and 18-11.)*

■ Growth hormone also directly exerts metabolic effects unrelated to growth, such as conservation of carbohydrates and mobilization of fat stores. *(Review Figure 18-11.)*

■ Growth hormone secretion by the anterior pituitary is regulated by two hypothalamic hormones, growth hormone–releasing hormone (GHRH) and growth hormone–inhibiting hormone (somatostatin). In negative-feedback fashion, IGF-I and GH both inhibit GHRH and stimulate somatostatin. *(Review Figure 18-11.)*

■ Growth hormone levels are not highly correlated with periods of rapid growth. The primary signals for increased growth hormone secretion are related to metabolic needs rather than growth, namely, deep sleep (during diurnal rhythm), exercise, stress, and low blood glucose.

Pineal Gland and Circadian Rhythms (pp. 685–687)

■ The suprachiasmatic nucleus (SCN) is the body's master biological clock. Self-induced cyclic variations in the concentration of clock proteins within the SCN bring about cyclic changes in neural discharge from this area. Each cycle takes about a day and drives the body's circadian (daily) rhythms.

■ The inherent rhythm of this endogenous oscillator is a bit longer than 24 hours. Therefore, each day the body's circadian rhythms must be entrained or adjusted to keep pace with environmental cues so that the internal rhythms are synchronized with the external light–dark cycle.

■ In the eyes, special photoreceptors that respond to light but are not involved in vision send input to the SCN. Acting through the SCN, the pineal gland's secretion of the hormone melatonin rhythmically fluctuates with the light–dark cycle, decreasing in the light and increasing in the dark. Mel-

*These factors all increase growth hormone secretion, but it is unclear whether they do so by stimulating GHRH or inhibiting GHIH somatostatin, or both.

atonin, in turn, is believed to synchronize the body's natural circadian rhythms, such as diurnal (day–night) variations in hormone secretion and body temperature, with external cues such as the light–dark cycle.

■ Other proposed roles for melatonin include (1) promoting sleep; (2) influencing reproductive activity, including the onset of puberty; (3) acting as an antioxidant to remove damaging free radicals; and (4) enhancing immunity.

Thyroid Gland (pp. 691–698)

■ The thyroid gland contains two types of endocrine secretory cells: (1) follicular cells, which produce the iodide-containing hormones, T_4 (thyroxine or tetraiodothyronine) and T_3 (tri-iodothyronine), collectively known as thyroid hormone; and (2) C cells, which synthesize a Ca^{2+}-regulating hormone, calcitonin. (*Review Figure 19-1.*)

■ Most steps of thyroid hormone synthesis take place on large thyroglobulin molecules within the colloid, an "inland" extracellular site within the interior of the spherical thyroid follicles. Dietary iodine is transported as iodide (I^-) from the blood into the follicular cells by the iodide pump, an energy-dependent symporter. From the follicular cells, I^- enters the colloid where it iodinates the amino acid tyrosine within thyroglobulin, yielding monoiodothyronine (MIT) and di-iodothyronine (DIT). Coupling of MIT and DIT produces T_3; coupling of two DITs produces T_4. Thyroid hormone is secreted by follicular cells phagocytizing a piece of colloid and freeing T_4 and T_3, which enter the blood. (*Review Figure 19-2.*)

■ Thyroid hormone is the primary determinant of the overall metabolic rate of the body. By accelerating metabolic rate, it increases heat production. It also enhances the actions of the sympathetic catecholamines and is essential for normal growth and for development and function of the nervous system.

■ Thyroid hormone secretion is regulated by a negative-feedback system between hypothalamic TRH, anterior pituitary TSH, and thyroid gland T_3 and T_4. The feedback loop maintains thyroid hormone levels relatively constant. Cold exposure in newborn infants is the only input for increasing TRH and thus thyroid hormone secretion. (*Review Figure 19-3.*)

Adrenal Glands (pp. 698–707)

■ Each adrenal gland (of the pair) consists of two separate endocrine organs—an outer, steroid-secreting adrenal cortex and an inner, catecholamine-secreting adrenal medulla. (*Review Figure 19-7.*)

■ Each steroid hormone is produced by stepwise modifications of cholesterol via specific enzymes present in a given steroidogenic endocrine gland. The adrenal cortex has enzymes to produce three categories of steroid hormones: mineralocorticoids (primarily aldosterone), glucocorticoids (primarily cortisol), and adrenal sex hormones (primarily the weak androgen, dehydroepiandrosterone). (*Review Figure 19-8.*)

■ Aldosterone regulates Na^+ and K^+ balance and is important for blood pressure homeostasis, which is achieved secondarily by the osmotic effect of Na^+ in maintaining the plasma volume, a lifesaving effect. Control of aldosterone secretion is related to Na^+ and K^+ balance and to blood pressure regulation and is not influenced by ACTH. Aldosterone is controlled by the renin–angiotensin–aldosterone system (RAAS) and by a direct effect of K^+ on the adrenal cortex. (*Review Figure 14-22, p. 535*).

■ Cortisol helps regulate fuel metabolism and is important in stress adaptation. It increases blood levels of glucose, amino acids, and fatty acids and spares glucose for use by the glucose-dependent brain. The mobilized organic molecules are available for use as needed for energy or repair. Cortisol secretion is regulated by a negative-feedback loop involving hypothalamic CRH and pituitary ACTH. Stress is the most potent stimulus for increasing activity of the CRH–ACTH–cortisol axis. Cortisol also displays a characteristic diurnal rhythm. (*Review Figures 18-3, p. 665; 18-7, p. 675; 19-9; and 19-13 and Table 19-2, p. 708.*)

■ Dehydroepiandrosterone (DHEA) governs the sex drive and growth of axillary and pubertal hair in females. It has no observable effect in males, in whom it is overpowered by testosterone. DHEA is under control of CRH/ACTH but negatively feeds back to the gonadotropin loop.

■ The adrenal medulla consists of modified sympathetic postganglionic neurons known as chromaffin cells, which secrete the catecholamine epinephrine into the blood in response to sympathetic stimulation. (*Review Figure 7-2, p. 239.*) Epinephrine reinforces the sympathetic system in mounting fight-or-flight responses and in maintaining arterial blood pressure. It also increases blood glucose and blood fatty acids. The primary stimulus for increased epinephrine secretion is activation of the sympathetic system by stress. (*Review Table 19-2 and Figure 19-13.*)

Integrated Stress Response (pp. 707–710)

■ The term *stress* refers to the generalized nonspecific response of the body to any factor that overwhelms, or threatens to overwhelm, the body's compensatory ability to maintain homeostasis. The term *stressor* refers to any noxious stimulus that elicits the stress response. (*Review Figure 19-12.*)

■ In addition to specific responses to various stressors, all stressors produce a similar generalized stress response: (1) increased sympathetic and epinephrine activity, which prepares the body for fight-or-flight; (2) activation of the CRH–ACTH–cortisol axis, which helps the body cope with stress by mobilizing metabolic resources; (3) elevation of blood glucose and fatty acids through decreased insulin and increased glucagon secretion; and (4) maintenance of blood volume and blood pressure through increased activity of RAAS and vasopressin. All these actions are coordinated by the hypothalamus. (*Review Figure 19-13 and Table 19-2.*)

Endocrine Control of Fuel Metabolism (pp. 710–726)

■ Intermediary or fuel metabolism is, collectively, the synthesis (anabolism), breakdown (catabolism), and transformations of the three classes of energy-rich organic nutrients—carbohydrate, fat, and protein—within the body. Glucose and fatty acids derived from carbohydrates and fats, respectively, are primarily used as metabolic fuels, whereas amino acids derived from proteins are primarily used for synthesis of structural and enzymatic proteins. (*Review Tables 19-3 and 19-4 and Figure 19-14.*)

■ During the *absorptive state* following a meal, excess absorbed nutrients not immediately needed for energy production or protein synthesis are stored to a limited extent as glycogen in the liver and muscle but mostly as triglycerides in adipose tissue. During the *postabsorptive state* between meals when no new nutrients are entering the blood, the glycogen and triglyceride stores are catabolized to release nutrient molecules into the blood. If necessary, body proteins are degraded to release amino acids for conversion into glucose (gluconeogenesis). The blood glucose concentration must be maintained above a critical level even during the postabsorptive state because the brain depends on blood-delivered glucose as its energy source. Tissues not dependent on glucose switch to fatty acids as their metabolic fuel, sparing glucose for the brain. (*Review Table 19-5.*)

■ Blood glucose concentration is controlled by factors that regulate glucose uptake by cells and glucose output by the liver. (*Review Figure 19-16.*)

■ The shifts in metabolic pathways between the absorptive and postabsorptive state are controlled by hormones, the most important being insulin. Insulin is secreted by the β cells of the islets of Langerhans, the endocrine portion of the pancreas. (*Review Figure 19-15 and Table 19-6.*)

■ Insulin is an anabolic hormone; it promotes the cellular uptake of glucose, fatty acids, and amino acids and enhances their conversion into glycogen, triglycerides, and proteins, respectively. In so doing, it lowers the blood concentrations of these small organic molecules. Insulin secretion is increased during the absorptive state, primarily by a direct effect of an elevated blood glucose on the β cells via excitation-secretion coupling. Insulin directs nutrients into cells during this state. (*Review Figures 19-17 through 19-21.*)

■ Glucagon, secreted by the pancreatic α cells, mobilizes the energy-rich molecules from their stores during the postabsorptive state. Glucagon, which is secreted in response to a direct effect of a fall in blood glucose on the α cells, in general opposes the actions of insulin. (*Review Figures 19-15, 19-20, and 19-21.*)

Endocrine Control of Calcium Metabolism (pp. 726–737)

■ Changes in the concentration of free, diffusible plasma Ca^{2+}, the biologically active form of this ion, produce profound and life-threatening effects, most notably on neuromuscular excitability. Hypercalcemia reduces excitability, whereas hypocalcemia brings about overexcitability of nerves and muscles. If the overexcitability is severe enough, fatal spastic contractions of respiratory muscles can occur.

■ Control of Ca^{2+} metabolism involves two aspects—regulation of Ca^{2+} homeostasis and regulation of Ca^{2+} balance—and depends on hormonal control of exchanges between the ECF and three compartments: bone, kidneys, and intestine. Regulation of *Ca²⁺ homeostasis*, maintenance of a constant free plasma Ca^{2+} concentration, involves rapid exchanges between bone and the ECF and to a lesser extent to adjustments in urinary Ca^{2+} excretion. Regulation of *Ca²⁺ balance*, maintenance of a constant total amount of Ca^{2+} in the body, is accomplished by adjustments in Ca^{2+} absorption from the intestine and urinary Ca^{2+} excretion.

■ Bone consists of an organic extracellular matrix, the osteoid, which is hardened by precipitation of calcium phosphate crystals. Bone constantly undergoes remodeling by means of bone-dissolving osteoclasts and bone-building osteoblasts. Entombed osteocytes are "retired" osteoblasts that have deposited bone around themselves. Osteoblasts and osteocytes are interconnected by long cytoplasmic arms that extend through the tiny canals that permeate the hardened bone, forming a continuous osteocytic-osteoblastic bone membrane. (*Review Figures 19-23 and 19-24.*)

■ Three hormones regulate the plasma concentration of Ca^{2+} (and concurrently regulate PO_4^{3-})—parathyroid hormone (PTH), calcitonin, and vitamin D.

■ PTH, whose secretion is directly increased by a fall in plasma Ca^{2+} concentration, acts directly on bone and kidneys and indirectly on the intestine to raise plasma Ca^{2+}. In so doing, it is essential for life by preventing the fatal consequences of hypocalcemia. PTH promotes Ca^{2+} movement across the osteocytic-osteoblastic bone membrane from the bone fluid into the plasma in the short term and promotes localized dissolution of bone in the long term by enhancing osteoclasts and suppressing osteoblasts. (*Review Figures 19-24 and 19-25.*)

■ Dissolution of the calcium phosphate bone crystals releases PO_4^{3-} as well as Ca^{2+} into the plasma. PTH acts on the kidneys to enhance the reabsorption of filtered Ca^{2+}, thereby reducing the urinary excretion of Ca^{2+} and increasing its plasma concentration. Simultaneously, PTH reduces renal PO_4^{3-} reabsorption, in this way increasing PO_4^{3-} excretion and lowering plasma PO_4^{3-} levels. This is important because a rise in plasma PO_4^{3-} would force the redeposition of some of the plasma Ca^{2+} back into the bone. (*Review Figure 19-28.*)

■ PTH facilitates activation of vitamin D, which in turn stimulates Ca^{2+} and PO_4^{3-} absorption from the intestine. The skin can synthesize vitamin D from cholesterol when exposed to sunlight, but frequently this endogenous source is inadequate, so vitamin D must be supplemented by dietary intake. From either source, vitamin D must be activated first by the liver and then by the kidneys (the site of PTH regulation of vitamin D activation) before it can exert its effect. (*Review Figures 19-26 and 19-27.*)

■ Calcitonin, a hormone produced by the C cells of the thyroid gland, is secreted in response to an increase in plasma Ca^{2+} and lowers plasma Ca^{2+} by inhibiting activity of bone osteoclasts. Calcitonin is unimportant except during the rare condition of hypercalcemia. (*Review Figure 19-25.*)

Uniqueness of the Reproductive System (pp. 741–748)

■ Both sexes produce gametes (reproductive cells), sperm in males and ova (eggs) in females, each of which bears one member of each of the 23 pairs of chromosomes present in human cells. Union of a sperm and an ovum at fertilization results in the beginning of a new individual with 23 complete pairs of chromosomes, half from each parent. *(Review Figure 20-3.)*

■ The reproductive system is anatomically and functionally distinct in males and females. Males produce sperm and deliver them into the female. Females produce ova, accept sperm delivery, and provide a suitable environment for supporting development of a fertilized ovum until the new individual can survive on its own in the external world.

■ In both sexes, the reproductive system consists of (1) a pair of gonads, testes in males and ovaries in females, which are the primary reproductive organs that produce the gametes and secrete sex hormones; (2) a reproductive tract composed of a system of ducts that transport and/or house the gametes after they are produced; and (3) accessory sex glands that provide supportive secretions for the gametes. The external genitalia are the externally visible parts of the reproductive system. *(Review Figures 20-1 and 20-2.)* Secondary sexual characteristics are the distinguishing features between males and females not directly related to reproduction.

■ *Sex determination* is a genetic phenomenon dependent on the combination of sex chromosomes at the time of fertilization: An XY combination is a genetic male, and an XX combination, a genetic female. *Sexual differentiation* refers to the embryonic development of the gonads, reproductive tract, and external genitalia along male or female lines, which gives rise to the apparent anatomic sex of the individual. In the presence of masculinizing factors, a male reproductive system develops; in their absence, a female system develops. *(Review Figures 20-4, 20-5, and 20-6.)*

Male Reproductive Physiology (pp. 749–758)

■ The testes are located in the scrotum. The cooler temperature in the scrotum than in the abdominal cavity is essential for spermatogenesis (sperm production), which occurs in the testes' highly coiled seminiferous tubules. Leydig cells in the interstitial spaces between these tubules secrete the male sex hormone testosterone into the blood. *(Review Figure 20-7 and 20-8.)*

■ Testosterone is secreted before birth to masculinize the developing reproductive system; then its secretion ceases until puberty, at which time it begins once again and continues throughout life. Testosterone is responsible for maturation and maintenance of the entire male reproductive tract, for development of secondary sexual characteristics, and for stimulating libido. *(Review Table 20-1.)*

■ The testes are regulated by the anterior pituitary gonadotropic hormones, luteinizing hormone (LH) and follicle-stimulating hormone (FSH), which are under control of hypothalamic gonadotropin-releasing hormone (GnRH). *(Review Figure 20-10.)*

■ Testosterone secretion is regulated by LH stimulation of the Leydig cells, and in negative-feedback fashion, testosterone inhibits LH secretion. *(Review Figure 20-10.)*

■ Spermatogenesis requires both testosterone and FSH. Testosterone stimulates the mitotic and meiotic divisions required to transform the undifferentiated diploid germ cells, the spermatogonia, into undifferentiated haploid spermatids. FSH stimulates the remodeling of spermatids into highly specialized motile spermatozoa. *(Review Figures 20-7, 20-8, and 20-10.)*

■ A spermatozoon consists only of a DNA-packed head bearing an enzyme-filled acrosome at its tip for penetrating the ovum, a midpiece containing the mitochondria for energy production, and a whiplike motile tail. *(Review Figure 20-9.)*

■ Also present in the seminiferous tubules are Sertoli cells, which protect, nurse, and enhance the germ cells throughout their development. Sertoli cells also secrete inhibin, a hormone that inhibits FSH secretion, completing the negative-feedback loop. *(Review Figures 20-7b and d and 20-10.)*

■ The still immature sperm are flushed out of the seminiferous tubules into the epididymis by fluid secreted by the Sertoli cells. The epididymis and ductus deferens store and concentrate the sperm and increase their motility and fertility prior to ejaculation. During ejaculation, the sperm are mixed with secretions released by the accessory glands. *(Review Table 20-2 and Figure 20-7.)*

■ The *seminal vesicles* supply fructose for energy and prostaglandins, which promote smooth muscle motility in both the male and female reproductive tracts to enhance sperm transport. The seminal vesicles also contribute the bulk of the semen. The *prostate gland* contributes an alkaline fluid to neutralize the acidic vaginal secretions. The *bulbourethral glands* release lubricating mucus.

Sexual Intercourse between Males and Females (pp. 759–764)

■ The male sex act consists of erection and ejaculation, which are part of a much broader systemic sexual response cycle. *(Review Table 20-4.)*

■ Erection is a hardening of the normally flaccid penis that enables it to penetrate the female vagina. Erection is accomplished by marked vasocongestion of the penis brought about by reflexly induced vasodilation of the arterioles supplying the penile erectile tissue. *(Review Figures 20-12 and 20-13.)*

■ When sexual excitation reaches a critical peak, ejaculation occurs. It consists of two stages: (1) emission, the emptying of semen (sperm and accessory sex gland secretions) into the urethra; and (2) expulsion of semen from the penis. The latter is accompanied by a set of characteristic systemic responses and intense pleasure referred to as orgasm. *(Review Table 20-4.)*

■ Females experience a sexual response cycle similar to that of males, with both having excitation, plateau, orgasmic, and resolution phases. Like the penis, the highly vascular clitoris undergoes erection (but not ejaculation). (*Review Figure 20-12.*) During sexual response, the outer portion of the vagina constricts to grip the penis, and the inner part expands to create space for sperm deposition.

Female Reproductive Physiology (pp. 764–796)

■ In the nonpregnant state, female reproductive function is controlled by a complex, cyclic, negative-feedback control system between the hypothalamus (GnRH), anterior pituitary (FSH and LH), and ovaries (estrogen, progesterone, and inhibin). During pregnancy, placental hormones become the main controlling factors.

■ The ovaries perform the dual and interrelated functions of oogenesis (producing ova) and secretion of estrogen and progesterone. (*Review Table 20-6, p. 794.*) Two related ovarian endocrine units sequentially accomplish these functions: the follicle and the corpus luteum.

■ The same steps in chromosome replication and division take place in oogenesis as in spermatogenesis, but the timing and end result are markedly different. Spermatogenesis is accomplished within two months, but the similar steps in oogenesis take anywhere from 12 to 50 years to complete on a cyclic basis from the onset of puberty until menopause. A female is born with a limited, nonrenewable supply of germ cells, whereas postpubertal males can produce several hundred million sperm each day. Each primary oocyte yields only one cytoplasm-rich ovum along with three doomed cytoplasm-poor polar bodies that disintegrate, whereas each primary spermatocyte yields four equally viable spermatozoa. (*Review Figures 20-14, 20-15, and Figure 20-8, p. 753.*)

■ Oogenesis and estrogen secretion take place within an ovarian follicle during the first half of each reproductive cycle (the follicular phase) under the influence of FSH, LH, and estrogen. (*Review Figures 20-16 through 20-20.*)

■ At approximately midcycle, the maturing follicle releases a single ovum (ovulation). Ovulation is triggered by an LH surge brought about by the high level of estrogen produced by the mature follicle. (*Review Figures 20-16, 20-18, and 20-21.*)

■ LH converts the empty follicle into a corpus luteum (CL), which produces progesterone and estrogen during the last half of the cycle (the luteal phase). This endocrine unit prepares the uterus for implantation if the released ovum is fertilized. (*Review Figures 20-16, 20-18, and 20-22.*)

■ If fertilization and implantation do not occur, the CL degenerates, withdrawing hormonal support for the highly developed uterine lining, causing it to disintegrate and slough, producing menstrual flow. Simultaneously, a new follicular phase is initiated. (*Review Figures 20-16 and 20-18.*)

■ Menstruation ceases and the uterine lining (endometrium) repairs itself under the influence of rising estrogen levels from the newly maturing follicle. (*Review Figure 20-18.*)

■ If fertilization does take place, it occurs in the oviduct as the released egg and sperm deposited in the vagina are both transported to this site. (*Review Figures 20-23 through 20-25.*)

■ The fertilized ovum begins to divide mitotically. Within a week it grows and differentiates into a blastocyst capable of implantation. (*Review Figure 20-26.*)

■ Meanwhile, the endometrium has become richly vascularized and stocked with stored glycogen under the influence of luteal-phase progesterone. (*Review Figure 20-18.*) Into this especially prepared lining the blastocyst implants by means of enzymes released by the trophoblasts, which form the blastocyst's outer layer. These enzymes digest the nutrient-rich endometrial tissue, accomplishing the dual function of carving a hole in the endometrium for implantation of the blastocyst while simultaneously releasing nutrients from the endometrial cells for use by the developing embryo. (*Review Figure 20-27.*)

■ After implantation, an interlocking combination of fetal and maternal tissues, the placenta, develops. The placenta is the organ of exchange between maternal and fetal blood and also acts as a transient, complex endocrine organ that secretes a number of hormones essential for pregnancy. Human chorionic gonadotropin (hCG), estrogen, and progesterone are the most important of these hormones. hCG maintains the CL of pregnancy, which secretes estrogen and progesterone during the first trimester of gestation until the placenta takes over this function in the last two trimesters. High levels of estrogen and progesterone are essential for maintaining a normal pregnancy. (*Review Figures 20-28, 20-30, and 20-31 and Table 20-5.*)

■ At parturition, rhythmic contractions of increasing strength, duration, and frequency accomplish the three stages of labor: dilation of the cervix, birth of the baby, and delivery of the placenta (afterbirth). (*Review Figure 20-33.*)

■ Parturition is initiated by a complex interplay of multiple maternal and fetal factors. Once the contractions are initiated at the onset of labor, a positive-feedback cycle is established that progressively increases their force. As contractions push the fetus against the cervix, secretion of oxytocin, a powerful uterine muscle stimulant, is reflexly increased. The extra oxytocin causes stronger contractions, giving rise to even more oxytocin release, and so on. This positive-feedback cycle progressively intensifies until cervical dilation and delivery are complete. (*Review Figure 20-32.*)

■ During gestation, the breasts are specially prepared for lactation. The elevated levels of placental estrogen and progesterone, respectively, promote development of the ducts and alveoli in the mammary glands. (*Review Figure 20-34.*)

■ Prolactin stimulates the synthesis of enzymes essential for milk production by the alveolar epithelial cells. However, the high gestational level of estrogen and progesterone prevents prolactin from promoting milk production. Withdrawal of the placental steroids at parturition initiates lactation.

■ Lactation is sustained by suckling, which triggers the release of oxytocin and prolactin. Oxytocin causes milk ejection (letdown) by stimulating the myoepithelial cells surrounding the alveoli to squeeze the secreted milk out through the ducts. Prolactin stimulates secretion of more milk to replace the milk the baby nurses. (*Review Figures 20-34 and 20-35.*)

ANATOMICAL TERMS USED TO INDICATE DIRECTION AND ORIENTATION

anterior	situated in front of or in the front part of
posterior	situated behind or toward the rear
ventral	toward the belly or front surface of the body; synonymous with anterior
dorsal	toward the back surface of the body; synonymous with posterior
medial	denoting a position nearer the midline of the body or a body structure
lateral	denoting a position toward the side or farther from the midline of the body or a body structure
superior	toward the head
inferior	away from the head
proximal	closer to a reference point
distal	farther from a reference point
sagittal section	a vertical plane that divides the body or a body structure into right and left sides
longitudinal section	a plane that lies parallel to the length of the body or a body structure
cross section	a plane that runs perpendicular to the length of the body or a body structure
frontal or coronal section	a plane parallel to and facing the front part of the body

WORD DERIVATIVES COMMONLY USED IN PHYSIOLOGY

a; an-	absence or lack	*epi-*	above; over	*osteo-*	bone
ad-; af-	toward	*erythro-*	red	*oto-*	ear
adeno-	glandular	*gastr-*	stomach	*para-*	near
angi-	vessel	*-gen; -genic*	produce	*pariet-*	wall
anti-	against	*gluc-; glyc-*	sweet	*peri-*	around
archi-	old	*hemi-*	half	*phago-*	eat
-ase	splitter	*hemo-*	blood	*-pod*	footlike
auto-	self	*hepat-*	liver	*-poiesis*	formation
bi-	two; double	*homeo-*	sameness	*poly-*	many
-blast	former	*hyper-*	above; excess	*post-*	behind; after
brady-	slow	*hypo-*	below; deficient	*pre-*	ahead of; before
cardi-	heart	*inter-*	between	*pro-*	before
cephal-	head	*intra-*	within	*pseudo-*	false
cerebr-	brain	*kal-*	potassium	*pulmon-*	lung
chondr-	cartilage	*leuko-*	white	*rect-*	straight
-cide	kill; destroy	*lip-*	fat	*ren-*	kidney
contra-	against	*macro-*	large	*reticul-*	network
cost-	rib	*mamm-*	breast	*retro-*	backward
crani-	skull	*mening-*	membrane	*sacchar-*	sugar
-crine	secretion	*micro-*	small	*sarc-*	muscle
crypt-	hidden	*mono-*	single	*semi-*	half
cutan-	skin	*multi-*	many	*-some*	body
-cyte	cell	*myo-*	muscle	*sub-*	under
de-	lack of	*natr-*	sodium	*supra-*	upon; above
di-	two; double	*neo-*	new	*tachy-*	rapid
dys-	difficult; faulty	*nephr-*	kidney	*therm-*	temperature
ecto-; exo-; extra-	outside; away from	*neuro-*	nerve	*-tion*	act or process of
ef-	away from	*oculo-*	eye	*trans-*	across
-elle	tiny; miniature	*-oid*	resembling	*tri-*	three
-emia	blood	*ophthalmo-*	eye	*vaso-*	vessel
encephalo-	brain	*oral-*	mouth	*-uria*	urine
endo-	within; inside				